VENUS

Edited by

D. M. HUNTEN, L. COLIN,
T. M. DONAHUE, and V. I. MOROZ

With the assistance of
MILDRED SHAPLEY MATTHEWS

With 69 collaborating authors

THE UNIVERSITY OF ARIZONA PRESS
TUCSON, ARIZONA

SPACE SCIENCE SERIES
Tom Gehrels, Space Science Consultant

PLANETS, STARS AND NEBULAE, STUDIED WITH PHOTOPOLARIMETRY, *T. Gehrels* Ed., 1974, 1133 pp.

JUPITER, *T. Gehrels* Ed., 1976, 1254 pp.

PLANETARY SATELLITES, *J. A. Burns* Ed., 1977, 598 pp.

PROTOSTARS AND PLANETS, *T. Gehrels* Ed., 1978, 756 pp.

ASTEROIDS, *T. Gehrels* Ed., 1979, 1181 pp.

COMETS, *L. L. Wilkening* Ed., 1982, 766 pp.

SATELLITES OF JUPITER, *D. Morrison* Ed., 1982, 972 pp.

VENUS, *D. M. Hunten, L. Colin, T. M. Donahue, and V. I. Moroz* Eds., 1983, 1143 pp.

THE UNIVERSITY OF ARIZONA PRESS

Copyright 1983
The Arizona Board of Regents
All Rights Reserved

This book was set in 10/12 IBM MTSC Times Roman
Manufactured in the U.S.A.

Library of Congress Cataloging in Publication Data

Venus.

(Space science series)
   Bibliography: p.
   Includes index.
   1. Venus (Planet)—Addresses, essays, lectures.
I. Hunten, Donald M. II. Series.
QB621.V46 1983   523.4′2   83-1064

ISBN 0-8165-0788-0

About the cover: On Nov. 20, 1978, 3.5 months after launch and some 13 million km or 19 days from encounter with Venus, the Pioneer Venus bus (foreground) released the three small probes in their pursuit of the Large probe, itself deployed pyrotechnically from the bus four days earlier. This dramatic moment was preserved on canvas through the original art of Paul Hudson, and is adapted here for the cover.

CONTENTS

COLLABORATING AUTHORS

R. E. Arvidson, *69*

V. S. Avduevskiy, *280*

A. T. Bazilevskiy, *137*

L. H. Brace, *779*

G. A. Burba, *137*

V. P. Chemodanov, *137*

Z. P. Cheremukhina, *280*

P. A. Cloutier, *941*

L. Colin, *10*

T. E. Cravens, *841*

D. P. Cruikshank, *1*

R. E. Daniell, Jr., *941*

T. M. Donahue, *1003*

A. P. Ekonomov, *131, 632*

E. Eliason, *69*

L. W. Esposito, *484*

K. P. Florenskiy, *137*

Yu. M. Golovin, *131, 632*

T. I. Gombosi, *779, 841*

A. Ya. Gorbachevskiy, *280*

K. I. Gringauz, *980*

D. M. Hunten, *ix, 650*

V. V. Kerzhanovich, *766*

A. J. Kliore, *779*

R. G. Knollenberg, *484*

W. C. Knudsen, *779*

V. A. Krasnopol'sky, *431, 459*

L. V. Ksanfomality, *565, 650*

Yu. N. Kulikov, *280*

S. Kumar, *299*

A. D. Kuz'min, *36*

M. C. Malin, *69, 159*

M. Ya. Marov, *280, 484, 766*

G. E. McGill, *69*

V. I. Moroz, *27, 45, 632*

B. Ye. Moshkin, *131, 632*

L. M. Mukhin, *1037*

A. F. Nagy, *779, 841*

M. K. Narayeva, *137*

H. Niemann, *299*

O. V. Nikolayeva, *137*

S. Nozette, *69*

A. S. Panfilov, *137*

V. A. Parshev, *431*

R. J. Phillips, *159*

I. M. Podgorny, *994*

J. B. Pollack, *1003*

R. Prinn, *299*

A. A. Pronin, *137*

R. D. Reasenberg, *69*

C. T. Russell, *873*

F. L. Scarf, *565*

G. Schubert, *681*

A. Seiff, *215, 1045*

A. S. Selivanov, *137*

V. P. Shari, *280*

Yu. A. Surkov, *154*

T. F. Tascione, *941*

F. W. Taylor, *650*

H. A. Taylor, Jr., *779, 941*

W. W. L. Taylor, *565*

M. G. Tomasko, *604*

O. B. Toon, *484*

R. P. Turco, *484*

G. R. Uspenskiy, *280*

O. Vaisberg, *873*

J. L. Warner, *69*

R. S. Wolff, *941*

U. von Zahn, *299*

PREFACE

The exploration of Venus by spacecraft began in 1962 and has continued at a fairly steady pace ever since. The end of 1978 was a landmark in these efforts, celebrated in this book. Six probes reached the surface, one entered the upper atmosphere, and three, an orbiter and two flybys, studied the environment. The Pioneer Venus orbiter is still in operation as this book goes to press. We now have data on the interior and surface, including a near-global map (Color Plate 1), the lower and upper atmosphere, and the interaction with the solar wind. The origin of the oven-like surface temperature is fairly well understood. Some elemental and isotopic abudances have been found to be remarkably different from those of Earth, and there is a ferment of discussion about the origin and evolution of the planets. All these topics, and many more, are treated in these pages. Most, but not all, of the authors are drawn from the Pioneer Venus and Venera communities.

The 30 chapters are almost equally divided between major surveys and the introductory and supplementary chapters. Many of the latter are by Soviet authors, and other Soviet contributions were integrated into the review chapters. Like other books in the Space Science Series of the University of Arizona Press, this one is built around a conference but is much more than a collection of conference papers. The International Conference on the Venus Environment, sponsored by the Ames Research Center and The University of Arizona, was held in Palo Alto, California, November 2–6, 1981. All the survey chapters were presented there in summary, and first drafts were submitted for review. Many of the other papers were submitted to the journal *Icarus,* and appear in two special issues, dated August and November 1982. The Soviet papers came in somewhat later, and were translated courtesy of NASA. A complete list of these Technical Translations may be found in the Appendix. Those not included in the book were transmitted to *Icarus* for further consideration. An attempt, only partially successful, was made to standardize the atmospheric nomenclature. The Editors recommend the names shown in the Figure below.

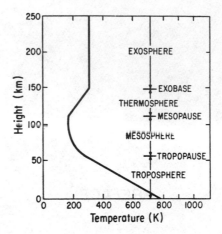

The difficulty in simply carrying them over from the Earth is that there is no distinction between the stratosphere and the mesosphere. The same is true of Mars, where a similar convention would be useful. The word *mesosphere* is recommended, partly because it is neutral, implying no more than being in the middle, and partly because it retains the useful term *mesopause*. Further discussion can be found in Chapters 11, 13, 20, and 21.

The production of this book was generously supported by the National Aeronautics and Space Administration through a grant from the Planetary Program Office, through translation services, and by the Ames Research Center. We are grateful to the University of Arizona Press, publishers of this book in their Space Science Series, to T. Gehrels, Series Consultant, to V. Adams for secretarial help, and to T. Smith who prepared much of the back matter for the book. Above all, we express our thanks to Assistant Editor M. S. Matthews, whose experience and hard work made the whole project possible.

Donald M. Hunten
Lawrence Colin
Thomas M. Donahue
Vassily I. Moroz

1. THE DEVELOPMENT OF STUDIES OF VENUS

DALE P. CRUIKSHANK
University of Hawaii

An historical summary is given of Venus studies before the space age. More recent groundbased work, some of it stimulated by the prospect of space missions, is also discussed.

That Venus is different from the bright stars because of its movements was recognized at least as early as the time of the Babylonians (Huber 1977) who recorded its appearances and influences on society just after 3000 B.C. It is clearly recorded in the astronomical chronicles of other civilizations, including those in China, Meso-America, and of course in Greece. Galileo turned his little telescope to Venus in 1610 and recognized from the lunar-like phases he saw that the planet is not self-luminous ("Cynthiae figuras aemulatur mater amorum" [Galilei 1610], translated as The mother of loves [Venus] emulates the figure of Cynthia [the Moon]).

Those who followed Galileo with the simple telescopes of the seventeenth and early eighteenth century confirmed the phases, and some detected spots on the nearly blank, white illuminated face. F. Fontana may have seen spots during his observations in Italy in 1645, and G. D. Cassini, while still in Italy, reported both bright and dusky spots in 1666 and 1667. In the eighteenth and nineteenth centuries many observers reported vague, dusky spots on Venus. As early as 1726, F. Bianchini of Rome attempted to time the rotation period of the planet by the variable appearance of the markings; he concluded that Venus had a period of 24 hr. J. Schroeter reported markings, as did W.

Herschel, W. Lassell, and E. M. Antoniadi; some of our contemporaries have seen them as well, though certain notable observers such as E. E. Barnard were never fully convinced of their reality. Markings similar to those detected visually were photographed by Wright (1927) and Ross (1928) with an ultraviolet filter. These cloud features have been dramatically confirmed by spacecraft photographs, and it is generally agreed that some visual observers are able to see with low contrast in the visual spectral region those markings that are so prominent in the violet and ultraviolet. Following the observations of the markings on Venus begin the studies which have led to solid physical insight as to the basic nature of Venus: its atmosphere, rotation, composition and character of its surface, and more.

In the middle of the eighteenth century the distance from the Earth to the Sun was unknown. Halley noted at the time of his unsuccessful measurement of the solar parallax during a transit of Mercury, that a transit of Venus observed from widely spaced stations would afford a greatly superior means of measuring this fundamental quantity. He also knew that the next transit of Venus on 6 June 1761 would occur after his demise (he died in 1742). In an address to the Royal Society in 1716 he urged his colleagues to ensure that the phenomenon be observed: "I recommend it therefore again and again to those curious astronomers who, when I am dead, will have an opportunity of observing these things...and I earnestly wish them all imaginable success" (Halley 1716).

Heeding Halley's words, J. N. Delisle organized a worldwide effort of 120 observers for the transit of 1761. Difficulties of timing the event, the determination of precise geographic location, and logistics resulted in only partial success in establishing the solar parallax. However, Lomonosov observing with an "optic tube" in St. Petersburg, noted a gray halo surrounding the planet as it was partially silhouetted against the Sun and correctly inferred that Venus has an atmosphere. Many other observers of the same event noted the halo and also that at second contact the small black disk remained connected to the Sun's limb with a black thread; when the thread broke, Venus already stood well on the disk. Lomonosov correctly concluded in the 1760s (see Lomonosov 1955) that these phenomena were caused by an atmosphere surrounding the planet, though others who saw the effects did not draw the same inference. Lomonosov's discovery was published in Russian in St. Petersburg and was not widely known. Yet his descriptions and inferences are clear, and he is rightly credited with the discovery of the Venus atmosphere (Vorontsov-Velyaminov 1956).

The second of the two eighteenth-century transits occurred on 3 June 1769 and was observed by, among others, Cook in Tahiti (which had itself been discovered only ~ 2 yr earlier) on a Royal Society expedition. Improvements in timing and position determination made the 1769 event a greater success, with the parallax determined to within a few tenths of an arcsecond of the correct value; the errors arose largely from the timing difficulties introduced by the planet's atmosphere. This method provided the most accurate parallax determination of the period. Captain Cook's expedition was the first of his three Pacific voyages of discovery; the islands now known as Hawaii were discovered on the third voyage in 1778.

Reasonable measurements of the diameter of Venus had been made at the transit of 1761, whereupon the concept of the Earth's twin planet arose. Until the atmospheric temperature and pressure at the surface were determined, the "twin" concept strongly supported the pervasive nineteenth-century belief in life elsewhere in the solar system, being overshadowed only by Lowell's hypothesis of life on Mars in the nineteenth and early twentieth centuries.

The brightness and nearly featureless appearance of the planet, as well as Lomonosov's observation of the halo effect, naturally led to the early conclusion that the atmosphere is cloudy, but the nature of the clouds and the detailed composition of the atmosphere have remained elusive up to the present time. Many scientists assumed that the clouds were dense water vapor clouds like Earth's, leading to the picture of Venus, common in the early twentieth century, as a rainy, carboniferous swamp.

Venus is often the brightest object in the sky, except for the Sun and Moon, and whenever a new astronomical technique is invented, Venus is among the first targets to be observed. This has led to many important discoveries. While visual and photographic spectroscopy had been applied to Venus many times, the first indication of spectral features different from the solar spectrum was found in 1932 with the high-dispersion spectrograph on the Mt. Wilson 2.5 m telescope. Adams and Dunham made spectrograms in a search for water vapor in the 8000 Å region at 5.6 Å mm^{-1} dispersion. Specially sensitized emulsions from Eastman Kodak had been provided by C. K. E. Mees for this spectral region, and a 4 hr exposure made in April 1932 showed two absorption bands with the characteristic structure of P and R branches. Few laboratory spectra existed for this spectral region so that Adams and Dunham could not make a direct comparison for quick identification. They measured the line positions and calculated the moment of inertia for the molecule and found it to be essentially the same as the known moment for carbon dioxide. Dissatisfied with the identification, they fabricated a 70 foot tube on Mt. Wilson, filled it with 10 bar of CO_2 and observed the spectrum using a lamp as a light source. The bands found on Venus appeared in the laboratory spectrum, thus confirming the identification beyond doubt (Adams and Dunham 1932; Dunham 1952).

Kuiper pursued the study of the spectrum of Venus further into the infrared, first with photographic emulsions that he used with the McDonald 2.1 m reflector in the 1940s, and soon after World War II with the newly developed lead-sulfide photoconductors. Kuiper and Chamberlain analyzed the photographic results at high spectral resolution for the rotational temperature of CO_2 at the cloud tops and made a study of the phase variation of the absorption band strength (Chamberlain and Kuiper 1956). Kuiper's first infrared results with spectral resolution of ~ 250 (Kuiper 1947) showed numerous bands of CO_2 in the region up to 1.8 μm and then as far as 2.5 μm at resolution 80. These early results were obtained with a prism spectrometer; eventually, good replica diffraction gratings became available and detector performance improved, so that by 1961 Kuiper (1962) was able to obtain resolution up to ~ 1500 over the same spectral region. He found nearly forty CO_2 bands, including hot bands and isotopic bands, permitting for the first time an estimate of isotopic abundances in the atmosphere of Venus.

Almost simultaneously, Gebbie et al. (1962) were obtaining the first spectra of Venus in the same spectral region (and with comparable resolution) with the Michelson interferometer spectrometer. After a few years, Kuiper also turned to the interferometer as the prime means of obtaining high-resolution spectra of the brighter planets, particularly of Venus and Mars (Kuiper et al. 1968). His success in this early high-resolution study of the CO_2 on Venus was due in large part to the extensive laboratory study of the carbon dioxide molecule that had just been completed by Courtoy (1959), and later to the fact that he made considerable use of NASA's airborne observatory, the Convair 990 "Galileo," for obtaining spectra without interfering telluric absorption features (Kuiper et al. 1967).

The composition of the clouds and the chemistry of the atmosphere vexed observers and theoreticians for decades, and even some years after several spacecraft had reached Venus, some entering the atmosphere, the fundamental problems were not solved. High-resolution spectroscopy from the ground, from balloons, and from aircraft gave conflicting results on a potentially vital constituent: water. Numerous investigators in the 1960s had reported a few hundred micrometers of precipitable water vapor above the clouds, but Kuiper's airborne measurements (at altitude 12 km) of the full disk showed only a five micrometer trace. This work, as well as the detection of ~ 50 μm H_2O by Belton et al. (1968) ruled out the possibility that the clouds were water vapor or ice crystals. The Belton et al. value has been supported by Pioneer Venus for small regions of the clouds. While the water vapor concentration above the cloud tops is variable, subsequent investigations have essentially corroborated Kuiper's value of the global mean of the mixing ratio to CO_2 of $\simeq 1.1 \times 10^{-6}$.

Microwave studies in the early 1960s showed an opacity source below the clouds, and a small quantity of water vapor was the favored candidate (Muhleman 1969; Pollack and Morrison 1969). Also, the early Soviet Venera entry probes detected small quantities of water, consistent with the amount required for microwave opacity, but quite insufficient to allow a water composition of the clouds. The cloud problem therefore remained a controversial issue as more details of the atmospheric composition were slowly sorted out by spectroscopy. The variability of the supracloud atmosphere complicated the problem, as R. A. Schorn, A. T. and L. D. G. Young, and colleagues reported the waxing and waning of the water vapor and the fluctuation of the carbon dioxide bands (e.g., L. Young et al. 1970). Belton and Hunten (1968) could not find oxygen in the A-band, though Prokofyev (1964) had reported a positive detection. Ultra high-resolution interferometric spectroscopy under the artistry of P. and J. Connes (Connes et al. 1967) revealed traces of HC1 and HF in the mid-1966s, and CO was also confirmed at about this time. The nature of the clouds remained obscure, however, until a superb study of the polarization of Venus at many wavelengths showed that droplets with refractive index $n = 1.45$ uniquely satisfied the obscrvations (Coffeen 1968). This soon led to Kuiper's (1969) proposal that the clouds were composed of dihydrated ferrous chloride. This model and most of the others were quickly abandoned when Sill (1972) and A. Young (1973) independently showed that concentrated sulfuric acid matches both the refractive index deduced from polarimetry and the spectral characteristics of Venus's clouds. A more complete analysis of the polarimetry in the light of the sulfuric acid discovery was made by Hansen and Hovenier (1974), and the Pollack et al. (1974) airborne infrared observations dramatically confirmed the sulfuric acid results. We still await a conclusive explanation of the ultraviolet markings in the bright clouds; they may be related to details of the sulfur chemistry of the planet's upper atmosphere. The discovery of SO_2 in the atmosphere from groundbased work (Barker 1979) and from an ultraviolet satellite (Conway et al. 1979) confirms the importance of sulfur chemistry on Venus.

The rotation period and the orientation of the spin axis of Venus eluded investigators until the advent of radar astronomy. Numerous attempts were made to determine the rotation period of the atmosphere from spectroscopy, but the Doppler shift in the Venus atmospheric lines was insufficient to be detected reliably indicating a period of many days or longer. The PEPSIOS experiments by Traub and Carleton (1974b) finally obtained a valid result. Those who could see or photograph the dusky markings tried to determine the period from the observed variations in the patterns, but this gave contradictory results. Moore (1959) lists some preradar determinations ranging from 20 hr to 60 d. An estimate of both rotation period and axis orientation was made by Kuiper (1954) who deduced from the variability of the markings photographed in ultraviolet that the rotation is "fairly rapid...no more than a few weeks at most, and definitely not in 225 days." From a detailed study of many ultraviolet photographs obtained over a broad time base, Boyer and Guerin (1966) correctly deduced the periodic repetition of patterns in the clouds and found that P ~ 5 d. This periodicity of the polar collar (which was measured from Earth) has been verified in spacecraft photographs, though it is now known to represent a characteristic period of atmospheric activity that is not directly related to the body rotation of the planet. Other periodicities at various latitudes are also on the order of a few days.

Radar measurements, begun in 1961, have provided the only reliable and quantitative determinations of the direction and rate of rotation of Venus (Shapiro 1968). The first successful radar experiments were conducted in 1961

at the Jet Propulsion Laboratory (JPL) and at Lincoln Laboratory, and while
they showed that the rotation was slow, they did not yield the period. The next
year, improved observations at JPL yielded two separate measurements of the
rotation rate (Carpenter 1964; Goldstein 1964) and also proved that the sense
of the rotation is retrograde with the axis nearly normal to the ecliptic. Further
improvements were made at the next several inferior conjuctions, and new
determinations were made from observations at the Arecibo Ionospheric
Observatory. As Shapiro (1968) recounts, it was soon (erroneously) concluded
that the retrograde rotation of Venus is locked in a resonance with the Earth
with a period of 243.16 d, such that Venus presents the same face to the Earth
at each inferior conjunction. Zohar et al. (1980) and others have since shown
that the resonance is not exact.

The orientation of the axis of rotation of Venus is almost perpendicular to
the ecliptic, as is the inclination of the pole of the planet's orbital plane.
Subsequent studies have shown that the ultraviolet markings are characteristically inclined several degrees to the planet's equator; this led early workers
astray when they tried to estimate the orientation from the shapes of the cloud
patterns. Kuiper (1954), for example, estimated the spin axis inclination at
~ 32° by this technique. Though the determination of spin rate and pole
position are coupled, the data are now accurate enough to show that the pole
tilt is ~ 3°.

The earliest deliberate studies of thermal emission of Venus were made in
daylight in 1923 and 1928 by Pettit and Nicholson (1955) with the Mt. Wilson
2.5 m telescope and vacuum thermocouples. They recorded the galvanometer
deflections on a moving photographic plate and used filters of rock salt and
glass. They were startled to find that the temperature on the illuminated side
was essentially the same as on the unilluminated side, ~ 240 K, and that it is
not strongly dependent on phase. This was confirmed with measurements
by Coblentz and Lampland (1925), and later by the more precise work of
Sinton and Strong (1960) with the 5 m telescope in 1953 and 1954. The
latter also obtained spectra in the thermal infrared (8–13 μm) as had Adel
and Lampland (1941) somewhat earlier, and instead of strong CO_2 absorptions, a more nearly graybody emission was revealed (Sinton 1961).

The microwave studies that emerged in the late 1950s (the first radio
detection was made by Mayer et al. [1958] in 1956 at 3.15 and 9.4 cm) and
early 1960s showed that the brightness temperature of Venus is ~ 600 K for
wavelengths ≳ 3 cm. Several theoretical investigations proceeded on the
assumption of a hot surface as responsible for the brightness temperature, but
other possibilities were also considered. Alternative explanations included
plasma instabilities in the ionosphere, atmospheric electrical discharges, and
thermal emission from a hot ionosphere (see Barath 1964).

The hot surface model was supported by the apparent thickness of the
atmosphere derived from the difference of some tens of kilometers in the
optical and radar radii. An assumed adiabatic lapse rate extrapolated from the

known temperature at the cloud tops to the surface yielded a temperature compatible with the radio measurements, and a surface pressure of several tens of bars (Sagan 1962).

In order to discriminate between a hot surface and an ionospheric (or atmospheric) origin of the high temperature, a microwave radiometer experiment was proposed (in 1960) for the Mariner spacecraft. The ionospheric theory predicted limb brightening, while limb darkening would occur in the case of a hot surface. Three Mariner 2 disk scans at 13.5 and 19 mm were made at encounter in December 1962, and revealed limb darkening, thus establishing the basic correctness of the hot surface model (Barath et al. 1963).

The new data base from all the Mariner 2 investigations, plus the erroneous groundbased detection of a microwave phase effect (variation of observed temperature with the phase angle of Venus) led to numerous theoretical investigations of the atmosphere and surface chemistry of the planet. Greenhouse models were given a major thrust by Sagan (1961) and Pollack and Sagan (1965b), though we should note that Wildt (1940) two decades earlier had recognized the potential role of CO_2 in heating the planet's atmosphere, in the course of his study of the possibility of photochemically-produced formaldehyde on Venus. Methods for understanding the physics of the atmosphere of Venus were also developing rapidly. The classic paper by Chamberlain (1965) showed, among other things, the importance of scattering geometry in the formation of spectral lines and the radiative properties of the clouds. Other important theoretical and observational work emerged with a spectroscopic study and model of the cloud tops by Belton et al. (1968) and an investigation of CO_2 photochemistry by McElroy (1968). An important discussion of the deep circulation of the atmosphere of Venus was published by Goody and Robinson (1966), and the early work on this topic culminated in the 1970 Fourth Arizona Conference on Planetary Atmospheres in Tucson (Gierasch 1970b).

A major concern in 1969 and the 1970s was the great stability of CO_2 in the photochemical region of the atmosphere. Much of the Fifth Arizona Conference on Planetary Atmospheres in 1971 was devoted to the subject of oxidization of CO, with various mechanisms emerging from considerations of the roles of hydrochloric acid and the unstable ClOO radical. As Hunten (1975b) has noted, these studies of the upper Venus atmosphere have direct application to the terrestrial stratosphere, including the threat to the ozone layer. Studies of the aeronomy of Venus stimulated the work of Wofsy and McElroy (1974) on the stability of terrestrial ozone against destruction by chlorine produced from chlorofluomethanes.

Throughout the 1960s Venus was a topic of major importance in the planetary sciences community. In the context of groundbased observation and planetary probes, it served as a nucleus of common interest to scientists in various fields and brought together experts in astronomy, atmospheres, aeronomy, particles and fields, geochemistry, and planetary interiors. It was a prime subject of the National Academy of Sciences report by Kellogg and Sagan (1961) and for several conferences: the Liège meeting on planetary problems in 1962; the Jet Propulsion Laboratory lunar and planetary conference of 1965; the Tucson planetary conferences of 1968 and 1970; and it figured heavily in the annual meetings of the newly formed (in 1968) Division for Planetary Sciences of the American Astronomical Society.

Certain peculiarities of Venus have emerged over the three centuries of telescopic observations, and one of the most intriguing is the Ashen Light, or the luminescence of the unilluminated hemisphere, best seen near inferior conjunction. Since the seventeenth century, some visual observers have reported occasions when the dark hemisphere can be seen faintly illuminated, usually with a slightly yellowish or coppery hue. W. K. Hartmann and I have seen this simultaneously and independently on an occasion when Venus was a few hours from inferior conjunction; I am convinced the phenomenon is real. A series of spectroscopic observations has been made (e.g. Kozyrev 1954; Newkirk 1959; Weinberg and Newkirk 1961) with no firm conclusion as to the nature of the nocturnal luminescence or its frequency of occurrence, though N_2^+ emission was a possible candidate. The discovery of oxygen emission on the night side by the Venera spacecraft (Krasnopolsky et al. 1976b; Lawrence et al. 1977) suggests that at times this molecular emission might be bright enough to be seen from Earth under especially favorable conditions (D. M. Hunten, personnal communication, 1980), but spectroscopic searches by W. K. Hartmann, D. P. Cruikshank, and D. Landman have so far been without success. Ksanfomaliti (1980b) suggests that the frequent lightning bolts discovered by the Venera 11 and 12 spacecraft (on the illuminated side) may illuminate the dark hemisphere as seen from Earth.

Another phenomenon is the appearance of the extended cusps which join to form a complete circle or halo when Venus is near inferior conjunction. Russell (1899) neatly explained this in terms of scattering in the upper atmosphere of Venus, and gave a means of deriving the height of the scattering layer above the opaque cloud deck from the observed extension of the cusps as they vary with phase angle. A height of a few kilometers is consistent with most of the visual and photographic observations of this phenomenon.

In an interesting study (as yet unpublished), A. B. Binder and I show that the unexpected illumination of the Venus halo at a time when the phase angle was insufficiently large to produce the usual joining of the extended cusps might be explained by the interaction of solar flare particles with the atmosphere of Venus. We discovered this peculiar phenomenon visually in 1959 with the Yerkes 1 m telescope and found a striking correlation with the occurrence of a series of class 3+ solar flares.

In contrast to the vague and diffuse markings on Venus seen by some observers, Lowell and his colleagues (e.g., Lowell 1897) drew thin, sharp features on the planet. These resemble his canals on Mars, Mercury, and the

Galilean satellites and are surely incorrect perceptions of whatever faint markings exist.

It would be hard to establish that particular dusky patterns reported by visual observers have been confirmed by spacecraft imagery. However, it is interesting to note that two of the most prominent and repeatedly reported features of prespacecraft visual observers were bright polar caps (called cusp caps) and a dark band bordering them (called the cusp collar or cusp band); those have been confirmed by the Pioneer images as permanent or semipermanent features of the ultraviolet cloud patterns. They may be dimly visible in the blue and should probably be regarded as discovered by visual telescope observers.

Anomalies in the apparent phase of Venus have been reported since the earliest telescopic observations. The Schroeter effect is the name given to the difference in observed and theoretical phase, or illuminated fraction of the planet. Specifically, dichotomy, or half phase usually appears to occur ~ 7 d early or late, depending on whether the phase is waning (evening apparitions) or waxing (morning apparations). Visual contrast perception (i.e. failure to see the faintest edges of the terminator) is probably the main cause of the Schroeter effect, along with the variability of dark cloud markings near the terminator.

Space-age studies are introduced in the next chapter by Colin and described in detail in the rest of the book. For the reader interested in further tracing the development of studies of Venus, the following additional references may be useful: Flammarion 1897; Sharonov 1965; Jastrow and Rasool 1969; Kuz'min and Marov 1975; *J. Atmospheric Sciences,* Vol. 32, 1975; Moroz 1981.

Acknowledgments. I thank D. Morrison and W. M. Sinton for information on their critical roles in early microwave and infrared studies of Venus, respectively. W. K. Hartmann, L. Colin, M. J. S. Belton, and J. Apt made helpful comments on an early draft of this chapter. I am grateful to the late Gerald P. Kuiper for involving me in a small way in his important work on Venus throughout the 1960s. Work on this chapter was supported in part by a grant from the National Aeronautics and Space Administration.

2. BASIC FACTS ABOUT VENUS

LAWRENCE COLIN
NASA Ames Research Center

Pre-space-age observations of Venus made from the Earth's surface produced a set of basic information about our mysterious sister planet. Since they were of necessity made from Earth, employing a variety of techniques (telescopy, radiometry, thermal mapping, spectroscopy, polarimetry, radar astronomy) and in different regions of the electromagnetic spectrum (visible, infrared, radio) looking at both emitted and reflected radiation, it is not surprising that much of the basic information collected was contradictory and confusing. The situation was not much improved when balloon and aircraft observations permitted the extension of observations to the ultraviolet, and studies in the visible and infrared could be made above the disturbing influences of the Earth's atmosphere. Major improvements in our understanding of Venus began only with the inauguration of the planetary exploration program in 1962 by Mariner 2; our understanding improved rapidly with data from the various Venera and Mariner spacecraft and culminated with the great influx of information from the Pioneer Venus and Venera 11 and 12 missions in 1978. This chapter focuses on generally accepted basic facts about Venus gleaned primarily from pre-space age observations but benefited quantitatively in many cases from a space age retrospective. Its purpose is to set the stage for the non-specialist reader as an introduction to the chapters that follow. Included also is a summary of the major features of the planetary missions already performed and those planned by the US and USSR.

During the month of December 1978, no less than ten separate, scientifically instrumented unmanned spacecraft assaulted the planet Venus. They were the Pioneer Venus orbiter, the component craft of the Pioneer Venus multiprobe: bus, Large (sounder) probe, three small probes (Day, Night, North) of the United States, and the two flyby and two descent lander craft of the Soviet Union Venera 11 and Venera 12 missions. The US and USSR missions were not coordinated, although each nation knew of the other's plans in some detail for many months prior to their respective launches. Not counting the Pioneer Venus orbiter, the scientific missions of each of the nine spacecraft were completed successfully within ~ 90 min of initial planetary encounter. The Pioneer Venus orbiter continues to collect valuable scientific data and should continue to have this capability through 1992 when the spacecraft will be captured and destroyed in Venus's upper atmosphere.

The Pioneer Venus missions represent a major milestone in the United States' program of Venus exploration: the first orbiter and first entry probe mission, all of the previous missions being remote-sounding flybys of the planet. The approach of NASA (National Aeronautics and Space Administration) of implementing a balanced, step-by-step program to exploring both the inner and outer planets and their satellites resulted in a relative neglect of our sister planet until the implementation of Pioneer Venus (Colin 1980). The USSR in contrast chose to concentrate their unmanned, scientific exploration program on the planet Venus, although they also conducted missions to the Moon and Mars. Thus their program consists of a series of small focused missions, the Venera (Venus) series, launched at nearly every opportunity (19-month centers), or every other opportunity, with each launch an improvement scientifically on the preceeding one (Moroz 1981). The Venera 11 and 12 missions were just two more in a series which will continue into the future. In fact, Veneras 13 and 14 both encountered Venus in March 1982; Venera 15 and 16 will be launched in 1984 and encounter Venus in 1985. The earlier pair again consisted of flybys and entry probe landers. The latter craft will deploy entry probe landers at Venus enroute to a Halley Comet encounter in 1986.

The Pioneer Venus and Venera 11, 12 missions provided a major increase in our knowledge about Venus and its gaseous environment. Unfortunately missions like these return ambiguous and even erroneous data along with the unambiguous data and results. Thus, in addition to the key scientific questions that are now answered, there remain some that were not. Many new questions have arisen to be answered by future missions. Early in 1980 it became clear that the time was ripe to take stock of where we are with Venus exploration. Accordingly, an international conference on the Venus environment was planned and held in November 1981. Written versions of the invited review papers at the conference form the major part of this book. (A special issue of the journal *Icarus* is devoted to some of the contributed papers given at this conference.) Soviet authors submitted a series of papers some of which appear in this book as separate chapters; material from others have been incorporated and duly referenced in chapters written by other authors. We are grateful to our Soviet colleagues for their contributions to the text.

For some time US and USSR scientists have attempted to establish cooperative arrangements in space exploration through the US/USSR Joint

Working Group on Near Earth Space, the Moon and Planets. Perhaps the most successful agreements to date involve those between Pioneer Venus and Venera 11, 12 scientists. Four meetings were held before and since planetary encounter in December 1978: Philadelphia, June 1976; Innsbruck, June 1978; Moscow, April 1979; and San Francisco, October 1981. The major accomplishments have been discussions of future plans and exchange of scientific results. One outstanding result is represented by the joint presentation of scientific results at the November 1981 Venus conference and in this book.

Because of the anticipated moratorium in planetary exploration by the US for the next decade or so, and in spite of the continuing Venera program, the editors are confident that the major knowledge about Venus described in this book will remain virtually valid and unchallenged for many years.

I. VENUS IN THE SPACE AGE

The Space Age was inaugurated in 1958 with the passage of the Space Act by the U. S. Congress and the formation of the National Aeronautics and Space Administration. The Act, among other things, called for the systematic exploration of space and the planets of the solar system. It was quickly decided that Venus would be a suitable target for mankind's first interplanetary spacecraft to be launched only four years later in 1962. It is interesting to recall what was known about Venus, or thought to be known, prior to that very first mission (Koenig et al. 1967; Moore 1959; Bronshten 1972). The reader is invited to return to this point upon completing this book to better appreciate the enormous gains in scientific knowledge about Venus gathered in just 20 years of space exploration compared to what was learned in the centuries before.

Man has been curious about our sister planet since antiquity (see Chapter 1). For thousands of years he had only his eyes for plotting the course of the planet as it wandered in the sky. With the advent of the telescope in the 17th century, he at last had a tool that opened new vistas in astronomy for the next 350 years. Two major facts resulted: Venus had a gaseous atmosphere and opaque clouds shrouded the planet completely obscuring the surface. The 1930s witnessed the application of infrared radiometry and spectroscopy to telescopic observations, and the first indication of an atmosphere of mainly carbon dioxide (above and thus by inference below the clouds) at Venus appeared. The birth of radio astronomy and radar astronomy in the mid-1950s permitted scientists to see below the clouds down to the surface for the first time. The picture evolved that the lower atmosphere was hot and dense, and that the surface was topographically smooth, compared to Earth's.

Let us now consider a survey of Venus concerning the available facts and theories existing in 1961, prior to the first spacecraft launch one year later (Lewis 1971). These facts were small in number: Venus was cloud-shrouded with a rich CO_2 atmosphere and the planet emitted a substantial flux at radio wavelengths. The prevailing theories led to qualitative descriptions of Venus which may be gathered into seven broad categories:

1. Moist, swampy, teeming with life, or
2. Warm, enveloped by a global carbonic acid ocean, or
3. Cool, Earth-like, surface water, dense ionosphere, or
4. Warm, massive precipitating clouds of water droplets, intense lightning, or
5. Cold, polar regions with 10-km-thick icecaps, hot equatorial region far above H_2O boiling point, or
6. Hot dusty, dry, windy, global desert, or
7. Extremely hot, cloudy, molten lead and zinc puddles at equator, seas of bromine, butyric acid, phenol at poles.

From this list it is not obvious that scientists were even talking about the same planet in 1961. For those who are impatient for the outcome, speculation (6) appears to represent most closely what we now think Venus is like. With all this knowledge, or lack of it, Mariner 2 was launched in 1962, inaugurating the age of planetary exploration. Facts about that mission and subsequent ones by the US and USSR are given in Table I. No attempt will be given to sort out the incremental gains in knowledge afforded by the individual missions since the subsequent chapters summarize our accumulated knowledge and remaining questions through mid-1982.

Prior to Pioneer Venus in 1978, the US program consisted of three flyby missions: Mariner 2, Mariner 5, Mariner 10 in 1962, 1967, 1973, respectively. Missions were thus launched on 5 or 6 year centers. The major difference between these missions was improved and/or new scientific instrumentation and varied flyby geometry. Pioneer Venus clearly marked a change in US planetary exploration philosophy (Colin 1980). The orbiter permitted repeated remote sounding compared to the flybys, and with the choice of orbit parameters permitted *in situ* sampling of the upper atmosphere, ionosphere and solar wind interaction. The multiprobe mission permitted *in situ* sampling within the clouds and lower atmosphere at widely spaced points in the day and night hemispheres, both in equatorial and polar regions.

The USSR inaugurated its Venus exploration program in 1967 with Venera 4 followed by Veneras 5 and 6; Venera 7; Venera 8; Veneras 9 and 10; Veneras 11 and 12; Veneras 13 and 14 in 1969, 1970, 1972, 1975, 1978, 1982, respectively. In fact, the USSR attempted 10 launches to Venus between 1961 and 1965, all of which were unsuccessful for one reason or another (Sheldon 1971). Three of these, launched on 12 February 1961, 12 November 1965, and 16 November 1965 were actually designated Venera 1, Venera 2, and Venera 3, respectively. Since launch opportunities to Venus occur every 19 months, it can be seen that since 1967 they have never failed to take advantage of at least every other opportunity. The Venera series has, with the exception of Veneras 9 and 10, consisted of flybys and entry probe landers. Venera 9 and Venera 10

TABLE I

Missions To Venus[a]

Spacecraft	Launch	Encounter	Type	Encounter Characteristics
Mariner 2	Aug. 27, 1962	Dec. 14, 1962	flyby	closest approach: 34,833 km
Venera 4	June 12, 1967	Oct. 18, 1967	bus	burn-up
			entry-probe	hard lander, nightside
Mariner 5	June 14, 1967	Oct. 19, 1967	flyby	closest approach: 4100 km
Venera 5	Jan. 5, 1969	May 16, 1969	bus	burn-up
			entry-probe	hard lander, nightside
Venera 6	Jan. 10, 1969	May 17, 1969	bus	burn-up
			entry-probe	hard lander, nightside
Venera 7	Aug. 17, 1970	Dec. 15, 1970	bus	burn-up
			entry-probe	soft lander, nightside
Venera 8	March 27, 1972	July 22, 1972	bus	burn-up
			entry-probe	soft lander, dayside
Mariner 10	Nov. 3, 1973	Feb. 5, 1974	flyby	closest approach: 5700 km
Venera 9	June 8, 1975	Oct. 22, 1975	orbiter	periapsis: 1560 km; apoapsis: 112,200 km; period: 48 hr, 18 min; inclination 34°10′
			entry-probe	soft lander, dayside
Venera 10	June 14, 1975	Oct. 25, 1975	orbiter	periapsis: 1620 km; apoapsis: 113,900 km; period: 49 hr, 23 min; inclination: 29°30′
			entry-probe	soft lander, dayside
Pioneer Venus 1	May 20, 1978	Dec. 4, 1978	orbiter	periapsis: < 200 km; apoapsis: 66,000 km; period: 24 hr; inclination: 105°
Pioneer Venus 2	Aug. 8, 1978	Dec. 9, 1978	bus	burn-up, dayside
			entry-probes	4 hard landers, dayside and nightside
Venera 11	Sept. 9, 1978	Dec. 25 1978	flyby	closest approach: 25,000 km
			entry-probe	soft lander, dayside
Venera 12	Sept. 14, 1978	Dec. 21 1978	flyby	closest approach: 25,000 km
			entry-probe	soft lander, dayside
Venera 13	Oct. 30, 1981	March 1, 1982	flyby	closest approach: unknown
			entry-probe	soft lander
Venera 14	Nov. 4, 1981	March 5, 1982	flyby	closest approach: unknown
			entry-probe	soft lander
Venera 15	1984	1985	flyby (Halley)	closest approach: unknown
			entry-probe	soft lander
Venera 16	1984	1985	flyby (Halley)	closest approach: unknown
			entry-probe	soft lander

[a]Table after Colin (1980).

had orbiters instead of flybys. Each of the Venera launches have produced modestly improved or new scientific instrumentation, modest changes in science emphasis or direction, and varied entry probe lander locations, in the day or night hemispheres.

Table II lists the impact locations of all the Venera and Pioneer Venus probe landers. The latitude and longitude coordinates are body-fixed given by the standard IAU 1970 convention. This is the same coordinate system used for the earth-based and Pioneer Venus orbiter radar mapper images and topographical maps shown in Chapter 6. Table II also lists the Venus local time and solar zenith angle of each impact point.

II. FACTS ABOUT VENUS

The important scientific data and knowledge of Venus gleaned from space exploration concerns the solid body, its shape and dimensions, the intrinsic magnetic field, the surface topography and composition, the atmospheric composition and structure, the cloud structure, particle size distribution and optical properties, the atmospheric dynamics, the ionospheric

TABLE II
Descent Probe Location[a]

Probe	Day	Latitude (deg)	Longitude (deg)	Local Solar Time (hr: min)	Solar Zenith Angle (deg)
Venera 4	Oct. 18, 1967	19	38	4:40	110
Venera 5	May 16, 1969	− 3	18	4:12	117
Venera 6	May 17, 1969	− 5	23	4:18	115
Venera 7	Dec. 15, 1970	− 5	351	4:42	117
Venera 8	July 22, 1972	−10	335	6:24	85
Venera 9	Oct. 22, 1975	32	291	13:12	36
Venera 10	Oct. 25, 1975	16	291	13:42	28
Venera 11	Dec. 25, 1978	−14	299	11:10	17
Venera 12	Dec. 21, 1978	− 7	294	11:16	20
Venera 13	March 1, 1982	− 7.5	303	9:27	38
Venera 14	March 5, 1982	−13.4	310.2	9:54	33
Pioneer Venus:					
Large Probe	Dec. 9, 1978	4.4	304.0	7:38	65.7
North Probe	Dec. 9, 1978	59.3	4.8	3:35	108.0
Day Probe	Dec. 9, 1978	−31.7	317.0	6:46	79.9
Night Probe	Dec. 9, 1978	−28.7	56.7	0:07	150.7
Bus	Dec. 9, 1978	−37.9	290.9	8:30	60.7

[a]Table after Moroz (1981).

structure and composition and the geometry of the solar wind interaction with
the ionosphere. These matters are the subjects of chapters 6 to 30 of this book.
There is a body of other facts about Venus which the serious student should
know and understand. These include facts about Venus as a planetary member
of the solar system, its orbital characteristics and axial rotation dynamics.
Also since mankind has and continues to view Venus from Earth, the facts
supporting these observations are of paramount interest. The remainder of this
chapter will be devoted to presenting this information. Some of these facts
have indeed resulted from space exploration, or at least the quantitative data
have been improved and refined. Since, however, the bulk of the data are
based on pre-space age observations, no attempt will be made to establish a
chronology or to differentiate sources.

Viewed from outside the solar system, Venus, along with Mercury, Earth
and Mars, are designated as the small inner, terrestrial, rocky planets (Abell
1969). These designations are used to distinguish them from the large outer
planets of the solar system that are by-and-large gaseous in nature, lacking a
solid surface: Jupiter, Saturn, Uranus and Neptune. All of these planets revolve in elliptical orbits in the same direction about the Sun, the latter located
commonly at one of the foci of each orbit. Mercury is closest to the Sun,
followed by Venus, Earth and Mars. Each orbit (with the exception of
Neptune-Pluto) is contained within that of its nearest neighbor i.e., the el-
lipses projected onto the ecliptic plane do not cross. Table III summarizes the
mean orbital features of the inner planets. Note that all of the inner planets
have small eccentricities so that their orbits are nearly circular. Venus, having
the smallest eccentricity (0.0068) is the most circular. In fact, the distance
of Venus from the Sun varies only from 0.718 AU (107.4 $\times$ 10^6 km) at peri-
helion to 0.728 AU (108.9 $\times$ 10^6 km) at aphelion and Venus's orbital velocities are 35.2 km s^{-1} and 34.8 km s^{-1} at those points. Note also that all
of the planets have orbits only slightly inclined from the ecliptic plane. Thus
for most practical purposes the nearly circular orbits of the inner planets
are nearly coplanar.

Let us now examine the physical characteristics of each of the inner
planets. Table IV summarizes the most important elements. Comparison of
mass, size, and density shows that Venus is the most Earth-like of the terrestrial planets, hence the designation of twin or sister planet; Venus being our
closest neighbor in the solar system (see Table I) strengthens this association.
Although Table IV indicates zero oblateness for Venus (implying a perfect
sphere) earth-based radar data indicate some measurable, yet small effects:
equatorial ellipticity of figure 1.1 $\pm$ 0.4 km, equatorial center-of-mass/
center-of-figure offset 1.5 $\pm$ 0.3 km (offset direction toward Earth at inferior
conjunction), moment of inertia difference (3 $\pm$ 3) $\times$ 10^{-4}. On the other
hand, Mars and Earth have strikingly similar rotational characteristics and are
quite dissimilar from Venus as can be seen by comparing sidereal rotation
periods and obliquity (spin axis orientation). In fact, Venus is the renegade of

TABLE III
Mean Orbital Elements of the Inner Planets[a]

	Mean Distance From Sun (AU)[b]	(10⁶ km)	Sidereal Period Orbital Period (d)	Sidereal Period (Tropical yr)	Mean Orbital Velocity (km s⁻¹)	Eccentricity	Inclination to Ecliptic (deg)
Mercury	0.387	57.9	87.969	0.24085	47.89	0.2056	7.00
Venus	0.723332	108.2	224.701	0.61521	35.05	0.006787	3.39
Earth	1.000	149.6	365.256	1.00004	29.79	0.0167	—
Mars	1.524	227.9	686.980	1.88089	24.13	0.0934	1.85

[a]Table after Beatty et al. (1981) and Hunt and Moore (1981).

[b]1 AU = 149,597,870 km

TABLE IV
Physical Elements of the Inner Planets[a]

	Mercury	Venus	Earth	Mars
Reciprocal Mass[b]	5,972,000	408,523.9	328,900	3,098,710
Mass[c] (Earth = 1)	0.0559	0.81503	1.0000	0.1074
Mass[c] (kg)	3.303×10^{23}	4.871×10^{24}	5.976×10^{24}	6.421×10^{23}
Equatorial Radius (km)	2.439	6051.3	6378	3398
Equatorial Radius (Earth = 1)	0.382	0.949	1.000	0.532
Ellipticity (Oblateness)[d]	0.0	0.0000	0.0034	0.0059
Mean Density (g/cm³)	5.42	5.24	5.52	3.94
Equatorial Surface Gravity (m s⁻²)	3.78	8.87	9.78	3.72
Equatorial Escape Velocity (km s⁻¹)	4.3	10.4	11.2	5.0
Sidereal Rotation Period	58.65 d	243.01d[e]	23.9345 hr	24.6229 hr
Obliquity, Inclination of Equator to Orbit	0°	−2°.6	23°.45	23°.98

[a]Table after Beatty et al. (1981) and Hunt and Moore (1981).

[b]Mass of Sun divided by mass of planet (including atmosphere and satellites).

[c]Satellite masses not included.

[d]$(Re-Rp)/Rp$, where Re and Rp are equatorial and polar radii, respectively.

[e]Retrograde.

the solar system with its retrograde rotational motion and very small spin
period as well as minuscule, if any, magnetic field, and no satellites. The best
estimate of the Venus spin-axis direction is given by the right ascension (RA
1950.0) of 273°3, declination (Dec. 1950.0) or 67°3 of the planetary north
pole (Ward and DeCampli 1979).

Venus's sidereal period of revolution about the Sun, 224.7 d, and its
sidereal rotation period about its axis, 243 d retrograde, combine to produce
a peculiar solar day on Venus. Figure 1 demonstrates that a Venus day is
equivalent to 116.75 Earth days. In other words a venusian would see
(through the clouds with his radio eyes) the Sun rise in the west, experience
58.37 Earth days of sunshine, watch the Sun set in the east and then witness
58.37 Earth days of night before the Sun rises again. Note also from Fig. 1
that a similar calculation for the Earth produces only about 4 min change in
24 hr for a day on Earth.

Now let us view Venus from the vantage point of the ancients, amateur
and professional astronomers, i.e., from Earth. Venus is the brightest of the
five naked-eye planets: Mercury, Venus, Mars, Jupiter and Saturn. (These 5

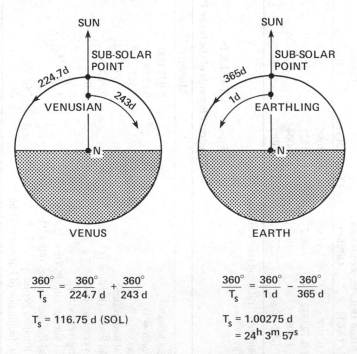

$$\frac{360°}{T_s} = \frac{360°}{224.7\,d} + \frac{360°}{243\,d}$$

$$T_s = 116.75\,d\ (SOL)$$

$$\frac{360°}{T_s} = \frac{360°}{1\,d} - \frac{360°}{365\,d}$$

$$T_s = 1.00275\,d$$
$$= 24^h\,3^m\,57^s$$

Fig. 1. Solar days on Venus and Earth. Venus revolves about the Sun in 224.7 Earth days and
rotates in a retrograde manner about its axis in 243 Earth days. These motions combine to
produce a Venus day (time between sunrises) of 116.75 Earth days (left). In contrast, the
rotation period of Earth (24 hr) is so short compared with the revolution period (365 days)
that a day is only 4 minutes longer than the rotation period (right).

planets along with the Sun and Moon were used to name the days of the week.) Venus is the most brilliant object in the sky, with the exception of the Sun and Moon; it is no wonder that it has probably been known almost from the dawn of history. In fact, a careful observer can see Venus in broad daylight and on a clear night it casts faint shadows.

The Sun and other stars, of course, generate their own light, but the planets are visible to us only because of sunlight reflected back to Earth. Venus's brilliance is due not so much to its size, but to its relative proximity to us and to its efficient reflection of visible solar radiation. The solar flux at Venus is 2.6×10^6 erg cm^{-2} s^{-1} (2600 W m^{-2}), about a factor of 2 $[1/(0.723)^2]$ greater than the solar flux at Earth (1.4×10^6 erg cm^{-2} s^{-1}). The most useful quantity describing reflection efficiency is the Bond albedo A (see Glossary for definition). The albedos of several solar system objects are listed in Table V. The relatively high value for Venus is due to its ubiquitous cloud cover, the material of which is highly reflective. The fraction of solar radiation absorbed by a planet and ultimately available to heat its atmosphere and surface and drive its winds is given by $1-A$. Venus absorbs rather weakly in the blue region of the spectrum, and strongly in the ultraviolet yielding a weak yellowish tint to its observed reflected light. There is also unconfirmed evidence for a weak Venus intrinsic glow (emitted light) in the visible from the dark hemisphere which has been called the Ashen Light.

Planetary scientists generally ascribe an average effective temperature T_e to the solar system bodies based on standard thermodynamic calculations. T_e is related to the Bond albedo by the equation of balance. For Venus $T_e = 231 \pm 7$ K. The effective temperature is related to the kinetic temperature at some level near Venus's cloud tops, but it took subsequent *in situ* spacecraft measurements to determine that level; atmospheric temperatures increase rapidly below this level in the clouds. At the surface the measured temperature and pressure is 735 ± 3 K and 90 ± 2 bar. Electromagnetic radiation from Venus (as from many of the other solar system bodies) is detectable from the ultraviolet through the radio. These observations sense different regions of the Venus atmosphere, from the exosphere to the surface. Visible and infrared observations are reviewed in Chapter 3 and the radio observations in Chapter 4.

The most useful quantity describing the apparent brightness of any celestial object (in either direct or reflected light) is the visual or stellar magnitude, an inverse logarithmic measure such that an increase of five magnitudes represents a hundred-fold decrease in the body's brightness. In other words, a difference in magnitude of 1 is equivalent to a brightness difference of 2.512 (since $2.512^5 = 100$). Some magnitudes are also listed in Table V. Note that the more negative (smaller) the magnitude is, the brighter the object. The objects are listed in decreasing level of brightness. The brightest stars have magnitudes of -1, 0, and $+1$, some of which are shown in Table V. There are ~ 20 stars in the brightness range -1.45 (Sirius, the brightest) to 1.26

TABLE V

Bond Albedo and Stellar Magnitude

Object	Bond Albedo	Stellar Magnitude
Sun	—	−26.91
Moon	0.06	−12.73 (max-full)
Venus	0.75	− 4.4 (max)
Earth	0.36	− 3.8
Mars	0.15	− 2.02
Jupiter	0.43	− 2.55
Sirius (1)[a]	—	− 1.46
Mercury	0.06	− 1.2 (max)
Canopus (2)	—	− 0.73
Arcturus (3)	—	− 0.06
Rigil Kentaurus (4)	—	− 0.01
Vega (5)	—	0.04
Saturn	0.61	0.7
Acrux (14)	—	0.9
Spica (15)	—	0.96
Antares (16)	—	1.0
Pollux (17)	—	1.15
Uranus	0.35	5.5
Neptune	0.35	7.8
Pluto	0.5	14.9

[a]The numbers in parentheses indicate a descending list of visual star brightness. In other words, Spica is the 15th brightest star in the sky.

(Mimosa); the faintest naked-eye stars are of 5th or 6th magnitude. Even the most casual stargazer realizes that there are thousands of stars with magnitudes equal to or greater than 6th. The telescope, of course, permits even fainter stars to be observed. The faintest star visible to the largest optical telescope on Earth at the Hale Observatories on Mount Palomar, is ~ 23rd magnitude, which is nearly 16×10^6 times fainter than the faintest stars easily seen with the naked eye.

Planets whose orbits lie closer to the Sun than Earth are termed inferior; those outside our orbit are designated superior (superior and inferior are derived from Latin words meaning higher and lower). The inferior planets, Mercury and Venus (and the Moon) exhibit phases to an observer on Earth (as seen through a telescope). Since the Sun illuminates only one half of every planet at a given time, as the Sun-planet-Earth geometry changes during a synodic year, various fractions of the sunlit hemisphere are visible at Earth. For a superior planet almost the entire illuminated disk is continually presented to us. Figure 2 is a sketch illustrating the phases to an observer fixed at Earth's orbit; it also demonstrates how the apparent angular size of Venus varies

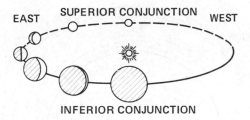

EAST   SUPERIOR CONJUNCTION   WEST

INFERIOR CONJUNCTION

Fig. 2. Changing apparent size and phase of Venus; a sketch of Venus observed from Earth at various phases. Near inferior conjunction, Venus is closest to Earth, and so is largest, but darkest. At inferior conjunction it is smallest but fully illuminated. Venus is brightest at east and west elongations, at an intermediate apparent size. (Figure after Baker 1958.)

(Baker 1958). Note that at the closest approach to Earth (called inferior conjunction), the planet appears largest but is dark since the illuminated disk is behind the planet (equivalent to the new Moon). At the farthest distance from Earth (superior conjunction), Venus is bright since we see the entire sunlit disk (equivalent to full Moon) but not its brightest because of its small size. Locations either side of inferior conjunction, termed elongation, present varying phase fractions and intermediate sizes to the observer on Earth. The greatest angular distance from the Sun achieved by Venus is ∼ 48°; this location is named maximum elongation. Maximum brightness (−4.4 stellar magnitude) occurs between maximum elongation and inferior conjunction. (Mercury's greatest elongation is 28° so it never is far from the Sun. For this reason and because its stellar magnitude is only −1.2 which is fainter than Sirius, Mercury is not easy to observe.)

This simplified picture is complicated by three factors: Venus's and Earth's relative motion, their non-circular, non-coplanar elliptical paths about the Sun, and the fact that the terrestrial observer is on a rotating planet. Figure 3 can be used to appreciate the first two factors. As a starting point, begin at the inferior conjunction marked Day 1. Following this each planet revolves about the Sun in a counter-clockwise direction (looking down from the north pole of the Sun) along their (idealized, circular) orbits. After 224.7 days Venus returns to its original position (sidereal year) but Earth has moved on completing ∼ 62% of its orbit (227.4/365.24) about the Sun. Venus continues its path gaining on Earth until they again meet at the next inferior conjunction. By performing a calculation similar to that given in Fig. 1, we can show that the time between conjunctions is 583.92 Earth days (1.6 yr or 19 mon). This time interval between inferior conjunctions is called the synodic period. Note that there are therefore five inferior conjunctions at Venus, spaced at equal intervals, every eight years (actually 8 yr−2 d). It is interesting to note in passing that Venus's axial rotation rate is almost, but not exactly, locked (synodic resonance implies a 243.16 d period; the actual period is

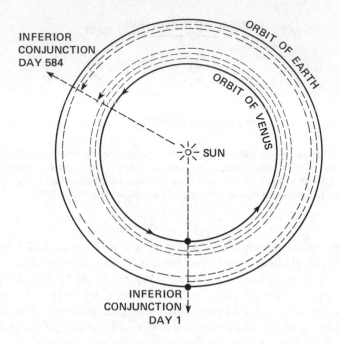

Fig. 3. Synodic period of Venus. Venus and Earth are closest at inferior conjunction. Starting at this point, Venus revolves about the Sun faster (period 224.7 d) than does Earth (period 365.25 d), taking 583.92 d (synodic period) to overtake Earth again.

243.0 ± 0.1 d) to its solar revolution rate so that at each inferior conjunction it presents the same face to Earth (Ward and DeCampli 1979).

However, the orbits of Venus and particularly Earth are not precisely circular but elliptical. Also, their orbital planes are tilted slightly with respect to one another (see Table III). Thus the geometry at each inferior conjunction, e.g. the Venus-Earth distance, differs each time. Precise relationships are repeated every 8 yr, or 5 synodic periods.

Let us digress here for a moment and return to the subject of planetary exploration launch opportunities. The optimum times to launch spacecraft from Earth to Venus are near the times of inferior conjunction when distances between the planets are minimum and the operation requires a minimum expenditure of energy (hence fuel) for launch. At minimum distance there is less time to wait and less time for things to go wrong between launch and encounter, and most important, communication distances are shortest, permitting the maximum data rate. Thus, the periods near inferior conjunction are termed launch opportunity periods. For Venus, they occur every 19 months. Table VI lists the dates of inferior conjunctions from 1959 to 1990 (Koenig et al. 1967). Listed also are all the US and USSR launch attempts to Venus, both accomplished and expected, to date. The Soviet plan for taking advantage of

TABLE VI

Space Age Inferior Conjunctions[a]

Date	Spacecraft launch
1. Sep. 1, 1959	
2. Apr. 10, 1961	Venera 1
3. Nov. 12, 1962	Mariner 2
4. Jun. 19, 1964	
5. Jan. 26, 1966	Venera 2, 3
6. Aug. 29, 1967	Venera 4, Mariner 5
7. Apr. 8, 1969	Venera 5, Venera 6
8. Nov. 10, 1970	Venera 7
9. Jun. 17, 1972	Venera 8
10. Jan. 23, 1974	Mariner 10
11. Aug. 27 1975	Venera 9, 10
12. Apr. 6, 1977	
13. Nov. 7, 1978	Pioneer Venus, Venera 11, 12
14. Jun. 15, 1980	
15. Jan. 21, 1982	Venera 13, 14
16. Aug. 25, 1983	
17. Apr. 3, 1985	Venera 15, 16
18. Nov. 5, 1986	
19. Jun. 12, 1988	
20. Jan. 18, 1990	

[a]Table after Koenig et al. (1967).

every launch opportunity in the mid-sixties and early-seventies and every second opportunity thereafter is apparent from this table. Comparison of the inferior conjunction dates in Tables VI with the launch/encounter dates in Table I shows that the latter always bracket the former. Such a compromise optimizes the launch opportunity benefits.

Let us return now to the Earth-bound observer. Figure 4 more clearly illustrates the Earth-Venus geometry and the development of the intermediate phases, between a "new" Venus and a "full" Venus. In this diagram Earth is allowed to remain fixed at its inferior conjunction position while Venus is supposed to revolve alone around the Sun (idealized, circular, coplanar orbits are used for simplicity). Table VII lists some key dimensions concerning this varying geometry. The values are typical and do not account for the variations due to ellipticity, inclination, etc. Note that Venus's apparent diameter in angular dimension shown in Table VII corresponds to a radius of 6121 km (cloud-top radius), some 70 km greater than the surface radius of 6051 km shown in Table IV. Visual observations from Earth do not, of course, see the latter. Starting at inferior conjunction in Fig. 4 and proceeding counter-clockwise around the Sun it takes Venus $\sim$ 36 d to reach the position of

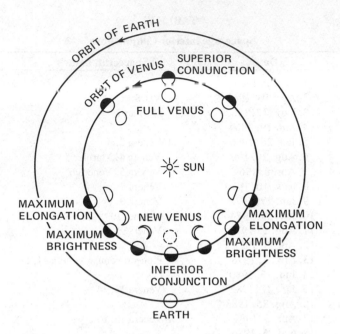

Fig. 4. Venus synodic period and phases. Concepts discussed with respect to Figs. 2 and 3 are combined to illustrate features observable at inferior and superior conjunction, and at the positions of maximum elongation and maximum brightness.

maximum brightness and another 36 days to reach maximum elongation. Venus is west of the Sun during this period and to the observer on a rotating Earth she appears as a morning star above the horizon before sunrise. Two hundred and twenty days elapse between maximum elongation and superior conjunction completing 292 d or one half the synodic period. The morning star disappears before superior conjunction in the brightness of the Sun. The 292-day period between superior conjunction and inferior conjunction repeats: 220 d to maximum elongation; 36 more days to maximum brightness and a final 36 d to inferior conjunction. During this half-synodic period, though, Venus emerges from the solar illumination as an evening star appearing above the horizon after sunset. A typical cycle then is as follows: 8-d period around inferior conjunction when Venus is not visible (Venus close to bright Sun, and new or dark), 263 d as a morning star, Venus again not visible for ~ 50 d around superior conjunction (Venus close to bright Sun), 263 d as an evening star. Since all this happens in 584 d, there is obviously no connection with the Earth's seasons, and events occur at different months year-to-year. For example, in 1982 Venus was visible in the evening sky briefly at the beginning of the year and then hovered in the morning sky from late January (after its 8-day disappearance) through October reaching its maximum brightness on

TABLE VII

Earth-Venus Relationships[a]

REVOLUTION TIMES	
Inferior conjunction to maximum brightness	36 d
Maximum brightness to maximum elongation	36 d
Maximum elongation to superior conjunction	220 d
Synodic period	583.92 d
VENUS-EARTH DISTANCE	
Inferior conjunction	41.9×10^6 km (0.28 AU)
Superior conjunction	257.3×10^6 km (1.72 AU)
ANGULAR DIRECTION FROM SUN	
Maximum elongation west (morning star)	48°W
Maximum elongation east (evening star)	48°E
VENUS APPARENT DIAMETER	
Inferior conjunction (new, dark)	64.5 arcsec
Maximum brilliance (large crescent)	39.0 arcsec
Maximum elongation (half)	37.3 arcsec
Superior conjunction (full, bright)	9.9 arcsec
VENUS APPARENT AREA	
Superior conjunction	77 arcsec^2
Inferior conjunction	3268 arcsec^2
APPARENT AREA OF SUNLIT SURFACE	
Superior conjunction	77 arcsec^2
Maximum brillance	188 arcsec^2
VISUAL MAGNITUDE	
Maximum elongation/brillance	−4.4
Superior conjunction	−3.5

[a]Table after Moroz (1981).

February 25. Venus returned to the evening sky for the last two weeks of the year, after 50 d disappearance (Murphy 1982).

Figure 5 is a sketch depicting how Venus moves in the western sky at sunset over several months (Aveni 1979). Never more than 48° from the Sun when the planet enters the picture at the left, Venus vanishes from view north of the sunset position at the point on the horizon indicated by the arrow. Venus is visible for ~ 3 hr after sunset at this point. Similarly as a morning star she is visible for a maximum of 3 hr before sunrise. Dotted lines connect the Sun and Venus at monthly intervals.

In conclusion, in our brief review of basic venusian facts we have discussed topics which were basically known and understood (some causes are still obscure, e.g., the slow, retrograde spin) before the space age. The follow-

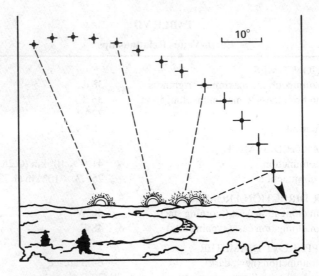

Fig. 5. The Evening Star. Another view of Venus as seen by an Earth observer in the western sky near sunset. Venus is shown at weekly intervals. (Figure after Aveni 1979.)

ing chapters discuss and summarize our current knowledge of other aspects of Venus: its surface, atmosphere, clouds, and solar wind interaction effects. Our current understanding of these regions are enormously improved from just a few years ago; however many enigmas and questions remain to be answered by a new generation of spacecraft experiments and experimenters.

3. STELLAR MAGNITUDE AND ALBEDO DATA OF VENUS

V. I. MOROZ

USSR Academy of Sciences

Magnitude and albedo data are summarized and tabulated. An analysis is given of the wavelength dependence, and optical constants are derived. Sulfuric acid cloud particles explain the data very well at red and longer wavelengths, but additional absorbers are needed for the blue and ultraviolet.

Irvine (1968) has made photoelectric measurements of the stellar magnitude of Venus in nine narrow ($\sim$ 100 Å) intervals of wavelength from 0.3 to 1.0 μm under all accessible phase angles and has determined for these intervals the phase functions, geometrical albedo, and spherical albedo. Later Barker et al. (1975) measured the relative distribution of energy in the spectrum of Venus, using a grating spectrometer. These observations covered phase angles from 40° to 100°. Figure 1 shows the curve of Barker, the results of Irvine for several phase angles, and also a series of other observations which are less systematic in nature. In the latter case the vertical scale was chosen so that the data were correlated with the Baker curve, while the measurements of Barker and Irvine were normalized with respect to the maximum.

It can be seen from Fig. 1 that there are considerable discrepancies between the results of Barker and Irvine. The Irvine data indicate a decrease in the reflectivity from $\lambda > 0.6$ μm and a step at $\sim$ 0.5 μm. These details are absent on the Barker curve. The procedure used by Barker is more reliable for measuring the relative energy distribution, even though the work of Irvine remains the single source of data for the phase functions in narrow spectral intervals. There are also some measurements of the phase function of the

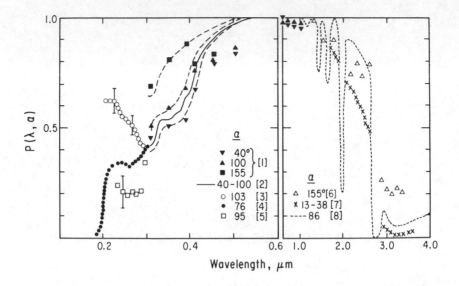

Fig. 1. The monochromatic albedo of Venus as a function of wavelength. The relative geometric albedo is given according to the data of various observers: [1] Irvine 1968; [2] Barker et al. 1975; [3] Anderson et al. 1969; [4] Wallace et al. 1969; [5] Evans 1966; [6] Sinton 1963*b*; [7] Moroz 1964; [8] Pollack et al. 1978. α is the phase angle. The dashes indicate curves of $C(\lambda)$, obtained by extrapolation of the data of Barker et al. (1975) to phase angles of 40°, 100° and 155° and the curve for the phase angle of 70° given by Barker et al. (1975) has been taken directly.

stellar magnitude in broad spectral intervals, including visual photometery, as well as photoelectric photometry in *UBV*.

This chapter attempts to represent the spherical (or Bond) albedo of the planet as a function of the wavelength in a broad range from 0.2 to 4 μm, starting with the following assumptions:

1. In the interval from 0.3 to 1.2 μm the relative dependence of the geometric albedo on the wavelength corresponds to the data of Barker et al. (1975).
2. The phase functions and phase integrals in the range of 0.3 to 1 μm correspond to the data of Irvine (1968).
3. Outside the interval 0.1 to 1 μm, the phase integrals behave as a certain empirical function of the size of the geometric albedo; this dependence can be found from the data of Irvine (1968), cf. below.

4. For the absolute calibration, we use a value of the visual spherical albedo, calculated from several determinations of the stellar visual magnitude V and phase function.

In the regions $\lambda < 0.3$ μm and $\lambda > 1.2$ μm there are few observations and data are only available for a few phase angles; nevertheless, an extrapolation is also possible for these regions under assumption (3) above.

We start by summarizing the formulas and definitions. The monochromatic geometric albedo at phase α is found from

$$\log p \, (\lambda, \, \alpha) = 0.4[m_\odot \, (\lambda) - m_\varphi(\lambda, \, \alpha)] - 2 \log \sigma \tag{1}$$

where $m_\odot$ is the magnitude of the Sun, m_φ has been adjusted to 1 AU Sun-Venus and Venus-Earth distances, and $\sigma \simeq \sin \sigma$ is the angular radius (in radians) of Venus at the same distance (de Vaucouleurs 1964). All logarithms are to base 10. Many writers reserve the term geometric albedo for the zero-phase value $p(\lambda,0)$; in the visual band

$$\log p = \log p \, (V, 0) = 0.4(V_\odot - V_\varphi) - 2 \log \sigma \, . \tag{2}$$

Conceptually, p is the ratio of the adjusted actual brightness to that of a Lambert disk of the same radius facing the Sun and viewed at zero phase. The phase integral is

$$q = 2 \int_0^\pi \frac{p \, (\alpha)}{p \, (0)} \sin \alpha \, d\alpha \tag{3}$$

and the spherical albedo is

$$A = p \, (0) \, q \, . \tag{4}$$

Compensating errors in p and q exist because of the unobservable region of phase near zero; this region occupies only a small solid angle and contains a very small part of the scattered radiation.

Table I summarizes the available results for the visual band (represented as 0.55 μm): the stellar magnitude $V(1.0)$ (adjusted to 1.0 AU), geometric albedo $p(V,0)$, phase integral $q(V)$, and spherical albedo $A(0.55)$ according to the results of four independent series of observations, along with the means and the root mean square error for the mean. The error in A is significantly less than that of the other three quantities for the reason mentioned above. For the stellar magnitude of the Sun, we used the value $V_\odot = -26.77$ mag, the mean of the measurements of Nikonova (1949), Kazjagina (1955), Stebbins and Kron (1957), and Gallemet (1964). To these was added the value $V_\odot = -26.79$ mag, found from absolute spectrophotometry of the Sun (Makarova

TABLE I

Stellar Magnitude V and Determination of $p(V, 0)$, $q(v)$, and $A(0.55)$

Author	Stellar Magnitude in Author's System	$V(1,0)$	$pV(0)$	$q(V)$	$A(0.55)$
Mueller (1893)	−4.16[a]	−4.50	0.736	1.194	0.878
Danjon (1949)	−4.27[b]	−4.33	0.629	1.296	0.815
Knuckles et al. (1961)	−4.76	−4.76	0.935	0.888	0.830
Irvine (1968)	−4.46	−4.46	0.709	1.149	0.819
Mean values		−4.51	0.752	1.132	0.836
rms error					
mean values		±0.09	±0.065	±0.087	±0.017

[a]System PD
[b]System HR

and Khazitonov 1972) and the stars (Kharitonov et al. 1978). The error of the mean value is $\Delta V_\circ = \pm 0.02$ mag, and therefore the accuracy of the absolute magnitude of A is somewhat less than that indicated in Table I. Moreover, in the value of A there will be introduced a slight correction after comparing the integrated spherical albedo with the effective temperature (cf. below).

Let us now consider the dependence of A (λ). We shall introduce the color function

$$C(\lambda, \alpha) = p(\lambda, \alpha)/p(0.55, \alpha) . \tag{5}$$

The monochromatic spherical albedo is

$$A(\lambda) = A(0.55) C(\lambda,\alpha) F(\lambda,\alpha) \tag{6}$$

where

$$F(\alpha,\lambda) = \frac{p(0.55, \alpha)}{p(0.55, 0)} \quad \frac{p(\lambda, 0)}{p(\lambda, \alpha)} \quad \frac{q(\lambda)}{q(0.55)} . \tag{7}$$

The general nature of the dependence of $C(\lambda, \alpha)$ on α is demonstrated by the data shown in Fig. 1. In the spectral segments with a more intense absorption, $C(\lambda, \alpha)$ rises more quickly with increase in α, than in regions with little or no absorption.

Comparing the data of Irvine (1968) and Barker et al. (1975), we may find an empirical dependence

$$F(C . \alpha) = \mathcal{A} + \mathcal{B} C(\alpha) \tag{8}$$

and, in Eq. (6), replace $F(\lambda, \alpha)$ by $F(C, \alpha)$.

Table II shows the coefficients $\mathcal{A}$ and $\mathcal{B}$, found in this manner for four phase angles, 40°, 70°, 100°, and 155°.

Figure 2 shows the dependence of $A(\alpha)$ in the range 0.2 to 4.0 μm, which was found by the above-described method from measurements, the results of which are given in Fig. 1. The solid line in the range 0.3 to 1.0 μm is the curve of $A(\alpha)$, based on the function $C(\alpha)$, obtained in the measurements of Barker et al. (1975); outside this region the result is the average of all the measurements shown in Figs. 1 and 2.

The integrated albedo is

$$A_I = \int_0^\infty A(\lambda) I_\odot(\lambda) d\lambda / \int_0^\infty I_\odot(\lambda) d\lambda = 0.74 \tag{9}$$

TABLE II

Coefficients $\mathcal{A}$ and $\mathcal{B}$ in Eq. (8)

α	40°	70°	100°	155°
$\mathcal{A}$	1.083	0.975	0.777	0.082
$\mathcal{B}$	−0.070	0.040	0.228	0.799

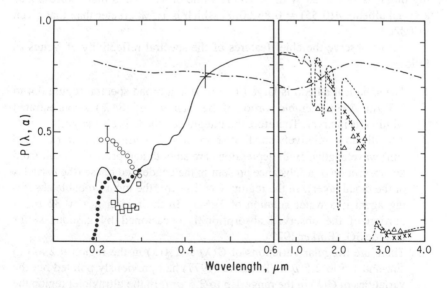

Fig. 2. The spherical albedo as a function of wavelength, calculated from the data of Fig. 1. Designations are the same as for Fig. 1, except the solid line beyond the limits of $\lambda < 0.3$ μm and $\lambda > 1.2$ μm, where this represents the mean of the data of different observers. The interrupted line is the asymmetry factor of the phase function for the upper portion of the cloud layer.

where $I_\odot(\lambda)$ is the energy distribution in the spectrum of the Sun. The uncertainty of the value is $\Delta A_I \sim 0.04$.

The integrated spherical albedo of the planet and its mean effective temperature T_e are related by the thermal balance equation:

$$4 \sigma T_e^4 = E_0 (1 - A_I) \tag{10}$$

where $E_0 = 2600$ W m^{-2} is the solar constant at the orbit of Venus. Thus, the measurement of the emitted infrared radiation theoretically enables the determination of A_I in an independent manner.

However, the determination of T_e is complicated by the fact that it is necessary to allow for the intensity of the thermal radiation as a function of the angle, latitude, and time of day. Using the data of the infrared radiometer of Venera 9 and 10 orbiters, Ksanfomality (1977) has found $A_I = 0.79^{+0.02}_{-0.01}$, which corresponds to $T_e = 222^{+20}_{-6}$°K. According to the data of the infrared radiometer on board the Pioneer Venus orbiter, F. Taylor et al. (1980) obtained a lower value of $A_I = 0.76^{+0.02}_{-0.03}$, for which $T_e = 228 \pm 5$ K. The results of Taylor et al. (1980) are similar to those obtained in this chapter. Another discussion is found in Chapter 20 by Taylor et al.

As a compromise, we may take the self-consistent pair of values: $A_I = 0.75 \pm 0.03$ and $T_e = 231 \pm 7$ K. The corresponding value of the mean outgoing flux F is 162 ± 20 W m^{-2}. To the value of $A_I = 0.75$ there corresponds a visual albedo: $A(0.55) = 0.85 \pm 0.03$ which is 1.5% greater than that given in Table I.

Let us observe the chief features of the spectral reflectivity of Venus as follows:

1. The spherical albedo is large ($A > 0.6$) in a broad spectral region 0.4 to 2.5 μm. A considerable portion of the solar energy (84%) is concentrated within this interval. Therefore the integrated albedo is also large.
2. The albedo is relatively small at short ($\lambda < 0.4$ μm) and long ($\lambda > 2.5$ μm) wavelengths. Both depressions are smooth and probably result from the absorption of a substance present in the condensed phase (the particles of the cloud layer). In the region $\lambda > 2.5$ μm, the most probable absorbing agent is a water solution of H_2SO_4. In the region $\lambda < 0.32$ μm a portion of the observed absorption is occasioned by gaseous sulfur dioxide SO_2 (Barker 1979).
3. There are irregular variations of $C(\lambda)$ and $A(\lambda)$ in the region 0.2 to 0.3 μm and 1.5 to 2.5 μm. Taranova (1977) had previously pointed out the variations of $C(\lambda)$ in the range 1.5 to 2.5 μm. In the ultraviolet region the variability is apparently somewhat greater.
4. Bands of CO_2 render the structure of the curve $C(\lambda)$ very complex in the range 1.2 to 2.5 μm. However, the influence of the bands on magnitude is slight, as only 20% of the solar energy is in the region $\lambda > 1.2$ μm.

Terrestrial spectroscopic and photometric measurements had already revealed that the cloud layer of Venus has a great optical thickness. Photometric and nephelometric measurements on board the probes confirmed this fact and provided detailed information on the macrostructure and microphysical characteristics of the venusian clouds. Their total optical thickness is ~ 30. The three main layers are described in detail in Chapter 16 by Esposito et al. The upper cloud contains particles of the so-called Modes 1 and 2 with mean diameters of 0.4 and 2.0 μm, respectively. The refractive index is between 1.44 and 1.46. This layer scatters a large portion of the luminous flux that leaves the atmosphere of the planet.

Two physical parameters affect the intensity of radiation in an optically thick scattering medium: the single scattering albedo a (often $\tilde{\omega}_0$) and the phase function. The latter is quantitatively described by the anisotropy factor g, which varies between zero (for isotropic scattering) and 0.8 to 0.9 (for great anisotropy). Large dielectric particles are strongly forward scattering. In terms of the size parameter

$$x = 2\pi r/\lambda \tag{11}$$

when $x > 1$, usually $0.6 < g < 0.9$. In the atmosphere for Venus, for Model 1 particles the condition $x > 1$ is satisfied (for $\lambda < 1\mu$m) and even more so for particles of Modes 2 and 3. Sobolev (1968, 1944), from phase curves of the brightness of Venus, has found that $g = 0.7$ and $a = 0.995$ for $\lambda = 0.55$ μm. Since this pertains to groundbase observations, these estimates may be assigned to the upper cloud. An analysis of photometric measurements within the cloud layer of Veneras 9, 10, 11, 12, and Pioneer Venus (cf. e.g., Moroz et al. 1979b, 1981) produces

$$0.99 < a < 1.00 \tag{12}$$

for the region 0.6 to 1.0 μm (outside the absorption bands of CO_2 and H_2O). When the optical thickness is >30 and a is close to 1, the clouds transmit radiation almost as well as in the case of conservative scattering, when the exact equality of $a = 1$ is attained.

The dependence of the albedo of Venus on the wavelength (Fig. 2) contains certain information on the chemical nature of the particles in the clouds of Venus (at least, in the upper portion). The curve in Fig. 2 distinctly reveals two regions with a reduced albedo ($0.32 < \lambda < 0.6$ μm and $\lambda > 2$ μm); the absorption in these regions is apparently due to a condensed phase. Attempts have already been made to determine quantitatively the absorbing properties of the material which comprises the particles (Moroz 1971; Sobolev 1972). Table III shows the results of similar calculations, performed especially for this chapter, with allowance for all the currently available supplementary information. The appropriate explanations are given below.

TABLE III

Spherical Albedo, Single-Scattering Albedo, Index of Refraction (Real and Imaginary Parts) in the Range 0.32-3.8 μm.

λ (μm)	$C(\lambda)$	$A(\lambda)$	$g(\lambda)$	$a(\lambda)$	Index of Refraction Real Part m_1	Imaginary Part m_2	Absorption Coefficient $k = \dfrac{4\pi m_2}{\lambda}$ (cm^{-1})	m_2 for 75% H_2SO_4
0.32	0.48	0.41	0.77	0.968	1.46	8×10^{-4}	300	$<2 \times 10^{-8}$
0.34	0.54	0.50	0.77	0.980	1.46	5×10^{-4}	200	$<2 \times 10^{-8}$
0.36	0.54	0.50	0.77	0.980	1.46	5×10^{-4}	200	$<2 \times 10^{-8}$
0.38	0.57	0.52	0.76	0.981	1.46	5×10^{-4}	200	$<2 \times 10^{-8}$
0.40	0.67	0.60	0.76	0.988	1.45	3×10^{-4}	100	$<2 \times 10^{-8}$
0.42	0.80	0.71	0.75	0.994	1.44	1.5×10^{-4}	50	$<2 \times 10^{-8}$
0.44	0.88	0.77	0.75	0.997	1.44	10^{-4}	30	$<2 \times 10^{-8}$
0.46	0.91	0.79	0.75	0.997	1.44	10^{-4}	30	$<2 \times 10^{-8}$
0.48	0.95	0.82	0.74	$\geqslant 0.998$	1.44	$\leqslant 10^{-4}$	<25	$<2 \times 10^{-8}$
0.50	0.97	0.83	0.74	$\geqslant 0.998$	1.44	$\leqslant 10^{-4}$	<25	$<2 \times 10^{-8}$
0.55	0.99	0.85	0.74	$\geqslant 0.999$	1.44	$\leqslant 10^{-4}$	<23	$<2 \times 10^{-8}$
0.60	1.00	0.85	0.72	$\geqslant 0.999$	1.44	$\leqslant 10^{-4}$	<20	$<2 \times 10^{-8}$
0.80	1.00	0.85	0.70	$\geqslant 0.998$	1.43	$\leqslant 10^{-4}$	<16	9×10^{-8}
1.00	1.00	0.85	0.69	$\geqslant 0.998$	1.43	$\leqslant 10^{-4}$	<12	1.5×10^{-6}
1.20	1.00	0.85	0.70	$\geqslant 0.998$	1.42	$\leqslant 10^{-4}$	<10	5×10^{-6}
1.50	0.95	0.82	0.72	$\geqslant 0.998$	1.40	$\sim 10^{-4}$	10	1×10^{-4}
1.75	0.90	0.78	0.75	0.997	1.39	6×10^{-4}	40	4×10^{-4}
2.20	0.79	0.70	0.80	0.995	1.38	1×10^{-3}	60	2×10^{-3}
2.40	0.71	0.63	0.81	0.992	1.36	1.3×10^{-4}	70	2×10^{-3}
2.85	0.080	0.080	0.80	0.847	1.29	8×10^{-2}	3500	6×10^{-2}
3.00	0.055	0.055	0.79	0.814	1.28	1×10^{-1}	4200	9×10^{-2}
3.20	0.035	0.035	0.78	0.781	1.28	1.2×10^{-1}	4700	1.3×10^{-1}
3.40	0.040	0.040	0.77	0.780	1.36	1.2×10^{-1}	4400	1.6×10^{-1}
3.60	0.055	0.055	0.76	0.793	1.39	1.2×10^{-1}	4200	1.4×10^{-1}
3.80	0.065	0.064	0.75	0.797	1.40	1.1×10^{-1}	3600	1.3×10^{-1}

For the conversion from a spherical albedo $A(\lambda)$ to the single-scattering albedo a, the formulas of Sobolev (1972) and Yanovitskiy (1972) are commonly used. Their applicability is confined to conditions of semiinfinite homogeneous atmospheres and low absorption ($1 - a \ll 1$). These restrictions do not apply in a 2-stream approximation. The approximate albedo R found in this way differs little from A. For a semiinfinite homogeneous atmosphere

$$R = \frac{\sqrt{1 - ga} - \sqrt{1 - a}}{\sqrt{1 - ga} + \sqrt{1 - a}}. \tag{13}$$

A comparison of R values, calculated by Eq. (3), and A values, calculated from the much more complicated equation of Yanovitskiy (1972), allows us to find a simple relation between R and A

$$1 - A = K(1 - R) \tag{14}$$

where K is a coefficient which is constant within limits of $\pm$ 2%. When $g = 0.7$ and $a > 0.95$, the coefficient $K = 1.127$.

In Table III we show as a function of wavelength the color function $C(\lambda)$ and the corresponding spherical albedo $A(\lambda)$. Anisotropy factors $g(\lambda)$ are calculated for a moderately narrow size distribution corresponding to the upper cloud, by use of the tables from Mie theory of Zel'manovich and Shifrin (1968). The single-scattering albedo $a(\lambda)$ is then obtained from Eqs. (3) and (4). The real part of the refractive index m_1 is assumed to be that of H_2SO_4, ~75% concentration by weight; the values at 0.36 and 0.55 μm are those of Hansen and Hovenier (1974). The imaginary index m_2 can then be found, again assuming the same size distribution and the Zel'manovich and Shifrin tables. As long as m_1 and the size distribution are fixed, and $(1-a) \ll 1$, $m_2 \propto 1-a$, approximately. The absorption coefficient k of the material in the cloud drops can then be found, since $k = 4\pi m_2/\lambda$. The last column in the table shows the values of m_2 for 75% H_2SO_4 (Palmer and Williams 1973).

Perhaps the most striking result is the intense absorption between 3 and 4 μm. Investigations of the center-limb effect in the thermal radiation of the planet indicate that $a(\lambda)$ is also much less than 1 at longer wavelengths, at least up to 40 μm (cf. e.g. Ksanfomality 1980). The cloud particles are apparently the most important absorbing agent affecting the transfer of thermal radiation in the altitude range of 48 to 70 km.

Comparing the last and third last columns of Table III, it is seen that the particles of 75% H_2SO_4 satisfactorily account for the drop in $A(\lambda)$ in the region of $\lambda > 1.5$ μm. This is not a novel result (Pollack et al. 1978). There is a close coincidence not only with respect to the absorption coefficients in the infrared range, but also the refractive indices in the visible range (Sill 1972; Young 1973).

However sulfuric acid does not account for the decrease of the albedo in the region of $\lambda < 0.5$ μm. At present there is no conclusive explanation of this fact. Two rather geochemically common elements may produce the yellow tinge: sulfur and iron. In connection with the discovery of chlorine in the cloud particles of Venus (Surkov 1979b), a compound such as ferrous chloride is worth attention, as a small admixture of this to sulfuric acid may apparently explain the observed behavior of the albedo in the region $\lambda < 0.5$ μm (Zasova et al. 1980).

4. RADIO ASTRONOMICAL STUDIES OF VENUS

A. D. KUZ'MIN
USSR Academy of Sciences

The first radio measurements of Venus in 1956 gave an indication of its high surface temperature. These and later measurements are summarized and placed on a uniform scale. Implications for properties of the surface and atmosphere, including the water vapor content, are discussed.

I. EMISSION MEASUREMENTS

Radio emission from the planet Venus was detected in 1956 by Mayer et al. (1958). Such emission from Venus was not unexpected since it is known that any body can emit thermal radio waves. The brightness temperature of thermal radio emission T_b is $\leqslant$ the temperature T of a body. In 1956 the planet Venus was considered a sister and even a twin of planet Earth and consequently Venus's surface temperature T_s would be nearly 300 K and therefore the brightness temperature of Venus's radio emission also would be nearly 300 K. However, Mayer et al. (1958) measured almost twice this value or $\overline{T}_{b\,\venus}$ = 560 ± 73 K at 3.15 cm wavelength. (The bar over T indicates the brightness temperature averaged over the visible disk of Venus.) If this emission was really thermal and its source was the surface, then it leads to the deduction that the temperature of the venusian surface is higher than 560 K. Such a high surface temperature drastically changed existing notions about Venus as a planet where conditions are similar to those on Earth. This fundamental discovery at once raised the questions: (1) is this emission thermal or nonthermal; and (2) does it originate on the surface, in the atmosphere, or in the ionosphere?

One of the tests for a thermal versus a nonthermal mechanism is the frequency spectrum of radio emission. The first attempt to obtain a spectrum was made by Mayer et al. (1958). Besides the above mentioned 3.15 cm observation, they had observed also at a longer wavelength of 9.4 cm with reduced accuracy; the two observations gave $\overline{T}_{b\varphi} = 430$ K and $\overline{T}_{b\varphi} = 740$ K. Gibson and McEwan (1959) observing at a shorter wavelength of 8.6 mm found a brightness temperature between 250 and 570 K; this fitted both the infrared temperature of 235 K (Pettit and Nicholson 1955) and the Mayer et al. value of 560 K. In 1959 more accurate observations at 8 mm were made by Kuz'min and Salomonovich (1960) whereby they found $\overline{T}_{b\varphi} = 350 \pm 70$ K, a value definitely lower than that at centimeter wavelengths. Kuz'min and Salomonovich proposed an explanation for this difference, suggesting that the lower brightness temperature at mm wavelengths is due to the absorption of these wavelengths in relatively cool atmosphere surrounding a hot surface. Independently, such a hypothesis had been developed by Sagan (1960a). He suggested that absorption is due to water vapor and high-pressure carbon dioxide and also proposed a high surface temperature caused by a runaway greenhouse effect in the Venus atmosphere; this model was called the greenhouse model.

Öpik (1961) proposed another mechanism for the heating of the surface by wind friction. Quite a different model for interpretation of experimental data was proposed by Jones (1961). He suggested that at millimeter wavelengths thermal radio emission from the surface was observed and therefore the surface was only moderately warm (400 K). On the other hand, the radio emission at centimeter wavelengths originated in a very dense ($N_e \sim 10^9$ cm^{-3}) ionosphere having a higher electron temperature than the surface. This model was called the ionosphere model.

During the next several years many observers measured Venus's radio emission in a wide range of wavelength from 2 mm to 70 cm. The results of these measurements are given in Table I. We have included in this table measurements in which the antenna was calibrated by reference horn or by reliably measured discrete sources; the resulting errors do not exceed 15%. An exception is made only for millimeter wavelengths, where measurements accurate to 20% are included.

With the data on Jupiter's brightness temperature (Wrixon et al. 1971) and flux densities of a series of discrete sources (Kellerman et al. 1969) becoming more accurate, we have made corrections to $\overline{T}_{b\varphi}$. The corrected values of brightness temperatures are given in the last column of Table I. Table I also gives the original values of $\overline{T}_{b\varphi}$ published by authors (column four) and the parameters of the reference sources accepted by the observers (column five) and later more accurate parameters (Wrixon et al. 1971; Kellerman et al. 1969). For Jupiter the brightness temperature is in degrees Kelvin; for the discrete sources the flux density is in Jansky.

TABLE I

Brightness Temperature Averaged Over Venus Apparent Disk

λ (cm)	Observers	yr	$\bar{T}_{b\,♀}$ Publ. (K)	T_b or S of Reference Source		$\bar{T}_{b\,♀}$ Corr. (K)
				Accepted by Observers	Calibrated Accuracy	
0.225	Yefanov et al. (1969)	1967	235 ± 40	Jupiter 150 ± 20	—	235 ± 40
0.31	Ulich et al. (1971)	1971	386 ± 33	reference horn	—	386 ± 33
0.32	Tolbert and Straiton (1964)	1964	300^{+57}_{-27}	reference horn	—	300^{+57}_{-27}
0.43	Tolbert and Straiton (1964)	1964	330^{+56}_{-36}	reference horn	—	330^{+56}_{-36}
0.8	Kuz'min and Salomonovich (1962)	1961	382 ± 65	Jupiter 140	157 ± 8	430 ± 85
0.8	Vetukhnovskaya et al. (1963)	1962	394 ± 30	Jupiter 140	157 ± 8	440 ± 35
0.8	Yefanov et al. (1969)	1967	375 ± 60	Jupiter 144 ± 23	157 ± 8	410 ± 65
0.82	Vetukhnovskaya et al. (1971)	1969	428 ± 70	Jupiter 144 ± 23	157 ± 8	465 ± 25
0.835	Thornton and Welch (1964)	1962	395 ± 60	Jupiter 144 ± 23	157 ± 8	430 ± 65
0.835	Welch and Thornton (1965)	1964	390 ± 45	Jupiter 144 ± 23	157 ± 8	425 ± 50
0.86	Tolbert and Straiton (1964)	1964	375 ± 58	reference horn	—	375 ± 58
0.97	Griffin et al. (1967)	1966	412 ± 55	reference horn	—	412 ± 55
1.16	Griffin et al. (1967)	1966	495 ± 50	reference horn	—	495 ± 50
1.25	Griffin et al. (1967)	1966	451 ± 53	reference horn	—	451 ± 53
1.31	Vetukhnovskaya (1969)	1967	470 ± 40	Jupiter 150	139 ± 6	436 ± 40

TABLE I

Brightness Temperature Averaged Over Venus Apparent Disk (continued)

λ (cm)	Observers	yr	$\bar{T}_{b♀}$ Publ. (K)	T_b or S of Reference Source		$\bar{T}_{b♀}$ Corr. (K)
				Accepted by Observers	Calibrated Accuracy	
1.35	Griffin et al. (1967)	1966	530 ± 45	reference horn	—	530 ± 50
1.46	Griffin et al. (1967)	1966	595 ± 50	reference horn	—	595 ± 50
1.6	Vetukhnovskaya et al. (1963)	1962	534 ± 60	Taurus 470	—	534 ± 60
1.65	Griffin et al. (1967)	1966	560 ± 51	reference horn	—	560 ± 51
1.9	Kellerman et al. (1969)	1966	500 ± 75	reference source	—	500 ± 75
1.95	Morrison (1969)	1967	495 ± 35	Virgo; 29 3C 123; 5,2	—	495 ± 35
2.07	McCullough and Boland (1964)	1963	500 ± 75	Cassiopeia Taurus	—	525 ± 75
2.7	McCullough (1972)	1969	612 ± 37	Virgo; 38.4	—	612 ± 37
3.3	Vetukhnovskaya et al. (1963)	1962	575 ± 30	Taurus 560	—	575 ± 30
3.75	Kuz'min and Dent (1966)	1964	659 ± 65	3C 123; 10	—	—
4.52	Dickel and Warnock (1968); Warnock and Dickel (1969)	1967	654 ± 35	3C 123; 12.3	—	669 ± 36
6	Hughes (1965)	1966	630 ± 30	Virgo 70; 3C 123; 16.9	—	630 ± 30
6	Dickel and Warnock (1968); Warnock and Dickel (1969)	1966	706 ± 45	3C 218; 13.0	—	730 ± 46
7.89	Warnock and Dickel (1972)	1970	686 ± 38	3C 218; 17.2	—	686 ± 38

TABLE I

Brightness Temperature Averaged Over
Venus Apparent Disk (continued)

λ (cm)	Observers	yr	$\bar{T}_{b\,♀}$ Publ. (K)	T_b or S of Reference Source		$\bar{T}_{b\,♀}$ Corr. (K)
				Accepted by Observers	Calibrated Accuracy	
9.26	Warnock and Dickel (1972)	1969	675 ±38	3C 218; 199	—	675 ±38
9.6	Kuz'min (1965)	1961	660 ±76	Taurus 760	—	660 ±75
10.0	Drake (1962a)	1961	583 ±50	3C 123; 23.7	3C 123; 25	615 ±60
10.0	Drake (1964)	1962	580 ±40	3C 123; 23.7	3C 123; 25	610 ±60
10.3	Vetukhnovskaya (1969)	1964	620 ±30	Taurus 760	—	620 ±30
10.6	Clark and Spencer (1964)	1962	580 ±60	3C 295; 10.8	3C 295; 11.4	610 ±60
10.6	Kuz'min and Clark (1965)	1964	580 ±60	3C 48; 7.8	3C 48; 8.6	640 ±65
11.1	Stankevich (1970)	1969	710 ±35	—	—	—
11.3	Kellerman (1966)	1964	605 ±30	—	—	—
12	Warnock and Dickel (1972)	1970	701 ±41	3C 218; 25.2	—	≃701 ±41
13	Boischot et al. (1963)	1962	712 ±80	—	—	—
14.3	Warnock and Dickel (1972)	1970	670 ±37	3C 218; 29.5	—	670 ±37
21	Boischot et al. (1963)	1962	674 ±70	—	—	—
21.3	Kellerman (1966)	1964	590 ±30	—	—	590 ±30
21.4	Drake (1964)	1962	528 ±33	3C 123; 50.1 / 3C 348; 44.7 / 3C 353; 56.2	3C 123; 45.9 / 3C 348; 44.5 / 3C 353; 54.9	505 ±50
49.7	Muhleman et al. (1973)	1964	510 ±50	—	—	510 ±50
70	Gordon et al. (1973)	1972	498 ±33	—	—	

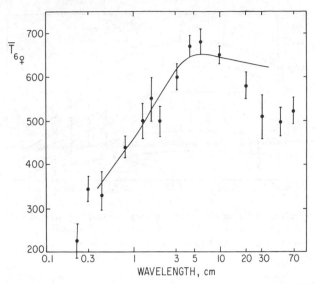

Fig. 1. The measured and calculated spectra of the averaged (for the apparent disk) brightness temperature of Venus $\overline{T}_{b\,\varphi}(\lambda)$.

Figure 1 shows the spectrum $\overline{T}_{b\,\varphi}(\lambda)$ plotted from data in Table I. The points are weighted averages. The brightness temperature of Venus has a maximum near $\lambda = 7$ cm. Such a spectrum does not represent classical thermal or nonthermal mechanism of radio emission and therefore the uncertainty between the greenhouse and the isospheric models still remains.

II. BRIGHTNESS DISTRIBUTION

Another approach to determining the model of the Venus radio emission is the study of the brightness distribution across the disk of the planet. One may expect that in the greenhouse model, which has the atmosphere cooler than the surface, the radio brightness will decrease from center to the limb of the planet. On the contrary, in the ionosphere model with ionosphere hotter than the surface, one may expect a limb brightening.

The measurement of the brightness distribution was carried out by Korolkov et al. (1963) by measuring the width of a response of the radio telescope produced by Venus transit relative to the original beamwidth of the radio telescope. The data showed that at a wavelength of 3 cm the radius of the emitting region does not exceed 1.07 radii of the visible disk r_φ: the form of the response fitted better the darkening of the limbs. However, at the same time Clark and Spencer (1962) came to the opposite conclusion. Their interferometric data at 9.4 cm do not fit the limb darkening nor even the uniform bright disk. The best fit was obtained for limb brightening or for a uniform

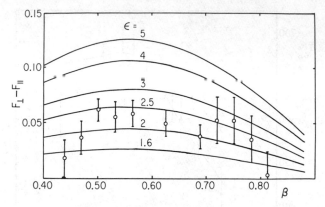

Fig. 2. The difference of the interferometer visibility function in polarization perpendicular $F_\perp$ and parallel $F_\parallel$ to the base of interferometer via resolution of interferometer $\beta = D/\lambda r_\female$. Here D is the length of the base of the interferometer.

disk 15% larger than the visible disk. Also at about this time Barath et al. (1964) performed measurements of brightness distribution by a space radio telescope on board the Mariner 2 space probe. At 1.9 cm wavelength they found limb darkening, but at the shorter wavelength of 1.35 cm where one would expect a stronger atmospheric effect, the limb darkening was not detected.

III. POLARIZATION

The definitive experiment for determining the emitting region and the temperature of the surface, based on the measurement of differential polarization, was proposed and realized in 1964 by Kuz'min and Clark (1965). The idea of the experiment was that thermal radio emission of the surface must be polarized at the limbs due to the difference in the Fresnel reflection coefficients for vertical and horizontal polarizations. On the contrary, an emission from the atmosphere or ionosphere, having no clear boundary, would be unpolarized. Observations were performed at Owen Valley Radio Observatory at 10.6 wavelength. The results of differential polarization measurements are shown in Fig. 2 as a difference of the interferometer visibility function in polarization perpendicular $F_\perp$ and parallel $F_\parallel$ to the base of the interferometer. One can see a clear excess of $F_\perp$ over $F_\parallel$ which proves that 10 cm radio emission of Venus is thermal emission from its surface. The value of the difference $(F_\perp - F_\parallel)$ depends on the dielectric constant ϵ of the surface allowing one to evaluate ϵ and the temperature of the surface T_s. Taking into account the attenuation in the atmosphere of Venus (Smirnova 1973) and correcting for the flux densities of the reference source (see Table I), we obtain $\epsilon = 3.6 \pm 1.0$ and $T_s = 760 \pm 80$ K from the Kuz'min and Clark

(1965) measurement. From the position of the zero of the interferometric visibility function of Kuz'min and Clark we also evaluated for the first time the radius of the surface of Venus, $r_{s\varphi} = 6057$ km. This value is in good agreement with the later more precise radar data, $r_{s\varphi} = 6050$ km.

IV. WATER VAPOR

An important objective of radio astronomical study of Venus was to estimate the water vapor content of its atmosphere. There were many attempts to detect water vapor from measuring the decrease in the brightness temperature at the wavelength of the water vapor absorption line $\lambda = 1.35$ cm. The most precise observations were made by Janssen et al. (1973). Their measurements at six wavelengths did not detect any decrease at $\lambda = 1.35$ cm; they estimated the upper limit of the water vapor abundance $f_{H_2O} \leq 0.4\%$. The second independent estimate of f_{H_2O} was made by the same group using interferometric observation of limb darkening at $\lambda = 1.35$ cm. They found $f_{H_2O} \leq 0.35\%$.

Smirnova and Kuz'min have estimated the water vapor content from radar data at 3.8 cm wavelength (Evans and Ingalls 1968) and found $f_{H_2O} = 0.1\% \pm 0.1\%$ at altitudes 15–30 km. Calculated spectra $\bar{T}_{b\varphi}(\lambda)$ for 0.1% H_2O and 97% CO_2 are plotted in Fig. 1 (solid line). One can see good agreement with experimental data in the wavelength range 2 mm to 10 cm. Good agreement between calculated and measured spectra obviates the need for any other components that absorb microwave radiation. Furthermore the content of these components cannot be large, and upper limits to their probable content can be estimated. Vetukhovskaya et al. (1971) and Kuz'min and Marov (1974) have estimated the upper limit of the moisture in the clouds of Venus, $M_0 \leq 0.4$ g m^{-3} showing that there are no strong water clouds.

At wavelengths > 10 cm the experiment showed a significant (~ 100 K) reduction of the measured brightness temperature. This reduction was not predicted theoretically within the framework of the existing data on the planet Venus. The discrepancy cannot be explained by an increase of ϵ at decimeter wavelengths, since the value needed for agreement, $\epsilon = 13$, is much larger than that measured in the same wavelengths by radar reflectivity $\epsilon = 5$. Additional atmospheric absorption of decimeter radio emission would reduce the reflection also, and this does not agree with the observations. This interesting feature of the spectrum requires further experimental testing.

All the above-mentioned measurements of Venus's radio emission were undertaken near inferior conjuction of the planet. Therefore they characterize the night side of Venus. For estimating a diurnal temperature variation, several observations near superior conjunction were undertaken. These results are given in Table II as a difference of brightness temperatures averaged over the illuminated $\bar{T}_{bo}$ and dark $\bar{T}_{b\bullet}$ sides of Venus. At 10 cm wavelength the radio emission originates basically on the surface of Venus. Therefore the absence

TABLE II

**Differences in Surface Temperatures of
Venus's Day and Night Sides**

λ (cm)	$T_{bo} - T_{b\bullet}$	Observers
11.3	24 ± 50	Kellerman (1966)
10.6	-30 ± 30	Kuz'min (1965)
10	26 ± 50	Drake (1962a, b)
4.52	2 ± 20	Dickel et al. (1968)
2.7	0 ± 9	McCullough (1972)
1.95	0 ± 25	Morrison (1969)
0.86	20 ± 8	Kalaghan et al. (1969)
0.82	4 ± 10	Vetukhnovskaya et al. (1971)
0.34	-22 ± 4	Epstein et al. (1968)
0.23	55 ± 60	Yefanov et al. (1969)

of a significant difference of $(\bar{T}_{bo} - \bar{T}_{b\bullet})$ at this wavelength is an indication that the surface temperature of the day side does not differ significantly from that of the night side.

This conclusion was confirmed also by the interferometric radio astronomy measurement by Sinclair et al. (1970). According to this measurement, the difference of the day- and nightside temperatures is 18 ± 9 K. Marov's (1974) calculation of the thermal regime of the venusian atmosphere showed that, because of the great heat content of the dense atmosphere, the diurnal temperature change in the near-surface atmosphere should not exceed 1 K. With such a massive atmosphere it is also difficult to expect significant temperature changes with latitude. Actually the Sinclair et al. (1970) measurement indicated that the reduction in surface temperature from the equator to the poles is no more than 12 K.

5. SUMMARY OF PRELIMINARY RESULTS OF THE VENERA 13 AND VENERA 14 MISSIONS

V. I. MOROZ
USSR Academy of Sciences

The parameters of the Venera 13 and 14 landing sites are given. Areas discussed are atmospheric, aerosol, and surface composition, water vapor, absorption of solar ultraviolet energy, electrical discharges, and imaging of the surface.

The Soviet probes Venera 13 and Venera 14 arrived on the planet Venus early in March 1982. Both missions were successful; however some time is necessary for processing the data and especially interpretation. Only preliminary results are available as this book goes to press. We give a short review of the first scientific publications that appeared for the most part in the July issue of the Soviet *Pisma Astron. Zh.* "Astronomical Journal Letters." The next and more complete set of papers will be in a special issue of Soviet *Kosmich. Issled.* "Space Research." English translations of both journals are available with a half-year delay.

I. OVERVIEW OF MISSIONS AND SCIENTIFIC EXPERIMENTS

Both missions are identical and belong to the so-called second generation of Soviet probes designed for studies of Venus; this generation started with Veneras 9 and 10 in 1975. Probes of the second generation have provided excellent opportunities for simultaneous use of many sophisticated instruments for investigations of atmosphere and surface. The total time of operation in Venus's high temperature medium is of course limited. To some extent it is possible to vary the time spent in the atmosphere and on the surface.

V. I. MOROZ

TABLE I

Landing Sites and Timings for Venera 13 and Venera 14[a]

	Venera 13	Venera 14
Launch date	30 Oct. 1981	4 Nov. 1981
Landing date	1 Mar. 1982	5 Mar. 1982
Opening of main parachute and start of receiving information	6:00:10	5:57:53
Jettison of parachute	6:09:05	6:06:46
Landing time[b]	7:01:00	7:00:10
End of receipt of information	9:08	7:57
Duration of activity on the surface	2:07	0:57
Landing site: latitude	$-7°\,30'$	$-13°\,15'$
longitude	305°	310° 10′
altitude[c]	$+0.1\pm0.2$	-0.6 ± 0.2
Solar zenith angle	36°0	35°5
Temperature of the atmosphere near the surface	738 K	743 K
Pressure near the surface	89.5 atm	93.5 atm

[a]Table from Sagdeev and Moroz (1982*b*). All times are Moscow time of receipt on the Earth.
[b]From the accelerometer signal.
[c]Above the reference level $R_0 = 6052$ km.

Veneras 13 and 14 spent a little less time in the atmosphere than Veneras 9 and 10, and a little more on the surface.

Information on timing, coordinates, etc. is given in Table I and the list of experiments in Table II. The reader interested in details of operational consequences of landing on Venus can find this in the description of the Venera 11 and 12 missions (Zaitsev et al. 1979). Here it is sufficient to say that the majority of the instruments are switched on at altitude $H_0 \sim 62$ km (see Table I) almost at the moment of the opening of the main parachute. At the same time the information was transmitted by the radiolink connecting the probe and the flyby spacecraft which retransmitted it to Earth. The parachute was jettisoned at ~ 47 km. The duration of the parachute descent was 9 min and the total duration of the descent between parachute opening and probe landing was a little more than 1 hr. Only one experiment, the accelerometer, was turned on before the opening of the parachute in order to measure the atmospheric parameters during the strong deceleration of the probe below 100 km. Table II shows the list of scientific experiments, their purposes and their altitude ranges. Letters (a,b,c . . .), explained in the table's footnotes, show the areas of planetary study that are related to the results of each experiment.

One of the most important and completely new experiments on Veneras 13 and 14 was the elemental analysis of rocks by the X-ray fluorescence spectrometer. Some difficult technical problems arose for this method, one of the most difficult being the introduction of the sample into a low-pressure volume inside the probe. All the problems were solved and the experiments on both probes were successful. Another new experiment was the first attempt to make seismic measurements. Many of the other instruments had prototypes in previous flights but some were redesigned and significantly improved (panoramic telephotometer, gas chromatograph, spectrophotometer). In other instruments capabilities were improved by relatively small changes. For example, the mass spectrometer is very similar to the prototype used on Veneras 11 and 12 but substantially more sensitive. All this improvement has expanded the range of the new results part of which are presented in this chapter.

II. ATMOSPHERIC CHEMISTRY: NEW DATA ON MINOR CONSTITUENTS

1. Mass Spectrometer (Istomin et al. 1982)

A description of the prototype of this instrument used on Veneras 11 and 12 has been given by Grechnev et al. (1979) and Istomin et al. (1979c). Due to improved laboratory preparation of the analytical systems and some changes in the electronics, the new version of this instrument is 10–30 times more sensitive. The basic unit of the instrument is a Bennett-type mass spectrometer initially designed for measurements in the upper atmosphere of the Earth, but Istomin and his collaborators have been able to adapt it for use inside very dense atmospheric layers. They used a very sophisticated inlet system which can be opened for 2 msec intervals to provide the intake of a small but precise quantity of gas into the high-vacuum part of the instrument. A small titanium sputter-ion pump was used. Some technical parameters of this instrument are given in Table III. The output of the instrument is analog with an automatic switching of ranges as each mass peak is registered. The 5 scales have gains of 0.1, 1.0, 10, 100, 1000. High peaks therefore are broken into several parts nested within each other and are somewhat difficult to read, but such troubles are amply compensated by the wide dynamic range and linearity.

Examples of original mass spectra obtained on Veneras 13 and 14 are given in Fig. 1. Only wings and central parts of peaks are shown in the cases of multiple switches of range. Peaks for the two neon isotopes ^{20}Ne (central part on the 2nd scale) and ^{22}Ne are easily visible on both spectra. Peaks for argon isotopes ^{36}Ar, ^{38}Ar, ^{40}Ar are measured on the 2nd and 3rd scales. The peaks for ^{36}Ar and ^{40}Ar are nearly equal (in contrast to terrestrial Ar; see Chapter 13). Also clearly visible on both spectra is the unresolved peak of Kr. A venusian xenon peak (the sum of isotopes ^{131}Xe and ^{132}Xe) may be seen on

TABLE II

List of Scientific Experiments and Instruments*

Experiment or Instrument	Measured Physical Parameters	Altitude Interval of Measurements (km)	Areas of Planetary Studies which can use Results	Principal Investigator
Pressure and temperature sensors	pressure P temperature T	62–0	(a)	V.S. Avduevsky
Accelerometer	dv/dt	110–63	(a)	V. S. Avduevsky
Doppler measurements	wind and turbulent velocity	62–0	(a)	V. V. Kerzhanovich
Mass spectrometer	chemical and isotopic composition of atmosphere	26–0	(b)	V. G. Istomin
Gas chromatograph	chemical composition of atmosphere	58–0	(b,d)	L. M. Mukhin
Spectrophotometer	spectral and angular distribution of solar radiation scattered by atmosphere and surface	62–0	(a,b,c,d,e)	V. I. Moroz
Groza-2	electrical discharges in atmosphere, acoustical events, electrical conductivity of aerosol particles, microseismical events	62–0 surface	(a,b,c) (f)	L. V. Ksanfomality

TABLE II (Continued)
List of Scientific Experiments and Instruments*

Experiment or Instrument	Measured Physical Parameters	Altitude Interval of Measurements (km)	Areas of Planetary Studies which can use Results	Principal Investigator
Nephelometer	volume scattering of aerosol medium	62–0	(c)	M. Ya. Marov
Humidity sensor	water vapor content	50–45	(b)	Yu. A. Surkov
X-RFS** for aerosol particles	elemental abundance of aerosol particles	62–45	(c)	Yu. A. Surkov
X-RFS for rocks	chemical composition of rocks	surface	(d,e)	Yu. A. Surkov
Telephotometer	panoramic images	surface	(d,e,a)	A. S. Selivanov
Sensors for measurements of physical properties of soil	rigidity and electrical conductivity of surface	surface	(e)	A. L. Kemurdzhan

*Table from Sagdeev and Moroz (1982b).
**X-ray fluorescence spectrometer.
(a) Structure and dynamics of the atmosphere.
(b) Chemical and isotopic composition of atmospheric gases.
(c) Physical properties and chemical composition of aerosol particles.
(d) Chemical composition of rocks.
(e) Geological structure and physical properties of the surface.
(f) Planetary internal structure.

TABLE III

Major Technical Parameters of Mass Spectrometers for Venera 13 and Venera 14

Mass range, a. m. u.	11–138
Resolution $M/\Delta M$ at level 0.1 (for ^{36}Ar)	30
Duration of scan, s	7
Duration of sampling, msec	2
Weight, kg	9.5
Power, W	17
Limit of sensitivity, ppb (10^{-9})	50

the upper spectrum of Fig. 1. The ^{136}Xe peak in this spectrum (Venera 13) is background and so is all xenon in the case of Venera 14.

The mass spectrometers operated in the altitude range from 26 km to the surface and gave nearly 250 mass spectra. Evaluation of data concerning neon isotopic ratios is in an advanced stage. For masses 20 and 22 a ratio

$$^{20}\text{Ne}/^{22}\text{Ne} = 11.8 \pm 0.7 \tag{1}$$

was found and also an upper limit for a ratio of masses 22 and 21

$$^{22}\text{Ne}/^{21}\text{Ne} > 10. \tag{2}$$

In Table IV a comparison of ^{20}Ne and ^{22}Ne ratios for Venus, Earth, and solar wind is given. These data show that the ratio of these two isotopes is a little higher on Venus than on Earth, but less than on the Sun.

There are no final conclusions concerning Kr and Xe. The range 10 to 40 ppb was given for both, values which will be narrowed down in the future. For krypton this range is not in accordance with the evaluation obtained earlier on Veneras 11 and 12 (600 ± 200 ppb; Istomin et al. 1980b) and

TABLE IV

Isotopic Ratios

	^{20}Ne/^{22}Ne	Reference
Solar wind	13.7 ± 0.3	Geiss et al. (1972)
Venus atmosphere	$14.3^{+5.7}_{-3.2}$	Hoffman et al. (1980a)
	11.8 ± 0.7	Istomin et al. (1982)
Terrestrial atmosphere	10.07 ± 0.35	Istomin et al. (1982)[a]
	from 9.24 to 10.31	earlier measurements by other authors (see refs. in Istomin et al. 1982)

[a]From calibration data of Venera 13 and Venera 14 mass spectrometers obtained before the flight.

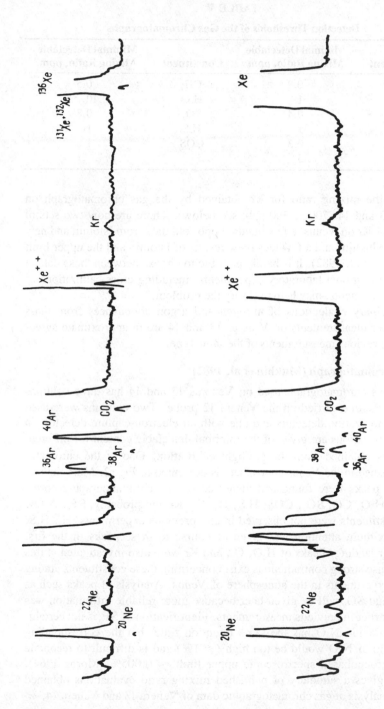

Fig. 1. Examples of original (unprocessed) mass spectra, obtained by Venera 13 (upper) and Venera 14 (lower) for the noble gases. Most of the peaks between Ne and Ar are instrumental background and the same for Xe on the lower spectrum. The large peak ^{136}Xe on the upper spectrum is background but close by it is the peak ^{131}Xe and ^{132}Xe which belongs to the venusian atmosphere (Istomin et al. 1982).

TABLE V

Detection Thresholds of the Gas Chromatographs

Constituent	Minimal Detectable Mixing Ratio, ppm	Constituent	Minimal Detectable Mixing Ratio, ppm
H_2	0.8	CH_4	0.5
N_2	1	H_2O	10
O_2	0.4	SO_2	0.8
Ar	1	H_2S	1
Kr	0.3	COS	1
CO	3		

also with the mixing ratio for Kr obtained by the gas chromatograph on Veneras 13 and 14 (700 ± 300 ppb; see below). There are now two sets of numbers for Kr on Venus: a few hundred ppb (old data from Istomin and new data from Mukhin) and a few tens (new results of Istomin and the upper limit of Donahue et al. 1982). It is hardly possible to choose between these data at this time. Additional laboratory experiments, including cross-calibrations of different instruments, may help to clarify the problem.

Preliminary evaluations of nitrogen and argon abundances from mass spectrometer measurements on Veneras 13 and 14 are in approximate agreement with previous measurements of the same type.

2. Gas Chromatograph (Mukhin et al. 1982)

The gas chromatograph used on Veneras 13 and 14 has three columns instead of the one carried on the Venera 12 probe. Two columns were used with neon ionization detectors and one with an electron-capture detector. In Table V evaluations are given of the minimal detectable content for different constituents obtained from the preflight calibration. One of the chromatograms obtained in the Venus atmosphere is presented in Fig. 2. More than 10 prominent peaks were found and identified on the chromatograph records: CO_2, N_2, H_2O, CO, SO_2, COS, H_2S, H_2, O_2, Kr and probably, SF_6. A few minor constituents were not observed in any previous experiments: H_2, H_2S, COS. Maximum attention was given of course to these peaks in the first weeks after landing. Peaks of H_2O, O_2, and Kr were also investigated at this time, because strong contradictions exist concerning these constituents among different experiments in the atmosphere of Venus. Analysis of peaks such as CO, N_2, and SO_2 will be given later because more reliable information was obtained earlier in previous measurements. Identification of SF_6 is not certain. Nitrogen dioxide N_2O has the same retention time, but the corresponding mixing ratio of N_2O would be too high ($\approx 1\%$) and is difficult to reconcile with the groundbased spectroscopic upper limit $\approx 0.005\%$ (Moroz 1964). Table VI gives a summary of published mixing ratio evaluations obtained from the analysis of gas chromatographic data of Venera 13 and Venera 14.

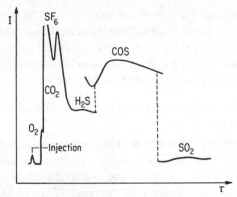

Fig. 2. Example of the chromatographic records obtained in the atmosphere of Venus with an electron-capture detector (Mukhin et al. 1982).

Many questions arise from a review of these remarkable numbers.

1. Why were COS and H_2S not observed in the Oyama experiment?
2. How can the simultaneous presence of strong oxidizing and reducing constituents, among O_2 and H_2 be explained?
3. What are the consequences of the relatively high abundance of H_2 for chemical models of the atmosphere?
4. What is the origin of SF_6?

Discussion of these and other questions in this field has not progressed far from the starting point. As usual new findings generated new questions.

3. Humidity Sensor

Surkov et al. (1982b) used an electrolytic sensor with LiCl salt for the measurements of water vapor content. This sensor is a small tube covered by a glass cloth impregnated with LiCl. An electrical current flows through the cover and heats it to a temperature higher than the atmospheric gas surrounding it. This temperature depends on the humidity of the gas, and the tempera-

TABLE VI

Some Results of Analysis of the Venusian Atmosphere by Gas Chromatographs

Constituent	Mixing Ratio, ppm	Altitude Range of Sampling
H_2	25 ± 10	58–49
O_2	18 ± 4	58–35
Kr	0.7 ± 0.3	49–37
H_2O	700 ± 300	58–49
H_2S	80 ± 40	37–29
COS	40 ± 20	37–29
SF_6	0.2 ± 0.1	58–3.5

ture difference between the cover and the gas was the parameter directly measured in this experiment. The authors gave a mixing ratio of water vapor in the narrow range of altitudes between 50 and 46 km as $(0.2 \pm 0.04)\%$, 10 times more than that obtained from spectroscopic measurements on the same probes and at nearly the same altitudes (see the next section)

4. Spectrophotometer: Comparison of H_2O Mixing Ratios Obtained in Three Experiments

Spectrophotometers (Moroz et al. 1982) on Veneras 13 and 14 measured spectral and angular distributions of solar radiation scattered in the Venus atmosphere. There are two spectral and one photometric channels in this instrument. The spectral channels scan the range 4500–12000 Å with resolution $\simeq 200$ Å and the photometric channel receives radiation with effective wavelength λ 3700 Å. The first spectral channel measures radiation from above constantly and has an almost hemispheric angular response. The second spectral and the photometric fields are relatively narrow (20° and 7°, respectively). Radiation is directed to these two channels by a step scanning device which provides measurements in 6 directions (including up and down). Spectral measurements were made from 62 km to the surface and on the surface itself. The ultraviolet photometric channel worked in the altitude range 62–45 km with a scan time of 2.5 s and an angular scan of 20 s. About 6000 spectra were obtained on both probes. An earlier form of this instrument was used on Venera 11 and Venera 12 in 1978 (Ekonomov et al. 1979). Two main improvements have been made in the new version: (a) measurements for several angles rather than one are provided, and this makes possible the direct evaluation of the volume scattering coefficient; (b) the ultraviolet photometric channel was added to measure the vertical profile of intensity in the ultraviolet, where a large quantity of the solar energy is absorbed by some unidentified absorber. A partial comparison of spectra for the up and down directions is given by a few examples shown in Fig. 3 (left, upper part).

Spectra from above are very similar to those obtained on Venera 11 and Venera 12 (Ekonomov et al. 1979; Moroz et al. 1981; see Fig. 27 in Chapter 13 of this book). Spectra from below show much deeper CO_2 and H_2O bands. We suppose that we are obtaining a more precise evaluation of the H_2O mixing ratio from these spectra than from earlier measurements, especially those inside the cloud layer where absorption bands from the direction above the probe are too weak. At this time, however, only the analysis of new measurements below 50 km and those for Venera 13 is completed. Synthetic spectra were calculated by the same method as used for interpretation of Venera 11 data (Moroz et al. 1979b). Some improvements in the program provide a more precise comparison of calculated and observed spectra. Four substantially different profiles of H_2O mixing ratio were chosen for calculation of synthetic spectra (Table VII). Some calculated spectra are presented (see Fig. 4) and conclusions obtained from comparison of observations and

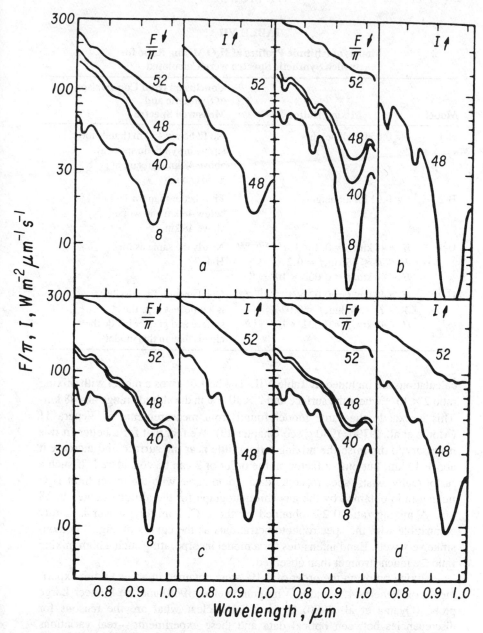

Fig. 3. Examples of scattered solar radiation spectra obtained by Venera 13. Spectra were registered in the wider range (0.45 to 1.2 λ). Here the part which was used for comparison with synthetic spectra calculated to find the best-fit vertical profile of the H_2O mixing ratio is shown. (a) *left*, measurements of the hemispherical intensity from above; *right*, measurements of the narrow angle intensity from below; (b) synthetic spectra for U-1 model profile (see Table VII); (c) the same for H-1 profile; (d) the same for U-2 profile (Moroz et al. 1982).

TABLE VII

Models of Altitude Profiles of H_2O Mixing Ratio for which Synthetic Spectra were Calculated

Model	Mixing Ratio, %	Conclusion from Comparison of Synthetic and Measured Spectra
H-1	$f = 0.05$ for all heights	No fit below 48 km (bands too strong in synthetic spectra), above 48 km agreement is good enough
H-2	$f = 0.01$ for all heights	Fit is better than for H-1 below 48 km and worse above 48 km
U-1	$H > 48$ km, $f = 0.2 \times 10^{-0.15(H-48)}$ $40 < H < 48$ km, $f = 0.2$ $H < 40$ km, $f = 0.002 \times 10^{0.05\,H}$	Nearly the same as for H-1
U-2	$H > 48$ km, $f = 0.02 \times 10^{-0.1(H-48)}$ $40 < H < 48$ km, $f = 0.02$ $H < 48$ km, $f = 0.002 \times 10^{0.025\,H}$	Good enough fit; possibly f is a little less near the surface and greater inside the clouds than in this model

calculations are included in Table VII. The best fit gives a model with mixing ratio $2 \times 10^{-3}\%$ near the surface and $2 \times 10^{-2}\%$ in the altitude range 40–48 km. This is exactly the same model found from measurements of Venera 11 (Moroz et al. 1979b, 1980a; see Chapter 13). We feel that for a better fit it is necessary to diminish the mixing ratio a little near the surface and increase it above 40 km, but only a factor on the order of 2 can be considered. If such a factor really exists, spectroscopic data would agree with the lower limit H_2O mixing ratio obtained by the gas chromatograph for the altitude range 49–58 km. A mixing ratio 0.2% obtained by the LiCl humidity sensor is hardly compatible with the spectrophotometric data as the curves of Fig. 3 demonstrate very well. Band intensities for a model incorporating such a high mixing ratio are much stronger than observed.

Mixing ratios on the order of 0.1% were obtained in a few earlier experiments: on Veneras 4, 5 and 6 (Vinogradov et al. 1968) and the Pioneer Large probe (Oyama et al. 1980). It is still not clear what are the reasons for discrepancies between optical data and these experiments—real variations (local or time-dependent) or some experimental errors in one or both sets of measurements. Variations do not however explain a discrepancy between instruments on the same probe. In any case it is clear that optical data are free from any effects of the local storage of water in the instrument or from danger of contamination of chemicals from other components than the water vapor. Our spectra of 1978 (Venera 11) were independently analyzed by L. and A.

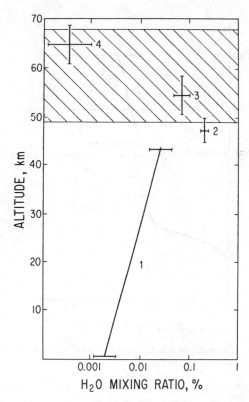

Fig. 4. Altitude profile of water vapor mixing ratios based on the data of recent measurements. (1) spectrophotometer of Venera 13; (2) LiCl humidity sensor; (3) gas chromatograph (Venera 14); (4) groundbased spectroscopy. Position of the cloud layer is marked by slanting lines.

Young, who spent 3 months in Moscow and had full access to the original records, calibrations, data, etc. Their first conclusion was that a constant mixing ratio $10^{-2}\%$ gives the best fit of Venera 11 spectra, but later they found that it is necessary to diminish the mixing ratio near the surface and to increase it near the lower boundary of clouds in the same manner as found in our interpretation made by a different method (Young et al. 1982).

Groundbased spectroscopic observations show that the mixing ratio is $<< 0.01\%$. The altitude determination of these measurements is not very accurate, but probably altitudes above 65 km give the main contribution to the observed intensities (see Chapter 13). If we add to this the recent information concerning H_2O content at the lower levels, we can conclude, in spite of still existing discrepancies, that profiles of the H_2O mixing ratio have a maximum somewhere in the range 40–60 km with substantial reduction above and below (Fig. 4). A profile of this type was first suggested from the Venera 11 spectral analysis and the new data confirm it.

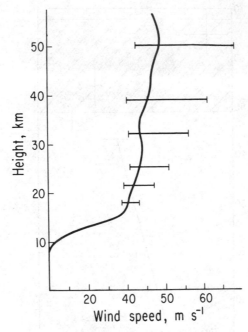

Fig. 5. Preliminary evaluations of the zonal wind velocity based on the Doppler measurements of Venera 13 (Kerzhanovich et al. 1982).

The steep gradient of mixing ratio above 50 km is reasonably explained by absorption of H_2O into the cloud particles of concentrated sulfuric acid (Chapter 16). The reduction between 50 km and the surface remains an unsolved problem. There must be a sink at low levels and a source near the cloud base. Some possibilities can be conceived, but they are too preliminary to be discussed here.

III. STRUCTURE AND DYNAMICS OF THE ATMOSPHERE

As the atmospheric structure has been well established by previous missions, no preliminary results have been published. The accelerometers and temperature and pressure sensors worked well. Surface values are shown in Table I. Preliminary evaluations of the zonal wind based on Doppler measurements (Kerzhanovich et al. 1982) are given in Fig. 5. The general properties of the zonal wind profile are the same as in previous experiments on Venera and Pioneer probes (see Chapter 21).

New evaluations of the wind velocity near the surface were obtained on Veneras 13 and 14. Doppler analysis is not sensitive enough for this purpose because the velocity is too small here. Rotary anemometers on Veneras 9 and 10 were the only previous source of information (Avduevsky et al. 1976b).

On Veneras 13 and 14 rotary anemometers were absent but the wind velocity near the surface was evaluated from the acoustic sensors (Ksanfomality et al. 1982a). During descent strong aerodynamic noise was recorded; after landing the noise continued but at a noticeably lower level. Laboratory simulation in the wind tunnel showed that this noise after landing could be generated by wind with a velocity from 0.3 to 0.6 m s^{-1}.

IV. CLOUDS AND ULTRAVIOLET EXTINCTION

Three experiments gave some information about clouds: the nephelometer, the X-ray spectrometer for aerosols, and the spectrophotometer. Only part of the spectrophotometric data is published at the time of the preparation of this chapter. Nevertheless, short communications about results of the other two experiments were given in the newspapers (see, e.g. Sagdeev and Moroz 1982a). The nephelometer (of backscattering type) showed on both probes fine altitude structure of clouds, especially complicated near the lower cloud boundary where very narrow ($\simeq$ 100 m) layers were recorded. A trap with special grids was used for collection of aerosol particles in the XRFS experiment. The X-ray fluorescence spectrum showed sulfur as the most abundant element. Chlorine was also identified but its abundance is less than that of sulfur. The opposite conclusion was obtained from the Venera 12 measurements: chlorine the main component and sulfur much less (Surkov et al. 1981). The new analysis is compatible with standard ideas concerning the composition of the clouds (see Chapter 16). However, there exists a big puzzle as to why identical instruments gave such different chemical contents of the clouds.

There are two sets of Venera 13 and Venera 14 spectrophotometric results (Moroz et al. 1982) connected with properties of the clouds: (1) altitude profiles obtained from records of spectra in the region 0.45–1.2 μm; (2) altitude profiles measured by the ultraviolet photometer at 0.37 μm. Profiles based on the spectral records from the up direction are qualitatively the same as those obtained by Veneras 11 and 12 (see Chapter 19). The spectral net flux data have not yet been analyzed. Some evaluations of the photometric properties of the clouds, the atmosphere as whole, and the surface (at 0.55 μm), obtained from the new spectra are presented in Table VIII. These data show that the optical thicknesses of the clouds are a little different at the two landing sites.

The ultraviolet profile is substantially new. The first attempt at such measurements was made on the four Pioneer Venus probes (Ragent and Blamont 1980) but was not very successful. Some undesirable pecularities of the spectral curve (windows in the region $\lambda > 0.7$ μm) were recognized too late. Moreover, the radiation was received in the horizontal direction only and would be practically uninterpretable in terms of the inverse problem of radiation scattering, even in the absence of trouble with the spectral curve.

TABLE VIII

Preliminary Evaluations of Photometric Properties
of the Atmosphere and Surface ($\lambda = 0.55 \ \mu$m)

	Venera 13	Venera 14
Solar zenith angle, deg	36	35.5
Normalized illuminance[a]		
$H = 60$ km	0.54	0.80
48	0.36	0.50
0	0.024	0.035
Clouds:		
height of lower boundary[b]	49	47.5
optical thickness (in the altitude range 47 to 62 km)	20	15
structure[c]	3 layers	3 layers
Surface:		
illuminance, W m^{-2} μm^{-1}	70	100
net flux, W m^{-2} μm^{-1}	5	7

[a]This is the value $E(h)/E_0$, where $E(h)$ is illuminance at altitude h, E_0 is illuminance outside the atmosphere at the orbit of Venus.
[b]Height above the level 6052 km.
[c]Structure as observed by Venera 11, Venera 12 and Pioneer probes: dense narrow (2–3 km) layer, relatively transparent middle layer, and again a denser layer above 57–58 km (see Chapter 16).

In our case, measurements of ultraviolet intensity were made at six angles of view: 0° (zenith), 45°, 135°, 180°, 225°, 315°. Measured intensities for 45°, 135°, 225° and 315° for Venera 14 are presented in Fig. 6. From these four measurements, flux from the upper hemisphere $F\downarrow$ and net flux ΔF were calculated. The zenith direction was not used at this time to avoid the problem of the influence of the parachute. $F\downarrow$ and ΔF are also shown in Fig. 6. Some irregular changes of intensity were observed in the altitude range 65–58 km for angles 0°, 45° and 315°. Irregularities were absent in the lower hemisphere. Two reasons for the upper hemisphere's irregularities are being considered: (1) inhomogeneities in the horizontal structure of clouds above 58 km; (2) a "halo" around the Sun direction could be visible at various angles owing to rotation and swinging of the probe. Such a halo could be noticeable to an optical depth $\tau \sim 8$ for a Henyey-Greenstein parameter $g = 0.8$. There were no large irregular changes in the intensities from either hemisphere below 58 km.

Changes in slope of the curves for $F\downarrow$ and ΔF suggest the presence of three regions: >57 km, 57–50 km, and 50–48 km. The same boundaries are found for visible radiation (see footnote c to Table VIII). Using the same inverse treatment of the problem as for the interpretation of the Venera 11 and 12 photometry in a longer wavelength region (Golovin et al. 1981), we calcu-

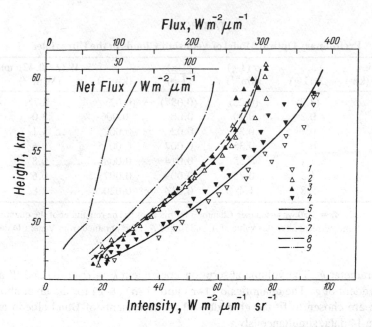

Fig. 6. Altitude profiles of ultraviolet radiation ($\lambda = 0.37 \ \mu$m) measured by Venera 14. The numbers 1, 2, 3, 4, represent intensities, measured for zenith angles 45°, 135°, 225° and 315°, respectively; 5, the intensity averaged for angles 45° and 315°; 6, intensity averaged for angles 135° and 225°; 7, the flux from upper hemisphere calculated from curves 5 and 6; 8, the flux from lower hemisphere calculated from curves 5 and 6; 9, the net flux.

lated aerosol extinction coefficient σ and single scattering albedo a in the altitude range 48-60 km. These quantities are independent of absolute calibration. The energy absorbed per unit volume was also found:

$$Q = dF/dh. \tag{3}$$

The measurements directly give Q_λ for $\lambda = 0.37 \ \mu$m. Assuming that in the range $\lambda < 0.42 \ \mu$m the optical properties of the atmosphere are the same as for $0.37 \ \mu$m, we obtained a rough evaluation of

$$Q \ (\lambda < 0.42 \ \mu\text{m}) = \text{const.} \ Q_\lambda \tag{4}$$

and total energy absorbed in the layer between altitudes h and $h + \Delta h$:

$$W = \int_h^{h + \Delta h} Q \, dh. \tag{5}$$

The quantities Q and W depend on the accuracy of the absolute calibration which can be further improved. Results of all these calculations are presented in Table IX. Values for $\sigma \ (1-g)$ and $(1-a)/(1-g)$ instead of for σ and a are presented here to remove any hypothesis concerning the phase function and

TABLE IX

Preliminary Optical Model of Venusian Clouds in the Ultraviolet

Altitude Range (km)	$\Delta\tau(1-g)$	$\sigma(1-g)$ (km^{-1})	$\frac{(1-a)}{(1-g)}$	$\bar{k}$ (km^{-1})	W ($\lambda<0.42\mu$m) (W m^{-2})
60–55	(3.7)	(0.25)	(0.087)	(0.02)	107[a]
58–60	0.2	0.10	0.06	0.006	3.0
56–58	0.7	0.35	0.02	0.007	1.1
54–56	0.7	0.35	0.007	0.0025	0.6
52–54	1.1	0.55	0.014	0.008	0.8
50–52	1.5	0.75	0.009	0.007	0.6
48–50	2.8	1.4	0.014	0.020	2.8

[a]Bond albedo $A_s = 0.50$ was used (see Chapter 3 in this book). The remainder of the quantities in this column correspond to this value of A_s and fluxes $F\downarrow$, ΔF obtained from Venera 14 data.

the parameter g. The volume absorption coefficient $k = \sigma(1-a)$ and W are independent of g. The quantities $\sigma(1-g)$ and $(1-a)/(1-g)$ for the layer above 60 km are chosen to fit the groundbased measurements of Bond albedo and Venera 14 data, simultaneously.

Table IX illustrates the main conclusions obtained from the analysis of the ultraviolet profile:

1. Most of the ultraviolet flux ($\approx 90\%$) is absorbed above 58 km. The rest is almost fully absorbed in the layer 10 km below.
2. The effective single scattering albedo is ≈ 0.99 for all the altitudes < 58 km, but substantially less above this level.
3. The volume absorption coefficient is a maximum in the upper (75–60 km) and lower (50–48 km) layers.
4. The effective optical thickness $\Delta t(1-g) \approx 7$ in the interval 48–60 km; more than half of this amount is in the lowest layer (48–50 km). If $g = 0.75$, the optical thickness is $\tau = 28$ in the range 48–60 km.

V. ELECTRICAL EVENTS IN THE ATMOSPHERE

Measurements of sporadic low-frequency radio emission observed for the first time on Venera 11 and Venera 12 were repeated by Ksanfomality et al. (1982$a,b,c,$) on recent probes as part of the experiment Groza-2 ("Groza" is thunderstorm in Russian). It is assumed that the source of these pulses is some electrical discharge similar to terrestrial thunderstorm but their real nature and the positions of the sources are a puzzle. Two possibilities are discussed in Chapter 17: discharge inside clouds and at some low level near the surface. In the latter case the phenomenon may be connected with volcanic eruptions.

For Veneras 13 and 14, altitude profiles of the averaged field strength at 10 kHz are published. Strong irregular variations of field strength are typical. There is a trend below 10 km for the emission to be systematically weaker with decreasing altitude. These properties were observed on Venera 11 and also Venera 12 but no emission was observed after landing. The authors of the experiment propose that the explanation is refraction of radiation from sources near the surface.

It is also conceivable that the whole phenomenon is due to electrification of the probe and is not natural. A thorough analysis of this possibility does not exist. As a first attempt to obtain relevant data, the Groza-2 experiment included a sharp electrode installed on the external wall of the probe to measure the corona discharge current. A smooth altitude-current curve was obtained and presented by Ksanfomality et al. (1982c) but not discussed.

VI. SURFACE IMAGES

The first optical images of the venusian surface (panoramas) (Selivanov et al. 1982; Florensky et al. 1982) were obtained in 1975 by Venera 9 and 10 probes. The experiment was repeated on Venera 13 and Venera 14 with a few improvements: (1) a full 360° field of view; (2) more digital brightness steps; (3) smaller pixel size; (4) use of color filters for some images. The quality of images is now much better and gives greater possibilities for geological analysis. A few versions of black and white and color panoramas were widely published (see Fig. 7 and Color Plate 5).

To provide a 360° field of view two cameras (telephotometers) were installed on each probe. They are mechanically scanning photoelectric devices with nominal field of view 37 × 180°. The angular resolution is 11 min and the height above the base of the probe is 90 cm. The axis of the "horizontal" scanning plane is 50° from the vertical. This is why the middle part of the panorama is a perspective image of the part of the surface which is near the camera while the borders are the images of more distant parts. The oblique line of the venusian horizon is visible more or less clearly in the upper left and right corners of every image. A few details of the probes visible in the field of view can be used for scaling: the ring base with triangular "crown" (the role of which is stabilization of the descent); a band with color test patterns (on the right half); the grating of the instrument for physical properties of the soil (PROP in Russian abbreviation); and covers of the telephotometers which were removed after landing and are lying on the surface. The distance between triangles of the crown is 5 cm; the length of the PROP-grating is 60 cm; the width of the test band is 9 cm; the semicylindrical covers of the telephotometers are 12 cm high and 20 cm wide.

The most common details of the surface structure visible in both panoramas are flat light stones, dark soil between them, and a quantity of small grains. There is more granulated substance around Venera 13 than Venera 14. The layered structure of the flat stones is noticeable in many cases. In general the neighborhood of each probe is reminiscent of the bottom of

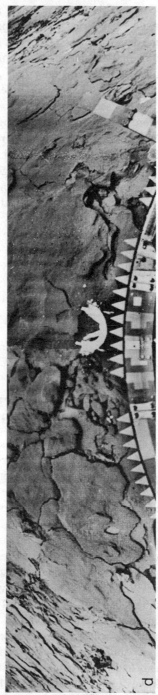

Fig. 7. (a) the panorama of the Venera 13 landing site, first half; (b) the second half of the Venera 13 panorama. (c) and (d) are the two halves of the Venera 14 panorama.

terrestrial oceans. This analogy is not senseless, because the atmospheric density is high (~ 0.1 g cm^{-3}) and the process (sedimentation) of the formation of surface details may be nearly the same as in our oceans. The simultaneous presence of flat stones and granulated material supports the idea of sedimentation. There is no doubt that grains can be moved by winds on Venus. Some evidence of such movements was obtained immediately from comparison of a few sequential images.

A few words are in order concerning the venusian horizon visible in the images. By normal refraction (when the refractive index rises with decreasing altitude to the surface) the venusian horizon must appear more distant than it geometrically is. The distance to the geometrical horizon is $\simeq 1$ km and Rayleigh scattering must smooth any sharp border even at this distance. A visibly sharp horizontal line means that the distance to the horizon is substantially less than 1 km, and may be only $\simeq 0.1$ km. Consequently there was anomalous refraction on both sides of the site. It is therefore necessary to have a temperature "break" near the surface (lapse rate $\geqslant 0°1$ m^{-1} in the lowest ~ 1 m layer). This phenomenon is very similar to a terrestrial mirage. The nearness of the visible horizon was first noted from the analysis of the Venera 9 and 10 panoramas (Moroz 1976). It may exist everywhere on Venus. In this case an observer on the Venus surface must have the impression that he is standing on a spherical body having a not very large diameter, much less than 1 km. The effect may depend strongly on the height from which it is observed.

VII. ANALYSIS OF THE COMPOSITION OF THE VENUSIAN ROCKS

This experiment was flown for the first time on Veneras 13 and 14 (Surkov et al. 1982a,b). A drilling device bored into the soil, got a sample ~ 1 cm^3 from depth ~ 3 cm and delivered it to a small chamber inside the probe. This chamber was pumped to a pressure of ~ 50 mbar which is low enough to avoid interference with the X-ray spectrometer. After this operation the sample is transferred to the XRS and is irradiated by radioisotopic sources, one of ^{238}Pu and two of ^{55}Fe. Simultaneous use of different types of source provides a wide enough range of measured elements from Mg to Fe.

Fluorescent X-ray emission of the soil was detected by four proportional gas counters with 40 micron beryllium windows. Spectra obtained on Venus at both landing sites may be seen in Fig. 8. The composition of the soil obtained from these spectra is given in Table X. At the Venera 13 site the composition of the soil is near to a terrestrial leucitic basalt with a high fraction of potassium. Such rocks are rare on Earth and found mainly on oceanic islands and in the continental rifts of the Mediterranian. A different type of rock was found at the Venera 14 site; here the rocks are similar to terrestrial tholeiitic basalt which is very common on our planet.

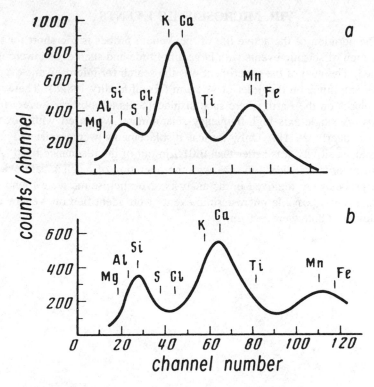

Fig. 8. X-ray fluorescent spectra of venusian rocks: a, Venera 13; b, Venera 14. Duration of measurement was 192 s.

TABLE X

**Composition of Rocks at the Landing Sites of
Venera 13 and Venera 14 (%)**

Elements (oxides)	Venera 13	Venera 14
MgO	11.4 ± 6.2	8.1 ± 3.3
Al_2O_3	15.8 ± 3.0	17.9 ± 2.6
SiO_2	45.1 ± 3.0	48.7 ± 3.6
K_2O	4.0 ± 0.63	0.2 ± 0.07
CaO	7.1 ± 0.96	10.3 ± 1.2
TiO_2	1.59 ± 0.45	1.25 ± 0.41
MnO	0.2 ± 0.1	0.16 ± 0.08
FeO	9.3 ± 2.2	8.8 ± 1.8
Σ	96	97

VIII. MICROSEISMIC EVENTS

The duration of the active life of the Venera probes is too short for the observation of seismic events with large amplitude and such events were not expected. The goal of the experiment was the search for microseismic events with an amplitude on the order of $\leqslant 1$ μm (Ksanfomality 1982c). There are many places on the Earth where such displacements are observed every few seconds. A single-axis high-frequency seismometer was used with a resonance of nearly 26 Hz. Only vertical displacements were registered and threshold sensitivity was better than 0.01 μm out of the resonance band. All parts of records coinciding in time with the operation of mechanical devices on the probes were removed in the analysis. No microseims were found on Venera 13; two possible microseismic events were identified on Venera 14. Amplitudes of both were ~ 1 μm.

6. TOPOGRAPHY, SURFACE PROPERTIES, AND TECTONIC EVOLUTION

GEORGE E. McGILL
University of Massachusetts

JEFFREY L. WARNER
NASA Johnson Space Center

MICHAEL C. MALIN
Arizona State University

RAYMOND E. ARVIDSON
Washington University

ERIC ELIASON
U.S. Geological Survey

STEWART NOZETTE and ROBERT D. REASENBERG
Massachusetts Institute of Technology

Despite the major bulk similarities between Venus and Earth, geologically interesting differences in atmospheric composition, atmospheric and lithospheric temperature, and possibly mantle composition suggest that the rock cycle on Venus will not be similar to the rock cycle on Earth. Exposed rock on the surface is indicated both by Venera images and by radar-derived dielectric constants. The materials sampled by Veneras 8–10 and 13–14 seem chemically similar to common igneous rocks on Earth. If differences of atmospheric pressure and temperature with elevation, and the probable wind transport of weathered regolith are taken into consideration, then it is thermodynamically possible for the minerals in these common rocks to be decomposed by reaction with the atmosphere. Radar data are not consistent with a thick, widespread, porous regolith as on the Moon. However, wind-transported regolith could be cemented into sedimentary rock that would be indistinguishable from other rocks in radar returns, so the surface of Venus may consist of weathered igneous rocks, or compacted and partially cemented sedimentary rocks, or both. The elevation spectrum of Venus is strongly unimodal, in marked contrast to Earth; only ~ 15% of the 93% of the surface mapped deviates > 1 km from the median elevation. Except for a few features very probably of volcanic origin, and troughs apparently analogous to continental rifts, most topographic features on Venus remain enigmatic. Some of the roughly circular forms previously inferred to be impact craters do not show the expected topography or shape, and thus may be of volcanic or tectonic origin. The overall rate of tectonic and volcanic activity on Venus seems similar to or slightly less than that on Earth, but the style is probably different. Considerations of internal thermal structure, topography, and gravity anomalies lead to two types of tectonic model: one involving a lithosphere too thick or too buoyant to participate in convective flow; and a second with a lithosphere that does participate in convective flow, implying the existence of venusian plate tectonics. Although trends implying structural control of topography are present, features consistent with Earth-like plate tectonics have not been recognized. Abundant depressions and elevations, some arranged in linear zones, suggest a form of hot-spot tectonics, but linear and arcuate troughs indicate that some type of plate tectonics may exist, at least in incipient form. Unless the high surface temperature of Venus is a geologically young characteristic, a thin lithosphere with dynamic mantle support of topography is more likely than static support. This suggests that the major topographic features seen on Venus are due to geologically recent tectonic or volcanic activity.

Clouds and hazes in the atmosphere of Venus prevent direct visual observation of its surface. During many years of optical observation, reports of breaks in the cloud cover often were accompanied by speculations on the nature of the surface materials and environment, which all proved untrue. In the early 1960s advances in passive radio astronomy permitted the reception of faint microwave signals due to thermal emission from the surface of Venus. Active microwave studies began in 1961 with the detection of radar signals reflected off the surface. Observations during the 1964 and 1967 inferior conjunctions focused on measurements of rotation rate, spin axis, and orbital motion, and on the initial observation of surface characteristics (e.g. Pettengill et al. 1967; Dyce et al. 1967; Goldstein 1965; Carpenter 1966; Zohar and Goldstein 1968). Beginning with observations made during the 1967 inferior conjunction, variations in radar reflectivity were mapped across a cap facing the Earth (Rogers and Ingalls 1969; Goldstein and Rumsey 1970). These first efforts were at low resolution, but subsequent observations provided data with radar cell dimensions approaching 10 to 20 km across (Rumsey et al. 1974; Campbell et al. 1976; Goldstein et al. 1976, 1978). In addition, average surface properties for the entire disk were derived from reflectivity determinations (reviewed in Kuz'min and Marov [1974] and Chapter 4).

As Earth-based radar observations reached a plateau and became somewhat routine, spacecraft observations began. Venera 8 landed on Venus in

1972, returning information on the radioactive components of its surface material through gamma-ray spectrometry (Vinogradov et al. 1973). Veneras 9 and 10 were even more successful, returning not only gamma-ray observations but also images of the surface (Surkov et al. 1974, 1976a; Florensky et al. 1977b,c). Veneras 13 and 14 provided additional images as well as the first chemical analyses for major elements. The Pioneer Venus orbiter accumulated a near-global radar altimetry map of moderate resolution, and near-global distributions of surface slopes and reflectivity, which allowed specific questions about the global-scale geology of Venus to be addressed.

Because Venus has nearly the same diameter and mass as the Earth, and the two planets are adjacent in the solar system, Venus has been referred to as Earth's twin. This similarity seems reinforced by similar amounts of two important atmospheric constituents, N_2 and CO_2, though most of Earth's CO_2 is not now in the atmosphere.

However, Venus has some intriguing differences from the Earth, of which the most geologically significant are:

1. Density and composition of the atmosphere;
2. High atmospheric and surface temperatures;
3. Bulk density slightly smaller than if Venus were exactly like Earth in composition.

Differences (1) and (2) affect the nature and rates of exogenic processes of weathering, erosion, transportation, and deposition; (2) and (3) affect the nature and rates of endogenic tectonic and volcanic processes.

Important first-order geological questions concern the tectonic evolution of a planet, and the processes that modify its surface. Internal heat causes poorly understood processes in the interior that, in turn, are manifested on the surface either by vertical displacement and lateral translation of surface layers (tectonics), by volcanic activity, or both. What are the nature and rate of Venus tectonics, especially as compared to Earth tectonics? Do the differences in mantle composition, thermal structure of the crust and upper mantle, and high surface temperature rule out an Earth-like system of plate tectonics on Venus? Does Venus transfer heat to the surface primarily by vertical tectonic activity and volcanism, rather than by a combination of these processes and lateral tectonic activity as on Earth?

Lack of data concerning the chemical and petrological composition of the Venus surface is a serious block to deciphering its tectonics. Major questions are:

1. What is the distribution of sial (rocks relatively rich in Si, Al, K, Na, etc; i.e. granitic in a broad sense) and sima (rocks relatively rich in Mg, Fe, Ca, etc; i.e. generally basaltic) on Venus's surface?
2. What is the range of concentrations of structurally bound H_2O in venusian rocks?

3. What are the systematics of isotopic and trace-element distributions in venusian rocks?
4. What is the distribution of igneous and sedimentary rocks on the surface of Venus?
5. Are metamorphosed rocks present on Venus?

Answers to such questions provide major constraints in modeling a planet's tectonic history.

The distribution and proportions of sial and sima are essential for tectonic analyses, because of their different physical properties; most importantly, the bulk density of sial is lower than that of sima. Sial is too buoyant to be recycled into the underlying mantle, but sima is not. Because sial is buoyant, it is concentrated on Earth by lateral motions of the sea floor into regions of limited extent and significant crustal thickness, called continents. Are there continents on Venus and if so, did they form by lateral tectonics like continents on Earth?

The presence or absence of structurally bound H_2O is important because it changes the physical properties of rocks. Rocks with structurally bound H_2O have lower melting points and lower viscosities at equivalent temperatures than anhydrous rocks. As little as 0.5% H_2O lowers the melting point of rocks as much as 100 to 200°C. Rocks saturated with structurally bound H_2O (which requires only a few % at low pressure) have melting points lowered by as much as 500°C. Furthermore, structurally bound H_2O at high pressure affects the chemical composition of the first partial melt, and thus affects the probable compositions of differentiated igneous rocks reaching the crust.

The distribution of igneous, sedimentary, and metamorphic rocks provides clues to a wide range of geological processes. Igneous rocks signify mass and energy transfer from the inside of a planet to its surface. Metamorphosed rocks imply two-way transfer of mass from the surface of a planet to some depth and back to the surface. Sedimentary rocks may imply topographic relief and erosion, although there are no strong constraints on whether the topography is due to volcanism, vertical tectonics, or lateral tectonics. The abundance, composition, and distribution of sediments and sedimentary rocks could provide key data concerning questions of exogenic processes:

1. What processes modify topographic relief, and what are their characteristic rates?
2. How active is weathering on Venus?
3. Can winds erode the surface effectively and transport debris to distant deposition sites?
4. Can solid-state creep of hot crustal rocks significantly modify landforms?
5. What are the chemical interactions between the atmosphere and the surface on an abiotic planet with a hot, dense atmosphere?

6. Do these reactions have any significant control over atmospheric composition?

This chapter will review our knowledge about the surface of Venus relevant to the basic questions discussed above, with reference to observations and speculations concerning the interior of the planet. Obviously, our knowledge is preliminary and sketchy, but we know far more now than when reviews were last prepared a few years ago (Hunten et al. 1977; Masursky et al. 1977). This chapter also will emphasize major questions to be addressed in the future.

I. PHYSICAL AND CHEMICAL PROPERTIES OF THE SURFACE

A. Radar Studies

Radio waves emitted or reflected from planetary surfaces contain information on: surface roughness, at about the scale of the wavelength involved; average surface slopes, at a scale $\geq$ 10 times the wavelength; surface brightness temperature; and the physical nature of surface materials (inferred from the reflectivity and dielectric constant). Excellent reviews of the principles of such studies may be found in Evans and Hagfors (1968), Evans (1969), Hagfors (1970), and Kuz'min and Marov (1974); the last also contains a thorough review of studies of the surface of Venus prior to 1974.

The total radar cross section of Venus may be represented as the sum of two components, a quasi-specular component and a diffuse component. The dependence of a specific region's relative cross section on the angle of incidence θ is complex, and cannot be represented by a single equation (e.g. Campbell and Burns 1980; Pettengill et al. 1980a). For $\theta \leqslant 20°$, the reflection is dominantly quasi specular and is dependent on the average slopes of the surface at scales large relative to the wavelength. Earth-based images of regions near the equator of Venus, and maps of the inverse slope parameter (Hagfors constant) appearing in the scattering equations used to model Pioneer Venus altimetric data, represent distributions of average (rms) meter-scale and larger slopes. At these low values of θ, the surface scattering properties are satisfactorily modeled by (Hagfors 1970)

$$\sigma_0(\theta) = \frac{\rho C}{2} \left(\cos^4 \theta + C \sin^2 \theta \right)^{-\frac{3}{2}} \tag{1}$$

where σ_0 is specific radar cross section for a given θ; θ, angle of incidence of the radar beam; C, constant approximately equal to inverse of mean-square slope (in radians), the Hagfors constant; and ρ, power reflectivity at normal incidence to surface, the Fresnel reflection coefficient defined in Eq. (2). For $20° \leqslant \theta \leqslant 60°$, diffuse scattering by surface roughness at a scale less than or equal to the wavelength is dominant; the areal variability of this property is represented by Earth-based images of most of Venus, and by the side-looking images from the Pioneer Venus radar. Campbell and Burns (1980) find that at intermediate values of the incidence angle the diffuse surface scattering varies according to $\cos^{1.5} \theta$, a relationship that falls between the Lommel-Seeliger ($\cos \theta$) and Lambert ($\cos^2 \theta$) scattering laws (Evans 1969). For $\theta \gtrsim 60°$ the reflectivity probably depends both on shadowing by very large-scale (> 1 km) surface slopes, and on diffuse scattering by small-scale roughness.

The total cross section measured from Earth is dominated by the quasi-specular component coming from the part of the disk closest to Earth (Kuz'min and Marov 1974) and thus at low incidence angles; hence the diffuse component generally is not important. For small regions at low θ, Eq. (1) may be used to determine the surface reflectivity ρ, if C can be independently estimated (Aleksandrov et al. 1980). Assuming a negligible diffuse component yields a minimum value for the reflectivity (Evans 1969). Surface reflectivity is related to the dielectric constant by

$$\rho = \left(\frac{1 - \sqrt{\epsilon}}{1 + \sqrt{\epsilon}} \right)^2 \tag{2}$$

where the complex dielectric constant ϵ consists of real and imaginary parts, the latter depending on volume conductivity. For most dry rocks and soils conductivity is negligible, hence ϵ is taken as equal to its real part.

Kuz'min and Marov (1974, pp. 56–68), using a weighted average of total cross section values from all observations at wavelengths long enough to free the results of significant atmospheric absorption, calculate a mean dielectric constant for Venus of 4.7 ± 0.8. Aleksandrov et al. (1980), using Eq. (1) plus Doppler filtering, show that the dielectric constant ranges from 2.7 to 6.6 along the track of the sub-Earth point, with weighted mean value 4.1 ± 0.9 Kolosov et al. (1981), using bistatic radar, find that the dielectric constant varies from 3.0 to 9.1 along two narrow ($\sim 2.5° \times 25°$) east-west strips $\sim 25°$ south of the equator. The mean values of the dielectric constant agree fairly well with values determined from emitted radio waves (Kuz'min and Marov 1974). Although the low extreme of 2.7 would be expected from a dry regolith, the mean values of $\geqslant 4$ are more consistent with dry rock (Keller 1966; Evans 1969; Campbell and Ulrichs 1969). These results suggest a much less widespread (or much thinner) unconsolidated and porous regolith than on the Moon, which is characterized by dielectric constants of 2.6 to 2.8 for radar of comparable wavelengths.

The dielectric constants obtained for Venus from Earth-based observations are, of course, average values for rather large areas. These areas lie near the equator, because that region was viewed by Earth-based radar with sufficiently low incidence angles for the quasi-specular component of the cross section to dominate.

Krotikov (1962) determined experimentally that the densities of dry, nonconducting, porous materials are related to the dielectric constant by

$$d = \left(\frac{\sqrt{\epsilon} - 1}{a} \right) \qquad (3)$$

where d is the bulk density in g cm^{-3}, and a is a wavelength dependent constant $\simeq$ 0.5 for the cm and dm wavelengths used to study Venus. The mean values of the dielectric constant determined for Venus from Earth-based radar yield densities $\sim$ 2.2 g cm^{-3}, with low extreme 1.3 and high extreme 3.1. If a non-negligible diffuse component exists, these values would be larger. The Rayleigh mixing formula

$$\frac{1}{d} \left(\frac{\epsilon - 1}{\epsilon + 2} \right) = \frac{1}{d_0} \left(\frac{\epsilon_0 - 1}{\epsilon_0 + 2} \right) \qquad (4)$$

where ϵ_0 and d_0 are the dielectric constant and density of solid rock, relates the dielectric constants and densities of solid and powdered rock as well as any model (Campbell and Ulrichs 1969). Assuming that weathered rock on Venus's surface is comparable in dielectric properties to laboratory powders, the average ϵ for Venus is consistent with a surface of weathered basalt with $\sim$ 20% porosity, or of weathered granitic rock with $\sim$ 5% porosity.

Determinations of reflectivity by the Pioneer Venus radar indicate a global mean dielectric constant of 4.5 $\pm$ 0.8, in excellent agreement with Earth-based results (Pettengill et al. 1982). However, dielectric constants higher than values characteristic of dry rocks are found locally, suggesting the presence of rocks containing significant proportions of minerals with non-negligible conductivities. Candidates include iron sulfides (Pettengill et al. 1982) or ferrimagnetic minerals such as magnetite and hematite.

B. Results from Venera Landers

The only direct observations of the small-scale characteristics of the surface of Venus were made by the Venera 8, 9, 10, 13, and 14 landers. Veneras 8, 9, and 10 used gamma-ray spectrometers/densitometers to determine surface material radioactivity and density; Veneras 13 and 14, using X-ray fluorescence (XRF), determined abundances of major elements. Density at the Venera 10 landing site, where the gamma-ray detector rested on rock, is reported as 2.8 $\pm$ 0.1 g cm^{-3}, typical of rock densities (Surkov et al. 1976b). Venera 8 measured a density 1.5 g cm^{-3} (Vinogradov et al. 1973), about half that expected for rock, but because Venera 8 did not return any images we do not know what the gamma-ray detector was resting on.

The abundances of K, U, and Th measured by the gamma-ray detectors on Veneras 8, 9 and 10 are more like those of the Earth than of the Moon or of meteorites (Fig. 1; Table I). Considering the difficulties inherent in obtaining these measurements, it is probably unwise to put too much weight on the precise abundances quoted. However, we may extrapolate from these data to infer rock compositions, by the correlation of enriched potassium with en-

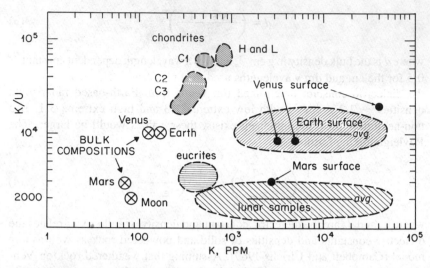

Fig. 1. Potassium abundances and K/U ratios for Venus, Earth, Mars, Moon, and selected meteorites. Estimated total planet bulk compositions were calculated by Taylor. Figure from Taylor (1982).

TABLE I

Venera Measurements of Uranium, Thorium, and Potassium Abundances[a]

Venera	U($\times 10^{-4}$ wt. %)	Th($\times 10^{-4}$ wt. %)	K(wt.%)	K/U($\times 10^{4}$)
8	2.2 ± 0.7	6.5 ± 0.2	4.0 ± 1.2	$1.82^{+1.65}_{-0.85}$
9	0.60 ± 0.16	3.65 ± 0.42	0.47 ± 0.08	$0.78^{+0.47}_{-0.27}$
10	0.46 ± 0.26	0.70 ± 0.34	0.30 ± 0.16	$0.65^{+1.65}_{-0.46}$

[a]From Surkov (1977).

riched Si and Al as observed among geological materials from Earth, the Moon, and meteorites. The enriched potassium abundances at the Venera 8 site are believed to represent sial, while the depleted potassium abundances at Venera 9 and 10 landing sites are interpreted to represent sima. The XRF analyses (Table II) from Veneras 13 and 14 suggest that the materials sampled were derived from basaltic rocks. All these results indicate that Venus is differentiated because they demonstrate that Venus surface rocks differ from primitive meteorite rocks in a way consistent with differentiation trends observed in Earth rocks (Fig. 1).

TABLE II

Preliminary Results of X-Ray Fluorescence Analyses by Veneras 13 and 14[a]

Oxide	Weight Percent	
	Venera 13	Venera 14
SiO_2	45 ± 3	49 ± 4
TiO_2	1.6 ± 0.4	1.2 ± 0.4
Al_2O_3	16 ± 3	18 ± 3
FeO	9 ± 2	9 ± 2
MgO	11 ± 6	8 ± 3
CaO	7.0 ± 1.0	10 ± 1.2
K_2O	4.0 ± 0.6	0.2 ± 0.1
MnO	0.2 ± 0.1	0.16 ± 0.08

[a]Modified from Surkov et al. (1982*b*).

Veneras 9, 10, 13, and 14 also acquired images of the surface. The image data for Veneras 9 and 10 were obtained with panoramic cameras 0.9 m above the surface, rotated about axes oriented 40° from the normal to the spacecraft equator (i.e. in the center of their travel the cameras looked downward at an angle of 50° from the equators of the landers). The nominal fields of view were 40° × 180° with angular resolution ~ 0.35°. The images extend from the base of the spacecraft to just over a meter away in the middle of the frames and to the horizon at the frame edges (Fig. 2) (see also Color Plate 5).

Venera 9 landed on a rocky surface sloping 15 to 25° (Florensky et al. 1977*a*). A number of discrete blocks can be discerned, ranging in width up to ~ 70 cm and up to 15 to 20 cm high. Most of the rocks are flat and slab-like, but some appear rounded. Rocks > 5 cm wide cover ~ 50% of the surface; the remaining surface area is covered by lower albedo material apparently composed of particles too fine to be resolved by the cameras. Several rocks display fractures and grooves suggestive of layering (Florensky et al. 1977*b, c*), but the image resolution is insufficient to definitely identify layers. Most of the rocks at the Venera 9 landing site appear to be massive boulders with little visible surface texture. However, one rock (lower right-hand part of the frame) shows light and dark splotches suggestive of varying texture. There is also a large number of small fragments, ≤ 10 cm in diameter, that may be partly derived from the boulders.

Venera 10 landed on a plain composed of flat, scattered outcrops separated by lower albedo, fine-grained material (Florensky et al. 1977*a*). The outcrops are typically 1 to 3 m in lateral dimensions and make up just over 50% of the surface area around the lander. The overall impression is that the surface is smoother at cm-dm scales than the surface at the Venera 9 site (Fig. 2). Within the regolith are smaller flat areas of rock, but it is not clear

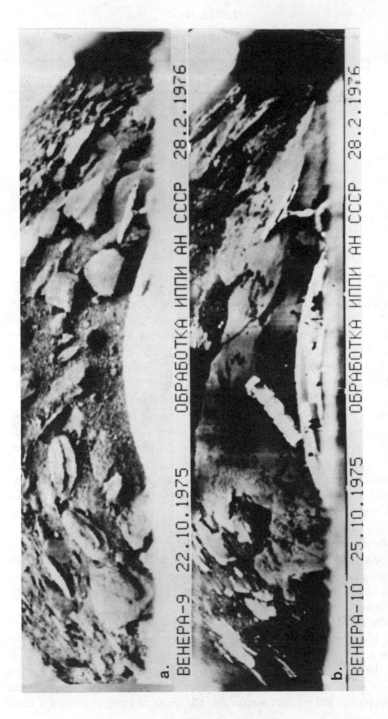

Fig. 2. Images of Venera 9(a), 10(b), 13(c) and 14(d) landing sites. The bright metallic feature in the right portion of the Venera 9 and 10 images is the gamma-ray spectrometer-densitometer head, ~40 cm long. Photographs courtesy of C. P. Florensky and V. L. Barsukov.

ВЕНЕРА-13 ОБРАБОТКА ИППИ АН СССР И ЦДКС

с.

ВЕНЕРА-14 ОБРАБОТКА ИППИ АН СССР И ЦДКС

d.

whether these are fragments broken off the larger outcrops (Florensky et al. 1977*b,c*) or bedrock partly exhumed or buried. The rock beneath the Venera 10 lander is fractured and grooved in places, but in general has a smooth, unmarked surface texture of similar appearance to that of the individual boulders at the Venera 9 site. Other patches of rock show rough, textured surfaces that seem to have trapped fine, dark material in pits and grooves.

The Venera 13 and 14 landing sites (Fig. 2) are more like the Venera 10 site than that of Venera 9, although the Venera 13 site includes a large area of dark regolith close to the spacecraft. The area viewed by the Venera 14 camera is almost entirely large, stepped platforms resembling outcrops of rock consisting of layers with thicknesses on the order of centimeters.

Mechanical properties of the regolith and rocks at the Venera 9 and 10 sites were estimated from the effects of the gamma-ray spectrometer/ densitometer head on these materials (Leonovich et al. 1977). Tests before launch allowed comparisons of landing site observations with simulations. From these comparisons it has been deduced that the strengths of individual stones at the two sites are on the order of tens of megapascals, and that the regolith beneath the Venera 9 rocks has a bearing strength nearly as large, suggesting a composition of well-compacted rock chips and fragments.

The surface albedos at both Venera 9 and 10 landing sites are low (Selivanov et al. 1976*a*). The dark, fine-grained materials of the regolith have albedos < 3%. Rocks at both sites have ~ 5% albedos though a few at the Venera 10 site have albedos approaching 12%.

Decreases in radiation immediately following landing were measured by photometers carried by both Veneras 9 and 10, a result interpreted as due to surface dust disturbed into the atmosphere by the landers (Ekonomov et al. 1980). This implies that some of the regolith must be poorly consolidated. Calculations by Garvin (1981) suggest that the dust particles raised into the atmosphere have diameters in the range 18 to 30 μm. The Venera 13 lander also raised dust, whereas the Venera 14 lander evidently did not (V. L. Barsukov, oral presentation, *Lunar and Planetary Science Conf.* XIII, 1982).

The Venera 9 and 10 imaging systems are similar to the facsimile cameras used on the Viking landers on Mars, but the Viking cameras are 1.3 m above the surface, rotate around axes perpendicular to the equators of the landers, and have better resolution (Table III). It is possible to compare the materials seen in Venera lander images to those in Viking lander images by digitally altering the latter to simulate Venera perspective and resolution (Fig. 3).

Figure 3 shows what the high-resolution mosaic of Viking lander 1, camera 1 would have looked like had it been acquired with the Venera system. The difference in information content is indicated by Fig. 4, a section of the Viking mosaic corresponding to the area outlined in white in Fig. 3. Note that the distorted Viking view in Fig. 3 extends much farther toward the horizon than the Venera views. In fact, the Venera image would have just reached the

TABLE III

Characteristics of Viking and Venera
Lander Facsimile Cameras

Parameter	Viking 1, 2	Venera 9, 10
Height	1.3 m	0.9 m
Tilt of camera (relative to spacecraft equator)	0°	50°
Angular resolution (pixel width)	0°.04	0°.35
Vertical field of view	65°	40°
Horizontal field of view	342°.5	180°

Fig. 3. Viking lander 1 high-resolution mosaic for ∼ 180° of azimuth, degraded to Venera 9 and 10 resolution. (a) normal perspective (white box defines area of Fig. 4). (b) distorted version to simulate Venera perspective.

Fig. 4. Area of Viking lander image shown within white box on Fig. 3. (a) full resolution. (b) degraded version to simulate Venera 9 and 10 resolution.

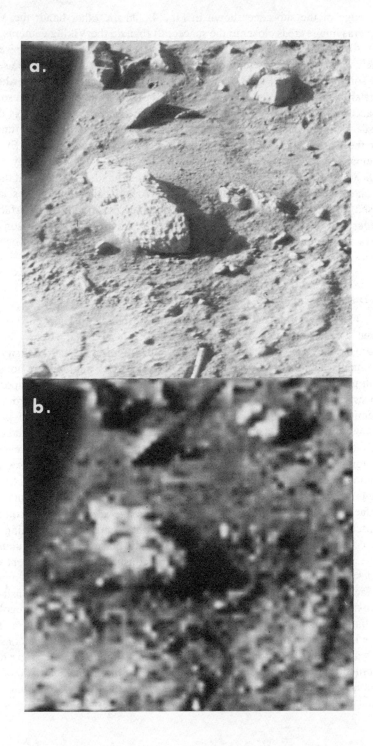

top edge of the subscene shown in Fig. 4. On the other hand, the Venera cameras image areas closer to the spacecraft than do the Viking cameras.

About 12% of the Viking 1 site is littered with blocks ranging in size from a few cm to ~ 1 m in width and height. Considerable texture can be seen in the soil exposures in Fig 3, although most of the information is lost when the data are degraded to Venera resolution. The degradation of angular resolution is accomplished by averaging adjacent pixels; thus, the technique is only a first-order approximation because differences in modulation transfer functions for the Viking and Venera imaging systems are not considered. Even so, comparison of the soil exposures in all three areas shows a clumpy texture indicating a finite cohesive strength. On the other hand, it is unclear if the lack of checkerboard brightness patterns on the rocks at the Venera sites, as opposed to the Viking site, is due to a dominance of smooth rock surfaces on Venus or to differences in the camera systems. At any rate, comparison of the surfaces seen at the three sites suggests that the Venera 9 site is roughest at scales ≤ 1 m, then the Viking site, and then the Venera 10 site.

C. Discussion

The Venera images of the surface clearly indicate the presence of rock, some as pieces detached from the bedrock, but some apparently exposed bedrock; they also indicate the presence of a regolith. Venera density measurements, gamma-ray spectrometer results, and XRF analyses are consistent with the visual observations of rock, because the densities and compositions correspond to those of igneous rocks abundant on Earth which, from basic chemical considerations, should be abundant on other Earth-like planets. Strength measurements are also consistent with solid rock, and they indicate as well that the regolith is strong enough to be consolidated; however, the photometer results and the lack of any data on regolith thicknesses at the landing sites suggest caution in interpreting this result. The dielectric constants of surface materials, as determined from radar measurements of the surface reflectivity, generally fall within the range appropriate for somewhat porous rocks (possibly due to weathering), but do not tell us anything about the composition of the rocks except perhaps in places where ϵ is unusually large. However, the values for the dielectric constant are not consistent with a widespread thick regolith as on the Moon. Dielectric constants for most dry surface materials like those on the Moon and Venus are very sensitive to porosity, so the Venus results are consistent with the presence of a regolith or aeolian sediments only if these materials have been consolidated and thus have lost some of their porosity. The data from all sources thus are consistent with any kind of low-porosity, consolidated rock—igneous, metamorphic, or sedimentary—with bulk compositions similar to those characteristic of common igneous rocks on Earth, and are also consistent with a porous, unconsolidated regolith constrained to be very thin or areally very restricted.

II. ATMOSPHERE-SURFACE INTERACTIONS

A. The Rock Cycle

Studies of process in the earth sciences commonly are organized by the concept of a rock cycle. This idea envisions a continuing dynamic balance between endogenic volcanic and tectonic processes that supply fresh rocks to the surface and create topographic relief, and cxogenic processes of weathering, erosion, transportation, and deposition that destroy both the fresh rocks delivered to the surface and the topographic relief created by endogenic processes. Almost all studies of planetary geology are attempts to understand some aspect of the rock cycle, and commonly evaluate the importance of specific processes and the overall rates of change in comparison to Earth. In the case of Venus, the differences from Earth are substantial: a much hotter and denser atmosphere; no liquid water; and much higher surface (and presumably crustal) temperatures. The first two characteristics, along with the specific composition of the atmosphere of Venus, are the basis for a comparative study of exogenic processes on Venus and Earth, the subject of this section. The last difference mainly affects endogenic processes and will be considered in Sec. IV.

The similar bulk properties of Earth and Venus, and the compositions of Venus surface materials determined by the Venera landers, indicate that igneous rocks delivered to the surface of Venus will contain the same major anhydrous rock-forming silicate minerals as on Earth. The questions then are:

1. What processes could weather these rocks, and what are the composition and characteristic grain sizes of the weathering products and the weathering rate?
2. What is the probability that these weathering products can be eroded and transported by the dense, dry atmosphere, and if so, at what rate?
3. What are the possible nature and extent of sediments and sedimentary rocks produced by deposition of transported weathering products?

This exercise is almost entirely speculative in the case of Venus, because we have no surface data at the scale and resolution needed to obtain the morphological clues used to understand the exogenic part of the rock cycles on Earth, the Moon, and Mars; nor for that matter do we know nearly enough about the chemistry and physical properties of the surface materials.

B. Chemical Weathering

Studies of chemical weathering on Venus essentially search for thermodynamically feasible reactions between the atmosphere (Table IV) and the minerals present in common rocks, taking into consideration the elevation-related ranges in atmospheric pressure and temperature and the probability that weathering products produced at high elevation (and thus at relatively low pressure and temperature) will be transported to low elevation (higher pres-

TABLE IV

Reactive Components of the Lower Atmosphere of Venus

Gas	Mole Fraction of Total Atmosphere	Observation
CO_2	0.96	Mariner, Pioneer Venus, Venera
CO	$10^{-4} - 10^{-5}$	Spectroscopic, Pioneer Venus
H_2O	$10^{-5} - 10^{-4}$	Pioneer Venus, Venera, spectroscopic
HF	10^{-8}	Spectroscopic
HCl	10^{-6}	Spectroscopic
SO_2	2×10^{-4}	Pioneer Venus, spectroscopic
COS	$< 2 \times 10^{-6}$	Pioneer Venus
H_2S	$\sim 3 \times 10^{-6}$	Pioneer Venus

sure and temperature). Models of weathering processes on Venus are very poorly constrained due to: (1) important uncertainties in our knowledge of abundances of H_2O and sulphur-bearing gases in the lower atmosphere (see Chapter 13); (2) the probability that not all important reactions have been anticipated; and (3) insufficient knowledge concerning reaction kinetics.

The problem of chemical weathering on Venus can be approached using chemical thermodynamics, if we assume that the surface temperatures are high enough to allow reaction equilibrium to be achieved rapidly relative to processes favoring disequilibrium (Mueller and Saxena 1977). If bulk solid-gas equilibrium is assumed, it is possible to determine which, if any, rock-forming minerals are stable on the surface. Disequilibrium will be favored over equilibrium: if fresh rock is not supplied quickly enough to keep pace with gas-gas reactions (Chapter 13; Orville 1975); if other processes, such as volcanic activity, supply gases too rapidly (Mueller and Saxena 1977); or if gases and solids do not communicate chemically under the appropriate pressure and temperature conditions, as might be the case if transport of weathering products from high to low elevation is too slow or burial of sediment at low elevation is too fast. On Earth, weathering reactions generally work within the medium of liquid water. Weathering on the hot, dry surface of Venus probably will be much slower, although even a minute amount of atmospheric H_2O may be more effective in promoting chemical weathering than is generally assumed (Warner 1982).

Various authors (Mueller 1963; Lewis 1968a, 1970; Khodakovsky et al. 1979) have attempted to deduce the likely mineralogy of Venus surface materials; their efforts were hampered by lack of knowledge of the exact abundances of H_2O, CO_2, and sulfur-bearing gases in the lower atmosphere. These authors also have suggested that the composition of the lower atmosphere may be regulated by gas-solid buffer reactions that limit the abundances of CO_2, H_2O, CO, HF, HCl, SO_2 and COS in the lower atmosphere. More recently, Nozette and Lewis (1982) have proposed a regional buffer by which wind transports carbonates produced at high elevations in silicate-CO_2 reactions to lower areas where they can react with quartz (SiO_2) to liberate CO_2. This provides an apparently reasonable model for the chemical weathering of common igneous rocks on Venus, but if almost all the CO_2 on Venus is in the atmosphere, buffering will not be an important global regulator (Orville 1975).

We can examine broad classes of reactions involving known atmospheric gases and mineral assemblages characteristic of common igneous rocks. The method of calculation of gas-solid equilibrium assemblages is given by various authors (e.g. Mueller and Saxena 1977). Reactions of the type: $\alpha M_1 + \beta G_1 + \gamma G_2 \rightleftarrows \delta W_1 + \epsilon G_n + \ldots$ are to be expected, where α, β, γ, δ, and ϵ are stoichiometric coefficients; M_1 a primary mineral; G_1 and G_2 atmospheric gases; W_1 a mineral product; and G_n residual gases. The equilibrium constant for this reaction is

$$K_T = \frac{(a_{W_1})^\delta (P_{G_n})^\epsilon}{(a_{M_1})^\alpha (P_{G_1})^\beta (P_{G_2})^\gamma} = \exp(-\Delta G_T^o/RT) \tag{5}$$

where P is partial pressure of each gas; a activity of the solid; and ΔG_T^o standard free energy of reaction at a specified temperature. The activity of each pure solid and the activity coefficients of each component are near unity under Venus conditions (Mueller and Saxena 1977). If ΔG_T^o for a reaction is known, K_T can be computed directly. Alternatively, K_T can be calculated if the abundances of the relevant gases are known. When the actual ambient pressures of the gases reacting with primary minerals exceed the partial pressure computed from the free energy data, the production of products is favored. Different reactions will tend to proceed toward the right or left at different places on the venusian surface depending on the ambient temperature, the pressures of the gases, and the actual activities of the minerals present. Free energy data are obtained from the JANAF (1971) tables, and from Robie et al. (1979). Thermodynamic data are available for a range of common igneous rock-forming minerals and for the gaseous components of the Venus atmosphere.

Current evidence is not adequate to determine whether the surface of Venus is oxidizing or reducing (Lewis 1970; Lewis and Kreimendahl 1980; Ringwood and Anderson 1977; Chapter 13). If Venus were initially Earth-like and lost an ocean of H_2O by H_2 escape, the crust should be highly oxidized. If Venus began its history with substantially less H_2O than Earth, a reducing lower atmosphere is expected. The question of oxidation state directly affects the chemistry of Fe and S. In a reducing lower atmosphere, COS and H_2S are expected to dominate over SO_2, and pyrite (FeS_2) would be produced by reaction with ferromagnesium silicates such as olivine. The observed CO/CO_2 ratio and the total abundance of sulfur gases (mole fraction $\sim 10^{-4}$) suggest a reducing lower atmosphere. However, experimental uncertainties caused by ingestion of an H_2SO_4 droplet into the PV Large probe mass spectrometer make these conculsions tentative (Hoffman et al. 1980a). If the lower atmosphere is oxidizing, then sulfates instead of sulfides would be produced by weathering of igneous minerals. To satisfactorily model chemical weathering on Venus, these uncertainties must be resolved.

We will assume that rocks on Venus contain the same major anhydrous rock-forming minerals found on Earth: olivines, pyroxenes, feldspars, and quartz. The earliest reaction investigated, proposed by Urey (1952) was

$$CaCO_3 \quad + \quad SiO_2 \quad \rightleftarrows \quad CaSiO_3 \quad + \quad CO_2 . \quad (6)$$
$$\text{(calcite)} \qquad \text{(quartz)} \qquad \text{(Wollastonite)}$$

Reaction (6) yields 90 bar of CO_2 at 742 K. This combination of $P(CO_2)$ and T is an almost perfect fit to the mean Venus surface. Wollastonite is a relatively uncommon mineral on Earth, and occurs mostly in deposits of metamorphosed limestones (rocks deposited in water, usually with organisms involved). But reaction (6) is thermodynamically feasible under Venus surface conditions if calcite can be produced by weathering of primary igneous rocks. Whether or not this implies buffering of the atmosphere remains controversial (e.g. Orville 1975; Nozette and Lewis 1982). Other weathering reactions involving CO_2 are:

$$Mg_2SiO_4 + CO_2 \rightleftarrows MgCO_3 \quad + \quad MgSiO_3 \quad (7)$$
$$\text{(forsterite)} \qquad \text{(magnesite)} \qquad \text{(enstatite)}$$

$$Mg_2SiO_4 + 2CO_2 \rightleftarrows 2MgCO_3 + SiO_2 \quad (8)$$

$$2Mg_2SiO_4 + CaMgSi_2O_6 + 2CO_2 \rightleftarrows CaMg(CO_3)_2 + 4MgSiO_3 \quad (9)$$
$$\text{(diopside)} \qquad\qquad \text{(dolomite)}$$

$$CaMg(CO_3)_2 + SiO_2 \rightleftarrows CaCO_3 + MgSiO_3 + CO_2 . \quad (10)$$

Reactions (7), (8), and (9) are most likely to proceed to the right at the top of Maxwell Montes, where atmospheric pressure and temperature are ~45 bar and ~650 K (Seiff et al. 1980). In reaction (10) weathering products from the previous reactions are used to form calcite. Pyroxenes and feldspars may weather to tremolite if the atmospheric mixing ratio of H_2O is sufficient (Fig. 5)

$$2CaMgSi_2O_6 + 3MgSiO_3 + SiO_2 + H_2O \rightleftarrows Ca_2Mg_5Si_8O_{22}(OH)_2 \quad (11)$$
$$\text{(tremolite)}$$

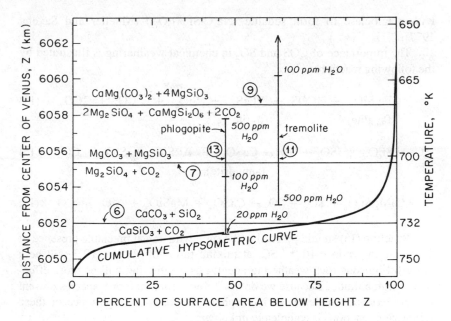

Fig. 5. Equilibrium altitudes for selected weathering reactions on Venus, plotted with the cumulative hypsometric curve of Venus topography. The equilibrium temperatures of reactions (11) and (13) are sensitive to the mole fraction of H_2O, and thus equilibrium altitudes are plotted for a range of X_{H_2O}. Figure modified from Nozette and Lewis (1982).

$$2CaAl_2Si_2O_8 + 5MgSiO_3 + SiO_2 +$$
(Ca-feldspar)

(12)

$$H_2O \rightleftarrows Ca_2Mg_5Si_8O_{22}(OH)_2 + 2Al_2SiO_5.$$
(andalusite)

Reaction (12) will proceed to the right at high elevations.

Hydrogen fluoride HF and hydrogen chloride HCl have been detected in the upper troposphere of Venus. If they also are present in the lower atmosphere, these reactions are possible:

$$KAlSi_3O_8 + 3MgSiO_3 + HF \rightleftarrows KMg_3AlSi_3O_{10}F_2 + H_2O + 3SiO_2 \quad (13)$$
(K-feldspar)   (phlogopite)

$$2NaAlSi_3O_8 + 2HCl \rightleftarrows 2NaCl + Al_2SiO_5 + 5SiO_2 + H_2O . \quad (14)$$
(Na-feldspar)   (halite)

Reaction (13) preferentially goes to the right at higher altitudes, assuming a mixing ratio of H_2O and HF constant with altitude. Halite will probably react with plagioclase to form sodalite ($Na_4ClSi_3Al_3O_{12}$) (Mueller and Saxena 1977).

The importance of COS and SO_2 in chemical weathering is illustrated by the following reactions:

$$Fe_2SiO_4 + 4COS \rightleftharpoons 2FeS_2 + SiO_2 + 2CO + 2CO_2 \quad (15)$$
$$\text{(fayalite)} \qquad\qquad \text{(pyrite)}$$

$$CaAl_2Si_2O_8 + SO_2 + CO_2 \rightleftharpoons CaSO_4 + Al_2SiO_5 + SiO_2 + CO \quad (16)$$
$$\text{(anhydrite)}$$

$$CaMgSi_2O_6 + SO_2 + CO_2 \rightleftharpoons CaSO_4 + MgSiO_3 + SiO_2 + CO . \quad (17)$$

Reaction (15) should proceed to the right with any appreciable presence of COS (mixing ratio $\sim 10^{-4}$). SO_2 at mixing ratios of 10^{-3} to 10^{-4} will be a powerful igneous rock weathering agent; this weathering will be more effective at high altitude. Because we do not know the amount of S species present nor the oxidation state of the atmosphere, the relative importance of these weathering reactions is completely unknown.

All of the reactions except (6), (10), and (13) involve minerals expected to exist in basalt and closely related igneous rocks (sima). Reactions (7), (8), (9), (12), (15), (16), and (17) predict that at high elevations on Venus the olivines, pyroxenes, and plagioclase feldspars characteristic of simatic rocks will weather to magnesite, enstatite, quartz, tremolite, andalusite, dolomite, and either sulfides or sulfates. Reaction (13) requires the presence of at least some sialic material at high altitudes. These minerals would constitute the regolith at high altitudes, the most probable source region for aeolian sediments. If this regolith is transported to lower elevations, reactions (6), (10) and (11) become possible.

Future determinations of the lower atmosphere composition, mineralogy of surface rocks, and major element abundances in the surface material of Venus will better constrain the processes of chemical weathering. Such experiments may be performed by a surface lander or on returned samples. Laboratory study of the proposed reactions to determine reaction kinetics is essential.

C. Mechanical Weathering

In addition to decomposition (chemical weathering), rocks on planetary surfaces are also subject to disintegration (mechanical weathering). On Earth, mechanical weathering results from freezing and thawing of water in pores and cracks, activity of organisms, volume changes associated with chemical weathering, sandblasting by wind-transported particles, and hydraulic effects of water waves and currents. On airless bodies such as Mercury and the Moon extreme surface-temperature fluctuations and meteorite impact are significant.

All of these must be insignificant or impossible on Venus, except for the volume changes associated with chemical weathering and possibly sandblasting effects, although the low near-surface wind velocities on Venus do not impart enough kinetic energy to moving particles to produce significant mechanical destruction of solid rock.

Two mechanical effects of chemical weathering might be important. If solid products have less densely packed crystal structures than the parent minerals (as in hydration reactions, for example), then the resulting volume increase in the surfaces of exposed rocks creates stresses that cause spalling. In addition, chemical weathering processes operate preferentially on certain minerals and along fractures and grain boundaries. This tends to convert solid, coherent rock into a mesh of strong and weak domains. In the absence of the more powerful disintegration agents present on Earth, this progressive weakening could proceed until wind-transported particles, or even gravity alone, dislodge fragments from weathered rock.

Unless the fine-textured dark material visible on the surface in the Venera images (Fig. 2) is volcanic ash, mechanical weathering must produce it. If winds are to transport solid particles, these particles must result from mechanical weathering of some sort. Although our information about the surface of Venus is not adequate to demonstrate that mechanical weathering occurs, it is reasonable to expect that it does, if chemical weathering occurs. However, it should be much slower than on Earth, and very likely constrains the overall weathering rate more than chemical weathering does.

D. Erosion and Transportation by Wind

In the absence of running water, wind is probably the dominant erosion and transportation agent on Venus. Wind action is controlled by the spectrum of wind velocities, the physical properties of the atmosphere, and the availability of particulate material on the surface.

Wind erosion occurs when the wind exerts lateral drag forces on particles exceeding the downwards and backwards forces that the immersed weights of the particles exert through their centers of gravity. The basic theory for this problem was investigated first by Bagnold (1941), and recently in a planetary context by Iversen et al. (1976a,b) and Greeley et al. (1980b). The critical parameter is the *threshold friction speed*

$$V*_t = A \left[\frac{(\sigma - \rho) gD}{\rho} \right]^{1/2} \tag{18}$$

where A is the threshold parameter; σ particle density; ρ fluid density; g gravity; and D particle diameter. Threshold friction speed is defined as the lowest friction speed at which the majority of exposed surface particles are set in motion. A second critical parameter is the *particle friction Reynolds number*

$$B = \frac{V*_t D}{\gamma} \qquad (19)$$

where γ is kinematic viscosity. Particle friction Reynolds number is a measure of surface roughness on the scale of individual particles. For $B < 5$, an aerodynamically smooth laminar sublayer exists adjacent to the surface. For $B > 70$, individual particles are large and the boundary layer is fully turbulent.

The threshold parameter A is a function of the threshold friction Reynolds number B; A is typically ~ 0.11 when $B > 3$, A increases as B decreases below 3. Values of A are dependent on the size-frequency distribution and particle diameter (Iversen et al. 1976b).

Iversen et al. (1976b) derived an empirical expression for A where $B < 0.22$ by considering the forces exerted on a particle by the aerodynamic flow. Their expression for equilibrium summation of drag, overturning, and lift moments, which takes into account particle weight, interparticle forces, and aerodynamic effects, is

$$A = 0.266 \left(\frac{1 + \dfrac{0.055}{\sigma g D^2}}{1 + 2.123 B} \right)^{\!\!1/2} . \qquad (20)$$

The coefficients 0.266, 0.055 (in units g s^{-2}), and 2.123 are experimentally determined. For A in the range $0.22 < B < 3$ (Iversen et al. 1976a)

$$A = \left(0.108 + \frac{0.0323}{B} - \frac{0.00173}{B^2} \right) \left(1 + \frac{0.055}{\sigma g D^2} \right)^{\!\!1/2} . \qquad (21)$$

The predicted threshold friction speed is plotted in Fig. 6 as a function of particle diameter for particles of two different densities; the following parameters were used for venusian near-surface conditions: $\rho = 0.1$ g cm^{-3}; $g = 888$ cm s^{-2}; $\gamma = 0.00443$ cm^2 s^{-1}. Surface wind speed differs from free-stream wind velocity because of surface drag and turbulence in the boundary layer. The free-stream wind velocity 1 m above a rocky surface (similar to the scene in the Venera 9 image) is approximately an order of magnitude higher than the surface wind speed.

The above expressions are based on observations of natural aeolian features on Earth, and experimental investigations in air and under martian conditions (6 mbar pressure). No experiments have yet been completed under Venus conditions (~ 96 bar pressure at median elevation) to test the validity of the extrapolation.

Near-surface, presumably free-stream wind velocities were determined directly by anemometers on the Venera 9 and 10 lander spacecraft. Local wind velocity at the Venera 9 landing site ranged from 0.4 to 0.7 m s^{-1}, and at the Venera 10 landing site between 0.8 and 1.3 m s^{-1} (Keldysh 1977). Near-

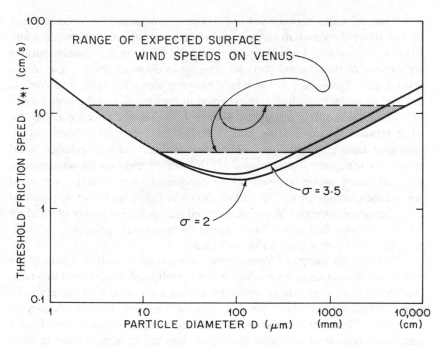

Fig. 6. Threshold friction speed ($V*_t$) as function of particle diameter D, under Venus conditions for particle densities σ of 2 g cm^{-3} and 3.5 g cm^{-3}.

surface wind velocity also was determined as the end point of a general study of atmospheric winds, by tracking the descent of the landing modules from Pioneer Venus and Venera spacecraft. These determinations revealed zonal winds in the venusian atmosphere that decrease in velocity with decreasing altitude. Counselman et al. (1980) used differential long-baseline interferometry to determine the three-dimensional velocity vector of each Pioneer Venus atmospheric probe. Doppler tracking of the Venera landers' transmitter signals was also used to determine the horizontal component of the lander velocity vector (Keldysh 1977). Counselman et al. (1980) show that below 55 km altitude in the Venus atmosphere the probe velocity vector will match the wind velocity to within 1 m s^{-1}. This justifies the use of the probe velocity vector as a measure of wind velocity.

All determinations near the surface indicate westward wind velocity of $\sim$ 1 m s^{-1} with a negligible north-south component. With increasing altitude the westward wind velocity increases, attaining $\sim$5 m s^{-1} at a 10 km altitude. The major difference among the wind velocity determinations from Pioneer Venus and the Veneras is that Venera data suggest day-night differences in wind velocity whereas the Pioneer Venus data suggest that the Venus winds are zonal and appear to flow completely around the planet with no significant day-night differences (Counselman et al. 1980).

Wind velocities between 0.4 and 1.3 m s^{-1} measured by Veneras 9 and 10 1 m above the venusian surface imply surface wind speeds between 4 and 13 cm s^{-1}. From Fig. 4 it is clear that the *in-situ* winds on the venusian surface are capable of transporting particles ranging in diameter from a few μm to several mm. Turbulence in the winds blowing along the surface of Venus, initiated by surface roughness, is expected in the presence of the large blocks seen in the surface field imaged by Venera 9. Turbulence affects capacity for wind erosion by introducing eddy currents and vertical motions in an otherwise laminar wind flow. Eddies and vertical motions enhance wind erosion by lifting particulate matter off the surface and into the windstream; thus turbulence makes a given set of conditions more effective for wind erosion than laminar flow. The curves shown in Fig. 4 are therefore regarded as conservative estimates of the erosion and transportation power of wind on Venus. Provided that particulate material of appropriate grain size is available, wind transport will be active on Venus.

However, the radar and Venera lander observations imply that most of the surface of Venus cannot be covered by unconsolidated wind-blown deposits; bulk densities of near-surface materials are not consistent with aeolian sediments, which typically have porosity approaching 30% and bulk density < 2.0 g cm^{-3}. Thus, present-day wind-blown sediments on Venus cannot form a continuous layer over the entire planet, but they could form patches in local depressions or in other regions that characteristically have lower mean wind velocities. Older aeolian sediments might be cemented and thus have lower porosity; if so, they could not be distinguished from weathered igneous rocks by density-dependent properties. Consequently, it is possible that a layer of cemented aeolian sediments covers much of the surface of Venus, especially those portions at low elevations (Warner 1980). For example, fresh basaltic rocks have densities $\sim$ 2.9 gm cm^{-3}, hence a regolith of basaltic composition with $\sim$ 30% porosity would have bulk density $\sim$ 2.0 gm cm^{-3}. Not much cementation would be required to increase this to the 2.2-2.3 gm cm^{-3} density (porosity $\sim$ 20%) implied by the radar data. Furthermore, the "bedrock platforms" seen by the Venera 10, 13, and 14 cameras may be cemented, layered aeolian sedimentary rocks rather than primary igneous rocks.

An extensive layer of aeolian sedimentary rocks could have important implications for general understanding of the planet. First, the smooth-floored, low-lying basins on Venus may be floored by sedimentary rocks rather than by basaltic lavas, as would be suggested by analogy with lunar basins. In addition, if the sediments were derived from the upper portions of Aphrodite, Ishtar, Beta, Phoebe, Alpha, and other elevated regions and deposited at lower elevations, they would mask the topography and chemical composition of the crust at those low elevations. A similar situation exists in the Earth's oceans. Deep sea sediments (below the carbonate compensation depth, where the sediments are not dominated by carbonates) contain quartz, feldspar, and clay minerals derived from weathering of continental rocks. The chemical composition of these sediments is more like the granitic continents than the underlying basaltic crust. Also, these sediments, as well as carbonate-dominated sediments above the carbonate compensation depth, form smooth surfaces that mask the ridges and valleys in the underlying basaltic crust.

Wind transports "dust" by suspension and "sand" by saltation. If a particle achieves a final velocity that equals or excceds the threshold friction speed, it will stay suspended in the atmosphere. Small particles, due to their larger surface area per unit mass, will achieve higher final velocities and will thus be carried in suspension. Larger particles will fall back to the surface after moving a finite distance. The division between suspension and saltation for Venus is predicted to be $\sim$ 20 to 50 μm particle diameter, slightly smaller than the $\sim$ 85 μm diameter at the minimum in the curves of V_{*t} versus D (Fig. 6). This minimum is the optimum particle size for transport by wind across the venusian surface, the diameter of the particle that may be eroded by the minimum wind velocity. Thus particles at the optimum diameter should be transported by saltation.

Greeley et al. (1980a) have calculated saltation parameters for Venus. They show that particles between 0.1 and 1 mm diameter will have path lengths ranging from a few to $\sim$17 mm and path heights ranging from a few to $\sim$13 mm. These values are about an order of magnitude lower than those for Earth under comparable dynamic conditions (friction wind speed slightly higher than the threshold value). This difference in saltation path length and height will cause a different geometry of aeolian geomorphic features. For example, ripples might be expected to have higher amplitudes and shorter crest-to-crest distances.

The possible presence of ripples and dunes on the venusian surface should be taken into account in interpretation of radar returns from spacecraft and Earth-based experiments. Ripples add to surface roughness at the scale of the wavelength of commonly used radars, and dunes should enhance the probability of specular reflections.

E. Lateral Creep of Crustal Rocks

Creep rates of rocks are strongly temperature dependent. If the high surface temperature of Venus implies a much hotter crust than on Earth, as seems likely, then the entire crust of Venus may be much less resistant to creep than the crust of Earth. This possibility is partly dependent on the relative H_2O contents of the two crusts. In a recent analysis Weertman (1979) has shown that a wet, hot crust on Venus will be incapable of supporting thick, isolated continental masses and high mountains for geologically significant lengths of time. This result is not sensitive to selection of any parameters other than temperature and H_2O content. Cordell (1981) and Solomon et al. (1982) have shown that large features such as impact basins will suffer essentially complete relaxation of their topographic relief in times on the order of 1 Gyr even if the crust is dry.

As a consequence, observed elevated regions probably are very young, or actively supported over long time periods. The tectonic significance of these possibilities is discussed in Sec. IV of this chapter, and in Chapter 10. Of interest here is that rock creep due to gravity may be a significant lateral transport process of surface material on Venus, especially if the crust is wet. This suggests that present elevated regions are young or continuously renewed, and that (unless we illogically assume that the relief of Venus is uncharacteristically high at present) elevated regions would have risen and spread to oblivion many times during the history of Venus. From the viewpoint of the rock cycle as known on Earth, this is a somewhat exotic concept. Because creep rate is so critically dependent on the content of bound H_2O in the crust of Venus, and because this parameter is so completely unknown, we cannot yet evaluate the contribution of this process to near-surface lateral transport of crustal material.

III. TOPOGRAPHY

A. Global Summary

The Pioneer Venus altimetric data include a large number of discrete elevations. From these one can interpolate elevation values at uniformly distributed points on the surface, or calculate average elevations of small surface areas containing several altimeter readings. In either case, the resulting array of numbers can be used to determine modal, median, and mean elevations on Venus. These statistical parameters have the following topographical significance:

1. The *mode* is the most common elevation: the peak on an elevation versus surface-area histogram;
2. The *median* divides the topography by surface area: half the surface area of the planet is below the median, half above it;
3. The *mean* divides the topography by volume; the volume of rock above the mean equals the volume of air below it.

On Venus, the modal radius is 6051.4 km, the median radius 6051.6 km, and the mean radius 6051.9 km. All topographic contour maps in this chapter are referenced to the median radius.

The most striking characteristic of the global topography of Venus is the pronounced unimodality of the elevation spectrum, in marked contrast to Earth (Fig. 7). The mean elevation on Venus is higher than the median elevation, because the high regions of Venus deviate from the median more than the low regions do. The near-equality of the median and modal elevations is especially interesting because it is so different from the situation on Earth (Fig. 7). The total relief measured by the Pioneer Venus radar altimeter is

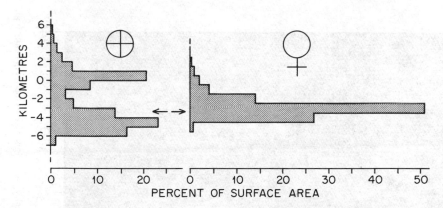

Fig. 7. Hypsometric histograms of Earth and Venus in 1 km elevation intervals. Scale at left is Earth elevation relative to sea level; the Venus histogram is matched to the Earth histogram at median elevations (arrows).

~13.5 km (see Color Plate 3), but this is almost certainly less than the true value because the large spacing and size of altimeter footprints result in systematic underestimation of heights and depths of limited areas of extreme high or low elevation.

Topographic data are available for ~ 93% of the surface of Venus; only the polar regions north of +74° and south of −63° remain unknown. Horizontal resolution of the topography is not constant because the footprint size varies from 7 × 23 km at periapsis (~17°N) to > 100 km in diameter at maximum range (Pettengill et al. 1980a), and because across-track footprint spacing varies from a maximum at the equator to 0 at +74°. The best resolution occurs in areas near periapsis where the altimeter tracks from three passes are neatly interleaved; in such places across-track footprint spacing can be <50 km locally. For most of the surface, the recognition resolution for roughly equidimensional features is between 100 and 200 km diameter (larger at high latitudes), but where footprint spacing and size are optimum it may be slightly better. Very long linear or arcuate topographic features can be resolved even if they are somewhat narrower than this recognition resolution, but their widths could be exaggerated on the resulting topographic maps (Arvidson 1981).

The global topography of Venus clearly does not resemble that of Earth (see Figs. 8 and 9). Furthermore, Fig. 9 demonstrates that the Pioneer Venus radar would be capable of defining many key elements of the Earth's topography, continents, ocean basins, major mountain belts, midocean ridges, and subduction troughs, if we were simply to remove the oceans without correcting for loss of load or for surface temperature differences. However, this obvious lack of topographic similarity does not necessarily imply that the two planets have equally dissimilar tectonic styles, as will be discussed in Sec. IV of this chapter (see also Chapter 10).

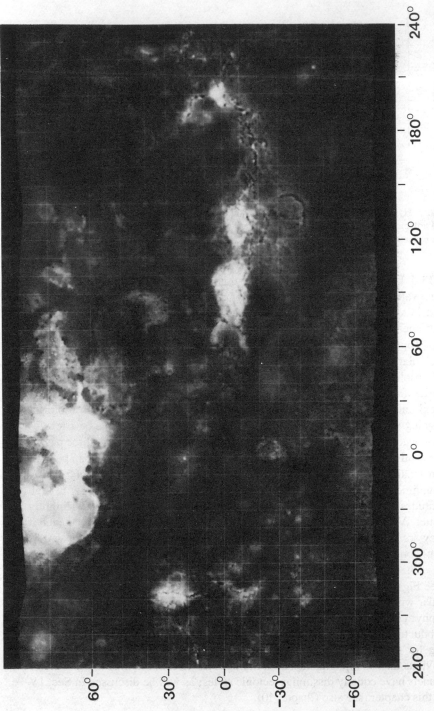

Fig. 8. Mercator projection of Venus topography from Pioneer Venus radar altimeter. Dark regions are topographically low, bright regions topographically high.

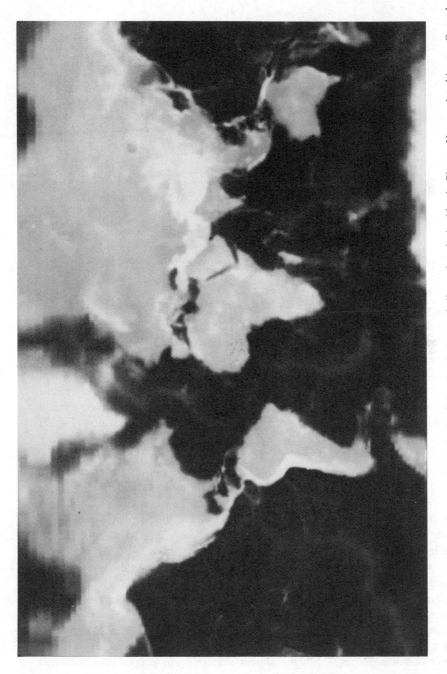

Fig. 9. Earth topography degraded to the resolution expected if elevation data were similar to that obtained by the Pioneer Venus radar altimeter. Ocean basins are assumed dry, but the topography has not been corrected for isostatic or temperature effects.

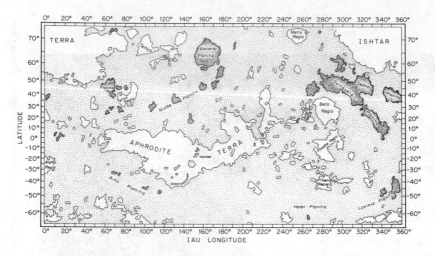

Fig. 10. Mercator projection of Venus topography showing distribution of regions with elevations
deviating > 1 km from the planetary median radius. Heavy stipple, radius > 1 km below
median; light stipple, radius = median ± 1 km; blank, radius > 1 km above median.

Most of the surface of Venus is characterized by a rolling topography at
intermediate elevations, with a horizontal scale of hundreds to thousands of
km and low local relief; only a few small regions stand > 1 km above or
below median elevation (Fig. 10). This extensive terrain has been called the
"upland rolling plains province" (Masursky et al. 1980). However, the upland rolling plains province was arbitrarily defined as lying between 0 and
2 km above median radius (Masursky et al. 1980), whereas Fig. 10 is simply a topographic map with contour lines 1 km above and below the median
radius. The plot of cumulative percent surface area versus elevation (Fig. 5)
demonstrates that the extensive terrain at intermediate elevations is more
naturally defined by contours ~ 1 km above and below the median than by
those selected by Masursky. Of course, the exact definitions of terrain levels
must inevitably be somewhat arbitrary, and nothing is to be gained by rede-
fining the elevation limits of named provinces. The important story is told
by the hypsography (Fig. 7); the areally dominant intermediate level, however defined, is characteristically rolling and also mostly without clearly
defined linear or arcuate topographic trends.

Of greater interest are those regions of Venus with elevations significantly higher or lower than the median (Fig. 10). Most of the lower terrain lies
within broad, shallow basins of irregular shape. Many of these are adjacent
to the regions of high elevation; this distribution produces a set of weak
linear trends in the topography (see Color Plate 3). In addition to these basins,
long, narrow, deep, linear-to-arcuate troughs occur on Venus. Because of
their narrowness, these troughs almost certainly are deeper than indicated
on our maps.

Two regions as large as small continents on Earth, plus several much smaller elevated regions rise significantly above the median elevation (Color Plate 3). As is clear from the hypsometric diagram (Fig. 7), the total area included in these elevated regions is not very large, but their relief above the median elevation is much more impressive than the relief of the low regions below the median.

Line-of-sight gravity anomalies correlate in position with many high and low regions of Venus (Color Plate 4). Anomalies of comparable wavelength on Earth do not coincide with topographic features, and generally are explained as due to lateral heterogeneity in the mantle (Phillips and Lambeck 1980). The amplitudes of the anomalies on Venus suggest that at least partial isostatic compensation of the larger topographic features occurs (Sjogren et al. 1980) or that, if compensation is complete, it is not shallow. Although gravity anomalies correlate areally with topographic features, the amplitudes of the anomalies do not show a simple relationship with the amplitudes of the corresponding topographic features (see Chapter 10).

Earth-based radar images of the accessible portion of Venus (Fig. 11) indicate that many both high and low topographic features seen in Color Plate 3 are also regions characterized by greater or less than average radar brightness. Side-looking radar images from Pioneer Venus (Fig. 12) confirm this observation. Generally, large regions that appear radar dark correspond to the large, irregular, topographically low areas. These areas are smooth at cm scale, a characteristic that has suggested a comparison with the lava-filled low areas on the near side of the moon (Rogers and Ingalls 1969; Schaber 1976). Conversely, the areas roughest at cm scale correspond to many of the most prominent positive topographic features: Maxwell Montes, Aphrodite Terra, Beta Regio, Akna Montes, Freyja Montes, and Vesta Rupes (the slope bounding the high-standing Lakshmi Planum). Also bright on Earth-based and Pioneer Venus side-looking radar are several less prominent elevated features: Alpha Regio, Phoebe Regio, Themis Regio, and Metis Regio. Many other radar-bright regions are evident on Figs. 11 and 12; these are generally less bright than the regions corresponding to identifiable elevations or slopes, and commonly take the form of elongated, wispy streaks. The apparent lack of correlation between slopes on the Pioneer Venus topographic maps and some of the areas of moderately enhanced cm-scale roughness may be at least partly due to the better horizontal resolution (by a factor of 5 to 10) of the Earth-based images: small-scale rough slopes resolvable on these images will not be seen on the Pioneer Venus topographic maps. It is also possible that some areas of anomalous cm-scale roughness are not slopes at all, but areas characterized by extensive rippling of wind-transported regolith, by fields of dunes (see Sec. II) or by extensive fracturing of rock along tectonically active zones.

Some of the wispy streaks define major linear zones (Schaber and Masursky 1981; Schaber 1982). Although Schaber suggests that these major linear zones of enhanced roughness are tectonic in origin, the global topo-

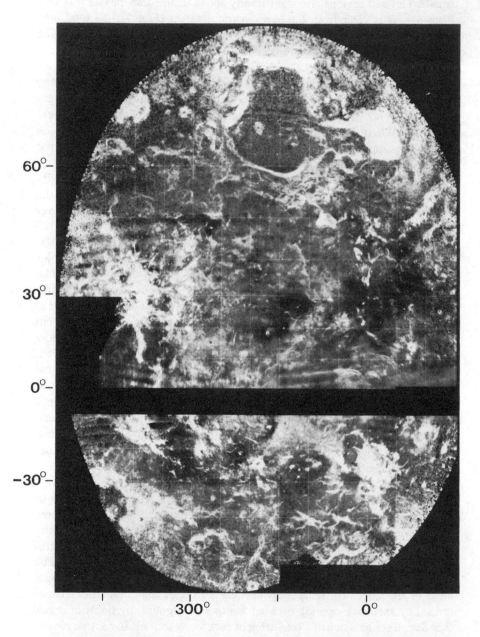

Fig. 11. Mercator projection of images obtained by Arecibo radar. Most of the brightness variability relates to differences in surface roughness at about the scale of the radar wavelength (12 cm); dark areas are relatively smooth at this scale, bright areas are rough.

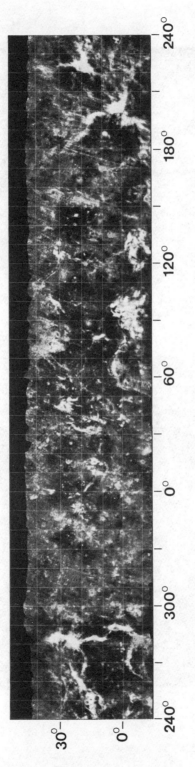

Fig. 12. Mercator projection of images obtained by the Pioneer Venus radar in side-looking mode. Brightness variability relates to differences in surface roughness at about the scale of the radar wavelength (17 cm); dark areas are relatively smooth at this scale, bright areas are rough.

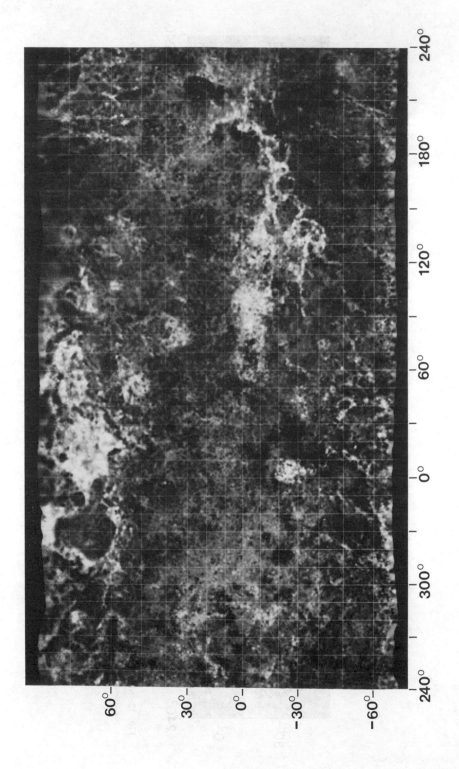

graphic maps (Color Plate 3; Fig. 8) are not of sufficient resolution to define their true nature. More detailed, hand-drawn contour maps show that much terrain within these linear zones is pitted and hilly, characterized by small, generally irregularly shaped high and low areas immediately adjacent to each other. The specific causes of such topography remain obscure, but we agree that the great length and continuity of these zones argues for an endogenic (tectonic) origin. Topographic details are presented later in this section.

The scattering law appropriate for the Pioneer Venus altimetric mode contains the Hagfors constant and Fresnel reflectivity coefficient introduced in Eqs. (1) and (2) (Hagfors 1970; Pettengill et al. 1980a). Figure 13 shows the global distribution of rms m-scale and larger slopes derived from Hagfors constant; Fig. 14 shows the global distribution of reflectivity. Many of the regions rough at cm scale also are rough at larger scales, but there are some interesting exceptions. Reflectivity variations probably relate primarily to the porosity of the surface material and secondarily to material composition (see Sec. II).

B. Characteristics of Specific Regions

The two elevated regions of continental dimensions, Ishtar Terra and Aphrodite Terra (Fig. 10), have been described in some detail by Masursky et al. (1980). Most of their topographic characteristics are well portrayed on the global topographic maps prepared by the U.S. Geological Survey (e.g. Color Plate 3; Fig. 8), but some local details are clearer in hand-drawn enlargements such as our Figs. 15–18. A third elevated region, Beta Regio, warrants detailed discussion despite its much smaller size.

Ishtar Terra. This region may extend northward beyond the limits of Pioneer Venus radar altimetry (Color Plate 3), but groundbased radar images (Fig. 11) suggest that its northern perimeter is very close to these limits (Campbell and Burns 1980; Masursky et al. 1980). The known extent of Ishtar includes an area about equal to that of Australia. It is dominated by Lakshmi Planum, a large plateau 2 to 3 km above the surrounding lowlands bounded by relatively steep slopes. Rising above this plateau are three massifs: Akna Montes and Freyja Montes along the western and northwestern edge of Lakshmi Planum, and Maxwell Montes along its eastern edge. The eastern part of Ishtar, east of Maxwell, is a complex pitted and hilly terrain ~ 1 km lower than Lakshmi Planum that lacks the steep, well-defined boundary slopes that characterize the plateau.

Fig. 13. Mercator projection of the distribution of Hagfors constant, derived from the Pioneer Venus radar in altimetric mode. Most of the brightness variability relates to differences in rms slopes at a horizontal scale of meters and larger, dark areas have relatively low mean slopes at this scale, bright areas have high mean slopes.

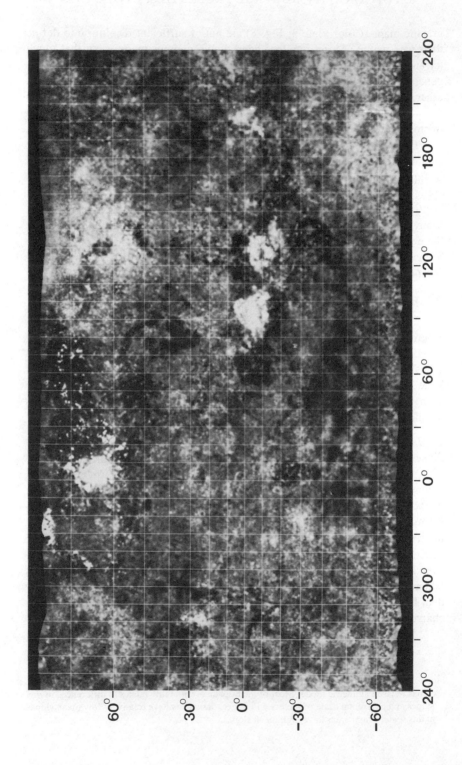

As illustrated in Figs. 15 and 16, there is a remarkable correlation between the distribution of cm-scale roughness variations and the detailed topography, and less correspondence between topography and the distributions of Hagfors constant and the Fresnel coefficient. Areas of steeper than average slope are especially likely to be rough at cm scale: the south margin of Lakshmi Planum (Fig. 16) and all of Maxwell Montes (Fig. 15) are notable examples.

The shape of Maxwell, and the parallelism between bands of varying cm-scale roughness and the length of the massif, suggest some sort of structural control. Maxwell has been inferred to be a fault-bounded massif (Masursky et al. 1980), but it also could be a large and complex domed or folded uplift, or a large, complex volcanic construct made up of a number of individual coalescing volcanoes. The large depression (Fig. 15) on its northeast flank resembles a volcanic caldera, but that tells us little about the origin of the entire massif. The absence of any resolvable depressions in the high areas of Maxwell is especially puzzling if its origin is purely volcanic. The pronounced banding in the Earth-based image of Maxwell has been inferred as due to differences in roughness between fault blocks and their bounding fault scarps (Masursky et al. 1980). Similar banding is present on images of Akna and Freyja Montes (Campbell et al. 1981). All these could be due to faulting, but the size and spacing of these bands are comparable to the dimensions of individual mountain ranges on Earth, whether defined by faulting or by folding. Unfortunately, the distributions of surface roughness and rms slopes do not permit unique solutions at the data resolutions available, and the Pioneer Venus altimetry is of insufficient resolution to determine if the roughness bands correlate with relief features.

The two large depressions Colette and Sacajawea on Lakshmi Planum are very interesting. They are irregular in shape; especially Sacajawea, which also occurs near the head of a large canyon-like cleft extending into the plateau from its southern rim (Fig. 16a). These characteristics suggest that Sacajawea is some sort of large volcanic-tectonic depression; it seems unlikely that it could be an impact crater. Although Colette's lack of roundness suggests that it also could be a volcanic-tectonic depression, the altimetric data are simply not good enough to characterize the plan shape very well.

The topography immediately south of Lakshmi Planum is very complex (Fig. 16a), including irregular closed depressions and an ill-defined arcuate elevated region. This region is characterized by wispy, moderately bright streaks against a dark background on Earth-based images (Fig. 16b), and by rms slopes almost equal to those on the scarp bounding the plateau (Fig.

Fig. 14. Mercator projection of the distribution of the Fresnel coefficient, derived from the Pioneer Venus radar in altimetric mode. Most of the brightness variability relates to differences in reflectivity (and thus dielectric constant) of surface materials; dark areas have relatively low reflectivities, bright areas have high reflectivities.

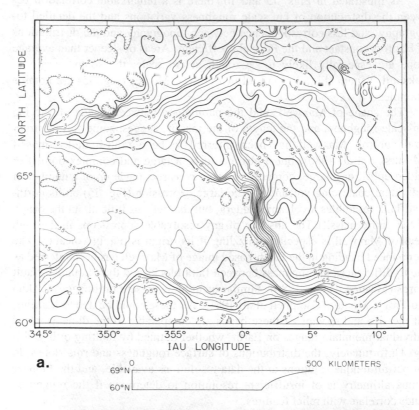

a.

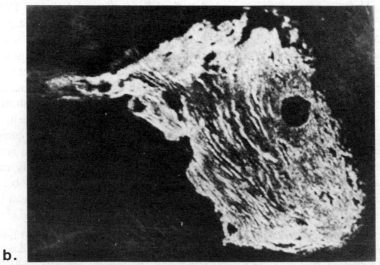

b.

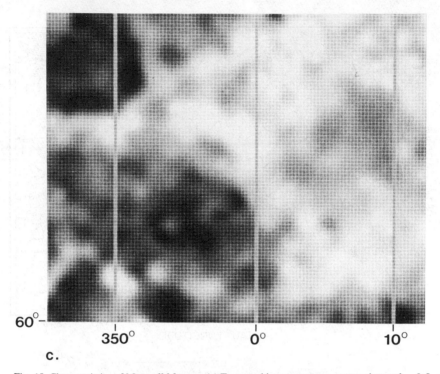

60° –

350°   0°   10°

c.

Fig. 15. Characteristics of Maxwell Montes. (a) Topographic contour map; contour interval = 0.5 km, datum is median radius (6051.6 km). (b) Distribution of cm-scale roughness from Arecibo image; bright areas are relatively rough. (Figure courtesy of D. B. Campbell, Arecibo, Puerto Rico.) (c) Distribution of Hagfors constant from Pioneer Venus radar; bright areas have relatively steep meter-scale and larger rms slopes.

16c). The nature of this sort of terrain is better defined in the Aphrodite Terra region.

Aphrodite Terra. This region covers an area approximately equal to that of South America, about twice as extensive as Ishtar Terra. However, it does not reach the extreme elevation that characterizes the Maxwell portion of Ishtar. In plan, Aphrodite is long and narrow, with several semi-isolated areas of maximum elevation (Color Plate 3). Most of the volume of Aphrodite is contained in the two largest of these elevated areas, both in the western half. With two stubby projections at its western end and the long, northwest curving tail at the eastern end (Color Plate 3), Aphrodite resembles a legless scorpion in shape. Aphrodite's most geologically interesting features are the remarkable troughs characterizing its eastern part (the base of the scorpion's tail), and the region south of the eastern of the two main elevated regions. These troughs are not well displayed on global topographic maps such as Color Plate 3 and Fig. 8, partly because they are too narrow to be seen easily

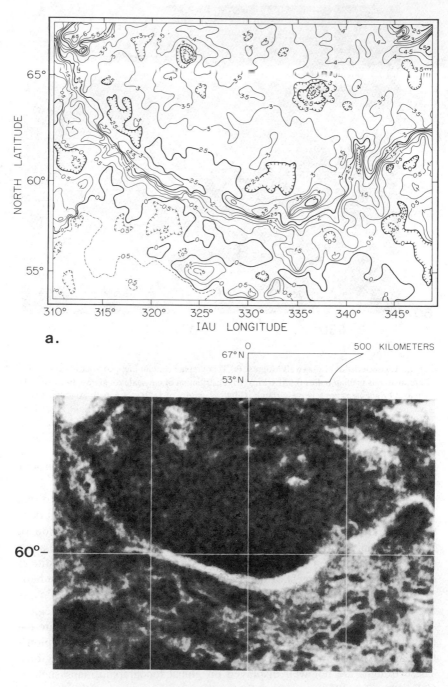

a.

b.

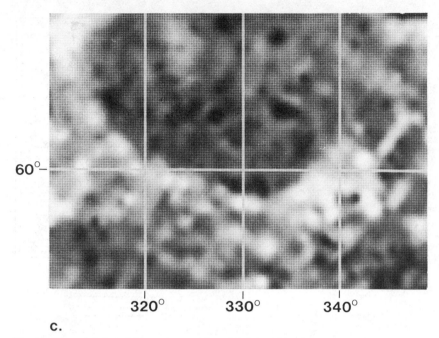

c.

Fig. 16. Characteristics of the southern portion of Lakshmi Planum and its boundary scarp. (a) Topographic contour map. (b) cm-scale roughness (Arecibo). (c) rms slopes (Pioneer Venus).

on such small-scale maps, and partly because the data smoothing involved in the preparation of these maps causes some loss of resolution. A more detailed map of the troughs in eastern Aphrodite has been published by Pettengill et al. (1980a, Plate 2).

The Aphrodite troughs, Artemis, Diana and Dali Chasmae, are truly impressive. They define a complex of troughs and parallel ridges 6000 km long; individual troughs and ridges are as much as 2500 km long, with trough-ridge relief up to 5 km (underestimated by Pioneer Venus) and trough or ridge widths on the order of 100 km (overestimated by Pioneer Venus). Figure 17, a detailed map of Dali and Diana Chasmae, provides an example of the characteristic topography of this trough-ridge system. This topography must be of tectonic origin, but the altimetry data alone do not resolve whether the troughs are analogous either to terrestrial rifts due to extension, as are found along divergent plate boundaries, or to terrestrial trenches due to compression, as are found along some convergent plate boundaries, or have no terrestrial analog.

Aligned with this trough-ridge complex and extending northeast towards the northern part of Beta Regio is a zone of pitted and hilly terrain enhanced in both cm-scale roughness and rms slopes (Fig. 18b and c). This is the most extensive of the linear tectonic zones defined by Schaber (1982). The peculiar

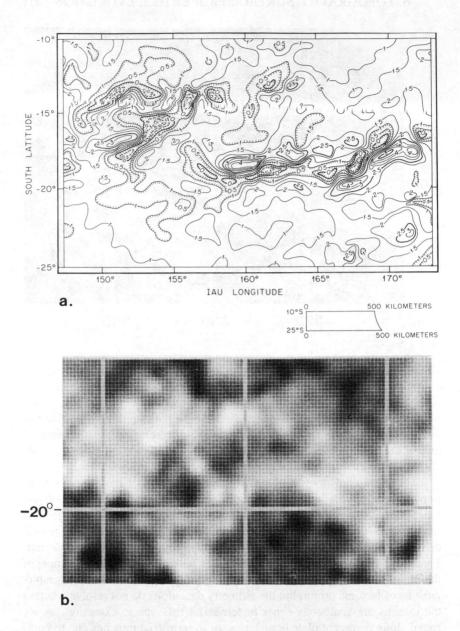

Fig. 17. Characteristics of Dali and Diana Chasmae. (a) Topographic contour map; half-km contour lines locally omitted where steepness of slope causes excessive crowding. (b) rms slopes (Pioneer Venus).

topography of pitted and hilly terrain is well illustrated by Fig. 18a. How real is this portrayal? Some of the high and low areas on the figure are defined by only one altimeter footprint and thus their true size, shape, and relief are not determinable. A topographic feature defined by only one point could of course be the artifact of an erroneous altimetry reading, but the abundance of pairs of readings, one higher than surrounding readings and the other lower, suggests that they all cannot be dismissed as artifacts. Their validity is also supported by the observation that the anomalous pairs can occur either in the same orbit or in adjacent orbits, and that in most places the high and low areas are in fact large enough to affect more than one altimetry footprint. There is some suggestion in the detailed maps (Figs. 17 and 18) that a continuum may exist from these small, irregular paired highs and lows, through the paired troughs and ridges of limited length included by Schaber and Masursky (1981) in their linear "disturbed zones," to the very large arcuate trough-ridge systems discussed above. If this is so, a common tectonic origin for all these features is possible, and the large arcuate troughs may well have a different tectonic origin on Venus than on Earth.

There is a remarkable change in topographic characteristics from the slightly elevated, rough, pitted and hilly terrain in eastern Aphrodite to the lower and smoother regions on either side (Fig. 18a). It is possible to interpret this relationship in two ways: (1) the pitted and hilly terrain represents a tectonic lineament, and the shallow crustal structure of the elevated region is somehow different from that of the flanking lower and smoother areas; or (2) there is no difference in shallow crustal structure, but the low areas are partly filled in by lava flows, aeolian sediments, or sedimentary rocks. As indicated earlier, the length and continuity of the linear zone favors the first interpretation. However, the possible effects of burial on the observable patterns imprinted into crustal rocks by tectonic processes must be carefully considered, especially because the second interpretation implies that the crustal structure responsible for the roughness is older than most surface materials, whereas the first interpretation suggests (but does not require) that the structural cause of the roughness is younger than most surface materials.

Beta Regio. This region is an elliptical elevated region with two areas of maximum elevation separated by a narrow "waist" at an intermediate elevation (Fig. 19a). Earth-based and Pioneer Venus side-looking radar show a very strong correlation between this region and cm-scale roughness (Fig. 19b and c). Beta Regio is not particularly distinctive with regard to rms slopes (Fig. 19d). Theia Mons, the southern elevated area, has very high reflectivity (Fig. 19e) (Pettengill et al. 1982). Extending nearly the full length of Beta Regio, and for another 2500 km southward to Phoebe Regio, is a major trough.

The trough in the Beta-Phoebe region differs from those in eastern Aphrodite in two respects: (1) it is not a more or less continuous arcuate low, but a system of generally linear depressions with many en-echelon offsets;

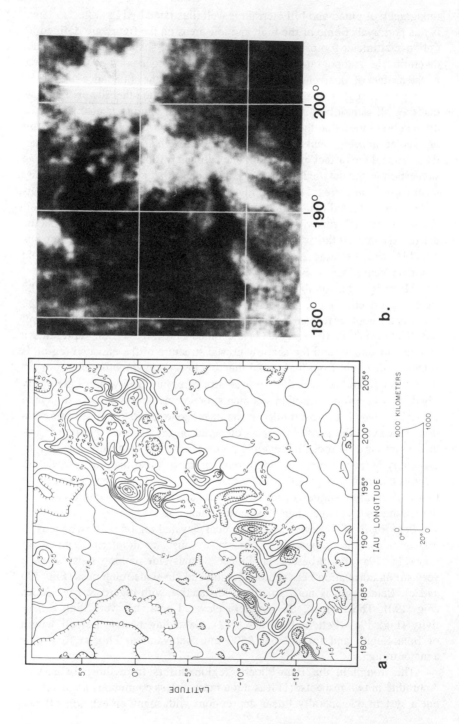

IAU LONGITUDE

LATITUDE

1000 KILOMETERS

a.

b.

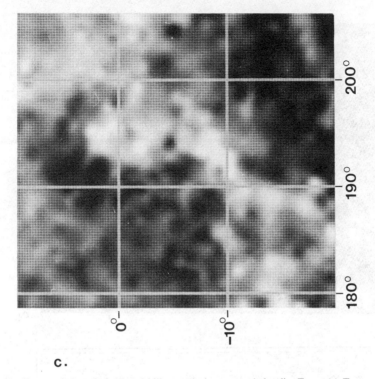

c.

Fig. 18. Characteristics of pitted and hilly terrain in eastern Aphrodite Terra. (a) Topographic contour map. (b) cm-scale roughness (Pioneer Venus). (c) rms slopes (Pioneer Venus).

and (2) for much of its length it lies within clearly defined elevated regions, comparable in relief and dimensions to the large crustal domes characteristic of major terrestrial continental rift systems (McGill et al. 1981). This trough has received greater attention than those near Aphrodite because it was discovered first (Goldstein et al. 1976; Malin and Saunders 1977) and because the topography of the trough and of Beta Regio is more amenable to interpretation by analogy with Earth. Part of Beta Regio was interpreted to be a large shield volcano by Saunders and Malin (1977); this idea was later supported by the basaltic K, U and Th abundances in surface rocks sampled by Venera 9 and Venera 10 (Surkov 1977). Subsequently, all of Beta Regio has been interpreted as a hugh volcanic construct (Masursky et al. 1980), or as a large crustal dome with one or several more modest volcanoes on it (McGill et al. 1981). The important difference between these hypotheses is that doming requires actual crustal uplift whereas building a large volcanic construct requires no crustal uplift. The topography as resolved by Pioneer Venus is more consistent with the crustal dome interpretation,

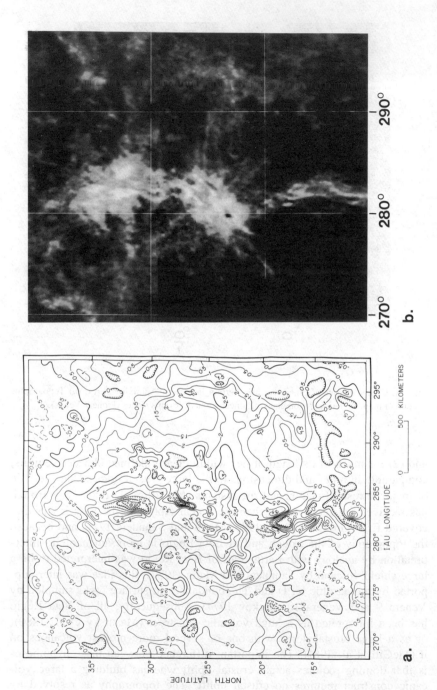

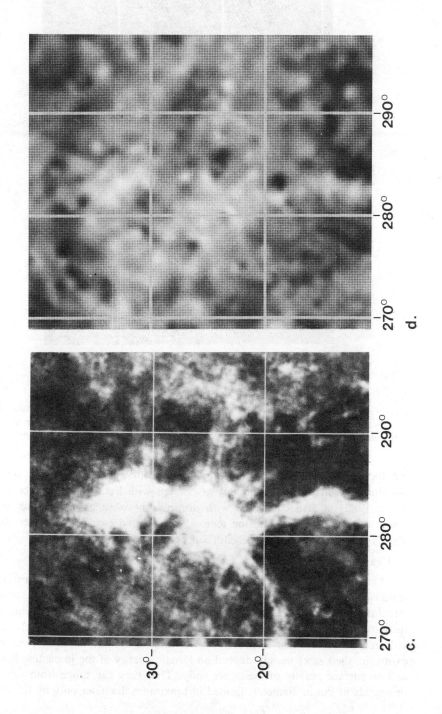

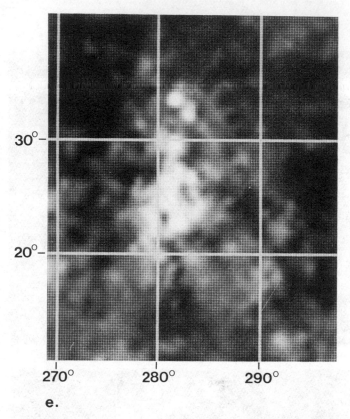

e.

Fig. 19. Characteristics of Beta Regio and associated rift. (a) Topographic contour map. (b) cm-scale roughness (Arecibo). (c) cm-scale roughness (Pioneer Venus). (d) rms slopes (Pioneer Venus). (e) Fresnel coefficient; bright areas have high reflectivity (Pioneer Venus).

but topography alone cannot unequivocally resolve the matter. The high-amplitude positive gravity anomaly associated with Beta Regio (Reasenberg et al. 1982) is also consistent with doming, because an anomalous and active upper mantle beneath the region is suggested (see also Chapter 10). This is exactly what is found beneath some rifted domes on Earth.

C. Craters

Craters are among the most abundant and easily recognized landforms on solid bodies in the solar system. A crater is a pit, cavity, or depression, often circular or nearly so, formed in a surface by various processes including meteoritic impact (impact crater), volcanic eruption (volcanic crater), and volcanic or tectonic subsidence with or without eruption. Impact craters are exogenic; their sizes mainly depend on kinetic energy of the impacting body and on surface gravity of the target body. Thus they can range from μm to thousands of km in diameter, limited in maximum diameter only by the size and velocity of impacting body that will fragment the target body. Because of the dense atmosphere, the minimum diameters of the impact craters on Venus should be on the order of 150 to 200 m (Tauber and Kirk 1976). Volcanic craters are endogenic and occur over a much narrower diameter range from a few m to perhaps a few km. Volcanic or tectonic depressions, also endogenic, can range in diameter from m (pit craters) to many tens of km (calderas). On Earth, there are a few depressions hundreds of km across whose origins are uncertain but presumed to be tectonic.

Crater-like forms have been proposed to exist on Venus, mostly on the basis of analyses of Earth-based radar reflectivity images. The criterion most often used to recognize craters is circular form; other criteria are required to interpret the mode of origin.

The resolution of early Earth-based radar brightness images was poor. Aside from a few radar-bright features interpreted as topographically high and rough, the largest forms are radar-dark features believed to be topographically depressed, by analogy to the lunar maria (Rogers and Ingalls 1969). Schaber (1976) and Schaber and Boyce (1977) followed this interpretation to explore the implications if such features were impact basins. Saunders and Malin (1977) examined altimetry of two such features as part of their study of the Beta region, and found them to be subtle depressions, supporting the basin interpretation. However, these basins could just as likely be of tectonic as of impact origin.

The difficulty in interpreting radar reflectivity images without the benefit of topographic data is demonstrated in the study of a large radar-dark area in the northern hemisphere of Venus, west of the prominent radar-bright feature Maxwell (Campbell et al. 1976). Using the lunar mare analogy and the observation that the dark area was surrounded by bright regions interpreted as rough rim deposits (again a lunar analogy), Campbell and co-workers interpreted the dark feature and bright surroundings as a large impact basin. It is now known from Pioneer Venus altimetry (Pettengill et al. 1979b) that the area in question, Lakshmi Planum, is an elevated plateau. Many of the large features discussed by Schaber (1976) and Schaber and Boyce (1977) have been examined in Pioneer Venus altimetry (Masursky et al. 1980); most show essentially no relief. The probability that rapid viscous relaxation, due to the high crustal temperatures on Venus, will erase all discernible relief by legitimate impact craters in geologically short times does not make studies of venusian craters any easier (Solomon et al. 1982).

Most of the discussion of craters on Venus concerns circular features between 30 and 600 km in diameter. The first report of small craters was made by Rumsey et al. (1974), using Earth-based observations of both reflectivity and altimetry at resolution $\sim$ 10 km per radar cell. They observed at least one 160 km diameter crater with its rim raised $\sim$ 500 m above its floor. Additional high resolution radar reflectivity and altimetry maps (Goldstein et al. 1976) provided sufficient numbers of craters for statistical examination. Using these images Schaber and Boyce (1977) created a crater size-frequency curve for Venus, from which they concluded that Venus underwent a bombardment of impacting objects not unlike that of both the Moon and Mercury, and that venusian craters had not been extensively erased by tectonic or volcanic phenomena. On the other hand, Malin and Saunders (1977) and Saunders and Malin (1977) note that at least half the radar images show no craters, and conclude that on Venus impact craters are preserved in some areas but not in others. Furthermore, volcanic or tectonic origin of some, and perhaps all, of the smaller craters cannot be ruled out.

Craters in Arecibo radar images were reported by Campbell et al. (1979) and Campbell and Burns (1980). Their principal identification criteria are circularity and radar-dark interiors surrounded by radar-bright rims, characteristics common to radar images of lunar impact craters. These craters are interpreted as mostly impact craters; the possibility of volcanic origin is acknowledged but not preferred. Some features display central bright spots interpreted as central peaks. Campbell and Burns (1980) analyzed the possible effects of the atmosphere on distribution of ejecta, and possible implications for the age of the craters if they are indeed impact structures. In this latter case, Campbell and Burns suggest that the area they examined is relatively young, being less than one fourth the age of Venus.

Some radar-dark areas surrounded by radar-bright boundaries show no relief, and some depressions have no distinguishing reflectivity markings. A few features (e.g. Colette and Sacajawea on Lakshmi Planum) show crater-like radar reflectivity patterns and are depressions (Fig. 16a). These are exceptions, not the rule, and of course Sacajawea is not circular. Overall, Pioneer Venus altimetry suggests that many circular features previously assumed to be due to impact in fact have topographic characteristics indicative of volcanic origin (Masursky et al. 1981b). This does not rule out the possible existence of preserved impact craters and basins, but it does indicate that attempts to determine surface ages for Venus by crater counting are premature.

Why do we care whether the features on Venus are volcanic or impact craters? If the craters are formed by impact, then we can learn a great deal about the age of surfaces on Venus, because observation of other planets indicates that craters on a surface accumulate as a function of time. Masursky et al. (1980) find that the craters reported by Campbell and Burns (1980) all lie on the lowland plains areas of Venus. If this observation is correct and if the craters are of impact origin, then the lowlands are older than the highlands. If on the other hand these craters are volcanoes, then the degree of volcanic activity might suggest that the lowlands are young. The volcanic interpretation of craters implies a very active interior at some time in the evolution of Venus, and preservation of the results of that activity.

Morphologic analysis of high-resolution images may be the only way to reduce this ambiguity of interpretation, and even then there may be problems. Examination of the radar-bright areas surrounding the craters is most likely to distinguish volcanic from impact structures. Radial abundance of secondary craters, morphology of lava flows, and characteristics of the surrounding surfaces may all help, but determination of the possible origins may not be possible short of *in situ* observations or sample return.

IV. TECTONICS

A. Introduction

The importance of Venus in planetary tectonics derives from its close similarity to Earth in such bulk properties as diameter, mean density, and presumably deep interior thermal structure. These similarities permit evaluation of the differences between Venus and Earth, which appear at first glance to be second-order: compositional differences implied by the small density deficiency of Venus; the probable orders-of-magnitude smaller amount of water; and the high surface temperature. The major significance of these differences is in the effects they may have on lithosphere thickness and buoyancy.

In Earth tectonics, "lithosphere" is generally defined as a cool, relatively strong thermal boundary layer lying on top of a slowly convecting weaker zone lower in the mantle (Turcotte 1979). The boundary between the lithosphere and the underlying less viscous asthenosphere is believed to be a zone of rapid change in effective viscosity, due either to partial melting or to highly nonlinear solid-state creep (Nur 1971; Mysen and Boettcher 1975; Weertman and Weertman 1975; Froidevaux and Schubert 1975; Schubert 1979). The depth to the lithosphere-asthenosphere boundary on Earth varies somewhat, but 100 km is a rough average (Wyllie 1971). This depth also corresponds to the average depth to the top of the seismic low-velocity zone.

Another definition of lithosphere is based on ability to statically support loads; a lithosphere so defined is called an "elastic lithosphere." Like the thickness of the thermal lithosphere, thickness of the elastic lithosphere depends on thermal gradient and composition; in addition it depends on dimensions of the load to be supported, time, and rate of loading. For time, loading, and length scales relevant to global tectonics on Earth, the elastic lithosphere is on the order of 50 km thick (Forsythe 1979); this corresponds roughly to the upper half of the thermal lithosphere.

The significance of lithosphere thickness to tectonic style and overall tectonic activity is still a matter of debate, but it is unlikely that very thick elastic shells could be broken or flexed as part of an Earth-like plate tectonics scheme. Thus, if Venus has a lithosphere significantly thicker than Earth's, as suggested by Schaber and Boyce (1977) and Warner and Morrison (1978), it would almost certainly not experience plate tectonics. However, the much higher surface temperature of Venus implies a thinner lithosphere (Chapter 10; McGill 1979*b*; Anderson 1980, 1981; Phillips et al. 1981; Kaula and Phillips 1981; Brass and Harrison 1982), because the lithosphere-asthenosphere boundary approximates an isotherm, and this temperature will almost certainly occur at shallower depths on Venus than on Earth. Consequently, it seems very unlikely that Venus has a lithosphere thicker than Earth's.

But a thin lithosphere does not guarantee plate tectonics. The downward motion of slabs of lithosphere into the underlying asthenosphere (subduction) requires that the subducting lithosphere have higher average density than the asthenosphere. Even though the lithosphere on Earth includes a low-density crustal cap its average density is high, except where the crust is very thick or the lithosphere very young, because the mantle portion of the lithosphere is colder than the mantle in the asthenosphere. Lithospheric density also can be enhanced by the high-pressure transition of basalt to eclogite, $\sim 15\%$ denser than basalt. This transition is inhibited by high temperature (Wyllie 1971). Subduction is thus inhibited by thick crust (continents) and by high lithospheric temperatures.

If the Venus lithosphere is thin and hot, then it is probably not dense enough to subduct, because the mantle to crust ratio within it will be too low and because the eclogite stability field will be well below its base. This has led some workers to predict that there cannot be plate tectonics on Venus at present (Anderson 1980, 1981; Smith 1980; Phillips et al. 1981). The question of lithospheric buoyancy is discussed in detail in Chapter 10.

It is intriguing that similar difficulties plague students of the early Precambrian era on Earth (pre-2.5 Gyr ago). Because of the decline in abundances of radioactive nuclides with time, heat generation must have been several times greater in the early Precambrian than it is now (Lambert 1976). For the same reasons as for Venus (insufficient cooling, low mantle/crust ratio in the lithosphere, eclogite stability field below the base of the lithosphere), the steep thermal gradient and thin lithosphere on the early Precambrian Earth would seem to preclude plate tectonics (Baer 1977; Lambert 1976; Hargraves 1978). Until we can solve the problem of early Precambrian hot-Earth tectonics here, interpreting Venus tectonics will be difficult, but perhaps having two examples will help solve the problem for both planets.

B. Observational Evidence

Observational evidence bearing on the tectonics of Venus includes its topography, the apparent presence of large craters, gravity anomalies, and the abundance of ^{40}Ar in the atmosphere. None of these provides definite answers to questions concerning rate and style of tectonic behavior, and all are subject to different and partly conflicting interpretations.

Tectonic and volcanic processes are manifestations of planetary heat loss. From its Earth-like size and bulk density, we would anticipate that Venus is tectonically active, for these similarities lead to thermal models predicting essentially Earth-like conditions in the interior of Venus. But a comparable level of tectonic and volcanic activity does not require that the specific mechanisms of heat loss, the tectonic style, be similar. Furthermore, differences in surface temperature, surface processes, and water abundance may well mean that similar tectonic and volcanic processes will produce different topography on Venus than on Earth. We will first discuss what is known about the level of tectonic activity on Venus, and then review critically the topographic evidence bearing on the question of tectonic style.

The amount of ^{40}Ar (in g/g of planet) in the atmospheres of planets can be used as a rough comparative index of tectonic and volcanic activity, because ^{40}Ar is a product of the decay of ^{40}K and so requires tectonic or volcanic activity to escape from the interior in any significant quantity. Venus has slightly less than one-third as much ^{40}Ar in its atmosphere as does Earth, but measurement errors are large enough for this ratio to range from one-fifth to one-half (Hoffman et al. 1980a; Chapter 13). This implies somewhat less tectonic and volcanic activity on Venus than on Earth, unless differences in abundance and distribution of K or differences in surface processes are significant. The abundances of K and the K/U ratio in Venus surface rocks (Vinogradov et al. 1973; Surkov 1977) are Earth-like, implying that the K contents of the mantle source regions on the two planets are similar (McGill 1979a); any significant difference would necessitate unrealistically high or low degrees of mantle partial melting to produce the Earth-like abundances measured in the surface rocks. Release efficiency on Venus will be lower than on Earth because of slow erosion in the absence of liquid water, but higher because of faster diffusion of gases out of the hot crust; the net effect is difficult to evaluate, but is probably less significant than the measurement error of the atmospheric ^{40}Ar mixing ratio. Thus Venus seems to have been a bit less tectonically and volcanically active than Earth, averaged over its entire history, but the difference probably is less than a factor of 3. In fact, the uncertainties are such that Venus might possibly be as tectonically and volcanically active as Earth.

The areal coincidence of gravity anomalies with major topographic features (Color Plate 4) is subject to more than one interpretation because of the inherent nonuniqueness of gravity solutions (Phillips and Lambeck 1980). The implications of these data for Venus tectonics are discussed by Phillips et al. (1981) and in Chapter 10 of this book. The correspondence of gravity and topography requires that either the elastic lithosphere is thick enough to support the topographic load, or the topography is the direct result of inhomogeneity in the mantle.

Because thermal models imply a thin thermal lithosphere on Venus, the probability of a thick elastic lithosphere is very low (Chapter 10). Comparisons of the amplitude of gravity anomalies and the topography within the rolling plains province, assuming elastic support, also lead to the inference of a very thin elastic lithosphere (Cazenave and Dominh 1981). Thus, the gravity anomalies mapped by Pioneer Venus imply a tectonically active planet with a dynamic mantle capable of maintaining significant lateral inhomogeneity despite high temperatures.

The most dramatic topographic difference between Venus and Earth is their contrasting elevation spectra (Fig. 7). The hypsometric diagram for Earth is bimodal, partly because plate tectonics continually creates thin, low-lying simatic oceanic crust while sweeping the thick, high-standing sialic crust into large continental masses. The unimodal Venus elevation spectrum is evidence against the existence of these processes (Masursky et al. 1980). However, two important uncertainties seriously weaken this argument: (1) if Venus had plate tectonics, it could exhibit an Earth-like, bimodal elevation spectrum only if its crust contained sufficient volumes of both sial and sima; and (2) the bimodal elevation spectrum for Earth is at least accentuated, and perhaps largely caused by, the oceans' effects on the erosion of continents and the distribution of the products of this erosion.

The present-day dryness of Venus suggests that it may have significantly less sial than Earth, because abundant water may be necessary for the differentiation of large volumes of sial (Anderson 1980, 1981). But there is good evidence that Venus once had more water than it does now (McElroy et al. 1982; Donahue et al. 1982), so it also is reasonable to expect that much sial evolved very early on Venus, as was apparently the case on Earth. Nevertheless, we do not know how much sial is present on Venus; this should be kept in mind during the following discussion, because all the arguments over the significance of unimodal versus bimodal hypsography assume that both sial and sima are of sufficient abundance to influence the topographic spectrum.

The presence of oceans on Earth must contribute to the strong bimodality of the topographic spectrum, but it is less clear whether oceans are essential for bimodality. Sea level provides a base level for the erosion of continents much higher than if there were no ocean water. In addition, the oceans tend to cause continent-derived sediments to be deposited on the flanks of the continents themselves; very little sediment reaches the deep ocean basins. If the oceans were to suddenly disappear, these controls on the erosion of continents and on the lateral transport of sediment also would disappear, and the sharp definition of the two elevation modes would be lost. But would the Earth's topography evolve all the way to a unimodal distribution like that of Venus? The most helpful approach to this question is not to assume, as have Brass and Harrison (1982), that an Earth without oceans would retain an unimpaired hydrological cycle, but to ask if erosion and transportation at rates characteristic of Venus could create a strongly unimodal topographic spectrum on a planet with Earth-like plate tectonics. Whether this hypothetical planet's topography will be unimodal or bimodal would depend on the relative rates of lithosphere spreading due to plate tectonics (tending to segregate sial and sima) and sediment transport (tending toward a globally uniform elevation). On Earth, sediment is transported from the continents to the ocean basins much too slowly to counteract the rate of plate spreading. Compared to subaerial processes on Earth, weathering, erosion, and transportation on

Venus are exceedingly slow. But the critical parameter is the rate at which these slow processes characteristic of Venus would transport continental material to ocean basins on our hypothetical waterless Earth; unless this rate is much faster than the actual rate on Earth (with oceans), lithosphere spreading rates must be slower than on Earth in order to maintain a unimodal topographic spectrum. We simply do not have the necessary quantitative data on rates of surface processes on Venus. A simple comparison with aeolian erosion rates on Earth (Brass and Harrison 1982) is inadequate because it ignores the great difference in the weathering rates, and wind cannot transport material any faster than it is supplied by weathering.

Plate tectonics on Earth results in a global system of diagnostic large-scale landforms and structures. A primary objective of both Earth-based and spacecraft radar mapping of Venus is to find evidence that a similar system of landforms and structures does or does not exist on Venus. Thus it is critically important whether or not the Pioneer Venus radar mapper could resolve the diagnostic landforms if they did exist on Venus. Confusion can be avoided by posing this question in two parts: (1) if Earth's oceans were drained, but its topography otherwise unchanged, could the Pioneer Venus radar mapper resolve the critical landforms? and (2) if the topography of Earth were corrected for removal of the load of ocean water, and for the much higher crustal temperature on Venus (and perhaps for other factors as well), would the critical landforms be so subdued as to escape detection? The answer to the first part is clearly "yes" (Fig. 9), despite the contrary impression given by some recent discussions (e.g. Brass and Harrison 1982). The answer to the second part is model dependent and open to controversy.

As a number of workers have pointed out (Arvidson 1981; Arvidson and Davies 1981; Masursky et al. 1981b; Phillips et al. 1981), the topography of Venus shows nothing comparable to the midocean ridges on Earth. Trench-like troughs do exist (Figs. 17 and 19), but these seem more like continental rifts on Earth (McGill 1980; Masursky et al. 1980; Schaber and Masursky 1981; McGill et al. 1981; Schaber 1982) than subduction-related trenches; i.e. they indicate divergence rather than convergence and, by analogy with continental rifts on Earth, only limited horizontal motion (McGill et al. 1981). Nothing like the global system of ridges, mountain chains, and trenches associated with plate tectonics on Earth is detectable in the topography of Venus.

A more meaningful comparison is possible if the simulated topography of Earth is corrected for isostatic response to removal of the oceans, and also for the fact that the high surface temperature of Venus imples more subdued topography from an otherwise Earth-like system of plate tectonics (see Chapter 10). These corrections significantly subdue the critical landforms, especially midocean ridges and subduction trenches, but they would still be resolvable in a simulated Pioneer Venus radar map (Fig. 20).

These topographic comparisons assume that plate tectonics on Venus

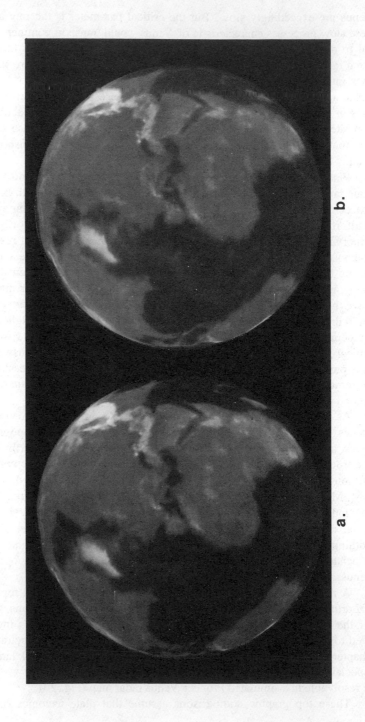

a.

b.

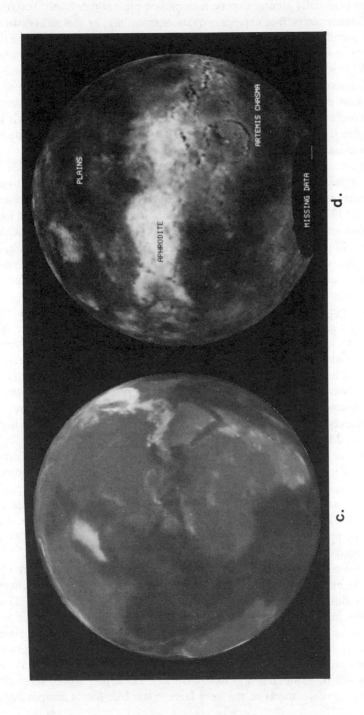

c.

d.

PLAINS

APHRODITE

ARTEMIS CHASMA

MISSING DATA

Fig. 20. Simulated Earth topography at resolution expected if elevation data were similar to Pioneer Venus altimetry; the hemisphere shown centers on mid-Atlantic ridge. High elevations are bright. (a) Oceans removed, but with no topographic corrections. (b) Corrected for isostatic response to removal of oceans. (c) Further corrected for differences in crustal temperatures. (d) Venus hemisphere centered on Aphrodite Terra.

would be essentially similar to plate tectonics on present-day Earth. If significant differences in surface expression, rate of spreading, or sediment distribution are postulated, models for plate tectonics on Venus can be devised that would not produce the diagnostic landforms we associate with plate tectonics on Earth, or would produce them in unrecognizable form (Brass and Harrison 1982; Solomon and Head 1982). These models are explicitly developed to explain how plate tectonics could occur without producing any recognizable topographic evidence; consequently, they cannot be tested against existing topographic data. Nevertheless, they remind us that the available data do not have the resolution needed to fully constrain the problem.

The crater-like forms visible on Earth-based radar images provide an interesting but potentially confusing clue to tectonic history. Attempts to use these craters to derive crustal ages have produced conflicting results (Schaber and Boyce 1977; Campbell and Burns 1980; Masursky et al. 1980), but most of the plots seem to indicate significant age (>1 Gyr). This age determination is valid if all or most of the observed features are impact craters, if venusian surface processes have not obscured significant numbers of similar features, and if the crater production curve for Venus is similar to the lunar production curve for large craters.

Experience from Mercury, the Moon, and Mars suggests that a crust pocked by a significant number of impact craters on the order of hundreds of km in diameter is an old crust. Thus, if the largest of the crater-like forms visible on radar images really are impact craters, at least some of the crust of Venus must be very old. Most of the very large crater-like forms occur on the areally extensive rolling plains province, so interpreting these features as impact craters leads to the conclusion that most of the crust of Venus must be old (Masursky et al. 1980) and that recycling by plate tectonics is not likely (Schaber and Boyce 1977).

However, as discussed earlier in this chapter (Sec. III), an impact origin for many of these crater-like forms is doubtful. If they are not due to impact, then these features do not imply an old crust; in fact they could be interpreted to imply youth. Many of the depressions on the surface of Venus are too irregular in plan to be due to impact (e.g. Figs. 16 and 18), and some areas seem to be pocked by very large, highly irregular depressions (Fig. 18) not quite like anything on Earth.

On the other hand, many of the smaller crater-like forms are more convincing as impact structures, because their appearance on radar brightness images is similar to that of lunar craters (Campbell and Burns 1980). These are found within a rather restricted area, but it is not clear if this restriction is real or an artifact of variable resolution and coverage. The apparent areal variation in density and types of crater forms may indicate age or tectonic contrasts from place to place, but more and higher-resolution imagery and altimetry will be required to test this possibility. However, it does seem clear that many (perhaps most) of the very large crater-like forms are not due to impact, and thus all past attempts to estimate crustal age by counting them are almost certainly invalid.

C. Tectonic Models

Because of the low resolution of the altimetry data and the scarcity of critical geophysical and geochemical data, venusian tectonic models are necessarily rather sketchy and insecure. Anderson (1980, 1981) argues that a thin lithosphere is virtually inevitable, for reasons already reviewed in this chapter, and that subduction of this thin lithosphere could not occur. Thus, all low-density differentiated igneous rocks that reach the lithosphere must stay there (many are recycled by subduction on Earth); this would progressively lighten the lithosphere by increasing the proportion of crust within it. The result would be a global crust permanently stable against subduction, with major elevation differences simply due to variations in the thickness of this crust. In Anderson's model, this global crust would be almost entirely simatic because of the assumed lack of the water needed to yield large volumes of sial. Such a crust would tend to have a unimodal elevation spectrum.

Phillips et al. (1981) present a similar model, except that they argue for the possibility of plate tectonics early in the history of Venus, if the present high crustal temperatures and associated loss of volatiles did not occur until significantly later than the planet's origin. A temporary retention of volatiles is supported (weakly) by the sialic K, U, and Th abundances at the Venera 8 landing site in the areally extensive rolling plains province (Vinogradov et al. 1973). This global sialic crust would also yield a unimodal elevation spectrum and would be stable against subduction, but a purely isostatic explanation for the large elevated regions is considered less likely than dynamic support, because the high temperatures in the crust and upper mantle would probably erase major differences in crustal thicknesses by creep in geologically short time periods.

On Earth, 60% of the total heat loss is related to the generation of new lithosphere at divergent plate boundaries (Sclater et al. 1980). If the lithosphere of Venus is too buoyant on the average to be subducted, then this lithosphere generation process must be much less important on Venus than on Earth. Two alternatives have been suggested: extensive volcanism; and local thinning of the lithosphere, and thus enhanced heat loss by conduction, over hot spots (Chapter 10; Kaula and Muradian 1982).

Some features in the topography of Venus are compatible with heat loss by localized volcanism plus conduction, but the term "hot spot" probably is not appropriate. The puzzling pitted and hilly terrain (Fig. 18) occurs extensively along linear, somewhat elevated zones such as the one extending from Aphrodite to Beta Regio (Color Plate 3; Schaber 1982); some of these zones include very large troughs (Fig. 17). These linear zones may well be incipient divergent boundaries, with high heat flow, active volcanism, anomalously thin lithosphere, and extensive crustal collapse producing large irregular depressions or very large linear to arcuate troughs. The rift and dome structure of the Beta-Phoebe region (Fig. 19; McGill et al. 1981) is also consistent with a model of heat loss by local volcanism and conduction, because it appears closely analogous to regions of continental rifting on Earth characterized by anomalously thin lithosphere, enhanced heat flow, and active volcanism. Our opinion is that these linear zones of enhanced heat loss and limited extensional tectonics would become divergent plate boundaries if subduction were possible elsewhere.

Gravity, topography and ^{40}Ar abundance suggest that Venus is tectonically active, and that it should lose heat at a rate roughly comparable to that for Earth. It is of great interest to determine the dominant mechanism(s) facilitating this heat loss, for they constitute the tectonic style of the planet. Insofar as we are able to interpret the available low-resolution radar imagery and altimetry, the tectonic style of Venus appears different from that of Earth; this is consistent with the predictions of most thermal models. The case is by no means closed, however; the thermal models are not well constrained, the composition of the crust is largely unknown, the ages of the crust and key topographic features are totally unknown, and available data are of insufficient resolution to resolve individual mountain ranges, yet alone specific structures within these ranges indicative of crustal processes. Any major improvement in our understanding of the geology and tectonics of Venus must await better data.

Acknowledgments. J. W. Head and G. H. Pettengill reviewed an early draft of this chapter and made a number of valuable suggestions. Extensive photographic and cartographic support was provided by the U.S. Geological Survey, Flagstaff, Arizona and by the Department of Geology and Geography, University of Massachusetts, Amherst.

7. SOME OPTICAL PROPERTIES OF THE VENUS SURFACE

YU.M. GOLOVIN, B.YE. MOSHKIN, and A. P. EKONOMOV
USSR Academy of Sciences

We review the results of the measurements by Soviet and U. S. probes of the optical properties of the surface of Venus.

The dense atmosphere and continuous cloud cover of Venus render its surface inaccessible to remote optical observations. Only direct measurements, carried out on board Soviet and U.S. probes, have been able to provide data on the optical properties of the surface of the planet. In this chapter we briefly survey these results.

The optical properties of Venus's surface have been carefully investigated by means of the probes of Veneras 9 and 10 (Golovin 1979; Ekonomov et al. 1980). Data on the surface albedo were likewise obtained by the probes of Veneras 11 and 12 (Ekonomov et al. 1979) and by the Large probe of Pioneer Venus (Tomasko et al. 1979*a*). Conclusions have been drawn concerning the limits on the range of values of the albedo from optical measurements by the probe of Venera 8 (Avduevsky et al. 1973*b*; Germogenova et al. 1977). The coordinates of the landing sites of all six probes, the spectral intervals (at the 0.5 amplitude level) in which the measurements were made, and the effective wavelengths λ_e for each spectral interval

$$\lambda_e = \int\limits_0^\infty S(\lambda)\lambda d\lambda / \int\limits_0^\infty S(\lambda)d\lambda \tag{1}$$

where $S(\lambda)$ is the spectral characteristic of the instrument, are given in Table I. The values of the surface albedo, allowing for the measurement error, are plotted in Fig. 1. The results and a description of the measurement procedures can be found in the references given in the last column of Table I.

[131]

TABLE I

Surface Albedo, Spectral Characteristics of Photometers, and Characteristics of Measurement Sites.

Descent Vehicle	Lat. (deg)	Long. (deg)	Spectral Interval, $\Delta\lambda$, μm	Effective Wavelength λ_e, μm	Surface Albedo A	Reference
Venera 8	10	12	0.52 –0.72	0.53	<0.6	Avduevsky et al. (1973)
					0.2 – 0.4	Gezmogencva et al. (1977)
Venera 9	32	231	0.5 –0.56	0.535	0.016 ± 0.008	
					0.042 ± 0.012	
					0.047 ± 0.015	
					0.057 ± 0.017	
Venera 10	16	291	0.52 –0.67	0.59	0.045 ± 0.01	Golovin (1979)
					0.058 ± 0.018	
			0.62 –0.7	0.67	0.076 ± 0.02	Ekonomov et al. (1980)
			0.7 –0.79	0.765	0.07 ± 0.02	
					0.105 ± 0.025	
			0.76 –1.06	0.87	0.16 ± 0.045	
Venera 11	–14	299	0.439–0.461	0.45	0.7 ± 0.07	Ekonomov et al. (1979)
			0.489–0.511	0.5	0.15 ± 0.08	
			0.624–0.646	0.635	0.23 ± 0.1	
Venera 12	–7	294	0.56 –0.68	0.64	$0.07^{+0.05}_{-0.02}$	Ekonomov et al. (1979)
Pioneer Venus	4	304	0.4 –1.0		<0.15	Tomasko et al. (1979a)

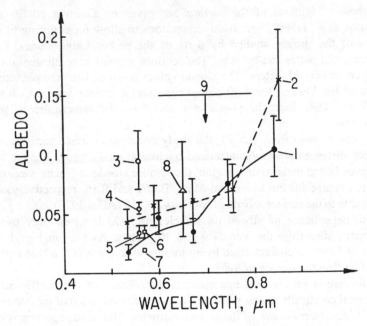

Fig. 1. Dependence of the surface albedo of Venus on wavelength. (1) results of the probe of Venera 9; (2) results of the probe of Venera 10 (Golovin 1979; Ekonomov et al. 1980). (3)–(7) albedo of characteristic sections on the panoramas in the work of Selivanov et al. (1976): (3) stone (Adams and McCord 1969), Venera 10; (4) slab (Tomasko et al. 1979a), Venera 10; (5) dark sections (Hovis 1965), Venera 10; (6) stone (Ekonomov et al. 1979), Venera 9; (7) rock fragments (Ross et al. 1969), Venera 9; (8) Venera 12 probe data (Ekonomov et al. 1979); (9) Pioneer Venus probe data (Tomasko et al. 1979a).

The photometers of the Venera 9 and 10 probes measured the intensity of radiation from the zenith, from the entire upper hemisphere, and from an angle 23° to the nadir. The measurements were carried out with five filters (cf. Table I). The scanning photometer of the Venera 12 probe measured the angular distribution of the intensity of radiation, from the atmosphere and from the surface of the planet before and after the landing. Scanning was performed in the vertical plane by a rotating mirror. One revolution of the mirror took ~ 1.3 s; the instrument's instantaneous field of view was ~ 20°.

At the landing sites of the Venera 8 and 11 probes, the value of the surface albedo was obtained indirectly, from measurements of the incoming fluxes alone. The albedo value was selected so that, for the lowest atmospheric layer (0–5 km), the measured and the calculated coefficients of transmission were identical. The layer was assumed to be gaseous and conservative.

Estimates of the ground albedo were also made during photometric processing of the surface images of Venus obtained by panoramic cameras on board probes of Veneras 9 and 10 (Selivanov et al. 1976b). The albedos of

characteristic segments of the surface are given by Golovin (1979) and
Ekonomov et al. (1980), who make corrections to allow for eclipsing of the
sections of the surface studied by parts of the probe. Large stones, rock
fragments, and plates can be seen. The sections studied have dimensions of
several cm to several meters. The albedo values obtained from the photometric data of the Venera 9 and 10 probes represent a surface segment of area
~ 300 m². Data from the photometer and from the camera are in good
agreement.

As can be seen from Table I, the study concerned surface sections considerable distances apart, characterized by low ground albedo. For the sections investigated in the visible region, the surface albedo of Venus was lower
than the average for the Moon and Mars, 0.067 and 0.16, respectively, and
comparable to the surface albedo of Mercury, 0.056 (Allen 1973).

The dependence of albedo on wavelength $A(\lambda)$ is given only by the
photometric data from the Venera 9 and 10 probes. As seen in Fig. 1, the
surface of Venus is characterized by an intense absorption in the blue region;
it is redder than the surface of Mars.

Information on the size and spectral distribution of the coefficient of
reflection theoretically makes possible a qualitative evaluation of the composition and characteristic size of the surface particles. Numerous specimens of a
variety of rocks were investigated, in order to obtain the dependence of their
spectral albedo on their dimensions and shape, the type of packing, and the
conditions of illumination (cf. e.g. Ross et al. 1969; Hovis 1965; Adams
and Filice 1967; Hunt and Ross 1967). Ross et al. (1969) investigated 22
specimens of igneous rocks and minerals in order to represent the complete
range of their characteristics: composition, mineralogy, origin, and size.
The study by Adams and Filice (1967) presents data for ~50 specimens of
rocks and minerals and provides a petrographic description of these rocks.

We have compared our values of dependences of $A(\lambda)$ with the spectra of
metamorphic rocks of the Earth, taking into account the fact that our results
were obtained with a much poorer spectral resolution than the spectra of the
Earth rocks. The relief greatly affects the reflective properties of the surface.
Numerous depressions may cause a lowering of the albedo. This effect is
evidently mainly responsible for the smaller albedo value at the landing site of
the Venera 9 probe, compared with the albedo determined from Venera 10 data.
In comparison with the spectra of reflection of terrestrial rocks, we used data
on the ground reflectivity at the landing site of the Venera 10 probe, where the
influence of the relief on the albedo was slight.

Adams and Filice (1967) have shown that, for identical rocks and identical estimates of the degree of dispersion, the determining characteristics are:
absolute magnitude of the albedo; dependence of the albedo on wavelength;
and the position of the absorption bands. These characteristics differ considerably for the three groups of silicate rocks: volcanic glass (obsidian,
pumice); acid rocks (tufa, rhyolite, granite); and basic and ultrabasic rocks

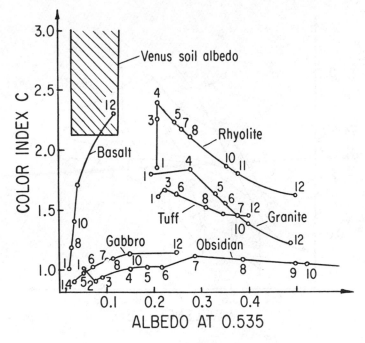

Fig. 2. Diagram of the color index and albedo, according to Adams and Filice (1967): (1) $d >$ 2.3 mm; (2) 1.68 $< d <$ 2.83 mm; (3) 1 $< d <$ 1.68 mm; (4) 0.5 $< d <$ 1 mm; (5) 354 $< d$ < 500 μm; (6) 250 $< d <$ 354 μm; (7) 150 $< d <$ 250 μm; (8) 100 $< d <$ 150 μm; (9) 75 $<$ $d <$ 100 μm; (10) 50 $< d <$ 75 μm; (11) 37 $< d <$ 50 μm; (12) 40 μm $< d$.

(basalt and gabbro). As a parameter to characterize the wavelength dependence we will use the ratio $C = A(0.87)/A(0.535)$.

Figure 2 is a diagram of the color index C and the albedo at 0.535 μm wavelength, for the various groups of silicate rocks. The particle size is the parameter. It is clear from Fig. 2 that the albedo of all metamorphic rocks, beyond a critical size d_c, begins to increase as the particle size decreases. The hatched region in Fig. 2 represents the data of the Venera 10 probe. Taking into account the measurement error, the color index C for Venus's surface is 2.1 to 10. We apparently may exclude volcanic glasses and acid rocks from consideration; for the former $C \leqslant 1.3$, and for the latter, although $1.2 < C <$ 2.4, still $A(0.535) > 0.2$ which is much larger than the surface albedo measured by the Venera 9 and 10 probes. Of the basic rocks, basalts do not contradict our measurements regarding the size of C (e.g. Pisgah and Reol Cinder). For these rocks, A is approximately twice as large as the measured value; this may be explained theoretically by the fact that the surface conditions (presence of relief) at the landing site of the Venera 9 and 10 probes differ from the condition of the specimens in laboratory measurements. The

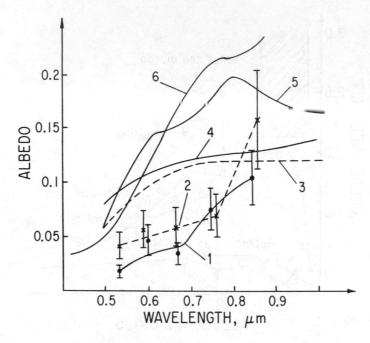

Fig. 3. Comparison of the measured surface albedo of Venus with the reflection spectra of
terrestrial metamorphic rocks and minerals: (1) Venera probe data; (2) Venera 10 probe data;
(3) hornblende, $d = 420–500 \ \mu m$; (4) biotite, $d = 420–500 \ \mu m$ (Ross et al. 1969); (5)
unpolished limonite (Hunt and Ross 1969); (6) Red Cinder basalt (Adams and Filice 1967).

entrainment of dust from the surface at the landing sites of the probes may also
cause a lowering of the albedo. Figure 3 shows the coefficients of reflection as
functions of wavelength for certain rocks and minerals that do not contradict
the results of our measurements.

Adams and McCord (1969) investigated the reflective properties of
basalts as functions of degree of oxidation. They found that with increasing
degree of oxidation, the coefficient of reflection in the shortwave region
decreases and the agreement between their measurements and ours of the
dependence of $A(\lambda)$ improves. We note that CO_2 is capable of oxidizing
basalt. Of the metamorphic rocks on Earth, only the basalts are similar to the
surface of Venus in regard to spectral characteristics, which agrees with the
results of an analysis of the content of K, U, and Th by Surkov et al. (1976a).

We note that, although the surface of Venus and certain types of basalts
agree with respect to A and C, the nature of the dependence of $A(\lambda)$ is different
for these (Fig. 3). The conclusions given here on the basis of optical data, that
basalt is the preferred composition of Venus's surface, should be considered
merely a plausible hypothesis.

8. PANORAMA OF VENERA 9 AND 10 LANDING SITES

K. P. FLORENSKIY, A. T. BAZILEVSKIY,
G. A. BURBA, O. V. NIKOLAYEVA, and A. A. PRONIN
USSR Academy of Sciences

A. S. SELIVANOV, M. K. NARAYEVA,
A. S. PANFILOV, and V. P. CHEMODANOV
The State Scientific Research Center

The panoramic images transmitted from the Venera 9 and 10 probes show the appearance of the surface of the planet Venus. The landscapes imaged are not distinguishable from the landscapes of Earth. The combination of radar observations and detailed phototelevision observations and determinations of the chemical composition of the landing sites of the descent probes permits a sophisticated approach to the study of the surface of Venus.

In October 1975 Veneras 9 and 10 transmitted to Earth the first images of Venus's solid surface. The probes of these missions landed near the extensive highlands of Beta which, judging by the general morphology of this structure, resemble twin shield volcanoes (Rhea Mons and Theia Mons), elevated above ancient gently rolling plains and younger lowland plains (Fig. 1) (Masursky et al. 1980). The landing site of the Venera 9 probe (32° N latitude, 291° E longitude) is on the eastern slope of Rhea Mons; the landing site of the Venera 10 probe (16° N latitude, 291° E longitude) is on a plain at the southeastern foot of Theia Mons (Abramovich et al. 1979).

I. DESCRIPTION OF THE PROBE

The layout of the television system and its mounting on the probe are shown in Fig. 2 (Selivanov et al. 1976*b*). The survey was performed by a

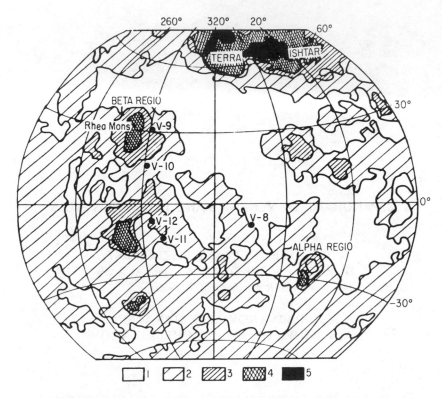

Fig. 1. Geological-geomorphological provinces of the western hemisphere of Venus (after Masursky et al. 1980). 1: lowlands (< 0 km); 2 and 3: rolling plains (altitude 0–1 km and 1–2 km); 4 and 5: mountainous regions (altitude 2–5 km and > 5 km). V8–V12 indicate landing sites of the Venera 8–12 probes.

panoramic camera (Abramovich et al. 1979), mounted inside an airtight heat-insulated container (Vinogradov et al. 1976), across a cylindrical illuminator (Gektin and Panfilov 1978). The scanning of the surface surrounding the probe through a nominal angular field of 40 × 180° was realized by two motions of a scanning mirror (Zasetskiy et al. 1979), a rotation around the panoramic axis and a stepping in the plane which passes through this axis. By means of a lens (Kemurdzhian et al. 1978), a pivoting mirror (Moroz 1979), and a diaphragm (Nepoklonov et al. 1979), the light was transmitted to a detector (Panfilov and Selivanov 1979) and converted into a video signal. A system of automatic regulation of the sensitivity by the level of the output signal was used to adapt the camera to the previously unknown actual conditions of illumination. The video signal from the camera, transformed into the digital form, was rebroadcast to Earth via the first orbiters of Venus. During the return path of the scan, the camera was calibrated from an internal light source.

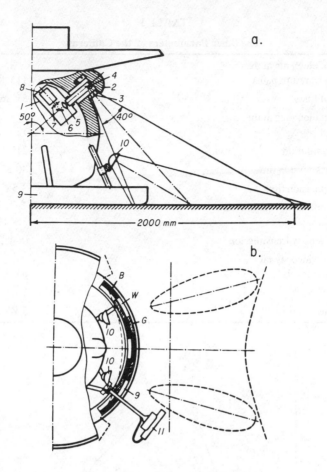

Fig. 2. Layout of the television system and its mounting on the lander: (a) side view; (b) top view. Zones of artifical illumination and trace of the panoramic zone on the surface are indicated by dashes.

In the field of view of the camera (Table I) seen in the central portion of the panorama, there was included the edge of the landing platform (Selivanov et al. 1976b), which had a test pattern painted in the form of white ($W - A = 0.44 \pm 0.02$), gray ($G - A = 0.17 \pm 0.02$), and black ($B - A = 0.07 \pm 0.02$) rectangles where A is albedo. In order to enhance the probability of reception of the image two floodlamps (Selivanov et al. 1976c) were mounted outside the probe, illuminating local zones of the surface (cf. Fig. 2b).

The spectral response of the cameras is shown in Fig. 3 and the modulation transfer functions are given in Fig. 4. The depth of field of the cameras was from 0.8 m to ∞. Figure 5 shows the response curves of the cameras,

TABLE I

The Basic Parameters of the Cameras

Number of elements in the line (without return path)	115
Number of lines	517 ± 13
Number of elements in the return path	13
Angular resolution	21′
Frame transmission time	30 ± 0.9 min
Line transmission time	3.5 s
Range of operating illumination, white surface	15–15,000 lux
Range of density transmission	0–1.2 (1.5)
Number of video signal quantization levels	64
Weight	5.8 kg
Power consumption	5 W

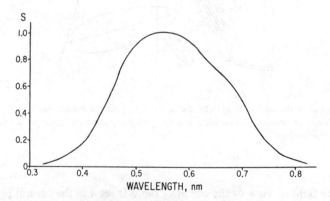

Fig. 3. Spectral response of the television cameras.

which were used to determine the optical densities corresponding to each element of the image. Using the difference between the optical densities ΔD of two segments of the image on the panorama, the contrast between them was determined: $K = 1 - 10^{-\Delta D}$. The reflection curves were determined from an analysis of the signal from flat extended features and a comparison with the test patterns on the landing platform (cf. Fig. 2). Thus, it

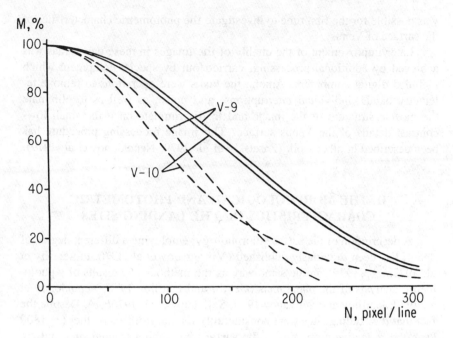

Fig. 4. Modulation transfer functions of the television cameras of Veneras 9 and 10. Dashes are curves plotted for a distance of 1 m from the lamp; solid line, for a distance ≥ 2 m. M indicates extent of modulation and N, the number of television elements in the line.

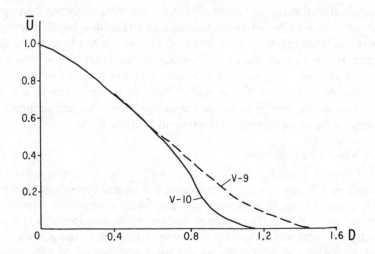

Fig. 5. Response curves of the Venera 9 and 10 television cameras. D is optical density and $U = U/U_{max}$ the relative signal size.

was possible for the first time to investigate the photometric characteristics of the surface of Venus.

Later improvement of the quality of the images in these panoramas was achieved by additional processing, carried out by special equipment which included digital computers. Among the tasks were operations to remove telemetry bands, individual interruptions, and noise, as well as to eliminate the mosaic structure of the image and to discriminate on it the small low-contrast details of the Venus surface. The image processing procedure has been described in other work (Zasetskiy et al. 1979; Nepoklonov et al. 1979).

II. THE MORPHOLOGICAL AND PHOTOMETRIC CHARACTERISTICS OF THE LANDING SITES

A description of the surface morphology, employing a different degree of detail, has been previously published (Vinogradov et al. 1976; Florenskiy et al. 1977a,b,c, 1979), in the same way as the methods and results of a photometric analysis of the panoramas (Gektin and Panfilov 1978; Nepoklonov et al. 1979; Panfilov and Selivanov 1979; Selivanov et al. 1976b, c). Despite the fact that the landing sites were considerably remote from each other ($\sim$ 1800 km), the reflective properties of the surface are similar in both sites and are characterized by an isotropic reflectivity and a low albedo: 0.03–0.12 (Selivanov et al. 1976b), peculiar to dark rocks with a rough surface (Tolchel'nikov 1974; Hapke and Van Horn 1963).

The contrasts between the most characteristic features in both images are related both to the differences between the reflective characteristics of the individual sections of the terrain, and to the difference between the slopes and the orientation of the relief surfaces in space (Panfilov and Selivanov 1979; Selivanov et al. 1976b). As a result of analysis, it has been found that the contrasts at both sites attain the following values: relief $\leq$ 0.5; albedo difference $\leq$ 0.7; both factors $\leq$ 0.9. Figure 6 shows typical histograms of the distribution of contrasts for individual lines of both panoramas, the angular positions of the lines from the origin of the panorama, and summary histograms for the entire panorama (Gektin and Panfilov 1978).

The Venera 9 Panorama

From the panorama it can be seen (Fig. 7) that the probe arrived on the surface in a region of rock detritus with a diameter between several cm and several tens of cm, in the spaces between which a surface of relatively fine-grained soil is visible. The image of the horizon at the upper edge is asymmetrically disposed which, in conjunction with the data of the probe's tiltmeter and the position of the gamma ray densimeter visible on the panorama, indicates that this rock detritus is situated on a slope $\sim$ 20° (Fig. 8). On the horizon at the right side projections are visible. Apparently these are also

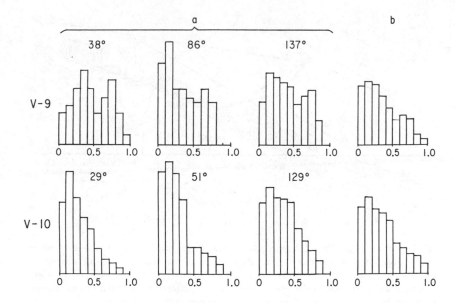

Fig. 6. Histograms of the distribution of contrasts for the panoramas of Veneras 9 and 10: (a) for individual lines; (b) for overall.

Fig. 7. Panorama transmitted by Venera 9 after digital processing (Zasetskiy et al. 1979).

stones. If we assume that their dimensions are the same as the stones near the probe (≤ 1 m), we arrive at the conclusion that the distance of the horizon is no more than several tens of meters at this location, which agrees with the conclusion that the probe landed on a slope. Above the horizon there is a bright image of the sky, which contains no other details. On the left side the image was cluttered with much noise. After processing by computer, in the left upper quadrant a bright field appeared, possibly an image of the sky. However, vague details, visible in this part of the image resembling clouds, are most likely artifacts.

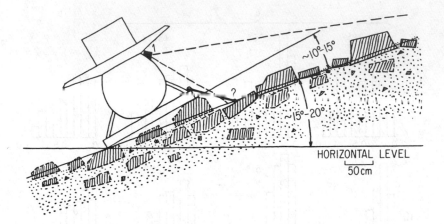

Fig. 8. Position of the Venera 9 spacecraft on the surface and structure of the near-surface layer, assumed as the basis for the analysis of the panorama. Hachures indicate stones, with the solid line showing contours of the stones observed on the panoramas; dots represent soil. Vertical and horizontal scales are identical. (1) indicates the television camera (dashes indicate the limits of its field of view) and (2) the sensor of the densimeter.

The stones in the detritus are rather large (with a diameter ranging from several to 50–70 cm) but relatively low, not more than 15–20 cm in height. The majority of these have a flattened slablike shape with a typical ratio between the height and the diameter in projection from ⅓ to ⅙. As a rule, these slabs lie on their broad face and only in isolated cases stand up, tilted toward another stone. On many stones there can be seen steplike cleavages which along with their slablike shape indicate a plane-parallel cleavage of the venusian rocks comprising these stones (Fig. 9). On the rectified image (Fig. 10) (Nepoklonov et al. 1979) it can be seen that, in cross section, the stones often have the shape of polygons, clearly formed by cleavages of the original rock along 2 or 3 primary directions, intersecting at angles of 50 to 70°. The edges of the slablike stones, as a rule, are sharp and rough, sometimes with a serrated contour. It appears that these stones have been formed by the disruption of rather strong rocks and have not yet been subjected to appreciable weathering by surface effects. Probably associated with this is the secondary maximum on the histogram of contrasts for Venera 9 (Fig. 6), due to the presence of a large number of gaps between the stones in the relief, depressions, and cracks (Gektin and Panfilov 1978). The stones have a uniform coloration, i.e. the size of their component grains is evidently less than the threshold of resolution (several mm). The stones are very dark; their albedo is ~ 0.03–0.04 or somewhat more. This estimate is not very reliable, as there are indications that a change in the characteristics of the calibration pattern took place on Venera 9 (Selivanov et al. 1976b). Only in certain cases, e.g. the

Fig. 9. Fragment of the panorama of Venera 9 (top, closeup view).

Fig. 10. Rectified plan view obtained from the panorama of Venera 9 (Nepoklonov et al. 1979).

stone at the left of the sensor of the gamma ray densimeter, can certain vague and even darker spots of 2–3 cm diameter be seen on the surface of the stones.

In certain cases the outline of several adjacent stones suggests their formation by fragmentation of an even larger stone. Thus, for example, we may reconstruct a single stone from three fragments with steplike cleavages, observed in the foreground on the left side of the panorama (Fig. 11).

In addition to the sharp-edged slablike stones in the field of view there are occasionally seen stonelike formations of different appearance. In particular, there is a stone of rounded and irregular shape with a diameter of roughly 40 cm, situated to the right and somewhat further away from the sensor of the

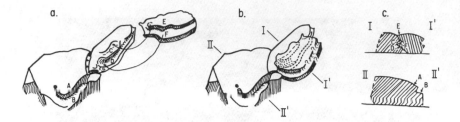

Fig. 11. Possible reconstruction of the large stone in the left portion of the panorama of Venera 9.

Fig. 12. Fragment of the Venera 9 panorama with a "strange" stone (to the right of the sensor of the gamma densimeter).

gamma densimeter (Fig. 12). The surface of this stone is colored more darkly than that of the slablike stones, and is distinctly mottled (with fine bumps?). Similar to this stone is a small (10–15 cm in diameter) stone of rounded isometric shape, near the center of the panoramic image, and several other stones. On the whole, these mottled stones give the impression of low-strength heterogeneous formations of the type of poorly-cemented breccia.

The surface which can be seen between the stones (Fig. 13) is much darker than the upward faces of the stones. The mean albedo of this surface is estimated between 0.02 (Selivanov et al. 1976b) and ~ 0.05 (Golovin 1979). It is important (and this estimate is reliable), that the contrasts between the stone and the ground in this location have a value between 0.29 and 0.5 (Selivanov et al. 1976b). On this surface there can be distinguished pebbles or protuberances with a diameter of several cm and even smaller (at the resolution threshold of the camera), spots of unclear morphology. The stones which comprise the detritus usually give the impression that they are freely lying on this surface, but sometimes they are partially embedded in it. The totality of features suggests that the surface between the stones is a fine-grain porous soil. The soil is probably loose, as it can be seen to be sprinkled on the

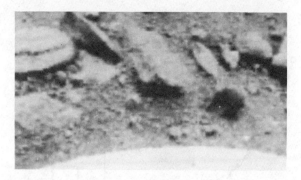

Fig. 13. Fragment of the Venera 9 panorama (closeup view). Soil is seen between stones. Visible at the bottom is a portion of the landing platform.

tops of some small (several cm) stones. It is characteristic that such sprinkling of soil is not seen on the surface of larger (10–20 cm in height) stones.

Using the Venera 9 image, an attempt was made to determine the size spectrum of the fragmented material near the probe. The calculations were performed in the central portion of the panorama ($\pm 45°$ to the right and left of the center). The diameter of the stones was determined from their angular dimensions and distance from the camera which, in its turn, was calculated from the known geometry for production of the image. The presence in the field of view of parts of the probe with known dimensions permitted an independent check. The results of the computation are shown in Fig. 14, from which it can be seen that the stones (more than 5 cm in diameter) in the detritus have a bimodal distribution. The prevalence of the small formations on the soil (3–5 cm in diameter) does not continue the dependence characterizing the stones, which is apparently related to their formation by a factor different from that of the stones comprising the detritus. On the whole, the location of the Venera 9 landing site evidently is a steep slope, actively worked by contemporary slope processes of the talus type.

The Venera 10 Panorama

It can be seen (Fig. 15) that the probe arrived on the surface in a region of a relatively level surface. The probe landed on a rock mass, not less than 3 m in diameter, slightly projected above the surrounding surface. At a distance of several meters from the probe at least three more such large blocks and several smaller blocks could be seen and in the spaces between the blocks a fine-nodular surface of darker soil was visible. The probe's tiltmeter indicated that it was listing by several degrees backwards and to the left (Fig. 16). The horizon was double in the right side of the panorama. The interface between the surface and the sky was not very distinct, which is natural because of the great optical thickness of Venus's atmosphere; however, $\sim 2°$ lower (closer to

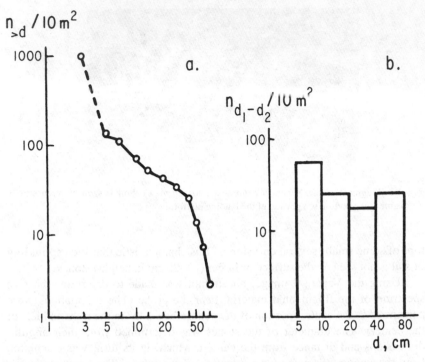

Fig. 14. Granulometric composition of the fragmented material at the landing site of Venera 9. (a) Cumulative dependence of the number of stones n with a size larger than that given as a function of their dimension d. (b) Dependence $n = f(d)$ in the form of a histogram. Area normalized for both figures is 10 m².

Fig. 15. Panorama transmitted by Venera 10 after digital processing (Zasetskiy et al. 1979).

the camera) a second more distinct line could be seen, parallel to the above, and evidently related to local inflections of the surface. Relief details could not be seen on the horizon. According to Moroz (1979), the mere visibility of the horizon in the environment of Venus indicates that its distance is no more than 1 km. On the whole, the general outline of the landscape on the image suggests that the landing site was located on a gently rolling plain.

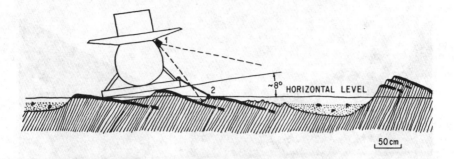

Fig. 16. Position of Venera 10 on the surface and structure of the near-surface layer, proposed as the basis for analysis of the panorama. Hachures indicate bedrock, more closely spaced as the stratified layers are more resistant to fracture; dots indicate soil. Vertical and horizontal scales are identical. (1) television camera (dashes indicate the limits of its field of view) and (2) the sensor of the densimeter.

The blocks of rock have a greatly flattened shape; their height above the surrounding surface is much less than their horizontal dimensions. In the right portion remote blocks can be seen to have a striated texture (stratification?). The striae are slightly inclined to the surface of the ground. Indications of stratification are also observed on the surface of the block on which the probe landed, in the form of a subhorizontal smoothed ridge, which is very prominent below the sensor of the gamma ray densimeter. The material of the rock mass has a rather high strength; the impact of the sensor of the gamma ray densimeter, weighting ~ 2 kg, at a speed of 7 m s^{-1}, did not produce a noticeable fracture of this rock which according to Kemurdzhian et al. (1978) corresponds to a compressive strength of more than 50–100 kg cm^{-2}.

The block on which the probe landed is characterized by a rectilinear shape in plan, especially prominent on the image which has been transformed into a rectified plan view (Fig. 17) (Nepoklonov 1979). The orientation of the rectilinear edges of this block clearly displays a system of 2 directions intersecting at an angle of 50–60°. One of these is followed by the above-mentioned extended ridge below the sensor of the densimeter, the other by several cracks which dissect the block. Apparently the formation of these cracks was associated with the relief of tectonic or crystallization stresses.

The rectilinear contour of the block is combined with a smoothness and a slight rounding of edges and corners. This smoothness also appears in the more distant rocks and generally resembles the phenomenon of sand erosion in terrestrial deserts.

The most probable values of the albedo of the smooth surfaces of the blocks are 0.05–0.06, while that of the brightest stone (at the left) is 0.07–0.12. These estimates are regarded as reliable (Selivanov et al. 1976b). On the surface of the blocks numerous darker spots can be seen. In the left portion of

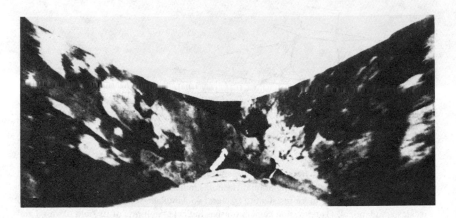

Fig. 17. Rectified plan view obtained from the panorama of Venera 10 (Nepoklonov et al. 1979).

the block these spots are numerous, but relatively small ($\leqslant$ 3–5 cm in diameter). Occasionally they merge into chains $\leqslant$ 10–15 cm long, the orientation of which is clearly indicated by the cracks which intersect this block. In the right portion of the block the dark spots have larger dimensions, but are scarce. It is unclear whether these spots are changes in the color of the surface of the stone, or whether they represent depressions in the stone. Perhaps here we are observing the effect of both these factors and the observed phenomenon has an affinity with the processes of honeycomb weathering, the predominant factor of which on Earth is the chemical disintegration of rock.

The soil between the blocks at the Venera 10 landing site basically resembles that between the stones at the landing site of Venera 9. It is characterized by albedo values which are lower than those for the stones (0.03–0.04). We point out that the contrasts between the stone and the soil are practically the same here as at the landing site of Venera 9 (0.38–0.5) (Selivanov et al. 1976b). At the Venera 10 landing site the surface of the soil is more distant from the camera ($\geqslant$ 2 m) and consequently is viewed with a poorer resolution (not better than 1–2 cm). In the left and right portions of the panorama it can be seen that the soil of the surface forms, as it were, a certain level, above which the soil does not extend. On the surface of the soil individual small (several cm) and relatively bright stones can be seen. On the whole it seems that the landing site of Venera 10 is in the form of a plain, covered by a thin shell of loose soil (regolith) with protruding outcrops of smoothed and dark rocks, on one of which the probe landed.

III. DISCUSSION OF RESULTS

A rather long time has passed since the Venera 9 and 10 missions. The results of measurements from Venera 11 and 12, the descent probes of Pioneer

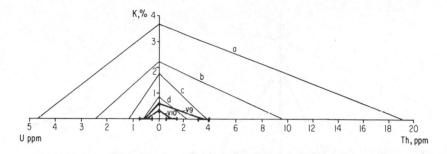

Fig. 18. Mean contents of K, U, and Th in terrestrial effusive rocks of the limestone-alkaline series and in rocks of the landing sites of Venera 9 (V-9) and Venera 10 (V-10) (Florenskiy and Nikolayeva 1981): (a) liparites; (b) dacites; (c) andesites; (d) basalts.

Venus, and, particularly, the radar sensing from the Pioneer Venus orbiter, obtained since that time, now permit a more profound interpretation of the results of the study of these panoramas.

The landscapes of the landing sites of Veneras 9 and 10 have both common and different features. The presence of a steep slope, covered by a detritus of sharp-cornered stones, suggests a recent geologic age for the landscape of the Venera 9 landing site, while the plainlike nature of the surface and the manifestations of weathering processes of the rock masses of the Venera 10 landing site testify to the comparative maturity of this landscape. But at both landing sites rocks and products of disintegration are observed, a loose soil (regolith) which is darker than the stones. The rocks of both landing sites have been formed of very dark, strong, and probably fine-grained material with a plane-parallel cleavage. According to the results of experiments (see Drake 1970), the primarily slablike shapes of the fragments are formed by the fragmentation of vitreous rocks, in contrast with the larger crystalline rocks which produce fragments of a more isometric shape (Garvin et al. 1981). In regard to the level of content of natural radioactive elements, the materials of the surface at both landing sites are similar to terrestrial basalts (Fig. 18) (Surkov et al. 1976a; Florenskiy and Nikolayeva 1981). From the sum of these features we may conclude with certainty that the rocks observed in the panoramas are basalts, possibly altered to a certain extent by surface processes (Florenskiy and Nikolayeva 1981).

Let us consider the geologic position of the landing sites. The Venera 9 probe reached the surface on the eastern slope of Rhea Mons, which is regarded as a basalt shield volcano (Masursky et al. 1980). The recent geomorphological age of the landing site of Venera 9 is in good agreement with the recent geologic age estimated for basaltic volcanism in this region (Masursky et al. 1980; Usikov et al. 1982). The landing site of the Venera 10 probe was on a plain bordering the southeastern base of Theia Mons, which is

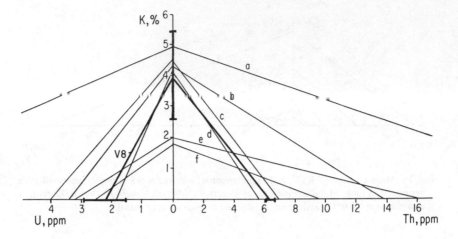

Fig. 19. Mean contents of K, U, and Th in terrestrial intrusive rocks of the alkaline series and in rock of the landing site of Venera 8 (V8) (Florenskiy and Nikolayeva 1981): (a) agpaite nepheline syenites; (b) syenites of the calc-alkaline series; (c) miascite nepheline syenites; (d) Na alkaline syenites; (e) diamond-bearing kimberlites; (f) melteigites, meymechites, ijolites, non-diamond-bearing kimberlites.

also probably a basaltic shield volcano. In terms of the hypsometric position, this area should be compared with the rolling plains which are regarded as the regions of development of the ancient crust (Masursky et al. 1980). However, the province of rolling plains is probably characterized by a material of a different petrological type. Venera 8, landing in this area, indicated that by the content of radioactive elements, the surface material of this location has affinities not with basalts, but with alkaline syenites (Fig. 19) (Florenskiy and Nikolayeva 1981). From spectrophotometric data (Golovin 1979; Ekonomov et al. 1980) it has been shown (Florenskiy and Nikolayeva 1981) that the material of the rolling plains (Veneras 8 and 11) is brighter than that of the basalt elevations (Veneras 9 and 12) and the landing site of Venera 10. It may be assumed that the plain of the Venera 10 landing site had been exposed to the influence of a superposed basalt volcanism, either associated with the shield volcanism of the Beta region or the concomitant processes of cleavage of the blocks of the ancient crust of Venus and their overflow by the basalts of the lowlands (Usikov et al. 1982). However, the landscape of the Venera 10 landing site does not resemble the fresh, untransformed surface of basaltic lava. The age of this landscape suggests either a relatively ancient formation or relatively high rates of weathering processes. In this connection it is interesting to note that the Venera 10 landing site is situated in a broad latitudinal-oriented saddle between the high ground of Beta and that of Phoebe to the south. Under the prevailing winds from the east to the west on the surface of Venus (Counselman et al. 1979), this circumstance may promote the more intense occurrence here of the processes of eolian weathering.

On both panoramas there are traces of exogenic working of the surface. Primarily, this concerns the physical fracturing of the original rocks along cracks, the product of which are the stones. This has been called a fracturing on a decimeter scale (Florenskiy et al. 1977b). Moreover, and this is especially apparent from Venera 10, the edges and corners of the stones and outcrops are smoothed out in the course of time and a change is seen on the surface, reminiscent of honeycomb weathering. This process has been called fracturing on a centimeter scale (Florenskiy et al. 1977b). It is evident that this process also acts on smaller dimensional fractions and it is conceivable that its development is due even more to the chemical interaction between the rocks and the atmosphere of Venus.

This assemblage of processes is a part of the chain of transformations of the rocks into fine-grained regolith. Probably the darker color of the regolith, in comparison with the rocks, is produced not so much by differences in the structure of the rough surfaces of the stones and the regolith, as by chemical changes. Since the coloration of the rocks, as a rule, is basically determined by the content and chemical form of iron, it is probable that one of these factors is different in the soil and in the rocks. But in addition to the formation process of the loose soil, there may also take place the opposite process, namely its lithification by chemical transformations, which is evidently promoted by the effect of high temperature in increasing the rate of diffusion in the solid phase. Possibly the mottled and rounded stones, observed in the panorama of Venera 9, are products of this process, a peculiar regolithic breccia, although this could also be a pyroclastic formation.

IV. CONCLUSION

The images transmitted from the Venera 9 and 10 landers have revealed to us a new world and for the first time shown the appearance of the surface of Venus. It is theoretically important that the landscapes do not display any special features that would distinguish them from the landscapes of Earth. On the contrary, they reveal analogies with very familiar terrestrial landscapes, e.g., the talus slopes of mountains in the arid zone and the peneplain with shallow occurrence of bedrock in the Central Kazakhstan steppes or the plateau sections of the Kola tundra. This circumstance instills confidence that the experience and methods of terrestrial geology in the broad meaning of this term are applicable to the study of the surface of Venus. The combination of radar observations, providing material for a global and regional geomorphological charting, with detailed phototelevision observations and determinations of the chemical composition at the landing sites of the probes, will enable a sophisticated approach to the study of the surface of Venus.

9. STUDIES OF VENUS ROCKS BY VENERAS 8, 9, AND 10

YU. A. SURKOV
USSR Academy of Sciences

The main results are summarized from an analysis by the probes of Veneras 8, 9, and 10, concerning the contents of natural radioactive elements and the density of venusian rocks. The characteristics obtained are compared with similar characteristics for rocks of the Earth's crust. Conclusions are drawn about the nature and origin of venusian rocks.

Experiments were conducted on board the probes of Veneras 8, 9, and 10 in 1972–1975 to investigate the composition and properties of rocks on Venus (Vinogradov et al. 1973; Surkov et al. 1973, 1976a, b, 1977; Surkov 1977). On the basis of the findings, theories have been developed about the nature, origin, and conditions of formation of the surface rock.

The severe climate on the surface of Venus greatly restricts the selection of methods for studying rock characteristics. Nuclear-physics methods were found to be most suitable in the initial investigations. Using scintillation gamma-spectrometers, the γ-radiation was measured and the content of natural radioactive elements in the venusian rock was determined (Vinogradov et al. 1973; Surkov et al. 1973, 1976a). Using a radiation densimeter, the density of the rock was also determined (Surkov et al. 1976b, 1977; Surkov 1977).

The gamma-spectrometers were based on a conventional design (Surkov et al. 1973). They used NaI(Ti) crystals as detectors, the signals from which were first amplified by photomultipliers, preamplifiers, and amplifiers, and then put into multichannel amplitude analyzers, from which the information was transmitted to Earth through the telemetry channel. Since the surface of

Venus is protected against cosmic rays by its dense atmosphere, and the γ-radiation of the rock is solely due to decomposition of the natural radioactive elements (the contribution to the γ-radiation of cosmogenic radionuclides, formed in the materials of the probe, is negligible), measurement of the spectra was performed in a narrow energy range of 0.3 to 3.0 MeV. The spectrometer was situated inside an airtight compartment of the probe and operated in normal climatic conditions. The measurement of the spectra of γ-radiation in each probe lasted for $\sim$ 1 hr.

Figure 1 is a contour map of the Venus surface obtained from the Pioneer Venus orbiter radar mapper (Pettengill et al. 1980b), indicating the landing areas of the Soviet space probes. The landing site of Venera 8 is in an abundantly cratered region of the ancient surface; the sites of Veneras 9 and 10 are on the slopes of the Beta highlands, probably an area of relatively recent shield basalt volcanism. Despite variance of 4 to 5 times for uranium and 10 times for potassium and thorium observed in the contents of rocks in different regions of the Venus surface (Surkov et al. 1973), the mean contents were found close to those of metamorphic rocks (basalts and granites) of the Earth's crust.

Figure 2 shows the K/U systematics of data for the major types of terrestrial rocks, and the points which correspond to the K/U ratios measured by Veneras 8, 9, and 10. At the landing site of Venera 8 the rock has the largest content of natural radionuclides, exceeding the average level in terrestrial basalts and approaching that of the granites. Possibly this ancient region where Venera 8 landed is similar in nature of its rocks to the continental portion of the Earth's crust. The data obtained by Veneras 9 and 10 classify the rocks studied with those of the transitional zone, tholeiitic and alkaline basalts, i.e. rocks produced by a relatively slight differentiation of the planet's primitive material. Apparently these rocks overlay more recent volcanic formations than the rock in the landing area of Venera 8.

The density measurements also give evidence for basaltic composition of the rocks investigated by Veneras 9 and 10. Radiation densimeters to measure the density of the rock were mounted on these probes (Surkov et al. 1976b, 1977; Surkov 1977). Densimeters work by measuring the intensity of incident γ-radiation scattered by the rock. The detector unit, containing a radioisotope γ-source and proportional counters, was placed outside the probe, directly on the rock surface, and operated in the severe climate of the planet. The electronic unit was housed inside the hermetic compartment of the probe and operated in normal climatic conditions. A density of 2.8 $\pm$ 0.1 g cm^{-3} was determined for the venusian rock. Material of this density corresponds to terrestrial basalts of massive texture and low porosity. Such rocks may be formed on Earth in conditions of slow cooling of basalt lava with a small release of gas.

Thus, data both on content of natural radioactive elements and on density suggests that rocks exist on Venus that are similar to the metamorphic rocks of

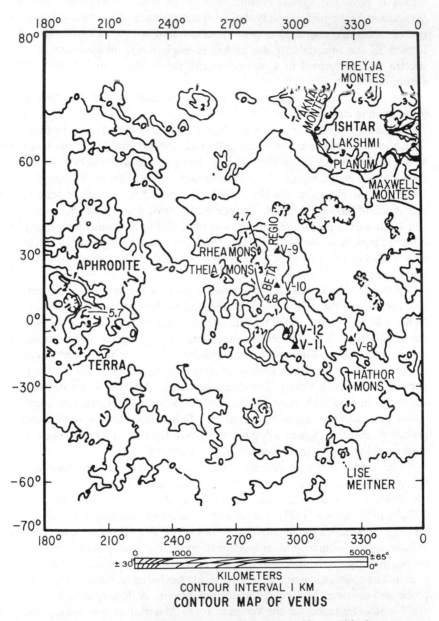

Fig. 1. Map of the surface of Venus obtained by radar by Pioneer Venus. Altitude contours are shown at intervals of 1 km. Landing areas of the Soviet Venera probes are indicated.

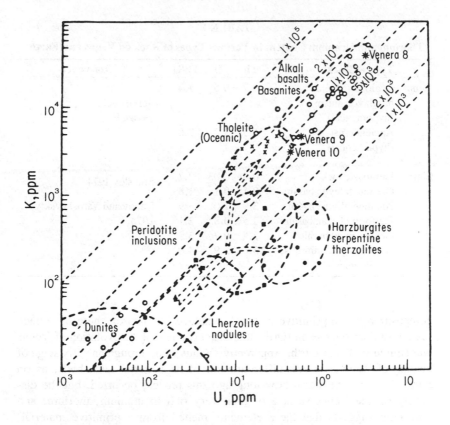

Fig. 2. K/U diagram for the content of potassium and uranium in the major types of Earth and Venus rock.

the Earth's crust. This suggestion also agrees with recent theories about the atmosphere of the planet.

If the great diversity of venusian rocks is generalized to two main types, rocks associated with ancient crust and with more recent areas of shield volcanism, and if their content of natural radionuclides is compared with that of rocks of the Earth's continental and oceanic crust, the types are easily seen to correspond. As shown in Table I, regions of ancient crust on Venus and continents on Earth have a higher content of radionulcides than regions of shield volcanism on Venus and the oceanic crust of the Earth. There is a larger variance on Earth than on Venus in the content of radionuclides in both types of rock, possibly indicating (without considering the poorly representative data from Venus) a greater degree of differentiation of terrestrial materials.

The K/U ratio ($\sim 10^4$) for the venusian rocks investigated is very close to that characterizing the majority of terrestrial magmatic rock. This suggests that if processes occurred on Venus which resulted in a change of the chemical

TABLE I

Thorium and Uranium Content in Various Types of Rock on Venus and Earth

	Type of Rock	Th	U	Th/U	Reference
Venus	Region of shield basalt volcanism (hypothetical)	0.7	0.5	1.4	Surkov et al 1976a, b
	Ancient crust (hypothetical)	6.5	2.2	3.0	
Earth	Ultrabasic rock	0.08	0.03	2.7	Smyslov 1974
	Oceanic toleyite basalts	0.18	0.1	1.8	
	Alkaline olivine basalts	3.9	1.0	3.9	Ronov and Yaroshevskiy
	Platform plateau basalts	2.5	0.8	3.1	1978
	Geosynclinal basalts	2.4	0.7	3.9	
	Granites, granodiorites, granite-gneisses	15.6	3.9	4.0	

composition of the primitive material (whatever its nature), these have been mainly similar to those on Earth. It is conceivable that the evolution of Venus was similar to that of Earth. Apparently, Venus went through an early stage of heating of its interior and differentiation of its material into shells, as on Earth. It cannot be stated how long ago this process occurred, but the discovery on the Venus surface of rock very rich in uranium, thorium, and potassium suggests that these elements melted from a primitive material, deposited in the interior of the planet. The large quantity of carbon dioxide in the atmosphere of Venus indicates that extrusion of magmatic rock onto the surface took place relatively recently or is even now continuing.

10. THE INTERIOR OF VENUS AND TECTONIC IMPLICATIONS

ROGER J. PHILLIPS
Lunar and Planetary Institute

and

MICHAEL C. MALIN
Arizona State University

The mean density of Venus, when corrected for compression, is ~3% less than that of Earth. This difference is smaller than the uncertainties of cosmochemical models. Since Venus undoubtedly has a separated core, its radial density profile is probably similar to Earth's. Because of the temperature dependence of viscosity, Venus's and Earth's interior temperatures should differ appreciably less than their surface temperatures do. However, the lithosphere of Venus is probably more buoyant than that of Earth because of higher lithospheric temperature, and so subduction, an essential process of plate tectonics, is less likely to occur. The major linear elevated regions on Venus, associated with Beta Regio and Aphrodite Terra, do not appear to have the necessary shape to be the divergent plate boundaries of seafloor spreading. Instead such tectonics would apparently be confined to the median plains and may not be resolvable in the Pioneer Venus altimetry data. The ratios of gravity anomalies to topographic heights on Venus indicate that surface load compensation occurs at great depths, e.g. ≥ 100 km under western Aphrodite Terra and ≥ 400 km under Beta Regio. At least some of this compensation is probably maintained by mantle convection. The shape of Venus's hypsogram is quite different from the ocean mode of Earth's hypsogram, and is evidence against seafloor spreading in the median plains. The topography more closely resembles intraplate areas on Earth. Ishtar Terra may be the only true continent on Venus. We propose that the tectonics of Venus is similar to intraplate basin-and-swell tectonics on Earth. Hot spot activity may be more vigorous on Venus than on Earth, if seafloor spreading does not remove heat.

The purpose of this chapter is both to review our knowledge of and to speculate about the interior of Venus. Our approach leads from compositional, structural, thermal, and gravitational modeling to the tectonics of the surface. We emphasize the relationship between thermal history and tectonics, because the tectonics express how a planet's lithosphere transfers heat from the interior. The Pioneer Venus mission has added a number of important data sets concerning the interior. Altimetry data have given a general idea of the physiographic provinces, and thus the tectonic style, of Venus. Gravity field mapping has provided clues to the relationships between surface features and interior processes.

The generalization is often made that planets of the same size and mean density, such as Venus and Earth, should undergo roughly similar evolution (Kaula 1975); it is natural that Venus should be studied by searching for similarities to or differences with Earth. With this in mind, various geophysical analyses for Venus are presented and the physiography is discussed, as objectively as possible; opinion and speculation are mostly confined to the final section of this chapter.

I. COMPOSITION AND STRUCTURE

Models for composition and density distribution within Venus are similar to those for Earth, because of its similar size and mean density. The two planets' present heliocentric positions suggest that their positions in the pre-accretionary solar nebula were also similar.

Venus is 18% less massive than Earth, while its mean density, 5.245 gm cm^{-3}, is 5% less than that of Earth; most of this contrast results from difference in self-compression. However, if we assume that Venus has the same structure and composition as Earth, and the same temperature for equivalent pressure, then its density should be 1.8% greater than actually observed (Ringwood and Anderson 1977), still 3% less than Earth's. The higher surface temperature of Venus leads to a slightly higher interior convecting temperature and, because of the coefficient of thermal expansion, to a decrease in the 1.8% difference in density (e.g. a 1.4% disparity using the convection results of Kaula and Phillips [1981]).

There are three general approaches to reconciling the remaining density difference between Venus and Earth: (a) bulk compositional difference (Basaltic Volcanism Study Project [BVSP] 1981, Chapter 4); (b) significant differences in internal temperature structure, which invalidate the analogy to Earth (Goettel et al. 1981); or (c) massive basaltic crust with depression of the basalt-eclogite transition (Anderson 1980). BVSP (1981, Chapter 4) has given an extensive review of possible fractionation processes within the pre-accretionary solar nebula that might explain the compositional differences among the terrestrial planets. Goettel et al. (1981) claim, however, that models based on these processes can all be adjusted to explain the density difference between Venus and Earth.

Equilibrium Condensation Models

In comparing Venus to Earth, it is convenient to place condensation/ fractionation models in four classes: equilibrium condensation, oxidation state, chondritic composition, and iron-silicate fractionation. Equilibrium condensation models (Lewis 1972, 1973, 1974) assume systematic decrease in temperature with distance from the protosun, and accretion of each planet from the specific solid condensates that were in temperature-pressure equilibrium with the nebular gas in the vicinity of the local accretion zone. Earth's greater density is explained by the presence of sulfur in the core; sulfur is condensable as FeS at the lower nebular temperature at Earth's distance from the protosun. Ringwood and Anderson (1977) have objected to the equilibrium condensation model on the grounds that: (1) the quantity of sulfur in Earth's core is insufficient to account for the density difference; (2) the model would indicate a higher density for Venus because the stronger reducing conditions of its higher nebular temperature would favor less oxidation of iron; and (3) Earth's potassium deficiency is not accounted for in the model. Goettel et al. (1981) have countered these arguments by estimating that the 1.8 ± 0.2 % difference between densities deduced by Ringwood and Anderson (1977) should really be 1.0 ± 0.4 %, assuming equivalent temperature with equivalent depth (not pressure) and considering a deeper basalt-eclogite transition. A 1 % difference can be accommodated by equilibrium condensation models, as these models also predict for Venus higher internal temperatures (see below) and higher pressures required for phase transitions. Both these effects lower internal density and, when corrected to standard conditions, reduce the density disparity between Venus and Earth. These results follow from the low to zero FeO content of the equilibrium condensation model of Venus (i.e. FeO does not condense at Venus's nebular equilibrium temperature). The assertion of higher pressure follows directly from experimental observation that the pressures of phase transformation from olivine to spinel and from spinel to perovskite are inversely proportional to the ratio FeO/ (FeO+MgO) (e.g. Ringwood and Major 1970).

The argument that lower FeO content leads to higher internal temperature follows from arguments regarding internal convection. Although the question of internal temperature state is discussed in detail in Sec. II, it is useful to examine briefly the interior temperature of a simple convecting system. Consider a spherical shell bounded on the inside by a core of radius R_c and on the outside by R_o, with temperature T_0. Convection is maintained by an internal heat source concentration Q. The average steady-state temperature (weighted with respect to density times specific heat), referenced to T_0, is given by

$$T - T_o = \Delta T \approx \left(\frac{Ra_{cr}}{Ra}\right)^\beta \frac{R_o^3 - R_c^3}{3kR_o^2} (R_o - R_c)Q \qquad (1)$$

where k is thermal conductivity in the boundary layer of the convecting region, Ra the Rayleigh number of convection, and Ra_{cr} the critical Rayleigh number to initiate convection. The term β is a parameter for vigorously convecting systems, found both observationally and theoretically to be between ⅓ and ¼. Since Ra is inversely proportional to the viscosity η of the convecting system

$$\Delta T \propto (\eta Ra_{cr})^{\beta} \tag{2}$$

where the viscosity can be expressed as

$$\eta = f(\sigma) \exp [(E + PV)/KT] \tag{3}$$

where E and V are activation energy and volume for creep, P is pressure, K is Boltzmann's constant, and $f(\sigma)$ represents a function dependent on stress. Specifying T_m as melting temperature, it has been suggested (Weertman 1970; Weertman and Weertman 1975) that

$$(E + PV)/K = gT_m \tag{4}$$

where g is a dimensionless constant. The relationship in Eq. (4) is not undisputed, and it is unclear whether T_m refers to solidus, liquidus, or some intermediate temperature. Yet the relationship, if correct, implies that

$$\Delta T \propto \exp[g\beta T_m/T]Ra_{cr}^{\beta} . \tag{5}$$

The melting temperature for olivine is inversely proportional to FeO content, and if all other factors were equal Venus, with a lower FeO content than Earth according to equilibrium condensation models, would sustain a higher internal convecting temperature.

Regardless of the FeO content of the Venus mantle, T_m for Venus could be high relative to Earth because of a relative lack of water in the interior. Of course, the low water content of Venus is only a supposition, based on equilibrium condensation models. If substantial water was initially present in Venus, the amount that now remains in the interior depends on the planet's degassing and recycling history. Assuming Eq. (4) to be true and T_m to represent the solidus temperature, equilibrium condensation models do predict a higher internal temperature for Venus from Eq. (5), on the basis of relative water content alone.

Goettel et al. (1981) brought out an additional point concerning plate tectonics and its effect on interior temperature. A free surface boundary condition, in which the lithosphere participates in convection (e.g. seafloor spreading and subduction), leads to a mantle more unstable to convection (i.e. a lower Ra_{cr}) than a rigid lithospheric boundary condition. This is because cooling of a convecting surface boundary layer is highly efficient in removing heat from the interior. We would anticipate higher interior temperatures with a rigid surface boundary condition, and indeed this is predicted by Eq. (5) using a higher Ra_{cr} (2772 [rigid] vs. 1308 [free]; Turcotte et al. [1979]). Goettel et al. (1981), quoting the results of Turcotte et al., conclude that a rigid venusian lithosphere could raise interior temperatures by 250 to 300°C, and if the compositional effects also work in the same direction Venus's mantle could be 500 to 600°C hotter than Earth's. The startling conclusion from all this is that the zero pressure or intrinsic density of Venus might be greater than that of Earth, and so the terrestrial planets would display densities monotonically decreasing with distance from the Sun.

Other Condensation-Fractionation Models

The oxidation state model is discussed in detail by Ringwood and Anderson (1977); it specifies that Venus and Earth accreted from matter of essentially the same composition, and that the difference in density arose solely from a difference in the oxidation state of iron, each planet containing 3.6 % sulfur. Specifically, the density difference between Venus and Earth is accounted for by 35 wt % of all iron in Venus being present as FeO. The Venus mantle contains 13.4 % FeO; the core (16 % sulfur) accounts for 23 % of the planet's mass. In contrast, Earth's mantle contains $\sim$ 9 % FeO, and its core (11 % sulfur) is 32.5 % of its mass.

For Venus to contain more FeO than Earth, it must have accreted at lower average temperatures than Earth. It is supposed that the nebular temperatures were approximately the same for both planets and that the temperatures imparted by conversion of gravitational to thermal energy during accretion were lower for Venus because of its 18 % lower mass. However, the accretion mechanism is complex and involves mechanical heat transfer by large-scale impacts into the growing protoplanet. According to Kaula (1981), the temperature of a planet the size of Earth would never greatly exceed the melting curve because of efficient heat removal by convection. The reduction of FeO to Fe is a volatilizing reaction which might take place on small pre-accretionary particles in the near-planet nebula, which in turn may be influenced by thermal radiation from the planet's surface. This difference in radiation energy would presumably provide more reduction of FeO for Earth than for Venus during their accretion. This concept has not yet been quantitatively examined and so the oxidation state hypothesis has not been rigorously tested.

Chondritic composition models (Wood 1962) assume that the planets are composed of the same material as chondritic meteorites, and can be simulated by varying the mix of key components of these meteorites. As a representative of this class of planetary models, the approach of Morgan and Anders (1980) is briefly reviewed. The approach relies on the hypothesis that in chondrites elements of similar cosmochemical properties are fractionated by similar factors and can be grouped together. If this is also true for planets, then only the abundances of five 'index elements,' one for each group, are required in the planet to calculate the abundances of all 83 (stable) elements. These five elements represent seven interdependent components of chondrites resulting from condensation, fractionation, and remelting. The model implies that groups tend to stay together during condensation because elements within each group have similar volatility. Physical mechanisms may be mostly responsible for the iron-silicate fractionation. The five key elements, expressed in terms of geophysically or geochemically measurable quantites, are: U, Fe, K/U, Tl/U, and FeO/(FeO+MgO). They respectively represent high-temperature refractory condensates, intermediate-high temperature metal condensates, intermediate-temperature volatile condensates, low-temperature volatile condensates, and low-temperature Fe-volatile reactions.

Applying the model of Morgan and Anders to Venus is only as accurate as are the estimates of the 5 key indicators. The amount of U and the K/U ratio can be extrapolated from Venera measurements to estimate bulk abundances. The value of the Tl/U abundance could be adopted from the Earth value. A gross estimate of Fe is obtainable from the mean density. However, there is no way to reliably estimate the FeO/(FeO+MgO) content of Venus; if there were, we could test both the oxidation state and equilibrium condensation hypotheses. Therefore, it does not seem possible to refine density estimates for Venus with the chondritic model.

Finally, we consider iron-silicate fractionation models, in which the silicate composition of the terrestrial planets is the same and the density differences are accounted for by different iron/silicate ratios (Urey 1952). Most models of this type involve physical fractionation processes and predict a systematic decrease in density with distance from the Sun. For example, nickel-iron planetesimals would possess inherently greater finite strength than silicate planetesimals, and given higher orbital velocities closer to the proto-sun we would anticipate the ratio of pre-accretionary metal to silicate planetesimals also to increase towards the Sun. Weidenschilling (1978) has presented an iron-silicate fractionation model based on orbital decay due to gas drag. The rate of orbital decay is slower for larger and /or denser bodies, due to smaller area-to-mass ratios.

Venus, of course, violates the monotonic decrease in density outward from the sun, and it is difficult to identify a collisional differentiation process with the type of selectivity required for such an effect. However, the discussion by Goettel et al. (1981) of a Venus denser than Earth, based on internal temperature arguments, would allow the iron-silicate fractionation process to be invoked.

Basalt-Eclogite Model

Anderson (1980) proposed the possibility that Venus and Earth are identical in composition, and that their difference in mean density is due to Venus's higher basalt-to-eclogite ratio.

On the basis of Earth's mantle composition, and an analogy with the

Moon, Anderson suggests that both Earth and Venus could have produced the equivalent of > 400 km thickness of basalt, along with extensive upward differentiation of heat sources. For Earth, subducted basalt would invert to eclogite and form a gravitationally stable layer. Anderson points out that the layer bounded by the 220 km and 400 km seismic discontinuities is well matched in both P-wave and S-wave velocities by eclogite. The eclogite between the 400 km and 670 km seismic discontinuities would be stable as garnetite.

For Venus, with higher near-surface temperatures and lower pressures, Anderson estimates a surface layer 100 to 170 km thick, composed of basalt and partial melt. For Earth, the oceanic basalt layer is only ~ 6 km thick, so this model suggests a much higher basalt-to-eclogite ratio for Venus. The model can account for most or all of the density difference between Venus and Earth, and therefore it may not be necessary to invoke differences in composition or oxidation state.

Radial Density Models

Zharkov (1981) has presented a detailed study of interior density models for Venus; the subject is briefly reviewed here. Any radial density model for Venus is heavily dependent on analogy with Earth, and less so on what is known about other terrestrial planets. The main assumptions are: (1) the excess density over that explained by common silicates is due to iron; and (2) most of the iron resides in a core. We do not actually have the data required to prove the existence of a venusian core: (1) moment of inertia, (2) seismic velocities and attenuation, or (3) free oscillations. To obtain the moment of inertia, either we would need to know the dynamical ellipticity (in addition to the known second zonal harmonic of gravity J_2), or the rotation rate of Venus would have to be considerably greater so that hydrostatic behavior could be assumed.

An intrinsic magnetic field for Venus would be strong evidence for the presence of a liquid iron core. Theoretical scaling arguments based on hydromagnetic flow (Busse 1976) suggest that a measurable field should exist for Venus. The absence of a measurable intrinsic magnetic field (Russell et al. 1980b) does not preclude the presence of a core, but this observation may have a bearing on the core's physical state, if it does exist. If the energy source of the geodynamo is solidification of the inner core, and if Venus's interior temperatures are at least as high as Earth's, then Venus will not have the energy to generate a magnetic field because its central pressure is lower than the pressure at the inner-core/outer-core boundary of Earth (Stevenson et al. 1980).

Various cosmochemical models, as discussed above, predict the composition of the Venus mantle, including the iron content of the silicates. Under reasonable assumptions about the composition of the core (e.g. Fe, Fe-FeS, Fe-FeO), core size is constrained solely by mean density.

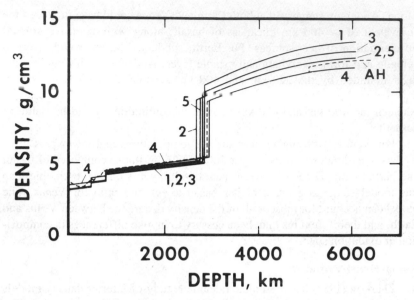

Fig. 1. Radial density models for Venus (Basaltic Volcanism Study Project 1981, Chapter 4).
Numbered profiles are for various cosmochemical models; dashed profile labeled AH is
for Earth.

Given the standard density of a particular composition, the actual density
in the planet is dependent on both temperature and pressure, including the
effects of phase changes. The approach to density modeling, as outlined in
BVSP (1981, Chapter 4), is first to convert the oxide chemistry to normative
mineralogy and then to apply known pressure and temperature corrections to
the estimated mineralogy of the planet. The uncertain temperature profile of
Venus is a dominant source of error in deriving density models. The pressure
effect on density ρ is given by the general finite strain equation:

$$P = \frac{3K_0}{m-n} \left[\left(\frac{\rho}{\rho_0} \right)^{\frac{1}{3}(m+3)} - \left(\frac{\rho}{\rho_0} \right)^{\frac{1}{3}(n+3)} \right] \tag{6}$$

where K_0 and ρ_0 are zero pressure bulk modulus and density and n is generally
1 or 2, m is 2 or 4, with $n < m$. The pressure at any depth z is

$$P(z) = \int_{R_0-z}^{R_0} \rho(r)g(r)\mathrm{d}r = \int_0^z \rho(z)g(z)\mathrm{d}z \tag{7}$$

where R_0 is planetary radius and $g(r)$ is planetary gravity. The effect of
temperature is included by using, for the appropriate mineralogy, the first and

second partial derivatives for the bulk modulus, and the coefficient of thermal expansion and its first partial derivative.

With the temperature and density corrections we have discussed, the density is integrated radially to derive planetary mass. Departures from the true mass (a few %) are accommodated by slight adjustments of the core radius.

Figure 1 shows density profiles based on various cosmochemical models. They are remarkably similar; the greatest variance among the models is in the core region. The mean core boundary is at ~ 3240 km radius, and this value is adopted for subsequent thermal modeling in this chapter. The effects of pressure on density, and the phase changes from olivine to spinel and spinel to higher pressure phases, are obvious in the models.

In summary, a wide variety of cosmochemical models permit construction of remarkably similar density profiles for Venus. This is because: (1) to first order, the abundances of Si, O, Mg, and Fe in models of Venus are similar to abundances in the solar atmosphere, carbonaceous chondrites, and Earth; (2) the mantle composition of Venus must be dominated by olivines and pyroxenes (or higher pressure equivalents); the various models give essentially minor variations among these minerals; and (3) only a fixed amount of Fe can be utilized to construct the mantle silicate mineralogy; the remainder must be relegated to a core.

II. THERMAL MODELS AND EVOLUTION

The thermal evolution of Venus depends on the content of radioactive heat sources and the initial temperature profile. Secondary effects include possible latent heat release of core freezing which apparently may not have happened in Venus (Stevenson et al. 1980) and the history of the surface temperature. The latter can also profoundly affect lithospheric tectonics. Both for simplicity and for the use of our calculations as bounding arguments, core contributions to the thermal state are omitted here.

The thermal evolution is obviously of fundamental importance for several major issues: (1) evolution of the atmosphere and general fate of interior volatiles; (2) differentiation and upward migration of basaltic melts; and (3) style of surface tectonics. We devote considerable attention to this last point in this chapter, for tectonics simply reflect the way the lithosphere is able to expel internal heat. Conversely, surface tectonics provide clues to internal thermal state and evolution.

Schubert et al. (1969) showed that the mantles of the terrestrial planets are unstable to solid state convection, and that planets remove most of their internal heat by convection with efficiency mainly dependent on planet size. Introducing the equations of motion into calculations of planetary heat transfer presents severe complexities due to strong nonlinearities in the advective term (temperature-velocity product) and to the exponential temperature dependence of the viscosity. The full convection equations have been solved by numerical techniques (e.g. finite difference), but the high Rayleigh numbers for Earth and Venus preclude accurate numerical simulations. However, because of these high Rayleigh numbers, certain asymptotic techniques for heat transfer in a convecting fluid may be applicable to predict reasonably accurately the surface heat flux, the average mantle temperature, and the general cooling history of the planet (McKenzie and Weiss 1975; Sharpe and Peltier 1978, 1979; Schubert et al. 1979, 1980b; Turcotte et al. 1979).

The parameterized convection approach adopted here uses the relationship for the heat flux, at the radius of the lithospheric boundary $r = R_L$, from a vigorously convecting mantle, with no core heat sources, as

$$q_0 = \frac{k(T - T_0)}{(R_L - R_c)} \, Nu \qquad (8)$$

where Nu is the Nusselt number, the ratio of total heat flux (convective plus conductive) to conductive heat flux. The Nusselt number is related to the Rayleigh number,

$$Nu \approx \left(\frac{Ra}{Ra_{cr}} \right)^{\beta} \qquad (9)$$

and the Rayleigh number is given by

$$Ra = \frac{\rho g \, \alpha(T - T_0) \, (R_L - R_c)^3}{\kappa \eta(T)} \qquad (10)$$

where ρ, g, and α are density, gravity, and coefficient of thermal expansion in the convecting region, and κ is diffusivity. Conservation of energy requires

$$\rho C_p (R_L^3 - R_c^3) \, \frac{dT}{dt} = -3R_L^3 q_0(t) + Q_m(t) \, (R_L^3 - R_c^3) \qquad (11)$$

where C_p is specific heat, and $Q_m(t)$ heat production rate per unit volume in the mantle. Successive substitution of Eqs. (3), (8), (9), and (10) into Eq. (11) yields an initial value problem for temperature which may be integrated in the time variable by standard techniques.

There are two basic results from parameterized convection modeling (Schubert et al. 1979, 1980b; Turcotte et al. 1979): that the terrestrial planets are still releasing a significant fraction of their primordial heat; and that this conclusion holds over a wide variety of modeling parameters and initial mantle temperatures. For equivalent mantle heat generation, Earth and Venus will evolve to a roughly similar convecting temperature T despite their different surface temperatures, because of the strong temperature dependence of viscosity (Fig. 2). Figure 3, from the parameterized convection results of Schubert

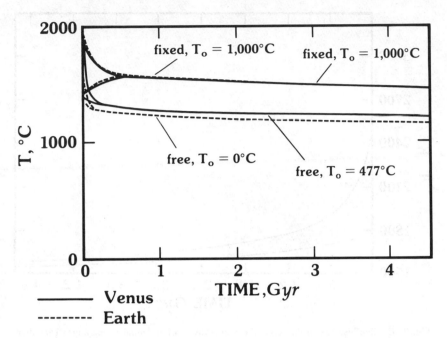

Fig. 2. Parameterized convection results for Venus and Earth (adapted from Turcotte et al. 1979). Interior convecting temperature is shown for fixed (T_o = lithosphere base temperature) and free (T_o = surface temperature) boundary conditions. Despite the large difference in surface temperatures for Venus and Earth, the convecting temperatures are nearly equal.

et al. (1979), shows that an initial temperature range as large as 1500°, in various Earth models, would evolve to a current maximum disparity of only ~ 10°C. The extensive parameter study for Earth by Schubert et al. (1980b) showed that secular cooling accounts for 15 to 35 % of the heat output of Earth over a wide range of the parameters Ra_{cr}, β, E, and f (σ). A similar result can be expected for Venus. Turcotte et al. (1979) reach similar conclusions, although they estimate the present-day secular contribution as only 7 to 12 %. The main point is that after the first 0.5 to 1.0 Gyr, the heat flux from the interior of Venus must be primarily controlled by the radioactive content of the interior, and the secular fraction is insensitive to uncertainty in the initial conditions.

The amount of primary heat producing elements (K, U, Th) present in Venus may be estimated by cosmochemical modeling, analogy with Earth, and extrapolation of surface measurements by the Venera landers. U and Th are highly refractory, and we might expect similar concentrations of them in Earth, Venus, and chondritic meteorites; we would certainly expect the Th/U ratio to be roughly constant because of similar volatilities. Determined from steady-state heat flux calculations and observed K/U ratios, the U concentra-

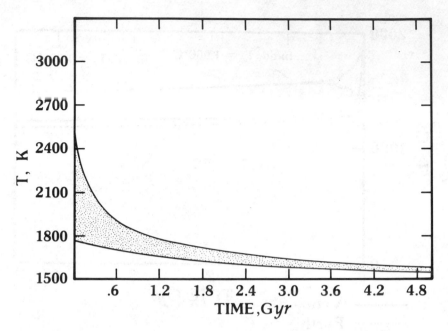

Fig. 3. Cooling histories of the terrestrial planets for various initial temperatures (Schubert et al. 1979), showing that present-day convecting temperature is insensitive to initial conditions.

tion in Earth is considerably greater than that found in chondrites (Wasserburg et al. 1964) as well as those values derived in cosmochemical models (BVSP 1981, Chapter 4). The convection calculations, as discussed above, show that secular cooling of Earth cannot be ignored; this in turn yields a bulk U concentration closer to chondritic.

The K concentration in Venus and Earth is difficult to constrain because of its relatively high volatility. The K/U ratio of Earth clusters closely about a value of 10,000 in crustal rocks. The main uncertainty in using this figure as representative of the bulk Earth is the possibility that potassium does not behave as a large-ion incompatible element at high pressures (Bukowinski 1976). Equilibrium condensation models of planet formation (Lewis 1972, 1973, 1974) predict essentially complete precipitation of K at Earth's distance, so the missing K is presumed to reside in the core. Ringwood and Anderson (1977) point out the almost complete lack of affinity of K with Fe-FeS systems at low pressures, but at high pressures K is predicted to be highly miscible in Fe-FeS melts (Bukowinski 1976).

In any partial melting of a mineral assemblage, K, U, and Th are partitioned into the melt with such efficiency that most of these elements enter the first few % of melt (Philpotts and Schnetzler 1970; Dostal and Capedri 1975). However, a significant fraction of potassium may be lost to the core before partial melting of the mantle, because of its miscibility with Fe-FeS melts. The Venera 8, 9, and 10 measurements of K/U ratios (Vinogradov et al. 1973; Surkov 1977) are on the average not much different from the terrestrial crustal value. Earth's crustal value, however, cannot be representative of the bulk Earth, for which the mean heat flow is reasonably well matched in parameterized convection models by assuming chondritic heat sources. In this chapter, chondritic values are assumed for Venus; specifically these are present-day abundances of 14 ppb for U, and Th/U and K/U ratios of 4 and 7×10^4, respectively.

Parameterized Convection Models for Venus

The goal of this chapter is to link interior thermal evolution with surface tectonics, on the basis of parameterized convection models. Three major questions are addressed here in terms of thermal state:

1. What is the buoyant stability of the lithosphere (i.e., is it now or was it ever subject to subduction, a prerequisite for plate tectonics);
2. Is there any evidence of seafloor spreading in the topography;
3. What are the constraints on interior heat flux in order to support the venusian topography by the finite strength of the lithosphere.

Equation (11) is the basis for our calculations, with $Q_m(t)$ describing the decaying K, U, and Th heat sources in present-day concentrations as described in the previous section. Schubert et al. (1979,1980b) introduced a second differential equation to describe evolution of a conducting lithosphere above a convecting interior. Specifically, the lithosphere boundary at radius $r = R_L$ and temperature T_L is modulated by the change in heat flux Δq across it;

$$-\rho C_p (T - T_L) \frac{dR_L(t)}{dt} = \Delta q \equiv q_L - q_0 \qquad (12)$$

where q_L is the heat flux immediately above the boundary and q_0 is the convective heat flux delivererd to the base of the lithosphere. Equation (12) has been interpreted as the thickening or thinning of the lithosphere with time according to the sign of Δq. Alternatively, Eq. (12) may be viewed as a description of migration of the T_L isotherm. Schubert et al. (1980b) identify the crust and lithosphere boundary depth as equivalent. In our formulation we separate the two boundaries. Both approaches describe a crust with heat source concentration Q_c:

$$Q_c(r) = Q_s \exp\left[(r - R_\venus)/D\right] \qquad (13)$$

where $R_\venus$ is the radius of Venus and Q_s and D are the surface concentration and spatial decay constants. In this analysis, the crust is bounded by $r =$

$R_{cr}(t)$, and the upper mantle lithosphere has a uniform distribution of heat sources at any given time, equivalent to the convecting mantle concentration Q_m. The conduction solution for temperature in the crustal portion of the lithosphere is given by

$$T(r) = A/r + B + \frac{Q_s D^2}{k_{cr}} (1 - 2D/r) \exp\left[(r - R_{\venus})/D\right] \tag{14}$$

and in the upper mantle portion by

$$T(r) = C/r + F - \frac{r^2}{6 k_{um}} Q_m \tag{15}$$

where k_{cr} and k_{um} are the thermal conductivities of the crust and lithospheric upper mantle and A, B, C, and F are constants determined by T_L, the surface temperature T_s (740 K), and continuity of temperature and heat flux across the crust-mantle boundary. The heat flux q_L is given by $-k_{um} \partial T/\partial r$ at $r = R_L$, using Eq. (15). The thickness of the crust increases with time according to an exponential increase in volume:

$$V_c(t) = V_{final}(1 - \exp[-\gamma t/t_{final}])/(1 - \exp[-\gamma]) \qquad t < t_{final} \tag{16a}$$

$$V_c(t) = V_{final} \qquad t \geq t_{final} \tag{16b}$$

where V_{final} is final crustal volume reached at $t = t_{final}$, and γ describes the rate of crustal buildup. Radioactive heat sources are depleted from the mantle into the differentiating crust according to

$$\frac{dC_c(t)}{dt} = \frac{dV_c(t)}{dt} c_m(t) f \tag{17}$$

where $C_c(t)$ is the heat source content of the crust, uncorrected for radioactive decay, and f is the enrichment factor of radioactive heat sources in the melted mantle component. The instantaneous concentrations of heat sources in the

$$Q_m(t) \equiv c_m(t) = [C_m(0) - C_c(t)] h(t)/V_m(t) \tag{18a}$$

and

$$c_c(t) = C_c(t) h(t)/V_c(t) \tag{18b}$$

respectively, where $V_m(t)$ is the instantaneous volume of the mantle, $C_m(0)$ is the initial heat source content of the mantle, and

$$h(t) = Q_{ch}(t)/Q_{ch}(0) \qquad (19)$$

is a fraction (< 1.0) giving the radioactive decay in heat production of the original chondritic source concentration in the mantle, $Q_{ch}(0)$. Thus, $C_m(0) = 4/3\pi(R_\varphi^3 - R_c^3)Q_{ch}(0)$, $Q_m(0) = Q_{ch}(0)$, and the average concentration in the crust and mantle combined is $Q_{ch}(t)$ but because of depletion into the crust for $t > 0$, $Q_m(t) < Q_{ch}(t)$. The term Q_s in Eqs. (13) and (14) is derived from $C_c(t)$.

Equations (11), (12), and (17) form a coupled set of first order differential equations that describe the thermal history of a planet with a convecting interior, an evolving lithosphere not participating in convective flow (as opposed to Earth's oceanic lithosphere), and a growing crust that is preferentially concentrating radioactive heat sources.

Comparative Lithospheric Stability of Venus and Earth

Hot thin lithosphere with attendant basaltic crust is formed on Earth at the midocean ridges and is carried away by the motions of seafloor spreading. At some point in the aging process of the moving plate, subduction may start and the lithospheric material may return to the interior, completing the flow process of mantle convection. The amounts of plate creation at the ridges and of plate loss at subduction zones and continental collision zones should be approximately equal.

Anderson (1981) and Phillips et al. (1981) raised the possibility that the higher surface temperature on Venus would lead to a more buoyant, and hence more difficult to subduct, lithosphere. To test this hypothesis quantitatively, the effects of negative buoyancy on both initiating and sustaining subduction must be considered separately. The two major factors in the buoyancy of unsubducted oceanic plate are compositional and thermal. Compositional stratification (for example, the 6 km thickness of tholeiitic basalt formed at the midocean ridges on Earth) leads to positive buoyancy. A positive temperature gradient will always lead to a negatively buoyant effect on the lithosphere, due to thermal expansion. Oxburgh and Parmentier (1977) calculated the effects of both compositional and thermal buoyancy on the oceanic lithosphere and concluded that overall negative buoyancy would not set in until the lithosphere had aged for ≥ 40 Myr. They implied that this condition was a prerequisite for initiation of subduction and that, while not all oceanic lithosphere that ages to the point of instability subducts, "subduction should occur if this unstable state is sufficiently disturbed."

Parsons (1982) showed that oceanic plate consumption is uniformly distributed with age, to a good approximation. He concludes that "... the positive buoyancy of the oceanic crust (Oxburgh and Parmentier 1977), which would tend to inhibit the consumption of young ocean floor, must represent only a small part of the total force balance." This is perhaps reasonable in that there is no net compositional difference between the lithosphere and the underlying undepleted mantle, since the basalt is a partial melt and the depleted mantle resides in the lithosphere. Any decrease in density must be due to conversion to low-pressure mineralogy, e.g. loss of garnet.

While subduction may be independent of age, there is no evidence to confirm the youngest age at which subduction is initiated. If Oxburgh and Parmentier's (1977) proposed age of 40 Myr is correct, we would have to explain the observation that some younger plates are presently subducting by concluding that the subduction must have started at a point on these plates where the age exceeded 40 Myr, and that the older portions have been consumed.

Once subduction is initiated, the negative buoyancy of the subducting plate may be important in maintaining seafloor spreading. Forsyth and Uyeda (1975) showed that plate motion is dominated by a balance between the forces of negative buoyancy of the descending plate (slab pull) and viscous resistance to its descent. However, this would suggest that the oceanic plates are in tension, whereas intraplate earthquakes suggest a compressional environment (e.g., Sykes and Sbar 1973). The major contribution of the descending slab may be to cancel much of the viscous resistance forces of the descent, and thus provide nearly a free boundary for other forces such as "ridge push" (see below).

Subduction is likely to start by thrust fault relief of compressive stresses in the oceanic plates. McKenzie (1977) has reviewed the force balances and other requirements to sustain subduction once thrusting has occurred. Of particular importance in application to Venus are the following:

1. Slab pull, proportional to the length of subducted plate, may be the dominant force in maintaining subduction, as pointed out by Forsyth and Uyeda (1975). There is a minimum required plate velocity, such that the subducted plate has not lost its negative buoyancy by cooling to the ambient mantle temperature before the plate has reached a critical depth for maintenance of subduction. If the lithosphere on Venus is more buoyant (i.e. hotter), then the minimum velocity requirement must be raised to achieve the same effect. In any case, absolute slab force diminishes as the buoyancy increases and vanishes for positive buoyancy.
2. Ridge push, i.e. the force due to the elevation of the divergent plate boundary above old seafloor, also contributes to maintaining subduction and could be the dominant force initiating thrusting. The force is proportional to the product of this elevation and the thickness of the oceanic lithosphere. According to arguments of boundary layer theory presented in the next section, ridge push forces on Venus might be less than half of their counterparts on Earth.
3. Subduction is resisted by shear stress on the fault plane at the subduction trench, but apparently uninhibited by coupling to the mantle. McKenzie suggests that the low value of the shear stress (100 bar) and the uncou-

pling due to a low viscosity layer are both due to the presence of water; thus, lack of water could lead to a different tectonic style.

4. Finally, the elastic resisting force of bending a plate into a trench increases approximately linearly with elastic plate thickness, but is decreased by the presence of an overlying material (e.g. ocean water on Earth). The boundary layer theory arguments imply that the differences between Venus and Earth of these effects approximately cancel, so the bending resistance may be about the same on both planets.

In summary, negative buoyancy of the oceanic lithosphere on Earth may be important in sustaining seafloor spreading. A decrease in negative buoyancy might tend to inhibit subduction, but this could be partially offset by increased plate velocity. Positive buoyancy of a plate might totally inhibit initiation and maintenance of subduction. If there are lower spreading ridge elevations and a thinner lithosphere on Venus than on Earth, ridge push forces that may play a strong role in subduction will be smaller. Finally, higher temperatures in the Venus lithosphere would move phase changes to greater depths and, on Earth, ". . . once cooling reaches a depth of 50 km, the garnet pyroxenite or eclogite transformations may make a substantial contribution to the [loss of] buoyancy of the lithosphere" (Anderson 1981).

Using the approach outlined in the previous section, a set of evolutionary models has been computed to test the stability of the venusian lithosphere at surface temperatures $T_s = 740$ K and 273 K. The latter temperature simulates an Earth-like surface condition; the former temperature assumes applicability of Venus's present surface conditions over geologic time.

While uncertainty exists in many of the modeling parameters, comparison of the two values of T_s is valid. A stability integral is defined in terms of the integrated density difference of the lithosphere relative to the density immediately beneath it (Oxburgh and Parmentier 1977)

$$S = \int_{R_{\venus}}^{R_L} \frac{[\rho_{SL} - \rho(r)]}{\rho_{SL}}\, dr \qquad (20)$$

where ρ_{SL} is sublithospheric density, and the variation in gravity is assumed negligible. If $S > 0$ then the lithosphere is positively buoyant; if $S < 0$ the lithosphere is unstable. It is convenient to normalize Eq. (20):

$$S' = 100 \left[1 - \frac{1}{(R_{\venus} - R_L)} \int_{R_{\venus}}^{R_L} \frac{\rho(r)}{\rho_{SL}}\, dr \right]. \qquad (21)$$

The normalized stability S' gives the percentage deviation from ρ_{SL} of the average lithospheric density. For a given (time-dependent) density profile, the

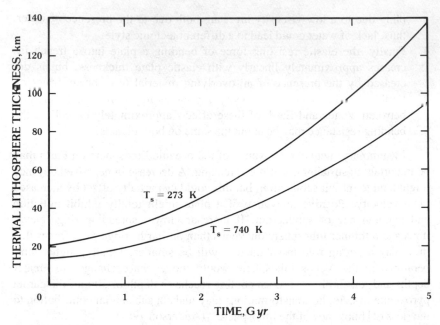

Fig. 4. Thickness of the lithosphere as a function of time for thermal model discussed in the text. Results are shown for two surface temperatures.

model temperature profile $T(r)$ is used to correct the densities for thermal expansion according to

$$\frac{\rho_0}{\rho(r)} = 1 + \alpha_1 T(r) + \frac{\alpha_2}{T(r)} + \alpha_3 T^2(r) \qquad (22)$$

where ρ_0 is density at $T = 0$; $\alpha_1 = 3.17 \times 10^{-5}$; $\alpha_2 = 1.05$; and $\alpha_3 = 7.1 \times 10^{-9}$ (Skinner 1962). The reference densities ρ_0 are 3.0 gm cm^{-3} for the crust and 3.3 gm cm^{-3} for the mantle. When a basalt-eclogite phase transition is allowed in the model, the eclogite density is 3.4 gm cm^{-3}, and from the phase diagram of Yoder (1976) a least-squares fit to the midpoint of the basalt-eclogite transition gives the condition for the eclogite stability field:

$$\frac{T(r) - 273}{145.4} - P(r) + 8.5 < 0 \qquad (23)$$

where $P(r)$ is pressure in kbar.

Lithospheric thickness as a function of time is shown in Fig. 4. The lithospheric base is defined by the temperature T_L, here taken as 1500°C in accordance with the boundary layer results of Kaula and Phillips (1981; see next section). This temperature defines the base of the thermal lithosphere; the

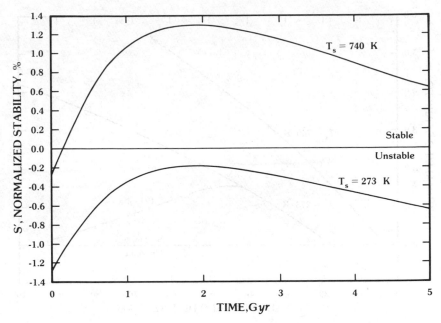

Fig. 5. Normalized stability as a function of time for thermal model discussed in the text. Results are shown for two surface temperatures and for crust allowed to grow exponentially to 10 km thickness in 4.6 Gyr. For $S' < 0$ the lithosphere is negatively buoyant.

elastic lithosphere discussed in Sec. III is much thinner than the thermal lithosphere. For both surface temperatures the lithosphere shows a monotonic increase in thickness as a function of time. At 4.6 Gyr the lithospheric thicknesses are 84 km and 121 km for $T_s = 740$ K and 273 K, respectively, in good agreement with the steady-state boundary layer results of Kaula and Phillips (1981).

In Fig. 5, normalized stability is shown as a function of time. In these two cases, the crust is allowed to build up exponentially to a final thickness of only 10 km, approximately simulating the present terrestrial oceanic crust. The buildup time is 4.6 Gyr and the exponential constant for crustal growth γ is 1.0 (see Eq. 16). In both models, the thin crust never enters the eclogite stability field. For both surface temperatures the evolution of lithospheric stability shows an increase in buoyancy up to 2 Gyr due to the growing low-density crust. After 2 Gyr, cooling of the lithosphere with the attendant increase in density dominates the evolution of the stability.

While the specific values of S' are dependent on the model parameters chosen, it is clear from Fig. 5 that the surface temperature of Venus leads to a lithosphere much more buoyant than that of Earth. Though it is not absolutely certain, it appears that Venus's lithosphere is much less likely to subduct than Earth's. This point is emphasized in Fig. 6, in which S' is plotted against final

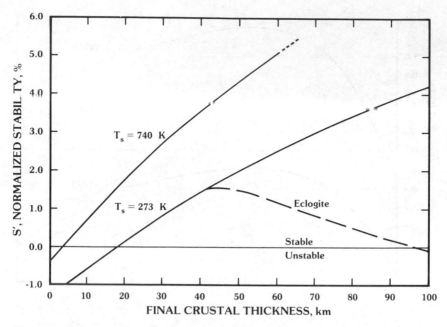

Fig. 6. Normalized stability as a function of crustal thickness at 4.6 Gyr in the evolutionary
model discussed in the text. Results are given for two surface temperatures. For $T_s = 740$ K
crustal thicknesses $\gtrsim 60$ km exceed the lithospheric thickness. Models were also run allowing eclogite to form, and for $T_s = 273$ K the lower crust is in the eclogite stability field for
thicknesses $\gtrsim 40$ km.

crustal thickness as an independent variable, for $t = 4.6$ Gyr. Since in these
models the intrinsic density of the upper mantle lithosphere is the same as that
of the convecting mantle, there is some maximum crustal thickness below
which the positive temperature gradient will render the lithosphere negatively
buoyant. For $T_s = 740$ K, this maximum value is ~ 3 km; for $T_s = 273$ K, $\sim$
18 km. There is no value of crustal thickness on the plot for which the $T_s = 740$ K model enters the eclogite stability field. For $T_s = 273$ K with crustal
thicknesses $\gtrsim 40$ km, the lower crust is in this field. If eclogite is allowed in
the model, Fig. 6 shows that it has profound effect on S'; negative buoyancy
occurs for crustal thicknesses $\gtrsim 100$ km. Thus, phase change is less likely to
contribute to subduction on Venus than on Earth.

The Quantitative Form of Seafloor Spreading on Venus

In addition to the problem of lithospheric stability, we can investigate the
quantitative forms that seafloor spreading might take on Venus and decide if
such forms could be resolved in the Pioneer Venus altimetric data. The importance of seafloor spreading cannot be overemphasized, because in Earth's
ocean basins the cooling of the oceanic lithosphere formed at the midocean
ridges accounts for nearly all the heat flow. Since the oceans occupy $\sim 2/3$ of

the surface area of Earth and oceanic and continental heat fluxes are approximately equal, the spreading process accounts for a large fraction of Earth's total heat loss (see Sclater et al. 1980, for a general review of Earth heat flow). If seafloor spreading does not occur on Venus, then a different mechanism for heat loss, and presumably different attendant tectonics, must be considered.

Kaula and Phillips (1981) examined the quantitative aspects of seafloor spreading on Venus, and included an estimate of the maximum possible length of midocean ridges on Venus. Their comparative analysis is outlined below.

In a vigorously convecting mantle there would be little temperature change (relative to the adiabat) across the convecting region, except for a thermal boundary of thickness d at the top (assuming no heat flux from the core). Here we assume that convection extends from the core radius R_c, to the surface of the planet R_0, so that the lithosphere is the boundary layer for thermal convection. The heat flux out of the convecting region can be equated to the temperature drop ΔT across the boundary layer; from Eqs. (8) and (9)

$$q_0 = \frac{k\Delta T}{d} = \frac{k\Delta T}{(R_o - R_c)}\text{Nu} = \frac{k\Delta T}{(R_o - R_c)}\left(\frac{\text{Ra}}{\text{Ra}_{\text{cr}}}\right)^\beta \tag{24}$$

and thus

$$d \propto \left(\frac{\text{Ra}}{\text{Ra}_{\text{cr}}}\right)^{-\beta} (R_o - R_c) \cdot \tag{25}$$

From Eq. (1), the average temperature difference between the convecting mantle and the surface shows a similar inverse proportionality to the Rayleigh number, and a direct proportionality to the heat source concentration Q

$$\Delta T \propto \left(\frac{\text{Ra}}{\text{Ra}_{\text{cr}}}\right)^{-\beta}Q = \left(\frac{\text{Ra}_q}{\text{Ra}_{\text{cr}}}\right)^{-\beta/(1+\beta)}Q \tag{26}$$

where Ra_q, the heat flux definition of the Rayleigh number, has been introduced $\left[\text{Ra}_q = g\,\alpha q_0\,(R_o - R_c)^4/(C_p\kappa^2\eta) = \text{Nu Ra} \approx \text{Ra}_{\text{cr}}^{-\beta}\text{Ra}^{1+\beta}$; see Kaula and Phillips 1981$\right]$. From Eq. (3), neglecting pressure effects

$$\text{Ra}_q^{-1} \propto \eta \propto \exp\left[E/KT\right] \tag{27}$$

so that if the boundary layer extends to the surface $(T_s = T_o)$, then

$$T = T_o + \Delta T = T_o + B\text{Ra}_q^{-\beta*} \tag{28a}$$

$$= T_o + A\,\exp\left[E\beta*/KT\right] \tag{28b}$$

where A and B are constants, and $\beta* = \beta/(1+\beta)$.

Since T can be estimated for Earth (Parsons and Sclater 1977) the constant A can be determined, utilizing the theoretical value for β of $\frac{1}{3}$. This value of A can be adopted for Venus; in doing so, heat source concentrations are approximately equated under the assumption that the other factors in A are similar for the two planets. Using the estimated value of A for Venus yields T only slightly higher than for Earth, 1755 K vo. 1630 K, consistent with time-dependent parameterized convection results. Thus the ΔT contrast for the two planets is still dominated by the surface temperature T_0 ($\Delta T = 1015$ K for Venus, 1350 K for Earth). Similarly, d equals 94 km for Venus, compared to 125 km for Earth. Under these conditions the thermal energy content of the venusian boundary layer (proportional to $d\Delta T$) is only $\sim 1/\sqrt{3}$ that of Earth.

However, in the steady-state assumption of this analysis, heat flux out of the planet can only depend on heat generation within the planet, and is independent of the value of Ra so long as the heat source concentration is fixed. For a fixed lithosphere, this is seen in Eq. (24), since both ΔT and d show the same proportionality to Ra. For a moving lithospheric boundary, as in seafloor spreading, heat is lost by removal of hot lithosphere from a divergent boundary at some velocity V, with the lithosphere ultimately cooling by conduction. In this case, the lower heat content of the boundary layer must be compensated for by a higher plate velocity since the rate of heat loss per unit length of divergent boundary is

$$q_b = \rho C_p V d \Delta T \qquad (29)$$

(Kaula and Phillips 1981). The heat loss q_b when integrated along the divergent boundaries (e.g., spreading ridges) is equivalent to the flux q_0 integrated over the surface area of the boundary layer in the nonspreading case. In Eq. (29) the Rayleigh number only serves to proportion the heat flux contributions between the heat content of the boundary layer $\rho C_p d \Delta T$ and the velocity V; the importance of the latter increases with increasing Ra.

The Rayleigh number for Venus is higher than for Earth because of Venus's slightly higher interior temperature (1755 K vs. 1630 K) and the exponential temperature dependence of viscosity. Specifically,

$$\mathrm{Ra}_{q_\venus}/\mathrm{Ra}_{q_\oplus} \approx \exp(E/KT_\oplus)/\exp(E/KT_\venus) \qquad (30)$$

and

$$\mathrm{Ra}_{q_\venus} \approx 3\mathrm{Ra}_{q_\oplus}. \qquad (31)$$

Since it can be shown that $V \propto \mathrm{Ra}_q^{1/2}/Z_c$, where Z_c is the depth extent of convection, then

$$V_\venus \approx \sqrt{3} V_\oplus \qquad (32)$$

for equivalent convection depths.

On Earth, the transverse topographic profile of the spreading, cooling oceanic lithosphere is well described theoretically (to 70 Myr age) by a conductively cooling, isostatically balanced slab moving away from a high temperature source region (Parsons and Sclater 1977):

$$h = h_0 - \delta h t^{\frac{1}{2}} \tag{33a}$$

$$h_0 = \frac{\rho d \alpha \Delta T}{2 \delta \rho} \tag{33b}$$

$$\delta h = \frac{2 \rho \alpha \Delta T}{\delta \rho} \left(\frac{\kappa}{\pi} \right)^{\frac{1}{2}} \tag{33c}$$

where $\delta \rho$ is the density differential between the lithosphere and the overlying fluid (in gm cm^{-3}, $\delta \rho_\oplus = \rho - 1.0$, $\delta \rho_\varphi = \rho$). If the shape of the cooling boundary layer is described observationally as

$$h = h_0 - \Delta h X^{\frac{1}{2}} \tag{34}$$

where X is measured transversely to the spreading axis, then

$$\Delta h^2 = \delta h^2 / V \tag{35}$$

(Kaula and Phillips 1981) where time and distance are related through the spreading velocity V. Using Eq. (26) with $\beta = \frac{1}{3}$,

$$\Delta h^2 \propto \frac{\Delta T^2}{V \delta \rho^2} \propto \frac{\mathrm{Ra}_q^{-\frac{1}{2}} Q^2}{(\mathrm{Ra}_q^{\frac{1}{2}} / Z_c) \, \delta \rho^2} \tag{36a}$$

or

$$\Delta h^2 \propto (\mathrm{Ra}_q \, \delta \rho^2)^{-1} Q^2 Z_c \tag{36b}$$

so that

$$\frac{\Delta h_\varphi}{\Delta h_\oplus} = \frac{\sqrt{\mathrm{Ra}_{q\oplus} Z_{c\varphi}} \, \delta \rho_\oplus Q_\varphi}{\sqrt{\mathrm{Ra}_{q\varphi} Z_{c\oplus}} \, \delta \rho_\varphi Q_\oplus} \tag{37}$$

and from Eq. (31), assuming $Q_\varphi = Q_\oplus$ and $Z_{c\varphi} = Z_{c\oplus}$

$$\Delta h_\varphi \approx 0.4 \Delta h_\oplus . \tag{38}$$

Because of the lower thermal content of the boundary layer and the difference in $\delta \rho$,

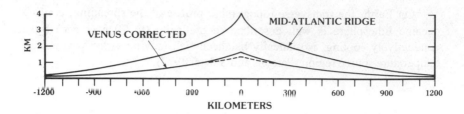

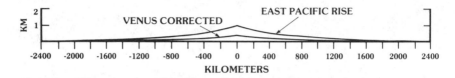

Fig. 7. Idealized ridge (divergent plate boundary) profiles for Venus and Earth. Examples for Earth are obtained by fitting Eq.. (34) to actual ridge profiles of a slow spreader (Mid-Atlantic Ridge) and a fast spreader (East Pacific Rise). Venus profiles are obtained by boundary layer theory relationships discussed in the text. The dashed line shows the smoothing that would be caused by the Pioneer Venus altimeter.

$$h_{0_{\venus}} \approx 0.4 h_{0_{\oplus}} \tag{39}$$

and this result is directly obtainable from the estimates of $d\Delta T$ given above, or from the proportionality $d\Delta T \propto Ra_q^{-\frac{1}{2}}$.

Thus, spreading ridges on Venus are predicted to reach crest heights h_0 only 2/5 as high as terrestrial crests, and decay (Δh) with distance from the crests at only 2/5 the terrestrial rate (i.e. 2/5 the slope at any given distance). The equivalency of the constant A (i.e. the equivalency of heat source concentrations) is explicit in this result; this leads to the numerical relationship between venusian and terrestrial Rayleigh numbers given in Eq. (31). Conversely, those venusian topographic features hypothesized as spreading centers either should satisfy the relationships in Eqs. (38) and (39), or heat source concentrations on Venus different from those of Earth must be considered.

Figure 7 shows two model terrestrial spreading ridge transverse topographic profiles and their venusian equivalents. Equation (34) has been fitted to a slow-spreading ridge (i.e. the Mid-Atlantic Ridge at ~ 28°N latitude) and to a fast-spreading ridge (the East Pacific Rise at ~ 17°S latitude). The profiles are converted to the venusian case by mutiplying h_0 and Δh by 0.4. A slow-spreading ridge clearly would be detectable with the Pioneer Venus altimeter (standard error of measurement 200 m; average footprint size $\approx$ 50 km by 50 km [Pettengill et al. 1980a]), provided that the longitudinal extent of the ridge was at least several hundred km. The difficulty of detecting a fast-spreading ridge such as the East Pacific Rise is equally obvious. This is illustrated in Fig. 8, in which the venusian equivalent of the East Pacific Rise

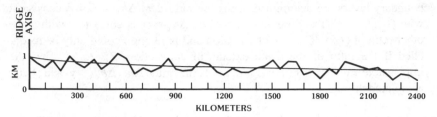

Fig. 8. One side of the Venus-corrected East Pacific Rise profile, along with a simulation of that profile when the standard error of the Pioneer Venus altimetry data is taken into account.

is plotted along with a profile derived on 50 km sampling intervals and plotted with random error obtained from a random number generator utilizing a 200 m standard error (Pettengill et al. 1980a). The published color topographic maps (Pettengill et al. 1980a; Masursky et al. 1980) obviously have too coarse a contour interval to detect spreading ridges of this type, even in the absence of error. With error, longitudinal integration of the transverse topographic profile for several thousand km would be required to detect the spreading ridges, even with the implausible assumptions that axis locations could be found and that the character of the spreading ridges does not change along the integration interval.

Kaula and Phillips (1981) asked if the major identifiable topographic lineaments on Venus could be spreading centers, and if so, how much heat could they deliver to the surface. They considered a ridge length of 16,700 km associated with Aphrodite Terra and its extensions, and 4100 km associated with Beta Regio. Topographically, these features satisfy two of four possible tests for seafloor spreading: (1) they are elevated with respect to a broad reference plain (i.e. the median plains; see Sec. IV) that could act as the boundary layer of mature ocean basin; and (2) the ridge shapes are predominantly concave. However, they fail two other important tests and are unlike the terrestrial ridge system, in that: (1) the ridge heights do not have a narrow distribution about a mode; and (2) they do not form a global interconnected system.

The topographic decay coefficient (Δh) can also be considered by fitting Eq. (34) to the Venus data. Values for this coefficient are derivable from Table 3 of Kaula and Phillips (1981). The mean of the ratio of all determinations of $\Delta h_{\varphi}/\Delta h_{\oplus}$ is ~ 1.0. Subsets of the ridge data in Table 3, based on various selection criteria, give about the same value. Earth and Venus Δh values appear indistinguishable, and are clearly inconsistent with the ratio 0.4 predicted on the assumption of equivalent heat sources.

In a steady-state convecting planet, once the surface temperature and heat source concentrations are specified (assuming the material constants and vertical dimension of convection are known), all properties of the convecting boundary layers are determined. Thus the ratio $\Delta h_{\female}/\Delta h_{\oplus} = 0.4$ is consistent with $Q_{\female}/Q_{\oplus} = 1$. The observed value $\Delta h_{\female}/\Delta h_{\oplus} \approx 1$ is consistent with some other ratio of $Q_{\female}$ to $Q_{\oplus}$. The observation and prediction could only be reconciled if the ratio of the depth extents of convection, $Z_{c\female}$ to $Z_{c\oplus}$, has the implausible value of ~ 6. For $Z_{c\female}/Z_{c\oplus} = 1$, the ratios $\Delta h_{\female}/\Delta h_{\oplus}(\equiv\gamma)$ and $Q_{\female}/Q_{\oplus}$ are related through

$$\exp\left[-E/KT_{\female}\right] = (0.5/\gamma^2)\exp\left[-E/KT_{\oplus}\right]Q_{\female}/Q_{\oplus} \tag{40}$$

and

$$T_{\female} = T_{0\female} + 25\left(\frac{Q_{\female}}{Q_{\oplus}}\right)^{3/4}\exp\left[E/4KT_{\female}\right]. \tag{41}$$

Solutions of these two equations for $T_{\female}$ and $Q_{\female}/Q_{\oplus}$ as a function of $\gamma = (\Delta h_{\female}/\Delta h_{\oplus})$ are shown in Fig. 9. The solution given by Kaula and Phillips (1981) is recoverable by a choice of $Q_{\female}/Q_{\oplus} = 1$. The observed value for $\Delta h_{\female}/\Delta h_{\oplus}$ of ~ 1 leads to $T_{\female} \approx 1400$ K and $Q_{\female}/Q_{\oplus} \approx 0.2$, implying a thicker lithosphere and slower plate velocities on Venus than on Earth.

Although application of a one-dimensional steady-state boundary layer theory should be viewed with caution, we have two qualitative choices: (1) If Venus has the same heat source concentration as Earth, then the Aphrodite and Beta ridge complexes are unlikely to be spreading centers, and we must examine the median plains for evidence of such physiographic features. As discussed above, fast-spreading ridges do not appear detectable in the Pioneer Venus altimetry. (2) Alternatively, if Aphrodite and Beta are spreading ridges, then the interior of Venus may be extremely depleted in heat sources. Both cosmochemical arguments and the Venera measurements of K, U, and Th suggest that this alternative is less likely.

Of course, the above application of boundary layer theory assumes totally free, or uninhibited, convection, which would be characterized by the full development of subduction zones. On Earth *inhibited* seafloor spreading, manifested by slow plate velocity, certainly does occur and is associated with those plates occupied by a significant fraction of continent and poor development of subduction zones. In the case of inhibited convection, velocity is not derivable from simple boundary layer theory, but is a dependent variable obtainable from Eq. (35) utilizing the results from one-dimensional cooling (δh) and the topographic decay coefficient (Δh). In this case, the results of Kaula and Phillips (1981) regarding the Equatorial Highlands are applicable, including the inference that this region of Venus could be a *very* slow spreading center of plate divergence delivering no more than 15% of the heat energy to the surface. Since the Equatorial Highlands are a global scale feature, it would be reasonable to assume that the median plains represent the cooler, older portions of these slow-spreading plates. It is thus difficult to imagine a second

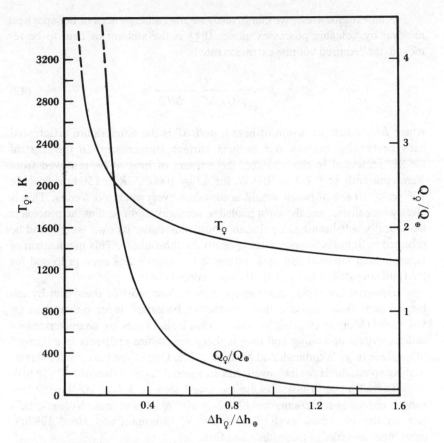

Fig. 9. Boundary layer theory results for the interior convecting temperature of Venus $T_♀$ and the ratio of heat source concentrations $Q_♀/Q_⊕$, as a function of the ratio of topographic decay coefficients $\Delta h_♀/\Delta h_⊕$.

mode of fast subduction (the one undetectible by Pioneer Venus) simultaneously occurring.

Heat Removal by Hot Spot Tectonics

As an alternative to heat removal by seafloor spreading processes, we investigate the mechanism associated with terrestrial hot spots. These phenomena are thought to represent the lithospheric expression of upwelling, hot convective plumes in the mantle (Morgan 1971). Hot spots are characterized by enhanced heat flow due to both lithospheric thinning and surface volcanism.

As an extreme view, we can investigate the consequences of hot spot heat removal by volcanic processes alone. If Q is the amount of heat to be removed, the required volume extrusion rate is

$$R = \frac{Q}{\mu(C_p \, \Delta T \; + \; \Delta H_f)} \tag{42}$$

where ΔH_f is heat of fusion of basalt, and ΔT is the temperature differential between basaltic magma and ambient surface temperature. In the thermal models presented in this chapter, the amount of heat to be removed from Venus presently is $\sim 2.3 \times 10^{13}$ W; for $\Delta T = 1000°C$, $R \approx 150 \text{ km}^3 \text{ yr}^{-1}$. In 10^8 yr, > 30 km of basalt would accumulate everywhere on Venus. This is clearly unrealistic, and the most plausible scenario might be that the process is isostatically self-limiting, i.e. basalt would sink under its own weight and be reheated to liquidus temperatures beneath the lithosphere. This mechanism of volcanic heat removal and mass return to the interior has been proposed for the Galilean satellite Io by O'Reilly and Davies (1981).

Terrestrial hot spots, however, appear to lose most of their heat by enhanced heat flow across a thin conductive boundary layer rather than by volcanism (Morgan and Phillips 1982). That is, hot spots are characterized by asthenospheric upwelling and thus higher temperature gradients in a thinned lithosphere (e.g. Wendlandt and Morgan 1982). One of the best studied terrestrial hot spots, the Hawaiian swell, has an anomalous heat flux of $\sim 7-12$ mW m^{-2} (Detrick et al. 1981). Taking the swell area as $2.7 \times 10^6 \text{ km}^2$, lithospheric conduction accounts for $19-32 \times 10^9$ W of heat loss. Volcanic heat loss for the Hawaiian swell, 2.5×10^9 W (Solomon and Head 1982a), contributes an order of magnitude less flux.

Morgan and Phillips (1983) investigated quantitiatively the thermal plausibility of the hypothesis (advanced in Sec. V) that much topography on Venus can be explained by the hot spot mechanism. In other words, the high topography results from thermal expansion of the lithosphere-asthenosphere column in regions of high temperature gradients associated with upwelling convective flow in the mantle. Analogously, topographic depressions are associated with low temperature gradients in the lithosphere. The maximum elevation of hot spot topography is dictated by the height of an asthenospheric column in isostatic balance with a reference lithospheric column. For the assumed model, the maximum hot spot elevation is at radius 6053 km, ~ 1.5 km above the mean radius of the planet. However, elevations below 6053 km comprise $\sim 93\%$ of Venus's surface area. The remaining 7% of surface area includes some of the most striking topography of the planet, such as Ishtar Terra which is capped by Maxwell Montes 11 km above the planetary mean radius. In the hot spot hypothesis, topography above 6053 km is due to buoyant crustal material. An attractive feature of this model is a lithosphere of variable thickness. Ishtar Terra, for example, could be composed of buoyant crustal material isostatically supported in a thick lithosphere, while the Equatorial Highlands (i.e. topography associated with Aphrodite Terra and Beta Regio [Phillips et al. 1981]) could be a region of high heat flow, thin lithosphere, and active volcanism.

Variable lithospheric thickness may well be necessary if any of the topography on Venus is to be supported isostatically. Assuming a globally uniform lithospheric boundary layer, Phillips et al. (1981) estimated a maximum elastic lithospheric thickness of 50 km; a plausible estimate is 25 km. This in turn severely constrains the isostatic support of the venusian topography; this aspect is discussed below.

General Requirement for Passive Support of Topography

As a final point in this discussion of thermal phenomena, we review the methodology for establishing requirements for passive support of the venusian topography. Passive support implies that topographic features are supported by the finite elastic strength of the lithosphere. Here the concept of isostasy must be considered. Isostasy, or the state of isostatic compensation, is usually defined as weight balance of surface loads by subsurface density anomalies distributed such that at some depth the pressure is everywhere constant and below this depth the planet is in hydrostatic equilibrium. Isostasy is synonymous with passive compensation, and assumes that the subsurface density distribution providing the compensation is also supported by the finite strength of the lithosphere, i.e. it is stable against creep or viscous deformation. Dynamic compensation is also considered, in which the compensating density distributions are maintained by thermal processes in the interior. In this section, we investigate requirements for passive compensation at some depth z^*; in Sec. IV the gravity data from the Pioneer Venus mission are used to estimate various values for z^*.

Given that passive compensation takes place at some depth z^*, the temperature at that depth is directly constrained. That is, since temperature is directly related to viscosity (Eq. 3), the viscosity must not be so low as to allow creep destabilization of the compensation zone. The critical temperature T^* is the upper bounding temperature at depth z^*, at which significant creep does not occur. To find T^*, the most creep resistant of the reliable flow laws is used, the dry olivine measurements of Kohlstedt and Goetze (1974).

There are several important characteristic times for the passive behavior of materials. One is the Maxwell time τ_m marking the transition in material behavior from elastic to viscous, measured from the time that stress is applied. A second characteristic time is the viscous relaxation time τ_η, in which imposed stresses and displacements relax to $1/e$ of their initial values. For the wavelengths considered in this chapter, $\tau_\eta \leqslant 10\tau_m$. Here we adopt the view that long-term geologic support requires Maxwell times $> 10^7$ yr, i.e. that isostatic support is stable against creep relaxation for $\geqslant 10^8$ yr. The Maxwell time is given by the ratio of viscosity to rigidity μ; adopting a value of 5 ×

10^{11} dyne cm^{-2} for the latter quantity, the viscosity is then required to be greater than some critical value η^*

$$\eta \geqslant \eta^* = \mu \tau_m \approx 1.5 \times 10^{26} \text{ poise} \cdot \tag{43}$$

Within a planetary lithosphere, the temperature at any depth z can be written as

$$T(z) = T_s + \frac{q_m z}{k_L} + f(Q_L, z) \tag{44}$$

where q_m is heat flux from the convecting interior; k_L is the lithospheric thermal conductivity; and $f(Q_L, z)$ is the temperature contribution due to heat sources in the lithosphere. For passive compensation, $T(z^*) \leqslant T^*$, or

$$\Delta T = T^* - T_s \geqslant \frac{q_m z^*}{k_L} + f(Q_L, z^*) \cdot \tag{45}$$

Since it can be shown that $f(Q_L, z)$ is non-negative, an upper bound q_m^* (i.e. no lithospheric heat sources) may be found on q_m that will satisfy the inequality of Eq. (45)

$$q_m^* \leqslant \frac{k_L \Delta T}{z^*} \cdot \tag{46}$$

Using Eq. (43) and the Kohlstedt and Goetze (1974) flow law, $T^* \approx 1000$ K and with $k_L = 4 \times 10^5$ erg cm^{-1} K^{-1} s^{-1} then in units m W m^{-2} km

$$q_m^* z^* \leqslant 1000 . \tag{47}$$

Equation (47) forms the basis for estimating an upper bound on allowable heat flux from the mantle for some given z^*. Analysis of gravity data leads to an estimate of compensation depth for a particular topographic feature, for which an assumption of isostasy implies that this depth is $\leqslant z^*$. This in turn yields an upper bound on q_m^* from Eq. (47). An alternate view of z^* is that, for the equality in Eq. (47), it may be taken as the depth extent of the elastic lithosphere. Equation (47) then gives a maximum allowable heat flux into the base of the lithosphere.

III. GRAVITY FIELD AND THERMAL IMPLICATIONS

The gravity field of a terrestrial planet reflects lateral variations in density due to surface topography and to various mechanisms in the interior that maintain density differences. The gravity field of Venus was first mapped by the Pioneer Venus spacecraft. Initial results were reported by Phillips et al.

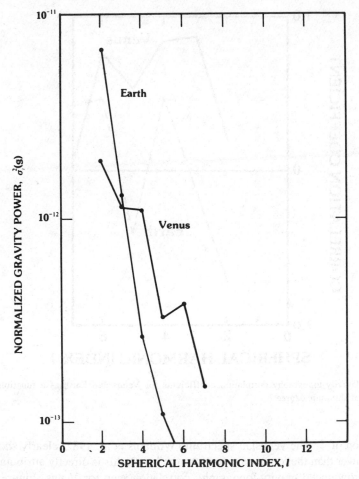

Fig. 10. Normalized gravitational power spectra of Venus and Earth as a function of spherical harmonic degree ℓ. Venus values are from Williams et al. (1982). The rotational effect is removed from the second degree terms.

(1979); systematic gravity anomalies were found for the limited region of Venus mapped at that time. Phillips and Lambeck (1980), and subsequently Reasenberg et al. (1981), reported a strong correlation of gravity to topography for all resolvable wavelengths (spherical harmonic degree $\ell \leq 20$). This is quite unlike Earth, where long wavelength anomalies ($\ell \leq 18$) receive little contribution from topography and crustal density anomalies (Phillips and Lambeck 1980). It is believed that the long wavelength anomalies on Earth are related to convective flow processes in the mantle, and that compensated topography contributes only $\sim 10\%$ of the power in this part of the spectrum. Figure 10 compares normalized gravity power spectra for Venus and Earth as

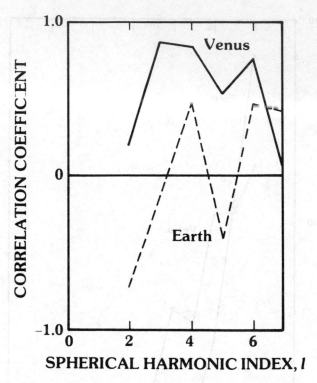

Fig. 11. Gravity-topography correlation coefficients for Venus and Earth as a function of spherical harmonic degree l.

a function of l. The venusian spectrum (Williams et al. 1982) clearly shows more power than the terrestrial spectrum for $l > 2$; this is directly attributable to the pronounced gravity-topography correlation seen for Venus. Figure 11 shows the gravity-topography correlation coefficients for Venus and Earth through the seventh harmonic; the higher correlation coefficients for Venus are quite evident (the seventh gravitational harmonic may be poorly estimated for Venus, leading to a low correlation coefficient).

Figure 12 shows the line-of-sight gravity component for Venus superimposed on a gray-level map of topography. The line of sight refers to the direction from the Pioneer Venus spacecraft to Earth, varying from vertical to horizontal relative to the local surface. The gravity is highly correlated to topography at all scales. Figure 13 is adapted from the gravity modeling results of P.B. Esposito et al. (1982) for Beta Regio. This solution results from simultaneous estimation of 191 surface disks and the Keplerian state of the numerous Pioneer Venus spacecraft orbits that passed near Beta Regio. In Fig. 13 the model has been evaluated for vertical gravity at 200 km elevation, and superimposed on a gray-level map of the Beta topography. The correlation of gravity to topography is striking, even though the gravity is less well resolved than the topography.

We find that the strong gravity-topography correlation persists at all regions of Venus mapped by the Pioneer Venus spacecraft. This relationship contrasts with the situation on Earth, where the long wavelength anomalies arise in the mantle and generally are uncoupled or only weakly coupled to the lithosphere, with correlation seen only in limited regions of the ocean basins.

Density Modeling

Various types of model can be matched to the gravity data in order to analyze subsurface density variations. Because of the high correlation of gravity to topography, the most important classes of model are those that interpret the gravity anomalies to arise from the topography and from mechanisms in the interior that support the topography.

It is convenient to analyze the gravity anomalies as resulting from the elastic flexural response of the lithosphere to a topographic load. The admittance function, the spectral ratio of gravity to topography (McKenzie and Bowin 1976), for flexural loading is given for spherical harmonic degree ℓ by its dimensionless form (Phillips et al. 1981)

$$Z_\ell = 1 + \frac{\Delta\rho}{\rho_c} \left(\frac{R_\female - t_c}{R_\female} \right)^\ell \frac{\alpha_\ell}{1 + \alpha_\ell}.$$ (48)

In this model, a layer of density ρ_c and thickness t_c is contained within the lithosphere of thickness t_L, and provides the major interior density contrast $\Delta\rho$ at its lower boundary. The expression α_ℓ is given by

$$\alpha_\ell = (c_1 f_\ell + c_2)/(f_\ell^3 + 4f_\ell^2 + c_3 f + c_4)$$ (49)

where

$c_1 = -\rho_c g_0 R^4/D; c_2 = c_1(1 - \nu^2); c_3 \equiv c_{31} + c_{32} = \Psi(1 - \nu) + (\rho_c + \Delta\rho) g_0 R^4/D; c_4 = 2c_{31} + c_{32}(1 - \nu); f_\ell = -\ell(\ell + 1)$. g_0 is planetary gravity; $R = R_\female - t_L/2$; ν is Poisson's ratio; $\Psi = 12 R_L^2/t_L^2$; and D is flexural rigidity, given by

$$D = Et_L^3/[12(1 - \nu^2)]$$ (50)

where E is Young's modulus. Two limits to Eq. (48) are given by $D \to \infty$, $Z_\ell \to 1$; the gravity signature is due to topography alone and $D \to 0$, for which

$$Z_\ell = 1 - \left(\frac{R_\female - t_c}{R_\female} \right)^\ell.$$ (51)

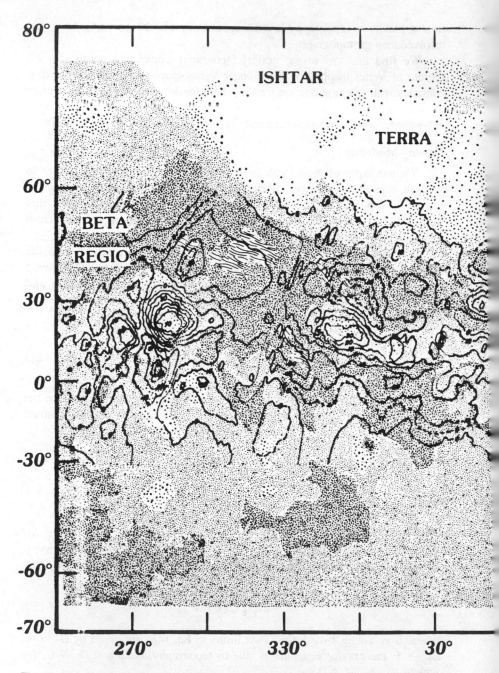

Fig. 12. Line-of-sight gravity map superimposed on generalized grey-level map of topography.
Shading is in 1 km intervals with elevations > 3 km shown in white. Countour interval is 10
mgal unless shown otherwise. (Gravity data courtesy of W. L. Sjogren.)

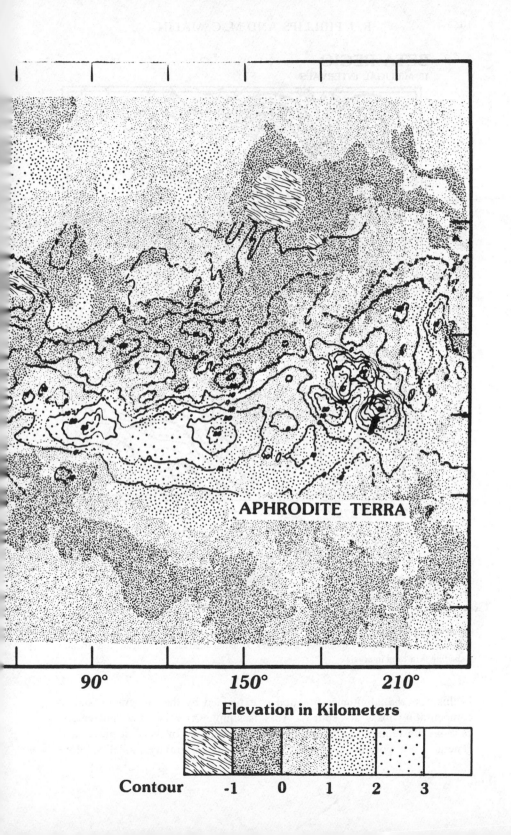

APHRODITE TERRA

90°   150°   210°

Elevation in Kilometers

Contour   -1   0   1   2   3

BETA REGIO
15 MILLIGAL INTERVALS

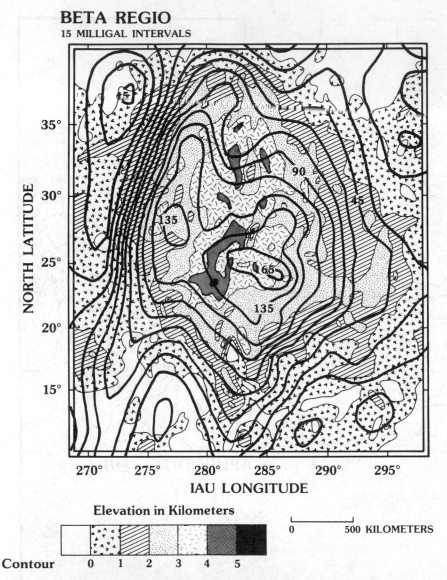

Fig. 13. Vertical gravity evaluated at 200 km altitude over Beta Regio, based on surface density model of P. B. Esposito et al. (1982). Contour interval is 15 mgal. Topographic shading is in 1 km intervals.

In this case the admittance function is explained by the topography and its compensating mass at depth t_c. This is simple Airy isostasy, but certain cautions are required in any compensation solution. In the first place, it is generally not possible to decide on the basis of gravity modeling alone whether the isostatic balance is due to a single surface (Airy isostasy) or to lateral variations of density in a layer (Pratt isostasy). The interpretation of a distributed density in terms of an equivalent single depth of compensation indicates that the single depth is less than the mean depth of the distributed density. Even more importantly, it cannot be determined whether the compensation is static (maintained by the finite strength of the lithosphere) or dynamic (topographic highs maintained by low density anomalies, thermal or compositional, associated with convective flow in the mantle).

Phillips et al. (1981) carried out a preliminary modeling analysis for the western region of Aphrodite and two of its northwestward extensions. From the admittance estimates, they generated curves for each region showing the trade-off between flexural rigidity and depth of compensation (Fig. 14). Because of the sharp rollover in the curves, the solutions seem to separate into two classes: (a) upper left branch, flexural support with flexural rigidity 10^{32} to 10^{33} dyne cm and density boundary at < 50 km; or (b) lower right branch, complete isostatic compensation (with $D \rightarrow 0$) at depth ≈ 100 km. Flexural rigidities in the range 10^{32} to 10^{33} dyne cm are in the same range as the seismically determined values on Earth, which correspond to a range of old oceanic lithosphere (100 km thickness) to continental lithosphere (200 km thickness). It has been argued, and supported by observation (Beaumont 1978; Watts et al. 1980), that the zero-age loading and seismically determined flexural rigidities are equivalent, and the former decreases with time due to stress relaxation. The values determined for Venus indicate either extremely high creep resistance (i.e. no stress relaxation) or extremely thick lithosphere. Using our nominal thermal model to calculate a value for q_m^*, $z^* \leqslant 20$ km from Eq. (47). The use of 20 km as the vertical dimension t_L of the elastic lithosphere generates a flexural ridigity of $D = 4 \times 10^{29}$ dyne cm, implying (1) that the upper left branch solutions of Fig. 14 with $D > 10^{32}$ dyne cm are implausible, and (2) that the lower right branch solutions suggest compensation at ≈ 100 km depth which cannot be passively maintained.

Alternatively, if lateral variations in heat flux are allowed as in the hot spot model, and $z^* > 100$ km, implying passive support of topography, then $D > 10^{32}$ dyne cm and neither branch of the solution curves can be eliminated. If z^* is set equal to 100 km, then q_m^* from Eq. (47) is 10 mW m^{-2}. What are the implications if this is a globally representative value? From the thermal model calculations discussed here, ~ 5 mW m^{-2} is secular heat loss, and so only ~ 5 mW m^{-2} could come from the steady-state flux of radioactive heat sources, implying the mantle must be ~ 90 % depleted by differentiation into a crust and/or by an initial depletion relative to the assumed heat source model. If the extreme case of differentiation into a crust is assumed, then this crust must now have an average heat source concentration about equal to that of granite, for a 30 km crust, and to that of primary basalt, for a 100 km crust. But Eq. (47) gives an upper bound to mantle heat flux, because it assumes no temperature contribution from lithospheric heat sources. However, if these

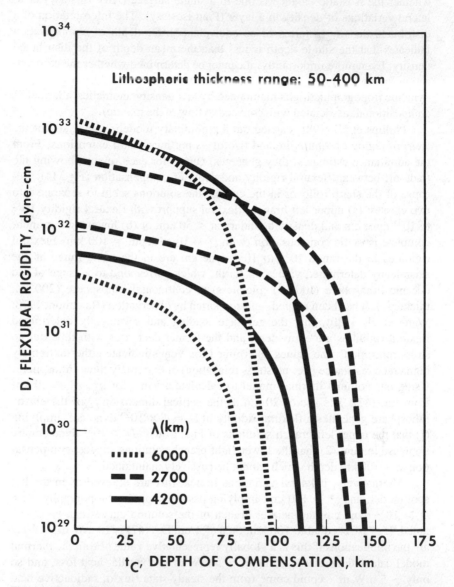

Fig. 14. Trade-offs of compensation depth t_c and flexural rigidity D, based on admittance estimates for three venusian topographic features of differing wavelengths (western Aphrodite Terra and two regions to the northwest). Each solution is bounded by arbitrary choices of lithospheric thickness: 50 km, lower curve; 400 km, upper curve. (From Phillips et al. 1981.)

sources are concentrated well above the 100 km depth, their contributions to the temperature at this level will be insignificant. Consider for example a mantle having undergone extreme differentiation and 90% depletion of heat sources, now residing in a lithosphere composed of a 30 km "granitic" crustal layer overlying a barren upper mantle. This lithosphere would provide a temperature low enough at 100 km depth to passively support topography. The nature of the density boundary at 100 km depth would of course be enigmatic, and so would the lack of significant density contrast at the base of the "granitic" layer. A more plausible view is that the boundary of the crustal layer, only tens of km deep, provides the major density contrast in the lithosphere. In this case the flexural support of topography is the correct solution branch of Fig. 14, since the flexural rigidity is 10^{32} dyne cm for a lithosphere 100 km thick, and the depth of the major density contrast is $\ll 100$ km.

In Fig. 15, the line-of-sight gravity observed by the Pioneer Venus orbiter over Beta Regio is shown. Also shown are the results of fitting isostatic models of various depths of compensation. This has been done by orbit simulation procedures described by Phillips et al. (1978). The best fit is for a compensation depth ~ 400 km.

For the Beta Regio region, P. B. Esposito et al. (1982) spatially sampled both the gravity field $g(x,y)$ of their multiple disk solution and the gravity field due to the topography alone $g_t(x,y)$, and produced a least-squares regression estimate of 0.44 for the ratio of $g(x,y)$ to $g_t(x,y)$. If Beta Regio can be characterized by a single wavelength λ, then

$$g(x,y)/g_t(x,y) \equiv \Psi = 0.44 = 1 - \left(\frac{R_{\venus} - d}{R_{\venus}} \right)^n \approx (1 - e^{-kd}) \qquad (52)$$

where $n = 2\pi R_{\venus}/\lambda$; $k = 2\pi/\lambda$; and d is the depth to an isostatically compensating surface. For $d = 400$ km, Eq. (52) yields a characteristic wavelength $\lambda \sim 4300$ km. A value of $\lambda = 4300$ km may be appropriate if the areal extent of Beta Regio is considered to be of half-wavelength dimensions.

By considering isostatic compensation by the density contrast at a single surface, an Airy mechanism, a minimum depth for compensation is realized. A vertically distributed (Pratt) compensation extends to greater depths and provides a more severe test for passive compensation. In terms of a topographic representation by a single circular disk compensated at depth d, for $\Psi = 0.44$

$$d = \frac{\Psi}{\sqrt{1 - \Psi^2}} \, w = 0.49 \, w \qquad (53)$$

where w is disk radius. For Pratt isostatic compensation distributed from the surface (to yield a minimum estimate) to a depth L

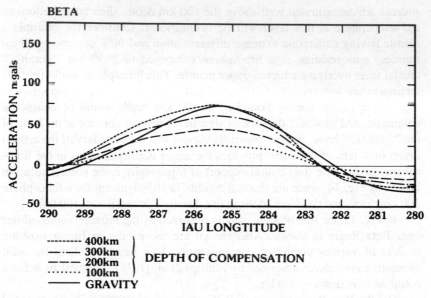

Fig. 15. Line-of-sight gravity (solid curve) in the Beta Regio area as measured by Pioneer Venus spacecraft on Orbit 566. Also shown are results generated from a model of compensated Beta topography with various depths of compensation.

$$L = \frac{2\Psi}{1 - \Psi^2}\, w = 1.09\, w \cdot \qquad (54)$$

Combining Eq. (53) and (54) yields a distributed compensation under Beta Regio to a depth of $L = (1.09/0.49)d \approx 900$ km, and deeper if the compensation zone does not extend to the surface.

A range of compensation depths of 400 to 900 km constrains, via Eq. (47), sublithospheric heat flux to a range of 1 to 2.5 mW m^{-2}. This is less than the secular heat flux from thermal models, and passive support then requires complete loss of heat sources from the mantle and a small secular cooling component.

Summary

The high correlation between gravity and topography on Venus is reviewed, and the results of interior density modeling of the gravity anomalies presented. The main goal, once depths of compensation of the topography are found, is to test whether the topography could plausibly be supported by the finite strength of the interior. If it cannot, then presumably the topography is geologically very young and/or is supported by dynamic thermal mechanisms, which in turn relate to lithospheric tectonics and the way Venus loses internal heat.

Assuming that the local heat flux is not significantly less than the global average, then for western Aphrodite and its extensions passive support of the topography appears possible, but only if there is extreme heat source depletion from the interior and concentration near the surface in rock that carries at least as much heat source energy as primary basaltic magmas. For Beta, complete differentiation of the mantle and almost total loss of the secular cooling contribution is required. Complete differentiation of Venus would suggest that ^{40}Ar should be able to escape easily into the atmosphere. If Venus has the same heat source concentrations as Earth, then the ratio $^{40}Ar_{\venus}/^{40}Ar_{\oplus}$ might be expected to exceed unity, in contrast to the observed value (Hoffman et al. 1980a) of ~ 0.3.

Earth still has an appreciable heat flux from its mantle, some tens of mW m^{-2}, due to both radioactive sources and primordial heat (Stacey 1980). For Venus to have removed most of its heat sources (Aphrodite interpretation), or all of its heat sources and virtually all of its primordial heat (Beta interpretation) may require implausibly efficient heat transfer at temperatures significantly below the solidus.

We find the alternative hypothesis more attractive, that Beta Regio and possibly other areas of the planet are dynamically maintained by thermal processes in the interior.

IV. TOPOGRAPHY AND PHYSIOGRAPHY

In discussing the interior of Venus and its manifestation in tectonics at the surface, the problem can also be turned around: What does surface morphology tell us about the interior and particularly the evolution of Venus? The purpose of this section is to review the form and distribution of the venusian topography in light of its interpretation in terms of interior processes.

Three points need to be stressed in addressing surface features. First, limitations in resolution of Pioneer Venus and earth-based telescopic radar observations severely restrict the unambiguous interpretation of landforms and topography. It is commonly held that data with resolution 100 km can be used to investigate global planetary tectonism (Carr et al. 1976); however, this is not always true, and the Pioneer data are of somewhat lower resolution than would apparently be needed to decipher certain equivalent terrestrial features (Arvidson and Davies 1981; Head et al. 1981). Second, we often approach Venus with a terrestrial bias acquired only in the past 20 or so years. This bias must be explicitly acknowledged when searching, for example, for plate tectonic landforms. Finally, specific processes do not necessarily produce unique landforms, and landforms need not reflect single or unique processes. Therefore, we admit great uncertainty and ambiguity in discussing the relationships between venusian landforms and internal processes.

Planetary Topography—Hypsograms

The topography of > 90 % of Venus was examined by the Pioneer Venus orbiter (Masursky et al. 1980; Pettengill et al. 1980a; Chapter 6). These three cited works contain detailed descriptions, documentation and interpretation of venusian topography and morphology. The topography of Venus has been divided into three physiographic provinces by these workers, based on height with respect to an average planetary radius. The lowest regions, termed lowlands province (Masursky et al. 1980), comprise ~ 27 % of the surface; high regions called highlands province occupy 8 %, and the remaining areas are called rolling plains province. These provinces are rather arbitrarily assigned by height relative to the mean surface elevation (planetary radius 6051.5 km). For our purposes, a somewhat simpler approach is warranted. We divide venusian topography into two zones: a distribution roughly symmetric about the peak (or mode) of the percent surface area-height graph of Venus (i.e. the hypsogram, Fig. 16) which we call the modal or median plains; and a much smaller fraction of the surface area at higher altitudes which we call highlands. The low latitude highlands, including Aphrodite Terra and Beta Regio, are called Equatorial Highlands (Phillips et al. 1981).

Much attention has been called to the difference between the hypsograms of Earth and Venus; Earth's is dramatically bimodal, while Venus's is unimodal (e.g. Phillips et al. 1981). Earth's topography clearly shows the continental landmasses and their difference from the ocean basins. The unimodal hypsogram of Venus suggests the absence of continents and ocean basins, and hence the lack of the principal physiography resulting from plate tectonics. Since such interpretations are derived at least partly from the hypsograms, we should perhaps ask what hypsograms tell us about planetary tectonics.

It appears to us that the gross form of hypsograms (i.e. the number, separation, and relative magnitudes of topographic modes) may reflect less the tectonics on a planet than the erosion, composition, and structure of the lithosphere. The bimodal terrestrial hypsogram most likely shows more the time-averaged level of ocean water (the "freeboard" concept proposed by Kuenen 1939; advanced by Hess 1962; and expanded upon by Wise 1972) than a primary result of plate tectonism, although interaction among erosion, sedimentation, and changing ocean basin geometry due to seafloor spreading clearly controls sea level.

In attempting to account for the effect of oceans on Earth's topography, Arvidson and Davies (1981) and Head et al. (1981) have subtracted from the ocean topography the effects of the weight of overlying water. The result is to reduce the separation of the two modes by slightly more than 1 km, leaving at least a 4 km difference. This difference can be attributed to the isostatic relationship between thicknesses of crustal units and their relief relative to one another, strongly modulated by the existence of an erosional baselevel.

Would Earth have a bimodal hypsogram if oceans did not create this

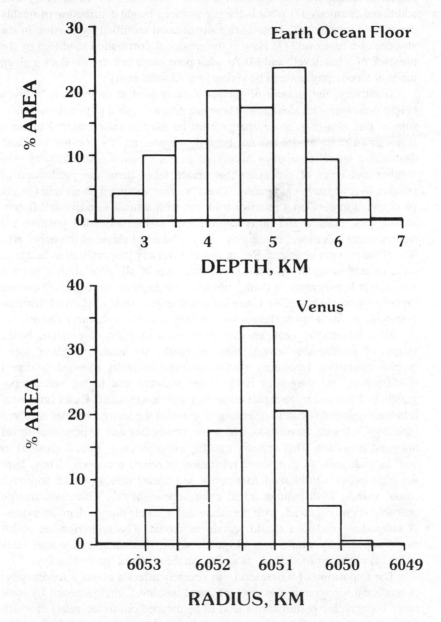

Fig. 16. Hypsograms for Earth's ocean mode and for Venus elevations < 6053 km. (From Morgan and Phillips 1983.)

erosional baselevel? This question must be examined by considering four additional questions: (1) What is the percent area-height distribution of freshly created topography? (2) How is this distribution modified by erosion in the absence of a baselevel? (3) How is the erosional distribution modified by the presence of a baselevel? and (4) At what percentage occurrence does a given mode of topography cease to be visible as a separate entity?

Intuitively, the amount of surface area created at or thrust to a given height decreases with elevation. There are probably good physical reasons to support this idea (e.g. more energy must be used to raise material higher), although an exact model has not been developed, mainly because on Earth destruction of relief always accompanies its creation. Additionally, the number and types of processes that create relief favor the production of positive over negative topography. Thus a surface starting with no relief might be expected to develop a relief distribution with a small negative tail (representing pits, craters, and other depressions) and an extended positive tail (representing volcanoes, mountains, etc.). The exact shape of the upper portion of the curve is not clear. Relationships inversely proportional to height or area, or related to an inverse exponential, would all yield similar percent area-height hypsograms. It is thus unlikely that hypsograms of newly created topography would resemble Guassian distributions; instead skewed distributions with modes at lower elevations than means would seem more likely.

How topography changes with time, as a function of erosion, is the subject of considerable investigation on Earth. Stochastic models of topographic evolution, laboratory studies of experimentally induced landform modification, and long-term field studies indicate that higher relief topography is destroyed more rapidly than that with lower relief. Exact functional relationships are difficult to determine because of the many variables involved (lithology, climate, vegetation, etc.), but power law and exponential forms are most prevalent. This appears true for exogenic (e.g. fluvial erosion) as well as endogenic (e.g. viscous relaxation of relief) processes. Thus, hypsometric peaks would tend to narrow and mean values would approach modal values. Distributions might even appear roughly Gaussian, though probably slightly skewed, with the means at higher elevations than the modes. A unimodal population would remain unimodal. The redistribution of bimodal topography would depend on the vigor, timescale, and spatial scale of the degradation process(es), as well as on the origin of the bimodality.

The imposition of a baselevel can severely affect a planet's topography. A unimodal topography has a geophysical baselevel, that imposed by isostasy. Topographic perturbations tend to be damped out to the relief at which finite strength of the lithosphere exceeds induced stresses, or at which isostatic uplift reaches equilibrium, which in itself is partially dependent on stress levels. An erosional baselevel can permit unimodal topography to become bimodal, given a benign geophysical environment, as is the case on Earth. The extreme peak of the continental mode at sea level suggests the strong and continued control of that mode by erosion to sea level. Beneath sea level, topography acted upon by submarine processes shows relief similar to but offset from the continental mode. Submarine topography is less subject to erosion; it may approximate freshly created relief complicated by sedimentary processes and cooling of the thermal lithosphere.

What would Earth's topography be like in a case without oceans and in which plate tectonic phenomena had ceased? As noted earlier, the weight of water accounts for ~ 1 out of the 5 km difference between the submarine mode and sea level. However, if erosion of the continents and high standing submarine topography proceeds rapidly and over extensive areas, then erosion, transportation, and isostasy might lead to a relatively smooth topography with only a few hundred m of relief. It is unlikely that the hypsogram of such topography would show distinguishable modes.

Finally, how does the fraction of surface that lies within a mode affect its detectability? The detection of modes involves their amplitudes and widths as well as their separation. A bimodal topography can seem unimodal if one mode has a much larger amplitude and/or is much broader than the other. Earth's partition of topography (67:33), separation of modes (5 km out of $\approx$ 15 km), and the full-width at half-power of each of its modes (≈ 1.5 km oceanic, ≈ 0.5 km continental) allow easy identification of its bimodality. Another planet might have a partition of topography (95:05), separation of modes (≈ 2 km out of 12 km) and full-width at half-power of two modes (both ≈ 2 km) that would not allow easy identification of bimodality. Venus may resemble this latter case.

In summary, what do the gross shapes of hypsograms of Earth and Venus tell us? On Earth, plate tectonic processes have generated enough continental crust to create a bimodal distribution of crustal thickness, reflected in part by low-frequency surface topography. Ongoing tectonic and volcanic processes have also created new high-standing topography superimposed on oceanic and continental crust, governed by poorly understood relationships that lead to the formation of progressively less area with increasing height. Working against the creation of relief are degradational mechanisms: sedimentation and cooling of the lithospheric boundary layer reduce relief on the seafloor, erosion reduces relief on the continents. The presence of the oceans and their surface as an erosional baselevel, and the feedback mechanisms that probably maintain this level, effectively separate Earth into two distinct topographic provinces. Removal of the erosional baselevel would very likely lead to a dramatic change in Earth's hypsogram.

On Venus, it is probable that no such baselevel exists. The distribution of topography can have several alternative explanations; none is unique. Venus may not have created a sufficiently large volume of acidic crust to produce continental-oceanic bimodality; it may also possess highly effective erosional processes that have leveled its topography.

The fine structure of hypsograms, e.g. the shape of individual modes, may also yield information about the tectonic style of a planet (Morgan and Phillips 1983). Figure 16 shows the dominant elevation contributions to Earth's oceanic mode and the single mode of Venus (the hypsogram for the median plains plus the lower elevations of the Equatorial Highlands is represented; i.e. elevations < 6053 km). The asymmetric shape of Earth's mode clearly shows the effects of seafloor spreading in two ways. The gradual slope to the left of the major peak is consistent with depth increasing with the square root of time (see Eq. 33), as predicted and observed for divergent, cooling plate boundaries (see Parsons 1982). The narrower range of elevations to the left of the peak in the venusian hypsogram is also consistent with the boundary layer results for seafloor spreading on Venus. That is, the smaller topographic height coefficient h_0 (see Eqs. 34 and 39) for Venus would yield this result. However, this interpretation provides no explanation for the lower elevations, to the right of the peak.

The rapid decrease in percent area to the right of the ocean-mode hypsogram's main peak strongly indicates that the ocean floor reaches a thermally mature baselevel, and that the topography at greater depths (e.g. trenches) contributes little to the surface area of the ocean floor. This is reflected in the well-known flattening at $\sim$ 70 Myr of the depth versus age curve for the oceanic lithosphere. The departure of both depth and heat flow from a $\sqrt{\text{time}}$ decay law led Richter and Parsons (1975) to propose a second smaller scale of convection (besides that of plates themselves) in the upper mantle. Heestand and Crough (1981) have advanced an alternative suggestion for the flattening of the depth curve, that the older a piece of ocean floor is, the more likely it is to have passed over one or more hot spots, showing the thermal imprint in terms of increased heat flow and decreased depth.

The hypsogram for Venus is very symmetric compared to Earth's ocean mode (and is even more symmetrical when plotted with more resolution in radius than in Fig. 16). There are at least two interpretations for the median plains, if seafloor spreading is operative: (1) The peak in the venusian hypsogram represents a time of excessive plate creation (this may cause the secondary peak in the ocean mode). If the venusian peak were removed then the shape of two hypsograms would be similar. (2) There is preferential consumption of older ocean floor on Venus (radii <6051 km), implying that a mature thermal boundary layer is not formed. This preferential consumption of older lithosphere could be due to enhanced lithospheric buoyancy on Venus. The total spread of elevations on both hypsograms is about the same. If the elevations represented in the venusian hypsogram mainly reflect the processes of seafloor spreading, then this equivalency in hypsogram width most likely requires an implausibly thicker lithosphere on Venus (i.e. heat flow significantly less than terrestrial) to compensate for the lack of ocean pressure on the topography and the higher ambient surface temperature. This is the same conclusion reached earlier in this chapter (see Fig. 9), using the observation that $\Delta h_{\venus}/\Delta h_{\oplus} \approx 1$ and Eqs. (40) and (41).

In the discussion of hot spot tectonics presented here and expanded by Morgan and Phillips (1983), the symmetrical distribution in the venusian hypsogram is due to roughly equal contributions of regions with greater and less than normal heat flow. The latter regions, cold spots, may not be well developed on Earth, because of episodic rapid rewarming of the lithosphere due to plate motion over the hot spots and because the plate motion itself tends to thermally homogenize the asthenosphere.

Topography of Plate Tectonics

To discover what venusian topography and landforms can tell us about interior processes, we first examine principal terrestrial tectonic features of known origins and then search our meager data for similar forms on Venus. This procedure obviously introduces considerable bias, since it attempts to set limits on how Earth-like Venus is. Acknowledging this bias is an important first step in this analysis. In discussing the relationships between global tectonism and landform genesis, it is convenient to divide the commonly involved plate tectonics into two phenomena, seafloor spreading and continental drift. Each has relatively distinctive related features, and thus the link between tectonic regime and landform is strongest.

Seafloor Spreading. There are three principal elements of the physiography of seafloor spreading: midocean ridges, deep-sea trenches and their associated island arcs, and fracture zones. Due to their topographic prominence (up to 4 km above surrounding seafloor), great integrated length (up to 56,000 km) and width ($\approx$ 3400 km mean total width), midocean ridges are the most striking seafloor features of terrestrial plate tectonics. They are also among the most important, since they represent the creation of new, hot oceanic crust that transports heat from Earth's interior once it is deposited at the surface. The average values for physical dimensions of midocean ridges conceal their considerable diversity. Although many form links in the continuous chain of ridges, some are clearly distinct. Their morphology is greatly dependent on velocity and cooling; slow-spreading centers create sharply peaked ridges, while fast centers create more subdued and broader ridges.

Considerable attention has been paid to factors that might work against detection of Earth-like midocean ridges on Venus, including the difference in plate velocities, the difference in weight of water overlying the ridge crest and ridge flanks, and the influence of surface temperature on cooling of the boundary layer (which accounts for the fall-off of topography with age away from the ridge crest). It is a strong comment on the inadequacy of our data that two groups working with the same data reach dissimilar conclusions. Arvidson and Davies (1981) performed corrections on a terrestrial global topographic data set with resolution comparable to that of the Pioneer Venus altimeter, and concluded that the continued visibility of terrestrial midocean ridges may prove that similar landforms do not exist on Venus. Head et al.

(1981) on the other hand conclude that the corrections for weight of water and differences in surface temperature modify the terrestrial topography enough that it is questionable whether such forms could be uniquely identified on Venus with the present data set.

Are there midocean ridge equivalents on Venus? Certainly none have been identified in the form seen on Earth. Ridge forms have been recognized in Earth-based radar images (Malin and Saunders 1977; Campbell et al. 1979, Jurgens et al. 1980), but they are much narrower than most terrestrial midocean ridges (although they do, for example, resemble the Ninety-East Ridge, which is not a spreading center). Phillips et al. (1981) have noted that the elevated (equatorial) regions of Venus tend to be elongated and included in very long systems of elevations roughly following great circles. Aphrodite is part of this trend. Phillips and co-workers compared three transverse cross sections of Aphrodite with typical transverse sections of terrestrial midocean ridges, and their qualitative assessment was that the venusian sections lacked the concave upward form characteristic of midocean ridge topography (as dictated by the cooling of the lithospheric boundary layer). However, Kaula and Phillips (1981), after a much more extensive and statistical treatment of a greater number of sections (51 on Venus, 60 on Earth), conclude that the venusian features in fact display a greater average concavity than do the terrestrial midocean ridges. As discussed in Sec. II, the major ridge forms on Venus (associated with Aphrodite and Beta) have topographic decay coefficient Δh values similar to spreading ridges on Earth; thus, under the assumption of uninhibited convection they might only be spreading ridges if the interior of Venus is extremely depleted in heat sources (Fig. 9). Alternately, these ridges have inhibited spreading velocities and act as an inefficient mechanism for heat removal (Kaula and Phillips 1981). Since both scenarios are considered unlikely, we conclude that the Aphrodite and Beta linear topographic trends are not spreading ridges, and the median plains must be examined for this type of tectonics. Terrestrial slow-spreading ridges, when corrected to venusian conditions of no ocean water load and higher temperatures and velocity, should be detectable in the Pioneer Venus data set, but none are seen. Fast-spreading ridges, the type of oceanic plate on Earth associated with significant subduction, seem beyond the detection capability of the Pioneer Venus altimeter.

Deep-sea trenches and their associated island arcs are the second main element of the physiography of seafloor spreading. Although the separation of the trench from its island arc may vary, the pairing of a narrow deep with an adjacent ridge or mountain chain creates a distinctive topography. Such paired topography is notably absent on Venus. However, Arvidson and Davies (1981) have demonstrated the difficulty of resolving terrestrial trenches at Pioneer Venus radar resolutions. Still, some features do resemble, at least in planimetric form and topographic relief, terrestrial island arcs or trenches. One good example resembling the former is Ut Rupes (55°N, 315°W), a ridge that parallels the southwest margin of Ishtar Terra (Malin and Paluzzi 1980). However, it does not have an associated trench (at least not one resolved in PV data). Masursky et al. (1980) and Arvidson and Davies (1981) find that Artemis Chasma (48°S, 135°W) resembles the Java and Aleutian trenches as seen in terrestrial data of comparable resolution. Their comparisons are based principally on the arcuate nature of Artemis; there does not seem to be an associated "island arc" nearby. The difficulty in observing these types of features using low resolution data makes such comparisons speculative at best.

Similarly, fracture zones or transform faults are difficult to see. Their existence can be inferred only by cases where they offset major topographic forms of other origins. No such offsets can be reliably inferred from the PV radar observations.

Continental Drift. Continental drift is also a major result of plate tectonics. Indeed, the search for a global tectonic pattern on Earth with mobile crustal plates was motivated by the geometric fit of continents, especially that between the west cost of Africa and east coast of South America. The map of Venus shows visual associations at least as good as those found on Earth. However, just as such associations were not especially rewarding until there was better and more quantitative evidence for continental drift, so would extending such analyses to Venus be fruitless. Thus, other aspects of continental drift must be examined for more definitive topographic and physiographic relationships; the two most noteworthy features are rifts and mountain ranges.

Rifts are the continental equivalent to midocean ridges, in that they represent the location of imminent creation of oceanic crust. However, the differences between continental and oceanic lithosphere cause a significantly different morphology to develop. Thermal uplift and associated extension create a broad, convex, elongate doming with a longitudinal trough or rift valley. Volcanism creates both relatively flat plains and large, central volcanic mountains. Due to spatial and temporal irregularities in these processes, continuous ridges and crustal troughs are rare; most continental rifts instead resemble a series or chain of elongate domes and basins.

Such features are very prominent on Venus, particularly associated with the highland areas of Aphrodite Terra and Beta Regio. McGill et al. (1981) have studied Beta Regio extensively; they note many similarities to continental rifts, especially those in East Africa. Masursky et al. (1980) have concentrated on the rifts of Aphrodite—Artemis, Dali, and Diana Chasmae. Head et al. (1981) also discuss these features. For a more complete examination of these landforms, see Chapter 6 by McGill et al. Again due to spatial resolution limitations in the Pioneer data, the origins of these large troughs cannot be unequivocally determined. As noted above, Artemis Chasma has some attributes of a deep-sea trench (a compressional tectonic phenomenon), and other attributes characteristic of continental rifting (extensional). Tectonic genesis (as opposed to impact genesis [Masursky et al. 1980]) is almost certain, but the nature of the tectonism is still debatable.

The second major feature of continental drift is the creation of mountain ranges, principally in collisions between continents. The Himalaya mountains result from the ongoing collision of India with Asia. Mountain ranges usually form along continental margins, but the suturing of two continents along a mountain range often places the mountains in the interior of a continental landmass.

True mountain ranges appear to occur only on Ishtar Terra; Akna, Freyja, and particularly Maxwell Montes are excellent examples of venusian mountain ranges. The similarity of the Maxwell Montes to the Himalayas has been noted, as has the strong resemblance of the associated portions of Ishtar east and west of the Maxwell Montes with those portions of India and Asia sutured together along the Himalayas. However, Masursky et al. (1980) interpret the Maxwell Montes as a fault-bound massif. The plate tectonic implications of any of the postulated origins are uncertain. Earth-based radar images of 6 km resolution that have been computer enhanced (J. W. Head, personal communication, 1982) clearly show the mountain ranges of Ishtar Terra to be composed of parallel radar backscatter bands that can be interpreted as ridges and valleys. They resemble compressional tectonic features on Earth, but could also be of extensional origin. Clearly, they show the importance of horizontal stress in the lithosphere at one time in the history of Venus. On Earth, significant horizontal stress is associated with plate tectonics.

Intraplate Phenomena. A final aspect of terrestrial topography, related to but not necessarily created by plate tectonic phenomena, is the features formed in intraplate locations, including volcanic and tectonic landforms produced by isolated mantle phenomena, usually called hot spots. These features can form inside the margins of both oceanic and continental plates, although they appear more prevalent in oceanic environments (probably due to the thinner lithosphere), in which they tend to create rises or subtle ridges surmounted by volcanic islands. Plate motions with respect to these hot spots tend to create chains of volcanoes, like the Hawaiian-Emperor chain of seamounts and islands. Within continents, hot spots create high standing plateaus, often topped by volcanoes, surrounded by large shallow basins. In Africa, where this phenomenon is most readily seen (probably because Africa is relatively motionless with respect to the hot spots), this form of relief is termed basin and swell topography (Burke and Wilson 1972). Tibesti in northern Chad is an example of a swell; south of Tibesti is the Chad Basin.

Venus seems to display considerable basin and swell topography. At all scales, from the limit of resolution to features the size of Aphrodite, equidimensional positive relief is surrounded by comparable lowland relief. There are many isolated topographic peaks on Venus, up to hundreds of km across and <2 km high, and down to the limit of earth-based radar observations (Malin 1981). There are nearly as many individual peaks per unit area on Venus as in the Pacific Ocean basin on Earth. Given the difficulty of creating isolated mountains except by volcanic or tectonic activity (see e.g. Malin 1978), these mountains clearly are an important element in venusian geology and geophysics.

Craters

Craters are prominent landforms on many planets and satellites, although on Earth degradation processes considerably reduce their numbers. They are important because each of the ways craters can form provides clues to various components of planetary evolution (see Chapter 6 by McGill et al.).

Venusian craters are especially controversial because of the scarcity of clear morphologic and topographic observations. They are most often interpreted as impact craters, on the basis of circularity, bright rims, randomly spaced locations, and abundance at diameters >75 km consistent with meteorite impact over 1 to 2 Gyr. However, many of the crater forms are not circular, and radar scattering properties are not particularly diagnostic of impact cratering landforms. Random locations are not unique to impacts; the true meaning of a time-related abundance resembling that of impact craters is obscure. Impacts could have made craters on Venus, but there are no conclusive data; it is premature to use such features of uncertain origin for impact cratering age determinations.

Volcanic craters could form on Venus as well, though this also is uncertain. However, we feel that craters at the tops of 1 to 2 km high mountains, which show rough surface textures at radar wavelengths, strongly suggest that some venusian craters are volcanic. For a more detailed discussion of venusian crater-like forms, see Chapter 6 by McGill et al.

Summary

Some important points to remember about the geology and topography of Venus, in attempting to decipher its geophysical history, are the following:

(1) Planetary hypsograms tell us (a) the relative fraction of crust of different densities; (b) the occurrence of erosional baselevels; and (c) the effectiveness of erosion. The absence of a clearly bimodal venusian hypsogram does not mean that Venus has not experienced plate tectonism. Instead, it suggests that (i) one crustal type is much more prevalent (which type is not known); (ii) if an erosional baselevel occurs, it is relatively near the topographic mode; and (iii) either erosion has been only partly effective in reducing old surface relief, or new relief and erosion have established an equilibrium dominated by erosion. Since a thinner lithosphere reduces the relief contrast produced by (a) above, the unimodal venusian hypsogram may indicate a relatively thin lithosphere.

(2) Surface features associated with seafloor spreading are problematic on Venus. The major ridge systems may not be spreading ridges, unless Venus is highly depleted in heat sources. The heights of these ridges also lack a well-defined mode and do not form a global system. Terrestrial-like slow-spreading ridges are probably detectable with Pioneer Venus resolution, but are not seen in the median plains. Fast spreaders, associated with well-developed subduction on Earth, seem unresolvable with the Pioneer Venus data. Strong evidence of features like deep-sea trenches and fracture zones is likewise missing, but may simply be below the limits of data resolution.

(3) Continental drift features, particularly rifts and mountains, are more easily discerned in the Pioneer data. Although these may have non plate-tectonic explanations, they clearly indicate that Venus is (or was) tectonically active.

(4) Topographically, Venus most resembles Earth's seafloor. On Earth, this topography reflects intraplate phenomena associated with isolated hot spots in the mantle beneath the crust. If this is also the explanation for such features on Venus, then Venus appears to be (or has been) extremely active over many length (and time) scales.

(5) Ishtar is remarkably different from the rest of Venus, having mountain ranges and flat plateaus. Its hypsogram resembles those of terrestrial continents. The geophysical explanation of Ishtar will probably not resemble that of other highland relief.

V. CONCLUSIONS: THE TECTONICS OF VENUS

In Sec. I we stated that Venus and Earth are similar in bulk composition and mean density, so similar that any reasonable cosmochemical or petrological model can explain the small discrepancy in density. It might then be expected that Venus and Earth are similar in other respects; indeed, we study Venus by trying to decide in what ways it is like and unlike Earth.

As we have said, the tectonics of a planet reflect the response of its lithosphere to removal of internal heat, and unless exogenic processes (e.g. impact cratering) dominate, the surface physiography of a planet is shaped by tectonic processes (and reshaped by erosional processes). In principle, given the thermal history of a planet the tectonic history can be predicted, and vice versa; in this chapter we have attempted to develop a framework to do just that for Venus.

The tectonic features of Earth clearly reflect its style of internal heat removal. Ocean-floor physiography is dominated by the cooling boundary layer of mantle convection. Away from the ridges, disorganized plume-like convective upwellings cause hot spot phenomena: lithospheric swells (e.g. Hawaii) and isolated volcanic groupings. The terrestrial continents themselves are enigmatic in origin, but their physiographic features clearly reflect plate-tectonic processes, i.e. continental sutures and rifts.

To decipher the tectonic history of Venus, the case for seafloor spreading has been closely examined as a mechanism for removal of internal heat, since that mechanism dominates on Earth. The higher buoyancy of the venusian than the terrestrial lithosphere suggests that seafloor spreading and subduction are likely to be less effective on Venus. The linear topographic trends of the Equatorial Highlands (e.g. those associated with Aphrodite Terra and Beta Regio), assuming Earth-like heat sources and uninhibited convection, do not appear to have the shape of divergent, cooling plate boundaries; they decrease much too rapidly with distance from their axial highs. If seafloor spreading takes place on Venus, we suggest that it must be confined to the median plains. We have demonstrated that divergent plate boundaries are not necessarily detectable with the resolution of the Pioneer Venus altimetry. In particular, fast-spreading terrestrial oceanic plates translated to the Venus environment cannot be resolved. However, if the convection is inhibited, then it is possible that the Equatorial Highlands represent a region of very slow plate divergence delivering only a small fraction of heat to the surface. Ridge push forces may also be diminished on Venus; this, combined with the greater lithospheric buoyancy, suggests that a different force balance might drive venusian plates. For example, coupling to the underlying mantle may be more important on Venus than on Earth in sustaining and/or inhibiting plate motions.

Physiographically, Ishtar is the most continent-like form on Venus (Phillips et al. 1981). It is a plateau with a distinct topographic boundary on much of its border. A flanking chain of topographic highs, Ut Rupes, stands off to the southwest in an island-arc relationship to the main mass of Ishtar. Three mountain ranges, Akna, Freyja, and Maxwell Montes, are contained on Ishtar; they are reminiscent of continental suture zones and can be interpreted from high resolution earth-based radar images as parallel ridge-valley topography.

The high-standing relief of Ishtar Terra most likely requires a buoyant crustal element in the lithosphere. If this is true, then some process must be invoked to put the original products of mantle differentiation through a secondary differentiation process. On Earth this is thought to occur mostly due to subduction of basaltic crust that is then subject, in a water-rich environment, to melting and differentiation of more acidic (granite-like) products. The acidic rocks rise to form island arcs behind subduction zones, and the fusion of these arcs may form the continents.

The physiography of Ishtar Terra is the most compelling evidence that there are or were Earth-like plate tectonics on Venus. At this writing, there unfortunately has been no definite test establishing even the roughest relative age of this feature. An interesting hypothesis is that before the present atmospheric greenhouse existed on Venus, surface temperatures were lower, water was more abundant, and terrestrial-like plate tectonics occurred. Ishtar would represent the continental mass produced by subduction and plate collision processes before the greenhouse and attendant high surface temperatures were imposed. Under this scenario, loss of water and/or increased surface temperature must be responsible for the cessation of plate tectonics. If this hypothesis is correct and the CO_2 greenhouse formed early in the history of Venus, then it must be shown to be rheologically plausible that Ishtar Terra was maintained over long periods of geologic time.

We believe that the Equatorial Highlands result from hot spot tectonics, and are led to this view by: (a) the prevalence of basaltic volcanism, as measured in the Venera geochemical results and inferred from earth-based radar observations; (b) the observation from altimetry that the topography of the Equatorial Highlands is characterized by roughly circular elevated regions; and (c) the gravity-anomaly interpretation that these elevated regions are supported by low-density roots that extend, at least in some cases, to depths well beneath the lithosphere, implying that the topography must be dynamically supported. Dynamic support means that the topography is supported by thermal buoyancy in the interior, through rising convective plumes that provide either direct buoyant support or indirect support by heating and expansion of the overlying lithosphere. This gravity interpretation defines these topographic regions to be hot spots.

It has been argued in Sec. IV that, aside from Ishtar, the closest physiographic analog to Venus found on Earth is the basin-and-swell and isolated hot spot terranes found in intraplate areas, mostly in ocean basins but also on continents (most notably Africa). This basin-and-swell topography, positive relief surrounded by comparable lowland relief, is seen at all scales on Venus from the limit of resolution to features the size of Aphrodite. Many isolated topographically high features exist, with areal density perhaps comparable to that of the Pacific Ocean basin. It is difficult to create isolated topographic highs other than by volcanic-tectonic processes. These positive relief features on Venus display phenomena similar to lithospheric swells on Earth, rifting (e.g. McGill et al. 1981) and volcanism (Saunders and Malin 1977).

The Beta Regio area of Venus would appear to be the youngest and most active hot spot. It has the deepest hot spot root, as interpreted from the gravity data, and well-defined volcanism and rifting. By the high ratio of gravity to topography, the elevated region at the eastern extremity of Aphrodite seems to be another youthful hot spot. Under this hypothesis, western Aphrodite is composed of two mature hot spots, each with distinct topographic and gravity highs. The depths of compensation are considerably less than that for Beta Regio. A possible interpretation is that the buoyant roots have spread out along the base of the lithosphere and this is reflected in the overlying topography, distributed over a larger area than Beta. Artemis Chasma may represent the boundary of a former hot spot in which the elevated topography collapsed. Smaller isolated topographic features presumably represent hot spots less vigorous than Beta or Aphrodite.

As discussed in Sec. II, the ocean basins on Earth account for most of the

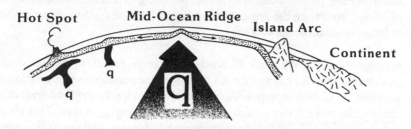

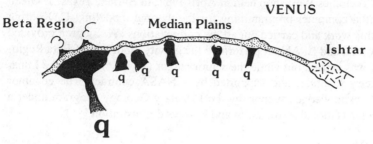

Fig. 17. Cartoon showing cross sections of ocean basin on Earth and hot spot model for Venus. Dark areas are regions of heat advection q; on Earth most of the heat is deposited at midocean ridges. In the model of Venus, heat is more nearly uniformly distributed at various hot spots.

planet's heat loss through seafloor spreading. How would the ocean basins remove Earth's heat if seafloor spreading stopped? We expect that the mechanism which is now secondary, the disorganized patterns of intraplate heat transfer through convective plumes (i.e. hot spots), would take over the heat transfer formerly carried out by seafloor spreading. The hot spots would grow in vigor, frequency, and magnitude to provide the required heat transfer. Regions such as Iceland, the North Atlantic Gravity High, and Hawaii would be more common, features larger than these might be expected, and there would be a denser background of small-scale hot spots (capped by individual volcanoes). This scenario provides the hot-spot tectonic model for Venus.

Could hot-spot tectonism be the primary way Venus expels its internal heat? While we hypothesize that it is a significant process, there is no way to judge the relative importance of hot-spot versus plate tectonism. If seafloor spreading exists, it would appear to be confined to the median plains, with attendant continental processes represented by Ishtar. Ishtar itself may transfer heat conductively, analogous to terrestrial continents.

Modest topographic features within the median plains all show associated gravity anomalies, where resolution is adequate. Modeling of these features will elucidate the style of tectonism in the median plains. Figure 17 shows a hot spot tectonic model for Venus, compared with the tectonic setting of Earth. Further progress can be expected in testing hypotheses for the tectonics of Venus, both from the present data sets and from new information acquired in future space missions to Venus.

Acknowledgments. We thank W. Kaula for rigorous review of the first draft of this chapter, and K. Goettel for his review of the section on composition and structure. S. C. Solomon provided an excellent review of the final draft of this chapter and pointed out a number of small but annoying errors. This chapter has also benefited from our discussions with P. Morgan and the participants in the Venus Tectonics Workshop held in April 1982, in Burnet, Texas. J. Green did the initial computer programming for the thermal modeling; C. Kohring extended this work and carried out all the computations. W. Sjogren provided the gravity data and E. Abbott performed the gravity modeling of Beta Regio. This work was carried out while the senior author (R. J. P.) was at the Lunar and Planetary Institute, and supported by a NASA contract. The coauthor (M.C.M.) acknowledges support by the Planetary Geology Program under a grant from the National Aeronautics and Space Administration.

11. THERMAL STRUCTURE OF THE ATMOSPHERE OF VENUS

A. SEIFF

NASA Ames Research Center

The available data sets determining the thermal structure are critically discussed and combined into a mean model. The implications of the observed contrasts for atmospheric motions, on small and large scales, are discussed. Greenhouse models which allow a convective atmosphere below 35 to 50 km give a reasonable explanation of the high surface temperature. Surprisingly, however, the deep atmosphere is generally stable. The radiative imbalance thus drives the general circulation rather than local convection. A distinct tropopause occurs at the cloud tops. Above this, the atmosphere is very stably stratified and not far from radiative equilibrium. Gravity waves are present. The upper atmosphere, with its very large day-night temperature contrast, must be in motion away from the Sun, but velocities are strongly limited to a sizable viscous dissipation.

The atmosphere of Venus may be divided into three natural vertical intervals, each with distinct characteristics: (1) the lower atmosphere, or troposphere which extends from the surface to the cloud tops; (2) the middle atmosphere, which is stably stratified extending from the cloud tops to 100 km; and (3) the upper atmosphere, above 100 km, the only region of Venus's atmosphere displaying large diurnal variability.

In the lower atmosphere, temperature and pressure increase continuously from earth-like conditions in the lower and middle cloud layers to ~ 737 K and 95 bar at the mean surface. Lapse rates vary with altitude, being strongly influenced or controlled by the presence of the cloud layers, an influence which extends well below the clouds. Steady diurnal variations in temperature are small at 30° latitude, < 1 K in the convective region above 20 km, but there are also oscillatory diurnal variations ~ 5 K which appear to be as-

[215]

sociated with gravity waves. Latitudinal variations in temperature at a given altitude are a few K below the clouds up to 60° latitude, the limit of observation. (There are indirect indications of possible temperature decrease on the order of a few K in even the deep atmosphere poleward of 60°.) Within and above the clouds (to 85 km), there are significant latitudinal pressure and temperature differences, which, from midlatitudes to 60°, are consistent with cyclostrophic balance of the zonal winds.

Much of Venus's lower atmosphere is statically stable. The mean lapse rate from the surface to 50 km is ~ 7.7 K km⁻¹, compared with the mean adiabatic lapse rate, 8.9 K km⁻¹. There is a 20 km deep stable layer just below the clouds with stability comparable to Earth's troposphere where waves in the diurnal thermal contrast have been observed. A second, less stable layer occurs just below 20 km. Convective layers are confined to the middle cloud, a 10 km layer below 30 km, and possibly within the lowest 10 km. Turbulent temperature fluctuations do not exceed 0.2 K in the convective layers near the morning terminator at 4° latitude.

Radiative-convective greenhouse models in which the convective layer is very deep, extending up to an altitude of at least 30 km, yield high surface temperatures comparable to those observed in Venus's lower atmosphere. However, the generally stable condition observed in the deep atmosphere implies that global circulation, rather than local convection, is responsible for the near-adiabatic state of the atmosphere. The radiative imbalance then drives the circulation.

Above the clouds, lapse rate decreases sharply, and from 60 to 100 km, the atmosphere is highly stable. It becomes essentially isothermal from 80 or 85 km to 100 km, and is isothermal above the cloud tops near the poles. Temperature increases poleward above the cloud tops. In this stably stratified region, diurnal variation is small, but temporal variations with amplitude ~ 10 K and periods of 5.3 and 2.9 d are present. Fluctuation amplitudes increase toward the poles. The variability appears to have two sources, gravity waves driven by diurnal temperature variation in the clouds, and the intermittent polar collar clouds and polar emission windows. (The latter rotate about the planet zonally with a period 2.9 d.)

Above 100 km, the upper atmosphere begins and there are major differences from the lower atmosphere in the diurnal contrast and the implied circulation patterns. The upper atmosphere is characterized by low temperatures relative to Earth's, and by surprisingly large diurnal changes, which begin just above 100 km. Exospheric temperatures are ~ 300 K (day side) and 100 K to 130 K (night side). Associated with nightside cooling is a major subsidence in flow across the terminators, accompanied by flow acceleration and downward transport of the bulk composition. There is indirect evidence that the circulation pattern changes from basically zonal below 100 km to basically subsolar-to-antisolar above 100 km. The global mean theoretical structure calculated by Dickinson is in first order agreement with the measured structure near the terminator. The Dickinson and Ridley dynamic models do not, however, predict the low nightside temperatures. A quantitative explanation for the low nightside temperatures is not yet established, but several possible contributing mechanisms have been advanced.

Over the last two decades the knowledge of Venus's atmosphere structure has advanced immeasurably; it has improved from the stage of controversial speculation to reasonably firm description. In this chapter, after a brief historical review, we will summarize the current state of knowledge. The discussion is organized by altitude intervals: Sec. I discusses the lower atmosphere (from the surface to the cloud tops), Sec. II the middle atmosphere (from the cloud tops to 100 km) and Sec. III, the upper atmosphere (from 100 km to the exosphere). We will in each case critically review the available data, discuss diurnal and latitudinal variability, static stability, convective regions, indications of wave phenomena, and theoretical understanding, where possible. Questions which remain unanswered, or which are raised by the data will be noted.

Historical Review

The period of astronomical observations of Venus's atmosphere, which began soon after the invention of the telescope, includes the following milestones:

~1660   Early telescopic observations of Christian Huygens show no evidence of visual features on Venus, leading him to pose the question, "is not all that Light we see reflected from an Atmosphere surrounding Venus?" (Huygens, Complete Works, 1967).

1761   Lomonosov, on observing the blurred edges of Venus in transit across the solar disk, concludes that the planet has an atmosphere "equal to, if not greater than...our earthly sphere" (Moore 1959, p. 64).

1792   Schröter infers the presence of a significant atmosphere from limb darkening and extension of the cusps near 90° phase (Moore 1959, p. 64).

1924   Cloud-top temperatures near 240 K measured by Pettit and Nicholson (1924).

1932   Detection of CO_2 in the atmosphere of Venus at abundance levels > 120 m atm by Adams and Dunham (1932).

1940   First suggestion that the greenhouse effect could be important to structure of Venus's CO_2 atmosphere by Wildt (1940a).

1953–5   Refinement of cloud-top temperatures of 225 ± 5 K, by Sinton (1953), and by Pettit and Nicholson (1955).

1953–5   Observation that day- and night-side cloud-top temperatures are nearly equal (same authors).

1958   Observation of high emission intensity at radio wavelengths, corresponding to blackbody temperatures ~ 600 K, by Mayer et al. (1960).

1960   Conclusion that high atmospheric temperatures can be sustained by the greenhouse effect if absorbers in addition to CO_2, in particular water vapor, are included, by Sagan (1960a).

1960   Inference from limb darkening of cloud-top thermal emission that atmosphere above the clouds is nearly isothermal, by King (1960).

1960   Indication from stellar occultation that the upper atmospheric temperature is ~ 300 K at the 100 km level and increases with altitude by ~ 3 K km^{-1} just above that level, by de Vaucouleurs and Menzel (1960).

1962   Observation that radio brightness temperature is relatively insensitive to local Venus time, by Mayer et al. (1962).

1965   Observations of small systematic cloud-top temperature differences which correlate with solar zenith angle, by Murray et al. (1963).

In 1967 with Venera 4 and Mariner 5, the exploration of Venus by spacecraft began. Starting with Venera 7, six consecutive Venera landers reached the surface and reported temperatures and pressures below the cloud levels. At the surface, temperatures and pressures ~ 750 K and 95 bar were measured, confirming that the microwave brightness temperatures were real atmospheric temperatures, and that the atmosphere was massive. Mariner 5 and 10 spacecraft measurements showed the detailed temperature structure both above and below the clouds, including inversions at cloud-top levels. (Not all of these indications were originally accepted as significant.) Mariner 5 and 10 also provided estimates of exospheric temperatures ~ 350 to 400 K, from interpretations of ultraviolet spectrometer data, while cloud-top winds were inferred from ultraviolet photographs obtained from Mariner 10.

The Pioneer Venus mission in late 1978 investigated the atmosphere by means of six coordinated spacecraft: four probes, a bus, and an orbiter. Both *in situ* and remote observations were used to define the atmospheric state properties and diurnal and latitudinal contrasts, below, within, and above the clouds, extending up to ~ 180 km. The findings of these and earlier spacecraft are incorporated in this chapter.

I. THE ATMOSPHERE BELOW THE CLOUD TOPS

A. Review of the *in situ* Measurements

A review of the data that presently define the structure of the atmosphere of Venus below the cloud tops must begin with the Venera landers. The temperature data from the five landers, Veneras 8 through 12, which obtained pressure and temperature data from the cloud levels to the surface are given in Fig. 1 (Marov et al. 1973b; Avduevsky et al. 1976a, 1979; Chapter 12). (The average profile of Veneras 11 and 12 has been given, following Avduevsky et al. [1979].) For comparison, Fig. 1 also includes $T(z)$ from the Pioneer Venus

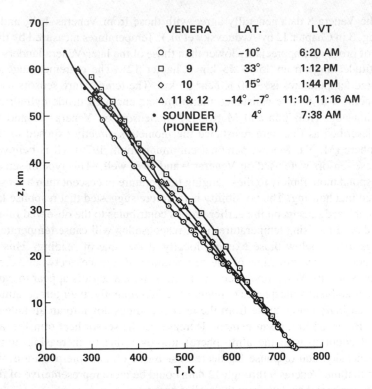

Fig. 1. Temperature data returned by the descent probes of Veneras 8 through 12 and Pioneer Large (Sounder) probe.

Large probe. To remove the effect of terrain elevation differences, all altitudes are referred to the 92 bar level, as explained below.

The Venera probes are very consistent in indicating a surface temperature of 740 K at 92 bar, within a few K. Measurement accuracies, including resolution uncertainty, are given as ± 8 K to ± 4 K, depending on the sensor range (Avduevsky et al. 1976a), and scatter of the measured data generally supports these estimates. Data points shown in Fig. 1 are from the best-fit curves to the collected measurements, rather than individual measurements, and the rms deviations of data about these curves are 2 K to 8 K (Marov et al. 1973b). (See also Chapter 12.) Altitudes were in all cases obtained from the measured pressures and temperatures, integrated in the equation of hydrostatic equilibrium. The mean lapse rate from the surface to 55 km is ~ 7.8 K km^{-1} for Veneras 10, 11 and 12 which is less than the adiabatic lapse rate, and temperature decreases monotonically from the surface to the cloud tops. These and the Venera 7 data definitely established that the near-surface atmosphere of Venus is as hot or hotter than had been indicated by the microwave emission interpreted as blackbody radiation.

The Venera 8 data generally agree with those from Veneras 4, 5, and 6 (see Fig. 3 in Chapter 12 by Avduevsky et al.). Temperatures measured by this group of probes are appreciably lower than those of the later Venera landers in the altitude range from 15 to 55 km (Chapter 12). The extreme range of measured temperatures is ~ 50 K near 35 km. The temperature sensors used on Veneras 4 to 8 were resistance thermometers enclosed inside cylindrical metal housings (Kuz'min and Marov 1975). Sensors on Veneras 11 and 12 were described as fine wire resistance thermometers directly exposed to the atmosphere (M. Ya. Marov, personal communication, 1979). (It is believed that these sensors were used on Veneras 9 and 10 as well.) This type of sensor will respond more rapidly to the changing temperature in descent than a sensor enclosed in a housing. The possibility is therefore suggested that response lag of the enclosed sensors on the earlier probes contributes to the observed range of the data. In a rising temperature field, response lag will cause temperature readings to fall below those existing locally at the time of reading. This is consistent with the direction of the temperature differences between, e.g., Veneras 8 and 10. When first uncovered, the Venera 8 sensors appear to have been near ambient atmospheric temperature. Subsequently, their temperatures fall increasingly below those from the later soundings down to an altitude of 35 km. Below 35 km, disagreement decreases as the sensor heat transfer rate (thermal coupling with the atmosphere) improves due to increasing atmospheric density, and response lag decreases. Based on this interpretation, we believe that the Veneras 9 through 12 data could be more representative of the local atmosphere than data from the earlier probes.

The Pioneer Venus (PV) Large probe data agree with those from Veneras 10, 11, and 12 within ~ 12 K below 55 km. (Differences are shown explicitly in the lower graph of Fig. 5 below.) Venera 9 measurements show higher temperatures than those from Veneras 10, 11, and 12 at 15 to 45 km. It has been suggested that this could be a result of error in one of the Venera 9 sensors (Chapter 12). In Secs. I.B and I.C, we will consider whether the observed temperature differences among the Venera soundings could indicate diurnal effects or effects of latitude.

Pressures measured during descent by the same Venera landers are shown as a function of altitude in Fig. 2. The points plotted are from curves that fit the data from multiple sensors. Sensor accuracy and resolution are given as ± 200 mbar below 3 bar to ± 4.5 bar at 100 bar. Scatter of the measurements about the best-fitting curves is 1.2 to 1.4 bar rms (Marov et al. 1973b).

Data from the several probes have been adjusted to a common landing elevation at the 92 bar level, where the vertical pressure gradient is ~ 5.7 bar km^{-1}. Thus, small terrain elevation differences will cause appreciable differences in surface pressure. Consider the probe that measured 85 kg cm^{-2} (83.4 bar) at touchdown (Venera 10). Its pressure reading at touchdown would correspond to a terminal elevation of 1.5 km relative to the 92 bar reference level. Based on this interpretation, all altitudes above touchdown are in-

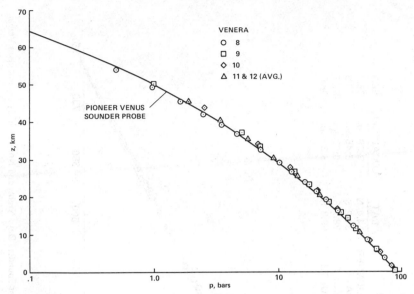

Fig. 2. Pressure data from Veneras 8-12, compared with Pioneer Venus Large (Sounder) probe data.

creased by 1.5 km to relate them to the 92 bar level. This adjustment for terrain elevation greatly reduces the indicated scatter in pressure data from the Venera landers (Avduevsky et al. 1976a). Veneras 9 and 10 with terminal elevations of 0.5 and 1.5 km, as defined by measured pressures at touchdown, landed in the foothills of Beta Regio, the 4.8 km high twin peaks at 280°E longitude, so the deduced elevations are reasonable. Similarly, Veneras 11 and 12, both with indicated landed elevations of 0.5 km, landed in the foot-hills of Phoebe Regio (maximum elevation ~ 2.5 km) which lies south of Beta Regio.

The Venera data generally lie close to the PV Large probe pressure profile. There is, however, a small systematic disagreement of up to 1 bar. The equivalent altitude discrepancy is ~ 1 km maximum. Since altitude is defined from the equation of hydrostatic equilibrium,

$$z = \int_{p}^{p_0} (RT/pg)\mathrm{d}p \tag{1}$$

a temperature difference leads to an altitude difference. Since Venera temper-atures were, for the later probes, somewhat higher than PV Large probe temperatures (Fig. 1), Venera altitudes at a given pressure will exceed the PV altitudes. However, this does not entirely reconcile the small disagreement in $p(z)$ in the altitude range from 10 to 40 km.

The temperature data below the clouds from the four Pioneer Venus probes are shown in Fig. 3, with altitudes referred to the 92 bar level (Seiff

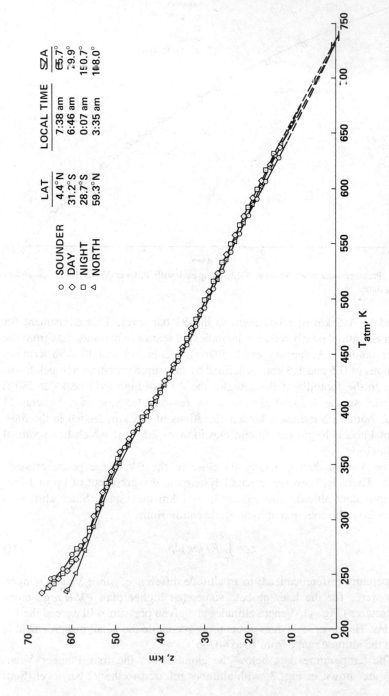

Fig. 3. Temperatures as a function of altitude from the four Pioneer Venus probes (Large or Sounder, Day, Night, and North) widely separated on Venus.

et al. 1980). A formal estimate of measuring accuracy for the sensors on these probes was 0.5 to 1.0 K. Disagreement of two independent sensors on each probe was typically 1 K and did not exceed 3.5 K. Hence the temperature differences measured should be significant to within 1 or 2 K.

Data from the four probes, widely separated over the earthward face of Venus, cluster together below 54 km, within a total span of 4 to 10 K (depending on altitude). The 54 km level is at the top of the middle cloud layer (Ragent and Blamont 1980). Thus, it is in the upper cloud layer that the temperatures measured at high latitude (60°) start to fall well below those measured at lower latitudes.

Temperature data were not obtained by the Pioneer probes below 12 km altitude. Extrapolation of the measured profiles to the surface is not greatly uncertain, however, and indicates surface temperatures from 731 to 735 K at the 92 bar level, which, like the temperatures at higher altitudes, is a little lower than T_0 from the Venera probes. The mean lapse rate up to 50 km is 7.7 K km^{-1}, compared to a mean adiabatic lapse rate of 8.86 K km^{-1} over this altitude range. This immediately indicates that there are regions of stable lapse rate below 50 km. Close examination of the profiles shows patterns of curvature which are relatively consistent from sounding to sounding and which indicate variation of stability with attitude. Stability profiles derived from these temperature data are presented in Sec. I.D.

Pressure measurements obtained during descent of the four Pioneer probes extend to the cloud tops at 67 km, where the pressure is $\sim$ 60 mbar (Fig. 4). After adjustment to the 92 bar reference level for terrain elevation differences of + 0.98 to − 0.65 km, data from the four probes define a single curve with no discernible pressure differences from the surface to 24 km. Above that level, the profiles gradually separate, primarily on the basis of latitude. The Day and Night probes at latitudes near 30° define essentially a single curve, while the other two probes, at latitudes of 4° and 60° define a slightly lower curve up to 58 km. Above 58 km, the near equatorial sounding moves toward those at 30° latitude. Measuring accuracies were estimated to be within ± 0.5 % of reading at pressures below 50 bar, and ± 1 % of reading above 50 bar, while pressure *differences* at 60 km are $\sim$ 10 %. Hence, differences found at the higher altitudes should be significant. Altitude accuracies were formally estimated to be $\sim$ 0.25 %, e.g., 150 m at 60 km (Seiff et al. 1980), which is $\sim$ 25 % of the altitude separation of the isobars (600 m) at the 150 mbar level.

B. Diurnal Variations

Diurnal temperature variations were sampled by the PV Day and Night probes at 30° latitude and results are shown in the top graph of Fig. 5. The local Venus time (LVT) difference at the landing sites of these two probes was from midnight to 6:46 AM. In the altitude range from 20 to 35 km, measured temperature differences were < 1 K. Above 35 km, the difference is oscilla-

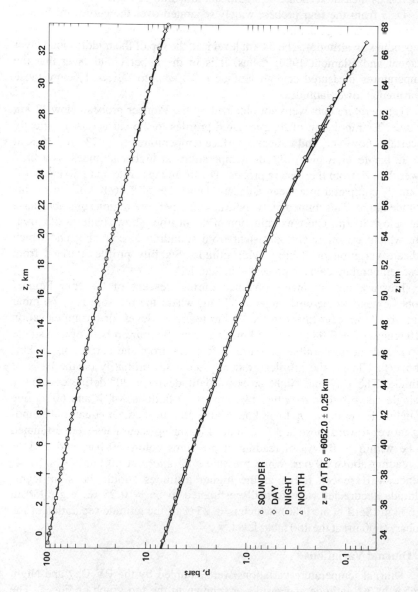

Fig. 4. Pressure data obtained by the four Pioneer Venus probes (Large or Sounder, Day, Night, and North) from the cloud tops to the planet surface. At the mean radius of Venus, 6051 ± 0.1 (Pettengill et al. 1980a), the pressure is 95 bar.

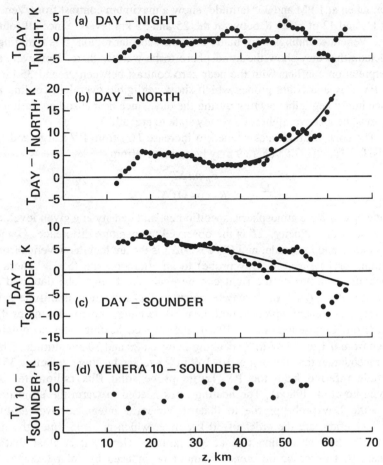

Fig. 5. Temperature contrasts from day to night and over latitudes from 4° to 60° from Pioneer
Venus probes. The bottom graph displays differences between Venera 10 and Pioneer Large
(Sounder) probe measurements.

tory with altitude, suggesting zero mean contrast in the presence of global
scale gravity waves. From 20 to 12 km, a diurnal temperature difference
increasing to 5 K was measured.

The Venera 11 and 12 soundings were made at near equatorial latitudes
(Fig. 1). While they were not simultaneous with those of the Pioneer probes,
they were obtained within 16 and 12 days, respectively, thereafter. Relative to
the Pioneer Large probe data, they seem to indicate a temperature difference
of ~ 6 K between 8 AM and 11 AM. They also agree closely with the Venera
10 sounding taken at 2 PM LVT > 3 yr earlier. The Venera 8 temperatures are
as much as 14 K cooler than those of the PV Large probe at almost the same
time of day, which could be a result of response lag, as noted earlier. Venera 9

data, taken at 1 PM and 33° latitude, show a maximum contrast from Veneras 10, 11, and 12 of ~ 25 K between the 25 and 35 km levels. Since all four of these Venera soundings were made near the subsolar region, it is tempting to conclude that they show the atmosphere to be warmer there. However, this interpretation conflicts with the near zero contrast between 25 and 35 km of the PV Day and Night probes which show negligible overnight cooling and hence imply negligible heating during the day, since heat absorbed during the day must be lost overnight for a steady state to prevail.

The solar heating rates needed to increase $T(z)$ from PV Large and Day probe levels to that of Venera 9 may be estimated from

$$\dot{Q} = \frac{c_p \, \rho u \, \Delta T}{(R_\venus \cos \Theta) \, \Delta \lambda} \tag{2}$$

where c_p and ρ are atmospheric specific heat and density at a given level, u is the zonal wind velocity, ΔT is the observed temperature difference, Θ and λ are latitude and longitude, and $(R_\venus \cos\Theta)\Delta\lambda$ is the arc length on Venus from a morning probe site (e.g. Day probe) to an afternoon site (e.g., Venera 9). Since all quantities on the right can be evaluated from probe data and gas property tables, $\dot{Q}$ (W m^{-3}) may be calculated as a function of altitude z. For the Day and Venera 9 probes, both near 30° latitude, and zonal wind profiles $u(z)$ (from Counselman et al. 1980), the implied heating rates so obtained range from 0.1 to 1.1 W m^{-3}, peaking between 20 and 30 km altitude. These are much larger than the ~ 4 × 10^{-4} W m^{-3} average heating rate below 35 km altitude inferred from the PV Large probe solar flux radiometer data (Tomasko et al. 1980a). The heating rates needed to warm the atmosphere from the Day probe profile to that of Venera 9 integrated over altitude, $\int_0^{60\text{km}} \dot{Q} \, (z)dz$, are the order of 30 kw m^{-2}, which is ~ 45 times the mid-day solar heat absorption rate at 30° latitude. Hence, it is clear that the Venera 9, Day probe differences cannot be induced by solar heating, but must be ascribed to other processes or to measurement uncertainties. This is also true for the Venera 10, 11, and 12 data relative to the Large probe data, for which necessary heating rates integrated over altitude are > 40 times the mean dayside solar input for the albedo of 0.71. Furthermore, the indicated temperature differences are comparable to stated rms measurement uncertainties. Thus, at moderate latitudes, diurnal temperature differences due to solar heating and nightside cooling are generally smaller than measurement uncertainties and are observed in one instance to be < 1 K below 35 km.

Above 50 km, where representative solar heating rates increase to 0.02 W m^{-3} at 60 km near the subsolar point (based on data presented by Tomasko et al. 1980a), the near equatorial atmosphere would, according to Eq. (2), be heated by ~ 5 K in a 90 m s^{-1} pass across the day side. At these altitudes then, moderate diurnal variations would be expected to result from solar heating.

Another source of diurnal temperature differences is adiabatic heating and cooling due to compression and expansion of the atmosphere in oscillatory vertical motion. There is an indication from Pioneer probe data that such processes occur above 35 km where stability determinations to be discussed below (Sec. I.D) show the presence of a stable layer which could support vertical oscillations, such as gravity waves.

If the day-night contrast shown below 20 km in Fig. 5 is real, and measurement error estimates suggest that it is probably due to dynamics, e.g., eddy motions. Both the Pioneer and Venera data indicate that the deep atmosphere is cooler near the equator than at 30° latitude, as shown by the Day probe-Large probe contrast below 40 km altitude in Fig. 5 and the Venera 9-Venera 10 contrast in Fig. 1. Large-scale horizontal eddies which intermittently transport the equatorial atmosphere to higher latitudes would under these circumstances lead to temperature inhomogeneities. This may be the source of the relatively large contrasts seen in the deep atmosphere. Other indications that suggest eddies in the deep atmosphere will be pointed out below.

C. Latitudinal Variations

Temperature variations with latitude shown by data in Fig. 3 from the four Pioneer probes are $\sim$ 5 K at fixed altitude below the clouds (Seiff et al. 1980). In Fig. 5, the Day probe-North probe temperature differences between latitudes of 30° and 60° are $\sim$ 5 K below the clouds and in the expected sense, i.e., cooler at the higher latitude. The Day probe-Large probe differences at 30° and 4° latitude show the equatorial region to be (surprisingly) cooler than midlatitudes below the clouds. Above 40 km, equatorial temperatures start to approach those at 30° latitude and rise above them in the upper cloud ($z > 57$ km). At these levels, the variation with latitude becomes larger and temperature decreases steadily from the equator to 60° latitude. The temperature contrast at 60 km between the equator and 60° latitude is $\sim$ 25 K.

This limited sample indicates that latitudinal variations below the clouds at latitudes below 60°, while small, are more significant than diurnal variations. Within the upper clouds, sizable temperature contrasts appear. Latitudinal variations poleward of 60° have also been measured below the cloud tops down to $\sim$ 45 km by the radio occultation technique (Kliore and Patel 1982; Yakovlev and Matyugov 1982) and will be discussed below.

The pressure data in Fig. 4 also group according to latitude. Below 25 km, pressure differences with latitude are smaller than measurement errors, which are $\sim$ 0.5 %. Above 25 km, the pressure differences generally follow the patterns described for temperature. Thus the two soundings at 30° latitude define essentially a single pressure-altitude curve up to 58 km and separate only a little above that. The 60° and 4° latitude soundings almost coincide below 55 km, above which the near equatorial pressures move toward those at 30° latitude, while the 60° sounding falls below the others. A graph of this upper region to an expanded scale is shown in Fig. 6.

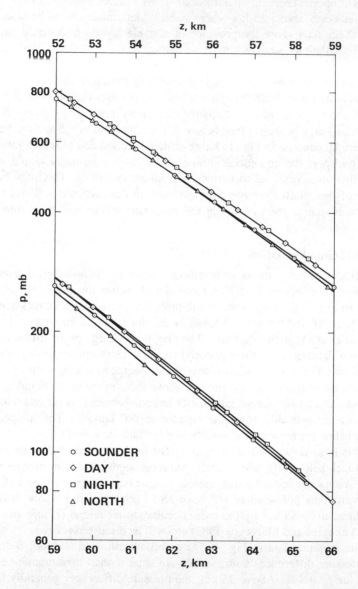

Fig. 6. Pressure differences measured among the four Pioneer Venus probes (Large or Sounder, Day, Night, and North) at altitudes from 52 to 66 km. The pressure differences between 30° and 60° latitude are consistent with cyclostrophic balance of zonal winds, as shown in Fig. 7.

For latitudes $> 30°$, the variation of pressure with latitude in the cloud layers has been shown to be consistent with cyclostrophic balance of the zonal winds (Seiff et al. 1979b, 1980; Chapter 21 by Schubert). This balance derives from the meridional momentum equation for steady frictionless atmospheric motion on a nonrotating planet

$$\frac{u^2 \tan\theta}{R_{\varphi}} + \frac{vw}{R_{\varphi}} = -\frac{1}{\rho}\left(\frac{\partial p}{\partial y}\right)_z = -g\left(\frac{\partial z}{\partial y}\right)_p \tag{3}$$

where u, v, and w are the zonal, meridional, and vertical wind velocity components, y and z are meridional and altitude coordinates, ρ and p are gas density and pressure, R_{φ} is the planet radius to the altitude z, θ is latitude, and g is the acceleration due to gravity. For $vw \ll u^2 \tan\theta$, which should hold for Venus not too close to the equator, this reduces to the equation of cyclostrophic balance

$$\rho u^2 \tan\theta = -R_{\varphi}\left(\frac{\partial p}{\partial y}\right)_z = -\left(\frac{\partial p}{\partial \theta}\right)_z. \tag{4}$$

Equation (4) may be integrated for a constant zonal wind velocity and density to obtain

$$u^2 = -\frac{\Delta p}{\rho \ln(\cos\theta_1/\cos\theta_2)}. \tag{5a}$$

Alternatively, Eq. (5a) can be written in terms of altitude difference at constant pressure

$$u^2 = -\frac{g\Delta z}{\ln(\cos\theta_1/\cos\theta_2)}. \tag{5b}$$

These equations describe a condition in which the meridional pressure gradient provides the horizontal component of centripetal force required to constrain the zonal flow circling the planet.

Given the measured wind profile, pressure differences for cyclostrophic balance can be calculated as a function of altitude. Or, with sufficiently accurate pressure differences, the winds may be estimated. Winds estimated from the pressure data of Fig. 6 for latitudes of 30° and 60° are shown in Fig. 7 at altitudes below 61 km (Seiff 1982; Seiff et al. 1979b, 1980). The comparison shown of winds estimated from pressure data with winds measured by differential long baseline (radio) interferometery (DLBI) (Counselman et al. 1980), leads to the inference that the principal latitudinal pressure variations in the cloud layers at latitudes $> 30°$ are associated with maintaining the zonal winds in cyclostrophic balance. However, this conclusion cannot be verified

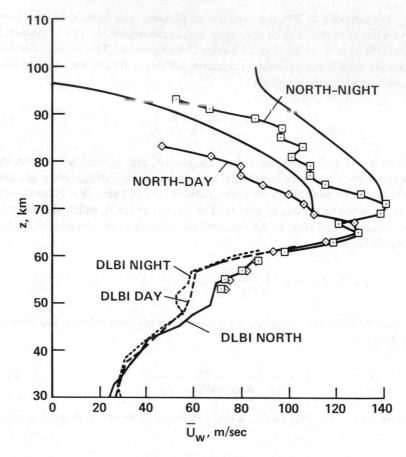

Fig. 7. Comparison of winds measured by radio tracking of probes below 60 km with winds
estimated from the measured pressures by means of the assumption of cyclostrophic balance. Above 60 km, the winds are inferred entirely from measured pressures.

very far below the clouds where pressure differences become comparable to
measurement uncertainty and the significant result is that pressure does not
vary measurably with latitude. Pressure differences at constant altitude between 30° and 60° latitude, predicted from measured wind velocities, are
given as a function of altitude in column 4 of Table I, and are compared with
measured pressure differences from Pioneer probes that entered near those
latitudes, in columns 6 and 7. When the pressure difference rises above the
level of ∼ 0.5 %, the predicted and observed differences agree in magnitude,
and the agreement improves with increasing altitude, i.e., as the measured
pressure difference becomes more accurate.

TABLE I

Comparison of Pressure Differences[a]

z (km)	Observed u (m s^{-1})	p (mbar)	Eq. (5a) Δp (mbar)	$\Delta p/p$	Δp North-Day (mbar)	Δp North-Night (mbar)
0	1	92,000	−0.36	3.9×10^{-6}	0	0
10	3	47,400	−1.9	3.9×10^{-5}	+200	+ 60
20	22	22,500	−54	0.0024	+ 40	−100
30	30	9,500	−50	0.0052	− 66	−140
40	38	3,450	−34	0.010	− 48	− 73
50	57	1,030	−28	0.027	− 30	− 40
60	83	220	−17	0.079	− 18	− 21

[a]Those deduced from measured winds (Eq. 5a) with those measured by two pairs of Pioneer small probes (last two columns).

The near-equatorial Large probe pressure data, however, do not follow the above pattern. Measured pressures at 4° latitude were, at all altitudes, lower than pressures near 30° latitude. Thus, Δp is of the wrong sign for cyclostrophic balance at these latitudes. At the higher altitudes, above 60 km, the North probe-Large probe pressure differences approach the expected magnitude but the Day probe-Large probe differences are still of the wrong sign; for the latitudes involved, these differences should be ∼ ¼ as great as the North-Day probe differences, i.e., ∼ −4 or −5 mbar at 60 km. The observed differences, which are ∼ + 10 mbar at 225 mbar, are large enough (5%) to be significant. They indicate that at latitudes below 30° other terms than the cyclostrophic balance term become significant in the equation of motion and dominate the pressure difference with latitude. Thus, e.g., unsteady flow or eddy motions may be present. The meridional velocities measured at altitudes above 40 km by the equatorial probe (Counselman et al. 1980) were biased equatorward, i.e., in the sense of the measured pressure differences.

Poleward of 60°, observations of the atmosphere structure below the cloud tops have been obtained down to the 4 bar level by the radio occultation technique (Kliore and Patel 1980, 1982; Yakovlev and Matyugov 1982) and down to the 800 mbar level from the Pioneer orbiter infrared spectrometer data of F. Taylor et al. (1980). If, following Kliore and Patel (1980) and Taylor et al. (1980), we compare these data sets on pressure-temperature coordinates as in Fig. 8, we eliminate the effect of altitude uncertainties. The altitude scale shown at the right is $z(p)$ from the North probe data. At altitudes < 67 km, we see evidence in both data sets of sizable temperature contrasts, up to 30 K in the occultation data and 40 K in the infrared data, at polar latitudes relative to 52° or 55° latitude. The occulation data show these contrasts to extend through and below the clouds but to be decreasing at 40 km altitude. The two occultation profiles at 76° and 84° latitude, compared on p, T coordinates, show no

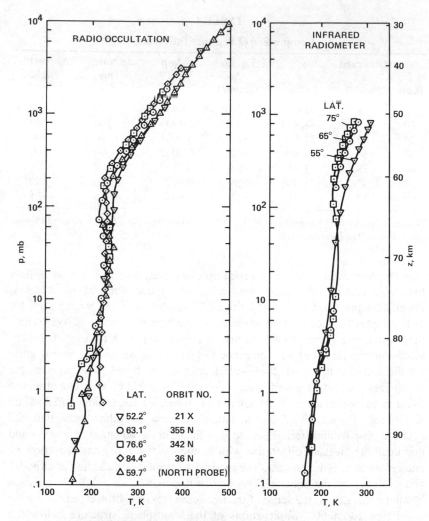

Fig. 8. Temperature contrasts at a given pressure in the polar atmosphere of Venus, defined by radio occultation and infrared soundings. The altitude scale is approximate (see text).

differences larger than a few K, although comparisons of other occultations at these latitudes show that a time dependent contrast exists there (see below). At lower latitudes (63° and 52°), the occultation data are generally comparable to the *in situ* profile from the North probe, which is included for comparison.

The infrared data at the right in Fig. 8 are temperature retrievals (rather than brightness temperatures) taken from Fig. 9, which shows zonally averaged global contours of constant temperature on log pressure, latitude coordinates. (These data were kindly furnished by F. W. Taylor and J. T.

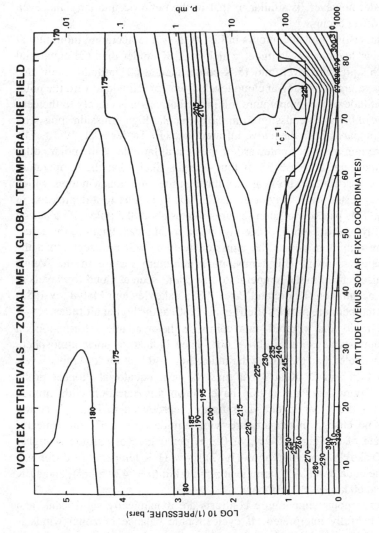

Fig. 9. Contours of constant temperature for altitudes above the 1 bar level on log pressure, latitude coordinates. These are global mean contours, averaging day side and night side, and averaging variability. The data are temperature retrievals from the Pioneer orbiter infrared radiometer (Taylor et al. personal communication). The line labeled $\tau_c = 1$ indicates the pressure level at which cloud optical depth = 1.

Schofield.) At the 800 mbar level, the infrared temperatures are offset by ~
− 30 K from the occultation data. The offset disappears at 200 mbar, and
at all lower pressures, the correspondence is quantitative. The offset is attributed to difficulties in modeling cloud effects and to the limited signal
contribution from these deeper levels. The relative variation of temperature
with latitude, however, is similar to that in the radio occultation data, even
where the offset occurs.

The interesting global contours in Fig. 9 show a temperature minimum at
the 200 mbar level at 75° latitude, which is 35 K colder than the equatorial
region at this pressure level, and 15 K colder than at 60° latitude. They also
show temperature increasing at cloud levels from 75° latitude toward the pole.

The latitude of the temperature minimum corresponds closely to the central latitude of brightest emission from the twin polar high-emission spots seen
in images prepared from the above infrared data (F. Taylor et al. 1980). The
temperature minimum lies poleward of the central latitude of the polar collar
clouds, which are at 65° ± 3°. It thus appears likely that the temperature
minimum is associated with heat loss through the emission windows which
has been suggested by F. Taylor et al. (1979b) to be due to a depression in
the clouds. This pair of emission features rotates about the pole with a period
2.7 d (F. Taylor et al. 1980). Since Fig. 9 is a zonal mean diagram, the actual
temperature minimum across the emission features will be deeper than the
average values shown. This temperature minimum feature in the Venus
clouds is also related to the temperature inversions seen at cloud-top levels in
both radio occultation (Kliore and Patel 1982; Yakovlev and Matyugov 1982)
and *in situ* temperature profiles (Seiff et al. 1980) at the higher altitudes.

These radio and infrared data comprise the available information on
thermal contrasts at latitudes above 60°. They indicate a polar atmosphere
appreciably cooler than that at low latitudes, by ~ 40 K at the 200 mbar level,
26 K at 1 bar, and 22 K at 3 bar than the near equatorial Pioneer probe
sounding. Kliore and Patel (1982) have shown this variation with latitude
explicitly for the 1 bar level. Their figure, reproduced in Fig. 10, shows
essentially no change in temperature with latitude up to 55°, and a nearly
constant rate of cooling $(\partial T/\partial \theta)_p$ of 0.9 K per degree of latitude at higher
latitudes. Yakovlev and Matyugov (1982) show 11 K temperature difference
between the equator and 75° latitude at $z = 56$ km ($p \simeq 420$ mbar), and 17 K
difference at 60 km ($p \simeq 210$ mbar).

The large polar temperature contrasts are dynamically significant, and
remain to be fully interpreted. If cyclostrophic balance of zonal winds is
assumed to be continued to the poles, with velocity decreasing as $\cos \theta$ from
93 m s^{-1} at 60° latitude to 0 at the pole, a further pressure difference of
~ 20 mbar would be expected poleward of 60° latitude at the 60 km level.
This corresponds approximately to a lowering of the 210 mbar isobar by
0.50 km, and to a mean cooling of the atmosphere below 60 km by
~ 1 %. The observed cooling at levels above 40 km is as much as 15%.

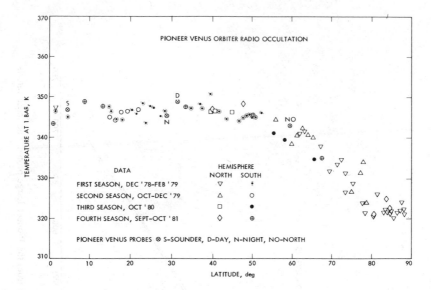

Fig. 10. Development of the polar temperature contrast with latitude, as defined at the 1 bar level by radio occultation data.

Furthermore, the altitude contraction of the atmosphere near the pole can be estimated from the temperature data for the interval from 40 to 60 km, and it is ~ 1.3 km by itself. Hence, it appears that simple cyclostrophic balance of zonal winds will not be consistent with these data. The rotating dipole configuration of the hot spots centered at 75° to 80° latitude also suggests a more complex dynamics. The symmetry of the temperature contours below the 100 mbar level ~ 75° latitude (Fig. 9) suggests that the twin hot spots may be a pair of vortices. Such vortex configurations are sometimes found on either side of the Earth's poles at midwinter (see, e.g., the isobar height contours given by Palmen and Newton [1969, Figs. 3.1 and 3.4]). In any case the contrasts poleward of 60° appear to be associated with a more complex dynamical configuration than the zonal flow prevalent below 60° latitude.

D. Thermal Fine Structure and Static Stability

There is a great deal of interesting fine structure in the temperature profiles obtained by the Pioneer probes during descent, especially by the near equatorial Large probe which descended slowly below 64 km and took a sample every 2 s. In Fig. 11, every third data point from this probe is plotted against altitude, and the profile is compared with the adiabat which intersects

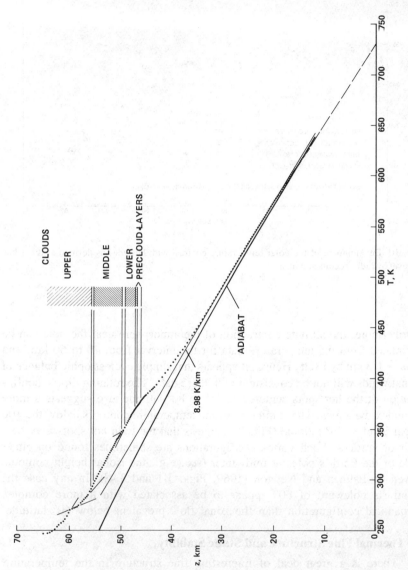

Fig. 11. Details in the temperature variation with altitude shown by data from the Pioneer Large probe. The lapse ra.e is generally stable, and exhibits changes at the cloud boundaries.

the data at 15 km altitude and with a constant lapse rate line having $dT/dz =$ -8.98 K km^{-1} which falls close to the measured points over the altitude range from 19 to 32 km. (An adiabatic lapse rate of 8.98 K km^{-1} corresponds to $T =$ 466 K or $z = 33$ km.) Locations of the cloud layers as determined from the PV Large probe nephelometer data by Ragent and Blamont (1980) are shown. This graph gives evidence that the lower atmosphere of Venus is generally statically stable except for limited altitude intervals. In the interval from 19 to 26 km, the measured lapse rate roughly parallels the adiabat, but is still slightly stable, while above and below this, it may be seen that the measured lapse rate is conspicuously stable (subadiabatic). Another interval which is approximately adiabatic occurs in the middle cloud (Seiff et al. 1979b). Outside of these two intervals, the data show an atmosphere which is stable rather than convective, whereas prior to measurement, a convective atmosphere had been generally expected.

Another way of viewing this profile is to observe that the cloud layers are warm relative to the extended temperature slope of the deep atmosphere (Seiff et al. 1979b). The warming, i.e., the local temperature difference from the -8.98 K km^{-1} line, is relatively large, ~ 28 K at 48 to 52 km. Phenomenologically, the cause of the warming appears to be absorption (trapping) of infrared radiation from the deep atmosphere. The radiatively warmed clouds then radiate downward and warm the atmosphere immediately below. This downward warming dies off with distance below the clouds, causing a stable lapse rate down to the 32 km level. The stable lapse rates also may be explained in the context of a 2-dimensional model of the global heat balance and circulation (Stone 1974). A brief discussion appears at the end of this section.

Changes in lapse rate occur near the independently determined boundaries of the cloud layers, and thus confirm the change in optical properties of the clouds across the layer boundaries. Thus, between the lower and middle cloud, there is a small nearly discontinuous change in temperature and slope. There also appears to be ~ 1 K discontinuity associated with the lowest precloud layer. At the lower boundary of the upper cloud a sharp change in lapse rate occurs.

Within the upper cloud where the temperature passes through 272 K, between the 58 and 60 km levels, the lapse rate decreases in a layer ~ 1 km deep to 2.3 K km^{-1}, and within this layer there is an interval 0.2 km deep in which the lapse rate is zero (there were three consecutive readings of 272.0 K). It has been suggested that this plateau may indicate a phase change reaction involving water or concentrated H_2SO_4 (Seiff et al. 1979b). The freezing temperature of 80% H_2SO_4, e.g., is 270 K, and the change from the liquid to the solid state should tend to hold temperature fixed at the freezing level until all cloud droplets are frozen. This may be the significance of the temperature plateau. This is also the altitude level above which the Mode 3 particles of Knollenberg and Hunten (1980) are not detected, while

Mode 1 increases sharply in number density; these changes also may affect the temperatures.

The comparison of the temperature data from Veneras 10, 11, and 12 with the adiabat through 737 K at 92 bar is shown in Fig. 12 and has also been given by Avduevsky et al. (Chapter 12). This figure, like Fig. 11, indicates a stable atmosphere below the clouds. Estimates of the degree of stability from Fig. 12 are given in Table II. The indication of general stability has one exception, in the Venera 10 probe data, in the layer from 30 to 35 km, where the temperature lapse rate is indicated to be unstable by ~ 0.1 K km^{-1}. (In Table II the lapse rate for the interval from 30 to 40 km altitude is from the Venera 11 and 12 data.) The stability near the surface may be characterized as near zero. Contrary to the Pioneer probe data, these data indicate the layer of the middle clouds is stable. However, in consonance with the Pioneer probe data, there is a highly stable layer just below the clouds. The stability inferences from Veneras 10, 11, and 12 are compared graphically with those from the Pioneer Large (Sounder) probe in Fig. 13a. The comparison is remarkably quantitative in spite of the sizable difference in local Venus time. However, the altitude resolution available from the data in Fig. 12 does not permit the comparison with detailed features shown by the Large probe data. Avduevsky et al. (Chapter 12) confirm that Veneras 10, 11, and 12 indicate a generally stable atmosphere, unlike Veneras 4, 5, 6, and 8, and suggest the possibility that measurement errors could have affected accuracy of lapse rates from the earlier probes.

Stability inferences were also made by Yakovlev and Matyugov (1982) from radio occultation measurements with Veneras 9 and 10. They found evidence in some temperature profiles for a stable layer beginning just below 50 km and extending to the lower limit of measurement, on the night side of the planet. In other profiles, this feature was less prominent, or absent. They also reported a slightly unstable (convective) layer in the clouds, extending from 51 to 58 km on the night side, and from 50 to 60 km on the day side at 62° latitude. At low latitudes on the day side ($\sim$15°), the convective layer was confined to lower altitudes, below 55 km.

Stability plots ($dT/dz - \Gamma$) (z) derived from the four Pioneer probe tem-

TABLE II

Stability Data Derived from Veneras 10, 11, and 12 Data in Fig. 12.

z (km)	$dT/dz - \Gamma$ (K km^{-1})
0–10	0.3
10–20	0.5
20–30	0.9
30–40	1.1
40–50	2.4
50–55	0.7

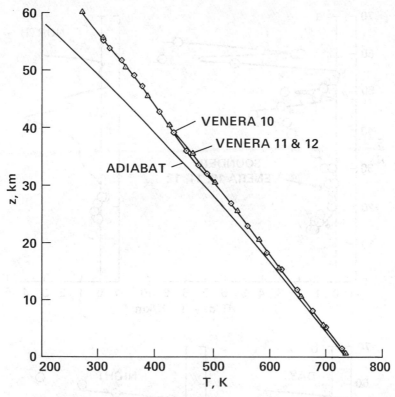

Fig. 12. Comparison of Venera 10, 11, and 12 data with the adiabat that passes through 737 K at the surface. The atmosphere is again seen to be generally stable, although essentially neutral in the lowest few kilometers.

perature profiles by Seiff et al. (1980) and shown in Fig. 13 exhibit strong similarities at the four widely separated sites, two at night and two in the Venus morning. Features found in all four profiles are:

1. The deep stable layer beneath the clouds;
2. A convective layer in the middle cloud;
3. A near-adiabatic layer found at ~ 20 to 30 km altitude, but displaced toward lower altitudes at night and at 60° latitude. The adiabatic layer is also indicated to be deeper at midnight than in the morning;
4. A rapid transition to strong stability in the upper cloud. This transition level we define as the tropopause (see below);
5. A tendency for stable lapse rates below 15 or 20 km to the end of the Pioneer probe soundings at 12 km. This tendency is also clearly apparent in Fig. 11 below 20 km from the change in slope of $T(z)$ relative to that of the adiabat. If this change is real, it may be a result of coupling of dynamics (vertical flow) with radiative transfer.

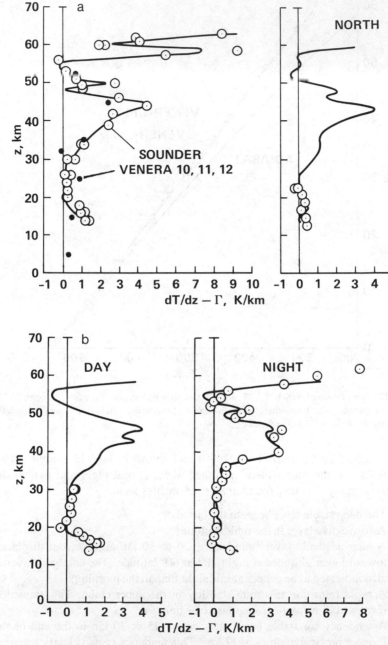

Fig. 13. Stability of the Venus atmosphere as a function of altitude determined from Pioneer Venus probe data at four widely separated locations: Large (Sounder) probe; North probe; Day probe; Night probe. Venera 10, 11, and 12 data from Table II show substantial agreement with the Large probe data in the graph at the top left.

The adiabatic lapse rates used in evaluating the stability (Seiff et al. 1980) are reproduced in Fig. 14. They allow for gaseous imperfections, and were calculated for both pure CO_2 and a mixture of 0.965 CO_2 and 0.035 N_2 from the equation

$$\Gamma = -\alpha T \frac{g}{c_p} \qquad (6)$$

where

$$\alpha = -\frac{1}{\rho} \left(\frac{\partial \rho}{\partial T} \right)_p \qquad (7)$$

and $\alpha T = 1$ for a perfect gas. The adiabat in Figs. 11 and 12 was obtained by integrating lapse rates from Eq. (6) and Fig. 14 for the CO_2-N_2 mixture, with $T_0 = 737$ K.

Returning to the temperature profile of Fig. 11, a close examination does not reveal any turbulent temperature fluctuations, i.e., scatter of points about a best fitting curve, greater than the resolution of 0.2 K. When all the points are plotted to an enlarged scale, the variation is still smooth. There is a region with some variation in slope, i.e., small amplitude waviness, between 44 and 50 km. This is in the layer of highly stable lapse rate, where convective turbulence would not be expected to occur. Wind shear turbulence requires that Richardson number

$$Ri = \frac{g}{\Theta} \frac{\partial \Theta}{\partial z} \Big/ \left(\frac{\partial u}{\partial z} \right)^2 \qquad (8)$$

be less than the critical value $\sim 1/4$. This is generally a layer of small wind shear except for locally high shear ~ -3 m s^{-1} km^{-1} between 46 and 48 km altitude (Counselman et al. 1980). The potential temperature Θ is also changing rapidly in this interval, as may be seen from Fig. 13, where $d\Theta/dz = dT/dz - \Gamma$ is plotted as a function of altitude. Richardson numbers calculated from these data are ~ 2 to 6. Hence, the shear layer should theoretically be stable against breakdown into turbulence. However, clear air (shear) turbulence is sometimes observed in the Earth's atmosphere with Richardson numbers of this magnitude. Woo et al. (1982) identify a lower turbulent layer on Venus near 45 km and attribute it to shear turbulence. They find the temperature fluctuation amplitude (rms) which is associated with this turbulence to be ~ 0.1 K for a turbulence scale of 100 m. Hence this finding is not inconsistent with the directly observed spatial variations of temperature. Yakovlev and Matyugov (1982) also detect turbulent fluctuations from radio occultation data peaking between 45 and 50 km on the night side. Both groups identify a second turbulent layer with peak fluctuating intensity just above 60 km.

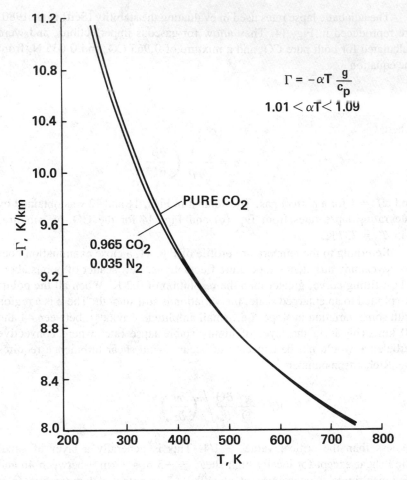

Fig. 14. Adiabatic lapse rates calculated from Eq. (6) for pure CO_2 and a 3.5% N_2 mixture. The adiabatic lapse rate varies strongly with temperature because c_p for CO_2 varies strongly.

In the enlarged scale graph of temperature in the interval below 50 km, there are two well-defined and nearly discontinuous slope changes and some small amplitude waves with a vertical wavelength of 0.5 to 0.8 km. The slope changes are found to correlate with cloud-layer boundaries. Thus, the largest slope change, accompanied by a small temperature offset, occurs in the gap between the lower and middle clouds, at 49.66 to 49.90 km, where the lapse rate falls locally to 3.94 K km^{-1}, producing the narrow stable peak seen just below 50 km in the Large probe stability profile (Fig. 13). Another slope discontinuity occurs at 46 km, near the bottom of the first detached cloud layer. The wavy structure continues below this, down to 44 km, with amplitude < 0.5 K. In this region, the cloud particle spectrometer detected

$\sim 4 \times 10^5$, 3 μm particles m^{-3}, a particle density lower by > 2 orders from that in the lower cloud, but still a sizable particle count. The wavy structure is possibly associated with these particles and gradual variations in their density (an inhomogeneous distribution) rather than with turbulence, since it is more ordered than random.

In the convective region of the middle cloud, there are again no temperature fluctuations > 0.2 K. In the near-adiabatic layer from 10 to 30 km, the same is true. There is some small amplitude waviness in $T(z)$ in the deeper layer, with amplitude ~ 0.2 K, and one wave peak of 0.5 K. These waves are irregular but ~ 1 km in length. The sensor response times used for these observations were ~ 1.25 s above 50 km and ~ 0.01 s at 19 to 30 km. These permit response to turbulent fluctuations at scales > 20 m in the cloud and 0.25 m in the deeper layer. The data thus appear to indicate that there are no turbulent temperature fluctuations > 0.2 K in the atmosphere of Venus below 60 km at the morning terminator. This conclusion supersedes that of Seiff et al. (1980).

E. Theoretical Models

Theoretical models to account for the high temperatures in Venus's deep atmosphere on the basis of the greenhouse effect date back to Sagan (1960a), soon after the high temperatures were detected from microwave emission. The models have been progressively improved in methodology (Pollack 1969b), and made to reflect the latest description of the chemical, particulate, and optical characteristics of the atmosphere. Sagan's analysis indicated a surface pressure of 4 bar, and found that a water vapor fraction of 0.05 % to 0.23 % was needed to close the windows in the CO_2 absorption spectrum sufficiently to maintain a surface temperature of 600 K. Pollack (1969a) assumed ice clouds and used Venera 4 and Mariner 5 data to select gas composition, but calculated the solar heating. Surface pressure was, at that time, still uncertain, because of the controversy about whether Venera 4 survived down to the surface, so several models with different p_0 were analyzed. Pollack found that ~ 0.5 % water vapor would maintain observed temperatures near the surface under the assumed conditions. Pollack and Young (1975) assumed 75 % H_2SO_4 clouds, a constant water vapor mixing ratio of 0.003, and used the Venera 8 data on downward component of the solar flux to define the solar deposition profile. The middle cloud layer was assumed to extend from the 1 bar level to the 6 bar level, whereas it is now known to lie between 0.45 and 1 bar. With these assumptions, and with a cloud particle diameter of 6 μm, a temperature profile was constructed which became superadiabatic below 16 km, where temperature was 600 K. Below this level, the atmosphere was assumed to be convective down to the surface. At higher altitudes, temperatures were generally within 50 K or better of the Venera 8 profile.

The calculations of Pollack et al. (1980a) reflect the Venus chemical and particulate environments and the solar heating measured by the Pioneer and

Venera 11 and 12 probes. There is still some thermally important uncertainty in water vapor concentration, solar heating, and particle distributions, so a variety of cases was considered. The calculation, like those of earlier papers, is one-dimensional and uses global average solar heating. The global net solar fluxes of Tomasko et al. (1980a) were used as an input to the calculation. The general conclusion is that combinations of the input parameters can be selected which give relatively close matches to the observed temperature structure and lapse rates. The combination is not, however, necessarily unique. This is illustrated in Fig. 15 which is based on the calculations of Pollack et al. (1980a). This figure differs from their Fig. 2 in which the altitude-pressure relation was assumed to be model independent. In Fig. 15, $z(p)$ is allowed to vary with $T(z)$. Nearly parallel adiabats are then obtained from the several models.

Four representations of the Venus environment are shown. In all four, the lapse rates for radiative equilibrium become superadiabatic below some level, indicated by the + symbols, and the atmosphere below this level is represented as adiabatic. In contrast, the data show a slightly stable atmosphere in this altitude range, as is evident from the Pioneer Large probe profile in Fig. 15 and as discussed earlier. Above 35 km, two of the four cases show a stable change in lapse rate, one of which follows the data closely from 44 km to 60 km and which therefore simulates approximately the observed stability.

The models that exhibit qualitatively the observed stability, models (a) and (b), both postulate the existence of a field of microparticles above the clouds, referred to as Mode 0 particles. These are given mean particle diameters of 0.0065 μm (65 Å) and an optical depth of 0.15 in the visible. They are assumed to extend from 64 km to the upper limit of the calculations, $\sim$ 85 km. The existence of particles of this size was first postulated by Suomi et al. (1980) to prevent the escape of more infrared radiation to space than is indicated by the overall radiative equilibrium of Venus. The calculations represented in Fig. 15 indicate that the Mode 0 particles are necessary to the observed warming of the atmosphere below the clouds and to the existence of the deep stable layer.

Two different levels of solar heating, and two vertical distributions of H_2O vapor mixing ratio are also represented in Fig. 15. Upper limit solar heating refers to the upper end of the uncertainty band in the measurements of Tomasko et al. (1980a), whereas nominal refers to the measured levels. The upper limit heating is $\sim$ 1.5 times the nominal in the clouds and $\sim$ 1.35 times the nominal below 40 km. The lower H_2O mixing ratio profile is based on the spectrophotometer data of Moroz et al. (1979c), while the higher level is from gas chromatograph measurements of Oyama et al. (1980a).

All of these combinations yield temperatures of the correct magnitude. In two cases, the entire atmosphere below the clouds is found to be in convective equilibrium rather than in radiative equilibrium. Hence, for these cases, the role of the greenhouse model is to establish conditions suitable for convection

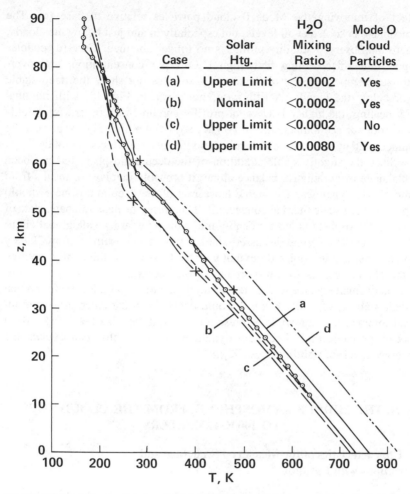

Fig. 15. Theoretical greenhouse models derived from Pollack et al. (1980*a*). These are radiative, convective equilibrium models in which it is assumed that the atmosphere is adiabatic
below the + symbols, where the radiative equilibrium state becomes superadiabatic.

to occur, and theoretically it is convection which controls the temperature
profile. A second important role of radiative transfer in such a case is that the
radiative imbalance is the heat source available to drive convection and/or
general circulation.

Case (a) best corresponds to the observations in and just below the clouds
(see Fig. 15). It represents the dry atmosphere with Mode 0 particles present
and upper limit solar heating. Case (b) changes only the solar heating level to
nominal, and this leaves the clouds and subcloud region cooler than observed,
while preserving the curvature seen in observed profiles. Case (c) shows the

effect of removing the Mode 0 cloud particles relative to case (a). The atmosphere is cooled at all levels, but especially in and just below the clouds, where the observed stability profile is no longer obtained. Case (d) couples upper limit solar heating with high H_2O vapor concentration, and over-estimates temperatures at all levels. It also does not show the deep stable layer below the clouds. A fifth case, not given in Fig. 15, with nominal solar heating, the higher H_2O vapor profile, and no Mode 0 particles, yields a too cool surface temperature of 699 K, so that the high H_2O vapor can be compensated by reducing solar heating and the thermal effects of clouds.

Since the slightly stable condition of the deep atmosphere has not been explainable from radiative balance alone, it probably involves or results from atmospheric dynamics. The stable lapse rate of the deep atmosphere should suppress small-scale thermal convection. Therefore, the near adiabatic state of the deep atmosphere is most probably maintained by large-scale global circu-lation. Stone (1974) has discussed stable lapse rates resulting from Hadley circulations, and has pointed out that a stable lapse rate is inherently required if the Hadley cell is to transport heat from the equator to the poles. For extremely small equator-to-pole temperature contrasts (0.02 K), he found that a static stability of ~ 0.1 K km^{-1} would occur. With the much larger merid-ional contrasts now known to be present ($\sim$25 K at the 1 bar level; Fig. 10), it appears probable that large-scale dynamics (Hadley cells) can explain the observed levels of stability of ~ 1 K km^{-1}.

II. THE MIDDLE ATMOSPHERE, FROM THE CLOUD TOPS TO 100 KILOMETERS

A. Characteristics of the Measured Structure and Variations with Latitude

The temperature structure of the atmosphere above the clouds to 100 km was measured by the probes and orbiter of the Pioneer Venus mission by use of three independent techniques. Representative data are shown in Figs. 16 and 17. The temperature data derived from *in situ* measurements of the atmo-spheric density structure (Seiff and Kirk 1982) are reproduced in Fig. 16, which also shows preliminary results from Veneras 11 and 12, obtained by a similar technique (Avduevsky et al. 1979). Data obtained by remote sensing from the Pioneer Venus orbiter by means of infrared radiometry (F. Taylor et al. 1980) and radio occultation (Kliore and Patel 1982) are shown in Fig. 17. Data obtained on Veneras 9 and 10 by radio occultation of Yakovlev and Matyugov (1982) are very similar to those of Kliore and Patel.

The infrared temperature retrievals shown were derived from Fig. 9 for latitudes of 30°, 65°, and 75° and therefore represent zonal mean temperature profiles, averaging day and night and also averaging the temporal variability.

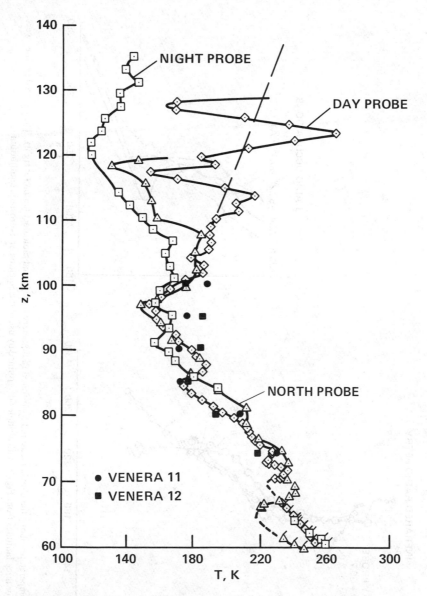

Fig. 16. Temperatures above the clouds derived from measurements of deceleration during entry of the three Pioneer small probes and Veneras 11 and 12. Flagged symbols show temperatures directly sensed during descent.

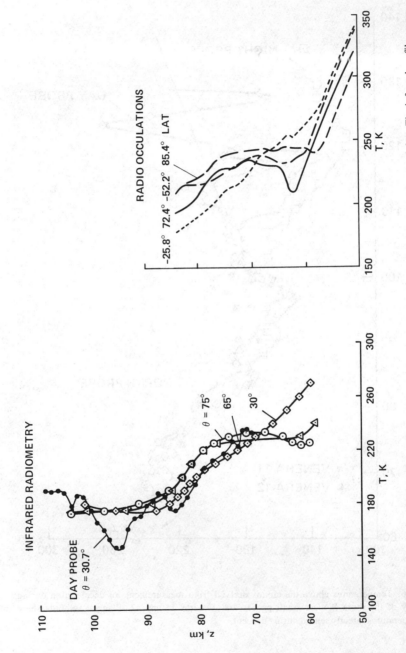

Fig. 17. Temperatures above the clouds measured by infrared radiometery (at left) and radio occultation (at right). The infrared profiles were derived for three latitudes from Fig. 9, with altitudes determined by use of the equation of hydrostatic equilibrium.

Altitudes shown were determined by graphical integration of the $T(p)$ data in the equation of hydrostatic equilibrium

$$z = z_{ref} + \int_{p}^{p_{ref}} \frac{RT}{g} \, d(\ln p) \, . \tag{9}$$

The altitude at the reference pressure level was taken from probe descent data discussed earlier (extrapolated for the highest latitude). The Day probe temperatures are included in Fig. 17 for comparison with the infrared temperature retrievals. The probe measurements generally agree within ~ 10 K with the zonal mean infrared profile at the same latitude, except at 95 km where a peak disagreement of ~ 30 K occurs. Close examination reveals the Day probe instantaneous profile to be roughly oscillatory about the infrared mean profile.

The occultation results, which display some variability, are four arbitrarily selected profiles from Kliore and Patel (1982) at latitudes near $-26°$, $-52°$, $72°$, and $85°$. Altitudes here were defined by the ray geometry and spacecraft position as radial distances from the planet's center and are believed to be accurate within ~ 0.5 km. A planetary radius of 6052 km was uniformly used for altitude reference. These four sets of occultation data show temperature profiles generally similar to the infrared and probe data, and indicate a distinct latitude dependence of the profiles in and above the clouds. Above 70 km, the polar atmosphere is warmer than that at low latitudes, a condition which is also evident at $p < 20$ mbar in Fig. 9. Very near the pole, the atmosphere above the clouds is almost isothermal at 243 K up to 73 km.

Temperature lapse rates in this altitude range are ~ 3 or 4 K km^{-1} from the 70 km level to ~ 85 km, at latitudes up to $\sim 60°$ (Fig. 16). At latitudes between 45° and 80°, inversions occur at the 60 to 70 km level which are variable in time, and are apparently associated with the intermittent polar collar cloud and the rotating pair of infrared emission windows centered at $\sim 75°$ latitude (F. Taylor et al. 1980; Kliore and Patel 1982), to be further discussed below. Above 85 km, the atmosphere is roughly isothermal up to 100 km (Figs. 16 and 17). Since the adiabat at these altitudes has a slope > 11 K km^{-1}, a lapse rate of 3 or 4 K km^{-1} is highly stable; the isothermal and inversion layers are even more stable. Because of this stability and the sharply stable transition in stability which occurs in the upper cloud layer, the overlying region is analogous to the Earth's stratosphere. The sudden stable change in the lapse rate in the upper clouds, seen in the stability diagrams (Fig. 13), is also visually evident in the temperature data of Fig. 18. It occurs near or just below the 60 km level, at the top of the middle cloud which may be identified as the tropopause level. A pressure level at the tropopause of ~ 200 to 400 mbar was determined from $T(z)$ measured by the four Pioneer Venus probes. Kliore and Patel (1982) studied a large number of occultation profiles and found the tropopause pressure level to decrease from ~ 450 mbar at the equator to an apparently double-valued minimum of ~ 200 mbar or ~ 125 mbar at 50° to 75° latitude (the region of the rotating polar collar clouds).

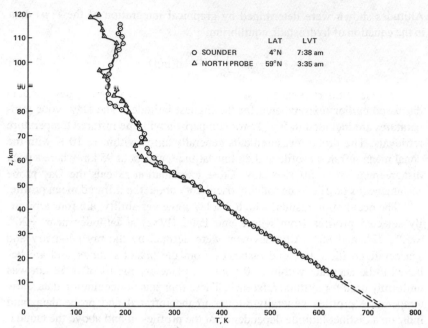

Fig. 18. $T(z)$ from the planet surface to 120 km. The pronounced change in lapse rate at the top of the middle cloud (near 58 km) marks the tropopause: O, Large (Sounder) probe; Δ, North probe. Compare also with Figs. 11 and 13.

The tropopause pressure increases again to $\sim$ 250 mbar above 80° latitude. Yakovlev and Matyugov (1982) give the tropopause pressure level of 300 $\pm$ 20 mbar on the night side at latitudes below 34°. The tropopause in Earth's atmosphere occurs at $\sim$ 225 mbar, and in the atmosphere of Jupiter, at $\sim$ 200 mbar (Hanel et al. 1979). Thus the pressure at the tropopause is around 200 mbar for all three planets.

The inversion layer just above 60 km is only faintly present at equatorial latitudes (Figs. 17 and 18), but becomes prominent with increasing latitude. Temperature rise in the inversion increases to $\sim$ 30 K near 70° latitude (Kliore and Patel 1980, 1982; Yakovlev and Matyugov 1982). Pressure at the lower boundary of the inversion is found from radio occultation measurements to decrease with increasing latitude to a minimum $\sim$ 90 mbar at 60° to 70° latitude (Kliore and Patel 1980), indicating movement to higher altitude. The *in situ* and radio occultation profiles show this inversion to be quite sharp with high positive lapse rates and rapid onset. Because of limited altitude resolution, the infrared temperature retrievals tend to smooth the inversion, and show it as a nearly isothermal interval at 65° latitude, while showing a shallow temperature minimum at 75° latitude (Fig. 17). However, the global contour diagram of the isotherms, obtained from the infrared radiometer (Fig. 9) is very instructive in showing the organization of temperatures in the middle atmosphere, and the temperature minimum $\sim$ 225 K occurring at the 150 mbar level at 75° latitude. It is clear that this minimum and the inversions are related.

It had been proposed that the inversions are associated with changes in radiative equilibrium that occur in the transition from the upper cloud to the "clear" atmosphere above (Seiff et al. 1980), and this may explain the mild inversions at the lower latitudes. The temperature minimum at 75° latitude now appears to have another explanation. As noted earlier, this is the latitude of maximum brightness of the infrared emission windows near the pole (see Plate 3 of F. Taylor et al. 1980). Hence, it appears that thermal emission to space may account for the temperature minimum near 60 km. The latitude of the temperature minimum is, at the same time, somewhat higher than the central latitude of the polar collar clouds, shown in Plate 3 of F. Taylor et al. (1980) to be at 65° ± 3°. These clouds may be in some way associated with the emission windows, but they do not directly correlate with the temperature minimum, being displaced in latitude. Therefore, it is suggested that the deep inversions at high latitudes, which reflect the presence of the temperature minimum, could be a result of cooling near the 60 km level effected by the rotating pair of infrared emission windows and radiative absorption or warming at higher levels. These emission windows have been attributed by F. Taylor et al. (1979b) to cloud depressions near the poles resulting from downward motion in Hadley cells. Yakovlev and Matyugov also found a "cold...layer...in the latitude interval of 66°–74° at heights between 54 and 80 km, [with] a temperature 10–15 K lower than the surrounding regions." This cold layer corresponds to Taylor's temperature minimum on its lower latitude side, and probably encompasses the intermittent polar collar clouds.

Above 70 km to $\sim$ 90 km, the middle atmosphere is shown to be generally warmer at high latitudes than near the equator (F. Taylor et al. 1979b) by the three sets of experiment data in Figs. 17 and 16. This is an unexpected direction of variation, since normally the equatorial regions would be expected to be warmer because of the greater solar heating there. The reasons for this observed behavior are not yet clear. Heating by the infrared emission from the lower atmosphere through the polar emission windows is one possibility. Another is that the temperatures are dynamically controlled.

Since the winds at the cloud tops are primarily zonal and cyclostrophically balanced by meridional pressure gradients (see above), some upward continuation of this condition might be expected. This possibility has been examined from the pressure data below 100 km which are shown in Figs. 19 and 20, the former from the entry probes and the latter from the orbiter infrared experiment. Pressure falls approximately exponentially from 200 mbar at 60 km to $\sim$ 0.025 mbar at 100 km, with a mean scale height of $\sim$4.5 km. (There is a small systematic deviation between the two data sets above 90 km, which leads to a 25 % disagreement in pressure at 100 km, the infrared data giving 0.030 mbar, and the entry data 0.024 mbar.) Within each data set,

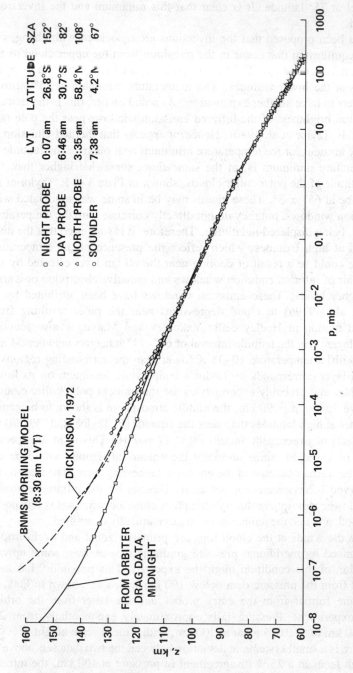

	LVT	LATITUDE	SZA
☐ NIGHT PROBE	0:07 am	26.8°S	152°
◇ DAY PROBE	6:46 am	30.7°S	82°
△ NORTH PROBE	3:35 am	58.4°N	108°
○ SOUNDER	7:38 am	4.2°N	67°

BNMS MORNING MODEL (8:30 am LVT)

DICKINSON, 1972

FROM ORBITER DRAG DATA, MIDNIGHT

Fig. 19. Pressure data above the clouds from three Pioneer probe, orbiter, and bus experiments. Below 100 km, pressure differences are small, but pressures sensed by the North probe at 60° latitude were consistently lower than those of the other three probes (Night, Day and Large or Sounder). Above 100 km, very large diurnal pressure contrasts develop.

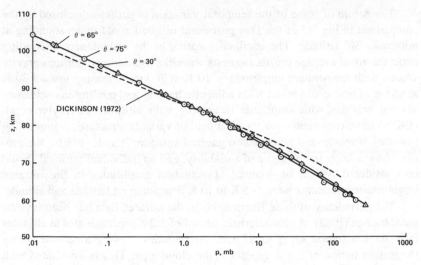

Fig. 20. Pressure data as a function of altitude from the infrared soundings. Below 80 km, pressure decreases with increasing latitude, consistent with the probe soundings.

pressure varies with latitude, the pressures at high latitude being smaller below ~ 85 km. At 70 km $\Delta p/p$ is ~ 20 % in both data sets, where Δp is pressure difference between latitudes of 30° and 60°. The pressure difference decreases with increasing altitude, and vanishes at 85 to 90 km. The pressure gradients are in the direction required to maintain cyclostrophic balance, and the zonal winds which they would balance were shown in Fig. 7 above 60 km. The symbols are derived from the probe pressure data (Seiff and Kirk 1982), and the solid curves above 70 km, from the infrared pressure data (Schubert et al. 1980a). The calculated winds peak near the cloud tops and decrease toward zero at ~ 90 km. The temperature increasing toward the poles thus is diagnostic of decreasing zonal winds above the clouds, just as temperatures decreasing toward the poles in the clouds are associated with the increasing zonal wind speeds there. Further discussion appears in Chapter 21 by Schubert.

B. Diurnal and Temporal Variations

Close examination of the Day and Night probe temperatures in Fig. 16 leads to the inference that temporal variability is more significant than diurnal variations, since day and night profiles do not separate cleanly below 100 km, but rather, they cross one another several times. The infrared temperature retrievals shown in Fig. 9 have not been separated to define the mean dayside-nightside variation, but infrared brightness temperatures from the same data have been used to examine day-night differences on individual orbits. These show differences no greater than a few K, and again, temporal variations appear larger than diurnal below 100 km (F. Taylor et al. 1979b, 1980).

The nature of some of the temporal variation is perhaps disclosed by the comparison in Fig. 17 of the Day probe and infrared zonal mean sounding at nominally 30° latitude. The oscillatory nature of the instantaneous sounding about the zonal average profile suggests wavelike oscillations, perhaps gravity waves, with temperature amplitude $\sim$ 10 K at 70 km increasing toward 30 K at 95 km. These could result from adiabatic heating and cooling in oscillatory vertical motions with amplitude increasing with altitude. F. Taylor et al. (1979b) have commented on "a great deal of variable structure . . . suggesting planetary waves . . . and other, less organized structure"; and, (1980) "the profiles show a fairly large degree of variability, and no individual retrieval should be considered typical of Venus." Fluctuation amplitudes in the infrared brightness temperatures were $\sim$ 5 K to 10 K depending on latitude and altitude.

The periodicity of these fluctuations in the infrared data has been investigated by Apt (1982). A characteristic period of 5.3 d predominates at altitudes from 65 km (cloud tops) to 87 km. At latitudes $> 70°$, a second strong fluctuation period of 2.9 d appears at the cloud tops. This is associated with the rotating polar dipole, the pair of infrared emission windows, which has been observed to rotate with a 2.9 d period (F. Taylor et al. 1979b). All fluctuation amplitudes diminish toward the equator, and they are just perceptible at latitudes of 10° to 20°, on the order of 0.25 to 0.1 of the amplitudes seen at 60° to 70° latitude. Apt has suggested that the oscillations may be thermal tides driven by temperature changes in the middle clouds at the level where the rotation period is 5.3 d. These temperature changes are attributed to solar heating of the atmosphere after it crosses the morning terminator.

Woo et al. (1980, 1982) and Yakovlev and Matyugov (1982) have found that the amplitude of the turbulent fluctuations which they observed in radio occultations just above 60 km increases at latitudes from 60° to 85° relative to those near the equator. This suggests a possible relationship between these turbulent fluctuations and Apt's oscillatory amplitudes, which also increase poleward. The latter are, however, of such long periods that they would not ordinarily be classified as turbulence. The long-period oscillations could conceivably give rise to shorter period disturbances, however; or, the radio occultations, by sampling over large planetary distances, could be sampling regions extending from peaks to adjoining troughs of the long-period oscillations which, if they are not perfectly regular, would thereby indicate spatial variability or turbulence.

C. Theoretical Models

Theoretical models for this altitude range and extending beyond 100 km through the mesosphere and thermosphere were published by Dickinson in 1972 and revised 4 yr later (Dickinson 1972, 1976). These models were for radiative-conductive equilibrium in a pure CO_2 atmosphere which was free of cloud or haze particles. The radiative model was detailed and thorough in its consideration of the molecular physics and radiative processes of the CO_2 molecule, including nonequilibrium aspects. Dickinson calculated a global mean model, which, in light of the small diurnal variations observed in the middle atmosphere, should be directly comparable to the temperatures measured at latitudes below 60°. This comparison has been made by Seiff and Kirk (1982) and is reproduced in Fig. 21. The model temperatures below 100 km are very similar to the measured profiles and there is little difference between the 1972 and 1976 models in this altitude range. Theoretical values agree closely with the data from 90 to 100 km and below 70 km. Between 70 and 90 km, the theoretical temperatures are ~ 25 K cooler than those observed. This may be due to Dickinson's neglect of the particles now known to exist in this region of the atmosphere (Kawabata et al. 1980), and also at higher latitudes, to neglect of the effects of the polar emission windows. Relative to the first of these possibilities, Tomasko et al. (1980*b*) have shown that a major fraction of the solar heating occurs in the altitude range above 66 km extending to ~ 90 km, and is associated with cloud particles or aerosols.

Pollack et al. (1980*a*) have modeled the altitude range from 60 to 85 km including aerosol absorption of solar radiation, and for cases (a) and (b) in Fig. 15, the model temperatures agree with observed temperatures to within 15 or 20 K.

A comparison of pressures from Dickinson's (1972) model with observed pressures is included in Figs. 19 and 20. The model pressures are in relatively close correspondence with the observations. The pressures are fully determined by the theoretical temperature profile through the equation of hydrostatic equilibrium, given the pressure at a single reference level. For the reference level, Dickinson chose the altitude of the Mariner 5 ionization peak of 141 km to which he assigned the pressure 2×10^{-6} mbar from his theoretical ionosphere model. This choice gives a reasonably close prediction of observed pressures above 80 km, and has been well supported by data at 140 km taken at the morning terminator, as we shall see in the next section.

III. THE UPPER ATMOSPHERE ABOVE 100 KILOMETERS

A. The Measured Structure and Diurnal Variation

In situ densities and density scale height data, from which upper atmosphere temperatures are inferred, were obtained by Pioneer Venus orbiter and bus experiments at altitudes from 130 to over 200 km (Keating et al. 1979*a,b*, 1980; Niemann et al. 1979*a*, 1980*b*; von Zahn et al. 1979*b*, 1980). Densities were measured by two independent techniques, from the aerodynamic drag force experienced by the orbiter, and from species number densities measured by two neutral mass spectrometers, one on the bus and one on the orbiter. While the bus neutral mass spectrometer (BNMS) sampled the atmosphere at a single local Venus time (LVT), the orbiter during the course of its mission measured densities over 24 hr of local solar time at a latitude ~ 16°N. The 24 hr temperature data from the orbiter drag experiment are

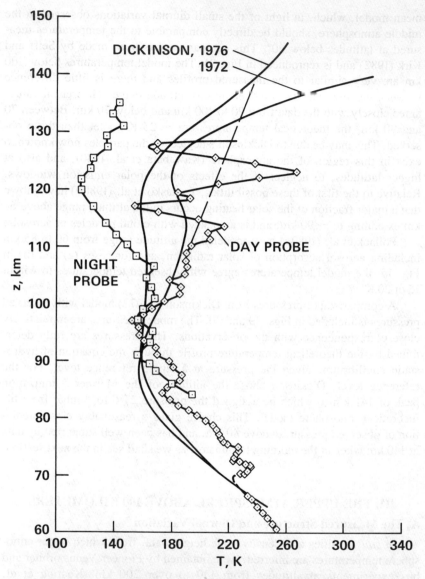

Fig. 21. Comparison of Dickinson's theoretical temperature models for the global mean atmosphere with temperature profiles above the clouds measured by the Pioneer small probes.

shown in Fig. 22 for altitudes > 140 km and those from the ONMS in Fig. 23 for exospheric altitudes > 180 km.

Two features of these data stand out: one is the low temperature in the exosphere, somewhat lower even on the day side than the 350 K to 400 K inferred from Mariner 10 ultraviolet data (Kumar and Hunten 1974; Broadfoot

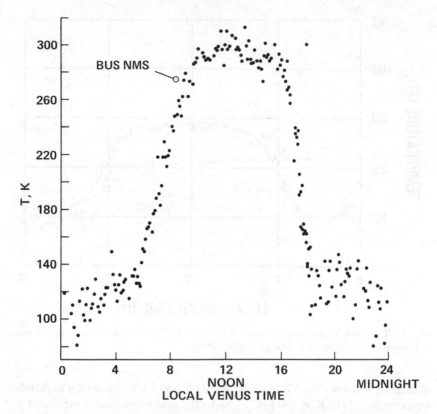

Fig. 22. Temperature deduced for the day and night hemispheres of Venus at 16° latitude from
the Pioneer orbiter drag experiment (Keating et al. 1980). The exosphere temperature derived from the bus NMS at −38° latitude, 8:30 AM, is also shown (von Zahn et al. 1980).

et al. 1974), and much lower than the ∼ 1500 K representative of Earth's
exosphere. The second feature is the strong diurnal variation, from 300 K on
the day side to the order of 100 K on the night side, which occurs sharply
across the terminator in 2 to 4 hr LVT. Fig. 22 includes the single exospheric
temperature ($z > 170$ km) derived from the BNMS data for 8:30 AM LVT at a
latitude of 38°S and a solar zenith angle of 61°.

While the temperatures in the thermosphere and exosphere show these
very large day-night differences, those in the lower atmosphere (below 100
km) are almost diurnally invariant except in the cloud layers (see above).
Thus, the altitude interval from 100 km to 140 km is an interval of transition
from one regime to the other. The nature of this transition has been measured
by accelerometers on the Pioneer Venus probes as they passed through these
altitudes at velocities > 11 km s^{-1} (Seiff et al. 1980; Seiff and Kirk 1982). Data
from the three small probes at these altitudes were included in Fig. 16. They
show a divergence of dayside and nightside temperatures beginning at an

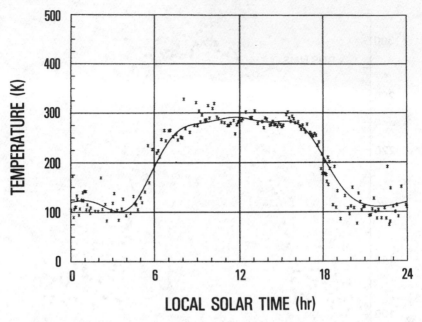

Fig. 23. Exosphere temperatures deduced from the PV ONMS at 16° latitude, shown as a function of local Venus time (Niemann et al. 1980).

altitude of 100 km. The Night probe at 12:07 AM LVT indicated a nightside temperature ~ 118 K at 120 km, close to the low temperatures measured at and above 140 km. The Day probe data, at 6:46 AM, yielded a steadily increasing mean temperature (dashed line) accompanied by a large amplitude oscillations. This oscillation may result from amplification of the gravity wave oscillation seen at lower altitudes in the middle atmosphere (Fig. 17, left). The temperature divergence thus occurs rapidly above 100 km, in the Venus lower thermosphere.

The combined data provided by the Pioneer Venus orbiter drag measurements and the Night probe entry measurements were analyzed and found to define upper atmospheric temperatures on the midnight meridian ~ 130 K, somewhat warmer than the midnight temperatures in Fig. 22 (Seiff and Kirk 1982). The PV ONMS temperatures at midnight are in near agreement with this, ~ 120 K according to the model fitted to the data (Fig. 23). This model indicates a temperature minimum 100 K at 4 AM, and one of ~ 110 K at 10 PM LVT, suggestive of possible recompression heating near midnight in the descending flow approaching the antisolar point.

Temperature data from four Pioneer Venus experiments are gathered in one graph in Fig. 24, which extends from the planet surface to 200 km. Upper atmospheric temperatures are shown for three LVT, midnight (± 1 hr), noon (± 1 hr), and 8:30 AM. The disagreement of the three measurements and

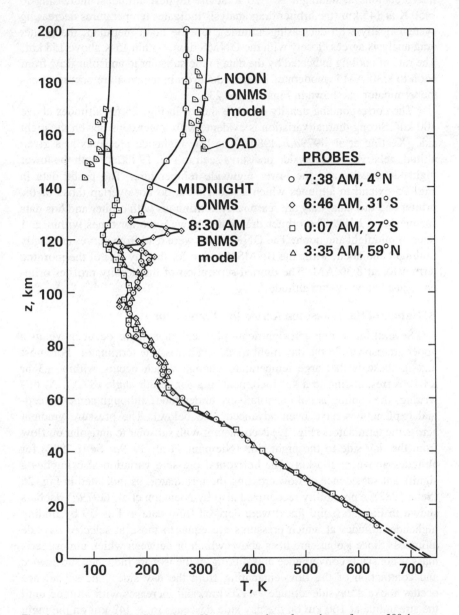

Fig. 24. Temperature profiles of the Venus atmosphere from the surface to 200 km, as indicated by four Pioneer Venus experiments. The different character of the temperature variation with altitude in the three altitude regimes is evident: (1) below the cloud tops; (2) from 60 to 100 km; (3) above 100 km.

interpretations at midnight is ~20 K at the highest altitudes, increasing to ~50 K at 145 km (the orbiter drag analysis indicates temperatures decreasing systematically with decreasing altitude). On the noon meridian, the orbiter drag analysis agrees closely with the ONMS data, within 15 K above 158 km. The rate of cooling indicated by the data in the subsolar to antisolar flow from noon to 8:30 AM is moderate. Most of the drop in temperature occurs across the terminator, as shown in Figs. 22 and 23.

The corresponding density data are shown in Fig. 25 for altitudes above 100 km. Strong diurnal variation is evident, with lower densities on the night side (Keating et al. 1979a,b, 1980). The low nightside pressures at a given altitude relative to the dayside pressures seen in Fig. 19 reflect both the lower nightside density and the lower nightside temperature. The probe data in Fig. 25 extend to altitudes which approach but do not overlap those of the orbiter and bus data; they are reasonably continuous with orbiter and bus data for the same LVT. The orbiter drag data shown were for times within ± 1 hour of midnight and noon. The ONMS data were read from curves at exactly midnight and noon, while the BNMS data are for the day side of the morning terminator at 8:30 AM. The diurnal separation of the density profiles originates just below 110 km altitude.

B. Nature of the Transition Across the Terminator

Several interesting and significant physical phenomena occur in Venus's upper atmosphere in the day-night transition across the terminator. Foremost among these is the large temperature change which occurs within ~3 hr LVT, corresponding to a 45° increment in solar zenith angle (SZA). At this writing, the cooling is not quantitatively understood, although some conceptual explanations have been advanced (see below). The pressure gradient across the terminators (Fig. 19) is consistent with subsolar to antisolar outflow from the day side to the night side (Niemann et al. 1979a; Seiff 1982) for altitudes above ~ 105 km. The horizontal pressure variation also implies a significant subsidence of flow crossing the terminator, as indicated in Fig. 26 (Seiff 1982), a possibility recognized also by Niemann et al. (1979a). Isobars shown in Fig. 26 (solid lines) were derived from data in Fig. 19 by finding nightside altitudes at which pressures are equal to those at selected dayside altitudes. Since isobars are lines above which or between which atmospheric mass is constant, convergence and sinking of the isobars indicates subsidence and contraction of the flow emanating from the day side. The subsidence occurs above a dayside altitude $\geq$ 120 km, and increases with altitude until the streamline at 160 km on the day side descends to ~ 142 km on the night side, descending 18 km. The mean angle of descent, however, is small, being ~ 0.005 radian at the evening terminator (Fig. 26). The descending flow transports dayside composition to lower altitudes on the night side as shown by the molecular weight contours across the terminator (dashed lines in Fig. 26). The molecular weights were derived from the ONMS and BNMS data

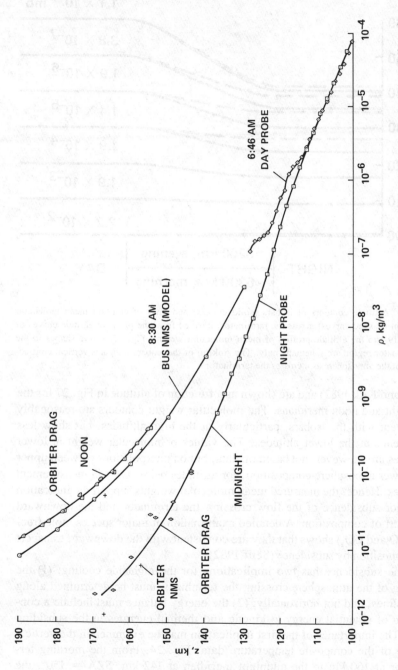

Fig. 25. Densities in Venus's upper atmosphere as a function of altitude. The data are from four Pioneer Venus experiments, and show the divergence of dayside and nightside densities, which begins at ~ 110 km.

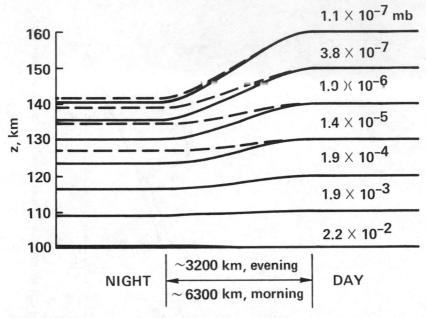

Fig. 26. Altitude contours of isobars (solid curves) and lines of constant mean molecular weight (dashed curves) across the terminator, derived from the pressure-altitude curves of Fig. 19 and the altitude profiles of mean molecular weight, Fig. 27. Curve shapes in the terminator region are schematic only. The sinking of the isobars indicates vertical contraction of the atmosphere as it crosses the terminator.

(Seiff and Kirk 1982) and are shown as a function of altitude in Fig. 27 for the midnight and noon meridians. The molecular weight contours are reasonably consistent with the isobars, particularly at the high altitudes, but show less subsidence at the lower altitudes. The values of molecular weight at lower altitudes are, however, not based on data, but on interpolations between upper and lower atmosphere composition, for altitudes below ONMS measurement altitudes. Hence, the measured mean molecular weights support the indication of major subsidence of the flow crossing the terminator and the downward transport of composition. A detailed examination of major species mole fractions (O and CO_2) shows that they are consistent with the downward transport of composition by subsidence (Seiff 1982).

The subsidence has two implications for the nightside cooling: (1) the cooling of the atmosphere crossing the terminator must be determined along streamlines, and not horizontally; (2) the energy balance must include a conversion of potential energy to kinetic and thermal energies in the subsiding flow. The importance of the first implication may be examined in the vertical profile of the composite temperature data, Fig. 24. From the morning terminator at 160 km to the midnight meridian at 142 km, SZA = 150°, the temperature drop may be as small as 128 K, rather than the 200 K implied by

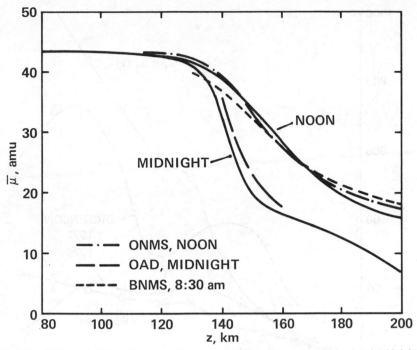

Fig. 27. Altitude profiles of mean molecular weight for the noon, morning, and midnight meridians, derived from PV orbiter and bus NMS data (after Seiff and Kirk 1982).

the more approximate generalization, 300 K to 100 K. The second effect, conversion of potential energy to heat and kinetic energy, counteracts this reduction by adding substantially to the thermal energy to be disposed of.

While complete models of these processes have not yet been developed, one-dimensional flow calculations which use measured pressures in the momentum equation indicate that for inviscid flow, the bulk of the potential energy would be converted to kinetic energy, producing supersonic flow velocities on the night side (Seiff 1982). However, the equation of continuity is not satisfied for supersonic nightside flow, given the measured densities and isobar contours. Applied to the data, the continuity equation indicates that the flow is essentially unaccelerated in crossing the terminator, and that the kinetic energy is dissipated into heat. This dissipation not only restricts the development of high velocities, but also adds to, and above 140 km dominates, the energy to be disposed of by the cooling process.

Cooling rates necessary to produce the nightside temperatures in the presence of subsidence and dissipation, estimated from the experimental data by use of one-dimensional calculations (Seiff 1982), are shown in Fig. 28. They are compared with global-mean 15 μm CO_2 radiative cooling rates calculated by Dickinson (1972, 1976). In the estimation of the cooling rates, velocity of the flow crossing the terminator is of first order importance. High

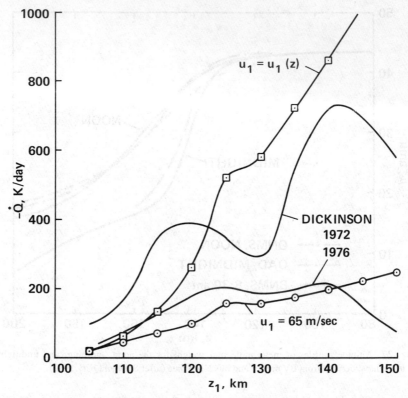

Fig. 28. Theoretical global mean radiative cooling rates in K/earth-day calculated by Dickinson (1972, 1976), plotted as a function of altitude and compared with cooling rates estimated
from measured structure data, with two assumptions for the dayside velocity profile. Estimated cooling rates include the effects of subsidence and kinetic energy dissipation in the
flow across the terminator.

velocity implies large cooling rates, and low velocity, smaller cooling rates.
The estimates shown were obtained for two extreme assumptions for the
dayside flow velocity, a low velocity of 65 m s^{-1} independent of altitude,
based on the velocity observed by Betz et al. (1977) at 109 km; and the much
higher velocities given by the dynamic model of Dickinson and Ridley (1977).
The latter, designated in Fig. 28 by $u_1 = u_1(z)$, assumes molecular viscosity
and gives $u \sim 300$ m s^{-1} at 140 km. (The importance of viscous dissipation in
the flow across the terminator [see above] may make the development of such
high velocities on the day side questionable.) Dickinson's global-mean radiative cooling rates are comparable in magnitude to the cooling rates estimated
from the analysis of the structure data and vary similarly with altitude up to
140 km. Hence it appears that the 15 μm radiation should contribute significantly to the cooling and the low nightside temperatures below 140 km,
particularly if flow velocities across the terminator are moderate, ~ 100 m

s^{-1}. Above 140 km, vertical diffusion with long mean free path, coupled with exchange of thermal energy for potential energy has been suggested as a process which can cool the higher levels (Seiff 1982).

Another mechanism which can contribute to the nightside cooling is downward turbulent heat transport i.e., large-scale overturning of the upper atmosphere, in the presence of a positive vertical gradient in potential temperature (Niemann et al. 1979a). The concept is here being applied at altitudes which extend well above the mixed region of the atmosphere, i.e. above $\sim$ 125 km (Fig. 27), which requires implicitly that diffusive separation proceed faster than turbulent mixing. Accompanying the turbulent heat transport, there is turbulent dissipation, and it is difficult to assess theoretically the relative magnitudes of the competitive processes of turbulent heat transport and dissipative heating (Hunten 1974). The analysis of the structure data (Seiff 1982) has indicated the magnitude of the dissipative heating which could be useful in future evaluations of this process.

A problem inherent in the radiative cooling mechanism is that the cooling decreases with decreasing atmospheric temperature. Alternatively stated, the requirement is that the radiating energy levels be repopulated by collisions rapidly enough to maintain the radiation. Equations given by Dickinson and Ridley (1977) indicate a decreasing, but still effective rate of cooling on the night side (Seiff 1982). Dickinson (1976) has pointed out that collisions with atomic oxygen are very efficient for exciting the CO_2 bending mode vibrations that radiate at 15 μm. This effect remains to be evaluated quantitatively for the observed levels of O concentration. It is probable that both radiative cooling and turbulent transport contribute to the production of the observed low nightside temperatures. Further discussion can be found in Chapter 20 by Taylor et al.

C. Further Comparisons with Theoretical Structure

Dickinson's theoretical models of the temperatures in the mesosphere and thermosphere are shown compared with Pioneer Venus probe entry data in Fig. 21. Dickinson placed the boundaries of the mesosphere at $\sim$ 90 and 122 km. The upper boundary, the mesopause is defined as the shallow temperature minimum which may be seen near 120 km in both models in Fig. 21. A minimum is also seen in the Night probe profile at 120 km. Above this lies the lower thermosphere, a region of relatively rapid temperature increase extending to 10^{-6} mbar at $\sim$ 145 km. There, Dickinson's upper thermosphere begins, and temperature asymptotically approaches exospheric temperature.

The theoretical curves shown in Fig. 21 are for the global mean atmosphere. Below 130 km, the 1972 global mean model corresponds closely with the Day probe temperatures obtained near the morning terminator. The Day probe temperatures are probably a little higher than the global mean atmosphere if the latter can be defined as midway between the extremes shown in Fig. 24. This would imply that the 1972 global mean model is somewhat warmer than Venus's global mean. The 1976 model, on the other hand, appears somewhat cooler than the experimental mean. Both models, however, are successful in estimating the magnitude and trends of variation in the observed atmosphere.

Above 130 km, the 1972 model goes off scale of Fig. 21 to an exosphere temperature of 475 K. The extreme ultraviolet (EUV) solar heating efficiency used to obtain this curve was 0.3. After the Mariner 10 indication that exospheric temperature was ~ 350 to 400 K, the EUV heating efficiency was revised downward to 0.1 for the 1976 model to give a dayside exosphere temperature of 400 K, and a global mean of 300 K. This produced a less satisfactory fit to the data below 135 km, which either suggests that EUV heating efficiency should vary with height from ~ 0.1 above 130 km to ~ 0.3 below, or that the radiative cooling model is somewhat in error. All things considered, these models were remarkably accurate predictions.

A comparison of theoretical with measured pressures was shown in Fig. 19 for the global mean atmosphere. It is likewise a remarkable anticipation of the Pioneer Venus measurements, at least up to 140 km.

The major deficiency in the dynamic theoretical models is that they did not anticipate the extreme coldness of the nightside upper atmosphere. Nightside tempertures as low as 160 K at 122 km increasing to 205 K at 140 km were obtained for SZA = 150° (Dickinson and Ridley 1977) from a model in radiative-dynamic equilibrium, thus overestimating the temperatures observed by 30 to 100 K. The models assume laminar flow, i.e., molecular viscosity. Since there are now several observational indications of turbulence in the upper atmosphere above 100 km (e.g., von Zahn et al. 1980; Seiff 1982), the assumption of laminar viscosity and conductivity at these levels may be the source of the disagreement of the Dickinson and Ridley models with the observations. The assumption affects vertical heat transport, dissipation, and flow velocities. All of these are factors in the nightside cooling. Hence it appears to be a promising direction for future research to explore effects of turbulent viscosity, dissipation, and heat transport on the nightside atmosphere. In an exploratory calculation, Dickinson and Ridley (1977) examined the effect of only one of these phenomena, turbulent vertical heat transport, on global mean temperatures of the upper atmosphere. The global mean temperatures were found to be lower by ~ 70 K above 125 km, when compared to the molecular conduction case. This calculation did not include dynamics or dissipation, but it suggests that large changes in the model will result from introducing turbulent processes.

IV. CONCLUDING REMARKS

Current knowledge of the thermal structure of the atmosphere of Venus is broadly summarized in Fig. 24, given also that the pressure at the mean surface is 95 bar. Below the clouds, steady, diurnal contrasts at midlatitudes appear to be <1 K, but oscillatory temperature differences with LVT as large as 5 K are present in the stable layer, and appear to signify the presence of gravity wave oscillations. Eddies may contribute to the contrasts ∼ 5 K seen in the deep atmosphere below 20 km. Contrasts as large as 50 K measured between probes of early Venera missions can be shown from energy arguments to most probably reflect measurement uncertainties.

The contrasts with latitude in the deep atmosphere are a few degrees Kelvin, but they become as large as 25 to 40 K in the cloud layers when the polar and equatorial atmospheres are compared at the 200 mbar to 2 bar level. The contrasts in the clouds between 30° and 60° latitude are consistent with cyclostrophic balance of the zonal winds. In the deep atmosphere near the equator, however, the pressures are not consistent with cyclostrophic balance, but could reflect large-scale eddy motions. Poleward of 60°, the large contrasts at cloud levels indicate more complex dynamics than simple zonal flow, possibly including a pair of vortices on opposite sides of the pole and moving around it with a 2.9 d period.

The temperature lapse rates from both the Pioneer Venus and Venera 10, 11, and 12 missions indicate a generally stable lower atmosphere, with limited regions of near neutral stability. The Pioneer probes showed stability profiles to be almost the same at four widely separated sites, with a deep stable layer just below the clouds extending down to 30 km, a shallow convective layer in the middle cloud, and an essentially neutrally stable layer from 15 or 20 km to 30 km. While this generally stable atmosphere is somewhat unexpected, it has a counterpart in the Earth's troposphere which is even more stable. A slightly stable lapse rate is a necessary condition for poleward heat transport in a Hadley cell (Stone 1974).

A range of greenhouse models by Pollack et al. (1980a) gives tropospheric temperatures close to those observed. The best fitting models yield a stable layer below the clouds. To obtain the stable layer, the models postulate the presence of microscopic haze particles 65 Å in diameter above the clouds. The several greenhouse models yield superadiabatic lapse rates at altitudes below 35 to 60 km. Where this occurs, the lapse rate is assumed adiabatic. Since the actual lapse rates are subadiabatic (stable), there should be no convective overturning. It would follow that the near neutral but slightly stable structure of the deep atmosphere is controlled by large-scale dynamics. The radiative balance calculations then can be used to determine the heat input driving the circulation.

Unanswered questions for the troposphere include whether some of the unexpectedly large contrasts and directions of temperature and pressure difference in the deep atmosphere are indeed evidence of eddy motions. If so, are the eddies horizontal or vertical, and what is their scale? A more complete description of the lower atmosphere in the polar region is needed to elucidate the polar dynamics and heat transport. More soundings in the subsolar region are needed to indicate whether the stability profile is similar to that observed on the night side and morning meridian, or, e.g., if the convective region at cloud levels might be deeper there than in the morning hours.

Even though the Pioneer mission was a very ambitious undertaking, its limitations should be recognized. The four probes were roughly equivalent to four sounding balloons in the Earth's atmosphere, from which one would not expect to completely characterize the atmosphere. More work is clearly needed before the findings relative to the deep atmosphere can be safely generalized.

In the upper cloud, there is a sharply stable change in the temperature lapse rate at $\sim$ 60 km, analogous to and occurring at roughly the same pressure level as the Earth's tropopause. Above this, the atmosphere is stable to 100 km, and approximately isothermal from 80 or 85 km to 100 km. At polar latitudes, the atmosphere is approximately isothermal above the cloud tops to $\sim$ 75 km. There is evidence for gravity waves in the middle atmosphere, perhaps thermal tides driven by diurnal temperature changes at cloud levels. The evidence is the variability, periodicity, and oscillations observed in the temperature profiles. A striking feature of the middle atmosphere is that it becomes warmer poleward. There is no confirmed explanation for this poleward warming, but it could result from radiative absorption at upper levels of the infrared emission through the two emission windows seen at cloud levels in infrared images. These are centered on 75° latitude and rotate about the pole. Associated with the poleward warming, there appears to be a slowing of the zonal winds, which peak in velocity at the cloud top and diminish toward zero at $\sim$ 90 km.

The 100 km level is a natural boundary in the atmosphere of Venus. Below this level, diurnal variations are generally small. Above it, they begin to increase to the very large contrasts in temperature, density, and pressure seen at 150 km. Below 90 km, a cyclostrophically balanced zonal circulation, which continues the flow pattern of the atmosphere at and below the clouds, is consistent with the pressure data. Above 100 km, the primary pattern suggested by the pressure contrasts is subsolar to antisolar circulation.

The altitude contours of isobars across the terminator in the upper atmosphere indicate a major vertical contraction or subsidence of the flow crossing the terminator. Streamlines descend by as much as 18 km from the 160 km dayside level, but at small angles of descent $\sim$ 0.005 radian. The subsidence also is indicated by lines of constant mean molecular weight connected across the terminator, which imply a downward transport of bulk composition as the atmosphere flows from the day side to the night side.

The dominant thermal characteristics of the Venus upper atmosphere above 100 km are its low exospheric temperature ($\sim$300 K on the day side) and its large diurnal variation. The low dayside temperatures can be explained by radiative-conductive equilibrium models with low efficiency of heating by the solar EUV. The nightside cooling to exospheric temperatures $\sim$ 100 to 130 K is not yet entirely understood. If the flow velocities are small ($\sim$100 m s^{-1}), the observed temperatures may be interpreted within the framework of a one-dimensional flow model to imply cooling rates generally comparable to the mean global cooling rates of the 15 μm CO_2 bands calculated by Dickinson. However, global temperature predictions from the dynamic models of Dickinson and Ridley do not include values as low as have been seen on Venus's night side, possibly because of high flow velocities in the models, i.e. molecular, viscosity and conduction, whereas there is increasing experimental evidence of turbulent processes in the atmosphere above 100 km. At the low temperatures of the night side, radiative cooling rates decrease, but this is moderated by the presence in the upper atmosphere of large fractions of atomic oxygen, an efficient collision partner for exciting the CO_2 15 μm bands. In addition to radiative cooling, downward turbulent transport of heat in the direction of the potential temperature gradient, may play a role in explaining the low nightside temperatures. Above 140 km, diffusion with long mean free path accompanied by exchange of thermal for potential energy may also be important. Future research should be directed toward quantitative understanding of these cooling mechanisms, more than one of which may be important.

Models of the thermal structure of the atmosphere of Venus are given in the Data Appendix at the back of this book. These models are for the three altitude regimes used to organize the discussion in this Chapter, and are heavily based on the experimental data reviewed herein. The rationale for selection of models is discussed in the Appendix below, which also includes comments on their limitations, estimated uncertainties, and expected variability of the atmosphere about the mean state which the models are intended to represent.

APPENDIX

Models of the Atmosphere of Venus

Models of the atmosphere of Venus have been previously proposed and published by Avduevsky et al. (1970), Kuz'min and Marov (1975), and Noll and McElroy (1972), among others. With the passage of time, as more experimental evidence on the structure is gathered, the models can become increasingly faithful to the actual atmosphere. The intent of this Appendix is to describe the rationale used to select models based on the state of knowledge reviewed in the Chapter. The models are given in tables in the Appendix at the back of this book for the lower atmosphere below the cloud tops (Table AI), the atmosphere above the clouds to 100 km (Table AII), and the upper atmosphere, from 100 km to 180 km (Table AIII).

1. The Lower Atmosphere: Below the Cloud Top

The models of the lower atmosphere, given in Table AI, were derived from data taken during descent of the four Pioneer Venus probes, which have high measurement resolution and small formal uncertainty. (These data are generally supported by the Venera probe data within measurement uncertainties.) Two models are given, one for low to middle latitudes up to 40°, and one for a latitude of 60°. The latitude boundary on the lower latitude model is based on data taken by radio occultation at the 1 bar level (Fig. 10), which show that the structure is essentially latitude independent up to 40°. Above this latitude contrast increases with increasing latitude. The Pioneer North probe defines the structure at 60° latitude. No data are available for the deep atmosphere poleward of 60° at pressures > 4 bar, and no polar model is presented.

The lower latitude model is a synthesis of data from three probes, two at ~ 30° latitude, and one near the equator at 4° latitude. There were systematic differences in structure with latitude observed among these three soundings. Whether the differences in the deep atmosphere are transient in nature or steady is presently unknown, but they represent too fine a level of discrimination for use in a general purpose model. Readers who are interested in these differences should refer to the tables of atmospheric state properties at the three locations given by Seiff et al. (1980). The temperature differences, illustrated in Fig. 5, indicate an altitude dependent offset between the mean curves at the equator and at 30° latitude plus an oscillatory component leading to temperature differences up to ~ 7 K. The equatorial sounding is cooler at altitudes below 55 km. The oscillatory differences have ~ 5 K amplitude with a vertical wavelength ~ 15 km. These are probably a result of global scale waves.

The model temperatures and pressures were chosen to be intermediate between those at 4° and at 30° latitude, but they are not a simple average. Rather, they were selected to give local temperature lapse rates compatible with the stability curves of Fig. 13, and the model presented preserves a first order representation of the observed atmospheric stability. It also satisfies the equation of hydrostatic equilibrium.

Temperatures are given to the nearest degree Kelvin, but because of the observed variability, could differ at some altitudes from temperature in the local atmosphere at a given time by ~ 5 K. Pressures and densities are given to three significant figures typically, and are probably known to within ~ 1 %. At altitudes above 60 km, where inversions which increase in strength with latitude are found, some additional variability in temperature could occur due to local differences in inversion structure.

The 60° latitude model is not significantly different from the lower latitude model at altitudes up to 40 km (pressures > 3.5 bar). Above this level, significant differences in temperature and pressure develop, as discussed in the text, while densities remain comparable to those at lower latitudes, suggesting that the cooling at higher latitudes occurs at constant volume. Below 20 km, pressure differences between 60° latitude and lower latitudes which appear in the models (i.e., in the data) are directed consistently equatorward, suggesting equatorward circulation near the surface. However, it is uncertain as to whether these differences are significant, since they are near the limit of measurement accuracy.

2. The Middle Atmosphere: Cloud Top to 100 km

Models of the middle atmosphere were based on data from three Pioneer Venus experiments: the orbiter infrared radiometer (OIR) (F. Taylor et al. 1980), the orbiter radio occultations (ORO) (Kliore and Patel 1982), and the Large and Small probe atmosphere structure experiment in its entry mode (LAS and SAS) (Seiff et al. 1980; Seiff and Kirk 1982). The OIR data were the zonal, temporal mean data shown by the temperature contours in log pressure, latitude coordinates (Fig. 9). These were converted to temperature-altitude plots, for comparison with the other two experiments, by the procedure described in the text. A reference altitude was selected near the 100 mbar level at which an altitude measured by one of the other two experiments was imposed. At latitudes up to 75°, these reference altitudes were taken from or extrapolated from probe descent mode data, and are estimated to be accurate within ~ 0.2 km. At latitude 85°, the reference altitude was from the radio occultation data, with estimated accuracy ~ 0.5 km. The $T(p)$ data from OIR were then integrated in the equation of hydrostatic equilibrium to define altitudes relative to the reference altitude. Comparisons of these three data sets are shown in Fig. 29 for latitudes $\leqslant 30°$, 45°, 60°, 75°, and 85°.

The latitude effects shown by these data, by the OIR temperature contour diagram, and the sequence of temperature profiles from ORO (Fig. 17; see also Fig. 1 of Kliore and Patel 1982) indicate that models of the middle atmosphere must be latitude dependent. Below a latitude of 30°, however, a single model can be used. Some changes from the low-latitude model start to appear at a latitude of 45°, as best illustrated in the ORO data referred to above. Major variations with latitude are found at $\theta \geqslant \sim 60°$ (Fig. 9). We therefore present a model for each of the latitudes represented in Fig. 29. Care has been taken to retain in the models thermal contrasts with latitude of the observed sense (warmer poleward above 70 km to 90 km) and magnitude. These contrasts are probably not accurate enough, however, to define the atmospheric dynamics exactly.

Diurnal variations are small in this altitude range (see text) so a single model suffices for all local times. The diurnal variations are, in fact, smaller than the temporal variability at a given local time. (This observation led Taylor and Schofield to define a single zonal, temporal mean contour diagram, independent of the local time as shown in Fig. 9.) The temporal variability is significant, and appears to be a result of waves in the atmosphere

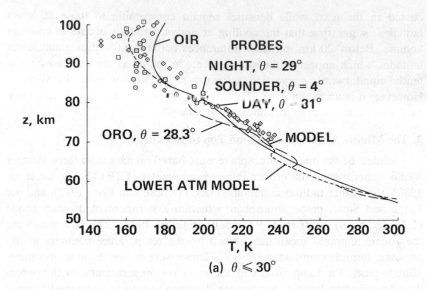

(a) $\theta \leqslant 30°$

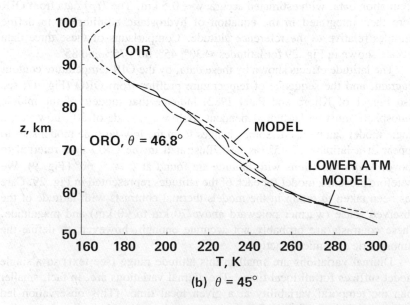

(b) $\theta = 45°$

Fig. 29. Comparisons at 5 latitudes of temperature data from three independent Pioneer Venus experiments in the middle atmosphere, and the selected models of $T(z)$.

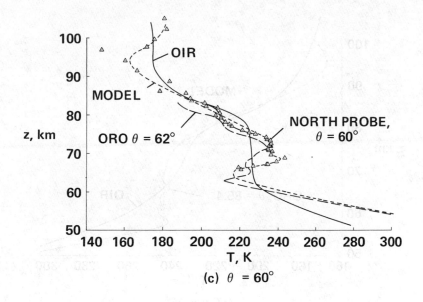

(c) $\theta = 60°$

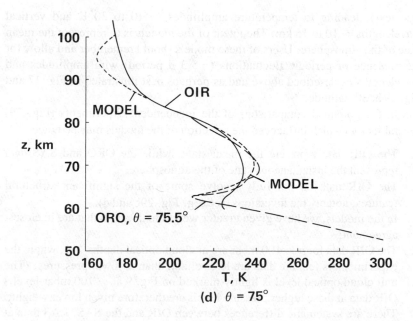

(d) $\theta = 75°$

Fig. 29. (Continued)

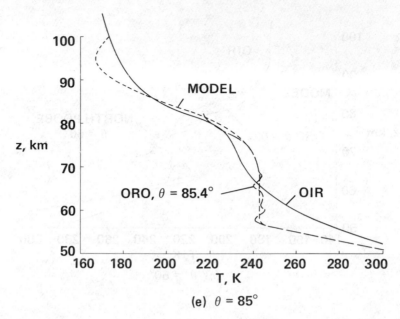

Fig. 29. (Concluded)

(see text), leading to temperature amplitudes ~ 10 to 30 K and vertical wavelengths ~ 10 to 15 km. The intent of the models is to represent the mean state of the atmosphere. Users of these models should remember and allow for the presence of periodic fluctuations (~5.3 d period) with amplitudes and wavelengths as described above and as perhaps best illustrated in Fig. 17 and Fig. 29 at 30° latitude.

In the graphical comparisons of the independent experiments (Fig. 29) several factors which influenced the selection of the models may be noted:

1. The OIR data represent the mean state, while the ORO and SAS data represent the instantaneous state of the atmosphere;
2. The OIR data do not fully resolve some of the significant structural features, notably the inversions (see e.g. Fig. 29c and d);
3. In the models, we have given greater weight to data sets that are in closest agreement;
4. The OIR data for $p > 100$ mbar are at cloud optical depths > 1, where the experimenters feel the data are less reliable than at lower pressures. (The unit cloud optical level is lightly marked on Fig. 9 at ~ 100 mbar level.) OIR data at these higher pressure levels are therefore given lower weight;
5. There are systematic differences between OIR and the SAS, LAS data at $z > 90$ km and at $68 < z < 80$ km which do not appear to be explainable as examples of instantaneous temporal variability, because data from all four probes deviate from the OIR mean profiles in the same direction.

(The probability that this would occur fortuitously due to temporal variability is $1/2^4 = 1/16.0$.) The models are based on the assumption that the vertical resolution limitation of OIR is responsible for this disagreement, and the selected model curves were heavily influenced in both intervals by the LAS, SAS data.

The selected models are continuous with the corresponding lower atmosphere models, except for discontinuities associated with the cloud-top temperature inversions. The models obey hydrostatic equilibrium and the equation of state. The mean molecular weight is taken to be 43.44 below 80 km, decreasing to 43.2 at 100 km (after Seiff et al. 1980), reflecting the higher concentrations of O and CO at the upper altitudes. Corresponding gas constants are 191.4 J kg^{-1} K below 80 km, increasing to 192.2 at 100 km, as given in the Data tables.

The temperature profiles very near the pole require comment. The Pioneer Venus radio occultations define a nearly isothermal structure above the 58 km level, as shown in Fig. 29e. Yakovlev and Matyugov (1982) obtain very similar polar profiles, also by radio occultation, from Veneras 9 and 10. The OIR profile, however, is not isothermal, but exhibits a small lapse rate at all levels. It comes into agreement with the occultation temperatures at 78 to 80 km altitude. The OIR profiles also show the fascinating symmetry about 75° latitude noted in the text (Fig. 9). It is not clear which of these differing patterns will prevail. We have chosen the isothermal profile for the polar model, since it has been seen repeatedly in occultation and by different observers.

The inversion layers which occur near the cloud tops are present in the model temperatures, but the altitude intervals used in the tables are not fine enough to define the inversions in detail. (To do so would probably be unwarranted, since they are somewhat time dependent.) The radio occultation data, which do define these inversions in detail, show that they are ~ 2 km deep at low latitudes, but increase in depth with latitude to ~ 7 km at 75° (Kliore and Patel 1982). The inversion in the model for latitudes $< 30°$ occurs between 66 and 68 km, and appears as a discontinuity between the model profiles above and below these levels, with a temperature offset of ~ 2.5 K. The offset increases with latitude as indicated by the ORO data in Fig. 29. At 60° latitude, e.g., the offset in the model is ~ 25 K, and the inversion layer altitude interval is from 65 to 69 km.

The temperature model for $\theta = 45°$ was guided at the 60 km level by the temperature contrasts with latitude indicated at that level by the OIR experiment. These were used as a guide for nonlinear interpolation of the probe temperatures at 30° and 60° latitudes. The lower atmosphere model temperatures given in Fig. 29b thus differ from those at $\theta < 40°$, reflecting the poleward cooling at cloud levels associated with cyclostrophic balance of the zonal winds. Similarly, the pressure at 45° latitude at the 60 km level was calculated from the equation of cyclostrophic balance (see text) for the zonal wind velocity at 60 km (82.9 m s^{-1}) which is implied by the pressure measurements at 30° and 60° latitudes. The model that results from these assumptions is intermediate in pressure and density to those for 30° and 60° latitude up to 82 km. Beyond that, it yields slightly higher pressures and densities than the models at the surrounding latitudes, probably indicating a slightly imperfect model temperature profile at 45° latitude.

Selection of the reference pressure level for $\theta = 75°$ also was based on the assumption of cyclostrophic balance, to obtain a pressure of 190 mbar at 60 km. This corresponds to a mean zonal wind velocity from 60° to 75° latitude of 83 m s^{-1} at 60 km altitude, which may be a little too large a velocity at the higher latitude. The pressure data of the ORO experiment (Kliore, personal communication, 1982) indicate that the 190 mbar level occurs at an altitude of 59.49 km, and thus agrees with the selected 190 mbar altitude (60 km) within the ORO experimental altitude uncertainty. The ORO pressure at 60 km, $\sim$ 171 mbar, in relation to the pressures at lower altitudes, would require a zonal wind velocity of 119 m s^{-1} for cyclostrophic balance, much higher than the observed velocities at 60 km. Hence, the selected 190 mbar pressure is judged to be more compatible with the models at lower latitudes, and to agree with ORO data to within the altitude uncertainty.

The temperature model below 80 km at 75° latitude was largely guided by the ORO profile in Fig. 29d. The selected inversion temperature offset maintains about the same temperature contrasts with latitude as are shown by the OIR data. It is also close to the peak temperature offset given in the Fig. 9 of Kliore and Patel (1982) for this latitude. Above 85 km, OIR provides the only temperature data for these latitudes. The tendency noted at 30° and 60° latitude for these temperatures to be higher than *in situ* data by about 10 K was assumed to apply here also. Accordingly, an adjustment of about this magnitude was made above 85 km in the model temperature profile at 75° latitude, as shown in Fig. 29d, to preserve similarity to the models at lower latitudes, and to preserve the temperature contrast with latitude shown by the OIR data.

The models derived from the data guided by the above considerations are given in Table AII.

3. The Upper Atmosphere

The data available on which to base a model of the thermal structure of the upper atmosphere are limited to: (1) the complete diurnal cycle at 16°N latitude in the altitude interval of 143 to 200 km, with most of the data between 150 and 180 km (Keating et al. 1980; Niemann et al. 1980b); (2) a sounding from 200 km to 130 km obtained by a neutral mass spectrometer during bus entry at 38°S latitude at 8:30 AM; (3) soundings from an altitude of 136 km (maximum) down to and through 100 km obtained by the four entry probes at latitudes of 4°, 30°, and 60° at LVT 7:38, 6:45, 12:07, and 3:35 AM LVT, respectively. These data are minimal for purposes of defining a global model. There are, however, theoretical and experimental indications that the upper atmosphere is approximatetly two dimensional, i.e., roughly axially symmetric about the subsolar-antisolar axis (see, e.g. Dickinson and Ridley 1977; Betz et al. 1977). By use of this assumption, a model based on the available data can be extended over much of the planet, with solar zenith angle as the independent variable. It should be cautioned, however, that there is also evidence for at least moderate deviations of the structure from global symmetry (see, e.g. Seiff 1982).

Data taken just below 100 km (preceding section) and above 150 km (Figs. 22 and 23) indicate that state variables are relatively invariant with solar zenith angle, for SZA < 50°, and a subsolar or noon model should be approximately applicable to this entire region. By similar reasoning, for SZA > 120°, the antisolar or midnight model should apply. The remaining terminator regions are regions of rapid variation with SZA. We will not attempt to model the terminator regions, other than to suggest that temperature structures similar to and intermediate between the subsolar and antisolar patterns should be used, guided by the exosphere temperatures indicated in Figs. 22 and 23. von Zahn et al. (1980) have given one model for the start of the terminator region, at SZA = 61°, 8:30 AM LVT, which indicates this type of intermediate structure. This model shows a structure relatively close to our subsolar model.

The antisolar model is based on a moderately smoothed midnight temperature profile, shown in Fig. 24, which was defined from Night probe and orbiter drag data near the midnight meridian (Seiff and Kirk 1982). This profile is reproduced with an expanded temperature scale at the left in Fig. 30. The model is obtained by integrating the equation of hydrostatic equilibrium upward from 100 km, by use of this temperature profile and a pressure at 100 km from the model of the middle atmosphere, averaged over latitudes up to 60°. Mean molecular weight is from the midnight curve of Fig. 27, and gas constant $R(z)$ is tabulated as part of the antisolar model, given in Table AIII. Pressures and densities are in near agreement with the measured midnight data. Accuracy of pressure and density is no better than 2 significant figures.

The subsolar or noon profile could not be similarly based on an experimental profile at noon, because of the lack of data between 100 and 150 km at SZA < 50°. Instead, the tabular density data of Keating et al. (1980) were used to derive $T(z)$ above 150 km. They show the exobase to be at ~ 162 km, and define a lapse rate between 160 and 150 km similar to that of von Zahn's morning model (right side of Fig. 30) (von Zahn et al. 1980). Integration upward from 100 km where $T \sim 170$ K (Table AII), was along temperature curves $T(z)$ which were iterated in initial slope at 100 km until the observed densities at $155 < z < 180$ km were obtained to within 15%. A noon $T(z)$ generally similar to von Zahn's morning profile resulted. The mean molecular weight profile for noon (Fig. 27) was used to define gas constants $R(z)$ which are very different above 125 km from those at the same altitude at midnight.

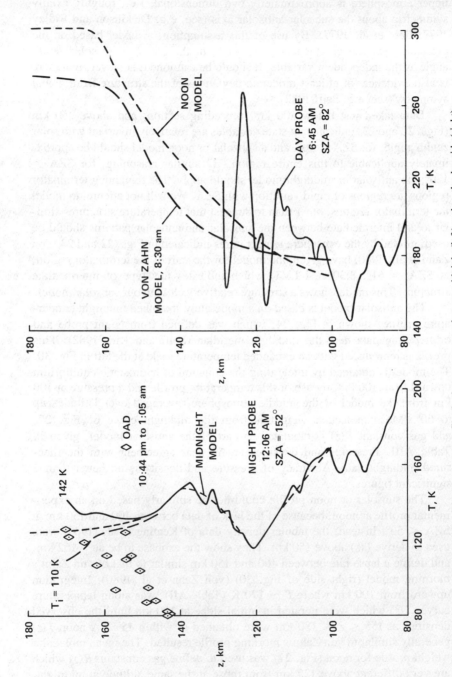

Fig. 30. Temperature data and model temperature profiles in the upper atmosphere at midnight (left) and on the day side (right).

Accuracy of this model of the noon or subsolar structure is probably lower than that for midnight.

The temperature structure measured by the Day probe below 129 km, shown at the right in Fig. 30, is highly oscillatory. However, it can be fitted by a mean temperature curve approximately parallel to the noon model below 130 km, but offset by +10 K. These data thus suggest that the noon model temperatures could be as much as 10 K low near the 100 km level, or that there is some variability at this level. The von Zahn model has made use of the Day probe data, by assuming implicitly that the difference in SZA between the bus entry and the Day probe entry sites is not critical near the 100 km level. Hence, the effect of this change in temperature on the models contributes to the differences in $\rho(z)$ and $p(z)$ between the noon and von Zahn models. As already noted, these differences are small, and in the expected direction. This consistency problem suggests the level of uncertainty which presently characterizes the dayside upper atmosphere.

Finally, it should be noted that the properties of the upper atmosphere of Venus, like those of Earth and Mars, would be expected to vary measurably through the solar cycle and with solar activity. Some efforts to investigate this kind of variation by Hedin et al. (1983) have shown density fluctuations in individual species which correlate with short-term variations in the 10.7 cm solar flux index. The associated temperature fluctuations are apparently smaller by a factor ~ 20 than in the Earth's atmosphere (see also Keating et al. 1980). The data on which the models are based were taken during a time of intermediate solar activity approaching solar maximum.

12. STRUCTURE AND PARAMETERS OF THE VENUS ATMOSPHERE ACCORDING TO VENERA PROBE DATA

V. S. AVDUEVSKIY, M. YA. MAROV, YU. N. KULIKOV,
V. P. SHARI, A. YA. GORBACHEVSKIY,
G. R. USPENSKIY, and Z. P. CHEREMUKHINA
USSR Academy of Sciences

Direct temperatures and pressure measurements, carried out for the first time on the entry probe of the Venera 4 mission in 1967, laid the foundation for systematic studies of the structure and parameters of the atmosphere of Venus. Measurements made during the Venera 4–12 missions cover the altitude range from 62 km to the surface of the planet on its night and day sides. Accelerometer measurements on Venera 8–12 pertain to the altitude region of 64 to 100 km. A thermodynamic analysis demonstrates that the temperature variation is almost adiabatic from the surface up to 50 km and subadiabatic up to ~ 85–90 km, or about the altitude of the mesopause. In the lower troposphere no significant temperature or pressure variations are apparent up to 20 km, whereas in the upper troposphere and stratosphere the variations in altitude profiles for atmospheric parameters apparently reflect actual space-time variations in the atmosphere of Venus. These data are generally in good agreement with the mean altitude profiles of the temperature and pressure given by a model (Kuz'min and Marov 1974; Marov and Ryabov 1974), which was constructed from measurements performed on the night side of the planet.

The atmosphere of Venus presents a striking contrast to the atmosphere of Earth. The study of its most important characteristics (temperature, pressure, chemical composition, optical and thermal characteristics, dynamics, properties of the clouds) has been made possible by space missions permitting direct measurements *in situ*. Venus was the first planet with an atmosphere to be examined, during the flight of Venera 4 in October 1967, by geophysical research techniques which later became common.

The realization of a regular program of flights to Venus by Venera spacecraft during the years 1967–1978 allowed repeated probing of the atmosphere and measurement of atmospheric parameters in various regions of the planet (Avduevsky et al. 1968, 1970, 1976a, 1979; Marov et al. 1973b). Basic information about these measurements is given in Table I. As can be seen, direct measurements conducted while the probes were descending on parachutes span the altitude range from 62 km to the surface. Four of the profiles were obtained on the night side (Veneras 4–7) and five on the day side (Veneras 8–12). If the venusian day is broken into 24 one-hour periods (see Table I), the direct measurements span roughly half the venusian day. The values of the atmospheric parameters on the surface were obtained in six regions of Venus.

Accelerometer measurements, performed as the Venera 8–12 probes entered the atmosphere, cover altitudes from 64 to 100 km. This region, as well as that down to ~ 40 km in which direct measurements were made while the probes were descending on parachute was also studied by radio occultation measurements from Venera 9 and 10 on both day and night sides of the planet.

The collection of all these data permitted a study of the structure of the troposphere and stratosphere of the planet, the thermodynamic state of the atmosphere and the nature and scale of variations of atmospheric parameters (Kuz'min and Marov 1974; Marov 1972, 1978, 1979).

I. RAW DATA AND DATA PROCESSING

A. Direct Measurements of Pressure and Temperature

For direct measurements of temperature T and pressure P from each Venera probe, arrays of resistance thermometers and aneroid manometers were used. These instruments possessed different ranges within the limits of 240 and 800 K in temperature, and 0.2 and 150 km cm^{-2} in pressure, with an error of ± 1.5% in the measurement range.

The experimental data points were represented by a polynomial approximation. In order to determine the functions which describe the variation of the temperature and pressure as functions of the parachute descent time (with aerodynamic shield after disconnecting the parachute for the Venera 9–12 probes) and which best approximate the experimental values, the method of least squares was used. The form of the function was specified by an n-th degree polynomial

$$\sum_{i=0}^{n} a_i t^i \tag{1}$$

when $n \leq 9$. The problem thus reduces to finding the coefficients a_i from measurement results for the temperature and pressure at various values of the

TABLE I

Summary of the Locations and Data Sets for the Venera Probes

Measurement Performed (LVT)	Spacecraft	Measurement Date	Coordinates of Landing Site[a]	Time of Day	Solar Zenith Angle	Measured Parameters	Altitude Range (km)
4:40	V 4	18.10.67	19° N 38°	night	—	T, P, ρ	51–24
~5:30	V 5	16.05.69	3° S 18°	night	—	T, P, ρ	55–16
~5:30	V 6	17.05.69	5° S 23°	night	—	T, P, ρ	49–16
~5:00	V 7	15.12.70	5° S 351°	night	—	T	55–0
6:30	V 8	22.07.72	10° S 335°	day	84°5	T, P, drag	55–0
							100–65
13:15	V 9	22.10.75	31°7 N 290°8	day	33°	T, P, drag	62–0
							110–76
13:45	V 10	25.10.75	16° N 291°	day	27°	T, P, drag	62–0
							110–63
~13:30	V 11	25.11.78	14° S 299°	day	20°	T, P, drag	61–0
							100–65
~12:00	V 12	21.11.78	7° S 294°	day	25°	T, P, drag	61–0
							100–65

[a]In the MAC system as per Transactions of the IAU (1971).

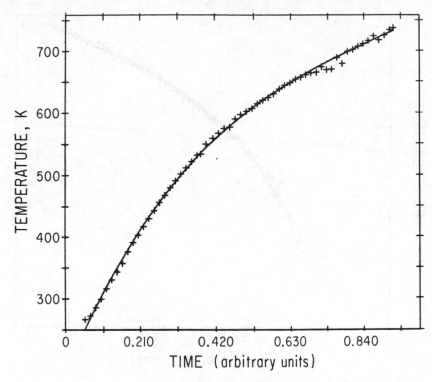

Fig. 1. Approximation of the experimental values of temperature, obtained by the Venera 11 probe, using a third-degree polynomial in time (solid line). The time is laid off along the abscissa in conventional units. For convenience in representing the initial data, a preliminary averaging was made for every 50 temperature values measured at consecutive moments in time. One of three averaged points corresponds to each mark.

time t, on the basis of the condition of producing the best approximation. Usually the value $n = 3$ corresponds to this condition. The finding of the coefficients of the polynomial took into account the statistical weights of the individual measurements. Examples of the resulting approximation curves for the Venera 11 spacecraft are shown in Figs. 1 and 2. The individual points are the result of a preliminary averaging of the first 50 measured values.

Using the obtained functions of $T(t)$ and $P(t)$ for each probe, the covered path z was determined and a corresponding altitude correlation of the measurement results achieved. When the gas composition of the atmosphere is known, by using the equation of hydrostatic equilibrium we obtain

$$dP = \rho g^{\varphi} dz \qquad (2)$$

$$\rho = \frac{P\mu}{RT} \qquad (3)$$

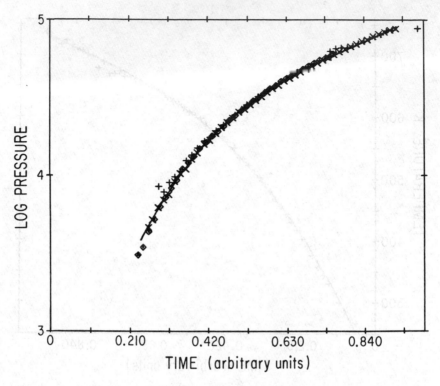

Fig. 2. Approximation of the logarithm of the measured pressure values, obtained by the Venera 11 probe, using a third-degree polynomial in time (solid line). As in Fig. 1, the time is measured in conventional units, and one of three points corresponds to each mark, obtained by preliminary averaging of 50 consecutive values of P for one of the three pressure sensors: +, data of sensor 1; ×, data of sensor 2; ◇, data of sensor 3.

from which

$$z = \frac{R}{\mu} \int_{P_0}^{P_x} \frac{T}{g^{\,\varphi}} \frac{dP}{P} \qquad (4)$$

where P_0 and P_K are the pressure values at the start and finish of the measurements t_0 and t_K, respectively; R is the universal gas constant; ρ is the gas density; μ is the mean molecular weight ($\mu = 43.4$). The acceleration of gravity on Venus at altitude h is

$$g^{\,\varphi} = g_0^{\,\varphi} \left(\frac{r_{\varphi}}{r_{\varphi} + h} \right)^2 \qquad (5)$$

where $g_0^{\,\varphi} = 8.81$ m s^{-2} at $h = 0$; r_{φ} is the radius of Venus ($r_{\varphi} = 6052$ according to *Transaction of IAU* [1971]).

On the other hand, if we know the aerodynamic characteristics of the probe and the parachute it is easy to calculate independently the distance covered with an allowance for the forces of buoyancy on the basis of the condition of a quasi-uniform descent (Kuz'min and Marov 1974)

$$mg^{\circleddash} \left(1 - \frac{\rho}{\rho_{am}}\right) = C_x S \frac{\rho V^2}{2} \tag{6}$$

which produces the formula

$$z = \left(\frac{2m}{C_x S}\right)^{1/2} \int_{t_0}^{t_x} \left[g^2 \left(\frac{1}{\rho} - \frac{1}{\rho_{am}}\right)\right]^{1/2} dt \tag{7}$$

where m and ρ_{am} are the mass and density of the descent probe; V is the speed of descent; C_x is the drag coefficient; S is the area of the parachute dome and/or of the maximum midsection of the probe and the aerodynamic shield.

The calculated heights above the surface by both methods are in satisfactory agreement. When a radio-altimeter is present (Veneras 4, 5, 6, 8), an additional independent correlation to the height above the surface is provided. For the landing probes, the moment of contact with the surface was plotted by the frequency recoil of the transmitter on the probe; from this moment, the altitude profile of the calculated distance covered $z(t)$ was reconstructed as the height above the surface $h(t)$ and the corresponding height $h(t_0)$ at the moment the measurements began was determined.

B. Measurements of Drag

On the segment of entry into the atmosphere of Venus, where the principal aerodynamic retardation of the descent probe takes place, direct temperature and pressure measurements are complicated by the complex pattern of supersonic flow and the high temperature of the heat-protection coating. More effective for this segment is the method of determining the atmospheric parameters by registration of the change in the longitudinal acceleration acting on the probe drag ratio (axial drag ratio $n_x = F/G$, where F is the drag force and $G = mg^{\circleddash}$ is the weight of the probe.

In a simplified version of the measurement, it is possible to record several intervals of time between a series of fixed values on the calculated curve $n_x(t)$, including the moment of attaining maximum value $n_{x\max}$. In this version, an experiment was realized for the first time on Venera 8. Using the equations of motion of the probe and assuming isothermal conditions for the stratomesosphere of Venus, estimates were obtained for the mean temperature and altitude scale (Cheremukhina et al. 1974).

On board the Venera 9 and 10 probes a single-range accelerometer ($n_{x\max}$ = 250) was used; on Veneras 11 and 12 a double-range meter ($n^{(1)}_{x\max}$ = 200; $n^{(2)}_{x\max}$ = 10) with a data storage and preliminary processing system transmitting on a segment of the parachute descent. In contract with Venera 8, in reconstructing the density profile no simplifications were assumed as to the temperature distribution, solving instead a nonlinear inverse boundary-value problem of ballistic descent. The set of equations of motion of the descent probe have the form

$$\frac{dV}{dt} = -n_x g^{\oplus} - g^{\female} \sin \theta$$

$$\frac{d\theta}{dt} = \left(\frac{V}{r_{\female} + h} - \frac{g^{\female}}{V} \right) \cos \theta \qquad (8)$$

where $dh/dt = V\sin\theta$, V is the speed of the probe, θ is the angle of entry in the atmosphere (the slope of the trajectory to the local horizon). This is supplemented by an equation for the density according to Eq. (6) in the form

$$\rho(t) = 2n_x G/C_x SV^2 \qquad (9)$$

as well as Eqs. (2) and (3), and the initial conditions for the moment of entry into the atmosphere t_0, (V_0, θ_0, h_0), which coincide with the conditions at the upper boundary of the integration. In view of the inaccuracy in the determination of θ, an additional boundary-value condition is used for the density at the lower boundary, using the entry altitude of the parachute system at the moment t_{pr}

$$\rho(t_{pr}) = \rho^*; h(t_{pr}) = h^*. \qquad (10)$$

Such a method, which uses an iteration procedure in order to refine the value of the angle θ with respect to the boundary value of the density ρ^*, was dictated by considerations of the best agreement between the results of the accelerometer measurements and the direct measurements, where these have become available.

It should be noted that in this particular method small deviations in the drag rates may lead to significant variations in the altitude dependence of the density (and, correspondingly, the temperature and pressure). Therefore a careful allowance was made for the influence of possible errors in the measurements of $n_x(t)$, especially on the descending branch of the drag ratio curve, in the altitude range of 72 to 65 km. Below 65 km, the solution is unstable, and the atmospheric parameters cannot be determined by this method. We also point out that in Eq. (9) the value of S is defined as the area of the protective sphere of the descent probe on the entry segment, to which the drag coefficient C_x is related. On Venera 8 the value of the axial drag ratio at an entry angle of $\theta_0 = -42°8$ was $n_{x\,max} = 378$; on Veneras 9 and 10 with entry angles $\theta_0 = -20°5$ and $-22°5$, $n_{x\,max} = 136\pm3$ and 167.5 ± 0.5, respectively. On Veneras 11 and 12, with $\theta_0 = -18°$ and $-19°5$, values of $n_{x\,max} = 153\pm3$ and 166 ± 3, respectively, were recorded.

II. ANALYSIS OF THE RESULTS

The assemblage of direct temperature and pressure measurement data, processed as described, from the Venera spacecraft (except Venera 7, which measured only the temperature, with a less definite altitude correlation [cf. Kuz'min and Marov 1974]), is given in Fig. 3 in coordinates of $\log P–T$ and in the form of the graphs $T(h)$ and $\log P(h)$ in Figs. 4 and 5, respectively. The results shown in Fig. 3 contain the most objective information, since they do not involve a subsequent processing for the purpose of an altitude correlation of the data. Thus, all the curves in Figs. 4 and 5 refer to a common reference level of the surface $h = 0$.

From a consideration of the curves in Figs. 3, 4, and 5, it is seen that the results of the measurements of the atmospheric parameters of Venus from the surface directly up to a level corresponding to a pressure $P \simeq 30$ bar are in satisfactory agreement. At the same time, above this level with smaller values of P there is a rather distinct tendency for the grouping of these data with respect to the minimum temperature values, produced by the nightside curves of Veneras 4 and 5, and the maximum values which pertain to the dayside measurements of Veneras 9–12. A pronounced deviation from the minimal toward the maximal values takes place in the altitude range 15 to 35 km for the curves of Veneras 5 and 6, which pertain to the equator. The largest deviation from all the measured temperature profiles is seen in the Venera 9 curve in the range 15 to 40 km, while the profiles of Veneras 10–12 display perfectly satisfactory agreement. In regard to the pressure, no significant deviations are observed in the curves of $\log P(h)$ from the surface up to 40 km.

The values of T_s, and P_s, measured at the surface, are summarized in Table II from data of all the spacecraft. The small discrepancies of these values may be explained by relative altitude variations from the mean level of the surface at the landing sites of the probes Δh. The values of Δh were found to lie within limits of ±1 km; unfortunately obtaining more accurate estimates is limited by the measurement errors.

Against this background of satisfactory agreement among the dayside profiles from Venera 10–12 data, the significant departure of the Venera 9 curve from these in practically the same altitude range with a maximum at an altitude ~ 28 km arouses reasonable doubts. Therefore these data were subjected to an additional comprehensive analysis, the results of which indicated the possible existence of a systematic error in the readings of one of the sensors of the temperature measurement system, operating on this segment in

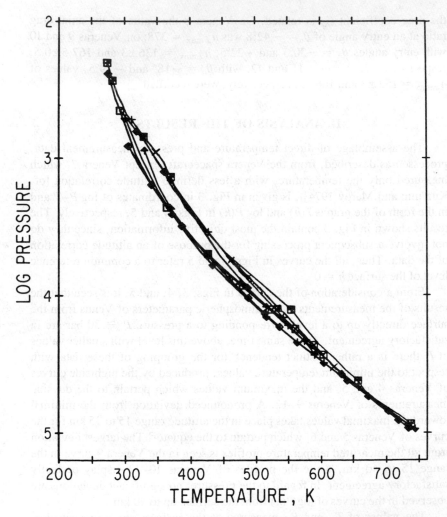

Fig. 3. Results of direct temperature and pressure measurements of Venera probes (V 4–V 12) in coordinates of log P–T. ◈, V4; ⊡, V5; ↑, V6; ↓, V8; ⊞, V9; ◈, V10; ×, V12; ⊞, V9 (processing did not take into account one of the five temperature sensors).

the atmosphere. In addition to the measured values of T, Fig. 4 also plots a corrected curve of $T(h)$, without taking into account the readings of this sensor, with a segment of interpolation by the measurement data of the remaining temperature sensors. This was found to be much closer to the other temperature profiles.

Discrepancies in the day and night profiles of $T(h)$ above 30 km definitely indicate the existence of a daily latitudinal variation of the temperature in the

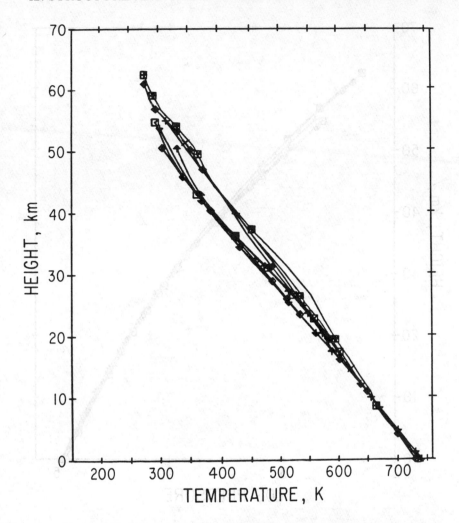

Fig. 4. Altitude profiles of the temperature of the lower atmosphere, according to data of direct measurements from Venera probes. Marks correspond to measurements in the same manner as for Fig. 3.

troposphere of Venus, attaining $\Delta T \simeq 40$ K at an altitude of 50 km. Temperature variations exert an appreciable influence on pressure changes, which comprise $\Delta P = 0.3$ kg cm^{-2} at this altitude.

Figures 6 and 7 show the values of $T(h)$ and log $P(h)$ according to the results of a processing of accelerometer measurements on Veneras 10, 11, and 12. On Venera 9 only the ascending segment of the curve $n_x(t)$ was obtained, and therefore the results are not presented. The temperature profiles obtained in the stratosphere are similar in nature for the data of Veneras 11 and 12,

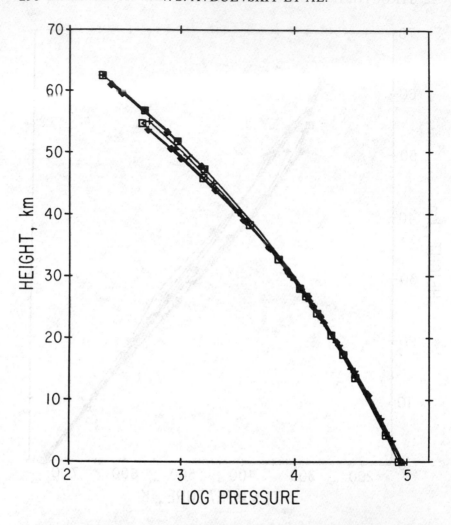

Fig. 5. Altitude profiles of the logarithm of the pressure in the lower atmosphere, according to the data of direct measurements from Venera probes. Marks and measurements as given in Fig. 3.

whereas the Venera 10 data suggest the existence of considerable variation. Above 65 km, at the upper cloud boundary, the temperature gradient is less than that of the lower lying troposphere, which is especially important for the entry segment of Venera 10. According to Venera 11 and 12 data, in the range 72 to 85 km the value of $\partial T/\partial h = (3.1–3.7)$ K km^{-1}, and the minimum value of the temperature of the stratomesopause is observed at 85 km: $T_{sm} = 160$ K (Venera 11) and $T_{sm} = 156$ K (Venera 12). At the same time, the Venera 10

TABLE II

Temperature, Pressure, and Density Values of the
Atmosphere of Venus at the Surface According to
Data of Direct Measurements

Space-craft	T_s (K)	P_s (kg cm^{-2})	$\bar{\rho}_s$ (kg cm^{-3})
V 7	747 ± 20	92 ± 15[a]	63[a]
V 8	743 ± 8	93 ± 1.5	64
V 9	728 ± 5	85 ± 3	60
V 10	737 ± 5	91 ± 3	63
V 11	734 ± 5	91 ± 3	63
V 12	735 ± 5	92 ± 3	64
Model with $R_{\venus} = 6052$ km	734	90	61

[a] Values calculated from the measured temperature and speed of descent.

data, with a smaller value of $\partial T/\partial h$, evidently testify to a temperature minimum situated at a greater height. In the stratomesosphere of Venus temperature variations and inversions are observed especially noticeable on the profile of Venera 12 between 85 and 100 km. Such a temperature variation, as well as the presence of inversions, is most likely explained by dynamic processes in the atmosphere at these heights (circulation, penetrating convection, influxes of heat).

A comparison has been made between the experimental data concerning the atmospheric parameters on the basis of direct measurements of P and T and an adiabatic temperature profile, obtained by numerical calculation, with an allowance for the deviation of the equation of state of the gas in the atmosphere of Venus from the ideal equation. $P = \rho(R/\mu)TF$, where F is the compressibility factor. A relation for the adiabatic temperature gradient was used in the form

$$\left(\frac{dT}{dh}\right)_{ad} = -\frac{g}{C_P}\left[1 - \rho\left(\frac{\partial i}{\partial P}\right)_T\right] \tag{11}$$

where C_P is the heat capacity at constant pressure and i is the enthalpy of the gas. Taking into account the hydrostatic equation, this relation can be rewritten as

$$\left(\frac{dT}{dh}\right)_{ad} = K\left(\frac{d\log T}{d\log P}\right)_{ad} \tag{12}$$

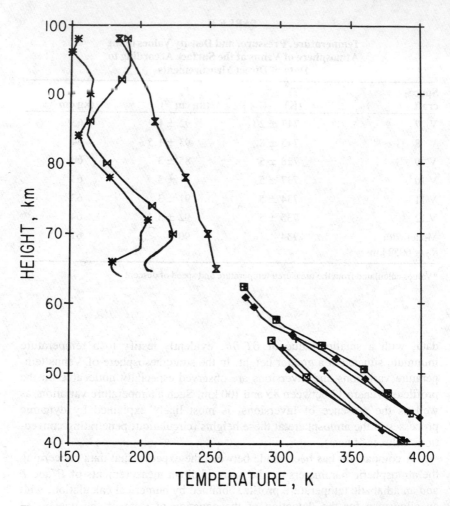

Fig. 6. Comparison of the data of accelerometer temperature measurements from Venera 10, 11, 12 probes with direct measurement data. Designations the same as given in Fig. 3. Upper curves correspond to accelerometer measurements, lower to direct: X, V10; ⋈, V11; *, V12.

where

$$K = \frac{g_0 \mu}{R}\left(F\,\frac{g_0}{g}\right)^{-1} \qquad (13)$$

and scarcely changes at all with height in the altitude range 0–50 km. The polytropic index is defined by the relation

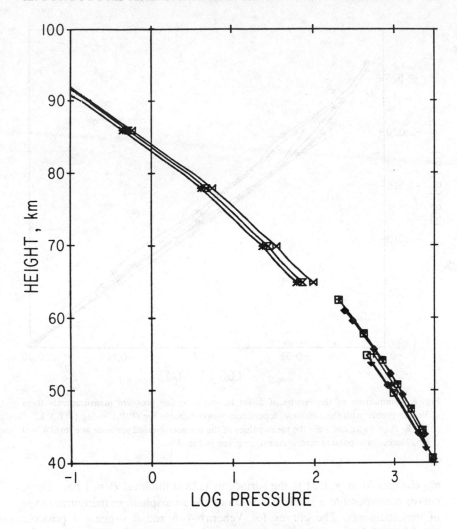

Fig. 7. Comparison of the results of accelerometer pressure measurements with direct measurement data from Venera probes. Designations the same as given in Fig. 3: X, V10; ⋈, V11; *, V12.

$$n = \left(1 - \frac{d \log T}{d \log P} \right)^{-1}. \tag{14}$$

Therefore, by comparing the experimental curves with the theoretical function of the adiabatic process, shown in Fig. 8 in coordinates $\log P/P_0 - \log T/T_0$, we can assess both the stability of the atmosphere and the polytropic index which corresponds to an altitude variation in the condition of the gas. As we see, the curves for Veneras 9–12 lead to temperature gradient values which are less than the adiabatic and to a value of $n < n_{ad}$, where the adiabatic index

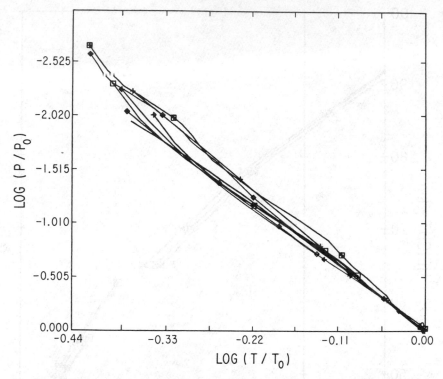

Fig. 8. Comparison of the results of direct temperature and pressure measurements from Venera probes with the adiabatic dependence in coordinates log (P/P_0) − log (T/T_0). T_0 = 735 K, $P_0 = 91$ kg cm^{-2} are the mean values of the temperature and pressure at a level $h = 0$ km. Marks correspond to measurements as given in Fig. 3.

n_{ad} changes from ≈ 1.21 at the surface to 1.28 at the level $P = 1$ bar. These curves correspond to a stable condition of the atmosphere in the entire range of measurements. The curves for Veneras 4–6 and 8 suggest a possible instability of the atmosphere at altitudes of 10–25 km (for Veneras 4 and 8), 25–37 km (for Veneras 5 and 6) where the temperature gradient is larger than the adiabatic. However this conclusion was made without allowing for measurement errors which are greatly reflected in the derivatives; therefore an additional confirmation is required.

III. COMPARISON WITH OTHER MEASUREMENTS

It is interesting to compare our profiles of $T(h)$ from the surface to an altitude of 100 km with other available measurements, as well as with the existing model distributions of the atmospheric parameters. Such a comparison is shown in Figs. 9–11. In addition to the data of direct and accelerometer measurements on the Venera spacecraft, these figures show the results of

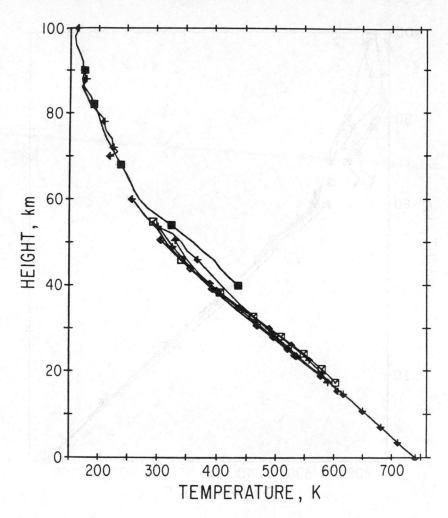

Fig. 9. Comparison of the altitude temperature profiles, obtained by measurement data from Venera probes and Pioneer Venus probes on the night side of the planet, near the morning terminator. Direct measurements: ◈, V4; ⊡, V5; ↑, V6; ↓, V8; ←, PV Day probe (Seiff et al. 1980). Radio sensing (Yakovlev et al. 1976,1978): ⊞, V9 and V10. Accelerometer measurements (Avduevsky et al. 1976a,1979): ←, PV Day probe.

direct measurements from the Pioneer Venus probes in the altitude range 65 to 13 km and data obtained during their entry in the atmosphere (Seiff et al. 1979, 1980; Chapter 11). On the whole, the variation of all the curves in the troposphere is rather similar, although definite differences have been discovered. The measurements of the Pioneer Venus probes, covering a period approximately from midnight to early morning in the latitude zone from − 30 to + 60°, i.e. virtually nightside values, when the atmosphere is coldest,

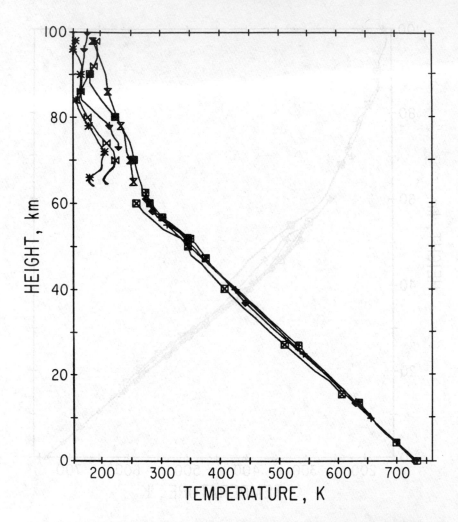

Fig. 10. Comparison of the altitude profiles of temperature, obtained by measurement data from Venera probes on the day side of the planet, with measurement data from Pioneer Venus near the morning terminator. Direct measurements: ⊞, V9 (without taking into account one of the temperature sensors); ⊕, V10; +, V11; ×, V12; ⊠, PV Large probe. Radio sensing: ⊡, V9 and V10. Accelerometer measurements: ⴵ, V10; ⴳ, V11; *, V12; ⊠, PV Large probe.

diverge little from 13 to 55 km. The Venera 4,5,6,8 data in Fig. 9, similar in time of day, are in good agreement with the Pioneer Venus measurements up to a height of 30 km. According to Fig. 10, the measured variation of the $T(h)$ curves by the Pioneer Venus probes is similar to the profiles obtained on Veneras 9 (corrected values), 10, 11 and 12, although the absolute values of the temperature according to these data, which concern the hours around

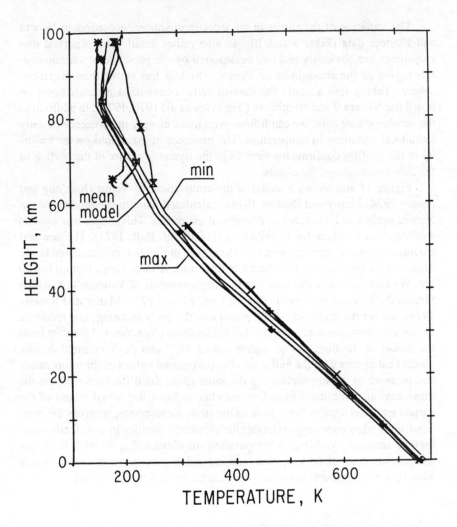

Fig. 11. Comparison of the altitude profiles of temperature, measured by Venera probes at the day and night side of the planet, with profiles from the model (Marov and Ryabov 1974): ↓, V8; ⊕, V10; +, V11; ×, V12; min, minimal model; mean model, basic model; max, maximal model. Accelerometer measurements: X̄, V10; ⋈, V11; *, V12.

noon, are 15 K higher, naturally explained by the different time of day. A latitude variation in addition to the diurnal may play a certain role in this case, since the latitudes of the landing sites of the Venera and Pioneer Venus spacecraft were different. A detailed comparison of the data is obstructed by the absence of a cross-calibration of the measurement instruments mounted on board the Venera and Pioneer Venus probes, as well as a certain discrepancy in the measurement precisions and, consequently, in the altitude correlation.

The variation of the curves in the stratomesosphere, according to Venera and Pioneer data (Figs. 9 and 10), is also rather small. The observed discrepancies are obviously real and occasioned by the presence of variations in this region of the atmosphere of Venus, which is less stable than the troposphere. Taking into account the data of radio occultation measurements on board the Venera 9 and 10 orbiters (Yakovlev et al. 1976, 1978), in addition to the accelerometer data, we can follow even more closely the presence of daily latitudinal variations in temperature. The presence of inversions on the majority of the profiles confirms the view as to the dynamic nature of the portion of the atmosphere above the clouds.

Figure 11 also shows a model of the atmosphere of Venus (Kuz'min and Marov 1974; Marov and Ryabov 1974), calculated from measurements of the Venera probes and a number of theoretical estimates. This model was used as a reference atmosphere for COSPAR (*COSPAR Inf. Bull.* 1973). The new and extensive series of measurements on the whole are well accommodated in the distributions provided by this model in the entire altitude range from 0 to 100 km. We should note in this case that the measurements of Veneras 4–8, which formed the basis of the model (Kuz'min and Marov 1974; Marov and Ryabov 1974), are for the night side of the planet and the early morning, and therefore the results of measurements in the day hemisphere (Veneras 9–12) differ from the model in the direction of higher values of T and P. A minimal-density model best agrees with the entire set of experimental values of the temperature and pressure in the troposphere; at the same time, the difference between the maximum and minimum models somewhat reduces the actual extent of the variations. This applies even more to the stratomesosphere, whereas the principal simulation curve approximates the measured profiles in a perfectly satisfactory manner, yielding a temperature minimum $T_{sm} = 160$ K in the stratomesosphere at 85 km. New and extensive measurement results which have become accessible in recent years can further refine this model.

13. COMPOSITION OF THE VENUS ATMOSPHERE

U. VON ZAHN
Universität Bonn

S. KUMAR
University of Southern California

H. NIEMANN
NASA Goddard Space Flight Center

and

R. PRINN
Massachusetts Institute of Technology

Spectroscopic observations and in situ measurements are utilized to study the composition and chemistry of Venus's atmosphere. We begin with a review of the historical development of the exploration of Venus's atmospheric composition (Sec. I) and the instrumental techniques employed for these studies (Sec. II). We next present a condensed list of recommended values for the mixing ratios of gases below 100 km altitude (Sec. III). Section IV presents a comprehensive review of the various constituent groups, their observations, related processes, and models. Included in this section is an outline of the mathematical background to current one-dimensional photochemical and transport models, a discussion on excited species and a list of references to upper limits for the abundances of unobserved gases. In Sec. V we review the available information on isotopic abundances. Finally, we outline a number of major open questions that need to be investigated and answered to further our understanding of the current state of the Venus atmosphere and to infer its origin and evolution.

I. OVERVIEW AND HISTORICAL DEVELOPMENT

Venus has by far the densest atmosphere of all the terrestrial planets. Below 80 km $> 99.9\%$ of its atmosphere is composed of only CO_2 and N_2. With respect to the composition of the Venus atmosphere, two basic issues come to our attention:

1. The trace gas content of the lower atmosphere: what is the composition of the remaining 0.1% in addition to CO_2 and N_2? What are the concentrations, reactions, sources, and sinks of these trace gases? What are their influences on the present state of Venus's atmosphere?
2. The composition of the upper atmosphere: what is the mean compositional state and its temporal and spatial variations? How can we understand the substantial abundance of O and CO in the upper atmosphere in combination with the extreme scarcity of O_2 and CO at the cloud-top level? What can we learn from the observed composition and its variations about dynamical processes acting in Venus's upper atmosphere? To what extent is the atmospheric composition responsible for the unexpectedly low exospheric temperatures found at Venus?

In addition to these broad issues there are a number of more specific problems of considerable interest, e.g:

a. Isotopic composition of the elements in the venusian atmosphere, in particular the abundance ratio D/H;
b. Exact mixing ratio of the two major constituents, CO_2 and N_2;
c. Coupling of the chemistry of the lower and upper atmosphere;
d. Density, composition, and temperature of the neutral particles in the exosphere.

Data on the atmospheric composition of Venus in conjunction with appropriate models provide us with important clues, sometimes crucial, to the type and strength of vertical and horizontal transport processes acting in the Venus atmosphere, to the infrared opacities of atmospheric gases like CO_2, H_2O, and SO_2 and their contributions to the greenhouse effect at Venus (see Chapter 18), and the identity and quantity of condensable vapors, such as water vapor and sulfur gases, from which to build haze and cloud particles (see Chapter 16). Knowledge of the upper atmosphere composition and in particular the oxygen abundance is essential for an understanding of the structure and behavior of the ionosphere (see Chapter 24). Furthermore, the measurements of the chemical and isotopic composition of Venus's atmosphere provide us with essential boundary conditions for theories and models describing the formation of the planet Venus and the origin and evolution of its atmosphere (see Chapter 29). Outstanding examples for such boundary conditions are the scarcity of water and the unexpectedly large abundance of nonradiogenic noble gases in the venusian atmosphere.

There exists, unfortunately, no generally agreed upon definition of the various altitude regions of Venus's atmosphere. It seems important, therefore, to define right at the beginning the nomenclature which we shall adopt in describing the regions of Venus's atmosphere:

lower atmosphere = the region between the surface (~ 100 bar) and the 0.1 bar pressure level.

middle atmosphere = the region between the 100 mbar and 0.1 mbar pressure level,

upper atmosphere = the region with total pressures < 0.1 mbar.

We note that the 100 mbar level falls close to 64 km altitude, the mean cloud top level. Also at 100 mbar a temperature inversion is commonly observed, which becomes particularly prominent at latitudes higher than 45° (Kliore and Patel 1980). The 0.1 mbar level falls close to 95 km altitude, the mean mesopause level. The lower and the middle atmosphere each comprise ranges in pressure a factor 1000.

A. Historical Development

Since it became established in the late 18th century that Venus possesses an atmosphere, many attempts have been made to unravel the mysteries about its composition. Early spectroscopic studies centered on a search for O_2, H_2O, and N_2. Many claims for the detection of O_2 and H_2O were made in the 19th century, but all were later shown to be unreliable. These measurements have many problems. To name a few: light from Venus passes through the Earth's atmosphere, hence absorption lines of the abundant terrestrial oxygen and water vapor appear prominently in the "Venus spectra"; also, we know today that there is in fact very little O_2 and H_2O above the reflecting cloud layer in Venus's atmosphere; furthermore, there are no N_2 lines in visible wavelengths suitable for remote sensing. For a comprehensive account of this earliest period of Venus research, the interested reader is referred to Moore (1961).

The first discovery on the atmospheric composition of Venus, which has withstood the test of time, was made by Adams and Dunham (1932) using the 100-inch reflector at Mount Wilson and photographic plates with improved sensitivity in the near infrared. They discovered three bands with heads at 782, 788, and 869 nm which they presumed to be due to Venus CO_2. This identification was verified theoretically by Adel and Dennison (1933) and through laboratory work by Adel and Slipher (1934). The latter were the first to estimate the CO_2 content above the reflecting layer to be ~ 3 km-atm (or km-amagat). For comparison, the Earth's atmosphere has a CO_2 column density of 2.6 m-atm.

Since then progress in our knowledge on the composition of Venus's atmosphere has come about in three phases. The last few years of the period 1932–1967 were characterized by continually improved earth-based spectroscopic measurements in the infrared. In the years 1967–1977, deep-space probes and a number of sounding rockets provided the major new discoveries. Also in this second phase earth-based studies succeeded in identifying the visible Venus clouds as composed of concentrated sulphuric acid, a discovery of far-reaching consequences for the atmospheric chemistry as well. The third phase started in December 1978 when no less than seven spacecraft entered the Venus atmosphere. One spacecraft was put in orbit around Venus, which is still in operation. Two additional spacecraft performed measurements during their flybys of Venus. Rather than the number of spacecraft, the vastly increased complexity and capability of the instruments employed in this latest series of spacecraft marks the start of this new phase of Venus studies. We summarize below the results obtained in these three phases.

By 1967 spectroscopic studies had clearly identified four gases in Venus's atmosphere: CO_2, CO, HCl, and HF (Connes et al. 1967,1968). Their abundances above the reflecting cloud layer were reasonably well established, as well as the effective pressure and temperature at the reflecting layer. For some time it was argued that CO_2 could only be a minor constituent of the Venus atmosphere. But arguments were put forward independently by Connes et al. (1967), Belton et al. (1968), and Moroz (1968) that CO_2 should be considered a major or even the dominant constituent. It was furthermore known that Venus's atmosphere above the clouds was extremely dry (Belton and Hunten 1966; Spinrad and Shawl 1966), but there was no general agreement as to the true water-vapor abundance (Owen 1967; Kuiper and Forbes 1967). Finally, upper limits for the abundances of many gases were established, most notably for O_2 (Spinrad and Richardson 1965). Numerous interesting details of these early observations are found in the historical account by L. Young (1974).

The second phase of Venus exploration encompasses the deep space missions of the Venera 4 through 10 spacecraft and of the Mariner 5 and 10 flybys. It also included earth-based studies of Venus's far ultraviolet (FUV) spectrum by means of sounding rockets, continued work in the visible and infrared on the elusive H_2O and O_2, and refinement of CO_2 observations. Microwave measurements at 1.35 cm wavelength of the H_2O content were improved, and similar studies at 2.6 mm of the upper atmospheric CO abundance were initiated. By the end of 1977, CO_2 was clearly established as the dominant gas with a mixing ratio larger than 93% (Vinogradov et al. 1971). The identity of the remaining 7%, however, was not known and was presumed to be mostly nitrogen and argon. The water-vapor content was recognized most likely to be variable, both in time and in space (Barker 1975a, b; Moroz et al. 1976). O_2 was finally identified, at least in the upper atmosphere, where it is responsible for prominent features of the nightglow (Lawrence et al. 1977). The upper limit of the O_2 mixing ratio close to the cloud deck was lowered nevertheless to 1 ppm (Traub and Carleton 1974). Ultraviolet and extreme ultraviolet (EUV) studies of the Venus airglow had yielded the exospheric density and altitude distribution of H (Barth et al. 1967; Kurt et al.

1968; Broadfoot et al. 1974; Bertaux et al. 1978) and of He (Kumar and Broadfoot 1975), as well as the upper atmospheric content of O (Rottman and Moos 1973). The latter remained controversial, however. The experimentally determined concentration amounted to $\sim 10\%$ at a CO_2 column density of 4×10^{16} cm^{-2}, whereas considerations of the hydrogen balance allowed theoretically for only $\sim 1\%$ atomic oxygen at comparable altitudes (Sze and McElroy 1975; Liu and Donahue 1975). The increased knowledge on the composition of the atmosphere led to the development of a number of models of the chemistry and of the transport of neutral constituents (e.g., Shimizu 1969; Lewis 1969,1970; Prinn 1971,1973b,1975; Sze and McElroy 1975; Liu and Donahue 1975; Dickinson and Ridley 1975,1977).

Phase three commenced in December 1978 with the arrival at Venus of the Pioneer Venus and Venera 11/12 spacecraft. They performed *in situ* and remote sensing measurements of Venus's atmosphere composition employing in the lower atmosphere 3 highly sophisticated mass spectrometers, 3 gas chromatographs and 2 optical spectrometers. In addition, the upper atmosphere was studied by these spacecraft using 2 mass spectrometers, 2 ion spectrometers, and 3 EUV spectrometers. New results of these latest spacecraft missions are: identification of N_2 as the other major constituent of Venus's atmosphere; measurements of the abundances of N_2, Ne, Ar, and Kr; improvement of the estimate for the He content of the lower atmosphere and determination of an upper limit for the Xe concentration; establishment of SO_2 (and not COS) as the major sulfur-bearing gas; measurements of the SO_2 and CO abundances below 30 km altitude; measurements of many new isotopic abundance ratios including surprisingly high ratios for $^{36}Ar/^{40}Ar$ and very likely for D/H; comprehensive measurements of density and composition of the upper atmosphere including their temporal and spatial variations; discovery of N_2, N, NO, and H_2 in the upper atmosphere; determination of the strength of vertical mixing in the altitude range 65 to 140 km; extensive studies of the densities and altitude distributions of exospheric H and O. Additional, but less than definite results were obtained for the concentrations of O_2, H_2O, and a number of other trace gases in and below the clouds. Earth-based observations made important contributions as well: the first discovery of SO_2 (Barker 1979), the detection of an intense airglow from $O_2(^1\Delta)$ on both the light and the dark sides of Venus (Connes et al. 1979) and measurement of the CO mixing ratio between 80 and 105 km (Kakar et al. 1976; Schloerb et al. 1980; Wilson et al. 1981). Since 1979 we have been faced with an enormous increase of information about Venus's atmospheric composition. More details and references will be given in Sec. IV.

II. INSTRUMENTAL TECHNIQUES

Our information about the composition of Venus's atmosphere comes from two sources: remote sensing observations and *in situ* measurements.

Remote sensing observations include earth-based experiments and those placed on board Venus orbiters and flyby spacecraft.

A. Remote Sensing Techniques from Earth

1. Observations of the Visible and Near-Infrared Spectrum. At the surface of Earth we can observe electromagnetic radiation from Venus either in the form of sunlight scattered from the Venus atmosphere and clouds, or thermal radiation from the surface and atmosphere of Venus. (There are a few examples of other sources of radiation from Venus observable at Earth, e.g., the emission of chemically excited atmospheric constituents like the 1.27 μm line of $O_2(^1\Delta_g)$.) During its passage through the Venus atmosphere radiation may suffer selective absorption, sometimes narrowband, at other times broadband, by various constituents of Venus's atmosphere. The resulting absorption features as observed at Earth allow us to identify the chemical nature of the absorbers. Narrow absorption lines are usually characteristic of specific absorbers and permit unambiguous identification. Depending on their oscillator strength, even rare trace constituents of Venus's atmosphere with mixing ratios well below 1 ppm may be detected in this way, e.g., HCl, HF, and SO_2. On the other hand, if atmospheric gases have no or only very weak optical transitions in the accessible wavelength range, then they are undetectable by remote spectroscopy. This latter situation applies, for example, to all noble gases, as well as to H_2 and N_2 so long as earth-based observations are restricted to the visible and near-infrared part of the spectrum. The only gases positively identified in this wavelength region are CO_2 (Adams and Dunham 1932), HCl, HF (Connes et al. 1967), CO (Connes et al. 1968), H_2O (e.g., Kuiper et al. 1969), and $O_2(^1\Delta)$ (Connes et al. 1979). In this connection it should be noted that spectral lines of H_2SO_4 in the gas phase have not yet been observed in the Venus spectra. Examples of the remarkable improvement since 1962 in the resolution of near-infrared Venus spectra, as presented by Connes and Michel (1975), are reproduced in Fig. 1.

2. Observations of the Ultraviolet Spectrum. The spectroscopic search for additional constituents has been extended to the FUV by means of Venus-pointing ultraviolet spectrometers aboard sounding rockets and Earth-orbiting satellites. Sounding rocket experiments led to the identification of O in the Venus upper atmosphere (Fig. 2) by Moos and Rottman (1971) and a remarkably accurate estimate of its column abundance by Rottman and Moos (1973). These experiments also confirmed the presence of H in the Venus exosphere, discovered earlier during the Venus flybys of Mariner 5 (Barth et al. 1967) and Venera 4 (Kurt et al. 1968). Furthermore, with a spectrometer of sufficient resolution to resolve the Lyman -α lines of H and D, Wallace et al. (1971) disproved the reality of a conjectured abundance ratio D/H ~10 at the exobase of Venus.

Two searches for minor constituents in the wavelength range 195 to 360 nm using the Orbiting Astronomical Observatory-2 by Owen and Sagan

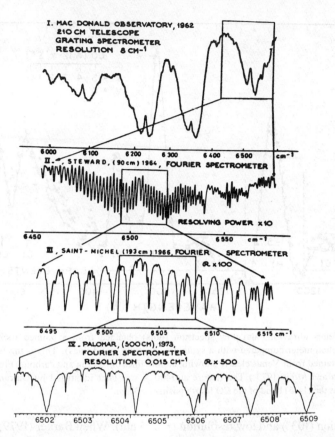

I. MAC DONALD OBSERVATORY, 1962
210 CM TELESCOPE
GRATING SPECTROMETER
RESOLUTION 8 CM⁻¹

II. STEWARD, (90cm) 1964, FOURIER SPECTROMETER

RESOLVING POWER x 10

III. SAINT-MICHEL (193cm) 1966, FOURIER SPECTROMETER

R x 100

IV. PALOMAR, (500CM), 1973,
FOURIER SPECTROMETER
RESOLUTION 0.015 CM⁻¹   R x 500

Fig. 1. Improvements in the resolution of near-infrared reflectance spectra of Venus. Four strong CO_2 bands are shown in trace I. Trace III displays a resolution of 1:80,000 as employed by Connes et al. (1967) in their discovery of HCl and HF absorption lines. Trace IV with a resolution of ~ 1:400,000 demonstrates the capabilities of a Fourier transform spectrometer in combination with the 5-m Hale telescope as used in the observation of the $O_2(^1\Delta)$ emissions from Venus by Connes et al. (1979). (Figure from Connes and Michel 1975.)

(1972) and Shaya and Caldwell (1976) did not succeed in firmly identifying new gases, although Owen and Sagan commented on a broad absorption feature centered at ~ 295 nm: "Sulfur dioxide has an absorption spectrum of this sort; ~ 5 × 10⁻³ cm-atm would produce a dip of approximately the observed extent.... But again because of the ultraviolet lability of this molecule (and vapor pressure problems?) we believe the Rayleigh scattering interpretation is more likely to be correct." Owen and Sagan proceeded to put an upper limit for the SO_2 mixing ratio of 0.01 ppm for the conditions of their observation. Their failure to identify SO_2 may be partially attributed to the comparatively large phase angle (angle between Sun, Venus, and Earth) of the

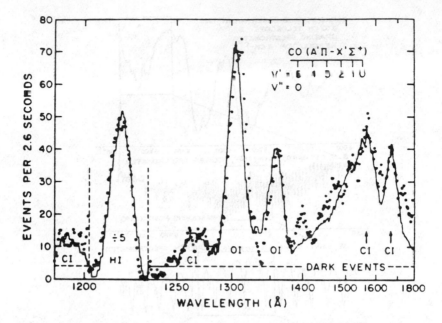

Fig. 2. Venus ultraviolet dayglow spectrum (dots) obtained from a pointed rocket-borne spectrophotometer compared with a synthetic spectrum (solid lines). This is the first spectrum obtained from Venus clearly showing atomic lines of oxygen and carbon. (Figure from Rottman and Moos 1973.) The feature at 1400 Å was later identified by Durrance et al. (1980) as the (14,5) band of the CO fourth positive system.

observation (103°) and low resolution ($\sim$ 2.5 nm). When Barker (1979) finally identified SO_2 he was able to detect it only at phase angles < 70°. He studied the 321 nm SO_2 vibrational band using the 2.7 m reflector at the McDonald Observatory with a resolution of $\sim$ 0.015 nm FWHM. Subsequently, Conway et al. (1979) confirmed the presence of SO_2 by observing with 0.01 nm resolution the reflectance spectrum of Venus at $\sim$ 215 nm using the International Ultraviolet Explorer (IUE) satellite. For their observations Conway et al. (1979) used the time of greatest western elongation of Venus in mid-January 1979. The same apparition of Venus was used by Feldman et al. (1979) to record low-dispersion spectra of the ultraviolet nightglow from Venus using the IUE satellite. These spectra, a portion of which is shown in Fig. 3, clearly revealed a number of the bands of the NO δ-system, thus leading to the discovery of nitric oxide in Venus's atmosphere.

The use of the ultraviolet spectrum as a hunting ground for minor constituents in Venus's atmosphere is certainly not yet over. The spherical albedo of Venus drops rapidly below 500 nm. Although SO_2 can explain detailed albedo features below 320 nm, the low albedo between 500 and 320 nm is not fully understood and calls for a continuing search for one or more as yet unknown absorbers, such as perhaps Cl_2 (Pollack et al. 1980b).

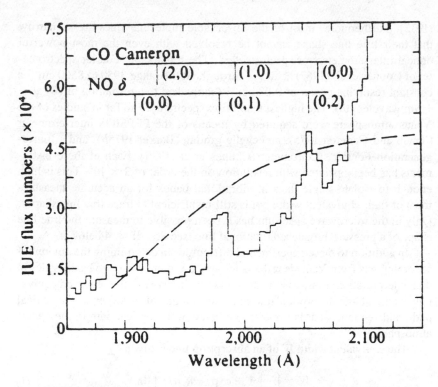

Fig. 3. Venus ultraviolet nightglow spectrum obtained from the IUE satellite, clearly showing a number of bands of the NO δ system and the CO Cameron system superimposed on a strong underlying continuum of scattered light. The dashed line gives the relative instrument sensitivity. (Figure from Feldman et al. 1979.)

3. Spectroscopic Abundance Measurements. So far we have emphasized the goal of identifying the chemical nature of gases in the Venus atmosphere. This is, however, only a first step. The second step is equally important, namely to determine the amount of gases present. This task is much more involved and complex than one would expect at first glance. Frequently there is an interference problem for genuine Venus absorption lines from the blending with either one of the numerous Fraunhofer lines or terrestrial absorption lines. The solar origin of the first can be identified by a comparison of a reflectance spectrum of the Moon or Mercury measured with the same instrument. The latter interference can be countered by making expedient use of the Doppler shift of venusian lines with respect to terrestrial ones due to the velocity of Venus relative to the observer on the Earth. There are many cases where experimental data for the oscillator strength of lines observed at Venus are totally lacking or are at least unknown for the environmental conditions of line formation in Venus's atmosphere. One historic example of this situation is the discovery of CO_2 in that planet's atmosphere by Adams and Dunham

(1932). Furthermore, most of the observable molecular lines are so narrow that their true line shape cannot be resolved with even the most powerful optical interferometers or spectrometers. The remarkable Fourier spectrometer of Connes et al. (1967) achieved throughout the range 3950 to 8500 cm^{-1} a constant resolving power of 0.08 cm^{-1} (equivalent to a resolution of 10^5 at $\sim$ 1 μm wavelength). The highest resolution spectra used so far in studies of the Venus atmosphere were acquired by means of the PEPSIOS interferometer (Traub and Carleton 1975), an echelle grating (Barker 1975b), and a fourth-generation Fourier spectrometer (Connes et al. 1979). Each of these instruments has been operated with resolution on the order of 3×10^5. This is high enough to isolate single lines in vibrational bands for an accurate measurement of their equivalent width but is still insufficient to trace true line shapes. Only in the microwave spectrum has it been possible to measure the true line shape of a pressure-broadened rotational line (see Sec. II. A.4 below).

In addition to these experimental difficulties in determining the amount of absorbing gas from Venus's reflectance spectra, there exist two basic issues in data interpretation: first, what is the relationship between the observed equivalent width of the absorption line and the amount of absorber in the optical path; and second, what is the true geometry of the optical path in the Venus atmosphere.

The equivalent width W of an absorption line is given by

$$W = \int_{-\infty}^{+\infty} \left[1 + \exp\left(-k_\lambda \rho l\right) \right] d\lambda \tag{1}$$

where k_λ is the absorption coefficient at wavelength λ, and ρ is the density of absorbing gas along the optical path of length l. W increases proportionally with the amount $u = \rho l$ of the absorbing gas only if W is much smaller than the actual pressure-broadened half width γ of the line (= weak line). Thus for $W \ll \gamma$

$$W = S u \tag{2}$$

where

$$S = \int_{0}^{\infty} k_\lambda d\lambda \tag{3}$$

is the integrated absorption coefficient, called the line intensity. If for pressure-broadened lines, W becomes much larger than γ (strong line), then for $W \gg \gamma$ the following asymptotic approximation holds:

$$W = 2 \left(S \cdot u \cdot \gamma\right)^{\frac{1}{2}}. \tag{4}$$

Here the equivalent width W increases only with the square root of the amount u of absorbing gas. In addition, W becomes dependent on γ which is proportional to the total pressure. Therefore the presence of any dilutant gas will affect the line width of the absorbing gas as well. In summary, before u can be determined from an observed equivalent width W, one also needs to know S and γ. On the other hand, γ is so small that in most cases it cannot be measured directly. Although a number of schemes have been devised to circumvent this problem (see e.g., L. Young 1972), one nevertheless realizes that the equivalent width of an absorption line is not a simple and direct measure of the amount of absorbing gas.

The latter statement becomes even more true if one takes into account the second issue: the geometry of the optical path in Venus's atmosphere. The simplest assumption is that the cloud tops act as a single reflecting layer so that all the absorption occurs in the "clear" atmosphere above the cloud deck. The situation, however, is not that simple. After traversing the clear atmosphere, solar photons in fact penetrate into the cloud. Only after multiple scattering by the cloud particles do some of these photons leave the clouds in the direction of the observing instrument, again traversing the clear atmosphere. In this way the optical path length is dependent on the scattering properties of the cloud particles; at least part of the absorption line, if not the dominant part, may be formed inside the clouds. Consequently, the optical properties of the clouds must be known (or assumed) before the amount of absorbing gas can be extracted from the observation of a Venus absorption line. Obviously at this point the quantitative analysis of remote sensing data becomes rather complex and this is not the place to go into further details. The state of the art has been reviewed in the mid-1970s by L. Young (1974). Other papers giving more emphasis to the problem of interpreting spectral lines from a scattering medium are by Chamberlain (1965), by Belton et al. (1968) and, in the context of Jupiter, by Wallace and Hunten (1978).

4. *Observations of the Microwave Spectrum.* Absorption lines have been detected also in the microwave part of the thermal radiation spectrum of Venus. The brightness temperature has been measured at many discrete wavelengths between 0.1 cm and 100 cm. It is ~ 650 K at 10 cm and decreases to 300 K at 0.1 cm. The atmospheric opacity causing the falloff of brightness temperature towards shorter wavelengths comes mostly from the far wings of the infrared CO_2 bands with potentially important contributions from SO_2, H_2O, and cloud particles (Janssen and Klein 1981). Pressure broadening, however, smears out all absorption lines formed below the cloud tops (or pressures above ~ 1 bar). Absorption by molecules in deeper atmospheric layers affects therefore only the level of the continuous spectrum. This applies also to the rotational transition of H_2O at 1.35 cm wavelength (see Fig. 4), which cannot be discerned from the continuum with presently available measurement accuracies. Nonetheless, careful comparison of the measured microwave opacity with that calculated for a model atmosphere consisting of CO_2, N_2, and one additional absorber can be used to infer upper limits for the

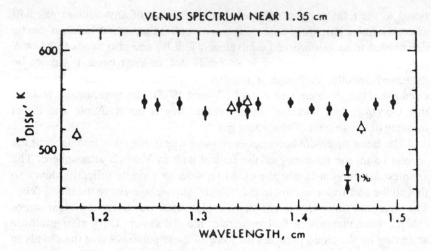

Fig. 4. Microwave measurements of the disk temperature of Venus near 1.35 cm wavelength. The dots represent Goldstone data and the open triangles Hat Creek measurements. The upper limit to any potential 1.35 cm H_2O absorption line is $\Delta T/T < 2\%$. (Figure from Janssen and Klein 1981.)

abundance of this additional absorber. Numerous upper limits derived by this method have been published, more recently e.g., those for H_2O by Pollack and Morrison (1971) and by Janssen et al. (1973), for H_2O and H_2SO_4 by Rossow and Sagan (1975), and for SO_2 and H_2O by Janssen and Klein (1981). With increasing altitude the line width due to pressure broadening decreases. It reaches the order of 1 MHz at the mesopause level ($\sim 10^{-1}$ mbar), making the absorption line clearly distinguishable from the continuum. It is in this altitude range that the CO line at 0.26 cm, initially discovered by Kakar et al. (1976), is formed. Subsequent observations by Gulkis et al. (1977) and Schloerb et al. (1980) established changes with phase angle of the obtained CO mixing ratio profile. Wilson et al. (1981) deduced profiles of the CO mixing ratio for the altitude range 80 to 105 km by using a numerical inversion of the measured profile of the CO absorption line. Fig. 5 shows examples of CO absorption lines observed on two occasions. To obtain reliable results from this method it is necessary to accurately measure the shape of a line which is itself exceedingly narrow ($\sim 1 \times 10^{-5}$ FWHM). Wilson et al. (1981) also searched for nine other molecular absorbers and determined separate upper limits for their concentrations above 70 km and above 95 km altitude. A systematic study of the detectability of millimeter wavelength spectral lines from planetary atmospheres has been published by White (1978), who showed that the most promising wavelength region is < 0.5 cm. For Venus White pointed out the possibility of detecting NO in the upper atmosphere. He also proposed a search of sulfur compounds in the atmosphere above 70 km altitude, a task later performed by Wilson et al. (1981) with negative result.

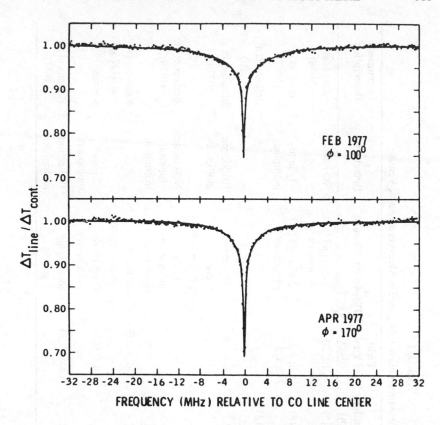

Fig. 5. CO absorption line at a frequency of 115 GHz (2.6 mm wavelength) superimposed on the Venus thermal spectrum. The line width is controlled by pressure broadening below 110 km altitude. The lines shown are formed by CO in the altitude range from 75 to 110 km. (Figure from Wilson et al. 1981.) For CO profiles deduced from these observations see Fig. 11.

Regarding the continuous spectrum White concludes that "precise continuum observations of Venus and Jupiter may yield otherwise unattainable information concerning the chemical composition of the denser layers of these two atmospheres. If the Venus continuum can be delineated with an accuracy approaching 1%, it may reveal the abundance of CO, OCS, H_2O, H_2S, NH_3, and possibly other molecules near or below the visible clouds."

B. Remote Spectroscopy from Space Probes

Remote sensing instruments aboard Venus flyby spacecraft and orbiters achieve much higher spatial resolution of the observed phenomena than do earth-based instruments. Their spectral resolution is, however, necessarily much more limited. Table I provides a summary of the experiments performed thus far. The pioneering observations of this kind were obtained from broad-

TABLE I

Remote Sensing Experiments for Atmospheric Composition Measurements at Venus

Spacecraft	Date of Observation	Wavelength Range (nm)	Resolution (nm)	Field of View (deg)	Dispersive Element	Detector(s)	Instrument Description
Venera 4 (flyby)	Oct. 18, 1967	105 to 134	(~15)	10; 20	2-channel photometer	2 NO counters	Kurt et al. (1968)
Mariner 5 (flyby)	Oct. 19, 1967	105 to 220	(~15)	3; 1.2; 1.2	3-channel photometer	3 photo-multipliers	Barth et al. (1967)
Venera 6 (flyby)	May 17, 1969	105 to 134	(~13)				Beljajev et al. (1970)
Mariner 10 (flyby)	Feb. 5, 1974	20 to 170	2	3 × 1/8	objective grating spectrometer	10 electron multipliers	Broadfoot et al. (1977a,b)
Venera 9/10 (orbiters)	Oct./Nov. 1975	122	—	0.4	H and D absorption resonance cells	3 NO counters for each instrument	Bertaux et al. (1978)
Venera 9/10 (orbiters)	Oct. 1975 to Feb. 1976	300 to 800	2	3.5 × 0.2	diffraction spectrophotometer		Krasnopolsky et al. (1976a)
Pioneer Venus (orbiter)	since Dec. 5, 1978	110 to 340	1.3	1.4 × 0.14	Ebert-Fastie spectrometer	2 photo-multipliers	Stewart (1980)
Venera 11/12 (flybys)	Dec. 21/25, 1978	30.4 to 165.7	2	1.1 × 0.33	holographic grating	10 electron multipliers	Bertaux et al. (1981)

band photometers aboard the Venus flyby spacecraft Venera 4 (Kurt et al. 1968) and Mariner 5 (Barth et al. 1967). Both experiments discovered the hydrogen corona about Venus by measuring the emission rate of Lyman -α radiation from atomic hydrogen (see Sec. IV. D and Sec. V. B). Both experiments also searched for emissions from atomic oxygen in the upper atmosphere but, interestingly enough, were unable to positively identify any atomic oxygen airglow radiation. These negative results can be explained according to Kurt et al. (1968) by the fact that Venera 4 approached Venus from the night side and observed the upper atmosphere below $\sim$ 300 km only in darkness. Kurt et al. derived for 300 km altitude an upper limit for the O abundance of n (O) $<2 \times 10^3$ cm^{-3} and concluded that "the nightside Venus is cold, consists primarily of molecular constituents, and undergoes a sharp transition to interplanetary space." In the light of the more refined Pioneer Venus measurements (see e.g., Niemann et al. 1980b), we may note that the first and third conclusion of Kurt et al., as well as their upper limit on nightside n (O) are correct. Barth et al. (1967), on the other hand, also observed the day side of Venus. However, all three channels were allowed to go off scale during the traversal of the line of sight across the bright disk; but when the field of view crossed the bright limb it took the oxygen channels $\sim$ 40 s to recover, at which time the line of sight had already passed over the atomic oxygen layer and only probed atmospheric layers above $\sim$ 450 km altitude. Hence again, no significant oxygen emissions were detected.

Mariner 10 carried the first optical spectrometer to Venus, an EUV+UV grating spectrometer having 2 nm resolution and employing a set of electron multipliers as detectors for the 10 selected wavelengths. The data obtained by Broadfoot et al. (1974) revealed for the first time the presence of significant concentrations of helium and carbon atoms in the upper atmosphere of Venus. The He I observations (see Fig. 6) were interpreted by Kumar and Broadfoot (1975) to indicate a dayside He density of $(2 \pm 1) \times 10^6$ cm^{-3} at 145 km altitude and a dayside exospheric temperature of 375 ± 105 K, which is very low compared to terrestrial conditions. Both results are in reasonable, although not exact agreement with later Pioneer Venus results (e.g., von Zahn et al. 1980). On the other hand, the helium became optically thick near 300 km altitude and thus the quoted density for 145 km represents an extrapolated value. If a somewhat lower exospheric temperature is chosen, within the error bar quoted by Kumar and Broadfoot, perfect agreement can be reached with the respective Pioneer Venus results at the 145 km level.

The Venera 9 and 10 orbiters each carried two remote sensing instruments: a hydrogen (and a deuterium) absorption cell placed in front of a NO counter (Bertaux et al. 1978) for continued study of the unique hydrogen distribution in the exosphere of Venus, and a spectrometer for the wavelength range 300 to 800 nm with a resolution of 2 nm (Krasnopolsky et al. 1976a) for investigation of Venus's nightglow. In the hydrogen absorption cell H$_2$ is thermally dissociated with the help of a heated tungsten filament, so that the

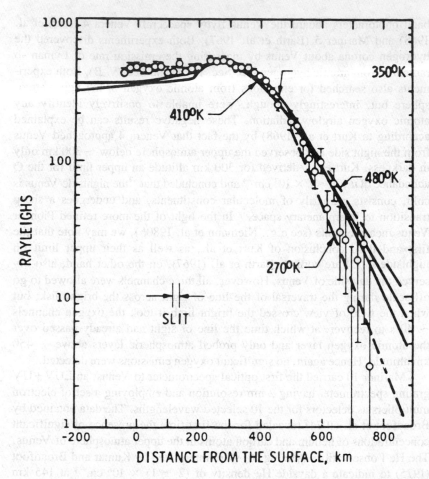

Fig. 6. The He I 58.4 nm dayglow brightness as a function of distance from the limb of Venus as derived by Kumar and Broadfoot (1975) from observations of the Mariner 10 ultraviolet airglow spectrometer. The solid lines are results of radiation transfer calculations including multiple scattering. The dashed lines represent brightness variations based on a single-scattering model. Also indicated are the exospheric temperatures inferred from the calculated helium distributions.

cell may obtain an optical thickness of 10 for Lyman -α when activated. Bertaux et al. (1978), from the analysis of one of the six counters employed, concluded that in addition to a "cool" exospheric hydrogen component there exists between $\sim$ 3000 km and 4500 km altitude an additional population of "hot" hydrogen atoms circulating the planet in satellite orbits. This notion and further results from this experiment are discussed in more detail in Sec. IV.D. The observations of the airglow spectrometers (see Fig. 7) led to the identification (Lawrence et al. 1977) of the strongest spectral features in the

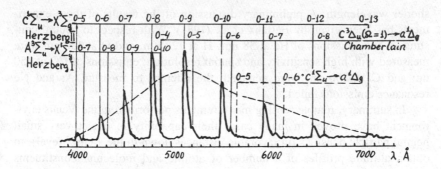

Fig. 7. Venus's visible nightglow spectrum as observed by the Venera 9 and 10 orbiters (Krasnopolsky et al. 1976*b*). The spectrum reveals a number of prominent O_2 band systems (see text). (Figure from Krasnopolsky et al. 1976*a*.)

visible portion of the Venus night airglow as due to the Herzberg II bands of molecular oxygen with some contribution from the Chamberlain system of O_2 by Slanger and Black (1978). Furthermore, Krasnopolsky and Tomashova (1980) obtained altitude profiles of the nightglow relative to the level of the cloud tops and mapped diurnal and latitudinal variations of its intensity.

So far, the most successful remote sensing instrument for atmospheric composition measurements is the ultraviolet spectrometer aboard the Pioneer Venus orbiter (Stewart 1980). Its main characteristics are listed in Table I. It is a highly versatile instrument capable of continuous and discrete spectral scans and of collecting limb profile data as well as providing images of the entire planetary disk with high spatial resolution. For imaging it makes use of the spinning and orbital motions of the Pioneer Venus orbiter. The experiment has in addition greatly benefited from its long operational lifetime in orbit about Venus, in 1982 approaching 3 full years. With respect to composition measurements this experiment provided, for example, crucial measurements of absorption features due to SO_2 in the reflectance spectrum from Venus's cloud layer (Stewart et al. 1979; Esposito et al. 1979; Winick and Stewart 1980). Furthermore, it mapped horizontally (Stewart et al. 1980) and vertically (Gérard et al. 1981) the recombination emission spectrum of NO on the night side of the planet.

The EUV and UV emissions from the upper atmosphere of Venus were again investigated by spectrophotometers aboard the Venera 11 and 12 flyby spacecraft (Table I). Each of the two instruments employed a mechanical collimator, a holographic grating, and electron multipliers placed behind the 10 exit slits (Bertaux et al. 1981). The slits were positioned at the wavelengths of the resonance radiation of He^+, He, Ne, O^+, Ar, H, O, CO, and C. Windows made of either MgF_2 or CaF_2 were placed in front of channels with wavelengths > 115 nm to give additional suppression of any unwanted light of shorter wavelength. A preliminary discussion of the results of these experiments has been given by Bertaux et al. (1981). With respect to neutral constituents the emissions of He at 58 nm, H at 122 nm, and O at 130 nm were measured with high sensitivity and spatial resolution; emissions of CO at 150 nm and C at 166 nm were detected; the attempt to measure Ar and Ne resonance emissions failed.

In summary, remote sensing measurements performed in the Venus environment have proven in many cases their capability to detect very small neutral particle densities (in favorable instances down to the 1 cm^{-3} level), to obtain altitude profiles of a number of atomic and molecular constituents (e.g., H, He, O, SO_2, NO*, and O*), and to map on a planetary scale certain airglow emissions (NO*, O* and O_2^*). Fig. 8 shows an example of the mapping capabilities of the Pioneer Venus ultraviolet spectrometer: circularized images of Venus at 130 nm, the resonance line of atomic oxygen. One would expect that continued analysis of the vast amount of data collected by the ultraviolet spectrometer aboard the Pioneer Venus orbiter should reveal many new features related to the composition of the Venus atmosphere. Nevertheless, past experience points out some shortcomings of the method and instruments. The determination of the density of ground-state atoms or molecules from the observed emission rates is generally rather complex. For resonant scattering of solar radiation the respective cross sections are so large that it takes only comparatively small column densities to produce an optically thick layer which in turn hides from observation all phenomena taking place at $\tau \gg 1$. In addition, the absolute value of the solar flux at the observed wavelength must be known. In some cases a number of excitation mechanism may contribute in uncertain proportions to the observed emission such as resonant and fluorescent scattering of solar radiation, direct photoelectron impact excitation, dissociative excitation of molecules, dissociative recombination of ions, and radiative association of atoms.

As for the instrument technology, first, the absolute spectral sensitivities at the various wavelengths still pose challenging problems for their calibration in the laboratory and for maintaining them constant throughout the lifetime of the instruments. Second, while the more advanced spectrometers allow us to study not only the few airglow lines with high intensity but also weaker emissions, the latter are rather susceptible to interference by unresolved neighboring lines. This problem is compounded in instruments without continuous spectral scanning capability (which applies so far to all instruments on Venus flyby spacecraft). Numerous examples can be found in the literature where difficulties in data interpretation were caused by implied strong emissions between the wavelengths actually monitored by the detectors. According to Bertaux et al. (1981), "many questions which are still open after this investigation could probably have been solved if a full EUV spectrum had been obtained.... This was not possible in this experiment for reasons independent of the will of the authors." The importance of obtaining

Fig. 8. Circularized images of Venus at 130 nm taken by the Pioneer Venus OUVS (L. Paxton and I. Stewart, personal communication). The orbit numbers (left to right, top to bottom) are: 477, 484, 491, 498, 505, 512, 519, 523, 533, blank, 547, 554, 561, and 568. The small-scale speckle is not significant, but the large organized features are. The typical brightness is ~12 kR.

at least one or two continuous spectra in the time period of critical observations is also borne out by the experience of the Pioneer Venus mass spectrometer studies of the Venus atmosphere. However, in spite of the difficulties of quantitative interpretation mentioned above, the remote sensing capability of orbiting spectrometers offers an unique opportunity for studying the two-dimensional morphology and temporal changes of composition, if the instrument has an imaging capability.

C. *In Situ* Measurements

1. Venera 4–8 Experiments. In situ measurements of the composition of the Venus atmosphere started with a series of experiments performed aboard the Venera 4, 5, and 6 entry spacecraft. These experiments were designed to determine the values or upper limits for the abundances of CO_2, N_2, O_2, and H_2O in the lower atmosphere below the level of the cloud tops. Because of its simplicity and technical reliability a manometric method was chosen for the CO_2 and N_2 measurements (Vinogradov et al. 1968). Ambient gas was introduced into two evacuated chambers one of which contained a chemical absorber with selective absorption properties for different gases. After filling the chambers with ambient gas and sealing them, a measurement was made of the pressure differential between the chambers that developed due to the selective pumping action of the absorbers. To determine an upper limit on the oxygen mixing ratio, the change of resistance of a 12 μm diameter, heated tungsten wire upon exposure to the ambient gas was observed. To determine the water content, two methods were used: (1) the manometric method employing $CaCl_2$ as absorber, and (2) an electrolytic method based on the change in electrical resistance of a P_2O_5 absorber. In this context we may recall that the discovery of HCl and HF in Venus's atmosphere by Connes et al. (1967) was published only about a year and a half before the launchings of Venera 5 and 6, not to mention the presence of H_2SO_4 in the clouds of Venus which was discovered much later. The composition experiments aboard Venera 4, 5, and 6 have been described in detail by Vinogradov et al. (1971) and by Surkov et al. (1977) who have also presented a summary of their results.

Venera 7 did not perform gas analysis experiments. On Venera 8 the absorber technique was used in modified form by Surkov et al. (1974) to search for NH_3 at the 2 bar and 8 bar level. The ambient gas was passed over tetrabromophenol-sulphaphthalein (bromophenol blue) which is known to turn dark blue on exposure to NH_3. This change of color was registered in an optical bridge arrangement. The instrument readings at both the 2 bar and 8 bar levels indicated a mixing ratio of NH_3 between 0.01 and 0.1% (Surkov et al. 1974); this result has met with considerable scepticism for various reasons (see Sec. IV. C.4).

2. *Mass Spectrometers.* These instruments are rather versatile, able to measure quantitatively the abundances of a great variety of different gases, to determine isotopic abundances, and to perform these measurements over a rather large dynamic signal range (up to 10^8). Furthermore, mass spectrometers with single ion counting detectors measure with such high speed even with high sensitivity, that their scan speed in most space research applications is determined more by the permitted output bit rate than by any other restriction. In spite of these advantages there are drawbacks one of which is the general instrumental complexity of a mass spectrometer which we need not discuss here. More specific to the Venus environment is the difficulty of measuring with mass spectrometers the reactive gases or radicals like H_2O, SO_2, or O. These constituents tend to adsorb or to chemically react to the ion source walls and once inside the instrument they are removed to a varying

degree from the ambient gas sample, as long as the ion source surfaces are not saturated with the respective gases. This problem is compounded for mass spectrometers operating in the lower atmosphere of Venus, because these instruments necessarily need gas inlet devices which reduce an ambient pressure of up to 100 bar down to the 10^{-5} mbar range inside the instrument ion source. This pressure reduction of more than *nine* orders of magnitude requires the use of leaks of extremely small pumping speed possibly having a high internal surface area. Initially when the inlet surfaces are clean they act as a sink for reactive gases. In the later portion of the mission, however, in particular when the inlet system is heated up along with the rise in ambient temperature, the adsorbed constituents may be desorbed again, now giving rise to additional and unwanted signals in the mass spectrometer.

There are various ways to enhance the measurement capabilities of mass spectrometers for reactive gases or separately for noble gases; a number of them have been employed in the neutral gas mass spectrometers flown into the Venus atmosphere (see Hoffman et al. 1980*c*). These measures were, however, only partly successful and not sufficient to obtain quantitative measurements of reactive gases at and below the 100 ppm level (see table 2 of Hoffman et al. 1980*c*).

The data analysis of mass spectra obtained in an unknown environment is greatly benefited by a continuous mass spectrum taken during the time of critical measurements. The mass spectrometers aboard the Venera 11 and 12 landers provided such continuous spectra at a data rate of 400 word/s. The Pioneer Venus orbiter mass spectrometer has continuous scan capability, though with a much smaller data rate (maximum of 56 word/s). It was not possible to take continuous spectra with the Large probe spectrometer of Pioneer Venus, basically because of its very high mass resolution in combination with a rather low data rate (4 word/s), nor was it possible with the spectrometer aboard the Pioneer Venus multiprobe bus because of the short time available for its measurements again in combination with a bit rate insufficient to transmit a continuous mass spectrum.

The first mass spectrometers dispatched to a planetary atmosphere other than Earth's were installed aboard the Venera 9 and 10 lander spacecraft (Surkov et al. 1978). The gas inlet systems of the instruments employed 3 porous plugs in series. Monopole spectrometers (von Zahn 1963) were used as mass analyzers. Both instruments were operating in the altitude range from 63 km ($\sim$ 130 mbar ambient pressure) down to 34 km ($\sim$ 6 bar). Unfortunately, neither instrument gave useful scientific data, due apparently to severe problems in the gas inlet systems. Surkov et al. (1978) report: "Analysis of these mass spectra indicated that for part of the descent time, the intensities of the spectral lines did not vary, and the spectra reflected the composition of residual gases of the instrument (background). The instruments began to trace...intensity variations with altitude approximately 15 minutes after being engaged." Surkov et al. (1978) published one of their mass spectra taken at an altitude of 47 km which, for example, shows an intensity ratio of $40^+/36^+ > 20$, totally disagreeing with the finding of later Venus missions of almost equal abundances of ^{40}Ar and ^{36}Ar in the Venus atmosphere. Furthermore Surkov et al. deduced from their mass spectra a nitrogen abundance of 1.8% which is a factor 2 below the mean value found during subsequent Venus exploration missions (see also Sec. IV. C. 1). These details lead us to believe that something catastrophic happened in the gas inlet systems of both the Venera 9 and 10 mass spectrometers: it would be interesting to know whether a connection has been established between the failures and the fact that the inlets were opened inside the upper cloud layer.

The Large probe mass spectrometer of Pioneer Venus (Hoffman et al. 1979a; 1980b) sampled the ambient atmosphere through two small microleaks, each consisting of a 3.2 mm diameter passivated tantalum tube whose outer end was forged to a flat-plate configuration having a conductance on the order of 10^{-7} cm^3 s^{-1} for the primary leak and 10^{-6} cm^3 s^{-1} for the secondary leak. A third leak was used for a special rare gas enrichment cell. These tubes were exposed directly to the ambient gas stream flowing around the Large probe, thus totally avoiding potential memory effects from the internal surfaces of inlet pipes, but at the same time accepting the risk of cloud particles impacting on the microleaks themselves. The latter event definitely happened to two and possibly all three of the leaks (Hoffman et al. 1980a). Mass analysis was achieved by a single-focusing magnetic spectrometer of high enough resolving power to fully resolve, for example, the mass doublet CH_4-O at mass 16^+. The instrument was opened to the ambient atmosphere at 62 km altitude and continued to operate until impact of the probe at the surface of Venus. The inlet leaks were blocked, however, between ~50 km and 29 km altitude by cloud particles.

The Venera 11 and 12 landers both carried mass spectrometers of identical design (Grechnev et al. 1979) using Bennett type RF spectrometers for mass analysis. The gas inlet system is unique in that it employs an electromagnetically operated microvalve which allows gas into the mass spectrometer in a series of short bursts lasting a few milliseconds each. No porous leaks are used, thus minimizing the effective surface area in the pressure-reducing components of the gas inlet system. Long pipes were employed, however, to pass a flow of venusian gas from the ambient atmosphere over the inlet valve and back out into the atmosphere. Ram pressure caused by the descending motion of the lander through the Venus atmosphere maintained the gas flow through the inlet and outlet pipes. The length of the entrance pipe from its inlet cone to the microvalve was on the order of 1 m, its diameter ~ 5 mm. Istomin (personal communication) notes, however, that at high altitudes and low ambient temperatures, cloud particles and condensable gases like water vapor may perhaps become attached to the internal surfaces of the inlet pipe. At low altitudes and much higher temperatures these species may detach again from the walls and be liable to introduce a bias to the low-altitude data.

The instruments of both landers took their first sample at an altitude of ~ 23 km (which is at an ambient temperature of ~ 550 K and well below the clouds) and operated until landing at the surface of Venus. None of the gases identified in the spectra by Istomin et al. (1979a) showed an altitude dependent mixing ratio except for the water peak at mass 18. It was thus possible for Istomin et al. (1980b) to reduce the electrometer noise by summing up 5 of their mass spectra obtained in the so-called "chemically-active components mode" and 28 mass spectra obtained in the "noble gas mode." In the first mode the high voltage was supplied to the ion sputter pump in standard fashion. In the noble gas mode, however, the high voltage to the ion sputter pump was switched off, thus reducing the pumping speed for noble gases to zero while maintaining at the same time an almost unattenuated pumping speed for CO_2 and N_2 due to prior sputtering of titanium inside the pump. Figure 9 shows the average of the 28 noble gas spectra obtained from the Venera 11 mass spectrometer, clearly outlining ^{20}Ne, the three argon isotopes and the unresolved krypton isotopes.

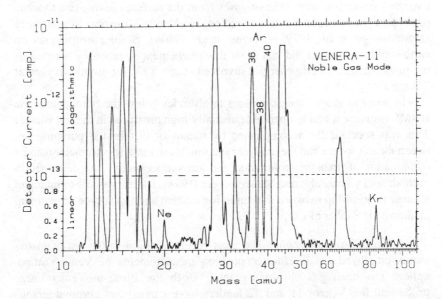

Fig. 9. Mean of 28 mass spectra obtained at altitudes below 23 km from the Venera 11 mass spectrometer in the so-called noble gas mode (see text) by Istomin et al. (1980b). The ordinate scale below 1×10^{-13} Å is linear, above this level logarithmic. Mass peaks of ^{20}Ne, three argon isotopes and of the krypton isotopes (unresolved) are clearly outlined. Note the near equality of the ^{36}Ar and ^{40}Ar peaks. This spectrum was computer processed by K. H. Fricke at the University of Bonn from data kindly supplied by V. G. Istomin.

Two mass spectrometers have performed measurements in Venus's upper atmosphere: the Pioneer Venus orbiter neutral mass spectrometer (ONMS) (Niemann 1977) and the mass spectrometer (von Zahn 1977) aboard the Pioneer Venus multiprobe bus (BNMS). Their basic design concepts are derived from similar instruments used in studies of the upper atmosphere of the Earth. Both instruments are capable of operating the ion sources with a retarding potential in order to measure selectively only those ambient particles which enter the analyzer without prior surface collisions in the ion source (also called the flythrough mode). This particular mode of operation is based on the high velocity with which the ambient particles enter the mass spectrometers, 9.8 km s^{-1} at periapsis for the ONMS and 11.6 km s^{-1} for the BNMS when it crossed the 150 km altitude level. The latter is so far the highest velocity at which mass spectrometers have been operated in upper atmospheres. It amounts to a kinetic energy of 0.7 eV amu^{-1} for the ambient particles entering the ion source, or e.g., 31 eV for CO_2 molecules. Although this energy is far above the threshold for dissociation and ionization of CO_2, no collision-induced processes of this kind have been noted in the flight data from the ONMS and BNMS instruments. On the other hand, the collisions of these high-energy particles with the ion source walls released larger than expected amounts of contaminant gases from the surfaces giving rise to some problems in measuring trace gases in Venus's upper atmosphere (Mauersberger et al. 1979; Niemann et al. 1980b). Similar instruments on earth-orbiting satellites did not show this phenomenon, probably because of the much smaller kinetic energies involved, e.g., 5 eV for atomic oxygen at 7.8 km s^{-1}.

In order to extend measurements to altitudes below the homopause, the BNMS instrument had to operate at unusally high pressures in the ion source. This was successfully accomplished by means of differential pumping between the ion source and the analyzer section of the mass spectrometer using a combination of titanium sublimation and ion sputter pumps. As described in more detail by von Zahn and Mauersberger (1978), it was possible to maintain a linear relationship between measured ion current and ion source pressure up to almost 10^{-2} mbar of CO_2.

3. Gas Chromatographs. The most recent addition to the list of instruments employed for *in situ* composition measurements of Venus's atmosphere has been gas chromatographs. Both the Pioneer Venus Large probe and the Venera 11 and 12 landers have carried gas chromatographs into the lower Venus atmosphere, details of which have been described both by Oyama et al. (1980b) and Gel'man et al. (1979a, b). Chromatographs by definition achieve the separation of substances by transporting a mobile phase containing the substances of interest over a stationary phase which has different degrees of affinity for each of the substances to be separated. For gas chromatographs the mobile medium is gaseous and consists of the sample plus carrier gas, whereas the stationary medium is solid and consists of a specially selected adsorber packed into a stainless steel capillary (column) of many meters length. The carrier gas is in principle streaming continuously through the column. The sample gas mixture is injected into the carrier gas stream at the head end of the column at discrete intervals. The retention time in the column is characteristic for each component of the sample gas (depending, for example, strongly on its vapor pressure). It can be measured in a way that not only the chemical identity of the component is determined, but also its absolute abundance in the original sample mixture.

Not surprisingly, the American Large probe gas chromatograph (LGC) and the Soviet Venera gas chromatograph (VGC) instrument designs showed considerable differences in detail; they can be explained best in a schematic of the two instruments (Fig. 10). Oxygen and argon are difficult to separate in gas chromatographs due to their similar retention times. Because they are considered very important in any analysis of Venus's atmospheric composition, Oyama et al. (1980b) in the design for their LGC provided for a long column of Porapak-N, which is the only sorbent presently known to achieve complete resolution of N_2, O_2, Ar, and CO (in that order), giving symmetric peaks at a column temperature of $\sim 20°C$. This performance, however, also dictated some rather long analysis times (> 5 min). A second, much shorter

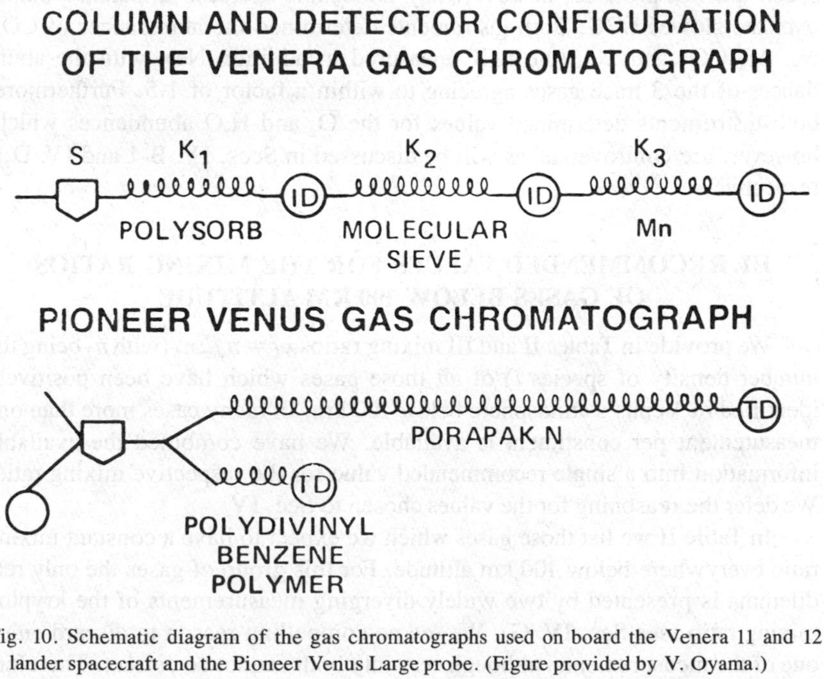

COLUMN AND DETECTOR CONFIGURATION OF THE VENERA GAS CHROMATOGRAPH

PIONEER VENUS GAS CHROMATOGRAPH

Fig. 10. Schematic diagram of the gas chromatographs used on board the Venera 11 and 12 lander spacecraft and the Pioneer Venus Large probe. (Figure provided by V. Oyama.)

column which separated CO_2, sulfur compounds, water and others for measurement, was designed around the parameters for the long column, matching its flow rate to give an approximately even sample-split and compatible analysis times. The approach of Gel'man et al. (1979b) to this problem has been to provide for their VGC two columns in series for the wide range of specific retention times encountered. Both are too short, however, to separate O_2 and Ar. Hence a third column was added in series just for the removal of O_2. A unique abundance measurement for O_2 and Ar was to be obtained by comparison of the signals from the ionization detectors (ID) Nos. 2 and 3. (For various reasons only an upper limit of 20 ppm O_2 was established.)

A second important difference between the two designs comes about by the detectors employed. The LGC used thermal conductivity detectors (TD) whereas the VGC used metastable ionization detectors. The latter have potentially 3 orders of magnitude higher sensitivity than thermal detectors and are not sensitive to temperature variations, thus eliminating the need for separate reference columns as used in the LGC. The difficulty with an ionization detector is that it requires extreme cleanliness in fabrication and operation of the instrument. Any type of contamination reduces the ionization efficiency, increases the level of standing currents, and changes in unpredictable manner the response characteristics of the detector.

During their actual operation in the Venus atmosphere, the Venera 12 VGC worked considerably faster than the Pioneer Venus LGC. This higher speed did not produce, however, any noticeable increase in science output over the slower LGC. Both instruments determined the abundances of CO_2, N_2, Ar, CO, and SO_2 (the LGC measured in addition Ne), with the abundances of the 3 trace gases agreeing to within a factor of 1.5. Furthermore, both instruments determined values for the O_2 and H_2O abundances which, however, are controversial as will be discussed in Secs. IV. B-1 and IV. D.1, respectively.

III. RECOMMENDED VALUES FOR THE MIXING RATIOS OF GASES BELOW 100 KM ALTITUDE

We provide in Tables II and III mixing ratios $x_i = n_i/\Sigma n_i$ (with n_i being the number density of species i) of all those gases which have been positively identified in Venus's atmosphere below 100 km. In many cases more than one measurement per constituent is available. We have combined the available information into a single recommended value for the respective mixing ratio. We defer the reasoning for the values chosen to Sec. IV.

In Table II we list those gases which we expect to have a constant mixing ratio everywhere below 100 km altitude. For this group of gases the only real dilemma is presented by two widely diverging measurements of the krypton mixing ratio (see Sec. IV. G). We see no compelling reason to discard either one of the measurements, although certainly both cannot be correct. Therefore

TABLE II

Recommended Values for Gases Measured below 100 km Altitude, with Mixing Ratios Assumed Constant with Altitude

Species	Mixing Ratio	Reference in Text
CO_2	96.5 ± 0.8 %	Sec. IV. B
N_2	3.5 ± 0.8 %	Sec. IV. C
He	$12^a \begin{smallmatrix}+24\\-8\end{smallmatrix}$ ppm	Sec. IV. G.1
Ne	7 ± 3 ppm	Sec. IV. G.2
Ar	70 ± 25 ppm	Sec. IV. G.3
Kr	0.7 ± 0.35 ppm	Sec. IV. G.4
	or 0.05 ± 0.025 ppm	Sec. IV. G.4

[a] Value is extrapolated into the lower atmosphere from measurements taken above 130 km altitude.

TABLE III

Recommended Values for Altitude-Dependent Mixing Ratios of Minor Constituents Measured below 100 km Altitude

Species	Mixing Ratio (ppm)	Altitude of Measurement (km)	Reference in Text
CO	350 to 1400[a]	100	Sec. IV. B
	180	90	Sec. IV. B
	< 10	75	Sec. IV. B
	50	64[b]	Sec. IV. B
	30	42	Sec. IV. B
	20	22	Sec. IV. B
H_2O	< 1 to 40[a]	cloud top	Sec. IV. D
	100	< 55	Sec. IV. D
SO_2	0.05	70	Sec. IV. E
	< 10	55	Sec. IV. E
	150	22	Sec. IV. E
H_2S	1	55	Sec. IV. E
	3	< 20	Sec. IV. E
HCl	0.4	64[b]	Sec. IV. F
HF	0.005	64[b]	Sec. IV. F
C_2H_6	2	(?)	Hoffman et al. (1980a)
O_2	(see text)		Sec. IV. B-1

[a] The quoted range of values encompasses the observed diurnal variation.
[b] For a discussion of the quoted altitude see text.

we refrain from recommending either one of the two published values for krypton.

In Table III mixing ratios for all those gases are given, which either have been observed to have a mixing ratio dependent on altitude (CO, H_2O, and SO_2) or which are expected on theoretical grounds to exhibit such altitude dependence (HCl, HF, H_2S, and C_2H_6) mainly because of their photochemical lability. Both vertical gradients of mixing ratios and diurnal variations of considerable strength have been observed in the 75 to 110 km altitude region for CO (see Sec. IV. B-2) and at cloud-top heights for H_2O (see Sec. IV. D.1). The earth-based observations of CO, HCl, and HF refer to mean pressures of line formation between 50 and 200 mbar. Because none of these constituents is expected to exhibit a steep gradient in mixing ratio at this pressure level, we prefer to quote here the altitude equivalent to 100 mbar, 64.4 km (see figure 26 of Seiff et al. 1980).

Molecular oxygen (see Sec. IV. B) has been identified in the 100 km altitude region by its emissions in the visible and infrared. These observations, however, do not help us derive an O_2 density or mixing ratio. O_2 mixing ratios were determined by the Pioneer Venus gas chromatograph at altitudes of 52 and 42 km. In Sec. IV. B.1 we shall discuss, however, why we feel unable to recommend these O_2 values for inclusion in Table III.

Besides the gases listed in Tables II and III, there exists a large body of observational and theoretical evidence, though indirect, for the presence of many additional minor constituents in Venus's middle and lower atmosphere. An example is gaseous H_2SO_4 which can be inferred from the existence of fluid cloud droplets of concentrated sulfuric acid. In addition, gaseous Cl_2 has been postulated by Pollack et al. (1980b) to explain Venus's decrease of albedo in the near ultraviolet. Similarly, from strong absorptions in the 400 to 600 nm region observed in Venus's lower atmosphere, Sanko (1980) inferred the presence of gaseous elemental sulfur S_3 at altitudes below 45 km. One last example is the observation of mass 2 ions in the Venus ionosphere which was interpreted by Kumar et al. (1981) to indicate the presence of 10 ppm H_2 in the middle atmosphere of Venus, whereas McElroy et al. (1982) used the same datum to alternatively infer large amounts of deuterium in Venus's atmosphere. As most of these inferences are either not unique or only qualitative, we will not dwell further on them here.

IV. CONSTITUENT GROUPS: THEIR OBSERVATIONS, RELATED PROCESSES AND MODELS

A. Vertical Transport in Atmospheres

The composition measurements discussed in the next sections were obtained over a wide range of atmospheric conditions, e.g., ambient pressures varying from $<10^{-15}$ bar to $\sim 10^2$ bar. We need therefore, mathematical models describing the physical and chemical processes acting in the atmosphere in order to make a comparison of observational data on one constituent obtained at different altitudes or to relate the results on different constituents. The behavior of planetary atmospheres is too complex, however, to allow complete calculation of their structure and variations from first principles with present-day knowledge. Nevertheless valuable insights into the operation of atmospheric processes can be gained by developing models which limit themselves to consideration of only a number of selected processes. Here we will not, for example, consider in detail the interactions between infrared radiation and the atmospheric constituents, thus bypassing altogether the heat balance of the Venus atmosphere (see Chapter 18 by Tomasko and Chapter 20 by Taylor et al. for treatment of this subject). On the other hand, realistic modeling of the atmospheric composition requires that we place primary emphasis on a detailed description of transport and photochemical processes. In the following we will outline some of the assumptions and approximations used in current models of Venus's atmospheric composition.

1. Molecular Diffusion. The largest force per unit mass acting in an atmosphere is due to the gravitational attraction of the planet. This dominant force causes planetary atmospheres to become strongly stratified, and therefore to be highly anisotropic. The gradients in state properties and chemical composition are much stronger along the local vertical than in a horizontal plane. We therefore will restrict ourselves in first approximation to the consideration of only those processes acting along the vertical. A more general case will be discussed in Sec. IV. B-3.

The major effect of gravity on a planetary atmosphere is to cause a gradient in total pressure p along the vertical direction z. In a static atmosphere

$$dp/dz = -\rho g \qquad (5)$$

where ρ is the total mass density, and g is the acceleration of gravity. According to the ideal gas law

$$p = nkT = \rho kT/m \qquad (6)$$

where n is the total number density, k is Boltzmann's constant, T is the absolute temperature, and m is the mean molecular mass of the gas. Combining Eqs. (5) and (6) we obtain

$$\frac{dp}{dz} = -\frac{mg}{kT}p = -\frac{p}{H} \qquad (7)$$

where $H = kT/mg$ is the (pressure) scale height. For the case where H can be taken independent of altitude, the integral of Eq. (7) becomes

$$p(z) = p(z_0) \exp\left(-\frac{z - z_0}{H}\right) \tag{8}$$

where $p(z_0)$ is the pressure at a reference level z_0. In the Venus atmosphere the scale height is ~15 km at the bottom of the atmosphere and ~4 km around 90 km altitude. Over extended altitude regions the total pressure therefore must be derived numerically from the formal solution of Eq. (7):

$$p(z) = p(z_0) \exp\left[-\int_{z_0}^{z} \frac{dh}{H(h)}\right] \cdot \tag{9}$$

A planetary atmosphere, of course, is composed of a mixture of gases. According to Dalton's law the partial pressures of the various atmospheric species add up to the total pressure

$$\sum p_i = p \tag{10}$$

where p_i is partial pressure of the ith species. In general the ratio of partial pressures of any two different gases will vary with altitude z, even for non-reactive gases, due to the process of molecular diffusion. An elaborate account of diffusion in gases has been given by Chapman and Cowling (1970). Their equations (18.2,6) and (18.3,13) can be combined to yield the basic equation of molecular diffusion in multiple gas mixtures

$$\nabla x_i + \left(x_i - \frac{\rho_i}{\rho}\right) \nabla \ln p - \frac{\rho_i}{p\,\rho}\left(\rho\, \mathbf{F}_i - \sum_j \rho_j \mathbf{F}_j\right) + k_{Ti}\, \nabla \ln T = \tag{11}$$

$$- \sum_j x_i x_j \left(\mathbf{C}_i - \mathbf{C}_j\right) / D_{ij}$$

where $x_i = n_i/n$, mixing ratio of the ith species, n_i is the number density of the ith species, ρ_i is the mass density of the ith species, $\mathbf{F}_i$ is the force per unit mass acting on the ith species, k_{Ti} is the thermal diffusion ratio for the ith species, $\mathbf{C}_i$ is the diffusion velocity of the ith species with respect to the mean mass velocity, and D_{ij} is the mean mass velocity diffusion coefficient of the ith species in the jth species. Rewriting for k_{Ti} we have

$$k_{Ti} = \sum_{j \neq i} x_i x_j \, \alpha_{ij} \tag{12}$$

where in the first approximation α_{ij} is the binary thermal diffusion factor (see Chapman and Cowling, equation 18.43, 4), and by considering motions only in the vertical, we obtain

$$- \sum_j \frac{x_i x_j}{D_{ij}} (v_i - v_j) = \frac{dx_i}{dz} + x_i \frac{m - m_i}{m} \frac{d\ln p}{dz} + \sum_{j \neq i} x_i x_j \alpha_{ij} \frac{d\ln T}{dz}$$

$$- x_i \frac{m_i}{m p} \left(\rho F_i - \sum_j \rho_j F_j \right) \qquad (13)$$

where v_i is the diffusion velocity along the z direction and m_i is the mass of the ith species; m is the mean molecular mass, equal to $\sum_j n_j m_j / \sum_j n_j$. We will dispose of the last term on the right side immediately by observing that the only external forces F_i per unit mass to be considered here are due to the force of gravity on the molecules. Because for all species

$$F_i = - g(z) \qquad (14)$$

the *direct* accelerative effects of gravity on the diffusive velocities are the same for all atmospheric constituents. Thus, the last term of Eq. (13) vanishes. Equation (13) can be rearranged to give

$$- \sum_j \frac{x_j}{D_{ij}} (v_i - v_j) = \frac{d\ln x_i}{dz} + \frac{m - m_i}{m} \frac{d\ln p}{dz} + \sum_{j \neq i} x_j \alpha_{ij} \frac{d\ln T}{dz} \cdot \qquad (15)$$

This equation shows most clearly that the velocity v_i of molecular diffusion comprises contributions from:

(a) gradients in composition x_i,

(b) a gradient in total pressure p,

(c) a gradient in temperature T.

The first term of the right side, which we will call composition diffusion, produces a diffusive flow as long as the gas mixture is inhomogeneous, that is, $dx_i/dz \neq 0$. It represents a tendency of the gas to attain a homogeneous state. In contrast, the second term, which describes pressure diffusion, and the third term, thermal diffusion, will turn a gas mixture into an inhomogeneous state as long as pressure or temperature gradients persist. In planetary atmospheres it is in particular the pressure diffusion or, more precisely, the pressure *gradient* diffusion which pushes the gas mixture towards spatial nonuniformity and causes the effect of diffusive separation of the various gaseous constituents in upper atmospheres.

For a one-dimensional static atmosphere subject to molecular diffusion the state of *diffusive equilibrium* is defined by

$$v_i = 0. \tag{16}$$

Under this assumption and noting that $d \ln p/dz = -m\, g/k\, T$, we obtain the diffusive equilibrium profiles for each species from

$$\frac{1}{n_i}\frac{dn_i}{dz} + \frac{m_i g}{kT} + \left(1 + \sum_{j \neq i} \frac{n_j}{n}\, \alpha_{ij}\right)\frac{1}{T}\frac{dT}{dz} = 0 . \tag{17}$$

Although Eq. (17) is of course easier to handle than Eq. (15) it must be said that the former lacks the physical insight which is offered by the latter. None of the three terms of Eq. (17) can be uniquely identified with either compositional diffusion, pressure diffusion, or thermal diffusion as is the case for Eq. (15). We see from Eq. (17) that even in diffusive equilibrium the density profile of each gas is dependent on all the neighboring gases. However, the coupling of the various constituent profiles through the term containing the weighted sum of the thermal diffusion factors α_{ij} turns out to be rather weak. For most applications this coupling can thus be neglected.

Realistic atmospheres are rarely, if ever, in diffusive equilibrium. This is because of the long times required to restore diffusive equilibrium after some arbitrary disturbance has been introduced into the atmospheric system. The characteristic recovery time is of the order (Mange 1957)

$$\tau_D \simeq H^2/D_{ij} . \tag{18}$$

If for an approximation we choose $H \simeq 7 \cdot 10^3$ m and $D_{ij} \simeq 5 \cdot 10^{20} n^{-1}$ m^2 s^{-1}, the recovery time in s is

$$\tau_D \simeq 1 \times 10^{-13} n \tag{19}$$

where n is in m^{-3}. Hence τ_D is longer than 1 day anywhere below the altitude level with $n = 10^{18}$ m^{-3}, which for Venus occurs at ~125 km altitude (von Zahn et al. 1980) and in the terrestrial atmosphere ~ 115 km altitude. We can safely assume that dynamic processes such as local turbulence, thermal updrafts, gravity waves, wind shears and global-scale circulation systems will all prevent the atmosphere from ever attaining a state of diffusive equilibrium in the middle and lower atmosphere.

2. *Eddy Diffusion*. In any atmosphere a considerable variety of processes act, as just mentioned, to mix the atmosphere and to homogenize its chemical composition much more efficiently than can be accomplished by molecular diffusion alone. Our knowledge about the sources, sinks, strength, and the temporal and spatial variations of these additional processes is, however, rather rudimentary even for the terrestrial atmosphere, not to mention that of Venus. For Venus our knowledge is insufficient to justify attempts of detailed modeling of individual dynamic processes and their effect on the atmospheric composition.

In order to make some headway towards a quantitative description of vertical transport processes beside molecular diffusion, we will introduce a generalization of the basic diffusion Eq. (15). We recall that it is the composition diffusion that represents a process of atmospheric mixing. Its strength, however, is constrained by the diffusion coefficient D_{ij} which is derived from kinetic gas theory and therefore cannot be considered a free parameter. In order to account for the contributions of all kinds of vertical mixing processes in one-dimensional models of the atmosphere, we will generalize Eq. (15) by adding a fourth term on the right hand side

$$-\sum_j \frac{x_j}{D_{ij}}(w_i - w_j) = \frac{d\ln x_i}{dz} + \frac{m - m_i}{m}\frac{d\ln p}{dz} + \sum_{j \neq i} x_j \alpha_{ij} - \frac{d\ln T}{dz}$$

$$+ \sum_j \frac{x_j}{D_{ij}} K \frac{d\ln(x_i/x_j)}{dz} \ . \tag{20}$$

This new term introduces a component of diffusion which, similar to composition diffusion, acts as long as there is a gradient in relative composition $d(x_i/x_j)/dz \neq 0$. It increases the *effective* strength of vertical mixing processes beyond the capability of molecular diffusion because K is considered a free parameter; it is called the eddy diffusion coefficient, or for short, eddy coefficient. K may vary with altitude, but it is presumed to be species independent. w_i is the diffusion velocity of the ith species resulting from the combined action of molecular diffusion and eddy diffusion; we will define the total diffusive flux of the ith species as

$$\phi_i \equiv n_i w_i \ . \tag{21}$$

The effect of K becomes particularly clear if we apply Eq. (20) to a *binary* gas mixture. Then $x_1 + x_2 = 1$ and we obtain, e.g., for the first constituent

$$(x_1 - 1)(w_1 - w_2) = (D_{12} + K) \frac{d\ln x_1}{dz} +$$

$$D_{12} \left[\frac{m - m_1}{m} \frac{d\ln p}{dz} + (1 - x_1) \alpha_{12} \frac{d\ln T}{dz} \right] . \tag{22}$$

Here the eddy coefficient K acts to increase the strength of only the composition diffusion term. $K >> D_{12}$ will turn the atmosphere into a homogeneous (mixed) state.

The total diffusive flux ϕ_i can be written as sum of the molecular diffusion flux $\phi_{m, i}$ and the flux $\phi_{e, i}$ due to eddy diffusion

$$\phi_i = \phi_{m, i} + \phi_{e, i} \tag{23}$$

hence

$$n_i w_i = n_i v_i + \phi_{e, i} . \tag{24}$$

Furthermore, after subtracting Eq. (15) from Eq. (20) and substituting Eq. (24) we obtain the eddy diffusion flux

$$\phi_{e, i} = - K \frac{n_i}{x_i} \frac{dx_i}{dz} = - K n \frac{dx_i}{dz} . \tag{25}$$

The expression (25) for the eddy flux is based on a gradient in $x_i = n_i / \sum n_j$. It agrees with the definition used by Banks and Kockarts (1973: see their equation 15.11). This definition differs slightly, however, from the one used by Hunten (1975a) and by others who define $\phi_{e, i} = - K n_a \frac{df_i}{dz}$, with $f_i = n_i / n_a$ and n_a being the number density of the dominant background gas (not total density). The latter definition appears useful only in case n_i is the density of a trace gas. As we develop equations here for a multiple gas mixture with no restriction to trace gases, the definition as represented by Eq. (25) is preferred in this context.

The eddy diffusion flux $\phi_{e, i}$ is proportional to the gradient of the mixing ratio x_i of the i of the species. In fact, it is this relationship which was presented by Lettau (1951) when he introduced the concept of eddy diffusion as a means to parameterize the mixing by any fluid motions. Nevertheless Eq. (22) shows more distinctly than Eq. (25) the close connection between the eddy coefficient K and the molecular diffusion coefficient D_{ij}. More in-depth discussions of the "essentially fictitious process" of eddy diffusion, as Lindzen calls it, can be found for example in articles by Lindzen (1971b, 1981) and Hunten (1975a). It may be helpful to point out that the word eddy in fluid dynamics and meteorology is used in a more general sense: there an eddy is any deviation $\mathbf{v}'$ from the mean flow $\bar{\mathbf{v}}$. In this context eddies occur over a wide range of spatial scales (10^{-2} to 10^{7} m) and temporal scales (seconds to months). The main interest in eddies rests in the fact that they provide transport or flux of properties in addition to the transport caused by the mean flow.

Modeling of vertical transport by eddy diffusion has been widely and successfully employed for the Earth's lower thermosphere and the stratosphere. It also forms the basis of most models of Venus's atmospheric composition. The part of an atmosphere where $K >> D_{ij}$ is called the homosphere, and the part with $D_{ij} >> K$ the heterosphere. The altitude level at which $K = D_i$ is called the homopause (formerly turbopause). Here D_i is the average molecular diffusion coefficient of the ith constituent given by

$$D_i = \left[\sum_{j \neq i} (n_j / n D_{ij}) \right]^{-1}. \tag{26}$$

The above given definition of the homopause accepts the (not particularly meaningful) fact that for each constituent the homopause occurs at a different altitude due to D_i being species dependent, whereas K is species independent. Usually the respective altitude differences amount, however, to at most a few kilometers.

3. *Photochemical Reactions.* For the case in which the ith constituent undergoes photochemical reactions we need to add to the system of differential Eq. (20) the continuity equations

$$\frac{\partial n_i}{\partial t} = P_i - L_i - \frac{\partial \phi_i}{\partial z} \tag{27}$$

where P_i and L_i are the photochemical production and loss rates, respectively, of species i at altitude z. Both steady-state or time-averaged calculations ($\partial n_i / \partial t = 0$) and time-dependent model calculations ($\partial n_i / \partial t \neq 0$) may be carried out.

B. Carbon and Oxygen Compounds

In this section we shall treat the observations and related chemical models of CO_2, CO, O_2, and O; each of these gases has been positively identified in the Venus atmosphere. In addition, O_3 is expected on theoretical grounds to exist in Venus's middle atmosphere, though only in minute amounts. CO_2 is the dominant constituent throughout the Venus atmosphere up to altitudes well above the homopause. We can consider that the other 4 gases mentioned above are derived from CO_2 through photochemical and thermochemical reactions. The interplay among the five constituents is greatly intensified by the presence of hydrogen, sulfur, and chlorine compounds in the atmosphere of Venus. Their specific roles in Venus's chemistry will be discussed below in Secs. IV. D–F.

B-1. Observations in the Lower Atmosphere

Since 1967, it has been firmly established that CO_2 is the dominant gas in the Venus atmosphere as outlined in Sec. I. There is only one other major constituent, nitrogen, the mixing ratio of which is (3.5 ± 0.8) %. Hence the mixing ratio of CO_2 is 96.5% with the same uncertainty of 0.8%. In the following we will assume that the abundance ratio $n(N_2) / n(CO_2)$ is constant throughout the lower and middle atmosphere. We should note that this assumption has little observational support. In fact, the data of the Pioneer Venus LGC indicate some altitude dependence of this abundance ratio at least in the lower atmosphere (Oyama et al. 1980a). Nevertheless, we shall postpone a discussion of this subject to Sec. IV. C.1 on nitrogen compounds.

The mixing ratio of CO in the lower atmosphere has been measured by the Pioneer Venus gas chromatograph at three altitude levels. The results of Oyama et al. (1980a) are:

$$32^{+62}_{-22} \text{ ppm at 52 km altitude,}$$

$$30 \pm 18 \text{ ppm at 42 km, and}$$

$$20 \pm 3 \text{ ppm at 22 km.}$$

The Venera 12 gas chromatograph performed 8 gas analyses between 42 km and the surface. Gel'man et al. (1979a) published only the average of all 8 analyses:

$$28 \pm 7 \text{ ppm below 42 km.}$$

Finally, earth-based near-infrared observations by Connes et al. (1968) yielded close to the top of the lower atmosphere ($\sim$ 100 mbar) a CO mixing ratio of

$$45 \pm 10 \text{ ppm.}$$

L. Young (1972) subsequently reanalyzed the Venus spectra taken by Connes et al. and deduced a CO mixing ratio of 51 ± 1 ppm. She inferred an effective pressure and temperature of line formation of 36 mbar and 249 K, respectively. The corresponding values derived by Connes et al. are 60 mbar and 240 K. In either case the so-called effective pressure is *not* synonymous with a mean pressure at the level of line formation. The latter is probably a factor $\sim$ 3 larger than the effective pressure (Connes et al. 1968).

These observations seem to establish that there exists a gradient in the mixing ratio of CO at least between 64 km altitude ($\sim$100 mbar pressure) and $\sim$ 40 km. Below 42 km none of the gas chromatograph experiments firmly identified a continuation of this gradient within the error bars of observations. On the other hand, the data taken by the Pioneer Venus LGC at 22 km are suggestive of such a trend. Furthermore, Moroz (1981: p. 25) reports: "Measurements on Venera 12 demonstrated some systematic trend of CO abundance between altitudes 36 and 12 km (a decrease of 40%)." The observed gradient in the mixing ratio implies a source of CO in the middle atmosphere, very likely photodissociation of CO_2, and a sink for CO below 40 km, for which the reactions with SO_3 and SO_2 reforming CO_2 are the most important candidates (see Sec. IV. E).

Our knowledge of the mixing ratio of O_2 in Venus's lower atmosphere is, unfortunately, in a less satisfactory state than for CO. For the top of the lower atmosphere Traub and Carleton (1974) have determined the very stringent upper limit of 1 ppm for the O_2 mixing ratio (no positive identification of O_2). On the other hand, the Pioneer Venus LGC (Oyama et al. 1980a) measured an O_2 mixing ratio of 44 $\pm$ 25 ppm at 52 km altitude, and 16 $\pm$ 7 ppm at 42 km. The Venera 12 VGC determined an upper limit of 20 ppm for the mixing ratio of O_2 anywhere below 42 km altitude (Gel'man et al. 1979a). Finally, from the spectrophotometer measurements aboard the Venera 11 and 12 landers Moroz (1981) determined an upper limit for the O_2 abundance of 50 ppm anywhere below 60 km altitude.

The substantial gradient in the O_2 abundance between 64 km and 52 km implied by the observations is rather difficult to explain. Furthermore, as pointed out by Oyama et al. (1980a), the amounts of O_2 and CO observed in the lower atmosphere cannot be in thermodynamic equilibrium, nor can the observed amounts of O_2 and SO_2 (Craig et al. 1981). Oyama et al. comment: "The coexistence of O_2 and CO in the lower atmosphere would seem to involve a distinctly nonequilibrium process. Only by postulating an elemental S/O ratio of $\sim$ 3 instead of 2 were we able to generate nontrivial amounts of O_2 in our thermodynamic calculations. Several tens of parts per million of O_2 near the surface were obtained only with models that also had SO_3 as the dominant sulfur-bearing species in this portion of the atmosphere and had a mixing ratio of about 1000 ppm."

The idea of saving the situation by allowing for the presence of high abundances of SO_3 in the lowest part of the Venus atmosphere has been studied in more detail by Craig et al. (1981). This study, however, presumes CO to be inert, which in our opinion is questionable in the light of the high reactivity of SO_3 towards CO. We therefore conclude, that either we have to accept a strong disequilibrium state among CO, SO_2, O_2, and H_2O in the lower atmosphere of Venus or discard at least one of the measurements in order to save the assumption of thermodynamic equilibrium. The latter course is our preferred one. It leads us to eliminate the O_2 measurements at 52 and 42 km altitude from the set of recommended observations, mainly because they are in fact based on a marginal signal and in addition are not supported by any other independent observation. This, however, leaves us with no positive O_2 abundance measurement anywhere in the Venus atmosphere, which is deplorable considering the importance of O_2 data for any model of Venus's atmospheric chemistry.

B-2. Carbon and Oxygen Compounds in the Middle Atmosphere

1. Observations. No measurement exists that accurately determines the mixing ratio of CO_2 in the middle atmosphere of Venus. Hence we will extrapolate the mixing ratio of CO_2 measured in the lower atmosphere (96.5 ± 0.8) % into the middle atmosphere by assuming that it remains constant up to at least the 0.1 mbar level. The total number density n at 0.1 mbar and 180 K is 4×10^{21} m^{-3}. At this density we obtain the time constant for establishing diffusive equilibrium from Eq. (19) as being $\tau_D \simeq 13$ yr. We can safely assume that eddy mixing acts on the atmosphere composition in much shorter time scales and will thus prevent any diffusive separation of gases in the middle atmosphere.

CO is the only constituent of the middle atmosphere for which a more elaborate set of measurements is available. As mentioned before, the CO abundance was first determined by Connes et al. (1968) to be 45 ± 10 ppm at the lower boundary of the middle atmosphere. Since 1976, detailed observations of pressure broadened microwave lines of CO by Kakar et al. (1976), Schloerb et al. (1980), and Wilson et al. (1981) have provided vertical profiles of the CO mixing ratio in the altitude range 75 to 105 km (see also Sec. II. A.4). These early measurements all agree that there exists a strong gradient of the CO mixing ratio within the altitude range of observations. The mixing ratio increases from roughly 10^{-4} at 85 km ($\sim$ 1 mbar) to 10^{-3} at 105 km ($\sim$ 0.01 mbar). Two subsequent studies (Wilson and Klein 1981; Clancy et al. 1981) extended the measurements to 110 km altitude and established the fact that the CO mixing ratio exhibits a significant diurnal variation. Two such profiles obtained by Wilson and Klein (1981) are shown in Fig. 11. The planetocentric phase angle $\phi = 320°$ (Fig. 11a) implies that the observed signals are coming preferentially from the late morning side of Venus, whereas at $\phi = 180°$ (Fig. 11b) mostly the midnight sector of Venus is probed. These and many other spectra imply strong day-night variations of the CO mixing ratio. They are particularly pronounced in the 75 to 85 km region where late morning CO abundances are a factor of ≥ 10 above those at midnight. The diurnal variation disappears at $\sim$ 90 km where the CO mixing ratio is found to be 180 ± 90 ppm. Above 95 km the phase of the diurnal variation reverses, with maximum CO mixing ratios occurring near midnight and minimum ratios near noon. In summary, the observations of Wilson and Klein indicate three significant facts:

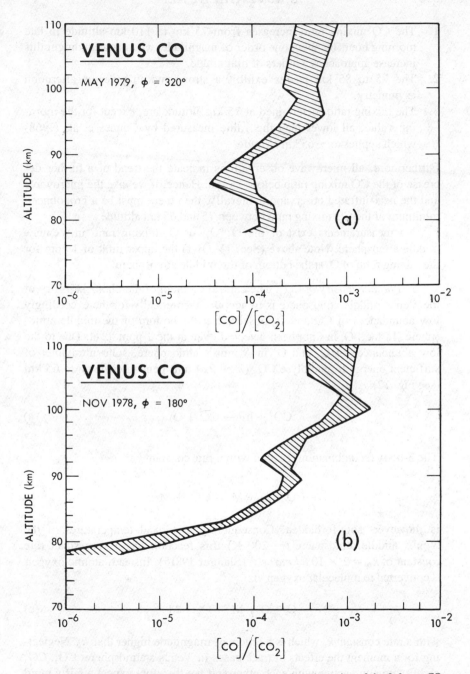

Fig. 11. Profiles of the abundance ratio CO/CO_2 obtained from inversion of the 2.6 mm CO absorption line by Wilson and Klein (1981). The phase angle of observation emphasizes late morning and midnight conditions in (a) and (b), respectively. Strong diurnal variations are observed, both above 100 km and below 85 km altitude.

1. The CO mixing ratio increases from 75 km to 110 km altitude in late morning hours by about one order of magnitude, whereas at midnight this increase approaches 3 orders of magnitude.
2. The 75 to 85 km results exhibit a strong morning versus afternoon asymmetry.
3. The mixing ratios determined at 75 km altitude are, except for the morning value, all lower than the value measured by Connes et al. (1968) which applies to ~ 65 km altitude.

Furthermore, all microwave observations indicate the trend of a further decrease of the CO mixing ratio below 75 km. Hence, if we take the microwave and the near-infrared observations literally, then there must be a pronounced minimum of the CO mixing ratio between 75 and 65 km altitude.

No measurements exist of the O, O_2, or O_3 mixing ratio in Venus's middle atmosphere. Note above (Sec. IV. B-1) the upper limit of 1 ppm for the mixing ratio of O_2 at the bottom of the middle atmosphere.

2. Photochemistry of CO_2. The central problem of the photochemistry of the Venus middle atmosphere is to account theoretically for the exceedingly low abundances of CO and O_2 observed at the bottom of the middle atmosphere. In fact, O_2 has not been detected even at the 1 ppm level. Due to the low abundances of O_2 and O_3 in Venus's atmosphere, solar ultraviolet of sufficient energy to photolyse CO_2 ($\lambda < 224$ nm) penetrates down to 65 km (see Fig. 12):

$$CO_2 + h\nu \rightarrow CO + O \tag{a}$$

The 3-body recombination reaction with a rate constant k_b

$$CO + O + M \rightarrow CO_2 + M \tag{b}$$

is, however, spin-forbidden. Consequently at typical temperatures of the Venus middle atmosphere (~ 200 K) this reaction has a very small rate constant of $k_b = 2 \times 10^{-37}$ cm^6 s^{-1} (Slanger 1981b). Instead, atomic oxygen is converted to molecular oxygen via

$$O + O + M \rightarrow O_2 + M \tag{c}$$

with a rate constant k_c which is 5 orders of magnitude higher than k_b. Neglecting for a moment the effects of trace gases in Venus's atmosphere, CO_2, CO, and O_2 are nonreactive with each other and we therefore expect a fairly rapid transition (on geologic time scales) of the CO_2 atmosphere to one dominated by CO and O_2. CO_2 would disappear from the upper atmosphere within a few weeks, and from the entire middle atmosphere in a few thousand years.

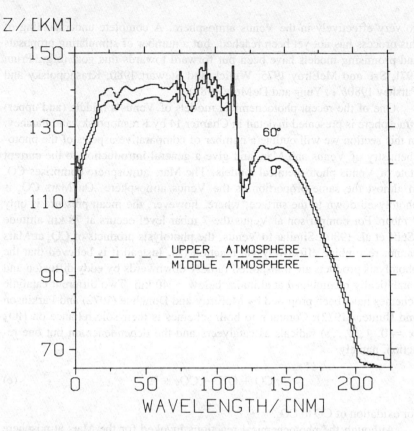

Fig. 12. Penetration depth ($\tau = 1$) of solar radiation into the Venus atmosphere for solar zenith angles of 0° and 60°, based on the upper atmosphere model of von Zahn et al. (1980) above 110 km, and the model of Yung and DeMore (1982) below 110 km.

Indeed, these arguments describe the situation correctly for the upper atmosphere of Venus, provided we take into account also the various dynamic processes exchanging gas between the upper and middle atmosphere. The above arguments, however, fall short in explaining the observed composition of the middle atmosphere which at least close to its lower boundary is characterized by an extreme dearth of the products of the CO_2 photolysis, that is, CO and O_2. The photolysis rate of CO_2 reaches 5×10^6 cm^{-3} s^{-1} near 70 km at middle and low latitudes (Krasnopolsky and Parshev 1980c). At this rate 1 ppm O_2 will be produced in merely 5 (Earth) days. Judging from the observed abundances of CO and O_2 (upper limit only), we must conclude that nature has found means to make the net reaction

$$CO + \tfrac{1}{2} O_2 \rightarrow CO_2 \qquad\qquad (d)$$

go very effectively in the Venus atmosphere. A complete understanding of this process has not yet been reached, but a number of stimulating proposals and promising models have been put forward towards this goal (e.g., Prinn 1971; Sze and McElroy 1975; Winick and Stewart 1980; Krasnopolsky and Parshev 1980b,c; Yung and DeMore 1982).

One of the recent photochemical models of Venus's middle (and upper) atmosphere is presented in detail in Chapter 14 by Krasnopolsky and Parshev. In this section we will outline a number of comparative aspects of the photochemistry of Venus and Mars and give a general introduction to the current state of Venus photochemical models. The Mars atmosphere comprises CO_2 in almost the same proportion as the Venus atmosphere. On Mars CO_2 is photolysed down to the surface, where, however, the mean pressure is only 7 mbar. For comparison at Venus the 7 mbar level occurs at 78 km altitude (Seiff et al. 1980). Similar to Venus, the photolysis products of CO_2 at Mars cannot recombine effectively via reaction (b). Instead it is believed that the photolysis products are transported rapidly downwards by eddy diffusion and catalytically recombined at altitudes below ~ 40 km. Two different catalytic schemes have been proposed by McElroy and Donahue (1972) and Parkinson and Hunten (1972). Common to both schemes is their sole reliance on HO_x ($x = 0, 1, 2, \ldots$) radicals as catalyzers and the dependence on but one reaction, namely

$$CO + OH \rightarrow CO_2 + H \qquad \qquad (e)$$

for oxidation of CO to CO_2.

Although the photochemical reactions invoked for the Mars atmosphere are in principle also at work in Venus's middle atmosphere, they alone do not offer a satisfactory explanation of the Venus atmospheric composition for the following reasons:

1. The mean solar ultraviolet flux at Venus is 4.4 times larger than on Mars;
2. In spite of (1), the observed mixing ratios of CO and O_2 are much lower at Venus (at least at the bottom of the middle atmosphere) than on Mars;
3. Venus's middle atmosphere contains much less water than does the Mars atmosphere;
4. There is hardly any $O(^1D)$ produced in the middle atmosphere of Venus which is capable of splitting up, for example, H_2O and H_2 to form OH. In the Earth and Mars atmospheres this excited $O(^1D)$ is primarily produced through photolysis of O_3. Ozone in turn is a side product of the photolysis of O_2 which is rather abundant in both atmospheres. Consequently the scarcity of O_2 on Venus reduces drastically the production of O_3 and $O(^1D)$ and limits the production of OH from H_2O to negligible amounts;
5. The most sophisticated model of the photochemistry of Mars (Kong and McElroy 1977; updated by McElroy et al. 1977) needs to invoke very

high eddy coefficients ($\sim 3\times10^7$ cm^2 s^{-1}) at 30 to 40 km altitude ($\sim 1\times10^{16}$ to 3×10^{15} cm^{-3} total number density) in order to match the model predictions with the available observations on the column abundances of O_2, H_2O, and O_3 in the Mars atmosphere. However, these high eddy coefficients could well be a model artifact due to an incomplete photochemistry scheme used in the model. The calculated altitude profile of the eddy coefficient would be much more constrained by measured altitude profiles instead of column abundances of critical constituents such as H_2O and O_3. Unfortunately, for the Mars atmosphere no such measurements exist. This situation is quite different for Venus. Here we have at least six independent inferences of the eddy coefficient including information on its altitude profile (see also Fig. 21). These refer to 60 km altitude for the analysis of radio signal scintillations by Woo and Ishimaru (1981), to 69 km altitude for the determination of the SO_2 profile by Winick and Stewart (1980), to $\sim$ 65 and 80 km altitude for the study of the cloud particle scale heights by Prinn (1974), to 70–95 km altitude for the aerosol scattering observations by Krasnopolsky (1979a), to 115 km altitude for the ultraviolet nightglow observations of Gérard et al. (1981), and to 130–145 km altitude for the dayside measurements of He, N_2, and CO_2 profiles of von Zahn et al. (1980). All observations obtained above 65 km altitude (except for Krasnopolsky's data above 80 km) are consistent with an altitude profile of the eddy coefficient $K = A \cdot n^{-1/2}$ (with n = total number density), where $A \sim 1 \times 10^{13}$ (within a factor of 2) if K is measured in cm^2 s^{-1} and n in cm^{-3}. This relationship, on the other hand, implies for the 10^{15} to 10^{16} cm^{-3} density range eddy coefficients at Venus which are > 2 orders of magnitude smaller than those invoked by McElroy et al. (1977) in their model of the martian photochemistry. Irrespective of the true situation on Mars, we see no obvious reason why at Venus there should exist a very limited altitude range with eddy coefficients as high as those inferred for Mars. This further illustrates the point that photochemical models which are successful at Mars do not necessarily also meet the constraints imposed by the observational facts on the Venus atmosphere.

6. The atmospheric region where photochemistry is driven by solar ultraviolet is laden with dust and in direct contact with the planetary surface at Mars. For Venus this region is far removed from the planetary surface, and it is interspersed by haze and cloud particles. These differences may account for many of the peculiarities via heterogeneous reactions taking place in the two atmospheres.

Is there a unique key to the CO_2 photochemistry of the Venus atmosphere? The answer is yes; the key is the presence of a few tenths of ppm HCl in the middle atmosphere. HCl has a considerable photodissociation cross section longward of 200 nm (see Fig. 13). Contrary to H_2O, HCl is therefore

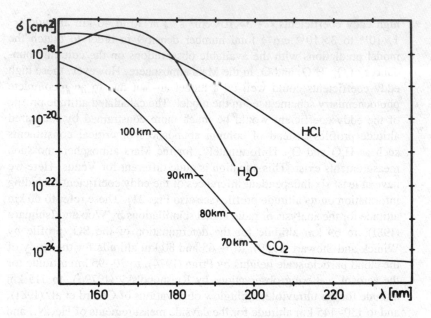

Fig. 13. Absorption cross sections of CO_2, H_2O, and HCl in the wavelength range between 150 and 240 nm. Altitudes are indicated where the vertical CO_2 column density is $1/\sigma$ ($\tau = 1$ for vertical incidence). HCl is photodissociated throughout the middle atmosphere, whereas H_2O is largely shielded by CO_2 from photodissociation below 90 km.

not shielded from photolysis by an extended layer of CO_2 as was first pointed out by Prinn (1971). Hence in the Venus atmosphere HCl is photodissociated right down to the cloud tops:

$$HCl + h\nu \rightarrow H + Cl \,. \tag{f}$$

On the other hand, HCl is reformed via the reaction

$$Cl + H_2 \rightarrow HCl + H \,. \tag{g}$$

Furthermore, the reverse process is also possible

$$H + HCl \rightarrow H_2 + Cl \,. \tag{h}$$

The net effect of the reactions (f, g, h) is that HCl catalyzes the dissociation of H_2 into H, as indicated in Fig. 14. As predicted by Prinn (1971) and confirmed in the elaborate model calculations of Krasnopolsky and Parshev (1980b,c) and Yung and DeMore (1982), HCl maintains in this scheme a near constant mixing ratio throughout the middle atmosphere. Further details of the HCl photochemistry are discussed in Sec. IV.F-1.

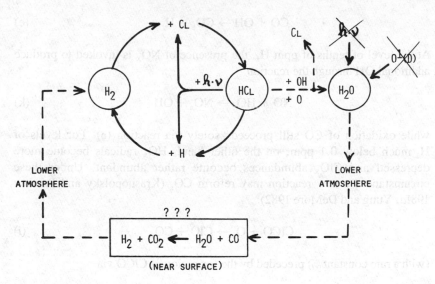

Fig. 14. Schematic diagram summarizing the transfer of hydrogen in the middle atmosphere from H_2 via HCl into H_2O. H_2 may be reformed from H_2O by thermochemical reactions near the surface.

H_2O becomes an almost irreversible sink for hydrogen, because it is neither photodissociated nor attacked by $O(^1D)$ in Venus's middle atmosphere. More precisely, the lifetime against photodissociation for H_2O becomes ~ 1400 yr at 70 km altitude at midlatitudes (Yung and DeMore 1982). It is produced, for example, via

$$HCl + O \rightarrow OH + Cl \qquad (i)$$
$$HCl + OH \rightarrow H_2O + Cl .$$

The hydrogen lost in H_2O production must be supplied either by H_2 or HCl. The resupply of H_2 in turn could be caused by the water gas reaction

$$H_2O + CO \rightarrow H_2 + CO_2 \qquad (j)$$

which at the pressures and temperatures present near the surface of Venus could perhaps transform H_2O back into H_2. Krasnopolsky and Parshev (1981a) first studied the other alternative, i.e., that hydrogen is in fact subtracted from HCl with the consequence of building up large amounts of Cl_2, at the same time strongly suppressing H_2. This model is discussed in detail in Chapter 14.

The balance between active hydrogen and chlorine radicals depends ultimately on the abundance of H_2 in the middle atmosphere. At and above levels of a few ppm of H_2, enough HO_x radicals will be produced to oxidize CO just as on Mars through only reaction (e) repeated below (Yung and DeMore 1982):

$$CO + OH \rightarrow CO_2 + H \,. \tag{e}$$

At the level of tenths of ppm H_2, the presence of NO_x is invoked to produce additional OH through the reaction

$$NO + HO_2 \rightarrow NO_2 + OH \tag{k}$$

while oxidation of CO still proceeds solely via reaction (e). For levels of H_2 much below 0.1 ppm, on the other hand, HO_x radicals become more depressed and ClO_x abundances become rather abundant. Under these circumstances a new reaction may reform CO_2 (Krasnopolsky and Parshev 1981a; Yung and DeMore 1982):

$$ClCO + O_2 \rightarrow ClO + CO_2 \tag{l}$$

(with a rate constant k_l) preceded by the formation of ClCO via

$$Cl + CO + M \rightarrow ClCO + M \,. \tag{m}$$

As discussed below in Sec. IV.D.3, we are still ignorant of the H_2 abundance anywhere in the Venus atmosphere. The only direct information we have is the upper limits of 200, 70, and 10 ppm at 52, 42, and 22 km altitude, respectively, obtained by the Pioneer Venus LGC (Oyama et al. 1980a). Lacking more substantial information, we cannot yet recommend adoption or elimination of either one of the two photochemical schemes, that is, the HO_x or the ClO_x as the dominant one. Also this is advisable because the important reaction rate k_l has only been indirectly obtained so far. Nevertheless Yung and DeMore put forward the following arguments in favor of the ClO_x scheme:

1. It is the only scheme capable of explaining the nighttime decrease of the CO mixing ratio in the 75 to 85 km altitude range;
2. It is the only model that predicts a significant abundance of gaseous Cl_2 at cloud-top heights, which could account also for the observed decrease of Venus's albedo in the wavelength range 320 to 400 nm, as suggested by Pollack et al. (1980b);
3. The ClO_x driven model does not require any tropospheric source of H_2, the identity and strength of which pose major questions for the HO_x driven model.

To this list of points in favor of the ClO_x driven scheme we may add one more: in these models the eddy diffusion coefficients are considered free parameters and used to optimize the comparison between model predictions and available measurements. For the ClO_x driven model, throughout the middle atmosphere, the eddy coefficient and its altitude profile can be chosen identical to the average terrestrial profile, which we consider a very reasonable result. On the other hand, for the HO_x driven model a steep increase of the eddy coefficient by a factor of 10 is required between 58 and 64 km, which we consider at odds with the observation of a rapidly increasing static stability just in this altitude range (Seiff et al. 1980). This need for high eddy coefficients is common to the Venus and the Mars HO_x photochemistry schemes. It is mandated by the requirement to assure fast downward transport of O_2 and O in order to avoid too large a build up of O_2 in the model of the middle atmospheres. Finally, we add that the suggestions of substantial amounts of deuterium in the Venus atmosphere (McElroy et al. 1982; Donahue et al. 1982) are contradictory to the conjectured H_2 mixing ratio on the order of 10 ppm (Kumar et al. 1981); they do not rule out, however, the presence of a few tenths ppm of H_2. Therefore a positive identification of the suggested deuterium does not help us much to decide between the HO_x and ClO_x driven schemes.

An introduction to the CO_2 photochemistry of Venus's middle atmosphere would be incomplete without at least a brief mention of the fate of the O and O_2 formed as consequence of CO_2 photolysis. Following Yung and DeMore (1982), for the HO_x-rich schemes (see Fig. 15a) $\sim$ 75% of all the O atoms formed enter subsequently the recombination scheme proposed by McElroy and Donahue (1972) for Mars (reaction (e) in Fig. 15). The remaining atoms recombine to O_2. Their O-O bond in turn is broken only below 70 km altitude in sulfur reactions such as

$$S + O_2 \rightarrow SO + O$$

$$SO + HO_2 \rightarrow SO_2 + OH \tag{n}$$

which were introduced by Winick and Stewart (1980) and Yung and DeMore (1982), respectively. Some of the oxygen is also consumed in the oxidation of SO_2 to SO_3 and H_2SO_4 near and within the upper-cloud layer (reaction (o) in Fig. 15a):

$$SO_2 + O + M \rightarrow SO_3 + M. \tag{o}$$

For the ClO_x-rich scheme the fate of O and O_2 is quite different. Atomic oxygen is almost completely recombined to O_2 which in turn is consumed in reaction (l) as indicated in Fig. 15b. The latter reaction achieves almost a local balance with the photolysis rate of CO_2 in the altitude range 65 to 90 km. Hence there is a greatly reduced need for vertical transport compared with the HO_x-dominated recombination scheme. The oxidation of SO_2 plays a minor role in the total oxygen balance and becomes noticeable only below 65 km.

As outlined above, the photochemical models of Krasnopolsky and Parshev (1981a) and Yung and DeMore (1982) are successful in matching model

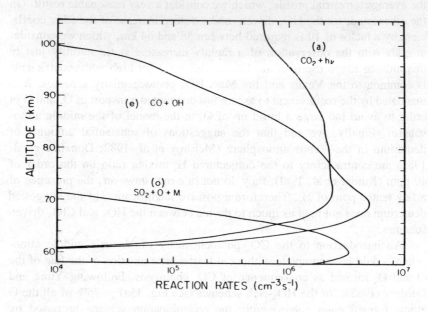

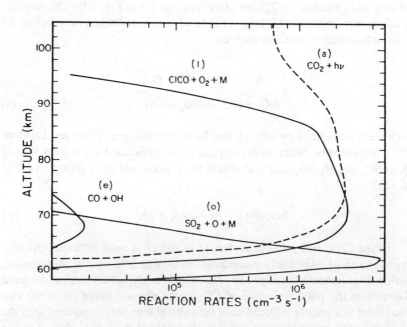

Fig. 15. The main oxygen-producing reaction (a) and consuming reactions (e, l, o) for the cases of a HO_x driven photochemistry (upper figure) and a ClO_x driven photochemistry (lower figure) of Venus's middle atmosphere. (Figures from Yung and DeMore 1982).

predictions with the (limited number of) relevant observations. It seems, however, that at least one observed feature is not covered by either model: the very low CO mixing ratios (< 10 ppm) measured at the 75 km altitude level by the microwave technique (see above). On the other hand, whether the microwave data for 75 km are totally reliable remains to be confirmed, as this is the lowest altitude for which such data have so far been published. In addition, further and improved measurements of the O_2, CO, and HCl abundances at 65 km altitude from earth-based, near-infrared observations would be highly desirable.

B-3. Composition of the Upper Atmosphere

1. Observations. The Pioneer Venus mission provided the first opportunity to measure the detailed composition of the neutral gas in the upper atmosphere of Venus. CO_2, CO, O, N_2, N, NO, He, and H have been positively identified and their absolute number densities determined, including many of their important temporal and spatial variations. Characteristic height profiles at equatorial latitudes are shown in Fig. 16 for the subsolar and

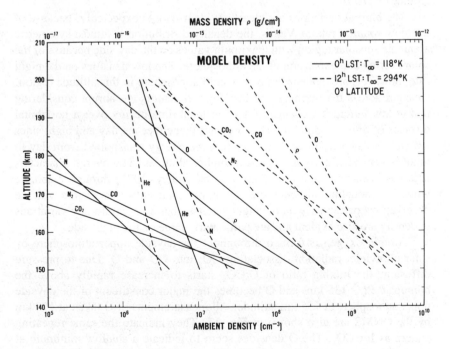

Fig. 16. Empirical model densities as a function of altitude at noon and midnight, 0° latitude based on the data of the Pioneer Venus orbiter neutral mass spectrometer. (Figure from Niemann et al. 1980*b*.) Here, as in Figs. 17 and 18, the density correction factor of 1.63 used by Hedin et al. (1982) is not included.

antisolar points (Niemann et al. 1980b). These results are derived mostly from the observations made with the neutral mass spectrometer (ONMS) on the orbiter spacecraft, whose orbital parameters allowed us to measure composition once every 24 hours at or near periapsis at all longitudes (Colin 1980). Periapsis latitude ranged from 13°N to 17°N. Complete diurnal observations were made in 225 Earth days, the sidereal period of Venus. The useful lifetime of the orbiter, i.e., with low-altitude periapsis for vertical composition profile measurements, extended through nearly three diurnal cycles.

The results for the density measurements of CO_2 and O at a reference altitude of 170 km are shown in Fig. 17, using local solar time (LST) as independent variable. The data were collected over a period of more than 600 Earth days. They reflect spatial, long-time, diurnal, and short-time variations in the upper atmosphere. The minimum time separation between original data points is 24 hours, the orbital period. Longer data gaps occurred during a few periods which were either dedicated to other specific mission objectives or were near superior conjunction when communication with the orbiter was difficult. Whenever necessary, for any given orbit, data points were interpolated between measurements made at altitudes adjacent to the reference altitude of 170 km.

The diurnal variations are clearly evident. Not unexpectedly, because of the slow rotation rate of Venus, the density distribution is almost symmetric about the subsolar region with maximum values on the day side because of the solar heating and expansion of the atmosphere. The low densities on the night side, however, are evidence of a very cold atmosphere in this altitude region. Changes across the morning and evening terminators are abrupt considering that at low latitudes a change of 1 hr in local time occurs over a horizontal distance of only ∼ 1500 km. This suggests that either the day and night sides are well isolated from each other or that the energy transported from day to night is very efficiently dissipated at higher altitudes. The overall characteristic has been predicted by Dickinson and Ridley (1977) through elaborate numerical modeling, although specific numerical results do not agree with the observations particularly on the night side (see below). Day-to-day variations in density are also evident and are particularly large on the night side.

Atomic oxygen is formed throughout the dayside upper atmosphere by solar ultraviolet radiation dissociating CO_2 into CO and O. Due to pressure diffusion, the mixing ratio of O/CO_2 starts to increase rapidly above the homopause (∼ 135 km) and O becomes the major constituent of the dayside upper atmosphere at 155 km altitude. All measurements of O taken at 170 km by the ONMS are also shown in Fig. 17. They indicate the same repeating pattern as for CO_2. The O densities seem to indicate a shallow minimum at approximately 15 hr LST, as do the CO_2 densities near 11 hr LST. It is not clear, however, whether these minima are genuine features of the diurnal variation or in fact a response to the variations in the solar 10.7 cm flux. This ambiguity arises because the repetition period for measuring the diurnal cycle

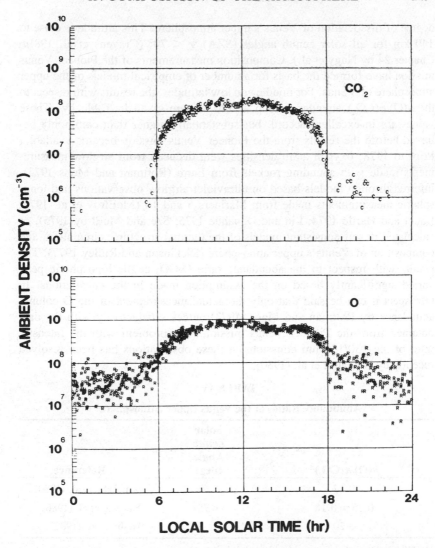

Fig. 17. Measurements of CO_2 and O number densities at 170 km altitude (see text) taken by the Pioneer Venus orbiter mass spectrometer over nearly three diurnal cycles.

is 225 d for the Pioneer Venus orbiter, which is nearly 8 times the 28 d period of solar activity as experienced at Venus. Although the 28 d variation exhibits now and then random phase changes and therefore cannot be considered a stable long-term period, the data sampling period of the ONMS (~ 3 diurnal cycles) is still too short to allow a reliable separation of the amplitudes of the diurnal and the 10.7 cm flux variations.

In the past, the abundance ratio $n(O)/n(CO_2)$ at the altitude of the day-side ionospheric peak has commonly been selected as an indicator for the

degree of dissociation of Venus's upper atmosphere. This altitude is close to
140 km for all solar zenith angles (SZA) $\chi < 70°$ (Cravens et al. 1981a;
Chapter 24 by Nagy et al.). Composition measurements of the Pioneer Venus
mission have formed the basis for a number of empirical models of the upper
atmosphere of Venus. For middle and low latitudes, the results with respect to
the $n(O)/n(CO_2)$ abundance ratio at 140 km are given in Table IV. These
values are in excellent accord, but substantially higher than commonly believed before the results from the Pioneer Venus mission became available.
Prior to 1978, oxygen atom densities were deduced from airglow measurements made with sounding rockets from Earth (Rottman and Moos 1973),
inferred through models based on ultraviolet airglow observations and ionospheric measurements made from Mariners 5 and 10 (McElroy et al. 1973;
Bauer and Hartle 1974; Liu and Donahue 1975; Sze and McElroy 1975), or
calculated in self-consistent models of the thermal structure, circulation, and
composition of Venus's upper atmosphere (Dickinson and Ridley 1977). The
results with respect to the abundance ratio O/CO_2 at the ionospheric peak
varied significantly based on the assumption made in the calculations. In
retrospect it can be said that only the actual measurement of the O column
abundance by Rottman and Moos (1973) agrees satisfactorily with the data
obtained from the Pioneer Venus mission. The problem with the intensity
ratio of the 130/136 nm emissions in these observations has been resolved
recently by Durrance et al. (1980).

TABLE IV

Abundance Ratios of the Venus Upper Atmosphere[a]

$n(O)/n(CO_2)$[a]	Solar Zenith Angle (deg)	Reference
0.17	61	von Zahn et al. (1980)
0.15 to 0.18	<75	Keating et al. (1980)
0.20	0	Hedin et al. (1982)

[a]At 140 km altitude for the solar zenith angles shown.

The diurnal variations of CO, N_2, and N follow a pattern similar to CO_2
and O. Helium and hydrogen, the two lightest constituents, are the exceptions. Similarly to Earth they are minor constituents and are carried along in
the flow of the heavy gases from day to night. The light weight and corresponding large scale height increases the transport efficiency with altitude
resulting in a diurnal density reversal, i.e., densities higher at night than
during the day. This is demonstrated in Fig. 18 which shows the He densities
measured by the ONMS at 170 km altitude over three diurnal cycles. Helium
shows a distinct bulge at night near the morning terminator with its maximum
at $\sim$ 5 hr LST and decreasing abruptly at the terminator to a nearly constant

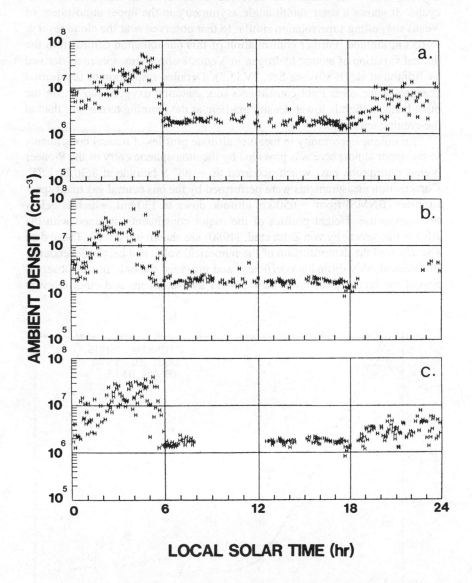

Fig. 18. Measurements of He number densities obtained at 170 km altitude by the Pioneer Venus orbiter mass spectrometer over 3 diurnal cycles (a,b,c).

daytime level. The bulge was observed at all three diurnal measurement cycles. It shows a solar zenith angle asymmetry in the upper atmosphere of Venus suggesting superrotation similar to that observed near the cloud level at ~ 65 km altitude. Further confirmation of this phenomenon comes from the diurnal variation of atomic hydrogen in Venus's upper atmosphere as derived by Brinton et al. (1980) (see Sec. IV.D.3). Detailed inspection of the diurnal changes of the other light constituents (e.g., atomic oxygen) shows, on the other hand, a slightly lower density gradient at the morning terminator than at the evening terminator.

An unique opportunity to measure altitude profiles of neutral constituents in the upper atmosphere was provided by the atmospheric entry of the Pioneer Venus multiprobe bus, which occurred at ~ 38° S latitude at 8:30 hr LST. Composition measurements were performed by the bus neutral gas mass spectrometer (BNMS) from ~ 650 km altitude down to 130 km, which is below the homopause. Height profiles of the major constituents obtained with the BNMS instrument by von Zahn et al. (1980) are shown in Fig. 19. These data also allowed the determination of the numerical value and height dependence of the local eddy diffusion coefficient and provided the only *in situ* observation of the homopause. From their BNMS measurements and other relevant

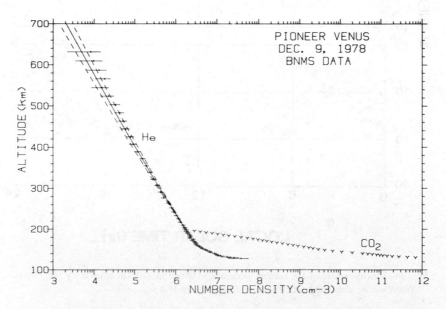

Fig. 19. Measurements of He and CO_2 number densities obtained by the neutral mass spectrometer on board the Pioneer Venus multiprobe bus. The gradient of He densities above 200 km indicates an exospheric temperature of 275 ± 15 K for this observation at 08:30 LST. The curvature of the He profile below 200 km has been used by von Zahn et al. (1979a) to calculate the eddy coefficient profile in the lower thermosphere.

Pioneer Venus data, von Zahn et al. (1980) determined the eddy coefficient K (in $cm^2 s^{-1}$) as dependent on the total number density n (in cm^{-3}):

$$K = 1.4 \times 10^{13} n^{-1/2}. \tag{28}$$

This places the homopause altitudes for N_2 and He at 136 km and 130 km, respectively. At the N_2 homopause $D_{N_2} = K \sim 4 \times 10^7$ cm^2 s^{-1}. This indicates a much more vigorous vertical exchange of gas at Venus than found in the Earth's atmosphere where we find at the homopause typically $D_{N_2} = K \sim 1 \times 10^6$ cm^2 s^{-1} (Hunten 1975a).

Total mass densities were not only measured by the ONMS and BNMS instruments, but were also derived by Keating et al. (1980) from the atmospheric drag affecting the Pioneer Venus orbiter near periapsis. However, the mass density values derived from drag measurements are generally 60% higher than those computed from the composition measurements made with the ONMS. Extrapolation of the orbiter results to the latitude of the bus entry further suggest that the ONMS results are low compared to the BNMS results. Ionospheric model calculations of Cravens et al. (1981a) also suggest that densities, needed to best fit the observed altitude of the ionospheric peak, should be higher than those measured by the ONMS by a factor 1.5. Therefore, for purposes of modeling Venus's upper atmosphere composition, Hedin et al. (1982) make the assumption that the ONMS densities previously published are too low by a factor of 1.63. Although the explanation for this discrepancy is likely to be found in uncertainties associated with gas dynamic effects in the ONMS instrument at high satellite velocities, the difference is not yet explained.

Kinetic temperatures have not been measured directly in the upper atmosphere but by observing the height variations of the densities, an effective scale height can be determined from which a kinetic temperature can be inferred. Although altitude and latitude change simultaneously along the orbit, it is the altitude changes that dominate (Colin 1980; Niemann et al. 1980b). Kinetic temperatures so derived from the ONMS measurements of number densities are shown in Fig. 20. Very similar temperatures have been inferred by Keating et al. (1980) from total mass densities derived from the atmospheric drag effects of the Pioneer Venus orbiter. Average values for both day and night are lower than predicted (Dickinson 1976; Dickinson and Ridley 1977) and, as already observed in the number densities, the changes across the terminators are abrupt. More remarkable still is the low temperature at night of ~ 110 k, which is considerably lower than the average temperature of 170 K observed near the mesopause at 95 km by the Pioneer Venus orbiter infrared radiometer (Schofield and Taylor 1982a). This differs considerably from conditions found on Earth where the upper atmosphere is warmer than the mesopause for all solar zenith angles. The explanation is most likely to be found in an efficient downward heat conduction on the night side caused by

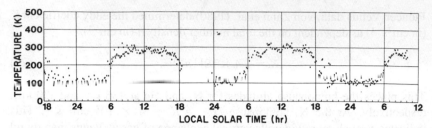

Fig. 20. Diurnal variation of the exospheric temperature as derived from the number density scale heights measured by the Pioneer Venus orbiter neutral mass spectrometer. The sampling of one diurnal cycle lasted 225 Earth days, the sidereal period of Venus.

turbulence and possibly lower-than-predicted horizontal energy transport across the terminator. Variations in densities, particularly on the day side, are also reflected in the inferred temperatures. A good many of these variations are probably caused by changes in the solar extreme ultraviolet (EUV) flux. The interesting implications of these data are treated in detail in Chapter 20 on the energy balance of the upper atmosphere of Venus.

Perturbations in the height profiles of the various gas constituents were observed by both Pioneer Venus upper atmosphere mass spectrometers. These pertubations were particularly large on the night side (see e.g., Niemann et al. 1980b). In addition strong perturbations in the total mass density profiles were measured by the Pioneer Venus Large, Day, and Night probes throughout their measurements in the upper atmosphere below $\sim$ 120 km (Seiff et al. 1980). Thus, there is ample evidence for turbulence or wave motion in the upper atmosphere of Venus.

2. Models. The photochemistry of the Venus upper atmosphere was first studied by Shimizu (1963). He calculated the vertical distributions of CO_2, CO, O, and O_2 under the influence of solar ultraviolet radiation, assuming photochemical equilibrium and neglecting vertical transport altogether. In this type of model, CO_2 is dissociated into O (and a thin layer of O_2 is formed) within a narrow altitude band of $<$ 5 km. Assuming relative abundances of 90% CO_2 and 10% N_2 in the middle atmosphere, Shimizu (1963) calculated this dissociation layer to occur at a particle density of 2.5×10^{13} cm^{-3}, later located by Pioneer Venus at an altitude of $\sim$ 113 km (von Zahn et al. 1980). In this model Shimizu used a rate constant $k_b \sim 7 \times 10^{-37}$ cm^6 s^{-1} for the 3-body recombination of CO and O to CO_2, only slightly higher than the currently accepted value of 2×10^{-37} cm^6 s^{-1}.

For the following 10 yr studies of the photochemistry of Venus's upper atmosphere (McElroy 1968, 1969; Shimizu 1969, 1971, 1973) were overshadowed by the misinterpretations of the Venera 4 and Mariner 5 observations concerning the O abundance and temperature in the upper atmosphere of

Venus. Neither Kurt et al. (1968) with their ultraviolet photometer on board Venera 4 nor Barth et al. (1967) with their similar instrument on board Mariner 5 were able to detect the resonance emission of atomic oxygen at 130 nm. Subsequently it was not sufficiently appreciated that for reasons given in Sec. II.A.2 these observations applied only to altitudes well above the exobase where indeed atomic oxygen is highly depleted. It was taken for granted that the oxygen deficiency observed at these high altitudes extended throughout the entire upper atmosphere of Venus, which in fact is not true (see above). In addition, the electron density profile of the dayside ionosphere obtained from the Mariner 5 occultation experiment by Kliore et al. (1967) was also taken as indicating unexpectedly low O abundances. This conclusion was based, however, on a rather incomplete understanding of the ion chemistry of Venus and on model calculations of the thermal balance which yielded an exospheric temperature much too high (McElroy 1968; R. W. Stewart 1968).

The search for causes of the inferred low O abundances led to a photochemical explanation (McElroy 1968) and an explanation in terms of upper atmosphere dynamics (Donahue 1968; Shimizu 1969). The former pathway was quickly rejected (e.g. Donahue 1968; Shimizu 1968; DeMore 1970). The conclusions drawn by Shimizu, on the other hand, remain valid, although in part derived from invalid assumptions. Shimizu (1969) writes: "By comparing our [model] results with the observed electron density at 250-500 km by Mariner 5, the deficiency of O atoms in the Cytherean upper atmosphere is concluded. This may be explained by the effect of the lower atmosphere dynamics on the upper atmosphere phenomena. We suggest $D_{eddy} = 1\text{-}3 \times 10^7$ cm^2 s^{-1} (or horizontal wind with the velocity of 100-300 m s^{-1}) in the upper atmospheric region." It is interesting to note, that even today there still exists the option to explain the composition of Venus's upper atmosphere either in terms of vigorous vertical eddy transport or by a global-scale circulation system. Most likely both mechanisms act together in proportions still unknown.

The concept of eddy diffusion was introduced to Venus aeronomy by Shimizu (1969). It became an essential part of all later one-dimensional photochemical models of the Venus upper atmosphere like those of Liu and Donahue (1975), Sze and McElroy (1975), and Krasnopolsky and Parshev (1981a). Considerably more realism was incorporated into these models when the assumed exospheric temperature was lowered from the previously used value of 700 K down to $\sim$ 350 K as first employed by Kumar and Hunten (1974). Although an altitude-dependent eddy diffusion coefficient has been used in studies of Venus's middle atmosphere chemistry since 1975 (Prinn 1975; Sze and McElroy 1975), for the *upper* atmosphere models only constant eddy diffusion coefficients were assumed until 1978 (see Fig. 21). This led to the much discussed inability of these upper atmosphere models (e.g. Liu and Donahue 1975; Sze and McElroy 1975) to describe with one eddy coefficient the available observations on both atomic hydrogen (requiring $K > 5 \times 10^7$

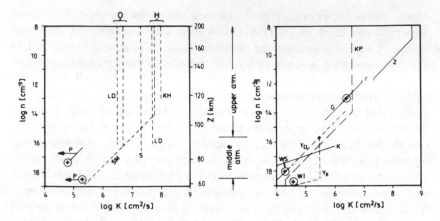

Fig. 21. Vertical eddy coefficient K versus total number density n in the middle and upper atmosphere of Venus. On the right-hand ordinate of the left frame an approximate altitude scale is given for dayside conditions. Pre-1978 photochemical models (left frame) include the profiles used by Shimizu (1969), S; Kumar and Hunten (1974), KH; Sze and McElroy (1975), SM; Liu and Donahue (1975), LD; and the upper limits of Prinn (1975), P. Post-1978 observations and models (right frame) are by von Zahn et al. (1979a), Z; Krasnopolsky (1979b), K; Krasnopolsky and Parshev (1980b,c), KP; Winick and Stewart (1980), WS; Gérard et al. (1981), G; Woo and Ishimaru (1981), WI; and by Yung and DeMore (1982), where Y_H indicates their photochemical model driven by HO_x and Y_{Cl} the one driven by ClO_x. Note that the data of Woo and Ishimaru pertain to the lower atmosphere which possesses a smaller static stability than the middle atmosphere.

$cm^2 s^{-1}$) and on atomic oxygen (requiring $< 1 \times 10^7 cm^2 s^{-1}$). As pointed out by von Zahn et al. (1980), this problem is considerably alleviated by using also in the upper atmosphere an eddy coefficient which increases with altitude. An overview of the eddy coefficients derived from observations and used in the various one-dimensional models of the Venus middle and upper atmosphere is given in Fig. 21a, b. The dayside upper atmosphere composition as calculated by Krasnopolsky and Parshev (1981a) is shown in Fig. 22. This model still exhibits a noticeable oxygen deficiency ($O/CO_2 = 0.08$ at 140 km instead of the observed 0.17; $O/CO_2 = 1$ at 162 km instead of 155 km) but otherwise it fits the available observations quite well, including the hydrogen Lyman-α measurements. However, the model does not incorporate the information available on nitrogen and helium densities.

An approach very different from the one-dimensional photochemical models was taken by Dickinson and Ridley who developed a two-dimensional model of Venus's upper atmosphere. Its two coordinates are altitude and solar zenith angle and it is therefore axisymmetric about the subsolar-antisolar direction. The latest version of this model (Dickinson 1976; Dickinson and Ridley 1977) is fully self-consistent and uses nonlinear perturbation theory to simultaneously calculate the thermal structure, global circulation, and com-

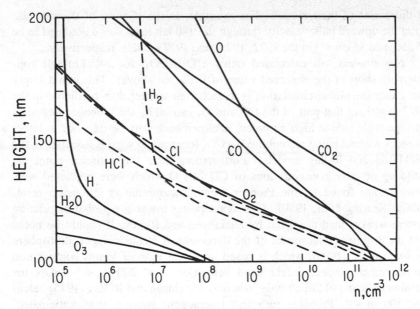

Fig. 22. Composition of the dayside upper atmosphere as modeled by Krasnopolsky and
Parshev (1980c). Concentrations indicated by dashed lines have been multiplied by a factor
of 1000 to show them more easily on the figure. Note the ratio HCl/CO_2 is nearly constant
well into the upper atmosphere.

position of Venus's atmosphere above the 0.1 mbar level. The atmosphere is
supposed to consist of only CO_2, CO, and O. The model considers the absorption of solar radiation and emission of thermal infrared, including radiative
transfer, and accounts for energy and mass transport associated with the
global-scale circulation. Dickinson and Ridley deliberately choose not to include in their models the effects of small-scale eddy mixing on composition
and thermal structure because of an acknowledged lack of a sound physical
description of this process. Dickinson and Ridley (1972) point out, however,
that with "an eddy mixing coefficient taken to be $\sim10^8$ cm² s⁻¹, we could
have derived for the venusian thermosphere global mean concentrations of
magnitude comparable to those obtained here."

Outstanding results of the Dickinson and Ridley models concern the
calculated exospheric temperatures and the wind system carrying upper atmosphere gas from the day side into the night side with velocities on the order
of 300 m s⁻¹. Composition was predicted to be almost independent of the
solar zenith angle on the day side. Dickinson and Ridley (1972, 1975, 1977)
made 3 calculations of the ratio $n(O)/n(CO_2)$ at the level of the ionospheric
peak ($\sim2 \times 10^{-6}$ mbar) on the day side obtaining 0.025, 0.045, and
0.056, respectively—the last being the low EUV heating efficiency case.
The increase is caused by a successive weakening of the dayside circulation

in this series of their models. For example, in the vicinity of the subsolar point the upward bulk velocity through the 140 km level was calculated to be 75, 50, and 35 cm s^{-1} in the 1972, 1975, and 1977 models, respectively.

Nevertheless, all calculated ratios $n(O)/n(CO_2)$ for 140 km fall considerably short of the observed value of 0.17 (see above). This could imply that either the global circulation is in fact even weaker than predicted in the 1977 model, or that part of the CO and O, carried by the global-scale winds into the night side at high altitudes, is blown back into the day side at lower altitudes without being recombined to CO_2 (contrary to what is assumed in the Dickinson and Ridley models). Furthermore, near the antisolar point the build up of large concentrations of CO and O which were predicted were however not found by the Pioneer Venus measurements (Niemann et al. 1980b; Keating et al. 1980). This again points towards a global circulation system weaker than predicted by Dickinson and Ridley. It should be noted that the 3-dimensional model of the dynamics of Venus's upper atmosphere by Mayr et al. (1980), which is based on actual Pioneer Venus composition measurements, indeed infers wind velocities of < 200 m s^{-1} across the terminator near 140 km altitude, whereas Dickinson and Ridley (1975) calculate 300 m s^{-1}. Probably turbulent interactions stronger than anticipated, between the middle and upper atmosphere also contribute to the observed diurnal distribution of atomic oxygen.

In summary, although the Dickinson and Ridley model has many attractive features such as its self-consistency, it cannot offer a satisfactory description of the observed Venus upper atmosphere composition, dynamics, and thermal structure. In particular a realistic description of the nightside conditions seems to call for basic modifications or extensions of the approach. Unfortunately, due to its highly structured sources and sinks, small-scale turbulence cannot be included in these models which are derived from first principles only. Its future incorporation therefore will automatically mean a loss of self-consistency for the model.

There is one more important feature which is not included in the Dickinson and Ridley (1975) model: pronounced deviations from an axisymmetric state of the upper atmosphere. Examples for these signatures are the observed nightside distributions of He (Fig. 18), NO, and the inferred diurnal variation of H (Fig. 31), all of which are strongly asymmetric with respect to a subsolar-antisolar axis. Less pronounced, but still noticeable deviations from symmetry between the dawn and dusk terminator are also recognizable in the diurnal variations of CO_2, O (Fig. 17), and the exospheric temperature (Fig. 20). These observations have been interpreted by Mayr et al. (1980); with the help of their previsouly mentioned model of upper atmosphere dynamics, they derive the strength of the day-to-night winds, the vertical eddy diffusion coefficient (taken to be altitude independent), exospheric flow, and the rate of superrotation of the upper atmosphere of Venus. These results are described in more detail by Schubert in Chapter 21 on atmospheric motions.

C. Nitrogen Compounds

1. Molecular Nitrogen. The attempts to accurately determine the mixing ratio of N_2 in the bulk atmosphere of Venus have had a colorful history. As described in Sec. II. C.1 the Venera 4, 5, and 6 landers carried no less than 8 instruments into the lower atmosphere of Venus which determined the volume abundance of the sum of nitrogen and all inert gases, thus in fact determining upper limits on the nitrogen mixing ratios. Because 3 of the individual measurements indicated upper limits for nitrogen < 2.5%, Vinogradov et al. (1971) felt justified to state: "the given measurements can be taken as reliable upper limits on the nitrogen content. Thus we can adopt a nitrogen content in the atmosphere of Venus of not more than 2%." Next came the mass spectrometers aboard the Venera 9 and 10 landers with their doubtful performance (see Sec. II. C.2). Preliminary results of these experiments were interpreted by Surkov et al. (1978) as indicating a N_2 abundance of 1.8%. Thus, in his review on results of Venus missions Marov (1978) was still describing the Venus atmosphere as having < 2% of nitrogen.

This picture was changed considerably by the more sophisticated *in situ* experiments of the Pioneer Venus probes and the Venera 11 and 12 landers. The respective results are shown in Table V. Three surprising results came out of these experiments: (1) all agreed on the fact that the nitrogen content is higher than was previously assumed; (2) the results of three instruments employed do not have overlapping error bars; and (3) the data of the Pioneer Venus gas chromatograph indicate that the nitrogen mixing ratio changes notably between 52 km and 42 km altitude (see Table V).

As to the first point we do not doubt the validity of the higher N_2 content in comparison with the Venera 4–10 results because all the later spacecraft were able to use considerably improved instrumentation. As to the spread in results we suspect that for the mass spectrometers so much CO from the ambient CO_2 and miscellaneous background is produced inside the ion sources that a reliable correction for these interference effects is difficult. Concerning the third point, there has been no explanation offered for either the unexpected difference in results between the two gas chromatographic experiments nor for the altitude gradient in the N_2 abundance observed by the Pioneer Venus LGC. Oyama et al. (1980*a*) comment on the latter result as follows: "This observation is not consistent with a simple model of a well-mixed atmosphere." We can only agree.

We summarize the results in Table V by stating that below 45 km altitude the nitrogen mixing ratio = (3.5 ± 0.8)%. Hence at the surface of Venus the partial pressure of N_2 is 3.2 bar, assuming a total pressure of 92 bar (Seiff et al. 1980). Furthermore, on Venus's surface the N_2 number density is 1.5 times higher than on the Earth's surface.

2. Atomic Nitrogen. Atomic nitrogen was first detected in the Venus thermosphere by the Pioneer Venus orbiter neutral mass spectrometer. On the

TABLE V
1978 in situ Measurements of N₂ Abundances

Spacecraft	Experimental Technique	Nitrogen Abundance in % at altitude				
		~ 0 km	~ 22 km	42 km	52 km	< 100 km
Pioneer Venus Large probe	mass spectrometer	4.0 ± 2.0[a]				
Pioneer Venus multiprobe bus	mass spectrometer					4.5 ± 1.3[b]
Venera 11 and 12 landers	mass spectrometer	4.0 ± 0.3[c]				
Pioneer Venus Large probe	gas chromatograph		3.41 ± 0.01[d]	3.54 ± 0.04[d]	4.60 ± 0.14[d]	
Venera 12 landers	gas chromatograph		2.5 ± 0.3[e]			

[a]Hoffman et al. 1980b, 0 to 22 km.
[b]Value extrapolated from in situ measurements taken > 140 km altitude; von Zahn et al. 1980.
[c]Istomin et al. 1980b.
[d]Oyama et al. 1980a, 0 to 42 km.
[e]Gel'man et al. 1979b.

surfaces inside its ion source, recombination of N with O to form NO appears to proceed almost quantitatively, helped by a large predominance of O over N. In this way ambient atomic nitrogen can be measured through the mass 30 peak in the mass spectrometer as demonstrated by Mauersberger et al. (1975) for atomic nitrogen in the terrestrial thermosphere. For the Venus upper atmosphere, Kasprzak et al. (1980) have used this method to derive number densities of atomic nitrogen in the altitude range from $\sim$ 150 km to 250 km. For 150 km altitude and a solar zenith angle of $70°$, $n(N)$ was found to be 1.5×10^7 cm^{-3}. When considering different altitudes and local times, a close proportionality was found between the atomic nitrogen and atomic oxygen densities. The average value for $n(N)/n(O)$ is 1.5% at 150 km.

Odd nitrogen densities in the upper atmosphere of Venus had been considered before by McElroy and Strobel (1969) and Cravens et al. (1978). It was, however, only through the information gained by the Pioneer Venus mission that model calculations of the nitrogen chemistry in the thermosphere of Venus attained a fair degree of realism. Rusch and Cravens (1979) have modeled the neutral and ion nitrogen chemistry in the daytime upper atmosphere. The steady-state continuity equations were solved for the density distribution of NO, $N(^4S)$ (the ground state), $N(^2D)$, and 8 charged constituents. All ion species were assumed to be in photochemical equilibrium and the odd nitrogen species were allowed to diffuse vertically. The eddy diffusion coefficient was chosen equal to 1×10^7 cm^2 s^{-1}. In the model of Rusch and Cravens $N(^4S)$ is produced primarily by

$$N_2 + e^* \rightarrow N(^4S) + N(^4S) + e$$

$$N(^2D) + M \rightarrow N(^4S) + M$$

$$NO^+ + e \rightarrow N(^4S; \,^2D) + O . \tag{p}$$

The major loss processes for $N(^4S)$ are:

$$N(^4S) + O_2^+ \rightarrow NO^+ + O$$

$$N(^4S) + NO \rightarrow N_2 + O . \tag{q}$$

The density of $N(^4S)$ at 150 km is predicted to be 8×10^6 cm^{-3} which we consider to be in fair agreement with the above-mentioned mass spectrometer value of Kasprzak et al. (1980). On the other hand, the calculated N^+ densities are smaller than measured densities (Taylor et al. 1980) by factors between 2 and 10. A similar, but less pronounced discrepancy between calculated and measured N^+ densities was found by Nagy et al. (1980) in their model calculations of the dayside ionic composition. It is not clear at the present time to what extent measurement inaccuracies on the one hand and incomplete or erroneous model assumptions on the other hand contribute to the noted discrepancies.

3. *Nitric Oxide.* The presence of nitric oxide (and therefore also atomic nitrogen) in the nightside upper atmosphere was discovered by Feldman et al. (1979) from observations of the ultraviolet nightglow from Venus made with the International Ultraviolet Explorer (IUE) satellite. In the wavelength band from 190 to 208 nm, three clearly visible features (see Fig. 3) were identified by Feldman et al. as the (0,0), (0,1), and (0,2) bands of the nitric oxide δ system. These observations were confirmed and extended to the γ bands of nitric oxide by Stewart and Barth (1979) using the Pioneer Venus orbiter ultraviolet spectrometer (PV-OUVS). The bands are produced by the rare process of an effective 2-body radiative recombination,

$$N + O \rightarrow NO + h\nu . \qquad (r)$$

The observed column intensity of the nitric oxide emissions allows one to obtain immediately a lower limit on the rate with which N and O atoms are recombining on the night side of the Venus upper atmosphere. Corresponding sources for the *in situ* production of N and O atoms on the night side are not likely to exist. In addition, nightglow emissions accompanying the dissociative production of N and O are known to be weak or absent (Stewart and Barth 1979). Therefore one is led to assume that the origin of the atoms is their production in the dayside thermosphere from where they are transported by a global circulation across the terminator into the night side (Feldman et al. 1979). The importance of the nitric oxide nightglow observations rests in the possibility of using the NO emissions as tracer for horizontal as well as vertical transport processes acting in the nightside cryosphere of Venus.

Planet-wide observations by the PV-OUVS (Stewart et al. 1980) revealed that the NO nightglow typically shows a bright spot just south of the equator at $\sim$ 2:00 LST which is $\sim$ 6 times brighter than the hemispheric mean brightness. We will compare in Sec. IV.H.2 the morphology of this NO nightglow with similar observations of the O_2 nightglow. Stewart et al. interpreted the brightness map as being also a map of downward flux of atomic nitrogen. In addition, they demonstrated that in a one-dimensional steady-state transport model the altitude of the emitting NO layer depends on the downward flux of O and on the strength of eddy mixing. This altitude z_p of peak emission intensity was derived by Stewart et al. to be 111 $\pm$ 7 km. For their detailed vertical transport calculations Stewart et al. chose at the 140 km level a downward flux of 1×10^{12} cm^{-2} s^{-1} for O and 1×10^{10} cm^{-2} s^{-1} for N. They further adopted for the vertical eddy coefficient K the form $K = A/M^{1/2}$, where M is the ambient number density and A is an empirical constant. Their calculated profiles of nightside O and N densities, and the volume emission rate of the NO δ bands, are shown in Fig. 23. Subsequent observations of Gérard et al. (1981) also using the PV-OUVS gave the improved value $z_p = $ 115 $\pm$ 2 km as mean over 3 selected orbits of the Pioneer Venus orbiter. Using the transport model of Stewart et al. (1980) and $z_p = 115$ km, Gérard et al.

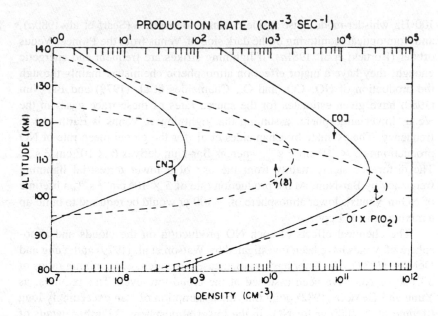

Fig. 23. Model profiles of nighttime densities of atomic nitrogen and atomic oxygen as derived by Stewart et al. (1980) from their observations of nightglow emissions of NO δ bands. Also given is the volume emission rate η (δ) of the NO δ bands which was measured to peak at 111 km altitude (subsequently updated to 115 km by Gérard et al. [1981]).

(1981) deduced the following parameters for the peak of the NO emission layer: $M_p \simeq 1.6 \times 10^{13}$ cm^{-3}, $A = 8 \times 10^{12}$ cm$^{1/2}$ s^{-1}, $K_p = 2 \times 10^6$ cm^2 s^{-1}. Because the molecular diffusion coefficient D is approximately 3×10^5 cm^2 s^{-1} for the total density M_p given above, we see that the NO emission layer occurs well below the homopause. If vertical transport occurred only through molecular diffusion, we would expect the maximum of the nightglow layer to be ~ 128 km (Gérard et al. 1981), which is 6 standard deviations from the altitude determined from the observations.

The strength of eddy mixing as expressed by the value of the constant A was deduced by Gérard et al. (1981) to be $A = 8 \times 10^{12}$ for deep nightside conditions, whereas for dayside conditions (SZA $\chi = 61°$) von Zahn et al. (1980) obtained $A = 1.4 \times 10^{13}$. This indicates an unexpectedly close agreement in turbulent transport properties for otherwise rather dissimilar geophysical environments.

Evidence for the frequent occurrence of thunderstorms in the lower atmosphere of Venus has been collected through *in situ* observation of noise bursts in the 10–80 kHz band (Ksanfomaliti et al. 1979) and by remote sensing observations of 27 MHz noise bursts emanating from Venus (Kraus 1957), of

100 Hz whistler-mode noise inside Venus's ionosphere (Scarf et al. 1980a), and from optical monitoring of the dark side of Venus from the Pioneer Venus orbiter (Borucki et al. 1981a). If lightning strikes are frequent and energetic enough, they have a major effect on atmospheric chemistry, mainly through the production of NO, CO, and O_2. Chameides et al. (1979) and Bar-Nun (1980) have given estimates for the source rates of these trace gases in the Venus lower atmosphere, assuming that lightning at Venus is Earth-like in frequency. The estimate by Chameides et al. for the global mean rate of NO production is 5×10^8 cm^{-2} s^{-1}, whereas Bar-Nun derives 6×10^7 cm^{-2} s^{-1}. The difference stems mainly from the use of a lower terrestrial lightning frequency by Bar-Nun. At the production rate of 5×10^8 cm^{-2} s^{-1}, a lifetime of NO in Venus's lower atmosphere of ~ 100 yr would be required to build up a mixing ratio of 1 ppb.

The chemical effects of such NO production on the clouds and atmosphere of Venus have been investigated by Watson et al. (1979) and Yung and DeMore (1982). The latter have studied cases in which, e.g., mixing ratios of 30 ppb of NO have been assumed at the cloud-top level. This requires, as Yung and DeMore (1982) point out, the assumption of "an exceedingly long lifetime of > 4000 yr for NO$_x$ in the lower atmosphere." Further details of these rather speculative studies may be found in the original literature. For future studies we suggest establishing more definite statistics on the thunderstorm frequency in Venus's atmosphere using the observational method of Kraus (1957) (but with currently available higher sensitivity) and by performing simultaneous observations from two, widely separated sites on Earth.

4. Ammonia. An instrument for the determination of the ammonia content in the lower atmosphere of Venus on board the Venera 8 lander (see Sec. II. C.1) carried out measurements at the 2 bar and 8 bar pressure levels of the Venus atmosphere, corresponding to 44 km and 32 km altitude, respectively. From the positive readings of the instrument, Surkov et al. (1974) estimated the ammonia mixing ratio to be between 10^{-3} and 10^{-4}. Goettel and Lewis (1974) examined the equilibrium chemistry of NH$_3$ in the Venus atmosphere and concluded, however, that the high NH$_3$ mixing ratios reported by Surkov et al. (1974) are inconsistent with the observation of rather low abundances of other gases in Venus's atmosphere, in particular H_2O and HCl. Florensky et al. (1978), on the other hand, argued that the upper part of the Venus troposphere can hardly be assumed to be in chemical equilibrium and that the tropospheric distribution of minor constituents can be strongly nonuniform, thus leaving open the possibility of a high NH$_3$ abundance in the lower cloud layer.

As to the method of NH$_3$ detection used by Surkov et al. (1974), it was pointed out by A. T. Young (1977) that bromophenol blue turns violet-red in concentrated H_2SO_4. He conjectured that the Venera 8 ammonia detector in fact responded to gaseous sulfuric acid.

None of the gas chromatographs and mass spectrometers on board the Pioneer Venus Large probe and Venera 11 and 12 landers detected ammonia, and regrettably none of the experimenters reports an upper limit for the ammonia abundance in the lower atmosphere of Venus. An upper limit of 1.6×10^{-5} was set by Smirnova and Kuz'min (1974) from microwave absorption measurements. Earth-based spectroscopic observations gave upper limits for the ammonia mixing ratio in the region of the Venus cloud tops of 3×10^{-8} (Kuiper 1968) and 1×10^{-7} (Owen and Sagan 1972).

D. Hydrogen Compounds

Hydrogen has been observed in the atmosphere of Venus in the form of H_2O, HCl, HF, H, H^+, and H_2^+ (or D^+). Hydrogen is also contained in the cloud droplets consisting of concentrated sulfuric acid. It is well known that Venus is severely deficient in water compared to Earth, yet the measured amounts of the above consitutents are significant enough to play an important role in the photochemistry of the middle atmosphere. Further, the escape of H is directly linked to the long-term balance of water vapor and the evolution of the Venus atmosphere. In this section we present a review of the available observations of hydrogen-bearing constituents in the atmosphere of Venus and discuss various models of hydrogen photochemistry and escape.

1. H_2O Observations. Many earth-based spectroscopic measurements have been made of the H_2O vapor content above the Venus clouds. In these observations interference by the strong telluric lines was a severe problem up to the mid-1960s. Since then two independent developments have led to a definite improvement in the quality of data on the venusian water abundance: first, the development of very high resolution interferometers and spectrometers (with resolving powers of better than 10^5) to separate the telluric from the venusian water absorption lines. Fig. 24 shows examples from Barker (1975*b*) of the separation of the two respective water lines and the observed line strength for ~10 ppm (left) and ~20 ppm (right) of venusian water mixing ratio. The second improvement is the development of airborne interferometers of sufficient quality to obtain the required Venus observations from stratospheric altitudes where the remaining column abundance of terrestrial water vapor is ~ 2 orders of magnitude lower than for an observatory at 2 km altitude. For these reasons we have listed in Table VI only the water vapor measurements obtained since 1967. Only Schorn et al. (1969) and Barker (1975*b*) observed Venus over a longer time span and with such a high spatial resolution that individual spectral scans covered only limited portions of the planetary disk. Both studies showed considerable variability with a preferential appearance of the larger H_2O abundances at intermediate phase angles 35° to 90° (see Fig. 25).

Barker (1975*b*) observed that H_2O total abundances for a 2-way absorption through the Venus atmosphere range from < 1 μm to 77 μm of precipi-

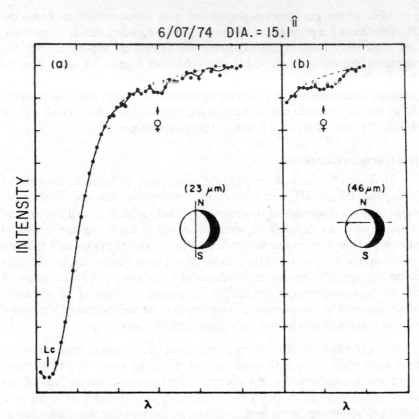

Fig. 24. Example of high-resolution observations of the Venus 819.7 nm water vapor line (♀), Doppler-shifted into the tail of the much stronger telluric water vapor line. Separation of data points is 1.18 pm. The left-hand diagram indicates the S/N ratio obtained for 23 μm of precipitable water vapor in Venus's middle atmosphere. The slit position across the Venus disk is shown in the small insert. (Figure from Barker 1975*b*).

table water, which according to Barker correspond to mixing ratios of < 0.5 ppm to 40 ppm. This range in turn encompasses almost all values published since 1967. We point out, however, that only 1 scan out of 206 values published by Barker corresponds to 40 ppm. All others are < 30 ppm; these measurements therefore suggest 30 ppm as a realistic upper limit for the H_2O mixing ratio above the clouds.

We consider the measurements collected in Table VI as a convincing proof for the fact that the Venus atmosphere above the clouds is rather dry. Nevertheless the question remains, how dry indeed is it and at what altitude? After all, the spectrometric measurements of Barker (1975*b*) indicate about an order of magnitude higher vapor abundances than the interferometric measurements of Kuiper et al. (1969) and Fink et al. (1972). Barker suggests that

TABLE VI

Earth-based Spectroscopic Water Abundance Measurements since 1967

Instrument	w^k	Water Bands (μm)	Phase Angle (deg)	Coverage of Venus Disk	Altitude of Observation (km)
Fourier interferometer[a]	<20	1.2 to 2.5			0.7
Spectrometer[b]	<16	0.82		partial	2.1
Fourier interferometer[c]	≤1	1.4/1.9		full	11.3 to 12.6
Spectrometer[d]	—[l]	0.82	66	partial	2.1
Spectrometer[e]	<16 to 40	0.82	52 to 92	partial	2.1
Fourier interferometer[f]	2	1.4/1.9/2.7	85/106	full	12.2
Fourier interferometer[g]	0.4 ± 0.1	1.4/1.9/2.7	87	full	12.2
Spectrometer[h]	<8	0.82	50/156	full	14.6
Fabry-Perot interferometer[i]	<1	0.82		full	2.3
Spectrometer[j]	<0.3 to 25[m]	0.82	21 to 162	partial	2.1

[a]Connes et al. 1967.
[b]Owen 1967.
[c]Kuiper and Forbes 1967.
[d]Belton et al. 1968.
[e]Schorn et al. 1968.

[f]Kuiper et al. 1969.
[g]Fink et al. 1972.
[h]Gull et al. 1974.
[i]Traub and Carleton 1974.
[j]Barker 1975b.

[k]Vertical column abundance w above the clouds in μm of precipitable water vapor (1 μm is equivalent to 3.35×10^{18} molecules cm^{-2}).
[l]Mixing ratio of $\leq 10^{-4}$ at cloud tops.
[m]Assumes mean air mass factor $\eta \sim 3$.

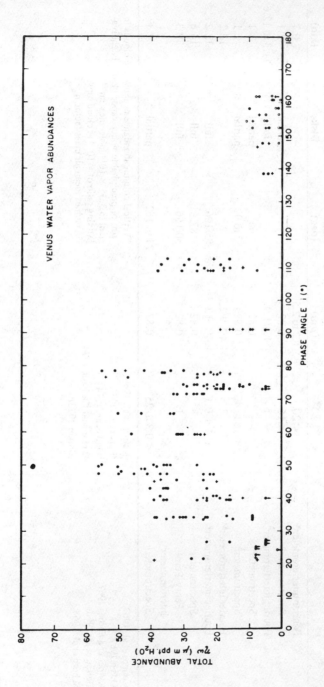

Fig. 25. Variability and dependence on the phase angle of Venus water vapor column abundances as observed by Barker (1975b). The observed H_2O abundances correspond to mixing ratios of 0.5 ppm to 40 ppm assuming a reflecting layer model and an average CO_2 abundance of 2 km atm above the clouds.

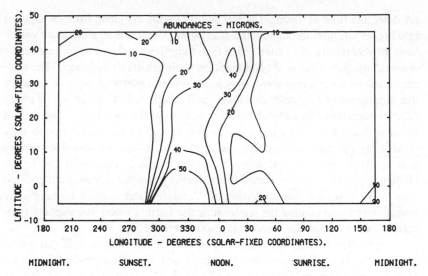

Fig. 26. Horizontal distribution of water vapor above the 90 mbar level as derived from the measurements of the Pioneer Venus orbiter infrared radiometer by Schofield et al. (1982). Column abundances are given in μm of precipitable water vapor (1 μm corresponds to a mixing ratio of ~2 ppm). The data extend from 10°S to 50°N at all longitudes. They are, however, averaged in bins of 10° latitude and 30° longitude. Thus the inner frame represents the boundary of the grid points and the outer frame the extent of the data.

he observes in fact local wet spots on the planet. At the same time, the full disk measurements of the airborne interferometer should measure lower abundance values because of the averaging over these wet spots. The statistics of H_2O abundances for Barker's 206 observations do not support this view, however; 80% of his scans yield mixing ratios > 1 ppm, the latter being the result of Kuiper et al. (1969). Thus, if wet spots exist, they must be rather abundant and full disk averages of 1 ppm and lower ought to be rather exceptional cases.

A new answer to the question about the global distribution of water vapor in Venus's middle atmosphere has been given recently by Schofield et al. (1982). They analyzed radiances measured by the infrared radiometer on the Pioneer Venus orbiter, in particular by the 45 μm channel, which is centered in the water vapor rotation band, and by the 11.5 μm channel, which is chiefly affected by cloud opacity. Fig. 26 shows the horizontal distribution of retrieved water-vapor abundance above unit cloud optical depth (which occurs at ~ 90 mbar level). Column abundances are indicated in μm of precipitable water vapor. The detection limit of the experiment was 3 μm, and 1μm vertical column abundance is according to Schofield et al. roughly equivalent to a mixing ratio of 2 ppm. The experiment clearly localizes a wet region at equatorial latitudes in the mid-afternoon sector, where mixing ratios exceed 100 ± 40 ppm. But mixing ratios throughout most of the night side are below the detection limit of 6 ppm. The wet region seems to extend from the equator into higher latitudes throughout the afternoon. The absolute abundance values have an uncertainty of a factor ~ 1.5 (see Schofield et al. for an in-depth error discussion). Schofield et al. comment on their results as follows: "The concentration of water vapor above the clouds on the equatorial day side suggests that it is generated by solar radiation, perhaps by selective heating of cloud aerosol particles or by photochemical processes at the cloud tops. The shift of the maximum from the expected subsolar region to the midafternoon may result from the mean atmospheric circulation. It is also possible that convection at deeper levels in the subsolar region coupled with uplift due to the Hadley circulation raises water from deeper, moister levels to the cloud tops." We conclude this discussion of Venus's water content in the middle atmosphere by reminding the reader that the Earth's lower stratosphere typically contains 4 ppm H_2O, hence it is also extremely dry.

Using the observed microwave spectrum of Venus near 1.35 cm wavelength, the H_2O abundance *below* the clouds has been studied by, e.g., Pollack and Morrison (1970), Kuz'min et al. (1971), Janssen et al. (1973), Rossow and Sagan (1975), and Janssen and Klein (1981). For their evaluation of the microwave opacity of Venus's lower atmosphere, only the last authors had the full benefit of Pioneer Venus mission results. They determined upper limits of 3000 ppm or 180 ppm for the mixing ratios of H_2O and SO_2, respectively, assuming uniform mixing beneath 50 km altitude and a scale height of 1 km above this level. If the SO_2 abundance in the lower atmosphere (see Sec. IV.E–1) and the cloud opacity were known more precisely, this upper limit of the H_2O mixing ratio could be lowered significantly.

In situ measurements of the H_2O content of the lower Venus atmosphere were first attempted by experiments on board the Venera 4–6 spacecraft. As described in more detail in Sec. II.C.3, two different but simple methods were used and the results of these experiments reviewed by Vinogradov et al. (1971) and Andreichikov (1978). These experiments were prepared at a time when the presence of H_2SO_4, HCl, and HF in the atmosphere of Venus were not yet known. Because of the quantitatively unknown effects of these trace gases on the results of Vinogradov et al. (1971), we will not consider further these early experiments.

Table VII gives a summary of the results of *in situ* measurements of H_2O abundances below the clouds since 1975. Moroz et al. (1976) used narrowband photometers on board Venera 9 and 10 to measure, from scattered sunlight, the brightness ratio of the 0.87 μm and 0.82 μm bands of CO_2 and H_2O, respectively. At an altitude of 40 km they determined an atmospheric water vapor abundance of 300 ppm. A new and much more sophisticated experiment was made on board Venera 11 and 12 spacecraft using scanning spectrophotometers for the wavelength range 0.43 to 1.17 μm (see also Sec. II. C.). Figure 27 shows samples of scattered daylight spectra taken with this instrument clearly delineating absorption bands of H_2O and CO_2 (the depres-

TABLE VII

In situ Measurements of Water Abundances below the Clouds since 1975

Spacecraft	Experimental Technique	Water Abundance in ppm at Altitude			
		0 km	22 km	42 km	52 km
Venera 9 and 10 landers	3 narrow-band photometers			300[a]	
Venera 11 and 12 landers	scanning spectro-photometers	20[b]	60[b]	150[b]	200[b]
Pioneer Venus Large probe	mass spectrometer	<1000[c]	<1000[c]	<1000[c]	<1000[c]
Pioneer Venus Large probe	gas chromatograph		1350[d]	5200[d]	<600[d]
Venera 11 lander	mass spectrometer	76[e]	67[e]		
Venera 12 lander	mass spectrometer	130[e]	52[e]		
Venera 12 lander	gas chromatograph	≤100[f]	≤100[f]	≤100[f]	

[a] Moroz et al. 1976.
[b] Moroz et al. 1979c.
[c] Hoffman et al. 1980a.
[d] Oyama et al. 1980a.
[e] Istomin et al. 1980b.
[f] Gel'man et al. 1980.

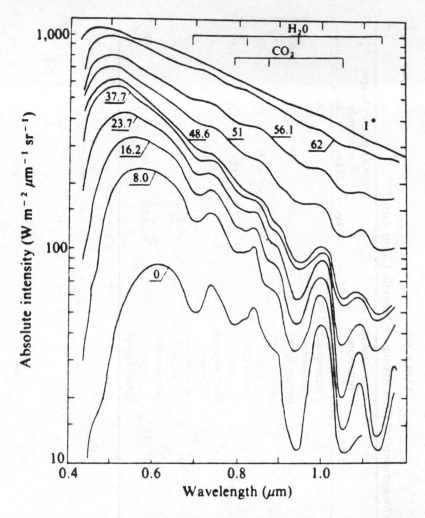

Fig. 27. Spectra of scattered solar radiation in the Venus atmosphere at different altitudes (in km) above the surface, obtained by the spectrophotometers on board the Venera 11 and 12 landers. The position of strong H_2O and CO_2 absorption bands are indicated at the top of the figure. (Figure from Moroz et al. 1980a.)

sion in the blue part of the spectrum below 30 km altitude will be discussed in Sec. E-1.3). Results of these remarkable observations have been published in a number of papers by Ekonomov et al. (1979) and Moroz et al. (1979c, 1980a). They indicate that the H_2O vapor mixing ratio varied from 200 ppm in the clouds to 20 ppm at the surface. This result is puzzling because no physical or chemical process has been identified so far that could explain the magnitude of this variation. A reanalysis of the same Venera data by

Young and Young (personal communication) shows that the two strong water bands (the 0.94 μm σ band and the 1.13 μm ϕ band) yield two different answers as to the H_2O mixing ratio in the Venus lower atmosphere. If they use only the absorption observed in the 1.13 μm band, they require a H_2O mixing ratio of 20 to 30 ppm for the entire atmosphere. However, if they use 0.94 μm band data and assume that it is entirely due to H_2O vapor, they get H_2O mixing ratios of 20 ppm at the surface, 200 ppm just below the clouds and 30 ppm just above the clouds, a result essentially identical to what Moroz et al. have published. Moroz (personal communication) does acknowledge the problem of fitting 1.13 μm data with the same model as that for 0.94 μm data. Young and Young are considering another possibility that SO_3 or some other sulfur compound below the clouds may be causing the abnormally intense 0.94 μm feature which may explain the discrepancy and in which case a relatively constant H_2O mixing ratio of 20 to 30 ppm would prevail.

Mass spectrometers were employed both in the Pioneer Venus Large probe and the Venera 11 and 12 lander spacecraft. The Pioneer Venus instrument (PV-LNMS) was opened to the ambient atmosphere at 62 km altitude, i.e., inside the upper cloud layer. Hoffman et al. (1980a) report that they observed at altitudes above 50 km altitude "a rather constant ratio H_2O to CO_2 of approximately 500 ppm. However, it is not clear if it is all due to ambient water vapor or partly to some instrumental source, although the residual before entry into the atmosphere was very low. Therefore, only a conservative upper limit of 1000 ppm will be reported at this time. In the region where the leak was blocked, the mass 18 peak grew rapidly (along with the SO_2 peak), indicating that the droplet on the leak was emitting water, a phenomenon that has been observed in laboratory studies of leaks coated with sulfuric acid. Below the altitude at which the leak reopened, the water peak again passed through a maximum, probably caused by heating of the inlet leak and a more rapid release of residual vapor into the mass spectrometer." The cloud droplet mentioned by Hoffman et al. also gave rise to a very interesting increase in the intensity ratio of the mass 19 over mass 18 peaks measured by the PV-LPMS, which is discussed in Sec. V. B.1.

The mass spectrometers of the Venera 11 and 12 landers, on the other hand, were opened to the ambient atmosphere at 23 km altitude, i.e., well below the cloud deck. By opening at this low altitude Istomin et al. (1980b) avoided any potential interference from cloud particles with their gas analysis. Istomin et al. deduced from the mass 17 and 18 peaks of their two instruments water abundances of ~100 ppm between 22 km and 4 km altitude (see Table VII). Due to the potential memory effects of the inlet pipe (see Sec. II.C.2) and to adsorption of water vapor at the internal surfaces of the mass spectrometer, a considerable uncertainty remains as to the quantitative relationship between the water abundance observed inside the instrument and the ambient water vapor mixing ratio as well as its altitude profile (Istomin, personal communication).

For the gas chromatographic measurements of the H_2O abundance we recognize a similar pattern. The Pioneer Venus instrument (Oyama et al. 1980a) took its first sample of ambient gas at 51.6 km altitude, when it was inside the middle cloud layer. During the subsequent analysis an anomaly was observed, in that a very large peak emerged from the short column 126 s after injection of this first sample. It could not be associated with any known constituent in the first injection and made an accurate measurement of the peak due to ambient water impossible. The abundances of water vapor measured in the second and third sample are shown in Table VII. The gas chromatograph on board Venera 12, on the other hand, took its first sample of ambient gas starting at 42 km altitude, well below the lower boundary of the cloud deck (Gel'man et al. 1979a). However, a previous control analysis had shown that the instrument contained some internal moisture and so only upper limits on the water abundance were derived (see Table VII), which are nonetheless much lower than the positive readings of the Pioneer Venus gas chromatograph.

In order to get a more unified picture from these sometimes conflicting data on the water abundance in the lower atmosphere, we should note that *within* the rather extended cloud deck of Venus there is but one positive identification of water vapor, the 200 ppm measurements of Moroz et al. (1979c). *Below* the cloud deck, i.e. beneath 48 km altitude, we are faced with a choice between (crudely stated) a few 1000 ppm water content as indicated by the Pioneer Venus gas chromatograph or an abundance on the order of 100 ppm as measured by all the other instruments using 3 different experimental techniques. As Oyama et al. (1980a) and Hoffman et al. (1980a) observe, both gas chromatograph and mass spectrometer measurements can be affected by the introduction of H_2SO_4 cloud droplets. In addition, the gas inlet systems and ion sources of mass spectrometers theoretically can both deplete the ambient gas sample of water vapor by adsorption to the inlet surfaces and can produce additional water by a large variety of chemical reactions at the inlet surfaces. Hence, neither a rather low nor a rather high value of water abundance measured in either type of instrument appears very reliable *a priori*. For the optical method employed by Moroz et al. (1979c) the situation apparently is otherwise. Although a high reading in this case could possibly be explained by a local contamination of the spacecraft environment with water (as sometimes is the case for balloon measurements of the terrestrial H_2O abundance in the stratosphere), a low reading of H_2O abundance can *only* be explained by a low column abundance of water vapor in the atmosphere. However, there is one qualification to be made. The data analysis of the optical experiment of Moroz et al. is far from straightforward and the solution may not be entirely unique as Young et al. (1981) claim. But much confidence can still be added to this analysis by laboratory studies of the optical properties of H_2O and CO_2 close to the environmental conditions encountered in the lower atmosphere of Venus. Barring any unforeseen development in this area, we argue that the experiment of Moroz et al. (1979c) is crucial in selecting the 100 ppm level of H_2O abundance as being more likely the correct one in preference over the 1000 ppm level. The 100 ppm level is also clearly the preferred case for the models of Venus's greenhouse effect and temperature structure by Pollack et al. (1980a) and for the study of the thermochemical fate of H_2SO_4 below the main cloud deck by Craig et al. (1981). To what extent the deviations of the various measurements from the mean 100 ppm level arc caused by natural variability or instrumental difficulties cannot be decided at this time.

2. HCl and HF Observations. HCl and HF were discovered in Venus's atmosphere by the earth-based observations of Connes et al. (1967). They obtained high-resolution spectra of Venus showing weak, narrow absorption lines which they identified as due to the lines of vibration-rotation bands of HCl and HF. Connes et al. (1967) estimated the abundance ratios HCl/CO_2 and HF/CO_2 to be 6×10^{-7} and 5×10^{-9}, respectively. A refined analysis of the Connes et al. spectra was performed by L. Young (1972), who derived an HCl mixing ratio of $(4.2 \pm 0.7) \times 10^{-7}$ or $(6.1 \pm 0.6) \times 10^{-7}$ depending on whether the HCl lines are formed at the level of 36 mbar or 50 mbar effective pressure, repectively, in the Venus atmosphere. She also argued that HCl appears to be nonuniformly mixed in the altitude regime under observation.

3. H and H_2 Observations. Lyman-α airglow observations of Venus from sounding rocket experiments, from Venera 4, Mariner 5, Mariner 10, Venera 9-12, and from the Pioneer Venus orbiter have provided the data on the hydrogen corona surrounding Venus; the salient features of these observations are summarized in Table VIII. A two-component Lyman-α distribution was observed from Mariner 5 (Barth et al. 1967). With increasing planetary distance the emission fell off exhibiting, unexpectedly, two different scale heights. The scale height observed at planetocentric distances > 9000 km was about twice as large as the one below 9000 km (Barth et al. 1968). During the 2nd Arizona Conference on Planetary Atmospheres this observation was interpreted in several different ways. Barth (1968) suggested that the Venus exosphere consists of an isothermal $H + H_2$ mixture; Donahue (1968) favored an isothermal $H + D$ mixture; A.I. F. Stewart (1968) conjected that a two-temperature distribution of H with 350 K and 700 K prevailed. The $H + H_2$ hypothesis was criticized by Donahue (1968) and McElroy and Hunten (1969) who showed that the required amount of H_2 led to excessive production of H which could not be disposed of. Donahue (1969), Wallace (1969), and McElroy and Hunten (1969) presented models of H and D distributions. Wallace also studied a two-temperature, nonspherical distribution of H about Venus. The deuterium hypothesis of a very large ratio D/H $\sim$ 10 at Venus's exobase had to be discarded when a rocket experiment by Wallace et al. (1971) failed to show the conjectured large D Lyman-α component. (Recent evidence for a much smaller ratio D/H $\sim$ 0.01 is further discussed in Sec. V. B.1.) The remaining evidence for a 700 K exosphere was the ionosphere data

TABLE VIII

Hydrogen Distribution in Venus Exosphere from the Lyman$-\alpha$ Data

	Mariner 5[b]	Mariner 10[c]	Venera 11, 12[d]
Thermal Component			
Dayside temperature (K)	275 ± 50	275 ± 50	300
Dayside density[a] (cm^{-3})	$(2 \pm 1) \times 10^5$	1.5×10^5	$4^{+3}_{-2} \times 10^4$
Nightside temperature (K)	150 ± 50	150 ± 25	—
Nightside density[a] (cm^{-3})	$(2 \pm 1) \times 10^5$	$(1.0 \pm 0.5) \times 10^5$	—
Nonthermal Component			
Dayside temperature (K)	1020 ± 100	1250 ± 100	1000
Dayside density[a] (cm^{-3})	1.3×10^3	$(5 \pm 1) \times 10^2$	1×10^3
Nightside temperature (K)	1500 ± 200	—	—
Nightside density[a] (cm^{-3})	1.0×10^3	—	—
Solar 10.7 cm flux	120	73	—
Solar zenith angle	0°	60°	80°

[a]All densities are referred to the critical level assumed at 255 km altitude.
[b]See Anderson 1976.
[c]See Takacs et al. 1980.
[d]See Bertaux et al. 1982.

which could be fitted with a 700 K exosphere model (McElroy 1969). Kumar and Hunten (1974) showed that a 'cool' exosphere at 350 K could well explain the Mariner 5 ionosphere data and proposed that the Mariner 5 Lyman-α data should be explained in terms of two-temperature distributions and that the required abundance of hot H atoms could be produced by ion-molecule reactions in the dayside ionosphere above the exobase (these reactions are discussed in detail below). A cool exosphere was subsequently confirmed by the Mariner 10 He 58.4 nm airglow measurements which showed that helium was distributed at a temperature of 370 ± 105 K in the exosphere (Kumar and Broadfoot 1975). The Mariner 5 Lyman-α data were reinterpreted by Anderson (1976) in terms of a two-temperature distribution of 270 K and 1020 K, respectively; this model with the Mariner 5 data is shown in Fig. 28. The Mariner 10 Lyman-α data again showed the presence of a two-temperature distribution (Fig. 29), though the hot component was suppressed by a factor of ~ 3 (Takacs et al. 1980). The hot component was observed on the night side as well as on the day side by both the Mariner 5 and 10 experiments. Dayside exospheric temperatures close to 280 K were obtained at once from *in situ* measurements of the Pioneer Venus bus mass spectrometer (von Zahn et al. 1979*b*) and the orbiter ultraviolet spectrometer (Stewart et al. 1979). This further confirmed that the Lyman-α data must be explained in terms of a two-temperature distribution.

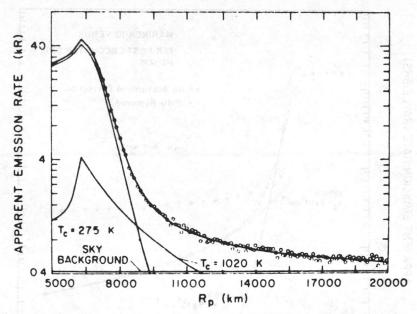

Fig. 28. Profile of Lyman-α dayglow brightness as observed by the Mariner 5 ultraviolet photometer (dots) and its decomposition by Anderson (1976) into three components: one caused by a population of 275 K hydrogen, another from 1020 K hydrogen, and a general sky background. The distance R_p is the planetocentric distance to the line of sight of the ultraviolet photometer.

The Venera 9 and 10 Lyman-α data were interpreted by Bertaux et al. (1978), who found that the dayside data could be fitted with a single component at 500 ± 100 K and n_c (H) = 1.5×10^4 cm^{-3} at 250 km, though the data were not inconsistent with a two-temperature distribution of 300 and 1000 K, respectively. These authors preferred a single-component explanation. However, it should be noted that the data used in their analysis were limited to altitudes below 4500 km, whereas the hot component (with > 600 K) shows through dramatically at higher altitudes only as evidenced by the Mariner 5 and 10 data (see Figs. 28 and 29). The Venera data, however, showed yet another nonthermal H population in the 3000–4500 km altitude range from the airglow line width measurement made with a hydrogen cell. The line width increased at 3000 km and then decreased at 4500 km, indicating the presence of a nonthermal H source in this region (Fig. 30). Bertaux et al. interpreted this signature as due to fast atoms circulating on satellite orbits created just behind the bow shock by charge-exchange collisions between the neutral atoms and the solar wind protons. We shall discuss this source further below.

Venera 11 and 12 observations of the Lyman-α airglow have led Bertaux et al. (1982) to the determination of exospheric temperatures at high latitudes (300 ± 25 K at 79°S and 275 ± 25 K at 59°S) which are virtually identical to

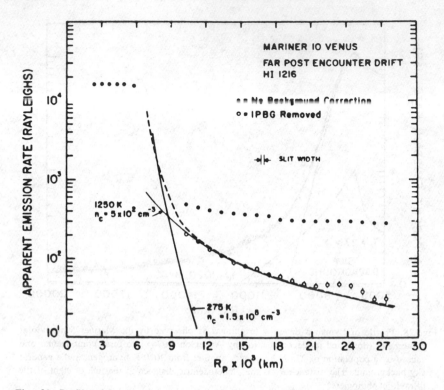

Fig. 29. Profile of Lyman-α dayglow brightness as observed by the Mariner 10 spectro-photometer (dots) and its decomposition by Takacs et al. (1980) into three components: one caused by a population of thermal hydrogen at 275 K, one from nonthermal hydrogen at 1250 K, and an interplanetary background (IPBG). Distance R_p is defined as in Fig. 28.

the values inferred from mass spectrometer measurements at 40°S (von Zahn et al. 1979*b*) and low latitudes (Niemann et al. 1980*b*). Bertaux et al. (1982) also find that a two-component exosphere provides the best fit to the Lyman-α data analogous to that obtained from Mariners 5 and 10. The densities and temperatures at the critical level (taken at 250 km) for the two populations are (see also Table VIII): $n_{1c} = 4^{+3}_{-2} \times 10^4$ cm^{-3}, $T_1 = 300$ K; and $n_{2c} = 1 \times 10^3$ cm^{-3}, $T_2 = 1000$ K.

Simultaneous *in situ* measurements of H$^+$, O$^+$, and O by the Pioneer Venus orbiter ion mass spectrometer (PV-OIMS) and neutral mass spectrometer (PV-ONMS) have been used by Brinton et al. (1980) to derive the global distribution of thermospheric atomic hydrogen by making use of the charge-exchange reaction of O$^+$ and H that controls the production of H$^+$,

$$O^+ + H \leftrightarrows O + H^+ . \qquad (s)$$

The results of Brinton et al. show a diurnal bulge of magnitude 400:1 in the hydrogen densities (Fig. 31). Such a bulge was predicted by Kumar et al.

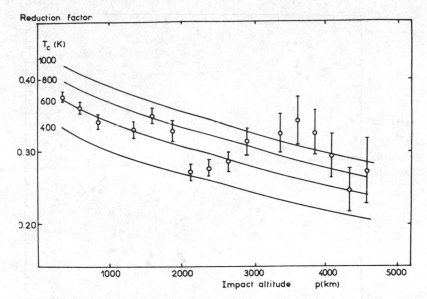

Fig. 30. Lyman-α line width measurement by the Venera 9 hydrogen absorption cell. Along the ordinate scale the reduction factor is given which is the ratio of measured emission when the H absorption cell is *on* over the intensity when this cell is *off*. The solid lines are model values with a hydrogen density of 1.5×10^4 cm^{-3} at 250 km altitude and various exospheric temperatures. A large increase of temperature is indicated between 2500 and 4000 km altitude. (Figure from Bertaux et al. 1978.)

(1978) and is well described by the transport model of Mayr et al. (1980). A similar bulge, but of smaller magnitude, has been observed for helium by Niemann et al. (1980b) and has been inferred also for H$_2$ by Kumar and Taylor (1981). The Lyman-α data obtained from the Pioneer Venus orbiter ultraviolet spectrometer (PV-OUVS) will provide a detailed morphology of the hydrogen bulge on the night side (I. Stewart, personal communication).

The observation of a mass 2 peak in the spectra of the PV-OIMS (H. Taylor et al. 1980) has been interpreted by Kumar et al. (1981) as the first direct clue to the H$_2$ abundance in the atmosphere of Venus. H$_2^+$ is produced in the dayside ionosphere in the photoionization of H$_2$ by solar ultraviolet and is destroyed by its reaction with O:

$$H_2 + h\nu \rightarrow H_2^+ + e$$

$$H_2^+ + O \rightarrow OH^+ + H. \tag{t}$$

Using the measured H$_2^+$ and O densities, Kumar et al. (1981) determined a mixing ratio of 10 ppm of H$_2$ in the Venus atmosphere. The resulting H$_2$ distribution in the thermosphere and the H distribution derived from the photochemistry of H$_2$ is shown in Fig. 32. A discussion of the hydrogen

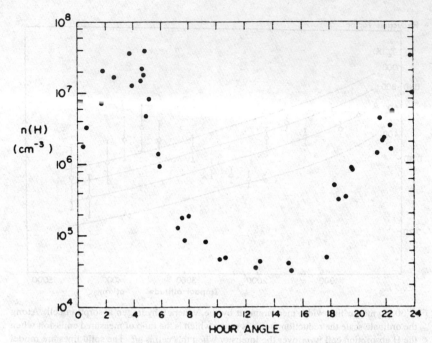

Fig. 31. Diurnal variation of the atomic hydrogen density near 165 km altitude as derived by Brinton et al. (1980) from *in situ* measurements of $n(H^+)$, $n(O^+)$, $n(O)$ and the neutral gas and ion temperatures.

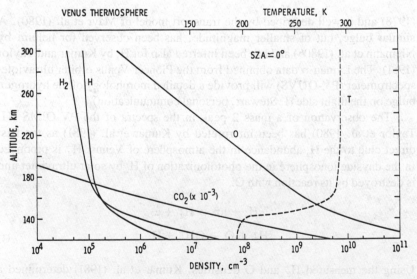

Fig. 32. Densities of H_2, H, O, and CO_2 in the dayside upper atmosphere of Venus. H_2 densities are derived by Kumar et al. (1981) from the observed abundance of ionospheric mass 2 ions under the assumption that the latter are in fact H_2^+ ions.

photochemistry will be given in Sec. IV. D.5. McElroy et al. (1982), on the other hand, have explained the mass 2 peak of the PV-OIMS as due to deuterium ions. This interpretation will be discussed further in Sec. V. B.1.

4. Sources of Nonthermal Hydrogen and Escape. The presence of a non-thermal hydrogen corona on Venus is unique among the terrestrial planets. Since the effective temperature of the observed hot H atoms is between 1000 and 1500 K, a similar corona on Earth would be obscured by the main thermal component at typical terrestrial exospheric temperatures of $\sim$ 1000 K. No such corona has been observed on Mars either. A study of this nonthermal corona is particularly important in order to understand the hydrogen escape processes on Venus. The exospheric temperature on Venus is only $\sim$300 K which allows a Jeans escape rate of hydrogen from Venus of only $\sim 10^5$ cm^2 s^{-1}, far below the Jeans escape rate of 1×10^8 H atoms cm^{-2} s^{-1} from Earth and 2×10^8 H atoms cm^{-2} s^{-1} from Mars (Hunten and Donahue 1976). However, the observed abundance of H atoms in the exosphere is too small to be consistent with 1 ppm of H_2O and 0.6 ppm HCl at the cloud tops if the escape flux is only 10^5 cm^{-2} s^{-1} and hence a nonthermal escape flux $> 10^7$ cm^{-2} s^{-1} is required (Liu and Donahue 1975; Hunten and Donahue 1976). Ever since the discovery of a nonthermal hydrogen corona on Venus, significant effort has gone into understanding the sources of nonthermal hydrogen in the Venus exosphere. Kumar and Hunten (1974) investigated a number of ion-molecule and charge-exchange reactions in the ionosphere that result in the formation of hot H atoms and suggested that the reaction of O^+ with H_2 and subsequent recombination of OH^+ could provide a potent source of hot H atoms,

$$O^+ + H_2 \rightarrow OH^+ + H_{hot}$$

$$OH^+ + e \rightarrow O + H_{hot} . \tag{u}$$

Acceptance of this model implies acceptance of an exospheric temperature near 350 K, presence of at least 2 ppm of H_2, and the presence of O^+ as a dominant topside ion. Encouraged by the Mariner 10 confirmation of a low exospheric temperature, Kumar et al. (1978) published an inventory of various possible sources of nonthermal hydrogen and their strengths (see also Sze and McElroy 1975):

(1) The ionosphere source; ion molecule reactions:

$$O^+ + H_2 \rightarrow OH^+ + H$$

$$OH^+ + e \rightarrow O + H$$

$$CO_2^+ + H_2 \rightarrow CO_2H^+ + H$$

$$CO_2H^+ + e \rightarrow CO_2 + H$$

$$He^+ + H_2 \rightarrow HeH^+ + H$$

$$HeH^+ + e \rightarrow He + H. \tag{v}$$

(2) The ionosphere source; charge exchange reactions:

$$H^+ + O \rightarrow O^+ + H$$

$$H^+_{hot} + H_{cold} \rightarrow H^+_{cold} + H_{hot} . \tag{w}$$

(3) Photodissociation of H_2:

$$h\nu + H_2 \rightarrow H + H . \tag{x}$$

(4) Charge exchange with solar wind protons:

$$H^+_{sw} + H_{exo} \rightarrow H^+_{cold} + H$$

$$H^+_{sw} + O \rightarrow O^+ + H . \tag{y}$$

Based on the available data, Kumar et al. (1978) concluded that the $O^+ + H_2$ reaction was still the strongest. The $He^+ + H_2$ reaction orginally proposed by Sze and McElroy (1975) was not significant in light of the Mariner 10 determination of the helium abundance in the Venus exosphere (Kumar and Broadfoot 1975). Ferrin (1976) showed that the hydrogen distribution resulting from $O^+ + H_2$ reaction could explain the Mariner 5 profile but he required a much higher H_2 abundance than assumed in the Kumar and Hunten (1974) model.

Pioneer Venus measurements have provided key data on O^+, O, H, and H_2^+ which have been used by Cravens et al. (1980b), Kumar et al. (1981), and Hodges and Tinsley (1981) to reevaluate various sources of nonthermal H. The results of these investigations are summarized in Table IX.

First, let us discuss the results on the $O^+ + H_2$ source. Cravens et al. (1980b) calculated the velocity distribution function of the nonthermal H atoms by a Monte Carlo technique and showed that on the day side the observed O^+ densities are so large that the reaction of O^+ with H_2 is the dominant source of nonthermal H and the assumed value of 2 ppm of H_2 is adequate to explain the Mariner 5 nonthermal component of hydrogen. Using the H_2 abundance of 10 ppm derived from the H_2^+ measurements, Kumar et al. (1981) have shown that the $O^+ + H_2$ source contributes a hot H flux of 10^8 cm^{-2} s^{-1} on the day side. The subsequent recombination of OH^+ produced an equal amount of super hot H atoms (energy 8.28 eV) in the exosphere. Half of these, directed upwards, would escape directly. A fraction of those hot H atoms striking the exobase will be reflected back up with little or no change in their velocity. Chamberlain and Smith (1971) calculated an albedo of $\sim 30\%$ for fast particles in the tail of the Maxwellian distribution exceeding the

TABLE IX

Production and Escape of Nonthermal Hydrogen from Venus

Sources of Nonthermal H	H Atom Energy (eV)	H Production Rate (cm⁻²s⁻¹) Day	H Production Rate (cm⁻²s⁻¹) Night	Global Event Rate	Global H Escape Rate (s⁻¹)
I. Ionosphere-exosphere source I					
$O^+ + H_2 \rightarrow OH^+ + H_{hot}$	$\leqslant 0.60$	1×10^8 (a)	1×10^8	5.0×10^{26}	3.3×10^{26}
$OH^+ + e \rightarrow O + H_{hot}$	8.28	1×10^8 (a)	1×10^8	5.0×10^{26}	
II. Ionosphere-exosphere source II					
$H + H^+_{hot} \rightarrow H^+ + H_{hot}$			1.3×10^8	3.3×10^{26} (b)	8.0×10^{25} (b)
$O + H^+_{hot} \rightarrow O^+ + H_{hot}$			1.4×10^7	3.5×10^{25} (b)	
III. Charge exchange with solar wind & escape (exosphere)					
$H + H^+_{sw} \rightarrow H^+ + H_{hot}$	<0.56	3.8×10^6		1.2×10^{25} (b)	7.0×10^{24} (b)
$O + H^+_{sw} \rightarrow O^+ + H_{hot}$	>0.56	1.4×10^6		7.0×10^{24} (b)	
		1.0×10^6 (c)			
IV. Loss by ionization and subsequent escape (exosphere)					
$H + h\nu \rightarrow H^+ + e$		8.4×10^6		2.1×10^{25} (b)	4.0×10^{24} (b)
$H + O^+ \rightarrow H^+ + O$			1.2×10^8	2.9×10^{26} (b)	
V. Ionosphere-thermosphere source					
$CO_2^+ + H_2 \rightarrow CO_2H^+ + H_{hot}$	1.17	0.9×10^8 (a)	0.9×10^8	4.5×10^{26}	1.0×10^{24} (a)
$CO_2H^+ + e \rightarrow CO_2 + H_{hot}$	7.97	0.9×10^8 (a)	0.9×10^8	4.5×10^{26}	
VI. Jeans escape	>0.56				2.5×10^{23} (c)
Total					4.3×10^{26}

(a) Kumar et al. (1981); (b) Hodges and Tinsley (1981); (c) Kumar et al. (1978).

escape velocity. Chamberlain (1977) stated that this albedo is not temperature dependent in the range of interest for nonthermal escape from Earth. Assuming a 30% albedo for particles directed downwards, we obtain an escape flux of 6.5×10^7 H atoms cm^{-2} s^{-1} from OH^+ recombination on the day side of Venus. On the night side, the H_2^+ observations from the PV-OIMS experiment show the presence of a predawn bulge which requires a bulge in II_2 den sity of ≥ 30 (Kumar and Taylor 1981). The O^+ density at night is $\sim 10\%$ of that on the day side, which implies that the nightside $O^+ + H_2$ source would be ~ 3 times stronger in the bulge region which covers $\sim 30\%$ of the night side. Hence on the average, the nightside $O^+ + H_2$ source magnitude is almost equal to the dayside value. This leads to a total hot H production rate of 1×10^{27} H atoms s^{-1}. The net escape rate is 3.3×10^{26} H atoms s^{-1}.

The detection of a large H^+ bulge on the night side requires a serious reconsideration of charge-exchange reactions of H^+ with H and O. Cravens et al. (1980b) showed that the contribution from these charge-exchange reactions at night is almost equal to that from the $O^+ + H_2$ source. Their model was based, however, on a uniform atomic hydrogen density of 2×10^5 cm^{-3} at the exobase over the globe (both day and night) after the results of Anderson (1976). This assumption is incorrect in light of the analysis by Brinton et al. (1980) who show that the H^+ bulge is actually due to a bulge of 400 to 1 in the atomic hydrogen density at the 160 km level (see Fig. 31). Hodges and Tinsley (1981) have allowed for a similar bulge by using a self-consistent model of global thermospheric hydrogen based on measured day-to-night temperature variation and by requiring no flow across the terminator. These authors find that the charge-exchange reaction,

$$H_{hot}^+ + H_{cold} \rightarrow H_{hot} + H_{cold}^+ \qquad (z)$$

is predominant on the night side and significant even on the day side leading to a total global nonthermal production rate of 3.3×10^{26} H atoms s^{-1}, of which 8.0×10^{25} s^{-1} escape. Additional escape of 1.1×10^{25} H atoms s^{-1} is provided by photoionization of H followed by subsequent pickup by the solar wind, and charge exchange of solar wind protons with atmospheric H.

Next let us consider charge exchange with solar-wind protons. The estimated global production rate from this source is 1.9×10^{25} s^{-1} (Hodges and Tinsley 1981) with $\sim 2/3$ of the atoms produced below the escape energy. This rate is factors of 50 and 17, respectively, smaller than the two sources discussed above. However, the particles produced could be in satellite orbits, as suggested by Bertaux et al. (1978), which would lead to a much longer lifetime than that for exospheric particles in ballistic orbits. Hence a smaller production rate may be adequate to populate the hot hydrogen corona. Bertaux et al. estimated that the density of the hot component resulting from this charge exchange would be $\sim 10\%$ of the total H density at 3500 km altitude. However, in the 2-temperature interpretation of the Lyman-α intensity data, the bulk of the H density at 3500 km altitude is due to the hot component. Hence it would appear that the solar wind charge exchange source is inadequate. Bertaux et al. (1982) have, however, revised this estimate to indicate that the 10% density value quoted by Bertaux et al. (1978) should be regarded as a lower limit because the lifetime of satellite particles is difficult to ascertain. In support of this argument, Bertaux et al. (1982) showed Venera 11 and 12 data indicating a kink or a sharp change in the scale height at 3000 km altitude in the Lyman-α intensity distribution which may be theoretically reproduced if the source of hot H is above this altitude. Hodges and Tinsley (1981) have argued against this source on the basis of their calculations which show that charge exchange with solar-wind protons is the dominant means of deorbiting satellite hydrogen, and thus unlikely to be a net source of orbiters. Obviously more work needs to be done to demonstrate the validity of this source; in the meantime we consider it a viable candidate.

Thus nonthermal H atoms that are observed in the hydrogen corona from the Lyman-α airglow can be produced by a combination of the ionosphere $O^+ + H_2$ source, the ionosphere H^+-H charge exchange, and the solar-wind charge exchange, with possibly appreciable contributions from each source. The escape of hydrogen is, however, dominated by the OH^+ recombination reaction. The total escape rate from all sources is 4.3×10^{26} s^{-1}.

5. Hydrogen Photochemistry. Photochemical models of the stratosphere and thermosphere including hydrogen photochemistry were published by McElroy et al. (1973) and Sze and McElroy (1975), whereas a thermosphere-only model was presented by Liu and Donahue (1975). Hydrogen photochemistry was also discussed by Kumar and Hunten (1974) with respect to an ionosphere model. Each of these models require some revision in light of Pioneer Venus measurements though several features are still valid. The salient features with respect to photochemistry are discussed below, and comments are made wherever a revision is required.

As introduced in Sec. IV. B-2.2 photodissociation of HCl proceeds right down to the cloud tops:

$$h\nu + HCl \rightarrow H + Cl. \qquad \text{(aa)}$$

The resulting chlorine atoms react with H_2 in the atmosphere forming HCl so that the HCl abundance is maintained:

$$Cl + H_2 \rightarrow HCl + H. \qquad \text{(bb)}$$

This is also the primary source of atomic hydrogen in the middle atmosphere and the primary sink for H_2. The H_2 density is determined by a balance between upward diffusion and chemical loss. McElroy et al. (1973) and Sze and McElroy (1975) proposed two schemes for recycling of CO_2. Odd hydrogen plays the key role in the following scheme in which ultimately OH reacts with CO to form CO_2,

$$H + O_2 + CO_2 \rightarrow HO_2 + CO_2$$

$$H + HO_2 \rightarrow 2OH$$

$$HO_2 + HO_2 \rightarrow H_2O_2 + O_2$$

$$h\nu + H_2O_2 \rightarrow 2OH$$

$$H + O_3 \rightarrow OH + O_2$$

$$CO + OH \rightarrow CO_2 + H .\qquad\qquad\qquad\text{(cc)}$$

The major source of water and the major sink of HCl is the reaction,

$$HCl + OH \rightarrow H_2O + Cl .\qquad\qquad\qquad\text{(dd)}$$

The second scheme involved formation of ClO which in turn reacts with CO to form CO_2,

$$Cl + O_3 \rightarrow ClO + O_2$$

$$Cl + HO_2 \rightarrow ClO + OH$$

$$CO + ClO \rightarrow CO_2 + Cl .\qquad\qquad\qquad\text{(ee)}$$

In the Sze and McElroy (1975) model, the two schemes were of comparable effectiveness but these authors used an H_2 mixing ratio of only 0.2 ppm, far below the Kumar et al. (1981) value of 10 ppm inferred from the PV-OIMS H_2^+ measurements. Yung and DeMore (1982) have reinvestigated the photochemistry of the middle atmosphere and shown that the observed O_2 upper limit of 1 ppm by Traub and Carleton (1974) is consistent with 10 ppm of H_2, and that with this large H_2 abundance, odd hydrogen controls the recycling of CO_2.

Photodissociation of HCl and H_2O leads to production of atomic hydrogen at a rate of 1×10^{12} cm^{-2} s^{-1} and 3×10^8 cm^{-2} s^{-1}, respectively. The reaction of Cl with H_2 would produce H at a rate $\geqslant 1 \times 10^{13}$ cm^{-2} s^{-1}. But diffusion limits the flow of H to $\leqslant 1 \times 10^8$ cm^{-2} s^{-1} above the mesopause. The primary source of atomic hydrogen in the upper atmosphere is the reaction of CO_2^+ with H_2 and subsequent recombination,

$$CO_2^+ + H_2 \rightarrow CO_2H^+ + H$$

$$CO_2H^+ + e \rightarrow CO_2 + H .\qquad\qquad\qquad\text{(ff)}$$

Kumar et al. (1981) estimate an H production rate of 3.3×10^8 cm^{-2} s^{-1} on the day side from this source. The first one of these reactions is also a potent sink for H_2 which does not attain diffusive equilibrium until $\sim$ 180 km (see Fig.

32). Since the escape rate is only 1×10^8 cm^{-2} s^{-1} from the nonthermal processes discussed above, the rest must be transported from day to night, greatly enhancing the hydrogen bulge. This is borne out by the transport models of Kumar et al. (1978) and Mayr et al. (1980) which are consistent with the observed bulge of 400 to 1 in atomic hydrogen.

Sze and McElroy (1975) assumed water-gas equilibrium to estimate the H_2 abundance at the surface:

$$CO + H_2O \rightarrow CO_2 + H_2 . \tag{gg}$$

However, for 20 ppm of CO and 20 ppm of H_2O, the H_2 mixing ratio is only 0.002 ppm if this equilibrium prevails (Kumar et al. 1981). Even for 100 ppm of H_2O, the revised value by L. Young (1981), the H_2 mixing ratio would be 0.01 ppm, well below the observed value of 10 ppm. Kumar et al. suggest that the S atoms, produced by visible light (Prinn 1978, 1979), may convert a fraction of the H_2O into H_2 and H_2S. Hoffman et al. (1980a) report only 3 ppm of H_2S near the surface, but such a reactive gas is difficult to measure with a mass spectrometer. The hydrogen chemistry near the surface is yet to be understood.

6. *Nonthermal Oxygen and Escape of H_2O*. Escape of oxygen is closely linked to the escape of hydrogen since Venus has most likely lost its original water content without a buildup of O_2. Production of nonthermal O atoms in the Venus exosphere has been investigated by Nagy et al. (1981) who find that the dissociative recombination of O_2^+ is the primary source,

$$O_2^+ + e \rightarrow O\,(^3P) + O\,(^3P) + 7.0\,eV$$
$$\rightarrow O\,(^3P) + O\,(^1D) + 5.0\,eV$$
$$\rightarrow O\,(^1D) + O\,(^1D) + 3.1\,eV$$
$$\rightarrow O\,(^3P) + O\,(^1S) + 2.8\,eV$$
$$\rightarrow O\,(^1D) + O\,(^1S) + 0.8\,eV. \tag{hh}$$

The product O atoms, which are energetic enough to escape the gravity on Mars and dominate the escape of oxygen from that planet, fall short of the escape energy of 8.86 eV required on Venus. They are, however, produced in such high abundance that a hot oxygen corona is produced which dominates the hydrogen density above 400 km until hot hydrogen takes over at much greater altitude. The O density distribution calculated by Nagy et al. is shown in Fig. 33. Prompted by these calculations, a search was made for a hot oxygen corona in the 130 nm airglow by the PV-OUVS experiment and indeed such a corona was found. The PV-OUVS data points are also shown in Fig. 33. They are a factor of 10 or so lower than the calculations of Nagy et al.

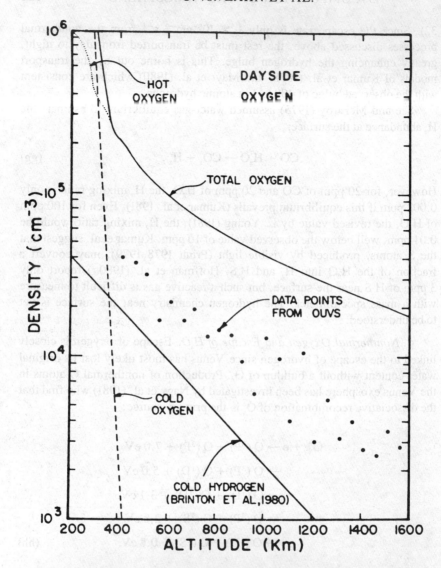

Fig. 33. Dayside hot oxygen densities ($\sim$ 6400 K) produced by dissociative recombination of O_2^+ as calculated by Nagy et al. (1981) and measured by the Pioneer Venus orbiter ultraviolet spectrometer. For comparison, the densities of cold oxygen and hydrogen ($\sim$ 300 K) are also shown.

who point out the discrepancy could be due to the following reasons: (a) >50% of the O atoms are produced in the ${}^1D$ state which has a lifetime of 110 seconds, whereas the 130 nm airglow emission is caused by the resonance scattering from the ground state; (b) significant differences exist in the electron and O_2^+ ion densities measured from various Pioneer Venus experiments and higher values would provide a better match with the 130 nm data of the PV-OUVS. An important effect of these hot O atoms would be to increase the charge-exchange rate of the ionosheath protons and thus increase the absorption of solar wind by the neutral atmosphere. The contribution to the escape of oxygen via hot O atoms is, however, relatively insignificant.

The most potent escape of oxygen from Venus is in the form of O^+ ions in the plasma clouds which have been observed above the ionopause with the Pioneer Venus orbiter electron temperature probe by Brace et al. (1982a). Brace et al. suggest that these clouds could originate at the dayside ionopause as wave-like structures, become detached from the ionosphere and swept downstream in the ionosheath flow, or they might be attached streamers analogous to cometary structure. Their estimate of total O^+ ion escape from this process is $\lesssim 7 \times 10^{26}$ ions s^{-1}, which is close to the H escape rate of 4.3×10^{26} atom s^{-1} listed in Table IX.

Comparable escape rates of H and O of $\sim 10^8$ cm^{-2} s^{-1} from Venus imply a loss of 2.5 m of water, if they have been constant over geologic time. In contrast, an average escape rate of 10^{11} cm^{-2} s^{-1} or 5×10^{29} s^{-1} is required to remove an ocean of water equal to the amount present on Earth today. This suggests that either Venus was deficient in water to begin with or the escape rate was much higher in the past.

E. Sulfur Compounds

There is now ample evidence that sulfur compounds are present and play vital chemical and physical roles in the atmosphere of Venus. Sulfur dioxide gas, concentrated sulfuric acid droplets, and possibly hydrogen sulfide and elemental sulfur gases have now been identified on Venus with the initial discoveries being made respectively by Barker (1979), Young and Young (1973) and Sill (1972), Hoffman et al. (1980a), and Moroz et al. (1979c). Chemical reactions involving sulfur species are intimately involved in the atmospheric oxygen balance and in the formation of clouds with their concomitant effects on the deposition of solar and infrared radiation in the atmosphere.

The chemistry of sulfur compounds on Venus appears to be driven by two essentially competing influences (Prinn 1978; Florensky et al. 1978). A photochemical region exists within and above the clouds where the chemistry is driven by solar ultraviolet radiation, an agent which produces oxidized, thermochemically unstable sulfur compounds. This photochemical region may also extend down conceivably to altitudes as low as 10 km with reactions driven by solar near-ultraviolet and visible radiation (Prinn 1979). In contrast, a thermochemical region exists at and near the surface where high temperatures and pressures enable the action of purely thermochemical reactions which produce reduced, thermochemically stable compounds. The observed composition of the atmosphere depends on the relative rates of reactions in these two regions and on the rate of vertical transport between them.

E-1. Photochemistry

In this section we discuss the photochemistry, thermochemistry, and chemical cycles of the principal sulfur compounds in the Venus atmosphere.

1. Sulfur Dioxide was first detected above the clouds by Barker (1979) using earth-based ultraviolet observations. It was later detected in both the middle and lower atmosphere by instruments carried by the Pioneer Venus orbiter and Large probe (Stewart et al. 1979; Oyama et al. 1979), and the Venera 11 and 12 landers (Gel'man et al. 1979b). A summary of these observations is given in Table X. An upper limit for the SO_2 mixing ratio published by Istomin et al. (1980b) has been withdrawn after laboratory tests with mixtures of CO_2 and SO_2 showed strong adsorption of SO_2 to the instrument surfaces in the early measurement phases. This reduced the effective instrument sensitivity for SO_2 initially by more than a factor of 100 (V.G. Istomin, personal communication).

Prior to the 1978 space missions, sulfur dioxide was first suggested as merely an intermediate in the photochemical formation of sulfuric acid from the precursors COS and H_2S, and as a product of decomposition of SO_3 vapor (Prinn 1971,1973b,1975). Later work suggested that if vertical mixing between the thermochemical regime near the surface and the photochemical regime within and above the clouds were sufficiently slow, then photochemically produced SO_2 and SO_3 could dominate thermochemically produced COS and H_2S throughout the upper photochemical regime (Prinn 1978). Under these circumstances SO_2 and SO_3 would be principal precursors of the sulfuric acid clouds rather than COS or H_2S.

Observations now show that SO_2 dominates over COS and H_2S above 22 km, and that the SO_2 mixing ratio has a value ~ 150 ppm between 22 km and the cloud base (50 km) but decreases within the clouds to a value ~ 0.1 ppm at 70 km. This is consistent with SO_2 being converted to H_2SO_4 above 50 km. As discussed later, this in turn demands that SO_2 be produced from H_2SO_4 below 50 km at the same global rate.

The gas-phase photochemistry of SO_2 and its conversion to H_2SO_4 have been most comprehensively discussed by Winick and Stewart (1980) and Krasnopolsky and Parshev (1980b,c and Chapter 14). These authors consider reactions involving chlorine and odd hydrogen in addition to sulfur compounds and include a number of chemical pathways not considered in the first studies of these three classes of compounds on Venus (Prinn 1971,1973b; McElroy et al. 1973). Winick and Stewart (1980) were able to successfully model the vertical distributions of SO_2 and cloud densities above 60 km but their chemical model produces O_2 and CO abundances respectively 6 and > 100 times the observed cloud-top values. Krasnopolsky and Parshev

TABLE X
Mixing Ratio of SO_2 (in ppm)

Reference	Mixing Ratios at Altitude of				Pressure of 40 mbar (69 km)	Column Density (cm^{-2})	Wave-Length Band (nm)	Experimental Techniques (or comment)
	0 km	22 km	42 km	52 km				
Barker (1979)	< 0.02 to 0.6[a]					2×10^{17} to 5×10^{18}[a]	321	Earth-based Spectroscopy
Conway et al. (1979)	0.1 to 0.8 for 3 to 26 mbar					4×10^{16}	212	IUE satellite
Stewart et al. (1979)						1×10^{17}	200 to 340	UV spectrometer PV orbiter
Esposito et al. (1979)				0.1				models of PV-OUV data
Winick and Stewart (1980)			4 ppm at 58 km	0.1				
Esposito (1981)				0.05				
Oyama et al. (1980a)		185±43	>176	<600				gas chromatogr. PV Large probe
Hoffman et al. (1980a)		<300		<10				mass spectrom. PV Large probe
Gel'man et al. (1979a)		130±35						gas chromatogr. Venera 12 lander

[a]Values subject to downward revision due to change in absorption cross section (see Wu and Judge 1981).

(1980b,c) were also able to model the vertical SO_2 distribution and in addition they predicted CO and O_2 abundances which agreed much better with observations. However, their scheme for recombining CO and O_2 utilized

$$ClO + CO \rightarrow Cl + CO_2 \tag{ii}$$

and assumed a rate constant for this reaction several orders of magnitude greater than that adopted by Winick and Stewart (1980). In addition, they included formation of COCl and the subsequent reaction

$$COCl + O_2 \rightarrow CO_2 + ClO \tag{jj}$$

which in fact dominates recombination of CO and O_2 in their model. The kinetics of COCl are at present poorly defined and therefore this latter reaction is speculative but certainly worthy of further investigation (see also Yung and DeMore 1982).

The solution-phase chemistry of SO_2, in particular the possible oxidation of SO_2 by photochemically produced H_2O_2, O_3, and chlorine oxides in solution in the sulfuric acid cloud droplets, has not yet been included in photochemical models. This is due to an almost complete lack of laboratory kinetic data on such reactions.

2. Hydrogen Sulfide and Carbonyl Sulfide. The models of Winick and Stewart (1980) and Krasnopolsky and Parshev (1980b,c) did not include any reduced sulfur gases, such as H_2S or COS. Although the importance of these gases is much less than proposed prior to the 1978 Pioneer Venus and Venera missions, they may nevertheless be playing important roles as elemental sulfur sources and O_2 sinks in the cloud region.

Hydrogen sulfide has been detected only by the PV-LNMS (Hoffman et al. 1979c). While the detection is not yet regarded as definite, the data suggest a mixing ratio of 3 ± 2 ppm below 20 km altitude (Hoffman et al. 1980a). For 22 km altitude, an upper limit of 2 ppm H_2S was established by PV-LGC (Oyama et al. 1980a). The mass spectrometer data, in addition, indicate a value ~ 1 ppm in the visible cloud region. This is consistent with production of H_2S in the lower atmosphere and destruction in the visible clouds. Carbonyl sulfide has not been detected in the atmosphere; the Pioneer Venus experiments suggest an upper limit of ~ 2 to 3 ppm above an altitude of 22 km (Hoffman et al. 1980a; Oyama et al. 1980a). For the region below 22 km little information on COS is available except for an upper limit of 500 ppm (Hoffman et al. 1980a) which is too high to put any serious constraints on the relevant models.

The photochemistry of H_2S has been discussed by Prinn (1978). Depending on the concentrations of O_2 and other oxidants in the cloud-top region, the SH radicals produced from H_2S photodissociation can lead either to sulfuric acid or to elemental sulfur production. Our current knowledge would suggest, however, that at least in the gas phase H_2S is much more important as an O_2 sink through the reactions

$$H_2S + h\nu \rightarrow SH + H$$

$$SH + SH \rightarrow S + H_2S$$

$$S + O_2 \rightarrow SO + O \tag{kk}$$

than as a source of sulfur or sulfuric acid. Production of elemental sulfur from H_2S is still possible if H_2S dissolved in acidic cloud particles is oxidized also by H_2O_2 in solution (Satterfield et al. 1954). Production of sulfuric acid in this reaction apparently requires alkaline conditions.

The photochemistry of COS has been discussed by Prinn (1971, 1973b, 1975). Like H_2S, photodissociation of COS can lead either to sulfuric acid or sulfur formation. Our current perspective suggests that, if present in the cloud region with a mixing ratio of 2 to 3 ppm, COS would be important as an O_2 sink and an elemental sulfur source through the reactions

$$COS + h\nu \rightarrow CO + S$$

$$SO + h\nu \rightarrow S + O$$

$$S + O_2 \rightarrow SO + O$$

$$S + COS \rightarrow S_2 + CO \tag{ll}$$

but that, like H_2S, it is not an important contributor to sulfuric acid formation. In contrast, COS photodissociation does appear to be an important source of H_2SO_4 in the Earth's stratosphere, where sulfuric acid concentrations are several orders of magnitude less than on Venus (Crutzen 1976).

3. *Gaseous Elemental Sulfur.* The Venera 11 and 12 spectrophotometers showed strong absorption in the 450 to 600 nm region between the altitudes of 10 and 30 km (see Fig. 27) which is interpreted in terms of absorption by gaseous elemental sulfur (Moroz et al. 1979c). The possible importance of S_n ($n = 1$ to 8) in the lower atmosphere was first pointed out by Prinn (1979). The light allotropes S_2 and S_3 (thiozone) are dominant at the high temperatures and low sulfur vapor pressures existing in the lowest three scales heights of the Venus atmosphere. Based on the thermodynamic equilibrium data of Mills (1974), abundance ratios of the various sulfur allotropes were calculated by Sanko (1980) as function of altitude above the Venus surface (see Fig. 34). To obtain also the absolute abundances, Sanko analyzed the observed absorption of scattered daylight near 450 nm (Moroz et al. 1979c). Of all the sulfur allotropes, only S_3 has a significant absorption cross section near 450 nm

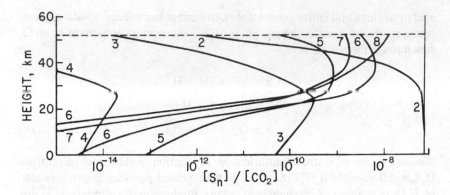

Fig. 34. The mixing ratio of sulfur allotropes S_n as function of altitude above the surface of Venus, calculated by Sanko (1980). The peak mixing ratio of S_3 is obtained from the scattered daylight measurements of Moroz et al. (1980a) and used to calculate the mixing ratios of the other allotropes from thermodynamic equilibrium data of Mills (1974).

wavelength (Meyer et al. 1972). Furthermore, as seen in Fig. 34, the calculated abundance of S_4 at Venus is rather low in comparison with the S_3 abundance. Hence from the observed absorption, Sanko (1980) determined the S_3 mixing ratio to peak at $\sim 4 \times 10^{-11}$ close to 23 km altitude. The mixing ratio of total sulfur then becomes $\sim 2 \times 10^{-8}$, dominated strongly by S_2 below 30 km. (The mixing ratios quoted are updates given by Sanko (1980) and somewhat lower than those shown in Fig. 34.) From a similar analysis of the 450 to 550 nm photometer observations on board Veneras 11 and 12, Moroz et al. (1981) determined the single scattering albedo $\widetilde{\omega}_0$ at the wavelength of 490 nm. As shown in Fig. 35, the parameter $(1-\widetilde{\omega}_0)$, which is proportional to the mass absorption coefficient, peaks ~ 1 scale height lower than inferred from the $S_3 + S_4$ profile calculated by Sanko (1980). To what extent this difference is due to the neglect in Sanko's calculations of the possibility of photodissociation and chemical reactions by S_3 (see below) remains to be studied. Finally, we note that Pollack et al. (1980b) also considered the absorption measurements by Moroz et al. (1979c); Pollack et al. state that "our calculations suggest that gaseous SO_2 may have been partly or wholly responsible for these results," but do not give any details of their respective calculation.

The spectroscopy and probable sources of S_2, S_3, and S_4, and the dynamical and chemical consequences of their presence in the lower atmosphere have been discussed by Prinn (1979). Of particular interest is the probable photodissociation of S_4 and perhaps S_3 by near-ultraviolet, blue, and yellow light and thus the existence of a photochemical regime below the visible clouds driven by solar radiation at much longer wavelengths than customarily considered by atmospheric chemists. The presence of photochemically produced

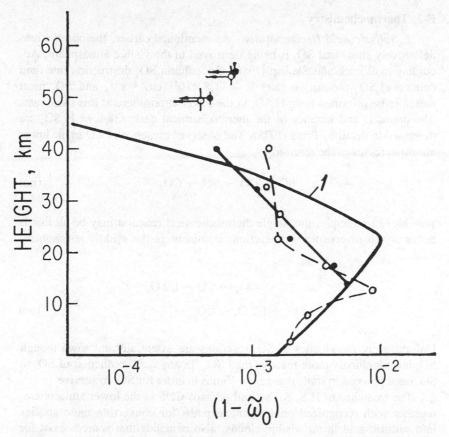

Fig. 35. The parameter $1-\tilde{\omega}_0$, with $\tilde{\omega}_0$ being the single scattering albedo at the wavelength of 490 nm, determined from data obtained by the spectrophotometers on board Venera 11 (open circles) and Venera 12 (dots). Solid line is a theoretical profile of $1-\tilde{\omega}_0$ based on the $S_3 + S_4$ mixing ratio derived by Sanko (1980). (Figure from Moroz et al. 1981.)

S and S_2 in the lower atmosphere would provide an important sink for COS through the chain reactions

$$S + COS \underset{\rightarrow}{\leftarrow} CO + S_2$$

$$S_2 + COS \underset{\rightarrow}{\leftarrow} CO + S_3$$

$$S_3 + COS \underset{\rightarrow}{\rightleftharpoons} CO + S_4. \qquad \text{(mm)}$$

Whether or not these photochemical S atoms can also attack H_2S, H_2O, CO_2, or H_2 at the high temperatures near the surface is not known. Laboratory experiments are required before the importance of this photochemical regime below the clouds can be further assessed.

E-2. Thermochemistry

1. Sulfuric Acid Decomposition. As mentioned earlier, the observations definitively show that SO_2 is being destroyed in the visible atmosphere. According to Winick and Stewart (1980), the column SO_2 destruction rate (and column H_2SO_4 production rate) is $\geq 1.8 \times 10^{11}$ cm^{-2} s^{-1}, and SO_2 must therefore be reformed from H_2SO_4 in the lower atmosphere at this same rate. The products and kinetics of the thermochemical destruction of H_2SO_4 are discussed in detail by Prinn (1978). The observed presence of CO in the lower atmosphere makes the reaction

$$SO_3 + CO \rightarrow SO_2 + CO_2 \tag{nn}$$

possible. In principle, this single thermochemical reaction may be all that is necessary to reverse the net reactions dominant in the middle atmosphere, namely

$$CO_2 \rightarrow CO + 1/2\,O_2$$
$$SO_2 + 1/2\,O_2 \rightarrow SO_3. \tag{oo}$$

Unfortunately the kinetics of SO_3 reactions are essentially unknown though SO_3 is undoubtedly more reactive than SO_2. In any case, reduction of SO_3 to SO_2 must proceed in some manner on Venus in order for SO_2 to survive.

The existence of H_2S, S_3, S_4, and perhaps COS in the lower atmosphere, together with recognized photochemical paths for converting these species into sulfuric acid in the visible clouds, also demands that sources exist for these species. As discussed by Prinn (1978), reduction of SO_2 by CO near the hot surface is suggested by laboratory experiments, with S_2, COS, and H_2S being the probable products. Continuity arguments would then demand that the total mixing ratio of the products of this SO_2 destruction (COS, H_2S, S_2, etc.) approach 150 ppm below 22 km. They also demand that SO_2 be reformed from COS, H_2S, and S_2 as air parcels rise above 22 km, which is possible but unproven (Prinn 1979). Indeed, an unambiguous answer to the question of the oxidation state of sulfur species below 22 km may have to await a definitive *in situ* measurement of the atmospheric composition close to the surface.

2. Equilibrium Versus Disequilibrium. In view of the fact that 80% of the atmospheric mass of Venus lies below 22 km, our present ignorance of the detailed composition of this lower region hampers our understanding of overall chemical cycles. A number of studies have been carried out which explicitly hypothesize chemical equilibrium in the lower atmosphere and between the atmosphere and surface minerals, and then explore the consequences of this hypothesis for atmospheric and surface composition (Lewis 1970; Florensky et al. 1978; Khodakovsky et al. 1979; Barsukov et al. 1980*a*,

b; Lewis and Kreimendahl 1980). The most recent studies indicate that the oxygen abundance in the atmosphere in chemical equilibrium is governed by the pyrite-anhydrite buffer (Lewis and Kreimendahl 1980; Barsukov et al. 1980b)

$$FeS_2 + 2CaCO_3 + 7/2\ O_2 \rightleftarrows 2CaSO_4 + FeO + 2CO_2. \qquad \text{(pp)}$$

In addition, the total mixing ratio of sulfur gases (SO_2 + COS + H_2S) decreases rapidly with increasing O_2, that is, with increasing abundances of FeO and $CaSO_4$ in the crust. In particular, a total mixing ratio for sulfur gases $\gtrsim$ 150 ppm requires FeO activities $\gtrsim$ 1/10 those on Earth (Lewis and Kreimendahl 1980). In other words even a mildly oxidized Venus crust is incompatible with significant amounts of any sulfur gases in the atmosphere. Finally, a total sulfur gas mixing ratio $\gtrsim$ 150 ppm requires COS and H_2S to be more abundant than SO_2. Indeed, the maximum SO_2 mixing ratio predicted under any conditions at the surface is 1/10 that observed (Lewis and Kreimendahl 1980).

These conclusions based on assumed chemical equilibrium between atmosphere and surface are clearly at odds with SO_2 remaining the dominant sulfur species at the surface. Thermochemical reactions at the surface would tend to either incorporate SO_2 in $CaSO_4$ until the SO_2 mixing ratio was decreased to $\leqslant$ 1/50 of that observed, or to transform SO_2 to COS and H_2S until the SO_2 mixing ratio was $\leqslant$ 1/10 that observed. Two possibilities appear to exist. The first is that COS and H_2S are indeed the dominant sulfur gases at the surface (i.e., equilibrium prevails). The second is that sulfur is first outgassed as COS and H_2S; then these species are oxidized to SO_2 (and H_2SO_4) by photochemical reactions, and finally the reaction times for incorporation of SO_2 into $CaSO_4$ and for reconversion of SO_2 to COS and H_2S are simply too long. Under these conditions SO_2 would be enhanced and COS and H_2S depleted relative to equilibrium (i.e., disequilibrium prevails). Which of these two possibilities prevails depends on the kinetics of the reactions involved.

While laboratory data on the kinetics of relevant gas-mineral reactions are nonexistent, there are some data on the kinetics of the gas reactions and these suggest the possibility of gas-phase equilibrium but only very close to the surface (Prinn 1978; Krasnopolsky and Parshev 1979a). Calculations can thus be carried out assuming gas-phase equilibrium which use elemental abundances derived from observations above 22 km rather than those determined by atmosphere-surface equilibrium. Calculations using data from Veneras 11 and 12 (Krasnopolsky and Parshev 1979a) and Pioneer Venus (Oyama et al. 1980a) predict SO_2, COS, S_2, and H_2S mixing ratios ~10^{-4}, 10^{-5}, 10^{-7}, and 10^{-6} to 10^{-7}, respectively. The abundances of SO_2 (and O_2) in these calculations are too large to be compatible with equilibrium with surface minerals but the SO_2, COS, S_2, and H_2S abundances are all roughly compatible with observations. This suggests that the slow or bottleneck reactions preventing

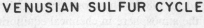

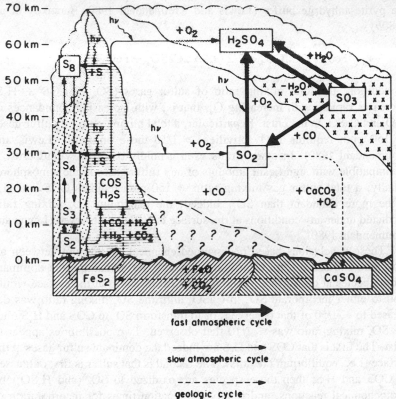

Fig. 36. The Venus sulfur cycle. The regions of stability and relative abundance of the major
sulfur compounds are qualitatively indicated by the position and size of the shaded area
assigned to each one.

complete equilibrium are the incorporation of SO_2 into $CaSO_4$ and O_2 into
$CaSO_4$ and FeO at the surface. These reactions may simply proceed too slowly
when compared with photochemical production of SO_2 and O_2 from COS and
CO_2.

E-3. Sulfur Cycles

Our present state of knowledge suggests the existence of three important
overall cycles for sulfur compounds on Venus as summarized in Fig. 36.
Although each of these cycles has several possible ramifications we present
only the essence of them here. The first cycle, which we refer to as the *fast
atmospheric cycle,* involves the net photochemical reactions in the middle and
upper atmosphere

$$CO_2 \rightarrow CO + 1/2 \, O_2$$

$$1/2 \, O_2 + SO_2 \rightarrow SO_3$$

$$SO_3 + H_2O \rightarrow H_2SO_4 \qquad\qquad (qq)$$

being reversed in the lower atmosphere by the net thermochemical reactions

$$H_2SO_4 \rightarrow H_2O + SO_3$$

$$SO_3 + CO \rightarrow SO_2 + CO_2 . \qquad\qquad (rr)$$

This fast cycle is thus intimately involved with the formation of the sulfuric acid clouds and at least in part with the recycling of photochemically produced CO and O_2. However, it does not explain the ultraviolet (300 to 400 nm) absorbing component of these clouds.

The second cycle which we call the *slow atmospheric cycle* involves the net photochemical reactions in the middle atmosphere

$$6CO_2 + h\nu \rightarrow 6CO + 3O_2$$

$$3/2 \, O_2 + H_2S \rightarrow SO_3 + H_2$$

$$3/2 \, O_2 + COS \rightarrow SO_3 + CO$$

$$2SO_3 + 2H_2O \rightarrow 2H_2SO_4$$

$$H_2S \rightarrow H_2 + (1/n) \, S_n$$

$$COS \rightarrow CO + (1/n) \, S_n \qquad\qquad (ss)$$

being reversed in the lower atmosphere close to and at the surface by the net thermochemical reactions

$$2H_2SO_4 \rightarrow 2H_2O + 2SO_3$$

$$SO_3 + 4CO \rightarrow COS + 3CO_2$$

$$H_2 + SO_3 + 3CO \rightarrow H_2S + 3CO_2$$

$$H_2 + (1/n) \, S_n \rightarrow H_2S$$

$$CO + (1/n)S_n \rightarrow COS . \qquad\qquad (tt)$$

Although this slow atmospheric cycle involves H_2SO_4 production, it is not an important contributor to the sulfuric acid clouds. However, it does involve production of elemental sulfur which is an attractive candidate for the ultraviolet absorbing component of the clouds (Prinn 1971, 1975; A. T. Young 1977). The photochemical regime in this slow cycle may be augmented by photochemical reactions instigated by S_3 and S_4 absorption in the lower atmosphere. Such reactions can convert COS and H_2S to elemental sulfur and perhaps to SO_2.

An important aspect of both the fast and slow atmospheric cycles is worth noting. Observations imply that the thermochemical reactions in the lower atmosphere in each cycle apparently proceed fast enough so that the SO_2/H_2S and SO_2/S_2 ratios, and by implication the SO_2/COS and SO_2/SO_3 ratios, are maintained approximately at the gas-phase equilibrium values expected when chemical interaction with the surface is negligible.

The third cycle which we refer to as the *geologic cycle* begins with outgassing of the reduced gases COS and H_2S from the crust as a result of mineral reactions such as those involving pyrite

$$FeS_2 + 2H_2O \rightarrow FeO + 2H_2S + 1/2\,O_2$$

$$FeS_2 + 2CO_2 \rightarrow FeO + 2COS + 1/2\,O_2 \,. \tag{uu}$$

These gases will then proceed many times through the fast and slow atmospheric cycles and in the process spend the majority of their time in the form of SO_2. In this manner continual outgassing will build up SO_2 in the atmosphere through the net photochemical reactions

$$2H_2S + 3O_2 \rightarrow 2H_2O + 2SO_2$$

$$2COS + 3O_2 \rightarrow 2CO_2 + 2SO_2 \,. \tag{vv}$$

Once SO_2 builds up to levels exceeding its thermochemical equilibrium concentration relative to surface minerals, several thermochemical reactions attempt to restore the system to equilibrium. These restoring reactions include formation of anhydrite

$$4SO_2 + 2O_2 + 4CaCO_3 \rightarrow 4CaSO_4 + 4CO_2 \tag{ww}$$

followed by conversion of anhydrite to pyrite to complete the cycle

$$4CaSO_4 + 2FeO + 4CO_2 \rightarrow 2FeS_2 + 4CaCO_3 + 7O_2 \,. \tag{xx}$$

Unlike the thermochemical reactions restoring equilibrium in the two atmospheric cycles, we have no observational evidence to suggest that the restoring reactions in the geologic cycle are proceeding fast enough to maintain atmosphere-surface equilibrium; nor do we have evidence that the geologic cycle is even in a steady state. Outgassing of sulfur gases may be dominated by sporadic volcanic activity or cometary impacts while their reincorporation into minerals is dominated by surface weathering and crustal recycling. Thus outgassing and reincorporation may only be rarely in balance, and the total sulfur content of the atmosphere including sulfuric acid cloud densities may experience oscillations or monotonic trends.

Laboratory experiments on the rates of various gas and gas-mineral reactions involving sulfur compounds are needed to further clarify the sulfur cycle on Venus. If we are ignorant of some of the details of atmospheric composition at the venusian surface, we are infinitely more ignorant of the nature of surface minerals. There are even recent suggestions that the compositions of the near-surface atmosphere and surface minerals are functions of the large-scale meteorological (Barsukov et al. 1980*b*) and topographical (Khodakovsky et al. 1979) patterns. These possibilities must be kept in mind in the future planning of the much needed *in situ* surface measurements on Venus.

F. Halogen Compounds

In the mid-1960s the development of a high-resolution Michelson interferometer enabled the remarkable discovery of the trace gases HCl and HF in the atmosphere of Venus using an earth-based telescope (Connes et al. 1967). The reported mixing ratios of HCl and HF in the cloud-top region were 0.6 ppm and 0.005 ppm, respectively. These mixing ratios were unfortunately below the sensitivity levels of the recent Pioneer Venus and Venera 11 and 12 gas-phase analyses, and thus only an upper limit of 10 ppm has been reported for HCl in the region below the clouds (Hoffman et al. 1980*a*). Chlorine in an unknown chemical form was, however, detected in the cloud particles themselves by the X-ray fluorescence experiment on Venera 12 (Surkov et al. 1981). The reported condensed chlorine density of 2×10^{-9} g cm^{-3} refers to atmospheric pressures ~ 1 bar and thus implies a Cl/CO$_2$ ratio $\sim 10^{-6}$, which seems reasonable. The reported particulate Cl/S ratio $\gtrsim 10$ is surprising but the analysis was apparently carried out at temperatures > 350 K so that any H$_2$SO$_4$ would have evaporated from the particles. Thus, as Knollenberg and Hunten (1980) suggest, the large Cl/S ratio may simply imply that Cl is in a less volatile form than H$_2$SO$_4$ in the Venus clouds. We note that no halogens other than Cl and F have yet been detected on Venus.

In this section we outline the photochemistry, thermochemistry, and probable chemical cycles of venusian chlorine and fluorine compounds, briefly discuss possible undetected bromine and iodine compounds, and suggest new candidates for chlorine species in the clouds.

F-1. Photochemistry

1. Hydrogen Chloride. The photochemistry of HCl on Venus was first studied by Prinn (1971,1972,1973*a*). This early work emphasized the importance of HCl photodissociation as a source of H$_2$ and the possible roles in the middle atmosphere of Cl, ClO, and ClOO reactions in the recombination of photochemically produced CO and O$_2$, and suggested photochemically produced Cl$_2$, HOCl, Cl$_3^-$, and OCl$^-$ in acidic cloud droplets are possible candidates for ultraviolet and blue absorbers in the visible clouds. Later McElroy et al. (1973) presented a chemical model for Venus's middle atmosphere which stressed the role of HCl and Cl in the production of odd hydrogen for recombination of CO and O_2, while Liu and Donahue (1975) presented a chemical-dynamical model of the venusian upper atmosphere emphasizing HCl as a source of H atoms.

The most comprehensive study of venusian chlorine chemistry prior to the Pioneer Venus and Venera 11 and 12 missions was that of Sze and McElroy (1975). They combined what appeared to be the most plausible chlorine and odd-hydrogen reactions from earlier papers in a chemical-dynamical model of the middle and upper atmosphere, which included vertical transport of the reactive chlorine species (Cl, ClO, ClOO, Cl_2). They concluded that reactions of CO with OH and to a lesser extent with ClO were the principal ways of recycling CO_2, and that HCl, Cl, and ClO were intimately involved in production and/or loss of middle atmospheric O_2, H_2, and H_2O. They also noted that the downward flux of O_2 into the clouds, and thus the maximum rate of production of H_2SO_4 from reduced sulfur gases, was a sensitive function of the rates assumed for vertical transport and for the important reaction

$$HO_2 + Cl \rightarrow HCl + O_2 \qquad\qquad \text{(yy)}$$

which destroys both odd hydrogen and reactive chlorine.

The photochemistry of chlorine compounds was readdressed by Winick and Stewart (1980) and Krasnopolsky and Parshev (1980b, c) after the 1978 U.S. and Soviet missions. Winick and Stewart (1980) deduced concentrations of odd hydrogen and reactive chlorine at least an order of magnitude smaller than those of Sze and McElroy (1975), because they used the latest estimates indicating a fast rate for the above odd-hydrogen and reactive-chlorine destruction reaction, and also because they omitted the reaction

$$HO_2 + Cl \rightarrow ClO + OH \qquad\qquad \text{(zz)}$$

for which laboratory evidence is lacking. The lower odd hydrogen predictions lead to much higher O_2 concentrations and downward fluxes and thus to much higher ClO/Cl ratios in the Winick and Stewart (1980) model relative to the earlier Sze and McElroy (1975) model. As mentioned in Sec. IV. E-1.1, even the inclusion of SO_2 oxidation as a sink for O_2 in the Winick and Stewart (1980) model fails to prevent the predicted O_2 concentrations above the clouds from greatly exceeding the observed upper limits for this gas. Since this latter model represented a serious attempt to study the problem using only those reactions whose kinetics are reasonably well defined, it is clear that additional reactions must be postulated and the necessary laboratory work carried out to substantiate them.

The study by Krasnopolsky and Parshev (1980*b,c*) indicated the same problems with overprediction of O_2 and CO which were noted by Winick and Stewart (1980) and, as discussed in Chapter 14, they postulated reactions of COCl with O_2, and ClO with CO, to serve as dominant O_2 and CO sinks. With these reactions included, the O_2 concentrations, O_2 downward fluxes, reactive chlorine concentrations, and ClO/Cl ratios become similar to those in the earlier work of Sze and McElroy (1975). However, the odd hydrogen concentrations, and consequently the oxidation rates of CO by OH, are much lower than those of Sze and McElroy (1975) and instead resemble those of Winick and Stewart (1980). The problem of how CO and particularly O_2 are kept at such low concentrations in the middle atmosphere of Venus is still not convincingly solved, despite over a decade of work on the subject. Four-center gas-phase oxidation reactions (e.g., between O_2 and COCl, SOCl, or SO_2Cl), and oxidation reactions in acidic solution (e.g., between dissolved H_2O_2, O_3, or chlorine oxides and SO_2 or H_2S) seem promising areas for laboratory investigations which might remove the present impasse. It is also apparent that the spectroscopic observation implying an O_2 mixing ratio < 1 ppm near the cloud tops (Traub and Carleton 1974) needs further corroboration in view of its great importance to our understanding of the upper atmosphere chemistry.

Vertical mixing rates in the middle atmosphere which are important parameters in the photochemistry of both chlorine and sulfur compounds fortunately seem better constrained than the chemistry. The available information has been presented in Sec. IV. B-2.2.

2. Hydrogen Fluoride. The much greater strength of the H-F relative to the H-Cl bond, together with the much lower abundance of HF relative to HCl on Venus, make reactive fluoride (F, FO, FOO, F_2) much less important than reactive chlorine in the atmosphere. The photochemistry should be dominated by the reactions

$$HF + h\nu \rightarrow H + F \qquad (\lambda < 213 \text{ nm})$$

$$F + H_2 \rightarrow HF + H \qquad\qquad\qquad\qquad \text{(aaa)}$$

and the ratio of HF to reactive fluorine on Venus much greater than the ratio of HCl to reactive chlorine. Parisot and Moreels (1981) studied the fluorine chemistry in Venus's middle atmosphere and concluded that fluorine concentrations are typically < 20 cm^{-3}. Indeed, the species F, FO, and FOO are so rare that their chemical roles in the middle atmosphere are expected to be negligible.

3. Hydrogen Bromine and Hydrogen Iodide. Since HBr and HI (or indeed any Br and I compounds) have not yet been detected on Venus, we mention these species only briefly. Sill (1975) suggested Br_2 produced from photodissociation of HBr, and dissolved in hydrobromic acid, as a coloring agent for the clouds but the required large Br_2 mixing ratio $\sim 8 \times 10^{-8}$ is incompatible with an upper limit of 2×10^{-10} derived from Venera 11 and 12 spectrophotometry (Moroz et al. 1981). If HBr is present above the clouds, an analogy with studies of HBr photochemistry on Earth (Wofsy et al. 1975) would suggest that the weakly-bonded HBr molecule will be largely converted to reactive bromine (Br, BrO) which could play some role in middle atmosphere chemistry if its mixing ratio $\gtrsim 10^{-7}$. Hydrogen iodide is even more weakly bonded than HBr and, if present in the middle atmosphere, should be essentially completely converted to oxides.

F-2. Chemistry of Cloud-forming Halogens

Halogen compounds have appeared prominently in the past as suggested candidates for the clouds of Venus. The list of compounds includes hydrochloric acid droplets, NH_4Cl and HCl hydrate crystals (Lewis 1968b), $FeCl_2$ hydrate crystals (Kuiper 1968), Hg_2Cl_2, Hg_2Br_2, and Hg_2I_2 crystals (Lewis 1969), chlorosulfonic acid (Prinn 1973b), hydrobromic acid (Sill 1975), Al_2Cl_6 crystals (Krasnopolsky and Parshev 1979a), and $FeCl_3$ hydrate crystals (Knollenberg and Hunten 1980). The discovery in 1973 that at least the upper portion of the clouds is composed of sulfuric acid droplets has tended to temper enthusiasm for other possible constituents. Nevertheless, the Pioneer Venus observations implying that crystals and not droplets are dominant in the lower portion of the clouds (Knollenberg and Hunten 1980), combined with the X-ray detection of Cl in the clouds mentioned earlier, forces us to consider involatile chlorine compounds as cloud constituents.

As reviewed by Knollenberg and Hunten (1980), none of the chlorine-containing candidates for the clouds suggested thus far seem satisfactory. Photochemically produced chlorine compounds have not, however, been seriously investigated. In particular, the photochemical production of reactive chlorine oxides from HCl suggests that more stable oxides (e.g., Cl_2O, OClO, Cl_2O_7) and oxyacids (e.g., HOCl, $HClO_3$, $HClO_4$) of chlorine may be produced as by-products. Of these possibilities, perchloric acid ($HClO_4$) appears most attractive. Like sulfuric acid, perchloric acid is a very strong acid with a low vapor pressure and a refractive index in the range ~ 1.38 to 1.42, depending on temperature and concentration. It also forms stable mixtures with concentrated sulfuric acid. In addition, as water is removed from the concentrated acid, or as the temperature decreases, a number of crystalline $HClO_4$ hydrates are formed with successively higher and higher melting points. Of greatest interest for Venus is the monohydrate $HClO_4 \cdot H_2O$ which forms colorless monoclinic or orthorhombic crystals with a melting point ~ 323 K.

Photochemically produced bromine and iodine compounds are also of interest. In particular, perbromic ($HBrO_4$), iodic (HIO_3), and periodic (HIO_4) acids, their respective crystalline hydrates, and some of their anhydrides, are reasonably stable and are worth exploring as cloud components. For example, H_5IO_6 ($\equiv HIO_4 \cdot 2H_2O$) forms colorless monoclinic crystals with a melting point $\sim$ 402 K. The various halogen oxyacids are also interesting because some of the halogen oxides formed when they thermally decompose (e.g., Cl_2O, $OClO$) are strong blue and ultraviolet absorbers. An expanded and more detailed discussion of the halogen oxyacids and oxides and their relevance to the Venus clouds is given by Prinn (1982).

F-3. Thermochemistry and Chemical Cycles

As for sulfur compounds discussed in Sec. IV. E, studies have been published which predict the concentrations of atmospheric halogen compounds in the situation where thermochemical equilibrium exists between surface and atmosphere (Lewis 1968a, 1970; Barsukov et al. 1980b; Khodakovsky 1982). In particular, the following equilibria at 750 K (Lewis 1970)

$$2NaCl + CaAl_2Si_2O_8 + SiO_2 + H_2O \rightleftharpoons 2NaAlSiO_4 + CaSiO_3 + 2HCl$$

$$2NaCl + Al_2SiO_5 + 3SiO_2 + H_2O \rightleftharpoons 2NaAlSi_2O_6 + 2HCl$$

$$2CaF_2 + MgSiO_3 + SiO_2 + 2H_2O \rightleftharpoons Ca_2MgSi_2O_7 + 4HF \qquad \text{(bbb)}$$

lead to predicted mixing ratios for HCl and HF of $\sim$ 10^{-6} and 2×10^{-8} respectively, which are only slightly greater than the values observed above the clouds. For the species $FeCl_2$, SiF_4, Cl_2, and F_2 the predicted mixing ratios were $\sim$ 7×10^{-11}, 8×10^{-20}, 2×10^{-20}, and 2×10^{-48}, respectively. Khodakovsky (1982) studied chlorine compounds of some of the heavier volatile metals and concluded the $AsCl_3$ mixing ratio is $< 10^{-4}$, the $SbCl$, $SbCl_3$, and $SbCl_5$ mixing ratios $< 10^{-6}$, the $SeCl_2$ mixing ratio $< 5 \times 10^{-6}$, the $TeCl_2$, $TeOCl_2$, and $TeCl_4$ mixing ratios $< 10^{-7}$ and the $HgCl$, $HgCl_2$, $PbCl$, and $PbCl_2$ mixing ratios $< 10^{-8}$. These are very strict upper limits since the thermodynamics of many possible stable minerals of these metals are not known and were therefore not included in the study. Crystals of $SbOCl$ were also suggested as a possible cloud constituent.

There is little doubt that the reason for the much higher atmospheric concentrations of HCl and HF on Venus relative to those on Earth is simply a result of the high surface temperature on Venus. The agreement between the predicted and observed mixing ratios for HCl and HF strongly suggest that thermochemical equilibrium is closely attained between the atmosphere and halite (NaCl) and fluorite (CaF_2) on the surface.

A definition of the chlorine cycle on Venus is hampered by the lack of data on gaseous chlorine compounds below the clouds. At present we have no reason to contradict the current belief that the principal chlorine compound outgassed from, and in equilibrium with, the crust is HCl. At least some of this HCl permeates into the region above the clouds where it is photodissociated to form reactive chlorine compounds. However, because it is also rapidly reformed *in situ* in this region by reactions such as

$$Cl + HO_2 \rightarrow HCl + O_2 \qquad \text{(ccc)}$$

its mixing ratio is not expected to decrease substantially in the middle atmosphere. The roles, if any, of heterogeneous HCl oxidation and chloride precipitation reactions within the clouds, and of photochemical HCl oxidation reactions involving S atoms below the clouds are not yet apparent. Thus the question of whether HCl is partially decomposed to cloud-forming chlorides, oxychlorides, and chlorine oxides in the cloud region has not been answered. Certainly any oxychloride and chlorine oxide particles formed will evaporate and thermally decompose (some explosively) as they settle toward the surface. For example, pure $HClO_4$ begins to decompose to H_2O, Cl_2, and O_2 at temperatures $\geqslant 473$ K (Levy 1962).

The fluorine cycle should bear some similarity to that of chlorine except that HF will be much more difficult to oxidize than HCl. As for bromine and iodine compounds, we know so little about them that discussion of their possible cycles is not useful at this time.

G. Noble Gases

1. Helium. Helium was measured extensively in the Venus atmosphere, although only at altitudes above 130 km. The first positive identification was provided by the observations of the He I resonance line at 58.4 nm wavelength during the Venus flyby of Mariner 10 (Broadfoot et al. 1974). Under the assumptions that the source of this emission is resonance scattering of solar radiation by helium and that the atmosphere is spherically symmetric, Kumar and Broadfoot (1975) derived from the Mariner 10 experiment a He density of 2×10^6 cm^{-3} at 145 km altitude and an exospheric temperature of 375 ± 105 K. The He 58 nm airglow was again observed from both the Venera 11 and 12 orbiters (Bertaux et al. 1981), but helium densities derived from these Venera experiments have not yet been published.

In situ measurements of He number densities were first performed by the mass spectrometers on board the Pioneer Venus multiprobe bus in the altitude range 700 to 130 km (von Zahn et al. 1979b) and on board the Pioneer Venus orbiter in the altitude range 260 to 150 km (Niemann et al. 1979b). Subsequent observations of the upper atmosphere helium densities over 3 diurnal cycles by the orbiter mass spectrometer (Niemann et al. 1980b) established clearly that on the day side the He densities show remarkably little temporal variations (see Fig. 18). Neglecting a possible, but unknown latitudinal dependence of He densities, we may therefore compare directly the three available and independent determinations of the absolute He densities on the day side (see Table XI). One notes a satisfactory agreement between the results of the two Pioneer Venus experiments. The comparison between the Mariner 10

TABLE XI

Helium Densities

Altitude km	n(He) cm^{-3}	Instrument	Reference
145	2×10^6	M10-EUV	Kumar and Broadfoot (1975)
150	5×10^6	PV-BNMS	von Zahn et al. (1980)
150	7×10^6	PV-ONMS	Hedin et al. (1982)
400	6.5×10^4	M10-EUV	Kumar (personal comm.)
400	1.1×10^5	PV-BNMS	von Zahn et al. (1980)

and Pioneer Venus results near the 150 km level could be improved by using the argument that the 58 nm brightness measurements of Mariner 10 already reach saturation near 300 km altitude due to the atmosphere becoming optically thick at lower tangent heights. For Mariner 10 the helium density (Table XI) at 145 km altitude therefore represents an extrapolation over ∼ 150 km using an exospheric temperature of 375 K. However, the 145 km density is increased a factor 1.8 if the extrapolation is based on an exospheric temperature of 275 K, as determined by von Zahn et al. (1980); this temperature determination is within the error bar for the exospheric temperature of Kumar and Broadfoot (1975). Therefore, if exospheric temperatures found by Pioneer Venus are indeed also representative for conditions during the Mariner 10 flyby, then all 3 independent measurements of helium densities yield for dayside conditions at 150 km altitude, n(He) = 5×10^6 cm^{-3}, within a factor of 1.5. On the night side helium densities are highly variable (see Fig. 18); for more information we refer the reader to the original literature (e.g. Niemann et al. 1980a; Hedin et al. 1982).

For the entire middle and lower atmosphere of Venus no measurement of the helium mixing ratio exists. To infer the lower atmosphere value we need to extrapolate the upper atmosphere measurements downwards into the well-mixed atmosphere. The lowest altitude, for which *in situ* measurements of helium densities were performed by the PV-BNMS, is close to the He homopause. The measurements yield a ratio He/CO_2 = 3.8×10^{-5} at 130 km altitude, but the profile above 130 km exhibits a considerable wave structure (see Fig. 1 of von Zahn et al. 1979a). By appropriately smoothing over this structure von Zahn et al. (1980) obtain for 130 km altitude He/CO_2 = 4.9×10^{-5}; Hedin et al. (1982) infer 5.4×10^{-5} from the PV-ONMS measurements. The extrapolation of this ratio into the lower atmosphere requires that one make assumptions about the strength of turbulent mixing and vertical bulk motions in the 100 to 140 km altitude region. For their respective one-dimensional model calculations von Zahn et al. (1979a) assumed that the net effect of turbulent mixing on the helium profile can be simulated by introduction of an eddy diffusion coefficient and that the mean vertical bulk motion is zero. Assuming an altitude *dependent* eddy coefficient of the form $K = A \cdot n^{-1/2}$ and fitting all helium data of the PV-BNMS up to 200 km altitude, von Zahn et al. (1980) obtained for the 100 km altitude level (and anywhere below) a ratio $He/CO_2 = 1.2 \times 10^{-5}$. If we assume an altitude *independent* eddy coefficient, the same lower atmosphere ratio is obtained for $K \sim 1.5 \times 10^7$ cm^2 s^{-1} (see Fig. 2 of von Zahn et al. 1979a). In fact, the helium mixing ratio extrapolated into the lower atmosphere turns out to be roughly proportional to the assumed eddy coefficient up to $K \sim 1 \times 10^8$ cm^2 s^{-1}. Although the altitude dependent eddy diffusion coefficient definitely gives the smallest least squares sum when comparing actual He (and N_2) observations with model predictions, it is clear that a large uncertainty remains for the true He/CO_2 ratio in Venus's lower atmosphere. This uncertainty was estimated by von Zahn et al. (1980) to be up to a factor 3.

Potential sources and sinks for Venus's atmospheric helium have been discussed by von Zahn et al. (1979a). Due to the severe unknowns in the global helium balance of Venus, we will not repeat these speculations here (see Chapter 29 by Donahue for the treatment of this subject).

2. Neon. The mixing ratio of Ne in the lower atmosphere of Venus has been measured by four instruments and the respective results are listed in Table XII. The data of the Venera 11 and 12 mass spectrometers have been combined into one value for the Ne mixing ratio by Istomin et al. (1980b). For the mass spectrometer experiments Hoffman et al. (1980a) and Istomin et al. (1980b) have in fact determined an abundance ratio of Ne/^{36}Ar. As discussed below we adopt here a recommended ^{36}Ar abundance of 31 ppm,

TABLE XII
Neon Mixing Ratios

Altitude (km)	Original Value (ppm)	Adopted[a] Value (ppm)	Reference
?	10 ± 7	10 ± 7	Hoffman et al. (1980a)
< 23	$12 \, {}^{+\,5}_{-\,3}{}^{b}$	8.6 ± 4	Istomin et al. (1980a)
52	< 8		
42	$10.6 \, {}^{+32}_{-10}$	4.3 ± 4	Oyama et al. (1980a)
22	$4.3 \, {}^{+\,5.5}_{-\,3.9}$		
		7 ± 3	this chapter

[a]Corrected, if applicable, for ^{22}Ne abundance and for normalized ^{36}Ar abundance (see text).
[b]Value represents ^{20}Ne mixing ratio.

TABLE XIII

Argon Mixing Ratios

Altitude (km)	Ar Mixing Ratio (ppm)	^{36}Ar Mixing Ratio (ppm)	Reference
< 25	$67 \begin{smallmatrix} +50 \\ -25 \end{smallmatrix}$	$30 \begin{smallmatrix} +20 \\ -10 \end{smallmatrix}$	Hoffman et al. (1980a)
< 23	110 ± 20	46.1 ± 9	Istomin et al. (1980b)
52	$60.5 \begin{smallmatrix} +40 \\ -47 \end{smallmatrix}$		
42	63.8 ± 14		Oyama et al. (1980a)
22	67.2 ± 2		
< 42	40 ± 10		Gel'man et al. (1979a)
(~ 140)	< 30[a]		Mauersberger et al. (1979)
	70 ± 25	31 ± 12	this chapter

[a]Value is extrapolated into the lower atmosphere from upper limit determined near 140 km altitude.

which is considerably lower than the value of 46.1 ppm published by Istomin et al. (1980b). Therefore we apply in the following a correction factor of $31/46.1 = 0.67$ to all noble gas measurements of Istomin et al. We combine the 3 values listed in Table XII for the Ne mixing ratio to a recommended mean of 7 ± 3 ppm.

3. Argon. No less than 3 mass spectrometers and 2 gas chromatographs have independently measured the argon mixing ratio in Venus's lower atmosphere. Their results and the upper limit extrapolated from the Pioneer Venus bus spectrometer are listed in Table XIII. It is obvious that some of the results do not overlap within their stated range of uncertainty. Hoffman et al. (1980c) have discussed reasons that may have contributed to this situation. It appears likely in addition, that the criteria for calculating the error bars have been chosen differently by the various authors. For these two reasons it is difficult to combine the results in an objective and quantitative way into a single recommended value for the Ar mixing ratio. We argue qualitatively that the gas chromatographic measurements of Ar are less affected by potential pumping problems and impurity peaks than the mass spectrometric observations. Furthermore, the mass spectrometers of Istomin et al. did not experience any problems from cloud particles blocking the gas inlet or problems in their special noble gas measurement mode as happened in case of the experiment by Hoffman et al. (1980a). With these somewhat vaguely defined arguments in mind we recommend combining the results listed in Table XIII into a mean value of the (total) argon mixing ratio 70 ± 25 ppm.

As mentioned before, the noble gas abundances determined by the Pioneer Venus and Venera 11 and 12 mass spectrometers operating in Venus's lower atmosphere are referred to the measured ^{36}Ar abundance. Assuming a relative isotopic abundance of 44.2% for ^{36}Ar (see Table XIII) we obtain a recommended value for the mixing ratio of ^{36}Ar 31 + 9 ppm. By comparing this ^{36}Ar abundance with the one published by Istomin et al. (1980b), we conclude that all noble gas abundances of Istomin et al. should be reduced by a factor 1.49 to make them agree with our recommended mean values.

4. Krypton. This gas was first discovered in the Venus atmosphere by the Venera 11 and 12 mass spectrometers. Preliminary results were reported by Istomin et al. (1979a); they indicated a mixing ratio of the isotope ^{84}Kr in the range of 0.5 to 0.8 ppm. A refined analysis led to the improved value for ^{84}Kr of 0.6 ± 0.2 ppm (Istomin et al. 1980b). Applying the normalization factor of 1.49 defined in the preceeding paragraph, this value reduces to 0.4 ± 0.14 ppm. The sum of 28 noble gas spectra obtained from the Venera 11 mass spectrometer is shown in Fig. 9. It demonstrates the excellent S/N ratio for the mass 84 peak. In addition, the intensities measured in the mass 80 to 86 range support the view that relative to ^{84}Kr the other isotopes of krypton are present in approximately the same abundances as in the terrestrial atmosphere (Istomin et al. 1980b).

Preliminary results of the Pioneer Venus Large probe mass spectrometer for Kr abundances were reported by Hoffman et al. (1979c) and indicated an upper limit of 1 ppm for the ^{84}Kr mixing ratio which was lowered to 0.2 ppm by Hoffman et al. (1980a). Finally, a refined analysis by Donahue et al. (1981) led to the claim that the mixing ratio of ^{84}Kr is in fact ~ 0.025 ppm with an upper limit of 0.028 ppm. Donahue et al. note: "It is not possible to reconcile such a result" with that of Istomin et al. (1980b).

To illustrate the implications of these different values of the venusian Kr abundance we compare in Table XIV the mixing ratios of the primordial gases Ne, ^{36}Ar, and ^{84}Kr in the atmospheres of the three terrestrial planets. If one

TABLE XIV

Mixing Ratios of Primordial Gases in the Atmospheres of the Terrestrial Planets (ppm)

Planet	Total Ne	^{36}Ar	^{84}Kr	Reference
Mars	2.5[a]	5.3	0.17[a]	Owen et al. (1977)
Earth	18	31	0.65	
Venus	7	31		this chapter
Venus			0.6	Istomin et al. (1980b)
Venus			0.025	Donahue et al. (1981)

[a]Calculated from original measurement assuming terrestrial isotopic abundances.

takes into account the true uncertainties of the values for the Mars and the Venus atmospheres, it is quite evident that the abundance ratios $Ne/^{36}Ar$ and $^{84}Kr/^{36}Ar$ are alike for all three atmospheres, provided we choose for the comparison the ^{84}Kr abundance measured by Istomin et al. Clearly the Kr value of Donahue et al. does not fit at all in this rather appealing picture. In particular, in light of the fact that there exists a fairly convincing measurement of a "planetary" krypton abundance at Venus, the divergent and rather low value of Donahue et al. looks suspicious.

We suggest the following instrumental arguments in favor of the Istomin et al. krypton abundance:

1. Istomin et al. have in fact measurements from 2 mass spectrometers, i.e., from Veneras 11 and 12. The performance and results of the two instruments are so similar that Istomin et al. were able just to add the spectra of the two instruments in order to improve the overall S/N ratio (see, e.g., the krypton data in Fig. 8 of Istomin et al. 1980*b*);
2. The instrument sensitivities for Kr were actually calibrated in the laboratory with mixtures of 2 ppm Kr in CO_2. During these calibrations no background or memory krypton was observed in the instrument (Istomin, personal communication);
3. The instruments were opened at 23 km altitude only and remained free from any interference by cloud particles;
4. During their descent in the atmosphere the ion getter pump was periodically switched on (so-called chemical analysis mode) and off (so-called noble gases analysis mode), each single mode lasting $\sim$ 200 s. After switching on the ion getter pump, all krypton peaks disappeared quickly in each of the five chemical analysis cycles;
5. The gas load introduced into the ion getter pump was considerably lower for the Istomin et al. spectrometers than for the Hoffman et al. instrument, because the Venera instruments opened their gas inlet valves only intermittently and during a shorter total measurement time (30 min vs. 54 min for the Pioneer Venus mass spectrometer);
6. The Venera instruments transmitted full mass spectra leaving no doubt about the position and height of the true peak maxima.

If the Donahue et al. (1981) analysis is correct then the high Kr abundances of Istomin et al. can only be explained by the presence of terrestrial Kr in both Venera mass spectrometers or by some unidentified background gas in the Venera instruments causing the observed mass 84 peak. If, on the other hand, Istomin et al. measured the correct Kr abundance, then the respective values of Donahue et al. are too small by a factor $0.4/0.025 \simeq 16$. This may have been caused by inadequate assumptions on the absolute sensitivity for Kr of the mass spectrometer which was not calibrated with Kr in the laboratory (Hoffman, personal communication). In retrospect this sensitivity is particularly difficult to establish due to the complex arrangements of various pumps and conductances in the instrument of Hoffman et al. (1980a). Furthermore, as this instrument had a very high resolution and used peak stepping only for the measurement of the Kr isotopes, there remains the possibility that for some reason the peaks of the Kr lines were missed and thus the count rates obtained were too low. Because of the severely limited bit rate it was not feasible to check this possibility by scanning a continuous mass spectrum inside the Venus atmosphere.

None of the potential failure modes which we have mentioned appears to us in any way convincing or even likely. We cannot make a clear choice between the two available krypton measurements.

5. *Xenon.* No positive identification of Xe exists in the Venus atmosphere. An upper limit of 0.12 ppm Xe was derived by Donahue et al. (1981) from the data gathered by the Pioneer Venus Large probe mass spectrometer. Judgement of the significance of this upper limit is obviously influenced by the conflicting results on krypton as discussed in the preceding section.

H. Excited Species

1. *O_2 Emissions.* The atomic oxygen, which is formed in copious amounts in the upper atmosphere through photodissociation of CO_2, recombines predominantly to molecular oxygen in the 85 to 100 km altitude region (see Sec. IV. B-2.2). In the course of these recombination processes a great number of excited oxygen molecules (and atoms) are formed, many of which in turn produce characteristic airglow emissions. The study of the intensity of these emissions and their spatial and temporal variations provides important information on the photochemistry and dynamics of Venus's middle and upper atmosphere.

Recombination of ground-state O atoms can lead to 7 bound electronic states of the O_2 molecule, 6 of which are experimentally known and 1 theoretically predicted (the $^5\Pi_g$ state) by Saxon and Liu (1977). Oxygen atom association can in addition raise O atoms to their lowest 2 excited states. The energy levels of these states (v = O for the O_2 states) and their nomenclature are given in Fig. 37. A peculiar property of oxygen is the fact that all excited states shown in this figure are metastable. Their radiative lifetimes range from ~0.2 s for the O_2 (A) state up to 3900 s for O_2 (a). Due to their reluctance to radiate, O_2^* and O^* represent a reservoir of readily available chemical energy which must be considered carefully in any scheme of the photochemistry of the atmospheres of Venus, Earth, and Mars.

On all three terrestrial planets the infrared atmospheric band of O_2 (a → X) at 1.27 μm is by far the strongest of all oxygen airglow emissions. Connes et al. (1979) discovered intense lines of this (0, 0) band on both the light and the dark sides of Venus, using a groundbased, high-resolution Fourier transform spectometer. The total band intensities were determined to be ~ 4.3 MR and 3.4 MR for the sunlit and the dark sides, respectively, with

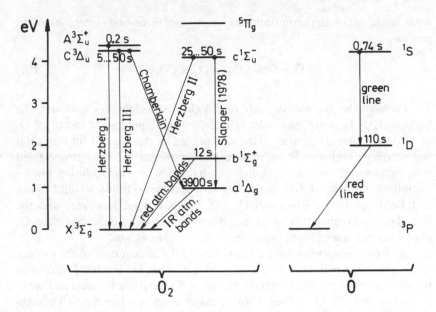

Fig. 37. Electronically excited states of O_2 and O which can be reached by recombination of groundstate oxygen atoms and the major interconnecting band systems and lines. All excited states shown are metastable. Their radiative lifetimes, if known, are also indicated.

an estimated uncertainty of $\pm$ 20%. The apparent brightness is, however, enhanced over the true emission by backscatter from the Venus clouds. After appropriate corrections, the true emissions become 1.5 MR and 1.2 MR for the dayglow and nightglow from Venus. This is to be compared with typical terrestrial values of (20 ± 10) MR for the 1.27 μm dayglow (Noxon 1968) and 0.1 MR for the nightglow (Llewellyn et al. 1973).

From a comparison of the height-integrated production rate of O atoms by photolysis of CO_2 with the observed column emission rate of O_2 $(a \to X)$, Connes et al. (1979) concludes: "...the production of O_2 $(a\,^1\Delta_g)$ on Venus must involve the major step leading from O atoms back to CO_2." This conclusion rules out the possibility that the recombination scheme for the net reaction $CO + O \to CO_2$ (which has been proposed by McElroy and Donahue [1972] for Mars) plays a dominant role in Venus's photochemistry. Connes et al. (1979) suggest that O_2 $(a\,^1\Delta_g)$ is formed predominantly in the catalytic cycle

$$Cl + O_3 \to O_2 + ClO \qquad \text{(ddd)}$$

followed by

$$ClO + O \to O_2 + Cl \qquad \text{(eee)}$$

with some secondary contributions from the three-body recombination process

$$O + O + CO_2 \rightarrow O_2 + CO_2 . \tag{fff}$$

Furthermore, the near equality of dayglow and nightglow indicates that O_3 photolysis by solar ultraviolet is relatively unimportant as source of O_2 (a $^1\Delta_g$) on Venus, in contrast to the role played by this process for terrestrial and martian dayglows. These implications were quantitatively investigated and, in general, confirmed in the low-hydrogen + high-chlorine photochemistry models of Krasnopolsky and Parshev (1980b, c) and Yung and DeMore (1982). Krasnopolsky and Parshev note, however, that the three-body recombination of O is probably even less important than O_3 photolysis as a source for the observed O_2 (a → X) emission.

As mentioned earlier the 1.27 μm band is the strongest of all the oxygen emissions in the nightglow of the terrestrial planets. On Earth the next most intense contributions come from the O_2 (A → X) Herzberg I system and the O green line (see Fig. 37). Thus it was a major surprise when the first spectra obtained from the visible nightglow of Venus by Krasnopolsky et al. (1976a) did not show either one of these two features but showed instead, between 400 and 700 nm, a strong molecular band system consisting of at least 8 spectral emissions of unidentified origin. This band system was subsequently identified by Lawrence et al. (1977) as being part of the O_2 (c → X) Herzberg II system. Closer scrutiny by Slanger and Black (1978) of the published Venera 9 nightglow spectrum revealed in addition minor contributions from the O_2 (C → a) Chamberlain system and the O_2 (c → a) system (see Fig. 7). The latter system was produced in the laboratory for the first time by Slanger (1978b) and may appropriately be called the Slanger system. Yet, the atomic oxygen green line remains totally absent from the Venera 9 spectra of the Venus nightglow.

What are the reasons for the big differences in the oxygen emissions from the atmospheres of Earth and Venus? Lawrence et al. (1977) conjectured that CO_2 played a special role in the three-body reaction

$$O + O + CO_2 \rightarrow O_2 + CO_2 + 5.12 \, eV \tag{ggg}$$

in that it allowed the O_2 molecule to become excited directly into the c-state, the v = 0 level of which has an energy of 4.08 eV (Saxon and Liu 1977). This direct line of argument had to be abandoned, however, when Slanger (1978a) was able to generate in the laboratory the Herzberg II emissions in the afterglow of both a pure O_2-He discharge and also a NO-titrated N_2-He discharge. Nevertheless, when adding CO_2 downstream from the atom generation point of the O_2-He afterglow. Slanger (1978a) observed clearly enhanced emissions from the O_2 c-state. He thus confirmed that CO_2 can play at least some peculiar role in the population of the c-state.

The same conclusions concerning the role of CO_2 were reached by Kenner et al. (1979) who studied the emissions emanating from a high purity O_2-Ar afterglow. They found that "CO_2 is not essential for the production of a spectrum which is predominantly made up of the Herzberg II bands." They also noted, on the other hand, that CO_2 is ~100 times as effective as Ar in producing the Herzberg II bands. Kenner et al. (1979) propose the following three-step mechanism

$$O + O + M \rightarrow O_2 (A) + M$$

$$O_2 (A) + M \rightarrow O_2 (c) + M$$

$$O_2 (c) \rightarrow O_2 (X) + h\nu \text{ (Herzberg II)} \qquad \text{(hhh)}$$

which is capable of explaining all their laboratory measurements, as well as the nightglow observations on Venus and Earth. After a number of simplifying assumptions Kenner et al. (1979) arrive at the result that the ratio of emission intensities for the Herzberg II and Herzberg I band should scale as follows

$$\frac{I \text{ (Herzberg II)}}{I \text{ (Herzberg I)}} = \text{const } \frac{n(M)}{n(O)} \qquad (29)$$

For a comparison of the above ratio in the venusian and terrestrial nightglow one needs to determine $n(M)$ and $n(O)$ at the emitting layer heights. Krasnopolsky and Tomashova (1980) have analyzed the altitude profiles of Herzberg II emissions for 5 cases of limb observations with the Venera 9 and 10 nightglow spectrometers. They refer their altitude scale to the height for which $\tau_{limb} = 1$. From Mariner 10 images O'Leary (1975) derived this height to be on the day side at the 4.1 mbar level, equivalent to 80 km altitude (Seiff et al. 1980). Assuming that during the night time $\tau_{limb} = 1$ occurs at 78 km, Krasnopolsky and Tomashova obtain the tangent heights of maximum emission as 90 ± 2 km for 3 observations and 113 km for another night (observations during a fifth night produced no useful altitude profile). Consequently, a big difference exists in the environment from which oxygen airglow emissions originate at Venus and at Earth. At Venus the total number density $n(M)$ is 1.4×10^{16} cm^{-3} at 90 km altitude, whereas on Earth it is 2.4×10^{13} cm^{-3} at the level of the maximum volume emission rate of the O green line near 96 km altitude (Offermann and Drescher 1973). At the respective altitudes the atomic oxygen densities $n(O)$ are for Venus 3×10^9 cm^{-3} (Yung and DeMore 1982; model C) and for Earth 4×10^{11} cm^{-3} (USSA 1976). Hence the ratio $n(M)/n(O)$ is almost five orders of magnitude larger in the Venus nightglow layer than in the corresponding Earth layer. The predominance of the Herzberg II bands in the Venus nightglow is further enhanced by the fact that there M = CO_2.

It appears that a successful convergence of venusian nightglow observations and laboratory simulations is rapidly approaching. Nevertheless the three-step mechanism of Kenner et al. needs further corroboration and study of the details. Krasnopolsky (1981) argues that "for M = CO_2 most recombination events lead to the formation of O_2 (A, C, c) states, and almost all O_2 (A) and O_2 (C) molecules turn into O_2 (c)." Thus we note a further complication of the basic three-step mechanism of Kenner et al. (1979). There remain in addition a number of quantitative unknowns, such as the yields for the production of O_2 (A), O_2 (C), and O_2 (c) in the atomic oxygen recombination reaction, many of the quenching rates for the three upper states by CO_2, O_2, N_2, O, O(^{1}D), and Ar, and the rate coefficients for the various reactions

$$O_2^* + O \rightarrow O_2 + O(^1S) \tag{iii}$$

which are precursor reactions for the emission of the atomic oxygen green line (Barth 1964)

$$O(^1S) \rightarrow O(^1D) + h\nu \, (558 \, nm) \,. \tag{jjj}$$

The weakly bound $^5\Pi_g$ state of O_2 calculated by Saxon and Liu (1977) has received little attention in the literature so far, perhaps because it is not yet entirely certain whether this state is indeed truly bound. Only Wraight (1982) has performed a study of the role of O_2 ($^5\Pi_g$) in the association of oxygen atoms. He concludes that this state, provided it has properties as calculated by Saxon and Liu, is of great importance for a number of oxygen association reactions and airglow emissions. It certainly deserves intensive laboratory and theoretical studies in the future.

The spatial and temporal variations of the oxygen airglow have been studied for the O_2 (a $\rightarrow$ X) emission by Connes et al. (1979) and for the case of the O_2 (c $\rightarrow$ X) emissions by Krasnopolsky and Tomashova (1980). For either case the observation of a fairly small variation with local time implies a transport time of O atoms from the day side into and across the night side comparable to or less than the chemical lifetime of these atoms. This transport most likely involves ascending motions on the day side, near-horizontal transport across the terminator and descending motions on the night side. In this way a substantial part of the transport occurs under conditions of reduced density with respect to the mean recombination height, thus increasing the effective chemical lifetime of the O atoms. From the photochemical model C of Yung and DeMore (1982) one can calculate the chemical lifetime of O atoms at 90 km to be $\sim$ 1 hr, and at 95 km to be $\sim$ 1 d. The lifetime of 1 hr appears too short to allow the maintenance of the airglow intensity throughout the night with only small temporal variations unless we invoke excessively

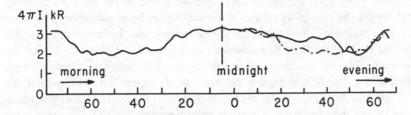

Fig. 38. Diurnal variation of the visible nightglow, primarily from the Herzberg II system, as derived by Krasnopolsky and Tomashova (1980) from observations of the Venera 9 and 10 orbiters. The dash-dotted line is the symmetric extension of the left part of the curve.

fast downward transport out of a higher altitude reservoir of O atoms. On the other hand, for an altitude of 95 km and above, the lifetime of O atoms predicted by the model of Yung and DeMore appears to be compatible with the lack of pronounced diurnal variations in the oxygen airglow and with transport times on the order of a few days.

In their studies of the visible nightglow, Krasnopolsky and Tomashova (1980) averaged 25 individual brightness scans across the dark side of Venus. Fig. 38 illustrates the result: the diurnal variation of the observed brightness versus hour angle. The characteristic features are a shallow maximum about 5° past midnight, two minima near 55° and pronounced increases in brightness from 55° to 70°, where the end of the measurements is reached. A comparison with the spatial distribution of the NO ultraviolet nightglow observed by the Pioneer Venus orbiter ultraviolet spectrometer will be given below. Krasnopolsky and Tomashova in addition find a significant hemispheric asymmetry of the visible nightglow intensity. The intensity exhibits a flat maximum between ∼ 28°S and 8°S and it falls off steadily into the northern hemisphere. At 20°N the observed intensity is a factor 1.8 lower than at 20°S.

2. NO Emission. In Sec. IV.C.3 we reviewed the observations and interpretation of ultraviolet nightglow consisting of bands of the NO γ and δ systems. Here we compare the spatial distributions of the NO and O_2 nightglows as observed by Stewart et al. (1980) and Krasnopolsky and Tomashova (1980), respectively. The NO nightglow reaches its maximum typically at ∼02:00 LST (local solar time) near 10°S. The O_2 nightglow reaches its maximum at ∼00:30 LST in the region of (18 ± 10) °S. Thus, both data sets agree on a noticeable asymmetry with respect both to local midnight and to the equator. Conditions are, on the other hand, not as repeatable as one may conclude from the above statements; the maximum of the NO airglow shows a rather wide scatter when observed only during single orbits of the Pioneer Venus orbiter. Gérard (1981) reports that these maxima occur over a range between 20:00 LST and 03:30 LST, and 40°S and 60°N. No corresponding analysis for the variability of the O_2 nightglow has been published. The latter task may be more difficult than for the NO nightglow, because of the smaller contrast if O_2 nightglow. Krasnopolsky and Tomashova obtained 3.3 kR for the near-midnight maximum and 1.9 kR for the early/late night minima, whereas for the NO nightglow Stewart et al. report the maximum to be ~ 5 kR and the hemispheric mean brightness 0.8 kR. The observed variations in the morphology and overall brightness of the NO emissions must result from irregularities in the global scale transport of N atoms from their source region on the day side into the sink region on the night side, as well as variations in the strength of downward transport in the 110 to 140 km altitude region on the night side.

3. *CO Emissions*. The ground state of $CO(X^1\Sigma^+)$ is connected to its first electronically excited singlet state $A^1\Pi$ via a strong band system, called the fourth positive system of CO. Since the work of Kassal (1975) it has been known that the (14,0) band of the CO 4+ system is accidentally in near-resonance with the H Lyman-α line. Approximately 12 rotational lines of the (14,0) band lie within the width of the very strong and broad solar Lyman-α line. The consequences of this resonance for the Venus dayglow were first pointed out by Durrance et al. (1980) and subsequently studied in depth by Durrance (1981). Fluorescent scattering from absorption of solar Lyman-α by the CO 4+ (14,0) band leads to dayglow emissions of many bands of the (14, v'') progression, the first members of which are given in Table XV.

It happens that the strongest emission comes from the (14,4) band which falls very close to the OI line at 135.56 nm. In low-resolution spectra, therefore, the CO 4+ (14,4) band will blend with the OI 136 nm line, as the (14,3) band may do with the OI 130 nm resonance lines. Before these near-resonances were recognized, there were considerable difficulties in reconciling the measured ratio of emissions close to 130 and 136 nm (Rottman and Moos 1973) with theoretical predictions of the ratio of OI emissions at 130 and 136 nm (Strickland 1973).

TABLE XV

Fluorescent Scattering from Absorption of Solar Lyman α by the CO 4+ Band

Dayglow Emission Bands	$\lambda\lambda$ (nm)
(14,0)	121.55
(14,1)	124.80
(14,2)	128.17
(14,3)	131.70
(14,4)	135.38
(14,5)	139.22

The (14,5) band is a prominent and unblended spectral feature in the Venus dayglow (Rottman and Moos 1973; Durrance et al. 1980). Since it is optically thin, the band intensity is directly proportional to the CO column density. On the other hand, the emitted 139.2 nm photons are partially reabsorbcd by the overlying CO_2. Therefore limb scans at 139.2 nm can be inverted to give a direct measure of the CO/CO_2 abundance ratio in the altitude range 120 to 150 km (Durrance et al. 1981).

I. Candidate Gases for Ultraviolet Absorption

A long-standing problem should be briefly mentioned here. When eventually solved, it may reveal interesting information on one or more yet unknown trace constituents of Venus's atmosphere. This problem is the chemical identification of the material that is responsible for the observed absorption of solar ultraviolet in the wavelength range 320 to 500 nm in the middle and lower atmosphere of Venus (see e.g. Wallace et al. 1972; Barker et al. 1975).

Formerly a problem of understanding Venus's albedo shortward of 320 nm also existed. This riddle was completely solved by the discovery of SO_2 in the Venus atmosphere (Barker 1979), extensive remote sensing and *in situ* measurements of SO_2 by the Pioneer Venus and Venera spacecraft (see Sec. IV. E-1.1), and new laboratory measurements of the absorption coefficient of SO_2 in the 208 to 800 nm spectral range (Pollack et al. 1980b; Wu and Judge 1981). The observed amounts and distribution of SO_2 fully explain the albedo of Venus between 200 and 320 nm (Esposito 1980). On the other hand, it is also certain that SO_2 cannot explain the absorption observed between 320 and ~ 500 nm, due to the rapid drop-off of its absorption coefficient longward of 320 nm. Thus, at least one more ultraviolet absorber is required, which dominates the opacity above 320 nm.

In any search for the chemical identity of this near-ultraviolet absorber we must keep in mind that its physical state is still unknown; is it a gaseous trace constituent, fluid haze or cloud droplet, or solid particulate material? Furthermore, much more information than just the spherical albedo of Venus has been collected, in particular by the Pioneer Venus spacecraft, which is relevant to this search. These additional data concern, for example, the radiative properties and vertical structure of cloud layers, the contrast measurements between ultraviolet cloud features and their dependence on wavelength and phase angle, and a considerable amount of new measurements pertaining to the composition of the Venus atmosphere. Pollack et al. (1980b) have synthesized these data in a radiative transfer calculation for the near-ultraviolet and visible. Their analysis led them to conclude "that the second UV absorber is located primarily within the upper cloud layer, that it is distributed throughout much of the upper cloud layer, but that it is largely absent from the top optical depth or two." Furthermore, "this absorber virtually disappears in a very abrupt manner at the base of the upper cloud layer." Pollack et al. propose as a candidate for this absorber gaseous Cl_2, of which ~ 1 ppm would be required inside the upper cloud layer. Gaseous Cl_2 has, however, not been identified by any other method in the Venus atmosphere. The hydrogen-poor + chlorine-rich photochemistry model of Yung and DeMore (1982) predicts a maximum of 0.07 ppm Cl_2 near 72 km altitude and only half of this value at 60 km and at 80 km. Therefore, neither the absolute amount of Cl_2 nor its predicted altitude profile appear to match satisfactorily the requirements for the near-ultraviolet absorber.

Zasova et al. (1981) propose as a possible absorber a 1% admixture of $FeCl_3$ in a solution of concentrated sulfuric acid. They base their proposal on laboratory measurements of the absorption coefficient of this solution and an appropriate cloud layer model (of which no details are given). They attempt to explain only the spherical albedo of Venus and recognize the difficulty of identifying a believable source for the $FeCl_3$.

The most recent model for explaining the near-ultraviolet albedo of Venus (and many other properties of its atmosphere and clouds) was advanced by Toon and Turco (1982). These authors suggest that elemental sulfur particles containing a few percent of S_3 are in fact the missing ultraviolet absorbers. Gaseous sulfur will be produced rapidly from SO_2 exposed to the solar ultraviolet provided that the O_2 abundance is significantly less than the SO_2 abundance. Toon and Turco assume that the sulfur vapor will condense to form particles when S_8 is supersaturated. These sulfur particles become quickly incorporated within H_2SO_4 droplets upon colliding with them. Once inside the H_2SO_4 droplets the sulfur particles would be prevented from further growth or evaporation.

Toon and Turco find that the sulfur vapor formed within the upper Venus clouds is enriched in the short allotropes S_3 and S_4. In particular S_3 absorbs strongly between 320 and 500 nm (see also Sec. IV. E-1.3) and could account for the albedo of Venus in this wavelength range. At the same time these allotropes are metastable and turn into S_8 with time constants ranging from hours to days. The well-known variability of the ultraviolet contrast thus would primarily be caused by a variable supply of SO_2 from the lower atmosphere to the upper cloud region. This model of Toon and Turco has many appealing features. A quantitative accessment of it must wait, however, for publication of the underlying photochemical and cloud-physical model.

J. Upper Limits for the Abundances of Unobserved Gases

Upper limits to the mixing ratio of many gases in the Venus atmosphere have been inferred from the lack of specific features in optical and radiowave spectra or from the absence of the appropriate detector signals in direct sampling experiments. In using these upper limits the reader must keep in mind that any one limit usually refers to a specific altitude range only, that this altitude range may depend on the peculiarities of the experiment (e.g., the phase angle of observation), and that there may exist strong natural variations, both in space and time, even for relatively stable gases like CO. Furthermore, in a few cases the determination of upper limits has involved the use of model assumptions which may or may not be entirely correct. The caution required in using published upper limits can be demonstrated by the fact that Barker (1979) discovered SO_2 in the Venus atmosphere with mixing ratios up to 5×10^{-7}, in spite of previously published upper limits much lower than this (down to 2×10^{-9}). The widely fluctuating upper limits and abundance measurements for the water-vapor content reflect, in our opinion, a combination of natural variability and considerable experimental difficulties in obtaining these results. Even for direct *in situ* measurements, upper limits on the mixing ratio of nitrogen have been obtained, which no longer reflect our current understanding of the atmospheric composition of Venus. With the above notes of caution the reader interested in upper limits to the mixing ratio of venusian gases is referred to the extensive compilations of such values and their references in the review papers of Moroz (1981), Wofsy and Sze (1975), and L. G. Young (1974). Shorter tables can also be found in Marov (1978), Marov (1972), NASA SP-8011 (1972), and Kuiper (1971).

Since publication of these review papers, the following new upper limits have been determined: on O_3, OCS, SO_2, H_2CO, N_2O, SO, CS, HCN, and H_2O by Wilson et al. (1981); on SO_2 and H_2O by Janssen and Klein (1981); on Xe by Donahue et al. (1981). In addition, we note the following upper limits which have not found their way into the compilations of some of the review papers: on CS_2 by Barker (1979); on NO, S, and SO by Krasnopolsky (1979a); on N_2O, H_2S, and SO_2 by Shaya and Caldwell (1976); on H_2O and H_2SO_4 by Rossow and Sagan (1975); on NH_3 by Goettel and Lewis (1974); and on SO_2, H_2S, C_3O_2, COS, O_3, and NO_2 by Anderson et al. (1969). Finally, we point out the compilation of theoretical detection limits of millimeter wavelength spectral lines by White (1978).

V. ISOTOPE ABUNDANCES

A. Historical Review

Isotope abundances in the Venus atmosphere have been determined so far by two methods: earth-based near-infrared spectroscopy, and *in situ* mass spectrometer measurements. In addition, the ultraviolet Venus dayglow has been searched for emissions of deuterium which, however, has not yielded a clear identification of this isotope.

Studies of the isotopic bands of CO_2 in the infrared spectrum of Venus started in 1948. With a newly designed prism spectrometer, Kuiper obtained on 1 April 1948 a spectrum of Venus with a resolution of 250 showing for the first time a clearly distinguishable, though poorly resolved isotopic band of $^{13}CO_2$ at 1.47 μm (Kuiper 1949; Herzberg 1949). Subsequent attempts by

Kuiper to improve on the 1948 results failed until a new grating spectrometer was built which allowed the recording of Venus's spectrum in the 1.0 to 2.5 μm region with a resolution of up to 1500. In Venus spectra taken in 1962, nearly 40 absorption bands were found (Kuiper 1962), all of which belonged to CO_2. Among them were several bands of $^{13}C^{16}O_2$ and one of $^{12}C^{18}O^{16}O$. Kuiper concluded that the isotopic ratios $^{13}C/^{12}C$ and $^{18}O/^{16}O$ are "equal to those on the Earth within the error of measurement," which Kuiper estimated to have a precision of $\pm$ 10%. Theoretical difficulties in relating the observed equivalent width ratios directly to ambient abundance ratios (see Sec. II. A.3) prevented Kuiper, from making a more quantitative determination of the desired isotopic abundance ratios.

A tremendous improvement in the resolution of infrared spectra of Venus was obtained by Connes and Connes (1966) by means of a Michelson interferometer. This instrument again covered the 1.2 to 2.5 μm region and led to the discovery of new isotopes in the molecular form of $C^{17}O^{16}O$, $H^{35}Cl$, and $H^{37}Cl$. However, the basic problem of deriving abundance ratios from equivalent width ratios remained. Finally in December 1978 the latest series of mass spectrometric measurements were performed by instruments on board the Pioneer Venus Large probe and the Venera 11 and 12 lander spacecraft. These experiments produced the first big surprise for Venus isotopic abundances, a $^{36}Ar/^{40}Ar$ ratio almost 300 times higher on Venus than on Earth (Hoffman et al. 1979b; Istomin et al. 1979a). Subsequently, strong evidence was found for a D/H ratio 2 orders of magnitude higher in the Venus atmosphere than on Earth (McElroy et al. 1982; Donahue et al. 1982).

B. Isotope Abundances of Various Elements

1. Deuterium. Deuterium was the first isotope in the venusian atmosphere in which scientists became keenly interested. This was not only because of the importance for astrophysics to obtain data on extraterrestrial deuterium abundances, but because of the peculiar altitude profile of Lyman-α emissions observed near Venus. As reported in Sec. IV. D.3, Donahue (1968) suggested explaining these observations by assuming a very large D/H ratio in Venus's upper atmosphere. For this deuterium interpretation one notes that the D Lyman-α line is shifted by 0.033 nm towards shorter wavelengths with respect to the H Lyman-α line. The solar Lyman-α line, on the other hand, is sufficiently broad to resonantly excite deuterium almost as efficiently as atomic hydrogen.

The conjectured deuterium distribution on Venus was subsequently modeled by Wallace (1969), Donahue (1969), and McElroy and Hunten (1969). All workers concluded that a ratio of D/H on the order of 10 at the critical level for escape (exobase) was required to fit the Mariner 5 Lyman-α observations. In this connection, McElroy and Hunten have shown that the effects of an escape flux could drastically alter the ratio of D/H between the bulk atmosphere and the exobase provided that eddy diffusion is much weaker in the upper atmosphere of Venus than in the same region on Earth and that hydrogen follows a limiting diffusion profile. They concluded that the Mariner 5 observations could be understood if the D/H ratio in the bulk atmosphere of Venus is ~0.1. Jointly with L. D. Kaplan, they made a search on the Connes spectra for the presence of DCl lines. From their failure to find any such lines, they set an upper limit of D/H < 0.1 by atoms for the lower atmosphere, still compatible with the H+D model of the Mariner 5 results (McElroy and Hunten 1969).

Wallace et al. (1971) were the first to search experimentally for deuterium in the upper atmosphere of Venus. For this they measured the D/H brightness ratio of the Venus dayglow by means of a rocket-borne ultraviolet spectrometer having sufficient resolution to separate the D and H resonance lines. The observed spectrum, though suffering from somewhat poor counting statistics, showed the H line, but not the D line. After subtraction of a general background, 32 ± 8 counts were left associated with H and -8 ± 8 counts with D. Of the 32 H counts 18 were ascribed to Venus, the remainder to the geocorona. Wallace et al. concluded "that, contrary to the deuterium interpretation, the bulk of the Venus emission is in the H line and not the D line...and that the deuterium interpretation of the Mariner 5 observations is not correct."

Another experimental search for deuterium in the Venus upper atmosphere was performed by Bertaux et al. (1978) who employed absorption-resonance cells filled with hydrogen and deuterium on board the Venera 9 and Venera 10 orbiters. Periodic activation of the hydrogen cells during pericenter passes of the two spacecraft gave clear absorption signals at the Lyman-α detector (NO counter), whereas activation of the deuterium cells produced no evidence of absorption. A later, refined analysis showed, however, that the deuterium cell did not function at all, for reasons still unknown. On the other hand, for some favorable viewing conditions, the incoming Lyman-α signal was almost totally absorbed by the hydrogen absorption cell from which an upper limit of D/H $\lesssim 4\times10^{-2}$ in the Venus upper atmosphere can be inferred (Bertaux, personal communication). This upper limit constitutes another proof that the deuterium interpretation of the Mariner 5 observations cannot be correct.

Recently McElroy et al. (1982) suggested that the mass 2 ion observed by the Pioneer Venus ion spectrometer (Taylor et al. 1980) is in fact D^+ rather than H_2^+. This interpretation implies a D/H ratio in the bulk atmosphere of $\sim 10^{-2}$, still much larger than the value for terrestrial seawater (see Table XVI). This D/H ratio could have been attained by selective, non-thermal escape of hydrogen from Venus's outer atmosphere over geologic times. As a source of the required nonthermal energy, McElroy et al. (1982) proposed collisions of thermal hydrogen atoms with fast oxygen atoms,

$$O_{fast} + H \rightarrow O_{fast} + H_{fast} (v \leqslant 10.5 \text{ km s}^{-1}) . \tag{kkk}$$

TABLE XVI

Isotopic Composition of the Elements
(in percent)

Element	Mass Number	Terrestrial Abundance[a]	Venus Abundance	Reference in Text
H	1	99.985		
(D)	2	0.015	1.6 ± 0.2	Sec. V. B.1
He	3	0.000138	< 0.03	Sec. V. B.2
	4	99.999862		
C	12	98.90		
	13	1.10	1.12 ± 0.02	Sec. V. B.3
N	14	99.634		
	15	0.366	0.366 ± 0.075	Sec. V. B.4
O	16	99.762		
	17	0.038		
	18	0.200	0.20 ± 0.01	Sec. V. B.5
Ne	20	90.51		
	21	0.27		
	22	9.22	7 ± 2	Sec. V. B.6
Cl	35	75.77	no significant	
	37	24.23	difference from	Sec. V. B.7
			terrestrial value	
Ar	36	0.337	44.2	
	38	0.063	8.6	Sec. V. B.8
	40	99.600	47.2	
Kr	78	0.35		
	80	2.25	7	
	82	11.6	23	
	83	11.5	14	Sec. V. B.9
	84	57.0	48	
	86	17.3	8	

[a]Values to be published in *Pure and Applied Chemistry*, 1982, as Part II of the Report of the International Commission on Atomic Weights (P. De Bièvre, Chairman "IUPAC Subcommittee on the Assessment of the Isotopic Composition of the Elements [SAIC]," personal communication).

The energetic oxygen atoms are produced in turn by dissociative recombination of O_2^+

$$O_2^+ + e \rightarrow O(^1D) + O(^3P) + \overline{5.0 \text{ eV}}. \tag{III}$$

McElroy et al. (1982) calculated from the above reactions a globally averaged escape rate for H on the order of 10^7 cm^{-2} s^{-1} at the current epoch, whereas the escape rate for D is negligible. If such a selective escape rate for H was maintained throughout most of Venus's history, as suggested by McElroy et al. a significant increase of the D/H ratio over its initial value 4 byr ago would have occurred. However, the linear backward extrapolation of this deuterium enrichment over geologic times must be considered rather speculative.

Kumar et al. (1982) have investigated various nonthermal and hydrodynamic escape mechanisms for past conditions on Venus when total H mixing ratio f_H was much higher and find that the hot oxygen impact source suggested by McElroy et al. would be unimportant in the past because the source saturates at only 3 times the present value and eventually ceases to operate once the hydrogen determined exobase is substantially above the ionosphere peak. Instead, Kumar et al. (1982) find that charge exchange between cold H atoms in the exosphere and high-temperature H^+ ions ($T_i \sim$ 2000 K) in Venus's plasmasphere is the strongest source. Charge exchange is also selective against escape of deuterium and the D/H ratio could be enhanced by a factor of 1100 over geologic time.

The deuterium interpretation of the observed mass 2 ion in the Venus ionosphere by McElroy et al. (1982) yields in fact an upper limit for the deuterium abundance in the Venus atmosphere, because contributions of H_2^+ to the measured signal cannot be excluded, small as they may be. Therefore, after the suggestion of McElroy et al. became known, efforts were intensified to unambiguously identify deuterium and its abundance in the Venus atmosphere. Attempts to derive the venusian D/H ratio from the data of the Pioneer Venus Large probe mass spectrometer were revived by Donahue et al. (1982). They studied in particular the behavior of the mass peaks found at 18.010 amu (caused by H_2O) and at 19.010 amu (appropriate for the detection of HDO). They showed that the ratio of 19 to 18 amu count rates increased greatly while the inlet leak of the mass spectrometer was clogged in the altitude range from 50 to 27 km, very likely with a venusian cloud droplet. Donahue et al. interpret this increase as indicating that in this altitude range venusian water from the cloud droplet was evaporated in the mass spectrometer and that this venusian water is deuterium-rich. Yet, this interpretation is not entirely self-evident. Besides HDO^+, the following compounds might contribute to the peaks observed at 19 amu: $H^{18}O^+$, F^+, $H^{37}Cl^{++}$, and $^{38}Ar^{++}$. Furthermore, the water observed in the mass spectrometer during its descent in the Venus atmosphere is always composed of two components, one of terrestrial and the other of venusian origin. On the other hand, the experiment collected a lot of supporting information, such as the signal strength of singly ionized ions ($H^{37}Cl^+$, $^{38}Ar^+$), of isotopic peaks ($H^{35}Cl^+$, $H^{16}O^+$), of dissociative peaks ($^{35}Cl^+$), on the time history of these various peaks, and the count rates in the three available mass 19 channels, which were located at 18.987, 18.998, and 19.007 amu (Donahue, personal communication). Putting this whole story together Donahue et al. (1982) come up with the conclusion that the venusian cloud droplet contained water with a ratio $HDO/H_2O = (3.2 \pm 0.4) \times 10^{-2}$.

Consequently the deuterium-hydrogen abundance ratio in the Venus atmosphere is inferred to be $(1.6 \pm 0.2) \times 10^{-2}$.

We view the arguments put forward by McElroy et al. (1982), as well as the data presented by Donahue et al. (1982) as strong evidence for the D/H ratio being highly enriched at Venus in comparison to that at Earth. Nevertheless, in our opinion, neither case can yet be considered firm proof for a dramatic deviation of Venus's isotopic composition with respect to terrestrial conditions. McElroy et al. interpret as D^+ the mass 2^+ ion observations in the Venus ionosphere which have also been considered to be H_2^+. In the case of Donahue et al. a full discussion of the data reduction leading to their result has not yet been published; when it is, we expect that some of the concerns raised above will be resolved. Therefore we suggest waiting for further corroboration and results of new experiments before considering firm a ratio of D/H $\sim 10^{-2}$ in the Venus atmosphere.

2. Helium. The isotope ^{3}He is exceedingly rare, as indicated in Table XVI. Mass spectrometers can in principle measure ^{3}He at the $m/e = 3$ peak. On the other hand HD^+ and H_3^+ also contribute to the peak intensity at mass 3. Here H_3^+ are strictly background ions formed in the ion source of the mass spectrometer, especially if there exists a sizable partial pressure of hydrogen and/or methane. A determination of ^{3}He and HD abundances via the mass 3 peak therefore requires separation of these 3 components.

The Pioneer Venus spacecraft carried 3 neutral gas mass spectrometers into the Venus atmosphere: one each on board the Large probe, the multiprobe bus, and the orbiter. The Large probe mass spectrometer showed the smallest relative peak intensity at mass 3 and therefore seems best adapted to derive upper limits on ^{3}He and HD abundances. In their analysis Hoffman et al. (1980a) compare the altitude profiles of the masses 1, 2, 3, 4, and 15 (CH_3 from background methane) peaks. The best fit for the altitude profile of the mass 3 peak is found using 90% methane-correlated H_3^+ and 10% HD^+. This fit does not require any ^{3}He. The above value of HD gives a HD/H_2 ratio of 3×10^{-4}, exactly the terrestrial value.

Since the venusian HDO/H_2O has been found to be on the order of 3% (Donahue et al. 1982), it is concluded that the mass 3 peak is heavily contaminated with terrestrial hydrogen, most likely from the methane calibration gas. Regarding ^{3}He, a maximum ^{3}He/^{4}He value of 3×10^{-4} is permitted from the uncertainties of the curve fitting. Therefore, a large excess ^{3}He in Venus's atmosphere over the terrestrial amount is allowable, but certainly not required by the data.

3. Carbon. Information on the venusian carbon isotopes was first derived from infrared spectra of Venus by Kuiper (1962). He found the abundance ratio ^{13}C/^{12}C equal to that on Earth within an estimated precision of $\pm 10\%$. The Pioneer Venus mass spectrometer of Hoffman et al. (1980a) gave the result ^{13}C/^{12}C $\leq 1.19 \times 10^{-2}$. The Venera 11 and 12 instruments of Istomin et al. (1980b) yielded (1.12 ± 0.02) × 10^{-2}. Wilson et al. (1981) measured absorption lines in the Venus spectrum of ^{12}CO and ^{13}CO at 115 GHz and 110 GHz, respectively. From their observations they calculated the abundance of ^{13}CO to be 1/(85 ± 15) that of ^{12}CO, again very close to the terrestrial ratio.

4. Nitrogen. Hoffman et al. (1979c) deduced from the 29/28 peak ratio observed in their mass spectrometer, that the venusian isotopic ratio ^{15}N/^{14}N is within 20% of that of terrestrial nitrogen, i.e. (0.366 ± 0.075) × 10^{-2}.

5. Oxygen. From infrared spectra of Venus Kuiper (1962) determined that the Venus ^{18}O/^{16}O ratio is equal to that on Earth within an estimated precision of ± 10%. Connes et al. (1967) remark: "The assumption, that the ^{18}O/^{16}O abundance ratio is the same on Venus as in the laboratory, appears confirmed by many intercomparisons of intensities of CO_2 lines." The Pioneer Venus mass spectrometer of Hoffman et al. (1980a) gave the result ^{18}O/^{16}O = (2.0 ± 0.1) × 10^{-3} matching perfectly the terrestrial value (see Table XVI).

6. Neon. The measurement of the isotopic composition of neon in the Venus atmosphere by means of mass spectrometers poses two general problems: first in the Venus atmosphere, the mixing ratio of Ne is only on the order of 10^{-5}, and second the mass 22 peak is also occupied by CO_2^{++}. To alleviate these problems for *in situ* measurements in Venus's lower atmosphere, both Veneras 11 and 12 and the Pioneer Venus mass spectrometers included special features. Istomin et al. (1980b) periodically put their instruments into a noble-gas analysis mode by switching off the high voltage of the ion sputter pump. In this mode the pumping speed for noble gases "falls to zero," whereas the pumping speed for chemically active gases like CO_2 and N_2 remains essentially unchanged. To minimize the contribution of doubly-charged ions without undue loss of sensitivity for singly-charged ions, Istomin et al. chose an energy (fixed) for the ionizing electrons of ~ 40 eV. With these features Istomin et al. obtained an abundance measurement for ^{20}Ne (see Fig. 9), but the instruments were not sensitive enough to measure the less abundant isotopes of neon. Thus no isotopic ratios were obtained.

Hoffman et al. (1980a) provided for their instrument a noble gas enrichment mechanism more elaborate than that employed by Istomin et al. (1980b). A single atmospheric sample was purified off chemically active gases through a combination of sorption and getter pumps. This purified sample was subsequently introduced into the mass spectrometer for a measurement of the isotopic ratios of the concentrated noble gases. The electron energy was 30 eV for this particular measurement. Hoffman et al. obtained a ratio ^{22}Ne/^{20}Ne = 0.07 ± 0.02 where the error bar is one standard deviation.

7. Chlorine. Connes et al. (1967) measured the equivalent width for unblended H^{35}Cl and H^{37}Cl lines in the spectrum of Venus. If the isotopic abundance ratio were the same on Venus as on the Earth, one would expect the equivalent width ratio of H^{35}Cl and H^{37}Cl line-pairs to be 3.13, provided

TABLE XVII

Isotopic Abundance Ratios

$^{40}Ar/^{36}Ar$	$^{38}Ar/^{36}Ar$	Reference
1.03 ± 0.04	0.18 ± 0.02	Hoffman et al. (1980a)
1.19 ± 0.07	—	Istomin et al. (1980a)
—	0.197 ± 0.002	Istomin et al. (1980b)

that the lines were unsaturated. Connes et al. obtained for the 3 least-blended pairs the ratio 2.4 ± 0.2. They attribute this to partial saturation of the lines, rather than to a difference in the chlorine isotopic ratio between Venus and the Earth. L. Young (1972) has extended this analysis to more HCl lines in the Connes et al. (1969) spectra. She finds an abundance ratio of 2.9 ± 0.3 and concludes "that the abundance ratio does not differ significantly from the telluric value, and that line saturation is occurring for the HCl lines."

8. *Argon*. Although there has been some confusion about the absolute abundance of argon in the Venus atmosphere (see Sec. IV. G. 3), the 1978 mass spectrometer experiments never left much uncertainty about the isotopic abundance ratios. The final values derived from 3 *in situ* mass spectrometer measurements in the lower atmosphere are given in Table XVII.

For a discussion of the remarkable difference in the ratio $^{40}Ar/^{36}Ar$ between Venus and Earth we refer the reader to Chapter 29 by Donahue and Pollack. The deviation found by Istomin et al. (1980b) of the venusian ratio $^{38}Ar/^{36}Ar$ from the terrestrial value of 0.187 is small, but possibly significant. It was independently measured by both the Venera 11 and 12 mass spectrometers. The smallness of the error bar of the ratio is due to adding and averaging ∼ 50 selected spectra of the Venera 11 and 12 instruments (Istomin, personal communication).

9. *Krypton*. We have reviewed in Sec. IV. G. 4 the krypton abundance measurements obtained by the mass spectrometers on board the Venera 11 and 12 landers and the Pioneer Venus Large probe. For the Venera 11 and 12 mass spectra it is the lack of appropriate resolving power which makes the calculation of isotopic abundance ratios for Kr rather difficult. We summarize these observations therefore with the semi-quantitative statement, that they do not give any evidence for a significant deviation of the Kr isotopic ratios of Venus from the terrestrial values. The analysis of Donahue et al. (1981) of the Pioneer Venus mass spectrometer gave the most probable mixing ratios for the Kr isotopes (see Table XVIII). Donahue et al. (1982) notes: "If these results do indeed represent observations of atmospheric krypton, the relative abundances of Kr isotopes on Venus are 0.07, 0.23, 0.14, 0.48, and 0.08" for the isotopes 80, 82, 83, 84, and 86, respectively. The isotope ^{78}Kr was not determined. The deviations from terrestrial abundances (see Table XV) are

TABLE XVIII

Mixing Ratios for Kr Isotopes[a]

Isotope	Ratio (ppb)				
^{80}Kr	0	<	3.6	<	10.2
^{82}Kr	6	<	12.2	<	17
^{83}Kr	0	<	7.2	<	19
^{84}Kr	7	<	25	<	28
^{86}Kr			4.3	<	9
total			52.3		

[a]Pioneer Venus mass spectrometer analysis (Donahue et al. 1981).

not significant in the light of the large error bars of the venusian isotopic abundances and the uncertainty surrounding the three independent Kr abundance measurements.

VI. MAJOR OPEN QUESTIONS

Selecting a limited number of questions and calling them major open questions can hardly be done in a purely objective manner. We strive instead to come up with a list of major questions whose answers appear to lie within reach of sophisticated earth-based measurements, or by the next generation of planetary probes to Venus. Photochemical models of Venus's middle atmosphere would greatly benefit from the following information:

1. What is the O_2 abundance (or improved upper limits) near the cloud tops?
2. What is the H_2 abundance in the middle atmosphere?
3. Are there any other chlorine-bearing gases beside HCl?
4. What are the NO_x abundances near the cloud tops?

For the thermochemical models of Venus's lower atmosphere of paramount importance is the question:

1. What is the atmospheric composition in the lowest scale height of the Venus atmosphere?

This query includes the question about the abundance of free O_2 in this atmospheric region. It also relates directly to the question about the degree of chemical equilibrium between the surface of Venus and its atmosphere.

To further the studies of the origin and evolution of Venus's atmosphere we would like to see answers to the following questions:

1. Is the abundance ratio D/H indeed 0.016 as claimed by Donahue et al. (1982)?
2. What are the abundances and isotopic composition of Kr and Xe?

This list reflects our belief that through decades of untiring efforts to unravel the mysteries of Venus's atmosphere, a fairly comprehensive picture of its complex chemistry has now been obtained. Yet, in the future it will be exciting to see some of the outstanding answers (and surprises) come in, further enhancing our understanding of Venus, the strange twin of Earth.

Acknowledgements. We gratefully acknowledge valuable discussions with J. L. Bertaux, G. C. Carle, J. H. Hoffman, D. M. Hunten, V. G. Istomin, V. A. Krasnopolsky, V. I. Moroz, J. T. Schofield, F. Woeller, A. T. Young, and L. G. Young. In preparing this manuscript U.v.Z. was assisted by K. H. Fricke and by a grant from the Bundesministerium für Forschung und Technologie, Bonn; S. K. was supported by grants from the NASA Planetary Atmospheres Program to the University of Southern California; H. N. acknowledges the assistance of A. E. Hedin and W. T. Kasprzak in preparation and analysis of Pioneer Venus ONMS data; R. P. was supported by the Atmospheric Chemistry Program of the National Science Foundation.

14. PHOTOCHEMISTRY OF THE VENUS ATMOSPHERE

V. A. KRASNOPOL'SKY and V. A. PARSHEV
USSR Academy of Sciences

The chemical composition of the lower atmosphere of Venus (0–50 km) can be explained by thermochemical equilibrium at the surface in the presence of a photochemical equilibrium among the different forms of sulfur S_n. Taking into account the law of constancy of elementary composition, it is very difficult to explain the measured altitude dependences of H_2O and O_2. At altitudes above 50 km, altitude dependences have been obtained for the concentrations of 27 components of the atmosphere and cloud layer by solving a set of equations which describe the processes of transfer and 102 chemical reactions in the atmosphere of the planet. The initial data for the calculation were the values of the concentrations of CO_2, H_2O, HCl, SO_2 at 50 km. In the list of chemical reactions there are catalytic cycles involving $COCl$ and $COCl_2$, which play an important role in the formation of CO_2 and the decomposition of O_2. An approximate method has been developed to describe the processes of transfer of radiation in the cloud layer, along with a procedure of assigning the boundary conditions for the secondary components, including as partial cases the photochemical and diffusional equilibrium. A good agreement is observed between the calculation and measurements of the chemical composition with $[H_2O]/[SO_2] \simeq 1.5$ at 50 km. Sulfuric acid and the products of its further transformation should comprise the major portion of the cloud layer in its upper and lower parts. The results of the calculation have been analyzed in detail for the measured components (CO, O_2, O, SO_2, H_2SO_4, H_2O). The appearance of O_2 inside and below the cloud layer is difficult to explain.

The photochemical analysis of the processes occurring in the atmosphere of the planet has been enlisted to provide a theoretical basis for the results of measurements of the chemical composition, which are inevitably incomplete and, in certain cases, contradictory. A photochemical model of the composition of the atmosphere produces vertical distributions of all components on the basis of data as to the quantities of the main parent molecules. This requires a knowledge of the rate coefficients of the chemical reactions k_i, the temperature T, and the eddy diffusion coefficient K. Since the k_i are not known for all the reactions, the number of components under consideration must be limited and it is necessary to employ an analysis for those regions of the atmosphere where these limitations apply. Therefore it is convenient to divide the atmosphere into two parts with a boundary at the base of the cloud layer at $\simeq 50$ km.

I. THE LOWER ATMOSPHERE BELOW 50 KM

Thermochemical Equilibrium

The high pressure and temperature (90 bar and 740 K at the surface), the absence of solar ultraviolet radiation, and the very limited data on the rate coefficients k_i permit us to use, as a first approximation, a calculation of thermochemical equilibrium, for which it is sufficient to know the thermodynamic characteristics of the particular substances. It is known that, for an equilibrium process

$$\alpha_1 A_1 + \alpha_2 A_2 + \ldots = \beta_1 B_1 + \beta_2 B_2 + \ldots \tag{1}$$

the equilibrium constant is

$$\log K_p = \log \frac{P^{\beta_1}(B_1) P^{\beta_2}(B_2) \ldots}{P^{\alpha_1}(A_1) P^{\alpha_2}(A_2) \ldots} = \sum_i \beta_i b_i(T) - \sum_i \alpha_i b_i(T) \tag{2}$$

where $b_i(T) = -10^3 G_i(T)/4.575\,T$; G_i is the free enthalpy in kcal/mole; P_i is the pressure in atm. Tables of the functions b_i are given by Mills (1974), and Barin and Knack (1973).

Chemical equilibrium is reached when the time for its establishment $\tau_c = (\sum_i k_i n_i)^{-1}$ is small in comparison with the mixing time $\tau_m = H^2/K$. Here, k_i and n_i are the rate coefficients of the reactions of destruction of the particular component and of the concentration of substances participating in these, and H is the scale height. A comparison of these times indicates the predominant role of mixing in the entire lower atmosphere with the exception of the surface, the catalytic effect of which should accelerate the chemical processes. For the estimates we used $K = 10^4 - 10^6$ cm^2 s^{-1}, $k_i \simeq 10^{-10}$ exp $(-25000/T)$ cm^3 s^{-1}, which is typical for reactions between different molecules (e.g. the reaction of CO and O_2, H_2 and O_2, H_2 and Cl_2), a mixing ratio $f_i \simeq 10^{-5}$, and an acceleration of the chemical reactions at the surface by 2 orders of magnitude. Consequently, thermochemical equilibrium is established at the surface and, above this, the chemical composition does not change with height. The results of the calculation of the chemical composition

TABLE I

Composition of the Lower Atmosphere of Venus Under the Condition of Thermochemical Equilibrium[a]

Gas	Veneras 11, 12 f_i	Pioneer Venus f_i
CO_2	0.96	0.96
SO_2	1.5×10^{-4}	1.5×10^{-4}
O_2	—	7.0×10^{-5}
CO	1.5×10^{-5}	—
S_2	1.0×10^{-7}	—
SO_3	—	1.3×10^{-3}
COS	2.0×10^{-5}	—
CS_2	—	—
H_2O	3.0×10^{-5}	1.35×10^{-3}
H_2	3.0×10^{-9}	—
H_2S	5.0×10^{-8}	—
N_2	3.4×10^{-2}	3.4×10^{-2}
NO	—	3.0×10^{-9}
NO_2	—	3.0×10^{-10}
NH_3	—	—
HCl	1.0×10^{-6}	1.0×10^{-6}
Cl_2	—	2.5×10^{-10}
HF	1.0×10^{-8}	1.0×10^{-8}

[a]Mixing ratios f_i of the components shown; values of $f_i < 10^{-10}$ not included.

of the lower atmosphere are given in Table I for components which have $f_i > 10^{-10}$; inert gases were not considered. The initial data for the calculation are the concentrations of CO_2, N_2 and SO_2, for which there is good agreement, and H_2O and S_2 (our estimate is based on spectroscopic measurements of Veneras 11 and 12 [Moroz et al. 1979b]), and HCl and HF from groundbased spectroscopic measurements in the infrared region (Connes et al. 1967). These calculations are simple; they were performed on the basis of the results of a preliminary processing of measurements of the chemical composition on Venera 11 and 12 and Pioneer Venus spacecraft, and certain conclusions have been confirmed by a more detailed processing of the measurements. This concerns in particular the data on CO and O_2.

Our assumption of a chemical composition of the atmosphere independent of height is merely an approximation which cannot apply to all the components. Let us now consider the substances for which experiment has found a dependence of the mixing ratio on the altitude.

The Law of Constancy of the Elementary Composition

The altitude dependences of the concentrations n_i of substances are described by the continuity equations

$$\frac{\partial n_i}{\partial t} + \frac{\partial \phi_i}{\partial z} = P_i - n_i L_i \tag{3}$$

$$\phi_i = -K \left(\frac{\partial n_i}{\partial z} + \frac{n_i}{H} \right) \tag{4}$$

where ϕ_i is the flux of component i, and P_i and L_i are the production and destruction of the component in the chemical reactions. When molecular diffusion prevails, in the expression for ϕ_i instead of K there appears the coefficient D_i and the mean scale height H is replaced by H_i. If diurnal variations are not considered (steady state), then $\partial n_i / \partial t = 0$. If we also sum the lefthand portions of the continuity equations for all the substances which contain the particular element, multiplied by the stoichiometric coefficients, then the overall production and destruction of the element should equal zero, i.e. its flux will be constant, and in the absence of inequality between the accumulation of the element at the surface and in the atmosphere or in the case of its dissipation into space, the flux should also equal zero. In the case of mixing, $\phi_i = -K \, [CO_2] \, df_i/dz$, from which it follows that f_i is constant. The law of constancy of the elementary composition is violated in the presence of molecular diffusion and condensation, since the expression for the flux is different in these cases.

CO, CO$_2$, SO$_2$, S$_n$. The decrease with height of f_{CO} from 3×10^{-5} at $h >$ 30 km to 1.5×10^{-5} below 25 km, discovered in both gas chromatograph experiments (Gel'man et al. 1979a; Oyama et al. 1980a), is accompanied by a decrease in f_{O_2} according to the data of Gel'man et al. (1979a) from 4×10^{-5} to 1.6×10^{-5} at 52 and 42 km. These variations do not correspond to the proportion of CO and O$_2$ in the formations of CO$_2$, and therefore we shall consider other possibilities to account for the altitude dependence of f_{CO}.

If we compare the measurements and the results of the calculation of thermochemical equilibrium (Table I), it is easy to see the approximate constancy of the sum $f_{CO} + f_{COS} = 3.5 \times 10^{-5}$. Then the altitude dependence of f_{CO} can be explained by the gradual conversion of CO to COS. Here the following expressions arise: in the first place, $f_{COS} \simeq (5-10) \times 10^{-6}$ at 22 km, which exceeds the upper limit of 2×10^{-6}, obtained in the third sample of the gas chromatograph on Pioneer Venus. In the second place, the free sulfur is not sufficient for the formation of such quantities of COS; if, however, sulfur is liberated by the consecutive reduction of CO$_2$ by means of CO, then it is necessary to expend three, and not one molecule of CO for the formation of each molecule of COS.

Our qualitative analysis of the problem has yielded the following solution. Due to the rapid photolysis of S$_3$ by visible light, the thermochemical equilibrium of the primary forms of sulfur at the surface S$_2$, S$_3$, and S is appreciably disturbed and transformed to a photochemical equilibrium. Optical measurements on board the Venera probes (Moroz et al. 1979b) have only been able to determine the S_3 concentrations which equal 6×10^{-9} cm^{-3} at the surface and which correspond to $f_{S_2} \simeq 10^{-3}$ at thermochemical equilibrium. The equilibrium quantities of S_2 and S now turn out to be 10 and 100 times greater than in the case of thermochemical equilibrium, and $f_{S_2} = 10^{-7}$ in accordance with our original estimate (Krasnopol'sky and Parshev 1979a). The change in the relative distribution of S, S_2, and S_3 affects the concentrations of COS by the processes

$$S + COS \rightleftarrows CO + S_2$$

$$S_3 + CO \rightleftarrows COS + S_2$$

$$S + CO + M \rightleftarrows COS + M. \qquad (a)$$

The second of these processes is most important, and $[COS]/[CO] = K$ $[S_3]/[S_2]$. Allowing for the photolysis of S_3, this ratio is 10 times less, and $f_{COS} \simeq 2 \times 10^{-6}$ at the surface. Thus, the probable cause of the diminution of f_{CO} is its transformation to CO_2 in reactions with SO_2 and then with SO. The variation of f_{CO} from 3×10^{-5} at 40 km to 1.5×10^{-5} at the surface corresponds to a diminution of f_{SO_2} by 7×10^{-6} or by 5% and to a growth of f_{S_2} by 3×10^{-6} and of f_{COS} up to 2×10^{-6}. And, in fact, an approximate calculation of the photochemical equilibrium at 12 km, where the maximum of $[S_3]$ is located, according to optical measurements (Moroz et al. 1979b), also indicates a rapid growth of f_{S_2} from the surface and $f_{S_2} \simeq 4 \times 10^{-6}$ at 12 km; however, these findings should be treated with caution, despite their agreement with the estimates of the anticipated values, in view of the approximate nature of the calculation.

H_2O and O_2. The measurement results of the Venera probes and Pioneer Venus are most contradictory in regard to water vapor and oxygen. We consider the spectrometric measurements of H_2O as the "purest", where the observed effect is produced in an atmosphere several km thick and not subject to the disturbances and contaminations introduced by the probe. Water can be absorbed on the surface of the instruments and distort the measurement results. Nonetheless, all the analysis methods of Veneras 11 and 12 yield $f_{H_2O} \simeq 10^{-4}$, which has also been assumed in Table I. If the higher values of f_{H_2O}, measured by Pioneer Venus, are due to local variations, then the data of the Venera probes are more representative, as it concerns two probes.

However, it is difficult to explain the variations of f_{H_2O} from 2×10^{-4} at 50 km to 2×10^{-5} at the surface. For this, in some way it is necessary to "fit" the considerable quantities of hydrogen and oxygen. The most obvious possibility

$$CO + H_2O \rightarrow CO_2 + H_2 \tag{b}$$

does not apply, as $f_{H_2} \simeq 10^{-4}$ at the surface contradicts the measurements; the equilibrium of the reaction is shifted to the left and, at thermochemical equilibrium, $f_{H_2}/f_{H_2O} = 6 f_{CO}$ at the surface, the variation of f_{CO} by 15 ppm is appreciably less than the change in f_{H_2O} by 180 ppm. It is possible that the errors of the optical measurements and the uncertainty of the initial data of the calculation for f_{H_2O} by these measurements do not completely exclude the absence of a dependence of f_{H_2O} on the altitude.

Greater difficulties occur in explaining the Pioneer Venus measurements of O_2 (Oyama et al. 1980a). It is interesting that $f_{O_2} = 44$ ppm at 52 km and 16 ppm at 42 km corresponds to a constant absolute content of $[O_2] = 8 \times 10^{14}$ cm^{-3}. This may occur in the presence of a source at the top and a sink at the bottom of the particular region; then the solution of the continuity equation at $h-h_0 > 2 H$ is equal to $n = \phi H/K$; here, h_0 is the height of the sink region. Substituting $[O_2] = 8 \times 10^{14}$ cm^{-3} at $K = 10^4$ cm^2 s^{-1} yields $\phi = 10$ cm^{-2} s^{-1}, i.e. twice as much possible production of O_2 under a photolysis of CO_2. The doubling does not produce problems, although in this case f_{CO} should exceed f_{O_2} by more than a factor of 2, which is not observed; moreover, the photolysis of CO_2 basically occurs above 65 km, and then f_{O_2} would be ~ 300 ppm at 65 km, instead of $f_{O_2} < 1$ ppm, in accordance with the spectroscopic measurements (Traub and Carleton 1974) that pertain to this altitude.

Thus, in order to explain the altitude variation of f_{H_2O} from the data of optical measurements (Moroz et al. 1979b) of the Venera probes and f_{O_2} from gas chromatograph measurements of Pioneer Venus (Oyama et al. 1980a), it is necessary for two gases containing H and O to be present in the lower atmosphere in quantities of $\sim 10^{-4}$. We have reviewed the objections to CO_2 and H_2; HCl in quantities of $\sim 10^{-4}$ at the surface would create the problem of an excess of chlorine in the cloud layer; NH_3 should not be formed in either photochemical or thermochemical processes.

II. THE ATMOSPHERE ABOVE 50 KM

The structure of a photochemical model for the composition of the atmosphere above 50 km should especially answer the following questions:

1. Why are the concentrations of the products of CO_2 photolysis so small?
2. How does the cloud layer of Venus form? Which processes lead to the formation of sulfuric acid which is absent in the evolution of gas of the lithosphere?
3. Why are the concentrations of gases containing sulfur that are the initial material and the intermediate links in the formation of H_2SO_4 so small in the upper portion of the cloud layer?
4. Why is f_{H_2O} in the upper portion of the cloud layer smaller by two orders of magnitude than in the atmosphere below the clouds?

5. How is the composition of the upper atmosphere determined? Are chlorine and sulfur involved?

These questions are related to the photochemistry of CO_2, H_2O, HCl and SO_2. Although other gases are also present in the atmosphere of Venus, they play a lesser role, either due to low chemical activity (nitrogen, inert gases), or due to a small relative content (HF); in order not to overburden the problem we shall not consider them.

Chemical Processes

The absorption of sunlight below 2075 Å leads to the photolysis of CO_2. In the absence of other gases, the reverse process is the reaction

$$O + O + CO_2 \rightarrow O_2 + CO_2 \tag{c}$$

which leads to an accumulation of CO and O_2. In the actual atmospheres of Mars and Venus this accumulation is prevented by catalytic reactions involving active forms of hydrogen H* (H, OH, HO_2, H_2O_2). These accelerate the formation of CO_2 and O_2 through the processes

$$CO + OH \rightarrow CO_2 + H$$
$$O + OH \rightarrow O_2 + H \tag{d}$$

and lead to an indirect photolysis of O_2

$$H + O_2 + CO_2 \rightarrow HO_2 + CO_2$$
$$HO_2 + h\nu \rightarrow OH + O$$
$$HO_2 + HO_2 \rightarrow H_2O_2 + O_2$$
$$H_2O_2 + h\nu \rightarrow OH + OH. \tag{e}$$

On Mars the presence of water vapor with a mixing ratio $f_{H_2O} \simeq 10^{-4}$ assures a quantity of H* which results in a decrease of f_{CO} and f_{O_2} down to $\sim 10^{-3}$. On Venus in the upper portion of the cloud layer, $f_{H_2O} \simeq 10^{-6}$, $f_{CO} \simeq 5 \times 10^{-5}$, and $f_{O_2} < 10^{-6}$. An important role is played by reactions involving active chlorine Cl* (Cl, ClO, ClO_2, Cl_2), which is formed in the photolysis of HCl. These reactions lead to the formation of CO_2 and O_2 in cycles such as

$$CO + ClO \rightarrow CO_2 + Cl$$
$$CO + ClO_2 \rightarrow CO_2 + ClO$$
$$O + ClO \rightarrow O_2 + Cl$$
$$O_3 + ClO \rightarrow ClO_2 + O_2. \tag{f}$$

In contrast with H*, these chlorine compounds cannot result in the breakdown of molecular oxygen and promote its accumulation. An important role in the photochemistry of the atmosphere of Venus should belong to compounds of chlorine with carbon monoxide, $COCl$ and $COCl_2$, which provide for the rapid combination of CO with O and O_2, reducing the quantity of free oxygen in the atmosphere

$$CO + Cl + CO_2 \rightarrow COCl + CO_2$$

$$COCl + O_2 \rightarrow CO_2 + ClO$$

$$COCl + O \rightarrow CO_2 + Cl$$

$$COCl + Cl_2 \rightarrow COCl_2 + Cl$$

$$COCl + Cl \rightarrow CO + Cl_2$$

$$COCl + CO_2 \rightarrow CO + Cl + CO_2$$

$$COCl + H \rightarrow CO + HCl$$

$$COCl_2 + O \rightarrow CO_2 + Cl_2. \tag{g}$$

We present here a more detailed list of the reactions of $COCl$ and $COCl_2$, because they have not been considered previously with respect to the atmosphere of Venus.

In addition to the photolysis of H_2O and HCl, a source of H* and Cl* is the reaction

$$O + HCl \rightarrow OH + Cl \tag{h}$$

as well as atoms of $O(^1D)$, formed in the photolysis of ozone

$$O(^1D) + H_2O \rightarrow OH + OH$$

$$O(^1D) + H_2 \rightarrow OH + H$$

$$O(^1D) + HCl \rightarrow OH + Cl. \tag{i}$$

The reactions of formation of CO_2 and O_2 do not involve the destruction of H* and Cl*; these disappear primarily in the processes

$$H + Cl_2 \rightarrow HCl + Cl$$

$$Cl + HO_2 \rightarrow HCl + O_2. \tag{j}$$

A source of sulfuric acid and free sulfur is the photolysis of SO_2 in the presence of water vapor

$$SO_2 + h\nu \rightarrow SO + O$$

$$SO_2 + O + CO_2 \rightarrow SO_3 + CO_2$$

$$SO_3 + H_2O \rightarrow H_2SO_4$$

$$SO + SO \rightarrow SO_2 + S. \tag{k}$$

If the drops of sulfuric acid have a solution concentration of 85%, the sum of the processes of their formation may be represented by the following relation

$$3\,SO_2 + 4\,H_2O \rightarrow 2\,H_2SO_4 \cdot H_2O + S. \tag{1}$$

It is clear from this that the quantity of aerosol sulfur is smaller than that of sulfuric acid by an order of magnitude in terms of mass. Moreover, a portion of the reactions between two molecules of SO leads to the formation of S_2O (we shall not consider this component), which entails a further reduction in the output of aerosol sulfur.

Besides $O(^1D)$, it is worthwhile to consider the excited states $SO_2\,(a^3B_1)$ and $SO_2\,(\tilde{A}^1B_1)$, which are formed by the absorption of light of 3200–3900 and 2200–3200 Å, respectively. Roughly 10% of $SO_2\,(\tilde{A}^1B_1)$ is turned into $SO_2\,(a^3B_1)$ (we shall designate this as $^3SO_2$) by radiation and quenching; for $^3SO_2$, in addition to extinction by collisions and radiation, the processes

$$^3SO_2 + CO \rightarrow SO + CO_2$$

$$^3SO_2 + SO_2 \rightarrow SO + SO_2$$

$$^3SO_2 + O_2 \rightarrow SO_3 + O \tag{m}$$

are possible. A complete list of the chemical reactions is much larger than the main processes described above and is given in Table II found in the Appendix at the end of this chapter.

Transfer of Radiation and Matter

At $h > 50$ km, the atmosphere of Venus is optically thick in both absorption (at $\lambda \simeq 2200$ Å, $\tau \simeq 1000$) and in scattering ($\tau \simeq 20$). The transfer of radiation in such an atmosphere is a rather complicated and independent problem, the solution of which is necessary for calculation of the rates of photolysis. We have avoided this difficulty by introducing an approximate description of the process of transfer of radiation by means of the factor of multiple scattering. It turned out in this case that the number of scatterings is close to unity when $\lambda < 3200$ Å and equals ~ 10 when $\lambda > 3200$ Å. Therefore the cross sections of absorption of substances in the cloud layer with $\lambda > 3200$ Å were taken to be 10 times larger than the actual. Another difficulty consists in the fact that, not only CO_2 but also SO_2 produces a strong absorption; the altitude distribution of the latter is not known, but determined as a result of the calculation.

The transport of aerosol particles is described by the formula of Stokes and Davis:

$$V = \frac{2}{9} \rho g \frac{r^2}{\eta} \left(1 + \frac{l}{r}\right) \left(1.257 + 0.4 \exp\left(1.1\, r/l\right)\right). \qquad (5)$$

Here ρ is the density of the material of the particles, g is the acceleration of gravity, η is the viscosity of the atmosphere, l is the mean free path of the molecules, r is the radius of the particles. It was assumed that $r \simeq 1\ \mu m$ at $h > 65$ km, $r = 4\ \mu m$ at $h < 60$ km, and $r(h)$ varies linearly at 60–65 km. Such a dependence approximately describes the results of the spectrometric measurements of particles from Pioneer Venus (Knollenberg and Hunten 1980). We assumed that a phase equilibrium exists between H_2O vapor and the water solution of H_2SO_4; it was assumed that sulfur passes into the aerosol phase after attaining the state S_2.

The processes of transport of the atmospheric gases are described by the coefficients of molecular and eddy diffusion D and K. For D, we used the standard values of $D \simeq AT^{0.75}/n$; the dependence of $K(h)$ is shown in Fig. 1, and is justified below.

The Boundary Conditions

At 50 km, the boundary conditions for the primary components CO_2, H_2O, HCl, and SO_2 are given in the form of their concentrations. Here, for CO_2 there are no problems; for SO_2 the gas chromatograph measurements on board Venera 12 (Gel'man et al. 1979a) and Pioneer Venus (Oyama et al. 1980a) are in good agreement and it has been assumed that $f_{SO_2} = 1.3 \times 10^{-4}$. The value of f_{HCl} was varied in the process of calculation in order to obtain 6×10^{-7} at the upper cloud boundary, in accordance with the spectroscopic measurements (Connes et al. 1967). A calculation revealed that f_{HCl} is constant between 50 and 100 km, and therefore $f_{HCl} = 6 \times 10^{-7}$ was assumed at 50 km. Taking into account the great uncertainty in the values of f_{H_2O}, we chose these so as to achieve the best agreement with the results of the measurements. In addition to these four components, we must include CO among the primary gases, as in the lower atmosphere CO is clearly nonphotochemical. It was assumed that $f_{CO} = 4 \times 10^{-5}$ at 50 km, in accordance with the spectroscopic (Connes et al. 1968) and the gas chromatograph (Gel'man et al. 1979a) measurements.

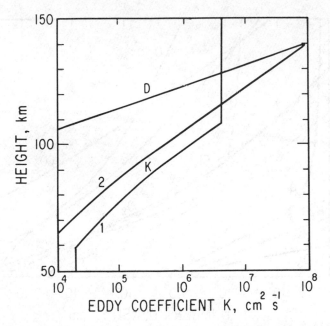

Fig. 1. The eddy diffusion coefficient K as a function of height. 1: assumed in the model; 2: proposed by von Zahn et al. (1979); D: coefficient of molecular diffusion.

Considerable difficulties occur in assigning the boundary conditions for the secondary components. Any of the ordinary forms, such as photochemical equilibrium, given concentration or flux, is inevitably arbitrary and poorly founded. We solved the problem by assuming that, for the secondary components, the conditions at 50 km are preserved directly down to the surface, where their concentrations were taken as zero. Then, in the region 0 to 50 km, the distribution of the particular component is determined by an ordinary differential equation, the solution of which for 50 km is

$$n(z_0) = \frac{P}{L} - \frac{2\,(\phi_0\,H/K + P/L)}{1 + \sqrt{1 + 4H^2 L/K}}. \tag{6}$$

It is easy to demonstrate that this corresponds to photochemical equilibrium when $\tau_m \gg \tau_c$ and to diffusional when $\tau_m \ll \tau_c$ and embraces the intermediate cases. This is an extremely convenient and universal form of notation of the boundary conditions for the secondary components.

At the upper boundary (200 km), the fluxes of all the components except H were taken to be zero. For atomic hydrogen, the flux of nonthermal escape was assumed to be 10^7 cm^{-2} s^{-1}, in accordance with the data of Kumar et al. (1978).

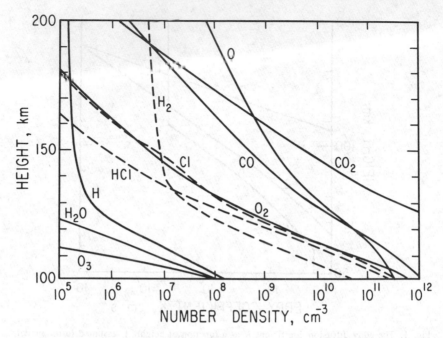

Fig. 2. Model of the composition of the upper atmosphere. For the dashes, the concentrations are a factor of 10^3 less than shown.

The Upper Atmosphere

Let us move on to a discussion of the results from our calculation. The resulting composition of the upper atmosphere, shown in Fig. 2, can be compared with the experimental results. The mass spectrometric measurements from Pioneer Venus (Niemann et al. 1980*b*) indicate very large variations of the atmospheric density as a function of the local time. It is therefore convenient to compare the calculated and measured composition of the atmosphere at altitudes where the density is identical. Such a comparison of data shows good agreement between calculation and observations (Niemann et al. 1980*b*) for CO_2, CO, H. For the other gases shown in Fig. 2, there is no measurement data. A correspondence between the calculation and the measurements of the photolysis products of CO_2 is achieved by selecting a value of $K = 4 \times 10^6$ cm^2 s^{-1} above 110 km (Fig. 1). By comparing the values of K at 90–140 km, used in our calculation and recommended by von Zahn et al. (1979), it may be noted that, at 110–140 km, K is larger, while at 90–110 km it is smaller than our values. As a result, the mixing effect is almost identical, and both models of the upper atmosphere are in sufficient agreement.

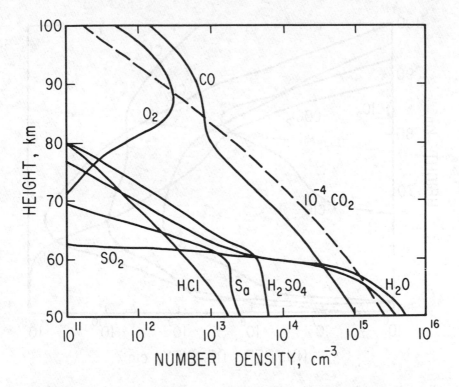

Fig. 3. Model of the composition of the atmosphere for 50–100 km (most plentiful components). Concentrations of S_a and H_2SO_4 are given for the aerosol phase.

The Composition of the Atmosphere at 50–100 km

In this region occur the most important chemical reactions determining the composition and structure of the atmosphere above the clouds and of the cloud layer. The results of the calculation of the atmospheric composition at 50–100 km are shown in Figs. 3 to 6. It is not difficult to see that, below the homopause and up to 90 km, $[CO] \simeq [O] + 2[O_2]$, from which it follows that the source of CO in the thermosphere is the photolysis of CO_2, and not a transport from the lower atmosphere. Below 80 km on the contrary, there prevails a transport from the lower atmosphere, because here a rapid formation of CO_2 after its photolysis takes place. Therefore f_{CO} increases from 5×10^{-5} at 75 km to $\sim 10^{-3}$ at 100 km. It should be noted that a similar dependence of f_{CO} has been obtained from measurements in the mm range (Schloerb et al. 1980). Among the processes which reverse the photolysis of CO_2 (Fig. 7), the most important is the reaction of COCl and O_2 (60%), followed by a cycle involving ClO (25%), and the hydrogen cycle (5.5%). In Fig. 7 curves 2–9 represent the following reactions:

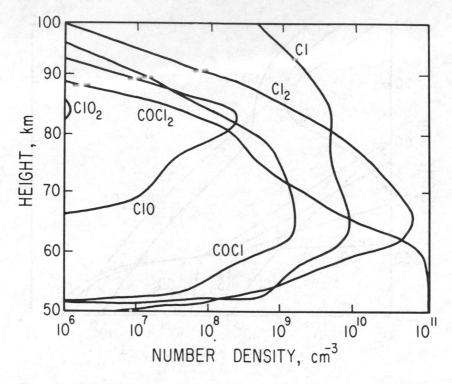

Fig. 4. Model of the composition of the atmosphere for 50–100 km (chlorine compounds).

$$(2)\ COCl\ +O_2\ \rightarrow CO_2+ClO$$
$$(3)\ CO\ +ClO\rightarrow CO_2+Cl$$
$$(4)\ CO\ +OH\rightarrow CO_2+H$$
$$(5)\ COCl\ +O\ \rightarrow CO_2+Cl$$
$$(6)\ COCl_2\ +O\ \rightarrow CO_2+Cl_2$$
$$(7)\ O+CO+CO_2\rightarrow CO_2+CO_2$$
$$(8)\ CO\ +{}^3SO_2\rightarrow CO_2+SO$$
$$(9)\ CO\ +ClO_2\rightarrow CO_2+ClO. \qquad (n)$$

The constant relative content of f_{HCl} at 50–100 km is explained, not by the predominance of the transport processes, but by the approximate equality of the rates of formation and destruction of HCl at any point of this interval. For example, at 65 km, the photochemical time of HCl is one third of the mixing time.

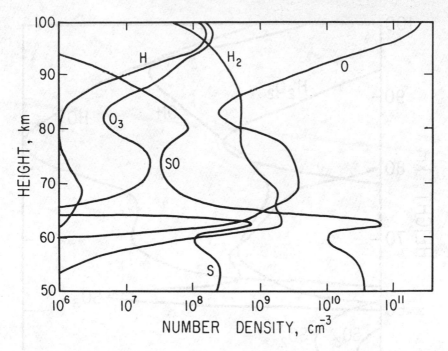

Fig. 5. Model of the composition of the atmosphere for 50–100 km (continued).

The vertical profile of SO_2 is determined by the equilibrium of transport from the lower atmosphere and of photolysis in the upper part of the cloud layer. The flux of sunlight in the region 2000 to 2180 Å, where practically each photon results in the photolysis of SO_2, is equal to 6×10^{13} quanta cm^{-2} s^{-1} and, with an allowance for the global averaging, the overall rate of photolysis of SO_2 in the atmosphere is 2.5 times larger than the photolysis of CO_2. Approximately ⅓ of this amount is composed of the reverse process

$$SO + O + CO_2 \rightarrow O_2 + CO_2 \tag{o}$$

and ⅔ forms H_2SO_4 and S. If we estimate the cross section of photolysis of SO_2 at 2000–2180 Å to be $\sigma_{SO_2} \simeq 10^{-18}$ cm^2 and the scale height of SO_2 to be ~ 1 km, then the maximum of the rate of photolysis should occur at altitudes where $[SO_2] \simeq 10^{13}$ cm^{-3}. Due to the small range of the cross sections, the photolysis of SO_2 takes place in a thin layer, below which the flux of SO_2 should be constant. Then by solving the continuity equation, it is not difficult to obtain an analytic expression for f_{SO_2}: $f(z) = f_0 - \phi_0 H/K$ $[CO_2]$ $(\exp(z/H) - 1)$. From this, the height of the photolysis layer, i.e. the level of $[SO_2] \simeq 10^{13}$ cm^{-3} or $f/f_0 \rightarrow 0$, is equal to $z_{max} = H\ln(K n_0/\phi_0 H + 1)$, with a conversion from the lower boundary. When $H = 8$ km and $K = 2 \times 10^4$ cm

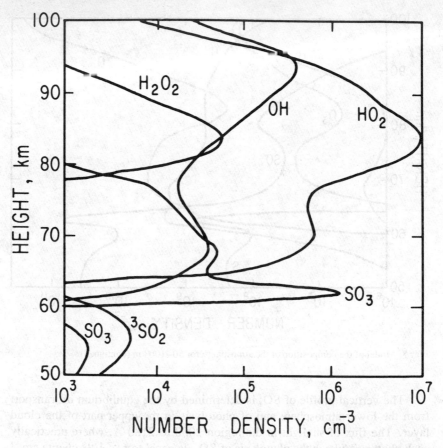

Fig. 6. Model of the composition of the atmosphere for 50–100 km (least abundant components).

s^{-1}, we obtain $z_{max} = 60$ km. For the best agreement with the experimental estimates of this height (64–66 km according to Esposito et al. [1979] and Zasova et al. [1981]), we required $K = (3-5) \times 10^4$ cm^2 s^{-1}, at which the calculated quantity of sulfuric acid is higher than the permissible. The numerical solution of Fig. 3 confirms our calculation. The excited states of SO_2 give $\sim 10\%$ of all the production of sulfuric acid.

It has already been noted above that SO_2 and H_2O are combined into drops of sulfuric acid and particles of aerosol sulfur in the proportion of 3:4. Thus it is not surprising that the profile of H_2O is similar to the profile of SO_2 and likewise characterized by a sharp decrease of f_{H_2O} at 55–65 km, in complete accordance with the experimental data. Approximately the same ratio should hold for SO_2 and H_2O at the lower boundary. If f_{H_2O} should decrease at the lower boundary, the atmosphere in the upper portion of the clouds would

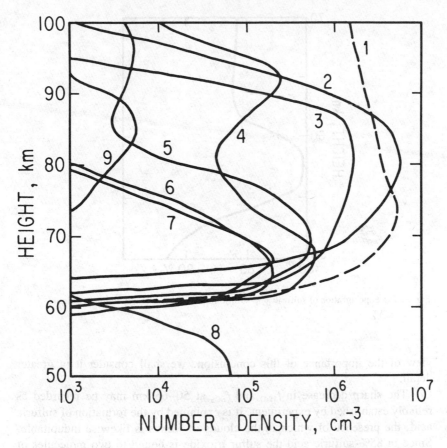

Fig. 7. Reactions of destruction (photolysis, curve 1) and formation of CO_2 (see text for reactions for curves 2 to 9).

be excessively "dry," the concentration of solution in the drops of sulfuric acid would be raised, and the photolysis of H_2O and the content of hydrogen in the exosphere would diminish. These effects would be reversed by an increase of f_{H_2O}. The sensitivity of the properties of the atmosphere to very small deviations of f_{H_2O} is exceedingly great; variations of several percent are sufficient for a complete discrepancy with the experimental data. The best agreement is observed when $f_{H_2O} = 1.9 \times 10^{-4}$ at 50 km, in excellent accordance with the spectroscopic measurements of Venera 11 and 12 (Moroz et al. 1979b). The results of the calculation of the quantities of H_2SO_4, its vertical profile (Fig. 3), the content of water vapor in the upper cloud layer, and the concentration of the solution of sulfuric acid (Fig. 8) are confirmed by observations, if we assume that the large particles of $r \simeq 4$ μm in the middle cloud layer consist of sulfuric acid or the products of its further transformation. In

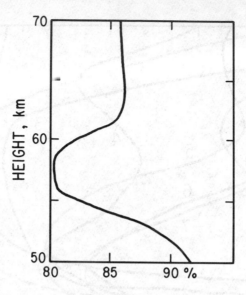

Fig. 8. The concentration of sulfuric acid solution in the aerosol as a function of height.

view of the importance of this conclusion, we shall consider it in greater detail.

The sharp decrease in f_{H_2O} and f_{SO_2} at 50–65 km may be regarded as reliably established by experiment. It is explained by the formation of sulfuric acid, the presence of which in the clouds of Venus is likewise indubitable. Since in 85% sulfuric acid the sulfur trioxide is bound to two molecules of water

$$\phi_{H_2O} + 2\,\phi_{H_2SO_4} = 0. \tag{7}$$

Considering that $\phi_{H_2O} = -K\,[CO_2]\,df_{H_2O}/dz$ and $\phi_{H_2SO_4} = -K\,[dn/dz + n/H1 - nV$ (where V is the Stokes velocity), and taking into account the experimentally measured weak dependence of the density of the aerosol particles on the altitude at 50–60 km, we may disregard the term dn/dz and obtain

$$n \approx -\frac{1}{2}\left[CO_2\right]\frac{df_{H_2O}}{dz}\bigg/\left(\frac{V}{K}+\frac{1}{H}\right). \tag{8}$$

If sulfuric acid is present only in the Mode 2 cloud particles, then $n = 5 \times 10^{12}$ cm^{-3} and, since all the other quantities are known, we obtain from this

$K \simeq 100$ cm^2 s^{-1}, which is absolutely unrealistic. If sulfuric acid and its products (we shall not consider the possible further transformations of sulfuric acid) comprise virtually the entire mass of the cloud layer in its upper and middle parts, then $n = 5 \times 10^{13}$ cm^{-3} and $K = 2 \times 10^4$ cm^2 s^{-1}. Upon further increase of K, the calculated mass of the cloud layer exceeds that of the particles, measured by spectrometer (Knollenberg and Hunten 1980). Values of $K = 2 \times 10^4$ cm^2 s^{-1} result in a discord between the calculated and measured profile of SO$_2$. We shall not discuss here the nature of the large particles in the lower portion of the cloud layer. The discovery of chlorine in the composition of the aerosol particles (Surkov et al. 1981) had suggested that AlCl$_3$ was the major component of the lower portion of the cloud layer (Krasnopol'sky and Parshev 1979a). Refined data as to the height of its lower boundary, density, and temperature (Knollenberg and Hunten 1980; Connes et al. 1967) excludes this hypothesis. However, there is yet another argument in favor of the presence of chlorides in the clouds of Venus; the ultraviolet spectrum of the clouds can be explained by the presence of a small ($\sim$1%) admixture of iron chloride FeCl$_3$ in the drops of sulfuric acid solution (Zasova et al. 1981) in the upper cloud layer. The presence of FeCl$_3$ in other portions of the cloud layer as well is not excluded.

Almost all the atomic oxygen produced by photolysis of CO$_2$ above 90 km forms molecules of O$_2$. The main processes of formation of O$_2$ are (Fig. 7)

$$O + ClO \rightarrow O_2 + Cl$$

$$O + O + CO_2 \rightarrow O_2 + CO_2$$

$$O + OH \rightarrow O_2 + H. \tag{p}$$

The destruction of O$_2$ is virtually complete in the reaction with COCl at altitudes of 65 to 90 km. As the source and sink of O$_2$ are situated at different heights, O$_2$ forms a layer with an overall content inversely proportional to the rate of mixing. In order not to exceed the spectroscopic limit of O$_2$ at 5×10^{19} cm^{-2} (Traub and Carleton 1974), K should be not less than 5×10^5 cm^2 s^{-1} at 90 km. Then the main elements of the dependence of $K(h)$ are obtained: 2×10^4 cm^2 s^{-1} from 50 to 60 km, 4×10^6 cm^2 s^{-1} in the upper atmosphere, and $K \geqslant 5 \times 10^5$ cm^2 s^{-1} between. At altitudes from 60 to 110 km, these values are in good correspondence with the dependence $K = A\, n^{-0.5}$, obtained by von Zahn et al. (1979).

Many processes of formation of O$_2$ produce excess energy, sufficient for excitation of the state of O$_2$ (a$^1\Delta_g$) which produces an intense emission of 1.27 μm (1.5 MR in the day and 1.2 MR at night [Connes et al. 1979]). A high production rate at altitudes above 85 km is possessed by the following reactions, which occur with spin preservation in the formation of O$_2$ (a$^1\Delta_g$)

$$O_3 + h\nu \rightarrow O_2 + O, k = 1.2 \times 10^{12} \, cm^{-2} \, s^{-1}$$

$$O_3 + Cl \rightarrow O_2 + ClO \qquad 1.3 \times 10^{12}$$

$$O + ClO \rightarrow O_2 + Cl \qquad 2.2 \times 10^{12}$$

$$O + O + CO_2 \rightarrow O_2 + CO_2 \qquad 3.5 \times 10^{11}$$

$$O + HO_2 \rightarrow O_2 + OH \qquad 3.2 \times 10^{11}$$

$$Cl + ClO_2 \rightarrow O_2 + OH \qquad 3.0 \times 10^{11}$$

$$O_3 + H \rightarrow O_2 + OH \qquad 1.4 \times 10^{11}$$

$$O + OH^* \rightarrow O_2 + H \qquad 2.5 \times 10^{11}. \qquad (q)$$

The last reaction is only possible for vibrationally excited molecules of OH*. Both processes of formation of OH result in its vibrational excitation, and a complete vibrational relaxation cannot take place during the lifetime of OH ($\tau \simeq 0.2$ s) at 100 km. The photochemical theory should account not only for the large intensity of the glow at 1.27 μm, but also for its small daily variation. Of all the enumerated reactions, only for the photolysis of ozone is the yield of $O_2 \, (^1\Delta_g) \, \gamma \simeq 1$ known. The large rates of transport of atomic oxygen from the day side to the night side in the upper atmosphere result in the fact that the concentrations of atomic oxygen and the rate of formation of ozone at ~ 100 km will be approximately identical at the night and day sides. It may be assumed that the sum of the yields of $O_2 \, (^1\Delta_g)$ in the second and third reactions is equal to unity, when each event of ozone destruction produces a single molecule of $O_2 \, (^1\Delta_g)$ on both the day and the night side. Furthermore, a selection of the constant of extinction of $O_2(^1\Delta_g)$ by molecules of CO_2 leads to agreement between the calculation and the measurements when $K_{CO_2} = 5 \times 10^{-20} \, cm^3 \, s^{-1}$. Taking into account the somewhat elevated intensities of sunlight assumed in our model from the measurements of Widing et al. (1970), we may recommend $k_{CO_2} = 3 \times 10^{-20} \, cm^3 \, s^{-1}$, which is in excellent agreement with the estimates of Noxon et al. (1976). The calculated profile of the glow of $O_2 \, (^1\Delta_g)$ is shown in Fig. 9.

For 50–100 km, the formation of active chlorine and hydrogen occurs as a result of the photolysis of HCl and the reaction of HCl and O, the destruction of which is largely produced by the processes of reactions 27 and 45 (see Table II in the Appendix following this chapter)

$$H + Cl_2 \rightarrow HCl + Cl$$

$$Cl + HO_2 \rightarrow HCl + O_2. \qquad (r)$$

It is interesting to note that, at identical rates of formation and destruction of H* and C*, their concentrations are [H*] << [Cl*]. The reason for this is the reaction 40

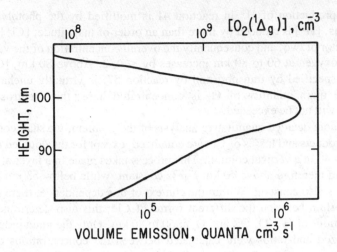

Fig. 9. Altitude dependence of the concentrations of O_2 ($^1\Delta_g$) and the 1.27 μm airglow.

$$OH + HCl \rightarrow H_2O + Cl \tag{s}$$

in which H* is destroyed and Cl* is produced. An excess of Cl*, equal to twice the rate of this process in the vertical column, is transported in the form of $COCl_2$ to the lower troposphere where at high temperature the reverse process will occur

$$COCl_2 + H_2O \rightarrow CO_2 + 2\,HCl. \tag{t}$$

The Influence of Some Reaction-Rate Constants on the Solution

After publishing our work on the photochemistry of Venus's atmosphere (Krasnopol'sky and Parshev 1980b,c, 1981b,c), we discovered that for certain reactions the values of the constants recommended in Baulch et al. (1980) differ from those used by us. The discrepancies are the most severe for reaction 40 $k_{40} = 3.0 \times 10^{-12} \exp(-425/T)$ instead of $2 \times 10^{-13} \exp(-310/T)$, reaction 41 $k_{41} = 10^{-11} \exp(-3370/T)$ instead of $1.8 \times 10^{-12} \exp(-2260/T)$, reaction 45 $k_{45} = 4 \times 10^{-11}$ instead of 10^{-13}.

It is interesting to analyze the influence of variations of k_{40}, k_{41} and k_{45} on the calculation results. Basically the situation can be described as follows: a reduction of k_{41} by an order of magnitude curtails the production of Cl* and H*, but an increase of k_{40} by $\sim$ 10 times preserves [Cl*], which is determined by the balance between the generation of Cl* in reaction 40 and its diffusion to the lower atmosphere. For the same reason, low concentrations of H* do not affect [Cl*] even when k_{45} increases by a factor of 400. The curtailed production of H* in reaction 41 is mollified by the photolysis of HCl. Thus, [H*] is reduced by more than an order of magnitude, [Cl*] by at least a factor of two, and consequently the overall concentration of the various forms of oxygen at 60 to 80 km increases by $\sim \sqrt{2}$. Above 80 km, [O_2] as basically specified by transport and by reaction 57, is virtually unchanged. Since the main quantity of O_2 is concentrated here, the spectroscopic threshold will not be exceeded.

In a more detailed quantitative analysis of the problem, we shall consider that the products and losses of Cl* are equalized, except for the doubled rate P of process 40 in a vertical column. This process takes place in a layer at 63 to 70 km, and therefore above 70 km f_{Cl*} is constant, while below 63 km [Cl*] $= P H/K$ is also constant. Within the context of this dependence, there exists an equilibrium between the different forms of Cl*; this also determines the concentrations of H*, O, O_2, and O_3 at 60 to 80 km. Only the main processes were utilized and ratios were calculated between the concentrations of the various components under the new and the old values of the reaction-rate constants. It turned out that, at heights of 60 to 80 km, the overall quantity of Cl* has been reduced by a factor of 1.75, that of O by 1.14, O_2 by 1.3, and O_3 by 1.5. Variations of [H*] are more complex in nature, as shown in Fig. 10.

The above remarks indicate that a change in the concentrations of chlorine and the oxygen-containing components is not significant, and thus the effectiveness of the catalytic and several other processes involving this change is preserved. Active hydrogen undergoes great variations, but its role in the atmospheric chemistry of Venus is small and our conclusions remain the same.

Are Considerable Quantities of O_2 Possible in the Cloud Layer of Venus?

The spectroscopic estimate of $f_{O_2} < 10^{-6}$ (Traub and Carleton 1974) pertains to a height of 62 to 65 km and does not contradict the gas chromograph measurements of Pioneer Venus (Oyama et al. 1980a), according to which $O_2 = 8 \times 10^{14}$ cm^{-3} at 52 and 42 km. If, for 50 km, we assign $f_{O_2} = 4 \times 10^{-5}$, then the calculation results will not contradict the spectroscopic threshold; the O_2 profile will be similar to the profiles of SO_2 and H_2O, and will possess a sharp drop at 55–65 km. The presence of molecular oxygen entails a reduction in the quantities of aerosol sulfur, due to the rapid oxidation of S and S_2. However O_2 cannot get into the lower atmosphere, where the concentrations of CO are large. A consideration of the photochemical processes in the clouds and atmosphere of Venus below the clouds likewise eliminates the upper atmosphere as a source of O_2 in the cloud layer. In order to explain the measurements of O_2, let us consider one possibility, no matter how unrealistic. Let us assume that the reaction of COCl with O_2 results, not in the formation of CO_2 and ClO, but in the formation of the complex $COClO_2$, the binding energy of which is sufficient to withstand collisions at $T \lesssim 200$ K, but not sufficient for $T \lesssim 300$ K, which is typical for the middle

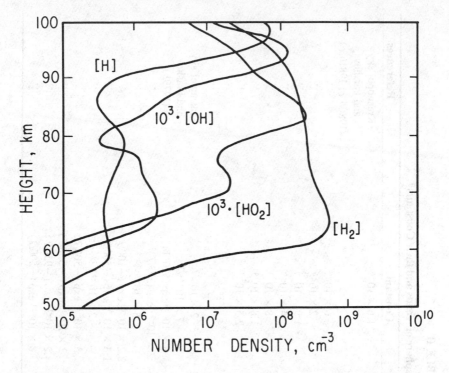

Fig. 10. Concentrations of H, H$_2$, OH, HO$_2$ when using the constants of the reactions from Baulch et al. (1980).

layer of clouds. Then a conservation of O$_2$ at 90–75 km would be possible, as well as transport and liberation in the middle and lower layers of clouds. This hypothesis is embarrassed by two circumstances: first, it is possible for O$_2$ to be liberated, not only in collisions, but also by the photolysis of a component like COClO$_2$; second, the capture of O$_2$ is accompanied by the capture of chlorine, which is only liberated in the lower atmosphere. A transport of chlorine atoms to the overlying layers would be accompanied by a transport of O$_2$, and then either O$_2$ would accumulate rather considerably at 65 km and the spectroscopic threshold would be exceeded, or chlorine would remain very scarce, since the overall rate of photolytic formation of Cl* is much less than the rate of photolytic formation of O$_2$. Thus this explanation is also compromised, and the appearance of O$_2$ in the cloud layer is not understood by us.

APPENDIX

TABLE II

Chemical Reactions in the Atmosphere of Venus and their Constants[a]

No.	Reaction	Constant	References
1	$CO_2 + h\nu \rightarrow CO + O$ (1675 $< \lambda <$ 2075 Å)	1.6×10^{-6}	Krasnopol'sky and Parshev (1980b,c; 1981b,c)
	$CO + O(^1D)$ ($\lambda <$ 1675 Å)		
2	$H_2O + h\nu \rightarrow OH + H$	6×10^{-6}	,,
3	$HCl + h\nu \rightarrow H + Cl$	2×10^{-6}	,,
4	$SO_2 + h\nu \rightarrow SO + O$	1.9×10^{-4}	,,
5	$O_2 + h\nu \rightarrow O + O$	4×10^{-6}	,,
6	$O_3 + h\nu \rightarrow O_2(^1\Delta_g) + O(^1D)$	8×10^{-3}	,,
7	$H_2O_2 + h\nu \rightarrow OH + OH$	1.8×10^{-4}	,,
8	$Cl_2 + h\nu \rightarrow Cl + Cl$	2.5×10^{-3}	,,
9	$SO_2 + h\nu \rightarrow {}^3SO_2$ (3200 $< \lambda <$ 3900 Å)	—	,,
10	$SO_2 + h\nu \rightarrow {}^1SO_2$ (2180 $< \lambda <$ 3200 Å)	—	,,
11	$CO + O + CO_2 \rightarrow CO_2 + CO_2$	2×10^{-37}	Krasnopol'sky and Prashev (1979b)
12	$O + O + CO_2 \rightarrow O_2 + CO_2$	$7 \times 10^{-31}/T$	,,
13	$O + O_2 + CO_2 \rightarrow O_3 + CO_2$	$3 \times 10^{-34} \exp(520/T)$	,,
14	$H + O_2 + CO_2 \rightarrow HO_2 + CO_2$	$2 \times 10^{-31} \exp(273/T)^{1.3}$	,,
15	$O + HO_2 \rightarrow OH + O_2$	2.5×10^{-11}	,,
16	$CO + OH \rightarrow CO_2 + H$	$2.2 \times 10^{-13} \exp(-100/T)$	,,
17	$H + O_3 \rightarrow OH + O_2$	$1.3 \times 10^{-10} \exp(-500/T)$	,,
18	$O + OH \rightarrow O_2 + H$	2×10^{-11}	,,
19	$HO_2 + HO_2 \rightarrow H_2O_2 + O_2$	$3 \times 10^{-11} \exp(-500/T)$	,,
20	$H + HO_2 \rightarrow H_2 + O_2$	$2 \times 10^{-11} \exp(-350/T)$	,,
21	$H + HO_2 \rightarrow OH + OH$	$4.2 \times 10^{-10} \exp(-950/T)$	,,
22	$H + HO_2 \rightarrow H_2O + O$	$8 \times 10^{-11} \exp(-500/T)$	,,

TABLE II (continued)

Chemical Reactions in the Atmosphere of Venus and their Constants[a]

No.	Reaction	Constant	References
23	$HO_2 + OH \rightarrow H_2O + O$	5×10^{-11}	"
24	$H + CO_2 \rightarrow H_2 + CO_2$	2.6×10^{-32}	Sze and McElroy (1975)
25	$Cl + CO_2 \rightarrow Cl_2 + CO_2$	2.6×10^{-32}	"
26	$H_2 + Cl \rightarrow HCl + H$	$8 \times 10^{-11} \exp(-2480/T)$	Kondrat'yev (1971)
27	$H + Cl_2 \rightarrow HCl + Cl$	$3 \times 10^{-10} \exp(-1000/T)$	Sze and McElroy (1975)
28	$HCl + H \rightarrow H_2 + Cl$	$10^{-11} \exp(-1600/T)$	Zahniser and Kaufman (1977)
29	$H + Cl + CO_2 \rightarrow HCl + CO_2$	2.6×10^{-32}	Sze and McElroy (1975)
30	$O_3 + Cl \rightarrow ClO + O_2$	$2.2 \times 10^{-11} \exp(-170/T)$	"
31	$Cl + O_2 + CO_2 \rightarrow ClO_2 + CO_2$	1.7×10^{-33}	Zahniser and Kaufman (1977)
32	$ClO_2 + Cl \rightarrow Cl_2 + O_2$	1.5×10^{-10}	Sze and McElroy (1975)
33	$ClO_2 + Cl \rightarrow ClO + ClO$	1.5×10^{-12}	"
34	$O + ClO \rightarrow O_2 + Cl$	$3.4 \times 10^{-11} \exp(75/T)$	"
35	$O_3 + ClO \rightarrow ClO_2 + O_2$	2×10^{-15}	"
36	$ClO + ClO \rightarrow Cl_2 + O_2$	2.3×10^{-14}	Sze and McElroy (1975)
37	$ClO + ClO \rightarrow ClO_2 + Cl$	10^{-31}	"
38	$ClO + CO_2 \rightarrow O_2 + Cl_2 + CO_2$	$1.2 \times 10^{-12} \exp(-1260/T)$	"
39	$O + Cl_2 \rightarrow ClO + Cl$	$10^{-20} \exp(-1560/T)$	"
40	$OH + HCl \rightarrow H_2O + Cl$	$2 \times 10^{-13} \exp(-310/T)$	"
41	$O + HCl \rightarrow OH + Cl$	$1.8 \times 10^{-12} \exp(-2260/T)$	"
42	$ClO_2 + CO_2 \rightarrow Cl + O_2 + CO_2$	$4 \times 10^{-12} \exp(-2500/T)$	"
43	$CO + ClO \rightarrow CO_2 + Cl$	1.7×10^{-15}	"
44	$CO + ClO_2 \rightarrow CO_2 + ClO$	10^{-15}	"
45	$Cl + HO_2 \rightarrow HCl + O_2$	10^{-13}	"

TABLE II (continued)

Chemical Reactions in the Atmosphere of Venus and their Constants[a]

No.	Reaction	Constant	References
46	$Cl + HO_2 \rightarrow OH + ClO$	10^{-13}	"
47	$OH + H_2 \rightarrow H_2O + H$	$7 \times 10^{-12} \exp(-2000/T)$	"
48	$O + H_2 \rightarrow OH + H$	$5 \times 10^{-11} \exp(-5100/T)$	"
49	$OH + O_3 \rightarrow 2\,O_2 + H$	$1.3 \times 10^{-12} \exp(-950/T)$	"
50	$HO_2 + O_3 \rightarrow 2\,O_2 + OH$	$1.3 \times 10^{-12} \exp(-820/T)$	"
51	$O + O_3 \rightarrow O_2 + O_2$	$1.3 \times 10^{-11} \exp(-2140/T)$	"
52	$Cl + H_2O_2 \rightarrow HCl + HO_2$	8×10^{-13}	Watson et al. (1976)
53	$CO_2^+ + H_2 \rightarrow HCO_2^+ + H$	10^{-9}	Sze and McElroy (1975) (estimate)
54	$CO_2^+ + HCl \rightarrow HCO_2^+ + Cl$	10^{-9}	
55	$HCO_2^+ + e^- \rightarrow CO_2 + H$	10^{-6}	Sze and McElroy (1975)
56	$CO + Cl + CO_2 \rightarrow COCl + CO_2$	3×10^{-33}	Kondrat'yev (1971)
57	$COCl + O_2 \rightarrow CO_2 + ClO$	2×10^{-14}	"
58	$COCl + O \rightarrow CO_2 + Cl$	10^{-13}	"
59	$COCl + Cl_2 \rightarrow COCl_2 + Cl$	$6 \times 10^{-13} \exp(-1400/T)$	"
60	$COCl + Cl \rightarrow CO + Cl_2$	$6 \times 10^{-14} \exp(-400/T)$	"
61	$COCl + CO_2 \rightarrow CO_2 + CO + Cl$	$10^{-12} \exp(-3200/T)$	"
62	$COCl + H \rightarrow CO + HCl$	10^{-13}	"
63	$COCl_2 + O \rightarrow CO_2 + Cl_2$	10^{-14}	"
64	$S + O_2 \rightarrow SO + O$	2.2×10^{-12}	Baulch et al. (1976)
65	$SO + OH \rightarrow SO_2 + H$	1.2×10^{-10}	"
66	$SO_2 + HO_2 \rightarrow SO_3 + OH$	3×10^{-16}	"
67	$SO_2 + OH + CO_2 \rightarrow HSO_3 + CO_2$	1.6×10^{-31}	"
68	$SO_2 + O + CO_2 \rightarrow SO_3 + CO_2$	10^{-32}	"
69	$SO + O \rightarrow SO_2$	$2.5 \times 10^{-16} \exp(300/T)$	"

TABLE II (continued)

Chemical Reactions in the Atmosphere of Venus and their Constants[a]

No.	Reaction	Constant	References
70	$SO + O_2 \rightarrow SO_2 + O$	$7.5 \times 10^{-13} \exp(-3250/T)$	''
71	$SO + O_3 \rightarrow SO_2 + O_2$	$2.5 \times 10^{-12} \exp(-1050/T)$	Kondrat'yev (1971)
72	$SO + O + CO_2 \rightarrow SO_2 + CO_2$	5×10^{-31}	''
73	$SO + SO \rightarrow SO_2 + S$	3×10^{-15}	''; Baulch et al. (1976)
74	$S + O + CO_2 \rightarrow SO + CO_2$	2.5×10^{-31}	(estimate)
75	$S + CO_2 \rightarrow S_2 + CO_2$	1.5×10^{-30}	Baulch et al. (1976)
76	$S + HO_2 \rightarrow SO + OH$	2.5×10^{-11}	(estimate)
77	$SO_3 + H_2O \rightarrow H_2SO_4$	10^{-12}	Heicklen (1976)
78	$OH + HSO_3 \rightarrow H_2SO_4$	10^{-13}	(estimate)
79	$S_n \text{ (aerosol)} + O \rightarrow S_{n-1} + SO \text{ (aerosol)}$	10^{-16}	Krasnopol'sky and Parshev (1980b,c; 1981b,c)
80	$H_2O_2 + SO_3 \rightarrow H_2SO_4 + O$	10^{-12}	(estimate)
81	$SO + O_2 + CO_2 \rightarrow SO_3 + CO_2$	$3 \times 10^{-32} \exp(-3250/T)$	Baulch et al. (1976)
82	$SO_2 + OH \rightarrow HSO_3$	5×10^{-13}	''
83	$SO + O_2 \rightarrow SO_3$	$10^{-12} \exp(-3250/T)$	''
84	$ClO + SO_2 \rightarrow SO_3 + Cl$	10^{-15}	(estimate)
85	$ClO_2 + SO_2 \rightarrow SO_3 + ClO$	10^{-15}	(estimate)
86	$HSO_3 + Cl \rightarrow HCl + SO_3$	10^{-13}	(estimate)
87	$S + OH \rightarrow SO + H$	2×10^{-11}	(estimate)
88	$O(^1D) + CO_2 \rightarrow O + CO_2$	1.8×10^{-10}	Krasnopol'sky and Parshev (1979b)
89	$O(^1D) + H_2O \rightarrow OH + OH$	3×10^{-10}	''

TABLE II (continued)

Chemical Reactions in the Atmosphere of Venus and their Constants[a]

No.	Reaction	Constant	References
90	$O(^1D) + H_2 \rightarrow OH + H$	3×10^{-10}	
91	$O(^1D) + HCl \rightarrow OH + Cl$	3×10^{-10}	(estimate)
92	$O(^1D) + CO_2 \rightarrow CO + O_2(^1\Delta_g)$	3×10^{-11}	Krasnopol'sky and Parshev (1980b,c; 981b,c)
93	$O(^1D) + CO \rightarrow O + CO$	8×10^{-11}	L. Anderson (1976)
94	$O_2(^1\Delta_g) + CO_2 \rightarrow O_2 + CO_2$		(cf. text)
95	$^1SO_2 \rightarrow SO_2 + h\nu$	$\gamma = 0.1$	b
	$^3SO_2 + h\nu$		
96	$^1SO_2 + CO_2 \rightarrow SO_2 + CO_2$	$\gamma = 0.1$	b
	$^3SO_2 + CO_2$		
97	$^1SO_2 \rightarrow SO_2$	$\gamma = 0.1$	b
	$^3SO_2$		
98	$^3SO_2 \rightarrow SO_2 + h\nu$	1.1×10^3	b
99	$^3SO_2 + CO_2 \rightarrow SO_2 + CO_2$	5×10^{-14}	b
100	$^3SO_2 + SO_2 \rightarrow SO_3 + SO$	$5 \times 10^{-11} \exp(-1370/T)$	b
101	$^3SO_2 + CO \rightarrow CO_2 + SO$	$8 \times 10^{-12} \exp(-2000/T)$	b
102	$^3SO_2 + O_2 \rightarrow SO_3 + O$	$5 \times 10^{-11} \exp(-1370/T)$	b

[a]Rates of photolysis given in s^{-1} for an optically thin atmosphere for average global conditions (half of the daily values). Constants have dimensions for the second and third orders, $cm^3\ s^{-1}$ and $cm^6\ s^{-1}$.

[b]Rao et al. (1969a,b); Otsuka and Calvert (1971); Jackson and Calvert (1971); Sidebottom et al. (1971); Strickler et al. (1974).

15. VENUS SPECTROSCOPY IN THE 3000–8000 Å REGION BY VENERAS 9 AND 10

V. A. KRASNOPOL'SKY

USSR Academy of Sciences

The grating spectrometers aboard Venera 9 and 10 orbiters provided data in several areas. Four band systems of O_2 were observed in the nightglow, and upper limits were set for other emissions, notably at 5577 and 5893 Å ([OI] and Na). Comparison with Earth, Mars, and laboratory data provides information on excitation processes and an upper limit to electron precipitation. Observations of the sunlit limb give information on high-altitude haze. Sporadic events included one interpreted as a large thunderstorm and a detection of a "dust ring" extending up to 700 km.

Veneras 9 and 10 carried grating spectrometers for the spectral region 3000–8000 Å. The instruments had a spectral resolution of 20 Å; the field of view was 3.6 × 12′. During the observations, 12′ on the disk of the planet corresponds to 15–20 km. The sensitivity threshold of the instruments at room temperature is 30 R (Rayleigh), which is equivalent to an albedo of $R = \pi I/I_0 = 4 \times 10^{-9}$. Unfortunately, the temperature of the instrument rose quickly from 20° C at the outset to 60–70° C at the end of the first month of operation of the orbiters, and the main portion of the observations were carried out with hot instruments, the sensitivity threshold of which was roughly 5 times worse. The scanning time was usually 10 s and sometimes 2.5 s. The motion of the field of view on the disk was 15–20 km per 2.5 s. The dynamic range of the instrument was 3×10^4 in the basic mode. For observations of the day side, the spectral range was 2700–7000 Å, the resolution 50 Å, the field of view 3.5, and the scanning time 2.5 s.

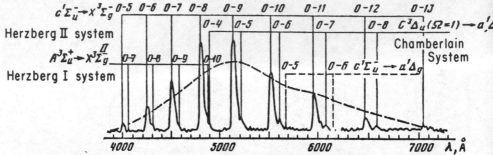

Fig. 1. Spectrum of the Venus nightglow. The positions of the four systems of O_2 bands are indicated. The dashed curve shows the instrument sensitivity as a function of wavelength.

I. THE NIGHTGLOW OF VENUS'S ATMOSPHERE IN THE VISIBLE REGION

The Spectrum

The spectrum of the glow of Venus is shown in Fig. 1. It was obtained by averaging 50 spectra, measured on the disk of the planet on 21 Feb. 1975 aboard Venera 9. During this series of observations, the zenith angle of the line of observation was small, and therefore the measurement conditions varied little from spectrum to spectrum.

Spectral identifications were made with the help of the laboratory experiments of Lawrence et al. (1977), Slanger and Black (1978), and Kenner et al. (1979). Four systems of O_2 bands are found, each forbidden and represented only by zero progressions, due to the rapid vibrational relaxation in the atmosphere of CO_2. The results are summarized in Table I, which also includes expectations for the undetected Herzberg III system. Relative intensities are from Slanger (1978a) and Degen (1977). The spectrum is dominated by the nine bands seen in Fig. 1 and two more (3562 Å and 3761 Å) seen only once. The other systems are more difficult to pick out, but seem to be present.

The ratio between the intensities of the systems $c^1\Sigma_u^- \to X^3\Sigma_g^-$ and $c^1\Sigma_u^- \to a^1\Delta_g$ is equal to 30 and should not depend on the conditions of excitation. According to Slanger (1978a), it equals 10, although the estimation of the intensity of $c^1\Sigma_u^- \to a^1\Delta_g$ in the Slanger spectra depends greatly on the adopted value of the continuous background, which varies with the wavelength. In the spectrum of Fig. 1 it is obvious that the ratio of $I_{6151 Å}/I_{5945 Å}$ is $< 1:7$, which would have been the case for a ratio of 1:10 of the intensity of the systems, and comprises $\sim 1:20$. If, in the Slanger spectrum, the background value is taken from the shortwave end of the band 5945 Å, then a value close to our result is obtained.

Various estimates have been made of the radiative lifetime τ of the Herzberg II system: 10^4 s (Krupenie 1972); 100 s (Krasnopol'sky et al. 1976b); 25–50 s for $v'' = 5$ or 0 (Slanger 1978a). Our results then give $\tau_0 = 1200$ s for $v' = 0$ of $c^1\Sigma_u^- \to a^1\Delta_g$.

TABLE I

O₂ Bands in the Nightglow of Venus in the Spectral Region 3000–8000 Å[a]

Transition Name τ_0, s	$c^1\Sigma_u^-\to X^3\Sigma_g^-$ Herzberg II 50			$c^1\Sigma_u^-\to a^1\Delta_g$ 1200			$A^3\Sigma_u^+\to X^3\Sigma_g^-$ Herzberg I 0.25			$A'^3\Delta_{u_1}\to a^1\Delta_g$ Chamberlain 50			$A'^3\Delta_{u_2}\to X^3\Sigma_g^-$ Herzberg III 100		
	v'v''	λ(Å)	4πI(R)	v'v''	λ(Å)	4πI(R)	v'v''	λ(Å)	4πI(R)	v'v''	λ(Å)	4πI(R)	v'v''	λ(Å)	4πI(R)
	0.3	3562	28							0.3	4553	2			
	0.4	3761	50				0.4	3457	4	0.4	4863	8			
	0.5	3980	145	0.5	5676	10	0.5	3641	8	0.5	5212	21		(Predicted)	
	0.6	4222	267	0.6	6151	16	0.6	3842	15	0.6	5606	31			
	0.7	4491	393	0.7	6695	20	0.7	4064	22	0.7	6055	36			
	0.8	4792	467				0.8	4308	25	0.8	6572	33	0.8	4427	
	0.9	5129	449				0.9	4579	24	0.9	7172	22	0.9	4713	
	0.10	5510	374				0.10	4880	19				0.10	5032	
	0.11	5945	276				0.11	5219	13						
	0.12	6445	186				0.12	5600	7						
	0.13	7026	65												
Total (R)			2700			100			140			200			30

[a] The radiative lifetimes of v' = 0 are shown as τ_0, and intensities are given in Rayleighs (R).

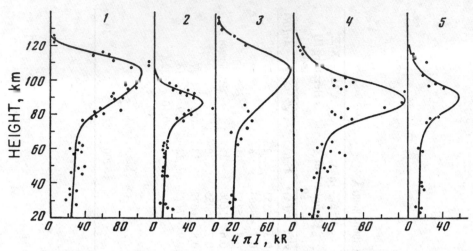

Fig. 2. Limb profiles of Herzberg II intensity for the five times indicated in Table II. Curve fits of Eq. (1) are shown by the lines.

The Chamberlain and Herzberg III systems emit from two nearby sublevels $\Omega = 1$ and $\Omega = 2$ of a single electronic state, differing in energy by $\Delta E = 120$ cm^{-1}. Under thermodynamic equilibrium, the ratio between the populations is $e^{-\Delta E/kT}$; allowing for $\tau_2 \simeq 2\tau_1$, we obtain ~ 30 R for the Herzberg III system. Its most intense band at 5032 Å should have an intensity of 5 R, below the threshold of detection.

As indicated in Fig. 1, the Chamberlain and Herzberg I systems can be made out in different wavelength regions (above and below 4700 Å, respectively) as small shoulders on the strong Herzberg II bands.

Altitude Profiles of the Emission

Figure 2 shows the intensity of the Herzberg II system as a function of the altitude above the limb. The instrument slit was parallel to the surface. Each point corresponds to a single measured band. The precision of the absolute height was ~ 20 km. It was found, as described below, from the cutoff on the bottom of the profile, attributed to extinction by atmospheric aerosols. This extinction was obtained from dayside measurements (Sec. III): optical depth is 1 at 83 km and the scale height is 3 km between 80 and 100 km. For the night side, unit optical depth was assumed to be at 78 km.

In calculating the profile of the emission on the nocturnal limb of the planet, it was assumed that $\tau = 1$ at 78 km. The effects of multiple scattering in the cloud layer were included by introducing a Lambert surface with albedo characteristic of Venus at 65 km. The phase function of the scattering of particles of the underlying haze with a dimension of 0.2 to 0.3 μm, was calculated from Mie theory and averaged over the half-space. The specific volume glow of the atmosphere is given in the form

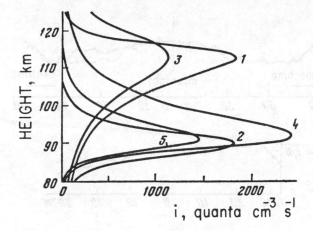

Fig. 3. The volume luminosities of the Herzberg II bands of O_2.

$$i(h) = \frac{2A}{e^{-\frac{h-h_0}{H_1}} + e^{\frac{h-h_0}{H_2}}} \tag{1}$$

and determined by four parameters: h_0 and A, the altitude of the maximum and the amount of glow in it; H_1 and H_2, the slope of the upper and lower portions of the layer. The solid lines in Fig. 2 indicate the best-fitting curves, with suitable averaging over the field of view. Figure 3 shows the resulting volume luminosities, while Table II gives the parameters of Eq. (1). The emission is concentrated at 85 to 120 km, the typical thickness of the layer is 15 km, and the zenith intensity is 2.5 kR. The variations in the height, shape of the layer and intensity of the emission are rather significant. Since the altitude was referred to the level of unit optical depth, which was assumed to

TABLE II
Parameters of Eq. (1) for the Herzberg II Band System

Date	Clock Angle t	Lati- tude φ	A Quanta $cm^{-3}s^{-1}$	h_0 (km)	H_1 (km)	H_2 (km)	Thick- ness, Δ (km)	$4\pi I$ (kR)	$4\pi I^{*a}$ (kR)
27 Nov. 1975	74	−19	1450	115	10	2	21	3.0	3.1
29 Nov. 1975	64	−33	1800	90	3	3	6	1.2	1.5
9 Dec. 1975	− 6	− 8	1100	115	10	5	24	2.4	3.4
11 Jan. 1976	13	− 4	2000	90	2	8	17	3.4	3.2
9 Feb. 1976	45	4	1300	90	2	5	11	1.5	2.1

[a] $4\pi I^*$ is the intensity of glow according to curves in Figs. 4 and 5.

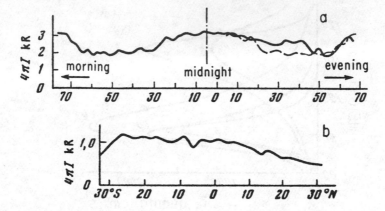

Fig. 4. (a) The intensity of emission as a function of the hour angle t. The dashed line represents the symmetrical continuation of the curve from the midnight maximum to the evening terminator. (b) Latitude dependence of the intensity of emission (curve normalized to unity).

equal 78 km, the actual deviations of this level from 78 km were incorporated in the error of determining the height.

The distribution of intensity over the night hemisphere was obtained by combining the data from 25 passes. It was assumed that the variations with solar hour angle t and latitude ϕ were independent. The results are shown in Fig. 4. In Fig. 4a there is a prominent variation with a period of 90°, displaced 5° toward the morning. The evening part of the curve has been folded over to the morning side for comparison. Atomic oxygen must be transported from the day side either by an upward continuation of the 4-day circulation, or by a symmetrical system of winds blowing from the terminator to the antisolar point. The first produces an asymmetrical component and the second a symmetrical component with maximum near midnight. As can be seen from Fig. 4, the transport is primarily across the terminator to the antisolar region.

We assume that the upper state of the Herzberg II bands is produced by association of two O atoms and quenched by collision with O. The emission is then proportional to [O] and variations of [O] at the maximum around 100 km should be a factor of 1.5 from $t \simeq 0$ to 45° with subsequent increase when $t > 45°$. These are less than the variations of [O] at 100 km calculated by Dickinson and Ridley (1977). A midnight value of [O] less than the calculated should lead to a colder atmosphere in this region, since there is less energy liberated here by adiabatic compression of the flow of oxygen and by the formation of O_2. In fact, the night atmosphere is much colder than predicted by the calculations.

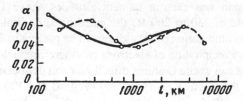

Fig. 5. The correlation of the O_2 emission. Dashed line represents analogous data for the atmosphere of Earth (Krasnopol'sky 1970).

The 5° displacement may be compared with 30° for the NO emission (Stewart et al. 1980), 15° for the temperature minimum from Pioneer Venus drag data (Keating et al. 1980), and the absence of a shift in the mass-spectrometric measurements (Niemann et al. 1980b). It should be noticed that the temperature measurement refers to altitudes of 150 to 160 km, while the NO emission is formed to 110 to 115 km and the O_2 emission at 100 km. This information is useful to clarify the variations of density and temperature in the interval 100–150 km.

The diurnal variation contains a noticeable wave structure with a period of 10 to 12°. Conceivably it is significant that the ratio between the wave period and 360° is close to the ratio between the period of circulation and the solar day (4/117). Within the observed latitude range of ±33° the variation is substantial (Fig. 5). It is asymmetric about the equator; the region of maximum values has a width of ∼30° with a center at 12°S.

Fluctuations in the O_2 emission can be described by the correlation function shown in Fig. 5, which contains also data for Earth from Kosmos-92 in the 2500–3150 Å range (Krasnopol'sky 1970). The measure for local fluctuations with a dimension between l_1 and l_2 is the mean relative size of the modulus of the difference between the intensity curve along the measurement route, averaged within limits l_1 and l_2. It should be proportional to $|I_1 - I_2|$ $(l_1 l_2)^{-1/2} \simeq \Delta I/I$. The coefficient of proportionality $\alpha(l)$ is a function of the size distribution of the fluctuations. The differences between the curves in Fig. 5 are slight; they demonstrate the closeness of the eddy mixing coefficients at 100 km for both planets. On Earth, $K \simeq 10^6$ cm^2 s^{-1} at 100 km (Hunten 1975a), and consequently the identical value should also be obtained on Venus. This value of K is in excellent agreement with our estimates from the aerosol distribution at 70 to 100 km (cf. Sec. III), from a photochemical analysis of a model of the atmospheric composition (Krasnopol'sky and Parshev 1980b, 1981c), and with the mass-spectrometer data of von Zahn et al. (1980).

Upper Limits for Other Emissions

The absence of detectable intensities at 5893, 5577, and 3914 Å set upper limits on the electron flux and other processes (Krasnopol'sky 1978, 1979a).

No 5577 Å emission was seen at tangent altitudes above 120 km. For a sensitivity threshold of 150 to 200 R, the corresponding limit on the zenith intensity is 4 R. Recombination of O_2^+, known to be present in the ionosphere, should give ~ 1 R. Precipitation of electrons can therefore yield no more than 3 R. Such electrons can excite O atoms directly, or can dissociate CO_2 into CO and excited O. If the electron energy is between 30 and 100 eV, maximum excitation occurs at 140 km, where CO_2 and O each have densities of 2×10^9 cm^{-3} (Niemann et al. 1980b). This is also the height of the nightside ionospheric peak (Chapter 23 by Brace et al.). The upper limit on the incident energy flux in the soft electrons is found to be 2×10^9 eV cm^{-2} s^{-1}, after allowance for the fact that $\sim 40\%$ are reflected. This value is 5 times greater than our original estimate due to the use of a more realistic model of the atmosphere, in which there are large quantities of atomic oxygen in addition to CO_2, and by taking into account more precise, recent data concerning the effectiveness of excitation of $O(^1S)$ (Fox and Dalgarno 1979).

If the maximum flux of electrons is realized, it may be sufficient to explain the upper peak of the nightside ionosphere. For electrons of 30 to 100 eV, the energy loss per ion pair is 40 eV (Fox and Dalgarno 1979), and when the thickness of the layer is 10 km and the coefficient of recombination is 2×10^{-7} cm^3 s^{-1} we obtain an electron concentration at the maximum of $n_e \lesssim 1.5 \times 10^4$ cm^{-3}. During the operation of Veneras 9 and 10, a typical value of n_e at the upper maximum was 7×10^3 cm^{-3} with variations in both directions.

Direct measurements from Veneras 9 and 10 gave electron fluxes of $\sim 2 \times 10^8$ cm^{-2} s^{-1}, or 10^{10} eV cm^{-2} s^{-1} for the mean energy of 50 eV (Gringauz et al. 1979; Chapter 27 by Gringauz). Such fluxes were shown able to sustain the nightside ionosphere. Our original estimate was smaller by a factor of 25, now reduced to 5. A direct comparison with Pioneer Venus data may not be valid because of the large difference in the levels of solar activity during the time of the experiments.

Upper limits were obtained for the twilight glow of N_2^+ 3914 Å (not more than 50 R for the solar depression angle $\beta = 3°5$) and 5577 Å (not more than 30 R when $\beta = 60°$). The interpretation incorporated a model of the atmosphere which differed greatly from reality ($T_\infty = 300$ K instead of 200 K in accordance with the measurements at the terminator by Pioneer Venus). Due to this, the estimation of the content of N_2 in the lower atmosphere turned out to be too small (a mixing ratio of $\lesssim 1\%$). It is not difficult to show that the upper limits given are not in contradiction with current information on the atmosphere.

From the absence of the corresponding emissions in the twilight glow, upper limits were also obtained for the content of certain atoms and ions of meteoric origin. The most interesting is the estimate for sodium: its content ($N < 4 \times 10^8$ cm^{-2}) and the nightglow of 5896 Å ($4\pi I < 20$ R) are smaller than in the atmosphere of Earth by an order of magnitude. The absence of the nightglow of sodium is explained by the smaller O density on Venus com-

TABLE III

Oxygen Nightglow in the Atmosphere of Earth, Venus and Mars

Emission	Earth $4\pi I$ (R)	Earth h_{max} (km)	Earth i_{max} quanta $cm^{-3}s^{-1}$	Venus $4\pi I$ (R)	Venus h_{max} (km)	Venus i_{max} quanta $cm^{-3}s^{-1}$	Mars $4\pi I$ (R)
O($^1S - {}^1D$) 5577Å Oxygen green line	120	97	120	5	—	—	$\leqslant 1$
$O_2(A^3\Sigma_u^+ \to X^3\Sigma_g^-)$ Herzberg I system	400	96	350	140	—	—	—
$O_2(A'\,^3\Delta_u \to a\,^1\Delta_g)$ Chamberlain system	100	—	—	200	—	—	—
$O_2(c^1\Sigma_u^- \to X^3\Sigma_g^-)$ Herzberg II system	100	—	—	2700	100	1800	$\leqslant 30$
$O_2(f^1\Sigma_g^+ \to X^3\Sigma_g^-)$ Atmospheric system	6000	94	6000	<200	—	—	—
$O_2(a^1\Delta_g \to X^3\Sigma_g^-)$ IR-atmospheric system	5×10^4	97	5×10^4	1.2×10^6	—	—	—

pared to Earth. The reason for the smaller quantities of free sodium on Venus is probably its rapid chemical bonding in the atmosphere of carbon dioxide.

II. EXCITATION OF OXYGEN EMISSIONS IN THE NIGHTGLOW OF EARTH, VENUS AND MARS

It is interesting to compare the results of measurements of oxygen emissions in the nightglow on Venus and similar data for Earth and Mars (Table III). An attempt of this type was undertaken (Krasnopol'sky 1981), but since that time new data have been published which should be included in the analysis. Especially important is the discovery of the Herzberg II bands in the nightglow of the Earth. An estimate has been made of their intensity and of the bands of the Chamberlain system (Slanger and Huestis 1981), and new information has appeared on the absence of quenching of O(^1S) by atomic oxygen (Slanger 1981a) (see Fig. 6).

In addition to the four systems of O_2 bands in our spectra, a glow of the infrared atmospheric system of O_2 ($a^1\Delta_g \to X^3\Sigma_g^-$) of 1.27 μm was detected on Venus and measured by Connes et al. (1979); it comprises 1.5 and 1.2 MR on the day and night sides of Venus. The glow of the atmospheric system of O_2 ($b^1\Sigma_g^- \to X^3\Sigma_g^-$), according to our data, does not exceed 200 R; it is subject to very intense quenching in CO_2 ($k \simeq 3 \times 10^{-13}$ cm^3 s^{-1}) (Wayne 1973). The upper limits of O_2 emission in the atmosphere of Mars were obtained

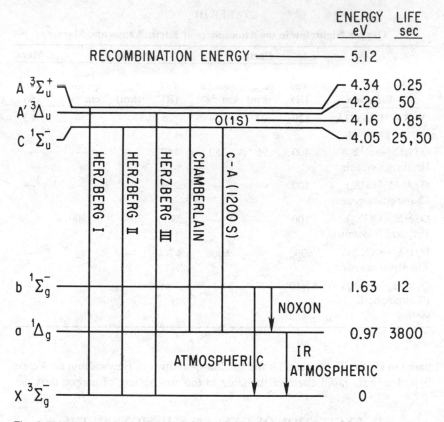

Fig. 6. The six lowest levels of O_2 and the band systems discussed.

from the results of measurements aboard the Mars 5 orbiter (Krasnopol'sky and Krysko 1976).

Data on the emission of the atmosphere of Earth has been cited from the works of Witt et al. (1979), Thomas et al. (1979), Evans et al. (1972), Slanger and Huestis (1981). The intensity of the Herzberg I bands was reduced by allowing for the presence of the emission of the Herzberg II and Chamberlain bands in the near ultraviolet region in a ratio of 4:1:1 (Slanger and Huestis 1981). The emission of $O_2(a-x)$ consists of two layers of roughly identical intensity (Evans et al. 1972); the lower layer at 87 km has a different source and is not considered here. Due to the lack of data on the altitude profile of the emission of $O_2(A)$ and $O_2(A')$ on Venus and $O_2(A')$ and $O_2(c)$ on Earth, we shall consider that they are similar to the profile of $O_2(c)$ and $O_2(A)$, respectively.

Table IV gives data on the compositions of the atmospheres at the maximum of the luminous layer. On Earth, the emission intensities and the concentrations of atomic oxygen have their maxima at similar heights, and

TABLE IV

Composition of the Atmosphere in the Luminous Layer[a]

Components	Earth, 96 km, 190K	Venus, 100 km, 170K	Mars, 35 km, 165K
$[M]^b$, cm^{-3}	2×10^{13}	1.5×10^{15}	5×10^{15}
$[O]$, cm^{-3}	7×10^{11}	2×10^{11}	2×10^{10}

[a]Data on Venus and Mars from Krasnopol'sky and Parshev (1979b, 1981c).
[b]$M = N_2 + O_2$ for Earth and CO_2 for Venus and Mars.

their difference can be determined only in the case of simultaneous measurements. The majority of the models of the atmosphere and most measurements yield $[O]_{max} = (5\text{-}10) \times 10^{11}$ cm^{-3}; we shall use a mean value of 7×10^{11} cm^{-3}. The variability of the profile of O at 90–100 km and its measurement errors do not, in our opinion, enable a comparison between the calculated and the measured profiles of the emission with precision better than 2 to 3 km in the absence of simultaneous measurements of [O].

Data on the five excited states of O_2, and on $O(^1S)$, are shown in Table V and Fig. 6. Some of the numbers are results of the following discussion. It is believed that the three states of O_2 near 4.5 eV are produced in the 3-body association of O atoms with yields α, β, and γ; the total rate coefficient is $k_1 = 4.7 \times 10^{-33}$ $(300/T)^2$ cm^6 s^{-1} (Campbell and Gray 1973). If the A state has a radiative life τ and is quenched with coefficient k_q, the volume emission rate is given by the Stern-Volmer formula

$$i = \frac{\alpha k_1 [O]^2 [M]}{1 + \tau k_q [M]} \tag{2}$$

where [M] is the concentration of third bodies. In the limit of strong quenching, $\tau k_q[M] \gg 1$, and

$$i \doteq \frac{\alpha k_i}{\tau k_q} [O]^2. \tag{3}$$

Herzberg I Bands, $A^3\Sigma_u^+ \rightarrow X^3\Sigma_g^-$

The coefficient in Eq. (3) is equal to 2.5×10^{-21} cm^3 s^{-1} (McNeal and Durana 1969; Young and Black 1966). In the Earth's nightglow, Eq. (2) is applicable and we obtain the values of α and k_q shown in Table V. Past work has assumed α to be 10 times larger and has therefore found k_1 to be larger also. In CO_2 the larger value shown for k_q follows McNeal and Durana (1969), and the Venus emission gives a consistent value of α. The upper limit for O follows Kenner et al. (1979).

TABLE V

Excited States of O_2 and O

State	$A^3\Sigma_u^+$	$A'^3\Delta_u$	$c^1\Sigma_u^-$	$b^1\Sigma_g^+$	$a^1\Delta_g$	$O(^1S)$
Energy ($v^1 = 0$), eV	4.34	4.26	4.05	1.63	0.97	4.16
Radiative life τ, s — air	0.18	50	25	12	3800	0.85
Radiative life τ, s — CO_2	0.25		50			
Recombination yield	$\alpha = 3.5(-3)$	β	$\gamma = 0.5$–0.7	small	small	small
Production coefficient[a]	—	—	—	5(−13)	3(−11)	7(−14)
Quenching coefficient $cm^3\,s^{-1}$ — air	2.6(−14)	6(−12)β	7(−14)	2(−15)	small	1.5(−12)$e^{-850/T/O_2}$
Quenching coefficient $cm^3\,s^{-1}$ — CO_2	1(−13)	(1−6)(−13)β	6(−15)γ	2(−15)	small	≤2(−14)
Quenching coefficient $cm^3\,s^{-1}$ — O	<2(−15)	<2(−12)β	6.5(−11)γ	8(−14)		

[a] Production coefficient is for transfer from O_2 (c); see Eq. (4).

Chamberlain Bands, $A'\,^3\Delta_u \to a\,^1\Delta_g$

On Earth, there is little doubt that the A' state is deactivated primarily by collisions rather than radiation; the radiative lifetime is long, and only one-quarter as many photons are seen as in the Herzberg I system. We therefore find the quenching coefficients shown in Table V, which are in the ratio 60 (air/CO_2) if the yield β is the same in both gases. A ratio of 10 is obtained from Slanger (1978a, Figs. 2 and 3). In this experiment introduction of CO_2 not only quenches the emission but also increases the rate of recombination of O. We also feel that quenching by O is negligible (Table V).

Herzberg II Bands, $c\,^1\Sigma_u^- \to X\,^3\Sigma_g^-$

Quenching of the c state by N_2 and CO_2 is negligible (Slanger 1978a), while O is an efficient quencher (Kenner et al. 1979). The Earth and Venus airglow intensities then give the quenching coefficient shown in Table V. The close agreement (15%) for the two planets again indicates that the distribution of excited states in recombination is the same in air and CO_2.

On Mars the Herzberg II bands are much weaker (Table III) and the CO_2 quenching coefficient given in Table V can be derived.

Oxygen Green Line, 5577 Å, $^1S - {}^1D$

It is now clear (Slanger and Black 1976) that the Chapman process (recombination of O atoms in the presence of a third atom) is insignificant. We therefore turn to the Barth process

$$O_2^* + O \to O_2 + O(^1S). \qquad (4)$$

O_2^* could be any of the three high states A, A', or c; lower states with high vibrational excitation are also conceivable, but seem to be absent on Earth. If τ^* is the radiative lifetime of O_2^* and i^* is the volume emission rate, the production rate of $O(^1S)$ is

$$k_3[O_2^*]\,[O] = k_3\,i^*\tau^*[O] \qquad (5)$$

where k_4 is the rate coefficient of reaction (4), shown in Table V as the production coefficient. The ratio of volume emission rates is

$$\frac{i_g}{i^*} = \frac{\tau^* k_4[O]}{\tau_g k_g[M]} \qquad (6)$$

where g refers to green line. Only the c state with some vibrational excitation satisfies all the constraints on O_2^*. On Earth, the observed band intensities show v' to be a maximum at 5–8, and k_4 is found to be 7×10^{-14} cm^3 s^{-1}. On Venus this vibrational excitation is absent, and the required value, $k_4 < 4 \times 10^{-15}$ cm^3 s^{-1} is readily met because of the 0.1 eV energy deficiency.

Concerning the other two candidates for O_2^*, the A state would need k_4 to be 3×10^{-12} cm^3 s^{-1} to explain the Earth airglow (this contradicts the much smaller quenching coefficient $<2 \times 10^{15}$ cm^3 s^{-1}) and for the A' state, a value $(1.7 \times 10^{-13}$ cm^3 s$^{-1})$ that explains the Earth airglow gives far too much intensity on Venus.

O_2 Atmospheric and Infrared Atmospheric Bands

These bands are at 0.76 μm $(b^1\Sigma_g^+\rightarrow X^3\Sigma_g^-)$ and 1.27 μm $(a^1\Delta_g\rightarrow X^3\Sigma_g^-)$. Quenching is small and so is the direct yield in recombination (Wayne 1973; Slanger and Black 1979). The large intensities of both systems suggest that they are excited by energy transfer from other metastable states of O_2. Due to the low yield of O_2(A), we only need consider two possibilities, A' and c. Llewellyn and Solheim (1980) and Greer et al. (1981) have shown that the intensities of the emissions of O_2(a) and O_2(b) in the atmosphere of Earth and their altitude profiles can be explained if virtually every event of oxygen recombination leads to the appearance of O_2(c), i.e. $\gamma \simeq 0.7$, and the quenching of O_2(c) produces the emission of O_2(a), O_2(b), and O(^{1}S) with an efficiency close to unity:

$$\left.\begin{array}{l} O_2(c) + O \rightarrow O_2(a) + O \\ O_2(X) + O(^1s) \end{array}\right\} k_7 = 3 \times 10^{-11} \text{ cm}^3 \text{ s}^{-1} \tag{7}$$

$$O_2(c) + O_2 \rightarrow O_2(b) + O_2 \quad k_8 = 5 \times 10^{-13} \text{ cm}^3 \text{ s}^{-1}. \tag{8}$$

In fact, the sum of the emissions of O_2(a), O_2(b), and O(^{1}S) at 90–100 km is equal to 35–70 kR; the rate of recombination in a vertical column is $(1–1.5) \times 10^{11}$ cm^{-2} s^{-1}; hence, $\gamma = 0.4–0.7$. Taking into account the measurements discussed above of O_2(c) in the atmospheres of Venus and Earth, $k_6 = 3 \times 10^{-11}$ cm^3 s^{-1} implies $\gamma \simeq 0.5$, which is consistent. The downward shift of the altitude profile of 0.76 μm is due to the rapid growth of [O_2] at lower altitudes. Although the experimental material is insufficient for a definite determination of the rate coefficients of the processes or even for the selection of the c state as the precursor of states a and b, there is some additional confirmation in two facts: (1) that the quenching of O_2(A') by atomic oxygen is small in the atmosphere of Earth, and thus O_2(A ') cannot be a precursor of O_2(a); and (2) that data on the emission of Venus definitely point to O_2(c) as the precursor of O(^{1}S).

It appears that reaction (7) does not account for more than one-third of the 1.27 μm emission from Venus (Krasnopol'sky and Parshev 1980b, 1981c). Moreover, although there is a major contribution on the day side from photolysis of ozone, the nightside emission is just as bright. We have suggested that the yield of O_2(a) is near unity in the destruction cycle

$$O_3 + Cl \rightarrow ClO + O_2$$

$$ClO + O \rightarrow Cl + O_2. \tag{9}$$

The required quenching coefficient in CO_2 is 3×10^{-20} cm^3 s^{-1}, in agreement with Traub et al. (1979).

The OH Meinel System

In the spectra of Venus there are no traces of the vibration-rotation system of OH bands. A comparison with the spectra of terrestrial nightglow (Broadfoot and Kendall 1968) is most favorable for the R_1, R_2 branches of the (7,2) band which produces on Earth an emission of 30 R with a width of 15Å. In the spectrum of Venus this emission does not exceed 40% of the intensity on Earth. We suggest that the reason is efficient vibrational quenching by CO_2, whose density at the peak of the production is 100 times greater than the corresponding air density on Earth. Any excitation of $O_2(a)$ from OH* is similarly suppressed.

III. LIMB MEASUREMENTS OF HIGH-ALTITUDE AEROSOLS

The spectrometers on Veneras 9 and 10 had a special mode for dayside observations (Krasnopol'sky 1979a, 1980). The wavelength range was 2700–7000 Å, the resolution 50 Å, and the field of view of the instrument 3'5, which corresponded to 7 km for observations on the limb. The scanning time was 2.5 s, during which the altitude changed by 20 km.

We shall discuss two data sets, in which the slit of the instrument was directed along the tangent to the limb of the planet. For Venera 9 (series 1 in Fig. 7a), the solar zenith angle was 72°, the latitude 31°, the hour angle of 250° (16:40 LVT), and the phase angle (the angle between the Sun and the optical axis of the instrument) 152°5. The analogous data for Venera 10 (series 2, Fig. 7b) are 74, 14, 253° (16:52), 148°; thus, both sets refer to virtually identical conditions, with the exception of a slight difference in latitude.

The readings changed very slowly during the tracking of the planetary disk and the gradual settling on its limb. Then, in the course of ~ 4 s, the intensity was abruptly reduced by three orders of magnitude, after which it remained almost unchanged for a long time due evidently to the light scattered in the instrument (and serving as a measure of this). The results (corrected for the scattered light) at 500 Å intervals are shown in Fig. 7. Apart from the spectral dependence, the altitude variation of the logarithm of the coefficient of brightness $R = \pi I/I_0$ (I_0 and I are the intensities of the solar and measured radiations) is well approximated by a line, and below a certain altitude level constant.

The scattering of light by the limb of the planet can be described with sufficient accuracy by the approximation of single scattering. This approxima-

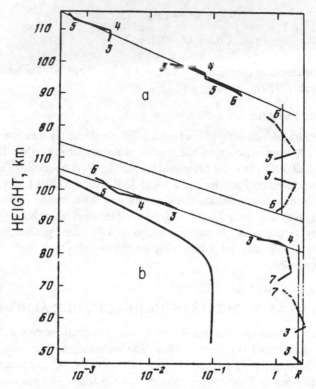

Fig. 7. The brightness coefficient $R = \pi I/I_0$ as a function of altitude for two series of observation on the limb of the planet. Wavelengths are taken at 500 Å intervals; numbers indicate the wavelength in 10^3 Å. The broken line corresponds to sections of the spectra beyond the scale. The solid line of Fig. 7b is the calculated brightness coefficient for a pure gas atmosphere.

tion is valid when the optical depth $\tau \leqslant 0.3$, i.e., when $h \geqslant 70$ km. The optical thickness of the limb, along the visual beam of the instrument, is 2 orders of magnitude greater in this case, i.e. $\tau_L < 30$. This approximation may be refined to a certain extent by allowing for the radiation reflected by the atmosphere, which increases the coefficient of brightness by a factor of $1+2a$, where a is the albedo of the planet. Here, the coefficient 2 is obtained in the following manner: if, for an incident flux of sunlight, the cross section of a spherical particle is πr^2, then for a flux reflected from a half-space, this section is equal to the area of the hemisphere, i.e. $2\pi r^2$. It can be shown that this coefficient is preserved for particles of arbitrary shape and random orientation. Then, the coefficient of brightness is

$$R = \frac{\gamma + 2a\bar{\gamma}}{4}\tilde{\omega}\left[1 - \exp\left(-\tau_L\right)\right]. \qquad (10)$$

Here, $\bar{\omega}$ is the albedo of single scattering, γ is the value of the phase function for the scattering angle (180° minus the phase angle, i.e. 27.5 and 32°), and $\bar{\gamma}$ is the value of the phase function averaged for a hemisphere. The optical thickness on the limb is

$$\tau_L = \tau(2\pi a/H)^{1/2} = t(2\pi aH)^{1/2} \tag{11}$$

where τ is the vertical optical thickness, t is the extinction coefficient, a is the radius of the planet, and H is the scale height.

Let us assume that t, and consequently τ and τ_L, vary exponentially with height. Then the dependence of $R(h)$ in Fig. 7 should be almost a broken line with an inflection point at a level of $\tau_L = 1$. This assumption, as can been seen from Fig. 7, is in good agreement with the data and will be used in further interpretation. The upper part of the broken line enables the determination of the scale height, equaling 3.6 and 2.9 km; the error is ± 0.3 km.

The pointing data from the spacecraft are not good enough to determine a useful height scale. Fig. 7 assumes that $\tau = 1$ at 68 km. The Rayleigh gas scattering provides a negligible contribution to the observed brightness, even in the shortwave portion of the spectrum (Fig. 7b). Here our data differ from the results of the observations on Mariner 10 (O'Leary 1975), in which the role of the gas scattering is comparable to that of the aerosol. In our experiment the scattering angles are small, and this leads to large values of γ for the aerosol scattering and, moreover, large values of the brightness coefficient $R \gtrsim 1$ when $\tau_L > 1$.

As we shall demonstrate below, the measured radiation is largely produced by scattering from particles with a radius of $r = 0.15$–0.3 μm. For such particles, an integration of the phase function over the hemisphere with axis perpendicular to the direction of scattering yields $\bar{\gamma} \simeq 0.1$. Therefore, despite the large value of the albedo, the effect of the scattering by the lower altitude cloud layer is negligible in our observations, and shall be disregarded.

Spectra of the Scattered Light, Particles Sizes, and Numbers

Figure 8 shows several spectra corrected for motion during the scan to the height that corresponds to the middle of the spectrum. The spectra contain numerous features, but not all are real. In dividing the measured radiation by the intensity of the solar spectrum (Makarova and Kharitonov 1972), errors arise due to the possible shifting of the wavelengths, and differing spectral resolutions and instrument functions. An additional source of errors is the inaccuracy of calibration and the logarithmic scale of intensity in the instrument, the spacing of which is 7%. Therefore, it is not worthwhile to discuss fine details of the spectra.

The major details are evidently produced by the wavelength dependence of the phase function. With the help of these details we shall attempt to determine the size of the scattering particles, assuming that they consist of a

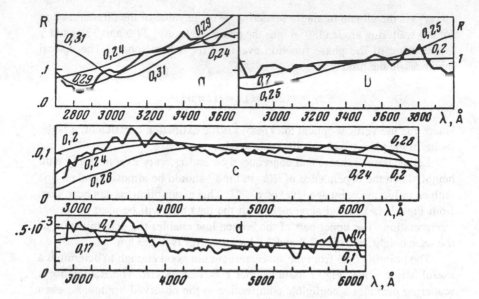

Fig. 8. Scattering spectra of the atmosphere at various altitudes h: (a) series 1, h = 70 km, $h(\tau_L$ = 1) = 77 km; (b) series 2, h = 85; (c) series 1, h = 94 km; (d) series 2, h = 99 km. Fine lines represent calculated scattering spectra of sulfuric acid particles; the value of the radius of the particles is indicated near the curves.

concentrated solution of sulfuric acid. Taking into account the temperature dependence of the refractive index n, we use n = 1.46 instead of 1.44.

Figure 9 shows the dependence of γ on $4\pi r/\lambda$, calculated for the two scattering angles of interest, in accordance with Mie theory, using a special program. The scattered spectrum for arbitrary r can therefore be obtained. Mie theory may only be an approximation, if the particles are solid at 180–200 K; it should be accurate if they are supercooled liquid.

Figure 8 shows the spectra of the particles that best correspond to the measured spectra. For the height scale of Fig. 7, at 94 km the size of the particles corresponds to 0.24 ± 0.04 μm, while at 70 km it is $0.29^{+0.02}_{-0.05}$ μm. Since τ_L is very large at 70 km, the result obtained should be assigned to an altitude of 77 km, where unit optical thickness is achieved at a sighting altitude of 70 km. For the profile of Fig. 7b at 99 km, $r \simeq 0.12$ μm; at 85 km, $r \simeq 0.2$ μm. To the values obtained we may add r = 1 μm at 68 km in accordance with earth-based polarimetric observations, and draw the curves of r as a function of height (Fig. 10a). For a known r, we can find the extinction cross section and then, knowing τ and t, we can determine the number of particles and their density. These quantities are shown in Fig. 10. It can be seen that the number of particles changes by 3 orders of magnitude from 70 to 100 km, the density by 5 to 6 orders. The minimum optical thicknesses, recorded at 100 to 110 km, correspond to a visual range of $\sim 2 \times$

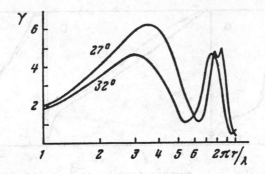

Fig. 9. The phase function of scattering γ as a function of the parameter $2\pi r/\lambda$ for two values of scattering angle.

10^6 km. The relative content of sulfuric acid is $\simeq 10^{-6}$ at 70 km and 5×10^{-9} at 100 km.

The Eddy Diffusion Coefficient

The vertical distribution of aerosol particles can be used to determine the eddy diffusion coefficient K. The photochemical formation of sulfuric acid takes place below 70 km, and therefore it can be assumed that the aerosol of $H_2SO_4 \cdot H_2O$ reaches altitudes higher than 70 km only by mixing. In this case, the vertical flux of aerosol should be compensated by a gradual dropping of the particles with a velocity V

$$\phi = -K\left(\frac{d\rho}{dz} + \frac{\rho}{H}\right) - \rho V = 0 . \tag{12}$$

Here, ρ is the density of the aerosol, H is the scale height of the atmosphere, and H_a is the scale height of the aerosol (cf. Prinn 1974). Hence,

$$\rho = \rho_0 \exp\left\{-z\left(\frac{1}{H} + \frac{V}{K}\right)\right\}, \frac{1}{H_a} = \frac{1}{H} + \frac{V}{K}, K = \frac{VHH_a}{H - H_a} . \tag{13}$$

The velocity of fall of the particles can be determined by the Stokes formula with the corrections of Davis, which extend its area of application to cover the case when the mean free path of the molecules is $l \geqslant r$:

$$V = \frac{2}{9} \rho_m g \frac{r^2}{\eta}\left(1 + \frac{l}{r}\right)\left(1.257 + 0.4 \exp^{-1.1r/l}\right) \tag{14}$$

where η is the viscosity of the atmosphere and $\rho_m \simeq 1.6$ g cm^{-3} is the density of the material of the particles.

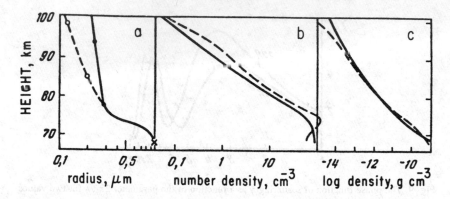

Fig. 10. The radius of particles r, number of particles n and density ρ as functions of altitude. Solid lines represent series 1, dashes series 2.

Now, by using the data of Fig. 10c, we can calculate $K(h)$ (Fig. 11). In consideration of our approximations, K is determined within roughly an order of magnitude. The error is especially large for the Venera 9 data at heights greater than 90 km, since here $H_a \simeq H$, and the denominator has a small value. The nature of the dependence of K on height is in agreement with the results of the photochemical calculations of Krasnopol'sky and Parshev (1980b, 1981c), with the estimate of $K \simeq 10^6$ cm^2 s^{-1} at 100 km, obtained from a comparison of the local variations of O_2 emission on Earth and on Venus (cf. above), and with the dependence of K obtained by von Zahn et al. (1980). However, the slope of the curve at 75 to 90 km turned out to be larger than that in the above-cited works.

Several spectroscopic limits shown in Table VI have been obtained from the absence of absorption bands in the spectra of Fig. 8.

IV. LIGHTNING ON VENUS

The observations of the low-frequency electromagnetic radiation aboard the Venera 11 and 12 (Ksanfomaliti 1979a) and Pioneer Venus (Scarf et al. 1980) spacecraft indicate the presence of rather powerful electric discharges in the atmosphere of the planet, probably in the cloud layer. In order to investigate this phenomenon, in this chapter we have made use of the results of optical measurements aboard Veneras 9 and 10 Krasnopol'sky 1979a, 1980).

Detection of Lightning from Spacecraft

Let us consider the attenuation of the light of a bolt by dispersion and absorption in the cloud layer. We shall assume that the lightning is created at its lower boundary. The situation is unchanged as far down as ~ 20 km, since the atmospheric attenuation between 20 and 48 km is much less than that in

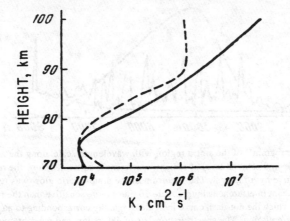

Fig. 11. The eddy diffusion coefficient K as a function of height from the data of two observation series.

the cloud layer. Light from within the clouds can escape even more easily. Because the single-scattering albedo of the cloud particles is near unity, we estimate that, between 4000 and 7000 Å, 15% of the light escapes. However, it is concentrated from 4π into π steradians, and the intensity is therefore reduced to only 60%.

We shall assume that the radius of the spot seen from space is equal to ½ the geometrical thickness of the cloud layer, 15 to 20 km. The field of view of the instrument at an average distance of 2500 km is 160×9 km². Then, if the center of the flash is located in the field of view of the instrument, approximately ½ of the entire energy is collected.

In evaluating the possibility of detecting lightning, we shall consider a value of 300 R at 60° C as an average threshold across the spectrum. A

TABLE VI
Upper Limits for Absorbing Gases

Substance	λ_{max} (Å)	6×10^{19} (cm²)	Maximum Relative Content		
			77 km	85 km	94 km
Cl_2	3300	2.6	4×10^{-7}	7×10^{-7}	1×10^{-5}
SO_2	2880	7	1.5×10^{-7}	3×10^{-7}	3×10^{-6}
CS_2	3150	0.025	4×10^{-5}	5×10^{-5}	1×10^{-3}
NO_2	4000	6.5	2×10^{-7}	3×10^{-7}	4×10^{-6}
O_3	2800	26	1×10^{-7}	1.5×10^{-7}	1×10^{-6}
Br_2	4200	2.1	1.5×10^{-6}	1.5×10^{-6}	1×10^{-5}
S_8	3500	—	1.5×10^{-7}	3×10^{-7}	3×10^{-6}

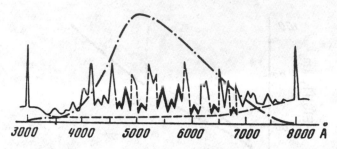

Fig. 12. One "spectrogram" of the storm region with wavelength scale along the zero of the instrument. Dashes indicate the level of radiant interference; dark lines show the reading for a scale less sensitive by a factor 8. The spectrum of the atmospheric glow was recorded in the absence of radiant interference along a scale 10 times more sensitive than the main scale in the figure, for which the dot-dash curve indicates a reading corresponding to an energy of 1 W/Å km² of the emission. The spectrum consists of 480 points and was obtained in 10 s; pulses at 3000 and 8000 Å indicate the beginning and end of the spectrogram.

conversion of this quantity for an isolated lightning flash, taking into account the above factor, yields 300 W/Å. In terrestrial lightning, the duration of the flash is ~0.25 s, and the portion of the entire energy of the lightning, radiated in the spectra region of 4000–11,000 Å, is 4×10^{-3} (Uman 1969). We assume that, on Venus, this quantity comprises 3×10^{-3} for the spectral region 4000–7000 Å. The total energy of the lightning, corresponding to the threshold value, is 7×10^7 J. The mean energy of Earth lightning is $\sim 3 \times 10^9$ J (Uman 1969; Chalmers 1967), i.e., it was possible to record much weaker lightning than is typical on Earth. The threshold increases by a factor of ~ 10 when the spacecraft is in sunlight.

The Observations

Observations were made on the spectra of the night side of the planet over an area of 3.5×10^6 km² at low latitudes, $\pm 32°$. Roughly half of the data was obtained in the shadow of the planet. Lightning is absent in the entire zone of measurements, except for a region of 450 km, at a latitude of 9° S with solar zenith angle 106° and an hour angle $t = 287°$, corresponding to 19:00 LVT (local time). The observations were carried out during the very first session of Venera 9, 26 October 1975. In this area there are sharp and chaotic variations of the signal size, which lasted for 70 s (Fig. 12). It should be noted that in precisely this session the largest contrasts were observed in the ultraviolet images of Venus and in the measurement data for 3500 Å (Ksanfomaliti and Selivanov 1978).

Two arguments favor the claim that the effect is real, and not the result of an imperfection of the instrument. The effect was never observed when the instrument was pointed away from the planet, even though the measurements at this position lasted 3 times longer than at the night side of Venus. Furthermore, the pulse amplitude diminished toward the edges of the spectrum, in accordance with the sensitivity curve of the instrument.

Energy, Spectrum, and Duration of a Lightning Flash

During the lightning observations, the scanning time was 10 s, and seven scans were taken during the passage through the thunderstorm region. The length of the flashes was much shorter than the scanning time, and the different wavelengths during the time of the measurements correspond to the different phases of the flash or to different flashes and pauses between these. A characteristic spectrogram is shown in Fig. 12. Of course, this is hardly any sort of spectrum analysis, but if we average the amplitudes of the flashes for the various wavelengths over all seven spectra, we may arrive at a rough distribution of the lightning energy in the spectrum. This distribution is rather uniform; it has a weak maximum, equal to 3×10^4 W/Å at 5500–6500 Å, and values of $\sim 2 \times 10^4$ W/Å at 4500 and 7500 Å.

The characteristic width of the peaks is ~ 0.25 s, much longer than the response time (5 ms) or the sampling interval (20 ms) but similar to the duration of a flash on Earth. However, the substructure typical of 4 short pulses of an Earth flash is missing. The luminous energy of a flash is 3×10^7 J, and the total energy 10^{10} J. These values would drop by a factor of 3 if the flashes were in the middle or upper cloud.

Sizes of Thunderstorm Regions and Frequency of Lightning Flashes

On Earth there take place 2000 storms at the same time, producing a total of 100 lightning flashes per second and taking up 3.6×10^{-3} of the surface of the globe (cf. Handbook of Geophysics 1960). Therefore the mean area of a terrestrial thunderstorm is 10^3 km^2 and its size is 30 to 40 km. The size of the storm on Venus is larger by an order of magnitude (450 km), at least along the line of roughly constant latitude that we observed. If the north-south extent is 150 km, then the area of a storm on Venus is 5×10^4 km^2, i.e., much larger than a storm on Earth. The ratio between the area of the storm and the area surveyed is 1.4×10^{-3}, which is close to the terrestrial value of 3.6×10^{-3} (cf. above), taking into account the statistical uncertainty of the first quantity.

In a 1000 km^2 field of view we saw ~ 2 flashes per second, ~ 40 times more frequent than the 3 per minute typical of an Earth storm. Assuming that our poor occurrence statistics are representative, we obtain the figures shown in Table VII. A summary of the sources of illumination of the Venus night side is given in Table VIII. The size of the thunderstorm region is close to the resolution of a good telescope; high-quality images of the night side of Venus, obtained by our telescopes, may have bright spots in the regions of thunderstorm activity and spectroscopy of these spots may be useful for investigating the chemical composition of the atmosphere of Venus. Several observations of the ashen glow of Venus might be associated with cases of a considerable increase in the storm activity.

TABLE VII

Characteristics of Lightning and Thunderstorms on Earth and Venus

Quantity	Earth	Reference	Venus[f]
Energy of lightning, J	3×10^9	a,b	$(3 - 10) \times 10^9$
Duration of a flash, s	0.25	c,d	0.25
Area of the storm region, km^2	10^3	c,d	5×10^4
Frequency of flashes in the storm region per $10^3 \, km^2 \, s^{-1}$	0.05	c,d	2
Portion of planetary surface in the zone of $\pm 32°$ latitude, occupied by storms	7×10^{-3}	c,d	1.4×10^{-3}
Frequency of flashes on the planet in the zone of $\pm 32°$ latitude, s^{-1}	100	c,d	400
Coefficient of transformation of solar energy into lightning energy	2×10^{-5}	e	5×10^{-5}

a: Uman 1969; b: Chalmers 1967; c: Tverskoy 1962; d: Handbook of Geophysics 1960; e: Bar-Nun 1979; f: Krasnopol'sky.

Lightning and the Chemistry of the Atmosphere

The idea of calculating the chemical products in lightning is due to Zeldovich and Reiser (1966). At high temperature in the discharge region a thermochemical equilibrium is rapidly established; as the temperature drops, the time for its establishment increases, and at a certain moment becomes larger than the cooling time. At this moment, the chemical composition stabilizes. The calculations of Chalmers et al. (1979) and Bar-Nun (1980) indicate products of CO, O_2, NO, and O equaling $(15, 5, 0.5, 5) \times 10^6$ molecules J^{-1}. Then, in accordance with Table VII, we obtain a global mean production of NO of $\sim 10^9 \, cm^{-2} \, s^{-1}$. The production of NO in the storm region is $3 \times 10^{12} \, cm^{-2} \, s^{-1}$; if the storm lasts one hour, the accumulated quantity of NO comprises $10^{16} \, cm^{-2}$ and yields an excess concentration of $[NO] \simeq 10^{10} \, cm^{-3}$.

The global mean concentration of NO may be estimated if the downward flux Φ is $10^9 \, cm^{-2} \, s^{-1}$ to a sink below the production level. For an eddy diffusion coefficient $K = 10^4 \, cm^2 \, s^{-1}$, the concentration is

$$[NO] \simeq \Phi H / K \simeq 10^{11} \, cm^{-3} \tag{15}$$

TABLE VIII

Illumination of the Night Side of Venus

Source	Brightness (stilb)
Atmospheric glow	6×10^{-8}
Light of the stars	7×10^{-9}
Light of Earth (at phase 90° of Venus)	2×10^{-9}
Lightning (average for the planet)	1.5×10^{-7}
Threshold of detection	2×10^{-5}
Lightning (average for the storm region)	1×10^{-4}

where H is the scale height. This is large compared to the maximum on Earth, $\sim 10^9$ cm^{-3} in the stratosphere. It is probably still too small to have a substantial effect on the cloud composition.

V. A DUST OBJECT NEAR VENUS

In one pass of Venera 10 an extended dust object was observed, which might be a cloud produced by a comet and captured by the gravitational field of Venus. The optical spectrum of scattering of the dust is an exponential function of the wavelength with an exponent of -2.1 at a sighting altitude of 100 km or -1.3 at 500 km. This corresponds to a mass spectrum of dust particles of $n = am^{-\alpha}$, where α varies from 1.37 to 1.1. Assuming an annular structure of the cloud, its mass is 5×10^{-3} g cm^{-1}; its maximum density at a thickness of 100 km is 10^{-17} g cm^3; its optical thickness along the radius is 7×10^{-6}, while along the normal to the plane of orbit at maximum it is 4×10^{-6}.

Only this short synopsis is given here, because a full description is readily available in Krasnopolsky and Krysko (1976, 1979). We originally supposed that the dust ring is a rather stable and long-lived object, in which the source of dust is a small satellite of Venus with a size of ~ 1 km, not visible from the Earth. This satellite should gather dust in the surrounding space, as well as produce dust by the gradual disintegration of its surface. However, we know of no observations from Pioneer Venus which support the presence of a dust ring, even though the ultraviolet spectrometer on that spacecraft should have a sensitivity perfectly capable of detecting the ring. Therefore it is most reasonable to assume a short-lived object, the source of which was transient in action. Possibly this was a dust swarm, created by a comet and captured by the gravitational field of Venus. Possibly this swarm does not form a continuous ring, but remains for a certain time as a cloud, rotating around the planet in an eccentric orbit.

16. THE CLOUDS AND HAZES OF VENUS

L. W. ESPOSITO
University of Colorado

R. G. KNOLLENBERG
Particle Measuring Systems, Inc.

M. YA. MAROV
USSR Academy of Sciences

O. B. TOON
NASA Ames Research Center

and

R. P. TURCO
R & D Associates

Clouds totally enshroud Venus horizontally, and have an enormous vertical extent, > 50 km. They are relevant to problems in Venus meteorology, geochemistry and evolution. With the data from Pioneer Venus and the Venera series, we have recently made remarkable progress in understanding the clouds. The Venus clouds consist of a main cloud deck at 45–70 km altitude, with thinner hazes above and below. Even the densest parts of the clouds are very tenuous, with visibility of several km. The microphysical properties of the main cloud allow further subdivision into a middle, upper, and lower cloud. Much of the cloud shows a multimodal particle size distribution. The mode most visible from the Earth is H_2SO_4 droplets with 2–3 μm diameter. At all altitudes we find small particles, though not necessarily of the same composition. The largest particles range up to 35 μm; these may either be nonspherical, incorporating crystalline cores, or merely be the tail of the H_2SO_4 droplet distribution. The upper region of the cloud visible from Earth shows both short and long term variations. Contrasts visible in the ultraviolet have time scales of hours to months. The haze above the main cloud varies on a time scale of years. However, we do not have a detailed enough record of the deeper regions of the clouds to describe their variation. Despite variations the vertical structure of *the clouds shows persistent features at sites separated by years and great distances. Recent models are able to explain the dominant chemical and growth processes in the clouds. Growth models can yield multimodal distributions consistent with the observations, if contaminants are invoked. The production of cloud droplets by photochemical oxidation of SO_2 at the cloud tops and condensation of H_2SO_4 near the cloud base seems plausible. Upward flowing SO_2 can explain the observed ultraviolet contrasts and planetary spectral reflectivity, at least at wavelengths < 320 nm. Another absorber is needed to explain the observations at longer wavelengths; amorphous sulfur and molecular chlorine are two candidates. The clouds are more strongly influenced by radiation than by latent heat release. The small particle size and weak convective activity in the observed clouds seem incompatible with lightning of cloud origin. Further advances may be expected from the long time base provided by the Pioneer Venus orbiter, future in situ measurements, and more detailed comparison and synthesis of data already in hand.*

Atmospheric particulates are an important recent concern of planetary sciences; planets with atmospheres invariably have particulate suspensions in the form of hazes or clouds. Differentiation of clouds from hazes or other suspended particulate matter is somewhat subjective; here, we will consider condensable material having optical depth greater than unity as cloud regardless of its tenuousness. Condensable material with optical depth less than unity we consider to be haze. Noncondensable particulate matter we refer to merely as suspended particulate. Mercury is the only planet in our solar system without an atmosphere and particulate suspension of some sort. Clouds have been identified on all the remaining planets, and of these only Earth and Mars have visible surfaces. Even Saturn's giant satellite Titan has a cloud-covered surface. Since aerosols may also exist above other satellites, in Saturn's rings, and in cometary tails, we see that cloud studies are broadly relevant.

The nature of the particulate matter in planetary atmospheres is still largely unknown. We do know of probable dust and ice clouds on Mars, and ammonia and methane clouds on the major planets. Nevertheless, the only cloud system that we basically understand, beside Earth's, is that on Venus. This chapter reviews our knowledge of Venus's clouds, their formation, growth, and morphology; where appropriate we compare venusian and terrestrial clouds and cloud processes.

Until the 1970s the Venus clouds were merely an obstruction to seeing the planet's surface. These clouds are unbroken and featureless to the eye, totally covering the planet. Our knowledge has increased as a result of directed ground observations and recent space missions (most notably Veneras 9 and 10; Pioneer Venus (PV) orbiter and probes; Veneras 11 and 12). For a review of our knowledge of Venus before the Pioneer mission, see Knollenberg et al. (1977) and Marov (1978a). Although studies of the new data are by no means complete, and remote observations continue on a daily basis from the Pioneer Venus orbiter, we now have enough knowledge for a general understanding of Venus's clouds.

We subdivide the clouds in the atmosphere of Venus into the main cloud deck and hazes above and below the main cloud. There may also be some undefined suspended aerosols. This is very similar to our terrestrial atmosphere in which there are boundary-layer hazes and fogs, and a variety of tropospheric cloud forms topped by a stratospheric aerosol layer (Junge layer). The most striking differences are the higher altitude of the Venus main cloud deck and the smaller amount of condensable vapor available for cloud mass on Venus, which has no oceanic liquid reservoir. The upper level haze is more pronounced on Venus, while subcloud particulate amounts are more enhanced on Earth.

The clouds and hazes of Venus can be further subdivided in terms of observed microphysical properties. We discuss five distinct regions of cloud and haze particles which have varying composition and microphysical properties. The top layer is the submicron aerosol-haze at 70 to 90 km altitudes. The temperature in this region is quite low, -50 to $-85°C$. Growth and evaporation processes are probably slow compared to the lower haze and cloud. This upper haze overlies the polar regions and has average optical depth as large as 1 (Kawabata et al. 1980). In the equatorial and midlatitude regions the haze is $\sim 1/10$ as dense and its particles mix with the 2 μm diameter H_2SO_4 cloud droplets at the top of the main cloud deck. The main cloud deck has three prominent regions referred to by Knollenberg and Hunten (1979) as upper, middle and lower cloud. A sharp base is evident at 47–50 km, but thin cloud layers are often found below the nominal cloud base (Ragent and Blamont 1979). The size distribution is generally multimodal throughout the main cloud deck, which has optical depth 20–30. Below the main cloud deck are small haze particles of low number density extending down to $\sim$ 30 km, where all particles essentially disappear. The cloud and haze properties we discuss are summarized in Table I.

In what follows our presentation is basically historical. Following the program of exploration of the Venus clouds itself, we examine the cloud system on Venus from the top layers to the planetary surface. We discuss groundbased observations first, then spacecraft remote measurements; both necessarily deal with the few uppermost optical depths, including the upper haze and upper cloud region. We then move downward into the main cloud deck and the most recent *in situ* observations. A number of special topics are discussed, and recent synthesis and modeling efforts dealing with the entire cloud system are presented. We conclude with a comparison of terrestrial and venusian cloud properties, and a summary of our state of knowledge.

I. THE UPPER CLOUDS AND HAZES OF VENUS: REMOTE OBSERVATIONS FROM EARTH AND SPACECRAFT

Observations of the Venus clouds prior to *in situ* probe measurements are restricted to the upper haze and cloud-top regions. Furthermore, the

TABLE I
Summary of Venus Cloud and Haze Properties

Region	Altitude (km)	Temperature (K)	Optical Depth τ (at 0.63μ)	Average Number Density (N cm^{-3})[a]	Mean Diameter (μm)	Proposed Composition[b]
Upper haze[c]	70–90	225–190	0.2–1.0	500	0.4	H_2SO_4 + contaminants
Upper cloud	56.5–70	286–225	6.0–8.0	(1)–1500 (2)–50	Bimodal 0.4 & 2.0	H_2SO_4 + contaminants
Middle cloud	50.5–56.5	345–286	8.0–10.0	(1)–300 (2)–50 (3)–10	Trimodal[e] 0.3, 2.5, & 7.0	H_2SO_4 + crystals (?)
Lower cloud	47.5–50.5	367–345	6.0–12.0	(1)–1200 (2)–50 (3)–50	Trimodal[e] 0.4, 2.0, & 8.0	H_2SO_4 + crystals (?)
Lower haze[d]	31–47.5	482–367	0.1–0.2	2–20	0.2	H_2SO_4 + contaminants
Precloud layers[d]	46 & 47.5	378 & 367	0.05 & 0.1	50 & 150	Bimodal 0.3 & 2.0	H_2SO_4 + contaminants

[a]Mode 1 number density as determined by log-normal fit to data in Sec. III; modes designated by numbers in parenthesis.
[b]See Sec. IV for interpretations of composition.
[c]From Travis et al. (1979a, b) and Kawabata et al. (1980).
[d]Observed only at PV Large probe site.
[e]Possibly only bimodal. See Toon et al. (1982a).

nonuniformity of these regions was not always recognized by earlier investigators. Observations from outside the planet's atmosphere "see" a mixture of haze and cloud particles. Fortunately, a critical period of ground observations occurred during the late 1960s, when the cloud-top region was less complicated and consisted mainly of uniform size (monodispersed) 2 μm diameter H_2SO_4 droplets.

A. Remote Observations of the Main Cloud Deck

Remote observations of Venus extend into antiquity, and scientific studies of the planet began shortly after the invention of the telescope. The lack of features at visible wavelengths and the overall brightness suggested a uniform cloud cover. At the resolution available to groundbased observations in the middle of this century, infrared imagery also showed no contrast. Only at ultraviolet wavelengths was it possible to observe contrasts suggesting changing cloud morphology, and to trace motions.

Throughout this century, candidates have been suggested for the cloud composition. Terrestrial analogs such as water and ice clouds were quite popular, and prior to planetary space probe exploration the list was expanded to include some unusual materials: mercury compounds (Lewis 1969; Rasool 1970), polywater (Donahoe 1970), and hydrated ferric chloride (Kuiper 1969). Eventually water clouds were ruled out because of the small amount of water vapor above (e.g. Barker 1975a) and below (Rossow and Sagan 1975) the clouds, and because derived cloud particle refractive indices were not compatible with water (see below). Two main types of data were then known to constrain the cloud composition. Spectroscopic observations showed that Venus had broad absorption features at ~ 0.3 μm, ~ 3 μm, and ~ 11.2 μm. Second, the observed polarization of the disk of Venus could not be matched by pure Rayleigh scattering (Horak 1950), demonstrating the effect of cloud particles. It appeared that the polarization characteristics could be definitive in determining the cloud microphysical properties, and thus possibly their composition. Studies by Coffeen (1969) and Dollfus and Coffeen (1970), giving the globally averaged polarization of Venus as a function of solar phase angle, provided the observational data set on which such studies could be based. Several authors (e.g. Coffeen 1968, 1969) were able to place limits on the possible composition by modeling the cloud particles as dielectric spheres (Mie scatterers). There were several difficulties with these early studies. First, all calculations assumed that the cloud particles were spherical. Second, the effect of multiple scattering on the polarization was ignored, i.e. it was assumed that only the single-scattered radiation contributed to the polarization. Third, (partly as a result of these assumptions) the range of refractive indices determined for the Venus cloud particles was broad enough to include almost all suggested constituents.

Resolution of these problems was contingent on conceptual and technical advances. The conceptual advance was that sphericity of the aerosols could actually be demonstrated (Hansen and Arking 1971); spheres have distinctive features in the angular variation of their polarization which were clearly seen in the Venus data. At the same time, numerical methods were developed to accurately include polarization effects in calculating multiple scattered light in planetary atmospheres by methods such as "doubling" routines (Hansen and Travis 1974). This technical advance allowed stringent limits to be placed on possible refractive index, size, and range in size of the particles.

The results of Hansen and Arking (1971) and Hansen and Hovenier (1974) showed the cloud particles to be spherical with radius $\sim$ 1 μm, and narrowed the allowable range of real refractive index to $\sim$1.45. With these new constraints, Sill (1972) and Young and Young (1973) independently proposed that the Venus clouds were composed of droplets of concentrated ($\sim$ 75% by weight) sulfuric acid. The support for this identification, summarized by Young (1973), is that:

1. Sulfuric acid has the right refractive index; furthermore, its dispersion relation yielded the correct refractive index at several wavelengths at which polarization observations had been analyzed to give refractive indices.
2. Sulfuric acid solutions would be liquid at cloud-top temperatures (240–250 K) and thus be in the form of spherical drops.
3. The high affinity of sulfuric acid for water would explain the extreme dryness of the Venus atmosphere at the cloud tops.
4. Similar aerosols of H_2SO_4 are formed in the Earth's Junge layer.
5. The near infrared observations of Venus could be well matched by sulfuric acid clouds, as shown by Pollack et al. (1975).

Thus, sulfuric acid has physical, chemical and optical properties that match many characteristics of the Venus clouds.

The analysis by Hansen and Hovenier (1974) established precise limits on the optical properties of the visible cloud particles, summarized as follows. The index of refraction of the aerosols is $n_r = 1.44 \pm 0.015$ at $\lambda = 0.55$ μm, and the index has a normal dispersion (see Fig. 1). The variation of polarization both with angle and with wavelength confirm that the cloud particles are spherical. The cross-section weighted mean radius of these spheres is 1.05 μm $\pm$ 0.10 μm (2.1 μm diameter). The particle size distribution has an effective variance in radius of only 0.07 $\pm$ 0.01; this narrow size distribution has a standard deviation of only 0.26 μm. The pressure at the cloud tops (optical depth of unity) is 50 $\pm$ 25 mbar; the cloud top is at 65–70 km altitude. Furthermore, all these properties must be quite uniform across the planet and to some depth into the clouds, because the analysis is based on globally-averaged polarimetry. The implication is that the observed global cloud of Venus is composed of aerosols of one size, shape, and composition: 2 μm diameter, 75 % concentration sulfuric acid droplets.

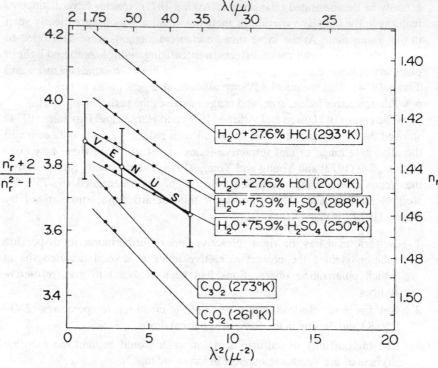

Fig. 1. Refractive indices of the Venus cloud particles (open circles) reduced from the polarization. The error bars represent the maximum uncertainty, not a probable error. The experimental values (dots) for the indicated liquids are based on laboratory measurements and, in some cases, interpolation formulas. (From Hansen and Hovenier 1974.)

B. Remote Observations of the Submicron Haze

The findings of Hansen and Hovenier on the narrowness of the particle size distribution were remarkable in ruling out other sizes and compositions of aerosols near the Venus cloud tops. Although there was some evidence for variability of the visible clouds, these results were strong enough that more complicated models seemed unnecessary until the Pioneer Venus cloud photopolarimeter images began to arrive (Travis et al. 1979b). The remote observations from the Pioneer Venus orbiter also support sulfuric acid particle composition, from wavelengths in the ultraviolet (Stewart et al. 1979) through the visible to the near infrared (Travis et al. 1979b), and over most of the planet. However, the polarimetry at 9350 Å, especially near the poles, shows strong evidence that the Venus clouds are more complicated than the uniform 2 μm diameter from earth-based observations. It now seems that Hansen and collaborators were fortunate in having a data set of Venus polarimetry that coincided with a period when the Venus cloud structure was relatively simple.

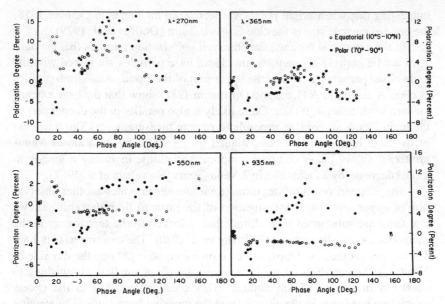

Fig. 2. Comparison of OCPP polarimetric data characteristic of the polar (solid circles) and equatorial (open circles) regions. The polar data are obtained by integrating over the 70° to 90° latitude region for each hemisphere and averaging, while the equatorial data correspond to an integration over the 10°S to 20°N zone. (From Kawabata et al. 1980.)

The Venus cloud microphysics shows variations with latitude and time not apparent in the earth-based polarimetry of the late sixties.

The anomalous polarization observed at 9350 Å cannot be due to excess Rayleigh scattering from CO_2. The Rayleigh scattering cross section is $\simeq \lambda^{-4}$, so that even greater anomalies would be evident at the shorter wavelengths, where in fact none was seen. The natural conclusion was that this positive polarization was due to particles substantially smaller in diameter than 9350 Å (so as to be in the Rayleigh-like regime at that wavelength), but comparable to or larger than the shortest wavelengths observed (2700 Å) (Travis et al. 1979a).

A definitive study of the haze particles overlying the main cloud was made by Kawabata et al. (1980, 1982). The basic data for this study are shown in Fig. 2, in which linear polarization is plotted against phase angle for regions near the pole and near the equator, for the four wavelengths at which the Pioneer Venus cloud photopolarimeter operates. The wavelength and phase coverage strongly constrain the aerosol properties in the haze. The haze has total optical depth over the poles $\tau = 0.6$–0.8; in the equatorial regions the optical depth is a factor of $\sim$10 less. This amount of haze exceeds any pre-Pioneer report; the time of Pioneer orbit insertion may thus represent the peak of a cycle decades long. The correlation is quite strong between the anomalous (haze-related) polarization and the bright polar regions seen in coincident images, so this haze is responsible for the bright polar caps. It is interesting that when bright polar caps were seen on Venus in October 1958, the derived particle size in the clouds was 0.3 μm (Dollfus et al. 1979).

In the equatorial regions, the observed polarization requires that the haze does not lie entirely above the main cloud; haze particles are mixed with the main cloud particles at least in the upper part of the cloud. Also, limb profiles at 6900 Å and 3650 Å (Lane and Opstbaum 1981) show that the haze extends upward to altitudes ⩾ 90 km. Quite likely it also persists in the deeper part of the main cloud, where it is invisible to remote measurements. The thickness of the haze layer has now been studied for ~ 900 d by the Pioneer Venus orbiter; it shows rather complex temporal variations, including a long-term global decrease by a factor of 5 in 2 Venus years (Kawabata et al. 1982).

The Pioneer Venus orbiter infrared radiometer provides another observation of upper level haze. Investigation of the infrared limb darkening shows that there are substantial high-altitude hazes locked to the terminators in the equatorial and middle latitudes (F. Taylor et al. 1980). The morning haze is 3000 km in extent at the cloud tops, centered on the equator 15° into the day side of Venus. At higher altitudes the haze region becomes smaller, diminishing to 1000 km in size at 85 km altitude. This could correspond to the excess submicron haze seen in the ultraviolet at the morning terminator. The evening haze is lower, denser, and less uniform than the morning haze, centered above 25°N latitude and extending well into the planet's night side. This may or may not be related to the morning haze, or for that matter to the submicron haze seen in the polarimetry.

C. Morphology of the Cloud Top Region

This section is a descriptive overview of the variability seen at the cloud tops in the ultraviolet; Sec. I.D discusses possible causes for the observed features. Dark markings on Venus visible only in the ultraviolet were first reported by Wright (1927) and Ross (1928). During the following fifty years these features were studied sporadically from Earth (Boyer and Camichel 1961; Smith 1967; Boyer and Guerin 1969; Kuiper 1969; Scott and Reese 1972; Dollfus 1975). Earth-based observations have severe limitations of resolution in space and time; small features cannot be seen, and there is danger of inferring spurious speeds and periodicities from the undersampled time record (Scott and Reese 1972; Beebe 1972).

Some early studies identified the following important properties of the large-scale cloud features (Dollfus 1975). Time scales for appearance and disappearance range from days to years. Discernible features such as the dark horizontal Y persist for weeks, reappearing at ~ 4 d intervals. The polar regions occasionally brighten relative to the rest of the planet on scales of up to one year. Many but not all features are symmetric about the equator; for example, the brightness of the two poles does not seem correlated.

A major advance in our understanding was provided by the television pictures from Mariner 10 in 1974 (Murray et al. 1974; Belton et al. 1976a, b;

Fig. 3. Mariner 10 mosaic of Venus. 6 February 1974. Source: NASA.

Anderson et al. 1978). (See Fig. 3.) This data set possessed both spatial and temporal resolution unobtainable from Earth. By tracking small features near the cloud tops, wind velocities could be inferred (Suomi 1974; Limaye and Suomi 1981). The investigators found features ranging in size from 30 km to the planetary-scale markings known from groundbased observations. The measured contrast of the features and their lifetimes increased with feature size. The inferred zonal wind speeds showed jets of faster rotation at 40 to 50° latitude, superimposed on a solid-body rotation at lower latitudes. The wind velocity was ~ -100 m s^{-1} near the equator, consistent with the four-day retrograde rotation of the cloud tops inferred from earth-based observations. The larger features were seen to be continuous across the midlatitude jets;

thus, these planetary-scale features (which travel at the mean wind velocity characteristic of the equatorial latitudes) must retain their coherence through some guiding mechanism, such as wave phenomena. The largest features seen from Earth were therefore explained by propagating planetary waves (Belton et al. 1976b).

A new and more extensive data set was provided (and continues to expand while the mission continues) by the Pioneer Venus orbiter cloud photopolarimeter (Travis et al. 1979a,b; Rossow et al. 1980; Del Genio and Rossow 1982). Although the resolution from this imaging is not as high as the best from Mariner 10, the daily systematic coverage provides a more uniform data set. A sample of some images is shown in Fig. 4. The brightness of the pole is not enhanced in these photos, as was done in the Mariner images (Limaye and Suomi 1977b). A uniformly reflective sphere would have dark polar regions. The images clearly show that the orbiter has been observing Venus in one of the periods of polar brightening. In this figure a common characteristic of Venus is evident, that the features can have different tilts with respect to the equator. The extremes are shown schematically in Fig. 5. The tilt of features seems to oscillate regularly between these extremes.

In addition to whorls, bands, caps, and streamers as shown in Fig. 5, cellular features were seen, similar to those observed by Mariner 10 (Murray et al. 1974). The majority of the Pioneer Venus images show these cells at low latitudes ($\pm 20°$), predominantly near and just downstream from the subsolar point. Both bright cells with dark rims and dark cells with bright rims were seen; both kinds are considered as evidence for convection (see Fig. 6). Pioneer and Mariner results agree that the lifetimes of the cells range from < 3 hr for the smallest up to one day for the largest.

The longer time of observation from Pioneer Venus has provided a better understanding of the Y features well known from groundbased and Mariner investigations. Rossow et al. (1980) believe that this feature is actually a composite structure formed by the correlated appearance of two or three separate features. This explains why its appearance is only quasi periodic. The regular evolution of the feature was interpreted by the Mariner 10 team as the shearing of an original bow-like wave downstream from the subsolar point. The more extensive data from Pioneer Venus require a view more symmetric in time; all the albedo features (such as the Y) evolve in a coordinated manner relative to latitude circles, to produce an oscillatory sequence (Rossow et al. 1980; Del Genio and Rossow 1982). Because they saw only part of this oscillation, the Mariner 10 investigators interpreted it instead as a systematic trend.

Perhaps a more important point concerns the permanence of a latitudinal shear in the zonal velocity. Both Mariner investigators (Suomi 1974; Limaye and Suomi 1977a,b, 1981) and Pioneer investigators (Rossow et al. 1980) calculated the wind velocities at the cloud tops by tracking small features. Comparison of these results (see Fig. 7) shows substantial differences.

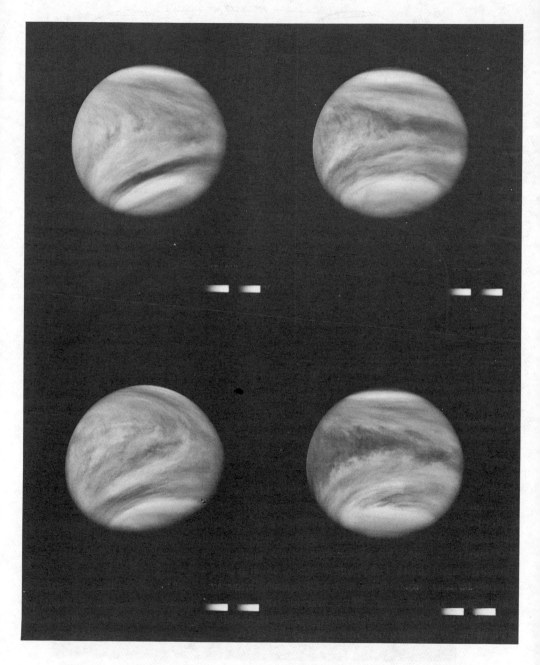

Fig. 4. Four representative images of Venus from the Pioneer Venus cloud photopolarimeter, taken 10, 11, 14 and 16 February 1979. The images show various levels of contrast and inclination of features with respect to latitude circles.

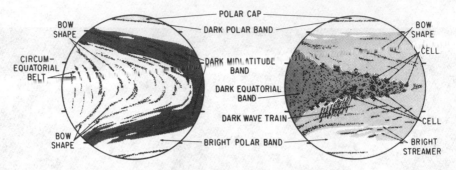

Fig. 5. Schematic diagram defining the basic types of cloud features observed in Venus ultraviolet images. The two views depicted here typically occur two days apart and represent the maximum and minimum tilt configurations, respectively. The tick marks on each circle indicate 20° and 50° of latitude in each hemisphere. (From Rossow et al. 1980a.)

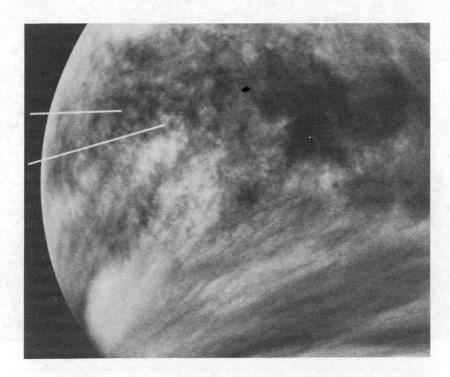

Fig. 6. Enlargement of equatorial region showing many cells suggestive of convective activity. The upper arrow points to a dark-rimmed cell, the lower arrow to a bright-rimmed cell.

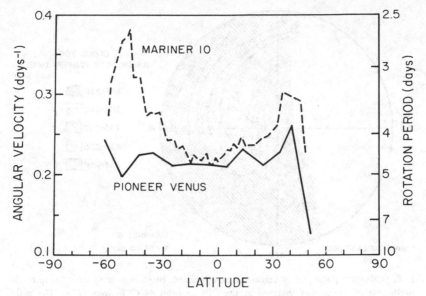

Fig. 7. A comparison of angular velocity profiles obtained from the OCPP cloud-tracked winds and Mariner 10 cloud-tracked winds measured by Limaye and Suomi (1980). (From Rossow et al. 1980*a*.)

Mariner 10 observations show the Venus atmosphere rotating as a solid body with strong midlatitude jets superimposed at ±45° latitude. Constant angular velocity is also evident in the Pioneer data, but only weak jets are seen at 35°S and 15°N, but since these results are based on only a five-week period (Rossow et al. 1980) the true average wind profile on Venus remains unclear. The differences with the Mariner 10 results show that even mean atmospheric motions can change significantly with time.

At the poles, some unusual horizontal features were seen by the PV orbiter infrared radiometer (F. Taylor et al. 1980). Above 70° latitude the investigators found a crescent-shaped collar around the pole consisting of anomalous and variable temperature and cloud structure (see Fig. 8). By combining the PV orbiter infrared radiometer data with radio occultation results from the same region, a consistent explanation has emerged for the polar anomalies (McCleese et al. 1981). The combined data show that the midlatitude collar (53°–70°N) is a region of intense thermal inversions, not elevated clouds as was previously concluded from infrared data alone. Temperatures in the collar region are up to 50 K colder than equatorial temperatures at the same altitude. In the polar hot spots the temperature profile does not show an inversion, but is nearly isothermal. Thus, these features appear as hot "eyes" with a dipole structure that rotates about the pole with a 2.7 d period.

All these polar phenomena provide some evidence for subsidence of the atmosphere in the polar region. There is a substantial overall lowering of

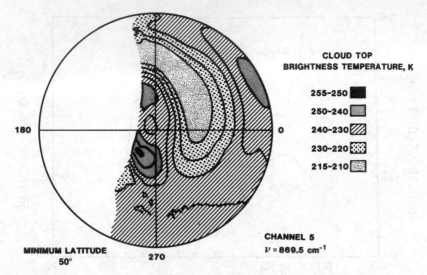

Fig. 8. OIR north polar plot of brightness temperature. Brightness temperature map of the
north pole of Venus was observed by the OIR on orbit 34 (7 January 1979). The radial
distance from the perimeter at 50°N is proportional to latitude. (From Knollenberg et al.
1980.)

the upper cloud north of 55° latitude. The cloud top at 65°N is at the 125
mbar pressure level, compared with 50 mbar at the equator. Thus, the mean
meridional circulation of the atmosphere might be a simple Hadley cell, with
warm air rising at the equator and streaming to the poles where it cools to
return to the equator at lower altitudes. The meridional circulation measured
by cloud-tracked winds (Limaye and Suomi 1981; Rossow et al. 1980) (see
Fig. 9) shows small but generally poleward velocities at all latitudes, and
is consistent with this simple interpretation (for more details see Chapter 21
by Schubert).

D. Nature of the Ultraviolet Contrasts

The nature and origin of the dark cloud-top features discussed in Sec. I.C
have caused lively debate. The Pioneer Venus mission and continuing earth-
based studies have provided a broad range of descriptive knowledge, but
important questions remain. What absorbers are responsible for the darkening? What physical and chemical mechanisms bring the absorbers selectively
into view? Is the same cause responsible for all km- to planetary-scale features? Any picture of the cloud-top chemistry and physics is incomplete if it
fails to explain the observed contrasts. Furthermore, studies of cloud-top
dynamics use the features as markers; understanding these markers is essential
to understanding the dynamics of the Venus atmosphere, which is evidenced
by their motion.

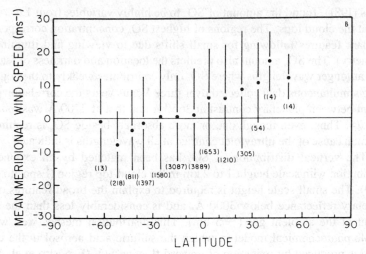

Fig. 9. Time and longitudinally averaged meridional wind speed as a function of latitude obtained from tracking ultraviolet cloud features. The number of wind vectors averaged to produce each value is indicated by the numbers in parentheses. The vertical lines indicate the rms deviation of the measurements about the mean. (From Rossow et al. 1980a.)

Of all the absorbing constituents suggested for the Venus atmosphere prior to the Pioneer Venus mission, the most popular was sulfur. Hapke and Nelson (1975) suggested that allotropes of sulfur had the correct color to explain the visible and ultraviolet albedo of Venus, and Prinn (1975) suggested that sulfur would result from COS photolysis occuring at the cloud tops. Hapke and Nelson (1975) required a large fraction of the total cloud mass to be composed of sulfur; however, Pollack et al. (1979, 1980b) and Tomasko et al. (1979) showed that orthorhombic sulfur alone, when identified with any of the size modes observed *in situ* by the PV cloud particle size spectrometer (Knollenberg and Hunten 1979), could not explain the known spectral albedo of the planet.

Coincident with the beginning of the Pioneer Venus mission, sulfur dioxide was found in the Venus atmosphere from earth-based observations (Barker 1979), PV ultraviolet observations (Stewart et al. 1979), the International Ultraviolet Explorer (Conway et al. 1979), and *in situ* gas chromatographic observations from the PV probes (Oyama et al. 1979). Using the SO_2 distribution determined by Esposito et al. (1979), Pollack et al. (1979) successfully matched the spectral reflectance of Venus over a range of wavelengths, and suggested that SO_2 alone might explain all the absorption in the Venus clouds. Later work by Esposito (1980) and Pollack et al. (1980b) showed that both the contrast of the dark features and the overall planetary reflectance could be matched by SO_2 only at wavelengths < 3200 Å.

The analyses of Esposito et al. (1979), Esposito (1980), and Esposito and

Gates (1981), found the amount of SO_2 to be highly variable, from 1 to $>$ 100 ppb at the cloud tops. The regions of highest SO_2 concentration correspond to the dark features (allowing for small shifts due to viewing and illumination geometry). The SO_2 amount also predicts the location and darkness of features seen at longer wavelengths where SO_2 only contributes weakly to the contrast; using simultaneous data collected over three Venus years the correlation coefficient between planetary contrast at 3650 Å and that at 2700 Å was found to be 0.74. Thus, even though one or more absorbers beside SO_2 is required, a common cause of the ultraviolet contrast at all wavelengths is indicated.

The vertical distribution of SO_2 has been matched by an exponential distribution with scale height 1 to 2 km in the cloud-top region (Esposito et al. 1979). The small scale height is required to explain the broad features of the planetary reflectance below 3000 Å, and is considerably less than the scale height of the ambient gas ($\sim$ 5 km). This distribution meshes well with a simple photochemical model in which the sulfuric acid aerosol in the upper cloud is produced by oxidation of upward flowing SO_2 (Esposito et al. 1979; Winick and Stewart 1980; also see Sec. IV).

The identity of secondary absorbers remains speculative. Pollack et al. (1980a) exclude orthorhombic sulfur, various iron compounds, and nitrogen dioxide gas because their absorption spectra do not match those derived for the absorber (see Fig. 10). They suggest Cl_2 as a second absorber, but there are two problems with Cl_2: first, the amount required is orders of magnitude larger than predicted by chemical models of the Venus atmosphere (Sze and McElroy 1975; Winick and Stewart 1980); and second there is no strong reason why it should physically be associated with the SO_2. Recent results by Yung and DeMore (1982) show that large quantities of Cl_2 might be produced above the clouds; this possibility deserves further investigation.

Recently Toon and Turco (1982) have shown that amorphous sulfur may be the ultraviolet absorber, citing as evidence that:

1. Sulfur could account for the high real refractive index required to explain the large backscattering observed in the upper clouds by the Pioneer Venus nephelometer (see Toon and Blamont 1982);
2. Sulfur, through formation of new particles by nucleation, provides an explanation for the bimodal size distribution and relative maximum particle concentration observed near 60 km by the PV Large probe cloud particle size spectrometer (LCPS);
3. A cloud physics and chemistry model indicates that sulfur will be produced in O_2-poor regions of the upper clouds. Upper cloud regions in which strong updrafts occur would be favorable for sulfur formation, as the correlation between SO_2 and ultraviolet features suggests;
4. Sulfur would be properly located in altitude since it cannot form above the cloud top, would be depleted in the lower clouds, and would have a natural correlation with SO_2;

5. The chemistry model shows that sulfur vapor in the upper clouds is greatly enriched in short chain allotropes such as S_3 and S_4 which are strongly absorbing with bands centered near 4000 Å and 5300 Å. Even a few percent of these allotropes in particles composed mainly of S_8 is sufficient to account for the albedo of Venus and the solar energy deposition profiles;

6. Due to thermal relaxation to S_8, the amorphous sulfur provides a mechanism to account for the short lifetime of the ultraviolet features;

7. Sulfur particles constituting a small mass fraction of the upper cloud do not violate any of the constraints imposed by Pioneer Venus observations (Pollack et al. 1980a, b; Tomasko et al. 1980b).

The high correlation between observations at 2700 Å and at 3650 Å requires some strong association between the absorbers. Esposito (1980) suggested that physical uplift or mixing might bring the absorbers into view. The small SO_2 scale height assures that a small upward transport would significantly increase the amount of SO_2. If some loss process likewise maintained a steep distribution of the second absorber, such updrafts would simultaneously increase this absorber's mixing ratio as well. This explanation postulates a physical connection between the two absorbers, because their vertical distributions are similar near the cloud top. Both Esposito (1980) and Pollack et al. (1980b) find that the amount of absorber increases with depth into the clouds. A second possibility is that there is a chemical connection between the absorbers. Secondary absorbers may be intermediates or reaction products in the sulfur cycle that produce the aerosols. One example is amorphous sulfur, discussed above; a further possibility is that the unknown absorber results from SO_2 combining with other reactive photochemicals at the cloud tops.

Several physical mechanisms could account for the nonuniform visibility of these absorbers, and thereby create the observed contrasts in brightness across the planet. Esposito's (1980) mechanism noted above accomplishes this by variations in the local effectiveness of vertical mixing. Where the mixing is strongest, more absorber is brought to the cloud tops. Pollack et al. (1980b) estimate that a factor of ~ 3 increase in the amount of absorber is sufficient to reproduce the observed contrast. Since the SO_2 has a scale height of only 1 to 2 km, a factor of 3 difference in abundance represents an altitude difference of only a few km; even small variations in vertical transport might be very effective over this short range in altitude.

Another possibility suggested by Pollack et al. (1980b) is that the bright ultraviolet regions are created by a layer of overlying haze similar to that seen over the Venus poles (Kawabata et al. 1980; see above). Where the optical depth of the haze layer is smaller, the main cloud deck and its included absorbers are more visible, forming an ultraviolet dark feature. The amount of haze necessary to match the contrast ($\tau \simeq 0.5$–0.8) is comparable to the

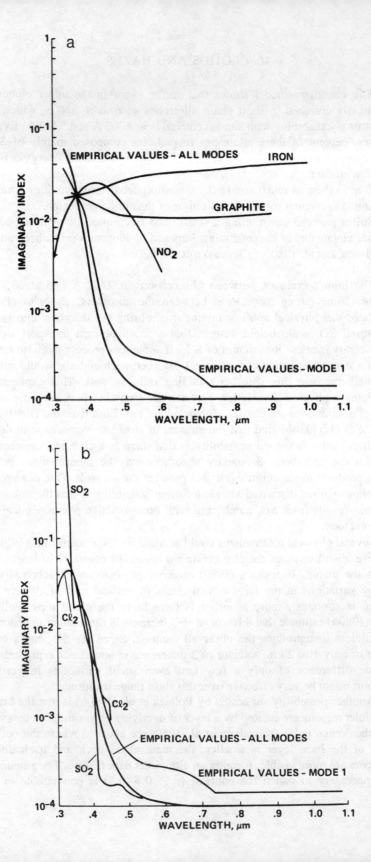

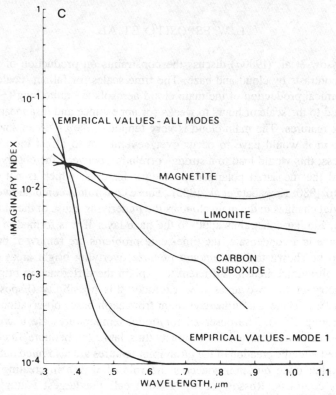

Fig. 10. Comparison of the empirically determined imaginary index of refraction of the ultraviolet absorber present alternatively in just Mode 1 cloud particles and all sized particles and the indices of refraction and absorption coefficients of several candidate solids and gases. The coefficients of each candidate have been multiplied by a scale factor so that it agrees with the empirical values at a wavelength of 0.37 μm. Scale factors of 8.0×10^{-3}, 2.0×10^{-2}, 3.0×10^{-2}, 2.1×10^{-2}, and 3.7 were used for metallic iron, graphite, limonite (hydrated ferric oxide), magnetite, and carbon suboxide, respectively. The absorption coefficients of Cl_2, SO_2, carbon suboxide, and limonite were obtained from Serry and Britton (1964), R. Nanes, R. Boese, and J. B. Pollack (manuscript in preparation), Smith et al. (1963), and Egan and Becker (1969), respectively, while those for the other materials are referenced in the work by Pollack et al. (1979). Information on the absorption of Cl_2 between 0.48 and 0.60 μm is given in Douglas et al. (1963). (From Pollack et al. 1980b.)

optical depth over the pole inferred by Kawabata et al. (1980). The ultimate cause for the variation in thickness of the overlying haze is temperature variations; lower temperatures could favor formation of haze above the main cloud.

Recently, a number of authors have suggested a chemical cause for the contrasts. Toon and Turco (1982) propose oxygen limitation in the dark areas, allowing the amorphous sulfur to exist there long enough to be seen. Esposito and Travis (1982) put forward a similar scheme in which oxygen and/or water is limiting; at these locations a buildup of SO_2 and a depletion of H_2SO_4 occurs, creating an ultraviolet dark area. Explicit calculations show that the mass budget of sulfur is balanced between excess sulfuric acid haze over the bright regions and excess SO_2 in the dark regions (Esposito and Travis 1982).

Rossow et al. (1980*a*) discuss the constraints on production of the observed contrasts by cloud and haze. The time scales for fallout, coalescence and chemical production of the main cloud aerosols are quite long ($\sim 10^7$ s) compared to the scale of hours to weeks for appearance and disappearance of the dark features. The main cloud is very tenuous (like an urban smog) and changes in it would have to occur over several km to affect the observed brightness; this would lead to a strong correlation between the observed contrast and the measured polarization of dark features, which is not evident (Esposito 1980; Kawabata et al. 1980). These constraints combine to exclude substantial changes in the main cloud as the underlying cause of the ultraviolet features. Similar constraints apply to the haze layer if it is formed of H_2SO_4. If the haze is a condensate, the time-scale problems are removed, but it remains to be shown that the amount of haze overlying bright areas relative to dark ultraviolet areas is sufficient to explain the difference in brightness. Polarimetry of the two areas now excludes this possibility (Esposito and Travis 1982). There is a further constraint from earth-based observation of CO_2 lines (Young 1975). The observed rotational temperatures are uncorrelated with the amount of ultraviolet contrast; thus large temperature fluctuations associated with the creation of the ultraviolet features are also ruled out.

Another open question concerns the role of waves in creating the ultraviolet contrasts. Rossow et al. (1980*a*) call the largest features quasi periodic; they are certainly reminiscent of waves. Covey and Schubert (1981*b*) interpret the contrasts as large-scale atmospheric waves propagating slowly with respect to the rapid cloud-top zonal winds. Planetary-scale gravity waves trapped at cloud levels have properties which could correspond to the observed contrast patterns. If they do, the next step is to show how these waves produce the proximate cause of the dark features (e.g. enhanced vertical transport, decreased haze thickness, or temperature alteration). Houben (1982) has shown that radiative-dynamic feedback involving the ultraviolet absorber and the cloud can drive a planetary wave which produces Y-shaped features consistent with the observations.

II. THE MAIN CLOUD DECK: RESULTS OF *IN SITU* MEASUREMENTS

The early remote observations that preceded *in situ* measurements offered little variety in microphysical problems. Consider a cloud of monodispersed 2 μm H_2SO_4 droplets: there are only a few microphysical processes to consider since the lack of larger particles obviously precludes precipitation mechanisms. This uniform cloud-top region covered the entire planet, and observations over several years showed little change. However, these observations were limited to the uppermost cloud regions; the bulk of the lower cloud structure remained to be discovered.

Lacis (1975) attempted to model the clouds to lower depths using remote observations and Venera 8 photometric data (Avduevsky et al. 1973b). He was able to place the nominal cloud base in the vicinity of 40 km, near the level of 35 km argued by the Venera 8 team (Marov et al. 1973a). A somewhat higher cloud base was inferred from Mariner 10 occultation data (Kliore, personal communication 1975). Multiwavelength photometers on board Veneras 9 and 10 confirmed the general location of the clouds (Avduevsky et al. 1976e). However, the details of the vertical structure of the main cloud deck were not apparent until entry probe experiments that specifically measured cloud particle characteristics began to return data.

A. Nephelometer Observations

The cloud structure and the microphysical characteristics of the aerosol component populating the clouds were first investigated in detail by nephelometric experiments on Veneras 9 and 10 (Marov et al. 1976a,b; in greater detail Marov 1978, 1980). More recently nephelometric measurements were made by Venera 11 (Marov et al. 1979a) and Pioneer Venus probes (Ragent and Blamont 1979). The most detailed analysis of the Venera nephelometric data processing is given by Marov et al. (1980). We will summarize the principal issues from this analysis and discuss some problems pertinent to the Venus clouds.

The nephelometer technique is based on determining microphysical properties of aerosols from the measured angular dependence of radiation scattered by the medium, taking into account experimental error. We note that such problems belong to the ill-conditioned class: unique results are not guaranteed. The Venera nephelometer measured the scattered radiation at scattering angles of 4°, 15°, 45° and 180°.

Detailed predictions of light-scattering characteristics expected from Venus cloud particles required extensive numerical experiments (Shari and Lukashevich 1977; Marov et al. 1979a). This work has shown that reconstruction of the phase function is possible by using scattering measurements taken at only a few scattering angles Θ_j, assuming a certain size distribution for aerosol particles to obtain sufficiently reliable estimates of aerosol parameters: particle mean size (effective radius) r_{eff}; concentration N; distribution width μ; and refractive index n. The investigation was carried out for a wide class of plausible atmospheric aerosols and for the angles 4°, 15°, 45°, and 180° chosen as the most informative. Mie theory for single scattering, generalized for an ensemble of particles, was used to describe the scattering behavior of particles with radius r at wavelength λ (where their size parameter $x = 2\pi r/\lambda$), and to find which of the parameters involved have the most influence at these angles. This theory is only rigorously correct for spherical particles, but a similar approach can be applied in some respects for irregularly shaped particles (Pollack and Cuzzi 1979). In the analysis of Marov et al.

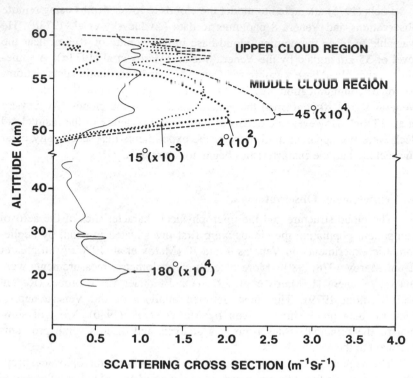

Fig. 11. Venera 9 nephelometer results. (From Marov 1978.)

(1980), a gamma distribution was assumed for the particle size distribution function: $n(r) = r^{\mu} \exp(-\mu r / r_m); 0 \leqslant r \leqslant \infty$.

The data were analyzed by comparison with a theoretical model. The altitude dependence of the observed efficiency yields the vertical structure of the clouds. Figure 11 gives the scattering cross section observed by Venera 9 at its four angles of observation. The results of the detailed analysis for the dependence of volume scattering coefficient from Veneras 9, 10 and 11 are shown in Fig. 12 (Marov et al. 1980). Table I summarizes the principal findings from nephelometric measurements for all Venera missions. The Venera measurements use a different surface reference altitude than Pioneer Venus's. We have adjusted all Venera altitudes downward by ~ 1.0 km at cloud altitudes to normalize with Pioneer measurements.

Marov defines three principal layers with different microphysical and optical characteristics found within the clouds:

Layer I, the upper layer from 60 km (where measurements began) to 57 km;

Layer II, the middle layer from 57 to 51 km;

Layer III, the lower layer from 51 to 47 km.

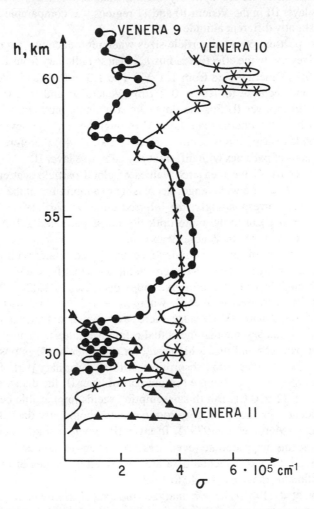

Fig. 12. The vertical dependence of the volume scattering coefficient from observations by
Venera 9, 10, and 11. (From Marov et al. 1980.)

This evidence clearly precludes considering the Venus clouds as uniformly
mixed. However, the layers are not isolated; their properties are interrelated.
Two intermediate layers: I′ between I and II; and II′ between II and III, can
also be distinguished. All layers in the clouds exhibit spatial and temporal
variations of microphysical properties. In the daytime the lower layer III
boundary is 48, 47, and 46.5 km above the planetary surface, according to
Veneras 9, 10, and 11 respectively. In the Venera 10 descent region, both

layer I and layer III are optically thicker than in Venera 9 measurements, whereas layer III in the Venera 10 and 11 regions was comparable in thickness but at a slightly different altitude.

The spectrum of cloud particle sizes when reproduced by a gamma distribution gives rise to the effective radius r_{eff}. These radii vary from 1.0 ± 0.05 to 1.35 ± 0.1 (Venera 9) and from 1.1 ± 0.1 to 1.3 ± 0.15 μm (Venera 10) in layer I, and increase to 1.9 ± 0.2 μm (Venera 9) and to 2.0 ± 0.5 μm (Venera 10) in layer II. For the upper layer a single-mode curve is satisfactory, but from the intermediate layer I' through layer II the deviation from a single-mode distribution is distinct. Marov et al. (1980) therefore argued that ≥ 2 fractions of particles with different r_{eff} comprise layer II.

In layer III all the measured values of cloud particle scattering change synchronously over a wide range, decreasing to a minimum at the very bottom of this zone. Correspondingly, the inferred effective particle size also varies little: 1.6 ± 0.1 μm in the more optically dense parts and 1.1 ± 0.2 μm in less dense parts, for both Veneras 9 and 10.

The behavior of refractive index n_r generally correlates with particle size distribution. In layer I the best-fit aerosol model is that for which $n_r = 1.46 \pm 0.01$. Through layer I' the refractive index decreases to 1.42 ± 0.01, then it dramatically decreases in layer II, where $n_r = 1.35 \pm 0.02$ was derived for both Veneras 9 and 10. Because there was evidence for two fractions of particles, we can assume that the actual refractive index is mainly determined by larger particles and that a higher n_r for the other mode is possible. Note also that if we assume some absorption, with an imaginary part of the refractive index $n_i \simeq 3 \times 10^{-3}$ in the bottom part of layer II, the data are consistent with $n_r = 1.42 \pm 0.01$. But this assumption seems implausible because there is no evidence for such an absorption in spectrophotometry data at ~ 0.9 μm wavelength (Moroz et al. 1979b). In layer III, where a single mode distribution also seems inappropriate (even assuming a very broad distribution with μ as low as 2), the experimental data are best fitted by values of the refractive index within the range $n_r = 1.40$ to 1.42.

Toon et al. (1982a) have reanalyzed the data of Marov et al. (1980) using a more general type of size distribution. They find that using size distributions of the type reported by Knollenberg and Hunten (1980), but replacing the largest size mode with a tail of sulfuric acid particles, yields optical parameters similar to those reported by Venera 9 if the particles are composed entirely of compounds with refractive indices typical of concentrated H_2SO_4 solutions.

Other parameters of aerosols that evolved from the inversion of experimental data are presented in Table II. The width of the distribution is parameterized by μ; the scattering coefficient, number density, and optical depth by $\sigma, N,$ and τ, respectively.

Below the clouds, whose bottom seems to vary in height, there is a subcloud haze between 46–48 and 31–34 km; its properties are also given in

TABLE II

Structure and Properties of Venus Clouds[a]

Designation of Cloud Layers	Height Range km	r_{eff} μm	μ	n_r	σ km^{-1}	N cm^{-3}	τ
Upper layer I	63–57	1.0–1.4	8–20	1.46 ± 0.01	2–7	200–350	7–20
Middle layer II	57–51	1.7–2.5	8–20	1.35 ±0.02 [b,d]	3–5	250–350	27–26
Lower layer III [c]	51–47	1.7–0.9	2–4	1.41 ± 0.01	1–5	50–150	6–12
Subcloud haze	47–31	0.15	2–12	1.7	0.1–0.5	2	3

[a]From Marov et al. (1980).
[b]Assuming nonabsorbing aerosols the lower limit is more probable.
[c]Variable in height especially near the bottom layer.
[d]Toon et al. (1982a) find refractive indices compatible with sulfuric acid.

Table II. The scattering pattern in this region was found to be produced mainly by finely dispersed aerosols ($r_{eff} \sim 0.15$ μm); a contribution of larger highly refractive particles is also possible (Marov et al. 1976a). This interpretation leads to a relatively small optical depth and agrees well with the results of Venera 8, 9, and 10 measurements of the attenuation of integral and quasi monochromatic solar radiation fluxes (Marov et al. 1973a; Avduevsky et al. 1973b, 1976e).

For layer I a single mode for the size distribution is satisfactory, whereas in layer II the features observed cannot be reconciled with the assumption of a single mode distribution. We therefore argue that there are $\geqslant 2$ fractions of particles with different characteristics of r_{eff} that compose layer II. There are also some indications for a multimodal distribution in layer III, though the inversion results are less definite here so we cannot confidently argue for this.

Some general trends should be noted: (1) the integrated optical depth reported by Marov (1978) was ~ 32; (2) cloud opacity generally increases with descent through the cloud system, reaching a maximum at ~ 50 km, and then decreases rapidly, vanishing at ~ 48 km; (3) only backscatter (180°) is observed below 47 km, and it increases with decreasing altitude. This region of backscatter-only signals led to a great deal of speculation about possible particle composition. Young (1979), realizing that only large particles of high index could create such a result, suggested that large particles of liquid sulfur were present down to 20 km altitudes. Also solid sulfur particles would explain many of the features associated with the ultraviolet absorption. Hapke and Nelson (1975) suggested a complete sulfur cycle, equivalent to the Earth's hydrological cycle, with liquid sulfur rain. We now believe that the backscatter measurements are not reliable, perhaps due to temperature effects on the probes; later Pioneer Venus results support this conclusion. It is noteworthy that Marov invoked multimodal properties, independent of actual size distribution measurements by Knollenberg and Hunten (1979) who found multimodality dominant throughout the cloud system.

About the same time that Marov's (1978a) synthesis results were available, Pioneer encountered Venus. The cloud *in situ* measurements included four nephelometers (Ragent and Blamont 1980) on board all four entry probes, and a particle size spectrometer (Knollenberg and Gilland 1980) on board the Large probe. The coordinates and locations of the four entry probe sites are given in Table III. We discuss the nephelometer results first because they are directly comparable to Marov's results.

The Pioneer nephelometers only measured backscatter (175°). The entry altitudes for the four probes varied from 66 km at the Day probe site to 62 km at the North probe site; all the nephelometers survived to the surface (the Day probe survived on the surface for nearly an hour). The measurements reported by Ragent and Blamont (1979, 1980) for all four nephelometers are summarized in Fig. 13. These nephelometer profiles reveal the same general

TABLE III

Locations of the Pioneer Venus Probes

Probe	Descent Time [a] (UT)	fps [b] (deg)	1950.0 Aries/Ecliptic (Celestial)		1970 IAU Pole (Body Fixed)		PV Pole, IAU Long (Body Fixed)		χ [c] (deg)	LT [d] (HHMM)
			CLAT (deg)	CLON (deg)	Lat (deg)	Long (deg)	Lat (deg)	Long (deg)		
Large	1845:54.31	−31.14	3.1	341.6	4.4	304.0	4.0	304.0	65.7	0738
North	1849:52.75	−68.74	58.8	40.4	59.3	4.8	60.2	3.2	108.0	0335
Day	1852:46.04	−23.78	−32.5	354.8	−31.2	317.0	−31.3	317.7	79.9	0646
Night	1856:21.58	−40.63	−28.1	94.9	−28.7	56.7	−27.4	56.6	150.7	0007
Sun direction			−1.2	276.1	−0.5	238.5	−1.8	238.5	0	1200
Earth direction			−2.2	39.3	−1.6	1.7	−0.7	1.7	123.1	0347

[a] At spacecraft—to obtain time of receipt of signal at earth add 3 min and 12 s.
[b] Relative to rotating planet.
[c] Solar zenith angle.
[d] Venus local time.

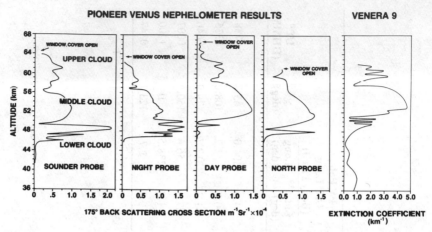

Fig. 13. Pioneer Venus nephelometer results. The remarkable similarity in the above nephelometer profiles is indicative of the planetary nature of this cloud system. This comparison extends to the Venera 9 results of Fig. 12 which preceded Pioneer Venus by three years. (From Knollenberg et al. 1980.)

structure reported by Marov et al. (1976a, b, 1978a), a fairly uniform central cloud region 6 to 8 km thick bounded by regions of reduced particle activity which separate upper and lower more structured regions, which we assume are similar to Marov's transition regions. Comparison of Figs. 11 and 13 shows striking similarities among the nephelometer results, though separated by almost 3 yr. The cloud base is at 45–47 km in all cases; the upper transition region varies in altitude from 56 to 59 km and the lower transition region from 49 to 51 km. The vertically integrated backscatter ranges from 0.5 to 2.5×10^{-4} m^{-1} sr^{-1} for both Venera and Pioneer nephelometers.

Although these similarities are striking we should not ignore the differences, which show considerable variability with time. Marov's layer II is very prominent while layer III is a layered region of weak backscatter which corresponds most closely to the PV Day probe nephelometer results. As discussed below, the middle and lower cloud defined at the PV Large probe site have similar characteristics to Marov's layer II.

B. Size Distribution Measurements

The only direct measurement of particle size spectrum in the Venus clouds is that of the particle size spectrometer (LCPS) on board Pioneer (Knollenberg and Hunten 1979, and in greater detail, 1980). This instrument was on the Large probe, and so provided detailed size measurements at only one location. Measurements of the particle size distributions by the LCPS included both single particle light scattering and shadowgraph imaging techniques. The light scattering method covered particles from 0.5 to 5 μm

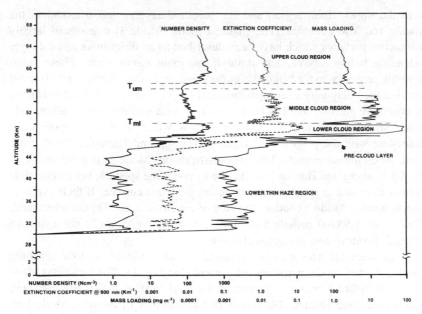

Fig. 14. Vertical structure of Venus cloud system. The data in this figure represent direct computations for the LCPS data. No allowance has been made for particles < 0.6 μm, the lower limit of size sensitivity or above the largest sizes actually measured. A density of 2 g cm^{-3} was assumed for mass loading computations. The thickness of T_{um} is ~ 1 km while T_{mi} is several hundred meters thick. (From Knollenberg and Hunten 1980.)

diameter; the imaging method covered larger particles from 5 to 500 μm. It is important to remember that all sizes reported from this instrument are diameters, not radii.

Data from the LCPS provide sufficient size resolution to characterize the microstructure of the clouds at altitude intervals as small as 50 m. While this extensive data set exists only at the Large probe location, similarities observed at all entry sites by the nephelometers suggest that this information is generally applicable to all entry sites. Figure 14 presents the vertical structure as observed by the LCPS for number density, integrated extinction coefficient, and computed mass loadings for particle sizes > 0.6 μm. The behavior of the individual particles is totally masked in the larger picture presented here. However, the same vertical structure observed by the Venera and Pioneer nephelometers is evident, i.e. a central cloud region separated from lower and upper cloud regions by transition layers.

On the basis of particle size properties, Knollenberg and Hunten (1979) defined a main cloud deck composed of upper, middle, and lower cloud regions (previously presented in Table I). A tenuous haze and two precloud layers lie beneath the main cloud deck. The size distribution is bimodal in both the upper cloud region and the precloud layers, and trimodal in the middle and lower regions. The smallest mode (Mode 1) consists of largely submicron particles which have been described as an ubiquitous aerosol layer extending below, above, and through the main cloud deck. These small aerosols probably extend into the upper haze region at altitudes of 70 to 90 km, but they may vary in composition and origin. Mode 2 of the trimodal set is extremely narrow, and has been identified with sulfuric acid at all altitude levels since the modal size and the narrow width of the distribution are consistent with the properties defined by Hansen and Hovenier (1974) from earth-based measurements. The largest size mode (Mode 3) is believed likely by Knollenberg and Hunten (1980) to be a crystalline species, but its chemical composition and exact physical morphology remain elusive. If their explanation is correct, Mode 3 contains the bulk of the cloud mass. On the other hand, Toon et al. (1982a) believe that Mode 3 is a tail to Mode 2, not a distinct peak; this controversy is discussed below.

The trimodal sizes may be separated at most altitudes and the resulting number density modal partitions are presented in Fig. 15. The measured Mode 1 aerosol includes only the fraction of the total aerosol > 0.6 μm diameter. Knollenberg and Hunten (1980) provided an estimate of the total Mode 1 by fitting log-normal plots to the aerosol size distribution data. However, the use of such functions and the resulting estimates of total aerosol assume that growth has evolved through coagulation of smaller aerosol. While this is probably true in the upper cloud region it may not be true elsewhere, as we will show. Thus we chose to only present the estimated Mode 1 total in the upper cloud region.

Upper Cloud Particle Size Characteristics. The upper cloud region is populated by a mixture of two particle sizes creating a bimodal size distribution of Mode 1 and Mode 2. The average size distribution for this region is shown in Fig. 16, after Knollenberg and Hunten (1979). They identified H_2SO_4 as the larger Mode 2 particles by matching the narrow size distribution of Mode 2 in the middle cloud region (where Mode 2 is clearly separated from Mode 1) with previous earth-based data. Mode 1 was then traced back up into the upper cloud region, where the two populations tend to merge. There are distinct altitudes in the upper cloud region where the bimodality is unmistakable, but the numbers of the smaller aerosol are sufficient everywhere to mask the narrow size distribution of Mode 2. Identification of these small particles is complicated because the LCPS only sees the tail of this underlying aerosol population, due to the 0.6 μm lower size limit.

We see that the upper cloud region has a maximum number density (and thus extinction coefficient and mass loading) in the vicinity of 60 km. A transition region between the middle and upper cloud regions is reached at 56.5 km, where there is a sharp decrease in particles mentioned above in our discussion of the nephelometer observations. In this transition region, particle

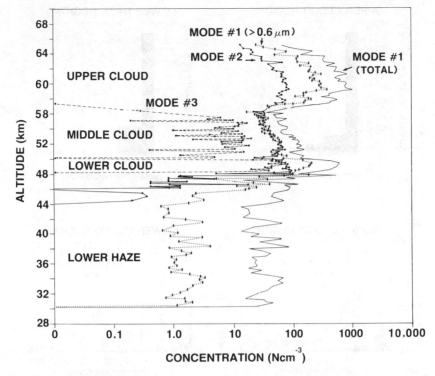

Fig. 15. Modal partitioning of number density. Mode 1 is shown for both the sizes measured by the LCPS and the upper cloud Mode 1 total expected from modeling efforts. (From Knollenberg and Hunten 1980.)

properties change from those characteristic of the upper cloud region to those characteristic of the middle region. The transition region is ~ 1 km thick and is at the same altitude as Marov's first transition region; it is designated T_{um} in Fig. 14.

Middle Cloud Particle Size Characteristics. In the middle cloud region the total number density gradually increases with decreasing altitude, while the extinction coefficient and mass loading remain essentially constant down to ~ 50.5 km. The greatly reduced particle activity at this altitude defines a second transition region between the middle and lower cloud regions, designated T_{ml}. The average size distribution for the middle cloud region is shown in Fig. 16.

The middle cloud region is characterized by three observed size distribution changes: (1) the truncated Mode 1 aerosol population decreases in number density by nearly an order of magnitude; (2) increased growth in the Mode 2 aerosol provides clear size separation between Modes 1 and 2; and (3) the

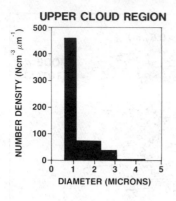

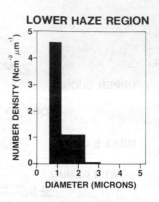

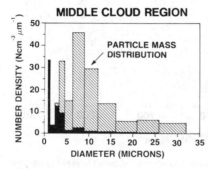

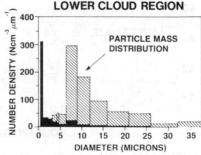

Fig. 16. Average size distributions measured by the LCPS. The multimodal character of the size distributions is even evident when vertically averaging over several kilometers. This is most apparent in the size distributions in the middle and lower cloud regions, but it is also apparent within the upper cloud and lower haze regions at precloud layer altitudes. The two precloud layers have been included with the lower haze and account for nearly all the particles with diameter > 1.2 μm. The mass distributions are relative distributions and assume all particles are spherical.

larger particles appear, resulting in the observed trimodal size distribution. The extinction coefficient and mass loading are higher due to the presence of Mode 3 particles, while the total number density is lower in comparison with the upper cloud region.

Lower Cloud, Precloud, and Lower Haze Particle Size Characteristics. In the lower cloud region the numbers of Modes 1 and 3 particles both increase by a factor of ~ 5, while Mode 2 is largely unchanged. However, the trimodal average size distribution as shown in Fig. 16 is still obvious. The lower cloud region contains roughly the same particle number densities as the upper cloud region, but much higher extinction coefficients and mass loadings due to the great number of large Mode 3 particles.

Below the lower cloud region there is a gradual decrease in aerosol particles down to 31 km; below this altitude essentially no particles were observed. Imbedded in this tenuous aerosol population (shown in Fig. 14) are two sharp cloud filaments at 46 and 47.5 km, previously designated as precloud layers. The size distributions for these two layers were also found to be strongly bimodal (see Knollenberg and Hunten 1980). Those in the upper precloud layer resemble Modes 1 and 2 in the lower cloud region; those in the lower precloud layer resemble Modes 1 and 2 of the middle cloud region.

The lower thin haze extends from 46 to 31 km. The lack of any discernible particles below 31 km is rather remarkable, although in the last few km above the surface zero particle activity is not entirely certain due to noise bursts that, though easily discountable as spurious, could mask real particles. However, the existence of particles > 2 μm is ruled out despite the noise; we favor the idea that the atmosphere is clear of particles > 0.6 μm all the way from 31 km altitude to the surface.

Behavior of Modal Number Densities. Knollenberg and Hunten (1980) analyzed the modal partitions of number density, extinction coefficient, and mass to gain insight into cloud particle origin, optical properties, and growth history. We will discuss optical and mass properties later in those sections dealing specifically with optical properties (Secs. II.D and E) and growth models (Sec. IV.B). Number density is of interest because it relates directly to origin and can be used as a tracer of mixing processes.

Knollenberg and Hunten found that the Mode 1 particles in the upper cloud region approached 1000 cm^{-3} at 60 km, the maximum anywhere in the main cloud deck. Above 61 km the particle number density decreases with increasing height faster than the gas concentration ($H_g/H_p \simeq 2$). Below 59 km the decrease with decreasing height is even more rapid. The Mode 2 H$_2$SO$_4$ droplets in the upper cloud region mimic Mode 1 at several altitudes, but their number density does not fall off nearly as fast with altitude above or below the 60 km maximum. It would appear that Modes 1 and 2 both have sources near 60 km but have somewhat different chemistry.

In the middle cloud region both Modes 1 and 2 track very closely in number density. Both sets of particles appear well mixed with the gas, having scale-height ratios $\simeq 1$. At depths below T_{um} in the middle cloud we first observe Mode 3 particles. (To be exact, one particle of Mode 3 size was observed just above T_{um} in the upper cloud region and a few more just after turn on near 65 km. Large particles seen just after turn on are thought to originate from the spacecraft during reentry.) The Mode 3 particles have number densities ~ 10 cm^{-3} throughout the middle cloud region but drop off rapidly near both T_{um} and T_{ml}. Mode 3 particles show considerable statistical scatter, as seen in Fig. 15.

In the lower cloud region all modes increase in number density, although Mode 2 seems to show a continuation of its number density in the middle region. Modes 1 and 3 both increase by a factor of $\sim$ 3. An interesting observation here is that the number density of Mode 2 continues to increase with descent through the lower cloud, while Modes 1 and 3 reach a maximum near 49.5 km and rapidly decrease below. Mode 2 shows no discontinuity throughout the lower cloud and into the upper precloud layer; we conclude that this layer may simply be a detached region of the lower cloud base.

From the behavior of the modal number densities Knollenberg and Hunten (1980) concluded that:

1. Modes 1 and 2 have a generating source near 60 km;
2. The modes cannot all be H_2SO_4;
3. The lower cloud region and especially the precloud layers result from new condensation;
4. Modes 1 and 3 are more strongly growth coupled than Modes 2 and 3 or 1 and 2 in the middle and lower cloud regions.

C. The Vertical Cloud Structure

The Venus clouds and hazes have enormous vertical extent, with a lower haze down to $\sim$ 30 km and an upper thick haze up to 90 km; the entire system covers a vertical depth of $\sim$ 60 km. For such an enormous amount of cloud and haze the average opacity must be rather low; in fact the average visibility in the Venus clouds is better than several km. The main cloud deck extends from $\sim$ 70 km (the level of unit optical depth in the ultraviolet) down to altitudes to 45 to 50 km. According to Cimino et al. (1980) radio occultation measurements imply cloud bases as low as 40 km (430 K) in high latitudes. This is somewhat puzzling considering that the North probe nephelometer found the base at 47.5 km, higher than all measurements except those of the Day probe, and higher than prior occultation results from Mariner 10 (Kliore et al. 1979a).

Steffes and Eshleman (1982) made a recent suggestion that may resolve this discrepancy. Their measurements of the microwave opacity of H_2SO_4 vapor show it may be the dominant microwave absorber in the 30 to 50 km altitude range. Thus, the radio occultation attenuation may be due to gaseous H_2SO_4 and not to a depressed cloud base.

Considering all the nephelometer results and the LCPS measurements, it appears that the middle and upper cloud structure are planetwide features. Furthermore, T_{um} is defined in all measurements by Venera and PV nephelometers. In all cases the opacity is higher in the middle than upper cloud, typically by a factor of 2. The lower cloud is less well defined and highly variable from location to location. Sharp layers are evident and these have the highest opacity at the Large and Night probe sites. Knollenberg and Hunten defined T_{ml} as the region separating the middle and lower cloud. A careful examination of Marov's definition of his lower transition region allows two interpretations, in the Venera results it is either a low density region similar to T_{um}, or the altitude where the largest particles disappear. The first interpretation places the lower transition region near the base of what corresponds to the Venera 50–56 km layer (layer II), since there is not a strong lower cloud feature. But in many respects the second interpretation is preferable, since it is tied to the loss of Mode 3 particle just as his upper cloud ends with the onset of Mode 3 particles. Using a definition of T_{ml} tied to multimodality places it at ~ 47.6 km; the lower cloud must be redefined to include only the sharply layered filaments. At the Large probe site this is near the base of the currently defined lower cloud region. The difficulty in defining T_{ml} at any site other than that of the Large probe is then due to the lack of particle size information. Another possibility is that this boundary is more transient than the others and not ubiquitous.

The clouds within the main deck would all be thin stratoform in terrestrial classification. Instabilities are slight and latent convection potential is negligible (see Knollenberg et al. 1980). Only the middle cloud region appears to have any potential for convective overturning. Above T_{um} an inversion is noted at a temperature of ~ 273 K. This is probably a planetary feature of large-scale circulation patterns, not produced by cloud processes.

The lower thin haze extending from ~ 47 km down to 31 km was detected only by the LCPS at the Large probe site, although the Day probe nephelometer saw another region of weak backscatter between 30 km and 49 km. The lower boundary is nearly identical to that found by the LCPS. That the nephelometer saw the thin haze implies that it is nearly two orders of magnitude stronger (in terms of 900 nm backscatter) at the Day probe site than at the Large probe site, since the haze at the latter was ~ 2 orders of magnitude below the nephelometer's minimum detectable signal level.

D. Particle Refractive Indices

A number of cloud optical characteristics cannot be inferred from any single experiment but instead require a synthesis of measurements. If such measurements are simultaneous we can avoid problems of correlating different time and spatial periods. The PV Large probe provided the most complete set of *in situ* measurements. Using the LCPS and nephelometer measurements, the solar flux parameters measured by the solar net flux radiometer (LSFR) allow calculation of optical depths and single scattering albedos. A synthesis report was given by Knollenberg et al. (1980) involving all PV cloud related experiments on probes. At the Large probe site the LSFR provided important cloud-related measurements and the neutral mass spectrometer (LNMS) and gas chromatograph (GC) aided in defining the gaseous parent vapor phase. A comparison of optical measurements by PV Large probe instruments is presented in Fig. 17. In this section we will concentrate on analysis from these PV data which together comprise a complete and unique set. We will make reference to Venera and prior data where comparisons are useful.

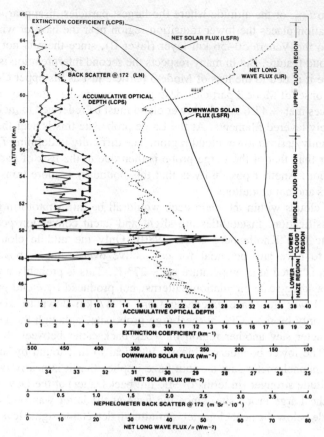

Fig. 17. Venus cloud system optical properties. This figure illustrates the complementary nature of the Pioneer Venus instruments providing cloud optical information. On the Large probe the LCPS and LN data show a great deal of similarity. The relative differences between the two profiles coincide with the changes in refractive indices. The LSFR net flux data show most of the absorption occurring in the upper cloud region. Changes in fluxes on the LSFR and LIR (from Boese et al. 1979) coincide with changes in the visible and infrared opacities, respectively. The sharp change in the LIR flux at 58 km and just above T_{um} results from the loss of infrared opacity, at least partially due to the disappearance of Mode 3 crystals. (From Knollenberg et al. 1980.)

Perhaps the most useful single particle optical property that can be deduced is the refractive index. The necessary data are the particle size distributions from the LCPS and the measured nephelometer backscatter. In essence the LCPS particle size measurements replace the forward scatter measurements of Marov's Venera nephelometers. Depending upon the accuracy of the particle size measurements, the computed refractive indices could be more accurate than those from the multiangle nephelometer of Marov; however,

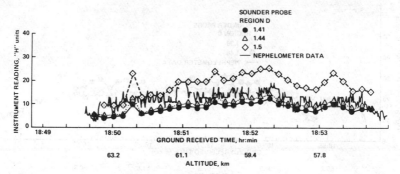

Fig. 18. Comparison of signals measured by the Large (Sounder) probe in the D region (upper cloud region) with values calculated using the particle size distributions measured simultaneously by the Large probe cloud particle size spectrometer (LCPS) and under the assumptions of spherical homogeneous particles, all having the same real index of refraction. Three cases shown are for real indices of refraction of 1.41, 1.44, and 1.50. (From Ragent and Blamont 1980.)

a possible source of error is that the LCPS and PV nephelometer operate at different wavelengths (see Knollenberg and Hunten 1980). Computations involving the LCPS and nephelometer data were presented by Ragent and Blamont (1980): the results of those computations for the upper, middle, and lower cloud regions are shown in Figs. 18, 19, and 20.

In the upper cloud region Ragent and Blamont used directly the LCPS data which truncated the Mode 1 size distribution at 0.6 μm. There is little cross section below 0.6 μm at visible wavelengths and even less at the 940 nm wavelength used by the nephelometer. Figure 17 shows computed values for three real indices 1.41, 1.44, and 1.5 assuming homogenous transparent spheres for Modes 1 and 2. As noted above, the best fit is for a value of 1.44 which is the value originally determined by Hansen and Hovenier (1974). These results are consistent with those of Marov in the same region.

Toon et al. (1982a) have reanalyzed the nephelometer data in the upper cloud region finding that a refractive index near 1.55 is needed at the highest levels seen by Pioneer Venus. The index declines to 1.45 at T_{um} (see Fig. 21). Toon et al. also show that Marov's data are consistent with this higher refractive index.

The middle and lower cloud regions are more difficult to model than the upper cloud region because the Mode 3 particles may be asymmetric. Because Mode 3 particles are very large, any such uncertainties have a profound effect on the computed backscatter at 940 nm. The number of Mode 3 particles observed by the LCPS is small enough that we must average over a number of altitude samples. This is a result of the relatively small sample volume isolated by the imaging probe of the LCPS for such sizes. On the other hand, the nephelometer views nearly one thousand times the LCPS volume and views

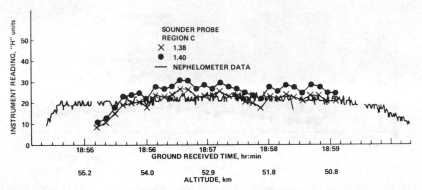

Fig. 19. Comparison of signals measured by the Large (Sounder) probe in the C region (middle cloud region) with values calculated using the particle size distributions measured simultaneously by the Large probe cloud particle size spectrometer (LCPS) and under the assumptions of spherical homogeneous particles, all having the same real index of refraction. Two cases shown are for real indices of refraction of 1.38 and 1.40. LCPS data are averaged over large intervals giving rise to lower calculated values near upper boundaries of the region. (From Ragent and Blamont 1980.)

an ensemble of Mode 3 particles at all times. In their analysis, Ragent and Blamont averaged over a number of LCPS samples using a running mean. Figure 19 shows the middle cloud comparison of the data with best fits assuming spherical particles with real refractive indices of 1.38 and 1.40. In the lower cloud region the relative weight of Mode 3 to Modes 1 and 2 increases substantially (as indicated in Fig. 20) and an inferred index < 1.33 is required.

We can explain the source of these low refractive indices by the following argument. Mode 3 particles dominate the cross section in both the middle and lower clouds. A direct integration assuming spherical Mode 3 particles indicates that the contribution of Mode 3 cross section in the lower cloud is at least four times that of Modes 1 and 2; in the middle cloud region it was only 50% more. Since the nephelometer backscatter responds more or less to r^2 (similar to the extinction cross section presented in Fig. 17), we would expect the nephelometer signal to be at least three to four times greater in the lower cloud than in the middle cloud region. However, as can be seen in Fig. 17, it is roughly only twice as strong. The lower nephelometer backscatter values yield lower values of computed refractive index. This is more clearly revealed in Fig. 19 where the refractive index was assumed to be 1.4 for Modes 1 and 2 and 1.33 for Mode 3.

We must conclude that the observations of Mode 3 are not consistent if it has the same composition as Modes 1 and 2. Furthermore, the index requirement for Mode 3 is unrealistically low for anything but water. Marov (see above) reached similar conclusions with regard to the lower cloud using his multiangle nephelometer data. However, Toon et al. (1982a) can match

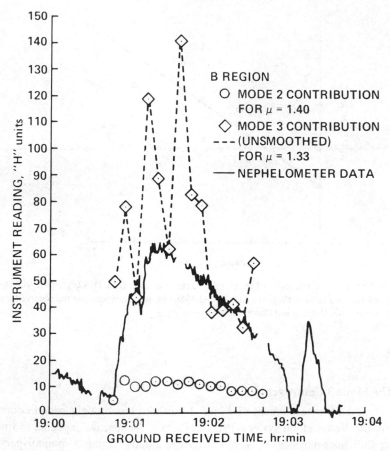

Fig. 20. Comparison of signals measured by the Large probe for the *B* region (lower cloud region) with values calculated using the particle size distributions measured simultaneously by the Large probe cloud particle size spectrometer (LCPS) and under the assumptions of spherical homogeneous particles. Calculated contributions are shown for particle size Modes 2 and 3 using refractive indices of 1.40 and 1.33 for Modes 2 and 3, respectively. (From Ragent and Blamont 1980.)

Marov's observations with more realistic refractive indices by considering more general particle size distributions. Near the base of the lower cloud region the large Mode 3 particles disappear, leaving only Modes 1 and 2. In this region a refractive index of 1.44 is again computed. Also, below the lower cloud region in the two precloud layers, computations by Ragent and Blamont again show values of refractive indices of 1.46 and 1.44, respectively. Knollenberg et al. (1980) discussed the problem presented by the inferred extremely low refractive indices in the lower cloud dominated by Mode 3 and concluded that only an asymmetric particle (crystal) could explain the results of all the measurements.

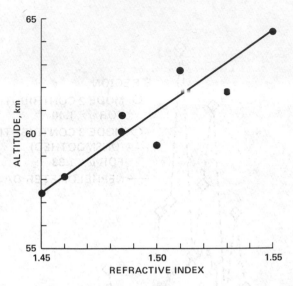

Fig. 21. The real refractive index of the cloud particles that Toon et al. (1982a) require to match the observed backscatter (Ragent and Blamont 1980) using Mie theory and the observed size distribution (Knollenberg and Hunten 1980).

E. The Mode 3 Controversy

The LCPS measurement of Mode 3 particles has brought added confusion to the Venus cloud system. Prior to Pioneer Venus, we expected to find rather dull microphysics on Venus with the clouds having a monodisperse population. Instead we found particles both larger and smaller than the expected 2 μm diameter H_2SO_4 droplets. In many respects the Mode 1 particles are as interesting as Mode 3; neither were anticipated and small particles are present at all altitudes. The primary fascination with Mode 3 comes from direct evidence for asymmetric (possibly crystalline) particles provided by Knollenberg and Hunten (1980). Knollenberg et al. (1980) further state that only crystals of high aspect ratio could satisfy the LCPS, LSFR, and LN results simultaneously. However, since the largest amount of mass (~80% according to Knollenberg and Hunten [1980]) is within the Mode 3 particles, it is extremely important to verify their existence and determine their composition.

The LCPS undoubtedly detected large particles, but the evidence for solid particles is indirect. There were internal inconsistencies in the LCPS measurements as well as inconsistencies between the LCPS measurements and the measurements made by other instruments. Some of these inconsistencies were:

1. Calculations employing LCPS size distributions do not give the back-scatter observed by the PV nephelometer in the lower clouds if reasonable refractive indices are used.
2. The LCPS size distributions do not yield the optical depths derived by the LSFR, assuming spherical particles.
3. Overlapping size ranges of the LCPS give conflicting measurements in the lower clouds.

In addition, independent Venera results show some oddities at the same altitudes:

1. Venera nephelometer phase function measurements are inconsistent with spherical particles having reasonable refractive indices in the lower cloud.
2. X-ray fluorescence measurements (Surkov 1979a) show about ten times as much chlorine as sulfur in the Venus clouds.

The various inconsistencies can be explained by the simple hypothesis that Mode 3 is composed of solid, nonspherical particles. However, this explanation requires an abundant gas-phase chemical in the clouds as the source for these particles. No such gas has yet been discovered.

Toon et al. (1982a) reexamine the evidence that solid particles form a distinctive size mode. They find that Mode 3 is defined by a discontinuity located between two size ranges of the LCPS. Although this could be real, it could also be the result of a small calibration shift of the PV instrument. A shift in the calibration removes the discontinuity, along with the internal inconsistency of the LCPS. The revised size spectrum is consistent with the Venera and Pioneer optical data in the lower clouds; all the modes are composed of sulfuric acid droplets without any solid particles. The only unexplained data are those showing large amounts of chlorine compared to sulfur in the clouds. We note, though, that the most recent Soviet measurements from Veneras 13 and 14 show a large sulfur to chlorine ratio, the opposite of Surkov's (1979a) findings.

From the data in hand, it seems impossible to disprove the existence of Mode 3. Two self-consistent, alternative interpretations of the data exist. Accepting the spacecraft observations at face value, we are led to the existence of a mode of large solid particles whose composition is unknown and whose source vapor has escaped detection. On the other hand, we may conclude that the large particle mode is merely the (mis-measured) tail end of the Mode 2 sulfuric acid droplets. This allows a simple understanding of the source of all the cloud particles, but at the cost of disbelieving some of the measurements.

If the large particles are solid, several possible compounds have been suggested. Thin chloride crystals have been proposed; these would explain the observations of Surkov (1979a). Watson et al. (1979) and Sill (1981) independently suggested that nitrosylsulphuric acid could be produced on Venus (Watson et al. in fact suggested the possibility independent of Knollenberg and Hunten's findings). Given that such crystals could form, it is not necessary for all of the mass of Mode 3 to be crystalline. It is much more likely that a quantity of Mode 3 mass could still be H_2SO_4 in the form of coated nitrosylsulphuric acid crystals. This may be a necessary assumption since H_2SO_4 is the only observed condensable having sufficient abundance to generate the measured mass loading.

F. Cloud Optical Depths and Albedos

Like estimates of refractive index, derived optical depths are also useful in constraining particle composition. Single scattering albedos are important in the upper cloud region as they influence the ultraviolet contrasts and limit options in selecting particle composition. The simplest way to determine optical depth of the various cloud particle modes is to integrate the size distributions or to calculate the extinction coefficient vertically using the LCPS data. A profile of cumulative optical depth assuming all the particles are spherical was presented in Fig. 17 and extinction coefficients are shown in Fig. 22. The partitioning of the optical depth among size modes is also a useful exercise and a summary of the resulting optical depths are presented in Table IV (Knollenberg and Hunten 1980). Of course it may be that Mode 3 is not spherical and thus the Mode 3 contribution would be too high owing to the overestimate in particle cross sectional area. To provide a more reasonable estimate of optical depth we have arbitrarily selected a value of refractive index for Mode 3 of 1.5 and computed the necessary reduction in cross section to provide agreement with the nephelometer results in terms of residual backscatter cross section given in Fig. 23. The reduced values of cross section are given in Table IV in the column headed Crystal Mode 3. Also shown in Table IV are the comparisons from Tomasko et al. (1980b) for the LSFR data. The agreement using revised values for Mode 3 is well within the known sources of error for the LCPS, LSFR, and nephelometer measurements. Table IV also gives the single scattering albedos derived from the combination of LSFR and LCPS data.

From the PV orbiter cloud photopolarimetry (OCPP) data we can estimate optical depth for the cloud top and the upper haze region above the *in situ* measurements. As previously mentioned the haze optical depth varies with location and time. In the early observations of Pioneer Venus the derived values ranged from upwards of one optical depth in the polar regions to perhaps as little as 0.1 in equatorial regions. Kawabata et al. (1980) found that the haze contributes $\sim$ 15% of the volumetric scattering coefficient at 550 nm. During two years of OCPP observations the amount of haze has rapidly decreased by a factor of five (Kawabata et al. 1982).

The amount of material above the first LCPS sample can be determined from LSFR observation of the dependence of sky brightness on azimuth. For

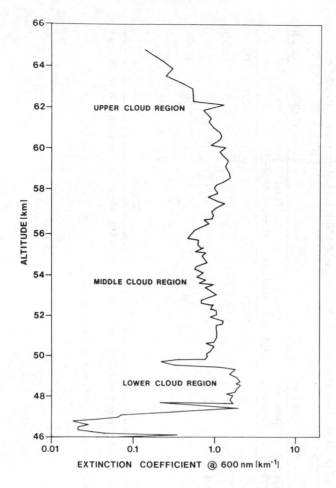

Fig. 22. Computed extinction coefficient for Modes 1 and 2 from LCPS particle size distributions. Comparing this figure with either the LCPS or nephelometer profiles in Fig. 17 illustrates
the contribution of Mode 3 particles.

the first several rotations of the Large probe, the direction of the Sun was
observed. Below that level, multiple scattering caused the sky brightness to be
independent of azimuth. This yields an estimate of optical depth of 2.8, which
is consistent with the extrapolated values from the LCPS. Below ~ 65 km the
upward and downward intensities from both the LSFR narrowband and broadband channels can be used to compute the optical depth and single scattering
albedo of the cloud particles in a particulate layer from an estimate of the
single scattering phase function consistent with the LCPS data. These values
are presented in Table IV.

TABLE IV

Summary of Optical Depths in Main Cloud Deck According to LCPS Size Modes & LSFR Inversions Optical Depth[a]

Region	(LCPS)						(LSFR)	$\bar{\omega}_0$ (LSFR)
	Mode 1	Mode 2	Mode 3 (Spherical)	Mode 3[b] (Crystals)	Total (Spherical Mode 3)	Total[b] (Crystal Mode 3)		
Upper cloud	2.15	5.12	0	0	10.77	10.11	13.7	0.9977
Middle cloud	0.34	3.53	6.2	2.2	11.07	7.07	8.3	0.9996
Lower cloud	0.48	1.82	9.9	3.2	12.20	5.6	4.8	0.9997
Precloud layers	0.06	0.24	0	0	0.30	0.3	1.3[c]	0.9992
Lower haze	0.20	0.05	0	0	0.25	0.25		
Totals	3.23	9.76	16.1		31.09	23.48	28.3	

[a] Approximately 3.5 optical depths are indicated above the first region of LCPS measurements by the LSFR which have been added to LCPS upper cloud values.

[b] These values are calculated from inferred aspect ratios which remove LCPS internal discrepancies.

[c] Includes CO_2.

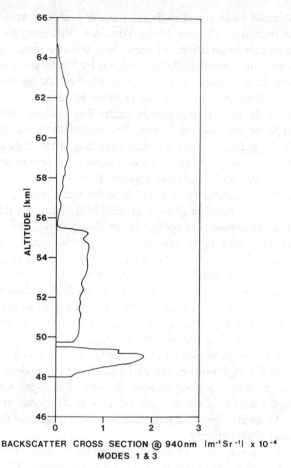

Fig. 23. Residual nephelometer backscatter attributed to Modes 1 and 3. The above profile was constructed by computing the nephelometer backscatter for Mode 2 as determined by the LCPS and subtracting it from the nephelometer measurements.

G. Cloud Dynamics

Considering the dispersed nature and low mass loadings of the particles in the clouds, cloud dynamics are more influenced by radiative exchanges than by the release of latent heat. Knollenberg et al. (1980) estimated that the maximum effect of latent heat released in the lower cloud region (coinciding with peak mass loading and steepest gradient of mass loading) would amount to only 0.5°C in excess temperature in a 1 km moist adiabatic ascent. If the mass of particles is less due to their asymmetry, a value closer to 0.1°C would be more appropriate. Thus, the cloud parcel stability is largely a function of radiative heating or cooling.

The thermal balance and radiative transfer problems associated with the clouds are treated in Chapter 18 by Tomasko. However we consider here several important implications of these heat transfer processes upon cloud dynamics and microphysics. As pointed out by Tomasko et al. (1980b) and Knollenberg et al. (1980) there are a number of inconsistencies between measured cloud structure and net flux profiles. In our own atmosphere we are very aware of the roles clouds play in controlling infrared opacity. This does not appear to be the case on Venus. Because of the tenuous nature of the Venus clouds and hazes coupled with abundant infrared gaseous absorbers (especially CO_2), the gases and not cloud particles appear to be the important sources of infrared opacity in most regions. Even near cloud base where there is the sharpest discontinuity in cloud material there is little observed effect upon the radiation net flux (Suomi et al. 1980), implying that the gaseous region below cloud base is as opaque as the cloudy region above. On the other hand, the upper cloud region appears to be higher in infrared opacity than the integrated particle size distributions can provide. Suomi et al. (1980) have proposed a region of highly concentrated ultrafine ($<0.2\mu$m) particles as an infrared opacity source for this region which could be called Mode 0. However, these suggested small particles must have a mass loading of 0.5 to 2.68 mg m^{-3} (Suomi et al. 1980). Even the lower value would produce number densities of 10^6 cm^{-3} at modal sizes of 0.1μm diameter. Taking a smaller particle size further increases number density ($\sim r^{-3}$), which would appear to be necessary if a high ratio of infrared/visible opacity is desired (e.g. >10). Coagulation is very rapid for number densities this high, with the number density falling by $1/e$ in a few minutes. This makes Mode 0 somewhat implausible. As an alternative, a molecular gas would achieve the highest ratio of infrared to visible opacity.

Particle size variability of the middle and lower cloud strongly influences the net flux profiles observed on board the Small probes. Comparisons with the nephelometer results indicate that while the middle cloud backscatter profiles are quite smooth, the net flux profiles are highly variable (factors of 2 within the clouds). Variability in Mode 3 might be a possibility here. Without particle size measurements we have difficulty in formulating definitive answers; however, it may also be true that gases and not particles are the controlling sources of infrared opacity. In fact, Pollack et al. (1980a) have determined that, overall, the sources of infrared opacity in order of decreasing importance are CO_2, H_2O (gas), cloud particles, and SO_2.

Only at the Large probe site do we have qualitative consistency between thermal models and cloud behavior. The base morphology of the lower cloud region is suggestive of condensation and thus one suspects local heating with convective motions. The sharp base results in a discontinuity in infrared opacity. Assuming that the CO_2 window regions are adequately blocked by the cloud particles, the cloud base would be heated from below, and the subbase region in turn would be cooled. Given enough time, this results in a subadiabatic region just below cloud base with a convective region in the lowest cloud layers. This one-dimensional analysis coincides with the measured structure (see Chapter 11 by Seiff) and the infrared flux discontinuities measured by the LIR shown in Fig. 17.

The observed temperature discontinuity (inversion) near 64 km on the North probe (Seiff et al. 1980) is suggestive of similar inversions observed at the top of terrestrial clouds. As in the terrestrial case, it suggests that the layer radiatively cools. This would be possible if there was a sharp increase in infrared opacity below this region with little infrared opacity above. The upper cloud region at the North probe site is the weakest of all those found in the nephelometer observations, while the middle cloud region is similar to the Large probe site. It would be necessary to have much larger particles (Mode 3) in the middle cloud and not in the upper cloud to produce the necessary radiative divergence. Furthermore, the proposed Mode 0 or gaseous absorbers must still allow a window for radiative losses by these larger blackbody radiators. Again gases, as opposed to Mode 0 particles, could more easily satisfy such conditions.

H. Lightning

There is little doubt that confirmation of a cloud origin for the Venus lightning observations would have tremendous implications on the cloud particle chemistry and microphysics. The measurements of lightning originate from acoustical (Venera), electrical (PV and Venera), and possibly even optical (PV) sensors. Ksanfomaliti et al. (1979) reported low-frequency noise bursts from 60 km to very near the surface. They interpreted this "thunder" as direct evidence for lightning. W. Taylor et al. (1979a) and Scarf et al. (1980) concluded that highly impulsive, intense radio signals consistent with whistler mode propagation through the ionosphere were caused by lightning. For the most recent review, see Chapter 17 by Ksanfomality et al.

Lightning has far-reaching consequences on cloud physics problems. As pointed out by Knollenberg et al. (1980), in addition to charging processes, there must be charge separation processes before lightning can occur. The only known means of generating the necessary charge separation is via falling large particles. Because charge relaxation processes are also active, charge separation generally benefits from both strong updrafts and large particles. On Earth, lightning involves precipitation; it is not observed until the ice phase is well established even though mechanisms for charge generation in liquid-liquid interactions appear plausible. Lightning is notably absent in clouds producing precipitation solely via coalescence processes. Likewise lightning is very rare in clouds producing only snow. By inference it is the accretion process that is actively involved in the charging process, since mixed-phase clouds are the preferred makers of thunderstorms. Regardless of the mix of ice crystals and water drops, precipitation-size elements are required. The difficulty on Venus is that we do not have confirmed observations of any large

particles. One additional problem with believing the existence of Venus lightning is the lack of optical sightings in repeated observations on the night side
by PV orbiter instruments. Terrestrial thunderstorms are easily visible for
hundreds of km from an airliner.

If lightning occurs in the Venus clouds, it is clear that cloud-to-ground
lightning would be rare due to the 40 to 50 km of travel required by such a
discharge. Cloud-to-cloud or even cloud-to-ionosphere lightning would be
more likely. The latter path, while equal to the cloud-surface distance, is less
resistive. A recent analysis by Scarf et al. (1982) indicates that the amount of
lightning is not as great as previously thought. Also, it appears that it is
strongly correlated with geographical locations, the highlands being a favored
location of lightning observations. This suggests that the lightning might be
associated solely with volcanic activity. Cloud electrical discharges, even if
not of sufficient magnitude to be considered lightning as we know it, are
extremely important to the chemistry of the gases and cloud particles. The
generation of NO_x is essential to the production of $NOHSO_4$, a possible Mode
3 crystalline candidate.

III. PARTICLE COMPOSITION

Physical descriptions of particles within a cloud have limited value without coexisting composition information. For the particles we have been discussing, we really only have compositional information on Mode 2 which we
have convinced ourselves is H_2SO_4. In addition, only one *in situ* experiment
has been conducted which purposely collected particles and analyzed their
chemical constituency (Surkov 1979*a*). We are thus left to identify the bulk of
the particulates through indirect methods, e.g., by analysis of the gaseous
vapors assumed to be in equilibrium with the particles.

On the basis of Surkov et al. (1981) analysis of X-ray fluorescence measurements of collected particles, we would conclude that chlorine and not
sulfur is the major particle chemical component. Surkov's data support a
chlorine-to-sulfur ratio of 10 and the amount of sulfur is not unreasonable if
we compute the total integrated sulfur within Mode 2 composed of H_2SO_4.
However, it is extremely difficult to imagine anything except Mode 3 that
could yield sufficient chlorine to generate a chlorine-to-sulfur ratio of 10.
While a great deal of speculation about possible chlorides exists, HCl and Cl_2
would seem to be eliminated on the basis of negative PV LNMS measurements. One could further argue that because the LNMS inlet was blocked by
what appears to be H_2SO_4, there is a strong possibility that the large Mode 3
particles contain H_2SO_4 and not copious chlorides. This conclusion is based
upon the fact that 2 μm droplets have vanishingly low collection efficiencies
and could not possibly wet a surface through impact collection.

In addition to Mode 3 we are also uncertain about possible identities
for Mode 1. Knollenberg and Hunten (1980) suggested that Mode 1 is a

background population with highly variable number densities but with little optical depth and suggested that free sulfur, extra-venusian materials, and pure H_2SO_4 were all candidates. From the above discussion it is clear that the identity of possible particulate constituents is a highly volatile issue.

A starting point in attempting to identify any condensable particulate is to assemble data on the available condensable gases throughout the atmosphere. This, however, is not as easy as it sounds. The only direct measurement of a condensable that we have is for H_2O. The projected concentrations of gaseous H_2SO_4 are so low that it would be unmeasurable. Its precursor SO_3 reacts so rapidly with H_2O that it would likewise be difficult to measure. One must therefore trace back through the reaction processes to SO_2 to find a surrogate condensable gaseous species. Fortunately in the case of SO_2 we have data from a number of sources including PV and Venera gas chromatography, PV mass spectrometry, and PV ultraviolet spectrometry. Figure 24, from a variety of sources, indicates available abundances of SO_2, HCl, and $H_2SO_4-H_2O$ at various levels within the cloud. Table V gives the columnar masses contributed by various size modes from Knollenberg and Hunten (1980). The H_2SO_4 values are estimated from $H_2SO_4-H_2O$ equilibria using the measured H_2O profiles from Moroz et al. (1979) and from Winick and Stewart's model results presented in Fig. 25.

Mode 2

It is preferable to start our discussion with Mode 2 simply because we are more convinced of its identity, i.e. H_2SO_4. It is somewhat disconcerting that, in spite of all of the entry probe measurements, the only positive identification we have for Mode 2 has resulted from earth-based measurements. However, Mode 2 is also the best behaved. The Mode 2 number density increases almost monotonically with depth through the main cloud deck and the size distribution remains narrow and well defined in almost all regions. We have been able to trace it throughout the upper, middle, and lower cloud deck and into the precloud layers as a singular particle population. While one may speculate on possible contaminants within Mode 2 (Tomasko et al. 1980b), it is doubtful that they amount to more than a few percent, even under the most highly contaminated conditions.

Mode 1

The Mode 1 particles comprise most of the total number of particles in the clouds and hazes. They may however be of different composition at the various altitudes. In the upper cloud region they could contain substantial amounts of sulfur; this would account for the observed refractive index and net flux profile. However, it would be necessary for the majority of this sulfur to be destroyed below cloud top so as not to dominate the optical cross section when viewed from space. Mode 1 has a minimum in number density in the middle cloud region. The Modes 1 and 2 populations track in number density

L. W. ESPOSITO ET AL.

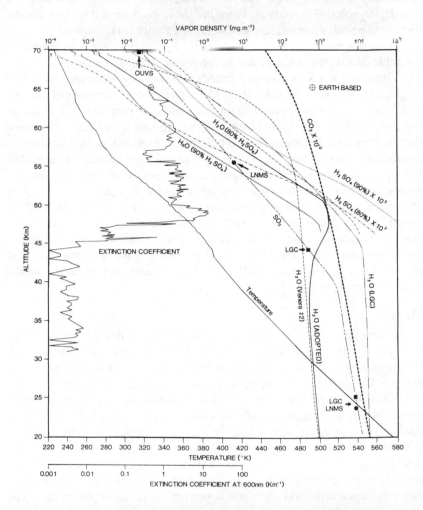

Fig. 24. Vertical structure of parent vapor inventory. Data presented in this figure are from a variety of sources. Measured H_2O: PV gas chromatograph (LGC) after Oyama et al. (1979). Venera 12 water vapor spectrophotometer (Venera 12) after Moroz et al. (1979d). Earth-based measurements after Hunten (1971). Measured SO_2: PV ultraviolet spectrometer (OUVS) after Esposito (1980). PV neutral mass spectrometer (LNMS) after Hoffman et al. (1980c). PV gas chromatograph (PGC) after Oyama et al. (1979). Temperature: PV atmospheric structure experiment after Seiff et al. (1979b). Our adopted H_2O profile used the earth-based measurements and Venera 12 data. It is consistent with 85 ± 5 % H_2SO_4 at all cloud levels. (From Knollenberg and Hunten 1980.)

TABLE V

Summary of Columnar Mass
According to LCPS Size Modes [a]

Region	Columnar Mass (g cm^{-2})		Mode 3 (Spheres)	Mode 3 (Crystals)	Total (Spheres)	Total (Crystals)
	Mode 1	Mode 2				
Upper cloud	1.1×10^{-4}	5.5×10^{-4}	0	0	6.6×10^{-4}	6.6×10^{-4}
Middle cloud	0.2×10^{-4}	4.7×10^{-4}	0.004	7.0×10^{-4}	0.0045	11.9×10^{-4}
Lower cloud	0.2×10^{-4}	3.8×10^{-4}	0.006	10.5×10^{-4}	0.0064	14.5×10^{-4}
Precloud layers	10^{-5}	0.6×10^{-4}	0	0	0.6×10^{-4}	0
Lower haze	10^{-5}	0.1×10^{-4}	0	0	0.1×10^{-4}	0
Totals	15×10^{-4}	18.7×10^{-4}	0.01	17.5×10^{-4}	0.012	0.0032

[a] No allowance has been made for cloud material above LCPS entry altitude of 65 km.

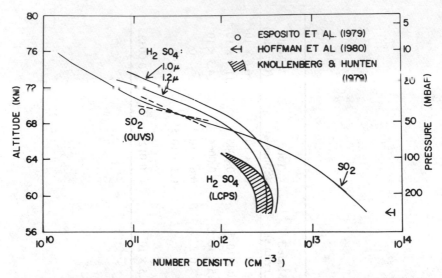

Fig. 25. Comparison of SO$_2$ and H$_2$SO$_4$ model profiles with experimentally inferred profiles. The SO$_2$ OUVS results with a scale height of 1 to 2 km are shown around 69 km. The model aerosol profile is shown for 1 μm and 1.2 μm radius particles, assuming all H$_2$SO$_4$ goes into these particles. The LCPS results (Knollenberg and Hunten 1979b) are obtained by assuming that the Mode 2 particles are 80 % H$_2$SO$_4$. (From Winick and Stewart 1980.)

in a similar manner with altitude change. However, subtle size differences suggest that these two particle sizes are not of the same chemical composition. In particular, one can site the region near the base of the middle cloud region where Mode 1 increases in size and number density at the apparent expense of Mode 2. Since Mode 1 in the middle region constitutes negligible cross section and mass, it may not be as important to identify its composition. In the lower cloud Mode 1 increases by a factor of $\sim$ 3 in number density while Mode 2 appears as a continuation of its profile in the middle region. However, the lower cloud region could more likely support populations of Modes 1 and 2 H$_2$SO$_4$ droplets since this is a region of condensation growth and supersaturations would be higher. We would thus favor Mode 1 as largely H$_2$SO$_4$ in the lower cloud as well as precloud regions. It cannot be entirely H$_2$SO$_4$ or it would rapidly grow to the size of Mode 2 (see Sec. IV below). Mode 1 also exists in the upper level haze; however, there is a variety of evidence suggesting that Mode 1 in this haze is different from Mode 1 in the upper cloud. In fact, from refractive index, size, and remote observations Mode 1 in the upper level haze is most compatible with sulfuric acid.

Mode 3

On the basis on Surkov's measurements it is tempting to suggest that Mode 3 is a chloride. Indeed, Barsukov et al. (1980a) have even suggested

HCl. While the low refractive index of HCl could explain the low value observed, HCl could not go undetected in the quantities required for vapor equilibrium. Solid chlorides have been suggested by Krasnopolsky and Parshev (1979a) and many others.

Of the candidates suggested so far none seems compelling. Al_2Cl_3 and $FeCl_3$ are difficult to accept for several reasons. The abundance of Cl in the atmosphere is less than a few ppm (Hoffman et al. 1980c), which probably precludes forming a high-mass cloud. $FeCl_3$ is colored, which may violate the observed low solar energy absorption in the lower cloud. $FeCl_3$ has a vapor pressure too low to condense and form a large-mass cloud. Al forms such stable compounds with oxygen that appreciable quantities should not occur in the atmosphere and furthermore, Al_2 was not found by the PV mass spectrometer. Various volatile metals might be suspected. However, mercury was not detected in the clouds by the Veneras (Knollenberg and Hunten 1980), nor by the PV mass spectrometer. No other volatile metals have been reported.

From the point of view of having adequate vapor reservoir, practically the only solid particle that appears remotely possible in the Mode 3 issue is nitrosylsulfuric acid ($NOHSO_4$) as suggested by Watson et al. (1979) and Sill (1981). $NOHSO_4$ may be considered a derivative of H_2SO_4 with $-NO$ replacing one hydrogen and is a product of the lead chamber process utilized in the production of H_2SO_4. A requirement for the production of $NOHSO_4$ is the presence of odd nitrogen in quantities of several ppm by volume. The production of NO_x is not easily envisaged in the equilibrium chemistry of the venusian atmospheric constituents. NO_x could be produced by lightning in the clouds but it is still difficult to imagine that this source is adequate in the oxygen-poor environment of the Venus clouds. There are strong spectroscopic upper limits on the various oxides of nitrogen at both cloud base and cloud top; the abundance of NO_2, for example, is less than 10 ppb in both locations (Owen and Sagan 1972; Krasnopolsky and Parshev 1979a). In addition, according to Sill (1981) the dissolution of $NOHSO_4$ in H_2SO_4 colored the solution; this may cause the lower cloud to absorb too much solar energy. Pollack et al. (1980a) show that NO_2 does not have the proper spectrum to explain the ultraviolet absorption on Venus as suggested by Sill (1981).

Several aspects of $NOHSO_4$ are particularly interesting. It forms small asymmetric feathery crystals in "midair." It forms ultraviolet-absorbing reaction by-products, develops with depleted H_2O, and its melting point (decomposition) is apparently dependent on contamination of H_2SO_4 with reported values from 74°C to 130°C (Sill 1981). Such temperatures are close to the temperatures at which Mode 3 disappears. It is, however, quite soluble in H_2SO_4 (~50% by weight in 98% acid) thus seeming to make it unlikely that $NOHSO_4$ crystals could coexist with H_2SO_4 droplets. It is likely that H_2SO_4 would coat such crystals and form saturated solutions, and this might be one way in which Mode 3 could coexist with Mode 2. However, in other respects it would be just as desirable if any Mode 3 crystalline component were insoluble as this would allow coexistence with Mode 2 at similar vapor pressures.

The composition of Mode 3 and therefore of the bulk of the Venus clouds, as well as the composition of the ultraviolet absorber, remain a mystery. Further remote studies of the chemistry of the Venus atmosphere may yet resolve the question of the composition of the ultraviolet absorber. However, the Mode 3 particles cannot be observed from outside the planet. To determine their composition and to confirm their existence will probably require future explorations using planetary probes.

Mode 0

Finally, we have Mode 0, a proposed particulate offered to explain the infrared opacity in the upper cloud region (Suomi et al. 1980; Tomasko et al. 1980*b*). In order for these particles to have remained undetected by remote and *in situ* methods it would be necessary that they have high infrared opacities and low visible opacities. The necessary opacity (approaching unity optical depth at 10 μm) is sufficiently high that a gaseous absorber would be easier to accept than a particulate absorber. The required number density is so high it would be unstable and normal coagulation would easily produce detectable sizes. The LCPS was capable of detecting absorbing particles just as easily as transparent ones.

IV. MICROPHYSICAL MODELING

The observed physical properties of the Venus clouds and the gases in its atmosphere provide clues to the processes which create the clouds and control their structure. Although many new ideas have come from the latest Pioneer Venus and Venera data sets, some physical constraints have long been understood.

Young (1973) and Prinn (1975) suggested that photochemical processes create sulfuric acid vapor near 70 km altitude. This vapor then nucleates and condenses at low temperatures in the upper atmosphere of Venus to form particles. Gravitational sedimentation prevents large numbers of particles from being carried above the region of chemical production by atmospheric motions. Therefore, the cloud-top altitude is determined by a balance between the production rate of sulfate by chemistry and the loss rate due to sedimentation. Although these early concepts involved COS as the precursor gas rather than SO_2, the basic idea concerning the location of the cloud top is still sound.

Young (1977) pointed out that the droplets at the cloud top would be relatively involatile due to the low vapor pressure of sulfuric acid. Hence, at the cloud top the particles cannot easily evaporate, which partly accounts for the lack of breaks in the cloud deck. Young (1973, 1975) and Wofsy and Sze (1975) also realized that the cloud would evaporate in the lower atmosphere because high temperatures would cause the vapor pressure of the droplet to exceed the ambient H_2O and H_2SO_4 abundances. Hence, the abundances of H_2O and H_2SO_4 in the lower atmosphere below the cloud deck control the location of the cloud base.

Knollenberg et al. (1977) drew an analogy between the Venus clouds and terrestrial smogs and stratospheric aerosols. These terrestrial aerosols can have particulate concentrations similar to those of the upper clouds of Venus, and they are also formed from photochemically-produced sulfuric acid. By analogy with the terrestrial aerosols, these authors suggested that the narrowness of the size distribution at the Venus cloud tops was probably due to the r^{-1} dependence of the condensational growth rate characteristic for particles large compared with the mean free path of the gas. This idea seems to be correct. Knollenberg et al. (1977) also suggested that Venus's clouds probably formed entirely from photochemical process at the cloud top rather than from condensation driven from below as is the case with terrestrial water clouds.

Rossow's (1977, 1978) numerical models of the Venus clouds incorporated the physical processes of coagulation, condensation, vertical transport by eddy diffusion, and particle sedimentation. Rossow showed that if the cloud particles near the cloud top were produced locally by photochemistry, then coagulation could not be responsible for the observed narrow size distribution at the cloud top. He, however, felt that the cloud could be produced by condensational processes near the cloud base and suggested that particles in the narrow size distribution near the cloud top might result from upward mixing of particles formed by coagulation of the particles that originated at the cloud base. Overall, Rossow believed the evidence was ambiguous as to whether the cloud forms chemically from its top or forms condensationally from its base. Recently Toon et al. (1982) proposed that the upper cloud is photochemically created while the lower one is condensational.

The great increase in our observational knowledge of the Venus cloud properties, as described earlier in this chapter, allows us to make more definitive statements about the mechanisms responsible for the cloud structure than were possible in earlier cloud modeling studies. We shall first review the evidence for vapor reservoirs for the cloud material. Then we shall discuss the fundamental physical processes controlling the cloud. Finally we shall describe a recent model which reproduces the observed cloud structure.

The chemical and physical processes that give rise to the cloud and haze aerosols have been greatly clarified in recent years largely through detailed measurements of abundant sulfur dioxide and other parent vapors. Previously, earth-based observations (Owen and Sagan 1972; Cruikshank and Kuiper 1967) had put strong upper limits on the mixing ratio of SO_2 near the cloud tops ($<10^{-8}$). However, the earth-based limit was much too low for several reasons:

1. The SO_2 mixing ratio decreases rapidly with altitude (Esposito et al. 1979);

2. The original observations were taken near quadrature so that only higher altitudes were probed by the spectroscopy (Young 1979);
3. SO_2 is horizontally highly variable above the cloud tops, with hemispherical mean amounts differing by a factor of 4 on successive days (Esposito and Gates 1981).

Previously, the most credible model for forming the cloud aerosols was due to Prinn (1973b, 1974, 1975), with COS as the sulfur-carrying gas. In a two-pronged cycle (or "bi-cycle"), COS arising from the deep atmosphere was reduced to yield solid sulfur, the then likely absorber (Young 1979) which under appropriate conditions was oxidized to form the H_2SO_4 droplets in the clouds. This model thus provided both the cloud aerosols and the ultraviolet absorber: the oxygen abundance at any point determined the relative abundance of clouds and absorber and thus the ultraviolet brightness. In addition, this chemistry provided a strong sink for O_2 at the cloud tops, consistent with the earth-based upper limit of 1 ppm (Traub and Carleton 1973). Both solid sulfur and acid droplets fell out of the photochemical production region into the hotter lower atmosphere where they were converted back to COS (Prinn 1979).

Although SO_2 has been found both above the cloud tops and in the deep atmosphere (Oyama et al. 1979), we have no reliable detection of COS anywhere in the Venus atmosphere (Hoffman et al. 1980c). SO_2-based models, with their ability to draw together a number of diverse individual observations, have therefore replaced the COS-based ones.

The most recent photochemical models (Yung and Demore 1982) are successful in matching the sulfur budget and also reproduce the observed cloud top mixing ratios of CO ($\sim$50 ppm) and O_2 ($\leqslant$1 ppm). Although the model of Winick and Stewart (1980) is successful in matching the sulfur budget, it fails on CO and O_2. A selective sink of O_2 is required to remove this discrepancy. Nevertheless, a lower mixing ratio of O_2 consistent with earth-based observations would not disrupt the sulfur cycle, since the photolysis of SO_2 itself provides the majority of the oxygen needed for its oxidation to sulfuric acid (Winick and Stewart 1980).

A. Reservoir of Cloud Material

While modeling has determined SO_2 to be the major gaseous precursor of the H_2SO_4 particles in the cloud deck (Winick and Stewart 1980; Turco et al. 1982a; Toon et al. 1982), the cloud does not directly interact with SO_2, but rather with H_2SO_4, SO_3, and H_2O vapors. The abundances of the former two gases have not been directly measured, while the abundance of H_2O is highly uncertain. Since the acid particles are probably only slightly supersaturated with respect to H_2SO_4 and are certainly in equilibrium with H_2O, the H_2SO_4 vapor mixing ratio can be determined from the measured or inferred concentrations of H_2SO_4 in the droplets.

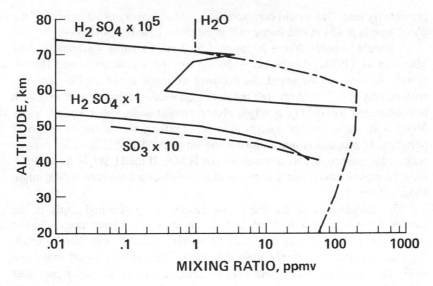

Fig. 26. H_2SO_4, H_2O, and SO_3 profiles.

Using the Gmitro and Vermeulen (1964) equations to find the acid concentration in equilibrium with the Venera water concentration observed, we find ∼ 96% acid solution at 46 km. A factor of 4 increase or decrease in the water vapor amount would only result in a concentration change from 92 to 98%. Using acid concentrations determined in this fashion we then calculated the acid vapor pressures shown in Fig. 26. Much smaller H_2SO_4 amounts are implied by the PV water vapor measurements which exceed the Venera values by orders of magnitude. An uncertainty of 4 in water abundance results in a 50% uncertainty in acid vapor pressure.

The vapor pressures in Fig. 26 were obtained from the Gmitro and Vermeulen equations but were divided by a factor of ten. Several workers have recently shown that these theoretical equations overestimate the measured vapor pressure by a factor ∼ 10 at high concentrations. Ayers et al. (1980) measured the vapor pressure above 98% sulfuric acid at the temperatures found at the cloud base in the Venus atmosphere confirming the factor of 10 error.

At the cloud base where the droplets evaporate, the H_2SO_4 gas concentration must equal the vapor pressure. Below that region the acid vapor must be uniformly mixed, or decline toward lower altitude since further sources of vapor would not be available. Ragent and Blamont (1980) report the cloud base is located at altitudes above 47 km, though the Large probe showed one cloud layer as low as 46 km. Even if one assumes that the cloud base is as low as 45 km, only 16 ppm by volume or 35 ppm by mass of H_2SO_4 could be present. By mass this would correspond to a cloud density of 90 mg m^{-3} if the cloud base is at 45 km and 10 mg m^{-3} if the cloud base is at 48 km.

It should be noted that ~ 20 mg m^{-3} total mass loading was measured by Marov et al. (1978a, 1980) and ~ 50 mg m^{-3} by Knollenberg and Hunten (1980). As we have discussed, the estimate we made above applies to H_2SO_4 with density $\rho \simeq 2.0$. Although the observed mass loading agrees well with that estimated for H_2SO_4, it might also represent some other volatile since Mode 3 is the dominant source of mass. Further, this assumes spherical particles. If Mode 3 is not H_2SO_4, some other condensible must be present with an abundance of 1 to 10 times that of H_2SO_4. If the H_2SO_4 in the cloud is only the mass of the Mode 2 particles then the observed mass of H_2SO_4 might be only 2 mg m^{-3}.

The calculations of the mass abundances just performed apply to the case of the average Venus cloud. In some locations on Venus, convection or other dynamical processes may lead to upward motions and thereby cause the condensation of a larger mass of H_2SO_4. The largest cloud mass that might be created in this manner may be estimated from the vapor pressures in Fig. 26.

If 35 ppm of H_2SO_4 vapor is present at 40 km and ascent of this air yields total condensation of the vapor, as is likely, then the condensed mass would be 4.2 kg m$^{-3} \times 35 \times 10^{-6} = 150$ mg m^{-3}. This might be increased by $<25\%$ by the mass of H_2O. This relatively low mass points to an interesting difference between Venus and Earth clouds. The base of many terrestrial clouds is located at temperatures from ~ 0 to 10°C. The mass of water vapor at 10°C and 100% relative humidity is ~ 10 gm m^{-3} and at 0°C it is ~ 5 gm m^{-3}. Hence, thick terrestrial clouds with these typical masses are to be found. On Venus the acid vapor mass is simply very low and so is the resulting cloud mass.

The above calculation suggests an H_2SO_4 cloud mass of ~ 150 mg m^{-3}, assuming that the only mass available was that in the air parcel containing the drops. However, since a rapidly falling particle does not move with the air, larger masses could result locally as the large drops grow by picking up vapor from rising particle-free air columns. The amount by which this process increases the condensed mass is not easily estimated but could be significant.

In addition to the H_2SO_4 vapor which may be present at the cloud base at ~ 20 ppm, SO_3 is also present. However, calculations of the SO_3 partial pressure suggest that its abundance is only $\leqslant 1$ ppm near the cloud base. Below ~ 30 km the acid vapor should be decomposed into SO_3 and H_2O. Hence tens of ppm of SO_3 should be present in the lower atmosphere.

B. Aerosol Physical Processes

The evolution of the aerosols is described by a continuity equation (e.g. Turco et al. 1982a):

$$\frac{dn(v, z, t)}{dt} = q - \frac{\partial}{\partial r}(gn) + \tfrac{1}{2}\int_0^v dv' K(v', v-v')n(v')n(v-v')$$

$$-\int_0^\infty dv' K(v,v')n(v')n(v) + \frac{\partial}{\partial r}\left(V_f n + NK_z \frac{\partial}{\partial z} n/N\right). \qquad (1)$$

Here n is the number of particles per unit volume, v is the particle velocity, q is the nucleation rate, g is the growth rate, K is the coagulation kernel, V_f is the fall velocity, N is the air number density, and K_z is the eddy diffusion coefficient.

The terms on the left hand side of Eq. (1) represent nucleation, growth, coagulation, sedimentation, and vertical diffusion. The rates of these physical processes are controlled by the dynamical regime, which is specified by the values of several nondimensional numbers (see Rossow 1978). For the Venus clouds the only nondimensional number which varies significantly is the Knudsen number Kn. Kn is the ratio of the mean free path between air molecule collisions to the particle radius. Near the cloud base the mean free path is $\sim 6 \times 10^{-2}$ μm but near the cloud top the mean free path is ~ 6 μm. Since we are interested in particles of size 0.1 to 10 μm we must consider the range of both large and small Kn.

Nucleation may occur homogeneously and involve only molecules of the condensing gas; it may occur on ions or on preexisting particles; or it may occur by other processes. These various mechanisms are assessed for the stratospheric aerosol layer of Earth by Hamill et al. (1982) who show it is not possible to accurately predict the rates of nucleation of sulfuric acid vapor. Fortunately, Hamill et al. (1982) also show that the characteristics of submicron size particles are not strongly dependent on the mechanisms causing nucleation.

Nucleation leading to new particles cannot occur (disregarding nucleation on small particles) unless the gas-phase molecular abundance exceeds the saturation vapor pressure. The vapor pressure over a spherical aerosol is strongly dependent on radius due to the Kelvin effect as shown in Fig. 27. Hence new nanometer size particles can only be created when there is a high supersaturation. The saturation ratio S is the ratio of the ambient vapor pressure to the vapor pressure over a flat surface and $S-1$ is the supersaturation. Large supersaturations can only occur at the cloud top where low temperatures cause low vapor pressures and where the cloud mass is too small to remove vapor efficiently from the gas phase.

Nucleation onto preexisting soluble particles may occur at saturation ratios less than unity because the vapor can interact with the molecules composing the surface of the preexisting particles, thus reducing the vapor pressure of the condensed phase. At the cloud base, nucleation must occur in this manner except possibly in regions having strong vertical motion which might create large supersaturations through adiabatic cooling. Once new particles are created, they are acted upon by several processes which can be roughly quantified.

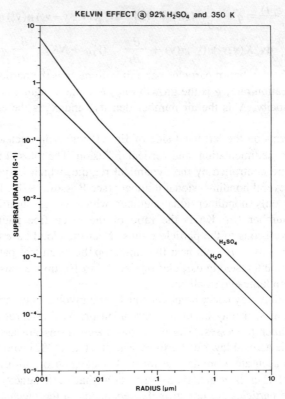

Fig. 27. Effect of radius of droplet curvature (Kelvin effect) on H_2O and H_2SO_4 supersaturation.

Sedimentation and eddy diffusion control the vertical distribution of the particles. Figure 28 presents the time required for 1 μm and 5 μm radius particles to fall a characteristic distance L of 5 km, $T_f = L/V_f$. The particle fall velocity is from the classical Stokes-Cunningham equation (Hamill et al. 1977) which bridges the Kn range of interest. When the mean free path is larger than the particle radius (Kn>1) the fall speed is proportional to the radius and inversely proportional to the air density. When the mean free path is less than the particle radius (Kn<1) the fall speed is proportional to the radius squared and is independent of the air density. These trends are evident in Fig. 28.

The diffusion times given in Fig. 28 assume that in the presence of a large mixing ratio gradient $\tau_d = L^2/K_z$. The vertical eddy diffusion coefficient is not well known. Woo and Ishimaru (1981) deduced $K_z \simeq 4 \times 10^4$ cm^2 s^{-1} at 60 km and Prinn (1974) suggested that the eddy diffusion coefficient increased toward higher altitudes in inverse proportion to the square root of the air density. Below 60 km we have simply chosen a curve whose shape resembles that of the atmospheric stability (Seiff et al. 1980) and whose magnitude is

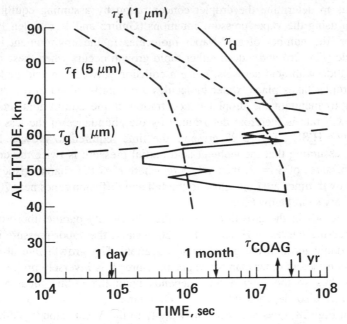

Fig. 28. Characteristic diffusional growth, and sedimentation times for particles of 1 μm and 5 μm radius as functions of altitude. The characteristic fall times τ_f are for 5 km distances. The growth times τ_g represent the times required to grow a 1 μm (radius) H_2SO_4 droplet with $S = 2$. The coagulation time τ_{coag} refers to the time required to reduce 100 cm^{-3} 1 μm (radius) H_2SO_4 droplets by one half at 60 km.

10^6 cm^2 s^{-1} near 52 km. This choice of coefficient is probably an upper limit and is uncertain by at least an order of magnitude. With these choices of eddy diffusion coefficient, typically particles fall about as fast as they are transported by diffusion.

The growth rate in Fig. 28 is taken largely from the work of Fuchs and Sutugin (1971), but has been modified to include the growth of a mixed composition aerosol (Hamill et al. 1977). Unfortunately, the growth process of sulfuric acid aerosols has been misunderstood by many previous workers, leading to serious conceptual errors. Since growth is the dominant process in the clouds, we shall discuss it at some length.

Rossow (1978) stated that the vapor over the droplets was entirely water vapor. Also he felt that droplets of different size were prevented from exchanging vapor because they were able to adjust their composition to compensate for vapor pressure differences due to droplet curvature. It is true that the pressure of water vapor over sulfuric acid droplets greatly exceeds that of sulfuric acid vapor (Fig. 26). Therefore, as discussed by Hamill et al. (1977), the droplets quickly come to equilibrium with the ambient water vapor. Indeed, knowing the temperature and the water vapor amount allows us to determine the droplet concentration by assuming equilibrium and then using the vapor pressure equations (Gmitro and Vermeulen 1964). However, the number of acid vapor molecules in the environment is not negligible (Fig. 26) and indeed rather rapid growth occurs when these molecules collide with acid aerosols. Once a collision between an acid molecule and a droplet takes place, water molecules are rapidly added to the droplet in order to maintain the droplet concentration at the equilibrium amount. The growth rate is therefore determined by the abundance of the less common vapor H_2SO_4. Figure 28 presents the time required to grow a 1 μm particle, assuming that the ambient acid partial pressure is twice the droplet vapor pressure, of $S = 2$, then $\tau \simeq r_0/g$ where r_0 is the characteristic size. These growth times are quite close to the fall and diffusion times near 70 km, and are very short below 60 km.

The trends in the growth curve of Fig. 28 closely parallel those of the vapor pressure curves in H_2SO_4 in Fig. 26 because the vapor pressure is the most variable term in the growth rate equation. The growth rate also depends upon factors other than the vapor pressure. It is inversely proportional to the radius so the growth time depends upon the radius squared. The growth rate also depends upon $S - 1$. Of course, $(S - 1) = 1$, as was assumed in Fig. 28, does not necessarily apply to the Venus clouds. Although models show that chemical processes are able to maintain large values of $(S - 1)$ above 70 km (Turco et al. 1982a; Toon et al. 1982), below 60 km $(S - 1)$ is certainly quite small. In order to attain a growth time for a 1 μm particle at 50 km equal to the removal time due to vertical mixing, $(S - 1)$ must be as small as 10^{-4} to 10^{-5}. Such small values of $(S - 1)$ are difficult to achieve, particularly if there are aerosols of different sizes present.

Rossow (1977, 1978) pointed out that due to the Kelvin effect, particles of different size should have different vapor pressures such that the smaller drop has the higher vapor pressure. For a pure component cloud, like terrestrial water clouds, this effect causes the small drops to evaporate and the large drops to grow when $S \simeq 1$. Rossow (1977, 1978) argued that, for acid droplets, the smaller drop would simply adjust its acid weight percent upward so that the decrease in water vapor pressure due to the higher weight percent would balance the increase in water vapor pressure due to the smaller size. This argument is correct, but as we have already discussed, it is the acid component that controls the growth rate. The smaller drop will have a larger acid vapor pressure than the bigger drop, both because it is smaller and because the small drop has a higher acid concentration. Rather than having no exchange of vapor, acid-water particles exchange acid vapor more rapidly than would pure droplets with the same vapor pressure. The loss of acid would induce a water loss, which would induce more acid loss, etc.: the droplets would evaporate.

Figure 27 illustrates the Kelvin effect for sulfuric acid solutions. The water vapor pressure over a 92% acid droplet at 350 K with 10 Å radius is

80% larger than that over a flat surface. The Kelvin effect causes a serious problem in understanding the multimodal structure of the Venus clouds. At 50 km, particles of radius 0.1, 1 and 5 μm all are present. Suppose that these are all droplets of H_2SO_4, and the supersaturation is so small that 5 μm particles are in equilibrium. Then, due to Kelvin effect, 1 μm radius particles would be undersaturated with $(S-1) \sim -1.5 \times 10^{-3}$ and, for 0.1 μm particles, $(S-1)$ would be $\sim -2 \times 10^{-2}$. The time to double the radius for $(S-1) = 1$ is ~ 30 s for a 1 μm particle, so the evaporation time on the Kelvin curve would be ~ 5 hr for the 1 μm drop and ~ 20 s for the 0.1 μm particle. Hence the observed size distributions are highly unstable; if the particles are all pure sulfuric acid, separate size modes are probably impossible. This problem is not as important for altitudes above 60 km where growth times are quite long even for large supersaturations.

In order to maintain multiple-mode size distributions in the lower cloud, it is necessary either that sulfuric acid compose only one of the modes or that some effect create small size-dependent changes in the vapor pressure that offset the Kelvin effect. The vapor pressure may be altered by the presence of dissolved materials in the droplet. This effect is very important in terrestrial cloud physics. Ambient aerosols of NaCl, $(NH_4)_2SO_4$ or other salts dissolve in water droplets causing a slight decrease in vapor pressure. Small droplets are more heavily enriched in the salts than large ones, so the smaller droplets have their vapor pressures depressed more than the large ones which compensates for the Kelvin effect. For strong sulfuric acid-water solutions we do not know that dissolved substances will lower the vapor pressure. Water and acid molecules are already strongly interacting so that the vapor pressure is severely depressed compared with that for the pure substances.

Figure 29 illustrates the Kelvin curve and two Kohler curves. The Kohler curves represent the vapor pressures for droplets containing dissolved material assuming that the material depresses the vapor pressure of sulfuric acid according to Raoult's law (Pruppacher and Klett 1978). In one case we assume that the mass of dissolved material is equivalent to a particle of radius 0.05 μm and in the other case to a particle of radius 0.1 μm. Figure 29 also illustrates a hypothetical case in which the supersaturation is assumed to be that needed for equilibrium with 5 μm droplets. If no material were dissolved in the droplets, all particles of radius < 5 μm would be undersaturated and would evaporate, while those > 5 μm in radius would grow. With dissolved material equivalent to 0.1 μm particles, droplets > 5 μm and < 1.5 μm would grow. However, the smaller particles could not become > 1.5 μm or they would begin to evaporate. Hence 1.5 μm is a stable size at which droplets would begin to collect. If less material were dissolved in the drop, the stable collection size would move to a smaller size. For example, with $r_c = 0.05$, the collection point would be at 0.25 μm. Hence, the presence of soluble material allows multiple modes to develop and be maintained. This process is observed in terrestrial fogs.

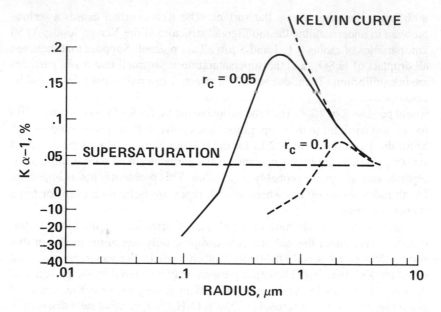

Fig. 29. Kohler curves for solution droplets. Two dissolved masses equivalent to 0.1 and 0.05 μm radius salt core particles are indicated with assumed complete dissociation and full vapor pressure reduction according to Rauolt's Law.

The final process which affects the Venus cloud particles is coagulation. While growth conserves particle numbers but changes the particle mass, coagulation conserves the mass but alters numbers. Coagulation of small particles is mainly due to their Brownian motion (see Hamill et al. 1977). However, for larger drops the differential fall velocity between particle sizes leads to coalescence. Coalescence can be neglected for particles $\lesssim 10$ μm radius because hydrodynamic forces prevent such small particles from actually touching each other. Hence, coalescence is not significant in the clouds which have so far been observed on Venus.

Figure 28 presents the time needed for coagulation to reduce the number of monodisperse 1 μm radius particles at 60 km by a factor of ½. $T_{coag} \simeq 2/n_o K$, where n_o is the number density of particles. In the calculation for Fig. 28, 100 particles per cm³ was chosen. The Venus cloud particles, particularly below 55 km, are not monodisperse, and the coagulation time is actually slightly shorter than indicated in Fig. 28. The particle coagulation kernel K is relatively independent of altitude or particle size within the cloud region (Fuchs 1964), and a typical value is 10^{-9} cm² s⁻¹.

C. Models of Cloud Structure

From the physical concepts just discussed, as well as experiments with numerical models of cloud physics (Turco et al. 1982a,b; Toon et al. 1982), we present one interpretation of the processes controlling the cloud structures. We first discuss the H_2SO_4 cloud and finally consider the possibility of solid Mode 3 particles.

The presence of large numbers of small particles on Venus indicates regions of particle nucleation: one at the cloud top, one at the cloud base, and one within the upper cloud region. The most obvious region of nucleation is at the cloud top. The possible presence of Mode 0 particles suggests a freshly nucleated sulfuric acid haze. Model studies (Turco et al. 1982a; Toon et al. 1982) show that large H_2SO_4 supersaturation can occur above the cloud top so nucleation is possible. In the model these nucleated particles slowly grow larger as they reach altitudes close to the cloud top. These particles probably correspond to the submicron particles observed by the orbiter at the cloud tops, especially at polar latitudes (Travis et al. 1979b). Due to the cold temperatures small growth rates occur above 70 km (Fig. 28) so the particles are relatively involatile and tend to be transported horizontally with little modification to their physical size. Figure 28 shows that 1 μm radius particles are not likely to be found above 75 km because the fall times are much shorter than the growth times or the assumed vertical mixing time.

The second most obvious region of nucleation occurs near the base of the cloud. If particles had formed within the cloud then coagulation and evaporation at the cloud base would lead to a decline in the number mixing ratio toward the base of the cloud. In fact, the number mixing ratio is high near the cloud base and declines slightly toward higher altitude. Therefore, a source of particles at the base of the cloud is implied.

Cloud models show that the supersaturation must be very small near the cloud base. Otherwise, as Fig. 28 indicates, very rapid growth would occur. Hence, the sulfuric acid particles must nucleate upon some preexisting particles. The LCPS discovered a thin haze of nonacid particles extending below the cloud deck to 30 km. The mass of this haze is very small so its precursor gas needs to constitute only ~ 1 ppb of the atmosphere. These haze particles may be the ones upon which the large cloud particles form. Continued condensation of sulfuric acid upon these particles and subsequent growth probably produces the Mode 1 particles observed near the cloud base. These particles may remain segregated from the Mode 2 particles due to the nature of the Kohler curve (Fig. 29).

The third region of cloud particle formation which is located in the upper cloud is the least obvious, but also the most interesting. The LCPS discovered two cloud particle size modes throughout the upper cloud. Furthermore, the number densities of both modes decrease rapidly going upward from 60 km toward 65 km (the cloud top is located near 70 km). Orbiter observations show that, in comparison to 65 km, there is a larger number density for Mode 2 and similar number densities for Mode 1 at the cloud top. These comparisons might represent temporal variation, or they may show that the cloud top as observed from space is actually a layer distinct from the one in which the

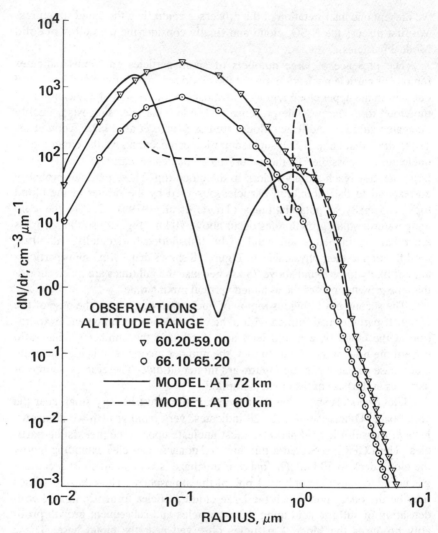

Fig. 30. Model predictions for the upper cloud region. (After Turco et al. 1982, Toon et al. 1982 and Toon and Turco 1982.)

initial measurements of the LCPS were made. Chemical models (Toon and Turco 1982) indicate that elemental sulfur can be created and can condense to form new small particles within the region of the LCPS upper cloud.

Figure 30 presents calculated size distributions in comparison with observed ones for the portion of the cloud located above ~ 56 km. The details of the calculation and analysis are presented by Turco et al. (1982a,b), Toon et al. (1982), and Toon and Turco (1982). In the model, SO$_2$ from the lower atmosphere is mixed into the upper cloud where it photodissociates, yielding both H_2SO_4 and elemental sulfur as end products. The H_2SO_4 is relatively stable against photodissociation and reaches altitudes near 80 km where the supersaturation is great enough to lead to nucleation. The freshly created nanometer-size particles coagulate and grow to larger sizes while they mix and fall to lower altitudes. Near 70 km a bimodal size distribution results from mixing together small sulfuric acid particles newly created at higher altitudes and slightly larger sulfuric acid particles which have resided for several months near 70 km. The narrow Mode 2 size distribution is created partly by the r^{-1} dependence of the growth rate and partly by the fact that particles > 1 μm are removed by sedimentation more rapidly than they can be created by growth.

Near 60 km the model also produces a bimodal distribution. However, in this case, the sulfuric acid particles dominate Mode 2 while sulfur particles compose the bulk of Mode 1. Sulfur is photochemically too unstable and is too rapidly oxidized by O_2 to form particles above the cloud top, which accounts for its general absence near 70 km. The fact that the sulfuric acid particles still have a narrow size distribution with a modal radius of 1 μm near 60 km is due to the near equivalence of the time needed to grow to 1 μm and the time for such particles to fall to lower altitudes.

A model simulation of the middle cloud is presented in Fig. 31. Small particles, which are assumed to condense near 30 km, enter the cloud at its base and dominate the Mode 1 cloud particles. Sulfuric acid nucleates on these particles to create the Mode 2 particles and the largest of the Mode 1 particles. The size to which the Mode 2 particles grow is strongly dependent upon the rate of supply of sulfuric acid vapor from below and upon the modification to vapor pressure caused by dissolved materials. In Fig. 31 the crosses represent a calculation in which the rate of supply of acid vapor from below is low and in which the solubility of the Mode 1 particles does not affect the acid vapor pressure. The sulfuric acid particles occur as a single size mode centered near 2 μm radius. Another case, presented in Fig. 31 by dots, has a higher rate of acid vapor supply from below and the Mode 1 particles affect the acid vapor pressure. In this case two sulfuric acid modes form with radii of ~ 1.5 μm and 4 μm.

An important point concerning the Venus clouds is that the lower clouds form by condensation and therefore have the potential for great temporal variability. The impression that Venus has an unbroken cloud deck with relatively little variability is based upon the upper cloud which is so cold that it is relatively involatile. However, the lower cloud could respond quickly to vertical motions which increase the supply of acid vapor locally, or carry particles downward toward higher temperatures. The variability is observed in the greatly differing structures of the lower cloud that were observed by the PV nephelometers on the various entry probes (Ragent and Blamont 1980) as well as by the Venera nephelometers (Marov et al. 1980).

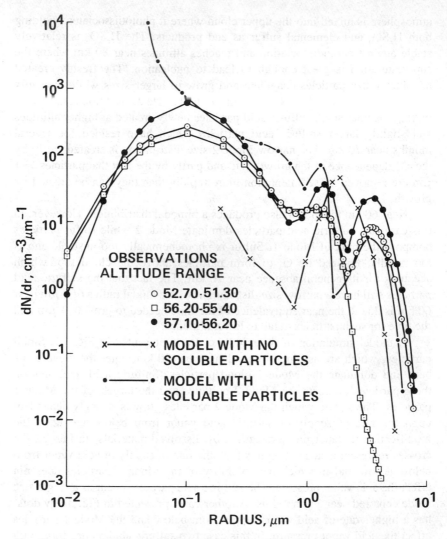

Fig. 31. Model predictions for the lower cloud region. (After Turco et al. 1982, Toon et al. 1982 and Toon and Turco 1982.)

The middle cloud calculations described above suggest that sulfuric acid clouds could form with or without Mode 3 particles of H_2SO_4 depending upon the details of the dynamics near the cloud base and upon the unknown physico-chemical properties of the hypothesized small haze particles below the cloud. The asymmetric Mode 3 particles suggested by Knollenberg and Hunten (1980) could be coated by and perhaps even largely composed of sulfuric acid. Another possibility is the suggestion of Toon et al. (1982a) that the Mode 3 particles may not be an independent collection of solid particles, but rather the large-particle tail of the H_2SO_4 Mode 2 particles. Since the Mode 3 particles dominate the mass of the entire cloud, determining their composition is of the utmost importance.

V. COMPARATIVE VENUSIAN AND TERRESTRIAL CLOUD PHYSICS

From our previous discussions it is obvious that the Venus clouds are not normal clouds in the terrestrial sense. The particles are simply too small to provide the optical density one normally associates with terrestrial clouds. It is only through their immense depth that they generate the optical depth one associates with clouds. Table VI summarizes average properties of Venus and Earth adapted from Knollenberg and Hunten (1980).

An examination of Table VI reveals several interesting contrasts. First, Venus has 100% cloud cover while the Earth has typically 40%. This is directly related to the greater volatility of H_2O compared to H_2SO_4. The average optical depth is 5 times greater on Venus than on Earth, yet mass loadings are at least 10 times less; this is due to the much smaller average particle size. The small aerosol populations are quite different with much more submicron aerosol mass on Venus at cloud levels. A further distinction here is that Earth clouds have reduced aerosol populations within cloud while Venus appears to have enriched populations. An explanation may be the lack of precipitation and therefore of effective scavenging mechanisms on Venus compared to Earth. The morphological consequences of the highly volatile H_2O terrestrial clouds versus the stable H_2SO_4 Venus clouds are manifested in dominant stratiform cloud masses in the visible part of the Venus atmosphere with low temporal variability and negligible latent instability and precipitation. Worthwhile comparisons are thus only possible at the microphysical process level; in the previous discussion we have drawn a number of comparisons.

One obvious counterpart to the Venus H_2SO_4 clouds and haze particles is the terrestrial stratospheric Junge layer (typical optical depth 10^{-3}) which has maximum mass mixing ratio near 120 mbar ($\sim$20 km). If this layer were much thicker, Earth would appear much as Venus does to the external observer. Sampling of this layer has increased in the late 70s and early 80s via balloon, aircraft, and satellite measurements. Earlier measurements (see Toon and Farlow 1981 for a review) determined size distributions with modal diameters of ~ 0.15 μm. Like the upper haze of Venus, the terrestrial Junge layer is highly variable in time and location; 1 to 2 orders of magnitude variations in number density are observed. Also, like Venus, this region of our atmosphere does not enjoy the cleansing effects of precipitation in reducing particle numbers, volatility is low, and sedimentation requires months to be effective.

TABLE VI

Comparative Properties of Venus and Earth Cloud Systems

Property	Earth	Venus
Percent coverage	40	100
Average optical depth	5–7	25–40
Maximum optical depth	300–400	40
Composition	solid & liquid H_2O	H_2SO_4 droplets, crystals, plus contaminants
Number density	100–1000 cm^{-3} (liquid) 0.1–50 cm^{-3} (ice)	50–300 cm^{-3} (liquid) 10–50 cm^{-3} (crystals)
Average mass loading (mass density)	0.3–0.5 g m^{-3}	0.01–0.02 g m^{-3}
Maximum mass loading (mass density)	10–20 g m^{-3}	0.1–0.2 g m^{-3}
Distribution function	normal—log normal bimodal (ice)	multimodal
Typical background aerosol at cloud base 0.5 μm	1 cm^{-3}	100–200 cm^{-3}
Condensation process	homomolecular	heteromolecular
Average precipitable mass	0.03–0.05 mm	0.1–0.2 mm
Mean scattering size (diameter) (in the visible)	10 μm	2–4 μm
Mean mass size (diameter)	30 μm	10 μm
Dominant optical cloud form	stratiform	stratiform
Dominant mass cloud form	cumulus	stratiform
Potential latent instability	high	low
Temporal variability	high	slight
Dominant cloud atmosphere heat exchange process	latent heat	radiation

Recently with the influence of Mt. St. Helens and other volcanic eruptions the stratospheric aerosols have been substantially enhanced. Figure 32
shows several characteristic vertical profiles over Panama from Knollenberg
et al. (1982) and Farlow (1981). An interesting feature of these measurements
is the decreasing size with increasing altitude (the reverse of observations
reported by Toon and Farlow 1981) and the lower altitudes (18 versus 20 km)
for maximum mass mixing ratios. Recently, measurements over Wyoming

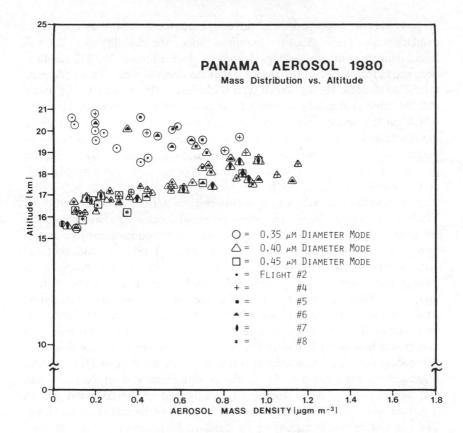

Fig. 32. Panama aerosol mass distribution.

(Rosen et al. 1982) confirm number densities > 50 cm^{-3} at 16 to 17 km. It would appear that the Venus upper haze in weaker periods is similar to the terrestrial Junge layer during enhanced periods and that both are highly variable.

In addition to the stratospheric Junge layers, low level haze clouds of strong anthropogenic influence can develop sufficient acidity levels to strongly influence microphysical properties. These precursors to the acid rain plaguing the industrialized areas of the world were referred to by Knollenberg (1977) as venusian-type clouds. From Braham's (1975) observations, such haze-clouds develop over statewide regions without any direct association with any one source region, but generally within a stagnating anticyclonic mixing layer capped by a subsidence inversion. These haze layers are characterized by 3 to 4 μm diameter H$_2$SO$_4$ droplets with 1 cm^{-3} > 10 μm. The precursor gases are H$_2$O and SO$_2$, as in the case of Venus. The H$_2$SO$_4$ droplets are concentrated by recycling within isolated scud layers near the haze top and form in a matter of hours as opposed to months in the stratosphere. The droplets survive large periods of advective transport as well as convective displacements. These droplets sometimes reduce the visibility to ~ 2 km in broad regions seemingly decoupled from the source region. In 1977 Knollenberg stated that we should be thankful that the Earth is blessed with adequate rainfall to clear the atmosphere of such acidic hazes. Today we are fully aware that the consequences of this atmospheric cleansing is acid rain, the effects of which on the surface ecosystem are all too apparent. As terrestrial cloud physicists seek to understand the processes incipient to acid rain, data should become available to provide detailed microphysical comparisons. For the present we can compare size spectral features of more typical terrestrial H_2O clouds with Venus H_2SO_4 clouds.

One worthwhile comparison is between the narrow size distribution of Mode 2 observed on Venus and terrestrial cloud spectra. We have previously stated that clouds do not develop narrow size distributions through coagulation processes. Coagulation processes generally produce broadened lognormal distributions. However, clouds do produce narrow distributions through diffusional growth. Diffusional growth occurs in all clouds given an updraft or a diabatic cooling process. Coagulation processes are slow at sizes characteristic of terrestrial clouds (5-50 μm) but coalescence processes can dominate at sizes > 25 μm. Where would one then look for a narrow size distribution in terrestrial clouds? The answer is in regions where the time scale is too short for coagulation to be effective and sizes are too small for effective coalescence; near a cloud base updrafts are significant and droplets are small and just starting to grow. Figure 33 is a typical terrestrial cloud droplet spectrum observed 300 m above cloud base in a continental cumulus cloud. The size distribution is narrow with a standard deviation of 2.0 μm. In other words, it is as narrow (comparing the mode to standard deviation) as any of the LCPS measurements indicate for Mode 2. More importantly, it is narrower for the same reasons: (1) diffusional growth dominates, (2) coagulation and coalescence are negligible, and (3) new nucleation is curtailed.

An interesting feature of the Venus clouds and hazes is the observed characteristic multimodal size distributions. Multimodal size distributions are also observed in terrestrial clouds and for similar reasons. There are several identifiable processes that generate multimodal size distributions in terrestrial clouds:

1. The activation process mentioned in Sec. V involving Kohler trapped small solution droplets coexisting with larger more dilute droplets;
2. Erosion of cumulus cloud parcels with ambient air;
3. Mixing of two cloud droplet populations;
4. Rapid coalescence;
5. Mixed phases.

The Kohler phenomenon and erosion of cumulus produce similar spectra though the sizes involved vary. Figures 34 and 35 are observed distributions in

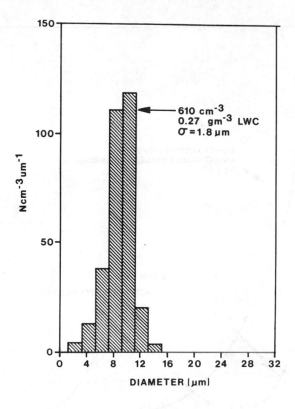

Fig. 33. Narrow droplet size distribution observed 300 meters above cloud base in continental cumulus in Colorado. The narrow size distribution results from rapid diffusional growth.

ground fog and cumulus, respectively, in Colorado. The ground fog observations reveal quite similar size distributions to those measured in the two precloud layers on Venus, the lower of which is shown in Fig. 36. Note that σ_g in the log-normal fit for the terrestrial fog corresponds to a narrower measured size distribution than σ for the Gaussian distribution in the Venus observations, but the comparison is otherwise very similar. Indeed, the underlying unactivated droplets have very similar σ_g. Figure 34 also shows fog with unactivated droplets where a log-normal distribution has been fitted to the observed large tail of the observed sizes. The σ_g of 3.1 is somewhat larger than for Mode 1 in Fig. 36.

As droplets become larger the effects of curvature (Kelvin effect) rapidly become negligible. Thus, while 0.1 and 1 μm diameter droplets may have vapor pressure differences of 1%, 10 and 100 μm droplets differ in vapor pressure by $< 0.01\%$. Since normal supersaturations in terrestrial clouds are above 0.1%, all droplets larger than a few microns can coexist rather easily. Thus when cloud parcels having differing droplet size mix, these populations

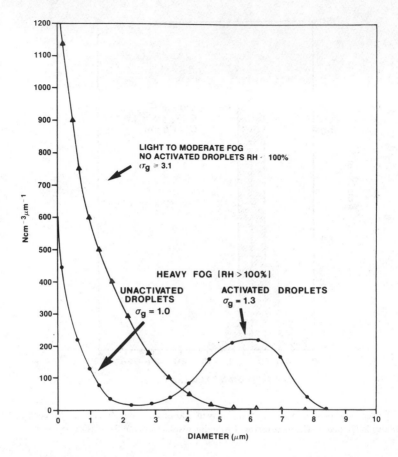

Fig. 34. Droplet size distribution observed in fog. The bimodal spectrum results from activation of a portion of the aerosol droplets when the relative humidity exceeds 100 % or $S-1 > 0$.

are stable, at least on the time scales that such clouds survive. On Venus supersaturations are lower so that 10 and 20 μm diameter Mode 3 droplets may not interact in competitive growth but Mode 2 and Mode 3 droplets would.

Once droplets become large enough to become involved in coalescence two effects occur. Depending on the narrowness of the original diffusion growth population, a second mode will develop with masses equal to or greater than twice that of the original mode. This stochastic growth process eventually results in the development of raindrop sizes. The secondary mass mode is easily traced in development but the bimodal size spectrum is only apparent with high size resolution in the onset coalescence stage and is a transient phenomenon.

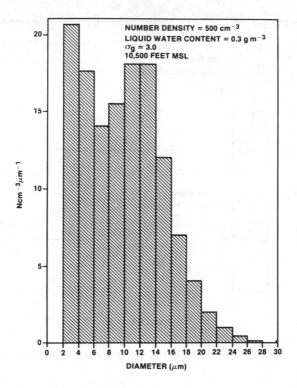

NUMBER DENSITY = 500 cm^{-3}
LIQUID WATER CONTENT = 0.3 g m^{-3}
σ_g = 3.0
10,500 FEET MSL

Fig. 35. Droplet size distribution within eroded continental cumulus. A bimodal size spectrum is often observed near the edges of terrestrial clouds.

Probably the commonest terrestrial case for multimodality is in mixed phase clouds. In terrestrial clouds the ice phase is produced under supercooled conditions where ice crystals grow at the expense of the water droplets. In a matter of minutes ice crystals are at least a factor of 10 larger than the surrounding water droplets. Depending on the availability of ice nuclei, the larger size mode may contain far fewer particles but rapidly consume all droplets resulting in a glaciated cloud. We are not suggesting that this is the case on Venus with Mode 3, since H_2SO_4 does not freeze at the temperatures in the middle or lower cloud. However, if Mode 3 is solid, it could be competing for certain vapor molecules with Mode 2.

VI. SUMMARY

Through spacecraft and earth-based investigations, our knowledge of the Venus clouds has advanced much in recent years. A number of earlier questions have achieved partial resolution. We now have a fairly good understand-

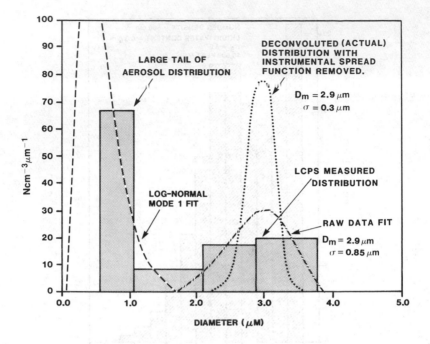

Fig. 36. The lower precloud layer size distribution resembles that in the middle cloud region.

ing of the vertical cloud structure and of the aerosols that form the clouds. We can count as major advances our present understanding of the cloud-top chemistry and physics, the studies of ultraviolet absorbers and the role of SO_2, the discovery of the mutlimodal aerosol size distribution, and the determination of the properties of the high-altitude submicron haze. The PV orbiter is providing us with an invaluable long-term record of the cloud-top region; the *in situ* measurements of the PV multiprobe mission have given us a measure of "cloud truth." Nevertheless, some tasks remain unfinished. The following summarizes our current state of knowledge on the important cloud problems with suggestions for continued effort.

A. Upper Haze and Cloud Top Region

Ultraviolet Absorption. This was the most frustrating question on Young's list in 1975 (Young 1975) and it is still with us. Sulfur dioxide is responsible for absorption at short ultraviolet wavelengths, but the identity of the absorber visible from Earth in the near ultraviolet remains unknown. Particulate sulfur (Mode 1) appears to be a possible choice as well as molecular chlorine. The related question of how the physical and/or chemical processes at the cloud top act to bring about the horizontal variability of these absorbing constituents is still a puzzle although several mechanisms have been proposed. This question is important also for the interpretation of cloud-tracked winds, which depends on implicit assumptions about the nature of the tracked features. The dark ultraviolet features are the most obvious characteristic of Venus to the external observer; the question of their nature is certain to haunt us while it remains unresolved. Some plausible lines of investigation that bear on this problem are high-resolution spectroscopy in the near ultraviolet, improved models of the cloud-top chemistry and physics, photometric and polarimetric studies of the bright and dark features, and laboratory work on the absorption cross sections and chemistry of plausible absorbers.

Submicron Haze. We now have a quantitative first understanding of the haze layer that covered the poles in the Pioneer Venus mission. Such a layer is probably the cause of earlier polar brightenings. These particles appear to be H_2SO_4, a counterpart to the terrestrial Junge layer. Much work remains: to determine the vertical distribution of the haze particles and to explain the morning/evening terminator asymmetry; to establish the microphysics of the haze formation. The time history of this phenomenon provides one basic piece of information to infer a steady or non-steady equilibrium. The study of the haze provides an insight into the weather (or perhaps climate?) at the Venus cloud tops.

Weather and Climate. What is the mean state of the Venus clouds and how does it vary? In the cloud-top height, in the zonal circulation, and in the latitudinal brightness distribution, Venus was significantly different as seen by Mariner 10 and Pioneer. These characteristics and others need to be observed in detail over long time periods. What is the equilibrium state and is it the same as the mean state? Are the clouds and the winds in a steady or nonsteady equilibrium? How does the regular oscillation in cloud morphology give rise to familiar average brightness distribution like the Y feature? The idea of propagating waves in the cloud-top region seems compelling, but how are waves related to the details of small- and large-scale morphology, circulation and transport, and to the mechanism for creating dark and bright regions? Do the cloud features just "go along for the ride" or do wave interactions help to determine the cloud features? What waves are involved, how are they created, how do they interact, how long do they last? Both deeper analysis of existing data and better theoretical formulations are required for advancement in this area. The continued data taking and analysis from the Pioneer Venus extended mission may answer some of these questions.

B. The Main Cloud Deck

In situ experiments have provided extensive microphysical information in the main cloud deck. Of greater importance is the evidence for the existence of two if not three separate particle constituents. The vertical distribution of these constituents generates the observed layered cloud structure evident planetwide in the cloud morphology. Planetary transition regions have been defined at altitudes coinciding with changes in microphysical properties and perhaps particle chemistry. The following list of summary findings highlights our current understanding of the clouds at greater depths.

Composition. The bimodal and trimodal size distributions observed indicate two or three different particle constituents. Mode 1 is present in all cloud regions as a background aerosol population with high number density but little optical depth except perhaps in the ultraviolet. Free sulfur, extra-venusian materials, and pure H_2SO_4 are candidates at various altitudes. In the upper cloud region the Mode 1 aerosols may have a higher refractive index and thus imply sulfur compounds. In the lower cloud region H_2SO_4 appears more plausible. Below the lower clouds, the lower thin haze which consists of Mode 1 aerosols may be desiccation products (cores) of cloud particle growth and decay. Mode 2 is almost surely sulfuric acid in all cloud regions and is highly concentrated. Mode 3 in the middle and lower cloud regions may be made up of asymmetric particles; however, these may also be coated with sufficient H_2SO_4 that it is the dominant form of mass. Such crystalline cores may be $NOHSO_4$. Nevertheless, it is also possible that these Mode 3 aerosols are merely large H_2SO_4 particles whose existence as a separate mode is somewhat artifactual. Resolving this controversy is quite important since Mode 3 provides the dominant mass in the Venus clouds. Both possibilities (asymmetric particles or observational error) seem open at this time. If the particles are not H_2SO_4, their composition and source vapor remain unknown. On the other hand, the theoretically more satisfying interpretation as H_2SO_4 requires discounting measurements by Pioneer Venus and Venera probes. Apparently further direct observations are required.

The Mode 2 H_2SO_4 droplets are probably several months old in the upper cloud region unless vertical exchange is faster than suspected, but perhaps only hours old in the lower cloud region. At upper altitudes H_2SO_4 droplets may be contaminated through scavenging of aerosols over their long lifetime. In the upper cloud region, H_2SO_4 vapor may be highly supersaturated, contributing to nuclei production, and supporting a bimodal population.

Particle Microphysics. The particle size distributions in the main cloud deck are inherently multimodal, while unimodal in the lower thin haze. In the lower and middle cloud regions the size distribution is observed to be trimodal with modal diameters to 0.1 to 0.5, 1.8 to 2.8, and 6 to 9 μm. Particles comprising the smallest size mode belong to an aerosol population extending throughout the main cloud deck and 15 to 20 km above and below it. The second size mode consists of droplets of H_2SO_4 with primary growth in the upper cloud region and a gradual increase in number density descending through the main cloud deck. This mode regenerates through condensation in the lower layers. Its size distribution is exceedingly narrow at any one altitude, and it is the dominant planetary size mode. The largest size mode could be asymmetric, implying crystalline cores. These particles may have sufficiently high aspect ratio to preclude accurate mass determination.

The particle size distributions are similar to those in terrestrial clouds where growth proceeds via similar processes. A quantitative model (Toon et al. 1982c) can produce the observed multimodality by competition between the Kohler and Kelvin effects, a process observed on Earth. However, the number of unknowns and assumptions in the model invites some scepticism. Clearly more *in situ* data are required to constrain such studies. Dominant but competitive droplet diffusional growth and sedimentation helps to explain the extreme narrowness of the size distributions in the Venus clouds. The uniformity of the Mode 2 distribution over most of the planet is indicative of extreme stability.

Optical Properties. The entire Venus cloud system has an optical depth of 20 to 35 at visible wavelengths. The particle real refractive index at visible wavelengths for Mode 2 is ~ 1.44 to 1.46, consistent with H_2SO_4. The refractive index of Mode 1 could be as high as 2 in the upper cloud region. The real refractive index of the suspected crystal within Mode 3 is unknown, but probably ranges from 1.5 to 1.7. The imaginary index of any Mode 3 crystals must be $< 10^{-3}$. Single scattering albedos at 0.63 μm range from low values of 0.995 in the upper cloud region to 0.999 in the lower cloud region. With regard to the ultraviolet absorber, gaseous absorption by SO_2 is clearly indicated, but an additional nonparticle contributor to ultraviolet contrasts at wavelengths longward of 3200 Å is required. No particulate absorber restricted to a single mode is satisfactory. The large Mode 3 particles are essentially nonabsorbing, but absorption in both Modes 1 and 2 is possible in the upper cloud region.

Other Cloud Processes. The main cloud deck is imbedded in the zonal circulation at altitudes of greatest wind velocity and vertical wind shear. The clouds are all of stratiform morphology. Cloud particle growth is not strongly influenced by either the large-scale circulation or latent heat released during condensation. These clouds absorb some heat from beneath via radiation. The H_2SO_4 droplet growth times range from months in the upper hazes to hours in the lower cloud region.

There is little evidence of potential latent instability in these clouds, nor for the existence of large precipitation particles. Known processes for the formation of lightning require both. If cloud processes generate the observed lightning, then much larger undetected particles must exist in clouds not observed by the handful of direct measurements.

C. The Future

How shall we go about finding the answers to the questions that remain? The extended operations of the Pioneer Venus orbiter will provide continued study of the Venus cloud top region. At least for this altitude range, we can hope to determine the long-term variability and stability of the cloud system. The answer is not as easy for the deeper regions. We need more *in situ* information for intercomparison and for settling such nagging questions as the composition of the deep clouds and the existence and shape of the Mode 3 cloud particles. Furthermore, the burst of new knowledge from the Pioneer and Venera missions has not yet run its course. Fruitful intercomparisons are still possible, along with deeper theoretical inquiry into the chemistry, dynamics, and microphysics of the clouds. We can hope that these efforts will not only advance our understanding of Venus, but of the Earth as well.

Acknowledgments. Much of our understanding of the Venus clouds has developed at meetings and activities of the Pioneer Venus Clouds Working Group, which has met regularly over the past few years to discuss, interpret and reconcile the diverse Pioneer Venus data. We have had many helpful comments on this manuscript from W. B. Rossow and B. Ragent and substantial assistance from C. A. Smith. The writing of this chapter was supported by the Pioneer Venus Project and the NASA Planetary Atmospheres Program.

17. THE ELECTRICAL ACTIVITY OF THE ATMOSPHERE OF VENUS

L. V. KSANFOMALITY
USSR Academy of Sciences

F. L. SCARF AND W. W. L. TAYLOR
TRW Space and Technology Group

Electrical discharges in the atmosphere of Venus produce electromagnetic radiation which can be detected by remote sensors. The lowest-frequency wave components (f ≤ 300 Hz) propagate through the ionosphere in the whistler mode when the magnetic field strength is high and the field points toward the planet; the signals are detected on the Pioneer Venus orbiter. More complete information on the total electromagnetic spectrum of Venus lightning comes from the direct atmospheric measurements of electromagnetic radiation on the Venera probes. The full analysis suggests that sources of electromagnetic radiation may be located much lower than the cloud layer and that unusual methods of space charge accumulation could be operative. The question of the optical radiation from discharges remains unclear.

Prior to the end of 1978, lightning represented one phenomenon for which we had direct observations only at Earth. There had been theories proposing existence of electrical discharges (lightning) in the atmospheres of other planets, primarily Jupiter, in order to account for certain unusual types of radio emission or unusual chemical characteristics of the atmosphere; however, even indirect measurements were unavailable.

When radio emissions from Venus were discovered at the end of the 1950s, Kraus (1956, 1957) initially theorized that the bursts of radio signals from Venus were related to lightning in the atmosphere of the planet. However, it is now known that the high-frequency radio emission proceeds from the intensively heated planetary surface and lower layers of the troposphere, with no relation to electrical discharges. Studies of the composition of the atmosphere by probes of the Venera series from 1967 through 1975 (and afterwards by Venera 11, Venera 12, and the Pioneer Venus probes) identified an additional problem of the formation of certain small gaseous components of the Venus atmosphere; there were conjectures that their characteristics involve electrical discharges, since it is known that the electrical activity of the terrestrial atmosphere appreciably affects the composition of the minor components. For instance, discharges produce nitrogen oxides NO, ozone O_3, nitrogen hydride NH, and even cyanogen CN. The nitrogen oxides formed during the lightning discharges react with the ozone and certain derivatives of chlorine (Chameides et al. 1979; Levine et al. 1980; Yung and McElroy 1979). Thus, investigations designed to detect lightning were placed on the Venera 11 and 12 probes, and the Pioneer Venus plasma wave instrument was also designed with a limited capability to detect whistler-mode signals from lightning.

On 21 and 25 December 1978, as the Venera 11 and 12 probes descended into the atmosphere of Venus, the Groza instrument detected a large number of electromagnetic pulses, basically very similar to the radio impulses or "atmospherics" produced by distant lightning flashes on Earth (Ksanfomality et al. 1979; Ksanfomality 1979a,b, 1980a). The Groza instrument responded to the magnetic component of the field. Shortly afterward, similar types of pulses were recorded by the plasma wave experiment in orbit on board Pioneer Venus; here, the electric component of the wave field is measured. These observations were also interpreted in terms of detection of Venus lightning (W. Taylor et al. 1979a,b; Scarf et al. 1980a, 1982).

The Venera and Pioneer Venus investigators interpreted these wave observations in terms of discharge sources on the basis of extensive terrestrial experience with similar measurements. For instance, while broadband impulsive electrostatic noise bursts are commonly detected by plasma wave instruments above the Earth's ionosphere, none of the very-low-frequency whistler mode electromagnetic waves that are known to originate within the magnetosphere (such as chorus and hiss) are impulsive on the time scales of interest. Similarly, in the terrestrial atmosphere, all the intense and impulsive low-frequency noise bursts are identified with lightning-like discharges in clouds, volcanic plumes, etc. To our knowledge, there are no other sources in a planetary atmosphere for radio bursts with these characteristics, and we use the general term lightning to describe the relevant Venera and Pioneer Venus observations, although direct observations of discharge phenomena (e.g., photographs) are certainly not available.

The first reports on the wave measurements stimulated additional work devoted to a search for venusian lightning flashes from orbiters. The first of these contained a discussion of spectrometric measurements of the night side of Venus, made in 1975 by the orbiters Veneras 9 and 10. Bursts were observed; however, the author conjectured that these were related to disturbances in the functioning of the instrument (Krasnopolsky 1980 and Chapter 15). The star sensor system of the Pioneer Venus orbiter was also used as a photodetector (Borucki et al. 1981), but negative results were obtained; this might have been due to the small amount of observation time and, also, to increased noise associated with penetrating energetic particles. The connection between lightning and generation of certain chemical components in the atmosphere of Venus became the object of many subsequent studies (Bar-Nun 1980; Levine et al. 1982; Chameides et al. 1979; Yung and McElroy 1979). Reports on the experiments aboard the Venera spacecraft and Pioneer Venus had apparently stimulated interest in this problem. Initially, all the papers on the electrical activity in the atmosphere of Venus were classified under the general heading of cloud research, since no one doubted that the charges are formed in the cloud layer where the lightning is also created. Serious doubts as to the validity of this proposition were advanced by Croft (personal communication, 1980) who noted that the refraction effects in the atmosphere of Venus would suggest a source location below 30 km. Scarf et al. (1982) recently speculated on sources very near the surface.

In December of 1978, reports were received on strange phenomena which may possibly relate to electrical activity at Venus. At altitudes of $\sim$ 12 km, certain detectors on board all the Pioneer Venus probes, even those independently installed on different vehicles, were damaged. Electrical discharges have been proposed as a probable cause (Polaski 1979). Unusual properties of the atmosphere of Venus near altitudes of 12 to 18 km also appeared in the Groza data and in the observations of the Doppler residuals obtained from Venera 12. In traveling through these altitudes, phase fluctuations were observed that were not related to the motion of the probe. This suggests the possibility that observed electrical activity may be related to the mysterious phenomena that develop at levels of 10 to 15 km. Ksanfomality (1979b) noted that the electrical activity of the atmosphere of Venus could possess merely a superficial similarity to lightning storms, manifested in the form of electric discharges, but possibly with a different underlying phenomenon. Apparently, an answer can be furnished only from the experimental data, to which we shall now refer.

I. EXPERIMENTAL RESULTS

In our survey of electrical activity at Venus, we use the results from several investigations, shown schematically in Fig. 1. The direct observations come from the Groza investigations on Veneras 11 and 12 and the plasma wave detector (OEFD) on the Pioneer Venus orbiter. These vehicles are shown in the lower left and upper right of the figure. Both experiments detected low-frequency electromagnetic waves. The search for related optical bursts was

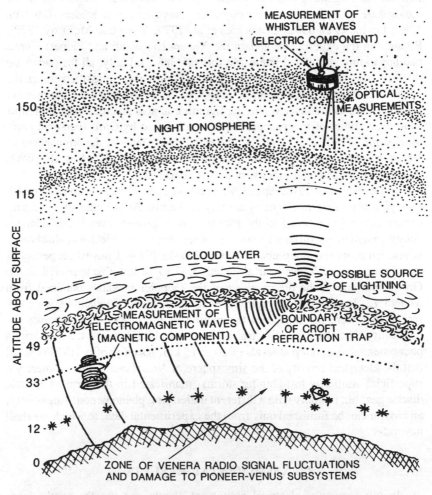

Fig. 1. Electrical discharges in the atmosphere of Venus can be recorded by a receiver on board the probe in the form of a low-frequency electromagnetic pulse, or by a receiver on board a low-flying spacecraft as a wave propagating through the ionosphere in the whistler mode along the fluctuating magnetic field. If the discharge takes place in the clouds, the scattered light may theoretically be recorded by the optical instruments of the spacecraft.

carried out using the star sensor on the Pioneer orbiter and the spectrometers on Veneras 9 and 10 (upper left corner of Fig. 1). We also discuss information on the fluctuations of the signal phase of the radio transmitter of Venera 12 and certain anomalies in the operation of the probes of Pioneer and Venera, as indicated at the bottom right of the diagram. (The nighttime ionosphere in Fig. 1 does not pertain to Groza.) As shown in the figure, there was free access of radio waves to the probes of Venera 11 and Venera 12, while the Pioneer Venus orbiter was shielded by the lower layers of the ionosphere. The Venera 9 and 10 spacecraft were located beyond the limits of the ionosphere and did not possess equipment for the detection of discharges. Pioneer Venus probes carried no equipment for detecting discharges.

II. THE LOW-FREQUENCY ELECTROMAGNETIC MEASUREMENTS

As seen from Fig. 1, the Groza measurements were obtained in an optimum location, since the observations were conducted in the actual surface-ionosphere waveguide. The instrument has four bandpass channels with center frequencies of 10, 18, 36, and 80 kHz and with bandwidths of 1.6, 2.6, 4.6, and 15 kHz, respectively (Ksanfomality 1979a). The instrument acts as a very long wavelength radio receiver with an external loop antenna of 250 mm diameter, reacting to the magnetic component of the field. Since there were no previous data available on the intensity of atmospheric disturbance fields, the instrument was tuned to the maximum possible sensitivity, consistent with the presence of other instruments of the spacecraft; the sensitivity is $\sim 0.3~\mu V$ at a frequency of 10 kHz and $\sim 1~\mu V$ at 80 kHz. An automatic gain control (AGC) system prevented overloading of the channels by large signals and operated with a signal in a broad-frequency band, 8-90 kHz. The instrument also had two pulse height discriminators with counters. The pulse counting rate f_p could also be evaluated independently by analysis of the nature of the recording, despite the relatively low frequency of sampling of the instrument output ($3~s^{-1}$) and the possible statistical counting errors. For a rather high frequency of impulses, the threshold signal level rises, as the averaged period of succession of the pulses $1/\overline{f_p}$ becomes much less than the time constant of the detector τ_d (0.24 s). The RC circuit of the detector does not have time to discharge between the frequent pulses, and if the average size of the maxima of the peaks is u_0, there appears a constant "pedestal":

$$u_s = u_0 \exp(-1/\tau_d f_p) . \qquad (1)$$

Thus, we can find the mean frequency of impulses independently of the counters, using

$$f_p = 1/\left[\tau_d \ln (u_0/u_s)\right] . \qquad (2)$$

For closely spaced discharges with rather intense acoustical fields (more than 78 dB), the sounds of the discharges could also have been recorded by means of a microphone and acoustic channel, but we attribute all of the sounds detected during the mission to aerodynamic noise (during the descent) or to spacecraft subsystems (on the surface).

The measurements began at an altitude of $\sim$ 60 km and continued down to the surface of the planet, and on the surface until contact was lost with the relay spacecraft. The time of operation of the Groza instrument on the surface was 110 min for Venera 12 (12 December 1978) and 76 min for Venera 11 (25 December 1978). The coordinates of the landing points were (Kurt and Zhegulev 1979):

	Venera 11	Venera 12
Latitude	$-14°$	$-7°$
Longitude	299°	294°
Solar Zenith Angle	20°	25°

The radio measurements were transmitted to Earth at a rate of 3 s^{-1} during the entire descent of the probes. The mean impulse counting rate was 16.5 s^{-1}, being reduced to 13 s^{-1} at an altitude of 20 km, while below 5 km it was 10 s^{-1}. During the preparations for the experiment, there was concern with respect to possible discharges involving the charging of the probes during their descent, even though such descent velocities are entirely insufficient for noticeable charging at Earth. Interference due to differential charging could be manifested in the form of short pulses at the outputs of the instrument, and therefore it was essential to demonstrate that these signals were absent.

During the descent, radio noise was recorded which was extremely similar to that found in the atmosphere of Earth during electrical thunderstorms. The first reports were given by Istomin et al. (1979*b*) and Ksanfomality et al. (1979*c*). As Venera 11 descended, the burst rate was high, whereas during the descent of Venera 12, the situation was more quiescent. A comparison of the development of individual bursts of radio noise during the descent of the two probes along identical routes in essentially the same region of the planet revealed appreciable differences in the nature of the recorded phenomena. This result would seem to preclude any relationship between the recorded radio bursts and the charging of the probe (Ksanfomality 1979*a*). A series of other facts also suggest that the radio bursts are unrelated to charging of the probe. For instance, one group of pulses was periodic, supposedly related to the aerodynamic rotation of the probe during the descent and to characteristics of the beam pattern of the Groza antenna. Finally, one burst was detected during operation on the surface of the planet, when the influence of charging could be completely excluded.

It is convenient to begin the discussion of the results of Groza with summary graphs of the averaged intensity of the electromagnetic field variation during the descent of the probes. The averaging was performed over 20 s

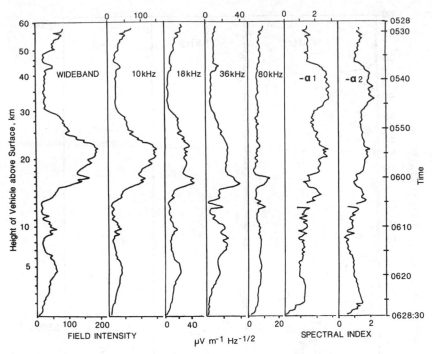

Fig. 2. The field intensity recorded on 25 December 1978 during the descent of the Venera 11 probe. Along the vertical: Earth-received time of the signal (right) and altitude of the probe above the landing point. Along the horizontal: the intensity in the channels 10-80 kHz (10, 18, 36, and 80 kHz), adjusted per unit of frequency interval and averaged over 20 s intervals. The two columns on the right show the spectral indices for the channels 18/10 and 36/10 kHz.

intervals for the entire period of descent, the influence of the AGC system being excluded. It makes no sense to represent the field intensity in this averaged form after the landing of the capsules, since only a single small group of pulses was recorded at the surface. Figure 2 shows the curves of the intensity $E(\mu Vm^{-1}Hz^{-\frac{1}{2}})$ in a broad band (8–90 kHz) and in four narrow spectral intervals (E_{10}, E_{18}, E_{36}, and E_{80}) for Venera 11. The curves were constructed as a function of Earth-received time at Moscow (right vertical scale) and as a function of the altitude of the probe above the landing site in km (left scale). The spectral density of the low-frequency pulsating atmospheric radio emission, detected by the experiment can be characterized by the spectral index α, with

$$E_1/E_2 = (f_1/f_2)^\alpha \qquad (3)$$

where f_1 and f_2 are the frequencies at which the radio noise was received, and the intensities are normalized with respect to the varying spectral intervals.

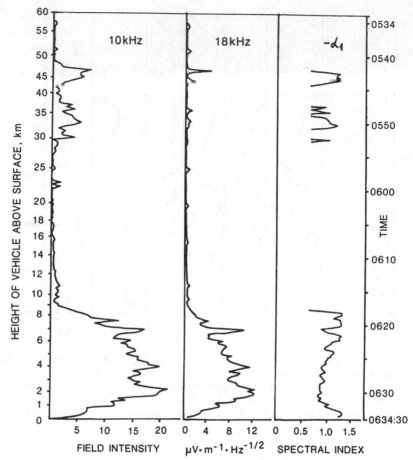

Fig. 3. The same as Fig. 2, for Venera 12, 21 December 1978. The signals in channels 36 and 80 kHz, except for the altitude interval of 0 to 9 km, were at the noise level.

From the ratios E_{18}/E_{10} and E_{36}/E_{10}, α_1 and α_2 were found, as shown in columns 6–7 of Fig. 2. Due to the relatively high noise level in the 80 kHz channel, the ratio E_{80}/E_{10} is not shown in the figure. Discrepancies in the spectral indexes of individual bursts of radio noise are attributed either to different distances of the sources, or to differences in the conditions or characteristics of the discharges at the sources.

The Venera 12 measurements, also in the form of graphs of the averaged field intensity, are shown in Fig. 3. Here, the signals were much weaker and the number of recorded pulses fewer. The region of maxima in Fig. 2 in the altitude interval of 32 to 12 km corresponds to an almost complete absence of signals in the case of Venera 12 (Fig. 3). On the contrary, in the interval of 43 to 30 km, only Venera 12 received pulses. This portion of the recording is shown in Fig. 4 in greater detail for 10 kHz. Along the vertical the field

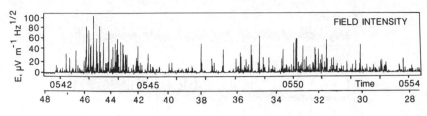

Fig. 4. High-resolution (unaveraged) profile of the pulses detected by Groza on Venera 12 in the altitude region of 28 to 48 km.

intensity is marked off (up to 100 μV m^{-1} Hz$^{-1/2}$). Fig. 4 and the subsequent diagrams were constructed for a compressed horizontal scale, but without averaging; and, therefore, due to the small time constant of the detector, the recorded pulses have the appearance of vertical spikes, each of which represents a single discharge. Along with pulses of considerable amplitude, there are a large number of minor bursts with amplitudes on the order of several μV m^{-1} Hz$^{-1/2}$. Basically, Venera 12 recorded only two large groups of radio noise. After the burst in the interval 5:42 to 5:54, shown in Fig. 4, there were practically no pulses for 24 min (Fig. 3). The signals reappeared only at 15–16 min prior to landing. Up to this time, the responses in the 36 and 80 kHz channels were very small, and are not shown in Fig. 3. The columns from left to right represent the field intensity in channels 10 and 18 kHz and the spectral index α_1.

The field intensities detected on Venera 12 (Figs. 3 and 4) were compared with terrestrial thunderstorm measurements at Moscow and at Frunze, in Central Asia, using a twin of the instrument on board Venera. A constant sensitivity was maintained in both cases by a built-in calibration. For the "threshold" portion at Venus, 25 to 12 km in Fig. 3, as well as at the surface of the planet, the pulse rates were similar to measurements on Earth during clear weather without nearby thunderstorms, especially in the range ≥ 18 kHz or higher. However, in the case of Venera 11, the mean frequency f_p of discharge was, on the average, an order of magnitude larger, and the threshold levels were only found at the surface of the planet.

The most intense bursts were observed on Venera 11 from 5:45 to 6:05, in the altitude range 30 to 13.5 km. The sequence consists of many thousands of individual pulses. The development of the initial portion of the sequence is shown in Fig. 5. The field intensity in each of the channels is plotted along the vertical, and the Earth-received time and the altitude of the probe above the level of 6052 km is along the horizontal. The averaged intensity reached 150 μV m^{-1} Hz$^{-1/2}$ at a frequency of 10 kHz, and 6 μV m^{-1} Hz$^{-1/2}$ at 80 kHz. A comparison of the frequencies f_p obtained from the counting rate and the level of the pedestal indicates that f_p attained 9 to 12 s^{-1}. We point out that the information given by Ksanfomality (1979a, 1980a) for the frequency of the

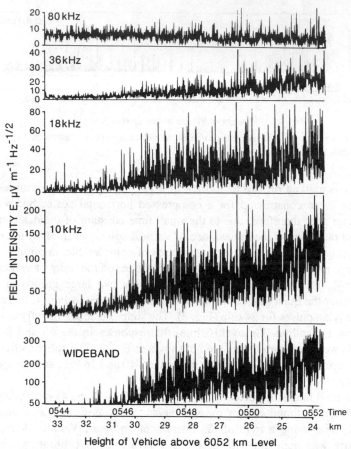

FIELD INTENSITY E, μV m^{-1} Hz$^{-1/2}$

Height of Vehicle above 6052 km Level

Fig. 5. The development of a large radio noise burst detected by Venera 11 between 5:46 and 5:59. The maximum intensity reached 300 μV m^{-1} Hz$^{-1/2}$ at a frequency of 10 kHz, while the frequency of the discharges reached 40 s^{-1}.

discharges cannot be directly considered as the number of lightning flashes. As a rule, a single lightning flash on Earth includes many separate radio impulses, on the average, 3 to 4 for return strokes, but sometimes as many as 20 (Uman 1969). The estimate of the parameter f_p did not depend on the sampling rate. In the interval 5:55–5:59, the frequency f_p increased by another factor of 3–3.5. The spectral index was ~ -2. The peak amplitudes reached 280, 100, 45, and 25 μV m^{-1} Hz$^{-1/2}$ at frequencies of 10, 18, 36, and 80 kHz. The burst shown in Fig. 5 was the longest, and lasted > 15 min.

Let us now consider the bursts of radio noise preceding the landing of the probes. These represent the only groups of pulses present in the identical altitude interval for both probes. Here, as in the entire descent phase (excluding the surface), the mean values of the field intensity were higher on Venera

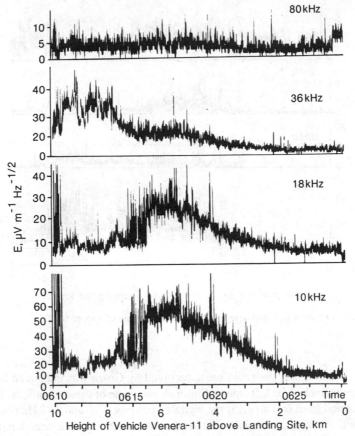

Fig. 6. Radio noise detected by Venera 11 prior to landing. The burst covers the altitude interval from 10 to 0 km. At the left of the figure there is a burst which is noted only at frequencies of 36 and 80 kHz.

11 than Venera 12 (Figs. 2, 3, 6, 7). The averaged intensity in this altitude interval reached 40 μV m^{-1} Hz$^{-\frac{1}{2}}$ at 10 kHz for Venera 11 and 20 μV m^{-1} Hz$^{-\frac{1}{2}}$ for Venera 12. The spectral index in both cases varied from -1.60 to -0.80. The parameters and abbreviations in Figs. 6 and 7 are the same as those in Fig. 5. It is noteworthy that there was a smooth decrease in the field intensity toward the surface, practically down to the noise level, in all channels. The development of these bursts began at an altitude of 8 to 9 km. The drop in the field intensity for Venera 11 was smooth in nature (Fig. 6), while that for Venera 12 was more abrupt, as shown in Fig. 3. In contrast with Fig. 6, Fig. 7 reveals several bursts. The similarity in the altitudes on both recordings may, of course, be accidental. The frequency f_p here is very high, rising to 30–55 impulses s^{-1}. The pulses on the recording of Fig. 7 in the 80 kHz channel basically represent the instrument noise.

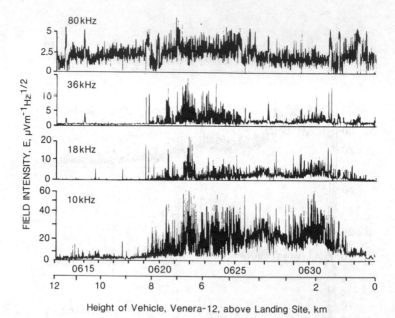

Fig. 7. A phenomenon similar to that in Fig. 6 for the case of Venera 12.

In certain cases, the radio noise received by Groza was enhanced in one or more channels. The left side of Fig. 6 shows some of these instances. Here, in the period from 6:11 to 6:15, the signals in channels 10 and 18 kHz changed little; but, at 36 kHz (and also at 80 kHz, as shown in Fig. 2) a large burst was recorded with field intensity up to 50 μV m^{-1} Hz$^{-\frac{1}{2}}$ and with f_p up to 30 s^{-1}. Obviously the spectral index of Fig. 2 has no meaning here. Similar anomalies can also be seen in Fig. 2 for the interval of 6:00 to 6:03. The origin of these high-frequency enhancements remains unknown.

One of the most interesting results from Groza is illustrated in Fig. 8. The central panel shows that the field intensity at 18 kHz has the form of pulses grouped in clusters with a noticeable periodicity. There are several hundred pulses in each cluster. The entire group consists of six large bursts, following each other with rising amplitude. This phenomenon was recorded between 6:04 and 6:11. The first three bursts are separated by intervals of 90 and 80 s, and the next by 50 s each. The height of the probe above the surface ranged from 13 to 9 km. The periodicity is even more evident in the pulse height data (lower portion of Fig. 8). At frequencies of 36 and 80 kHz, the phenomenon is much less pronounced. We have established that this fact is related to the angular response characteristics of the instrument antenna. The entire sequence of bursts was detected only until 6:11, and the abrupt termination is still quite evident in a high resolution plot of the counting rate in the large-

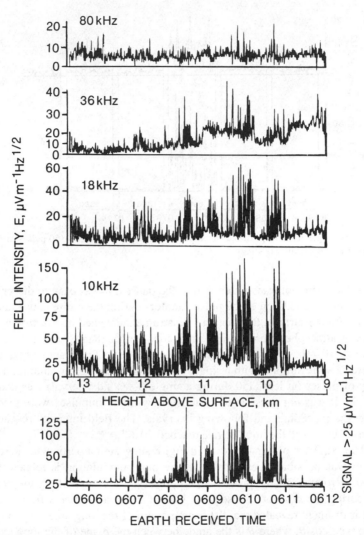

Fig. 8. The field intensity in the altitude interval 9 to 13 km for Venera 11 had the appearance of periodic pulse packets. The sequence is abruptly interrupted near 6:11. The decrease in modulation with frequency rise is related to distortions of the beam pattern at 36 and 80 kHz.

amplitude pulse height analysis channel, averaged over intervals of 10 s (see Ksanfomality 1979a). A similar periodicity, but less pronounced, was detected in the final portion of a large burst (observed between 5:59 and 6:03, when Venera was descending from 18 to 16 km above the surface).

Let us now consider the period of operation at the surface of the planet. During this time the telemetry sampled the outputs of Groza intermittently, for 8-second-long intervals, separated by 6- and 3-min pauses. Although the

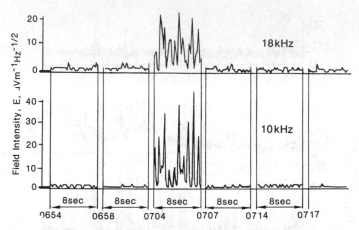

Fig. 9. The single group of pulses detected on the surface of the planet (150 pulses in 8 s) by Venera 12, 30 min after landing.

Groza data were not transmitted during the pauses, it was easy to discern any changes in the condition of the pulse counters during the entire pause, and we find that on the surface, the Venera 11 instrument detected no electromagnetic pulses of atmospheric origin with intensities higher than 5 μV m^{-1} Hz$^{-\frac{1}{2}}$ in the range 10 kHz (where the intensity was greatest). The results obtained by Venera 12 were similar, with a single exception; 30 min after landing, a large group of pulses (at least 150 during a single 8-second interval) was detected (Fig. 9). It was not possible to establish how many impulses were recorded during the preceding and following intervals. The field intensity reached 40 μV m^{-1} Hz$^{-\frac{1}{2}}$, and the impulse rate reached 20 s^{-1}.

The sequence of bursts shown in Fig. 8 suggested a particular interpretation. It was possible to establish that the burst modulation is related to the period of rotation of the probe, which preserved a course along the vertical axis during the descent. Subsequent laboratory measurements from a flight spare instrument revealed that the beam pattern of the loop antenna, normally of the type cos θ, where θ is the angle between the plane of the loop and the direction toward the signal source, was somewhat distorted by the influence of the probe (Fig. 10). As a result, during the rotation of the antenna, the signal modulation was only appreciable at low frequencies. If the azimuth angle of a point source is θ, and the angular velocity of rotation of the probe around the vertical axis is ω_x, the voltage at the output of the loop antenna with effective height H when the field intensity is E is given by

$$u = EH \cos (\omega_x t + \theta). \qquad (4)$$

While an actual emitter cannot be a point source, from the amplitude of the signal at the minimum of the sequence it is possible to find its angular

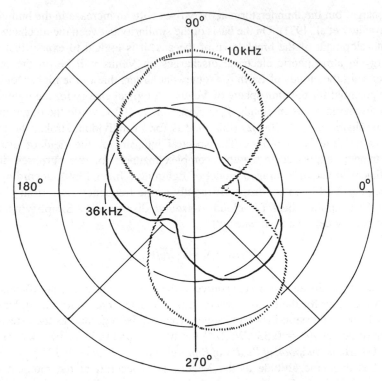

Fig. 10. Distortions of the beam pattern of the Groza antenna, produced by the influence of the vehicle.

dimension and to determine that it is close to 5°. Thus, it was established that the source is, in fact, small (but the accuracy of this determination is poor). The next step involved estimating the distance to the source, and here we assumed that certain information concerning lightning in the atmosphere of Earth could be used to interpret the Venus data. During the instrument design period for the Groza experiment, Hara (1976) and Ksanfomality independently expressed opinions about possible lightning activity on Venus. Hara considered the conditions of formation of electric discharges, and concluded that they are satisfied in the atmosphere of Venus. Hara's analysis is discussed by Ksanfomality (1979a), and his results are summarized there. The analysis shows that the Stokes velocities, which are no more than fractions of mm s^{-1} for the aerosols of Venus (Ksanfomality 1980b), can be disregarded as sources for charging in comparison to the considerable velocities of turbulent movements.

It is known that an aerosol environment merely entails an accumulation of the charges in the clouds (Imyanitov et al. 1971). The actual mechanism of electrical breakdown is identical in both air and carbon dioxide. The conductance of the aerosol particles is not directly related to the development of the discharge, but the thunderstorm activity rises with an increase in the humidity (Imyanitov et al. 1971). On the basis of the similarity between the mechanisms of development of the breakdown of the gas, it is logical to expect that the energy in atmospheric electrical discharges on Venus will be on the same order as in the clouds of Earth (or somewhat larger, due to the higher ionization potential for the atmosphere of Venus). A typical energy for the lightning in a discharge between clouds W_0 is close to 2×10^9 J, while the duration of the discharge is $t_p \simeq 100$ μs (this time is for an individual stroke; the total discharge time is ~ 0.5 s). The spectral intensity of the field of distant lightning is expressed in a rather complex manner; for our purposes, it is sufficient to use an averaged model of lightning (Uman 1969), according to which the field intensity drops off rapidly as the frequency rises from $f = 5$ kHz (this applies best for cloud-to-ground discharges). Simplifying the problem, we derive an estimate with

$$E_L \propto \sqrt{\gamma W_0/t_p}\big/r \qquad (5)$$

where r is the distance from the source to the receiver and γ is a coefficient of proportionality for the radio emission from the discharge. With this lightning model, the expression inside the square root can be regarded as the effective radiated power. ($P = 0.33$ kW Hz^{-1} for 10 kHz, and 0.33×10^{-2} kW Hz^{-1} for 100 kHz [*Handbook of Basics of Radioelectronics Theory* 1977].)

Based on the altitude of the Venera 11 spacecraft at the moment the sequence shown in Fig. 8 was recorded, the minimum distance to the source could be several tens of kilometers (although it is highly unlikely that the spacecraft was directly under the source). However, for the measured field strength, up to 110 μV m^{-1} Hz$^{-\frac{1}{2}}$ at 10 kHz, the energy in the discharge, according to Eq. (5), is close to 10^{-3} W_0. Such low-energy lightning is not encountered in the Earth's atmosphere, since for the development of a discharge, one needs high field gradients, which arise only for a sufficiently large charge buildup in the clouds; the discharge energy is then also correspondingly large. For example, a potential difference of 3×10^8 V over a path of 3 km is created by a spherical charge of 40 C with a radius of 1 km. We also note that for Fig. 8, the mean discharge rate was ≥ 20 s^{-1}, which, for an object of small linear dimensions (several km), would apparently rule out a thunderstorm origin. There are, besides thunderstorms, at least two known natural phenomena in the troposphere and on a planet's surface which produce electromagnetic fields similar to those in thunderstorm discharges. These are the discharges near an erupting volcano and radiation from the Earth's crust when it is in a state of mechanical stress. Both of these sources are localized, as a rule, with the second giving a field strength significantly weaker than the electromagnetic fields of a thunderstorm.

We considered the most natural source of the Venus signals to be associated with a thunderstorm front. For large distances, we use an empirical

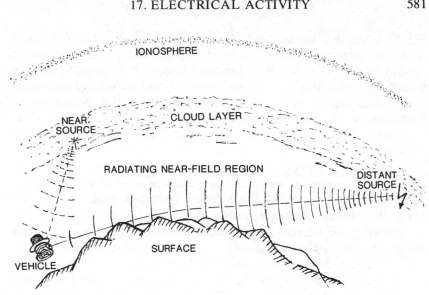

Fig. 11. Diagram illustrating the probable nature of propagation of radio waves from a remote source during the recording of the sequence in Fig. 8. Upon radio eclipse, the region essential for free propagation of radio waves is intersected by the probe on a transit of 70 to 100 min. This greatly exceeds the time during which the signal vanished in Fig. 8.

formula of Austin (Chernyshov and Sheinman 1973) to relate the field intensity to the wavelength and distance

$$r = \frac{300\sqrt{P}}{E_L} \sqrt{\frac{\Delta}{\sin\Delta}} \exp\left(-\frac{1.4 \times 10^{-3}}{\lambda^{0.6}} r\right) \tag{6}$$

where Δ is the half-angle formed by the source, the center of the planet, and the receiver. With a frequency of 10 kHz, Eq. (6) gives a distance of 1250–1350 km. The source of angular dimensions given above suggests a horizontal extent of 120 to 150 km, which corresponds to an extremely extensive thunderstorm front on Earth. According to the measurements at other frequencies, r may reach 1600 to 1900 km. A diagram showing these interpretations for near and far sources is contained in Fig. 11.

Thus the natural hypothesis concerning the similarity in the mechanisms of development of electrical discharges in the clouds of Venus and the Earth leads to the conclusion that the sequence of bursts in Fig. 8 proceeded from an extended thunderstorm front, at a distance of 1200 to 1500 km from the landing site of the probe. In this region, there were 10 to 30 discharges per second, and the spectral index of the field intensity was close to −1. The relatively small dispersion of the pulse amplitudes indicates that all the discharges apparently occurred between separate portions of the cloud layer (or a different charge-bearing medium). The source of the radio noise shown in

Fig. 5 was probably located much closer, at a distance of 700 to 1000 km, if we assume identical parameters for the discharges. The spectral index here is close to -2. The frequency-dependent burst observed by Venera 11 from 5:59 to 6:02 had a spectral index from -1.6 to -1.0. The last burst of radio noise preceding the landing (Fig. 6) more or less coincided with the decrease in the field intensity calculated for the shadow zones and penumbra by the diffraction formulas (Feinberg 1961). The attempt to deduce the Venus soil dielectric permeability and conductivity from these data has not been successful, because the curves of the field intensity for a remote source virtually coincide for different combinations of the ground parameters at super-long waves. However, the dielectric permeability deduced by the radar method (Kuz'min and Marov 1975) has previously been estimated to be $(3–3.6) \pm 0.4$.

Therefore, the Venera 11 observations are consistent with a state of high thunderstorm activity in the equatorial region of the planet. During the descent, at least 5–6 bursts were recorded, apparently related to different sources. Four days prior to this, in the identical region, Venera 12 recorded much weaker thunderstorm activity. The solar activity was essentially identical in both cases, since the entire period from 16 through 29 December was characterized by low activity; the flux at radio frequencies of 100–200 MHz was near 7–30 units, and the number of sunspots from 19 through 25 December was very small (*Solar Data* 1978). Thus, we may conclude that the development of the thunderstorm activity took place simultaneously over a large area, during a period of no more than four Earth days (roughly 1/30 of the venusian solar day).

Let us once again return to the right side of the bottom two recordings for the sequence in Fig. 8. The signal terminated abruptly at 6:11, and there was only a single pulse at 6:11:25, when the next maximum should have occurred. In a report by Ksanfomality et al. (1979c), it was proposed that the cessation of the signal was due to the source passing below the horizon. The basis for this conclusion was the close coincidence between the range of the radio horizon and the previously obtained estimate of $r = 1250–1600$ km

$$r_h = \sqrt{2R_v}\,(\sqrt{h} + \sqrt{z}) = 1200\ \text{km}. \tag{7}$$

These values can be reconciled if we allow for the strong refraction of the beams which can be evaluated using an altitude model for the density distribution. In this case, the ratio between the angles of refraction δ_1 and δ_2 for an inclined beam is given by

$$\sin\delta_2 = \frac{m_1 R_1}{m_2 R_2}\,\sin\delta_1 \tag{8}$$

where m_1, m_2 are the refraction coefficients in layers located at levels with radii R_1, R_2. We use a 30-layer atmosphere model and find a maximum angle of refraction of 3°. This leads to an increase in the range of the radio horizon up to 1400 km, which is relatively close to the result from Eq. (7).

The propagation of radio waves in the super-long wavelength range over distances of 1000 km or more is determined by diffraction above the spherical surface of the planet. The deduced distance to the source was close to 50λ at 10 kHz and 180λ at 36 kHz. Considering the radiating near-field region, it has been shown that the abrupt break in the sequence of Fig. 8 at 6:11 cannot be explained by the radio eclipse of the source by the horizon, as was considered in the preliminary communication (Ksanfomality et al. 1979). In actuality, the radius of the n-th Fresnel zone at a distance of ρ_1 from the source and ρ_2 from the receiver is

$$r_n = \sqrt{\frac{\rho_1 \rho_2}{\rho_1 + \rho_2} \lambda n}. \tag{9}$$

For the first zone at $\rho_1 = \rho_2 = 500$ km, $r_1 = 86$ km for 10 kHz, and 45 km for 36 kHz. Since the speed of descent of the Venera 11 probe at 6:11 was close to 0.6 km min^{-1}, the characteristic time for the probe to travel from the penumbra zone to the shadow zone exceeds by a very large amount the 1–2 min during which the signal disappeared. We also point out that the periodicity of the bursts, distorted by the random distribution of the pulses, does not correspond to the classical diffraction conditions observed above a half-plane or a spherical planetary surface (Feinberg 1961). The descent speed of the probe did not increase, and therefore, the period of succession of diffraction maxima could only increase, whereas in the experiment it diminished from 90 to 50 s.

Let us turn to the averaged graphs of the field intensity. It follows from Figs. 2 and 3 that almost all the events detected by the Groza experiment were observed when the probes were located below the level of 30 km. A striking explanation for the measured dependence of the field intensity has been proposed by Croft and Price (1983), who conjecture that a dominant role in the long-wave propagation from remote atmospheric sources is played by a peculiar refraction waveguide, which is discussed further below. The "roof" of this waveguide is found to be situated somewhat above the level of 30 km. However, the "quieting" of the atmosphere as observed from the planetary surface cannot be explained by this single mechanism. The most probable cause of the low field intensity is the strong absorption of waves at the lower boundary of the ionosphere. It is known that low-frequency waves undergo attenuation in the ionosphere, depending on the frequency ν of collisions between electrons and neutral molecules or ions; as a result, a portion of the incident-wave energy is dispersed in the collisions. The losses are determined by the coefficient of absorption $\omega\kappa/c$, where κ is the index of absorption in the complex coefficient of refraction: $m = n - i\kappa$

$$\kappa = \left\{ \frac{\left[\epsilon^{*2} + (4\pi\sigma/\omega)^2\right]^{1/2} - \epsilon^*}{2} \right\}^{1/2} \qquad (10)$$

The complex dielectric permeability of the ionosphere is

$$\epsilon^* = \epsilon - i4\pi\sigma/\omega \qquad (11)$$

where ϵ is the dielectric permeability

$$\epsilon = 1 - 4\pi e^2 N_e/m(\omega^2 + \nu^2) \qquad (12)$$

and σ is the conductivity as determined by collisions

$$\sigma = \nu e^2 N_e/m(\omega^2 + \nu^2). \qquad (13)$$

Here, e and m are the charge and mass of the electron.

The collision frequency ν depends on the concentration of neutral molecules N_m. As the wave frequency f is lowered, the absorption associated with the collisions is increased, with the main losses coming from the lower layers of the ionosphere. The largest absorption at a frequency f will be observed when

$$1/f \cong 2\pi/\nu. \qquad (14)$$

Over a wide frequency range ($f_{max}/f_{min} \approx 10^2$), the absorption curve has a broad maximum (at the level 0.5). In order to compute these losses, it is necessary to know the vertical profile of the electron concentration in the lower layers of the ionosphere of Venus, but this is poorly known for altitudes near 90–110 km. However, instead of this, using Groza data, we can find the altitude at which the required wave attenuation should occur. For $f = 20$ kHz, the required concentration N_m can be estimated as

$$N_m = 2\pi f N_0 C \sqrt{T_0/T} = 2.4 \times 10^{13} \text{ cm}^{-3} \qquad (15)$$

where N_0 is the Loschmidt number, $T_0 = 300$ K, C is a constant, and T is the measured temperature. Since

$$\rho = N_m m_H \mu \qquad (16)$$

where m_H is the mass of the hydrogen atom and μ, the molecular weight of the atmosphere, was determined to be 42, ρ could be evaluated as 1.8×10^{-8} g cm^{-3}. Seiff et al. (1980) found that $T = 190$ K and $= \rho\ 1.8 \times 10^{-8}$ g cm^{-3} corresponds to an altitude of 117 km. However, the result of the calculation is strongly affected by the electron profile. The validity of this statement and the actual value of the C coefficient depend on how close the electron density profile in the lower part of the Venus ionosphere is to the terrestrial profile.

Along with the conclusion of absorption in the ionosphere, the experimental results seem to require that the scattering of waves by reflection from the surface in the radio-refraction channel be diffuse in nature, or that the coefficient of reflection should not be very high. Then, the low field intensity observed from the surface can be understood. The absorption (or scattering) in the refraction channel implies that the observed sources were not very remote, probably not more distant than 2000 km. In the case of Venera 12, the distance r could be larger. The mean amplitude of 10 kHz pulses in Fig. 7 is $\sim$ 35 μV m^{-1} Hz$^{-\frac{1}{2}}$. We obtain a mean distance to the source of about 3000 km using the Austin formula, Eq. (6).

If we assume that sources 10% more distant than $r = 3000$ km were still accessible to observations in the limiting case (since these bursts of radio noise were still not at the threshold intensity), we then find that $r^2/4R_v^2 =$ 7.5% of the area of the planet was under observation; over this territory there were two (in the case of Venera 12) or five to six (in the case of Venera 11) sources of discharge. This model then implies that the mean number of simultaneously acting sources of discharge on the planet is $\sim$ 50, if they are uniformly distributed over all latitudes (and if the 7–8 Venera 11 and 12 examples provide a sufficient statistical reliability). The mean frequency of discharges in each source region was within the limits of 10 to 30 s^{-1}. Since the total duration of the thunderstorm phenomena is, of course, unknown, we cannot draw a realistic conclusion as to the average number of discharges per unit of area or time. According to the *Handbook of Geophysics* (1960), the number of simultaneous thunderstorms on Earth is close to 2000, and they occupy an area of $\sim 3.6 \times 10^{-3}$ of the total surface area.

The wave instrument on the Pioneer Venus orbiter (Scarf 1977; Scarf et al. 1980c) provides additional basic information on electrical activity at Venus. The instrument, called the OEFD (Orbiter Electric Field Detector), was primarily designed to study the interaction of the solar wind with the ionosphere of Venus, but a longstanding secondary goal involved the search for whistler mode noise bursts associated with atmospheric lightning. The spacecraft is in a 24-hr orbit with apoapsis of 12 R_{φ}; for the first few years, the orbit elements were controlled to keep the periapsis altitude down near 150 to 180 km, so that the orbiter penetrated to the level of the ionosphere peak or even just below it. However, since Orbit 600, the orbit characteristics have not been controlled, and the periapsis altitude has been steadily rising.

Figure 1 shows that while Groza measured wave components within the neutral atmosphere, the OEFD was screened by the Venus ionosphere. Other important differences involve the wave sensors (OEFD uses a short electric dipole) and the frequency coverage. The Pioneer Venus bandpass channels are centered at 100 Hz, 730 Hz, 5.4 kHz, and 30 kHz, and each filter has a bandwidth equal to 30% of the center frequency. Since the ionosphere essentially blocks all electromagnetic waves with frequencies between the electron plasma frequency

$$f_p = 9 \times 10^3 \sqrt{N_e}$$ (17)

and the electron cyclotron frequency

$$f_c = |e|B/2\pi mc \approx 28 \, B$$ (18)

most of the radiation from lightning cannot reach the orbiter (here, N_e is the electron density in cm^{-3}, and B is the magnetic field strength in gammas). Specifically, since the local plasma density at the position of the Pioneer Venus orbiter must generally be sufficiently high so that $f_p \gg 30$ kHz (i.e., $N_e \gg 12 \, cm^{-3}$), only whistler mode signals can reach the OEFD.

The initial Pioneer Venus orbits had periapsis in sunlight, and no lightning signals were detected. Scarf et al. (1979, 1980a) and W. Taylor et al. (1979b) noted that the dayside passes generally have turbulent magnetic fields (with the average value for $f_c \leqslant 100$ Hz) along with rapid variations in field direction. This leads to strong Landau damping near the ionopause for whistler mode turbulence from the ionosheath, and for this situation, whistler mode noise bursts from the atmosphere would also be strongly absorbed within the ionosphere. A final complication associated with dayside measurements arises because the illuminated solar array itself couples low-frequency noise from the spacecraft subsystems to the antenna, and the photoelectrons emitted from the spacecraft surface produce an asymmetric plasma sheath which further modulates the local noise.

The nightside measurements were clearly more promising because of the much greater sensitivity achievable in the absence of solar array-photoelectron noise phenomena, and during the first Venus night, W. Taylor et al. (1979a) reported that at low altitudes the OEFD detected impulsive signals with $f < f_c$. These bursts were interpreted as whistler-mode noise signals, and it was further assumed that the sources were atmospheric lightning. Accordingly, an extensive study of the OEFD measurements of Venus whistlers was initiated. During the early nightside orbits, when these bursts were first observed, the Pioneer Venus Temperature probe (OETP) detected very low local electron densities ($N_e \leqslant 10^2 \, cm^{-3}$) with much irregularity, and it was conjectured (W. Taylor et al. 1979a,b) that the lightning could only reach the orbiter when such ionospheric conditions were present. A more comprehensive study soon showed that this initial conclusion was wrong; signals that we classify as whistlers are readily detected even when the ionospheric density approaches $10^5 \, cm^{-3}$, although it turns out that the magnetic conditions favorable for whistler detection are frequently associated with depressed local density values. Figure 12, from Scarf et al. (1980a), shows an example with all of these characteristics. The impulsive lightning signals appear in the 100 Hz channel,

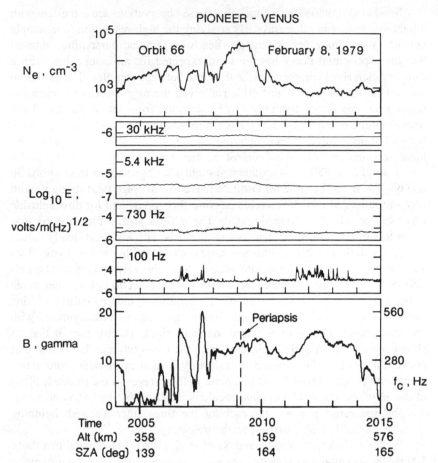

Fig. 12. Intensity of the whistlers associated with discharges in the atmosphere, recorded by the OEFD instrument on the Pioneer Venus orbiter on 8 February 1979. The vertical scales give the logarithm of the electromagnetic field intensity, the magnetic field intensity (bottom), and the electron density N_e (top). The whistlers were detected in the 100 kHz channel. At the bottom right is a gyrofrequency scale.

and they are clearly associated with high magnetic fields ($f_c \gtrsim 360$ Hz). In this case, the lightning pulses are also detected only when N_e is relatively low ($N_e \lesssim 10^3$ cm^{-3}). We interpret these bursts as signals that propagate up to the orbiter in the whistler mode, with the index of refraction $\hat{m}$ given by

$$\hat{m}^2(f,\psi) = 1 + f_p^2/f_c\,(f_c\cos\psi - f) \qquad (19)$$

where ψ is the angle between the magnetic field and the direction of propagation for the wave. Since $f_c/f \gtrsim 3$, $\hat{m}$ is real for small ψ values, and the wave can propagate through the ionosphere essentially along the magnetic field.

Scarf et al. (1980a) determined that these observations are consistent with a lightning source in other ways. By studying the high-resolution (one sample per 0.5 s) amplitude versus time profiles for individual bursts, they showed that the exponential decay has the form expected for an input signal with a time duration short compared to the 0.5 time between samples. The extensive data set for the Pioneer Venus orbiter allowed the investigators to determine under what conditions lightning can be detected from the spacecraft. They found that the relation $f_c \gtrsim 300$ Hz is a necessary but not sufficient condition for OEFD detection of Venus lightning, and it also clearly suggests that the local electron density does not control the detectability.

Scarf et al. (1980a) also contrasted nightside observations from Orbits 36 and 68. The B-field profile for Orbit 36 had a low average field strength, with short-duration pulses that resemble dayside flux ropes, and for this nightside traversal, no whistlers were detected. The B-field profile for Orbit 68 was much less disordered, there were many intervals with high and steady values for f_c, and 100 Hz whistler bursts were associated with some of these. This study suggested that the conditions leading to a relatively well-ordered night-side field configuration are also necessary for the detection of whistler mode noise bursts. A great step forward in achieving understanding of this phenomenon came when the times for lightning detection were compared with the changing orientation of the local magnetic field. The top part of Fig. 13 shows some of the E and B field data from the Orbit 68, and the bottom part contains a drawing of the spacecraft trajectory in solar cylindrical coordinates, along with projections of 1 min magnetic field averages in the (rotated) plane of the drawing. The field magnitude scale is indicated by the bar on the right, and the spacecraft position for each of the three intervals with lightning signals (regions 1, 2, 3) is marked on the trajectory plot.

Figure 13 is a typical case, and Scarf et al. (1980a) concluded that those 100 Hz bursts which had already been assumed to be due to a lightning source (on the basis of their impulsive amplitude variations and the fact that $f_c \gg$ 100 Hz) are only detected on the Pioneer Venus orbiter when the local B field is oriented parallel or antiparallel to a line that intersects the planet. Thus, the waves do have the characteristics of parallel propagating whistlers, and since Landau damping disappears for $\mathbf{k}$ parallel to $\mathbf{B}$, these waves can reach the Pioneer Venus orbiter without significant attenuation by wave-particle interactions in the plasma. The possibility of a local source mechanism was discounted because internally generated whistler mode signals, such as hiss or chorus, are not impulsive on these very short time scales. The frequent detection of these impulsive whistler signals in the shadow of Venus is attributed (a) to the fact that the nightside B field generally has a quasi radial configuration with high f_c, and (b) to the decreased threshold noise level for the OEFD in eclipse. It is also significant that the quasi radial field configuration can allow good matching between the upcoming wave vector incident on the lower boundary of the ionosphere and the parallel whistler mode signal of

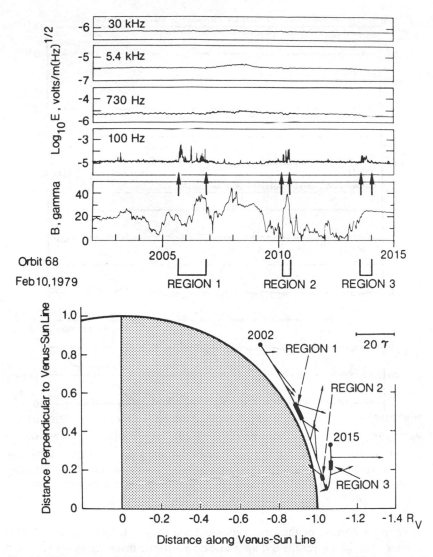

Fig. 13. Whistlers associated with discharges were recorded in those cases when the direction of the field coincided with the direction of propagation of a wave rising from the planetary atmosphere. Arrows at various segments of the orbit, shown at the right in the bottom half of the figure, indicate the measured direction of the field for regions 1-3, marked off on the upper graphs.

interest within the plasma. Scarf et al. (1980a) tried to identify sources on the surface or in the cloud layer by tracing rays along the B field from the orbiter down to 60 km or the surface itself. Essentially they assumed that the whistlers are generated on the field lines intersecting the instantaneous position of the spacecraft, and they presented results for $\sim$ 30 nightside orbits,

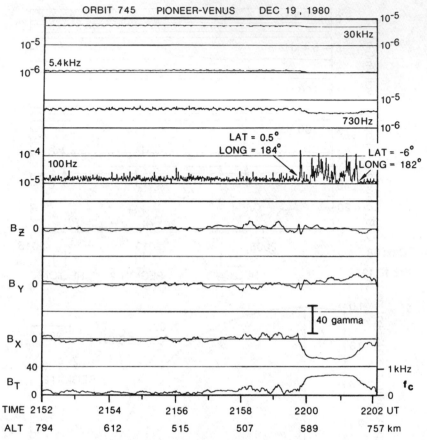

Fig. 14. A more recent example of lightning detection during the period when the Pioneer Venus orbiter periapsis altitude was no longer being controlled. The signals are again detected when f_c is high and steady, with a field orientation pointing down to the planet.

covering a portion of the first Venus eclipse season for the Pioneer Venus orbiter. The initial results did not indicate a strong clustering of sources, but such a conclusion could not be ruled out.

The OEFD investigators have now analyzed all of the data from the first 750 orbits covering almost four full eclipse seasons. In a recent report, Scarf et al. (1982) demonstrate that this extensive data set strongly indicates a clustering of signal sources, and Fig. 14, taken from this study, shows an example of whistler detection during the fourth Venus night season, when periapsis had climbed to almost 500 km. The intense signals attributed to lightning were detected between 22:00 and 22:01:30, in a region with a very high f_c and with $|B_x| \gg |B_y|$, $|B_z|$, so that $\mathbf{B}$ did point down to the atmosphere. However, Fig. 14 also shows that in adjacent intervals with $f_c \gtrsim$

300–500 Hz and large B_x (e.g., the period near 21:59:10 and the period after 22:01:30), bursts were not detected, although the parallel whistler mode conditions were still satisfied. Extensive analysis of the entire data set supports the conclusion that there is a phenomenon which restricts the OEFD ability to detect the whistlers, independent of B field magnitude and orientation. However, these comments should not be interpreted to imply that the detection of intense whistler noise bursts is continuous; during the first eclipse season, the OEFD detected these impulses during only 2% of the time when the spacecraft was in darkness.

Figure 14 has Venus-centered longitude and latitude for the spacecraft positions at the indicated times; these parameters have been used by Scarf et al. (1982) to make maps showing where the lightning is detected. Figure 15 contains a radar map of Venus highlands (from Masursky et al. 1980) along with data for *all* whistler bursts through Orbit 750. Here, the instantaneous spacecraft coordinates at the time of detection and the simultaneous measurements of the vector magnetic field are used to deduce a hypothetical source point on the surface, assuming propagation strictly along the extended magnetic field direction. This task is not a simple one for Venus because the local field, which originates in the solar wind, fluctuates in magnitude and direction. Despite this variability in "look direction", Fig. 15 shows a remarkable degree of order, with most of the deduced sources found to be clustered near the Beta and Phoebe Regios; there is an additional significant cluster near the Atla Regio at the eastern edge of Aphrodite Terra.

The curve labeled Limit of Analysis marks the boundary of the nightside region sampled during the first four Venus nights (up through Orbit 750); it can be seen that there were extensive opportunities to search for whistler mode bursts over a thick near-equatorial belt covering a large range of longitudes. Because the B field is inherently variable, it is not possible to present firm statistics, but it seems clear that the clustering is not random. Scarf et al. (1982) considered several tentative explanations, but no firm conclusion was arrived at. Since the lightning sources were associated with highlands (the gray areas in Fig. 15 represent areas with altitudes > 2 km above the mean), Scarf et al. (1982) considered the possibility that mountainous terrain produced strong cloud lightning activity. This conclusion seems unacceptable because the Thetis Regio on the eastern edge of the main Aphrodite massif ($120°–140°$ longitude) is as mountainous as Atla, but it is clearly not a source of this lightning. Scarf et al. (1982) also speculated that volcanic activity produces lightning near the surface. This conjecture has support in several ways: (1) Masursky et al. (1982) have analyzed the terrain shapes and surface features and concluded that the highlands of the Beta, Phoebe, and Atla Regios are of volcanic origin; (2) Phillips and Malin (see Chapter 10) find that these same locations have anomalously high gravity fields, suggesting local stresses that could be characteristic of volcanic activity. It should be noted, however, that this interpretation of the data may also have problems. Borucki

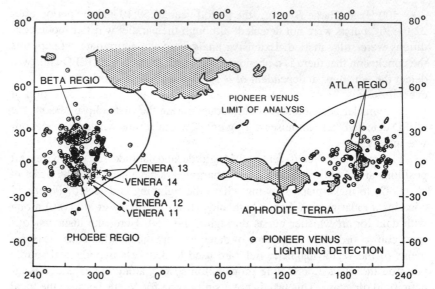

Fig. 15. A Pioneer Venus orbiter radar/lightning map. The gray areas show the highlands, as defined by Masursky et al. (1980), and the smooth curve bounds the region where the OEFD was able to search for lightning through Orbit 750 (almost four Venus eclipse seasons). The dots show the field-aligned projections on the surface for the ray paths, assuming parallel whistler mode propagation, and the locations for the Venera 11, 12, 13, and 14 landings are marked. Masursky et al. (1982) have argued that the Beta, Phoebe, and Atla regions, where the clusters occur, are of volcanic origin, and Phillips and Malin (Chapter 10) note that these regions have significant gravity anomalies.

(1982) pointed out that the Pioneer Venus probes did not find evidence of dust or ash which might be expected if the Venus volcanoes were presently active.

III. THE SEARCH FOR LIGHTNING FLASHES BY SPACECRAFT

Certain types of optical instruments on board spacecraft can detect discharges in clouds. We start our discussion of Venus observations by estimating the possible contribution of lightning to the overall luminous flux emitted from the planet. After an analysis of the initial results from the wave instruments, there was a strong temptation to try to explain the ashen glow of the planet by lightning. Since the ashen glow is detected from Earth, this implied a rather large optical brightness for the source. In the early work of Ksanfomality (1979a), the following argument was used to estimate the rate of lightning strokes necessary for the ratio between the fluxes from the night side and day side of Venus in quadrature to be 10^{-4}. The coefficient 10^{-4} is arbitrarily chosen, but it is probably close to the maximum capabilities of terrestrial photometry. This flux ratio is rather large. The luminosity of the day side is magnitude -2.7, giving a corresponding nightside magnitude of 7.3. The flux of solar radiation F_d, reflected in the direction of Earth by the sunlit portion of the planet, is

$$F_d = \zeta E_s p f(\phi) R_c^2 / a^2 \tag{20}$$

where E_s is the solar constant, ζ is that part of the solar constant in the spectral interval of 400–720 nm, p and $f(\phi)$ are the known geometric albedo and phase function of Venus for quadrature, R_c is the radius of Venus at the upper boundary of the cloud layer, and a is the major semiaxis of the orbit. Ksanfomality (1979a) used

$$F_n = \bar{\epsilon} N_L \eta W_0 e^{-z_0} \chi \tag{21}$$

where N_L is the number of lightning flashes per second over one fourth of the planet, η is the conversion efficiency of transformation of lightning energy W_0 into light, z_0 is the optical depth of location of the lightning, χ is the limb darkening, $\bar{\epsilon}$ is an allowance for the conservativity of scattering of the light in the clouds. It was assumed that the lightning is situated not too deep in the cloud layer, $z_0 = 0.5$, and that $\bar{\epsilon} = 1.6$ and $\chi = 0.8$. With the coefficient of 10^{-4}, it follows that

$$N_L = 10^{-4} \frac{\zeta E_s p f(\phi) R_c^2}{\bar{\epsilon} a^2 W_0 \eta e^{-z_0} \chi} \tag{22}$$

The coefficient η for lightning on Earth in the range of 400 to 1100 nm is 4×10^{-3} (Uman 1969). The maximum radiation is in the ultraviolet, since the temperature in the lightning column reaches 2.5×10^4 K. An attempt at experimental measurement of the luminous efficiency of the discharge in an environment similar in composition to the atmosphere of Venus at first provided a very high value for η, ~ 0.1, which led to the estimate $N_L = 6 \times 10^3$ (Ksanfomality 1979a). It was later discovered that η is much lower. Let us assume that for Venus this coefficient is the same as that of Earth. Then, in order to satisfy the conditions $F_n/F_d = 10^{-4}$, for $W_0 = 2 \times 10^9$ J (the value for Earth), there should be 1.38×10^5 discharges per second. From the above estimate (20 to 30 discharges in a region with a linear dimension of 150 km), Ksanfomality (1979a) deduced that roughly 1.4×10^{-3} discharges km^{-2} s^{-1} should take place on the night side. Therefore the total power of the lightning would be 2.76×10^{14} J s^{-1} or 4×10^{-3} of the solar radiation absorbed by the planet which is dependent on the ratio chosen for F_n/F_d. According to Bar-Nun (1980), in the tropical regions of Earth this ratio is 2×10^{-5}. Over the entire Earth, it is estimated that there are 100 lightning flashes per second; therefore the value of the discharge frequency obtained above, $\leq 1.4 \times 10^5$ s^{-1}, appears much too large. This point has also recently been made by M. Williams et al. (1982).

After the discovery of electromagnetic pulses at Venus, the data from previous photometric observations were reexamined for evidence of lightning flashes. A check of all nightside data obtained with ultraviolet photometers on Venera 9 and Venera 10 (Ksanfomality and Petrova 1978) yielded no results, since the instruments, with a single exception, were operating in the low-sensitivity mode for observations of the day side of Venus. A similar check of the data from the spectrometer on board the same orbiters revealed that, in a single case, on the night side on 26 October 1975, the spectrometer observed chaotic variations of the signal amplitude in the form of pulses with a duration of ~ 0.3 s. The entire event lasted for 70 s. No such event was observed on other observation days. The threshold sensitivity of the instrument was 500 R. An analysis of the data led Krasnopolsky (1980 and Chapter 15) to the conclusion that the instrument was recording lightning flashes in an area with dimensions of ~ 450 km, at 9° N latitude and $\sim 190°$ E longitude. The pulses appeared while the instrument was operating in the normal spectral scanning mode. The scanning time was 10 s, and the pulses were observed in 7 consecutive scans. Kransnopolsky has noted that the envelope of the pulses roughly follows the curve of spectral sensitivity of the instrument, but statistical processing demonstrated the presence of a weak maximum in the region 550–650 nm.

The pulses detected by Krasnopolsky have two peculiarities. First, despite the high time resolution of the instrument (5 ms), the peaks are continuous in nature, whereas on Earth lightning produces many short consecutive bursts with intervals on the order of 50 ms. A second peculiarity is a rise in the intensity during a time of ~ 150 ms, with an identical fall time. This is not typical of lightning on Earth, where more than half the bursts consist of several pulses. For cloud-to-ground discharges, the luminosity decreases with the number of the pulse, with the first pulse usually being the brightest and the front edge of the burst very short, on the order of 50 μs (Uman 1969). The average total duration of the strokes for all four stroke flashes on Earth does not exceed 300 μs (Uman 1969); the remainder of the time, 250 ms, is occupied by the intervals between the strokes. Krasnopolsky finds that the duration of the burst is a total of 250–300 ms, which is more than the active time of lightning on Earth by a factor of 10^3. Therefore, we cannot agree with Krasnopolsky (1980) that the "typical duration of a lightning flash is the same for Venus and Earth." If their peak powers coincide, then according to Fig. 20, the lightning on Venus should carry 10^3 times more energy, or 10^{12} to 10^{13} J. On the other hand, if the energies are the same, then for an overall duration of 0.3 s for the burst, the power (and brightness) will be 10^3 times smaller than that for identical phenomena on Earth.

Krasnopolsky (1980) supposes that lightning is formed in the lower cloud layer (49–52 km). Given an optical thickness of the cloud layer $z_0 \simeq 20$, an albedo of the atmosphere beneath the clouds 0.85, and a single scattering albedo $\bar{\omega} \simeq 0.999$, the calculated attenuation of the radiation passing through the cloud layer from the bottom to the top is 6. Krasnopolsky points out that the outgoing radiation was concentrated within solid-angle limits close to π, while the ratio of the outgoing luminous radiation is close to 0.6 of the original. It was further assumed that the radius of the bright spot on the surface of the clouds is 7 to 10 km, i.e., half the thickness of the cloud layer. The field of view of the instrument, adjusted to the cloud surface, was 160 × 9 km; it was assumed that this covers half the area of the bright spot. The luminous energy of lightning, calculated in this manner, was 3 × 10⁷ J, but when taking into account the coefficient η, the total energy of the lightning flash should reach 10^{10} J. The latter estimate is an order of magnitude higher than that assumed by Ksanfomality (1979a) for a cloud-cloud discharge and coincides with terrestrial lightning of the cloud-Earth type. It was further assumed that the latitudinal extent of the thunderstorm region is 3 times less than the longitudinal. The ratio between this area and the entire area covered by the observations is 1.4×10^{-3}, which is close to the identical parameter for Earth: 3.6×10^{-3} (*Handbook of Geophysics* 1960).

The conclusions drawn by Krasnopolsky (1980) are the following: The lightning energy on Venus is 10^{10} J; the coefficient of transformation of solar into lightning energy is 5×10^{-5}. Given the absence of flashes on the other days of observation, the average number of lightning flashes was found to be moderate, 400 s⁻¹ in the band between latitudes of ± 32°, while the ratio between the brightnesses of the day side and night side of the planet is 10^{-7}. Observations of this ashen glow may require a very appreciable intensification of the thunderstorm activity of Venus. We add that even an increase of the discharge energy up to $W_0 = 10^{10}$ J reduces the estimate of N_L in Ksanfomality (1979a) down to only 2.8×10^4 s⁻¹, which is 70 times higher than given by Krasnopolsky (1980). The brightness B_c of the clouds is not mentioned in the article, although it is easy to calculate from the data. For a peak energy $W_0 = 10^{10}$ J, an overall duration $\tau_p = 0.3$ s for the flash, $\eta = 4 \times 10^{-3}$, a solid angle Ω of π ster, a diameter $d = 20$ km for the light spot at the level of 0.5 of the maximum brightness, and a ratio $\phi = 0.6$ between the generated and the original powers

$$B_c = \phi\eta W_0/\Omega\pi d^2\tau_p = 2 \times 10^{-6}\text{W cm}^{-2}\text{ster}^{-1}. \qquad (23)$$

This brightness is in the range 400 to 700 nm. For a duration of a normal return stroke on Earth ($\tau_p \simeq 300~\mu$s), the brightness B_c under the identical conditions would be 10^3 times higher. The measurements in Krasnopolsky (1980) occurred on the same day on which the largest contrasts of the ultraviolet details were observed (Ksanfomality and Selivanov 1978).

Borucki et al. (1981) used the star sensor on the Pioneer Venus orbiter to search for flashes in the optical range during the period from 29 September through 14 November 1979. They investigated a region with coordinates from

70° N to 50° S latitude and from 240° to 60° E longitude. Since the sensitivity of the detector was very high, only the light scattered in the lenses was measured. The detector was converted into a wide-angle photometer with a threshold sensitivity of 2×10^{-9} W cm^{-2} for a signal-to-noise ratio of 5. The spectral range of the instrument of 550 to 800 nm was determined by the silicon light detector. The specific operating features of the instrument imposed a number of boundary conditions. First, in 36 periapsis passes, no more than 450 s were available for the search for lightning. Second, the 18 pulses observed are statistically indistinguishable from the pulses produced by high-energy particles. Borucki et al. assert: ''If lightning on Venus has an amplitude distribution or a rate much larger than the upper estimates for the lightning of Earth, the stellar detector on Pioneer Venus would have discovered it.'' Nonetheless, the recorded events are not statistically distinguishable from observations of charged particles, and the results are essentially negative. It has also been reported that two series of observations of the night side of Venus in 1980, performed at observatories in the U.S., did not detect lightning flashes (E. Nye, personal communication, 1981). The data presented by Borucki et al. (1981), unfortunately, do not permit a direct comparison of the effective threshold sensitivity of the Pioneer Venus star sensor and the Venera spectrometer. The power threshold of the spectrometer was 1.8×10^{-10} W, and that of the star sensor (when the lens area is taken into account) is 2.5×10^{-8} W. For a correct comparison, it is necessary to know the instrument sensitivity to the brightness of an extended source, which must be expressed in W cm^{-2} ster^{-1}.

In summary, it appears that the optical data and the low-frequency electromagnetic data disagree, and several possible explanations can be contemplated. For example, electrical discharges on Venus may have a low light yield (small coefficient η). However, this is apparently contradicted by the data obtained with the Venera spectrometer. In general, this work (Krasnopolsky 1980) appreciably complicates the situation: on the one hand, the electrical discharges exist and there are many of them; on the other hand, their optical radiation exists but there are few bursts. It might be easier to find an explanation if there were no flashes of light at all, as in Borucki et al. (1981). Then, we could suggest that the flashes are created not at the base of the cloud layer, but deep within the gas atmosphere, say at a height of 12 km above the surface, or even much lower. In this case, the overall luminosity of the lower cloud boundary is reduced by $\sim 10^3$ times in comparison with the position of the flash at 1 km from the boundary of the cloud layer, but the light spot on the outer surface of the clouds would have at least an order-of-magnitude larger area and low surface brightness (by rough estimate $\sim 7 \times 10^{-8}$ W cm^{-2} ster^{-1}). However, M. Williams et al. (1982) note that this large attenuation applies only for blue light. Since Borucki et al. used a red-sensitive detector, they note that 5% of the photons should have escaped, even with a source at the surface.

IV. WHERE DO THE DISCHARGES TAKE PLACE?

In order to understand how discharges occur in the atmosphere of Venus and what is the mechanism of charge accumulation, it is essential to know the source altitude. Scarf et al. (1982) have studied this by analyzing Pioneer observations in connection with terrain features. Here, we will discuss again the Groza observations in terms of ray tracing.

In Figs. 2 and 3, the field spectral density is shown as a function of the descent time of the probe and its altitude above the surface of Venus. But how well do we know the altitude of the discharges? In Istomin et al. (1979*b*), Ksanfomality et al. (1979), and Ksanfomality (1979*a*), it was assumed that the discharges occur in the cloud layer. Although we have no direct proof, they proceeded from the following basis: it is well known that on Earth, large extensive charges and their associated lightning discharges occur almost exclusively in thunderstorm clouds. In certain cases, an accumulation of charges is observed in winter clouds (winter lightning). There are also known to be strong lightning discharges in dust storms and over erupting volcanoes. According to Latham and Stowe (1969), on Earth the largest charges are observed in clouds with drops of supercooled water, and in clouds of ice crystals. If, however, the cloud consists solely of liquid droplets, the charge will be small. Regarding charging mechanisms, contact between substances with differences in dielectric permeability, electrochemical potentials, densities, or surface hardnesses under laboratory conditions invariably results in charge separation (Loeb 1953). It is, therefore, presumed that the presence of several types of particles in the middle of the cloud layer of Venus (Marov et al. 1976*a*, 1979*a*; Knollenberg and Hunten 1980; Ragent and Blamont 1980) probably creates conditions for charge separation. However, Knollenberg and Hunten (1979*b*) and Knollenberg et al. (1980) express doubt that accumulation of a spatial charge in a medium which contains an aerosol of a strong electrolyte, sulfuric acid, is possible. Generally speaking, this situation is no impediment to charging. Imyanitov et al. (1971) showed, in laboratory experiments, that the presence of electrolytes or surfactants resulted in an increase in the potential difference up to several hundred volts. According to this same source, electrical conduction is smaller in the clouds of Earth than in the free atmosphere due to ion capture by drops. The electrification upon breaking contact between two particles is determined by the difference between their chemical potentials and the charge which is released by the breaking (Imyanitov et al. 1971)

$$q = V_{1,2} C_{1,2} (1 - e^{-t/\tau}) \tag{24}$$

where $V_{1,2}$ is the contact potential difference, $C_{1,2}$ is the capacitance between the particles at the moment of separation, and τ is the relaxation time, given by the product of the dielectric permeability and the electrical conduction of the particles. Knowing the properties of all types of particles of the cloud layer of Venus will allow a more definite position as to the possibility of charge accumulation.

However, the electrical charges may occur at altitudes well below the cloud layer. The principal bursts of radio noise were observed when the detectors were below an altitude of 30 to 32 km, and not in the cloud layer. This may be because the signals were received from a remote source. Changes in the field intensity produced by the source itself are extremely difficult to distinguish from other changes due to peculiarities associated with receiving the signal by the antenna on the probe, discontinuities in the region through which the signal propagates, or refraction. Croft (personal communication, 1980) arrives at this same conclusion. Furthermore, there are events which appear similar in data from the two probes. These are bursts of comparable amplitude and similar shape in the altitude interval of 0 to 8 km, low field intensities in the bottom portion of the cloud layer (50–60 km), and an absence of signals at the surface of the planet. We shall assume that these recurrent features are typical of the atmosphere of Venus. Then, as Croft and Price (1983) point out, a more important role for refraction should be examined than the role postulated by Ksanfomality (1979a). It follows, from the analysis of Croft, that with a source at an altitude of 70 km and at a distance on the order of 1000 km from the probe, there results high electromagnetic energy density on the surface (about half of all the radiated power), while the zone of low density lies beyond 2000 km. Therefore, Croft asserts that the conclusion that the source of discharges is in the cloud layer may be erroneous. It also follows from his computations that there exists a peculiar "radio refraction trap" which concentrates the electromagnetic field between the surface and the critical level of the atmosphere (33 km).

Croft analyzes the large burst of radio noise observed by Venera 11 between the altitude of 31 and 13 km (Figs. 2 and 5). He concludes that the formation and development of the burst may have been determined not by the start of a distant thunderstorm, but by the probe's crossing the critical level of 33 km, the source being situated below 33 km and more than 1000 km away. According to his computations, all the waves from such a source, propagating nearly horizontally, will remain below 33 km. If, however, the source is located above 33 km, the density of the atmosphere will be insufficient for critical refraction.

All waves with elevation angles up to $7°5$ from a radiator on the surface of the planet will reach heights of up to ~ 30 km, after which they are refracted toward the surface. The effect is reminiscent of reflection from the ionosphere, but differs in its low height. The propagation distance in such a situation depends on the coefficient of reflection of the waves from the surface of the planet. As the altitude of the source increases from zero, the maximum elevation angle for reflection decreases (e.g., in the computations of Ksanfomality [1979a], down to 3° for an altitude of 11 km). The effect van-

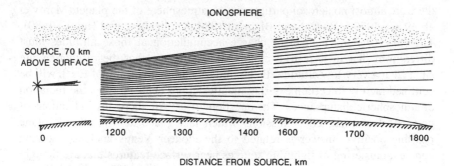

Fig. 16. A location of the signal source at a height of 70 km and a distance of 1200 to 1800 km, according to the analysis of Croft (personal communication, 1980), produces an altitude distribution of the field intensity which differs considerably from the position of the maximum obtained in the experiment (below 30 km). The density of the beams on this diagram represent the field intensity of the source. The drop in intensity toward the surface is displayed with special clarity on the fragment for the range interval of 1700 to 1900 km or more. The best correspondence is provided for a source at an altitude below 30 km.

ishes for horizontal beams at 33 km, while at a level of 40 km, according to Croft, only beams with an angle of elevation of −1° can reach the surface. But, after reflection from the surface, they no longer return to the surface. The low field intensity in the region above 33 km, in the case of Venera 11, should be observed (according to Croft) due to the effect of defocusing, but only under the condition that the source is remote and low. Reflections from the ionosphere may substantially complicate the altitude dependence of the field of intensity. Croft mentions that an important question is still the degree of absorption of low-frequency waves in the lower layers of the ionosphere.

The data in Figs. 2 and 3 are consistent with the view of Croft. Figure 16 shows several ray paths from Croft and Price (1983). The density of radio paths reflects the field intensity at the particular point. Like any real phenomenon, the altitude changes in the field intensity, measured by Veneras 11 and 12, are more complex than the model. Ionospheric reflection may nonetheless create a low field intensity and low field intensity on the surface can be accounted for by absorption of low-frequency waves by the surface.

Let us generalize from these conclusions. According to Croft, there is reason to suppose that the discharges are localized not in the clouds, but in the lowest layers of the atmosphere, below 33 km. Optical observations from satellites (Krasnopolsky 1980; Borucki et al. 1981) do not show optical lightning in the clouds with rates as high as those observed from the low-frequency electromagnetic instruments. A possible explanation of this apparent contradiction may be the fact that the region where the discharges are formed is far below the clouds. It is known that below the level of 48–49 km, there are almost no aerosol particles in the atmosphere of the planet (Marov et al. 1976a, 1979a; Ragent and Blamont 1980; Knollenberg and Hunten 1980). Therefore, the mechanism of charge accumulation at those altitudes must differ from that in terrestrial clouds. If we assume that the discharges are situated at a very low altitude, their similarity to the lightning of Earth will be confined merely to certain external manifestations such as, e.g., the radiation of atmospheric disturbances. The question arises: Are there any indications of unusual electromagnetic emissions which take place at low altitude? One possible positive answer is related to the Pioneer Venus analysis showing apparent clustering of lightning sources near surface features that are thought to be volcanic (see Scarf et al. [1982] and the discussion of Fig. 15 above).

In this connection, it is also of interest to note that other anomalies, perhaps related to electrical activity, were detected at relatively low altitudes. On 22 December 1978, during the descent of Venera 12, the Doppler frequency shift was measured by a signal which was rebroadcast from the probe via the relay spacecraft to Earth. The data obtained were used to determine the wind speed and turbulence of the atmosphere. Kerzhanovich et al. (1979c) pointed out that for 15 min prior to landing, variations in the signal frequency were observed which were not related to the action of the wind on the probe. In the cited work, this segment was excluded from the discussion. Kerzhanovich thought that the unusual fluctuations may have a relation to our problem and graciously made available the omitted data, three fragments of which are shown here in Fig. 17. The average slope of the line reflects the Doppler frequency shift produced by the relative movement of the orbiter (transponder). Deviations from the mean lines represent phase fluctuations. A signal of "tranquil" character was recorded for a long time (e.g., in the left part of the figure). At an altitude of 22 km, there appeared slight fluctuations in the signal phase which continued down to the level of 16 km and then vanished. At an altitude of 8.5 km, these reappeared and achieved many hundreds of degrees (see fragments of the recording in the middle portion of Fig. 17). The fluctuations continued until the actual landing, after which they vanished entirely. No malfunctions were discovered in the operation of the equipment on board. The signal after the landing is shown at the right. During the landing of Venera 11, three days later, in the same region, at the eastern slope of the highlands of Phoebe, no phase fluctuations were observed. It remained unclear whether this event was related to the electrical activity of the atmosphere; however, in time, it fully coincides with the burst of radio noise in the lower portion of Fig. 3 and in Fig. 7.

Tyler (personal communication, 1980) also noted that observations of the telemetry signals from the Pioneer Venus Day probe *on* the surface of Venus showed slow variations in amplitude, with typical changes of $\pm$ 0.5 db in a 500 to 1000 s period. These variations were judged to be real because they were observed to be the same with three of four receivers employed at the Deep Space Net Stations 14 and 43. It is possible that the Pioneer Venus

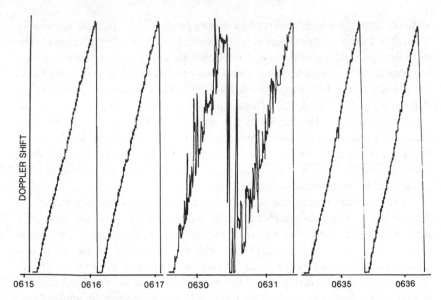

Fig. 17. Phase fluctuations of the radio signal of Venera 12 shown by deviations from the mean
sloping lines with appearance of the signal at altitudes of 10 km (left), 2 km (center), and at
the surface (right).

telemetry fluctuations could be related in some way to those observed
on Venera.

Colin (1980), Seiff et al. (1980), and Suomi et al. (1980) have reported
even stranger phenomena on board the Pioneer Venus probes. On all the small
probes (North, Night, and Day) in the deep layers of the atmosphere, there
occurred malfunctions of the temperature and total flux sensors. The temperature sensor also failed aboard the Large probe. It is noteworthy that the probes
were completely independent, with enormous distances separating them;
nevertheless, the disturbances of these sensors were observed at the identical
altitudes of 12 to 12.5 km, the malfunctions in the temperature and flux
sensors being separated by no more than several minutes. The circuits of both
types of sensors were electrically insulated. The sensors were mounted on
rods protruding from various sides of the probes, and the layouts of the
sensors were entirely different. In one case, a platinum filament was used; in
the other, a thermistor was used. The authors cited above confirm that all the
instruments passed through the full array of essential tests. Laboratory exposure of the twin instruments to corrosive media, high temperatures and pressures, as well as possible simulated defects, could not produce any disturbances in their operation. Even though we cannot completely exclude certain
defects, Colin states that ''certain research forces us to assume that the failure
could be related to the action of an external electrical discharge on all of the
probes.'' It is also stated that, at the same time as the breakdown of the

sensors, anomalies were observed in other systems of the probes, apparently also related to an electrical action on the vehicle. Polaski (1979) has made the most detailed analysis of these malfunctions. He also noted anomalies involving considerable current surges in the power bus circuits. These investigations led Polaski to the following conclusion: "The only remaining possibility is that the probe was enveloped in a plasma. But where this plasma could come from is unknown." He concludes his article with the words: "...Each of these disturbances [in the operation of the instrument] could be the result of a discharge of static electricity... An analysis... allows us to assume that these disturbances are the result of an unexpected electrical interaction between the probes and the atmosphere of Venus."

Thus, it would seem that we have an indication that the atmosphere of Venus may be electrically active at great depths. But the existence there of a plasma with high concentration, while the number density of the neutral gas is 4.66×10^{20} cm^{-3}, is inconceivable; and the formation of electrical discharges, not provoked by the probe, is incomprehensible. In order to achieve full understanding, we require additional information as, for instance, the altitude dependences of the electric field and conductance of the atmosphere, which are now being analyzed theoretically (Levin and Tzur [1981], electrical field; Borucki et al. [1982], conductance), as well as the localization of the discharge sources by taking bearings from the descent probes, or by other means. Careful photometric measurements from orbiters would also be extremely useful, even if the results were negative.

From the discussions in this chapter we arrive at the following conclusions:

1. Below the ionosphere of Venus, there are numerous sources of impulsive electromagnetic waves. The maximum impulse rate from a single source extends up to 20 or more per second.

2. When the interplanetary magnetic field is captured in the layers of the night ionosphere above the electron maximum, so that f_c is large, and B points down to the planet, the lowest frequency components of the radiation from these disturbances (on the order of 100 to 300 Hz) travel in the whistler mode and are detected by the orbiter in eclipse.

3. The ionosphere-surface channel does not, apparently, function at frequencies of 10 to 80 kHz. This is consistent with the idea that the electron concentration is low directly down to altitudes on the order of 115 km, where there can be an attenuation of the low-frequency waves by collision. But even this conclusion presently refers only to the day side of the planet.

4. There are good reasons to believe that the sources of electromagnetic radiation (discharges) may be located much lower than the cloud layer, and that the mechanism of accumulation of space charges may differ from that on Earth.

5. The clustering of apparent whistler mode wave sources near Phoebe, Beta, and Atla Regios suggests a possibility of volcanic activity.

6. Thanks to refraction, a radio frequency signal is able to travel over distances of 3000 km or more, gradually weakening by virtue of the diffuse nature of reflection from the surface and possible absorption by the latter. However, these nearly horizontal wave propagation conditions refer to signals that will not escape beyond the ionosphere.

7. The question as to the optical radiation of the discharges remains very unclear.

In summary, we address the fundamental question: Does lightning occur on Venus? The wave measurements of electrical activity would seem to provide a resounding positive answer to this question, since we know of no other sources for such intense impulsive low-frequency radio waves of atmospheric origin. Nevertheless, there are many fascinating open questions on this topic, and future missions to Venus should be designed to provide new information.

18. THE THERMAL BALANCE OF THE LOWER ATMOSPHERE OF VENUS

MARTIN G. TOMASKO
University of Arizona

Since the late 1950s, the temperature near the surface of Venus now established as 730 K has been known to be remarkably high in view of Venus's cloud cover which causes the planet to absorb even less sunlight than does Earth. Early attempts to understand the thermal balance that leads to this unusual state were hindered by the lack of basic information regarding the composition, temperature-pressure structure, cloud properties, and wind field of the lower atmosphere. Beginning in the 1960s, a series of successful space missions has measured many of the above quantities that control the transfer of heat in Venus's lower atmosphere. In this chapter, the relevant observational data are summarized and the attemps to understand the thermal balance of Venus's atmosphere below the cloud tops are reviewed. The current data indicate that sufficient sunlight penetrates to deep atmospheric levels and is trapped by the large thermal opacity of the atmosphere to essentially account for the high temperature observed. Nevertheless, several features, particularly those due to atmosphere dynamics, remain to be understood.

In order to understand the thermal balance of the lower atmosphere of Venus, we must evaluate the rates of various processes that can transport heat in this region. To know these rates, we need to have detailed information regarding the composition, temperature-pressure structure, clouds, and dynamical state of the atmosphere. For the lower atmosphere of Venus, loosely defined here as the region below the cloud tops, this information was especially meager before the advent of space missions to the planet. In the past decade or so, however, a series of remarkably successful Mariner, Venera, and Pioneer missions has advanced our knowledge to the point where separate review chapters in this book are devoted to each of these topics. Accordingly, while our understanding is still incomplete, we have finally measured or are in a position to quantitatively evaluate some of the heat transport rates we believe should be most important in this no longer inaccessible portion of the Venus atmosphere.

The earliest available data bearing on the thermal balance of Venus concerned its size, distance from the Sun, and albedo. The data suggested that while Venus is closer to the Sun than is Earth, it also has a more complete cloud cover than Earth's, with the result that Venus absorbs an amount of sunlight that is somewhat less than that absorbed by the Earth. Judging by conditions on Earth, Venus's extensive cloud cover suggested the presence of a considerable amount of water, and for some years the popular conception of Venus was of a planet with a climate similar to that of Earth, a view gradually revealed to be far from true.

Section I of this chapter reviews our growing knowledge of the conditions in the lower atmosphere of Venus which bear on its thermal balance. These conditions include the temperature profile, cloud structure and optical properties, composition, and wind field. Section II relates the heating and cooling rates to several of the observed atmospheric parameters (radiative flux profiles, cloud and atmospheric structure), and briefly reviews some of the major developments in attempts to model the thermal balance of Venus's lower atmosphere. Present problems and areas requiring further work are indicated.

I. SUMMARY OF RELEVANT OBSERVATIONS

A. Temperature Structure

An estimate of the temperature in the cloud-top region of Venus's atmosphere can be obtained from the relative strengths of absorption lines arising from different rotational levels in the near-infrared region of Venus's spectrum, using observations made from Earth. Results using a homogeneous scattering model give temperatures of 240 to 270 K and a pressure of ~ 0.2 bar as typical of cloud-top conditions (Belton 1968), values not grossly different from cloud-top conditions on Earth.

However, the characteristics of the microwave emission from Venus, first detected by Mayer et al. (1958), greatly changed our view of conditions on Venus (Chapter 4). In the spectral range between 2 and 20 cm, Venus was seen to emit radiation at a roughly constant brightness temperature of ~ 600 K. In 1962, limb darkening observations at 1.9 cm by Mariner 2 indicated that this emission indeed originated from the surface of the planet and not from some unusual phenomenon in the upper atmosphere. In 1967 the combined radio occultation measurements of Mariner 5, direct entry measurements from Venera 4, and earth-based radar observations of the radius of Venus were sufficient to yield a surface pressure of $\gtrsim 70$ bar and a surface temperature of

~ 700 K. Direct measurements by a series of entry probes beginning with Venera 7 have confirmed the remarkably high temperature and pressure near the surface of Venus (Marov 1972, 1978).

Among the most accurate measurements of the temperature-pressure structure of the lower atmosphere of Venus are those made on the four Pioneer Venus (PV) probes (Seiff et al. 1980; Chapter 11 in this book). The probe entry locations are shown in Fig. 1 and vary in latitude from 30°S to 60°N and in solar zenith angle (SZA) from 66° to 150°. The close agreement between the atmospheric structures observed at these sites is apparent in the figure. The Day and Night probe temperature profiles are essentially equal between 20 and 35 km altitude, while their differences show a wave with an amplitude of a few degrees K between 35 and 65 km. The mean temperature profiles at these altitudes are very nearly equal at the Day and Night probe sites. The wave appears also in the temperature difference profiles of the Day-North and Day-Large probe sites, but superimposed on a general trend that has the North probe profile becoming 20 K cooler and the Large probe a few degrees warmer than the Day probe at 65 km altitude (see Seiff et al. 1980; Chapter 11).

The profiles of static stability in the lower atmosphere at the four PV probe sites are shown in Fig. 2. The atmosphere exhibits a subadiabatic (stable) temperature gradient above the clouds, and becomes neutrally stable in the middle cloud. At all four sites, the atmosphere exhibits a markedly stable profile from the cloud bottom at ~ 50 km down to ~ 30 km altitude. The adiabatic profile is then followed down to altitudes of ~ 20 km, below which several of the profiles again indicate a return to stable conditions before the data from the temperature sensors on all four probes terminate at ~ 13 km altitude. These data regarding static stability are especially useful for constraining models of the thermal balance of Venus's lower atmosphere.

B. Absorption of Sunlight

There are at least two separate problems concerning the absorption of sunlight: the variation of absorbed energy with wavelength and the variation with height. The variation with wavelength of the spherical albedo of Venus was obtained by Irvine (1968) from observations at many phase angles at each of a variety of narrow spectral bands from the near ultraviolet to the near infrared. From these data the bolometric spherical albedo was estimated as 0.77 ± 0.07 with the uncertainty largely due to the uncertainty in the magnitude of the Sun. The high value of the albedo A implies a small and relatively uncertain value for the amount of absorbed solar energy, which is proportional to $(1 - A) = 0.23 \pm 0.07$. If Venus is in equilibrium with absorbed sunlight, it should emit 150 ± 45 W m^{-2} corresponding to an effective temperature of 227^{+15}_{-20} K. The albedo of Venus is very high in the visible, but drops in the ultraviolet and is quite low at wavelengths longward of 1 μm due to absorption by carbon dioxide and the cloud particles. Since a considerable

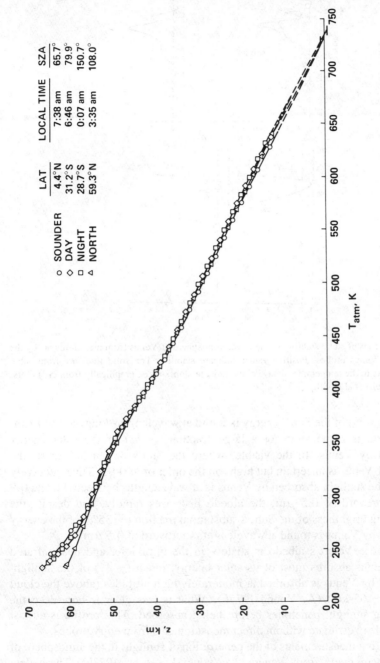

Fig. 1. Comparison of temperature profiles from the four Pioneer Venus probes. SZA is the solar zenith angle. (From Seiff et al. 1980.)

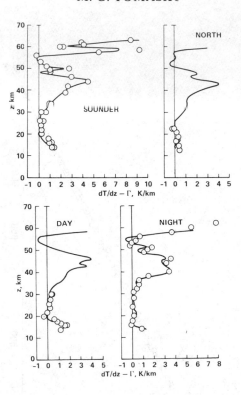

Fig. 2. Profiles of static stability of the lower atmosphere of Venus from measurements by the
Pioneer Venus probes. Positive values indicate stability. The solid lines are from cubic
spline fits to the temperature data. Points indicate slopes taken graphically from $T(Z)$ plots.
(From Seiff et al. 1980.)

fraction ($\sim \frac{1}{3}$) of the Sun's energy is found at wavelengths longward of 1 μm,
this region is responsible for a large fraction ($\sim \frac{1}{2}$) of the solar energy
absorbed by Venus. In the visible, where the Sun's output is largest, the
albedo of Venus is uncertain but high, on the order of $\geqslant 0.90$. Thus, relatively
little of the sunlight absorbed by Venus is at wavelengths between 0.6 and 0.9
μm. Shortward of 0.5 μm, the albedo decreases rapidly, and despite the
decreasing brightness of the Sun, a substantial fraction ($\sim 35\%$) of the energy
absorbed by Venus is found at wavelenghts shortward of 0.5 μm.

Because Venus's albedo is so low in the ultraviolet and near infrared
where Venus absorbs most of its solar energy, much ($\sim \frac{1}{2}$) of the sunlight
absorbed by Venus is absorbed at moderately high altitudes (above the cloud
tops at ~ 65 km) (A. Young 1975b). What is less clear is how deep the
remaining sunlight penetrates before being absorbed. This profile is almost
impossible to estimate without direct measurements from entry probes.

The first measurements of the penetration of sunlight in the atmosphere of
Venus were made from Venera 8 (Avduevsky et al. 1973b). These data

consisted of measurements of the downward flux of sunlight between ~ 0.5 and 0.8 μm from 48 km altitude to the surface. While only $\sim 10\%$ of the solar energy absorbed by Venus is contained in this wavelength range, it is at these wavelengths that solar radiation penetrates most deeply into Venus's atmosphere. Despite uncertainties in the exact solar zenith angle and the ground reflectivity, the data from this experiment indicated that roughly 1% of the solar flux incident on Venus at a solar zenith angle of $\sim 85°$ is absorbed at the surface of the planet (Lacis 1975).

This experiment was followed by improved experiments on Venera 9 and 10 which measured upward as well as downward flux between 0.5 and 1.05 μm at 28° and 33° SZA (Moshkin et al. 1978), on Veneras 11 and 12 (which included spectral measurements) at $\sim 20°$ SZA (Moroz et al. 1979b,1980c), and on Pioneer Venus at 66° SZA in the wavelength range from 0.4 to 1.8 μm (Tomasko et al.1980a). Near the surface, the measurements made by these various missions are relatively consistent with the expectations of forward scattering cloud models (Tomasko et al. 1980b) in showing greater penetration of sunlight at small solar zenith angle (see Fig. 3). The measurements indicate that some 2.5% of the sunlight incident on Venus is absorbed at the surface of the planet (which would require a net radiative flux directed upward at

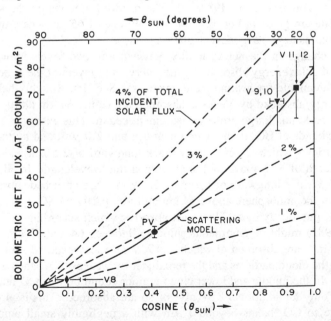

Fig. 3. Net solar flux at the surface of Venus as a function of the cosine of the solar zenith angle. Dashed lines represent constant fractions of the incident solar flux as indicated. The solid line represents the variation computed for a forward scattering cloud model adjusted to pass through the Pioneer Venus measurement at the Large probe site (labeled PV). This measurement and those measured at the Venera 8 (labeled V8), Venera 9 and 10 (V9, 10) and Venera 11 and 12 (V11, 12) sites are quite consistent. (From Tomasko et al. 1980b.)

thermal wavelengths of ~ 17 W m^{-2} over the entire planetary surface for thermal balance). While this is a small fraction of the sunlight incident at the top of the atmosphere, it is important in the thermal balance of Venus's lower atmosphere.

At visible wavelengths, the ground reflectivity is fairly low, $< 15\%$ (Ekonomov et al. 1980) and at solar wavelengths the downward flux exceeds the upward flux at the surface by a large factor, making net flux measurements relatively easy. At increasing altitudes, however, both upward and downward fluxes increase dramatically in a way that makes the net flux a moderately small fraction (15%) of either one by the time the cloud level is reached. In these circumstances, the measurement of net flux can be made most reliably if the same detector is used to meaure the upward and downward flux. Such measurements were made on Veneras 11 and 12, and are reported by Moroz et al. (1980c; see also Chapter 19). The Venera 11 and 12 narrowband net flux profiles remain to be incorporated in a model that yields an estimate for the globally averaged bolometric net solar flux with altitude.

The Pioneer Venus Large probe measurements of net solar flux used different detectors for upward and downward flux measurements; but have been modeled to yield an estimate for the globally averaged bolometric net solar flux (Tomasko et al. 1980a, b). The resulting bounds on the solar net flux profile are given in Fig. 4. Above an altitude of 65 km where the probe measurements begin, a model calculation is shown.

The change in the net solar flux between any two levels indicates the amount of solar energy absorbed by the intervening layers of the atmosphere, which is available for heating these layers (see Sec. II). Roughly half of the solar energy absorbed by Venus is absorbed above the 64 km altitude where the Pioneer Venus Large probe measurement began. This energy is absorbed in strong bands of CO_2 in the region near 1.6 and 2.0 μm, and by the H_2SO_4 cloud particles which are essentially black longward of 2.5 μm and have an optical depth of ~ 2 above 64 km altitude at this wavelength. In addition to the ~ 35 W m^{-2} longward of 1 μm, ~ 22 W m^{-2} are absorbed shortward of 0.5 μm in the atmosphere above 64 km altitude, partly by SO_2 (shortward of 0.35 μm) but mostly by an ultraviolet absorber which according to Toon and Turco (1982) might be amorphous sulphur. Between 64 and 57 km altitude, ~ 20 W m^{-2} are absorbed of which ~ 15 W m^{-2} are absorbed shortward of 1 μm by the cloud particles and the remainder by weaker CO_2 bands longward of 1 μm. In the middle and lower cloud regions found between 57 and 48 km altitude, only an additional 8 W m^{-2} or so are absorbed. This is almost entirely due to CO_2 bands beyond 1 μm with a negligibly small amount absorbed by the cloud particles in this region (which have a visible optical depth of ~ 17). Between the cloud base at 48 km and ~ 40 km very little solar energy is absorbed. Between 40 and 15 km altitude an additional 15–20 W m^{-2} are absorbed, largely by an ultraviolet absorber shortward of 0.6 μm (suggested by Moroz to be a form of sulphur vapor) and partly by weak CO_2

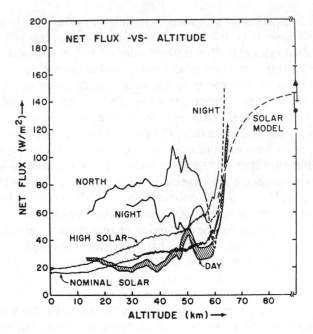

Fig. 4. The thermal net flux measurements as published by Suomi et al. (1980) at the three Pioneer Venus small probe entry sites (labeled Night, North, and Day) compared with estimates of the globally averaged bolometric net solar flux based on measurements at the Pioneer Venus Large probe entry site. The solar flux measurements scaled up to include energy outside the passband of the instrument and averaged over the planet using a forward scattering cloud model are labeled "nominal solar." Also shown is the highest solar net flux profile (labeled high solar) consistent with instrument calibration uncertainties. The long dashed curve above 60 km altitude is a model solar flux calculation. The triangle and dot correspond to estimates of the net thermal and solar flux (respectively) at the top of the atmosphere by the PV orbiter. The magnitudes of the net fluxes are plotted; the solar net flux is directed downward, and the thermal net fluxes are directed upward. (From Tomasko et al. 1980*b*.)

and water bands longward of 0.7 μm. Little solar energy is absorbed in the lowest 15 km of the atmosphere. At the surface, some 17 W m^{-2} are absorbed.

C. Emission of Thermal Radiation

From the outside of the atmosphere, the total bolometric thermal emission can be measured rather directly. The measurements require integration over wavelength, emission angle, and longitude and latitude on the planet. The analysis of the measurements made by the Pioneer Venus orbiter infrared radiometer (OIR) experiment give a total emitted flux of 157 ± 6 W m^{-2} (corresponding to effective temperature T_e = 229.4 ± 2.2 K) from measurements made in the northern hemisphere (Schofield and Taylor 1982*b*). Presumably the southern hemisphere would yield a similar value. For the whole planet to be in equilibrium with absorbed sunlight, the bolometric spherical albedo would have to be 0.76, which is in good agreement with Irvine's (1968) measured value of 0.77 ± 0.07. A value of 0.80 ± 0.02 was given by F. Taylor et al. (1980) in a preliminary interpretation of OIR data in a broad solar channel, but the uncertainty in this value may be larger than the value quoted due to calibration uncertainties (see Chapter 20 by Taylor et al.) Schofield and Taylor (1982b) also give the thermal emission of Venus averaged in 10-deg wide bands of latitude. The emission is remarkably constant with latitude as already known from earlier work at a variety of wavelengths from Earth (see Tomasko et al. 1977). By contrast, the total amount of absorbed sunlight falls to zero at the poles somewhat more rapidly than the cosine of the latitude, μ_l (roughly as $\mu_l^{1.4}$, [Tomasko et al. 1980b]) due to the increasing reflectivity of the forward scattering clouds at glancing incidence. The difference between the profile for the absorption of sunlight and the emission of thermal energy with latitude provides a basic drive for atmospheric dynamics.

The direct measurement of net thermal fluxes within the atmosphere is much more difficult than measurements from outside, and has only been attempted on the recent Pioneer Venus mission (Boese et al. 1979; Suomi et al. 1980). The measurements by the Large probe infrared radiometer (LIR) were seriously affected by a window misalignment at altitudes below 50 km, but seem reasonable above this altitude when the factor π is removed from the caption of the figures in Boese et al. (1979) as revised by the author (Boese, personal communication, 1982). 'In particular, the Large probe thermal net fluxes are highly correlated with cloud structure measurements made on this probe, indicating the significant role of cloud thermal opacity (Boese et al. 1979).

The only thermal net flux measurements below the clouds are those made on the three PV small probes (Suomi et al. 1980). The instrument used a wide field of view and a wide spectral range (0.2 μm to 150 μm wavelength) to measure net flux. Two of the small probes entered in darkness, and measured thermal radiation only. The third (the Day probe) entered at a solar zenith angle of 79°9 and contains solar as well as thermal contributions. A model solar flux profile consistent with the Large probe solar flux measurements has been used to generate the thermal flux profile shown in Fig. 4 for the Day probe. For the other probes, the values as originally obtained by the experimenters are shown along with estimates of the globally averaged solar net flux from the LSFR experiment for comparison.

The thermal flux profiles are surprisingly variable from site to site in view of the great similarity in temperature profiles measured at these sites. In addition, at both the Night and North probe sites they are much greater than the globally averaged solar net flux profile at low altitudes, implying a substantial radiative imbalance in the lower atmosphere. In view of the large and variable nature of these flux measurements, the investigators have searched for instrumental problems which could have affected the measurements, and

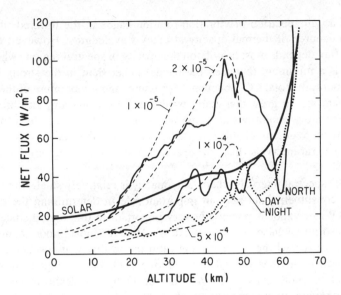

Fig. 5. Thermal net flux measurements from the three small Pioneer Venus probes after correction for instrumental effects (from Revercomb et al. 1982). The fluxes are compared to radiative-transfer calculations for the indicated values of the water mixing ratio. Also shown is the "high" global average solar net flux profile from Fig. 4. The uncertainty of the corrected thermal fluxes depends on the uncertainty in the model computed fluxes at 14 km. This effect gives an estimated uncertainty of 10 W m^{-2} at 14 km with decreasing uncertainty at increasing altitude.

have found one that could have systematically increased the measured thermal net fluxes (Revercomb et al. 1982). The authors believe that they understand the vertical dependence of the flux errors, and by adjusting the fluxes to reasonable values at low altitudes, they have derived corrected thermal fluxes as shown in Fig. 5. The fluxes of both the Day and Night probes (near latitude 30°S) are now in rough agreement with each other and with the solar net flux profile estimated to be typical of globally averaged conditions. The North probe thermal net flux profile is still somewhat larger than the others, but is consistent with model calculations.

It is important to be able to compute the net thermal radiative fluxes expected in Venus's lower atmosphere for the measured temperature profile. This requires knowledge of the composition and thermal opacity sources in the atmosphere (see Eq. 4 in Sec. II). While CO_2 was long known to be an important constituent in Venus's atmosphere from its many absorption lines, the complexity of line formation in a scattering atmosphere prevented finding an accurate value for its mixing ratio until entry of the Venera probes (Marov 1972). These measurements, together with the gas chromatograph measurements on PV have indicated its mixing ratio to be ~ 96.5% (Oyama et al. 1980a). Because this gas is the primary component of the Venus atmosphere and has a rich spectrum of vibration-rotation bands in the infrared, it is the dominant source of thermal opacity at many wavelengths. However, the net thermal flux depends most heavily on the opacity in spectral regions where the opacity is a minimum (between the bands) rather than in the strong bands. Furthermore, the total CO_2 abundance, pressure, and temperature of the lower atmosphere are so great that many weak transitions that are not observable under laboratory conditions are expected to occur. Thus it has been particularly difficult to obtain reliable information on the net thermal flux expected in certain spectral intervals (like, for example, the 3–4 μm region) due to CO_2 opacity alone.

In the spectral intervals where CO_2 opacity is relatively small, the opacity of trace constituents can play an important role in determining the thermal opacity. Water vapor is potentially very important in this regard since it has strong absorptions in several CO_2 windows. Here the most important question has long concerned the water mixing ratio in the lower atmosphere. Earth-based spectroscopy has yielded water mixing ratios in the range from 10^{-4} to 10^{-6} (Belton 1968; Barker 1975b; Chapter 13 by von Zahn et al.) in the cloud-top region, though measurements on several Venera probes gave values of a few tenths of a percent (Marov 1972). However, if water makes up a part of the cloud particles, many types of direct sampling analyses may be prone to yield too high a value for the mixing ratio of the atmospheric water vapor due to contamination by water from cloud droplets. This is known to have happened during the PV mass spectrometer experiment when the inlet was actually plugged by a cloud droplet during a significant portion of the descent (Hoffman et al. 1980a) and may have contributed to the relatively high water mixing ratios (up to 0.5%) measured by the PV gas chromatograph experiment (Oyama et al. 1980a).

On the other hand, the optical spectrometer experiment on Veneras 11 and 12 would not be affected by this problem, and these data have been interpreted (Moroz et al. 1980a; see also Chapter 13) as indicating a local mixing ratio of water vapor which varies from $\sim 2 \times 10^{-4}$ in the region of the cloud bottom to $\sim 2 \times 10^{-5}$ near the surface. While smaller than some other estimates, these water vapor values are sufficiently large to provide an important source of thermal opacity in Venus's lower atmosphere (Pollack et al. 1980a).

Also observed in the lower atmosphere were SO_2 (at ~ 200 ppm) and CO (at ~ 20 ppm) (Oyama et al. 1980a), two gases which are significant sources of thermal opacity. Gases that have no allowed transitions in the infrared such as the noble gases provide negligible thermal opacity. Other gases like HCl which may be present in mixing ratios of $\leqslant 1$ ppm are also ineffective thermal opacity sources.

In addition to the gases listed above, significant thermal opacity can be provided by the clouds covering the planet. The effectiveness of the cloud particles for blocking thermal radiation of a particular frequency depends on the composition of the particles (through the complex index of refraction) as well as the size distribution and the total mass of cloud material. Of particular interest is the opacity in the thermal infrared compared to that in the visible. This ratio is an important parameter in determining the effectiveness of the clouds in aiding the greenhouse effect.

Knowledge concerning the cloud structure is reviewed by Knollenberg et al. (1980) and by Esposito et al. (Chapter 16 in this book). Briefly, earth-based observations of the infrared spectrum by Pollack et al. (1975), and of the index of refraction of the cloud particles derived from polarization measurements by Coffeen (1969), Hansen and Hovenier (1974), Sill (1972), and Young and Young (1973) indicate that the clouds are composed of strongly concentrated sulfuric acid ($>$ 70% by weight) in the cloud-top region. The earth-based polarimetric measurements indicate that optical depth unity in the visible occurs at $\sim$ 50 mbar pressure level, and that the cloud particles are spherical with a narrow size dispersion about a mean effective radius of $\sim$ 1 μm (Hansen and Hovenier 1974). The cloud particle size spectrometer (LCPS) on the PV Large probe made measurements of the size distribution and number density of the cloud particles at pressure levels from $\sim$97 mbar (above which the visible cloud optical depth is $\sim$ 4) to the surface. In addition to confirming the presence of the 1 μm radius size mode (termed Mode 2), a distinct mode (Mode 1) of smaller particles whose mean effective radius is $\sim$ 0.5 μm was also seen. Both size modes were present in the three rather distinct cloud layers whose lower boundaries were at 48, 50, and 57 km altitude at the Large probe entry site. The Mode 1 particles were seen in all three cloud layers as well as in an optically thin (at visible wavelengths) haze that extended below the main cloud layers to an altitude of $\sim$ 30 km, and also above the main cloud layers.

In the middle and lower clouds, significant number densities of cloud particles up to 35 μm diameter (Mode 3) were also seen. The presence of these larger particles is especially important for providing thermal opacity. If the sulfuric acid cloud particles have diameters of $<$ 3 μm, their absorption cross section at wavelengths longward of 10 μm will be nearly an order of magnitude smaller than their extinction cross section in the visible (see Fig. 6). Particles with diameter $d > 6$ or 7 μm, however, will have cross sections at thermal wavelengths which are generally within a factor of $\sim$ 2 of their visible extinction cross sections. Thus, the Mode 1 particles are quite ineffective in providing thermal opacity. The Mode 2 particles ($d \simeq 2$–3 μm) with a total visible optical depth of $\sim$ 11 can provide a thermal optical depth on the order of unity. The Mode 3 particles had a visible optical depth of $\sim$ 5 at the Large probe site (Tomasko et al. 1980a) and can provide thermal optical depths of 2 to 3, but only within the middle and lower clouds where they are found.

The above estimates for the thermal opacity of the Mode 3 particles follow Fig. 6 and thus assume the particles are composed of concentrated sulfuric acid. There has been some question about the validity of this assumption arising from the fact that the PV nephelometer measurements of back-scattered

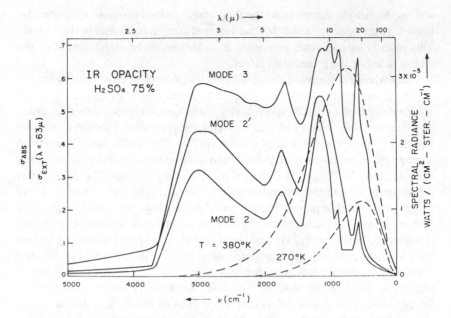

Fig. 6. The ratio of infrared absorption cross section to extinction cross section in the visible (0.63 μm) as a function of wavelength for sulfuric acid droplets having the size distributions measured for Mode 2 (effective particle diameter $\bar{d} = 2.1$ μm) Mode 2' ($\bar{d} = 2.8$ μm) and Mode 3 ($\bar{d} = 6.7$ μm). Also shown are the blackbody emission curves (dashed lines) for 270 K and 380 K, the temperatures at the base of the upper and lower cloud layers, respectively. (From Tomasko et al. 1980b.)

light, as well as the observed rate of decrease of upward and downward flux through the middle and especially the lower cloud, are not in agreement with the expected values calculated by assuming all the cloud particles measured by the LCPS experiment are spherical sulfuric acid droplets. Essentially no disagreement is seen in the upper cloud where no Mode 3 particles are observed by the LCPS. The LCPS investigators have suggested that the problems can be resolved if the Mode 3 particles are elongated crystals rather than spheres, and cite some features in their raw data to support this suggestion (Knollenberg and Hunten 1980). However, no completely satisfactory composition has yet been suggested for the solid cloud material.

Recently, Toon and Blamont (1982) have suggested that the Mode 3 particles may not be a separate size mode but simply an extended tail of the Mode 2 sulfuric acid spheres which reaches to quite large sizes. These authors suggest that a slight miscalibration of one size range in the LCPS instrument may have led to the impression that these particles formed a separate size mode. If this interpretation proves correct, the composition of the largest particles would be sulfuric acid, and the visible optical depths deduced from the Large probe solar flux radiometer (LSFR) experiment could safely be converted to thermal opacities along the lines outlined above.

One other possibly important source of thermal opacity was suggested by Suomi et al. (1980). If aerosol particles are sufficiently small for a given real refractive index, their scattering cross section can be made small in the visible as well as in the infrared. However, their absorption optical depth is separately determined by their imaginary index of refraction and the mass of aerosol per unit surface area in the clouds. For sulfuric acid, the imaginary index is large in the infrared and near zero in the visible. Thus it is possible to adjust the aerosol mass per unit area to achieve any desired absorption optical depth in the infrared, and to choose a sufficiently small particle size to make the scattering optical depth negligibly small at wavelengths longer than any specified wavelength in the visible. Therefore the particles might not have been seen by the LSFR experiment, which is sensitive in the visible, or by the LCPS experiment, because the particles were too small, but they could still provide an important contribution to the thermal opacity.

One motivation for suggesting the presence of these particles is the fact that the orbiter infrared (OIR) instrument on PV measured brightness temperatures near 11 μm (where gaseous opacity is a minimum) which implied that over much of the planet an aerosol optical depth of unity at 11 μm is reached at a pressure of $\sim$ 100 mbar (Schofield and Taylor 1982c). At 0.63 μm in the visible, the LSFR data (Tomasko et al. 1980a) suggest that the aerosol optical depth above the 100 mbar level is $\sim$ 4. By extrapolating the measured LCPS partition between Mode 1 and 2 type particles to higher altitudes, these authors assumed that the optical depth in the visible above the 100 mbar level would be $\sim$ 2.3 due to Mode 2 and $\sim$ 1.7 due to Mode 1. The corresponding absorption optical depth at 11 μm would be only $\sim$ 0.1 for Mode 1 and $\sim$ 0.5 for Mode 2, approximately half the value required by the OIR measurements. The proposed mode of submicron particles (dubbed Mode 0) could be responsible for the rest.

However, the partition between Modes 1 and 2 between 97 mbar (where the visible optical depth is 4) and $\sim$ 40 mbar (where the visible optical depth is 1) is rather uncertain because neither the orbiter nor probe experiments measured this region well. If nearly all (instead of ½) of the visible optical depth in this region were caused by Mode 2 particles, the 11 μm aerosol optical depth above 100 mbar would be nearly 1 as observed. Thus the need for a significant population of Mode 0 particles is questionable. Furthermore, it is worth noting that the Mode 1 particles in models E and F in Tomasko et al. (1980a) were mislabeled as being H_2SO_4 when in fact their optical properties were computed using a real refractive index of 1.93 (more appropriate for sulfur). Thus, the value of the scaling factor (1-g), which should be used to scale the Mode 1 particles optical depths to effective isotropic optical depths, is 0.515 rather than 0.265 as given in Table 6 and shown in Fig. 12 of that paper. That is, the entries in the second column of Table 6 giving the effective isotropic optical thickness of the Mode 1 particles in each layer of model F should be nearly twice as large as shown. Even if only half the optical thickness of the Mode 1 particles between 38 and 97 mbar in model F were replaced by an equivalent amount of Mode 2 particles, the thermal opacity of the clouds above 100 mbar would nearly double, without changing the visible opacity of the clouds or adding Mode 0. Despite this labeling error, none of the other conclusions of Tomasko et al. (1980a), particularly concerning the scaling of the measured solar flux profile to conform to globally averaged conditions or to include wavelengths outside the LSFR bandpass, are affected. Indeed, model F provides some support for the suggestion of Toon and Turco (1982) that amorphous sulfur may be the ultraviolet absorber, and for Toon and Blamont (1982), who require a component with high refractive index to explain the Pioneer Venus nephelometer results in the upper cloud. It now also seems that the particle number densities measured by the LCPS in the upper cloud do not provide sufficient visible optical depth to be consistent with the decrease in solar flux through this cloud, unless a component with high refractive index is present.

The horizontal variability in the thermal opacity produced by the clouds is difficult to estimate. The OIR measurements indicate that aerosol optical depth unity is reached at about the same pressure level ($\sim$ 100 mbar) except in the polar regions. Also, there are strong similarities in the visible cloud structure as revealed by the PV nephelometers and the Venera and PV solar flux profiles over much of the planet. On the other hand, the thickness of the lower cloud (which contained the greatest contribution of the Mode 3 particles at the Large probe site) appears to be rather different at the 4 PV probe sites in the nephelometer data. Also, the thermal net flux measurements in the clouds made by the Small probe net flux radiometer (SNFR) instrument are quite different at the three Small probe entry sites. Nevertheless, in view of the relative uniformity of the planet in the thermal infrared, the thermal opacity of at least the upper portions of the clouds varies relatively little over most of the planet.

D. Wind Field

The winds in the lower atmosphere have been measured by the differential long baseline interferometry (DLBI) experiment on the 4 PV probes (Counselman et al. 1980), and also on the Venera 8 through 12 probes (Marov et al. 1973a; Antsibor et al. 1976; Kerzhanovich et al. 1979b). Figure 7a from Schubert et al. (1980a) summarizes the zonal wind profiles measured below the clouds. The PV measurements for the Day and Night probes, which landed at nearly equal latitudes near 30°S, are almost identical and are shown as a single curve. The zonal winds are westward at all altitudes. High velocities of $\sim$100 m s^{-1} are measured at cloud-top altitudes decreasing to $\leqslant$1 m s^{-1} (about the measurement uncertainty) at altitudes below $\sim$ 7 km. Below $\sim$ 30 km altitude the zonal wind speed decreases with increasing latitude. Many of the profiles show alternating layers of high and low wind shear. Measurements using anemometers on Veneras 9 and 10 while on the surface gave wind speeds of 0.3 to 1 m s^{-1}, consistent with these results (Avduevsky et al. 1976b).

Figure 7b shows the DLBI measurements of meridional winds from the 4 PV probes (Counselman et al. 1980); the data show the noise level ($\sim$ 0.5 m s^{-1}) superimposed on a complex structure. In the lowest 30 km the meridional velocities are quite small but seem to show significant variations which may be indicative of eddy motions (Schubert et al. 1980a). The small meridional motions (comparable to the measurement uncertainties) make it impossible to determine whether or not a mean Hadley cell exists at these atmospheric levels, or to estimate the meridional heat transport due to such a cell or to the eddies.

While the meridional winds vary significantly in magnitude and direction at altitudes above 40 km, they are equatorward on all 4 PV probes at altitudes between $\sim$ 50 and 55 km. Near the cloud tops (65–70 km altitude) the winds measured by tracking ultraviolet cloud features (Rossow et al. 1980a) give small ($\sim$ 10 m s^{-1}) meridional motions directed toward the poles in each hemisphere. Taken together, these measurements imply the existence of a weak Hadley cell superimposed on the strong zonal winds operating at cloud levels. Since most of the solar radiation absorbed by Venus is absorbed at cloud levels and above, the ability of this Hadley cell to transport heat pole-

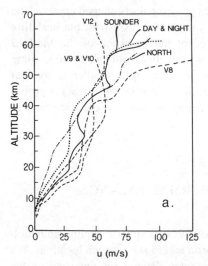
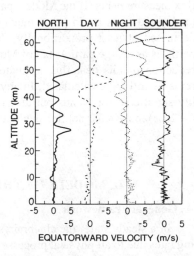

Fig. 7. (a) A comparison of zonal wind velocity profiles from Pioneer Venus and Venera probes. (From Schubert et al. 1980a.) (b) Meridional wind profiles from Pioneer Venus differential long baseline interferometry experiment (DLBI) and Doppler tracking data. (From Counselman et al. 1980.)

ward can be expected to be important in limiting the equator-to-pole temperature contrasts to the modest values observed at these altitudes.

E. Other Processes

Several other processes have been suggested from time to time as contributing in important ways to the thermal balance of Venus's lower atmosphere. For example, it has been suggested that wind-blown dust could cause significant thermal opacity or possibly even contribute to heating the surface through frictional effects (Öpik 1961). The observed virtual absence of aerosols in the lowest 30 km of the atmosphere by the PV nephelometer and LCPS experiments indicates that these effects are negligibly small.

Heat generated by radioactive decay in the interior of the planet has also been suggested as an important energy source for the lower atmosphere (Hansen and Matsushima 1967). However, current estimates (Toksoz et al. 1978) place the flux of internal heat for Venus $\sim$ 2 orders of magnitude less than the solar flux that has been measured to reach the surface.

One potentially important heat transfer process involves the vertical transport due to heats of formation and latent heat release during the vertical cycling of cloud vapor. Assuming that all the particles measured by the LCPS are spherical led Knollenberg and Hunten (1979a, b) to estimate an upward heat flux of 8 W m^{-2} for this effect, $\sim$ 20% of the globally averaged downward net solar flux at cloud levels. However, most of the cloud mass was contained in the Mode 3 particles, and the assumption that these particles are spherical leads to problems in the interpretation of the nephelometer and solar flux measurements. If the Mode 3 particles are irregularly shaped crystals, the total cloud mass could be as much as a factor of 2 less, with the heat flux decreased at least proportionally. The recent interpretation of the Mode 3 particles as a large particle tail of Mode 2 presumably would also lead to some reduction of this heat flux estimate, but probably by a smaller factor that remains still to be calculated. In any case, this heat transfer process is probably significant for Venus but at a level comparable to current uncertainties ($\sim$ 20%) in the solar net flux at cloud levels.

II. MODELS OF THE THERMAL BALANCE

A. General Framework

In a steady state, the algebraic sum of the atmospheric heating and cooling rates due to all physical processes should add to zero at each location. At a given location, the heating and cooling rates depend on the state of the atmosphere: its temperature, pressure, composition, radiation and wind fields. A successful steady-state model for the thermal balance of Venus's lower atmosphere will include the relationships of the heating and cooling rates to the atmospheric state, and show that the atmospheric structure leading to zero net heating or cooling at each location is equal to the observed structure.

For radiative processes, these rates are defined and related to the atmospheric structure as follows. Let the diffuse intensity of radiation at altitude z in the atmosphere

$$I(z, \lambda, \theta, \varphi) = dE/(dt, dA_\perp \, d\Omega \, d\lambda) \tag{1}$$

be defined as the rate at which radiant energy dE between wavelength limits λ and $\lambda + d\lambda$, traveling within a differential cone of solid angle $d\Omega$ about the direction specified by zenith and azimuth angles θ and φ respectively, crosses a unit area $dA_\perp$ oriented perpendicular to the direction of the beam. The units of I are, for example, W m^{-2} ster $^{-1}$ μm^{-1}. The net rate at which sunlight of wavelength λ crosses a horizontal surface of unit area at altitude z is given by the net flux $\mathscr{F}_n(\lambda, z)$, evaluated by integrating over solid angle:

$$\mathscr{F}_n(\lambda, z) = \int_0^\pi \int_0^{2\pi} I(\lambda, z, \theta, \varphi) \cos \theta \sin \theta \, d\varphi d\theta \tag{2}$$
$$+ \cos \theta_0 \, \mathscr{F}_0 \exp \left[-\tau(\lambda, z \, \theta_0) \right].$$

Here the $\cos \theta$ projection factors in each term convert $I(\theta)$, energy per unit area perpendicular to each beam direction, to the energy per unit area of horizontal surface. The second term represents the reduced incident solar flux that survives to altitude z without being absorbed or scattered into the diffuse intensity field. Note that when $I(\theta = 0)$ represents intensity traveling upward this equation gives the net solar flux (directed downward) as negative. (With this convention, θ_0 the zenith angle toward which the incident solar flux is traveling is $\theta_s + \pi$, where θ_s is the local zenith angle of the Sun.)

The magnitude of the net solar flux at any altitude is the amount of solar energy absorbed below that altitude. The difference in net flux at any two levels is the amount of energy absorbed by the intervening layer of atmosphere, and available for heating that layer. Thus, when the atmosphere can be approximated by plane parallel layers (as when the thickness of the atmosphere is much less than the planetary radius), the radiative heating rate $\Gamma(z, \lambda)$ at altitude z for radiation of wavelength λ is given by

$$\Gamma(z, \lambda) = -\frac{1}{\rho(z) \, c_p} \frac{d \mathscr{F}_n(z, \lambda)}{dz} \tag{3}$$

where $\rho(z)$ is local atmospheric density, and c_p is specific heat capacity per unit mass at constant pressure, dependent on atmospheric composition. The local solar heating rate also depends strongly on longitude and latitude, through its dependence on the zenith angle of the Sun.

In addition to being heated by absorbing sunlight, the atmosphere can be cooled (or heated) by emitting and absorbing thermal radiation. The upward and downward intensities of thermal radiation at altitude z are given by

$$I(z, \lambda, \theta) = \int_z^0 B\left[T(z'), \lambda\right] \exp\left\{-\left[\tau(z', \lambda, \theta) - \tau(z, \lambda, \theta)\right]\right\} d\left[\tau(z', \lambda, \theta)\right]$$
$$+ I_s(z=0, \theta, \lambda) \exp\left\{-\left[\tau(0, \lambda, \theta) - \tau(z, \lambda, \theta)\right]\right\} \tag{4a}$$

for $0 \leq \theta < \pi/2$, and

$$I(z, \lambda, \theta) = \int_z^0 B\left[T(z', \lambda)\right] \exp\left\{-\left[\tau(z, \lambda, \theta) - \tau(z', \lambda, \theta)\right]\right\} d\left[\tau(z', \lambda, \theta)\right]$$

for $\pi/2 \leq \theta \leq \pi$. $\tag{4b}$

The upward intensity includes a term for the contribution of emission from the planet's surface that survives to altitude z. Here $B\left[T(z), \lambda\right]$ is the Planck function, giving the thermal emission of a blackbody at temperature T and wavelength λ. Note that the optical depth τ at wavelength λ depends on the slant path through the gas at zenith angle θ. In the thermal infrared, special techniques are often required to deal with the rapid variation of the opacity of the gases with wavelength (see Pollack 1969b).

The net thermal flux is given by integration over solid angle as in Eq. (2), except that I is independent of azimuth angle, and the second term which represented the reduced solar flux is omitted at thermal wavelengths. Positive net thermal flux is directed upward, and the local cooling rate due to any increase of net thermal flux with altitude is given by

$$\Lambda(z, \lambda) = \frac{1}{\rho(z)\,c_p} \frac{d\,\mathscr{F}_n(z, \lambda)}{dz}. \tag{5}$$

This function can also vary with longitude and latitude, due to changes in opacity sources (clouds or trace constituents) or to changes in the temperature structure of the atmosphere.

In principle, if the temperature and pressure structure and the composition of the atmosphere are known, the optical depth between any two levels can be calculated at each wavelength and the thermal radiation field, net flux, and local radiative cooling rate can be computed from Eqs. (4) and (5). Of course, the opacity due to aerosols must also be included. Likewise, if the optical properties and spatial distribution of the aerosols are known in sufficient detail, the solar radiation field can be obtained by solving the equation of radiative transfer including multiple scattering (see e.g., Chandrasekhar 1960); the solar net flux profile and heating rate follow from Eqs. (2) and (3). In practice, comparison of measured and computed flux profiles is useful in providing checks on the gaseous composition and the distribution and nature of aerosols.

At any location on the planet, the difference from zero of the sum of the radiative solar heating and thermal cooling rates, integrated over all wavelengths for an appropriate time, must be balanced by other processes such as atmospheric dynamics to yield a long-term steady state. Because of the difficulties of correctly including a three-dimensional treatment of atmospheric dynamics, some extreme simplifications are commonly made. For example, the problem can be reduced to one dimension by assuming that horizontal motions cause the globally averaged heating and cooling rates to balance at each altitude. By iterative techniques it is possible to find a temperature-pressure structure for the given composition and cloud properties, for which the globally averaged radiative heating and cooling rates balance. This temperature structure is then tested for stability against convective overturning by comparing the temperature gradient in the model with the adiabatic temperature gradient. Wherever the magnitude of the model temperature gradient exceeds the magnitude of the adiabatic gradient, convection is assumed to occur, and the temperature profile of the model in radiative equilibrium is replaced by the adiabatic profile. Because of the efficiency of atmospheric convection, it is assumed that the temperature gradient needs to be only very slightly superadiabatic for convective transport to make up for the excess of the solar heating rate over the radiative thermal cooling rate (see, e.g. Goody and Walker 1972).

Despite their simplicity, such one-dimensional radiative-convective models have been important in developing theories of the thermal balance of Venus's lower atmosphere, as shown in Sec. II B. As indicated in Sec. II C, the extension of these models to include a more realistic treatment of atmospheric dynamics remains a challenge for the future.

B. Historical Developments

Advance beyond the "Earth's twin" notion of the Venus atmosphere, and the beginning of attempts to understand the state of Venus's atmosphere date from the discovery of the high brightness temperature of Venus at centimeter wavelengths (Mayer et al. 1958). Early attempts to explain these observations were hampered by the general uncertainty regarding the state of Venus's lower atmosphere including its composition, surface pressure, and cloud structure, and range widely—from nonthermal ionospheric emissions to heating by wind-blown dust (Öpik 1961) to greenhouse models driven by internal heat or absorbed sunlight (Sagan 1960b). With the Mariner 5 and Venera 4 missions (in October 1967) the surface temperature and pressure as well as the high mixing ratio of carbon dioxide were clearly established, removing any doubt that the high microwave brightness temperatures were due to thermal emission from the hot surface of the planet. Still, several possible explanations of the hot lower atmosphere remained. Because the effective temperature seen from outside the atmosphere was nearly 500 K lower than the surface temperature, it was clear that the thermal opacity of the atmosphere must be great. Debate concerned how to achieve the required large thermal opacity without violating spectroscopic, microwave, or Venera spacecraft constraints on the amount of water present. Clouds could always be assumed to provide a large thermal opacity, but the thicker the clouds were made in the infrared, the thicker they were in the visible and the less sunlight would penetrate to the surface.

After the Venera 4 and Mariner 5 flights, Pollack (1969) used nongray calculations to show that a composition consistent with available constraints ($\sim$ 90% CO_2, 10% N_2, 0.5% H_2O) and a cloud of optical thickness $\sim$ 20–40 in the visible (corresponding to $\sim$ 3 times less in the thermal infrared) would provide sufficient thermal opacity to give the measured effective temperature from the observed temperature structure. Equally significant, Pollack calculated models of the penetration of solar energy through these clouds (assumed to be composed of water spheres) and showed that sufficient sunlight would penetrate and be absorbed beneath the clouds to maintain the high surface temperature.

Nevertheless, the water vapor abundance measured from the Earth was too low to be consistent with clouds composed of water or ice; the composition of the clouds therefore still was unknown (the properties of dust clouds were being actively pursued [Samuelson 1967]), and no firm information on the optical thickness of the clouds in the visible was available. It was still unclear whether the actual cloud structure resembled Pollack's model and whether sufficient sunlight could penetrate the actual clouds to replace even the small amount of thermal flux which was seen to be leaking out. If sufficient sunlight did not reach the surface, it was suggested that very large amounts of dust in the atmosphere (Hansen and Matasushima 1967) might provide an extremely large amount of atmospheric thermal opacity such that the small amount of heating due to radioactive decay in the planet's interior could replace the small escaping thermal flux and support the hot lower atmosphere. But again, the mechanism for raising the dust (particularly after the dust became thick enough to prevent sunlight from heating the lower atmosphere) remained unclear.

At about this time, Goody and Robinson (1966) considered the possibility that the nonuniform solar heating over the planet could cause a deep circulation of the atmosphere even if all the solar heating occurred at cloud levels and none penetrated to the ground. They suggested that the difference in heat input between the subsolar and antisolar points would result in a narrow descending column of atmosphere near the antisolar point and a slowly rising motions over the rest of the globe. Using the Boussinesq approximation (neglecting density variations except for buoyancy terms) they found the circulation would penetrate essentially to the surface and cause an adiabatic temperature gradient throughout the lower atmosphere and a high surface temperature.

The ability of this indirect circulation to lead to an adiabatic profile in the atmosphere was soon questioned by other authors (e.g. Hess 1968). Stone (1975) pointed out that the thermal time constant for the lower atmosphere was sufficiently large so that even for the slow rotation period of Venus, the diurnal temperature difference expected was negligibly small. Thus, the Goody and Robinson circulation would be expected to occur between the equator and the poles rather than the subsolar and antisolar points. Numerical calculations by Kalnay de Rivas (1975) showed that the Goody and Robinson circulation did indeed occur also when she used the Boussinesq approximation. However, when she included the effects of the variation of density (by a factor of $\sim$ 100) between the surface and the cloud levels, the circulation did not penetrate more than a few km ($\sim$ a scale height) below the levels where the solar heating occurred. The lower atmosphere remained at rest and followed an isothermal, not an adiabatic, temperature profile. She concluded that penetration of sunlight to low levels (or some other deep heat source) was required to explain the observed vertical temperature structure of the lower atmosphere.

Of course, even with the deep penetration of sufficient sunlight, mass motions would be required to transport heat meridionally to give the relatively small latitudinal thermal contrasts observed in the face of the large variation in solar heating with latitude. The long thermal time constant of the lower atmosphere which could smooth diurnal effects would not suppress latitudinal temperature differences without atmospheric dynamics, but with deep penetration of sunlight to provide a drive for the circulation, a variety of types of circulation from eddies to a smooth Hadley cell could be conceived which might provide the transport.

C. Recent Models

By the time of the Venera 8 entry probe to Venus, sulfuric acid had been identified as an important constituent of the clouds of Venus (Young and Young 1973; Sill 1972; Pollack et al. 1975). Nevertheless, among the great uncertainties in understanding the thermal balance of Venus's lower atmosphere was the unknown thickness and optical properties of the clouds which would control the penetration of solar energy. The Venera 8 measurements showed for the first time that a significant amount of sunlight penetrated to the surface of the planet. Also, they indicated that the clouds extended down to an altitude of $\sim$ 35 km at the Venera 8 entry site, occupying a substantial vertical extent of the atmosphere below their tops at $\sim$ 70 km altitude. For the first time, solar deposition profiles could be calculated for comparison with direct observations.

Pollack and Young computed a globally averaged, one-dimensional radiative-convective equilibrium model, subject to the new observation constraints; the H_2O vapor pressure was taken to be that in equilibrium with concentrated sulfuric acid cloud particles in the cloud (and 0.3% below them), and the cloud optical depths were consistent with the Venera downward visible flux meaurements. The results of greatly decreasing the water vapor abundance in the altitudes above 35 km suggested that the thermal opacity of the cloud particles would have to play an important role for the greenhouse to work. When 3 μm radius droplets were used instead of the 1 μm radius droplets deduced from polarimetry of the cloud-top region, the radiative temperature profile became steeper than the adiabatic profile at the $\sim$ 15 bar pressure level, leading to a high surface temperature, though not quite as high as the observed level. The authors felt that sufficient uncertainty still existed regarding several atmospheric parameters and that reasonable adjustments could allow the model to produce essentially the observed high surface temperature. For example, useful as the Venera 8 optical measurements were, they measured only downward flux and did not uniquely determine the visible cloud optical depths. Also, uncertainties in the particle size distribution deep in the clouds still produced large uncertainties in the thermal opacity of the clouds. (See Tomasko et al. [1977, 1979b] for a review of the uncertainties concerning the thermal balance of the Venus atmosphere at this time.)

Many of these uncertainties have been considerably reduced by the Venera 9-12 and Pioneer Venus missions. Data from the nephelometers, upward and downward viewing optical radiometers, and the cloud particle size spectrometer have indicated the global similarities in the cloud structure with a well-defined lower boundary at 48 km altitude from measurements at 8 additional entry sites. The particle size distributions have been precisely measured at one site and strongly constrained by the nephelometers at all the sites. The net solar flux profile has been measured at several sites, and some estimates of the globally averaged profile are beginning to appear. Measurements of the water vapor abundance beneath the clouds have been made in a way that is not susceptible to contamination by the liquid water content of the cloud droplets. Direct measurements of the abundance of trace constituents which can also provide thermal opacity (such as SO_2) have been made in the lower atmosphere by several techniques. Equally important, very accurate profiles of the temperature and thus the static stability have also been determined at several sites.

One-dimensional globally averaged radiative-convective models have been computed incorporating these new observational constraints. Even in the presence of some simplifying assumptions, early calculations (Tomasko et al. 1979) showed that the stably stratified layer observed under clouds (from $\sim$ 30 to 50 km altitude) and the high surface temperature could be reproduced by models incorporating the solar net flux profile and the Venera water vapor mixing ratio ($\sim$ 2 $\times$ 10^{-4}). When a water vapor mixing ratio larger by a factor of 10 was used below the clouds, the profile became adiabatic in the clouds and remained so to the ground.

Calculations have been made by Pollack, et al. (1980a) that include opacity due to SO_2, CO, HCl, as well as some additional pressure-induced transitions in CO_2. The water vapor mixing ratio in their base case varied from 2 $\times$ 10^{-5} at the surface to 2 $\times$ 10^{-4} in the clouds before rapidly decreasing to 2 $\times$ 10^{-6} at the cloud tops. Models with the nominal and high solar net

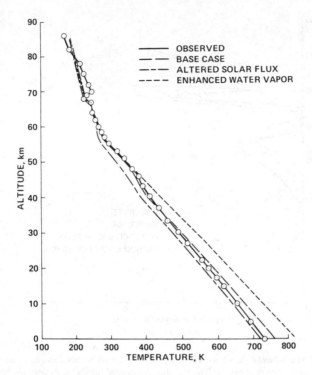

Fig. 8. Comparison of the observed temperature structure in Venus's lower atmosphere with that of several one-dimensional radiative-convective models. The "base case" uses the Venera 11 water vapor profile and the PV "high solar" net flux profile. The "altered solar flux" model uses the somewhat lower "nominal solar" net flux profile from PV. The "enhanced water vapor" model uses the water vapor measured by Oyama et al. (1980a). (From Pollack et al. 1980a).

flux profile from PV and the Venera 11 water profile yield surface temperatures of 720 K and 760 K, bracketing the observed surface temperature of 730 K (see Fig. 8). Furthermore, these models contain an extended stable layer beneath the clouds and have static stability profiles qualitatively similar to the observed profiles (Fig. 9). When a water vapor mixing ratio corresponding to the PV gas chromatograph measurements (a factor ~ 50 greater) is used, this stable layer disappears from the model and the surface temperature increases to ~ 820 K (nearly 100 K higher than observed).

In the base case model, the second stable region observed between 15 and 20 km altitude does not form, but the authors point out that the shift of the blackbody curve to shorter wavelengths at the high atmospheric temperatures in the lowest 30 km leads to an increase in the net thermal flux due to a relative window in the 2.1 to 2.6 μm region. As a result, the net thermal radiative flux carried by the adiabatic profile is nearly as large as the solar net

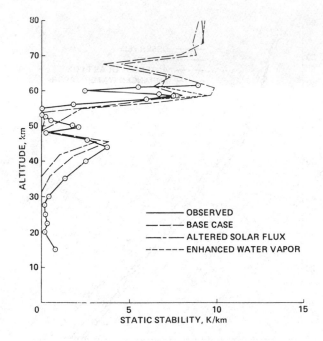

Fig. 9. Comparison between the observed stability structure of Venus's lower atmosphere and
that of the models shown in Fig. 8. The stability parameter equals the difference between the
adiabatic and actual lapse rates. At all altitudes from 0 to 80 km, where a given theoretical
curve is not explicitly illustrated, the stability parameter equals zero. (From Pollack et al.
1980*a*).

flux. Small changes in the model might lead to the reappearance of a stable
region at low altitudes.

Considering the simplifications inherent in a one-dimensional radiative-
convective model, the agreement with the observed temperature profiles is
remarkable. Nevertheless, significant uncertainties remain even in these
simplified models. For example, the total thermal flux emerging from the top
of the model can exceed the observed flux by a larger factor (nearly ⅓) when
the nominal cloud model is used and no Mode 0 aerosols are added. If
sufficient mass ($\sim$0.1 mg cm^{-2}) of Mode 0 aerosols is included above the
middle cloud to provide the required thermal opacity ($\sim$ 1 at 10 μm), the
diameter of these particles must be $\lesssim$ 0.07 μm to give less than a scattering
optical depth of unity near 0.32 μm in the ultraviolet (Tomasko et al. 1980*b*).
Their number density would be $\sim$ 2 $\times$ 10^5 cm^{-3} if they were distributed over
a 30 km range of altitudes above 57 km. The time for coagulation to remove
these particles (convert them to significantly larger sizes) would be several
hours. It remains to be seen whether mechanisms can be proposed to replace

this aerosol mass in an equally short time. On the other hand, it may be that the optical depth of Mode 2 particles has been underestimated in the nominal cloud model (as suggested in Sec. I) in a way that might provide the bulk of the missing opacity by the Mode 2 particles alone. Interestingly enough, Pollack et al. (1980a), remark that the existence and nature of the stable layer found just beneath the clouds in their models depends on the amount and distribution of the Mode 0 absorbers. The infrared opacity of the clouds is apparently still a matter for further investigation.

Also somewhat unsettled is the treatment of the thermal opacity produced by the gases in the high temperature, high pressure lower atmosphere of Venus. Pollack et al. (1980a) show much of the thermal flux at altitudes $\leqslant 30$ km being carried at rather short wavelengths (< 3 μm). While these authors relied in part on theoretical calculations in an attempt to include a fairly complete set of transitions, numerous transitions arising from excited states (hot bands) and from very weak overtone and combination bands, undoubtedly occur in this spectral region, even though they have not yet been observed in the laboratory. Thus, it is difficult to be sure of the completeness and accuracy of the gaseous opacities used in the model computations.

If the gaseous opacities are inaccurate, they are probably somewhat too low due to the limited number of transitions included, rather than too high. Thus, while the measured thermal net fluxes might well have been a bit lower than computed fluxes, it is surprising to find the first measured thermal fluxes significantly greater than the computed fluxes in the lower atmosphere. The "corrected" measured fluxes at low latitudes given by Revercomb et al. (1982) are now in agreement with fluxes computed using water abundances in the range of values measured at low latitudes by the Venera 11 and 12 probes. If it were available, a separate measurement of lower water abundance at high latitudes, as suggested by the revised North probe flux profile (see Fig. 5), would be an important confirmation of both the revised flux profiles and the opacity tables currently in use.

It could be argued that since the thermal net flux produced by the adiabatic temperature profile is close to the solar net flux in the models and a slightly stable region is seen in the lowest portions of the PV descent profiles, the thermal gaseous opacities are close to correct in the models. Small adjustments, for example, in the average solar flux profile or even a slight decrease in the thermal opacities would produce a marginally stable temperature profile as observed at low altitude ($\sim$ 15 km). However, no effects due to global circulation have yet been included in the model other than assuming that horizontal motions distribute the heat as required to produce the small horizontal contrasts observed. On Earth, global circulation (aided by latent heat effects) produces a mean lapse rate that is only ⅔ of the adiabatic value despite the presence of convection (Stone 1975). It may be unreasonable to expect the one-dimensional models to reproduce all the features in the vertical temperature profile.

In this regard we should comment briefly on the state of including the dynamics in a comprehensive picture of the thermal balance of Venus's lower atmosphere. (For a more complete discussion, the reader is referred to Schubert et al. [1980a] and Chapter 21 in this book by Schubert.) As outlined by Stone (1975), comparison of the radiative and dynamical time constants as functions of height in the atmosphere can be instructive. He points out that below ~ 56 km altitude, the dynamical time constant is much less than the radiative time constant, and the dynamics are expected to determine the thermal structure of the atmosphere. He also points out that the radiative time constant is greater even than the length of one Venus day, so that the thermal inertia of Venus's lower atmosphere is sufficient to eliminate diurnal temperature variations even without zonal winds. Meridional heat transport by mass motions would be expected to occur, and should be driven by the strong latitudinal variations in solar heating. These motions could take the form of a slow Hadley cell with rising motions at the equator, flow toward the poles, and a deeper slow return flow. Substantial heat flux rates can be carried by even very slow motions in the lower atmosphere due to the great density; the predicted wind speeds near the surface are $\lesssim 1$ m s^{-1}. Eddy motions could also transport the heat. Stone points out, though, that a Hadley type circulation transports heat poleward only if the poleward branch is at a higher potential temperature than the equatorward return flow at lower levels. In other words, this type of motion transports heat poleward only if the temperature gradient is slightly subadiabatic. It seems possible that improvement in the thermal opacity tables for the lower Venus atmosphere will increase the opacity and make the one-dimensional models no closer to the slightly stable measured temperature profile. The details of the modifications to the temperature field caused by the dynamics will have to await a successful global circulation model, but such modifications could conceivably be responsible for the low-altitude stable layer observed.

Substantial work remains to be done before the global circulation is modeled, despite the great body of observational data recently collected. For example, many features of the wind field measured for Venus are qualitatively as expected, including the low wind speeds observed at depth, and the direction and magnitude of meridional flows in the clouds. However, the most striking feature of the circulation remains the rapid zonal flow (~ 100 m s^{-1}) near the cloud tops. This dramatic aspect of the flow still remains to be understood and reproduced in a detailed calculation.

In summary, instruments on entry probes have now measured many of the properties of the lower atmosphere of Venus which control the transfer of solar and thermal radiation. One-dimensional radiative-convective models indicate that sufficient sunlight penetrates to low levels and is trapped by the large infrared opacity of the atmosphere to yield essentially the high temperatures observed. Nevertheless, several features in the temperature structure, such as the relatively small latitudinal contrasts and the existence of a deep stable region, indicate that atmospheric motions also play an important role. While we know that atmospheric dynamics is responsible for some of these effects, the complexities of the processes are such that they have yet to be modeled successfully in detail.

Acknowledgments. It is a pleasure to acknowledge the benefit of many enlightening discussions with numerous colleagues from the Pioneer Venus community during the course of my participation in that program. This work has been supported by a grant from the National Aeronautics and Space Administration.

19. SOLAR SCATTERED RADIATION MEASUREMENTS BY VENUS PROBES

A. P. EKONOMOV, YU. M. GOLOVIN,
V. I. MOROZ, and B. YE. MOSHKIN
USSR Academy of Sciences

The fluxes of scattered sunlight in the deep layers of the atmosphere of Venus were measured in several series of experiments. Results show the structure of the aerosol medium through photometric data. Absorption of solar energy by the atmosphere and surface of Venus is also discussed.

Between 1972 and 1978 several experiments were made by the Venera spacecraft measuring the fluxes of solar radiation, that penetrate the deep layers of Venus's atmosphere. The purpose of this chapter is to survey the information acquired, which is important for the problems of thermal balance and structure of the aerosol medium.

The need to measure the fluxes of scattered sunlight in the deep layers of Venus's atmosphere was evident before the first space flights to the planet. In fact, the early hypothesis of the runaway greenhouse effect capable of maintaining the high temperatures at the surface of Venus required two conditions to be fulfilled: the atmosphere had to be (a) very opaque in the infrared, and (b) partially transparent to solar radiation all the way down to the surface. The possibility of the first condition in an atmosphere rich in CO_2 was obvious; concerning the second condition, measurements were indispensable.

I. BASIC SPECIFICATIONS OF THE INSTRUMENTS

Veneras 4, 5, 6, 7 (1967–1970) entered the atmosphere of the night side of the planet. This choice of landing site was dictated by the method of data transmission used for this first generation of Venera-type spacecraft, direct transmission from the probe to Earth. Nonetheless, it was possible to land the last of this group, Venera 8 in 1972, near the terminator in the illuminated portion of the planet. A photometer was included in the Venera 8 payload, and the first data were obtained concerning the flux of sunlight in the deep layers of the atmosphere.

This first experiment (Avduevsky et al. 1973a,b) was relatively simple. Measurements were made in a single and comparatively broad wavelength band (0.52–0.72), and radiation was received only from the upper hemisphere. The proximity to the terminator caused a large error in the value of the air mass. Nevertheless, a qualitatively new result was obtained, that an amount on the order of 0.015 of the incident solar flux reaches the surface of the planet. This estimate, refined in subsequent experiments, became the main argument in favor of the greenhouse hypothesis.

Veneras 9 and 10 (1975) and Veneras 11 and 12 (1978) were greatly improved, and far more complex optical experiments were carried out. The two identical photometers on Veneras 9 and 10 (Avduevsky et al. 1976c,d; Ekonomov et al. 1978) received radiation in five broad (0.1–0.3 μm) wavelength bands, covering a region from 0.5 to 1.1 μm. The radiation was recorded from above and below the descending probes in relatively narrow fields of view as well as from the entire upper hemisphere. Measurements were made all the way down to the surface. Moreover, on Veneras 9 and 10 a narrowband photometer operated in the 40–60 km interval of altitudes with three filters $\sim$ 0.005 μm wide, one of which was centered on the CO_2 band at 0.78 μm, another on the H_2O band at 0.82 μm and the third in a nearby "background" region (Moroz et al. 1976).

The next stage of measurements was completed in 1978 by Veneras 11 and 12 (Ekonomov et al. 1979; Moroz et al. 1979a,b, 1980a). The instrument recorded a continuous scan of the radiation spectrum in the range 0.45 to 1.2 μm ($\Delta\lambda \simeq 0.03$ μm) from the zenith plus an angular scan in the vertical plane for four broad spectral intervals. Investigations of the solar radiation were also carried out on Pioneer Venus (Tomasko et al. 1979a,b, 1980a; Suomi et al. 1979,1980; Blamont and Ragent 1979; Ragent and Blamont 1979,1980), but there were no spectral measurements.

Table I gives a summary of the quantitative characteristics and performance conditions of all the Soviet and U.S. photometric experiments *in situ* on Venus. This table also presents results such as the normalized illumination at several levels and some conclusions as to the structure of the clouds.

The basic design of all the Soviet photometers except the narrowband instruments on board Veneras 9 and 10 was generally identical. The optical unit was mounted on the outside of the spacecraft, the electronic unit on the inside. The optical unit had walls which could withstand 100 atm pressure and thermal insulation to assure proper operation for the entire descent from 60–64 km altitude to the surface. In other words, the optical unit was a miniature and almost independent probe. The radiation inside the optical unit

TABLE I

Photometric Measurements in Deep Layers of the Venusian Atmosphere[a]

Space Vehicle	Zenith Distance of Sun	Altitude Range (km)	Wavelength Range (μm)	Spectral Resolution (μm)	Viewing Directions[b]	Angular Field of View[c]	Normalized Illumination[d] 60 km	50 km	0 km	Conclusions on Cloud Structure	References
V8	84.5±2.5	50 0	0.52-0.72	≃0.2	0	~60°	—	0.15	0.015	e	Avduevsky et al. 1973a,b.
V9	33°	64 0	0.5-0.56	0.06	0, 157of	40°,120of	0.8	0.4g	0.05	Lower boundary	Avduevsky et al. 1976a,b.
			0.52-0.67	0.15							
V10	27°	63 0	0.62-0.70	0.08						~49 km	Ekonomov et al. 1973.
			0.70-0.79	0.09							
			0.76-0.106	0.30							
V9	33°	64 0	0.78	0.005	45	~15°	0.8	0.5	—	Lower boundary ~50 km; local condensations	Moroz et al. 1976.
			0.80								
V10	27°	63 0	0.82								
Pioneer	65.7°	62 0	0.4-1.0	0.6	27,60,83,102,142	5°	0.24	0.17h	0.02	Lower boundary ~48 km; several layersi	Tomasko et al. 1975a,b.
Large probe			0.4-1.8	1.4	27,60,83,102,142						
			0.63	0.07	60,142						
PL	65.7°	66 0	0.365	0.04	~90	11.5°	0.16	0.093	?	Lower boundary ~48 km,i	Ragent & Blamont 1979, 1980.
PD	79.9°	64 0	0.365	0.04	~90	11.5°	0.06	0.03	?j	several layers	Blamont & Ragent 1979.
PD	79.9°	64 0	0.530	0.07	~90	11.5°	0.11	0.07	?		
V11	20°	64 0	0.45-1.2k	0.22-0.035	0	15°	0.90	0.55	0.08m	Lower boundary	Moroz et al. 1979ab, 1980;
V12	20°	64 0	0.4-0.6	0.2	scan	15°				~49 km, 3 cloud layersj	Ekonomov et al. 1979.
			0.6-0.8	0.2	0-360°						
			0.8-1.3	0.5							
			1.1-1.6l	0.6							

Notes to Table I

[a]Table taken from the survey by Moroz (1981).

[b]Angle between the optical axis and the vertical (0° is axis oriented toward the zenith).

[c]Full width at half maximum response.

[d]Downward flux F at the indicated altitudes divided by the solar flux F_s (in the same spectral region) evaluated outside the atmosphere across a surface perpendicular to the solar beam. For measurements using narrow fields of view, the radiation field is assumed isotropic in the evaluation of F.

[e]Apparently the measurements began at the lower cloud boundary, and at 35 km a change was noted in the altitude derivative dF/dz, possibly related to the presence of a more dense aerosol medium below the clouds discovered in other experiments.

[f]One wide field of view (120°) centered at the zenith and viewing essentially the entire upper hemisphere, and two narrower fields of view (40°) centered at the zenith and 23° from the nadir.

[g]F/F_s for a channel from 0.62 to 0.70 μm.

[h]F/F_s for a channel at 0.63 μm.

[i]A laminar structure revealed by variations of dF/dz.

[j]No reliable estimates of F at the surface, due to the presence of additional peaks in the infrared range in the transmission curve of the filters.

[k]Spectral measurements with continuous spectral scan.

[l]Measurements with wide filters and continuous angular scan.

[m]F/F_s for $\lambda = 0.7$ μm.

---

was transmitted along a lightguide 1 m long. This enabled the optical unit to be mounted near the hull of the main spacecraft, improved the thermal regime, and at the same time reduced obscuration of the field of view by the probe structure.

The typical relative measurement error in the photometric experiments was from 1 to 10%; the absolute error was ∼ 20%. The calibration problems are discussed in Golovin et al. (1977).

Of great interest from the viewpoint of the thermal regime and dynamics of the atmosphere is the net flux ΔF, downward minus upward flux. The solar heating is

$$\frac{dQ}{dt} = \frac{d}{dH} (\Delta F) \cdot \tag{1}$$

To determine dQ/dt with meaningful accuracy it is necessary to measure $d(\Delta F)/dH$ with a high enough precision, which is difficult because $d(\Delta F)dH$ is the derivative of a small difference between two large quantities. On Veneras 9 and 10, separate measurement channels were used to measure the upward and downward fluxes and so the data on the differential fluxes were very poor. To solve this problem, a continuous angular sweep was used on Veneras 11 and 12; this is an essential improvement because the radiation from above and below is gathered by a single receiver. Yet even in this case the measurement precision for the differential fluxes was not satisfactory.

II. MEASUREMENT RESULTS: VERTICAL PROFILES OF THE ILLUMINATION, INTENSITY, AND NET FLUXES

Let us define the quantities measured in the photometric experiments. With a hemispherical or nearly hemispherical direction pattern oriented upwards, the illumination of a horizontal surface (downward flux $F\downarrow$) is measured. With a narrow direction pattern, the intensity from the top $I\downarrow$, from the bottom $I\uparrow$, or as a function of the zenith angle $I(\Theta)$ is measured. In an optically thick atmosphere (see e.g. Sobolev 1972), the angular distribution of intensity is well approximated by

$$I(\Theta) = I\downarrow (1 - b + b \cos \Theta) \cdot \qquad (2)$$

In this case, the fluxes and intensities are related by

$$F\downarrow = 2\pi \int_{0}^{\frac{\pi}{2}} I(\Theta) \sin\Theta \cdot \cos \Theta \, d \Theta = \pi I\downarrow \frac{3 - b}{3} \qquad (3)$$

$$F\uparrow = \pi I\downarrow \frac{3 - 5b}{3} = \pi I\uparrow \frac{3 - 5b}{3(1 - 2b)} \cdot \qquad (4)$$

The net flux is

$$\Delta F = F\downarrow - F\uparrow = \frac{4}{3} \pi b I\downarrow = \frac{2}{3} \pi \Delta I \qquad (5)$$

where $\Delta I = I\downarrow - I\uparrow = 2bI\downarrow$. It is convenient to introduce a quantity called the internal albedo

$$A = \frac{F\uparrow}{F\downarrow} = \frac{3 - 5b}{3 - b} \cdot \qquad (6)$$

The measurements by Veneras 11 and 12 revealed that the angular distribution of intensity in the atmosphere of Venus closely corresponded to the law of Eq. (2). Therefore the use of the above equations for the transition from intensities to fluxes is justifiable.

Figures 1–5 show the photometric profiles obtained in each experiment. On Venera 8, the measurements spanned an altitude interval from 50 km to the surface; on the other probes measurements began at 62 to 64 km and went down to the surface. The overall data for Venera 8 contained 27 point measurements. On Veneras 11 and 12 a spectrum from the zenith and a complete angular sweep cycle (with four filters) was transmitted every 10 s ($\sim$ 300 points in 10 s). In all $\sim$ 500 spectra and angular scans were obtained from Veneras 11 and 12.

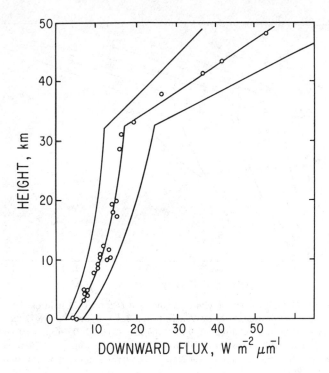

Fig. 1. Flux $F\downarrow$ as a function of altitude from measurements by Venera 8 with $F\downarrow$ (in W m^{-2} μm) along the abscissa, altitude along the ordinate. Effective wavelength is 0.63 μm (Avduevsky et al. 1973b), with modification of the scale of $F\downarrow$ to W m^{-2} μm^{-1} instead of W m^{-2}.

III. STRUCTURE OF THE AEROSOL MEDIUM FROM PHOTOMETRIC DATA

All the measurements started when the probes were already within the cloud layer. The first measurement points by Veneras 9, 10, 11, 12, corresponding to the maximum altitudes, indicated an irregular behavior in certain cases, possibly related to effects of direct sunlight or to interruptions in the clouds. At $\sim$ 60 km altitude the effects of direct solar radiation are imperceptible, and all the measured radiation is scattered. A rough estimate of the optical depth at this altitude is $\tau \lesssim 5$.

The profile obtained by Venera 8 (Fig. 1) has a break at $\sim$ 33 km altitude; before new data was received this height was assumed to correspond to the lower boundary of a dense cloud layer. None of the profiles from the later experiments have any sharp feature at 33 km altitude; instead, they display a

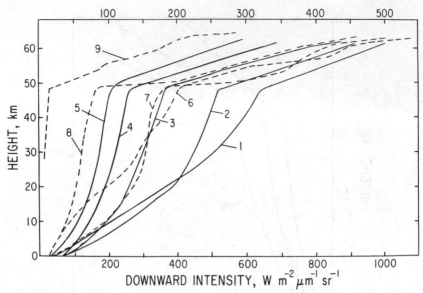

Fig. 2. Intensity $I{\downarrow}$ from the zenith with a narrow (20° at level 0.5) field of view and wide filters, according to measurements by Venera 10 (curves 1–5) and Venera 12 (curves 6–9): (1) wavelength range 0.5–0.56 μm; (2) 0.52–0.67 μm; (3) 0.62–0.7 μm; (4) 0.7–0.79 μm; (5) 0.76–1.06 μm (Ekonomov et al. 1978); (6) 0.45–0.55 μm; (7) 0.56–0.68 μm; (8) 0.9–1.14 μm; (9) 1.15–1.55 μm. Lower scale is for curves 6 and 7, upper for curves 8 and 9 (Moroz et al. 1981).

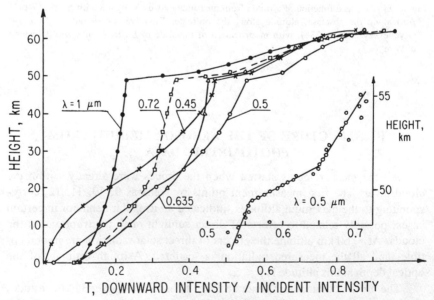

Fig. 3. $I{\downarrow}$ from the zenith according to measurements by Venera 11. Spectral measurements have resolution 0.02–0.03 μm. Intensity outside the atmosphere is taken as the unit of measurement (Ekonomov et al. 1979).

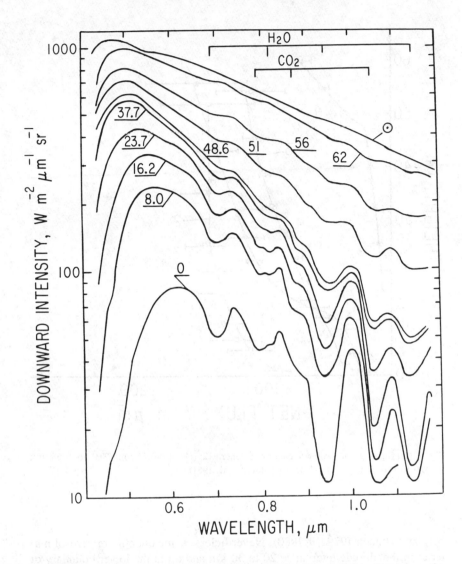

Fig. 4. Spectra of sunlight from the zenith region according to Venera 11 data. Numbers on the curves represent altitude above the surface in km (Ekonomov et al. 1979).

break at altitudes of 48 to 49 km. It is now clear that the lower boundary of the dense clouds is at these higher altitudes. The same conclusion is supported by experiments with nephelometers (Marov et al. 1976a; Blamont and Ragent 1979; Ragent and Blamont 1979, 1980) and with a particle counter (Knollen-

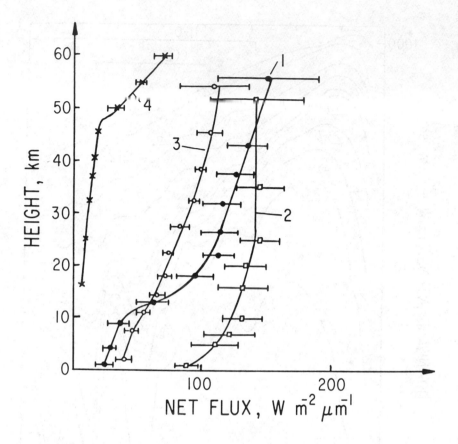

Fig. 5. Net fluxes ΔF according to data from Venera 12: (1) 0.45–0.55 μm; (2) 0.56–0.68 μm; (3) 0.9–1.14 μm; (4) 1.15–1.55 μm (Moroz et al. 1981).

berg and Hunten 1979a,b, 1980). Nevertheless, some quantity of aerosol material in the altitude interval ~ 20 to 30 km and up to the lower boundary of the main clouds was detected in almost all measurements after Venera 8, though the absolute magnitudes of the scattering coefficients were 1 to 2 orders of magnitudes less than that within the main cloud layers. For Venera 8, an anomalously high density of aerosol in this region was detected, possibly related to effects of condensation at the terminator. All the photometric experiments indicate that below ~ 10 km the atmosphere of Venus is practically free of aerosol (Avduevsky et al. 1973a,b; Golovin et al. 1978; Ekonomov et al. 1979).

The structure of the main cloud layer located above 48 km is also somewhat variable. On the profiles from Venera 11 and 12 data, lower B, middle

C, and upper D layers are identified. The derivative dI/dz is a maximum in layer B and a minimum in layer C. There is no such structural division on the profiles of Veneras 9 and 10. From the observed profiles we can obtain quantitative characteristics of the scattering medium such as optical thickness $\Delta\tau$ of individual layers of the clouds or haze, mean volume coefficient of scattering α in them, and single scattering albedo ω_0. These quantities are determined with accuracy down to a factor of $1-g$, where g is a factor of anisotropy equal to 0 for an isotropic or Rayleigh scattering and to 0.6 to 0.9 for large particle scattering ($2\pi r/\lambda > 1$). Estimates of this factor are taken from other data (ground observations or *in situ* nephelometric observations). The laws of light propagation in a medium with multiple scattering are used for the interpretation. Several different methods are employed.

(a) One such method is solution of the direct problem by calculation of a series of model profiles; from such profiles α, ω_0 are chosen from the models that most nearly reproduce the measurements. A solution of the direct problem is obtained either by means of asymptotic formulas (Lukashevich et al. 1974; Moroz et al. 1976; Germogenova et al. 1977; Ekonomov et al. 1978,1979; Dlugach and Yanovitzky 1978), or by the method of addition of layers in a two-stream approximation (Lacis and Hansen 1974a,b; Moroz et al. 1979), or by the method of doubling and addition (Tomasko et al. 1979a,b, 1980a).

(b) Another method is solution of the inverse problem (Ustinov 1977; Golovin et al. 1981; Moroz et al. 1981; Golovin and Ustinov 1981), which most easily produces estimates of α and ω_0

$$\alpha(1 - \omega_0) = \frac{1}{3(I{\downarrow} - I{\uparrow})}\frac{d(\Delta I)}{dH} \tag{7}$$

$$\alpha(1 - g) = \frac{1}{\Delta I}\frac{d}{dH}(I{\downarrow} + I{\uparrow}) \cdot \tag{8}$$

Dividing Eq. (7) by Eq. (8) we obtain

$$\frac{1 - \omega_0}{1 - g} = \frac{d(\Delta I)}{d(I{\downarrow} + I{\uparrow})}\frac{\Delta I}{3(I{\downarrow} + I{\uparrow})} \cdot \tag{9}$$

Equations (7) and (8) make possible the determination of α and $1-\omega_0$ with accuracy down to the factor $1-g$, directly from the altitude profiles of $I{\downarrow}$ and $I{\uparrow}$.

The optical thickness of the main cloud layer (in the interval 48 to 49 km) is practically independent of the wavelength, indicating scattering by large particles. Other data from the nephelometers and particle counter agree with this, indicating that the main contribution to scattering comes from particles of sizes on the order of 1 μm.

The values of the optical thickness of the clouds according to data of Veneras 9, 10, 11, and 12 are 20, 25, 29, 38 respectively where $\Delta\tau$ is given for the 48 to 62 km interval and g = 0.7 is assumed (Ekonomov et al. 1978; Golovin et al. 1981). The optical thickness of the clouds at the Pioneer Venus Large probe entry site was estimated as $\sim$ 25 by Tomasko et al. (1980a). Figure 6 shows the profile of the quantity $3\alpha\ (1-g)$ in the cloud layer, according to measurements of $I\downarrow$ and $I\uparrow$ by Venera 12. Optical thicknesses of 25 to 35 are typical for the cloud layer of Venus.

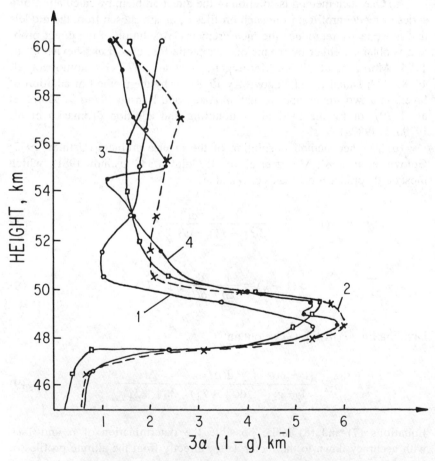

Fig. 6. Profile of the quantity $3\alpha\,(1-g)$ in the cloud layer according to measurements of $I\downarrow$ and $I\uparrow$ by Venera 12 (Golovin et al. 1981). Numbers on the curves indicate the spectral intervals which are the same as for Fig. 5.

TABLE II

**Optical Thickness of the Aerosol Component
of the Layer Below the Clouds[a]**

Spacecraft	λ_e (μm)	ΔH (km)	τ_a	ΔH (km)	τ_a
Venera 9	0.535		1.0		<2.8
	0.6		0.7		<1.77
	0.67	45-25	0.5	25-0	<1.14
	0.76		0.37		<0.68
	0.87		0.3		<0.39
Venera 10	0.533		1.66		<3
	0.58		1.1		<2.26
	0.67	45-25	0.87	25-0	<1.2
	0.77		0.55		<0.68
	0.87		0.4		<0.42
Venera 11	0.49		1.27	32-22	1.83
	0.715	42-32	0.47	32-10	1.17
	1.0		0.13	32-25	0.067
Venera 12	0.49		1.53	30-20	<0.25
	0.64	45-30	0.83	30-0	<0.5
	1.0		0.4	30-20	<0.15

[a]Based on data from Golovin et al. (1981) and Golovin and Ustinov (1981).

Aerosol scattering contributes to the Rayleigh scattering below the clouds as well (Figs. 7 and 8, and Table II). The overall aerosol optical thickness here is on the order of unity, and in contrast with the main cloud layer depends greatly on the wavelength. This suggests scattering by smaller particles (with radius on the order of 0.1 μm); the corresponding estimates of dimensions and concentration are given in Tables III and IV. An important parameter in calculating the radiation field in the atmosphere is the surface albedo, which was measured by Veneras 9, 10, and 12, giving a value of $A_s \simeq 0.1$.

IV. ABSORPTION OF SOLAR ENERGY BY THE ATMOSPHERE AND SURFACE OF VENUS

The first estimates of the absorption of solar energy by the atmosphere and surface were made from photometric measurements by the probes of Veneras 9 and 10 (Moshkin et al. 1978). During the day in the equatorial region the surface absorbs ~3%, and the atmosphere ~18%, of the solar

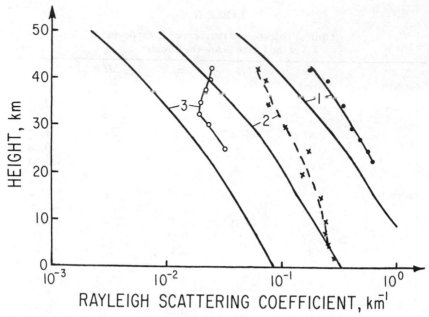

Fig. 7. Altitude dependence of the coefficients of Rayleigh scattering σ_R (solid curve) and of the parameter α $(1-g)$ (dashed curve), obtained from Venera 11 data for three effective wavelengths: (1) 0.49; (2) 0.715; and (3) 1.0 μm (Golovin et al. 1981).

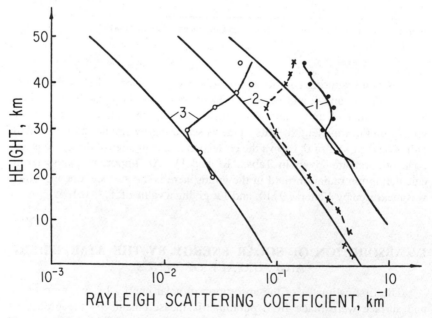

Fig. 8. The same as for Fig. 7, from the data of the Venera 12 probe for (1) 0.49, (2) 0.64, and (3) 1.0 μm (Golovin et al. 1981).

TABLE III

Dimensions of the Particles of the Aerosol Below the Clouds[a]

H (km)	Veneras 9 and 10 r (μm)	Venera 11 r (μm)	Venera 12 r (μm)
45	≤0.1	—	0.2-0.3
40	≤0.1	0.08-0.1	≈0.2
35	≤0.1	0.05-0.07	≈0.1
25	≤0.1	≤0.04	—

[a]From Golovin and Ustinov (1981).

energy arriving at the upper boundary of the atmosphere. This result agrees well with the hypothesis of the greenhouse effect in Venus's atmosphere. To obtain these estimates from the filter data it was necessary to calculate the smoothed-out spectral composition of the radiation. The first recordings of the spectrum at the surface were obtained by Venera 11.

The amount of energy absorbed in the layer of atmosphere with upper boundary at altitude H_1 and lower boundary at H_2 is

$$Q_{1,2} = \Delta F_1 - \Delta F_2 \tag{10}$$

the influx of heat due to absorption of solar energy in this particular spectral range. In the 0.45 to 0.55 μm range the maximum amount of energy is absorbed in a layer of ~ 10 to 35 km, where $Q \simeq 10$ W m^{-2}. In the layer from 10 km to the surface the absorption is several times less (Fig. 5). There is little or no absorption in the clouds, at least in the lower (48–51 km) and middle (51–58 km) portions of the cloud layer. Despite the large optical thickness of the clouds ($\Delta\tau = 29$ and 38 for the landing sites of Veneras 11 and 12), the net flux is virtually unchanged in the 48 to 58 km interval of altitudes. The rigid upper limit here is $Q < 7$ W m^{-2}. In the spectral range 0.45 to 0.55 μm, and incidentally also in that of 0.56 to 0.68 μm, the attenuation of radiation by the aerosol of the cloud layer is mainly determined by the scattering, which is essentially conservative scattering.

TABLE IV

Relative Content by Mass $\bar{\rho}_a$ and the Concentration N of Aerosol Particles at the Landing Sites of Venera 9, 10, 11, 12[a] probes[b]

	Venera 9	Venera 10	Venera 11	Venera 12
N (cm^{-3})	6400	7750	9260	8400
$\bar{\rho}_a \cdot 10^{-8}$	0.4	0.485	0.58	0.52

[a]$H = 35$ km, $r = 0.1$ μm, $m = 1.5 - 0i$, $\rho = 2$g cm^{-3}.
[b]From Golovin and Ustinov (1981).

For the spectral range 0.56 to 0.68 μm there is characteristically a very weak absorption or none at all, at altitudes $H > 15$ km. At lower altitudes the absorption is considerable; according to Venera 12 data, ~ 5 W m^{-2} is absorbed. The absorption takes place in the H_2O and CO_2 bands, arriving in the long-wave branch of this filter.

A more intense absorption is observed in the spectral interval 0.9 to 1.14 μm, especially in the atmosphere below the clouds ($Q \approx 15$ W m^{-2}). Here the H_2O and CO_2 bands are also absorbed. The absorption is slight in the clouds. Solar energy in the spectral interval 1.15 to 1.55 μm hardly reaches the surface of the planet at all. Most of the energy (~ 15 W m^{-2}) is assimilated in the cloud layer by absorption in the H_2O and CO_2 bands.

Table V gives data on solar energy fluxes, absorbed at various altitudes and averaged over the spectral bands. Ustinov et al. (1979) calculated a model for the solar heating based on results from Veneras 9 and 10, with the addition of a series of supplementary data. This model should be reviewed in the light of new measurements, though the qualitative nature of the curve shown in Fig. 9 (for the subsolar point) can hardly have changed. Indeed, Ustinov's curve for the solar net flux profile Q' at the subsolar point (given in terms of the solar constant at Venus) is in good agreement with the Pioneer Venus solar flux measurements scaled to the subsolar point using a forward scattering model (Tomasko et al. 1980a). Despite the small absorption in the visible range, maximum solar heating can be expected to occur at the cloud layer (due to the ultraviolet and infrared segments of the spectrum that are only partially covered in our measurements).

Using the altitude dependence of the net flux, the single scattering albedo was determined at various altitudes by a formula similar to Eq. (7). Table VI lists the values of ω_0 for the spectral bands 0.45 to 0.55 and 0.56 to 0.68 μm.

TABLE V

Solar Heating (W m^{-2}) at Various Heights in Four Spectral Ranges[a]

Altitude (km)	Spectral	Ranges (μm)		
	0.45-0.55	0.56-0.68	0.9-1.14	1.15-1.55
Venera 11				
60-48	<7	<10	<15	15±5
48-10	12±2	<5	13±3	8±1
10-0	—	—	—	—
Venera 12				
60-48	<6	<8	<10	16±3
48-10	10±2	<4	13±5	14±1
10-0	2±1	4±2	3±3	—

[a]From Moroz et al. (1981).

TABLE VI

**The Parameter $(1-\omega_o)$ for the Cloud Layer
and the Atmosphere below the Clouds[a]**

Altitude (km)	Venera 11		Venera 12	
	$\lambda_e = 0.49$ (μm)	$\lambda_e = 0.726$ (μm)	$\lambda_e = 0.49$ (μm)	$\lambda_e = 0.634$ (μm)
52-58[b]	$<7 \times 10^{-4}$	$<8 \times 10^{-4}$	$<7 \times 10^{-4}$	$<4 \times 10^{-4}$
48-52[b]	$<5 \times 10^{-4}$	$<5 \times 10^{-4}$	$<4 \times 10^{-4}$	$<3 \times 10^{-4}$
15-48[c]	$(5\text{-}50) \times 10^{-4}$	$<1.4 \times 10^{-3}$	$(1.5\text{-}9) \times 10^{-3}$	$<3.1 \times 10^{-3}$

[a]From Moroz et al. (1981); for the cloud layer it is assumed that the absorption takes place either in the middle or lower portions of the clouds.
[b]The parameter g is from Table 3 of Golovin et al. (1981).
[c]The parameter $g = 0$.

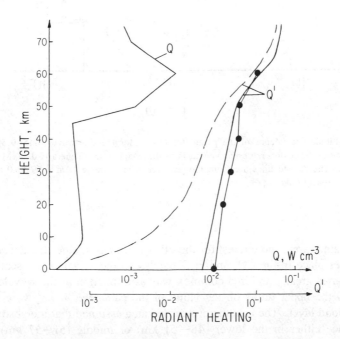

Fig. 9. Dependence on height of radiant influx of heat per unit volume $Q(H)$ and dependence of the radiant influx of heat below the level H, $Q'(H)$, evaluated at the subsolar point. Solid lines, atmosphere + surface; dashed line, atmosphere alone. $Q(H)$ is in W cm^{-2}; $Q'(H)$ is in units of the solar flux outside the atmosphere. The solid lines and dashed line are from a model by Ustinov et al. (1979). The solid dots are from a model for Q' based on the Pioneer Venus data by Tomasko et al. (1980a). This model was given at a solar zenith angle $\Theta_o = 19°3$ and Q' has been evaluated by dividing by (cos Θ) times the solar constant at Venus.

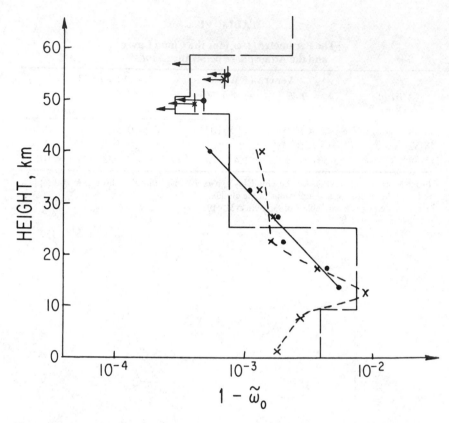

Fig. 10. Altitude dependence of the parameter $(1-\omega_0)$ for the shortwave (0.45–0.55 μm) spectral range. Solid dots represent Venera 11 probe data; crosses, Venera 12 data (Moroz et al. 1981). The dashed lines are from Pioneer Venus measurements at 0.59 to 0.67 μm (Tomasko et al. 1980a).

The ω_0 values are of no interest for the other two ranges, since the absorption there takes place in the CO_2 and H_2O bands and the actual single scattering albedo varies greatly and in a complex way as a function of the wavelength.

Only the upper limits on the value of the parameter $1-\omega_0$ were found for the cloud layer. The calculations in this area assumed that the absorption takes place either in the lower (48−51 km) or middle (51−57 km) layer alone. If the absorption occurs in the entire thickness of the clouds, the values of the parameter $1-\omega_0$ given in Table VI will be reduced by roughly a factor of 2. The upper limits of $1-\omega_0$ for the clouds were found to be rather low (from 3×10^{-4} to 8×10^{-4}). Tomasko et al. (1980a) obtained comparable values for ω_0 at 0.63 μm from the Pioneer 11 Large probe. When the optical thickness of the clouds is $\Delta\tau \approx 30$, such low values indicate that the scattering in the cloud layer at altitudes < 57 km may be regarded as conservative for many purposes.

Moroz et al. (1981) compare the albedo of the atmosphere, measured at 60 km altitude, with data concerning the spherical albedo of Venus. To make our results agree with groundbased measurements it is necessary to assume that there is an actual absorption $1 - \omega_0 \approx 10^{-2}$ for $\lambda = 0.7$ μm in the upper portion of the clouds (≥ 60 km), though alternative explanations are possible.

For the spectral band 0.45 to 0.55 μm, a vertical profile was obtained for the parameter $1 - \omega_0$ as shown in Fig. 10. In the calculations for this profile $g = 0$ was assumed below the clouds. Note that a minimum is observed in the profile of the single scattering albedo between 10 and 20 km, and consequently the volume coefficient of absorption has a maximum here. This behavior of $1 - \omega_0$ can be explained as absorption in gaseous sulfur, the relative content of which is 2×10^{-8} in a conversion for S_2 molecules.

20. THE THERMAL BALANCE OF THE MIDDLE AND UPPER ATMOSPHERE OF VENUS

F. W. TAYLOR
Oxford University

D. M. HUNTEN
University of Arizona

and

L. V. KSANFOMALITI
USSR Academy of Sciences

The various physical processes thought to be important in controlling the energetics of the atmosphere of Venus above the clouds are reviewed. Their ability to account for the observed thermal structure is discussed. Existing models account reasonably well for the global mean structure but fail markedly to predict or explain the diurnal variations and various high-latitude phenomena.

The temperature structure of the venusian atmosphere, and its variability, have now been measured in some detail. The availability of these data raises the question of whether our current understanding of the processes at work (radiative heating and cooling, dynamical transport of heat energy, etc.) are adequate to explain the observations. One might expect that the basic similarities between the atmosphere of Earth, and that of Venus above the 1 bar pressure level, would mean that the approaches developed in considerable detail for calculating radiative transfer and dynamics in the terrestrial case could be applied successfully with relatively minor modifications to predict or at least model conditions on Venus. This is indeed the case, provided only the global mean structure is considered. However, the diurnal variations, the equator to pole gradients, and the planetary wave modes, are proving more difficult to explain. These will be considered in some detail below.

The nomenclature developed for the different vertical regions of the atmosphere (see Chapter 1) is based on the mean thermal structure. The lowest 45 km or so (the troposphere) is in a state near to radiative-convective equilibrium and hence has a lapse rate (vertical gradient) close to the adiabatic value $\Gamma = -g/c_p$ (Houghton 1977, p. 3). On Venus the acceleration due to gravity g and specific heat c_p are nearly the same as for Earth, so Γ is approximately the same, at ~ 10 K km^{-1}. The energy balance of this region has been covered in the preceding chapter. Above the tropopause the venusian mesosphere is a region in which the lapse rate gradually decreases, as predicted by models of radiative equilibrium. Sources of heat for this region are thermal infrared flux from below (primarily quasi-blackbody radiation from the cloud tops), and absorption of solar flux in the near-infrared vibration-rotation bands of carbon dioxide and water vapor. These are balanced by thermal emission from the low-frequency bands of the same species, directly or indirectly to space. Above the mesopause, additional heating by solar extreme ultraviolet (EUV) produces a region where the temperature increases with height (the thermosphere), which on Venus extends from ~ 100 to 140 km above the surface. At higher levels still, the mean free path of molecules is greater than the local scale height and escape to space by light particles (e.g. hydrogen atoms) is possible. In this region, the exosphere, the temperature is constant with height.

Before considering the energetics in each of these regions in more detail, we will review the relevant measurements.

I. MEASUREMENTS OF TEMPERATURE, RADIATIVE FLUXES, AND WINDS

The most useful measurements for studies of energetics are those of atmospheric temperature as a function of pressure and height, location on the planet, and time. The location, composition, and variability of clouds and the composition of the atmosphere itself are also important parameters. These measurements are of two main kinds, remote and direct; the former offering good coverage in space and time, and the latter higher vertical resolution and greater ultimate precision. Quite large sets of both kinds of data are now available for our region on Venus (see, in particular, the special Pioneer Venus issue of *Journal of Geophysical Research*, December 1980), and we shall discuss each briefly in turn. More details may be found in Chapters 11, 13, and 16 on atmospheric structure, composition, and cloudiness. Fundamental information has also been obtained from the Pioneer and Venera orbiters on the fluxes of reflected solar radiation (which relate to the total energy absorbed by the planet), and on planetary thermal radiation at the top of the atmosphere. These, of course, relate to the planet as a whole rather than to individual layers of the atmosphere. For such layers it is necessary to take the seemingly less direct approach of working with the local temperature and computing the absorption and emission of radiation. Fortunately, above the tropopause we can usually neglect the additional complications stemming from convective heat transport. An exception may be the thermosphere-cryosphere, where eddy conduction is suspected to be important.

A. Remote Sensing Measurements

Table I and Fig. 1 provide an overview of typical equatorial temperatures as obtained by earth- and spacecraft-based radiometers and spectrometers sampling the thermal infrared spectrum of the planet (see references given in Table I). There is considerable variance between individual measurements but a fairly consistent picture can be drawn nevertheless. The dense clouds on Venus behave much like a blackbody at a temperature ~ 235 K. In the strong CO_2 bands, particularly ν_2 near 670 cm^{-1}, radiation from the clouds is strongly attenuated and emission from the colder upper atmopshere is observed. At wavenumbers less than 300 cm^{-1} the temperature starts to increase as the clouds become less opaque, in accordance with Mie theory. This trend is arrested at far infrared wavelengths (~ 10 to 100 cm^{-1}) by the pressure-induced band of CO_2 and the rotational bands of H_2O, CO, HCl, and others, while at very long (radio) wavelengths, the atmospheric opacity finally becomes negligible and brightness temperatures approaching that of the surface are observed. It was by measurements at radio and infrared wavelengths that the high surface temperature of Venus and the approximate height of the clouds were first determined (Chapter 4).

Earth-based instruments on large telescopes or spacecraft-borne radiometers can obtain sufficient spatial resolution to map features and their evolution, an essential step. If good spectral resolution is available, vertical discrimination can also be obtained because the radiation measured, according to the radiative transfer equation

$$I_{\Delta\nu} = \int_0^\infty B_{\Delta\nu}(z) K_{\Delta\nu}(z)\, dz \tag{1}$$

depends for its altitude of origin only on the *weighting function* $K_{\Delta\nu}(z) = dT_{\Delta\nu}(z)/dz$, where $T_{\Delta\nu}(z)$ is the transmission in the special interval $\Delta\nu$ from level z to space. A set of weighting functions for different $\Delta\nu$, all close together in the wing of the 15 μm band of CO_2, span the altitude range from the cloud tops to 115 km with a mean vertical resolution of ~ 10 km (Fig. 2). From a spacecraft in a near-polar orbit, coverage of all latitudes is possible as required for a study of the global system. Measurements of this kind were made from the Pioneer Venus orbiter in 1978–1979 (Taylor et al. 1980).

TABLE I

Summary of Measurements of the Brightness Temperature of Venus

Spectral Interval (μm)	Brightness Temperature (K)		Reference
	local	disk	
8–13		225	Sinton and Strong 1960
8–13	241		Pettit 1961
8–13		226	Sinton 1963a
7.8–9.0	240		Chase et al. 1963
10.1–10.7	240		Chase et al. 1963
8–14	233		Pollack and Sagan 1965a
8–13.5		225	Gillett et al. 1968
8–14	233		Sagan and Pollack 1969
8–14	230	220	Moroz 1971
8–14	234[a]		Kunde et al. 1977
8–14	238		Kunde et al. 1977
8–14	230		Logan et al. 1974
8–14	235		Chase et al. 1974
35–55	255	245	Taylor 1975
7.4–13.3 + 17.7–30	240[b]		Ksanfomaliti 1980b
	244[c]		Ksanfomaliti 1980b
	237[d]		Ksanfomaliti 1980b
11.3–11.6	238–244[e]		Taylor et al. 1980
14.6–14.8	170–185[e]		Taylor et al. 1980
45–110		245	Ward et al. 1977
31–38		244	Courtin et al. 1979
47–67		255	Courtin et al. 1979
71–94		238	Courtin et al. 1979
114–196		282	Courtin et al. 1979
450		277	Whitcomb et al. 1979
750		323	Whitcomb et al. 1979
700–2500		276	Werner et al. 1978

[a]Resolution not better than ⅓ of the disk
[b]Averaged over the planet.
[c]Nocturnal hemisphere.
[d]Dayside hemisphere
[e]Range includes real variability over a 72 day period and zenith angles of emission from $\cos^{-1}(0.85)$ to $\cos^{-1}(0.3)$.

Simpler measurements in one channel only were obtained earlier by the Venera 9 and 10 orbiters in 1975–76. A single spectral band (7.4 to 30 μm, with the 13.3 to 17.7 μm region blocked out) was employed, the wavelengths being chosen to exclude most of the CO_2 absorption so that cloud opacity makes the dominant contribution to $T_{\Delta\nu}(z)$. The weighting function

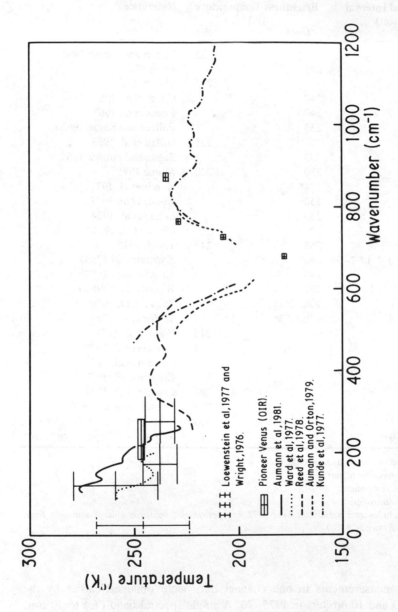

Fig. 1. Brightness temperature versus wavenumber for Venus: a summary of the available groundbased and spacecraft data excluding those with very low spectral resolution.

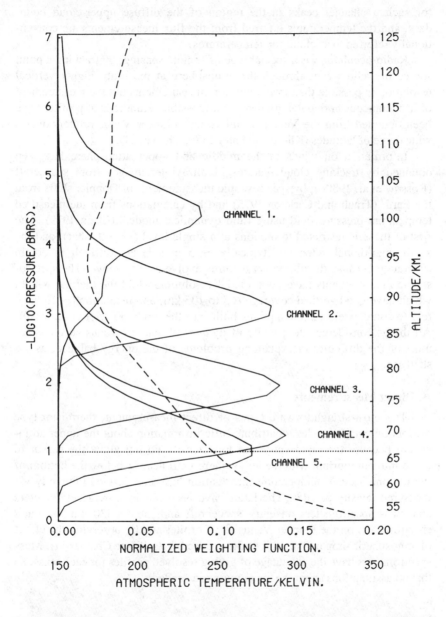

Fig. 2. Weighting functions for the Pioneer Venus orbiter infrared radiometer, superimposed on a global mean temperature profile. The weighting functions describe the height distribution of the emitted thermal radiation measured by the instrument and subsequently converted into temperature-height profiles (figure after Taylor et al. 1979c).

for such a channel peaks in the region of the diffuse upper-cloud boundary, and the temperatures inferred from the flux measurements are conventionally referred to as cloud-top temperatures.

Radio occultation is a special type of remote sensing; by tracking a point source of radio waves through the atmosphere at the limb, higher vertical resolution is possible than from emission measurements but the measurement of instantaneous horizontal gradients is impossible. Hundreds of profiles have been obtained from the Venera 9 and 10 and Pioneer Venus orbiters over a wide range of latitudes (Kliore and Patel 1980; Chapter 11).

Information on winds in the middle and upper atmosphere has been obtained by tracking cloud features in ultraviolet images from spacecraft (Rossow et al. 1980a), by spectroscopic measurements of Doppler shifts from the Earth (Traub and Carleton 1975) and by calculations from the measured temperature-pressure field using fluid-dynamical models (Elson 1979). The first of these is restricted to motions at a single level ($\sim$ 100 mbar), and the small, meridional velocities (typically $\sim$ 5 m s^{-1}) are difficult to obtain accurately because the ultraviolet features evolve as they move. The spectroscopic measurements have poor spatial resolution while the thermal winds, which have good vertical coverage (65 to 105 km) and resolution ($\sim$ 10 km), rely to some extent on assumptions built into the model (Taylor et al. 1980). All things considered, the paucity of dynamical data on Venus accounts for most of the difficulty in explaining problems of the energy balance, as we shall see.

B. Direct Measurements

The most straightforward types of direct measurement, thermistor-type sensors on entry vehicles, contribute little information about the upper atmosphere because the rapid entry and low density do not allow the sensor to come into equilibrium with the atmosphere until near or below the bottom of the region. Instead, temperature information has been derived indirectly via measured density profiles. The latter have been obtained by accelerometers and mass spectrometers on entry spacecraft and, above 130 km, by mass spectrometers on the Pioneer Venus (PV) orbiter and by observing the effect of atmospheric drag on the orbit of the spacecraft itself (see Chapter 11). Mass spectrometers have the advantage of giving resolved profiles for each mass, so that no assumptions about the mean mass are needed.

II. THE GLOBAL ENERGY BUDGET

Before embarking on a discussion of the internal energetics of the atmosphere it is convenient to consider here the energy budget of the planet as a whole. While not strictly an upper atmospheric problem much of the exchange of energy between the Sun, Venus, and space directly involves upper atmospheric processes and the behavior of the atmospheric system as a whole depends crucially on whether Venus has any internal sources or sinks of energy. More pragmatically, the components of the incoming absorbed and outgoing emitted power as a function of latitude are key boundary conditions on the energy balance of the upper atmosphere. The important parameters defining these components at a point are the *albedo* and the *bolometric temperature* (also called *effective temperature*). The latter is obtained from the energy flux of the radiation at all wavelengths λ leaving a spherical surface surrounding the planet and is defined by

$$\Theta_b = \frac{1}{\sigma} \left[\int_0^{2\pi} \int_0^{\infty} I(\lambda, \Omega) \, d\lambda d\Omega \right]^{\frac{1}{4}}$$ (2)

where Ω is solid angle, I thermal radiance, and σ Stefan's constant. Similarly, the albedo (strictly the spherical bolometric albedo) is given by

$$A = \frac{\int_0^{2\pi} \int_0^{\infty} I_\uparrow(\lambda, \Omega) \, d\lambda d\Omega}{\int_0^{2\pi} \int_0^{\infty} I_\downarrow(\lambda, \Omega) \, d\lambda d\Omega}$$ (3)

where $I_\uparrow$ is the reflected and $I_\downarrow$ the incoming solar radiance. The denominator of Eq. (3) is known quite well from a knowledge of the solar constant and the elements of the solar system; $I_\uparrow$ and I (for the thermal part, Eq. 2) can be measured from an orbiting spacecraft. The wavelength integrals in Eqs. (2) and (3) are to be taken over the regions occupied by the planetary and solar spectra, respectively.

The integrals in Eq. (2) have been evaluated using Pioneer Venus data for 108 latitude-longitude bins covering the whole northern hemisphere by Schofield et al. (1982). They find an area-weighted mean of 229.4 ± 2.2 K for the bolometric temperature of the hemisphere, corresponding to a mean outgoing flux of 157.0 ± 6.0 W m⁻². Ksanfomaliti (1977, 1980b) obtained a bolometric temperature of 221.9 K from the measurements of Veneras 9 and 10 in 1975. The difference is mostly explained by the more limited wavelength coverge of the Veneras and by the fact that they occupied low-inclination orbits which afforded no coverage of the polar regions. To overcome the latter restriction, it was necessary to assume that the mean bolometric temperature of the equatorial regions is characteristic of the whole planet. Pioneer measurements show that this is a fairly close but not exact assumption (Tomasko et al. 1980b).

Measurements of albedo are more difficult to calibrate than those of thermal flux because of the problem of obtaining an accurate reference source. Using earth-based measurements Irvine (1968) calculated a value for A of 0.77 ± 0.07, which was later revised upward to ~ 0.80 ± 0.07 by Travis (1975). The Pioneer Venus infrared radiometer had a 0.4 to 4.0 μm channel, calibrated by a lamp, from which Tomasko et al. (1980b) obtained a preliminary albedo for Venus of 0.80 ± 0.02.

Another approach to determining the albedo is simply to assume that the atmosphere is in net radiative balance, whence the equation

$$\sigma \,\Theta_b^4 = \frac{(1-A)E_o}{a^2} \tag{4}$$

should apply. Here E_o is the solar constant and a the distance from the Sun. This expression allows the albedo to be calculated from thermal measurements alone. In this way a value of $0.79^{+0.02}_{-0.01}$ has been obtained from Venera radiometry (Ksanfomaliti 1977, 1980b) and of 0.76 ± 0.006 from Pioneer Venus emission measurements (Schofield et al. 1982).

Clearly, the Pioneer measurements of emission and reflection are not consistent with each other if net radiative balance applies. A source inside Venus equal in magnitude to 20% of the solar input (i.e., accounting for the difference between $A = 0.76$ and 0.80) is very unlikely since Venus is thought to have an Earth-like makeup which would imply internal heat sources several orders of magnitude less than this. Also, even if such sources were postulated, it is difficult to construct a model in which these fairly large amounts of heat can be transported from the core to the atmosphere via a rocky crust without the latter becoming sufficiently plastic to collapse the observed surface relief. This could only be avoided if the transport was very localized, i.e., via a relatively small number of giant volcanoes. Although large, fresh-looking volcanoes do appear to exist on Venus (see Chapter 6) and the composition of the atmosphere is consistent with vigorous output from these, a simple comparison with terrestrial volcanism shows that the volcanic activity on Venus would have to be on an awesome scale to account for the missing 5×10^{15}W or so of power. A more acceptable alternative is that the preliminary estimate of 0.80 ± 0.02 for the albedo from the PV measurements is too high, since the uncertainty limit is now known from further work to be too conservative (J. V. Martonchik, personal communication). A fuller analysis of the PV albedo data—still the best, in terms of wavelength, spatial, and phase coverage, and radiometric precision, which is likely to be obtained for the foreseeable future—is likely to resolve this puzzle. In conclusion, then, the best thermal measurements of Venus, with the assumption of global energy balance, yield a value for the albedo of 0.76 ± 0.01; this is the most probable value.

III. STRUCTURE AND PHENOMENA

We now examine the mean structure and the principal phenomena in the upper atmosphere of Venus as revealed by the extensive set of measurements outlined in Sec. I.

A. Diurnal Temperature Structure

The diurnal variability of the cloud-top temperature was observed during a period in 1975–76 by radiometers on Veneras 9 and 10. Fig. 3 shows a diagram which summarizes the data, for a latitude band from 40°S to 50°N. Wherever possible, the curve was constructed using nadir observations (i.e., looking vertically downwards); in regions where such measurements were not obtained, measurements at different zenith angles were corrected for normal viewing using the discovered limb-darkening function. This procedure introduces some uncertainties since the limb-darkening law is not the same everywhere on the planet, and is especially variable on the day side. The regions where nadir measurements were available are indicated on the diagram.

From Fig. 3 it can be seen that the Venera results imply an asymmetry amounting to 10% in units of flux in the average thermal emission from the equatorial zone, the night side being generally warmer. This confirms the pioneering groundbased observations of Pettit and Nicholson (1924). Higher nightside than dayside temperatures are also apparent in the earth-based infrared maps of Apt and Goody (1979).

A partial explanation for the Venera observations of lower morning and evening fluxes from the cloud tops (which approximates the net flux from the planet, since emission from the clouds is the dominant contributor) can be obtained by examining an equatorial cross section of the mean (time-averaged) temperature structure and cloud-top morphology. Fig. 4 shows such a plot, as determined from PV measurements made in 1978–79 (Taylor et al. 1981; Schofield and Taylor 1982b). From this it can be seen that the diurnal variability observed by the Veneras and numerous groundbased observers is in fact due to a combination of two effects. The first is the variability with local time of day in the height of the cloud top (defined as the level at which the vertical optical depth of the cloud is equal to 1 at some reference wavelength, in this case 11.5 μm). The second is the actual variation of temperature at constant height (or, more conveniently, constant pressure) with local time. The latter, the dependence of temperature on the position of the Sun in the sky, is called the solar thermal tide.

Near the cloud tops the day-night temperature difference is small, with an amplitude of only ∼ 2 to 3 K, but with remarkable tidal structure which changes from predominantly wavenumber one at the cloud tops to a pronounced wavenumber two in the middle atmosphere (70–90 km) and then back to wavenumber one at higher levels. The phase of the tide is seen to vary fairly smoothly as it propagates upward.

Above 95 km the diurnal contrast increases and is ∼ 15 K at 105 km and ∼ 200 K at 150 km (Fig. 5). The daytime high in the thermosphere is ∼ 300 K, the night-time low ∼ 100 K. This surprisingly low value has led to the coining of the term cryosphere for the night-time thermosphere on

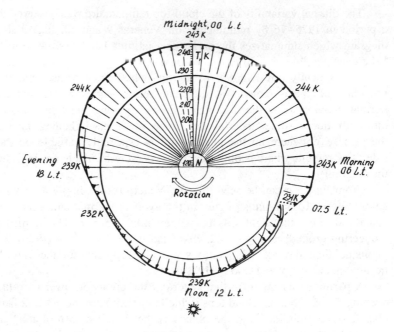

Fig. 3. Diagram of the distribution of fluxes of net outgoing thermal radiation from the equatorial region, as observed by Veneras 9 and 10. View is from the north pole. The radius at each point is proportional to the net flux. Semicircles inside the curve mark the regions where it was possible to obtain nadir measurements.

Venus. The sharpness of the transition at the terminator is also very marked. Light elements in the cryosphere show a concentration, or bulge, at ~4 AM (Chapter 13). There is little or no sign of a corresponding feature in the temperature. At the base of the thermo/cryosphere, at the highest level for which global-scale maps have been obtained so far, there occurs a pronounced local temperature maximum near the antisolar point, the spatial extent of which is smaller than that expected of a thermal tide feature (see Color Plate 6).

B. Equator to Pole Structure

Before Pioneer Venus there was little useful information on the latitude dependence of the thermal radiation since the earlier Veneras were restricted to near-equatorial orbits. PV infrared data at five different wavelengths is plotted as a function of latitude in Fig. 6. Channel 5 is a window channel (11.5 μm) and so gives the latitudinal variation of the cloud-top temperature. In this, and at other wavelengths corresponding to emission from levels near the cloud top, there is a pronounced minimum in the flux at latitudes near 70°N

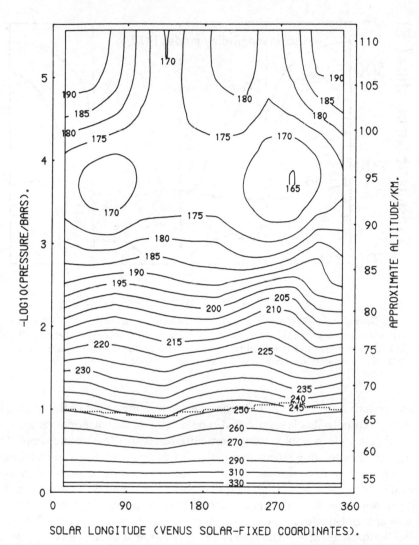

Fig. 4. Equatorial cross section of the mean thermal structure in solar-fixed coordinates, from Pioneer Venus data. 0° longitude is local noon (Schofield and Taylor 1982*b*).

with the equatorial and polar values roughly equal. At wavelengths corresponding to higher altitudes, the flux has a maximum value over the polar regions and is ~35% smaller at the equator at 15 μm (channel 2). The minimum around the pole had been noted in early earth-based maps (e.g. Diner and Westphal 1978) and is now known as the polar collar. The maximum at the pole is associated with two bright spots of elevated brightness temperature, sometimes joined by a bridge-like filament of high brightness

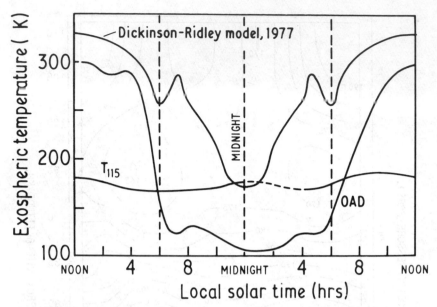

Fig. 5. Diurnal variation of upper atmospheric temperature, according to Pioneer Venus measurements (Keating et al. 1980) and a predictive model (Dickinson and Ridley 1977). The temperature at 115 km derived from Pioneer Venus infrared soundings (Taylor et al. 1980) is also shown (figure from Keating et al. 1980).

temperature some 2000 km in length. The flux in the center of the brightest parts of this polar dipole is estimated to be ~220% of the equatorial value.

Again it is possible, using vertically-resolved PV measurements, to examine how the flux relates to the detailed thermal structure and cloud-top morphology as a function of latitude. Figure 7 shows these quantities averaged over the same period of time (December 78 to February 79) as for Fig. 4. It can now be seen that the appearance of relatively low cloud-top temperatures at high latitudes in earth-based thermal maps (Murray et al. 1963; Diner and Westphal 1978; Apt and Goody 1979) is due to structure associated with the polar collar rather than the expected global trend of lower temperatures associated with reduced solar heating towards the poles. In fact, the temperature increases with latitude over most of the mesosphere and the atmosphere, for example, is typically 10 to 15 K warmer over the pole than over the equator on a 1 mbar surface (see Color Plate 7) above the 0.1 mbar level the gradient reverses. This state of affairs has important consequences for the general circulation, which are discussed below and in Chapter 18.

The strong, localized feature centered near 75°N in Fig. 7 is the polar collar, now seen to be a ribbon of very cold air surrounding the pole, roughly 1000 km wide and 10 km deep. Temperatures in the middle of the inversion are up to 40 K colder than their equatorial counterparts at the same level. The

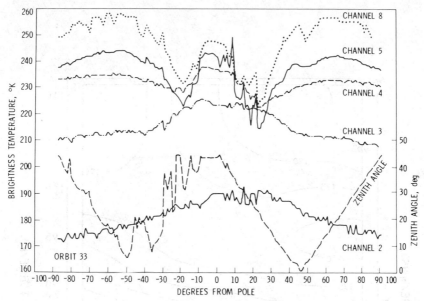

Fig. 6. Latitude dependence of the flux (expressed as brightness temperature) reaching space from different levels of the atmosphere. See Fig. 2 for key to channels (from Taylor et al. 1980).

collar is not zonally symmetric, however; it has a strong wavenumber-one component with minimum temperatures in midmorning (see Color Pate 7).

C. Waves and Transient Phenomena

Waves have been observed on Venus in ultraviolet images (Belton et al. 1976a,b; Rossow et al. 1980a), satellite drag data (Keating et al. 1980), Doppler wind measurements (Traub and Carleton 1975) and thermal infrared maps (Taylor et al. 1980). Apt and Leung (1982) have undertaken a study which reveals the amplitude of waves of various periods as a function of latitude and height as present in the PV infrared data (Fig. 8). They conclude that the dominant mode has a period of 5.3 d, is most active at latitudes of 60 to 70°, and that this and other modes have a minimum amplitude at altitudes near 80 km.

The most prominent quasi-permanent wave structures in the Venus upper atmosphere are the solar tides (which include the diurnal variation of temperature in the thermosphere), the Y-shaped wave seen in ultraviolet images of the clouds, and the high-latitude polar collar and polar dipole features. None of the ultraviolet features, including the Y, correlates significantly with features in the thermal maps; it therefore seems that their role in the energy budget of the atmosphere is either small or perhaps confined to levels well

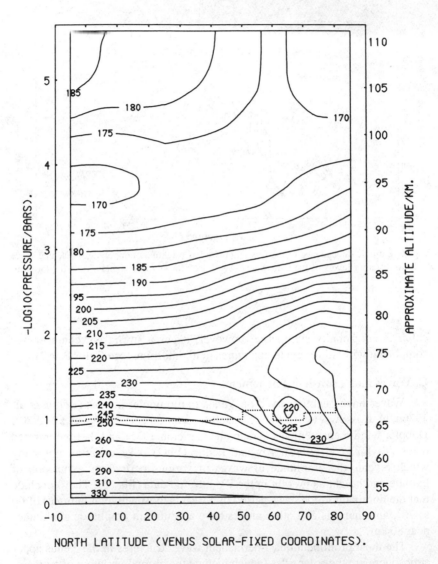

Fig. 7. Meridional cross section of the mean thermal structure in solar-fixed coordinates, from Pioneer Venus data (Schofield et al. 1982).

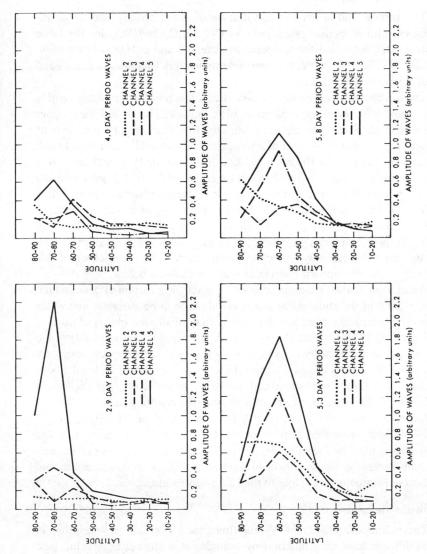

Fig. 8. Amplitudes of free wave modes as a function of period and altitude as observed by the Pioneer Venus orbiter infrared experiment (OIR). Channel numbers are related to altitude approximately as: Channel 2, 90 km; Channel 3, 80 km; Channel 4, 70 km; Channel 5, 65 km (see Fig. 2). (From Apt and Leung 1982.)

below 60 km. The tides, collar, and dipole, on the other hand, strongly modify the thermal structure, cloud morphology and dynamics of the region and explaining them is a key aspect of understanding the upper atmospheric energetics.

D. Cloud and Composition

The vertical and horizontal distribution of cloud opacity and the abundances of infrared active gases such as CO_2, H_2O, and SO_2 are the basic quantities that determine the temperature structure and energy balance of an atmosphere. These properties are reviewed in detail in Chapters 16 and 18 and a brief summary will suffice here.

The top boundary of the main clouds lies in the middle atmosphere and is fairly laterally homogeneous except at high latitudes. It has a diffuse upper boundary, with a scale height only slightly less than that of the gaseous atmosphere, ~ 4 km. The level of unit optical depth for 11.5 μm wavelength radiation lies close to the 100 mbar (65 km) level over most of the planet. It is 3 or 4 km lower in the vicinity of the polar collar (70°–80° N) and 2 or 3 km higher north of 80°N, forming a lip to the eye of the polar vortex, the so-called dipole. The latter is a region of reduced cloud cover of unknown depth at latitudes within 10° of the pole. Its appearance (see Color Plate 8) suggests either 2 cusp-shaped holes on either side of the pole or a single elliptical hole with the center obscured by a patch of higher cloud. The significance of this structure for the atmospheric energetics will be discussed below.

Apart from clouds, the main sources of opacity are carbon dioxide, which forms $\sim 96\%$ of the atmosphere and is not thought to be variable, and water vapor. The latter has a controversial history in groundbased observations (see e.g. L. Young 1972; Chapter 16) but PV orbiter mapping tends to confirm the 'dry' results and is broadly consistent with the most recent spectroscopy by Barker (1975b). The highest concentration occurs shortly after local noon, near the equator, but does not exceed a mixing ratio of 100 ppmv (Schofield et al. 1982). Over most of the planet the above-cloud mixing ratio is <6 ppmv, too small to play a major role in the radiative energy budget.

It has been suspected (Taylor et al. 1980) that CO_2 condenses at the mesopause at the dawn terminator. There is evidence in the form of anomalous limb darkening for an enhanced high-altitude haze at this location, and the observed temperature is close to the CO_2 condensation point.

E. Albedo Variations

The bolometric albedo of Venus (integrated over all wavelengths) is remarkably constant as a function of latitude, at a current best value (see above) of 0.76 ± 0.01. Figure 9 shows the zonal mean value as a function of latitude as measured by the PV orbiter infrared experiment (Taylor et al. 1980). At discrete wavelengths, strong latitude dependences do occur and for the two intervals for which data exist, 0.365 μm and 2.0 μm, we

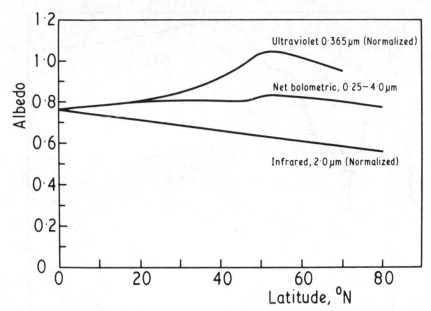

Fig. 9. Smoothed fits to the zonally- and time-averaged values of the albedo of Venus as a function of latitude at wavelengths in the ultraviolet, the infrared, and for total solar energy. The scale is absolute for the net albedo and arbitrarily normalized for the two single wavelength curves (after Rossow et al. 1980 and Taylor et al. 1980).

see (Fig. 9) that the ultraviolet albedo increases towards the pole while the infrared albedo decreases. The ultraviolet increase suggests a thickening in the mesospheric submicron haze with latitude, but the infrared data contradict this. The 2.0 μm wavelength was chosen to be in a weak CO_2 band, so the reflectivity at this wavelength is a strong function of cloud height or haze thickness. Probably, some presently unexplained wavelength- or height-dependent property of the haze and cloud particles will be necessary to reconcile the data. Other information on albedo as a function of wavelength is summarized by Travis (1975), from which Fig. 10 is taken.

It is also of interest to use the measurements of albedo and thermal flux variations to compare the rates of net heating and cooling of the atmosphere as a function of latitude (Fig. 11). These show net heating in tropical regions and net cooling at high latitudes, qualitatively similar to Earth's. The implication, of course, is that the atmosphere transfers energy polewards, by a combination of advection and eddy transport with radiation and conduction probably playing minor roles. The details of this process, in particular at which height level it mostly occurs and the relative contributions of bulk and wave motions, remain unknown and largely unstudied.

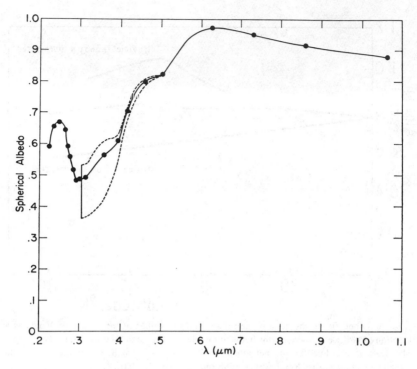

Fig. 10. The wavelength dependence of spherical albedo. The envelope contained by the dashed lines represents the maximum range of variability between the light and dark ultraviolet markings (from Travis 1975).

IV. LOCAL ENERGY BUDGETS: SOURCES, SINKS, AND TRANSFER

A. Solar Radiation

The solar constant at Venus is nearly twice that at Earth. Although this is more than offset by the higher albedo in terms of the input of energy to the lower atmosphere, the reflection takes place near or below the base of the middle atmosphere, giving intensities within the region that are double their terrestrial counterparts. Thus, of the 2621 W m^{-2} incident at Venus, $\sim$ 80% is reflected to space, $\sim$ 10% is absorbed above 64 km altitude and the rest below, 2.5% at the surface itself (Tomasko et al. 1980b). The upper-atmospheric portion is mostly absorbed a few km inside the cloud tops, with a small fraction due to solar EUV (wavelength $\leqslant$ 1080Å) radiation absorbed above 100 km in the thermosphere (Fig. 12). Relatively little solar absorption takes place in the upper mesosphere, the small amount which does being mostly due to the near-infrared absorption bands of carbon dioxide.

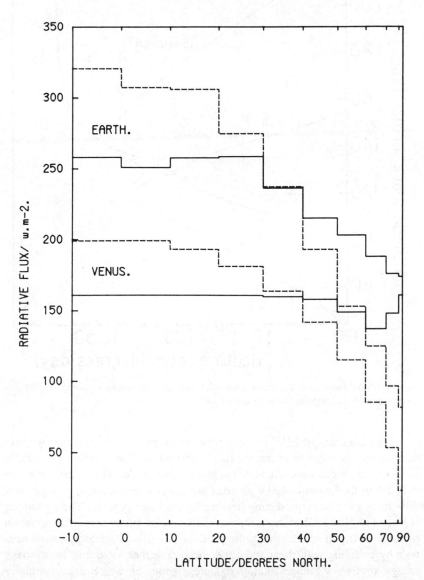

Fig. 11. Net flux in and out of the northern hemisphere of Venus as measured by the Pioneer Venus orbiter (Schofield et al. 1982). The comparable terrestrial curves are from Von der Haar and Suomi (1971).

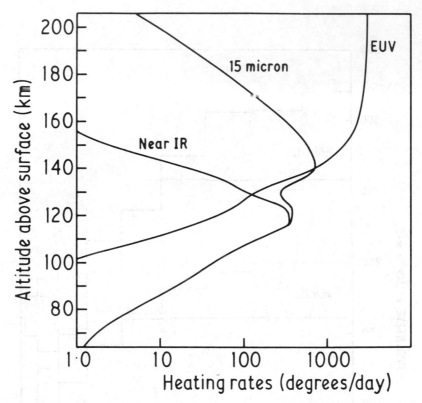

Fig. 12. Global mean rate of extreme ultraviolet heating, near-infrared heating, and 15 μm cooling, all in K/Earth day (from Dickinson 1972).

The efficiency of EUV heating depends on two terms, one due to direct heating and the other to excitation of vibrational-rotational states, principally in CO_2, which can be quenched. Dickinson and Ridley (1977) estimate that only 2% of the incoming EUV photons are directly thermalized, with perhaps 40% more eventually producing heating by the latter process. The remainder is lost as airglow. All of these mechanisms involve ionization or dissociation of some of the molecules, with the products contributing energy to the neutrals by colliding with them and either becoming thermalized or transferring energy to CO_2 vibration-rotational states, some of which are eventually quenched. A calculation of the probability of whether an upper vibration-rotation state of CO_2 will transfer its energy to molecular motion or emission of a quantum is crucial to quantifying this process, as well as the heating rate by absorption of near-infrared radiation. It also, of course, determines the rate of cooling at the same levels. This so-called non-LTE (local thermodynamic equilibrium) problem is complicated because numerous relaxation paths are available, and since collisions are involved the process is dependent on temperature, pressure and composition. The best treatment for Venus to date is by Dickinson (1972,1976).

B. Thermal Radiation

The boundaries of the upper atmosphere, so far as thermal radiation is concerned, are cold space (effectively a blackbody at absolute zero) and the cloud tops (to a fair approximation, a blackbody at ~ 240 K). The atmosphere between exchanges thermal energy with these surfaces and between its different regions by radiative transfer in its vibration-rotation bands, of which the most important by far is the ν_2 band of CO_2 at 15 μm, which lies near the maximum of the Planck function of $T = 240$ K and is very strong. Other CO_2 bands near 10 μm, 4.3 μm and in the near and far infrared play a smaller role while the contributions of bands of less abundant species such as H_2O, HCl, CO and SO_2 may be neglected for most practical purposes.

The rate of thermal infrared cooling to space can be obtained to a good approximation by neglecting radiative exchange between levels and considering only cooling to space (Rodgers and Walshaw 1966). In this limit $Q(z)$, the cooling rate per unit volume is given by

$$Q(z) = \frac{1}{2} \rho(z) \frac{hc}{\lambda} r T(\lambda, z) \tag{5}$$

where $\rho(z)$ is the number density of molecules, hc/λ the energy per photon, r the rate of photon emission per molecule, and $T(\lambda, z)$ the transmission from level z to space at wavelength λ. Note that according to Kirchhoff's law, r is zero when T is unity so cooling, like heating, occurs only at those wavelengths inside spectral bands of CO_2. At such wavelengths T will tend towards its maximum value of unity at high altitudes where $\rho(z)$ is low, and vice versa. Thus cooling at any particular wavelength has a maximum value at some level determined by the opacity of the atmosphere at that wavelength (cf. Fig. 1). As noted earlier, the atmosphere is nearly transparent at most wavelengths (except in the CO_2 bands) down to the cloud tops. This is the reason why the cloud-top temperature and the bolometric temperature of the planet have nearly the same value. The rate r is a function only of temperature at a given wavelength, so long as LTE (local thermodynamic equilibrium; see Houghton 1977, p. 10) applies, i.e. below ~ 120 km. Above this level it has a dependence of the form

$$r = \frac{r_0}{1 + \frac{1}{2} \left(\frac{A}{\sigma_r}\right) T(z)} \tag{6}$$

where A is the Einstein coefficient for spontaneous emission and σ_r is the collisional relaxation rate. For the 15 μm band, Dickinson (1972) finds that the second term in the denominator is essentially zero below $\sim$ 120 km and is on the order of 10^3 near 150 km, indicating the low efficiency of infrared cooling at exospheric levels.

C. Advection

The transport of heat by vertical or horizontal motions is an important process whenever the velocities are greater than the characteristic distances divided by the radiative time constant of a parcel of air. It is easy to find, following Stone (1975), how the ratio ϵ of the dynamical to the radiative time constants varies as a function of height in a pure CO_2 atmosphere provided the orders of magnitude of the three components of the wind velocity can be assumed. The dynamical time constant is defined by

$$\tau_{\text{dyn}} = \frac{R}{v(z)} \tag{7}$$

where R is the radius of Venus and v a typical horizontal velocity, while the radiative time constant is

$$\tau_{\text{rad}} = \frac{\gamma p(z) H}{(\gamma - 1) \sigma \Theta(z)^4} \tag{8}$$

which is simply the time constant for the atmosphere to cool if all heating is removed. In Eq. (8), γ is the ratio of specific heats (equal to 1.3 for CO_2), p the pressure, H the scale height, $\Theta(z)$ temperature at height z, and σ Stefan's constant. Inserting suitable values (from PV data for winds and temperatures), we find that $\tau_{\text{dyn}} \simeq \tau_{\text{rad}}$ for zonal motions when p is on the order of one tenth mbar (Fig. 13). This is consistent with the pressures at which diurnal temperature variations are observed to become large. For the meridional direction the radiation-dominated regime extends to higher pressures (some tens of mbar) than for the zonal direction, and again this is consistent with the appearance of larger equator-to-pole temperature gradients above than below the clouds.

D. Waves and Eddies

There is ample evidence for organized waves and random eddies on all spatial and time scales on Venus. As on the Earth, these will certainly play a major role in the transport of energy, momentum, and minor constituents. The inert constituents of the atmosphere are observed to be uniformly mixed up to a level, the homopause, at $\sim$ 130 km (Chapter 13). At greater heights, each one takes on its own scale height. It is conventional and convenient to describe the mixing by an eddy diffusion coefficient K; the molecular diffusion coefficient D is inversely proportional to density and exceeds K above the homopause. Accompanying the eddy mixing is eddy heat transport, and the coefficients should be approximately equal. The thermosphere, and even the

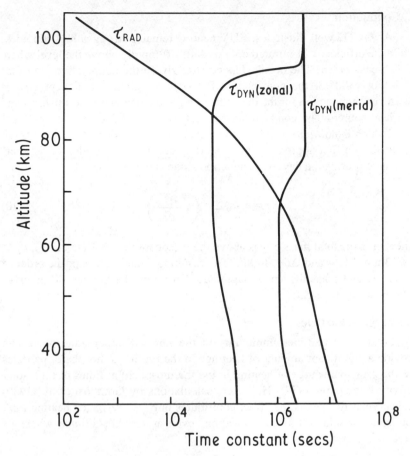

Fig. 13. Radiative and dynamical time constants from simple theory, shown as a function of altitude.

cryosphere, is statically stable; the vertical mixing must therefore be externally driven. The source of mechanical energy is presumably the troposphere, although on Venus the horizontal winds from the thermosphere to cryosphere may contribute. In any case, energy is dissipated, and the resulting heat may offset the cooling due to the downward eddy conduction (Hunten 1974; Izakov 1978). Although the question remains unsettled, the existence of a cryosphere suggests that the cooling effect dominates. The heat flux is

$$F = \kappa c_p \rho \left(\frac{d\Theta}{dz} + \Gamma \right) \tag{9}$$

It can be downwards even if the temperature gradient $d\Theta/dz$ is negative so long as its magnitude is less than the adiabatic lapse rate Γ.

E. Conduction

As Fig. 12 well illustrates, EUV heating remains strong at high altitudes, while the efficiency of infrared cooling falls off rapidly above the level where LTE breaks down. The details depend strongly on the temperature, and Fig. 12 is not valid for the cryosphere. The only remaining cooling process is downward conduction for the day side; at night the conduction is upwards and is a heat source. The conductivity for CO_2 is $\kappa = 0.82 \, \Theta^{1.28}$ erg cm^{-1} s^{-1} K^{-1}, which is not quite valid for the actual mixture of CO_2, CO, and O, but adequate as a first approximation. The time constant for cooling of the thermosphere (during the day) by conduction alone is on the order of

$$\tau = m \, c_p \, \Delta T \, \left(\kappa \, \frac{dT}{dz} \right)^{-1} \tag{10}$$

where m is the total mass of gas above the mesopause ($\simeq 0.1$ gm cm^{-2}), $c_p \simeq$ 0.67 J gm^{-1} K^{-1} and $\Delta T \simeq 100$ K. The resulting value for τ is of the order of one year, and τ has the same magnitude if we consider heating of the cryosphere during the night.

F. Energetic Particles

Solar wind particles impinging on the rarefied upper atmosphere can provide a significant amount of heating. In the Earth's atmosphere, particles are the principal source of heating in the thermosphere at times of high solar activity (Houghton 1977). However, calculations by Gombosi et al. (1980) indicate that for Venus the effect contributes only $\sim 1\%$ of the heating rate, compared to solar ultraviolet radiation, even at 128 km altitude where its effect is greatest.

V. LOCAL ENERGY BUDGET: MODELS AND DISCUSSION

A. Mean Thermal Structure

The mean temperature versus height of the mesosphere can be computed by evaluating explicitly the energy absorbed at each level from the incoming solar spectrum, that absorbed and emitted at thermal wavelengths from the lower boundary and all of the other layers, and the temperature at which they all balance. The most detailed treatment of this straightforward but tedious problem is by Dickinson (1972). His results for the temperature profile are plotted in Fig. 14 along with the envelope of a planet-wide set of measurements. It can be seen that the model agrees tolerably well with the Pioneer Venus data, so that the mean structure of the region can be said to be reasonably well understood. The smallness of the diurnal variations in the model calculations for the mesosphere is consistent with the small ratio of the dynamical to the radiative time constant, but is too small because of course the models do not include the effects of the polar collar or the solar tides.

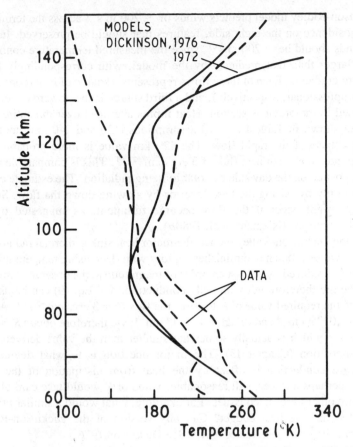

Fig. 14. A comparison between the model temperature profiles of Dickinson (1972, 1976) and the envelope of measured temperature profiles for the middle and upper atmosphere (Schofield and Taylor 1982b).

The agreement is much poorer at altitudes above 120 km where the model calculations are made more difficult by, among other things, the onset of non-LTE radiative processes and the increased relative significance of eddy heat transfer. The most sophisticated model, that of Dickinson and Ridley (1977), fails to predict correctly the mean temperature of the thermosphere or its day-night variation, although there are some qualitative similarities (Fig. 5). Dickinson (1976) had earlier shown that very low night-time exospheric temperatures are predicted by models in which fairly large eddy diffusion coefficients are assumed. It remains to be shown that the Dickinson-Ridley model can reconstruct the measurements in detail with an appropriate selection of parameters. In the meantime we must try to adapt their results to the much lower temperature, following Niemann et al. (1979a, 1980b). The

Dickinson-Ridley model predicts winds of ~ 200 m s^{-1} across the terminator and subsidence on the night side, features that should be preserved. In fact, the winds should be > 200 m s^{-1}, since the measured temperature contrast is much larger than that predicted by the model, with correspondingly larger pressure gradients. Each of these flows represents a source of heat (convective and compressional, respectively), and a third source is the upward conduction discussed in the previous section. Heat inputs averaged over the entire night side are shown in Table I for two assumptions (120 and 140 km) about the effective base of the rapid flow. The 120 km value is more probable, and corresponds to a mean heat flux of 3 ergs cm^{-2}s^{-1}. This is comparable to the heat deposited on the day side by solar ionizing radiation. The estimate can be reduced only by raising the base level or by slowing down the flow. Such a slowing might occur if the flow became turbulent, as suggested by the Reynolds criterion (Niemann et al. 1980b).

In the Dickinson-Ridley model, the major heat sink is thermal radiation, a process whose efficiency diminishes rapidly with decreasing temperature and is therefore reduced by at least an order of magnitude at the observed temperature. We are therefore left with eddy conduction, and Eq. (9) can be used to estimate the required value of K to carry the flux $F = 3$ ergs cm^{-2} s^{-1}. At 120 km $\rho = 10^{-9}$ g cm^{-3} and dT/d$z + \Gamma = 1°$ km^{-1}; we therefore obtain $K = 2 \times 10^7$ cm^{2}s^{-1} which is actually somewhat smaller than the value derived from the composition (Chapter 13). The major question is to what degree this downward conduction is offset by the heat from dissipation of the eddy energy; perhaps a larger, still reasonable, value of K would take care of this. There are other unresolved issues. For example, what would a similar process do to the dayside heat budget? Can any version of the Dickinson-Ridley solution exist with these values of K? (See Dickinson 1976.)

B. Thermal Tides

The thermal tide is the name given to the daily rise and fall of temperature at a given level in the atmosphere in response to the apparent motion of the Sun across the sky. The familiar afternoon maximum and predawn minimum in the mean air temperature near the surface is one manifestation of atmospheric tides on the Earth. Particularly at higher atmospheric levels, the terrestrial tides contain large components of both wavenumber one and wavenumber two (i.e., one and two complete cycles per day, respectively). This occurs partly because of the fact that the solar forcing function, although primarily wavenumber one, also contains sizable components of wavenumber two and higher orders. Also, the conditions in the atmosphere can be such that wavenumber two propagates more efficiently than the fundamental in the vertical direction, so that the amplitude of the latter, although more weakly forced, can dominate at high enough altitudes (i.e., in the mesosphere). In spite of this, it is somewhat surprising to find that on Venus wavenumber two

TABLE II[a]

Heat Sources for the Nightside Thermosphere

Boundary (km)	Convective (erg cm^{-2} s^{-1})[b]	Compressional (erg cm^{-2} s^{-1})[b]	Conductive (erg cm^{-2} s^{-1})[b]
120	2.6	0.52	0.017
140	0.03	0.006	0.0088

[a]After Niemann et al. (1979a).
[b]Averaged over the hemisphere.

is the dominant mode from the cloud top to the mesopause (Taylor et al. 1980). Mathematical theories of atmospheric tides (Chapman and Lindzen 1970) which work quite well for the terrestrial case, do not appear to fit the Venus observations at all in a preliminary study (L.S. Elson and R. Zurek, unpublished report). A more sophisticated analysis may reveal that the vertical heating profile due to deposition of solar radiation inside the clouds fits the vertical wavelength of $\nu = 2$ better than $\nu = 1$. Alternatively, perhaps the $\nu = 2$ mode propagates more efficiently through the region of strong vertical shear in the zonal winds which exists near the cloud tops. As a third possibility, $\nu = 1$ may be more prevalent than the available measurements reveal if its vertical wavelength is short compared to the vertical resolution of the data (5–10 km).

The rapid dissipation of the tides at an altitude of 90 km or thereabouts is a predictable consequence of radiative cooling to space which becomes very efficient near those levels. The development of relatively high dayside and low nightside temperatures around the 100 km level is also understandable in terms of near-infrared heating by the Sun, as discussed above, although the small but definite displacement of the temperature maximum to 20° or 30° of longitude *before* local noon is rather disconcerting and presently unexplained. Similarly, the compact warm feature at local midnight may be a high-altitude remnant of the upward propagating wave two, or it may be a symptom of compressional heating in the descending arm of a thermospheric subsolar to antisolar circulation cell.

Note that the cloud-top height also exhibits a tidal variation (Fig. 4). This is distinctly wavenumber one with the maximum in late afternoon and the minimum near dawn and is most readily interpreted as being due to rising warm air near the subsolar point and descending cool air near the antisolar point taking cloud particles along in the bulk motion. The phase shift from noon (where the heating is maximum) to midafternoon would be expected since the tides are embedded in the rapid zonal flow, ~ 100 m s^{-1} at cloud-top level. This behavior is evidence for a substantial subsolar to antisolar flow component of the general circulation of the atmosphere, even at the cloud-top level, since it cannot be explained by a zonally-rotating Hadley cell alone.

C. Polar Heating Mechanisms

Three mechanisms have been advanced so far to explain the unexpected equator-to-pole temperature gradients in the 70 to 90 km region. These are:

Mechanism 1. Compressional heating in the descending branch of an axially-symmetric, equator-to-pole Hadley cell. There is a considerable amount of observational evidence for the existence of such a circulation (Chapter 11), at least at the cloud tops some 10 km below the lowest level which unambiguously displays the polar warming phenomenon (Fig. 7). It is straightforward to show (e.g. Houghton 1977) that a parcel of air transferred adiabatically from pressure p_1 and temperature Θ_1 to p_2 will increase its temperature according to

$$\Theta_2 = \Theta_1 \left(\frac{p_2}{p_1} \right)^{1 - 1/\gamma} \tag{11}$$

where γ is the ratio of specific heats. Inserting typical values for Venus of Θ_1 = 180 K at p_1 = 1 mbar gives a value of $\Theta_2 > 300$ K at 10 mbar, where the mean temperature is only ~ 200 K. In practice, of course, the compression will not be perfectly adiabatic, especially if the descent is not very rapid. A parcel would take two days to descend from 1 to 10 mbar at a rate of 5 cm s^{-1}, and the vertical velocity is unlikely to be much larger than this.

A problem with this mechanism is that if the cloud-top Hadley cell extends into the region of the polar warming it becomes difficult to explain how the cell is driven, since Hadley's mechanism requires cool, sinking air at the pole. One possibility is that a thermally indirect cell, i.e., one moving against the pressure gradients, is driven by a more powerful direct cell below the clouds, coupled by viscosity. This might happen if the solar heating rate is maximum well inside the upper cloud boundary, which is likely, and if the main cell driven by this is somehow constrained in its vertical extent. Otherwise this heating mechanism must be discarded in favor of one of the others (below), and a divergence of the flow over the pole into an upward branch above the clouds and a downward branch (down into the clouds, creating the dipole) is implied (Elson 1979).

Mechanism 2. Solar heating of the 70 to 90 km altitude range may be greater at the pole due to geometric effects and density variations in the optically thin, partially absorbing haze which has been detected in this part of the atmosphere (Chapter 16). If the atmosphere is well mixed zonally, and the haze is optically thin, twice as much heating is obtained at a given altitude near the pole than at the equator, since all longitudes are illuminated. This effect is further enhanced if the haze is more dense at high latitudes, as has been suggested (Kawabata et al. 1980). The heating rate in a haze of optical thickness for absorption of only 10^{-6} is still several hundreds of degrees per day at a mean pressure of 1 mbar. The measured optical thicknesses quoted by Kawabata et al. are much larger than this but apply, of course, mostly to scattering with absorption playing a much smaller role.

Mechanism 3. The polar dipole may warm the mesospheric atmosphere above it by allowing thermal radiation from the hot lower atmosphere to pass through the cloud-free regions and become absorbed in the CO_2 atmosphere above. The base temperature in a polar hotspot, for example, is probably $\sim$ 290 K (Taylor et al. 1979c) compared to $\sim$ 240 K for a typical equatorial cloud-top temperature. In the optically thin approximation (e.g. Goody 1964) the mesopause temperature is given by

$$\Theta_m = 2^{-\frac{1}{4}} \, \Theta_b \qquad\qquad (12)$$

where Θ_b is the effective temperature as defined in Eq. (2). This effect could, therefore, raise the temperature at 80 to 90 km altitude (where the lapse rate is close to zero) by as much as $2^{-\frac{1}{4}} \times (290 - 240)$, or $\sim$40 K, over the small area occupied by hot spots.

One problem in discriminating among these mechanisms is that all of them may be contributing to some degree. Whatever the origin or origins of the polar warming, it has an important effect on the height dependence of the zonal wind field. According to the principle of cyclostrophic balance a positive equator-to-pole temperature gradient produces pressure forces which must be balanced by a vertical shear tending to reduce the zonal wind speed (Schubert et al. 1980a; Chapter 21). A model by Elson (1979) has been used to infer from the actual measured temperatures on Venus that the zonal wind should fall essentially to zero, from its cloud-top value of $\sim$ 100 m s^{-1}, by a height of 80 to 90 km. If the cloud-top wind is $>$ 100 m s^{-1}, then the zonal wind may persist to higher levels (see Chapter 16) but, conversely, if the model is made more sophisticated by including a frictional dissipation term, the upper bound on the wind is lowered again (Taylor et al. 1980). Considering the likely range of values of the various parameters, it seems unlikely that the zonal wind is significant at 100 km altitude, although there is some evidence from the distribution of certain exospheric species that it may build up again by the 150 km level.

D. Polar Wave Phenomena

No models exist yet to show plausible mechanisms for generating and sustaining the polar collar or the dipole nature of the polar vortex. However, the fact that ultraviolet cloud-tracked winds suggest a state near to solid body rotation at latitudes up to at least 60° (Rossow et al. 1980a) provides a possible clue. The rotation rate is $\sim$ 4.8 d at those latitudes; at 80°N it is 2.7 d (Taylor et al. 1980). This implies a transition to a state closer to angular momentum conservation somewhere between 60° and 80°. If the transition zone is narrow, it will occur at $\cos^{-1}$ (2.7 cos 80°/4.8) = 72° which is very close to the central latitude of the collar. Thus the collar may be a manifestation of the breakdown of solid body rotation at high latitudes. The dipole, on the other hand, is presumably a wavenumber two instability in the eye of the polar vortex caused in some way by its rapid rotation. Wavenumber two patterns often surround the poles on Earth, but so do other orders at other times. The dipole varied greatly during the 72 d for which it has been observed, but always retained its elongated double character. It has been suggested (R. Hide, personal communication) that this shape may transport angular momentum downwards more efficiently than a simple vortex, and if so it may be required to conserve angular momentum and mass simultaneously as both converge on the pole in large quantities.

VI. SUMMARY

The overall picture of the upper atmosphere of Venus is now reasonably complete. What is needed at this stage is a phenomenological model of the region in which the parameters governing the various processes are adjusted to match more closely the structure and behavior which have been observed since the pioneering work of Dickinson and his co-workers. Much of the groundwork exists for accounting satisfactorily for the global thermal structure and mean circulation (given a 4-day zonal rotation driven from below the lower boundary), although difficulties are likely with accounting for the diurnal variability of the thermosphere without resort to unreasonably large eddy conductivities. The main untackled problems have to do with the formation and propagation of planetary-scale wave phenomena and a quantitative evaluation of their role in transferring heat and momentum. For the present and the immediate future, we have the healthy situation where the data base is more complete than the interpretative theory, and considerable progress is possible in the next few years.

21. GENERAL CIRCULATION AND THE DYNAMICAL STATE OF THE VENUS ATMOSPHERE

GERALD SCHUBERT

University of California at Los Angeles

Because of numerous and striking differences between Venus and Earth with regard to atmospheric structure and composition, manner of solar energy deposition, and planetary rotation rate, there is little similarity in their global atmospheric circulations and dynamics. The principal mode of atmospheric circulation on Venus is a zonal retrograde superrotation of the entire atmosphere, from just above the lowest scale height to ≥ 100 km. There may also be a retrograde superrotation of the thermosphere-cryosphere superimposed on a strong subsolar-antisolar circulation. The angular momentum of the atmospheric superrotation is ~ 0.15% of the solid planet angular momentum. The substantial amount of angular momentum in the atmosphere suggests the possibility of significant angular momentum exchanges between the two reservoirs, resulting in changes in the length of day on Venus, perhaps on the order of hours. Such changes might be detectable with Earth-based radar. Approximate cyclostrophic balance prevails throughout the nonequatorial lower atmosphere of Venus wherever the westward winds are large. Eddies, the mean meridional circulation, and planetary-scale waves may all be involved in the upward transport of retrograde angular momentum to maintain the atmospheric superrotation. The weak mean meridional circulation of the lower atmosphere may consist of Hadley cells at cloud heights and the surface, separated by an indirect cell located in the planetwide stable layer below the clouds. Eddies have been observed in the lower atmosphere and they may also transport significant amounts of heat and momentum latitudinally and vertically. The zonal wind and cloud level Hadley cell combine to form a polar vortex. When viewed from above the south pole, the vortex rotates counterclockwise and the flow is poleward and downward at the cloud tops. The dark horizontal Y and other large-scale patterns in the ultraviolet photographs are planetary-scale waves drifting slowly with respect to the atmospheric superrotation. Waves are present throughout the atmosphere over a broad range of spatial scales. The cloud level region is probably a center of wave generating activity by prominent mesoscale convection cells. The mean zonal and meridional circulations may not be symmetric about the equator. A detailed explanation of Venus's global atmospheric circulation remains a major challenge to the theory of the dynamics of planetary atmospheres.

I. SUMMARY OF ATMOSPHERIC PROPERTIES AND STRUCTURE

There now exists an observational data base, including *in situ* measurements of wind velocities, which makes it possible to confidently describe certain aspects of Venus's global atmospheric circulation. These data have been compiled from decades of observations of Venus made from Earth and from a series of Mariner, Venera and Pioneer Venus spacecraft. The spacecraft missions to Venus are listed in Table I of Chapter 2; included are flybys, orbiters, entry probes, and landers. The orbiters have provided extensive horizontal and temporal coverage of regions of the atmosphere above the clouds. The probes have explored the atmosphere from $\sim$ 65 km, just below the cloud tops, to the surface at the widely separated locations shown in Fig. 1. The Pioneer Venus probes reveal latitudinal and longitudinal contrasts on the planet at a single time. Although the data set for Venus has grown enormously, especially during the intensive exploration in the 1970s, it is nevertheless incomplete and limited, both spatially and temporally. Some significant features of the large-scale atmospheric motions are unknown; this is particularly true of the circulation above the clouds and in the lowest 10 km adjacent to the surface. It is hardly surprising that we do not have a complete picture of Venus's general atmospheric circulation, given the nature of the Venus data base. Compare the situation of Venus with that of the Earth's oceans; even with more comprehensive observations, there are still gaps in our understanding of oceanic circulation. For example, the role of mesoscale eddies or rings in the mean meridional transports of heat and salt is only now being clarified (The Ring Group 1981). It is perhaps surprising that we could determine as much as we have about Venus's global atmospheric circulation; we have had some success because the general character of Venus's large-scale atmospheric circulation has not changed dramatically over the many years of observations.

The atmosphere of Venus is so different from the Earth's atmosphere that our terrestrial experience is of little help in understanding how it circulates. Table I compares basic information on Venus and Earth and their atmospheres. The mean orbital radius of Venus is only 72% of that of the Earth. This accounts for the much larger solar constant S at Venus; $S_\female/S_\oplus = 1.9$. Although the mean incident solar flux is nearly twice as large at Venus as at the Earth, Venus absorbs only $\sim$ 55% as much solar energy as Earth does. Because of its complete cloud cover, Venus reflects 80% of the incident solar flux (Irvine 1968; F. Taylor et al. 1980; Chapters 18, 20) compared with $\sim$ 30% for the Earth (Vonder Haar and Suomi 1971).

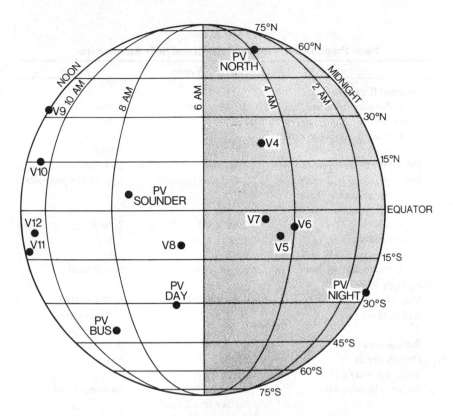

Fig. 1. The entry locations of the Pioneer Venus (PV) and Venera (V) probes. All the probes penetrated the deep atmosphere except the PV bus which burned up at 110 km. Except for the PV North probe which entered at 60°N, the entry locations are all in the equatorial region between ~ 30°S and ~ 30°N. Almost all the probes have entered between local midnight (the antisolar meridian) and local noon (the subsolar meridian). The V11 and V12 probes landed shortly after noon.

Latitudinal, zonal, and vertical variations in solar energy absorption provide the drive for atmospheric motions on Venus and Earth. Venus's orbital eccentricity and the tilt of its equator to its orbital plane (Shapiro et al. 1979) are both small. Therefore, in contrast to the Earth, seasonal variations in insolation are negligible for Venus. The dynamical state of Venus's atmospheric circulation should also be very nearly symmetric about its equator because its obliquity is so small. The nearly uniform level of Venus's surface (Masursky et al. 1980; Pettengill et al. 1980a) and the absence of any land-sea contrasts strengthen the case for an equatorially symmetric atmospheric circulation. While topographic forcing of atmospheric motions is expected to be unimportant because of Venus's mild topographic variations, the planet does have three major elevated regions of continental scale. The largest of these,

TABLE I
Basic Properties of Venus and Earth and their Atmospheres

	Venus	Earth
Orbital and Rotational Data		
Mean distance from Sun (km)	1.082×10^8	1.496×10^8
Eccentricity	0.0068	0.0167
Inclination of equator to orbit, obliquity	177°	23°45
Orbital (sidereal) period (d)	224.701	365.256
Rotation period (sidereal)	243 d retrograde	23.9345 hr prograde
Rotation rate (rad s^{-1}) and Coriolis parameter at 30° latitude (s^{-1})	2.99×10^{-7}	7.29×10^{-5}
Length of solar day, τ_{day}	117 d	1 d
Solar constant S (kW m^{-2})	2.62	1.38
Overhead motion of Sun	west to east	east to west
Solid Body Data		
Mass (kg)	4.870×10^{24}	5.973×10^{24}
Radius (km)	6051.5	6378 (equator), 6357 (pole)
Surface area (m^2)	4.6×10^{14}	5.1×10^{14}
Density (kg m^{-3})	5250	5515
Surface gravity g (m s^{-2})	8.87	9.78
Surface characteristics	nearly uniform surface level; few continental-scale highlands	land-sea contrasts
Angular momentum (kg m^2 s^{-1})	2.1×10^{31}	5.86×10^{33}
Atmospheric Data		
Composition (percentage by volume)	96.5% CO_2, 3.5% N_2, <0.1% SO_2, H_2O, Ar, CO, Ne	dry air—78% N_2, 21% O_2, 1% Ar, <0.1% CO_2, Ne, He; H_2O highly variable up to a few percent
Mean molecular weight (kg kmol^{-1})	43.44	28.96 (dry air)
Gas constant R (J kg^{-1} K^{-1})	191.4	287.1
Surface temperature T_s (K)	730	288
Surface pressure p$_s$ (MPa)	9.2	0.1
Surface density ρ_s (kg m^{-3})	65.0	1.225
Pressure scale height at surface H_s (km)	15.75	8.43
Mass (kg)	4.77×10^{20}	5.30×10^{18}
Excess angular momentum (kg m^2 s^{-1})	3.4×10^{28}	1.6×10^{26}

TABLE I (continued)
Basic Properties of Venus and Earth and their Atmospheres

	Venus	Earth
Exospheric temperature (K)	300 (day), 100 (night)	1000 K (mean solar conditions)
Albedo	0.8	0.30
Absorbed solar flux (W m^{-2})	132	242
Cloud cover	100%	40%
Cloud composition	concentrated sulfuric acid	solid and liquid water
Altitude of solar energy deposition	mostly within and above the clouds; EUV above 100 km; some near the surface and at the ground	mostly at the ground; ultraviolet absorption by ozone in stratosphere; EUV above 100 km
Approximate mean lapse rate in lower atmosphere (K km^{-1})	8	6.5
Approximate adiabatic lapse rate near surface (K km^{-1})	8	10 (dry), 6 (wet)
Specific heat at constant pressure near surface c_p (kJ kg^{-1} K^{-1})	~1.2	1.0 (dry)
Specific heat at constant pressure/specific heat at constant volume γ	~1.2	1.4
Sound speed at surface c_s (m s^{-1})	412	340

Aphrodite Terra, lies mostly in the southern hemisphere close to the equator, between about 0 and 30°S and 60 and 210°E. The second largest region, Ishtar Terra, containing the highest mountain on Venus, Maxwell Montes, lies mostly between about 60 and 75°N and 315 and 45°E. The smallest highland region on Venus, Beta Regio, lies between about 15 and 35°N and 275 and 295°E. These highland areas, which comprise ~ 8% of the surface, stand as much as 11 km (Maxwell Montes) above the mean topographic level. It is possible that these elevated regions force nonsymmetric motions and disturbances in the form of internal gravity waves.

Because of Venus's extremely slow rotation, winds in its atmosphere are subjected to only a weak Coriolis force, while on a rapidly rotating planet such as the Earth, the Coriolis force exerts a strong control on atmospheric motions. Winds in the Earth's atmosphere are in approximate geostrophic balance, a dynamical state in which parcels of air move under the dominant influence of nearly equal and opposite horizontal Coriolis and pressure gradient forces. In the absence of a strong Coriolis force, nonlinear inertial forces must balance pressure gradient forces. Therefore, the dynamical state of the Venus atmosphere is fundamentally different from the quasi-geostrophic regime that describes the large-scale wind field in the middle and high latitudes of the Earth's atmosphere.

The relative unimportance of the Coriolis force in Venusian atmospheric motions can be demonstrated quantitatively by evaluating the Rossby number

$$Ro = \frac{U}{fL} \tag{1}$$

where U is the horizontal wind speed, L is the horizontal length scale of the circulation, and f is the Coriolis parameter

$$f = 2\Omega \sin\phi \tag{2}$$

(where Ω is the planetary rotation rate and ϕ is the latitude). The Rossby number is the ratio of the inertial force to the Coriolis force. If $Ro \ll 1$ the motion is quasi geostrophic; if $Ro \gg 1$ the Coriolis force is negligible. Typical values of U and L for large-scale motions are 10 m s^{-1} and 10^6 m. With $\phi = 30°$ and f from Table I, we find $Ro = 33.4$ and 0.14 for Venus and Earth, respectively. The Rossby number based on Venus's planetary rotation rate is large; the Venus atmosphere cannot be in a quasi-geostrophic dynamical state. We will see later that the cloud level Venus atmosphere is in a quasi-cyclostrophic state (Leovy 1973) in which the centrifugal force of the rapid cloud level atmospheric superrotation is approximately balanced by the meridional pressure gradient. (At cloud heights the Venus atmosphere rotates ~ 60 times faster than the planet.) However, there is no significant superrotation of the deep Venus atmosphere, so in the lowest scale height the atmosphere is neither quasi cyclostrophic nor quasi geostrophic.

The slow retrograde rotation of the planet combines with Venus's orbital motion to produce a solar day lasting 117 days, about half a Venus year. Comparison of the duration of the Venus day τ_{day} with other characteristic time scales gives information on the importance of different processes in the atmosphere (Stone 1975). For example, the Rossby number can be written

$$Ro = \frac{U}{fL} = \frac{U}{2\Omega \sin\phi\, L} = \frac{\tau_{day}/2\pi}{2 \sin\phi\,(L/U)}$$

$$= \left(\frac{1}{4\pi \sin\phi}\right)\frac{\tau_{day}}{\tau_{dyn}} \tag{3}$$

where the dynamical time scale τ_{dyn} is L/U. The Rossby number is thus a measure of the ratio of length of day to characteristic dynamical time. Since

Ro is large for Venus, the dynamical time is less than one Venus day and Coriolis effects should be relatively unimportant. As an additional consequence of Venus's combined orbital and rotational motions, an observer on the surface would see the Sun rise in the west and set in the east. The direction of the Sun's overhead motion plays an important role in one of the proposed explanations of the generation and maintenance of Venus's atmospheric superrotation (Schubert and Whitehead 1969). Thermal tides follow the motion of the Sun in Venus's sky.

While Venus absorbs less solar energy than the Earth, the amount it does absorb has profound consequences for its massive atmosphere. Although most of Venus's solar energy absorption takes place within and above its highest clouds, $\sim$ 2.5% of the sunlight incident on the planet is absorbed at the ground (Tomasko et al. 1980*a,b*). This is apparently enough to account for the enormously high surface temperature by the greenhouse mechanism (Sagan 1962; Pollack 1969*a*; Pollack and Young 1975; Pollack et al. 1980*a*; Chapter 18); there need be no downward transport of heat by a large-scale circulation, as originally proposed by Goody and Robinson (1966). The main absorbers of infrared radiation in Venus's atmosphere are CO_2, H_2O, cloud particles, and SO_2 (Pollack et al. 1980*a*).

Solar energy absorption is the drive for atmospheric motions. Unlike the Earth's atmosphere, which is mainly driven from below by absorption of sunlight at the surface, Venus's atmospheric circulation is largely driven from above by the absorption of sunlight within and above the upper cloud layer. Venus's ubiquitous cloud cover influences atmospheric circulation by controlling the amount and location of solar energy absorption. However, the latent heat associated with the condensation and evaporation of the H_2SO_4 cloud droplets is not an important energy source or sink for motions, simply because the mass loading of the clouds is small (Rossow 1978; Knollenberg et al. 1980). This is in contrast to the importance of cumulus convection in determining the thermal structure and large-scale circulation of the Earth's tropical atmosphere. Minor constituents of Venus's atmosphere such as H_2O and SO_2 also influence the thermal structure and motions, because they contribute to the absorption of long-wavelength planetary radiation.

Figure 2 shows how the pressure scale height

$$H = \frac{RT}{g} \qquad (4)$$

where T is temperature, varies with altitude in the atmospheres of Venus and Earth. Here R is the gas constant (universal gas constant divided by mean molecular weight), T is temperature, and g is the acceleration of gravity. Because of the high surface temperature T_s, the scale height at the surface is nearly twice that of Earth; $\sim$ 65% of the entire mass of an atmosphere lies within the lowest scale height. The vertical profile of H basically reflects the

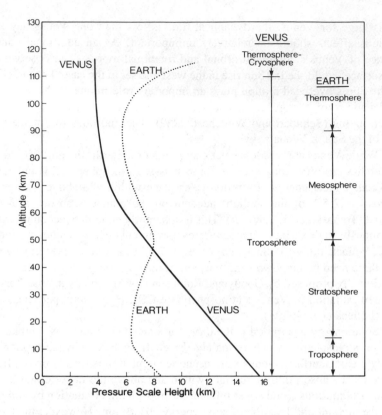

Fig. 2. Pressure scale height H versus altitude for Venus and Earth. H for Venus is based on the Pioneer Venus Large probe temperature measurements (Seiff et al. 1980); H for Earth is based on the U.S. Standard Atmosphere (1976). The altitude profile of H for Venus has been smoothed above 65 km. The variation of g with altitude is taken into account for the Earth but not for Venus.

thermal structure because R and g are essentially constant in the lower atmosphere. Thus the altitude dependence of H reveals the troposphere, stratosphere, mesosphere, and thermosphere of the Earth's atmosphere, and the troposphere of the Venus atmosphere. We have arbitrarily referred to the region of Venus's atmosphere between the surface and ~ 100 km as the troposphere, encompassing the ''mesosphere'' defined in the Preface, Fig. 1. Temperature increases approximately monotonically with height throughout this altitude range (Seiff et al. 1979a,b, 1980; Kliore and Patel 1980; Taylor et al. 1981; Seiff and Kirk 1982). However, pronounced inversions occur at heights of ~ 60–70 km for latitudes $\geqslant 50°$ (Kliore and Patel 1980) and a different terminology or classification might be more appropriate for this high latitude thermal structure. Venus's upper atmosphere above the troposphere

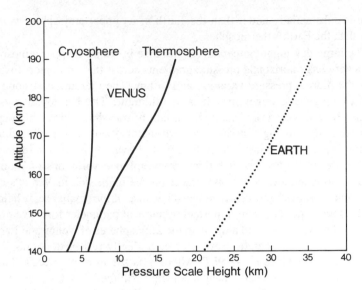

Fig. 3. Pressure scale height dependence on altitude in the Earth's thermosphere and in Venus's thermosphere/cryosphere. H for the Earth is based on the U.S. Standard Atmosphere (1976). H for Venus is computed from the model upper atmosphere of Keating et al. (1980). The variations of g and the mean molecular weight with height are accounted for. In computing the mean molecular weight for Venus's upper atmosphere only CO_2 and O are considered. This is only an approximation at certain altitudes, but is generally acceptable since Venus's upper atmosphere is CO_2 dominated below ~ 145 km (night side) or ~ 160 km (day side), and mainly O above these heights for the altitudes considered.

has the structure of a dayside thermosphere in which temperature increases with height to an exospheric temperature of ~ 300 K (von Zahn et al. 1979b, 1980; Niemann et al. 1980b; Keating et al. 1979b, 1980) and a nightside cryosphere (Schubert et al. 1980a) in which temperature decreases with height to an exospheric temperature of ~ 100 K (Keating et al. 1979a, 1980; Niemann et al. 1979a, 1980b; Seiff and Kirk 1982) (see Fig. 38 in Sec. VII). The altitude profiles of H in the Venus thermosphere-cryosphere and the Earth's thermosphere are shown in Fig. 3. Scale heights in Venus's upper atmosphere are much smaller than they are in the Earth's thermosphere, because of the relatively low temperatures in Venus's exosphere. The increase of H with altitude in the thermospheres of both Venus and Earth is a consequence of the increase of temperature and the decrease of mean molecular weight and g with height. The smaller scale heights in Venus's cryosphere than in its thermosphere are due to the very low temperatures in Venus's nightside upper atmosphere. Since temperature decreases with height in the Venus cryosphere, the small increase in H with altitude is due mostly to the decrease in mean molecular weight with height. While Venus's lower atmosphere is substantially hotter than the Earth's, its upper atmosphere is much colder than the Earth's thermosphere.

The large day-night temperature difference in Venus's upper atmosphere creates very large horizontal pressure gradients across the terminator. Figure 4 shows the diurnal pressure variation at 190 km. The pressure p at noon is 3 orders of magnitude larger than it is at midnight. The horizontal pressure gradient force arising from this huge variation in p should drive high-speed winds from day to night across the terminator (Seiff 1982). The diurnal variation in p is more extreme than the day-night variations in either temperature T or density ρ because it reflects the combined variations of T and ρ; $p = \rho RT$. Temperature is not only lower on the night side in Venus's cryosphere, but ρ is also significantly reduced because of the smaller scale height.

The sound speed c_s is an important dynamical parameter for wave propagation, and its variation with altitude in the atmospheres of Venus and Earth is shown in Fig. 5. The height profiles of c_s basically reflect the atmospheric thermal structures since $c_s^2 = \gamma RT$, where γ is the ratio of specific heat at constant pressure c_p to specific heat at constant volume c_v. Below ~ 40 km, c_s is larger for Venus than it is for Earth because of Venus's much higher surface temperature, but above this height Venus's atmosphere is colder than Earth's and accordingly its sound speed is lower. By combining c_s and H according to

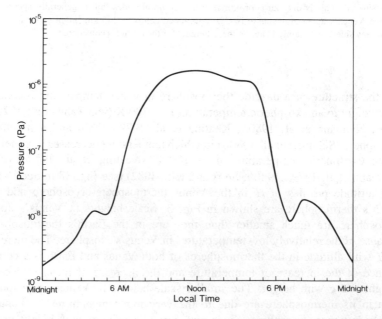

Fig. 4. Diurnal pressure variation at 190 km in the upper atmosphere of Venus. Pressures are computed from the model atmosphere of Keating et al. (1980) using the temperature and atomic oxygen number density.

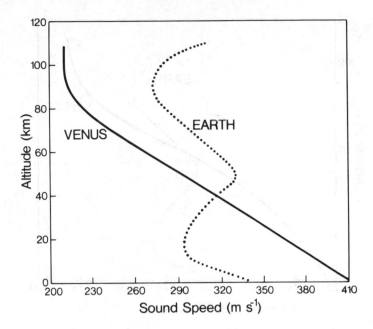

Fig. 5. Altitude profiles of sound speed c_s in the atmospheres of Venus and Earth. For Earth, c_s is based on γ and R from Table I and the U.S. Standard Atmosphere (1976) temperature profile. For Venus, c_s is computed from the thermal structure used in Fig. 2 and the value of R in Table I. The ratio of specific heats for Venus is calculated from the adiabatic lapse rate given in Seiff et al. (1980), by equating this lapse rate to g/c_p. Thus the computed sound speed for Venus accounts for the variation of γ with temperature and pressure. The altitude profile of c_s for Venus has been smoothed above 65 km.

$$\omega_a = \frac{c_s}{2H} \tag{5}$$

we obtain the acoustic cutoff frequency ω_a; ω_a is the lowest frequency at which sound waves can propagate in an isothermal, nonrotating atmosphere (Houghton 1977). The variation of ω_a with height in the atmospheres of Venus and Earth is shown in Fig. 6. Sound waves can propagate with lower frequency below cloud heights in Venus's atmosphere than they can at comparable altitudes in the Earth's atmosphere; the reverse is true for the altitude range 60 to 110 km. There is relatively little variation in ω_a with height in the Earth's atmosphere, whereas there is a factor of 2 variation in ω_a between the surface and 100 km on Venus, a direct consequence of the drop in temperature from ~ 730 K at the surface (Seiff et al. 1980) to ~ 180 K at 100 km altitude (Taylor et al. 1980). Atmospheric structure also has an important influence on dynamics and wave propagation, through a parameter known as the Brunt-Väisälä frequency N:

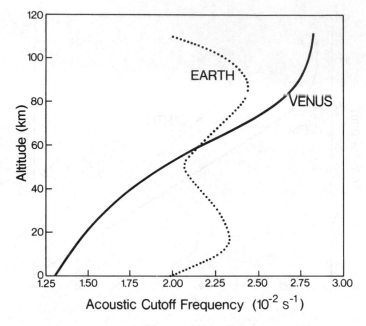

Fig. 6. Acoustic cutoff frequency $\omega_a = c_s/2H$ versus height in the atmospheres of Venus and Earth. c_s and H are obtained from previous figures.

$$N = \left[\frac{g}{T}\left(\frac{dT}{dz} - \Gamma\right)\right]^{1/2} \qquad (6)$$

where dT/dz is the vertical temperature gradient and Γ is the adiabatic temperature gradient. The Brunt-Väisälä frequency is a measure of atmospheric stability; the larger N the more stable the atmosphere. In an isothermal, nonrotating atmosphere N is the maximum frequency of internal gravity waves. Altitude profiles of N for Venus and Earth are shown in Fig. 7. N is small in the Earth's troposphere where the temperature gradient is close to the wet adiabat. In the Earth's stratosphere and above, N is relatively large and the atmosphere is highly stable. The lowest one or two scale heights of Venus's atmosphere are close to being adiabatic, and the Brunt-Väisälä frequency is accordingly small. Between $\sim$ 30 km and the base of Venus's clouds at $\sim$ 50 km there is a stable layer in which N reaches a maximum of $\sim 10^{-2}$ s^{-1} at $\sim$ 45 km. The stability of this region of the atmosphere has important consequences for the mean meridional circulation and the vertical transport of zonal momentum. There is a nearly neutral layer between $\sim$ 50 and 55 km within Venus's clouds; at still greater heights Venus's atmosphere is strongly stable. Internal gravity waves can propagate with higher frequency in the Earth's lower atmosphere than they can in Venus's lower atmosphere.

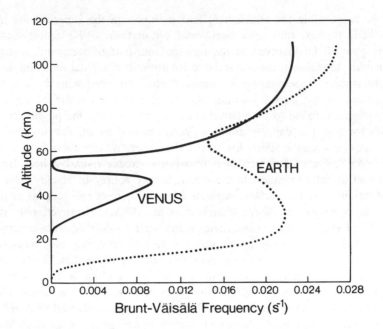

Fig. 7. Brunt-Väisälä frequency as a function of height in the lower atmospheres of Venus and Earth. The profile for Venus is computed from the static stability and temperature data of the Pioneer Venus Large probe (Seiff et al. 1980). The static stability above 65 km in Venus's atmosphere has been smoothed.

Circulation and Dynamics

Having completed this overview of the properties and structure of Venus's atmosphere, let us proceed to discuss its circulation and dynamics. The principal mode of atmospheric circulation on Venus is a zonal retrograde superrotation; the entire atmosphere above the lowest scale height, extending upward to ≥90 − 100 km, participates in this global superrotation (Schubert et al. 1980a). Zonal winds increase in speed from ≲ 10 m s^{-1} at 10 km to 100 m s^{-1} at cloud heights (Counselman et al. 1979, 1980). The zonal retrograde superrotation may also be the dominant mode of circulation in the lowest scale height and at even higher altitudes, but there are insufficient data to reliably establish the nature of the large-scale circulation in these regions. There is indirect evidence for a superrotation of the thermosphere-cryosphere at 150 km (Schubert et al. 1980a).

We begin with a review of the observations of the global atmospheric superrotation (Sec. II). We then discuss the implications of this dominant circulation mode for the dynamical state of the atmosphere, and conclude that approximate cyclostrophic balance prevails in nonequatorial regions where the zonal winds are large (Sec. III). In Sec. IV the physical mechanisms that may be responsible for establishing and maintaining the superrotation are described. Eddies, the mean meridional circulation, and planetary-scale waves may all be involved in the upward transport of westward angular momentum to maintain the atmospheric rotation. The angular momentum of the atmospheric superrotation is estimated and compared with the angular momentum of the slowly rotating planet. The two reservoirs of angular momentum are found to be relatively comparable, raising the possibility of significant length of day variations on Venus caused by angular momentum exchanges between the atmosphere and solid planet. We discuss in Sec. V the meridional circulation which also transports excess radiative heat from equatorial to polar latitudes. In the mean, it may consist of Hadley cells at cloud heights and the surface, separated by an indirect cell located in the stable layer below the clouds (Schubert et al. 1980a). However, only the cloud-level Hadley cell has been reasonably well established from observations. Eddies occur in the lower atmosphere and they may also transport significant amounts of heat and momentum latitudinally and vertically. The zonal wind and poleward flow at the cloud tops combine to form a polar vortex (Suomi and Limaye 1978). When viewed from above the South pole, the vortex rotates counterclockwise and the flow is poleward and downward at the cloud tops. Section VI deals with waves and emphasizes the planetary-scale variety which forms the dark horizontal Y and other ultraviolet albedo patterns of global extent. Finally, Sec. VII discusses the circulation of the thermosphere-cryosphere. Although this is probably dominated by a strong subsolar-antisolar circulation which is axisymmetric about the Venus-Sun line, there may also be a retrograde superrotation of the upper atmosphere; the thermosphere-cryosphere superrotation may be the upward continuation of the superrotation of the lower atmosphere (Schubert et al. 1977).

In view of the numerous and striking differences between Venus and Earth in atmospheric structure and composition, manner of solar energy deposition, and planetary rotation rate, it is hardly surprising that there is little similarity between their large-scale atmospheric circulations. A number of these differences and their consequences have been discussed in previous reviews of the dynamics and circulation of Venus's atmosphere by Stone (1975), Schubert et al. (1977, 1980a), Schubert and Covey (1981), and Moroz (1981).

II. ZONAL RETROGRADE SUPERROTATION

The first observations to reveal the dominant east-to-west rotation of Venus's atmosphere were the Earth-based ultraviolet pictures of the planet. While Venus is essentially featureless when observed in visible light, contrast markings are apparent in ultraviolet photographs. Earth-based ultraviolet pictures of Venus have been available for more than 60 years (Quénisset 1921; Wright 1927; Ross 1928). However, it was not until the 1960s that a prominent large-scale albedo feature known as the dark horizontal Y was observed to reappear with a period of ~ 4 d, indicating a retrograde zonal rotation of the cloud level atmosphere with equatorial wind speeds of $\sim 2\pi \,(6050 \text{ km})/4 \text{ d} = 110 \text{ m s}^{-1}$ (Boyer and Camichel 1961; Boyer 1965; Smith 1967). Intensive observation extending into the 1970s revealed other planetary-scale ultraviolet features (C or ψ shapes) and long-term variations in these markings (Dollfus 1975). Because these recurring large-scale albedo markings might possibly be manifestations of propagating planetary waves, the Earth-based ultraviolet photographs did not conclusively establish the existence of a bulk rotation of Venus's atmosphere.

The higher resolution ultraviolet pictures of Venus taken by Mariner 10 (Murray et al. 1974; Belton et al. 1976a,b; Anderson et al. 1978) and Pioneer Venus (Travis et al. 1979a, b; Rossow et al. 1980a) spacecraft have made it possible to separate actual atmospheric rotation from the motions of planetary waves, by the differences in movements of small-scale and large-scale markings. The motions of the small-scale ultraviolet features reflect the bulk velocity of the atmospheric winds (Suomi 1974) while the displacements of the large-scale markings reveal the propagation of planetary-scale waves relative to these winds (Belton et al. 1976a,b). The longitudinally averaged zonal wind speeds deduced from tracking of small-scale features in the spacecraft ultraviolet images are shown as a function of latitude in Fig. 8. The solid curve is the zonal wind velocity for solid body retrograde rotation with an equatorial speed of 92.4 m s^{-1}. During the Pioneer Venus nominal mission, the Venus atmosphere at cloud level was rotating with a nearly constant angular velocity (Rossow et al. 1980a). However, at the time of the Mariner 10 flyby, there was a prominent midlatitude jet; mean zonal wind speeds in the middle latitudes distinctly exceeded the velocities corresponding to solid body rotation of the cloud level atmosphere at the period of the equatorial flow (Suomi 1974; Limaye 1977; Limaye and Suomi 1981). This difference between the Mariner 10 and Pioneer Venus latitudinal profiles of the zonally averaged east-to-west wind shows that the zonal wind varies with time in a manner that may help explain how this global atmospheric rotation is maintained, a point we return to in Sec. IV. The cloud-tracked zonal wind speeds in Fig. 8 are not symmetric about the equator. Other data discussed later also suggest the possibility of a hemispherical asymmetry in Venus's mean circulation, even though solar forcing is expected to drive only symmetric motions. Alternatively, the mean circulation may indeed be symmetric about the equator, and these measurements may only reflect the occurrence of antisymmetric wave modes or eddies during observation.

Because the large-scale markings maintained their overall coherence despite the nonuniform atmospheric rotation at the time of Mariner 10, it must be concluded that global patterns such as the dark horizontal Y are manifestations of planetary-scale waves propagating with respect to the winds (Belton et al. 1976a,b). Thus, the Earth-based observations of periodically recurring global

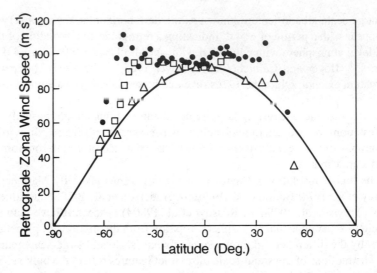

Fig. 8. Longitudinally averaged retrograde zonal wind velocities versus latitude, inferred from the tracking of small-scale features in spacecraft ultraviolet images of Venus: △, Pioneer Venus (Rossow et al. 1980a); ●, Mariner 10 (Limaye 1977; Limaye and Suomi 1981); □, Mariner 10 (Travis 1978). The solid curve is the zonal wind velocity for solid body retrograde rotation with equatorial speed 92.4 m s⁻¹ corresponding to a period of 4.8 d.

markings did not actually reveal the atmospheric rotation; instead, the velocities inferred from the periodicities of Y, ψ and reversed C patterns are the sums of the wind velocities and the wave speeds relative to the winds. However, these planetary-scale waves move rather slowly with respect to the winds, and the combined movement of waves and winds is not very different from that of the atmospheric rotation alone, fortunately for those who interpreted the Earth-based ultraviolet observations as bulk atmospheric motion.

Earth-based spectroscopic (Traub and Carleton 1975, 1979) and heterodyne (Betz et al. 1976, 1977) measurements of Doppler-shifted CO_2 lines also show that Venus's atmospheric circulation at cloud heights is mainly an east-to-west rotation with equatorial speeds of 100 m s⁻¹. The heterodyne data indicate that the rapid zonal circulation extends from the cloud top to up to altitudes of at least 70 to 80 km in the equatorial region. They also suggest a different circulation regime at still higher altitudes, namely a predominantly subsolar to antisolar flow at 115 km with 100 m s⁻¹ winds across the terminator. The heterodyne results are the only direct measurements of wind speeds above Venus's clouds. However, there may be large uncertainties in the wind velocities and circulation patterns inferred from these data, as indicated by the heterodyne measurements of 35 to 45 m s⁻¹ average vertical wind speeds over periods of weeks in the 75 to 115 km region. Wind velocities can also be inferred from temperature and pressure measurements between ~ 70 and 110 km, but the speeds obtained are also uncertain because of assumptions in the dynamical modeling. Thus we cannot be sure of the height to which the zonal retrograde superrotation dominates the atmospheric circulation.

Doppler tracking of Venera probes and landers (Marov et al. 1973*b*; Antsibor et al. 1976; Keldysh 1977; Kerzhanovich et al. 1979*c*; Moroz 1981) and interferometric tracking of the Pioneer Venus probes (Counselman et al. 1979, 1980) have resulted in the altitude profiles of the zonal wind shown in Fig. 9. When compared with vertical profiles of meridional wind velocities from the Pioneer Venus interferometric tracking experiment (see Fig. 28 in Sec. V), these data show that east-to-west rotation is the predominant mode of atmospheric circulation at all altitudes between ~ 10 km and cloud levels. In the lowest 10 km of the atmosphere both zonal and meridional wind velocities are only a few m s^{-1}. Surface winds measured by anemometers on Veneras 9 and 10 (Avduevskii et al. 1976*b*) were in the range 0.3 to 1 m s^{-1}. The observed motions in the lowest scale height of the Venus atmosphere are sluggish and mainly neither meridional nor zonal. Therefore, the few measurements of wind velocity we have are inadequate even for guessing at the circulation pattern from these data alone. Zonal wind speeds increase essentially monotonically with height. Significantly, there is no indication of an eastward zonal wind at any level of the atmosphere. This is an important constraint on atmospheric circulation theories and the mechanisms for the maintenance of the retrograde zonal superrotation. There is a gradual increase in westward zonal wind velocity with height for some of the profiles (Pioneer Venus North probe, Veneras 9 and 10), while others show layers of relatively constant wind speed alternating with regions of large increase in velocity with height (Pioneer Venus equatorial probes, Veneras 8 and 12). The layering in the zonal wind velocity profiles correlates qualitatively with layered structures in the vertical profiles of atmospheric static stability (Schubert et al. 1980*a*); the vertical shear in the zonal winds is large in very stable layers. This agrees with our dynamical intuition that high static stability strongly inhibits the vertical mixing which tends to reduce the wind shear.

The zonal wind velocities at cloud heights are on the order of 100 m s^{-1} for all Pioneer Venus probes and for Venera 8. However, wind speeds at cloud level are only ~ 50 m s^{-1} for Veneras 9, 10 and 12. These latter probes all entered the atmosphere near local noon (see Fig. 1); their measurement of smaller wind speeds at cloud level suggests that the zonal winds at these heights may be slower near the subsolar region. This would imply a deceleration in atmospheric rotation rate at cloud level between morning and noon and an acceleration after noon (times of day are with respect to retrograde rotation). Both Earth-based ultraviolet observations (Boyer 1973) and the spectroscopic data of Traub and Carleton (1975) indicate that the zonal wind velocities at cloud height are larger in the afternoon than in the morning. This is also supported by the analysis of Mariner 10 ultraviolet images reported in

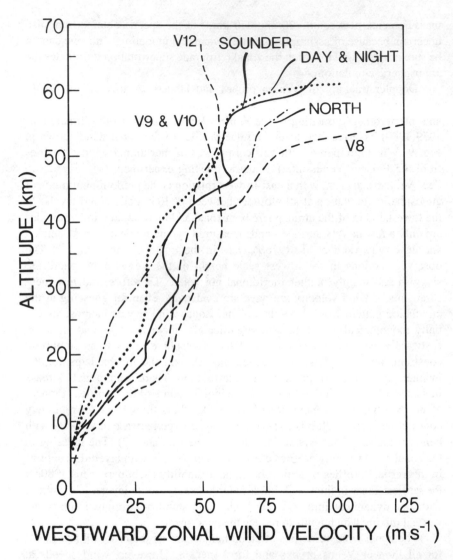

Fig. 9. Vertical profiles of east-to-west wind speed from Doppler tracking of Veneras 8, 9, 10, and 12 (V8, V9, V10, V12) and interferometric tracking of Pioneer Venus probes (from Schubert et al. 1980a).

Fig. 3 of Schubert et al. (1977) (see Limaye and Suomi [1981] for a more comprehensive study of these data). An apparent acceleration in the zonal wind between morning and afternoon has also been found in the wind speeds inferred from tracking of features in the Pioneer Venus ultraviolet images (Limaye et al. 1982). If morning to afternoon acceleration in the cloud level zonal wind exists, it has important implications for theories of the retrograde

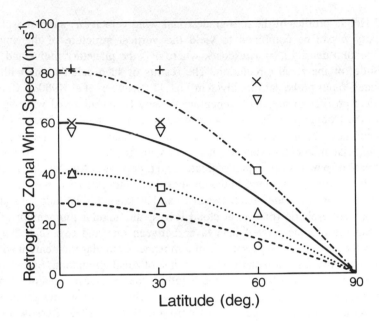

Fig. 10. Latitudinal variation of retrograde zonal wind speeds measured by interferometric tracking of Pioneer Venus probes. The symbols refer to different altitudes: 20 km, (O); 30 km, (Δ); 40 km, (□); 50 km, (∇); 55 km, (×); and 60 km, (+). The curves represent solid body rotation at different rates. It is assumed that the zonal circulation is approximately symmetric about the equator, so the wind speeds for the Day and Night probes can be plotted at 31°N.

superrotation; it suggests that direct solar heating plays a significant role in the maintenance of the zonal circulation. The close similarity of zonal wind velocity profiles from the Day and Night Pioneer Venus probes shows, however, that there is little change in atmospheric rotation between midnight and 6 AM.

The latitude dependence of Venus's atmospheric rotation below the clouds can be ascertained from the wind profiles in Fig. 9. Retrograde zonal wind speeds for the Pioneer Venus probes are plotted as a function of latitude in Fig. 10. At 20 km the data are approximately consistent with solid-body rotation of the atmosphere, but not at higher altitudes. The wind speeds at 60° latitude are substantially higher than they would be if the atmosphere above 20 km were in solid-body rotation with the angular velocity of the equatorial zonal circulation. Temporal changes in atmospheric rotation rate can reconcile the probe results for 60 km with the approximate solid-body rotation of the cloud tops at slightly higher levels implied by the Pioneer Venus ultraviolet cloud tracking. The cloud-tracked winds refer to a time 2 mo after the probes entered the atmosphere.

Height profiles of the retrograde zonal wind velocity u_λ and atmospheric density ρ can be combined to yield the vertical structure of the angular momentum density $L = \rho a u_\lambda \cos\phi$, where a is the planetary radius and ϕ is latitude, of the zonal circulation. The results of such calculations with the Pioneer Venus probe data are given in Fig. 11 (Schubert et al. 1980a). Similar vertical profiles of angular momentum density based on zonal wind speeds from the Doppler tracking of Venera probes are presented in Chapter 22. At low altitudes the increase in u_λ with height dominates the decrease of ρ, and the angular momentum density increases with height. Above $\sim$ 20 km the decrease of ρ with altitude predominates and L decreases with height. There is therefore a broad maximum in the angular momentum density, with the peak $\sim$ 20 km at low latitudes and between 20 and 30 km at 60° latitude. In spite of the fast atmospheric rotation at cloud levels, the angular momentum of the atmosphere is concentrated in a layer between one and two scale heights above the surface. Possible sources of atmospheric angular momentum will be considered in the section on the maintenance of zonal circulation. Since one of these sources is the solid planet, the height of the angular momentum density peak may characterize the vertical scale over which efficient upward transfer of angular momentum occurs (Del Genio and Rossow 1982; Rossow 1982). The scale could be associated either with the thickness of a surface Hadley cell or with the height of eddies driven by diurnal variations in heating.

Figure 11 shows that angular momentum density decreases with latitude at all heights between $\sim$ 10 km and the clouds; the same is true for the angular momentum per unit mass $a u_\lambda \cos\phi$. Although u_λ increases with ϕ between 30° and 60° latitude above $\sim$ 30 km (Fig. 10), L and L/ρ still decrease with ϕ because of the decrease of $\cos\phi$ with latitude. Since L/ρ decreases with ϕ, poleward meridional motions transport retrograde angular momentum to higher latitudes. Meridional velocities measured by the Pioneer Venus interferometric tracking experiment (see Fig. 28, Sec. V) change direction with height, suggesting poleward advection of retrograde angular momentum at some altitudes and equatorward advection of relatively prograde momentum at other heights. The latitudinal variation in angular momentum density tends to decrease with height between $\sim$ 20 km and the base of the clouds of $\sim$ 50 km.

At the top of the clouds, $\sim$ 65 to 70 km, ultraviolet cloud-tracked winds show that L/ρ can either decrease with ϕ or remain approximately constant. When the cloud-top atmosphere is in approximate solid-body rotation with frequency ω, as it was at the time of Pioneer Venus, L and L/ρ decrease with latitude proportionately with $\cos^2\phi$ since

$$L = \rho a u_\lambda \cos\phi = \rho \omega a^2 \cos^2\phi. \qquad (7)$$

At the time of the Mariner 10 flyby, approximate solid-body rotation of the cloud-top atmosphere prevailed poleward of midlatitudes (Fig. 8) and L and

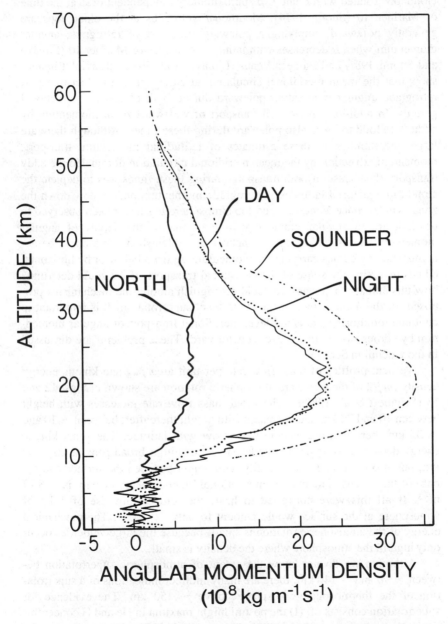

Fig. 11. Retrograde angular momentum density of the atmosphere versus height, from the zonal wind velocity and density measurements of the Pioneer Venus probes (Schubert et al. 1980a).

L/ρ decreased with ϕ in this region. In equatorial latitudes, however, the cloud tops rotated with L and L/ρ approximately independent of ϕ at the time of Mariner 10 (Suomi 1974). Meridional velocities at the cloud tops are generally poleward, implying a poleward transport of retrograde angular momentum when L decreases with latitude. Analyses of Mariner 10 (Limaye and Suomi 1981) and Pioneer Venus (Limaye et al. 1982) ultraviolet images show that the mean meridional circulation at cloud top did in fact transport retrograde angular momentum poleward during both of these observational periods. In addition, meridional transport of westward zonal momentum by eddies at cloud top was also poleward during these times. Although there are large uncertainties in these estimates of latitudinal momentum transport, momentum advection by the mean meridional circulation dominated the eddy transport. The sense of both transports during these times was to deplete the cloud-top equatorial latitudes of retrograde momentum and to slow down the zonal winds. Since Venus's cloud-top atmosphere has never been observed to be deficient in angular momentum at the equator, the supply of angular momentum to this region must somehow be replenished. This could be accomplished by eddies carrying the momentum up from below or by horizontal eddies reversing the sense of the meridional transport. If we could determine how this actually happens, we could distinguish among the mechanisms proposed for the maintenance of the cloud-level superrotation. It is important to continue monitoring the cloud-level meridional transport of angular momentum by eddies, to see if it is ever equatorward. These problems are discussed in more detail in Sec. IV.

Vertical profiles of mass flow rate per unit area ρu_λ and kinetic energy density $\rho u_\lambda^2/2$ of the westward atmospheric rotation are shown in Figs. 12 and 13 (Schubert et al. 1980a). The zonal mass flow rate increases with height between 10 and 20 km and decreases with height thereafter; between ~ 10 and ~ 35 km there is a decrease in mass flow with latitude. The zonal kinetic energy density at equatorial latitudes shows a similar broad peak at 20 to 30 km, but at 60°N it increases steadily with height between the surface and the base of the clouds. The maximum rotational kinetic energy density is ~ 8 kJ m^{-3}. If all this were converted to heat, the temperature rise of 1 kg of atmosphere at the surface would amount to only ~ 0.1 K. The mechanical energy of the atmospheric rotation is small, because the large velocities occur only high in the atmosphere where the density is small.

Although we do not know the extent of atmospheric superrotation between ~ 90 and ~ 140 km, there are many indirect indications of a superrotation of the thermosphere-cryosphere above ~ 150 km. The evidence for superrotation consists of: (1) the postmidnight maxima in He and H concentrations (Niemann et al. 1979c, 1980b; Brinton et al. 1979); (2) the phase of the diurnal temperature variation (Mayr et al. 1980; Keating et al. 1980); and (3) the postmidnight enhancement in intensity of the ultraviolet night airglow, probably associated with subsidence (Stewart and Barth 1979; Stewart et al.

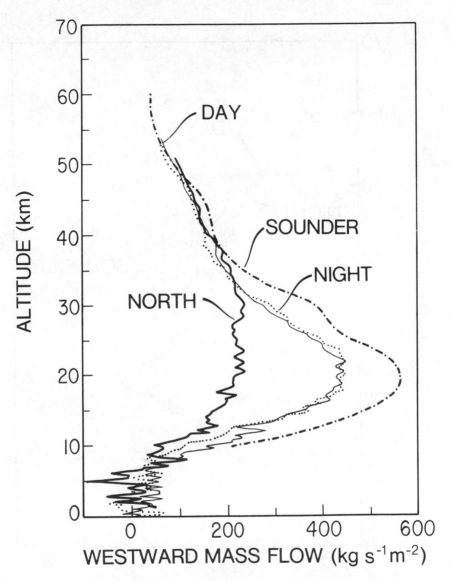

Fig. 12. Vertical profile of the westward mass flow rate per unit area ρu_λ, from the wind and density measurements of the Pioneer Venus probes (Schubert et al. 1980a).

1980). The nighttime He and H bulges imply retrograde zonal winds at altitudes > 150 km in excess of 50 m s^{-1} at the equator (Mayr et al. 1980). The phase of the diurnal temperature variation indicated by mass spectrometer and drag measurements is consistent with theoretical superrotation models with 5 to 10 d rotation periods (Elson 1977; Mayr et al. 1980). Large 5 to 6 d

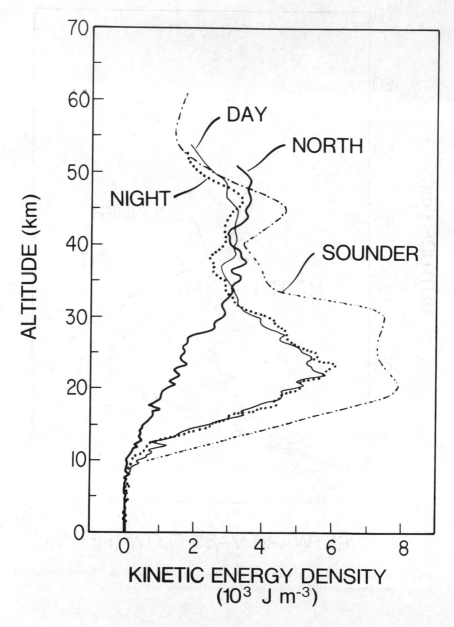

Fig. 13. Kinetic energy density $(\frac{1}{2})\rho u_\lambda^2$ of the atmospheric rotation as a function of altitude, from the wind and density measurements of the Pioneer Venus probes (Schubert et al. 1980a).

oscillations in temperature on the night side (Keating et al. 1979*a*) may also be indicative of a 5 to 6 d rotation period of the upper atmosphere. Even if the superrotation of Venus's upper atmosphere is real, we do not expect it to be the dominant form of circulation. The large day-night horizontal pressure gradients (Fig. 4) should drive a strong subsolar-antisolar circulation with high-speed cross-terminator winds (Niemann et al. 1980*b*; Mayr et al. 1980; Seiff 1982).

III. CYCLOSTROPHIC BALANCE

Venus's nonequatorial atmosphere between $\sim$ 10 km and the cloud tops is in approximate cyclostrophic balance, a dynamical state in which the equatorward component of the centrifugal force on a zonally rotating atmospheric parcel is balanced by a poleward meridional pressure gradient force (see Fig. 14). Leovy (1973) first suggested that cyclostrophic balance might occur in the rapidly rotating Venus atmosphere; measurements of wind speeds and horizontal temperature and pressure contrasts by the Pioneer Venus probes have since provided qualitative evidence for this balance (Seiff et al. 1980; Schubert et al. 1980*a*). The equatorward latitudinal component of the centrifugal force per unit mass due to zonal motion with westward velocity u_λ is $u_\lambda^2 \tan\phi/a$. The poleward meridional pressure gradient force per unit mass is $- (\partial p/\partial y)/\rho$, where y is a locally northward cartesian coordinate. Cyclostrophic balance thus takes the form

$$\frac{u_\lambda^2 \tan\phi}{a} = -\frac{1}{\rho}\frac{\partial p}{\partial y} \, . \tag{8}$$

With the introduction of the geopotential Φ (g times altitude z) and the hydrostatic approximation

$$\frac{\partial p}{\partial z} = -\rho g \tag{9}$$

Eq. (8) can be written

$$\frac{u_\lambda^2 \tan\phi}{a} = -\frac{1}{\rho}\left(\frac{\partial p}{\partial y}\right)_z = -g\left(\frac{\partial z}{\partial y}\right)_p =$$

$$-\left(\frac{\partial \Phi}{\partial y}\right)_p = -\frac{1}{a}\frac{\partial \Phi}{\partial \phi} \, . \tag{10}$$

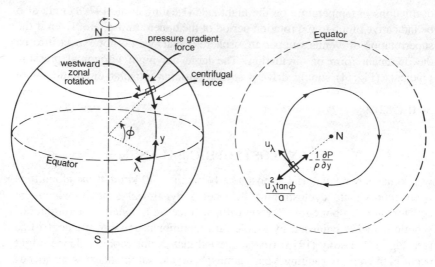

Fig. 14. Sketch of the forces acting on a zonally rotating atmospheric parcel in cyclostrophic balance.

In terms of the logarithmic pressure coordinate ζ:

$$\zeta = -\ln \frac{p}{p_{\text{ref}}} \tag{11}$$

where p_{ref} is a reference pressure, the hydrostatic equation becomes

$$\frac{1}{g} \frac{\partial p}{\partial z} = -\rho = \frac{\partial p}{\partial \Phi} \tag{12}$$

$$p \frac{\partial \Phi}{\partial p} = -\frac{p}{\rho} = -RT = -\frac{\partial \Phi}{\partial \zeta} \tag{13}$$

$$\frac{\partial \Phi}{\partial \zeta} = RT. \tag{14}$$

The perfect gas law $p = \rho RT$ is used in Eq. (13). By differentiating Eq. (10) with respect to ζ and substituting Eq. (14) we obtain

$$2u_\lambda \frac{\partial u_\lambda}{\partial \zeta} = -\frac{R}{\tan\phi} \frac{\partial T}{\partial \phi}. \tag{15}$$

Equation (15) is the thermal wind equation for cyclostrophic balance. The vertical derivative of the zonal kinetic energy per unit mass is directly proportional to the latitudinal temperature gradient on a constant pressure surface. If T increases poleward then u_λ decreases with height; if T decreases poleward the zonal wind velocity increases with height. Note that ζ increases with z according to Eq. (11).

Altitude profiles of zonal wind velocity can be inferred from measurements of T as a function of latitude and height by integrating Eq. (15) over ζ. Derived wind speeds can be compared with measured zonal wind velocites to test the validity of the cyclostrophic balance approximation. Also, since there is only one direct determination of wind speed above the clouds (Betz et al. 1976, 1977), model-derived velocities provide valuable insights into the nature of the circulation at these heights. Latitudinal temperature contrasts at different altitudes are the basic inputs to the cyclostrophic zonal wind calculation; these are known from a number of experiments including *in situ* probe measurements (Seiff et al. 1979*a,b,* 1980; Seiff and Kirk 1982), orbiter infrared radiometry (F. Taylor et al. 1979*a,b,* 1980), and radio occultations (Yakovlev et al. 1976, 1978; Kolosov et al. 1980; Chub and Yakovlev 1980; Kliore and Patel 1980). Below 70 km, T decreases poleward on constant pressure surfaces (see Fig. 27, Sec. V) and u_λ increases with height, but between 70 and 100 km, T increases poleward (see Fig. 27, Sec. V) and u_λ decreases with height.

Zonally averaged temperatures from the Pioneer Venus orbiter (PVO) infrared radiometer experiment were used to construct the altitude profiles of cyclostrophic zonal wind velocity shown in Fig. 15 (Schubert et al. 1980*a*; Taylor et al. 1980). As the wind speed at 70 km is unknown, the temperature data do not determine a unique velocity profile. While all the profiles show a decrease in u_λ with height, as required by the poleward increase in T, the magnitude of the zonal wind at 100 km depends very sensitively on the speed at 70 km. At 33° lat, for example, zonal wind speeds at 100 km are between 40 and 80 m s^{-1} for u_λ, at 70 km between 100 and 130 m s^{-1}. If the wind speed is as low as 60 m s^{-1} at the cloud tops, on the other hand, strong zonal winds vanish as low as 85 km. A similar state of affairs exists at 45°N, while at 65°N, where u_λ at 70 km is probably < 80 m s^{-1} (see Fig. 8), the zonal superrotation may be confined to altitudes $\lesssim 90$ km. F. Taylor et al. (1979, 1980) and Elson (1979) have used a dynamical model, which includes a parameterization of frictional effects by an eddy diffusivity, to infer mean zonal wind speeds from averaged infrared brightness temperatures. Their models predict a more rapid reduction in zonal wind velocity with height than do the cyclostrophic balance models. More complex modeling does not necessarily yield more reliable zonal wind estimates unless all relevant physical processes are included; the models of F. Taylor et al. (1979*b,* 1980) and Elson (1979) allow for eddy diffusion of mean zonal momentum but not for eddy production of mean momentum, the process which probably maintains the strong zonal winds in the first place.

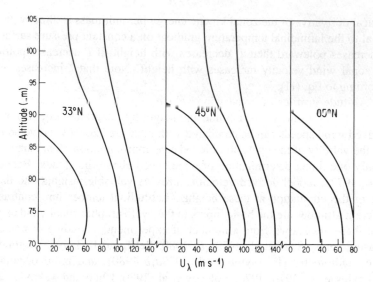

Fig. 15. Retrograde zonal wind velocities u_λ, derived from a cyclostrophic balance model and the zonally averaged brightness temperatures of the PV orbiter infrared radiometer experiment. Altitude profiles of u_λ are shown at 3 different latitudes for several choices of the *a priori* unknown value of u_λ at 70 km.

Figure 16 shows cyclostrophic zonal wind velocity profiles derived from PV Day, Night and North probe temperature and pressure data (Seiff et al. 1980; Seiff 1982). The derivation assumes that the data from the Night and Day probes (at 31°S) represent conditions at 31°N. The inferred zonal wind speeds increase with altitude up to heights of 65 to 70 km; maximum speeds at the cloud tops are between 130 and 140 m s^{-1}. Zonal winds increase with height below the cloud tops because of the poleward decrease of temperature on constant pressure surfaces below $\sim$ 70 km. The derived zonal winds decrease with height in the altitude range 70 to 95 km because of the poleward increase of temperature on constant pressure surfaces in this region of the atmosphere. Extrapolation of the model profiles in Fig. 16 to greater heights suggests that the vertical shear in the zonal winds above the cloud tops reduces westward wind speeds to very small values at altitudes of 90 to 100 km. The dominant superrotation of the lower atmosphere would then be confined to heights < 100 km, and a different circulation regime, probably a subsolar-antisolar flow, would prevail above 100 km (Seiff 1982). However, simple extrapolation of the probe-derived cyclostrophic wind profiles to higher altitudes is not possible because the latitudinal temperature gradients responsible for vertical shear in the zonal wind between 70 and 90 km disappear between 90 and 100 km.

Vertical temperature profiles from radio occultations have also been used to deduce cyclostrophic zonal wind velocities (Chub and Yakovlev 1980).

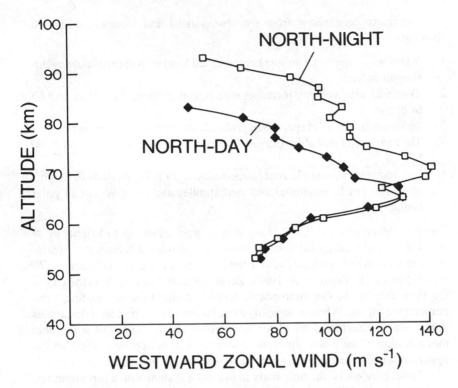

Fig. 16. Altitude profiles of cyclostrophic zonal wind velocities, deduced from Pioneer Venus
probe temperature and pressure measurements (Seiff et al. 1980; Seiff and Kirk 1982; Seiff
1982). It is assumed that the Night and Day probe temperatures and pressures represent
conditions at 31°N.

These data also show a poleward decrease in temperature below ~ 70 km,
implying an increase of rapid zonal winds with height and a poleward increase
in temperature above 70 km, implying a decrease in atmospheric rotation rate
with altitude. Inferred wind speeds are ~ 100 m s⁻¹ at 65 km.

    While the Pioneer Venus probes measured conditions consistent with
cyclostrophic balance at cloud levels in midlatitudes (30° to 60°), their measurements of temperature and pressure at cloud heights near the equator indicate a breakdown of this dynamical state. The pressure measured by the Large
probe (at 4°N) at a given height in the clouds was smaller than the pressures
measured by the Day and Night probes at the same altitude (at 31°S, assumed
to be representative of 31°N) (Seiff et al. 1980). These observations imply an
equatorward pressure force at cloud levels between 4° and 30° latitude which
is inconsistent with cyclostrophic balance. The breakdown of cyclostrophic
balance at cloud heights near the equator suggests that nonaxisymmetric motions contribute significantly to the meridional force balance in this region of
the atmosphere (Seiff et al. 1980).

The major conclusions from the observations and modeling discussed above are:

1. Venus's nonequatorial atmosphere at cloud level is in approximate cyclostrophic balance;
2. The zonal wind velocity increases with height between the surface and 65 to 70 km;
3. Maximum zonal wind speeds are probably attained at the cloud tops;
4. The zonal wind probably decreases with altitude between $\sim$ 70 and $\sim$ 90 km;
5. The dominant retrograde zonal superrotation probably extends to heights of $\geqslant$ 100 km in equatorial and midlatitudes and to $\geqslant$ 90 km in polar latitudes.

The persistence of reasonably strong east-to-west winds up to heights of $\geqslant$ 100 km is supported by the relative smallness of diurnal temperature variations in comparison with latitudinal temperature differences between $\sim$ 70 and $\sim$ 100 km (F. Taylor et al. 1980). Zonal advection of heat from the day to the night side by the fast atmospheric rotation reduces the longitudinal temperature variations. While a subsolar-antisolar circulation probably dominates the general circulation of the upper atmosphere above 100 km, it is possible that a portion of the lower atmosphere's superrotation impresses itself on the upper atmosphere, causing it to superrotate with speeds of several tens of m s^{-1}.

Venus may not be the only body in our solar system with a superrotating, cyclostrophically balanced atmosphere; Titan may have one also. Based on measured latitudinal contrasts in infrared brightness temperature, together with the observed absence of longitudinal variations in thermal structure, Flasar et al. (1981) have suggested that Titan's stratosphere may have cyclostrophically balanced 100 m s^{-1} winds.

IV. MAINTENANCE OF THE RETROGRADE ZONAL SUPERROTATION

From the dynamical viewpoint, the major challenge of the Venus atmosphere is to explain its bulk superrotation. We have seen that this is westward at all latitudes and at altitudes between $\sim$ 10 and $\sim$ 100 km. Zonal wind velocity increases steadily with height above the surface and probably reaches a maximum in the vicinity of the cloud tops at 65 to 70 km. The angular momentum density of the atmosphere's rotation increases with height from the surface to $\sim$ 20 km and decreases with height above this level. Most of the atmospheric angular momentum is contained in the layer between 10 and 40 km. The lowest 10 km of the atmosphere has relatively little excess retrograde angular momentum (relative to corotation with the solid planet). The problem facing the dynamicist is to find the source of the excess atmospheric angular momentum, and the processes which transport this momentum both vertically and horizontally.

The Earth's atmosphere also superrotates with an excess prograde or eastward angular momentum of $\sim 1.6 \times 10^{26}$kg m^{2}s^{-1} (Hide et al. 1980). This is only 2.73×10^{-8} times the angular momentum of the solid Earth (5.86×10^{33}kg m^{2}s^{-1}). The Earth's excess atmospheric angular momentum per unit mass of the atmosphere is 3.0×10^7 m^{2}s^{-1}. Venus's total atmospheric angular momentum can be obtained by integrating the angular momentum density (see Eq. 7) over the volume of the atmosphere

$$L_{TOT} = \int_{-\pi/2}^{\pi/2} \int_{o}^{2\pi} \int_{a}^{\infty} \rho u_\lambda r^3 \cos^2\phi \, dr \, d\lambda \, d\phi \, . \tag{16}$$

Assuming that $\rho = \rho(r)$ and $u_\lambda = u_\lambda (r, \phi = 0) \cos\phi$, i.e. that each spherical shell of the atmosphere is in solid body rotation, the integral expression for Venus's atmospheric angular momentum becomes

$$L_{TOT} = \frac{8\pi}{3} \int_{a}^{\infty} r^3 \rho u_\lambda (\phi = 0) \, dr \, . \tag{17}$$

The integration over altitude can be approximated by setting r equal to a and using the PV Large probe data in Fig. 11 for ρu_λ ($\phi = 0$). Venus's total atmospheric angular momentum is $\sim 3.4 \times 10^{28}$kg m^2 s^{-1}. This is probably a low estimate because the atmospheric superrotation in polar latitudes is faster than that of solid body rotation with the equatorial rotation rate (Fig. 10). An estimate of Venus's excess atmospheric angular momentum can also be made by treating the atmosphere as a thin shell of mass (p_{surf}/g) per unit area, rotating with constant angular velocity ω. The atmospheric angular momentum in this approximation is

$$L_{TOT} = \frac{8\pi}{3} a^4 \frac{p_{surf}}{g} \omega \, . \tag{18}$$

Using this formula, the parameter values given in Table I, and a 12 d rotation period appropriate to the atmospheric layer containing most of the angular momentum, we obtain 3.6×10^{28} kg m^2 s^{-1} for the atmospheric angular momentum, close to our previous estimate.

Venus's total atmospheric angular momentum exceeds the excess angular momentum of the Earth's atmosphere by a factor of $\geqslant 200$. Per unit atmospheric mass, Venus's specific excess atmospheric angular momentum of $\sim 7.1 \times 10^7$m^{2}s^{-1} is more than twice that of the Earth's atmosphere. Venus's excess atmospheric angular momentum is a relatively large fraction, 1.6×10^{-3}, of the planet's solid body angular momentum, 6×10^4 times as large as the corresponding fraction for the Earth. Atmospheric angular momentum fluctuations on Earth cause changes in the length of the day on the order of milliseconds (Hide et al. 1980). Because Venus's atmosphere and solid body have very much more comparable amounts of angular momentum, it may be that atmospheric angular momentum fluctuations on Venus produce much larger changes in the length of its day (l.o.d.). Assuming conservation of solid-body plus atmospheric angular momentum and introducing ϵ for the fractional change in atmospheric angular momentum, we can write

$$\text{change in l.o.d} = \epsilon\,\tau_{\text{day}}\,\frac{L_{\text{TOT}}\,(\text{atmosphere})}{L_{\text{TOT}}\,(\text{solid planet})}$$

$$= 1.6 \times 10^{-3}\epsilon\,\tau_{\text{day}} \qquad (19)$$

$$= 4.5\epsilon\,(\text{hr})\ .$$

Thus, changes of 20% or more in the atmosphere's angular momentum will lead to l.o.d. changes on Venus on the order of hours. The angular momentum of the Earth's atmosphere is known to change by several tens of percent on short time scales (Hide et al. 1980); Venus's atmosphere may undergo short-term angular momentum variations of similar magnitude, as suggested by differences in zonal wind speeds among measurements made at different times. Length-of-day changes on Venus on the order of hours are on the edge of detectability by Earth-based radar (Shapiro et al. 1979). Changes of this magnitude could also be measured by some future long-term experiment on an orbiter or on the surface. Golitsyn (1982) has independently reached similar conclusions regarding the potential significance of angular momentum exchanges on Venus.

Not only does Venus's atmosphere contain significantly more excess angular momentum than the Earth's atmosphere, but its momentum is distributed in a fundamentally different way than the Earth's. The excess prograde momentum of Earth's atmosphere is concentrated in its middle and high latitude eastward winds; in low latitudes the winds in the Earth's atmosphere are westward and there is a local deficiency in relative atmospheric angular momentum. The Earth's equatorial atmosphere does not superrotate; on the contrary, it subrotates. On Venus, however, the mean atmospheric rotation is unidirectional and even the equatorial regions superrotate. Thus the dynamical processes that distribute excess angular momentum in Venus's atmosphere are able to transfer this momentum to low latitudes, whereas on Earth these processes, if they operate at all, are ineffective. The major difficulty in understanding Venus's atmospheric superrotation therefore is not in explaining the simple fact of a net excess atmospheric angular momentum, but rather in elucidating the reasons for equatorial atmospheric superrotation. A closely related problem is explaining the westerly equatorial jets in the atmospheres of

Jupiter and Saturn (Smith et al. 1979*a,b*, 1981, 1982). The processes supporting the equatorial superrotation probably involve feeding of angular momentum to the low latitude mean zonal flow by nonaxisymmetric eddies, because
an equatorial superrotation cannot be generated by a zonally symmetric circulation with angular momentum advection by meridional winds balanced
against down-gradient vertical diffusion of momentum (Hide 1969; Held and
Hou 1980).

The source of Venus's excess atmospheric angular momentum must lie
outside the atmosphere; the solid planet or the Sun are possibilities. Retrograde angular momentum of the solid planet could be transferred to the atmosphere through the net effect of frictional and atmospheric pressure torques
acting over the surface. The angular momentum of the Sun could be deposited
directly in Venus's atmosphere by the action of solar torques on thermally
induced diurnal atmospheric mass variations. Once angular momentum has
been imparted to the atmosphere, it could be redistributed both horizontally
and vertically by the mean meridional circulation and by the net transport of
eddies. On Earth, both the equatorial Hadley circulation and the midlatitude
baroclinic eddies are important in the meridional transport of angular momentum (Palmen and Newton 1969). Venus's mean meridional circulation can
transfer retrograde angular momentum upward at some latitudes and altitudes
and downward at others; similarly, it can transport angular momentum either
equatorward or poleward depending on altitude and latitude. Eddies can also
transfer angular momentum up or down, or equatorward or poleward, depending on location in the atmosphere. In addition, because different types of
eddies occur over a broad range of spatial scales, there can be several reinforc-
ing or opposing contributions to the net eddy transport of angular momentum.

The ways eddies and mean circulations transport angular momentum in
Venus's atmosphere depend on the source of the momentum. If the Sun were
feeding westward momentum directly into Venus's atmosphere, and if the
atmospheric rotation were basically steady over long intervals of time, then
eddies and mean motions would have to transfer the momentum downward
into the solid planet. This might be accomplished by small-scale turbulent
eddies diffusing retrograde momentum down the gradient in zonal wind velocity, a direction that is downward throughout the atmosphere below the clouds.
Although the solid planet would be receiving retrograde angular momentum
from the Sun via the atmosphere, the solid body need not speed up since other
prograde torques acting on it could balance the atmospheric torque. If the
solid planet were the source of the excess atmospheric angular momentum, then momentum would have to be transferred upward by atmospheric
motions during the initial spinup of the atmosphere. The retrograde angular
momentum of the solid body would be reduced during this process. After the
establishment of a quasi steady atmospheric superrotation, subsequent net
interaction of solid planet and atmosphere could be negligible over long time
scales. Instead, there might be a balance between upward and downward

and equatorward and poleward fluxes of angular momentum already in the atmosphere through different mean circulations and eddy transports. Yet there could also be a significant net exchange of angular momentum between the atmosphere and solid planet, if not in the long term, then perhaps in some alternating cycle on short time scales. We have already pointed out the remarkable fact that the atmospheric angular momentum is only 1.6×10^{-3} of the solid body angular momentum, making the possibility of short-term momentum exchanges between the atmosphere and solid planet highly significant. Thus, the value of understanding the mechanisms that maintain Venus's atmospheric superrotation goes beyond atmospheric dynamics; the rotational state of the solid planet may be intimately tied to the rotation of the atmosphere. We now discuss these mechanisms and the connection between atmospheric and solid body rotations in more detail.

The idea that periodic solar heating could impart a net angular momentum to an atmosphere dates back to Halley (1686) who suggested it for the Earth. Figure 17 shows how eddies in planes parallel to the equator can transport retrograde angular momentum upward. There must be a westward tilt to the eddies along the rising branch of the circulation and an eastward tilt along the descending branch. The eddies will transport retrograde angular momentum upward and prograde angular momentum downward. Eddies of this type can be induced by the eastward overhead motion of the Sun, as shown schematically in Fig. 18. This is the moving-flame mechanism (Fultz et al. 1959; Stern 1959; Davey 1967; Hinch and Schubert 1971; Young et al. 1972; Whitehead 1972; Douglas et al. 1972) proposed by Schubert and Whitehead (1969) and Schubert and Young (1970) to explain the 4 day rotation of the Venus atmosphere at cloud heights. Heating due to the Sun diffuses vertically away from the level of solar energy absorption by small-scale turbulent mixing and radiative transport. In the time this takes, the Sun moves horizontally overhead, thereby creating a phase lag or tilt to the induced motions. The direction of the tilt is westward for rising motions when heat diffuses upward from the absorption level. Thus, eddies which transfer westward momentum up can be induced in the lower atmosphere by absorption of sunlight near and at the surface, and induced in the atmosphere above cloud level by sunlight absorbed within the upper cloud deck. Even downward diffusion of heat from sunlight absorbed above the clouds can lead to eddies tilted similarly to those in Fig. 18, because of the strong stable stratification of the region (Young and Schubert 1973).

Tilted convection cells in a zonal plane can also arise from an instability mechanism suggested by Thompson (1970). Figure 19a shows a steady convection pattern produced by a stationary periodic heat source. This could be a subsolar-antisolar circulation in the equatorial plane, viewed in a coordinate system fixed to the Sun. Suppose that a small westward mean velocity is somehow established as shown in Fig. 19b. The westward corotation of the atmosphere with the solid planet could be a source of such a perturbation

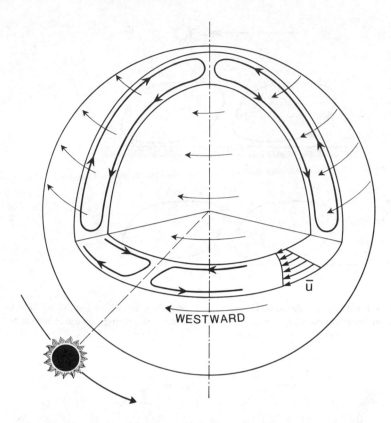

Fig. 17. Sketch of motions that could maintain the atmospheric superrotation: westward tilted
eddies in the equatorial plane and a mean meridional circulation.

mean flow $\bar{u}$. The convection cells will be slightly tilted by the flow as shown
in Fig. 19c. The tilt in the cells creates a correlation of the velocity components resulting in an upward transport of westward momentum, as discussed
above. This amplifies the initial mean velocity and could eventually lead to
buildup of a high-speed westward flow. The initial perturbation mean flow
could conceivably be eastward; the instability mechanism would then produce
a large eastward flow. If this feedback process does play a role in establishing
Venus's atmospheric superrotation, then a westward perturbation provided
by Venus's retrograde rotation or a small westward flow generated by the
moving-flame mechanism would be essential to explain the sense of the
present-day atmospheric rotation, assuming that direction is not fortuitous.

The eddies we have been discussing are driven by diurnal variations in
heating. A dimensionless ratio which can be used to estimate the size of
diurnal effects is τ_{rad}/τ_{day} where τ_{rad} is the radiative time constant (Gierasch
et al. 1970; Stone 1975). If $\tau_{rad}/\tau_{day} \gg 1$, the atmosphere will not cool

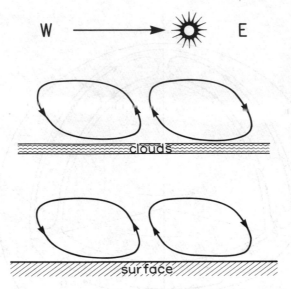

Fig. 18. Vertical eddies induced by the eastward overhead motion of the Sun, tilted so as to transport westward angular momentum upward. This is the moving-flame mechanism for driving the atmospheric superrotation.

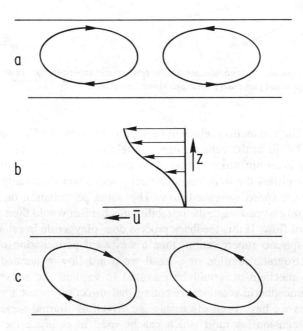

Fig. 19. A symmetric convection pattern (a), induced by a periodic heat source, which leads to tilted convection cells (c) when perturbed by a mean flow $\bar{u}$ (b).

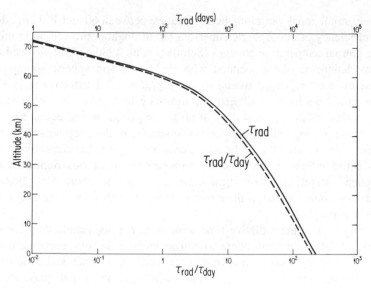

Fig. 20. The radiative time constant τ_{rad} and the ratio τ_{rad}/τ_{day} as functions of altitude in the Venus atmosphere.

appreciably by radiation during the night and the radiative drive for day to night zonal circulations should be small. The radiative time constant and the ratio τ_{rad}/τ_{day} are shown as functions of altitude in Fig. 20; τ_{rad} is from the calculations of Pollack and Young (1975). The radiative time constant decreases with height from a maximum value of $\sim 120\ \tau_{day}$ at the surface to $10^{-2}\ \tau_{day}$ at 75 km; $\tau_{rad}/\tau_{day} = 1$ at ~ 60 km. Thus, diurnal effects and day-to-night zonal circulations might be anticipated only at cloud heights and above (Stone 1975). Accordingly, we might also expect the moving-flame mechanism to function only in and above the clouds. Gierasch (1970a) used a radiative-dynamical model to show that the moving-flame mechanism could generate strong zonal winds in Venus's clouds. His calculations predicted wind magnitudes by a balance between advective and radiative time constants. The flow greatly reduced the amplitude of the diurnal thermal oscillation.

Temperature measurements generally confirm the above idea of where in the atmosphere diurnal effects are important, but there are notable differences (Seiff et al. 1980; F. Taylor et al. 1980; Kliore and Patel 1980; Seiff and Kirk 1982). Longitudinal variations in temperature are smaller than anticipated between 60 and 100 km, while temperature contrasts among the Pioneer Venus probes show that there are significant diurnal variations in thermal structure in the deep atmosphere. The temperature difference between the Day and Night probes was ~ 5 K at 12 km; the Large probe temperature was 11.5 K hotter than the Night probe temperature at 13 km (Seiff et al. 1980). The

relatively small zonal variations in temperature between 60 and 100 km attest
to the efficiency of atmospheric superrotation throughout this region in mod-
erating diurnal temperature changes (Schubert et al. 1980a). Had we used the
effective length of day associated with the actual atmospheric rotation to
estimate the ratio τ_{rad}/τ_{day} instead of using $\tau_{day} = 117$ Earth days as in Fig.
20, we would not have concluded that diurnal effects might be important at
cloud heights. With $\tau_{day} \approx 4$ days at 60 km, τ_{rad}/τ_{day} would equal 30. The
relatively large longitudinal temperature contrasts in the deep atmosphere at
the time of the various PV probe measurements suggest that diurnally forced
eddies could influence the structure and circulation of this region of the
atmosphere, despite the above time-scale argument (Schubert et al. 1980a).
Indeed they would have to, if the moving-flame mechanism were to operate in
the lower atmosphere.

    It may be that the radiative time scale does not adequately describe the
response of the lower atmosphere to diurnal changes in solar energy absorption because of dynamical heat transports. Instead, a dynamical time scale
τ_{dyn} might be appropriate if τ_{dyn} were sufficiently short that quasi steady
day-night circulations could be established within a Venus solar day. The time
scale τ_{dyn} is

$$\tau_{dyn} = R_{\venus}/v \qquad (20)$$

where v is a characteristic horizontal velocity. Stone (1975) computed τ_{dyn}
using $v = (gH)^{1/2}$, the typical speed of an external gravity wave or sound wave.
Such large velocities (200 to 300 m s^{-1} from Fig. 5) probably seriously
underestimate the time scale required to establish quasi steady diurnal circulations. If we use a conservative estimate of $v = 1$ m s^{-1}, a typical velocity of
winds in the deep atmosphere and of meridional winds at cloud levels, we
obtain $\tau_{dyn} = 70$ d, only 0.6 Venus day. A velocity of 1 m s^{-1} is also
reasonable for the estimate of τ_{dyn} for diurnally forced motions, because it is
the speed a subsolar to antisolar circulation would attain in the lower atmosphere of a nonrotating Venus (the speed of such a flow should be comparable
to the velocity of a surface Hadley cell, 1 m s^{-1} according to Stone [1975] and
Golitsyn [1970]). If a subsolar-to-antisolar circulation in the nonrotating case
has speeds such that τ_{dyn} is only a fraction of the planet's day, it seems
possible that diurnally forced motions can occur. While there may not be
enough time to establish a mean subsolar-to-antisolar circulation in Venus's
lower atmosphere, we should not rule out the possibility that diurnally driven
eddies are present. If we applied this reasoning to the Earth, again taking $v =$
1 m s^{-1}, we would compare $\tau_{dyn} \approx 70$ d with $\tau_{day} = 1$ d, and we would
conclude that diurnal forcing could not establish any quasi steady day-night
circulations, in agreement with observations. This is one reason that the
moving-flame mechanism might work on Venus even though it is not effective on Earth.

The stability structure of the Venus atmosphere, especially the highly stable layer below the clouds, makes it difficult to imagine moving-flame eddies extending all the way from the surface to cloud levels. It seems more likely that if such eddies existed they would be vertically layered, and it is by no means obvious that the required upward transfer of momentum could then occur. This same criticism applies to meridional circulation cells which can also transfer angular momentum upward. We will consider this problem in more detail when we discuss the meridional circulation mechanism for the maintenance of Venus's superrotation.

The motions driven by diurnal variations in solar heating could be cellular as already discussed, or they could be thermal tides. With tilted Hadley-like zonal cells, near surface winds blow toward the subsolar point, but the tidal winds blow away from the region of heating. Thermal tides simply transfer mass between heated and cooled regions of the atmosphere, resulting in low atmospheric pressure where the atmosphere is hot and high pressure where it is cold. The tides propagate around the planet from west to east following the overhead motion of the Sun. As a consequence of atmospheric mass redistribution in the thermal semidiurnal tide, the Sun might impart retrograde angular momentum directly to Venus's atmosphere. If the maximum of pressure (and hence atmospheric mass per unit surface area) occurs before noon, as on Earth, the Sun's gravitational field will exert a torque on the atmosphere which will tend to increase the atmospheric rotation rate (Fig. 21). Gold and Soter (1971) proposed that this mechanism is responsible for the cloud-level atmospheric superrotation (see also Teitelbaum and Cot 1981). The phase of the mass wave is most readily understood in terms of the phase of the thermal wave. Because of radiative and diffusive heat transport in the atmosphere, it is reasonable to expect that the phase of the thermal wave might lag the Sun, i.e. the maximum atmospheric temperature might occur after noon. The second harmonic mass wave should be $\sim 90°$ out of phase with the second harmonic thermal wave and, as shown in Fig. 21, the phase of the mass wave would then lead the Sun. However, these phase relationships could be significantly altered by dynamical effects (Gold and Soter 1979), a point discussed in more detail below. In the case of the Earth, observations show that the mass wave of the semidiurnal tide leads the Sun by ~ 2 hr (Haurwitz 1964), thereby producing a prograde torque, i.e. one tending to spin up the atmosphere.

The effectiveness in producing retrograde zonal winds of a solar torque acting on a Venus semidiurnal atmospheric thermal tide depends not only on the magnitude of the torque, and the magnitude and phase of the mass wave relative to the Sun, but also on the eddy kinematic viscosity ν of the atmosphere (Young and Schubert 1973). The atmospheric torque must be able to account for the observed vertical shear of the zonal wind. The torque is related to the shear stress on the atmosphere by

$$\text{torque on atmosphere} \approx \pi^2 R_\circ^3 \, (\text{shear stress on atmosphere}) . \qquad (21)$$

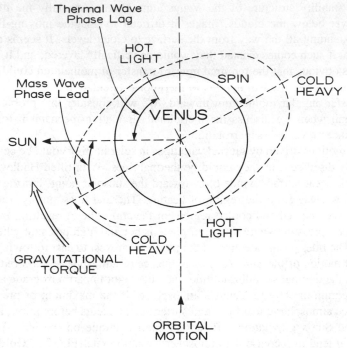

Fig. 21. Retrograde spinup of the Venus atmosphere by a solar torque acting on a thermal semidiurnal tide.

The shear stress on the atmosphere is in turn related to the vertical shear of the zonal wind du_λ/dz by

$$\frac{du_\lambda}{dz} \approx \frac{\text{shear stress on atmosphere}}{\rho\nu}. \tag{22}$$

By combining Eqs. (21) and (22) we obtain

$$\frac{du_\lambda}{dz} \approx \frac{\text{torque on atmosphere}}{\pi^2 R_\venus^3 \rho\nu}. \tag{23}$$

We can estimate an upper bound on the shear by using $\rho = 0.2$ kg m^{-3} (the density near the cloud tops) and 2×10^{16}J for the torque on the atmosphere (Dobrovolskis and Ingersoll 1980)

$$\frac{du_\lambda}{dz} \approx \frac{5 \times 10^{-5}}{\nu} \tag{24}$$

with ν in m^2s^{-1}. The observed shears are $\geq 10^{-3}s^{-1}$, requiring ν to be $\leq 10^{-1}$ m^2s^{-1} for the mechanism to work. This is at the lower limit of turbulent eddy viscosities found in the Earth's atmosphere (Houghton 1977), and less than the estimate of 4 m^2s^{-1} for the 60 km level of the Venus atmosphere from radio scintillation measurements (Woo and Ishimaru 1981). Other estimates of ν in the Venus atmosphere generally exceed 10^{-1} m^2s^{-1} (Prinn 1974, 1975; Sze and McElroy 1975; Belton et al. 1976b; Kumar et al. 1978; Winick and Stewart 1980; Covey and Schubert 1981a). Thus the estimated torque is not large enough to produce the observed shear, because the atmosphere is fully turbulent.

The solar torque on the atmospheric tide could also be transmitted to the solid planet by friction at its surface; it could then balance a torque in the opposite direction caused by the solid body tides raised on Venus by the Sun's gravity (Gold and Soter 1969, 1979; Kundt 1977; Ingersoll and Dobrovolskis 1978; Dobrovolskis and Ingersoll 1980). The torque due to solid body tides, if acting alone, would force the planetary rotation to become Sun-synchronous within $\sim 10^8$ yr (Goldreich and Peale 1970). Thus, the present slow retrograde rotation of Venus could be an equilibrium state between solar torques on the atmospheric and solid body tides. Torques exerted by Earth could also play a role. The solar torque on the thermal semidiurnal atmospheric tide offers a single explanation for the unusual rotation states of both the cloud-level Venus atmosphere and the planet itself. If the Sun were not feeding retrograde angular momentum into the atmosphere, the mechanism for maintaining Venus's planetary spin by an atmospheric torque on the solid planet would not be tenable.

Observations of the solar thermal tide in Venus's atmosphere are too limited to resolve the key first-order question of whether the phase of the tide is consistent with the theories outlined above. Earth-based observations of thermal infrared emission from Venus are primarily restricted to the night side (Ingersoll and Orton 1974; Apt and Goody 1979a; Apt et al. 1980). These observations generally show a local maximum in equatorial temperatures shortly before midnight at 9–10 PM. A temperature maximum after midnight would be expected if the phase of the semidiurnal tide were as assumed by the theories. However, interpretation of the Earth-based data is complicated by the broad band of wavelengths used. The measured brightness temperatures include a component due to cloud-top temperatures. Thus an increased infrared emission could indicate a shift in cloud heights rather than a temperature increase at constant altitude. Data from the Pioneer Venus infrared radiometer indicate a maximum temperature occurring before noon (F. Taylor et al. 1980), again in conflict with theoretical expectations.

It is important to note that Pioneer Venus infrared observations cover a time span of only 10 weeks, <1 Venus solar day (117 Earth d), while Earth-based observations represent up to several years' data. Thus, the temperatures seen by the Pioneer Venus orbiter could indicate a long-term eddy rather than the actual solar tide (Schubert et al. 1980a). This would be possible if the dynamical time scale τ_{dyn} were short enough to set up quasi steady day-night circulations in <1 Venus d. It is argued above that this could be the case even in the deep atmosphere. Therefore, it is possible that quasi steady, day-night cellular motions dominate tidal oscillations in the Venus atmosphere. This is not the case on Earth because an Earth day is too short for diurnal forcing to set up day-night circulations. The validity of Dobrovolskis and Ingersoll's (1980) quantitative estimate of the atmospheric tidal torque is therefore open to question. Although simple tidal theory might be inadequate to describe the diurnally induced circulation of the lower atmosphere, the mass asymmetries associated with these motions could still provide an atmospheric torque on the solid planet capable of balancing the solid body torque (Gold and Soter 1979).

Mean meridional circulation, as sketched in Fig. 17, can also transport westward momentum up. If the angular momentum at low latitudes, where ascending flow occurs, exceeds the angular momentum at high latitudes, where the atmosphere descends, the Hadley circulation will transport net angular momentum upward. Gierasch (1975) proposed that this mechanism could be at work in the Venus atmosphere to sustain its rapid superrotation. In addition to transporting momentum vertically, a meridional flow would also carry westward momentum poleward if retrograde angular momentum decreases with latitude (see Fig. 11). From the tracking of features in Mariner 10 and Pioneer Venus ultraviolet photographs it is clear that poleward transport of westward zonal momentum by a mean meridional flow occurs at the top of Venus's clouds (Rossow et al. 1980a; Limaye and Suomi 1981; Limaye et al. 1982). It has been suggested (Rossow et al. 1980a) that this is the process which led to the buildup of midlatitude jets (see Figs. 8 and 22) in the zonal flow at the time of the Mariner 10 flyby. Because the poleward branch of a Hadley cell extracts angular momentum from the equatorial region, meridional circulation by itself cannot explain Venus's atmospheric superrotation at low latitudes. The tendency of the mean meridional circulation to deplete equatorial latitudes of excess angular momentum is apparent in the Earth's atmosphere, which actually subrotates at low latitudes although it superrotates on the whole. The equatorial Hadley cell dominates the meridional transport of angular momentum in the Earth's atmosphere at low latitudes. Therefore, in the case of Venus another process must counteract this tendency of the Hadley cell by advecting momentum to lower latitudes. Gierasch (1975) suggested that horizontal eddies redistribute the momentum to low latitudes. However, if the eddies that mix the momentum latitudinally also diffuse heat with similar efficiency, the diffusion would reduce the latitudinal temperature gradients driving the meridional flow and shut off the entire process (Kálnay de Rivas 1975). Thus, the meridional transport mechanism will explain Venus's equatorial atmospheric superrotation only if there are eddies that mix momentum without diffusing heat.

Rossow and Williams (1979) have argued that the requisite motions could

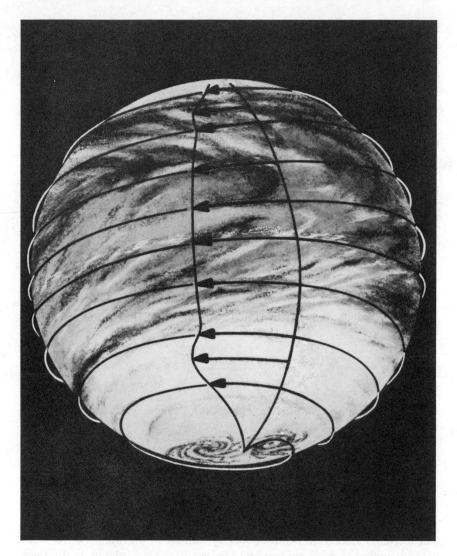

Fig. 22. Venus's atmospheric rotation at the cloud tops at the time of the Mariner 10 flyby exhibiting midlatitude jets.

take place at cloud heights if the atmosphere were essentially barotropic at these levels, as a consequence of the relatively slow rotation and high static stability. The horizontal exchanges could then be almost nondivergent and the two-dimensional motions would transport little heat. The nonlinear interactions of the horizontal eddies with each other and the mean zonal flow would resemble a field of two-dimensional turbulence. The absence of midlatitude jets in the latitudinal structure of the cloud-level zonal wind at the time of the

Pioneer Venus observations, and their presence at the time of the Mariner 10 flyby, could reflect dynamical states of the atmosphere at opposite ends of a cycle of buildup and redistribution of retrograde angular momentum (Rossow et al. 1980a). We could view the Mariner 10 state of atmospheric rotation with its midlatitude jets as the culmination of zonal wind buildup due to an angular-momentum-conserving poleward transport by the mean meridional circulation. The Pioneer Venus state of essentially solid-body atmospheric rotation would then represent the end product of latitudinal redistribution of angular momentum by barotropic eddies. While the Mariner 10 and Pioneer Venus rotational states are consistent with this dynamical scenario, the actual processes of buildup and redistribution have not been observed.

The equatorward mixing of the excess angular momentum in the midlatitude jets could be triggered by a dynamical instability of the mean zonal wind. Barotropic instability is one possibility; it is an instability associated with the latitudinal shear $d\bar{u}_\lambda/dy$ of the mean zonal wind $\bar{u}_\lambda$. Amplifying disturbances extract energy from the zonal flow and redistribute momentum by broadening and weakening the jet. Rayleigh's criterion provides a necessary condition for barotropic instability (Houghton 1977). It requires the latitudinal gradient of absolute vorticity of the mean flow to vanish somewhere i.e.

$$\beta - \frac{d^2\bar{u}_\lambda}{dy^2} = 0 \tag{25}$$

where β is the latitudinal gradient of the Coriolis parameter, or the planetary vorticity

$$\beta = \frac{2\Omega\cos\phi}{a} . \tag{26}$$

Planetary vorticity tends to stabilize zonal jets, but the effect is small on a slowly rotating planet like Venus. Travis (1978) and Elson (1978) have shown that Rayleigh's criterion could be met by latitudinal profiles of zonal wind velocity similar to those observed by Mariner 10. These velocity profiles not only meet this criterion but in fact support growing barotropic disturbances; perturbations do exist with the proper horizontal tilt to transport momentum out of the jet and feed on its zonal kinetic energy.

The picture of a barotropically unstable midlatitude zonal wind triggering two-dimensional eddies that transport angular momentum equatorward is consistent with the results of power spectral analyses of the brightness fields of Mariner 10 ultraviolet images (Travis 1978). The brightness power spectra in midlatitudes have an average slope of ~ -1.8 for zonal wavenumbers $\geqslant 3$, while in equatorial regions the slope is ~ -2.7 for wavenumbers > 5. These spectral characteristics can be interpreted dynamically if the ultraviolet brightness spectra reflect the eddy kinetic energy spectra of the motions. A slope of -3 describes the inertial subrange of two-dimensional turbulence, and since this applies at wavenumbers > 5, the energy input is at smaller wavenumbers or larger scales. The slope of -1.8 may characterize the growth of barotropically unstable disturbances in midlatitudes.

The explanation in terms of meridional circulation and eddy mixing for the maintenance of Venus's atmospheric superrotation is plausible for the cloud levels because: (1) there is probably a cloud-level Hadley cell extending from the equator to high latitudes that can carry excess angular momentum upward; and (2) there are, at times, cloud-level zonal winds capable of generating barotropic eddies by latitudinal shear instability to replenish the equator with angular momentum extracted by the Hadley cell. However, the horizontal eddies transporting angular momentum equatorward need not be forced by barotropic instability of the zonal wind. Diurnal heating at cloud heights, for example, might also drive the horizontal eddies that feed angular momentum back to the equator. Another such eddy mechanism may be required to explain the equatorial atmospheric superrotation at cloud heights if the state of near-solid-body rotation observed by Pioneer Venus persists for long enough times. The cloud-level Hadley cell on Venus might be more efficient than the Earth's equatorial tropospheric Hadley cell in transporting angular momentum upward and poleward, because the Venus cell probably extends to much higher latitudes.

It is relatively straightforward to understand how a mean meridional flow transports angular momentum upward in a single direct cell, but it is more difficult to imagine how the mechanism could work throughout the thick Venus atmosphere, whose meridional circulation is undoubtedly more complex and probably consists of Hadley cells and indirect cells stacked upon each other (the meridional circulation is discussed in the next section and illustrated in Fig. 24). The processes maintaining Venus's zonal circulation must be effective everywhere between the surface region and cloud levels and above, because all this atmosphere superrotates except perhaps the lowest scale height. In addition, if the solid planet is the source of the excess atmospheric angular momentum, then the dynamical processes must carry the momentum upward from the surface to cloud heights and above. Could a layered configuration of alternating direct and indirect cells accomplish this? How is the momentum transported from one cell to the next higher one? The indirect cells transfer angular momentum downward. If the mean meridional circulation consists of near-surface and cloud-level Hadley cells with an indirect cell in between, as discussed later, then upward momentum transport by the meridional flow might be effective in the two direct cells with some other mechanism transporting the momentum across the layer of atmosphere occupied by the indirect cell. Not only is it unnecessary for a single momentum transfer process to be effective throughout the entire atmosphere, but more than one mechanism is quite likely to be involved, given the complexity of

Venus's vertical atmospheric structure and mode of heating. On the other hand, Rossow (1982) has argued that even a complex meridional circulation consisting of direct and indirect cells alternating with height can by itself accomplish a net upward transport of angular momentum. This may be correct despite the fact that the process did not produce a significant atmospheric superrotation in Rossow's (1982) numerical calculations.

The mechanism of the meridional circulation is not the only one to be challenged by the difficulty of transporting angular momentum upward through the thick atmosphere to cloud heights. Could the tilted eddies of the moving flame reach from the surface to the clouds? Probably not, for they would tend to be layered by the same atmospheric structural and heating characteristics responsible for the likely breakup of the meridional circulation into multiple cells, e.g. the highly stable layer just below the clouds and the strong cloud level heating. The peak in angular momentum density at ~ 20 km (Fig. 11) may indicate the height to which the near-surface upward dynamical transports of momentum are effective, independent of which process is at work. In the past, theoretical modeling of the generation of Venus's rapid zonal circulation was focused on the cloud-level winds. Future efforts should concentrate on how the entire atmosphere, with realistic vertical variations in structure and heating, is driven to superrotation.

We discussed above how horizontal eddies might replenish the equatorial regions with angular momentum carried away from low latitudes by the meridional circulation. It was also noted that the eddies should behave barotropically to avoid reducing the latitudinal temperature gradient that drives the meridional flow. How likely is it then that the eddy motions are barotropic? After all, the cloud-level atmosphere is in a fundamentally baroclinic, cyclostrophically balanced state. How do the meridional temperature gradients of this state contribute to the behavior of the eddies? In a rapidly rotating atmosphere the importance of baroclinic effects can be assessed by comparing the length scale of the eddies with the Rossby radius of deformation L_R

$$L_R = \frac{NH}{f} . \tag{27}$$

If L_R is large compared to the size of the eddies, they behave approximately barotropically; if on the other hand the horizontal dimensions of the eddies are similar to L_R, baroclinic effects influence their behavior. Because of Venus's very slow rotation it may not be appropriate to use L_R as a characteristic length scale to estimate the potential importance of baroclinicity on eddy behavior. However, at cloud heights where the atmosphere is rotating with a ~ 4 d period, L_R may be a relevant parameter if we base f on the 4-d period.

In general, we can conclude from Eq. (27) and the above discussion that barotropic eddy behavior requires a high stability or large Brunt-Väisälä frequency. From Fig. 7 we see that N is large above 60 km and in the stable layer below the clouds, but the atmosphere is nearly neutrally stable, i.e. N is very small, in the middle cloud region between ~ 50 and ~ 55 km and near the base of the stable layer between ~ 20 and ~ 25 km. If we evaluate L_R near 60 km using $N \approx 10^{-3}$ s^{-1}, $H \approx 6$ km and $f = 2\left[2\pi/(4 \times 24 \times 3600)\right]$ (1/2) we get $L_R \approx 330$ km, suggesting that baroclinic effects could be important for large-scale eddies in this part of the atmosphere. Above 60 km, where the Brunt-Väisälä frequency is more than an order of magnitude larger, L_R would be $\geqslant 3000$ km and the behavior of large-scale eddies should be more nearly barotropic.

The numerical calculations of Young and Pollack (1977) also suggest that the meridional circulation mechanism can generate rapid retrograde zonal winds at least at cloud levels. Mean meridional circulation is the principal means by which zonal momentum is transported vertically in their model. Horizontal planetary-scale eddies are found to be important for the latitudinal transport of momentum. These processes generate high-speed westward winds in the model only above ~ 50 km, while the entire atmosphere below 30 km rotates in a prograde sense. There are no eastward zonal winds in Venus's atmosphere and we have seen that its westward zonal momentum is concentrated below 30 km (Fig. 11), where the model atmosphere of Young and Pollack (1977) rotates eastward. The computations therefore do not realistically simulate the dynamical processes generating the superrotation of the entire Venus atmosphere. It is possible that the model atmosphere has no net retrograde angular momentum and that its cloud levels are accelerated westward only at the expense of an equal and opposite eastward acceleration of the lower atmosphere. In that case the model atmosphere fails to transfer westward zonal momentum from the solid planet to all levels including the clouds. The inability of the model to represent all the significant dynamical processes in Venus's atmosphere may be due to the few spherical harmonic modes retained in the model and to the particular form of parameterization of vertical momentum diffusion for all scales not explicitly included (Rossow et al. 1980b; Young and Pollack 1980).

Atmospheric waves, particularly those associated with the planetary scale ultraviolet markings, may also be important in maintaining the superrotation by transporting angular momentum upwards (Belton et al. 1976b; Young and Pollack 1977; Schubert et al. 1980a). Large-scale waves are significant carriers of energy and momentum in the Earth's atmosphere (Holton 1979) and there is abundant evidence for their ubiquitous presence in Venus's atmosphere (see Sec. VI on waves). From a statistical analysis of ultraviolet temporal variability over a period of about two months of Pioneer Venus imaging, Del Genio and Rossow (1982) concluded that the dark horizontal Y is probably formed by two distinct waves, one with a prominent amplitude in midlatitudes and the other dominating in the equatorial region. These waves could be a midlatitude Rossby wave and an equatorial Kelvin wave (Belton et al. 1976b; Covey and Schubert 1981b,1982a). Both waves propagate upward; the Kelvin wave should transport westward momentum upward and tend to accelerate the high-altitude zonal winds (Covey and Schubert 1982a).

Fels and Lindzen (1974) have discussed a wave transport mechanism for generating retrograde velocities in atmospheric layers that absorb solar energy. Internal gravity waves generated by diurnal heating variations carry prograde momentum out of the absorbing layers, thereby accelerating them in the retrograde direction. Other regions of the atmosphere would absorb the prograde momentum and move in that direction. This specific mechanism could only produce thin layers of high-speed retrograde flow, and not the observed bulk retrograde rotation of the entire atmosphere. Although this particular wave generation and transport process can be ruled out as the sole drive of atmospheric superrotation, upward transport of zonal momentum by planetary waves may still be involved in supporting the westward winds.

In summary, the atmospheric retrograde superrotation is maintained by dynamical processes that transport westward momentum vertically upward. The vertical transport may be accomplished by mean meridional circulations, diurnally driven eddies, or upward propagating planetary-scale waves. A single mechanism may not work throughout the atmosphere; different processes may be effective at different levels. In assessing the likelihood that any of these mechanisms are at work we should keep in mind the following. There is no observational evidence for the existence of the moving-flame eddies. There are numerous observations of planetary-scale waves, but no direct evidence for their involvement in maintaining the zonal winds. There is evidence for a cloud-level Hadley cell and strong theoretical reasons to expect a Hadley circulation near the surface. The net vertical transport of momentum by such direct mean meridional circulations must be upward, and accordingly this process certainly occurs in at least some regions of the atmosphere. However, these same Hadley cells carry momentum away from the equator, and horizontal eddies must transport it back. There are no observations of this equatorward transport of westward zonal momentum by horizontal eddies. All observations of cloud-level motions so far show only a poleward transport of retrograde angular momentum by eddies (Limaye and Suomi 1981; Limaye et al. 1982). While there is no actual evidence of indirect meridional cells, there are strong theoretical reasons to expect that they exist; the net vertical transport of momentum by these cells is downward, further complicating the role of the meridional circulation in maintaining the superrotation. Of all the mechanisms discussed here, the meridional circulation/eddy mixing process is the least speculative because we are reasonably certain that there is at least a Hadley cell at cloud levels. However, our current knowledge does not permit the definitive identification of this process as the one actually at work even at cloud heights, and there are problems, enumerated above, with how it could work throughout the atmosphere. A strong piece of evidence for it would be observation of equatorward transport of momentum by horizontal eddies in future analyses of ultraviolet imaging data. On the other

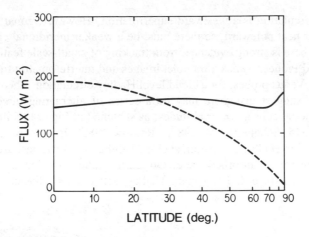

Fig. 23. Latitudinal variation of absorbed solar energy (dashed curve) and emitted infrared energy (solid curve) for Venus (after Tomasko et al. 1980*b*).

hand, if ultraviolet observations continue to show only poleward eddy transport of retrograde angular momentum, then the evidence would be strongly against this mechanism, and would imply instead that the equatorial region was being supplied with angular momentum from below by one of the other mechanisms.

V. MERIDIONAL CIRCULATION

The basic drive for atmospheric circulation is latitudinal imbalance between absorbed solar radiation and emitted infrared radiation, as illustrated for Venus in Fig. 23. In equatorial latitudes, more solar energy is absorbed than is reradiated to space. The opposite is true in polar latitudes. The atmosphere must redress this imbalance by transporting heat poleward. Meridional transport of heat in the Earth's atmosphere is accomplished by a Hadley circulation in low latitudes and by baroclinic eddies in midlatitudes. In the rapidly rotating atmosphere of the Earth, the sloping convection associated with midlatitude baroclinic eddies is a more efficient heat transfer mechanism than advection by the mean meridional circulation of a Hadley cell (e.g. Houghton 1977; Holton 1979). However, baroclinic heat transports would not be expected to dominate in the midlatitudes of an atmosphere on a slowly rotating planet like Venus, and a Hadley circulation on Venus would be expected to extend to high polar latitudes. Thus, it would be anticipated that the primary atmospheric circulation mechanism on Venus would be a pair of Hadley cells symmetric about the equator, with rising motions over the equatorial latitudes, poleward flow at high altitude, sinking over the polar regions, and equatorward flow near the surface. It is therefore remarkable that Venus's dominant atmospheric circulation is a westward superrotation. However, zonal winds cannot transfer heat poleward, so there must be a weaker meridional circulation as well. There is strong evidence, from tracking of small-scale features in Mariner 10 and Pioneer Venus ultraviolet images and interferometric tracking of the Pioneer Venus probes, for a cloud level Hadley circulation. Theoretical considerations suggest that there may be a series of alternating direct and indirect meridional cells at lower altitudes, as sketched in Fig. 24 (Kálnay de Rivas 1973, 1975; Schubert et al. 1980a; Rossow 1982). It is not surprising that there is a main Hadley circulation at cloud heights on Venus because most of the solar energy is absorbed there. On Earth, solar radiation is mostly absorbed at the surface and the main Hadley cell occurs adjacent to the

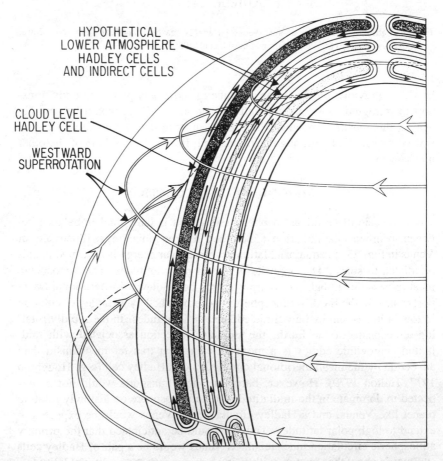

HYPOTHETICAL
LOWER ATMOSPHERE
HADLEY CELLS
AND INDIRECT CELLS

CLOUD LEVEL
HADLEY CELL

WESTWARD
SUPERROTATION

Fig. 24. Cloud level Hadley cell carrying the excess radiative energy deposited at high altitudes in the equatorial region to polar latitudes. Below the clouds there is probably a series of alternating direct and indirect meridional cells, including a ground-level Hadley cell.

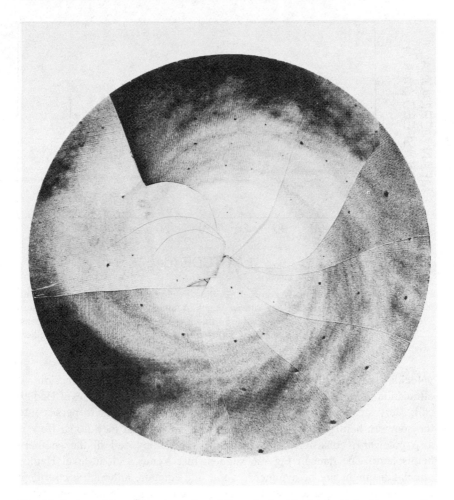

Fig. 25. A view of the south polar region of Venus in the ultraviolet, obtained by remapping and piecing together Mariner 10 photographs taken at different viewing angles and times (Suomi and Limaye 1978). The outer circle of the composite picture is the planet's equator. The atmosphere is rotating counterclockwise and the spiral pattern of the polar vortex is readily visible in the cloud-like markings.

ground. The cloud-level Hadley cell and the atmospheric superrotation combine to produce the polar vortex seen in the Mariner 10 ultraviolet image, Fig. 25 (Suomi and Limaye 1978).

Average meridional wind speeds from tracking of small-scale ultraviolet features in the Mariner 10 (Limaye and Suomi 1981) and Pioneer Venus (Rossow et al. 1980a) images are shown as a function of latitude in Fig. 26. The cloud-top north-south winds are an order of magnitude slower than the winds of the dominant east-to-west circulation; average meridional wind

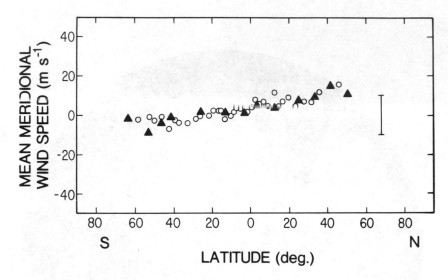

Fig. 26. Time and longitudinally averaged meridional wind speeds from tracking of small-scale features in Mariner 10, ○ and Pioneer Venus, ▲ ultraviolet photographs. The vertical error bar is a representative rms deviation of the measurements about the plotted means. Positive speeds in northern latitudes are northward motions; negative speeds in southern latitudes are southward motions.

velocities are ≤10 m s^{-1}. In the northern hemisphere, the mean meridional circulation is northward at all latitudes, consistent with a cloud-level Hadley cell. Very small northward velocities (1 or 2 m s^{-1}) appear to persist into the southern hemisphere, but south of ∼ 25°S the mean meridional flow is also poleward, consistent with a cloud-level Hadley cell in the southern hemisphere. The data in Fig. 26 suggest that Venus's cloud-level Hadley circulation might not be symmetric about the equator, although a confident determination of equatorial asymmetry is precluded by the estimated errors in the cloud-tracked wind vectors and the location of Venus's pole. Indeed, there is no obvious cause in the orbital or rotational motion of Venus for an equatorial asymmetry in the mean meridional circulation of the magnitude suggested by Fig. 26. The northern hemisphere cloud-level Hadley cell might not only be larger in latitudinal extent but also faster; the wind speeds in the northern hemisphere in Fig. 26 are somewhat larger than the south-ward wind velocities south of 25°S, although this observation is again subject to uncertainties. The extension of the northern hemisphere cloud-level Hadley cell into the southern hemisphere could help explain the observed pressure differences among the Large, Day, and Night PV probes (Seiff et al. 1980). The pressure at cloud heights at the Day and Night probe locations (∼ 30°S) exceeded the pressure at the Large probe location (near the equator). The sense of these pressure contrasts is consistent with a high-altitude northward flow in low southerly latitudes.

Certain features of atmospheric thermal structure are consistent with a downwelling motion at high latitudes that would be expected for a cloud-level Hadley cell that extends far poleward. These include the dipolar infrared hot spots which straddle the north pole and rotate around it every 2.9 d (F. Taylor et al. 1980; Apt and Leung 1982). The hot spots may be lowerings in the cloud tops associated with descending motion near the center of the polar vortex (F. Taylor et al. 1980). However, other prominent thermal features of the cloud-level polar atmosphere are not so obviously related to the descending motions of a polar vortex, although they may nonetheless have a dynamical origin. Most important among these are the strong cloud-level temperature inversions that occur at latitudes $\geqslant 50°$ (Kliore and Patel 1980; Kolosov et al. 1980). The deepest inversions are found between latitudes of 60° and 70°, the approximate location of the cold collar seen in the infrared soundings of the Pioneer Venus orbiter (F. Taylor et al. 1980). The pressure at the inversion temperature minimum is least and the temperature at the inversion minimum is lowest between 60° and 70° lat. The latitude variation of temperature along a constant pressure surface shows a strong minimum between 60° and 80° at cloud heights (Fig. 27). No explanation has been offered for these low cloud-level temperatures at 60° to 80° lat. The thermal structure of the cloud-level polar atmosphere suggests that the dynamical regime at high latitudes may be more complicated than in the equatorial region.

The height to which the cloud-level Hadley cell extends above the cloud tops is uncertain. The thermal structure of the atmosphere again supplies some clues. Temperatures between 70 and 90 km altitude over the polar regions are higher than those over the equator at comparable altitudes (F. Taylor et al. 1980; Fig. 27). An analogous situation may exist in the Earth's atmosphere, where at the solstices the temperatures over the winter pole near the mesopause are much higher than the temperatures at the same heights over the summer pole (Holton 1975). This situation occurs in the presence of a zonally averaged stratosphere-mesosphere meridional circulation that involves upwelling over the summer pole, northward meridional flow at high levels, and downwelling over the winter pole. This circulation is driven by strong radiative heating at the stratopause over the summer pole. There, at ~ 50 km, ozone absorbs solar ultraviolet radiation, heats the atmosphere, and drives the large-scale circulation that persists to the mesopause (~ 80 km) and above. Thus, at the altitude of the mesopause, there is upwelling over a relatively cold summer pole and downwelling over a relatively hot winter pole. The large-scale meridional circulation at the mesopause is a direct circulation driven by radiative heating below the mesopause. The warm winter pole at the mesopause has been attributed to adiabatic compressional heating in the descending flow over the winter pole together with diabatic heating during the northward motion of the air from the summer pole (Leovy 1964). However, it is not certain that this explains the high temperatures over the winter pole at the mesopause. Dynamical transports of heat to this region are doubtlessly required, but other

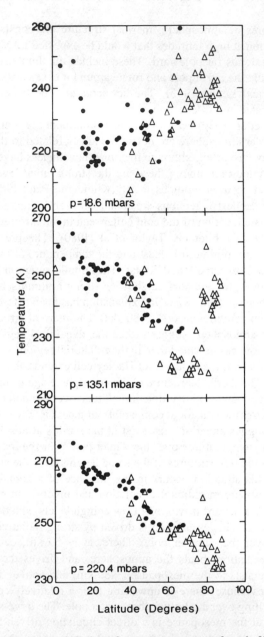

Fig. 27. Temperature versus latitude on 3 constant pressure surfaces in Venus's atmosphere. The approximate altitudes of the pressure levels are 73, 63, and 60 km. Temperature decreases with latitude at 220.4 mbar (60 km) but increases with latitude at 18.6 mbar (73 km). There is a strong temperature minimum between 60 and 80° latitude at 135.1 mbar (63 km), corresponding to deep inversions in the atmospheric thermal structure at these latitudes. Data are from the Pioneer Venus radio occultation experiment, courtesy of A. Kliore. ●, southern hemisphere occultations; Δ, northern hemisphere measurements.

mechanisms may be involved, e.g. the deposition of heat by upward propagating large-scale gravity waves. The analogy with Venus is in the possibility that the Hadley cell driven by radiative heating at cloud levels extends to heights of ≥ 90 km and results in descending motions over the poles in the 70 to 90 km range. Radiative heating along the poleward branch of this circulation along with adiabatic compressional heating in the descending flow over the pole could then produce the warm pole at 70 to 90 km observed on Venus (Schubert et al. 1980a). It is not at all certain however that the requisite diabatic heating along the poleward portion of the Hadley cell actually occurs.

F. Taylor et al. (1979b,1980) and Elson (1979,1982) have used the brightness temperatures measured by the infrared radiometer on the PV orbiter to infer the circulation above the clouds. Their models predict a poleward flow below 90 km, but this motion is the return branch of a mean meridional circulation with upwelling over the polar regions, equatorward flow above 90 km, and descending motion over the equatorial regions. The direction of motion along the upper branch of this circulation agrees with heterodyne observations (Betz et al. 1976, 1977). However, the model wind speeds on this branch are larger than those observed, and very large eddy diffusivities must be invoked to limit the flow velocities predicted by the model. The sense of the meridional circulation high above the clouds depends on the direction of the meridional pressure gradient, which can be determined from infrared temperatures by integrating Eq. (14) over height. However, the result depends on the unknown value of the meridional pressure gradient at the lower boundary. Therefore, the direction of mean meridional flow is uncertain for the same reason that the persistence of the mean zonal wind with height is uncertain (refer to the discussion in Sec. III on cyclostrophic balance). A more physical description of this uncertainty is that although the observed temperature gradients between 70 and 90 km by themselves tend to drive a direct meridional circulation with rising motion at the pole, heating below 70 km tends to drive the opposite circulation and this may predominate. If the mean zonal wind persists to high altitudes, it would be because the heating at cloud levels provides a strong enough poleward pressure force to overcome the equatorward pressure force associated with the warm pole at higher altitudes, thereby also tending to maintain cyclostrophic balance. This could further imply that the cloud-level drive for the meridional circulation is strong enough to impose a downwelling motion in the 70 to 90 km altitude range over the pole (Schubert et al. 1980a).

In view of the uncertainty in the direction of meridional circulation at 90 km, we must consider other explanations for the warm pole above the clouds, such as enhanced radiative heating from below. The descending flow of the polar vortex near 65–70 km could lead to a lowering of the cloud tops and clearing at high altitudes. This would allow a greater flux of infrared radiation to penetrate into the atmosphere above the clouds, which would cause a warming over the pole if an infrared absorber were located there. Upward propagating planetary-scale waves may also deposit heat in Venus's polar atmosphere above the clouds. This mechanism is thought to be import-ant in producing polar warming of the Earth's stratosphere. The leakage of ultra-long planetary-scale waves into the lower stratosphere causes the eddy heat transport in the stratosphere to be against the temperature gradient, main-taining the reversed gradient against radiative damping (Holton 1975, 1979).

Altitude profiles of meridional wind speeds from interferometric tracking of the Pioneer Venus probes are shown in Fig. 28 (Counselman et al. 1980). North-south velocities are $\lesssim 10$ m s^{-1} at all altitudes below the clouds. Winds change direction with height and increase in magnitude with proximity to the clouds. Meridional wind velocities below the clouds do not exceed ~ 5 m s^{-1} and are generally $\leqslant 1$ or 2 m s^{-1}. Because these profiles reflect the altitude variations of north-south winds at only a few locations on Venus at a single time, they may not be indicative of mean meridional circulations; they could simply be snapshots of temporally evolving eddies. Undoubtedly, some of the structure in the profiles of Fig. 28 is due to eddies, but features of the profiles common to all the widely spaced probes may reflect the temporally and zonally averaged meridional circulation. It is therefore instructive to interpret the measured meridional winds in this way, and see how far the hypothetical circulations are both self-consistent and consistent with theoretical expecta-tions based on general atmospheric structure and mode of heating. The direc-tions of the meridional winds, and the cellular circulations implied by a strict interpretation of these winds as mean meridional cells, are summarized in Fig. 29. All the wind direction profiles are consistent with the existence of a cloud-level Hadley cell marked (a) in the figure; this is particularly clear for the North and Large (Sounder) profiles. The Night probe, with equatorward flow above 60 km, may have measured the southward extension of the north-ern hemisphere Hadley cell as discussed above. The upper limb of the cloud-level Hadley cell at the Day probe site may have been above 60 km; the data for this probe show strong equatorward flow between 60 and 45 km, which is consistent with the lower limb of the cloud-level Hadley cell and the upper limb of an underlying indirect cell. All the wind directions are consistent with the base of the cloud-level Hadley cell being approximately coincident with the base of the clouds, and the upper branch of the cell extending at least to the cloud tops at 65–70 km. Meridional wind directions at all the probes also show consistent evidence for an indirect cell immediately below the clouds. Figure 29 suggests that this indirect cell (b) might be contained between the base of the clouds and ~ 40 km, the height of maximum static stability in the stable layer beneath the clouds (see below). At lower altitudes there are no cells common to all the probes. Clearly, at least some of these low-altitude meridional winds are associated with eddies and are not part of a consistent mean meridional circulation. This interpretation of the meridional wind ve-locities measured by the PV probes in terms of the mean meridional circula-tions shown in Fig. 29 is only speculative, and is guided by other observations

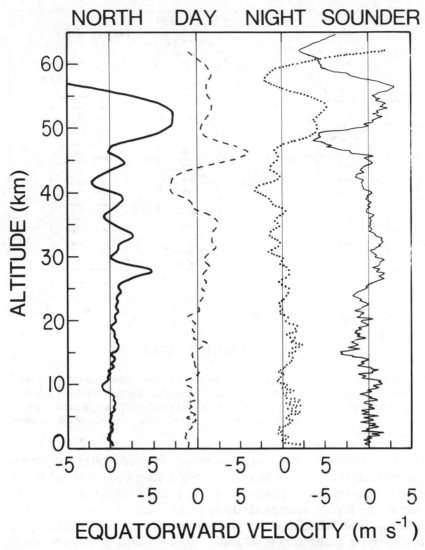

Fig. 28. Meridional wind velocity profiles from interferometric tracking of the Pioneer Venus probes (Counselman et al. 1980).

of wind speeds and atmospheric structure and theoretical ideas about the nature of the mean circulation. Even a complete set of wind measurements at a single instant certainly cannot define a temporally averaged circulation.

Theoretical considerations (Stone 1974, 1975; Kálnay de Rivas 1973, 1975; Schubert et al. 1980a; Rossow 1982) suggest a surface Hadley circulation to latitudinally redistribute solar energy absorbed in the lower atmosphere and at the ground. A direct meridional circulation in the deep atmosphere is

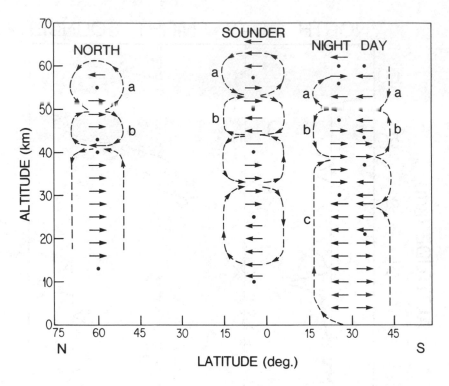

Fig. 29. Directions of meridional winds versus height at the sites of the Pioneer Venus probes. The wind direction profiles of the Night and Day probes are displaced slightly in latitude to facilitate the plotting. The meridional cells suggested by each profile are indicated schematically. Cells with the same designation may correspond to the same part of a mean global circulation pattern.

also needed to balance the observed increase in infrared cooling with latitude (Suomi et al. 1979, 1980; Schubert et al. 1980a; Revercomb et al. 1982). It is expected that the surface Hadley cell would extend to high latitudes, in contrast to the Earth's tropospheric Hadley cell, because sloping convection associated with baroclinic eddies, the heat transfer mechanism replacing the Hadley cell in midlatitudes in the Earth's rapidly rotating atmosphere, would not be expected to occur in the slowly rotating lower atmosphere of Venus. However, the latitudinal extent of a Hadley cell depends on eddy momentum transports as well as planetary rotation rate (Held and Hou 1980; K. E. Taylor 1980), and we cannot be certain that direct meridional circulations extend all the way to the pole on Venus. Because most of the solar energy absorbed by Venus is deposited in and above the clouds, we also expect a cloud-level Hadley cell. Thus, because of the way the Sun heats the atmosphere and the thickness of the atmosphere, we expect a vertically layered mean meridional circulation with indirect cells frictionally driven by adjacent direct circulations (Kálnay de Rivas 1973, 1975; Schubert et al. 1980a; Rossow 1982).

One possibility, suggested by the thermal structure of the lower atmosphere, is a 3-layer mean meridional circulation: a cloud-level Hadley cell; a Hadley cell between the surface and ~ 40 km; and a weak indirect cell between 40 and 50 km. Temperature measurements on the Pioneer Venus probes (Seiff et al. 1980) and the Venera 10 lander (Avduevskii et al. 1976a) reveal a strongly stable region of the atmosphere between ~ 30 km and the base of the clouds at ~ 50 km. The static stability in this layer peaks at heights between ~ 40 and ~ 45 km (Figs. 30 and 31). The planetwide stable layer just below the base of the clouds is probably due to enhanced heating in the clouds by absorption of long wavelength radiation from the lower atmosphere (Pollack et al. 1980a). The large static stability of this layer might cause it to act as a lid on the underlying atmosphere, preventing any mean meridional circulation from extending upward from the lower atmosphere (Schubert et al. 1980a), just as the highly stable stratosphere largely confines the Earth's equatorial Hadley cell to the troposphere. Therefore, a surface Hadley cell on Venus could be restricted to heights of ≤ 40 km.

The PV Night probe wind directions (Fig. 29) are consistent with 3 major mean meridional cells. In addition to the cloud-level Hadley cell and the indirect cell already discussed (cells a and b in Fig. 29), the equatorward flow between the surface and ~ 30 km and the poleward flow between ~ 30 and ~ 40 km (cell c in Fig. 29) are in the respective flow directions in the lower and upper limbs of a surface Hadley cell. This consistency may only be fortuitous, however, since a sampling of the latitudinal winds at a single time and location will not necessarily reveal the temporally and zonally averaged mean flow. In addition, as discussed below, the wind directions measured by the other Pioneer Venus probes are not indicative of a surface Hadley cell.

We can use a simple mass balance argument to check the plausibility of a surface Hadley circulation extending to 40 k. If $\bar{v}_u$ and $\bar{v}_\ell$ are the magnitudes of the mean meridional wind speeds in the upper and lower limbs of the Hadley cell, then

$$\frac{\bar{v}_u}{\bar{v}_\ell} = \frac{p_s - \bar{p}}{\bar{p} - p_{40}} \tag{28}$$

where p_s is surface pressure; p_{40} the pressure at 40 km; and $\bar{p}$ the pressure at the level separating the upper and lower branches of the cell. Figure 32 shows how $\bar{v}_u/\bar{v}_\ell$ depends on the altitude of the $\bar{p}$ level. The mean poleward velocity in the upper limb of the surface Hadley cell is < 10 times larger than the mean equatorward velocity in the lower limb, so long as the level separating the branches is ≤ 30 km. If this level were 25 to 30 km, the mean meridional wind speeds near the surface could be fractions of 1 m s^{-1} while the mean winds at 25 to 40 km would be only several m s^{-1}. The Night probe meridional wind speeds in Fig. 28 are consistent with this possibility, although they do not necessarily represent the mean north-south winds. The main point of

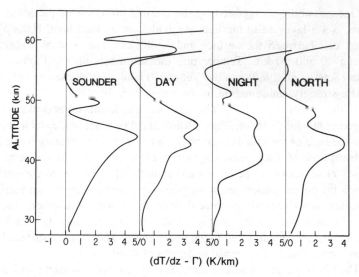

Fig. 30. Altitude profiles of atmospheric static stability at the Pioneer Venus probe sites. Γ, adiabatic lapse rate; dT/dz, vertical temperature gradient in the atmosphere. A positive value of $dT/dz - \Gamma$ indicates stability.

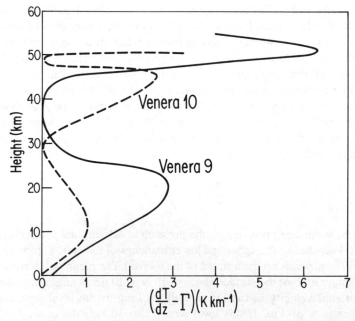

Fig. 31. Altitude profiles of atmospheric stability deduced from Venera 9 and 10 temperature measurements (Avduevskii et al. 1976b). $dT/dz - \Gamma$ is calculated from reported values of γ and n, the polytropic exponent ($p \propto \rho^n$), using the formula $dT/dz - \Gamma = |\Gamma| (\gamma - n)$ $(\gamma - 1)^{-1} n^{-1}$. This equation is valid for a perfect gas in hydrostatic equilibrium. The stability profiles are approximate: variations in the reported values of n have been smoothed.

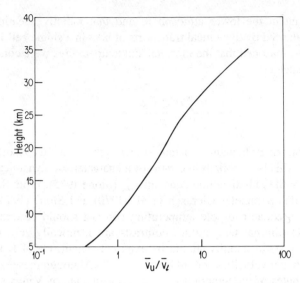

Fig. 32. The ratio of mean velocity $\bar{v}_u$ in the upper branch of a surface Hadley cell to mean velocity $\bar{v}_\ell$ in the lower branch, as a function of the level separating the two branches. The Hadley cell extends to 40 k.

this discussion is that a surface Hadley cell could extend to the stable layer beneath the clouds, with a lower branch 25 to 30 km thick and with relatively weak meridional wind speeds throughout, consistent with the data in Fig. 28. The surface Hadley cell could of course also be confined to lower altitudes.

As noted above, the meridional winds below 40 km at the other Pioneer Venus probe sites (Fig. 28 and 29) are not consistent with a 3-layer mean circulation pattern and probably represent eddies (Counselman et al. 1980; Schubert et al. 1980a). A conspicuous example of this is the poleward sense of the meridional winds throughout the lowest scale height at the Day probe location. This is consistent with diurnally forced eddies in Venus's lower atmosphere. The flow below 55 km at the North probe location is equatorward at almost all altitudes above 15 km; it is significantly poleward only between ~ 40 and ~ 43 km. While the equatorward flow is consistent with the lower limb of a surface Hadley cell, the poleward flow between 40 and 43 km does not transport enough mass to be the upper branch of the cell. The many reversals of wind direction with height at the Large probe location also indicate eddy motions.

The structure and meridional circulation of Venus's lower atmosphere at any given time is evidently more complicated than the temporally and zonally averaged state expected theoretically. Gierasch et al. (1970), Golitsyn (1970) and Stone (1974, 1975) have used scaling arguments to predict the nature of the mean state of the lower atmosphere. They concluded that diurnal effects were unimportant in the lower atmosphere and that radiative heating and cooling were balanced by dynamical transports of heat in a single cell Hadley circulation. They reasoned that the motions and temperatures were controlled by a single parameter

$$\frac{\sigma^{3/8}\, q_{abs}^{5/8}\, Rg}{c_p^{3/2}\, p_s}$$

where σ is the Stefan-Boltzmann constant, and q_{abs} the solar flux absorbed by the atmosphere. This is essentially the ratio of a characteristic dynamical time scale $R(gH_s)^{-1/2}$ to the radiative time constant τ_{rad} (Stone 1975). From the very small value of this parameter Gierasch et al. (1970) and Stone (1974) concluded: the mean equator-to-pole temperature contrast should be extremely small, O(0.1 K) (diurnal temperature contrasts are identically zero in this model); the mean static stability should be extremely small, O(0.1 K km^{-1}); and mean meridional velocities should be $\sim$ 1 m s^{-1}. Although these could in fact be the attributes of the temporal and zonal mean state of Venus's lower atmosphere, conditions were quite different at the time of Pioneer Venus. Horizontal temperature contrasts both latitudinal and longitudinal were as large as 5 to 10 K, one scale height above the surface (Fig. 33). The temperature near 30°S at $\sim$ 15 km was $\sim$ 5 K hotter than the equatorial temperature at

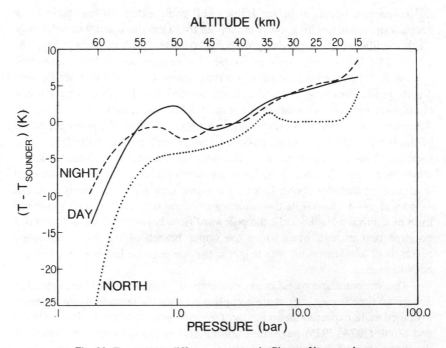

Fig. 33. Temperature differences among the Pioneer Venus probes.

the same height. There were similar temperature differences between day and night, and equatorial latitudes and 60°N. The static stability of the lower atmosphere was very large in certain altitude ranges. We have already described the stable layer between 30 and 50 km. Venera 9 measured a very stable lower atmosphere beneath 30 km with a stability peak at $\sim$ 20 km (Fig. 31); the Pioneer Venus probes also detected stable layers at $\sim$ 15 km (Seiff et al. 1980). Any planetwide stable layers of the atmosphere below 30 km could significantly modify the meridional circulation by their tendency to subdivide the surface Hadley cell into several vertically layered cells. The lower atmosphere meridional winds shown in Figs. 28 and 29 suggest the existence of eddies. Even with the small value of the Coriolis parameter on Venus (Table I), only tens of days would be required for Coriolis forces to produce 1 m s^{-1} east-west winds from north-south motions of 1 m s^{-1}. Zonal winds of O(1 m s^{-1}) have been measured in the lowest scale height of the atmosphere (Fig. 9) and, of course, higher velocity zonal winds dominate the circulation at all levels between $\sim$ 10 and $\sim$ 100 km. Thus, while there is probably a temporally and zonally averaged surface Hadley cell in the lower atmosphere, the motions there at any instant of time are not merely those associated with the mean circulation. Instead, eddies are probably involved in a temporally evolving pattern that transports heat poleward and retrograde zonal momentum upward, to build up the dominant westward superrotation.

VI. WAVES

Waves are undoubtedly present in the Venus atmosphere over a wide range of spatial scales. Figure 34 is a Pioneer Venus ultraviolet image of what is probably a train of internal gravity waves with horizontal spacing $\sim$ 200 km (Rossow et al. 1980a). These waves have phase fronts inclined to constant-latitude circles. Other likely trains of internal gravity waves known as circumequatorial belts are nearly parallel to latitude circles over distances of thousands of km. A group of these waves is visible in the high-pass, spatially-filtered composite Mariner 10 ultraviolet image shown in Fig. 35 (Belton et al. 1976a). The circumequatorial belts are separated by $\sim$ 500 km, and they have only been observed to travel southward with low phase speeds of $\sim$ 20 m s^{-1} (Belton et al. 1976b). The global ultraviolet patterns visible from Earth and recorded in detail in spacecraft images (Murray et al. 1974; Belton et al. 1976a,b; Anderson et al. 1978; Travis et al. 1979a,b; Rossow et al. 1980a) are probably manifestations of planetary-scale waves propagating slowly with respect to the bulk superrotation of the atmosphere (Belton et al. 1976b). The best known of these features is the dark horizontal Y shown prominently at two times 4 d apart in Fig. 36. A single Y completely encircles the planet. Besides the Y, similar zonal wavenumber 1 features such as a horizontal ψ and reverse C are often seen. All these features include dark midlatitude bands that tilt westward; i.e. the more westward portions are

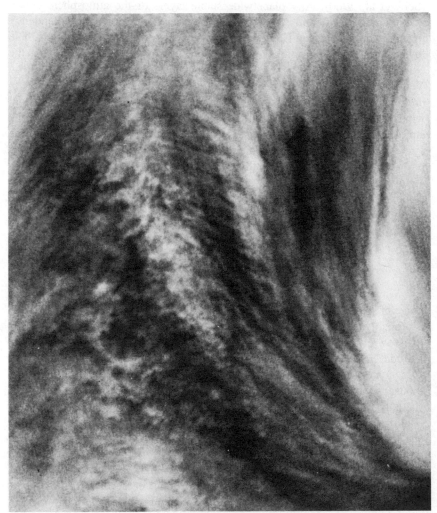

Fig. 34. A train of gravity waves in a Pioneer Venus ultraviolet image.

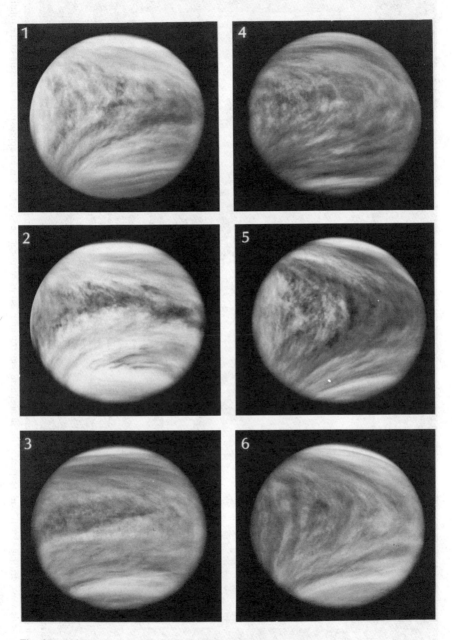

Fig. 35. A high-pass, spatially filtered, composite Mariner 10 ultraviolet photograph showing a southward propagating train of circumequatorial belts.

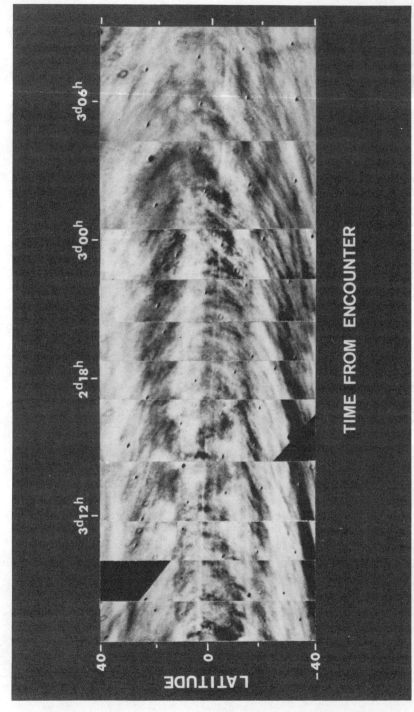

Fig. 36. Series of Pioneer Venus ultraviolet images at intervals of 1 d, showing the dark horizontal Y in (1) and 4 d later in (5). The Y pattern represents a planetary-scale wave drifting slowly with respect to the atmosphere superrotation, so its reappearance after 4 d is an approximate measure of the rotation period of the atmosphere.

located at higher latitudes so that the midlatitude bands produce a pattern that opens outward in the direction of atmospheric rotation. The Y and ψ, but not the C, include a dark band aligned with the equator. The impression gained from studying a time series of ultraviolet images is that the midlatitude and equatorial components appear and disappear, usually with the same relative phase, and thereby produce the changing ultraviolet features (Rossow et al. 1980a). The midlatitude and equatorial components are therefore somehow interrelated, perhaps by nonlinear dynamical interactions (Covey and Schubert 1982b).

A complete interpretation of these ultraviolet contrasts is not possible because the chemical identities of all the ultraviolet absorbers are not known (Esposito and Travis 1982). Sulfuric acid, the major component of the upper-cloud particles, does not absorb in the ultraviolet. Therefore the ultraviolet dark areas must be produced by trace quantities of ultraviolet-absorbing substances. Sulfur dioxide gas is the likely absorber for wavelengths < 0.32 μm; another as yet unidentified absorber is required to explain the absorption at longer wavelengths (Pollack et al. 1980b; Esposito 1980). The absorbers are probably formed in the lower atmosphere and destroyed photochemically upon reaching the cloud tops. At high latitudes the dark areas may be produced when the polar haze overlying the cloud tops decreases in optical depth, perhaps because of increased temperature. In equatorial regions, ultraviolet contrasts may result from variations in upward transport which either deplete or enhance the concentration of absorbers near the cloud tops (Pollack et al. 1980b). The important point here is that the ultraviolet albedo is probably correlated with some dynamic or thermodynamic property of the atmosphere such as vertical velocity or temperature, and so the contrasts serve as tracers of waves, convection, and other aspects of atmospheric dynamics.

The global scale ultraviolet markings are known to be traveling waves for two reasons. First, the midlatitude or spiral bands, essential elements of the large-scale patterns, cannot be streamlines of the atmospheric cloud-top circulation. If they were, they would be nearly aligned along circles of latitude, because the zonal flow speeds are 1 to 2 orders of magnitude greater than the meridional wind velocities. Second, and most important, the global ultraviolet patterns cannot be composed of features passively advected by the zonal wind, because at the time of the Mariner 10 flyby there were small but significant differences between the phase propagation speed of the planetary-scale pattern and the zonal wind speed. The Y pattern exhibited solid body rotation, maintaining itself over two successive atmospheric rotations, while it was deduced from tracking of small-scale ultraviolet markings that the zonal wind contained a high-speed midlatitude jet (see Fig. 8). If each component of the Y had traveled at the local zonal wind speed, differential shear would have destroyed the pattern in one or two rotations (Belton et al. 1976b). The midlatitude component of the Y must have been traveling more slowly than the high-speed winds of the jet. Belton et al. (1976b) estimated that large-scale midlatitude ultraviolet features were propagating upstream (eastward) at a velocity of $\sim 20–30$ m s^{-1} relative to the winds. The equatorial part of the Y was thought to be moving slightly faster, $5–15$ m s^{-1}, than the atmospheric winds.

Even though the atmosphere rotated nearly as a solid body at the time of Pioneer Venus, there are indications that the phase velocity of the planetary-scale features differs from the zonal wind speed. Time series of ultraviolet brightness at low latitudes show a 4 d periodicity, whereas a 5.2 d period is detected at middle latitudes (Del Genio and Rossow 1982). Together with the cloud-tracked wind measurements, this implies that the planetary-scale features are traveling ~ 15 to ~ 20 m s^{-1} faster than the zonal wind at low latitudes and ~ 5 to ~ 10 m s^{-1} slower than the zonal wind at midlatitudes, in qualitative agreement with the results from Mariner 10. The 4 d period at low latitudes is associated with the zonal motion of the dark equatorial band, while the 5.2 d midlatitude period is connected with oscillations in the latitude of the bright polar bands (Del Genio and Rossow 1982). The 5.2 d periodicity is most prominent in the latitude range $45°–55°$. This is the farthest equatorward location of the ultraviolet bright polar bands, which undergo quasi periodic changes from tilted to parallel in their alignments with latitude circles (Belton et al. 1976b; Rossow et al. 1980a). Both the equatorial and midlatitude components of the large-scale ultraviolet patterns are zonal wavenumber 1 features (Belton et al. 1976b; Del Genio and Rossow 1982), although other interpretations have ascribed wavenumbers 2 and 3 to these structures (Dollfus 1975; Suomi and Limaye 1978).

Wavelike features in polar regions have both wavenumber 1 and wavenumber 2 zonal variations. The wavenumber 2 feature is a pair of infrared brightness maxima which straddle the pole (F. Taylor et al. 1980) and rotate around it in a retrograde sense with a 2.9 d period (Apt and Leung 1982). Since no wind measurements exist at these high latitudes, we do not know how the phase speed of the dipolar infrared feature compares with the atmospheric rotation. However, the 2.9 d period is significantly smaller than the atmospheric rotation periods of > 4 d deduced from cloud-tracked winds at lower latitudes, suggesting that the dipolar infrared feature is a wave traveling westward relative to the local zonal winds. Near $60°–70°$ N there is a crescent shaped band of cold temperatures at cloud heights and above (F. Taylor et al. 1980; Kliore and Patel 1980; Fig. 27). This cold infrared collar is primarily wavenumber 1, and a component of its motion follows the Sun (F. Taylor et al. 1980).

There is direct evidence of planetary-scale waves in Venus's atmosphere from oscillations of the zonal wind velocity. The spectroscopic measurements of Traub and Carleton (1979) reveal a sinusoidal variation of zonal wind speed between limits of 40 and 120 m s^{-1} with a period of 4 days. This period is consistent with a wave drifting slowly with respect to the cloud-top atmospheric superrotation. Cloud tracked zonal wind velocities from Mariner 10 ultraviolet images correlate with the phase of the Y (Limaye and Suomi 1977*a*,1981), further supporting this picture. However, the amplitude of the wind variations obtained from both Mariner 10 and Pioneer Venus cloud-tracked winds is only 10–15 m s^{-1}, significantly less than that observed by Traub and Carleton (1979).

Oscillations of temperature as well as velocity have been observed by the PV orbiter. Five to six day oscillations of infrared brightness temperature have been detected in the 70–90 km altitude range (F. Taylor et al. 1980; Apt and Leung 1982). These fluctuations have an amplitude of 1–10 K; maximum amplitudes occur at 60°–70° lat. The waves appear to propagate vertically with a vertical wavelength of ~ 20 km. Five to six day oscillations are also seen at higher altitudes (155 km) in the exospheric temperature deduced from the atmospheric drag experiment on the PV orbiter (Keating et al. 1979*a*). Thus, both sets of observations may be detecting the same upward propagating wave.

Altitude profiles of temperature and wind velocity, obtained by the 4 Pioneer Venus probes as they descended below the cloud tops, reveal wavelike variations with wavelengths of 5–10 km (Seiff et al. 1980; Counselman et al. 1980). Similar vertical variations are seen in infrared temperature soundings (F. Taylor et al. 1980), radio occultation temperatures (Kliore and Patel 1980; Kolosov et al. 1980), and number densities of upper-atmosphere neutral species (Niemann et al. 1980*b*; von Zahn et al. 1980). All these observations are snapshots of vertical wave structure, with no information about horizontal structure or period. It is difficult to judge whether these altitude profiles reveal the vertical structures of planetary-scale waves or of much smaller scale waves. The latter seems to be the case for the wavelike variations in temperature and wind velocity recorded by the probes, for there appear to be no correlations among the probes even though they entered the atmosphere simultaneously.

The planetary-scale traveling waves apparent at cloud-top levels in the global ultraviolet patterns could affect the mean structure and circulation of Venus's atmosphere by transporting heat poleward and zonal momentum upward. Evidently, because of their 4 to 6 d periods they are not directly forced by stationary topographic features of Venus's surface or by the diurnal and higher harmonics of the solar heating that give rise to atmospheric tides. Solar heating variations in a coordinate system rotating with the atmosphere at cloud heights could produce waves, but these would have a 4 d period in the corotating system. The observed waves have a very long period with respect to this system. One possible explanation for Venus's planetary-scale waves is that they are normal-mode oscillations of the atmosphere. Resonant responses of the Earth's atmosphere have been identified in observations of pressure and temperature fluctuations (Eliasen and Machenhaur 1965; Rodgers 1976; Madden 1978, 1979). These include westward propagating waves with zonal wavenumber 1 and periods of 5 and 16 d. The 5 d wave is the lowest symmetric normal mode oscillation of the atmosphere with zonal wavenumber $s = 1$, and the 16 d wave is the second lowest normal mode (Madden 1979; Walterscheid 1980). These waves are Rossby-Haurwitz modes. Free Rossby-Haurwitz waves on Venus would travel eastward, due to Venus's retrograde rotation, with phase speed ~ 20 m s^{-1}. This is consistent with the midlatitude upstream velocity of the Y inferred from Mariner 10 (Belton et al 1976a,b) and Pioneer Venus (Rossow et al. 1980a; Del Genio and Rossow 1982) observations. At the equator, free gravity waves would have phase velocities > 200 m s^{-1}, much larger than the observed 15 to 20 m s^{-1} westward speed of the Y relative to the mean zonal flow. However, a combination of two oppositely moving but otherwise similar gravity waves (gravity modes can propagate both east and west) could form a quasi standing wave relative to the atmospheric circulation, and drift with it. Belton et al. (1976b) have shown that it is possible to simulate the Y on Venus by a combination of the free Rossby-Haurwitz and gravity modes, with the proper phase relations. However, their analysis was carried out for a model atmosphere with uniform 4 d rotation and constant static stability. The normal modes of a realistic Venus model atmosphere whose static stability and zonal wind velocity varies with height have yet to be determined.

Alternatively, Venus's planetary-scale traveling waves might result from an instability of its mean circulation. The barotropically unstable Mariner 10 midlatitude jet stream is discussed above. Barotropic instability transfers the kinetic energy of a latitudinally varying zonal flow to the growing waves. However, this type of instability cannot be responsible for the 4 d equatorial wave on Venus, because the wave propagates significantly faster than the zonal flow at almost all latitudes (Del Genio and Rossow 1982). While barotropic instability could possibly produce the 5.2 d midlatitude component of the Y, this seems unlikely, because the PV latitudinal wind profiles showed no pronounced jet streams over a five-week period and were barotropically stable (Rossow et al. 1980a). Despite this, the Y and other wavelike features were quite apparent in the PV data (Schubert et al. 1980a). However, the midlatitude wave at the time of Pioneer Venus might have been the remnant of an earlier barotropic instability. There is some support for this view in a recent report of a prominent, barotropically unstable midlatitude jet which existed during a 7 d period of PV ultraviolet observations in Fall 1979 (Rossow and Kinsella 1982). PV ultraviolet observations of zonal winds at earlier (Spring 1979) and later (Spring 1980) times revealed a cloud-top state of nearly solid body rotation. With low dissipation, barotropic instabilities might persist long after the zonal mean profile showed no jets. Del Genio and Rossow (1982) have noted a similarity between the morphology of the bright polar bands and the streamlines of a zonal flow with a superimposed advection wave, which could have evolved by two-dimensional interactions from an initial barotropic disturbance (Rossow and Williams 1979).

Another type of instability that could give rise to the traveling waves is baroclinic instability, which utilizes the potential energy of the equator-pole temperature contrast. Young (1981) has shown that the PV altitude profiles of zonal wind and temperature satisfy the necessary condition developed by Charney and Stern (1962) for baroclinic instability under geostrophic conditions. No one has yet formulated a similar criterion for baroclinic instability under the cyclostrophic conditions that are more appropriate for Venus. A related form of instability, symmetric baroclinic instability, may also occur on Venus (K. E. Emanuel, personal communication). We note, however, that the necessary conditions for these types of baroclinic instability—high wind shear and low static stability—are met only in a thin layer of the atmosphere at 50 to 55 km. Young et al. (1982) have recently calculated the growth rates of baroclinic unstable modes in a three-dimensional, spherical, primitive equation model of the Venus atmosphere. The momentum transport by the baroclinic modes tends to create a jet at middle or high latitudes.

It is also possible that the traveling waves are neither resonant oscillations nor the result of instability. Instead, they could be the preferential response of the atmosphere, at 4 to 6 d periods, to forcing over a broad range of frequencies. This situation apparently occurs in the Earth's equatorial stratosphere, where observations reveal westward traveling waves with a 4 to 5 d period (Yanai and Maruyama 1966) and eastward traveling waves with a 15 d period (Wallace and Kousky 1968). From the phase speeds observed over a limited number of stations, the westward waves appear to have zonal wavenumber 4 while the eastward waves have wavenumber 1 or 2. The former have been identified as mixed Rossby-gravity waves and the latter as Kelvin waves (Wallace 1973; Yanai 1975). These are, respectively, the lowest westward propagating and lowest eastward propagating modes of oscillation in a rotating atmosphere or ocean (Longuet-Higgins 1968). The Rossby-gravity and Kelvin waves are definitely not resonant oscillations of the Earth's atmosphere. The equivalent depths implied by the observed frequencies and wavenumbers (see Figs. 2–5 in Longuet-Higgins 1968) are on the order of 100 m, far smaller than the 10 km equivalent depth for resonant oscillations. Furthermore, the small equivalent depths lead to the conclusion that the Rossby-gravity and Kelvin waves are vertically propagating. They transport energy and momentum upward, indicating that they are forced in the troposphere (Wallace 1973; Yanai 1975).

Monsoons and convection at the Intertropical Convergence Zone are obvious candidates for the forcing of these waves. However, no source of excitation traveling with the phase speed of either Rossby-gravity waves ($20-30 \text{ m s}^{-1}$ westward) or Kelvin waves ($20-30 \text{ m s}^{-1}$ eastward) has been observed in these disturbances. Instead, Wallace (1971) found a 4 to 5 d standing-wave oscillation in cloud brightness patterns observed over the equatorial Pacific by satellites. This can be viewed as a superposition of eastward propagating waves, each with a 4 to 5 d period. Theoretical modeling by Holton (1972, 1973) indicates that, for certain assumed mean zonal

wind profiles, the stratosphere preferentially responds to the westward propagating component, producing a Rossby-gravity wave with a 4 to 5 d period as
observed. For the Kelvin waves there is no distinct excitation source with a 15
d period, but according to Holton's (1972, 1973) model the stratosphere
preferentially responds to forcing with a 20 to 40 m s^{-1} eastward phase speed
to produce Kelvin waves. The reason for the preferential response is that this
phase speed is associated with a vertical wavelength that matches the vertical
scale of the diabatic heating that forces the wave. Thus the Kelvin wave can
be explained as a preferential response to forcing over a broad range of
frequencies. Latent heat release in the tropics plays a crucial role in generating
both the Kelvin and mixed Rossby-gravity waves (Hayashi and Golder 1978).
Both types of equatorial stratospheric waves interact with the mean flow and
thus affect the general circulation of the tropical stratosphere. The Kelvin
waves transport eastward momentum up while the Rossby-gravity waves
transport westward momentum up, according to calculations by Lindzen
(1971a). Holton and Lindzen (1972) have successfully modeled the alternation
of eastward and westward mean zonal winds in the equatorial stratosphere (the
quasi biennial oscillation) as arising from the momentum transports of Kelvin
and Rossby-gravity waves.

  The forcing for the planetary-scale Venus waves could be convective
activity at cloud heights, or turbulence arising from small-scale shear
(Kelvin-Helmholtz) instability of the zonal flow or the breaking of small-scale
gravity waves (Woo et al. 1980). The observed phase speed of the waves
would be due to a preferential atmospheric response to forcing with 4 to
6 d periods, in analogy to the forcing of the Kelvin wave in the Earth's equatorial stratosphere. Convective activity is a particularly plausible driving
mechanism because of the cellular disturbances observed in ultraviolet photographs. Figure 37 shows bright-rimmed and dark-edged cellular structures
visible in the Mariner 10 images of the subsolar region (Belton et al. 1976b).
The cells, suggestive of vigorous convection at cloud heights (Murray et al.
1974; Tritton 1975; Belton et al. 1976b; Rossow et al. 1980a; Covey and
Schubert 1981a), are also seen in the ultraviolet images taken by the Pioneer
Venus (Rossow et al. 1980a) and Venera 9 (Selivanov et al. 1978; Kerzhanovich et al. 1979a) orbiters. The large horizontal dimensions of the cells,
O(100 km), compared with the probable depth of convection (a scale height)
might be understood either as a consequence of radiative effects and anisotropic diffusivity (Covey and Schubert 1981a) or as the lateral spreading of
upwelling regions confined by a stable overlying atmosphere (Rossow et al.
1980a). The trains of gravity waves, including the circumequatorial belts,
seen in ultraviolet photographs may also originate in convective activity
within the clouds. Cloud-level convection may be a principal wave generating
mechanism in the Venus atmosphere.

  Because solar energy absorption is strongest in the subsolar region, we
might expect that cloud-level convection and wave-generating activity would

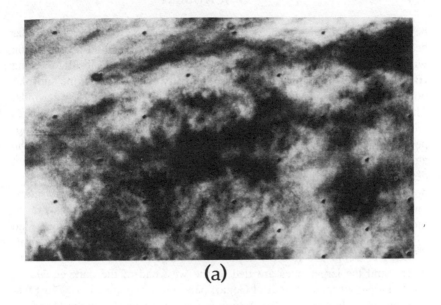

(a)

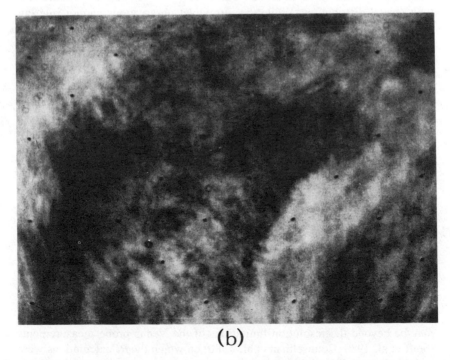

(b)

Fig. 37. Bright-rimmed (a) and dark-edged (b) cellular structures observed by Mariner 10 in the subsolar disturbance.

be strongest there. Disturbances could be generated not only by direct solar heating but also by the interaction of the fast atmospheric cloud-level rotation with a temperature or pressure obstacle provided by the subsolar heating (Belton et al. 1976b). Disturbances preferentially generated near the subsolar point would be carried westward or downstream by the cloud-level winds. Thus, any obscrved association of cellular or wavelike ultraviolet features with locations near to and downstream of the subsolar point would support a concentration of activity in the subsolar cloud-level region of the atmosphere. The limited ultraviolet observations of Mariner 10 could not determine if convection and wave propagation were enhanced near to and downstream of the subsolar point. However, the long series of PV ultraviolet images (Rossow et al. 1980a) suggests that these processes are no more vigorous near the subsolar region than elsewhere. The cells seem tied to dark material, not the subsolar point (Belton et al. 1976b; Rossow et al. 1980a; Del Genio and Rossow 1982). Bow shaped features are seen everywhere, and the larger ones are tied to the west end of the dark equatorial band, not the subsolar point (Del Genio, personal communication). Most features appear fully formed in the morning and are advected into the afternoon quadrant (Travis, personal communication). Circumequatorial belts and other wave trains have not been observed often enough to characterize their spatial distribution. While the ultraviolet observations cannot rule out a concentration of convection and wave production in the subsolar cloud-level region, they offer no support for it.

Turbulent activity has been identified in 2 layers of Venus's atmosphere centered at ~ 45 and ~ 60 km, by radio scintillation measurements of small-scale temperature fluctuations (Woo et al. 1974, 1980; Woo 1975; Timofeeva et al. 1978; Kolosov et al. 1980). These turbulent layers are located in statically stable regions of the atmosphere where small-scale gravity waves can propagate and be trapped, and where large shears in the zonal wind can lead to Kelvin-Helmholtz instability (Woo et al. 1980; Woo and Ishimaru 1981). The potential for shear or Kelvin-Helmholtz instability is measured by the Richardson number Ri:

$$Ri = \frac{N^2}{(du_x/dz)^2} \cdot \qquad (29)$$

Instability requires $Ri < 0.25$. The Richardson number of Venus's atmosphere is small at 60 and 45 km, though it appears to exceed the maximum value of 0.25 for shear instability to occur (Schubert et al. 1980a). However, as Woo and Ishimaru (1981) have emphasized, measured values of Ri in the Earth's atmosphere are generally as low as 0.25 only in very thin layers which may be beyond the resolution limitations of the Venus probe measurements (Seiff et al. 1980; Counselman et al. 1980) on which the Ri calculations were based. Thus, Kelvin-Helmholtz instability appears to be a plausible explanation for the observed turbulence, especially in the 60 km layer. Although there may be considerable turbulence associated with activity in the clouds, such layers are nearly neutrally stable and would not be identified as turbulent by a radio scintillation experiment that is sensitive only to fluctuations in potential temperature (Woo and Ishimaru 1981).

There have been several interpretations of Venus's Y and other large-scale ultraviolet patterns in terms of particular types of forced waves. Some of these assume a barotropic state in which the mean zonal wind velocity and static stability are independent of altitude. Belton et al. (1976b) used this type of model to show that the Y could be a combination of a midlatitude Rossby wave and a westward propagating equatorial Kelvin wave. The Y morphology could result if the Rossby wave were shifted westward in phase by an appropriate amount with respect to the Kelvin wave. Other ultraviolet patterns such as the ψ and reversed C shapes could also be obtained with appropriate values of the phase shift. However, if these waves were completely uncoupled, their relative phase would change with time as they drifted with respect to each other at different speeds. The relative phase would vary from 0 to 2π within one or two rotations of the cloud level atmosphere, and changing ultraviolet patterns, including ones that are never seen, would unfold (Covey and Schubert 1982a,b). The Y and other planetary-scale features are actually observed typically to persist for several 4-d rotations of the cloud-level atmosphere (Belton et al. 1976b; Rossow et al. 1980a), and the components of the patterns do not drift in and out of phase (Rossow et al. 1980a). Thus there must be some nonlinear coupling between the midlatitude Rossby wave and the equatorial Kelvin wave, which selects and tends to preserve the particular relative phases that produce the observed global-scale ultraviolet patterns (Covey and Schubert 1982b). The midlatitude and equatorial waves forming the Y must be locked in phase, independent of their identification as Rossby and Kelvin waves. Statistical analyses of temporal variations in the ultraviolet brightness of Pioneer Venus images show that there certainly are distinct midlatitude and equatorial waves making up the Y (Del Genio and Rossow 1982). Moreover, the observed relative phase velocities of these waves require that their phases be tightly coupled to avoid rapid pattern modification.

Analysis of cloud-tracked winds in the Mariner 10 ultraviolet photographs (Limaye and Suomi 1981) suggests that the identification of the equatorial wave as a Kelvin wave might not be correct. The Kelvin wave is symmetric about the equator, and the meridional velocity perturbation of the wave must therefore vanish at 0° lat. The north-south velocities in the Mariner 10 ultraviolet images, when averaged with respect to a coordinate system moving with the Y, are not zero at the equator (Limaye and Suomi 1981). However, the measured speeds are small, there are large uncertainties in the estimates, and the length of time covered by the images analysed was only 3.5 d. Additional, similar analyses with the more extensive PV ultraviolet images are necessary before a definite conclusion about the Kelvin wave can be reached.

Rossow and Williams (1979) have generated a single planetary-scale advective wave that has some features reminiscent of the Y, with a nonlinear, two-dimensional barotropic model. The advection wave is a nonlinear wave on a slowly rotating planet, analogous to the Rossby wave but advecting mostly relative vorticity rather than planetary vorticity. The single advection wave cannot constitute the entire Y, which must have distinct equatorial and midlatitude components, but it could be the midlatitude part of the Y (Del Genio and Rossow 1982). Houben (1982) has produced a feature similar to the Y, using a model of radiative instability in an atmosphere whose zonal wind velocity is constant with height. The model (Leovy 1966; Gierasch et al. 1973) assumes that atmospheric parcels displaced upward near the cloud tops are radiatively heated, because they carry ultraviolet absorbers to greater height. This enhances their buoyancy during the ascending part of a wave cycle and leads to instability. We do not know if this physical mechanism actually operates on Venus, but it does seem plausible. However, the process generates a zonal wavenumber 2 pattern (Houben 1982) which does not agree with the observed wavenumber 1 of the Y.

None of the wave models discussed so far have accounted for the observed height variations of both the zonal winds and static stability. The actual atmosphere is essentially motionless at the surface and rapidly rotating at cloud levels, and there are large vertical variations of static stability. These characteristics of Venus's atmosphere must profoundly affect the properties of any planetary-scale waves that propagate over large vertical distances to the regions we see in the ultraviolet. The influence of realistic variations of basic state zonal wind and static stability on the vertical structure and horizontal propagation speeds of global-scale waves has been studied by Covey and Schubert (1981b, 1982a), who determined the preferred modes of oscillation of Venus's atmosphere by finding the frequencies of enhanced atmospheric response to random forcing over a broad range of periods. Preferred responses of the atmosphere were found at frequencies comparable to the observed frequencies of midlatitude and equatorial components of the Y. The preferred modes of oscillation correspond to Rossby waves which are prominent at midlatitudes and propagate somewhat slower than the cloud-top zonal wind, and Kelvin waves which are prominent at the equator and travel somewhat faster than the zonal wind. Both these wavenumber 1 oscillations are symmetric about the equator and propagate vertically at large altitudes. The amplitude of the Kelvin wave is enhanced by partial trapping in the region of high static stability below the clouds. The Rossby wave must be forced at cloud heights because of critical-level absorption. Upward-propagating Rossby waves forced at the surface or deep in the atmosphere will be absorbed at a critical level below the clouds, because the westward zonal phase speed of the wave is smaller than the cloud-top zonal wind velocity. However, if the forcing occurs in the clouds (by convective activity, for example) at a level where the wind speed exceeds the phase speed of the wave, then the oscillation cannot encounter a critical level as it propagates farther upward, assuming that the winds do not decrease with altitude in the region of interest. As noted above, the Rossby and Kelvin waves can be combined, with the appropriate phase, to give a Y-shaped pattern reminiscent of the planetary-scale cloud pattern seen on Venus. The earlier remarks about the necessity of phase-coupling between the waves apply here as well.

The results of Covey and Schubert (1981b, 1982a) for the preferred responses of a realistic Venus atmosphere support the original suggestion by Belton et al. (1976b) that the Y is superposed Rossby and Kelvin waves. There are some difficulties with the Kelvin wave, one of which has been discussed. In addition, the phase speed of the Kelvin wave in the model of Covey and Schubert (1982a) is somewhat too fast relative to the cloud-top zonal wind to agree completely with the ultraviolet observations. This might be due to the particular forcing used in the model, as opposed to the real forcing in Venus's atmosphere. As already mentioned, further analysis of PV ultraviolet cloud-tracked winds from the perspective of the moving Y could clarify whether a Kelvin wave is relevant to its equatorial component. Concerning the midlatitude component, the linear Rossby wave and its nonlinear generalization, the advection wave, are the best candidates to be the actual wave in Venus's atmosphere. The work of Covey and Schubert (1982a) also shows that the response of the Venus atmosphere to random forcing depends strongly on major features of its basic state, such as the stable layer below the clouds and the rapid zonal winds at cloud heights. Thus, large static stability variations at and above the clouds and a decrease in zonal wind speed above the clouds would significantly modify the character of preferentially forced planetary-scale waves. This offers the possibility of further constraining the mean zonal wind and stability structure of the atmosphere above the clouds, by observation and modeling of the waves. As an example, if the 5–6 d oscillations of infrared brightness temperature detected in the 70–90 km altitude range propagate upward to produce fluctuations of similar period in the exosphere temperature, then the zonal wind could not decrease substantially above the cloud tops, because the oscillations would suffer critical-level absorption in a region of strong shear. If there is a large wind shear above the clouds, then the temperature oscillations of the exosphere could not be connected with those of the cloud-top atmosphere. This, of course, assumes westward propagation of the waves.

The static stability and zonal wind variations with height will also strongly influence the vertical propagation of small-scale gravity waves. Many of these waves will not be able to reach cloud levels, if they are generated in the lower atmosphere. Waves traveling westward with horizontal speed $\leqslant 100$ m s^{-1} will be absorbed by critical layers as they propagate upwards. Vertically-propagating waves moving east-west, with Doppler shifted horizontal phase velocites $\geqslant 10$ m s^{-1}, will be attenuated in the region of low static stability near the base of the clouds (see Fig. 30). Since the zonal wind velocity u_λ greatly exceeds 10 m s^{-1} at $\sim$ 50 km (the approximate height of the static stability minimum), all eastward vertically propagating waves are damped in this region, as are all westward vertically propagating waves with horizontal phase speeds $\gtrsim$ 100 m s^{-1}. Vertically propagating waves traveling north-south with horizontal phase speeds $\geqslant$10 m s^{-1} will be attenuated in the low stability layer in the clouds. Thus, essentially all vertically propagating small-scale gravity waves forced in the lower atmosphere will be either attenuated or completely absorbed before they arrive at cloud heights.

Our estimates of gravity wave attenuation due to static stability variations are based on the properties of waves in an atmosphere with constant static stability and horizontal wind velocity. The dispersion relation for waves with horizontal wave number k and Doppler shifted horizontal phase velocity c_D in an isothermal atmosphere with pressure scale height H, Brunt-Väisälä frequency N, and sound speed c_s is (e.g. Houghton 1977, pp. 88–93)

$$m^2 = \frac{N^2}{c_D^2} - \frac{1}{4H^2} - k^2 + \frac{k^2 c_D^2}{c_s^2} \tag{30}$$

where the vertical structure of the wave is given by exp (imz) exp $(z/2H)$. Equation (30) assumes east-west wave propagation; for north-south propagation it is simply necessary to interpret c_D as the actual phase velocity of the wave. Attenuation occurs when m^2 is negative; from Eq. (30), this happens when

$$\frac{1}{2} - \frac{1}{2} \left\{ 1 - \frac{4N^2 k^2}{c_s^2 (k^2 + 1/4H^2)^2} \right\}^{\frac{1}{2}} < \frac{c_D^2/c_s^2}{(1 + 1/4H^2 k^2)} < \frac{1}{2} + \frac{1}{2}$$

$$\left\{ 1 - \frac{4N^2 k^2}{c_s^2 (k^2 + 1/4H^2)^2} \right\}^{\frac{1}{2}} \tag{31}$$

If

$$k^2 \ll \frac{1}{4H^2} \quad \text{and} \quad \frac{4N^2 k^2}{c_s^2 (k^2 + 1/4H^2)^2} \ll 1$$

then the condition for attenuation simplifies to

$$4H^2 N^2 < c_D^2 < \frac{c_s^2}{4H^2 k^2} - 4H^2 k^2 N^2 \tag{32}$$

The largest value of H for the region of the Venus atmosphere below the clouds is the surface value 15.75 km (Table I and Fig. 2). Thus, k^2 will be $\ll (4H^2)^{-1}$ for waves with horizontal wavelength $>$ 188.5 km. At cloud heights H is only 6 km (Fig. 2), and waves with horizontal wavelength > 72 km will satisfy $k^2 << (4H^2)^{-1}$. From the altitude profiles of H, c_s and N in Figs. 2, 5 and 7, the quantity $4N^2k^2c_s^{-2} (k^2 + 1/4H^2)^{-2} \leqslant 0.4$ at both cloud levels and the surface, provided the horizontal wavelength is $\geqslant 100$ km. (We have assumed that N at the surface is $\leqslant 2 \times 10^{-3}s^{-1}$.) Thus the simplified attenuation criterion Eq. (32) is valid for waves with horizontal wavelength $\geqslant$ 100 km. The attenuation criterion implies that waves with Doppler shifted horizontal phase velocity $c_D > 2HN$ will undergo some decay in amplitude while propagating upward in the atmosphere below the clouds. From Figs. 2 and 7, we obtain $2HN \approx 150$ m s^{-1} at the height of the stability maximum near the base of the clouds, and $2HN \approx 10$ m s^{-1} for $N \approx 5 \times 10^{-4}s^{-1}$ in the low-stability layer in the clouds. Almost all vertically propagating small-scale gravity waves will be attenuated in any nearly neutrally stable layer in the clouds or near the surface. Only waves with very small horizontal phase speeds ($\leqslant 10$ m s^{-1}) have a chance of passing with little attenuation through regions of low static stability in Venus's atmosphere. Because of the sensitivity of wave transmission to altitude variations in static stability and zonal wind velocity, the possibility exists of using wave observations in ultraviolet images of the cloud-top levels to study atmospheric structure below the clouds.

Small-scale acoustic-gravity waves propagating in the thermosphere-cryosphere (von Zahn et al. 1980) will be affected by wave-induced diffusion (Del Genio et al. 1979). The same wave-induced diffusion effects that determine relative amplitudes and phases of neutral upper atmosphere species also occur in the lower atmosphere, for example, if there are sharply defined layers of individual gases produced photochemically. Such layers do not share the scale height of the background atmosphere and will therefore exhibit amplitude and phase differences relative to the background when a wave passes through. Chiu and Ching (1978) were able to explain observations of large fluctuations in the terrestrial ozone and sporadic-E layers this way. The particle scale height in the Venus clouds is only 2–3 km (Lacis 1975; Knollenberg et al. 1980) while the atmosphere has a scale height of $\sim$ 6 km (Fig. 2). The SO_2 scale height inferred from the wavelength dependence of the strength of its ultraviolet absorption is 1–2 km (Esposito et al. 1979; Esposito 1980). Such gaseous species might fluctuate during wave passage 4–5 times as much as the CO_2 wave amplitude. This result would apply for planetary-scale waves as well as localized gravity waves. For a typical CO_2 wave density perturbation of 10%, the corresponding cloud-layer species fluctuations would then be 40–50% of the ambient density of these species. These sharp fluctuations could greatly enhance or suppress the visibility or perhaps even the formation rate of the dark ultraviolet clouds (Travis 1975) over different parts of the wave cycle, by the preferential advection of large amounts of dark cloud-forming material up to the visible cloud tops over half of each cycle. This might help explain why such well-defined wavelike ultraviolet contrast features are observed on Venus (Del Genio et al. 1979).

VII. THERMOSPHERE AND CRYOSPHERE

There are no direct measurements of wind velocity in Venus's upper atmosphere above the turbopause. The circulation of this region of the atmosphere must therefore be inferred from observations of its structure, the most prominent feature of which is the contrast in temperature between day and night. Temperatures on the day side of Venus increase from ~ 180 K at the 100 km level (F. Taylor et al. 1979b, 1980; Seiff et al. 1980) to near 300 K ~ 150 km altitude (von Zahn et al. 1979b, 1980; Niemann et al. 1979c, 1980b). The structure of the dayside Venus atmosphere at these altitudes is similar to that of the Earth's thermosphere, though the temperatures are considerably lower. However, there is no nightside thermosphere. The temperature at ~ 150 km on the night side of the planet is as low as 100 K (Keating et al. 1979b, 1980; Niemann et al. 1979a, 1980b; Seiff and Kirk 1982), while the nightside temperatures at 100 km are substantially the same as the dayside, ~180 K. Thus, in the altitude interval where temperature increases with height on the day side, temperature decreases with height on the night side. This coldest region of Venus, the nightside upper atmosphere, is called the cryosphere (Schubert et al. 1980a). The approximate vertical temperature structures of the dayside thermosphere and nightside cryosphere are shown in Fig. 38.

The temperature difference between night and day gives rise to horizontal pressure gradients (Fig. 4) which should drive a strong subsolar-antisolar circulation in Venus's upper atmosphere (Seiff 1982). The simplest possible flow pattern would be axisymmetric about the Venus-Sun line and would involve dayside upflow centered on the subsolar point, cross-terminator flow from day to night, and nightside downflow centered on the antisolar point.

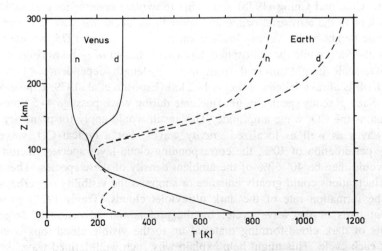

Fig. 38. Altitude profiles of temperature in the atmospheres of Venus and Earth.

Return flow from night to day at lower levels in the atmosphere would complete the circuit. This is the glóbal upper atmospheric circulation pattern hypothesized in a series of numerical models by Dickinson and Ridley (1972, 1975, 1977). While this type of circulation must occur, the actual motions in Venus's upper atmosphere are more complex. Postmidnight maxima in the He and H densities (Niemann et al. 1979c; Brinton et al. 1979), displacement of the minimum in the diurnal temperature variation toward morning (Mayr et al. 1980; Keating et al. 1980), and a postmidnight airglow maximum (Stewart and Barth 1979) all suggest a westward superrotation of the upper atmosphere with zonal wind speeds > 50 m s^{-1} at the equator (Mayr et al. 1980). The circulation of Venus's thermosphere-cryosphere may not be axisymmetric about the Venus-Sun line; it may be fully three dimensional with a significant zonally symmetric component. Future models of the Venus upper atmosphere should not only allow for nonaxisymmetric flow, but should also generate the superrotation, assuming that the phenomenon is distinct from the superrotation of the lower atmosphere. However, the two regions of the atmosphere may be coupled and the lower atmosphere superrotation may drive a zonal circulation of the upper atmosphere (Schubert et al. 1977).

The changes in density, temperature, and pressure between day and night in the Venus upper atmosphere occur very suddenly at the terminator (Keating et al. 1980; Niemann et al. 1980b). Diurnal variations in the Earth's thermosphere are much more gradual (Jacchia 1977). The horizontal pressure gradient across the evening terminator in Fig. 4 is $\sim 2.5 \times 10^{-13}$ Pa m^{-1} and is somewhat smaller, $\sim 1.5 \times 10^{-13}$ Pa m^{-1}, across the morning terminator. The average horizontal pressure gradient between day and night at a comparable altitude in the Earth's thermosphere is $\sim 3.5 \times 10^{-12}$ Pa m^{-1} (Jacchia 1977), a factor of ~ 10 larger. The horizontal pressure gradient is larger in the Earth's thermosphere than in Venus's upper atmosphere, because pressures are a factor of > 100 greater at comparable altitudes and these more than compensate for the sharpness of Venus's cross-terminator pressure variations. Pressures in Venus's upper atmosphere are less than on Earth because Venus's low exospheric temperatures result in relatively small scale heights and more rapid reduction in pressure and density with altitude. Although the horizontal pressure gradient in Venus's upper atmosphere is smaller than in the Earth's thermosphere, it is still large enough to drive high-speed cross terminator winds from day to night. The velocities of these winds need not be > 200 to 300 m s^{-1}, because viscous forces associated with a vertical shear of the winds can balance the pressure forces (Mayr et al. 1980). The viscosity μ in the upper atmosphere of Venus, at altitudes where atomic oxygen is the dominant species, is given by

$$\mu = 4 \times 10^{-7} T^{0.7} \tag{33}$$

where μ is in Pa s and T is in K (Banks and Kockarts 1973). At the average temperature of 200 K, Eq. (33) gives 1.63×10^{-5} Pa s. This can be compared with a viscosity of 5.04×10^{-5} Pa s in the atomic oxygen dominated part of the Earth's thermosphere at an average temperature of 1000 K. The vertical scale of the shear h in the horizontal wind u, required to balance the cross terminator pressure gradient $|\nabla_t p|$ in Venus's upper atmosphere, is therefore

$$h \approx \left(\frac{\mu u}{|\nabla_t p|} \right)^{\frac{1}{2}} \approx \left(\frac{1.63 \times 10^{-5} \times 200}{2 \times 10^{-13}} \right)^{\frac{1}{2}} \text{m} \approx 130 \text{ km} . \tag{34}$$

Since this is such a large distance, we conclude that viscous shear forces can easily limit the cross-terminator wind speeds to values of $\leqslant 200$ m s^{-1}. Viscous forces in Venus's upper atmosphere are basically as effective in limiting wind velocities as they are in the Earth's thermosphere. Although the viscosity in Venus's upper atmosphere is a factor of ~ 3 smaller than in the Earth's thermosphere, the horizontal pressure gradients on Venus are at least an order of magnitude smaller, even across the terminator region.

The values of the pressure gradients given above show that circulation is asymmetric with respect to the noon-midnight meridian plane (see also Seiff 1982). The pressure gradient is larger across the evening terminator, where the superrotation reinforces the winds of the subsolar-antisolar circulation, than it is across the morning terminator where the superrotation opposes the subsolar-antisolar winds.

Dynamical processes may be required to explain the very low nightside temperatures in Venus's upper atmosphere and the characteristic thermal structure of this region, i.e. decrease of temperature with height, responsible for its classification as a cryosphere. The models of Dickinson and Ridley (1977) do not predict the low nightside temperature or the nightside decrease of T with height. They also show gradual day-night transitions similar to those in the Earth's thermosphere, but contrary to the Venus observations discussed above. The models do show reasonable dayside exospheric temperatures for sufficiently low values of extreme ultraviolet heating efficiency. The low nightside temperature must be maintained against a number of processes that transfer heat to the cryosphere: advection and conduction of heat across the terminator from the dayside thermosphere; compression of descending flow on the night side; and upward molecular heat conduction from the relatively warm 100 km level (Chapter 20). Radiative cooling alone does not appear adequate (Keating et al. 1980; Niemann et al. 1980b), although Seiff (1982) argues that CO_2 15 μm cooling rates are sufficient at altitudes below 140 km.

Downward eddy heat transport against the vertical temperature gradient has been suggested as a means of cooling the cryosphere (Niemann et al. 1979a; Keating et al. 1979a). Forced vertical mixing of the nightside upper atmosphere is possible because the decrease of temperature with height does not exceed the adiabatic lapse rate of ~ 10 K km^{-1}. The external forcing could result from breaking of solar thermal tides or planetary-scale waves as they propagate upward from the lower atmosphere (Lindzen 1970; Prinn 1975; Leovy 1979). However, the same eddy processes that force heat downward in stably stratified atmospheric layers also heat these layers by viscous dissipation (Hines 1960, 1965). It is not known which effect, heating by frictional dissipation of eddies or cooling by forced eddy mixing, might dominate in the Venus cryosphere, or for that matter in the Earth's thermosphere (Hunten 1974; Johnson 1975; Dickinson 1976; Izakov 1978). Walterscheid (1981) has recently shown how upward propagating gravity waves can induce a downward transfer of sensible heat and induce cooling of regions of the upper atmosphere. Recent improvements in our knowledge of the solar extreme ultraviolet heating efficiency in the Earth's thermosphere (Torr et al. 1980) eliminate the wave energy dissipation as a necessary source of net heating.

Several other features of Venus's upper atmosphere structure also seem to require large eddy diffusivities on the night side. Transport of atomic oxygen to the night side by the subsolar-antisolar circulation was predicted to built up a nighttime bulge in O (Dickinson and Ridley 1977) similar to the winter atomic oxygen and helium bulges on Earth (Keating et al. 1971, 1972; Keating and Prior 1968). However, in Venus's upper atmosphere, the atomic oxygen bulge is on the day side in spite of the day-to-night transport (Niemann et al. 1980b). Enhanced downward transport of O by eddy processes on the night side (Keating et al. 1980) could remove the atomic oxygen from the upper atmosphere. The superrotation of the thermosphere-cryosphere and the low nightside temperature would also reduce the amplitude of atomic oxygen buildup on the night side (Mayr et al. 1978, 1980; Keating et al. 1980). The same process would also limit the amplitude of the nightside He bulge, which is observed to be smaller than predicted (Mayr et al. 1978; Niemann et al. 1980b; Keating et al. 1980). A large eddy diffusivity also helps to explain the high altitude of the homopause on the day side (von Zahn et al. 1979b, 1980; Chapter 13). The temperature of the dayside upper atmosphere could be maintained against enhanced eddy cooling by relatively efficient solar extreme ultraviolet heating.

Qualitative explanations have been given for a number of striking features of Venus's upper atmospheric structure and circulation. Among these are the nightside cryosphere, the sharpness of day-night property variations across the terminator, the absence of a nighttime atomic oxygen bulge, and a possible zonal retrograde superrotation. Quantitative and detailed explanations of these observations should be a goal of future research.

VIII. DIRECTIONS FOR FUTURE RESEARCH

The major unanswered questions about the dynamics and general circulation of Venus's atmosphere concern the nature of the secondary mean meridional circulation and the maintenance of the dominant retrograde zonal superrotation. The two problems are closely related, because the mean meridional flow could play an important role in the angular momentum balance of the atmosphere by transferring westward zonal momentum upward. The latitudinal transport of heat by the mean meridional circulation is also a crucial element in the thermal balance of the atmosphere, although the major question here, namely the maintenance of the high surface temperature, seems to have been satisfactorily answered as long as dynamics can be assumed to redistrib ute the heat globally (Pollack et al. 1980*a*; Tomasko et al. 1980*b*).

We cannot say for certain that we know the temporal and zonal mean meridional circulation at any level in Venus's atmosphere, even at the cloud tops. Our present best guess, guided by observations, theoretical considerations, and numerical modeling, is that the mean meridional circulation is vertically layered with a Hadley cell adjacent to the surface, one at cloud levels, and an indirect cell in between. However, this still needs to be confirmed or modified by observations. Thus, a primary focus of any future spacecraft exploration of Venus's atmosphere should be defining the mean meridional circulation. This is a tall order, and can only be accomplished by observations over extensive time periods and locations on the planet.

From a practical viewpoint, we are likely to succeed first at doing this at cloud levels, where long-term global ultraviolet imaging by an orbiter is currently feasible. The ultraviolet imaging by the Pioneer Venus orbiter is a first step, and every effort should be made to extend the lifetime of this experiment. Measurements of quantities such as temperature, which can be made from a long-lived orbiter, would also be useful in defining the mean meridional circulation above the clouds, although the motions could only be inferred with the aid of a dynamical model.

Our only hope for determining the mean meridional circulation below the clouds is some day to establish long-lived surface experiment stations to monitor wind velocity, pressure, and temperature and to release balloons into the atmosphere at different levels to measure atmospheric state properties and reveal wind velocities by their motions over long time periods. The actual circulation in the lowest scale height of Venus's atmosphere is completely unknown, and the dynamics of this region, which contains 65% of the atmospheric mass, is probably crucial to the buildup of the large zonal winds at greater heights because it is likely that the atmospheric angular momentum must be transported upward from the surface. The comparable angular momenta of solid planet and atmosphere, and the possibility of significant exchanges of angular momentum between them greatly enhance the potential importance of long-lived surface meteorological stations.

Determination of temporal and zonal mean meridional circulation from long-term global observations also would give the eddy velocities, and from these one could calculate the net eddy transports of angular momentum (and heat, if temperature data were also available). Estimates of the meridional eddy flux of angular momentum at the cloud tops have already been made from Mariner 10 and Pioneer Venus ultraviolet cloud-tracked wind velocities

(Limaye and Suomi 1981; Limaye et al. 1982). Information on eddy fluxes of angular momentum at all levels of the atmosphere is needed to distinguish among the proposed mechanisms for the maintenance of Venus's supperrotation. There is no alternative to accumulation of meteorological data over long periods of time, if eddy contributions to the angular momentum balance of the atmosphere are to be determined. Again, cloud-top imaging from an orbiter holds the best promise for progress along these lines in the near future. Ascertaining if a mechanism such as the moving flame is at work at any level in Venus's atmosphere requires the determination of eddy zonal velocities and the associated momentum transports. A quantitative estimate of the importance of eddy zonal motions at cloud levels, and an assessment of the significance of diurnal effects at cloud heights, could be obtained from long-term Pioneer Venus ultraviolet imaging, but the data have not yet been analyzed from this perspective. If such an analysis were to show that diurnal effects were unimportant at cloud heights, then a mechanism like the moving flame could be ruled out. However, a positive result concerning diurnal variations would need support from quantitative determinations of the vertical eddy flux of zonal momentum before we could establish the actual operation of the moving-flame mechanism in Venus's atmosphere.

Since the type of data needed to determine the mean meridional circulation of Venus's atmosphere and the eddy contributions to the atmospheric angular momentum balance will probably not be available in the near future (except perhaps for the cloud tops), our current best hope lies in theoretical analysis and numerical modeling. We have enough general knowledge of atmospheric structure and solar heating to make realistic calculations feasible. Computations should be in the form of fully 3-dimensional general circulation models, to insure that the nonlinear processes probably occurring in Venus's atmosphere are represented. Numerical modeling must play a central role in our understanding of the circulation of Venus's upper atmosphere, where motions will always have to be inferred from measurements of variables such as density, temperature, composition, etc. Fully 3-dimensional models of Venus's thermosphere/cryosphere should be developed to account for its possible superrotation, which, if actual, would destroy any axisymmetry of upper atmosphere motions about the Sun-Venus line.

Acknowledgments. This research was supported by grants from the National Aeronautics and Space Administration. I benefited greatly from colleagues who generously shared their research results and ideas. They include J. Apt, C. C. Covey, A. D. Del Genio, A. J. Kliore, S. S. Limaye, W. B. Rossow, A. Seiff, J. M. Straus, V. E. Suomi, L. Travis, R. L. Walterscheid, R. Woo, and R. E. Young. I am especially indebted to C. C. Covey for permission to quote portions of his Ph. D. thesis in the section on waves, and to L. Travis and A. D. Del Genio who provided detailed and thoughtful reviews of a preliminary version of this chapter.

22. THE ATMOSPHERIC DYNAMICS OF VENUS
ACCORDING TO DOPPLER MEASUREMENTS
BY THE VENERA ENTRY PROBES

V. V. KERZHANOVICH and M. YA. MAROV
USSR Academy of Sciences

The Doppler-tracked wind data from the Venera series of probes are summarized. The observed winds, vertical motions, and turbulence are discussed and compared with theoretical expectations.

Before the Venera spacecraft, very little was known about the dynamics and circulation of the atmosphere of Venus. Ultraviolet light images indicated a possible rotation of the atmosphere at great altitudes with period ~ 4 d (Boyer and Camichel 1961; Smith 1967). Spectroscopic measurements concerning the level of cloud layer yielded a broad range of wind speeds corresponding to either direct or retrograde rotation of the atmosphere (Richardson 1958; Guinot and Feissel 1968). Theoretical estimates led to a simple single-cell circulation model with wind speeds on the order of several m s^{-1} (Goody and Robinson 1966; Stone 1968; Golitsyn 1973).

The first direct estimates of the wind speed were obtained by Venera 4, and similar measurements were conducted on all the subsequent Venera spacecraft (Kerzhanovich et al. 1969, 1972, 1980; Andreyev et al. 1974; Antsibor et al. 1976). The method employed is a variation of a classical method of measuring the speed of a probe moving under the action of wind; when the descent speed is not too great, the inertia of the probe can be disregarded and the horizontal component of the probe's velocity will correspond to the wind speed.

On board Venera spacecraft which conducted a vertical probing of the atmosphere, the Doppler method was used to directly measure one component of the wind speed, along a radius directed toward the point of signal reception (to Earth for Veneras 4–8 and toward the orbiter or flyby for Veneras 9–12). The quantity measured includes a radial component of the speed of the reception point relative to the probe; for Veneras 4–8 this component was accurately calculated from the ephemerides of Earth and Venus, then subtracted from the measurements to obtain the planetocentric radial speed of the probe (Kerzhanovich et al. 1969, 1980). The vertical speed of the probe was found by several procedures based on measurements of the temperature or pressure of the atmosphere. The difference between the planetocentric Doppler speed and the radial projection of the vertical speed can be used to find a particular component of the wind speed. The speed measurements were implemented by a nonsampling method, using the frequency shift of the received signal with respect to a reference value corresponding to the probe at rest with respect to the planet. The signal of the transmitter on board came from a thermostatically controlled quartz oscillator with a short-term stability of several parts in 10^{10}.

On board Veneras 4–8 the reference value of the frequency was found from measurements during calibration communication sessions on a segment of the interplanetary flight. For Veneras 7–8 the reference frequency was also taken to be the frequency of the signal after the landing of the craft; these two values matched well. On board Veneras 9–12, where the change in the Doppler shift was produced by the movement of the orbiter or flyby relative to the probe, and no frequency calibration was performed, the wind speed was found by solving a combined problem in which the reference frequency and the coordinates of the probe's landing site were ascertained along with the wind speed. The calculations were performed by means of a numerical model, taking into account the structure of the atmosphere, the aerodynamics and wind drift of the probe, and the movement of the orbiter (Antsibor et al. 1976; Kerzhanovich et al. 1972).

The systematic errors in determining the wind speed were ± 2.8 m s^{-1} for Venera 4, ± 1.5 m s^{-1} for Venera 7, and ± 0.2 m s^{-1} for Venera 8, in terms of the radial component. The slowly changing errors, due to inaccuracies in allowing for the frequency drift of the master oscillators and in determining the vertical and radial speed of the orbiters (for Veneras 9–12), did not exceed several m s^{-1}. The procedure of processing and evaluating the measurement errors is given in greater detail in Kerzhanovich et al. (1980); Antsibor et al. (1976); and Kerzhanovich and Marov (1977).

The Venera spacecraft have probed the atmosphere in the equatorial and subequatorial zone: Veneras 4–7 at night, Venera 8 in the early morning, and Veneras 9–12 around noon (local time) (Fig. 1). For Venera 4, the measured component of the wind speed was closer to the meridional, while for Veneras 7 and 8 it was closer to the zonal. Veneras 5 and 6 probed the atmosphere near the surface, and did not allow a direct estimate of the horizontal speed.

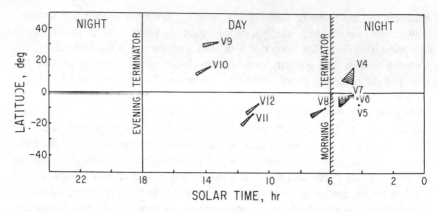

Fig. 1. Landing sites of Venera spacecraft, and directions of the wind speed components measured.

I. THE WIND SPEED AND ITS VARIATIONS

Vertical distributions of the wind speeds from Venera data are shown in Fig. 2. The altitude was determined by simulation of the atmosphere (Kuz'min and Marov 1974), with which the data of Pioneer agrees well if the zero levels are lined up (Seiff et al. 1980).

The wind speed profiles obtained in the various experiments conducted over roughly one and half decades, reveal a basic similarity. It is especially evident that in all the experiments except perhaps that of Venera 7, the measured component of the wind speed does not change direction. Although the Venera data do not allow an unambiguous interpretation (only a single component was measured), if we assume that the general nature of the circulation remains unchanged, then by allowing for the direction of the measured components of the wind speed we can coordinate the measurements of all the Venera spacecraft only by assuming that the wind speed direction remains close to zonal. Thus a wind in the same direction as the rotation of the planet, toward the west, corresponds to the positive values of the speed in Fig. 2. This hypothesis has been confirmed by the Pioneer measurements (Counselman et al. 1980). The similarity of the Venera 4–7 profiles becomes clearer if, in view of the systematic measurement errors and uncertainty of the landing site coordinates, we shift these profiles as a block and convert to the zonal component (dashes in Fig. 2).

We now note certain features of the atmospheric circulation which follow from our data. The Venera results indicate that a large part of the troposphere, beginning at altitudes of 15–20 km all the way up to 60–65 km, is involved in an intense zonal movement, the speed of which attains 100–140 m s^{-1} in the cloud layer. Although these data initially caused debate (Young 1975), the more recent Venera and Pioneer measurements have confirmed this

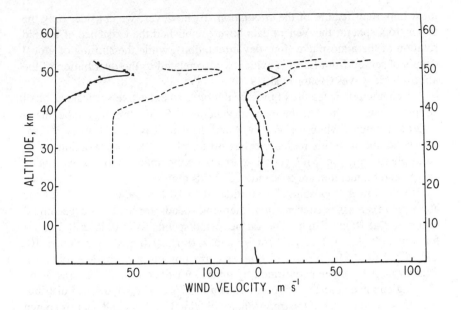

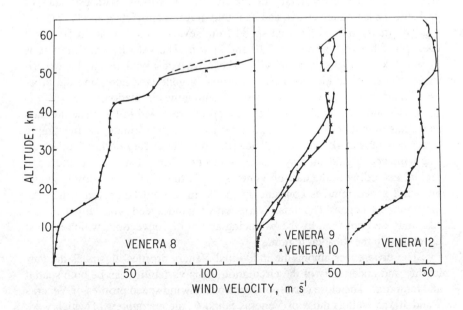

Fig. 2. Profiles of the zonal wind speed obtained by Doppler measurements on board Venera spacecraft. Positive values correspond to a westward wind, i.e. in the direction of Venus's rotation.

most important feature of the circulation. If, however, we consider only the upper troposphere, the Venera data have established the existence of a rapid rotation of the atmosphere (four-day circulation), while the shifting of global details observed from Earth is evidently also related to the distribution of the atmospheric waves (Belton et al. 1976).

A characteristic feature of the circulation is an extensive vertical region of almost constant wind. In the data of Veneras 4, 8, 12, this region begins at ~ 20 km height, while for Venera 9 and 10 data it begins at 30 to 35 km. The wind speed in this region reaches 40 to 50 m s^{-1}, corresponding to an atmospheric rotation with period 8 to 10 d; the main amounts of kinetic energy and momentum are concentrated in this region.

In the lower troposphere at altitudes 0 to 10 km, where ~ 40% of the atmosphere's mass is concentrated, the wind speeds are low, not exceeding 2 to 3 m s^{-1} at 10 km height. The surface wind speeds were 0.25 ± 0.5 m s^{-1} for Venera 8, 1.2 ± 1.0 m s^{-1} for Venera 9, 0.6 ± 1.0 m s^{-1} for Venera 10, and < 1.2 m s^{-1} for Venera 11 (values correspond to a retrograde rotation). These values have been confirmed by measurements on board Veneras 9 and 10 using cup anemometers (Avduevsky et al. 1976f). The wind speed distribution in the lower layer of the troposphere, obtained by Venera 8 data, is shown in Fig. 3 on a large scale. Here the error bars indicate a less reliable estimate of the change in wind speed at low altitudes, obtained by Venera 7. Discussion of these and other details of the illustrated profiles, with estimates of probable measurement errors, can be found in Kerzhanovich et al. (1977).

The main shift in the wind speed from several m s^{-1} to 40 to 50 m s^{-1} takes place between altitudes of 10 and 20 km. The velocity gradient here is 3 m s^{-1} km^{-1}. This zone of wind shear was observed in all the experiments. Most of the kinetic energy is probably generated here, and the processes in this zone may play a key role in forming the circulation.

A second zone of shear in the wind speed, revealed in the Venera 8 data as well as in that of Veneras 4–7, begins at 41–43 km altitude; the most rapid increase occurs at 48–49 km, i.e. near the lower boundary of the cloud layer. The wind speed gradient may attain 10–15 m s^{-1} km^{-1} here. In the afternoon region, according to the data of Veneras 9–12, this shear is not observed and the wind speed remains constant up to the maximum heights at which the mesurements began. The small variation of wind speed with height in this case may be explained by the balancing action of convection, which is most developed in the subsolar region.

The differences among the individual Venera profiles illustrates the variability and uncertainty of the circulation. This variability can be both spatial and temporal. The close correspondence of the wind speed profiles of Veneras 9 and 10, as well as those of Veneras 5 and 6, the encounters of which were separated by four and two days, respectively, indicates that the characteristic time of the changes is larger than several Earth days. Furthermore, from comparison of data of Veneras 4–8, which probed the atmosphere of Venus

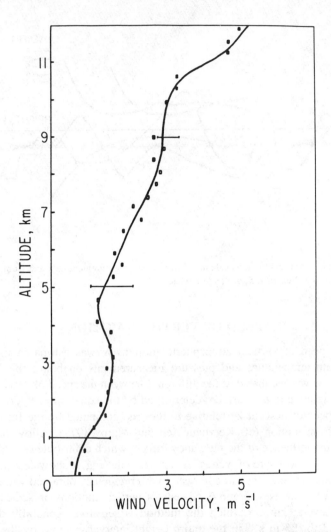

Fig. 3. Wind speed profile in the lower troposphere according to the measurements by
Venera 8.

during the morning and forenoon hours, it is evident that substantial changes
may take place on time scales <1.5 yr. The differences between the morning
and night profiles on the one hand, and the afternoon profiles on the other, are
apparently consistent, if we further take into account that the profiles obtained
at the same time by Pioneer probes over a time interval from midnight to early
morning are almost identical (Counselman et al. 1980).

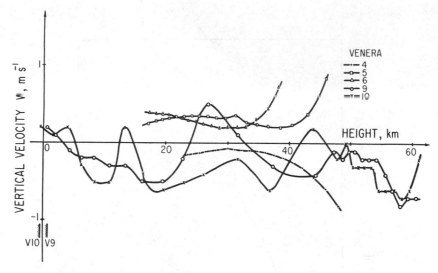

Fig. 4. Speeds of vertical motions calculated from direct measurement data and from descent aerodynamics of Venera 4, 5, 6, 9, and 10 spacecraft.

II. SPEED OF VERTICAL MOTIONS

The speed of vertical atmospheric motions w was estimated independently from temperature and pressure measurements on board the Venera spacecraft. w was estimated as the difference between the speed of descent V of the probe relative to the surface, determined by the equation of hydrostatics, and the speed of descent V' relative to the gas, computed by the formula of quasi-uniform motion (cf. Kerzhanovich and Marov 1977). Allowance was made for the influence of the buoyancy forces, which contribute considerably in the lower atmosphere of Venus. As an example, Fig. 4 shows estimates of w from Venera 9 and 10 data, contrasted with previously obtained values of w from that of Veneras 4, 5 and 6. Speeds of vertical motions on scales of the order of several km, to which the method is sensitive, generally did not exceed 0.3 to 0.5 m s⁻¹ in the entire height interval covered by the atmospheric temperature and pressure measurements; values outside these limits at the starting segments of the probe's descent were produced by procedural errors. On the main portion of the curve, the possible errors are less than the adjusted values of w, and therefore they should be regarded as estimates of the speeds of convective motions in the atmosphere of Venus. Change of sign may reflect actual changes in vertical currents from ascending ($+$) to descending ($-$).

Note the reasonable correlation between the curves of w from Venera 9 and 10 data, appearing at measurement ranges 20–30 km and 50–60 km. Like the measured profiles of the zonal wind themselves, this also suggests that the characteristic time of the changes in atmospheric movements is greater than the 2 to 4 Earth days separating the landings of each pair of probes. At the same time there is apparently no definite dependence on time of day because, though descending motions prevail for the daytime values, during the night both ascending (Veneras 5 and 6) and descending (Venera 4) motions are observed. The behavior of the w curves according to the Venera 11 and 12 data is similar to that indicated by the values from Veneras 9 and 10. The Venera 8 estimates produce mainly negative values in the entire altitude range, except for the interval between 0 and 5 km where the w curve changes sign.

A theoretical treatment of the nature of convection in the Venus atmosphere allows us to estimate the intensity of vertical movements and to compare these with the w values determined from an analysis of the descent dynamics of the Venera probes (cf. Kuz'min and Marov 1974). Using the Jeffreys condition for stability of a cellular convection in a viscous gas heated at the bottom, we can write

$$\frac{w}{a_1}(K) \simeq 0.6\,(1-K) \tag{1}$$

in the range of values $(1-K) = 1-10^{-2}$, where K is the ratio between adiabatic temperature gradient and actual gradient, and a_1 is the speed of sound, chosen as the velocity reference. Unfortunately, the small difference between the actual and the adiabatic temperatures $\Delta T = T - T_{ad}$ is difficult to determine reliably from our direct measurement, due to considerable errors in determination of T. In the intervals of most reliable measurement, $\Delta T \lesssim 0.5\,K$, from which $1-K = \Delta T/T = 10^{-3}$ and $w \lesssim 0.3$ m s^{-1}. This is apparently an upper limit, and is in good correspondence with previous estimates. On the other hand, an estimate on the order of magnitude of ΔT can be found from an approximation for the convective heat flux q_K, using calculations of the radiant heat fluxes q_T from the measured temperature profile (cf. Kuz'min and Marov 1974; Shari 1976), which agree with the measurements (Suomi et al. 1980). Supposing that $q_K \simeq q_T \simeq 10^2$ W m^{-2}, we obtain

$$(\Delta T)^{3/2} = \frac{q_K}{\rho C_p\,(2gH/T)^{\frac{1}{2}}} \tag{2}$$

where ρ, C_p and g are gas density, heat capacity, at constant pressure, and acceleration of gravity. For mean values of these quantities corresponding to the height of a convective cell assumed equal to the scale height H, we obtain $\Delta T \simeq 3 \times 10^{-3}\,K$, from which $1-K \simeq \Delta T/T \simeq 10^{-5}$, and $w \simeq 0.01$ m s^{-1}. This estimate should be regarded as a lower limit, and so we can write

$$0.01 \text{ m s}^{-1} \lesssim w \lesssim 0.3 \text{ m s}^{-1}. \tag{3}$$

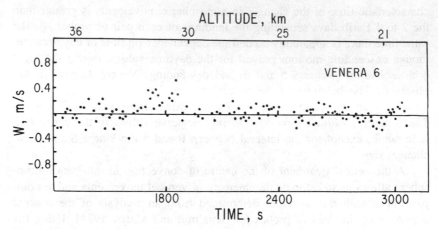

Fig. 5. Fluctuations of the vertical wind speed component according to measurements by
    Venera 6.

III. TURBULENCE

The comparatively short interval of wind drift data points of the probes t_p
$\sim$ 1–10 s for all Veneras, enabled us to estimate the fluctuations of wind
speed, i.e. the dynamic turbulence, from measurements of the speed variations of the probes. Although the specific variations of probe speed differs
from the variations of the wind speed due to distortions by the aerodynamics
of the probe, for time scales $>t_p$ the probe's speed variations may be regarded
as estimates of the turbulence. Although the division of the motion into
average and turbulent is somewhat arbitrary, especially in the case of vertical
probing, the scale height H can be used as an upper scale of the turbulence in
the free atmosphere, since fluctuations of wind speed with scales $<H$ are
regarded as turbulence. Such an interpretation is more natural than treating
small-scale variations as changes in the mean wind speed associated with
general circulation processes, or as the manifestation of atmospheric waves,
as was hypothesized by Schubert et al. (1980a).

The Doppler measurement during the descents of all the Venera probes
revealed that below 40 km the turbulence of the atmosphere is slight, despite
the large value of both the wind itself and its gradients; the recorded speed
variations lay within the limits of measurement error for scales from tens of m
up to several km. The measured speed fluctuations for Venera 6 are shown in
Fig. 5; their rms value did not exceed 0.1 to 0.2 m s^{-1} (with 10 s averaging) or
0.2 to 0.3 m s^{-1} (with 1 s averaging). Similar to this value, 0.3 to 0.5 m s^{-1},
is the upper limit of the amplitude of turbulent fluctuation at these altitudes in
other Venera experiments ($\sigma_w = 0.20$–0.35 m s^{-1}).

If we consider that in these particular scales the turbulence is nearly isotropic (these scales obviously lie within the inertial range [Kerzhanovich and Marov 1974]), then, even though Veneras 4–6 and 9–12 measured fluctuations of the vertical speed, approximately the same limit is obtained for the fluctuations of the horizontal component ($\sigma_u \approx 0.85\ \sigma_w$ [Monin and Yaglovi 1967]). This hypothesis is confirmed by the Venera 8 data in which the speed fluctuations had the same magnitude while the radial direction made a 38° angle with the local vertical since its vertical and horizontal components contributed almost identically to the measured Doppler speed.

This small magnitude of wind speed fluctuations persists in both the lower region of the wind shear and in the boundary layer. Direct measurements of the wind speed in the near-surface layer have produced fully consistent values of ~ 1 m s^{-1}. The published graph of Doppler measurements of the Pioneer North probe (Seiff et al. 1980) also agrees with the Venera data: $\sigma_u \sim 0.4$ to 0.5 m s^{-1} below 23 km. Thus, we may conclude that weak turbulence in the lower atmosphere is a characteristic feature of the circulation on Venus.

Above 40 km, the turbulence as a whole is much more developed. Its maximum is in the subsolar zone, where it was recorded by all four probes of Veneras 9–12; the turbulence here is apparently convective and related to radiative heating, which is most intense in this region. The mean amplitude of wind speed fluctuations was 1 to 1.2 m s^{-1} in the experiments of Veneras 11 and 12 with maximum variations $\leqslant 3$ m s^{-1}; for Veneras 9 and 10 the wind speed fluctuations were 2 to 2.5 times less. On the night side, turbulence above 40 km is not a characteristic phenomenon. It was distinctly recorded only by Venera 4, for which the rms fluctuations of the speed were ~ 0.75 m s^{-1} while the characteristic spatial scale was ~ 200 m. However, in Venera 5 and 6 data no turbulent fluctuations were detected in the wind speed.

In this region of 40 to 60 km altitude turbulence, when present, is distributed rather evenly and does not show distinct stratification. Also, an analysis of amplitude fluctuations of radio signals during radio occultations of the Mariner and Pioneer probes suggests concentrated turbulence in layers near 49 and 60 km (Woo et al. 1980; Woo 1975); the level of fluctuations can be correlated with the Venera data. The observed discrepancies are possibly related to the time variability of turbulence and the influence of temperature stratification on temperature fluctuations; also, it cannot be excluded that the amplitude fluctuations of the signals may have a different origin, for example refraction by random layers.

IV. SEVERAL EVALUATIONS OF THE CIRCULATION

We now can estimate several characteristic parameters of the circulation. Figures 6 and 7 show vertical distributions of the kinetic energy of zonal motion and angular momentum per unit atmospheric volume, constructed

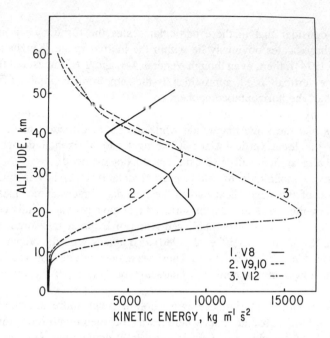

Fig. 6. Vertical distribution of the kinetic energy of zonal motion.

from Venera data. Most of the kinetic energy occurs at altitudes from 15 to 45 km; the maximum is at a height of ~ 20 km for Veneras 8–12 and 35 km for Veneras 9 and 10. The overall kinetic energy of a unit column of atmosphere averaged for four experiments is 3.0×10^8 J m^{-2}, while the energetic mean wind speed is

$$\bar{u} = \left[\int_0^h \rho u^2 \, dh / \int_0^h / \rho \, dh \right]^{1/2} \approx 21 \text{ m s}^{-1}. \tag{4}$$

The angular momentum has a maximum at 20 to 25 km height. The angular momentum of a unit column of atmosphere M averaged for Veneras 8–12, is 9×10^{13} kg s^{-1}. The variance in M among the individual experiments is $\pm 15\%$.

Using the measured wind speed fluctuations it is possible to estimate the dissipation rate of kinetic energy ϵ (Monin and Yaglom 1967):

$$\epsilon = \frac{(\sigma u)^3}{C L} \tag{5}$$

where $C \sim 1.9$ is a universal constant, and L is the characteristic scale of the fluctuations. Because of intense variations in the characteristic curves

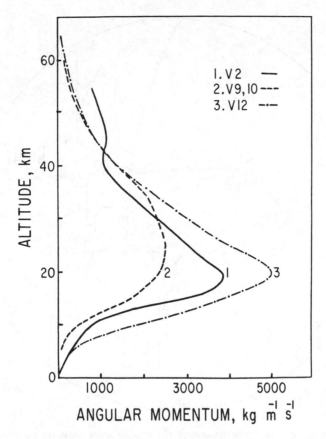

Fig. 7. Vertical distribution of the angular momentum per unit volume in the Venus atmosphere.

of turbulence in the upper troposphere (above 40–45 km), the range of estimates was very broad, from 1 to 2 cm^2 s^{-3} for Veneras 5 and 6 (night side) to 50–180 cm^2 s^{-3} for Veneras 11 and 12 (afternoon region). Below 40 km the fluctuations of wind speed with scales up to several km did not exceed 0.5 m s^{-1} in any of the experiments; assuming $L \sim 1$ km, we obtain $\epsilon \sim 0.7$ cm^2 s^{-3}.

We can also estimate the characteristic time to stop the circulation by friction when the sources of kinetic energy are extinguished, $T_u = Eu/\epsilon$. Then for the upper troposphere ($u \sim 70$ m s^{-1}) we obtain 1.6 d; for the region of maximum kinetic energy (20–40 km) $T_u > 100$ d; and for the lower troposphere $T_u > 0.5$ d. These values may characterize the variability of the circulation, and in this sense they agree with the available experimental data. The differences in T_u may possibly indicate different mechanisms of circulation at the corresponding altitude regions.

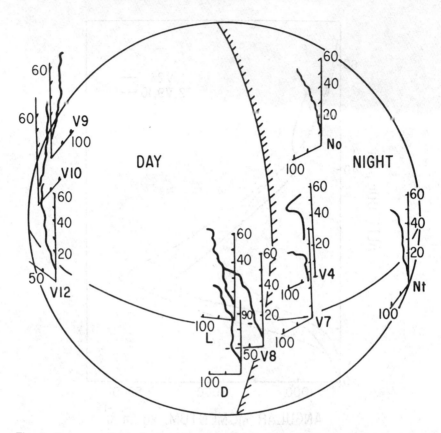

Fig. 8. Summary of data for determination of the wind speed on board the Venera and the Pioneer Venus probes. D, Nt, No, and L designate Day, Night, North and Large Pioneer probes.

The wind speed measurement results of the Pioneer Venus probes provide excellent confirmation of the Venera data (Fig. 8). The zonal nature and retrograde direction of the principal atmospheric motion at all heights, the low wind speed in the lower troposphere and at the surface, the wind shear zone at 10 to 20 km and the region of virtually constant wind, and the motion in the upper troposphere at speeds up to ~ 100 m s^{-1} have been confirmed. The discrepancies among the results of individual experiments may result from the space and time variability of the circulation.

23. THE IONOSPHERE OF VENUS: OBSERVATIONS AND THEIR INTERPRETATION

L. H. BRACE and H. A. TAYLOR, JR.
NASA Goddard Space Flight Center

T. I. GOMBOSI
Central Research Institute for Physics of Hungarian Academy of Science

A. J. KLIORE
Jet Propulsion Laboratory

W. C. KNUDSEN
Lockheed Palo Alto Research Laboratory

and

A. F. NAGY
The University of Michigan

United States and Soviet observations of the density, temperature, composition, motion, and magnetic structure of the Venus ionosphere are reviewed and their implications discussed. Although the ionosphere is created primarily by photo-ionization of the neutral atmosphere by solar extreme ultraviolet, nearly all ionospheric parameters are strongly influenced by interactions with the solar wind. The interplanetary magnetic field conveys solar wind pressure to the ionosphere, compressing, accelerating, heating and removing plasma, forming the ionopause and inducing a nightward convection of plasma. The products of the interaction are observed just above the ionopause in the form of superthermal ions and electrons and ionospheric plasma clouds or streamers which extend well downstream of the planet. The ionopause altitude varies from 290 *km at the subsolar point to 1000 km near the terminator, with large short-term variations that are well correlated with solar wind pressure. The ionopause expands on the night side and its altitude is more variable than on the day side. Within the ionosphere the main peak of the electron density is found at ~ 140 km on the day side and is believed to be formed by local production and loss, analogous to the Earth's E region. The nightside peak, which forms at a slightly higher altitude, appears to be produced primarily by the subsidence of ions transported from the day side and is more like an F_2 region. Throughout most of the ionosphere this nightward ion flow is driven primarily by the day-to-night plasma pressure gradient. Electron precipitation also contributes to the nightside ionization, particularly to the formation of a secondary peak occasionally observed below the main peak. The important ions are O^+, O_2^+, CO_2^+, NO^+, H^+, C^+, N^+, CO^+ (or N_2^+), He^+, N_2^+, O^{++} and $^{18}O^+$. O^+ is the major ion in the upper ionosphere, except in the predawn bulge, where H^+ has similar densities. The lower ionosphere is dominated by O_2^+, except at the lowest altitudes at night where NO^+ and CO_2^+ become significant ions. O^+ becomes the dominant ion above 200 km on the day side and 160 km on the night side. The ionospheric electron and ion temperatures are higher than those found in the Earth's ionosphere, owing primarily to solar wind energy deposited at the ionopause and propagated downward by conduction and to the night side by both conduction and convection. The magnetic field in the dayside ionosphere is generally too low to impede the horizontal transport of ionospheric plasma, except perhaps at low solar zenith angles where the magnetic pressure may exceed the thermal plasma pressure at times of high solar wind pressure. Large-scale holes which protrude radially into the antisolar ionosphere also have strong magnetic fields. These holes are believed to be signatures of an interaction of the magnetotail with the ionosphere.*

The interest of mankind in the exploration of the planets and their environments is deeply ingrained in the spirit and intellect of man. No doubt this urge to explore neighboring worlds has been reinforced recently by the realization that our own environment is in constant evolution in ways that are not well understood, the result of a complex interaction of anthropogenic and other natural processes. We sense that the new perspectives afforded by the exploration of the atmospheres of other planets will broaden our understanding of where our own fits into the spectrum of the possible.

Owing to the close proximity and similarities of Venus to Earth in size, mass, and distance from the Sun, the atmosphere of Venus has been explored more extensively than that of any planet other than Earth. This is particularly true of the ionosphere of Venus which has been visited intermittently by spacecraft since 1967 when Mariner 5 provided the first radio occultation measurements of the electron density N_e (Mariner Stanford Group 1967; Kliore et al. 1967, 1969). Since that time a series of United States and Soviet missions have provided many additional altitude profiles of N_e by the radio occultation method. The pace of exploration picked up markedly in 1975 when the Venera 9 and 10 spacecraft were injected into Venus orbit to provide many more occultations near the noon and midnight meridians (Ivanov-Kholodny et al. 1979).

These early results inspired a great amount of theoretical analysis and speculation on the nature of the upper atmosphere and ionosphere of Venus, some of it highly insightful, but most of it flawed by our almost complete lack of knowledge of the composition and temperature of the neutral upper atmosphere and ionosphere (McElroy and Strobel 1969; Gringauz and Breus 1969). Without this knowledge such observed characteristics as the ionospheric scale heights and the altitudes of the electron density peak could be reproduced theoretically by many assumed models of the neutral atmosphere, ion composition, and plasma temperatures, not to mention the magnetic field structure of the ionosphere. The theoretical tools developed over the years to explain the observations of the Earth's ionosphere, quite naturally were modified and extended to predict what we might find when it became possible to explore Venus. These theoretical codes (e.g. Chen and Nagy 1978), described extensively in Chapter 24 (Nagy et al.), included photoionization, photochemistry, photoelectron production, local heating and cooling of electrons and ions, and ionization by particle precipitation. Solar-wind interactions with the ionosphere were also treated theoretically (e.g. Cloutier and Daniell 1973) to predict the induced ionospheric current system and describe the scavenging of photoions by the solar wind. Thus, although there was extensive theoretical speculation about the ionosphere of Venus, there was little empirical knowledge about anything but the electron density. It was not until the arrival at Venus of the Pioneer Venus Orbiter (PVO) on 4 December 1978 that the first observations of the other parameters became possible.

The goals of the PV mission, the spacecraft, and its instrumentation are described in special issues of *Space Science Reviews* (1977a, b) and the *IEEE Transactions on Geoscience and Remote Sensing* (1980). An onboard propulsion system permitted periapsis to be lowered deeply into the ionosphere and maintained there, usually at altitudes in the range of 140 to 160 km. This allowed the parameters of major importance to the formation of the ionosphere to be measured repeatedly, sometimes down to the main peak of N_e. Complementary observations by radio occultation extended the N_e measurements through the main peak and the lower ionosphere. The occultation measurements also provided N_e profiles of the high-latitude ionosphere which could not be measured *in situ* because of the low latitude of periapsis ($\sim 15°N$).

The purpose of this chapter is to review the observational knowledge that has been gained in these early explorations of the Venus ionosphere and to recall some of the interpretations offered by the various observers at the time, as well as to trace the evolution of our empirical picture of the ionosphere as new data appeared. Some of these results were included in a review of the theory and observations of the ionospheres of the terrestrial planets by Schunk and Nagy (1980). For a more comprehensive theoretical framework in which to view these measurements we refer the reader to that paper and to Chapter 24 on ionospheric theory. The measurements of the composition and temperature of the neutral atmosphere, which are of great scientific relevance to the ionosphere, are covered in Chapter 13 (von Zahn et al.).

Much of the great complexity of the Venus ionosphere arises from the absence of an intrinsic magnetic field which would shield it from the solar wind. The magnetic fields that are found at Venus are induced by the interaction of the solar wind and the interplanetary magnetic field (IMF) with the ionosphere itself, an interaction resulting in compression of the IMF at the ionopause as it wraps around the planet (Russell et al. 1979a) and in the induction of currents within the ionosphere (Cloutier and Daniell 1973). Since these magnetic field phenomena are the subjects of Chapters 25 and 26, we will only consider the effects of the magnetic field on the configuration and behavior of the ionopause and on the transport of ionization and energy within the ionosphere.

I. EARLY OCCULTATION MEASUREMENTS

Prior to 1967 little was known about the ionosphere of Venus. On October 19 of that year the Mariner 5 spacecraft flew behind Venus and provided radio occultation measurements which led to the first N_e profiles of both the dayside and nightside ionosphere, shown here in Fig. 1 (Mariner Stanford Group 1967; Kliore et al. 1967, 1969; Fjeldbo and Eshleman 1969). The day side exhibited a main peak of 5×10^5 cm^{-3} at ~ 140 km and a sharp decline of N_e above 500 km, a feature that was recognized as having great significance for the understanding of the interaction of the solar wind with Venus. This upper boundary of the ionosphere was called the plasmapause, in analogy with a similar feature in the Earth's ionosphere; however it has since become known as the ionopause in recognition of the presence of shocked solar wind plasma above this boundary.

The nightside ionosphere also exhibited a peak of N_e at 140 km having an electron concentration of 1×10^4 cm^{-3}. The existence of such a dense nighttime ionosphere was a considerable surprise because the long Venus night (58 Earth days) and the short time for ion recombination near the peak (~ 100 s) lead one to expect the ionosphere to disappear shortly after solar extreme ultraviolet (EUV) radiation was removed. The Mariner Group ascribed the nightside peak to local ion production by cosmic rays, scattered solar Lyman-α, or meteoric bombardment. Alternatively, they suggested that the nightside ionosphere might be produced by horizontal transport of plasma from the day side, an insightful remark which has been supported by later results and analysis.

An almost constant shelf of N_e above 1000 km at night, leading to a sharp ionopause near 3700 km, was interpreted as either an instrumental effect or a population of H$^+$ ions with mean ion and electron temperatures between 625 and 1100 K. A transition region between this shelf and the main peak was interpreted as He$^+$ ions at 620 to 970 K. Later results from PVO have

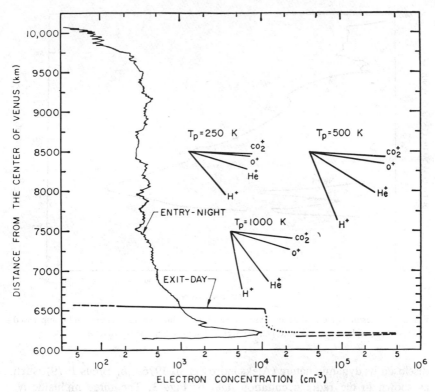

Fig. 1. Profiles of N_e based on radio occultation measurements from Mariner 5 showing a dense dayside ionosphere and a substantial nightside ionosphere extending to very high altitude. Distinct ionopauses were present on both the day side and night side (Mariner Stanford Group 1967).

confirmed the existence on many occasions of an extended ionosphere on the night side, but have also shown that the major ion is usually O^+ at much higher temperatures.

In 1974, the Mariner 10 radio occultation measurements (Fig. 2) provided a more or less similar picture, with a dayside N_e peak of 3×10^5 cm^{-3} at 142 km and a double peak on the night side (Howard et al. 1974; Fjeldbo et al. 1975). The upper nightside peak at 142 km had an N_e of 9×10^3 cm^{-3} and the lower peak at 124 km had an N_e of 7×10^3 cm^{-3}. The dayside ionopause was found at 335 km, and the region immediately below was inferred to be irregular and turbulent. The ionopause was again recognized as the probable site of solar wind interaction with the ionosphere.

These early flyby missions provided "snapshots" of Venus which confirmed the existence of a dayside ionosphere and surprisingly revealed a substantial nightside ionosphere. Well defined ionopauses were identified, particularly on the day side. Late in 1975 the first Venus orbiters, Venera 9 and 10, began to provide "moving pictures" of the Venus ionosphere which

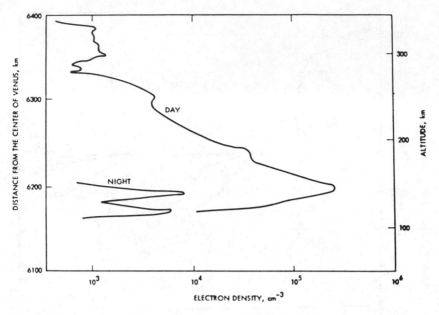

Fig. 2. Profiles of N_e from Mariner 10 radio occultation showing an irregular and compressed dayside ionosphere and a double-peaked nightside ionosphere (Howard et al. 1974).

disclosed its dynamic nature (Aleksandrov et al. 1976a, b; Breus 1979), such as shown in the radio occultation data of Fig. 3. The three nightside N_e profiles, in spite of their large orbit-to-orbit variability, clearly reveal the double-peaked nature of the ionosphere first evident in the Mariner 10 data. The upper peak was still present at 140 km, and a lower peak was present at 120 km $\sim$ 50% of the time. Gringauz et al. (1979) attributed the nightside peaks to the local production of ionization by low-energy electrons observed repeatedly far downstream in the planets shadow by the wide-angle analyzers and the Faraday cylinder on Veneras 9 and 10.

The dayside peak was much less variable in the Venera 9 and 10 data, as shown in Fig. 4. Ivanov-Kholodny et al. (1979) examined its density as a function of solar zenith angle (SZA) and concluded that the N_e variation was consistent with a Chapman layer, i.e., a layer in photochemical equilibrium. However, they noted that the height of the peak varied only weakly with SZA, an effect inconsistent with the behavior of a Chapman layer which should move upward as the sun sets. Gavrik et al. (1980) showed that the absence of a rise in the peak at sunset could be explained if there were a suitable decrease in the neutral gas temperature which would lower the altitude of maximum ion production. Such a drop in neutral thermospheric temperature had already been deduced from the neutral mass spectrometer results of PVO (Niemann et al. 1979a). Moving to higher altitudes, Ivanov-Kholodny found that the ionopause height changed with SZA, increasing only little up to 65°, reaching

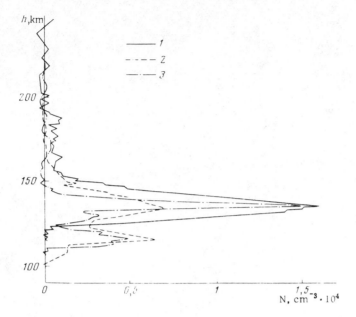

Fig. 3. Venera 9 and 10 radio occultation profiles of the nightside ionosphere (Aleksandrov et al. 1976*a*).

a maximum at ~85° and decreasing at larger SZA. We will see that the later PV data reveal a much greater increase in ionopause height with SZA and much greater variations with UT. In Sec. II we will outline the early results of the PV mission described primarily in two issues of *Science* (1979*a*, *b*).

II. INITIAL PIONEER VENUS RESULTS

The arrival of PVO on 4 December 1978, signaled the beginning of an intensive exploration of the Venus ionosphere. The radio occultation measurements of the nightside ionosphere (Kliore et al. 1979*b*) continued to show the great variability evident earlier in the Venera 9 and 10 data, as evidenced by an average density at the peak of 1.67×10^4 cm^{-3} with a standard deviation of 0.72×10^4 cm^{-3}. Figure 5 shows 12 nightside N_e profiles which illustrate this great orbit-to-orbit variability of the peak density. The peak altitude was more stable, however, at 142.2 ± 4.1 km. The major features of the Venera data are also present; however, the appearance of double-peaked profiles is much less common in the PV data (4 of 36 profiles). About half of the Venera profiles exhibited such peaks. It is not clear whether this difference is real or statistical.

The early PV occultation measurements of the day side which occurred near the terminator (Fig. 6) also revealed few surprises (Kliore et al. 1979*c*). These profiles showed two distinct regions of nearly constant scale height

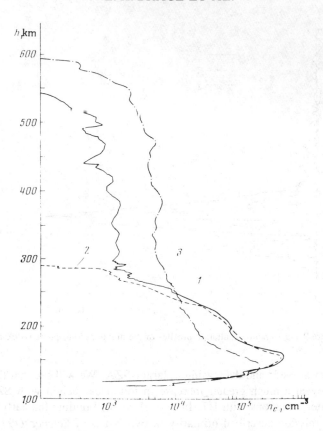

Fig. 4. Dayside N_e profiles from Veneras 9 and 10 (Aleksandrov et al. 1976*b*).

above the peak, a lower region in which the scale height ranged from 13.5 to 37 km, attributed to the predominance of O_2^+, and an upper region having a scale height between 32.4 and 70.8 km, representing the major ion O^+. The peak was observed at an average altitude of 143 km. Thus the PV occultation data were generally consistent with the Venera 9 and 10 measurements and those obtained on the early flyby missions. The electron densities at the peak were generally higher but this might have been expected because of the higher levels of solar activity in 1979.

PV Orbit

Before reviewing the early *in situ* results, a few words about the nature of the PV orbit are in order because it has had a direct bearing upon what could be learned about the ionosphere using the *in situ* measurements. The initial injection placed the spacecraft in a near polar orbit (105°), with apoapsis at 11 Venus radii and periapsis at 17° N lat. The altitude of periapsis, initially 378 km, was quickly lowered to ~ 150 km using onboard propulsion. An orbital

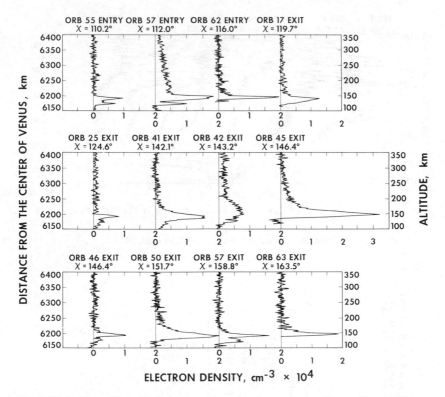

Fig. 5. Nightside profiles of N_e from the radio occultation method on Pioneer Venus (Kliore et al. 1979b).

period of 24 hr was also quickly established and maintained. Figure 7 shows the changes in periapsis altitude and local time through the first 600 orbits while propulsion was being used to maintain the desired orbit. During this interval the latitude of periapsis moved very slowly from its original latitude of 17° N to 14° N.

During the 600 Earth days in which periapsis was actively controlled, the annual movement of Venus about the Sun caused the orbit plane to sweep out all local times repeatedly, moving at the rate of 0.1 hr of local time per orbit. At this rate, periapsis moved through the night side three times and through the day side twice before beginning to spiral upward through the equatorial ionosphere, rising at a rate of ~ 400 km per Venus year.

Because of the great orbital eccentricity and low periapse latitude, the PV *in situ* measurements refer primarily to the low-latitude ionosphere. Only the radio occultation method permitted the high-latitude ionosphere to be observed. A second limitation, one that is inherent in direct measurements from eccentric orbit, is the difficulty of separating the altitudinal structure of the ionosphere from the horizontal structure. Thus, although the profiles of N_e

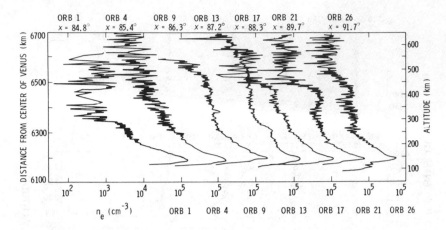

Fig. 6. Dayside N_e profiles from PV radio occultation (Kliore et al. 1979c).

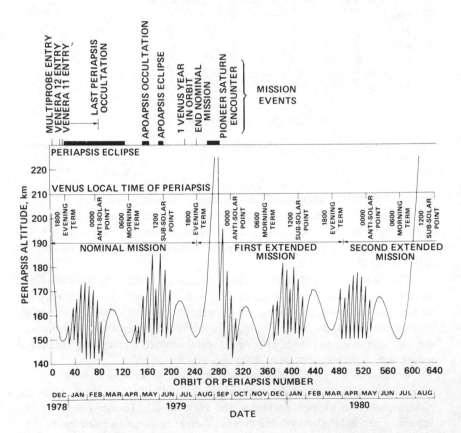

Fig. 7. Pioneer Venus periapsis altitude and local time during the first 600 days while propulsion was used to maintain periapsis deep in the ionosphere (Colin 1980).

along the orbit are usually dominated by altitude structure, important variations with latitude, longitude, and local time structure may be superposed. This is particularly true on the night side where the ionosphere is irregular and dynamic. The local time behavior itself is obtained only statistically by averaging through the data obtained during the 224 day period required for the Sun to rotate through 360° relative to the inertially fixed orbit plane of PVO. These factors must be kept in mind when discussing the PVO data.

Early Dayside Results

The first *in situ* observations of the Venus ionosphere were made by the PVO instruments in the afternoon ionosphere (at SZA 70 to 90°). H. Taylor et al. (1979a, b), using an ion mass spectrometer (OIMS), identified an extensive ion composition, rich in oxygen, nitrogen, and carbon chemistry as shown in Fig. 8a. The lower ionosphere was found to be dominated by O_2^+, identified as a photochemical layer peaking near 150 km, but with orbit-to-orbit variability in density and in height distribution. The molecular ions NO^+, CO^+, and CO_2^+ were found to contribute <10% of the ionization near the peak. At higher altitudes, the ionosphere was found to be dominated by O^+, with C^+, N^+, H^+, and He^+ determined to be important secondary ions. Trace amounts of O^{++}, $^{18}O^+$, Fe^+, and Mg^+ were also tentatively identified. Above the O^+ peak near 200 km, day-to-day differences in the ion scale heights were interpreted as resulting from variations in ion convection stimulated by the solar wind interaction. An ion spectrometer instrument identical to the OIMS was also carried on the PV entry bus spacecraft, which entered the dawnside ionosphere, during the first week of the Orbiter operation at dusk. These results, shown in Fig. 8b, revealed a great similarity between the composition and concentration of the dawn and dusk ionospheres, on the same day (H. Taylor et al. 1979c).

Knudsen et al. (1979b), using a retarding potential analyzer (ORPA), found that the ion composition was consistent with diffusive equilibrium above 220 km, though Bauer et al. (1979) found the upper ionosphere to be perturbed sometimes by plasma convection. Knudsen et al. (1979b) also found that the ion temperature T_i was elevated above the neutral gas temperature T_n (Niemann et al. 1979a) reaching values of $\sim$ 2000 K at 500 km (Fig. 9). The variations of T_i with altitude at altitudes below 250 km was clear evidence of a layered heat source which was attributed to Joule heating since heating by energetic O_2^+ ions was estimated to be too small to supply the needed heating rate. The electron temperature T_e was also very high, rising from < 1000 K at 150 km to > 4000 at 500 km. Similar T_e results were reported by Brace et al. (1979b) using the electron temperature probe (OETP). In addition, Brace et al. found very small-scale vertical wave structure in T_e and N_e on an orbit which followed a solar flare (Fig. 10), indicating that the Venus dayside ionosphere could respond dynamically to solar events. In general, the plasma

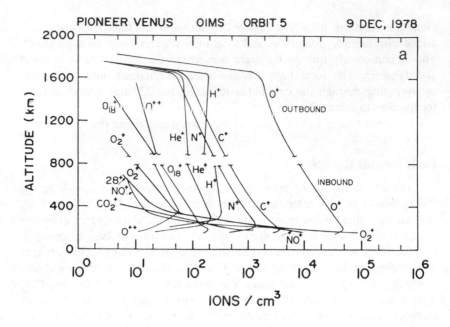

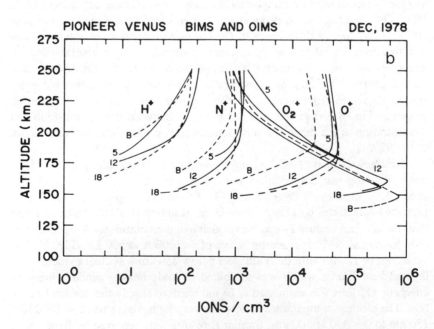

Fig. 8. (a) Ion composition measured in the late afternoon, and (b) a comparison of ion composition measurements from the PV Orbiter in the dusk sector (solid) with similar measurements in the dawn ionosphere (dashed) from the PV entry bus (H. Taylor et al. 1976*b, c*).

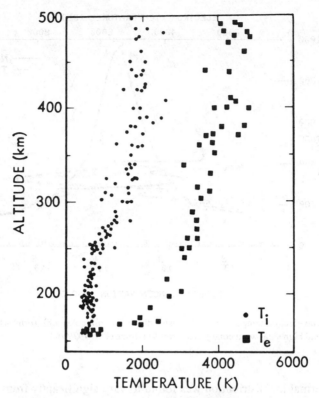

Fig. 9. The electron and ion temperatures in the afternoon ionosphere at 70° to 80° SZA measured on 6 of the first 10 PV orbits (Knudsen et al. 1979b).

temperatures were much higher than would result from heating by solar EUV photoelectrons and local cooling of electrons to ions (cf. Chen and Nagy 1978; Chapter 24 by Nagy et al.). Brace et al. (1979b) and Knudsen et al. (1979a) concluded that additional heating must be occurring at the top of the ionosphere, a result that was not totally unanticipated (cf. Cravens et al. 1979). A downward electron heat flux of $\sim 3 \times 10^{10}$ eV cm^{-2} s^{-1} and an ion heat flux of $\sim 2 \times 10^{8}$ eV cm^{-2} s^{-1} would be needed to maintain the observed T_e and T_i profiles, assuming that no magnetic field was present to inhibit vertical heat conduction. The total heat conducted through the electron gas was estimated to be $\simeq 10\%$ of the solar wind energy incident on Venus (Knudsen et al. 1979a).

All three of the ionospheric plasma instruments observed distinct ionopauses in every passage. The ionopause altitude varied widely from orbit to orbit. H. Taylor et al. (1979b) detected the presence of superthermal ion flow near the ionopause, and interpreted the effect as evidence for the acceleration and sweeping away of the ambient ions by the solar wind. The energy imparted to the dominant ionospheric constituent O$^+$ and the structure of the

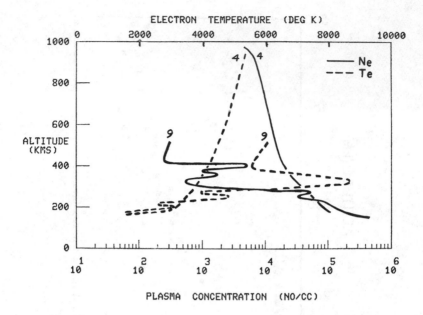

Fig. 10. The electron temperature and density observed on a quiet orbit (orbit 4) and an orbit which had high solar wind conditions (orbit 9) (Brace et al. 1979*b*).

superthermal ion distribution was found to vary significantly from day to day in response to changes in the solar wind and the IMF. Brace et al. (1979*b*) and Knudsen et al. (1979*a*) found that the plasma pressure at the ionopause was approximately equal to the magnetic field pressure just above. They concluded that the field must play an important role (probably the dominant role) in transferring the pressure of the solar wind to the ionosphere. This would imply that most of the solar wind plasma does not directly interact with the ionosphere but is largely excluded by the magnetic field. Knudsen et al. (1979*b*) noted that the spacecraft occasionally encountered patches of detached plasma above the ionopause.

H. Taylor et al. (1979*b*) and Scarf et al. (1980), noting that whistler mode waves observed by the electric field detector (OEFD) tended to disappear at the ionopause, showed that this could be explained by an interaction with ionospheric electrons. Figure 11 illustrates this effect on orbit 4. They identified whistler wave energy as a possible source of the high T_e observed throughout the dayside ionosphere.

A highly complex magnetic structure was evident in the early dayside magnetometer (OMAG) measurements (Russell et al. 1979*a*). Figure 12 is a high time resolution plot of the total magnetic field B encountered during an early passage through the afternoon ionosphere. The large fields observed in the region just above the ionopause are the ones mentioned above which are

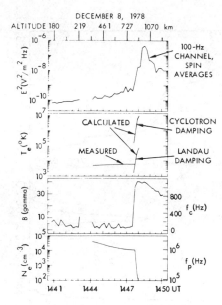

Fig. 11. Observations of plasma waves (100 Hz channel), the electron temperature, the magnetic field strength, and the electron density during an exit from the ionosphere (Scarf et al. 1980*b*).

thought to apply the solar wind pressure to the ionosphere. The field nearly disappeared within the ionosphere except for small-scale features at low altitudes which Russell et al. called flux ropes, in analogy to apparently similar features on the Sun. These ropes, which were the major surprise in the dayside magnetic structure, have dimensions along the orbit on the order of tens of km. They were observed in all transits of the dusk sector.

In summary, the early dayside results answered several of the first-order questions about the ionosphere that inspired the mission (Bauer et al. 1977). The primary ions were O_2^+ and NO^+ at low altitudes and O^+ at high altitudes. The ion scale height above the peak was consistent with diffusive equilibrium. The ionosphere was much warmer than had been expected ($T_e \approx 4500$ K and $T_i \approx 2000$ K at 500 km). A layered ion heating source centered near 200 km altitude and attributed to Joule heating was discovered. The ionopause was a distinct feature whose altitude varied from orbit to orbit, and an approximate pressure balance existed at the ionopause between the magnetic field and the plasma. Whistler wave damping was identified as a possible source for the elevated T_e. Magnetic flux ropes on the order of tens of km in diameter were discovered in the lower ionosphere.

Nightside Ionosphere

Early in 1979, during the first excursion of periapsis into the night side, the PV data revealed a number of startling discoveries. As noted earlier (Fig.

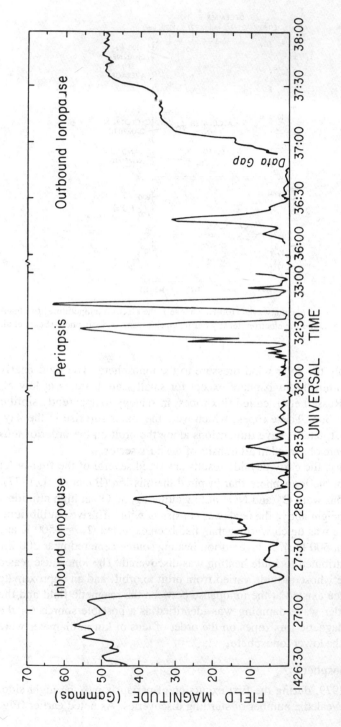

Fig. 12. High time resolution magnetic field strength across the afternoon ionosphere showing the enhancement of the magnetic field above the ionopause and narrow enhancements (flux ropes) within the ionosphere (Russell et al. 1979a).

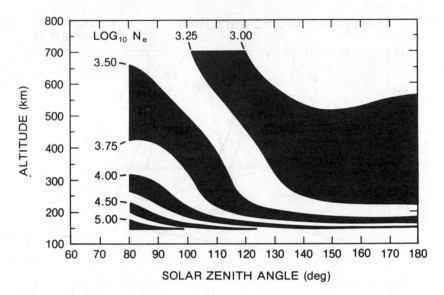

Fig. 13. Contour plot of an empirical model of N_e based on PV measurements between 80° and 180° SZA (Brace et al. 1979c).

5), the PV occultation data confirmed the fact that a substantial ionosphere existed on the night side (Kliore et al. 1979b). This was also evident in the *in situ* measurements. Brace et al. (1979c) used the OETP measurements of N_e to model the average variation of N_e with altitude between 80° and 180° SZA (Fig. 13). In this empirical model, the electron density below 200 km declined by more than a factor of 10 across the terminator but remained nearly constant beyond 120° SZA. N_e was $\sim 1 \times 10^4$ cm^{-3} at 150 km, a value consistent with the average occultation value of 1.5×10^4 cm^{-3} for the main peak at ~142 km (Kliore et al. 1979b). H. Taylor et al. (1979b) found that, like the day side, the major ion at 150 km was O_2^+, giving way to O^+ above 160 km (Fig. 14). Thus the nightside ion composition at low altitudes was not unlike the dayside.

The source of the nightside ionosphere remained uncertain. H. Taylor et al. (1979a) calculated that the transport of O^+ above the exobase (~170 km) from the day side at a velocity of 100 m s^{-1} would be adequate to produce the observed nightside O_2^+ concentration through charge exchange in the reaction $O^+ + CO_2 \rightarrow O_2^+ + CO$. Brace et al. (1979c) performed a similar calculation and were able to match the height profile of the OETP N_e model (Fig. 13) by an ion transport velocity of 300 m s^{-1} from the day side which led to a nightside downward flux of O^+ of 6×10^7 cm^{-2} s^{-1}. Alternatively, Brace et al. (1979c) also were able to match the N_e model equally well by invoking an isotropic flux of 30 eV electrons or 500 eV electrons with fluxes of 4×10^8 cm^{-2} s^{-1} and 1×10^8 cm^{-2} s^{-1}, respectively. Electrons with energies > 30 eV

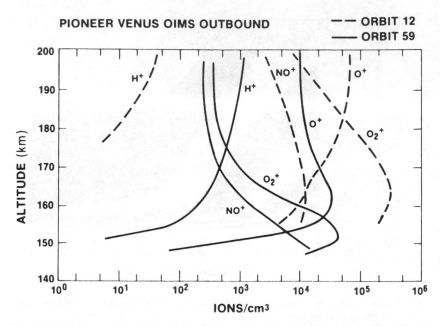

Fig. 14. Comparison of predusk and nightside ion composition profiles measured on orbits 12 and 59, respectively (H. Taylor et al. 1979a).

tended to produce ionization at altitudes below the observed main peak but could be responsible for the secondary peak at 120 km. Similar calculations on the effects of electron precipitation were performed by Kliore et al. (1979b) who reached similar conclusions, i.e., that low-energy electrons in sufficient fluxes would be capable of producing the observed nightside ionosphere. Intriligator et al. (1979), using the plasma analyzer (OPA), reported fluxes of suprathermal electrons within the nightside ionosphere. Preliminary estimates of the fluxes on one nightside orbit showed that the flux between 50 and 250 eV may have been sufficient to produce the amount of O_2^+ observed at 150 km. Thus, the early nightside data were consistent with a nightside ionosphere produced either by the transport of O^+ from the day side or by the precipitation of low-energy electrons ($\sim$ 30 eV). Higher energy electrons may have produced the lower of the two nightside peaks. These attempts to account for the existence of a nightside ionosphere are discussed in greater depth in Chapter 24 on ionospheric theory (Nagy et al.).

The nightside ionosphere exhibited more complexity than the dayside, both in terms of concentration and composition. Unlike the dayside, nightside ion concentrations were observed by Taylor et al. (1979a) to range from abundant levels or "well-established" cases to "depleted" events, for which the thermal ions essentially disappeared down to the periapsis height near 160 km, as shown in Fig. 15. Such extreme variations were interpreted as

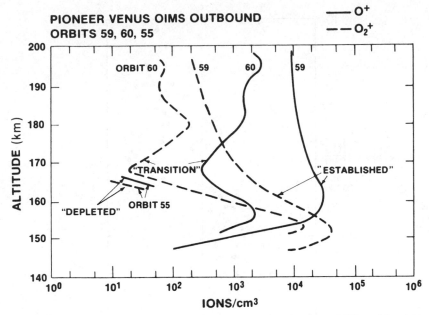

Fig. 15. Profiles of O^+ and O_2^+ on three nightside orbits showing the variability of the night-side ionosphere (H. Taylor et al. 1979b).

interruptions of the day-to-night ion convection process, which in the extreme resulted in the sweeping away or depletion of the night side, as indicated by orbit 55.

Another significant example of the nightside complexity was the discovery by H. Taylor et al. (1979a) of the increasing prominence of H^+ with increasing SZA as shown in Fig. 16. The H^+ to O^+ ratio at 250 km increased from 10^{-2} at dusk to about unity at dawn, with H^+ forming a distinct bulge in the predawn region near SZA 140°. Within this bulge, the H^+ densities were sometimes observed to exceed those of O^+, normally the dominant topside ion at all other local times. Brinton et al. (1980) combined these ion measurements with simultaneous atomic oxygen measurements from the neutral mass spectrometer (Niemann et al. 1979a, 1980b) to show that the dawn-dusk asymmetry was associated with a corresponding composition asymmetry in neutral hydrogen in the upper atmosphere, an effect believed to be induced by global circulation patterns in the thermosphere (Mayr et al. 1980). The asymmetrical buildup of H^+ was interpreted as resulting from lateral transport and subsidence combined with atmospheric superrotation. An additional important composition change observed at night was an increasing value of the ratio NO^+/O_2^+ with increasing SZA at the lowest altitudes, implying that NO^+ may become the dominant constituent of an E layer, peaking generally below the periapsis altitude of PVO.

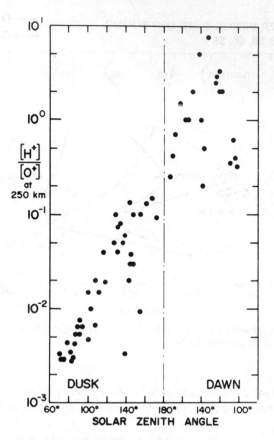

Fig. 16. The nightside variation of H⁺/O⁺ with SZA at 250 km altitude (H. Taylor et al. 1979a).

Another surprising discovery of this first excursion through the night side was that the plasma temperature did not decrease across the solar terminator. In fact, T_i was observed to be higher by a factor of 2, particularly enhanced in the antisolar region where T_i reached values comparable to T_e (Knudsen et al. 1979a). Figure 17a shows this T_i enhancement at SZA > 150°. T_e was found to remain high on the night side, perhaps slightly increased relative to the day side, although much more variable from orbit to orbit (Fig. 17b). Knudsen et al., in attempting to model the elevated ion temperatures, found it necessary to supply distributed heating to the ions which they presumed to be Joule heating. The total heat conducted through the electron gas over both the day and night hemispheres was estimated to be at most 10% of the solar wind energy incident on the dayside hemisphere. The whistler-mode damping observed at the dayside ionopause was not present at the nightside ionopause (W. W. Taylor et al. 1979a).

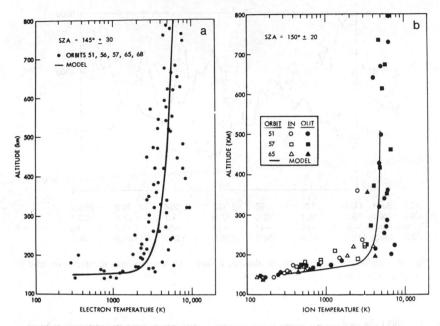

Fig. 17. (a) Electron temperature from five nightside orbits, and (b) ion temperatures from three nightside oribts. The solid lines represent theoretical models (Knudsen et al. 1979a).

Unlike the dayside ionosphere, the plasma on the night side is highly structured in both density and temperature as illustrated in the single transit shown in Fig. 18 (Brace et al. 1979c). This small-scale variability was to become even more evident in data discussed later. To provide a basis for examination of this extreme variability of T_e, Brace et al. (1979c) employed the OETP measurements to model average behavior as functions of altitude and SZA, as shown in Fig. 19. This empirical model showed that the average T_e was nearly independent of SZA below 300 km and actually declined slightly on the night side at great altitudes.

Nightside Magnetic Structure

The magnetic structure of the nightside ionosphere turned out to be even more complex than its dayside behavior. Figure 20 shows a series of orbits which reveal this complexity. Russell et al. (1979b) had hoped to be able to isolate the component arising from the intrinsic field of the planet, if any. However, the field directions in the nightside ionosphere appeared to track the changes in the solar wind field, suggesting that the magnetic field in this region was entirely induced by solar wind interactions. Especially strong and variable fields (~30γ) were found in the middle of the wake region within the ionosphere which were clearly not of planetary origin. Russell et al.

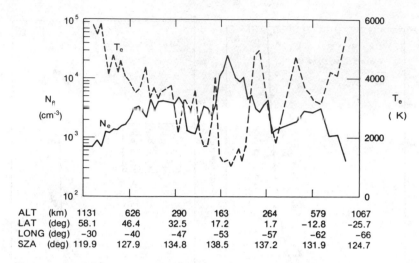

ALT	(km)	1131	626	290	163	264	579	1067
LAT	(deg)	58.1	46.4	32.5	17.2	1.7	-12.8	-25.7
LONG	(deg)	-30	-40	-47	-53	-57	-62	-66
SZA	(deg)	119.9	127.9	134.8	138.5	137.2	131.9	124.7

Fig. 18. Small-scale structure in the electron density and temperature, measured in orbit 98 through the night side (Brace et al. 1979c).

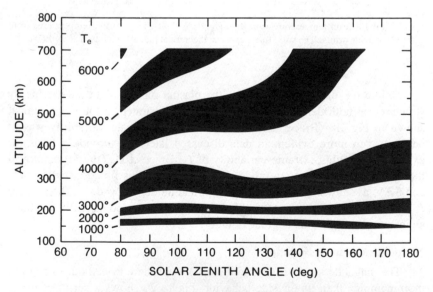

Fig. 19. Contour plot of an empirical model of the electron temperature based on measurements between 80° and 180° SZA (Brace et al. 1979c).

suggested that they might be remnants of an earlier magnetosheath field that had penetrated the ionosphere and had been convected into the night side. Elsewhere within the nightside ionosphere the field was nearly horizontal and appeared to bend around the planet converging into the wake.

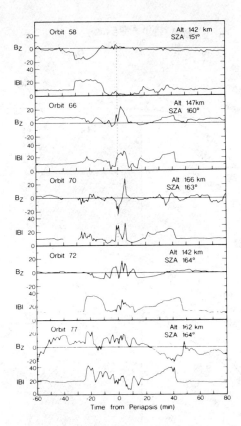

Fig. 20. The average magnetic field strength $|B|$ and the component parallel to the spacecraft spin axis B_z for 5 nightside orbits of PV. The great complexity and variability of the nightside magnetic fields is evident (Russell et al. 1979b).

Nightside Ionopause

The ionopause altitude on the night side was even more highly variable than in the late afternoon. Brace et al. (1979a) showed that the ionopause beyond the terminator often tended to hug the planet, at times having altitudes as low as 200 km (Fig. 21). The formation of an ionopause so close to the planet was attributed to ion scavenging by the interplanetary field as it was dragged tailward into the wake by field line tension. Perez de Tejada (1980a) suggested that viscous interactions of the shocked solar wind with the ionosphere near the terminator deflected solar wind plasma into the wake and produced the ionopause compression observed in the near-wake region. It should be recalled that the low altitudes of periapsis did not permit the ionopause to be observed when it was far downstream, so its true configuration in the wake region could not yet be measured directly. The rise of periapsis occurring throughout the PV extended mission is expected to pro-

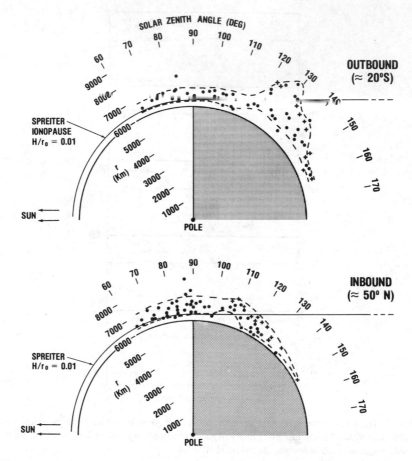

Fig. 21. The inbound and outbound ionopause altitudes plotted against SZA (Brace et al. 1979a).

vide more complete resolution of this near-wake region between 1981 and 1985 (Colin 1980).

III. RESULTS FROM THE PRIMARY PV MISSION

The measurements reviewed thus far were obtained from the early Venus flyby missions, the Venera radio occultation data, and the preliminary PV results obtained in December 1978 and in early 1979. The latter were reported primarily in the two *Science* issues featuring the early PV data (*Science* 1979a, b). In this section we discuss the more complete picture obtained in the period from early 1979 through mid-1980 in which fuel was used to maintain periapsis deep within the Venus ionosphere at an altitude of

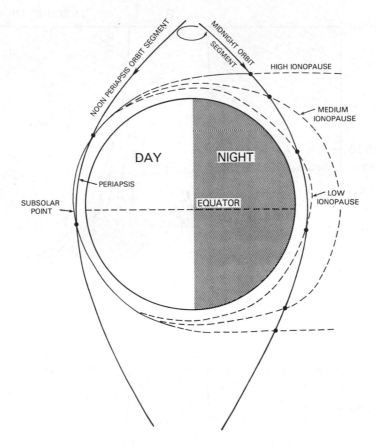

Fig. 22. Noon and midnight PV orbits showing how the orbit encounters the ionopause of Venus. Three hypothetical ionopause cases are shown to demonstrate the sampling configuration.

~ 150 km. During this period the orbit swept through all local times more than twice, allowing study of the diurnal variation of the low-latitude ionosphere. The accumulation of data in this interval also created a data base for later studies of the dynamic nature of the ionosphere.

Ionopause

Since the solar wind is so important to the structure and behavior of the Venus ionosphere, we begin the discussion at the ionopause where this interaction occurs. A few words are in order about how various measurements define the location of the ionopause. The ionopause is a region in which many parameters change drastically, including the ion and electron temperatures and densities, the magnetic field magnitude and direction, and the amplitude of

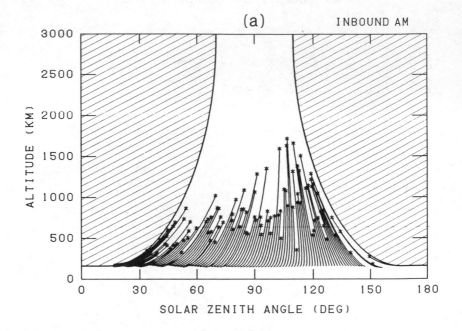

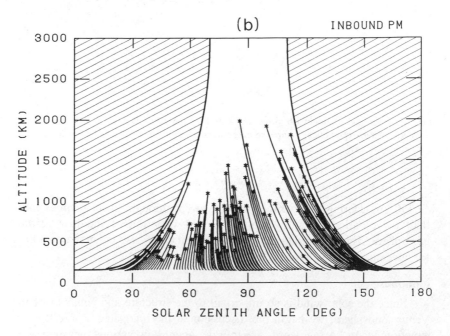

Fig. 23. The SZA variation of the ionopause altitude for inbound crossings (a) and (b), and outbound crossings (c) and (d) of the ionopause. Dawn and dusk sides are plotted separately. The hatched regions were inaccessable to the PV orbiter (Brace et al. 1980).

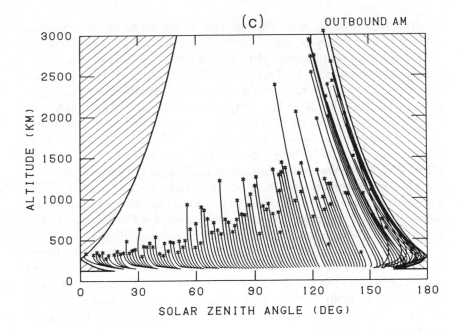

(c) OUTBOUND AM

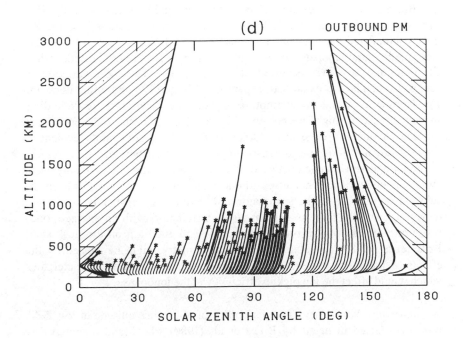

(d) OUTBOUND PM

plasma wave emissions. Since the ionopause as defined by all these parameters has a finite thickness, the assignment of a specific ionopause altitude differs somewhat with the particular parameter employed and various instrumental sensitivity factors. H. Taylor et al. (1980) have selected the ionopause altitude as the boundary between the thermal and suprathermal ion components observed by the OIMS at the top of the ionosphere. This transition is thought to mark the region of solar wind interaction with the ionospheric plasma. Knudsen et al. (1979b) and Brace et al. (1979a) use the altitude in the steep gradient in N_i or N_e where the density passes through 1×10^2 cm^{-3}. Elphic et al. (1980b) have used the point in the ionopause gradient of **B** where the plasma pressure equals the magnetic field pressure. Naturally the ionopause altitudes defined independently by these methods have the potential for significant disagreement, although in the majority of PV ionosphere transits the ionopause signature is sufficiently unique that the differences are small compared to its orbit-to-orbit variability. However, Cloutier et al. (see Chapter 26) have pointed out some rather important exceptions that may occur in regions of the ionosphere at low SZA where large-scale magnetic fields exist (Luhmann et al. 1980). With these qualifications concerning what is meant by the term ionopause, we move on to consider the characteristics of this upper boundary of the ionosphere.

Brace et al. (1980) expanded upon the early OETP ionopause study (Fig. 21) using data from the first full Venus year. Figure 22 shows how the PV orbit encountered the ionosphere during the first 3 Venus years. On the day side, the inbound ionopause crossings were at middle northern latitude, while the outbound crossings occurred at low latitudes. The SZA variation of ionopause height was examined, separating AM and PM and midlatitude and low-latitude crossings to attempt to isolate any dawn-dusk or latitudinal asymmetries that might be present. Figures 23a,b,c,d show the ionopause altitudes at these 4 crossings. Although the orbital constraints evident in Fig. 22 did not allow the downstream ionopause surface to be examined, several important characteristics of the ionopause were immediately clear from this study. The ionopause altitude was lower and less variable in the subsolar region than in the terminator and beyond. A statistically significant dawn-dusk asymmetry was evident at the terminator, with an average ionopause altitude of 700 km on the dawn side and 1000 km on the dusk side. Brace et al. (1980) speculated that this asymmetry might be caused by the effects of the spiral angle of the IMF upon ionosheath flow characteristics which would affect the ion pickup efficiency at the ionopause.

Ionopause Variability. The variability of ionopause altitude at low SZA was investigated in detail by Brace et al. (1980) who found a remarkably reliable inverse relationship between the ionopause altitude and the magnetic field pressure just above the ionopause (Fig. 24a). They also found that the magnetic field pressure, after correction for cos^2 SZA effects was $\sim 80\%$ of

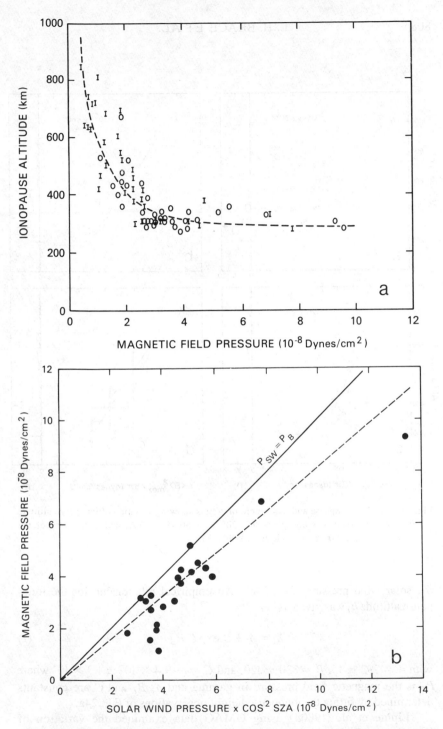

Fig. 24. (a) The variation of ionopause altitude near the subsolar point with magnetic field pressure, and (b) the correlation of magnetic field pressure at the ionopause with solar wind pressure measured just outside the bow shock. The dashed line in (a) represents a best least-squares fit to the measurements. The dashed line in (b) is a best fit which has been forced through the origin (Brace et al. 1980).

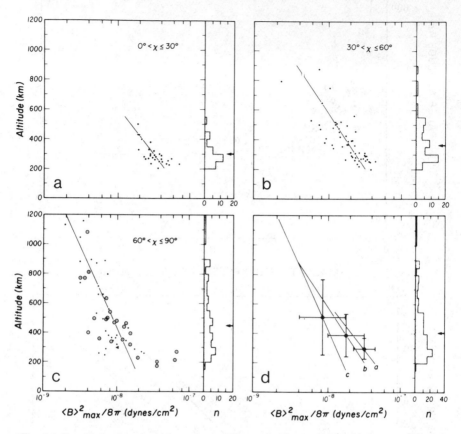

Fig. 25. Dayside ionopause and ionosphere of Venus showing variation of ionopause altitude with magnetic field pressure at 0°–30°, 30°–60°, 60°–90° SZA, and a summary of the behavior for these 3 ranges (Elphic et al. 1980a, b).

the solar wind pressure (Fig. 24b). An empirical relationship for the iono-pause altitude h_i was given as

$$h_i = A + B \exp C\, P_B \tag{1}$$

with $A = 292 \pm 15$, $B = 830 \pm 120$, and $C = -9.4 \times 10^7 \pm 1.3 \times 10^7$ where P_B is the magnetic field pressure in gamma and A, B, and C are constants determined by fitting the ionopause height observations of Fig. 24a.

Elphic et al. (1980a) using OMAG data examined the variation of ionopause altitude with magnetic field pressure for a wide range of dayside SZA as shown in Fig. 25. They concluded that the increasing ionopause

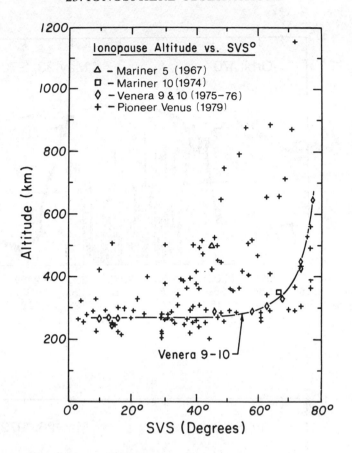

Fig. 26. Variation of ionopause altitude with SZA from Mariner, Venera (Ivanov-Kholodny et al. 1979) and PV observations (Elphic et al. 1980a).

height with SZA tended to obey a quasi-Newtonian approximation to a gas dynamic pressure balance at the obstacle surface of the form

$$B_{max}^2/8\pi = K\rho_0 v_0^2 (\cos^2 SZA)^v + p_0 \qquad (2)$$

where B_{max} is the maximum magnetic field strength just outside the ionopause, ρ_0 and v_0 are the undisturbed solar wind mass density and flow speed, v and K are constants that depend on solar wind flow conditions at the shock, and p_0 provides a finite ionosheath thermal pressure at the terminator. The OMAG ionopause height data are consistent with the Newtonian approximation in which v and K are unity, although a great deal of variability from this relationship was evident.

The variations of the ionopause altitude with SZA observed in 1979 by

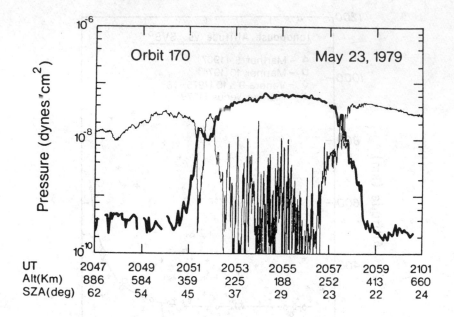

UT	2047	2049	2051	2053	2055	2057	2059	2101
Alt(Km)	886	584	359	225	188	252	413	660
SZA(deg)	62	54	45	37	29	23	22	24

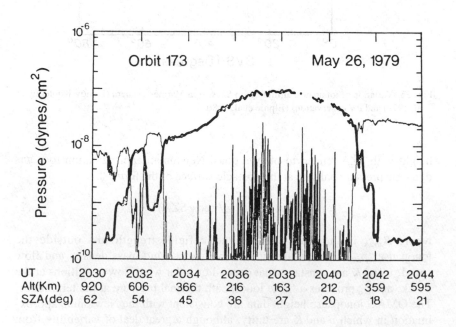

UT	2030	2032	2034	2036	2038	2040	2042	2044
Alt(Km)	920	606	366	216	163	212	359	595
SZA(deg)	62	54	45	36	27	20	18	21

Fig. 27. Comparisons of the magnetic field pressure (light lines) and the ionospheric plasma pressure (heavy lines) on 4 PV passages through the midday ionosphere (Elphic et al 1980a).

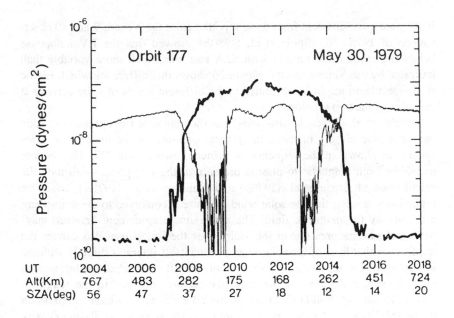

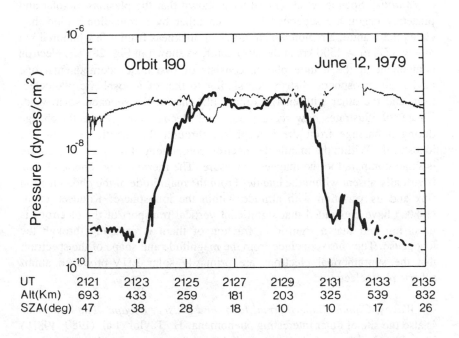

PVO were quite different from those reported from the Venera 9 and 10 observations of 1975–76. Elphic et al. (1980b) showed that the PV ionopause height increased more rapidly with SZA and was much more variable than indicated by the Venera results. Figure 26 shows this difference which Elphic et al. speculated might be attributed to the different levels of solar activity at the times of the two missions.

Elphic et al. (1980a,b) also found that the magnetic field pressure outside the ionopause closely matched the plasma pressure of the underlying ionosphere, as shown in the sequence of orbits shown in Fig. 27. The smooth transition from magnetic to plasma pressure at the ionopause confirmed the calculations of Brace et al. (1979b) and Knudsen et al. (1979a), based on preliminary results, that the solar wind pressure is conveyed to the ionosphere primarily by the magnetic field. The approximate agreement between magnetic and plasma pressure in the vicinity of the ionopause also carries the implication that little of the solar wind plasma itself interacts directly with the ionospheric plasma, though Vaisberg et al. (1980) concluded that hot plasma may contribute ¼ to ⅓ of the total plasma pressure within the magnetic barrier. Zwan and Wolf (1976) have discussed a process in which compression of the IMF against the ionosphere diminishes the solar wind plasma density adjacent to the ionopause. Gombosi et al. (1980) calculated that < 10% of the solar wind protons are absorbed by the atmosphere by charge exchanged before reaching the ionopause.

Mantle. Spenner et al. (1980) have shown that the plasmas of solar and planetary origin are separated from each other by a transition region they called the mantle, a region which grows in thickness from a few hundred km at low SZA to > 1500 km at the terminator, as shown in Fig. 28. The electron component of the mantle plasma consists of two major components, one having an ionospheric distribution similar to that of ionospheric photoelectrons and the other with energies more like that of the shocked solar wind. Figure 29 illustrates how the suprathermal electron energy spectra change during a passage from the ionosphere, through the mantle and into the ionosheath. Within the mantle the electron component of the particle pressure is small compared to the magnetic pressure. The thermal ionospheric plasma is virtually absent within the mantle. From the magnitude of the photoelectron flux and its variation with altitude within the ionosphere, Knudsen et al. (1980a) have concluded that significant vertical transport of the electrons is occurring and that a significant fraction of them is escaping through the ionopause. They also conclude from the magnitude and shape of the spectrum that the suprathermal electrons are primarily solar EUV-produced atmospheric photoelectrons.

Waves, Clouds, Superthermal Ions, and Plasma Escape. The ionopause is also the site of other interesting phenomena. H. Taylor et al. (1980, 1981b) investigated certain of the characteristics of the superthermal ion layer de-

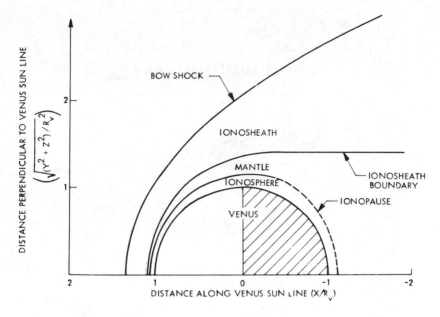

Fig. 28. Schematic location of the ionosheath, mantle, and the ionopause (Spenner et al. 1980).

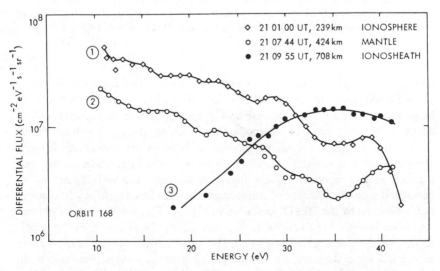

Fig. 29. Differential energy spectra of suprathermal electrons in the ionosphere ①, in the mantle ②, and the ionosheath ③ (Spenner et al. 1980).

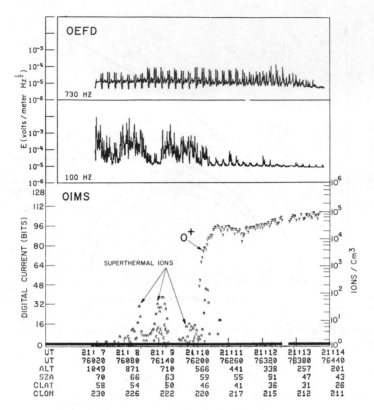

Fig. 30. An example from orbit 165 (18 May 1979) of a superthermal ion layer overlapping the ionopause gradient in O^+. Low-frequency plasma waves are evident in the 100 Hz channel of the OEFD instrument in the region of superthermal ions (H. Taylor et al. 1981*b*).

tected earlier in the mission. The superthermal ions, with energies in the range 10 to 90 eV, were detected on many dayside and terminator-region orbits, beginning just below the ionopause and extending for tens to hundreds of km above. The ion layer was interpreted as resulting from a combination of accelerated ionospheric ions and photoions born in the exosphere through photoionization of neutrals extending above the ionopause. The structure and location of the superthermal ion layer on the day side were found to be correlated with the structure of enhancements in the low-frequency ac electric fields detected by the OEFD, as shown in Fig. 30. The low-frequency plasma waves are thought to be generated by plasma instabilities driven by the acceleration of the exospheric photoions. The superthermal ion buildup and the associated wave structure were found to be generally more pronounced on inbound passes at high latitudes than found on outbound passes at low latitudes (see Fig. 22). At high latitude, the average convective flow was more

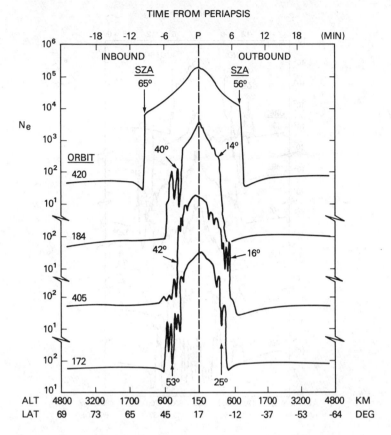

TIME FROM PERIAPSIS

Fig. 31. Wave-like irregularities observed at the dayside ionopause. The waves are common
during times of high solar wind pressure when the ionopause is driven to lower altitude (e.g.
orbits 184, 405, 172) (Brace et al. 1982a).

nearly perpendicular to the magnetic field, resulting in more efficient coupling
between the solar wind and the ambient ions. This is the same mantle region
in which Spenner et al. (1980) observed the transition from superthermal
electrons and ionosheath electrons (Fig. 29).

Wave-like irregularities often form in the ionospheric plasma at the dayside ionopause. Brace et al. (1980, 1982a) have interpreted these as ionopause
surface waves. Figure 31 shows several examples of these waves, which are
envisioned as arising from multiple partial transits of a wavy ionopause surface as the spacecraft sliced almost horizontally through the top of the ionosphere. Another interesting feature, that may be related to the ionopause
waves, is the plasma clouds observed above the ionopause (Knudsen et al.
1979b; Brace et al. 1980, 1982a). Figure 32 shows a series of examples of such
clouds observed by the OETP. Brace et al. (1982a), noting that the clouds

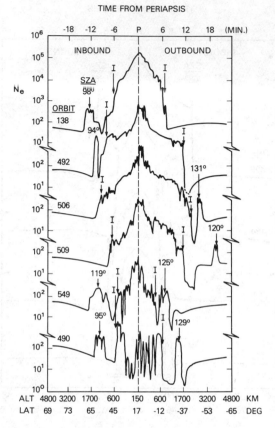

Fig. 32. Plasma clouds, identified by their SZA, observed above the nightside ionopause (I) (Brace et al. 1982*a*).

were composed of ionosphere-like plasma and that they tended to be found downstream from the wave-like structures, suggested that these two features might represent different stages in the process of ionospheric plasma removal by the solar wind. Figure 33 is a sketch showing the zone in which the clouds were observed (Brace et al. 1982*a*). This zone appears to correspond well with the downstream portion of the mantle (Fig. 28). Since the N_e in the clouds is generally too low to permit flow velocity measurements, they are not actually known to be escaping Venus. In fact, the measurements do not preclude the possiblity that the clouds are actually attached streamers analogous to cometary tails. Whatever their configuration, the clouds are observed within the near ionosheath or mantle region, and the possibility that they may be detached and flowing downstream is an interesting one. Brace et al. (1982*a*) estimated an upper limit to the planetary escape flux of 7×10^{26} ions s^{-1}, based on the OETP observation of cloud size, density, occurrence frequency,

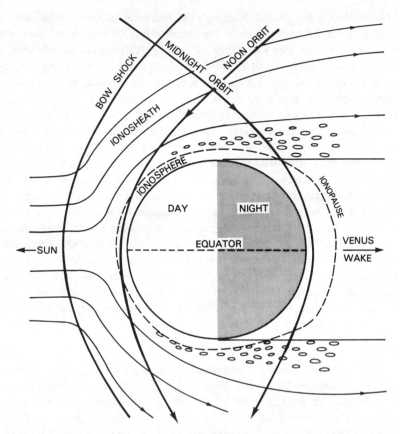

Fig. 33. Conceptual drawing of the plasma clouds observed near the terminator and as far downstream as the midnight orbit. The actual shapes of the clouds are unknown (Brace et al. 1982a).

spatial distribution, and cloud transit times. A planetary ion escape flux of this magnitude would require an upward flux of ions out of the dayside ionosphere that is near the diffusion limit of O^+ through O_2^+ ($\sim 3 \times 10^8$ cm^{-2} s^{-1}). Thus the dayside ionosphere is capable of supplying an ion flux equal to the upper limit of ion escape flux in the form of clouds. This upward flux would tend to leave the dayside ionopause depressed in altitude and the underlying ionosphere heavily eroded. Both of these features are typically observed on the day side, as will be seen below. The physical process which forms the clouds is not yet entirely clear. Wolff (1980) used Kelvin-Helmholtz instabilities, induced by shear flows at the ionopause, to form ''plasma bubbles'' that were subsequently picked up and carried downstream by the solar wind. This process could be the source of the observed plasma clouds.

The Nightside Ionopause. Although the ionopause altitude variations on the day side can be understood as the direct effect of solar wind pressure changes, the even greater variability of the nightside ionopause remains a mystery. The very existence of a well-defined nightside ionopause was something of a surprise. No process for its formation was known, and the earlier occultation measurements were not sufficiently sensitive (except for Mariner 5) to detect the nightside ionopause reliably because N_e was too low. Gasdynamic models of the ionopause (Spreiter and Stahara 1980) had simply assumed that the nightside ionosphere extended indefinitely downstream along the last flow lines which cleared the terminator ionopause. Perez de Tejada (1980b) invoked a viscous boundary layer above the ionosphere at the terminator to deflect the shocked solar wind into the wake region where an ionopause might be formed by a pressure balance between ionosheath and ionospheric plasmas. However, this process fails to explain the many occasions, evident in Figs. 21 and 23, when the nightside ionopause was found within a few hundred kilometers of the surface. Electrodynamic forces in the wake, induced by solar wind interactions, may ultimately explain the great variability of ionopause height on the night side. As periapsis rises, future PVO measurements in the upper ionosphere and near-wake region will allow further investigation of this suggestion.

Ionosphere

As noted earlier, the PV Orbiter passed through the Venus ionosphere ~ 600 times between December 1978 and July 1980, while periapsis was being maintained at altitudes in the vicinity of 150 km. Figure 34 a and b show a selection of N_e profiles from the OETP data from the first Venus year. These profiles illustrate the relatively smooth nature of the dayside ionosphere (Fig. 34b) in contrast to the extreme complexity of the nightside ionosphere (Fig. 34a). But before examining the structure and complexity of the ionosphere we will discuss its average conditions.

Average Conditions. Theis et al. (1980) employed the full set of OETP data from the first Venus year to devise spherical harmonic models of the average behavior of N_e and T_e as functions of SZA and altitude. They assumed that the latitudinal variations could be neglected and that local time variations were symmetrical about the noon-midnight plane. These models are shown in Fig. 35a,b in the form of T_e and N_e contour plots. They are consistent with the earlier models of Brace et al. (1979c), Figs. 13 and 19, which were based only on the early duskside data from PV. Similar T_e and N_e diurnal behavior were reported by Knudsen et al. (1979a) and Miller et al. (1980) from the ORPA measurements as shown in Fig. 36a,b.

The sources of the high electron temperatures were reexamined in the light of this more complete data set (Cravens et al. 1980). Two primary heat sources for the dayside ionosphere are known to exist: suprathermal electrons

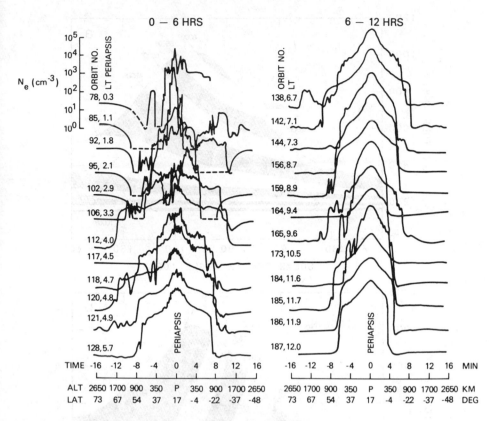

Fig. 34. Montage of N_e profiles between 0–6 hr local time (a), and 6–12 hr local time (b), illustrating the relatively smooth nature of the dayside ionosphere and the highly structured nature of the nightside ionosphere (Brace et al. 1980).

produced by photoionization of the major neutrals, and solar wind energy deposited at the ionopause. Earlier theoretical analysis of PV data (Knudsen et al. 1979a) suggested that the photoelectrons might supply enough heating to explain the observed T_e. However, examination of the ORPA suprathermal electron data within the ionosphere showed that these fluxes are at least a factor of 5 too small (Knudsen et al., 1980a), nor could the photoelectrons be responsible for the high electron temperatures observed across the entire night side. Therefore it appears that the high electron temperatures arise primarily from solar wind interactions at or near the ionopause. Hoegy et al. (1980) arrived at similar conclusions by comparing two-dimensional solutions of the heat conduction equation with the global model of T_e by Theis et al. (1980) (Fig. 35a). Miller et al. (1980) presented evidence that the variation of T_e with altitudes was independent of ionopause altitude. They suggested from the

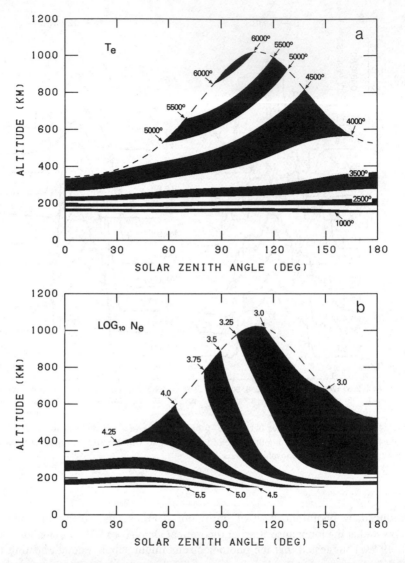

Fig. 35. Empirical models of T_e (a), and N_e (b), versus altitude and solar zenith angle. The upper limit of the models is determined by the ionopause altitude and the sampling limits of the orbit (Theis et al. 1980).

observations that the boundary condition for electron temperature at the ionopause is a constant net flux and not a constant temperature.

The variation of T_i with altitude and SZA, illustrated in Fig. 37, reveals several processes which require understanding. Above 300 km altitude, T_i remains constant with SZA up to a SZA of 150°. Between 150° and 180° SZA

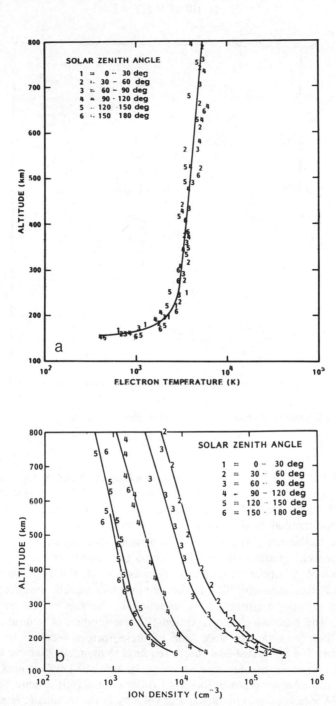

Fig. 36. (a) Median electron temperatures, and (b) densities for various ranges of SZA (Miller et al. 1980).

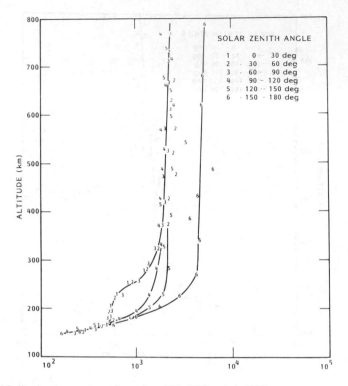

Fig. 37. Median ion temperatures for various SZA (Miller et al. 1980).

the median ion temperature then doubles, reaching the value of approximately 5000 K. Knudsen et al. (1980c) and Miller et al. (1980) suggest that this increase in T_i beyond 150° SZA results from the conversion of ion bulk kinetic energy into thermal energy as the flow of ions from the day side converges on the antisolar axis. The ion flow is driven predominantly by the large plasma pressure gradient which occurs across the terminator as a result of the sharp drop in density at the terminator (Knudsen et al. 1981). Below 300 km altitude, T_i increases with SZA as the terminator is passed, an effect which Miller et al. (1980) attribute to a reduction in the rate of ion cooling to neutrals. The rate, which is proportional to the product of neutral and ion density, decreases rapidly since both of these factors decrease across the terminator. The integrated heat source required to maintain T_i above the neutral temperature on the day side between 150 and 250 km is approximately 1.5% of the solar wind power incident (Miller et al. 1980). Figure 38 shows the ORPA measurements of the ion drift velocity in the terminator region. Ion velocities directed toward the night side with magnitudes of 2 to 3 km s^{-1} are typical. The velocity increases from < 1 km s^{-1} at periapsis to a few values as

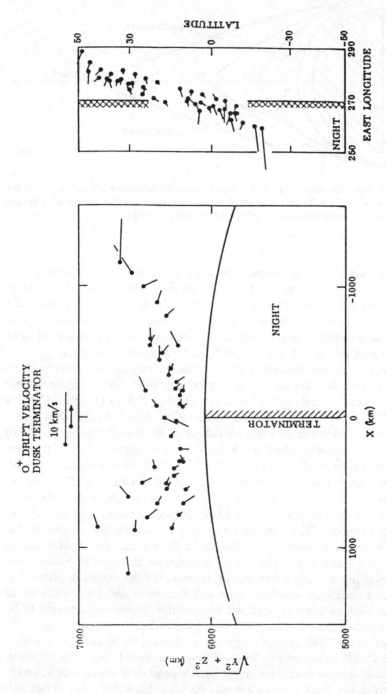

Fig. 38. O$^+$ drift velocities projected into two different planes for several orbits near the dusk terminator showing a nightward and downward ion flow having typical velocities of a few km s^{-1} (Knudsen et al. 1980b).

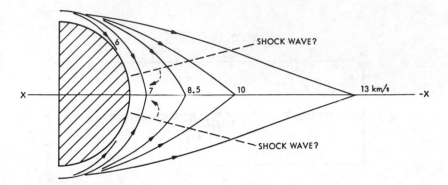

Fig. 39. Ballistic trajectories of particles (ions) crossing the terminator at a radius of 7000 km
with velocities of 6 to 13 km s^{-1}. A shock may form in the near wake, with most of the ion
flow returning to the nightside ionosphere (Knudsen et al. 1980b).

large as 8 km s^{-1} at the ionopause and behind the terminator. Since the ion
thermal velocity is typically > 1 km s^{-1}, the flow is supersonic. As shown in
Fig. 39, Knudsen et al. (1980b) suggest that a shock wave forms in the wake
to slow the flow as it converges on the antisolar axis. The directed flow kinetic
energy is converted to heat manifested by the large ion temperature. Most of
the ions were assumed to descend and form the nightside ionosphere.

The composition measurements from the primary mission completed the
coverage of the diurnal variation (H. Taylor et al. 1980, 1981b). Figure 40a,b,c
show the measurements at 200 km of H$^+$, O$^+$, O$_2^+$, NO$^+$, CO$_2^+$, He$^+$, CO$^+$ (or
N$_2^+$), C$^+$, and N$^+$. The full data set continued to show the dawn-dusk asymmetry in H$^+$, evident in the early PV data (Fig. 16), which Mayr et al. (1980)
attributed to neutral composition changes driven by superrotation of the thermosphere. Balancing this asymmetry of H$^+$, O$^+$ was more concentrated near the
dusk terminator. Thus the total ion density did not exhibit a significant dawn-
dusk asymmetry. Taylor et al. (1981a) also looked for systematic differences
between the ion composition at middle latitude (inbound leg) and at low
latitudes (outbound leg). Finding none, they concluded that, relative to the
Earth's ionosphere, where the influence of the geomagnetic field is strong
(Taylor and Walsh 1972), the ion composition of the dayside Venus ionosphere was not strongly dependent on latitude. On the night side, however,
significant variability with both time and location is observed. Differences
between inbound and outbound ion composition profiles are thought to be
caused by plasma convection driven by the solar wind.

Theis et al. (1980) arrived at similar conclusions for N_e and T_e by comparing the OETP measurements to their empirical model (Fig. 35a,b) which
represented average conditions. Figure 41a,b are plots of the ratio of N_e and T_e
to the respective models, plotted versus the unmodeled parameter of latitude.

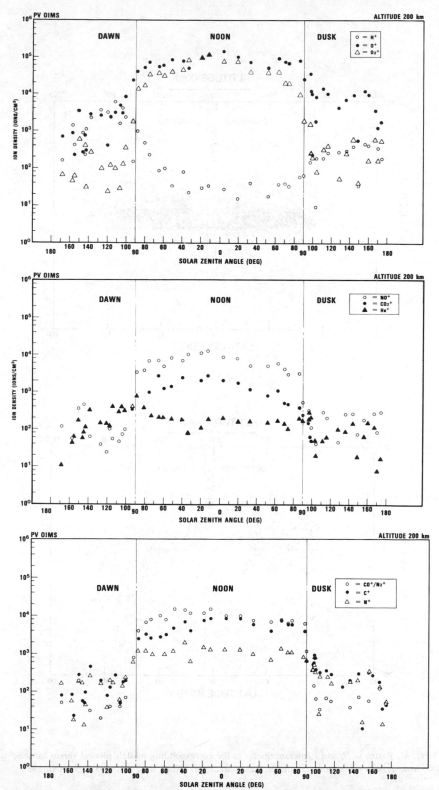

Fig. 40. Ion composition measurements from the first two Venus years showing the full diurnal variation of (a) H^+, O^+, O_2^+; (b) NO^+, CO_2^+, He^+; and (c) CO^+/N_2^+, C^+, N^+ (H. Taylor et al. 1980).

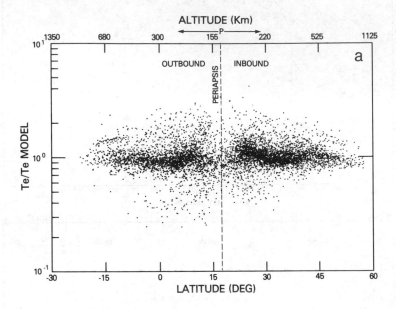

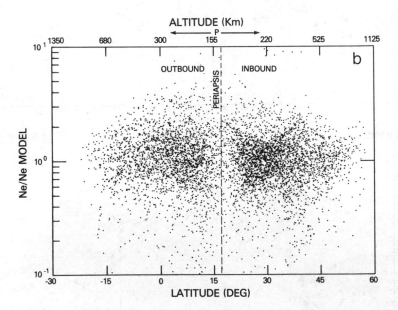

Fig. 41. Ratios of N_e and T_e measurements to the corresponding models plotted versus latitude to detect any latitudinal variation in these parameters. None were evident (Theis et al. 1980).

All local times are represented in the data. The scatter of the ratios from unity is due to the variability of the ionosphere, primarily on the night side. However, the absence of a systematic deviation from the model with latitude shows that the ionosphere is remarkably free of latitudinal structure, except of course that related to SZA.

A similar search for local time asymmetry in T_e and N_e found none, but did turn up systematic deviations of N_e from the model on the day side (Theis et al. 1980). The ratios of the measurements to the model are shown in Fig. 42a,b. The solid line is a free-hand fit of the N_e ratio. Since 224 Earth days are required to obtain data at all local times, it was not possible to determine whether these deviations of N_e from the N_e model reflected real local time effects not described by SZA, or were caused by changes in solar wind or solar EUV radiation. H. Taylor et al. (1981a), however, showed that a similar dayside variation in the concentrations of the ions correlated well with the 27 day variation in the $F_{10.7}$ cm flux, an index of solar EUV radiation. Bauer and Taylor (1981) showed that the same 27 day variation in the ion densities was directly correlated with the variations in the solar Lyman-β and He II lines measured in Earth orbit by Atmosphere Explorer and transposed to Venus, allowing for their different orbital locations. This correlation, shown in Fig. 43, clearly tied the variations in the Venus dayside ionosphere density to solar EUV changes. The analysis of Bauer and Taylor also showed that the changes in ion density are mainly due to changes in EUV ionization and to a much lesser extent the result of changes in the neutral atmospheric density caused by heating of the neutral atmosphere. Further analysis of the ion chemistry is reviewed by Nagy et al. in Chapter 24.

An interesting change in the altitude of the main peak of N_e occurs in the vicinity of the terminator (Fig. 44). As noted earlier, Kliore et al. (1979b) reported that the average altitude of the peak on the night side was ~ 143 km. Using PV radio occultation data, Cravens et al. (1981a) reported that the altitude of the peak declines beyond 70° SZA, reaches a minimum at ~ 137 km at 82° SZA and then rises to over 150 km just beyond the terminator at 90° SZA. This decline in peak altitude near sunset is probably caused by the dusk decline of neutral density and temperature. Garvik et al. (1980) used this effect to explain the absence of a rising peak in the Venera 9 and 10 data (also evident in Fig. 44). Apparently the preterminator decline of the altitude of the peak was absent at the lower levels of solar activity of that period. The PV peak height measurements do show the rise predicted by Chapman theory beyond 82° SZA.

The PV *in situ* measurements of the nightside ionosphere (H. Taylor et al. 1982) found that the average altitude of the peak was ~ 5 to 10 km higher than the 143 km reported from occultation (Kliore et al. 1979b). Both data sets exhibit orbit-to-orbit variations of the peak which exceed this difference. Part of the difference arises from a systematic bias in the *in situ* average peak location introduced by the behavior of periapsis. Since periapsis did not often

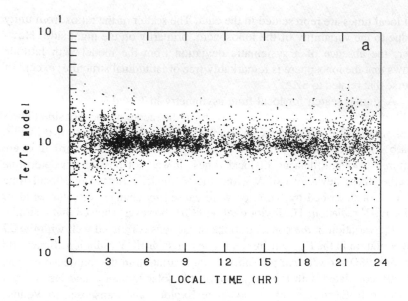

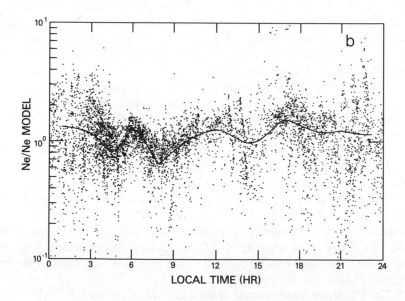

Fig. 42. Ratios of N_e and T_e measurements to the models plotted versus local time to examine any local time structure not accounted for in the model by SZA. The N_e measurements (plot b) showed significant variations from the model during daytime as illustrated by the solid line (Theis et al. 1980).

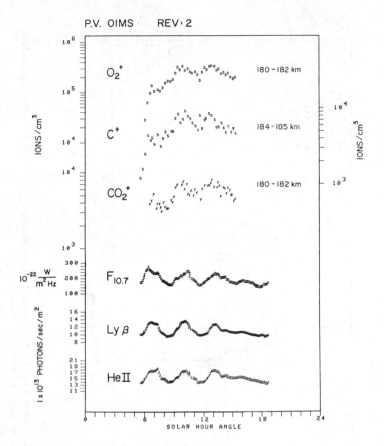

Fig. 43. Variations in ion composition with $F_{10.7}$ (solar flux at 10.7 cm) and the Lyman-β and He II lines observed by Atmosphere Explorer E. This result showed that the 27 day variation in solar EUV was responsible for the changes in ion concentration (Bauer et al. 1981).

go below 143 km, the peak could only be observed if it were above its average height. On the other hand, one cannot discount the possibility that the peak actually forms at a lower altitude at the higher latitudes where most of the occultations occur. The peak in the *in situ* data, if one was observed at all, could be seen only near 15° north latitude where periapsis was located.

Whatever the average altitude of the peak turns out to be, the wide range of orbit-to-orbit variability is a reality whose cause is yet to be determined. This variability is probably associated with the time varying roles of ion transport from the day side and local ion production in the formation of the nightside peak.

The magnetic structure of the ionosphere appeared more and more complex as periapsis swept through new regions of the ionosphere during the first PV year. Figure 45 shows a variety of magnetic field profiles observed by

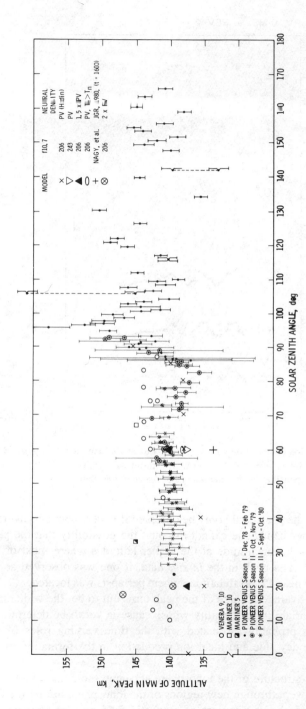

Fig. 44. Variations in the height of the main ionospheric peak observed by PV, Veneras 9 and 10, Mariners 5 and 10, and theoretical location for a variety of assumed solar fluxes and neutral thermospheric models (Cravens et al. 1981a).

OMAG as periapsis reached the subsolar ionosphere for the first time (Elphic et al. 1980a; Luhmann et al. 1981c). The ionosphere is usually field free except for the flux ropes (Fig. 45a); however large-scale currents are often found within the ionosphere, primarily at times of high solar wind pressure (Fig. 45b,c). These internal fields tend to be smooth and horizontal and approximately parallel to the magnetic field above the ionopause. Sometimes a diamagnetic layer forms between the ionopause and 200 km (Fig. 45c). Hartle et al. (1980) examined the effects of this magnetic field on the ion composition, ion density, and electron temperature at low SZA. They found that the portion of the ionosphere above 200 km is often compressed by $j \times B$ forces where the ion flow is expected to be vertical. Luhmann et al. (1980) interpreted these large-scale ionospheric fields as representing a horizontal belt whose latitudinal extent is greatest at small SZA, in qualitative accord with the theoretical ionospheric current system of Cloutier and Daniell (1979). The belt is believed to grow in area as the solar wind pressure increases, allowing the field to be observed sometimes at large SZA on the day side.

Elphic et al. (1980a) examined the plasma density for any correlation with the magnetic rope structure to determine whether $j \times B = 0$ in the vicinity of a rope. That is, were the ropes wound so tightly that their external field was negligible, thus requiring no gradients in thermal plasma pressure? The latter would appear in the N_e measurements near the ropes. Figure 46 shows a pair of $|B|$ and N_e comparisons which suggest that the ropes are not accompanied by reductions in N_e, except perhaps at higher altitudes where the ropes may not be so tightly wound.

The magnetic fields within the dayside ionosphere do not appear to play a dominant role in determining its density and composition, though the field may be strong enought to affect electron heat conduction and therefore T_e. The plasma pressure everywhere dominates the magnetic field pressure, except perhaps below 170 km at times of high solar wind pressure when the induced magnetic field pressure approaches or exceeds the plasma pressure at low SZA. Elsewhere on the day side, except within the flux ropes, the field appears to be too weak to affect the transport of plasma. The same is not true everywhere on the night side, as will be discussed below.

In summary, it appears that the Venus dayside ionosphere is simpler than that of the Earth in the sense that the average global variations of plasma temperature, density, and composition can be described as functions of SZA. However, we will see that the nightside ionosphere is much more dynamic than that of Earth because it is formed in large part by transport from the day side. This transport, although driven primarily by plasma pressure gradients at the terminator, is almost continually responding to the effects of solar wind changes on the nightside flow.

Nightside Structure. The nightside ionosphere is highly structured spatially and is highly variable on time scales which are probably much less than

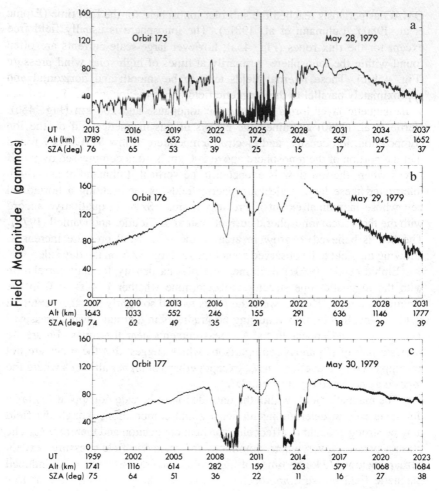

Fig. 45. Three examples of the magnetic field strength profiles observed during passages through the subsolar ionosphere (Elphic et al. 1980a).

the 24 hour orbital period of the PVO. Figures 15, 34, 41, and 42 illustrated this nightside variability. Since the ion transport time from the terminator to the antisolar point is on the order of 1 hr, individual consecutive daily passages of PVO represent entirely independent snapshots of an ionosphere which is evolving on a much shorter time scale. However, the structure encountered along any given nightside passage is likely to be primarily spatial because the satellite velocity greatly exceeds the ion transport velocity essentially everywhere in the ionosphere. Brace et al. (1980) called attention to a number of distinctly recognizable spatial structures which were evident repeatedly in nightside transits during the first Venus year. Two of these

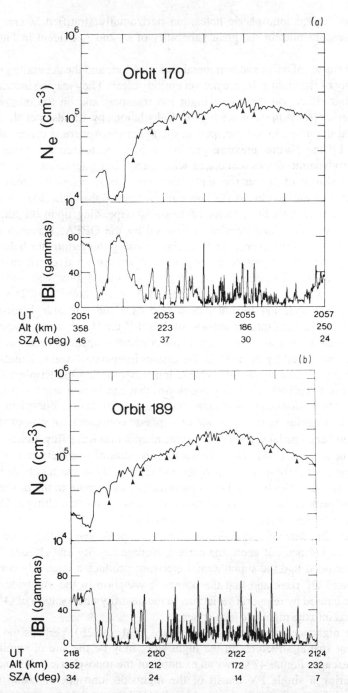

Fig. 46. Magnetic field strength B and N_e in orbits 170 (a), and 189 (b). N_e exhibits a slight inverse correlation with B in the flux ropes at higher altitudes (Elphic et al. 1980a).

structures, called ionospheric holes and horizontally stratified layers, were largely responsibile for the great variability of N_e and T_e evident in Figs. 41 and 42.

The causes of holes and horizontal stratification, and the generally greater variability of the night side, are not yet entirely clear. They may be induced by solar wind effects on the day-to-night ion transport and on the removal of ionospheric plasma in the wake region. Calculations by Knudsen et al. (1981) show that day-to-night ion transport in the lower ionosphere is driven almost entirely by the plasma pressure gradient across the terminator rather than by electrodynamic forces associated with solar wind interactions. If so, the great variability of N_e on the night side must arise primarily from electrodynamic control of plasma flow once it has reached the night side where the plasma pressure falls by a factor of 10 to 50 depending upon the altitude. Using ion velocities and densities measured by the ORPA, Knudsen et al. (1980b) estimated the averge total ion flux crossing the terminator below the ionopause into the nightside hemisphere. When assumed distributed uniformly over the nightside hemisphere, the estimated flux was shown to provide a column ionization source rate of 3×10^8 cm^{-2} s^{-1}. They also estimated the total ion recombination rate over the nightside hemisphere which yielded a hemispheric average of 2×10^8 cm^{-2} s^{-1}, a value approximately equal to the source rate from convection of ionization. From early ion flux rates measured by the OPA in the distant ionospheric wake, Knudsen et al. argued that most of the ionospheric ion flux crossing the terminator was returning to the planet. Thus they concluded that convection was a substantial and possibly predominant ionization source for the nightside ionosphere. The observed variability in the nightside ionospheric concentration was attributed to temporal and spatial variations in the terminator and wake flow fields.

Spenner et al. (1981) have shown that the suprathermal electron energy distribution is nearly uniform spatially and temporally in the nightside hemisphere and that suprathermal electrons primarily contribute to the production of the O_2^+ peak at 140 km and not to the O^+ peak at 160 km altitude. On the other hand, a downward flux of O^+ ions of 1 to 2×10^8 cm^{-2} s^{-1} is able to reproduce the density and altitude of the O^+ peak and at the same time produce an O_2^+ peak of about the correct average density and altitude. They have concluded that the suprathermal electrons produce a relatively constant background O_2^+ peak and that the observed variation in the nightside ionosphere is caused by temporal variations in the quantity and location of O^+ ions convected into the nightside hemisphere.

The nightside ionospheric holes (Brace et al. 1982b), perhaps the most consistent spatial structures in the night side, may be the site of significant plasma escape. Figure 47 shows an example of the ionospheric holes encountered during a single PV transit of the nightside ionosphere. Using many examples of these features measured by the OETP over 3 Venus years, Brace et al. have interpreted the holes as common, if not permanent, features which

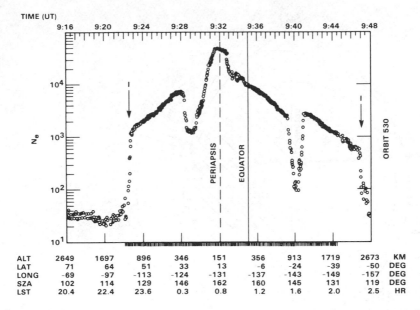

TIME (UT)

ALT	2649	1697	896	346	151	356	913	1719	2673	KM
LAT	71	64	51	33	13	-6	-24	-39	-50	DEG
LONG	-69	-97	-113	-124	-131	-137	-143	-149	-157	DEG
SZA	102	114	129	146	162	160	145	131	119	DEG
LST	20.4	22.4	23.6	0.3	0.8	1.2	1.6	2.0	2.5	HR

Fig. 47. An example of a pair of ionospheric holes observed in orbit 530. The inbound hole is less depleted than the outbound hole because of the difference in crossing altitude (Brace et al. 1982*b*).

penetrate the antisolar ionosphere from the wake region in north-south pairs. The holes are large-scale features having typical north-south dimension of 1000 km and average east-west dimensions of ~ 1800 km, and they penetrate down to as low as 160 km altitude. Noting that they contain strong radial magnetic fields, Brace et al. (1982*b*) concluded that the hole configuration was as shown in Fig. 48. The radial fields, shown conceptually within the holes, were of the same type as reported by Luhmann et al. (1981*b*) and shown in Fig. 49. Oribt 530 displays the field which corresponds to the N_e profile of Fig. 47.

Detailed study of the holes (Brace et al. 1982*b*) revealed some interesting characteristics of the dynamics of the nightside ionosphere. The strong radial magnetic fields within the holes exclude the entry of ions and electrons from the surrounding ionosphere, while an unspecified process removes plasma from within the holes. Since little plasma can enter through the sides of the holes, the removal of plasma must occur either by ion subsidence into the lower ionosphere and recombination, or by ion loss to the magnetotail through electrodynamic processes. Grebowsky and Curtis (1981) have argued that the holes are created by parallel electric fields which originate as crosstail electric fields in the magnetotail. Their connection to the magnetotail is supported by the presence of very high electron temperatures in the holes where the electron mean free path is great enough to permit the entry of hot magnetotail electrons.

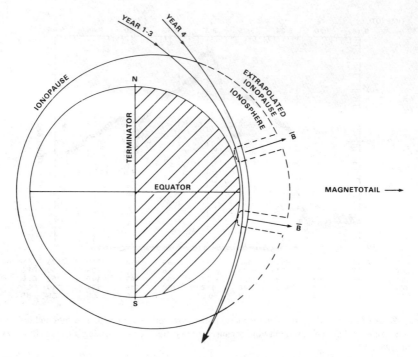

Fig. 48. Drawing of possible global configuration of ionospheric holes. Strong radial fields within the holes may have either polarity. The crossings in Venus year 4 show that the fields remain vertical in the upper ionosphere, although the orbit does not permit their configuration further downstream to be observed (Brace et al. 1982*b*).

The holes also appear to affect the dynamics and energetics of the nightside ionosphere. Brace et al. (1982*b*) calculated from PVO measurements that the pressure on the antisolar side of the holes was less than that on the terminator side. This pressure difference across the holes may arise because they act as a barrier to the antisolar ion flow which, in the process, must be diverted around the holes. Thus the holes are somehow involved in determining the ion flow patterns in the nightside ionosphere. The diversion of ion flow around the holes also converts some of the kinetic energy of flow into thermal energy and may contribute to the elevation of T_i in the antisolar region (Knudsen et al. 1980*b*).

IV. SUMMARY

The observations of the Venus ionosphere have revealed it to be a highly complex and dynamic region. Figure 50 is a sketch which is intended to

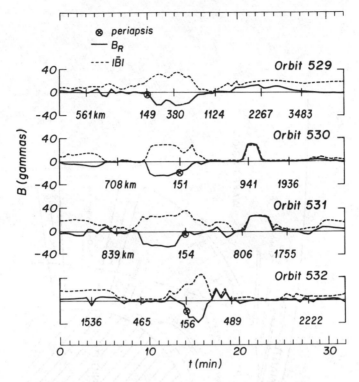

Fig. 49. Total magnetic field strength $|B|$ and its radial component B_R measured on four passages through the antisolar ionosphere. The orbits which exhibit ionospheric holes (orbits 530 and 531) exhibit strongly enhanced radial fields in the holes. At low altitudes the field may remain radial outside the hole (Luhmann et al. 1981b).

summarize the global configuration of the ionosphere and illustrate the major observed features and some of the processes currently thought to be responsible for its structure and behavior. The ionosphere has been expanded relative to the other features for clarity.

Solar EUV radiation falling on the dayside thermosphere creates an ionosphere which is in photochemical equilibrium near its peak at $\sim$ 142 km where the major ions are O_2^+, CO_2^+, and NO^+. The major ion above 200 km is O^+, except in the predawn sector where H^+ has comparable concentrations owing to changes in the neutral gas composition. The upper ionosphere is nearly in diffusive equilibrium much of the time, although ion transport to the night side and the loss of plasma at the ionopause frequently causes departures from diffusive equilibrium. The ionosphere is heated internally by photoelectrons and heated from above by the downward conduction of solar wind energy deposited at the ionopause.

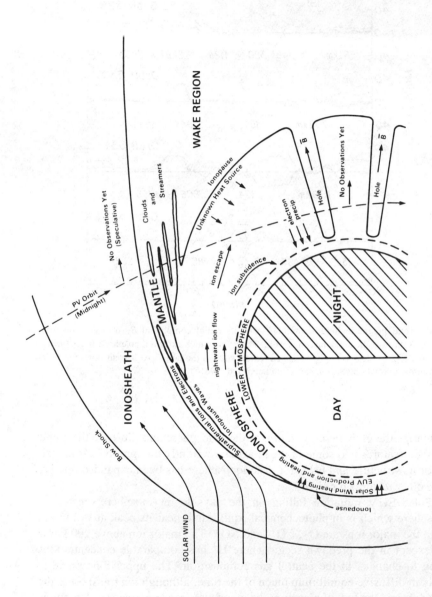

Fig. 50. Sketch summarizing the ionospheric observation and the processes inferred from them. The scale of the ionosphere has been expanded by a factor of about 3 relative to the other regions.

The day-to-night plasma pressure gradient across the terminator drives a nightward flow of ions, usually at supersonic velocities. Some of these ions subside to contribute to the formation of the nightside ionosphere, and other ions may be convected into the wake by electrodynamic forces. The great variability of the nightside ionosphere is believed to arise from changes in the proportion of the flow which is lost to the wake. Even without the nightward ion flow, a nightside ionosphere would be formed by the precipitation of electrons with energies up to a few hundred eV, though the measured electron fluxes do not appear adequate to constitute the only source of nightside ionization.

Large-scale radial holes extend downward to nearly the peak of ionization in the antisolar region. These holes, which occur in north-south pairs, enclose regions of strong radial magnetic fields which are believed to originate in the magnetotail of Venus. The holes act as a barrier to the flow of ionization into the antisolar region and may be related to the greatly elevated ion temperatures observed there.

The ionopause itself is a highly dynamic and complex surface which extends from an average altitude of 290 km at the subsolar point to $\sim$ 1000 km at the terminator. The ionopause altitude varies from 200 to $>$ 3000 km on the night side, although its configuration over most of the night side cannot be determined from the current PV orbit. The increase in ionopause altitude toward the terminator reflects the decline of solar wind pressure with SZA. The ionopause altitude is highly variable, probably with a time constant of $<$ 1 hr. On the day side these variations are caused by solar wind interactions which continuously remove just enough plasma from the top of the ionosphere to maintain a static balance between the solar wind and ionospheric plasma pressures. Thus the ionopause altitude varies inversely with solar wind pressure. The solar wind pressure is conveyed to the dayside ionosphere primarily by means of an enhanced magnetic field which is approximately parallel to the ionopause surface, though it is possible that some of the solar wind plasma interacts directly.

Just above the ionopause is the mantle, a transition region between the ionospheric plasma and the flowing shocked solar wind plasma. A variety of solar wind interaction products are observed in the mantle, including suprathermal ions and electrons, ionospheric photoelectrons, enhanced plasma wave activity, and clouds or streamers of ionospheric plasma. These features are probably signatures of the plasma acceleration processes that form the ionopause and divert the solar wind about Venus; however, the specific processes have not been identified.

Although these early exploratory missions to Venus thus far have provided an exciting glimpse of a dynamic new planetary ionosphere, the wealth of data which have yet to be processed and digested will surely turn up errors in our current conceptions. Therefore this chapter should be considered only a progress report on initial results and early speculations on their meaning.

Acknowledgments. Those of us who are Principal Investigators on the Pioneer Venus Orbiter thank the Pioneer Venus Project for their highly professional efforts in the development and launch of the spacecraft and for its continuing operation, data acquisition and processing. We also thank H. Brinton of NASA and C. Russell of UCLA for reviewing this chapter and for helpful comments and discussion.

24. BASIC THEORY AND MODEL CALCULATIONS OF THE VENUS IONOSPHERE

A. F. NAGY, T. E. CRAVENS, and T. I. GOMBOSI
The University of Michigan

A review of the theoretical understanding of the chemical and physical processes controlling the behavior of the ionosphere of Venus is presented. The level of this understanding has advanced immeasurably during the last few years, both because of the considerable data base now available from the Venera and Pioneer Venus missions and because of the broadly based theoretical studies stimulated by these data. The basic theory currently used in the general studies of planetary ionospheres is reviewed, including a discussion of the equations which describe the controlling physical and chemical processes of the ionosphere. The present status of the various models of the ion composition, density, and thermal structure that have been developed to reproduce the basic observed ionospheric behavior is discussed. Comparisons of calculated and measured dayside ion composition and density values have shown an overall agreement, but have also indicated the need for more sophisticated models which, for example, would include metastable chemistry. At this time no truly comprehensive and successful model of the nightside ionosphere has been published; the present status of nightside models is reviewed and directions for further improvements indicated. Dayside energy balance calculations lead to electron and ion temperature values in reasonably close agreement with measured values; however the energetics of the night side is poorly understood.

The flyby of the Mariner 5 spacecraft of Venus in 1967 (Kliore et al. 1967) provided great impetus to quantitative calculations of the basic characteristics of the ionosphere of Venus. These basically photochemical models published shortly after the encounter (e.g. McElroy 1969) were relatively successful in reproducing the measured electron density profiles, assuming that the major ion is CO_2^+. The observed presence of a significant nighttime ionosphere on

Venus, where the night is very long ($\sim$ 58 Earth days) was difficult to explain; McElroy (1968), McElroy and Strobel (1969), and Banks and Axford (1970) suggested that horizontal transport of light ions may be responsible for the maintenance of the nighttime ionosphere. More comprehensive dayside models were presented by Whitten (1969,1970) and Herman et al. (1971) who still had CO_2^+ as the major ion; but they also calculated the ion composition and structure of the upper ionosphere and found He^+ to be the dominant ion above $\sim$ 300 km. These authors solved the ion and electron energy equations using a variety of assumptions on the heat sources and boundary conditions, and predicted electron and ion temperatures significantly in excess of the assumed neutral gas temperature. The major advance in modeling the Venus ionosphere occurred in 1974, just before the Mariner 10 radio occultation data became available (Howard et al. 1974), with the publication of the work of Kumar and Hunten (1974). The two major changes in this model were the assumption of an exospheric temperature of only 350 K and an atomic oxygen mixing ratio of $\sim$ 1% at the homopause. The ion chemistry in the presence of even such a small amount of oxygen leads to molecular oxygen as the major ion near the ionization peak. This Kumar and Hunten (1974) model agreed reasonably well with the Mariner 5 radio occultation data; a very similar model by Nagy et al. (1975) reproduced most of the observational features of both Mariners 5 and 10. In the time period between Mariner 10 and Pioneer Venus further daytime electron density profiles were obtained from Veneras 9 and 10 (Aleksandrov et al. 1976a); a number of more comprehensive and detailed ionospheric models were published (Butler 1975; Chen 1977; Chen and Nagy 1978). Bauer and Hartle (1974) attributed the apparent ionospheric ledges and topside scale heights observed by Mariner 10 to ionospheric compression by the solar wind. Cloutier and his coworkers at Rice University published a series of articles dealing with the general problem of solar wind interaction with a nonmagnetic planet in which they calculated the current, potential and flow velocity characteristics inside the ionosphere (e.g. Cloutier and Daniell 1973; Daniell and Cloutier 1977). The presence of a significant nighttime ionosphere was reconfirmed by the Mariner 10 (Howard et al. 1974) and Venera 9 and 10 radio occultation (Aleksandrov et al. 1976b) observations, which led to a number of additional suggestions for the necessary nighttime source of ionization. Butler and Chamberlain (1976) suggested that the nighttime ionosphere consists of metallic ions, which are either transported from the day side or produced locally by meteor induced ionization. Electron impact ionization was proposed as a potential nighttime source by two groups independently (Gringauz et al. 1977; Chen 1977; Chen and Nagy 1978; Gringauz et al. 1979).

The instrument complement aboard the Pioneer Venus (PV) orbiter began, on 4 December 1978, to carry out the most comprehensive investigation ever attempted of a planetary upper atmosphere-ionosphere system other than Earth's (Colin and Hunten 1977). Cravens et al. (1978) presented a summary of the pre-Pioneer Venus understanding of the Venus ionosphere; the main purpose of this chapter is to summarize in some detail our present-day understanding of the physical and chemical processes controlling the Venus ionosphere. In Sec. I the basic theory underlying planetary ionospheres is discussed. The latest general models of the ion composition and density are discussed in Sec. II. Section III treats the energetics of the ionosphere, and in Sec. IV. a brief outline of some of the special phenomena observed in the ionosphere is given. A critical review of the shortcomings of our present-day understanding, and suggestions for future work conclude this chapter.

I. BASIC THEORY

A. Controlling Equations

The Pioneer Venus ion mass spectrometer (H. Taylor et al. 1979c) detected the presence of over a dozen ion species in the dayside ionosphere of Venus. Therefore, a detailed ionospheric model should include all these species, requiring the solution of a large complex set of nonlinear partial differential equations and the consideration of over a hundred chemical reactions. The relevant conservation equations that need to be solved to describe the temporal and spatial variation of the concentration, drift velocity, and temperature of the different ionospheric species have been extensively discussed and reviewed (Schunk 1975,1977; Schunk and Nagy 1980; Barakat and Schunk 1982). Only first order transport processes have been included in past theoretical studies of the Venus ionosphere; processes such as supersonic plasma flows, diffusion thermal heat flow, and anomalous resistivity, which have been found to be important in the terrestrial ionosphere have not yet been considered quantitatively and self-consistently. In this chapter only the first-order processes and resulting equations will be reviewed but it should be remembered that some of the neglected effects may be important, e.g., supersonic flows and shocks in the nighttime ionosphere (Knudsen et al. 1980b), and diffusion thermal heat flow (Schunk and St-Maurice 1981). The continuity, momentum, and energy equations (cf. Schunk and Nagy 1980) for a given ionospheric species s are

$$\partial n_s/\partial t + \nabla \cdot (n_s \mathbf{u}_s) = P_s' - L_s' n_s \tag{1}$$

$$n_s m_s \frac{D_s \mathbf{u}_s}{Dt} + \nabla p_s + \nabla \cdot \tau_s - n_s m_s \mathbf{G} - n_s e_s \left[\mathbf{E} + \frac{1}{c} \mathbf{u}_s \times \mathbf{B} \right] = \frac{\delta \mathbf{M}_s}{\delta t} \tag{2}$$

$$\frac{D_s}{Dt} (3/2 p_s) + 5/2 p_s (\nabla \cdot \mathbf{u}_s) + \nabla \cdot \mathbf{q}_s + \tau_s : \nabla \mathbf{u}_s = \frac{\delta E_s}{\delta t} + Q_s - L_s \tag{3}$$

where $D_s/Dt = \partial/\partial t + u_s \cdot \nabla$ is the convective derivative of species s, $p_s = n_s k T_s$ is the partial pressure, n_s is the number density, m_s is the mass, e_s is the charge, T_s is the temperature, u_s is the drift velocity, q_s is the heat flow vector, τ_s is the stress tensor, P_s' is the ionization production rate, L_s' is the ionization loss frequency, Q_s is the heating rate, L_s is the cooling rate, $\mathbf{G}$ is the acceleration due to gravity, $\mathbf{E}$ is the electric field, $\mathbf{B}$ is the magnetic field, $\partial/\partial t$ is the time derivative, ∇ is the coordinate-space gradient, c is the speed of light, and k is Boltzmann's constant. The double-dot operator in Eq. (3) corresponds to the scalar product of the two tensors (cf. Chapman and Cowling 1970). The quantities $\delta \mathbf{M}_s/\delta t$ and $\delta E_s/\delta t$ represent the rate of momentum and energy exchange, respectively, between species s and all other species in the plasma, due to elastic collisions.

These equations become a closed set only if appropriate expressions for the heat flow and stress tensor are used. A discussion and specific numerical values of the appropriate choice of these terms, along with those for the rate of momentum and energy exchange are given by Schunk and Nagy (1980) and will not be repeated here.

The most commonly used and greatly simplified one-dimensional form of Eqs. (1–3) are

$$\frac{\partial n_s}{\partial t} + \frac{\partial F_{sz}}{\partial z} = P_s' - L_s' \tag{4}$$

$$F_{sz} = n_s u_z = -D_s n_s \left[\frac{1}{n_s} \frac{\partial n_s}{\partial z} + \frac{m_s g}{kT_i} + \frac{T_e/T_i}{n_e} \frac{\partial n_e}{\partial z} \right.$$

$$\left. + \frac{1}{T_i} \frac{\partial}{\partial z} (T_e + T_i) + \frac{\alpha_s}{T_i} \frac{\partial T_i}{\partial z} \right] \tag{5}$$

$$\frac{3}{2} n_m k \frac{\partial T_m}{\partial z} - \frac{\partial}{\partial z} \left(K_m \frac{\partial T_m}{\partial z} \right) = Q_m - L_m \tag{6}$$

where F_{sz} is the vertical diffusive flux of sth ion, D_s is the diffusion coefficient of sth ion, g is the component of gravitational acceleration along z, T_i is the ion temperature, T_e is the electron temperature, α_s is the thermal diffusion coefficient of sth ion, and m is the index indicating either electrons or ions. In Eq. (6) the subscripts m denote either ion or electron parameters depending on whether the ion or the electron energy equations are being considered; K_m is the thermal conductivity.

In the case of the terrestrial ionosphere, which has a strong intrinsic magnetic field, these one-dimensional equations are usually solved along a field line because of the preferred direction of the various transport processes parallel to magnetic field lines. Venus does not have such a strong and steady magnetic field (see Chapter 25); therefore these equations are solved normally as a function of altitude, using various simplifying assumptions concerning horizontal transport and transport coefficients. Multidimensional models are generally necessary to calculate some of the ionospheric parameters in cases and regions where horizontal transport processes are important (e.g. terminator), but simpler approaches are also possible. A set of equations similar to Eqs. (4–6) but which include horizontal transport effects can be derived from Eqs. (1–3). The continuity equation which includes horizontal transport terms, and assumes the variables are a function only of altitude z and solar zenith angle χ ($=x/R$), can be written as

$$\frac{\partial n_s}{\partial t} + u_x \frac{\partial n_s}{\partial x} + n_s \frac{\partial u_x}{\partial x} + \frac{n_s u_x}{R} \, \text{ctn} \left(\frac{x}{R}\right) + \frac{\partial F_{sz}}{\partial z} = P_s' - L_s' \tag{7}$$

where $n_s(x,z)$ is the density of species s, x is the horizontal arc length from the subsolar point, R is the radial distance from the center of the planet and u_x is the horizontal velocity, which is generally a function of x and z. This equation can be solved rather easily for steady-state conditions if u_x and $\partial u_x/\partial x$ are known. In principle Eq. (2) can be used to calculate u_x, but quantitative information on many of the driving mechanisms responsible for horizontal ion motion is not generally available. Whitten et al. (1982) have, however, used a simplified form of this equation to derive horizontal velocities on the night side in the vicinity of the terminator.

An energy equation, analogous to Eq. (6) but which also includes the effects of horizontal transport can be derived from Eq. (3). Such an equation for the temperature $T_m(x,z)$ can be written, using the assumptions employed in obtaining Eq. (7), as

$$\frac{3}{2} n_m k \, \frac{\partial T_m}{\partial t} + \frac{3}{2} n_m k \, \left(u_x \frac{\partial T_m}{\partial x} + u_z \frac{\partial T_m}{\partial z}\right)$$

$$+ n_m k T_m \, \left(\frac{\partial u_x}{\partial x} + \frac{u_x}{R} \, \text{ctn} \left(\frac{x}{R}\right) + \frac{\partial u_z}{\partial z}\right) + \nabla \cdot \mathbf{q}_m = Q_m - L_m \tag{8}$$

where the heat flux, $\mathbf{q}_m$, is given by

$$\mathbf{q}_m = - K_m \nabla T_m \tag{9}$$

and u_z is the vertical velocity. This equation can also be solved rather easily for steady-state conditions if the horizontal contribution to the divergence of the heat flux can be neglected.

The ambipolar diffusion coefficient usually selected for use with Eq. (5) is the one derived by Conrad and Schunk (1979) for a fully ionized plasma which is given by

$$D_s = \frac{kT_i}{m_s \nu_{sj}} \left(\frac{1}{1 - \Delta_{sj}}\right) \tag{10}$$

where ν_{sj} is the momentum transfer collision frequency between ion species s and j, and $\triangle_{sj}$ is a corresponding diffusion correction factor. Expressions for the collision frequencies and correction factors may be found in Conrad and Schunk (1979). The thermal diffusion effect is highly simplified if it is expressed by the simple form given by the last term in Eq. (5). Bauer et al. (1979) presented values for the thermal diffusion coefficient α_s based on earlier work of Schunk and Walker (1969), which gives a reasonable approximation for typical ionospheric conditions on Venus. Note that the sign of the numerical values given by Bauer et al. must be reversed to be consistent with Eq. (5).

As mentioned earlier, magnetic fields as well as collisonal processes influence the motion of charged particles. Diffusive transport takes place preferentially along rather than perpendicular to strong magnetic fields. This means that in the terrestrial ionosphere, which is permeated with a strong and steady magnetic field, the transport equations are usually solved along magnetic field lines; however for Venus, which has a weak and fluctuating magnetic field the problem is much more complex.

Collisions tend to decouple charged particles from magnetic field lines and facilitate transport perpendicular to the field. The coefficients for transport perpendicular to the field, for a uniform and steady magnetic field can be obtained by multiplying the standard noninhibited coefficients by the factor F (Hochstim and Massel 1969; Johnson 1978),

$$F = \frac{\left(f\frac{\nu}{\Omega}\right)^2}{\left(f\frac{\nu}{\Omega}\right)^2 + 1} \tag{11}$$

where Ω is the gyrofrequency and f is a correction factor of order unity. This factor F, for the case of Venus where the magnetic field is weak and thus the gyrofrequency is small, is close to unity for ions. However, even for Venus, this factor is much less than unity for electrons and consequently electron transport on Venus is basically constrained along magnetic field lines.

The average magnetic field direction in the dayside ionosphere of Venus is horizontal (see Chapter 25). The presence of a strictly uniform and horizontal field means that, while vertical ion transport would not be seriously impeded, the vertical transport of electrons and electron heat would be severely restricted. However, the magnetic field in the dayside ionosphere is not at all uniform but highly structured; even though the average field tends to be horizontal, the field at any given time and location may be in any direction. A possible simplifying assumption, which was made by Cravens et al. (1980a), is to postulate a magnetic field which is completely random (or fluctuating) with a specified correlation length l. If the mean free path of the charged particle in this random field is λ then the thermal conductivity can be approximated by

$$K_m = \frac{3}{4} A \, n_m k \bar{v} \, \lambda \qquad (12)$$

where n_m is the particle density, $\bar{v}$ is the mean thermal velocity and A is a correction factor of order unity. For electrons the gyroradius (typically 0.2 km for thermal electrons) is smaller than the correlation length and consequently an electron will move along a field line. Since the field line performs a random walk with correlation length l, the electron mean free path will be equal to l. The ion gyroradius (typically 20 km for thermal ions) is usually larger than the correlation length (which is on the order of a few km) so the ion mean free path tends to be considerably larger than the correlation length and the ions do not really "feel" the magnetic field.

The actual magnetic field on the day side is really neither entirely uniform nor entirely random. Throughout most of the ionosphere are located complex magnetic structures called flux ropes (Russell and Elphic 1979) and between the flux ropes there exists a very weak and random background field. Perhaps a more accurate description of charged particle transport in such an ionosphere would involve diffusion or conduction towards a flux rope followed by rapid diffusion or conduction along the flux rope.

B. Ionization Processes and Photochemistry

Solar extreme ultraviolet (EUV) radiation is the main source of ionization in the dayside ionosphere. The calculation of the photoionization rate requires a knowledge of the number densities of the neutral constituents n_n as a function of altitude z, the absorption $\sigma_n^a(\lambda)$, and ionization $\sigma_n^i(\lambda)$ cross sections of these constituents as a function of wavelength λ, the branching ratios of the various excited ion states $p_n(\lambda, E_i)$, and the spectrum of solar radiation incident upon the top of the atmosphere $I_\infty(\lambda)$. In terms of these quantities, the primary photoelectron (ion) production rate P_e' as a function of energy and altitude is given by

$$P_e'(E,z) = \sum_i \sum_n n_n(z) \int_o^\infty d\lambda I_\infty(\lambda) \sigma_n^i(\lambda) p_n(\lambda, E_i) \exp\left[-\tau(\lambda, z)\right] \qquad (13)$$

where the optical depth τ is given by

$$\tau(\lambda, z) = \sum_n \sigma_n^a(\lambda) n_n H_n \, \text{ch}(R_n, \chi) \qquad (14)$$

and where

$$H_n = kT_n/m_n g \qquad (15)$$

$$R_n = (R + z)/H_n . \qquad (16)$$

In Eqs. (13–16), $E = E_\lambda - E_l$, E_λ is the energy corresponding to wavelength λ, E_l is the ionization energy of a given excited ion state l, R is the planetary radius, χ is the solar zenith angle, and $ch(R_n, \chi)$ is the Chapman grazing incidence function (Chapman 1931). Approximate expressions for this Chapman function valid for both large and small solar zenith angles have been presented by F. L. Smith and C. Smith (1972); for $\chi \leq 80°$ the Chapman function can be replaced by sec χ. As mentioned earlier, a knowledge of the neutral densities, the absorption and ionization cross sections, and the incident solar flux is required. The neutral composition and its dependence on solar zenith angle is described in Chapter 13. The absorption and ionization cross sections for the neutral constituents of importance in the Venus upper atmosphere are given in Table I. The solar extreme ultraviolet radiation spectrum has been measured at Earth by spectrophotometers carried aboard the Atmosphere Explorer (AE) Satellites C, D, and E (Hinteregger et al. 1973,1977). A reference spectrum, based on these measurements and appropriate for low solar cycle conditions, was compiled by Hinteregger for use by the AE investigator team. Some of the EUV spectra measured by Hinteregger during solar cycle maximum are available in the open literature (Torr et al. 1979; Hinteregger et al. 1982) and also daily values of 15 representative wavelength group covering the time period from mid-1977 to mid-1981 are available through the AE team. Of course, these solar flux values must undergo the appropriate r^{-2} scaling ($\sim 1 \cdot 93$) to make them appropriate for Venus; also, if time-dependent comparisons are attempted, the relative location of Venus and Earth with respect to solar longitude must be considered.

On the night side of the ionosphere, electron, and maybe even ion, impact ionization is believed to play an important role. Calculations of the impact ionization rates are significantly more complex than those for the photoionization case because particle transport as well as elastic and inelastic collision processes must be taken into account. The most commonly used form of electron (or ion) transport calculations is based on the so-called multistream approach. As an illustration, a simplified form of this formulation for the case of monoenergetic electrons in a single constituent atmosphere is given below:

$$\mu \frac{d\Phi(\mu,z)}{dz} = -n_n(z) \left[\sigma_e + \sigma_a \right] \Phi(\mu,z) + \frac{1}{2} n_n(z)\sigma_e$$

$$\cdot \int_{-1}^{+1} \rho(\mu',\mu)\Phi(\mu',z) \, d\mu' + P'(\mu,z) - L'(\mu,z) \qquad (17)$$

TABLE I

Weighted Photoionization and Photoabsorption Cross Sections[a]

$(\sigma_{eff}/10^{-18} cm^2)$

Interval	$\Delta\lambda$,Å	O+(4S)	O+(2D)	O+(2P)	Total O+	He+	N2(abs)[b]	N2(ion)[b]	CO(abs)[b]	CO(ion)[b]	CO2(abs)[b]	CO2(ion)[b]
								Species				
1	50–100	0.32	0.34	0.22	1.06	0.21	0.60	0.60	1.92	1.92	4.42	4.42
2	100–150	1.03	1.14	0.75	3.53	0.53	2.32	2.32	3.53	3.53	7.51	7.51
3	150–200	1.62	2.00	1.30	5.96	1.02	5.40	5.40	5.48	5.48	11.03	11.03
4	200–250	1.95	2.62	1.70	7.55	1.71	8.15	8.15	8.02	8.02	14.98	14.98
5	256.3	2.15	3.02	1.95	8.43	2.16	9.65	9.65	10.02	10.02	17.88	17.88
6	284.15	2.33	3.39	2.17	9.26	2.67	10.60	10.60	11.70	11.70	21.21	21.21
7	250–300	2.23	3.18	2.04	8.78	2.38	10.08	10.08	11.01	11.01	20.00	20.00
8	303.31	2.45	3.62	2.32	9.70	3.05	11.58	11.58	12.52	12.52	23.44	23.44
9	303.78	2.45	3.63	2.32	9.72	3.05	11.60	11.60	12.47	12.47	23.44	23.44
10	300–350	2.61	3.98	2.52	10.03	3.65	14.60	14.60	13.61	13.61	23.88	23.88
11	368.07	2.81	4.37	2.74	10.84	4.35	18.00	18.00	15.43	15.43	25.70	25.70
12	350–400	2.77	4.31	2.70	10.70	4.25	17.51	17.51	15.69	15.69	25.81	25.81
13	400–450	2.93	4.75	2.93	11.21	5.51	21.07	21.07	18.01	18.01	27.52	27.52
14	465.22	3.15	5.04	3.06	11.25	6.53	21.80	21.80	19.92	19.92	28.48	28.48
15	450–500	3.28	5.23	3.13	11.64	7.09	21.85	21.85	20.09	20.09	29.27	29.27
16	500–550	3.39	5.36	3.15	11.91	0.72	24.53	24.53	21.61	21.44	31.61	31.61
17	554.37	3.50	5.47	3.16	12.13	0.00	24.69	24.69	22.28	22.31	33.20	33.20
18	584.33	3.58	5.49	3.10	12.17	0.00	23.20	23.20	22.52	21.38	34.21	34.21
19	550–600	3.46	5.30	3.02	11.90	0.00	22.38	22.38	22.41	21.62	34.00	34.00

TABLE I (Continued)

Weighted Photoionization and Photoabsorption Cross Sections[a]

$$(\sigma_{eff}/10^{-18}\,cm^2)$$

Interval	$\Delta\lambda$,Å						Species					
		$O^+(4S)$	$O^+(2D)$	$O^+(2P)$	Total O^+	He^+	$N_2(abs)$[b]	$N_2(ion)$[b]	$CO(abs)$[b]	$CO(ion)$[b]	$CO_2(abs)$[b]	$CO_2(ion)$[b]
20	609.76	3.67	5.51	3.05	12.23	0.00	23.10	23.10	18.42	16.93	25.31	20.16
21	629.73	3.74	5.50	2.98	12.22	0.00	23.20	23.20	18.60	16.75	25.86	21.27
22	600–650	3.73	5.50	2.97	12.21	0.00	23.22	23.22	19.78	17.01	25.88	21.14
23	650–700	4.04	5.52	0.47	10.04	0.00	29.75	25.06	25.59	17.04	25.96	21.72
24	703.36	4.91	6.44	0.00	11.35	0.00	26.30	23.00	24.45	16.70	21.76	17.71
25	700–750	4.20	3.80	0.00	8.00	0.00	30.94	23.20	25.98	17.02	22.48	17.02
26	765.15	4.18	0.00	0.00	4.18	0.00	35.46	23.77	26.28	12.17	52.96	50.39
27	770.41	4.18	0.00	0.00	4.18	0.00	26.88	18.39	15.26	9.20	26.48	20.00
28	789.36	4.28	0.00	0.00	4.28	0.00	19.26	10.18	33.22	15.44	21.79	17.07
29	750–800	4.23	0.00	0.00	4.23	0.00	30.71	16.75	21.35	11.38	31.83	21.53
30	800–850	4.38	0.00	0.00	4.38	0.00	15.05	0.00	22.59	17.13	12.84	10.67
31	850–900	4.18	0.00	0.00	4.18	0.00	46.63	0.00	37.64	11.70	45.06	19.66
32	900–950	2.12	0.00	0.00	2.12	0.00	16.99	0.00	49.44	0.00	70.89	0.00
33	977.62	0.00	0.00	0.00	0.00	0.00	0.70	0.00	28.50	0.00	23.91	0.00
34	950–1000	0.00	0.00	0.00	0.00	0.00	36.16	0.00	52.90	0.00	34.41	0.00
35	1025.72	0.00	0.00	0.00	0.00	0.00	0.00	0.00				
36	1031.91	0.00	0.00	0.00	0.00	0.00	0.00	0.00				
37	1000–1050	0.00	0.00	0.00	0.00	0.00	0.00	0.00				

[a]Data are taken from M. R. Torr et al. (1979).

[b]Photoabsorption is denoted by (abs), photoionization by (ion).

where $\Phi(\mu,z)$ is the electron flux with $\mu = \cos \alpha$, α is the angle of electron with respect to reference direction, $n_n(z)$ is the number density of atmospheric species, σ_e is the total elastic scattering cross section at energy E and σ_a the total inelastic scattering cross section at energy E, $1/2\rho(\mu',\mu)$ is the probability that an elastic scattering of an electron at μ' will result in an electron at μ, $P'(\mu,z)$ is the differential production rate of electrons in $cm^{-3}s^{-1}sr^{-1}$, and $L'(\mu,z)$ is the differential loss rate of electrons in $cm^{-3}s^{-1}sr^{-1}$. The first term on the right-hand side of Eq. (11) represents scattering out of μ, while the second term corresponds to scattering from all other angles μ' into μ. A large number of papers (e.g., Nisbet 1968; Nagy and Banks 1970; Stolarski 1972; Cicerone et al. 1973; Oran and Strickland 1978; Mantas and Hanson 1979) have dealt with such transport calculations of auroral electrons and ionospheric photoelectrons, and discussed the equivalence and/or relationship among the various techniques; a comprehensive review paper on this subject is in progress (Oran 1983).

There are myriads of chemical reactions that are potentially important in the Venus ionosphere. A summary of the most important ion-neutral reaction and dissociative recombination rates is given in Tables II and III, respectively, along with the best available values of these rate coefficients. Figure 1 is a schematic diagram, that attempts to illustrate the ion chemistry dominating the O_2^+, O^+, CO_2^+, CO^+, C^+, N_2^+, NO^+, N^+ and He^+ abundances in the chemically controlled region of the ionosphere. For example, one can see from this figure that the major primary ions, CO_2^+ and O^+, are very rapidly transformed to O_2^+ which is the dominant ion in this region, via reactions N2 and N23. Reactions involving metastable species are not listed in Tables II and III and have not yet been included in any published models; Fox (1982) has shown that certain metastable reactions play an important role in the chemistry of some of the observed species, and that their inclusion improves the agreement between the observed and calculated values. A more detailed discussion of the general ion chemistry is left for Sec. II.

C. Heating and Cooling Processes

The main source of energy for the mid- and low-latitude terrestrial ionosphere is the EUV radiation from the Sun; however, for Venus, processes associated with the solar wind interaction also appear to be important. The absorption of EUV radiation by the neutral atmosphere results in both photoionization and excitation of the neutral gases. The resulting excited atoms and molecules lose their energy in quenching collisions with electrons and other neutral particles and by radiation. Photoionization produces energetic photoelectrons since the energy carried by the ionizing photons exceeds, in general, the energy required for ionization. The initial photoelectron energy depends not only on the energy of the ionizing photon and the identity of the neutral species but also on the ionization state of the photoion. Typically,

TABLE II
Ion-Neutral Reaction Rates[a]

Reaction Number	Reaction	Rate Constant ($cm^3 s^{-1}$)
N1	$CO_2^+ + O \rightarrow O^+ + CO_2$	9.6×10^{-11}
N2	$CO_2^+ + O \rightarrow O_2^+ + CO$	1.64×10^{-10}
N3	$CO_2^+ + NO \rightarrow NO^+ + CO_2$	1.2×10^{-10}
N4	$CO_2^+ + H_2 \rightarrow CHO_2^+ + H$	1.4×10^{-9}
N5	$CO_2^+ + H \rightarrow CHO^+ + O$	5.0×10^{-10}
N6	$CO_2^+ + H \rightarrow H^+ + CO_2$	1.0×10^{-10}
N7	$CO_2^+ + N \rightarrow products$	$\leqslant 1.0 \times 10^{-11}$
N8	$CO^+ + O \rightarrow O^+ + CO$	1.4×10^{-10}
N9	$CO^+ + CO_2 \rightarrow CO_2^+ + CO$	1.0×10^{-9}
N10	$CO^+ + NO \rightarrow NO^+ + CO$	3.3×10^{-10}
N11	$CO^+ + H_2 \rightarrow COH^+ + H$	1.8×10^{-9}
N12	$CO^+ + N \rightarrow NO^+ + C$	$\leqslant 2.0 \times 10^{-11}$
N13	$O_2^+ + NO \rightarrow NO^+ + O_2$	4.5×10^{-10}
N14	$O_2^+ + N \rightarrow NO^+ + O$	1.2×10^{-10}
N15	$N_2^+ + CO \rightarrow CO^+ + N_2$	7.4×10^{-11}
N16	$N_2^+ + CO_2 \rightarrow CO_2^+ + N_2$	7.7×10^{-10}
N17	$N_2^+ + NO \rightarrow NO^+ + N_2$	3.3×10^{-10}
N18	$N_2^+ + O \rightarrow NO^+ + N$	1.4×10^{-10} [b]
N19	$N_2^+ + O \rightarrow O^+ + N_2$	1×10^{-11} [c]
N20	$C^+ + CO_2 \rightarrow CO^+ + CO$	1.1×10^{-9}
N21	$O^+ + N_2 \rightarrow NO^+ + O$	1.2×10^{-12}
N22	$O^+ + NO \rightarrow NO^+ + O$	6.4×10^{-13}
N23	$O^+ + CO_2 \rightarrow O_2^+ + CO$	9.4×10^{-10}
N24	$O^+ + H_2 \rightarrow HO^+ + H$	1.7×10^{-9}
N25	$O^+ + H \rightarrow H^+ + O$	$2.5 \times 10^{-11} T_n^{1/2}$
N26	$N^+ + CO \rightarrow CO^+ + N$	4.0×10^{-10}
N27	$N^+ + CO \rightarrow NO^+ + C$	5.0×10^{-11}
N28	$N^+ + NO \rightarrow NO^+ + N$	9.0×10^{-10}
N29	$N^+ + CO_2 \rightarrow CO^+ + NO$	2.5×10^{-10}
N30	$N^+ + CO_2 \rightarrow CO_2^+ + N$	7.5×10^{-10}
N31	$He^+ + CO_2 \rightarrow CO^+ + O + He$	8.7×10^{-10}
N32	$He^+ + CO_2 \rightarrow CO_2^+ + He$	1.2×10^{-10}
N33	$He^+ + CO_2 \rightarrow O^+ + CO + He$	1.0×10^{-10}
N34	$He^+ + CO \rightarrow products$	1.68×10^{-9}
N35	$He^+ + CO \rightarrow C^+ + He + O$	1.4×10^{-9}
N36	$He^+ + NO \rightarrow N^+ + He + O$	1.25×10^{-9}
N37	$He^+ + N_2 \rightarrow N^+ + He + N$	9.6×10^{-10}
N38	$He^+ + N_2 \rightarrow N_2^+ + He$	6.4×10^{-10}
N39	$H^+ + CO_2 \rightarrow CHO^+ + O$	3.0×10^{-9}
N40	$H^+ + O \rightarrow O^+ + H$	$2.2 \times 10^{-11} T_i^{1/2}$
N41	$H^+ + NO \rightarrow NO^+ + H$	1.9×10^{-9}

[a]Data are taken from Albritton (1978), D. G. Torr and M. R. Torr (1978), and Schunk and Raitt (1980).
[b]$300/T^{0.44}$ for $T \leqslant 1500$ K.
[c]$300/T^{0.23}$ for $T \leqslant 1500$ K.

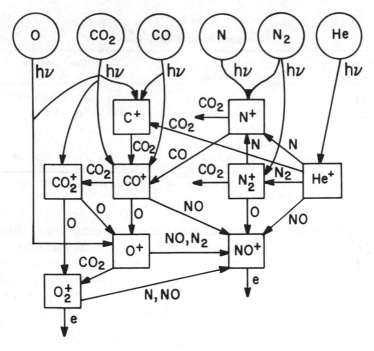

Fig. 1. Ion chemistry scheme.

photoionization produces photoelectrons with initial energies of some tens of electron volts.

Only a relatively modest amount of the initial photoelectron energy is deposited directly in the ambient electron gas. Most of the excess kinetic energy is lost in both elastic and inelastic collisions with the neutral particles and in Coulomb collisions with the ambient electrons. If the photoelectrons lose their energy at an altitude near where they are produced, the heating is said to be *local*, while if the photoelectrons lose their energy over a distance greater than a neutral scale height, the heating is termed *nonlocal*. Nonlocal heating effects occur mainly at high altitudes, where ambient densities are low, and at high photoelectron energies.

The heat gained by the ambient electron gas from the photoelectrons and from superelastic collisions with the neutrals acts to raise the electron temperature above the ion and neutral temperatures. The hot ambient electrons then lose energy in Coulomb collisions with the ambient ions and in elastic and inelastic collisions with the neutral atmosphere. The energy gained by the ion gas is generally sufficient to raise the ion temperature above the neutral gas temperature when $n_i \gtrsim 10^{-3} n_n$. The extent to which the electron and ion temperatures are elevated depends on the relative importance of the various heating, cooling, and energy transport processes.

TABLE III

Dissociative Recombination Rates[a]

Reaction Number	Reaction	Rate Constant cm^3s^{-1}
R1	$O_2^+ + e \rightarrow O + O$	$1.6 \times 10^{-7} (300/T_e)^{0.55}$
R2	$CO_2^+ + e \rightarrow CO + O$	$3.8 \times 10^{-7}(300/T_e)$
R3	$CO^+ + e \rightarrow C + O$	$\sim 2.0 \times 10^{-7}$
R4	$NO^+ + e \rightarrow N + O$	$4.2 \times 10^{-7}(300/T_e)^{0.85}$
R5	$N_2^+ + e \rightarrow N + N$	$1.8 \times 10^{-7}(300/T_e)^{0.39}$

[a]Data are taken from D. G. Torr and M. R. Torr (1978), Weller and Biondi (1967), and M. A. Biondi (personal communication, 1979).

Comparisons between observations and model calculations (e.g. Cravens et al. 1979, 1980a) have indicated that heating processes other than solar EUV play an important role in the energetics of the ion and electron gases for both the day and night ionospheres of Venus. The main proposed heating mechanisms for the ions are chemical heating and frictional (Joule) heating (Cravens et al. 1979; Knudsen et al. 1979b) and the transformation of kinetic energy into thermal energy (Knudsen et al. 1980b). Landau damping of low-frequency whistler mode waves in the ionopause region is a likely source of heat for the electrons in the dayside ionosphere (Scarf et al. 1979). The pertinent energy exchange rates among the electron, ion, and neutral species have been discussed in detail and tabulated in two reviews by Schunk and Nagy (1978, 1980) and will not be repeated here.

II. ION DENSITY AND COMPOSITION MODELS

The first step in any calculation of ionospheric densities is the determination of the ion production and loss rates. Photoionization is the major source of ionization in the dayside ionosphere; representative values of calculated photoionization rates for a solar zenith angle (SZA) of 60° are shown in Fig. 2. The primary ion production indicated in the figure refers to direct photoionization, while the secondary ion production denotes secondary and tertiary production rates due to photoelectrons. The photoelectron fluxes were calculated using the two-stream method (Nagy and Banks 1970). The ionization rates were calculated (Nagy et al. 1980; Cravens et al. 1980a) by using solar EUV intensities appropriate for 1979 conditions, neutral density values based on the Pioneer Venus Orbiter (PVO) neutral mass spectrometer results (Niemann et al. 1979b) and the cross-section values shown in Table I.

It has been established by a number of authors (e.g., Chen and Nagy 1978; Cravens et al. 1981a) that in the dayside ionosphere, photochemical equilibrium conditions hold in the altitude region below $\sim$ 170 km. The most

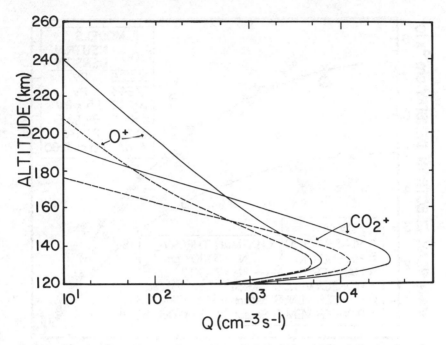

Fig. 2. O^+ and CO_2^+ primary photoionization and photoelectron impact ionization rates for SZA $\chi = 60°$. Solid line indicates primary ion production, dashed line secondary ion production. (From Nagy et al. 1980.)

recent and comprehensive calculations of the total electron and ion density values in this region are those of Cravens et al. (1981a). These authors were interested in a detailed comparison of the absolute value and SZA dependence of the measured and calculated electron densities at the dayside electron density peak region. Their model calculations, based on solutions of the appropriate photochemical equations, included all the reactions listed in Tables II and III, which involve O_2^+, O^+, CO_2^+, N_2^+, NO^+, or CO^+. However as mentioned earlier, in order to understand the electron density behavior in the vicinity of the ionospheric peak, only the major ion O_2^+ is important. CO_2^+ is the main ion produced near the peak (see Fig. 1), but reactions N2 and N1, N23 quickly convert it to O_2^+, which in turn almost always recombines dissociatively (R1). A reasonably accurate expression for the electron density in terms of the total ionization rate P_e', and the dissociative recombination rate for O_2^+ (Cravens et al. 1981a) is

$$n_e \approx 520 \sqrt{P_e'} T_e^{0.275} . \tag{18}$$

Figure 3, reproduced from the work of Cravens et al. (1981a), shows the comparison between measured and calculated peak electron densities as a

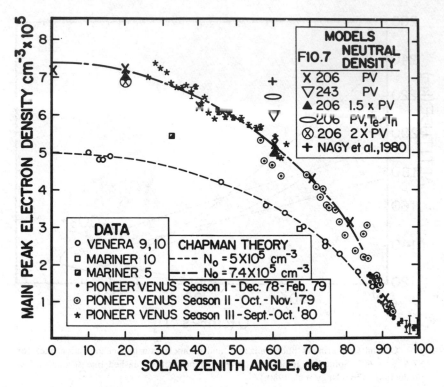

Fig. 3. Electron number densities at the ionospheric peak versus SZA. $T_e = T_n$ for all the models except for the model labeled $T_e > T_n$ and the Nagy et al. (1980) model. PV indicates that the Hedin neutral densities (multiplied by the factor shown) were used. The F10.7 for each model is indicated. (From Cravens et al. 1981a.)

function of zenith angle. Excellent agreement was found between the calculated and measured peak densities, as well as the altitude of the peak, if the neutral gas densities were increased by a factor of 1.5 from the model values, which were based on the PVO neutral mass spectrometer results (A. Hedin, personal communication). Since the publication of the work of Cravens et al. (1981a) these neutral model density values have been increased, for various reasons, by ~ 75% (Hedin et al. 1983); therefore, the behavior of the daytime electron and total ion densities in the chemically controlled region is well understood.

In order to model the structure and composition of the dayside ionosphere above ~ 170 km, where transport processes become important, the full set of continuity and momentum equations must be solved. The most recently published set of such model calculations are those of Nagy et al. (1979, 1980) who solved the coupled set of Eqs. (4) and (5) for O_2^+, O^+, CO_2^+, C^+, N^+, He^+, and H^+. In these equations the terms for all horizontal transport and vertical bulk flows are omitted. Therefore the results obtained by this model

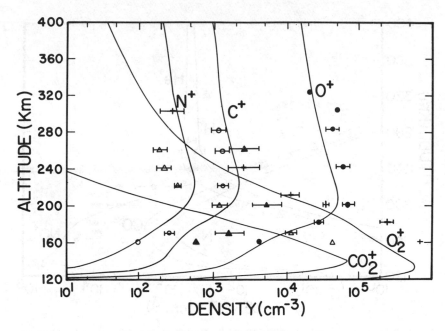

Fig. 4. Calculated ion density profiles of CO_2^+ (Δ), O^+ ($\bullet$), O_2^+ (+), C^+ ($\blacktriangle$), and N^+ (o) for SZA $\chi = 60°$. The measured ion density values are also shown for comparison; the horizontal bars are an indication of the data spread between the two selected orbits. (From Nagy et al. 1980.)

are not applicable near the terminator or the ionopause where the neglected terms are expected to be important. Chemical equilibrium solutions, describing conditions below ~ 200 km were also obtained by Nagy et al. (1980) for N_2^+, NO^+, and CO^+. The results of these calculations for a zenith angle of $60°$ are shown in Figs. 4 and 5, along with ion densities measured by the PVO ion mass spectrometer for corresponding conditions. The source and loss processes controlling O_2^+ behavior have been discussed earlier and will not be repeated. The second major ion O^+ is produced directly by photoionization, electron impact, or by the ion-atom interchange reaction N1. Numerous processes, including N22, N23, and N25 destroy O^+. C^+ is produced by dissociative photoionization and reaction N35 and is lost via reaction N20. The amount of neutral atomic carbon in the upper atmosphere is unknown at present and so the relative importance of photoionization is uncertain. Doubly ionized atomic oxygen can create both C^+ and N^+, through reactions with CO_2 and N_2 respectively (Fox and Victor 1981); however this source appears to be significant only above ~ 250 km even if rate coefficients as large as $10^{-9} cm^{-3} s^{-1}$ are assumed. N^+ can be produced by dissociative photoionization and the He^+ reactions N36 and N37; the contribution of direct photoionization to the N^+ production rate is negligible below ~ 200 km but becomes significant at higher altitudes.

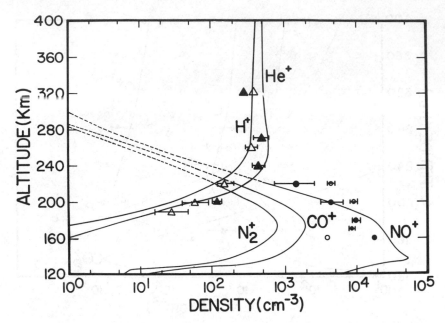

Fig. 5. Calculated ion density profiles of He$^+$ (▲), H$^+$ (Δ), N$_2^+$ (o), CO$^+$ (o), and NO$^+$ (●) for SZA $\chi = 60°$. The N$_2^+$, CO$^+$, and NO$^+$ values were calculated assuming photochemical equilibrium and are shown with dashed lines above 200 km, where this assumption breaks down. The measured ion density values are also shown for comparison; the horizontal bars are an indication of the data spread between the two selected orbits. (From Nagy et al. 1980.)

The major N$^+$ loss processes are reactions N29 and N30. NO$^+$ is produced by a variety of reactions but the most important source is the reaction of O$_2^+$ with atomic nitrogen (N14). The major source of He$^+$ is direct photo-ionization, while H$^+$, similar to the terrestrial case, comes from charge exchange with O$^+$.

The calculations of Nagy et al. (1980), presented in Figs. 4 and 5, were carried out with an early version of the PVO neutral density model, based on the mass spectrometer measurements (A. Hedin, personal communication); as mentioned earlier, the most recent version of that model has density values greater by 75%. These uncertainties associated with the model neutral atmosphere and reaction rates used in the calculations, and the uncertainties attached to the measured ion densities, mean that only disagreements by factors significantly greater than 2 between the calculated and measured ion densities cannot be ignored at this time. Among the major ions the best agreement by far is for O$^+$. The worst agreement is for the measured mass 28, which is the sum of both N$_2^+$ and CO$^+$. Since the calculated CO$^+$ densities are considerably larger than the N$_2^+$ ones, only CO$^+$ can be meaningfully compared with the measurements. The calculated C$^+$ densities are on the low side and, if they were higher, the CO$^+$ agreement would improve because CO$^+$ is produced from C^+ via reaction N20. The agreements for O_2^+, CO_2^+, NO^+, He^+, H^+, C^+, and N^+ are reasonable considering the many uncertainties. The discrepancies and the factors that need to be considered for a better agreement are discussed in more detail by Nagy et al. (1980) and will not be repeated here; however, it is appropriate to repeat some of their general observations. During the last few years it has been established that excited neutral and ion species play an important role in the chemistry of the terrestrial ionosphere (cf. D. G. Torr and M. R. Torr 1982); it is possible that many of the present difficulties in modeling the Venus ionosphere arise because only cursory consideration has been given to the excited state chemistry. The neglect of many of the potentially important transport processes (e.g. Schunk and St. Maurice 1977) may also contribute to difficulties in a quantitative treatment of the topside ionosphere. As summarized by Nagy et al. (1980) "model calculations indicate that while a basic understanding of the chemical and physical processes controlling the composition and vertical distribution of the dayside Venus ionosphere, well below the ionopause, has been achieved, there are many important details requiring further investigations."

The mechanisms responsible for the maintenance of the nighttime ionosphere have been, as mentioned earlier, the subject of much debate ever since the first Mariner 5 measurements. The two favored source mechanisms, electron impact ionization and horizontal transport, were first evaluated quantitatively using new PVO results by Brace et al. (1979c) and Kliore et al. (1979b). These authors showed that either a precipitating low-energy (~30 eV) electron flux of ~2–4×10^8 cm^{-2}s^{-1} or a downward thermal O^+ flux of $\sim 6 \times 10^7$cm^{-2}s^{-1} can account for the measured peak electron densities ($\sim1.6 \times 10^4$cm^{-3}) at the observed altitudes (~142 km). Initial results from the PVO plasma analyzer (OPA) (Intriligator et al. 1979) and the retarding potential analyzer (RPA) (Knudsen et al. 1980b) indicated the presence of nighttime precipitating electron fluxes and thermal ion fluxes across the terminator of about the required order of magnitude, thus providing support for both of the candidate maintenance processes. Spenner et al. (1981) have carried out detailed model calculations of the nightside ionosphere using precipitating electron fluxes and downward O^+ fluxes consistent with values measured by the PVO retarding potential analyzer. These calculations show again that a downward flux of O^+ ions of between 1 and 2×10^8cm^{-2}s^{-1} results in O^+ and O_2^+ ion density profiles comparable to the measured median values (see Fig. 6); furthermore, the PVO retarding potential analyzer results also show that the divergence of the horizontal O^+ fluxes on the night side is consistent with the required value (W. C. Knudsen personal communication). Spenner et al. (1981) have also calculated the ionosphere resulting from typical precipitating suprathermal electron fluxes, measured by the RPA, and find that the calculated O_2^+ density profile has a peak density of about one half of that typically observed and that the calculated O^+ densities are about an order of magnitude too small. They conclude that the "transport

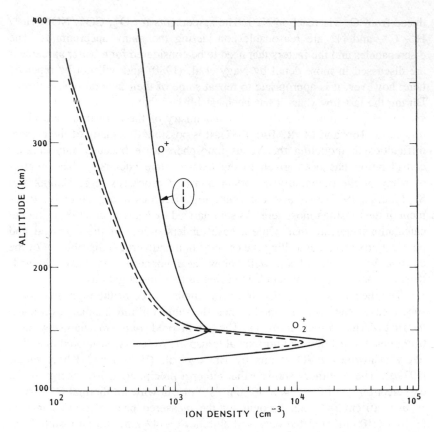

Fig. 6. Comparison of calculated O^+ and O_2^+ ion profiles for a 1×10^8 cm^{-2}s^{-1} downward flux of O^+ ions at 90° magnetic dip angle with (—) and without (--) the ionization produced by the S1 suprathermal electron spectrum. The O^+ profile for the two cases are indistinguishable. The suprathermal electron spectrum contributes primarily to the O_2^+ profile. Model O_2^+ density above $\sim$ 150 km altitude is overestimated because of model limitations. (From Stenner et al. 1981.)

of O^+ ions from the dayside ionosphere is responsible for most of the ionization required to maintain the nightside ionosphere."

Implicit in models seeking to explain the maintenance of the nightside ionosphere by a downward flux of O^+ is the assumption that there is a high-altitude horizontal flow of O^+ across the terminator sufficient to generate the desired downward diffusion from high to low altitudes deep in the night side. A more self-consistent approach is to solve the two-dimensional Eq. (7), together with Eq. (5), for both the downward flux and the density rather than just assume a downward flux. Cravens et al. (1981b) followed this approach, assuming horizontal flow velocities on the order of 0 to 9 km s^{-1}, consistent with the measurements of Knudsen et al. (1980b) and H. Taylor et al. (1980).

Their results confirm that horizontal transport can, for the most part, maintain the observed nightside ionosphere (see Fig. 7), at least out to a zenith angle of $\sim 140°$. The calculations shown in Fig. 7 were for an ionopause height of 1870 km. Horizontal transport can maintain an ionosphere quite deep on the night side if reasonable horizontal velocities and ionopause heights are used. Below 200 km, electron precipitation will certainly play a role, as discussed above. Whitten et al. (1982) have also shown that horizontal transport is a viable source for the nightside ionosphere, in the vicinity of the terminator.

III. THE ENERGETICS OF THE IONOSPHERE

A comparison between model calculations and direct measurements of the daytime ion temperatures in the ionosphere on Mars indicated that solar EUV radiation is not the only major ionospheric heat source (Chen et al. 1978). This led Cravens et al. (1978) to postulate that in the Venus ionosphere, similar to Mars, heating mechanisms other than those due to solar EUV radiation must be considered in any study of the ionospheric energy balance. The choice of appropriate transport coefficients is also more complex than for the terrestrial ionosphere because of the very weak and variable magnetic field (see Sec. II). Temperatures presently recorded by the PVO instrument complement (see Chapter 23) confirm these prior suggestions concerning the presence of nonsolar EUV energy sources, and the need to consider some mechanisms capable of modifying classic transport processes.

To calculate the amount of energy received by the thermal electron gas from solar EUV radiation via the photoelectrons, it is necessary to calculate the equilibrium photoelectron flux. Butler and Stolarski (1978) established that even very small quasi-horizontal magnetic fields will inhibit the vertical transport of photoelectrons and force them to deposit their energy locally to fairly high altitudes. Cravens et al. (1980) also considered the effect of fluctuating magnetic fields on photoelectron transport. Photoelectrons will scatter and at least partially decouple from the magnetic field lines if the amplitude of the fluctuations is large enough. If the gyroradius of a typical photoelectron (~ 3 km for a 10 γ field and 15 km for a 2 γ field) is equal to or smaller than the correlation length of the magnetic fluctuations (~ 10 km), then photoelectrons will travel along magnetic field lines and, because we are assuming that the field lines are randomly oriented, the electrons traveling along them will perform a random walk. Typical characteristic distances for photoelectron transport in the ionosphere are on the order of a few hundred km (approximately the vertical extent of the ionosphere) while the mean free path due to the fluctuating magnetic field is on the order of the correlation length (~ 10 km). This results in a sufficiently large number of scattering events to make this approach justifiable. To simulate crudely the effects of the increased total

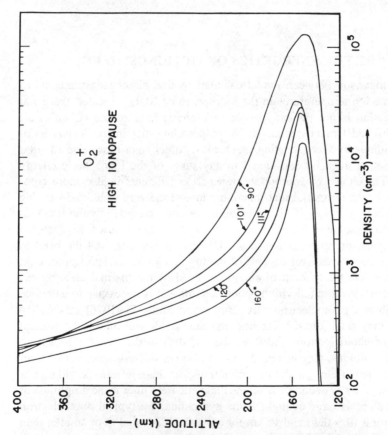

Fig. 7. O_2^+ density profiles are shown for a number of solar zenith angles. The horizontal velocity used increased linearly from 0.05 km s^{-1} at 120 km to 9 km s^{-1} at the assumed ionopause height of 870 km.

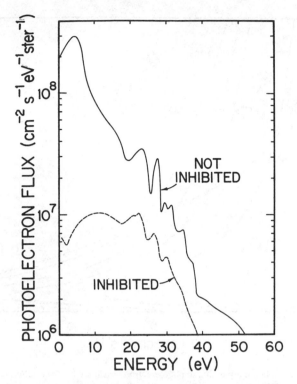

Fig. 8. The photoelectron flux at one altitude (304 km) is shown as a function of energy. SZA $\chi = 0°$. The flux per steradian is found by assuming that the calculated fluxes were isotropic over the upward (or downward) hemisphere. Both the inhibited and uninhibited transport cases are shown. (From Cravens et al. 1980*a*).

photoelectron path length due to such scattering events, the two-stream transport method with a dip angle of 2° was employed. The photoelectron flux calculated for this inhibited vertical transport case is shown in Fig. 8. If the magnitude of magnetic fluctuations is quite large and if the photoelectron gyroradius is considerably larger than the correlation length of the fluctuations (or if the field is very weak), the photoelectrons will not "feel" the magnetic field and their vertical transport will not be inhibited. In this case the two-stream method with a dip angle of 90° can be used; the results of such calculations are also shown in Fig. 8. The limited published data on photoelectron fluxes supports the case of "uninhibited" transport (Knudsen et al. 1980*a*). The calculated heating rate of thermal electrons due to photoelectrons is shown in Fig. 9 for 4 zenith angles and for no inhibition of vertical transport. The heating rate for the subsolar inhibited case is also shown for comparison. The major difference between the heating rates calculated using the different assumptions on photoelectron transport is at high altitudes; while the differences in the heating rates can be as large as an order of magnitude above

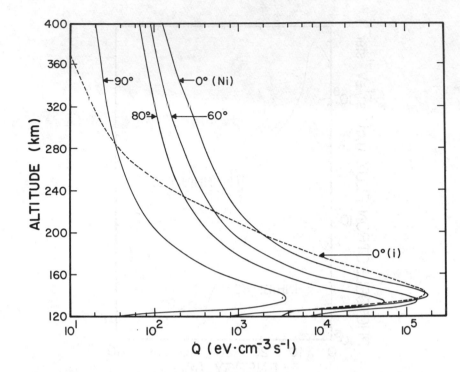

Fig. 9. The electron heating rate due to photoelectrons is shown for four values of the SZA for
the case of uninhibited transport. The heating rate appropriate to inhibited transport is only
shown from the subsolar point, labeled by the letter (i). (From Cravens et al. 1980a.)

300 km, the calculated temperatures are not, in general, significantly affected
because energy transport processes rather than local energy deposition dominate the heat balance at these altitudes.

The major sources of heating for the ion gas are believed to be (1) energy
transfer from the electron gas via Coulomb collisions, (2) exothermic ion
neutral chemical reactions, and (3) frictional (Joule) heating. The energy
transfer rates via Coulomb collisions are well known but large uncertainties
are associated with those due to chemical reactions and frictional heating. In
the case of the chemical reactions, the difficulties are mainly due to the
uncertainty of the fraction of the exothermic energy going into kinetic energy
(Rohrbaugh et al. 1979; Cravens et al. 1979). Calculations of the frictional
heating rates are uncertain due to our lack of knowledge of the neutral wind
velocities and the limited data base on the ion velocities.

The various uncertainties associated with the heating rates, and in the
proper choice of transport coefficients and topside heat input due to solar-wind
interaction processes, leave a modeler with a practically unlimited choice
of free parameters. The available data base on the electron and ion temperatures provides some constraints, but still does not lead to a unique answer.

A number of papers dealing with dayside ionospheric energy balance (Cravens et al. 1979; Knudsen et al. 1979a) were published shortly after experimental data on the electron and ion temperatures became available from the RPA (Knudsen et al. 1979b) and Langmuir probe (Brace et al. 1979c) carried by PVO. These model calculations, which were based on the solution of Eq. (6) for the electron and ion gases, obtained a reasonable fit to the observed daytime temperatures by (1) assuming a topside and/or distributed heat input not associated with solar EUV (Knudsen et al. 1979a), (2) modified transport coefficients (Cravens et al. 1979), and (3) both of the above (Cravens et al. 1979, 1980); an example of such a comparison from Knudsen et al. (1979a) is shown in Fig. 10. The most recently published comprehensive model calculations of the ionospheric energetics is that of Cravens et al. (1980), which is reviewed in some detail below.

The calculations of Cravens et al. (1979) demonstrated the fact that the presence of a weak, steady, and near-horizontal magnetic field reduces the necessary heat inputs; their work indicated that in order to match the measured temperatures using a reasonable magnetic field, topside heat inputs into the electron and ion gas on the order of a few times 10^9 and 10^7 eV cm^{-2}s^{-1} respectively, are needed. The electric field experiment aboard PVO (Scarf et al. 1979) has detected whistler wave signals in the dayside magnetosheath region which are strongly damped at the ionopause. The fraction of the wave energy that is reflected is not known; therefore only an upper limit on the energy deposited into the electron gas, via Landau damping, is available. The value of this upper limit is 3×10^{10} eV cm^{-2}s^{-1}, well above the energy necessary to explain the dayside electron temperature observations. The required topside energy input for the ion gas is 2 orders of magnitude less, and it is reasonable to expect such energy to be available through some process associated with solar wind-ionosphere interactions. However, no suggestions have been made for a specific mechanism. The measured daytime ion temperature profile shows a clear bump in the altitude region between $\sim$ 160 and 200 km, which implies the presence of some specific heat source(s). Cravens et al. (1979) found that exothermic chemical reactions can account for most of the required heat, while Miller et al. (1980) favor frictional heating as the dominant source. Schunk and St. Maurice (1981) have predicted that an ion temperature anisotropy will be present in this altitude region caused by large ion flow velocities. Their calculations have also indicated that diffusion-thermal heat flow is more important than ordinary ion thermal conduction in this lower ionospheric region.

The effect of magnetic field fluctuations on thermal conductivity, as discussed in Sec. I, was included in some of the model calculations of Cravens et al. (1980); results of one set of such calculations are shown in Fig. 11. For the case shown, a magnetic field strength of 10 γ and no inhibition on photoelectron transport was assumed; various values of effective mean free path λ were used to approximate a potentially wide range of magnetic fluctua-

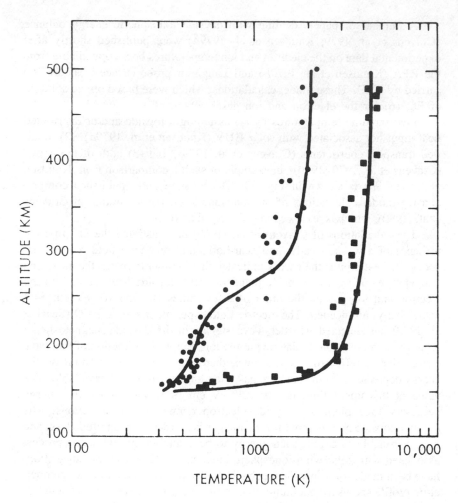

Fig. 10. Dayside ion (●) and electron (■) temperature profiles. A column heat input of 3×10^{-3} erg cm^{-2}s^{-1} uniformly distributed between altitudes of 150 and 250 km and presumably representing Joule heating was required to raise the model ion temperature to the measured temperature in this altitude interval. The model represented in the figure by a solid line included in electron heat input at the top of 3×10^{-2} erg cm^{-2} s^{-1}. The SZA was $72° \pm 11$. (From Knudsen et al. 1979a.)

tions. The calculated ion temperatures are almost independent of the value of λ chosen, because in a 10 γ field the ion collision frequencies are of the same order as or larger than the ion gyrofrequency for all altitudes less than a few hundred km. As indicated in Cravens et al. (1979), only when the magnetic field strength is considerably larger than 10 γ is the ion thermal conductivity significantly affected. Figure 11 also shows that reasonable electron temperatures result with an assumed λ of $\sim$ 3 km. As mentioned earlier, a large number of free parameters are present and no unique choices can be

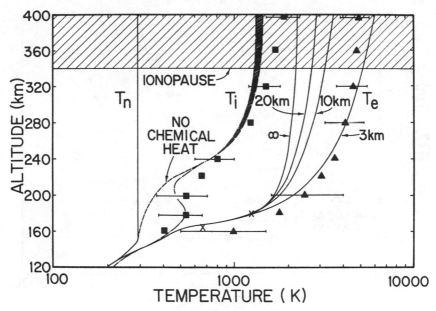

Fig. 11. Electron, ion, and neutral temperature profiles are shown for the subsolar point. The magnetic field is 10 γ and temperature profiles for a range of λ are shown. Uninhibited photoelectron transport is assumed. Electron temperature data from the Langmuir probe (OETP) (▲) and ion temperature data from the retarding potential analyzer (ORPA) (■) are shown. Some ORPA electron temperatures (x) are also shown. The SZA $\chi = 0°$. (From Cravens et al. 1980a.)

made in fitting the observed data. As a final example of such temperature calculations, which fit the data well, Fig. 12 shows the results obtained by Cravens et al. (1980a) assuming a 10 γ field, a mean free path of 10 km due to fluctuations, and topside heat inflows into both the electron and ion gas.

The energetics of the nightside ionosphere are even less well understood than those for the day side. Examples of the added difficulties for the nightside energy balance are (1) the role of precipitating particles, (2) the energetics associated with supersonic flows, and (3) the absence of a topside electron heat source due to whistlers. Only a few groups have so far attempted to carry out quantitative calculations of the nightside energetics. Hoegy et al. (1980) solved the two-dimensional electron energy equation between 80° and 170° SZA, using the observed mean temperature values as constraints. The free parameters in their calculations were the magnetic field direction and the topside heat inflow. They find that a good agreement between the calculated and measured values of T_e can be achieved if the magnetic field and corresponding heat flow directions are assumed as shown in Fig. 13. The assumed field configuration is not inconsistent with the observed mean values and implies that the lower nightside ionosphere is heated by conduction of heat from the

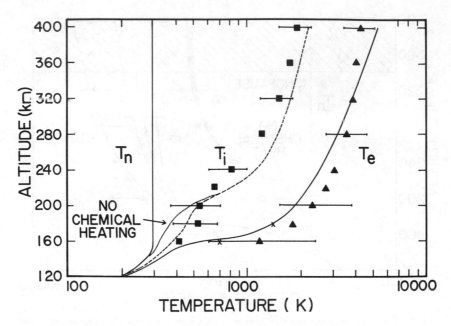

Fig. 12. As for Figure 10 except that the magnetic field is 10 γ and the mean free path is 10 km, and different heat inputs are used. $\Phi_e = 5 \times 10^9$ eV cm^{-2}s^{-1}; $\Phi_i = 3 \times 10^7$ eV cm^{-2}s^{-1}. The SZA $\chi = 60°$. (From Cravens et al. 1980a.)

day side, while the upper ionosphere receives the necessary heat from the ionosheath. Cravens et al. (1981b) solved Eq. (8) for electron and ion temperatures, though they ignored the horizontal heat conduction terms (but included the other horizontal terms), thus restricting the validity of their calculations mainly to the ions. They found that adiabatic compression on the night side helps somewhat to maintain the nightside ion temperature but not nearly enough to explain the very high temperatures measured by the Pioneer Venus RPA (Miller et al. 1980). Knudsen et al. (1980b) suggested that the heat necessary to maintain the high nightside ion temperatures is derived from the kinetic energy of the plasma flowing rapidly past the terminator. Merritt and Thompson (1980) pointed out that there is an upper limit to the rate at which heat can be transported by thermal conduction in a plasma, and that this limit is probably reached for electrons on the night side of Venus.

IV. MISCELLANEOUS IONOSPHERIC PHENOMENA

The discussion in this chapter has been restricted, so far, to the general or average behavior of the ionosphere; however the observations, discussed in Chapter 23, indicate the presence of many unique and interesting spatial and/or time varying phenomena such as detached plasmas, ionospheric holes,

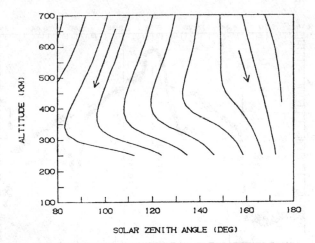

Fig. 13. Heat flux lines or magnetic field lines that satisfy electron energy balance by using the empirical T_e over SZA of 80° to 175° and over altitudes of 160 to 700 km. The arrows indicate the directions of heat flow. (From Hoegy et al. 1980.)

and disappearing ionospheres, to name only a few. No quantitative theoretical or model descriptions have been published of any of these features. As an example of our present understanding of these ionospheric phenomena, the mechanisms that may be responsible for the formation of ionospheric holes are reviewed briefly below.

The electron densities measured by the electron temperature probe carried by PVO through one of its periapsis passes, during which typical ionospheric holes were observed, are shown in Fig. 14. The observed general characteristics of these holes can be summarized as follows:

1. The densities in the holes are $\sim$ 2 orders of magnitude lower than in adjacent regions;
2. The holes are typically found at altitudes above 200 km and in two broad zones. These zones extend from about midnight to 3 AM local time and are $\sim$ 20° in latitude extent, centered at 30°N and 12°S;
3. The latitude extent of individual holes is on the order of 1000 km;
4. The magnetic field in the holes is approximately vertical and is enhanced compared to adjacent regions;
5. The electron temperature in the holes ($\sim$ 2000 K) is substantially lower than in the surrounding ionosphere, except in the lowest density regions where highly elevated temperatures ($\sim$ 20,000 K) are observed.

One of the first questions to be examined was whether there is a static pressure balance across the hole boundaries. Brace et al. (1982b) have shown that such a pressure balance does exist and thus the holes can be quasi-static

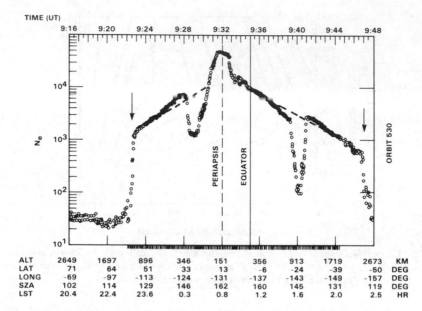

Fig. 14. A 32 minute periapsis profile of N_e illustrating a pair of holes encountered in an otherwise undisturbed nightside ionosphere. The spacecraft crossed the ionopause (I) at 1170 km, traversed a hole centered at 29°N and 240 km altitude, passed periapsis at 151 km, encountered a second hole centered at 24°S and 940 km, and left the ionosphere at 2500 km. The satellite altitude (ALT), latitude (LAT), longitude (LONG), solar zenith angle (SZA), and local solar time (LST) are given along the abscissa. (From Brace et al. 1982b.)

features. Two potentially likely mechanisms responsible for hole formation, using terrestrial analogies, can easily be dismissed. The near vertical magnetic field lines in the holes leading back in the tail suggest that polar wind types of outflow may be present; however, the relative enhancement of H^+ over O^+ inside the hole region (H. A. Taylor, personal communication) is contrary to the general predicted behavior of the polar wind. Horizontal neutral wind velocities of 200 to 400 m s^{-1} are believed to be present in the thermosphere of Venus, and such velocities in conjunction with appropriately tilted near-vertical field lines could drive ionization downward to regions of rapid recombination and result in significantly decreased densities (R. W. Schunk, personal communication). However, an examination of the magnetic field data has indicated no meaningful correlation between the field tilt and the presence of holes (J. G. Luhmann, personal communication), thus removing this mechanism from among the viable ones.

Grebowsky and Curtis (1981) propose that the observed plasma depletions in the holes are caused by electric fields parallel to the observed near-radial magnetic fields. They do not present a specific and/or quantitative mechanism for the formation of this electric field but suggest, using terrestrial analogy, that neutral sheet acceleration processes will result in an upward directed $E_\parallel$ distribution with a potential drop on the order of kilovolts. Given such a field, electrons from the tail region will accelerate down into the ionosphere becoming a source of nighttime ionization and a heat source for the ionospheric thermal electron gas. This electric field would also accelerate thermal ions out of the ionosphere into the tail region. Using rough estimates of densities and field strength, Grebowsky and Curtis (1981) indicate that their proposed mechanism is not inconsistent with the data base presently available. Hartle (1981) has examined the abundance and altitude behavior of the various minor and major ions inside the holes and has concluded that the observed ''height'' profiles (there is always some ambiguity between horizontal and vertical variations) are consistent with the presence of $E_\parallel$ and upward flowing ions.

This brief summary of our present understanding of the nightside ionospheric holes clearly indicates, as mentioned earlier, the lack of any quantitative theories of this well-documented ionospheric feature. There are intensive efforts under way to develop a detailed explanation for the various observed special ionospheric phenomena. Significant progress is expected as the available data base increases.

V. CONCLUSION

Our general understanding of the chemical and physical processes controlling the behavior of Venus's upper atmosphere and ionosphere has advanced greatly due to the significant observational data base now available together with broadly based theoretical studies. This chapter has reviewed briefly the controlling ionospheric chemical and transport processes and outlined the present status of the various general models of the ion composition, density and thermal structure developed to reproduce the basic observed behavior of the ionosphere. Comparisons of calculated and measured dayside ion composition and density values have indicated a basic and general agreement but they have also shown the need for (1) laboratory measurements of some of the rate coefficients, (2) inclusion of metastable chemistry (Fox 1982) and possibly certain nonconventional transport effects into the models, and (3) measurements of the ionospheric metastable populations. No truly comprehensive models of the nightside ion composition and density have yet been published and compared with data. Further advances in this area must wait for (1) the development of multidimensional models that include both the horizontal ion transport terms and particle precipitation effects, and (2) more data on ion (and maybe neutral gas) velocities as well as on precipitating particle fluxes.

It has been possible to reproduce, using model calculations, measured dayside ion and electron temperatures; however the controlling heating and transport mechanisms cannot be established uniquely, because of too many free parameters. In the future, more sophisticated models of charged particle transport in realistic ionospheric magnetic fields will be required. Our understanding of the nightside energetics is very limited; we know the amount of heat required to explain the observations but we are not yet able to establish the mechanisms responsible for the heating and/or for transporting that heat. As mentioned in Sec. IV, the bulk of this chapter has only dealt with the general, gross, behavior of the ionosphere, and not with any of the detailed time and/or spatially varying phenomena, which the observations are now beginning to reveal. Attempts to understand the processes responsible for these phenomena have just begun and it will be sometime before any definitive models become available. The recent intensive exploration of the aeronomy of Venus has greatly expanded our knowledge, but at the same time exposed new areas of ignorance. As the Pioneer Venus data base grows and more sophisticated models evolve, we expect further and significant advances in our understanding of the ionosphere of Venus.

Acknowledgments. The authors of this chapter wish to express their appreciation to L. H. Brace and W. B. Hanson for many helpful suggestions. This work was supported by grants from the National Aeronautics and Space Administration and the National Science Foundation.

25. THE INTERACTION OF THE SOLAR WIND
WITH VENUS

C. T. RUSSELL
University of California, Los Angeles

and

O. VAISBERG
Space Research Institute, Moscow

Venus has been more frequently studied by space missions than any other planet. Consequently, we know more about the solar wind interaction with Venus than with any planet except the Earth. Unlike the Earth, Venus's iono-sphere, not a magnetosphere, deflects the solar wind flow. However, as on the Earth, this deflection is accomplished with the formation of a bow shock, which heats and compresses the solar wind flow. The shock is both closer to the planet and weaker than would be expected for an ideal gas dynamic interaction with a perfectly reflecting obstacle. The ionized flow of the magnetosheath can interact directly with the neutral atmosphere through charge exchange and photoioniza-tion. The former process removes momentum from the flow; both processes add mass to the solar wind, since the high altitude neutral atmosphere is mainly hot oxygen, not hydrogen. Finally, Venus, like Earth, has a magnetotail but not for the same reason. The mass loading of the flow in the magnetosheath slows the transport of magnetic flux tubes past the planet, while the ends of the tubes continue to travel rapidly in the solar wind. Thus the planet accretes interplane-tary magnetic flux. This process is the dominant source for the magnetotail flux, not unipolar induction, although the latter process is present at least when the solar wind dynamic pressure is high. On the whole, the solar wind interaction with Venus is more comet-like than Earth-like.

[873]

Venus is the Earth's closest planetary neighbor, both in distance and in physical properties. The closeness in space has invited frequent exploration: the Venera series of spacecraft, three Mariner flybys and the Pioneer Venus orbiter and probe mission. The similarity in size, density, and a dense atmohere has invited speculation that Venus is similar to the Earth in other ways as well, but we find many differences. The most important difference in interaction of the solar wind is the weakness or absence of a planetary magnetic field on Venus, which in this respect is more akin to a comet than to the Earth. We would expect Venus's magnetic field to be weaker than the Earth's simply because Venus's rotation rate is much lower (Busse 1976). Venus rotates on its axis with respect to the stars once in 243 d; the earth's rotational period is only 23 hr 56 min 4 s. However, Venus's field is much weaker than would be extrapolated from comparison with the terrestrial field, implying that there are probably important differences between the interiors of Earth and Venus, which affect Venus's ability to generate a self-sustaining magnetic dynamo.

Venus orbits the Sun every 225 days; the Earth orbits the sun every 365 days. The beat between these two periods is 586 d or 19.3 mon, so there are nearly two equi-spaced launch opportunities every three years. The first of these to be used was by Mariner 2 in 1962 which flew by Venus with a distance at closest approach of 6.6 Venus radii (R_φ). The next Venus data were not obtained until three launch opportunities later, when in October 1967 both Venera 4 and Mariner 5 probed the planet. At the next opportunity, duplicate missions Veneras 5 and 6 were launched. Veneras 7 and 8 did not study the solar wind interaction and Mariner 10, whose primary mission was to explore Mercury, passed by Venus at too large a distance to provide new insight into the nature of the interaction. Hence four opportunities passed before new results on solar wind interaction were obtained in October 1975 with the Venera 9 and 10 orbiters. Repeated sampling with orbiters was a great advance over previous missions, and simultaneous data from two spacecraft provided the capability to unambiguously separate temporal from spatial effects. The next opportunity in May 1977 was skipped, but in December 1978 Venus was visited by a flotilla of spacecraft: the Pioneer Venus orbiter and probe, which consisted of a bus and four atmospheric probes, and Veneras 11 and 12, each having a flyby and a lander. Of the six U.S. spacecraft, however, only the orbiter provided data on the solar wind interaction and there have been no new results on the interaction from the four Soviet spacecraft, although the flyby craft did contain sophisticated plasma instrumentation for solar wind studies. These missions and their arrival dates and trajectory characteristics are listed in Table I.

The number of missions devoted at least in part to studying the interaction of the solar wind with Venus would suggest that much is now known about the interaction of the solar wind with Venus. While our understanding of some aspects of the interaction has in fact become quite extensive, in other aspects

Table I
Missions Investigating the Solar Wind
Interaction with Venus

Mission	Encounter/Orbit Injection	Orbit/Trajectory
Mariner 2	14 December 1962	Flyby Closest approach: 6.6 $R_{\venus}$
Venera 4	18 October 1967	Entry probe Lowest transmission: 200 km
Mariner 5	19 October 1967	Flyby Closest approach: 1.7 $R_{\venus}$
Venera 6	17 May 1969	Entry probe Lowest transmission: 32,000 km
Mariner 10	5 February 1974	Flyby Closest approach: 1.96 $R_{\venus}$
Venera 9	22 October 1975	Periapsis: 1545 km Apoapsis: 18.5 $R_{\venus}$ Period: 48 hr 18 min
Venera 10	25 October 1975	Periapsis: 1665 km Apoapsis: 18.8 $R_{\venus}$ Period: 49 hr 23 min
Pioneer Venus Orbiter	4 December 1978	Periapsis: 140–180 km Apoapsis: 12 $R_{\venus}$ Period: 24 hr

we have only rudimentary knowledge. Despite all this activity, each mission to Venus has involved some sort of compromise from the ideal mission for solar wind studies, leaving gaps in our knowledge. For example, the Pioneer Venus orbiter depends on the stability of solar wind conditions and the repetition of features to separate temporal from spatial variations. The Venera 9 and 10 orbiters permitted this separation of factors but had a limited number of simultaneous transmissions and no *in situ* ionospheric data. Throughout this review we caution the reader as to the limits of our present understanding in each area.

The basic features of the interaction are now well understood, as shown in Fig. 1. The planetary ionosphere presents an obstacle to the magnetized solar wind flow. The supersonic, or more correctly, supermagnetosonic, solar wind is deflected about the planet by a collisionless bow shock in much the same way as the solar wind is shocked and deflected about the terrestrial magnetosphere. Finally, in apparent analogy with the Earth's magnetosphere but more correctly in analogy with cometary ion tails, Venus has its own magnetotail.

In this chapter we will consider first the nature of the obstacle to the solar wind flow, and review the limits on the possible size of any planetary mag-

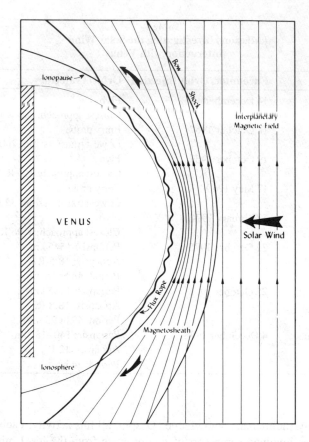

Fig. 1. Basic features of the solar wind interaction with Venus. The solar wind is deflected
around Venus by the planetary bow shock. The obstacle is the planetary ionopause; the
interface between the magnetic field which piles up in front of the planet and the ionosphere
is called the ionopause. At times of low solar wind dynamic pressure the ionosphere is
unmagnetized except for the occurrence of twisted filaments of field called flux ropes.
Behind the planet a tail is formed by flux tubes which become filled with plasma on the day
side and convect around the planet together with some flux tubes which penetrate the
ionosphere and become hung up there.

netic moment. We will also examine the early models of the interaction to
compare with our present understanding, and discuss the physical processes
occurring at the ionopause, the interface between the shocked solar wind and
ionospheric plasmas. Then we will examine the wake and tail region behind
Venus, including the question of ion pickup by the solar wind and mass loss.
Finally, we will move out from the planet to consider questions such as the
location and shape of the bow shock and the effects of Venus on the solar
wind beyond the bow shock. Other viewpoints on some of these matters are
represented in Chapters 26 and 27 by Cloutier and by Gringauz.

I. THE NATURE OF THE OBSTACLE

At the Earth the solar wind is deflected far above the atmosphere by a strong intrinsic magnetic field generated by a dynamo deep in the interior of the Earth. This shielding by the Earth's magnetic field does not prevent solar wind energy from coupling to the Earth's atmosphere. The aurora is one manifestation of this coupling. However, the solar wind does not directly interact with the atmosphere and current rates of loss to the solar wind are thought to play a minor role in the evolution of the terrestrial atmosphere. A similar type of obstacle to the solar wind is found at Mercury, Jupiter and Saturn; we can generally understand the behavior of their magnetospheres in terms of our terrestrial experience.

The earliest measurements suggested that Venus's magnetic field was far weaker than Earth's and possibly even absent, although the latter suggestion evoked some controversy. If the magnetic field were so weak that the solar wind interacted directly with the atmosphere, then a qualitatively different interaction would exist, one more like the interaction of a comet with the solar wind. The fact that the controversy on whether Venus had an intrinsic or induced magnetosphere raged for close to 20 yr shows the limitations of flyby missions for gathering *in situ* data, especially in situations involving variations in latitude, longitude, altitude, and time.

A. The Search for a Magnetic Moment

Mariner 2 was the first successful mission to another planet, passing within 6.6 radii ($R_{\venus}$) of the center of Venus in December 1962. It carried a triaxial fluxgate magnetometer (Smith et al. 1963, 1965) a solar wind probe (Neugebauer and Snyder 1965) and an energetic particle experiment (Frank et al. 1963). None of these instruments detected any evidence of the solar wind interaction with the planet, implying that the magnetic moment of Venus was <1/20th that of the Earth.

On 18 and 19 October 1967 Venera 4 and then Mariner 5 reached Venus. Venera 4 carried a triaxial fluxgate magnetometer (Dolginov et al. 1968) and four charged particle traps (Gringauz et al. 1968). Venera 4 transmitted ion and magnetic field data until it reached an altitude of 200 km. A bow shock was reported in both the magnetic field and ion data. Assuming that their data contained no offsets and noting that there was little change in the magnitude of the field with altitude, Dolginov et al. used the last field measurement to calculate an upper limit to the Venus moment of 1/3000th of the terrestrial moment. Later Dolginov et al. (1969) compared their entry data with simultaneous Mariner 5 data obtained 80 $R_{\venus}$ away in the solar wind. Figure 2 shows this comparison. Both spacecraft are in the solar wind simultaneously at the beginning of this figure. The traces have been aligned for the expected delay time from Venera to Mariner. The fields at the two spacecraft are quite different. Such differences are rare but not unknown over similar baselines at

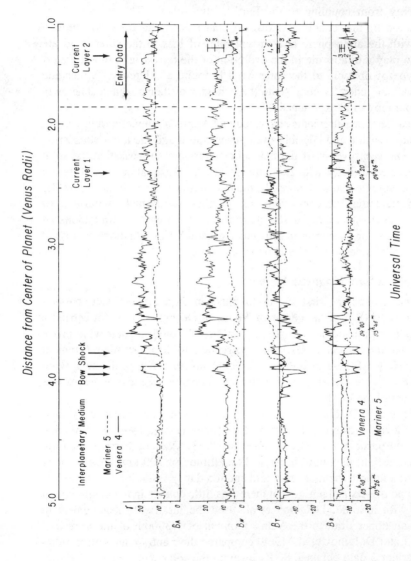

Fig. 2. The magnetic field measured by Venera 4 and Mariner 5 during the Venera 4 encounter with Venus (Dolginov et al. 1969).

1 AU (Russell et al. 1980c), where the time delay is quite variable also. Thus we should not be surprised at these differences, but they could also be instrumental in origin. The principal point of this comparison is that the magnitude of the interplanetary magnetic field as measured by Mariner and the magnitude of the shocked magnetosheath field near the planet both stayed nearly constant as Venera 4 approached within 200 km of Venus. Thus Dolginov et al. reaffirmed their conclusions on the weakness of the magnetic moment of Venus. Note, however, that the vector components of the field change significantly as the planet is approached.

The next day Mariner 5 flew by Venus, coming within 1.4 $R_{\venus}$ of the axis of the optical shadow and within 1.7 $R_{\venus}$ of the center of the planet. Mariner 5 carried a triaxial vector helium magnetometer, a solar wind probe (Bridge et al. 1967) and an energetic particle detector (Van Allen et al. 1967). Figure 3 shows the trajectory of Mariner 5 and the plasma and magnetic field magnitude observed during the encounter period. Superimposed on these data are dashed lines corresponding to the numerical solution of the problem of solar wind interaction with an impenetrable object the size of the ionosphere when the interplanetary magnetic field is aligned with the flow (Rizzi 1971). Between points 2 and 3 the velocity observed is less than that predicted by the nondissipative hydrodynamic model, suggesting the presence of a viscous boundary layer. Such viscous boundary layers have been proposed by Perez-de-Tejada and coworkers (Perez-de-Tejada and Dryer 1976; Perez-de-Tejada et al. 1977). However, momentum is removed from the solar wind protons in other ways. The first is charge exchange, in which a fast ion becomes a fast neutral and dumps its momentum into the upper atmosphere; another example is photoionization of neutral oxygen in the boundary layer; the oxygen ions become accelerated while the protons become decelerated. These processes will be discussed at greater length in Secs. I.B and II. C below. Later more sophisticated analysis of the solar wind data reaffirms these conclusions (Shefer et al. 1979).

The magnetic field strength at point 3 in Fig. 3 increases above the value of the field in the surrounding regions and becomes quite steady for a few minutes. This point is also the closest approach to the axis of the optical shadow. Figure 4 shows the Mariner 5 and Venera 4 magnetic field plotted along the trajectory in a coordinate system that assumes cylindrical symmetry about the Sun-Venus line (Russell 1976a). Near closest approach to the tail axis the field lines are stretched out in a tail-like fashion both on Mariner 5 and Venera 4. The boundaries that one would identify as magnetopause crossings into a magnetotail-like field are consistent in location with the boundary of the tail seen in Pioneer Venus data (Russell et al. 1981). The direction of the tail field at Venera and Mariner is opposite. Mariner 5 passed through the northern lobe of the Venus wake. However, the Venera 4 bus also entered in the northern hemisphere (Dolginov et al. 1978). Thus these data argue against the existence of an intrinsic planetary magnetic field, and are not consistent with

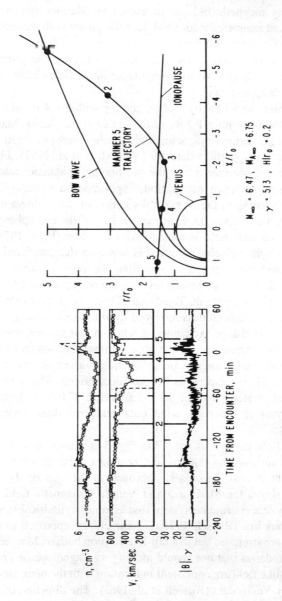

Fig. 3. The number density, solar wind velocity, and magnetic field strength during the Mariner flyby. On the right is the Mariner 5 trajectory in solar cylindrical coordinates, in which the distance from the center of the wake is plotted versus the distance along the Sun-planet line (Rizzi 1971). The dashed lines show the hydrodynamic solution for a specific set of ionospheric and solar wind parameters.

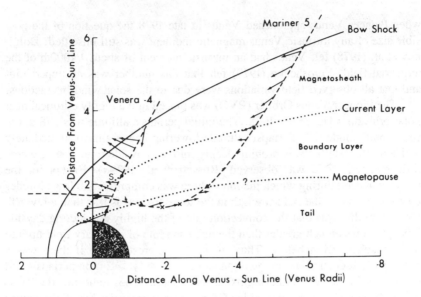

Fig. 4. The magnetic field measured by Mariner 5 and Venera 4 plotted along their trajectories in solar cylindrical coordinates. For Mariner 5 the 3-dimensional trajectory information is used. For Venera 4 we have only the ecliptic plane projection. Closed triangles mark magnetopause traversals; dots indicate the boundary of a second current layer, possibly the rarefraction wave; the S marks the location of the Mariner 10 shock crossing.

the existence of an intrinsic field as large as the upper limit of 6.5×10^{22} Gauss-cm^3 which had been advocated by Russell (1976b).

The next measurements, on Venera 6, returned data only on the distant bow shock (Gringauz et al. 1970). The next set of data came from the magnetometers on the Venera 9 and 10 orbiters which arrived at Venus on 22 and 25 October 1975. Their orbits had periods of slightly > 48 hr, periapsides of ~ 1500 km and inclinations of $\sim 35°$. These satellites made frequent passes through the wake region, crucial to probing for weak fields. The initial results suggested the intrinsic field to be weak (Dolginov et al. 1976).

Veneras 9 and 10 clearly observed a magnetic tail extending to large distances from the planet. The question to be settled was whether this long magnetic tail was due to sources intrinsic to the planet as concluded by Dolginov et al. (1978), or whether it was a manifestation of the interaction of the magnetized solar wind with the planet as concluded by Yeroshenko (1979). As laboratory simulations showed, it is certainly possible to have a long magnetic tail behind an unmagnetized obstacle (Dubinin et al. 1978). Despite the fact that simultaneous data from Veneras 9 and 10 could help distinguish the always vexing temporal and spatial variations, the amount of simultaneous coverage was limited. Furthermore, the Venera spacecraft were not triaxially stabilized in most cases when they were far from the planet; they were allowed to spin slowly about an axis close to the solar direction. Hence, when Pioneer Venus approached Venus in late 1978 the question of the possible size of any intrinsic Venus magnetic moment was still unsettled. Dolginov et al. (1978) felt Venus had an intrinsic moment of about 1/4000th of the terrestrial field. Yeroshenko (1979) felt that this number was an upper limit and that all observed field variations were due to the solar wind interactions.

The Pioneer Venus Orbiter (PVO) was placed into a highly elliptical near polar orbit on 4 December 1978. The initial periapsis altitude was 378 km at solar zenith angle 63°. Periapsis altitude lowering began almost immediately and soon the satellite was probing deeply into the dayside ionosphere, where little magnetic field was observed (Russell et al. 1979a). Except for the occasional orbit during which the ionosphere was compressed to low altitudes by the solar wind, the field strength in the ionosphere was less than a few nT. One possibility was that the convection rate of the highly conducting dayside ionosphere was much greater than the diffusion rate of planetary field up into the ionosphere from below. Thus, it was of interest to see if there was a buildup of magnetic flux on the night side. In slightly > 2 mon periapsis had moved to the center of the nightside ionosphere. The nighttime field was generally similar to the day side. Often there were very low field values throughout the night ionosphere as in Fig. 5a; sometimes the field was moderately high (tens of nT) and steady as in Fig. 5b. Occasionally there were limited regions of radial field extending apparently through the base of the ionosphere, but the field mainly wrapped around the nightside ionosphere as if the solar wind magnetic field had draped around the planet and was closing behind the obstacle (Russell et al. 1979c, 1980b). There was no obvious sign of a planetary magnetic field. However, it was possible that there was some net flux out of the planet masked by the external currents and their fluctuations. Thus, the average magnetic field over the nightside ionosphere was calculated and inverted to find the best fit for magnetic moment. Using data from two nightside passes the upper limit to the magnetic moment was found to be 4×10^{-5} of the terrestrial moment or $<3 \times 10^{21}$ Gauss-cm^3 (Russell et al. 1980a).

The magnetic field strength necessary to deflect the solar wind at 0.72 AU is $\sim$ 100 nT. If the solar wind doubles the planetary field by compression in the interaction, this deflection can be accomplished above the surface of the planet by a magnetic moment only 1/360th of the terrestrial moment. Small as this may be, the actual measured moment is < 1/25000th of the terrestrial moment, and we can conclude that at the present time intrinsic magnetic fields play no significant role in the solar wind-Venus interaction. Venus presents a very different type of obstacle to the solar wind than has been probed at the other planets.

B. Non-Magnetic Barrier Models

If a superconducting ball is placed in a uniform external magnetic field, the magnetic field lines bend so as to exclude the volume occupied by the ball

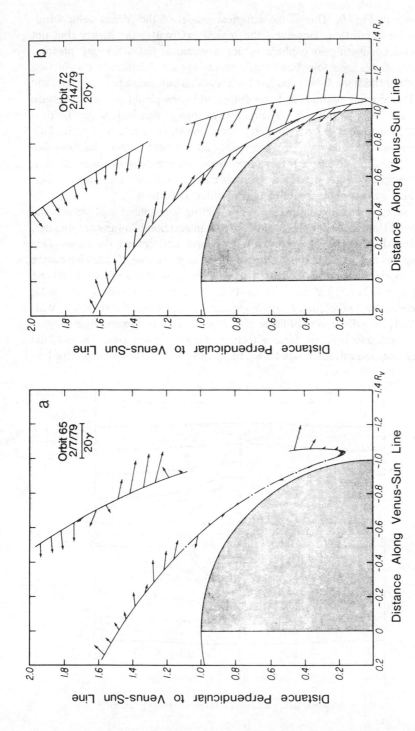

Fig. 5. Magnetic field in Venus wake plotted in solar cylindrical coordinates. (a) PVO orbit 65 2/7/79; (b) PVO orbit 72 2/14/79.

as shown in Fig. 6. This is the simplest model of the Venus solar wind-ionosphere interaction. However, the region external to the highly (but not infinitely) conducting ionosphere is not a vacuum, but a flowing plasma. Thought of in magnetohydrodynamic terms (as an electrically conducting magnetized fluid) the solar wind has both mass and momentum and an electric field. This flowing plasma has to be deflected by the planet by some pressure wave. The flow introduces day-night asymmetry, and unless the field is flow-aligned there will be asymmetry about the flow direction as well. This superconductor analogy is useful but is only an approximation to the condition of the Venus ionosphere. Furthermore, there are the complications of non-magnetohydrodynamic processes such as charge exchange and photoionization which we ignore for the moment, but will discuss later.

A superconducting or infinitely conducting ionosphere will deflect the solar wind flow around it. The nature of this interaction is quite complicated, involving the creation of a bow shock to heat and deflect the flow. This problem has not been solved analytically, although computer codes have long existed (Spreiter et al. 1966) which treat well the unmagnetized solar wind and the case of field-aligned flow (Rizzi 1971). The magnetization of the solar wind creates two problems. First, magnetic stresses affect the flow (cf. Walters [1964] for a discussion of these effects at the Earth); Spreiter et al. (1966) did not take into account these effects in their computations. Second, the flowing magnetized solar wind has an electric field. This was realized as

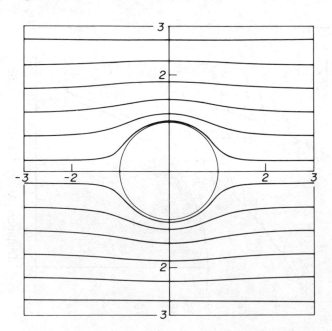

Fig. 6. Magnetic field lines around a superconducting sphere.

important by all early investigators (Dessler 1968; Sonett and Colburn 1968; Cloutier et al. 1969; Cloutier 1970) but the effect of this electric field depended critically on the nature of the obstacle. It is important at this point to distinguish between two types of induction. The most familiar type of induction is that associated with a time varying magnetic field, as seen in our case in the reference frame of the planet. This can be caused by propagating waves or by convected structure in the solar wind. The second type of induction is unipolar induction, in which currents are driven in the planetary conductor by the solar wind electric field associated with the moving magnetized highly electrically conducting solar wind. The former type of induction is very important on the moon whose surface, being nonconducting, prevents the latter type from occurring. But it may be important at Venus, and the former is a complication to be avoided in our present observational state.

We would not need to worry about unipolar induction if the ionosphere were either perfectly conducting or nonconducting. In the former case, the surface of the obstacle would be an equipotential and all flow would be diverted around the planet. In the latter, the flow would all crash into the planet as it does on the lunar surface. In the real world of Venus, the true case is somewhere between. Early investigators proposed a variety of possible scenarios for the solar wind interaction with Venus. Michel (1971) has classified these as the direct interaction model, the tangential discontinuity model, and the magnetic barrier model. These are illustrated in Fig. 7 with altitude profiles of the electron number density, magnetic field strength and horizontal flow velocity. In the direct interaction model (Fig. 7a) the inflowing post-shock solar wind depresses the ionosphere until a rather rapid transition occurs from flow to diffusive dispersion of the inflowing plasma. The solid line in Fig. 7a shows a case in which flow rate is entirely controlled by photoion pickup. Collisional stagnation above the peak of the ionospheric density would be expected to modify this profile as sketched by the dotted line. In the tangential discontinuity model illustrated in Fig. 7b, the ionospheric pressure equals the solar wind pressure across the ionopause and the plasmas are noninteracting. In the magnetic barrier model of Fig. 7c, a tangential discontinuity occurs between the flowing solar wind and the magnetic field. The magnetic field then transmits the pressure to the ionosphere, decreasing with decreasing altitude so there is net pressure balance in the ionosphere. None of these models agrees well enough with observations that we could call it a successful prediction. Nevertheless each of these models has some correct features.

The first clues to the correct nature of the planetary obstacle to the solar wind came from the Mariner Stanford Group's (1967) radio occultation data. The interesting feature of these data was the sudden drop in density with increasing altitude which has been called the ionopause. The name "anemopause" was suggested instead by Bridge et al. (1967) because it was supposed that the ionopause was where the solar wind stopped, but never

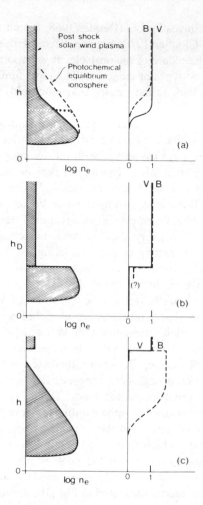

Fig. 7. Summary of Venus solar wind interaction models by Michel (1971). Profiles of electron concentration, n_e, magnetic field B, and horizontal flow velocity V are shown as a function of height for the three models. Incident solar wind velocity and field are normalized to unity. (a) Direct interaction model; (b) tangential discontinuity model; (c) magnetic barrier.

became popular. This lack of popularity had nothing to do with the fact that the anemopause concept, as proposed, was incorrect.

Mariner 10 provided another dayside occultation with somewhat lower ionopause and not as sharp as Mariner 5. Venera 9 and 10 followed with many more. Figure 8 shows three altitude profiles of the number density obtained by the Venera orbiters (Aleksandrov et al. 1976b) at solar zenith angles of (1) 14°, (2) 46°, and (3) 85°, illustrating the variability of the ionopause position.

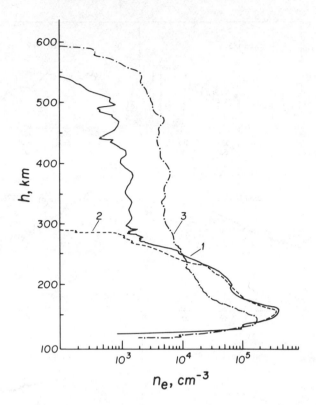

Fig. 8. Dayside electron density profiles from Venera 9 and 10 occultation measurements at solar zenith angles of 14° (profile 1), 46° (profile 2) and 85° (profile 3) (Aleksandrov et al. 1976b).

These provided support for the existence of an ionopause, but until *in situ* data were obtained the physical nature of this boundary was uncertain. As discussed in more detail in Chapter 23 by Brace et al., the ionopause can generally be considered as a tangential discontinuity with ionospheric thermal pressure on the inside balancing magnetic pressure on the outside (Elphic et al. 1980b). There is little plasma pressure immediately outside the ionopause. This is apparent from the near perfect pressure balance between the magnetosheath field and the ionospheric plasma (Elphic et al. 1980b) and the near equality of these two pressures with the preshock solar wind dynamic pressure (Elphic et al. 1980a; Brace et al. 1980; Vaisberg et al. 1980). One might wonder what happens to the postshock solar wind flow. As Fig. 9 shows there is a gradual buildup in magnetic field strength as the planet is approached in the subsolar region, though the strength of this magnetic barrier varies from day to day. The data suggest that the postshock solar wind is mainly lost from the flux tubes of the magnetic barrier. Zwan and Wolf (1976) have suggested

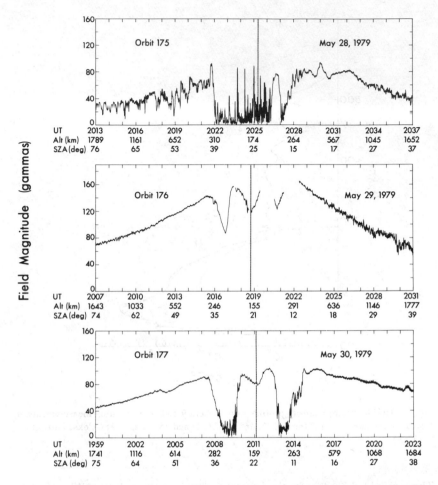

Fig. 9. Magnetic field strength as a function of time through periapsis on three successive
Pioneer Venus orbits at low solar zenith angle (Elphic et al. 1980a).

that this occurs because the hot plasma escapes along field lines as the tube
slowly convects against the planet. Observationally, we obtain the picture
shown in Fig. 10 as the post-Pioneer Venus magnetic barrier model.

The first noticeable difference between Fig. 10 and Fig. 7c is that there is
an abrupt increase in field strength with increasing altitude at the ionopause
and a gradual decay, rather than the reverse. In conjunction with the abrupt
increase in field strength the cold plasma density of the ionosphere abruptly
decreases so as to maintain pressure balance with the magnetic field in the
barrier. The density of the hot plasma decreases with decreasing altitude
mainly due to the Zwan and Wolf effect. Outside the ionopause charge exchange occurs at a rate dependent on both the neutral atmospheric density and

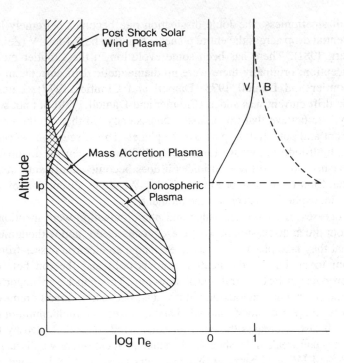

Fig. 10. Post-Pioneer Venus magnetic barrier model. See legend to Fig. 7 for further explanation.

the postshock solar wind density. This further depletes the hot component and adds mass, if the charge exchange is between a light ion (e.g. H^+) and a heavy neutral (e.g. O). Photoionization also occurs here as it does throughout the atmosphere, at a rate dependent on the extreme ultraviolet (EUV) flux and the neutral density. This further adds mass to the postshock solar wind flow. The mass accretion slows down the flow; hence the velocity decreases as the planet is approached as does the electric field (not shown). In this model the anemopause (where the solar wind stops) is a gradual transition whereas the ionopause is a sharp transition, albeit the cold plasma density increase is somewhat spread out due to charge exchange and photoionization in the slow-moving magnetosheath just exterior to the ionopause.

The key assumption in this post-Pioneer magnetic barrier model is the equipotential nature of the ionopause, in contrast to the assumptions of the Cloutier model (cf. Cloutier et al. 1981 and Chapter 26). However, at the observed average altitude of the ionopause, and with the observed thickness of the ionopause current layer (Elphic et al. 1981) it is necessary to take into consideration the gradient terms in the generalized Ohm's law. When this is done the current is found to be overwhelmingly diamagnetic drift currents which are dissipationless, the Joule dissipation rate becomes extremely low and the potential drop across the entire planet is found to be 0.1 to 1 V (Zeleny and Vaisberg 1981). There has been some evolution in the Cloutier model since its inception; originally there were no diamagnetic drift currents in the model (Cloutier and Daniell 1973; Daniell and Cloutier 1977). Later a diamagnetic drift current was added (Cloutier and Daniell 1979) but not self-consistently. We believe that this non-self-consistency led them to the larger Ohmic current and potential drop across the planet. This discussion concerns only the high-altitude ionopause (altitude $\geq$ 300 km); when the ionopause current is pushed down to lower altitudes it does become significantly resistive and the Cloutier model becomes viable. We defer discussion of the low-altitude ionospheric obstacle for the moment.

The processes of charge exchange and photoionization were mentioned above without much discussion of what these processes do, and without much evidence that they take place. In a charge exchange an electron passes from a neutral atom to an ion. If the neutral atom is cold and the ion hot, the exchange produces a fast neutral and a slow ion. This can be an important process in the solar wind interaction with Venus, because it directly removes momentum from the postshock solar wind and deposits that momentum in the atmosphere. It also decreases the need for solar wind to be deflected by the planet and thus can weaken the shock and change its location, as we discuss in Secs. IV. A and IV. B. Charge exchange occurs most readily, i.e., has the largest cross section between members of the same species, such as $H-H^+$. However, proton-oxygen charge exchange is accidentally resonant and Venus has a significant hot neutral oxygen exosphere (Nagy et al. 1981). This charge exchange reaction also adds mass to the postshock solar wind.

Photoionization, by extreme ultraviolet emissions from the Sun, adds mass to the postshock solar wind flow. Photoionization of oxygen is very important because the exosphere is dominantly oxygen in the region of interest and because oxygen ions are much heavier than protons. These processes are important in the formation of the plasma mantle and magnetotail of Venus and will be discussed again in Sec. II. C.

C. The Low-Altitude Barrier

There is a noticeable change in the ionopause at low altitudes $\leq$ 300 km. This change can be seen in Fig. 11, which shows the thickness of the ionopause current sheet as a function of altitude. The ionopause is typically $\sim$ 30 km thick above 300 km but increases to $\leq$ 90 km thick at $\sim$ 250 km (Elphic et al. 1981). We attribute this to the change in conductivity and Joule dissipation with altitude. When the ionopause is low, currents can be driven in the ionosphere as in the Cloutier model. In fact, as shown in Fig. 12, the low-altitude ionosphere is magnetized during these periods of low ionopause altitude, which of course are associated with periods of high solar wind dynamic pressure (Luhmann et al. 1980). These high field regions are also

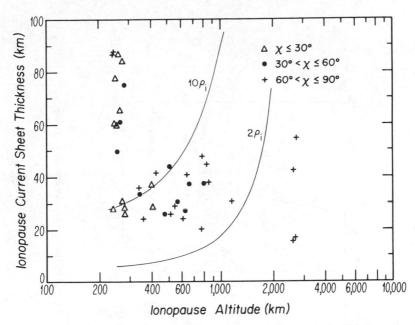

Fig. 11. Thickness of the ionopause current sheet as a function of altitude. Thin lines show 2 and 10 gyroradii (Elphic et al. 1981).

more prevalent at low solar zenith angles; this fact is consistent with these high field regions lying along a belt and not covering the entire ionosphere, and is also in accord with Cloutier's model. Suffice it to say that at times when the solar wind dynamic pressure is high the nature of the solar wind interaction with Venus may be significantly different than under typical solar-wind conditions.

D. Summary

We now know that for all practical purposes Venus is a nonmagnetic planet and the ionosphere is responsible for deflecting the solar wind flow. At times when the solar wind dynamic pressure is low and the ionopause altitude is above ~ 300 km, a magnetic barrier forms which deflects the solar wind before it directly encounters the ionosphere. At higher solar wind pressures, the ionopause moves to low altitudes, the current layer thickens, and a more direct interaction seems to occur in which currents are driven in the ionosphere by the solar wind electric field, i.e., by unipolar induction.

II. THE MAGNETOSHEATH

The region of postshock solar wind flow at Mercury, Earth, Jupiter, and Saturn is called the magnetosheath. As the physics of the postshock flow at

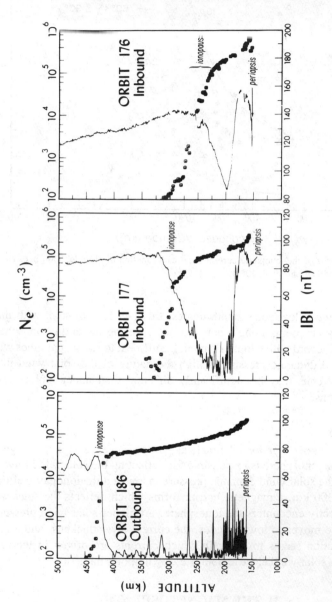

Fig. 12. Altitude profiles of the magnetic field strength and electron density measured by the electron temperature probe for the three passes shown in Fig. 10.

Venus is basically the same as at the other planets, we will call this region the magnetosheath even though the term ionosheath has at times been popular. In this section we examine what is known about the flow properties of the magnetosheath, especially near the ionopause in the region known as the plasma mantle. Are the properties of this region those of a viscous boundary layer transferring momentum from the magnetosheath to the magnetosphere, or are they the properties of a mass accretion layer as sketched in Fig. 10?

A. Magnetosheath Flow

In the terminator regions of the magnetosheath, plasma measurements show that the plasma flow direction is satisfactorily described by the hydro-magnetic model. Figure 13 shows Venera 10 RIEP analyzer measurements of the flow direction across the terminator compared with gas-dynamic calculations (Vaisberg et al. 1976). Note that the solar wind flow does not penetrate the wake region but the cold component does. Note also the heating of the cold component as the center of the wake is approached. Similar measurements have been obtained from a much larger data set on Pioneer Venus by Mihalov et al. (1980). As shown in Fig. 14, over the terminators where the vast majority of the data have been obtained, the flow near the ionopause is tangential to it, as expected. There are only three reported measurements in the range of solar zenith angles (SZA) 40–60°. Of these three, two have substantial components of the flow into the ionopause.

The postshock plasma parameters have been compared with theory by Mihalov et al. (1980), Mihalov (1981), and Smirnov et al. (1981). While there is rough agreement between the observed and expected decrease in velocity across the shock, Mihalov et al. find that the observed decrease in velocity is too small in 80% of the cases for solar zenith angles <70°. Furthermore, the observed magnetosheath ion temperature is close to a factor of two too low everywhere. These observations are supported by detailed case history models using the full gas dynamic modeling of Spreiter and Stahara (1980) in the study by Mihalov (1981). Deeper in the magnetosheath the observed ion temperature is found to be up to an order of magnitude too low. This observation is consistent with the significant decrease in ion temperature in the same region found earlier by Venera 10 (Vaisberg et al. 1976a), shown in Fig. 13. This could in part be due to the lack of information on the electron temperatures. However, in the terrestrial magnetosheath the electron temperature is seldom equal to the ion temperature, let alone greater than it. Furthermore, the ion densities are found to be less than predicted. Smirnov (personal communication, 1981) has reported that the observed ion temperature jumps were equal or somewhat lower than theoretical ones.

B. The Boundary Layer

A principal feature of the venusian magnetosheath is the boundary layer. This layer was characterized on different occasions by an unusually flat ion

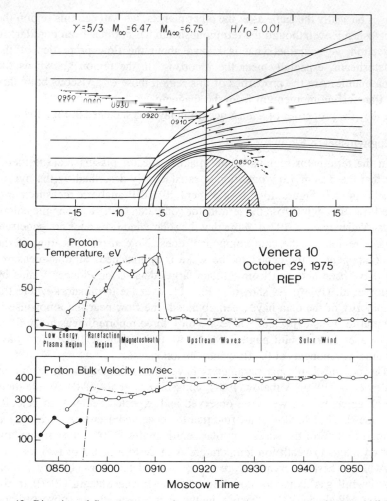

Fig. 13. Direction of flow (arrows on the upper part), ion temperature (in the middle) and velocity (below) as measured by Venera 10 RIEP on 29 October 1975. Location of the shock and the boundary of obstacle as well a flow direction (solid lines in the upper part) are from HD model of Spreiter et al. Dashed arrows show the low-energy ion flow, and closed dots (below) their parameters. Drafting errors in the originally published figure have been corrected. In particular, there is no component of solar wind flow into the wake anywhere on this pass.

distribution function, by the depletion of the energetic tail of the ion spectra, and sometimes is seen as a rarefaction region (Vaisberg et al. 1976a; Romanov et al. 1978). Depletion of the plasma by the loss of energetic (1.5-3.0 keV) ions is clearly seen on successive spectra as measured on two revolutions of Venera 10 (Fig. 15). Magnetosheath spectra (12-5 for 10/31/75 and 10-6 for 11/6/75) show a gradual loss of the energetic tail of the ions while

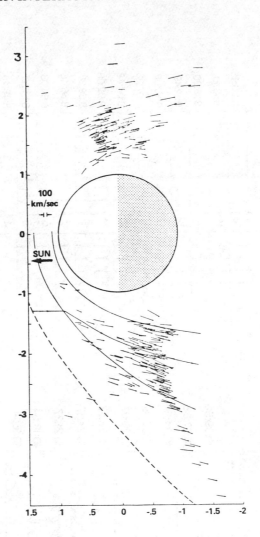

Fig. 14. Magnetosheath flow around Venus as measured by the Ames plasma analyzer in solar cylindrical coordinates. Vectors measured north of Venus are plotted in the upper half of the figure while the remaining vectors were measured south of Venus's orbital plane. Model ionopause and shock surfaces for Mach 8 and Mach 2 (dashed) are shown together with a Mach 8 streamline.

approaching the obstacle. This depletion is apparently due to the loss of energetic ions in the planetary atmosphere. The flow lines on which depletion occurs pass close enough to the atmosphere of Venus (comparable to twice the gyroradii of the energetic ions) for this depletion to take place.

On the night side of the planet, quite far downstream, the boundary layer is particularly apparent in the kinetic parameters of flow. Velocity and ion

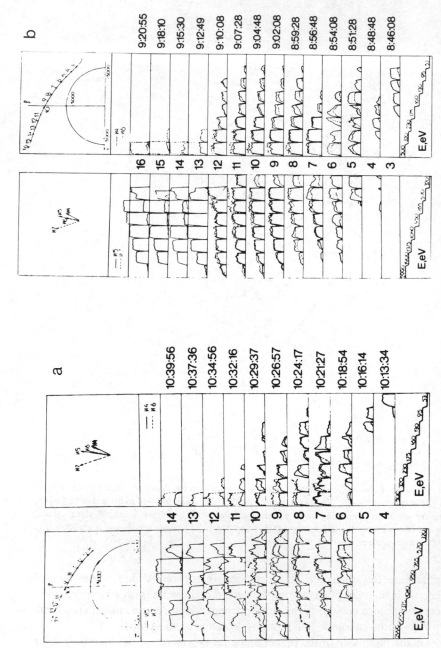

Fig. 15. Successive ion energy spectra as measured by 4 electrostatic analyzers of RIEP on (a) 29 October 1975: and (b) 31 October 1975. Insert in the upper part shows the orientation of analyzers. Numbers on the orbit show the location of spectra. Scale on the bottom shows the energy intervals; time is given on the right.

temperature profiles obtained during the crossing of the venusian interaction region from the bow shock to the central part of the wake, at distances 5-8 R_{φ} from the planet, are shown in Fig. 16 (Romanov et al. 1978). The boundary layer is seen as a decrease of velocity and rise of temperature, as in a viscous boundary layer. The velocity profile, calcuated from the ion energy spectra under the assumption that the flow consists of protons everywhere, changes monotonically across the wake boundary that was crossed at 0255 Moscow time. The temperature drops at the boundary of the wake. As we will see later, this seems due to change of the ion composition across the boundary of the wake.

The viscous interpretation of the solar wind interaction with Venus has been championed by Perez-de-Tejada and colleagues (Perez-de-Tejada and Dryer 1976; Perez-de-Tejada et al. 1977; Perez-de-Tejada 1979; 1980a, b).

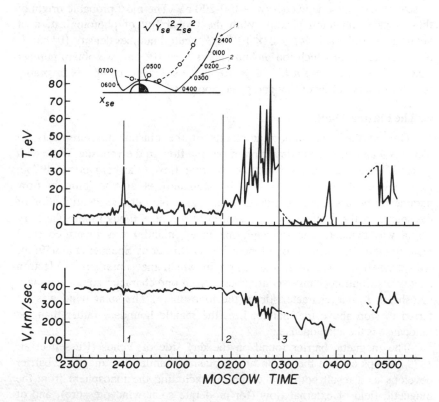

Fig. 16. Flow velocity and ion temperature along the orbit of Venera 10 on 18, 19 April 1976. Orbit of Venera 10 (solid line) is shown above orbit of Mariner 5 (dashed line). Boundary layer is clearly seen by velocity decrease and temperature increase. Temperature drops upon entry into magnetotail plasma. Circles on Mariner 5 trajectory correspond to events 1 through 5 in Fig. 3. Number 1, 2 and 3 on Venera 10 trajectory and time series indicate similar events on Venera 10 pass.

This viscous interaction transfers momentum to the ionosphere causing a decrease in velocity and an increase in temperature as the planet is approached; Fig. 16 agrees with this view. In the region of velocity decrease the temperature rises. On the other hand, these viscous boundary layer treatments of the interaction sweep under the rug the underlying plasma instabilities and wave-particle interactions responsible for the observed plasma behavior. The analogy may be valid, but it does not show how the momentum transfer or the heating occur.

Two-component flow structure near the planet is often evident in the energy spectrum; in addition to the broad spectrum of shocked solar-wind plasma a narrow peak appears in the low-energy part of the spectrum. This peak can be seen in Fig. 15a (spectra 6-8) and Fig. 15b (spectra 5-6). This component is sometimes seen at larger distances from the planet. The number flux of these particles reaches $\sim 10^6$ cm^{-2} s^{-1}, with temperature on the order of several eV and kinetic energy $\sim$ 100-200 eV. The most probable origin of these ions is photoion pickup. With the probability of photoionization of venusian exospheric ions $f \sim 2 \times 10^{-7}$ s^{-1}, neutral number density 10^4 cm^{-3}, and path length in which ion pickup occurs $L \sim 10^9$ cm, we obtain number flux of photoions $F = f n L \sim 2 \times 10^2$ cm^{-2} s^{-1}. This compares favorably with the observed flux of low-energy component.

C. The Plasma Mantle

One of the most important results of the plasma measurements on Veneras 9 and 10 was the detection of plasma flow in the venusian wake with quite different parameters from the external flow (Vaisberg et al. 1976c; Romanov et al. 1978). As shown by Romanov et al. the jump in flow parameters occurs at the boundary of the wake. No definite identification of the source of these ions was suggested until recently, though their planetary origin was evident. The same plasma layer (mantle) was identified from measurements of electrons on Pioneer Venus Orbiter by Spenner et al. (1980), who defined the mantle as the region in which the suprathermal electron energy distribution changes continuously from one characteristic of the magnetosheath to one characteristic of the ionosphere. The solar wind was inferred to stop above the mantle, i.e., the mantle boundary rather than the ionopause is the anemopause.

The magnetic barrier found on the day side of Venus (Russell et al. 1979a) helps explain the origin of the plasma mantle. The magnetic barrier develops as a result both of the current excluding the ionosphere from the magnetic field of external flow (for moderate solar wind pressure), and of magnetohydrodynamic (MHD) effects in the external flow. This barrier transfers the solar wind pressure to the ionosphere (Russell et al. 1979a).

Consideration of the pressure balance across the venusian ionopause shows that the magnetic pressure of the magnetic barrier is slightly smaller than both the solar wind pressure and the ionospheric pressure (Vaisberg et al.

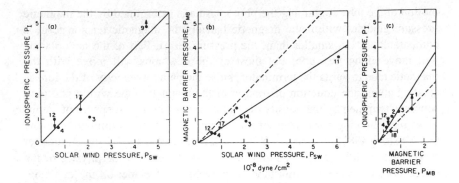

Fig. 17. Comparison of solar wind pressure with (a) ionospheric pressure; with (b) magnetic pressure in the magnetic barrier ; as well as (c) between the magnetic pressure and ionospheric pressure, according to PVO measurements on first 18 revolutions. Solid lines show the best least-square fit. The dashed lines show the variation if the pressures were equal.

1980) (see Fig. 17). This pressure deficiency could be provided by plasma pressure in the barrier (Vaisberg et al. 1980). An even smaller pressure deficiency was found in the analysis of data of Pioneer Venus orbiter (PVO) (Elphic et al. 1980a). In fact, the different estimations of the ratio of plasma pressure to magnetic barrier pressure are probably all consistent with unity within the calibration uncertainty of the plasma detectors.

Observations of the magnetic barrier allow us to determine the topological connection of the venusian wake with the dayside structure. Since the boundary of the wake is the internal boundary of the shocked solar wind and since wake plasma is much hotter and dilute than ionospheric plasma, the magnetic barrier is evidently the dayside counterpart of the wake (Zeleny and Vaisberg 1981). The similarity of the electron components of the mantle on the day side and on the night side of the planet according to PVO data (Spenner et al. 1980) confirms this supposition.

There are three possible sources of plasma in the magnetic barrier: relative diffusion of the ionospheric plasma and the magnetic field; photoionization; and charge exchange in the barrier. To understand some of the properties of the plasma mantle, it is not important how these ions enter the magnetic barrier region, but only that they accrete in the barrier at a specific rate. Zeleny and Vaisberg (1981) have considered a model of the plasma mantle using the assumption that the source of the ions found there is photoionization, together with some simplifying approximations about the dynamics of the magnetic tube in the solar wind approaching ionosphere. Part of the hot plasma filling the flux tube that decelerates on the day side will move along the tube, leading to its depletion (Zwan and Wolf 1976). At the same time the new photoions will fill the tube, and while it sinks into the magnetic barrier the solar wind plasma is replaced by new cold photoplasma. Subsequent motion of the tube containing the photoplasma is determined by the magnetic pressure gradient within the magnetic barrier (the magnetic tension term is comparable to, but smaller than, the pressure term). Part of the new plasma will move along the tube, but most of the new ions will move with the magnetic tube towards the terminator, since the parallel velocity of the ions is small. Solving the equation of motion of the magnetic tube with a photoion source determined by the density of neutral oxygen in the exosphere of Venus gives the following parameters of mantle plasma near the terminator (Zeleny and Vaisberg 1981): velocity ~ 60 km s^{-1}, number density 4-5 cm^{-3}, number flux $\sim 2.5 \times 10^7$ cm^{-2} s^{-1}. Figure 18 shows a sketch of the physics of the interaction. On magnetic flux tubes approaching the planet on the day side, the external plasma is replaced by planetary plasma within the magnetic barrier. This new plasma is convected in the wake along with the magnetic flux tubes. The wake is formed by magnetic field lines loaded with cold planetary plasma stretched out along the wake by the velocity shear in magnetic flux tubes, the ends of which are immersed in the solar wind.

Observations of wake plasma at the distance $\sim 0.5\text{-}1$ R$_{\venus}$ behind the terminator give a number flux smaller by an order of magnitude. This does not contradict the model, as the field tube should expand in the wake by the factor of the ratio of wake radius to magnetic barrier thickness, i.e., by a factor ~ 10. Yet the energy of the directed motion of oxygen ions with velocity ~ 60 km s^{-1} obtained in the model of Zeleny and Vaisberg is almost twice as large as the energy of motion of the tail ions, ~ 150 eV (Vaisberg et al. 1976b; Romanov et al. 1978). Taking the exospheric helium photoions as the source of the mantle, we obtain an even larger discrepancy between the model and the observed energy of wake ions. This discrepancy suggests that anomalous ionization processes investigated recently by Formisano et al. (1982) may be important in the formation of the mantle. The anomalous ionization process will lead to the earlier loading of the magnetic field tube by photoions and diminish the final velocity of plasma. Using the anomalous ionization criteria by Formisano et al. (1982); Zeleny and Vaisberg (1981) were able to obtain agreement between the model of Venus's mantle and observations. On the other hand, recent measurements of the critical ionization velocity phenomenon in weak magnetic fields (Brenning 1981) show that the critical ionization velocity effect disappears when the electron gyro frequency is much less than the electron plasma frequency, as it is everywhere in the vicinity of Venus. Fortunately, there are other candidates for increased ionization. Charge exchange certainly takes place. Furthermore, the photoionization rate is greater than that which was assumed, since the authors used the neutral atmosphere of Niemann et al. (1979a) rather than the more recent Nagy et al. (1981) calculations of the oxygen exosphere. Finally, recent studies by Mihalov and Barnes (1982) of the Pioneer plasma data suggest that O$^+$ is moving at speeds close to but less than that of the ambient magnetosheath plasma near to and sunward of the terminator plane. The most intense fluxes occur at low altitudes, as do

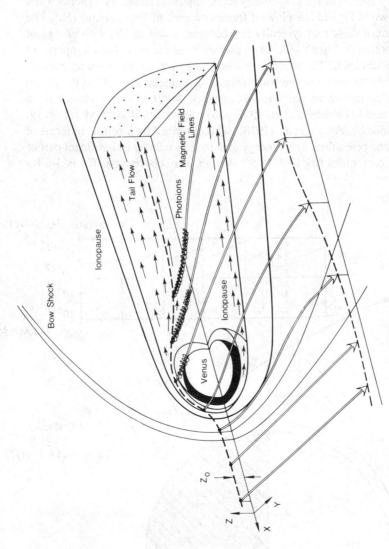

Fig. 18. Model of solar wind-Venus interaction (Zeleny and Vaisberg 1981). Subsequent locations of magnetic field tube are shown. Spirals are trapped photoions; arrows show the flow of low-energy ions in the wake.

the lowest speeds. At altitudes above a few thousand km, the two speeds
are comparable.

D. The Venusian Wake

Energy spectra of the low-energy ion component found on Veneras 9 and
10 are shown in Fig. 15 (two lowest frames on each of Fig. 15a and 15b). The
temperature of these ions typically lies between 1 and 10 eV, with energy of
directed motion ~ 150 eV. Closer to the center of the wake both temperature
and energy diminish. The velocity of low-energy motion has a strong component towards the center of the wake (Fig. 13). Immediately behind the planet
there is a cavity where no ions with energy ≥ 50 eV were observed on
Veneras 9 and 10 (Vaisberg et al. 1976c). Similar results illustrated in Fig. 19,
were obtained by Verigin et al. (1978), who described their results in terms of
an umbra and penumbra. Low-energy ions usually occupy a significant part of
the wake, yet regions free of $E \geqslant 50$ eV ions were observed as far as 5-6 $R_{\female}$
downstream.

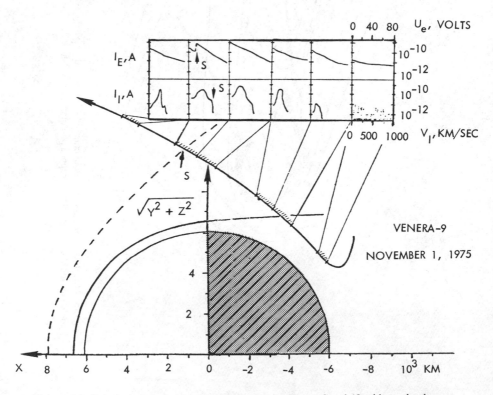

Fig. 19. Ion flux spectra obtained 1 November 1975 by Venera 9 and 10 wide angle plasma
detectors through the wake, into the magnetosheath, and through the shock (Gringauz et al.
1976a).

The layer of plasma in the wake is thick at the high magnetic latitudes, where the vector to the spacecraft measured from the center of the wake is nearly orthogonal to the interplanetary magnetic field direction projected on the terminator plane. The layer of wake plasma is thin at low magnetic latitudes, where the spacecraft position vector relative to the center of the wake is almost parallel to the interplanetary magnetic field direction (Vaisberg 1980). Measurements of the layer of plasma in the wake at different magnetic latitudes are shown in Fig. 20. The cross section of the mantle at a distance $\sim 1\ R_{\varphi}$ behind the terminator is shown in Fig. 21. These observations agree with the model of Zeleny and Vaisberg (1981) in that high-latitude plasma is convected within the flux tubes. Stronger deviation of plasma flow towards

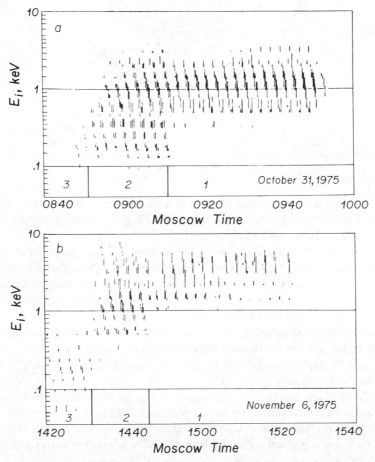

Fig. 20. Dynamic spectra of ions for two passes of Venera 10: (a) 31 October 1975 and (b) 6 November 1975. Length of the bar is proportional to logarithm of the RIEP counting rate in specific energy interval (scale is on the left). (1) solar wind, (2) magnetosheath, (3) wake; (a) low magnetic latitudes, (b) high magnetic latitudes.

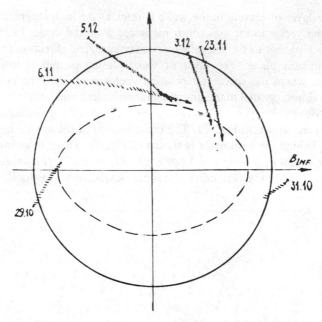

Fig. 21. Projections of the parts of orbits in which low-energy wake plasma was observed. Every orbit was rotated until YZ-component of interplanetary magnetic field was coincident with horizontal (axis). Ring is Venus; dashed oval is suggested boundary of the cavity.

the center of the wake is due to the magnetic tension of bent magnetic-field lines. Low-latitude plasma is squeezed out along a relatively narrow layer of field lines (Zeleny and Vaisberg 1981).

Comparison of the directions of flows near the boundary of the wake, where two plasma components (external and internal) mix, allows us to estimate the mass of the ions in the wake flow. From the measurements of directions of flow and energies of mass motion of these components the velocity of internal flow can be calculated, taking into account that the axis-directed convection velocities of the two components are equal. This estimate gives an ion mass in the internal flow between He^+ and O^+, closer to O^+. In this case the typical velocity of O^+ ions in the wake is ~ 40 km s $^{-1}$ with typical number density ~ 0.5 cm^{-3}.

Cases of quasi-periodic acceleration of ions were observed by Romanov et al. (1978); the periodicity of these events is 5 to 10 min. These events were interpreted by Romanov et al. (1978) as venusian substorms, the existence of which was suggested earlier by Russell (1976c) from the analysis of magnetic data.

The transition from external to internal flow at the wake boundary is quite distinct, although sharper and more gradual transitions were observed. The thickness of the transition layer ranges from 100 to 500 km. Strong fluctuations in the magnetic field with $\Delta B \simeq B$ are usually recorded near the wake boundary. An example of a wake boundary crossing (Romanov et al. 1978) is shown in Fig. 22, in which a sharp transition from one flow regime to another one is seen, as well as strong magnetic field fluctuations. A burst of ions with energy ~2 keV is also observed at the boundary. Bursts of accelerated ions were observed at almost all crossings of the wake boundary. They usually lasted from 10 to 20 s in the spacecraft time frame. It appears that the acceleration of ions to an energy of several keV occurs permanently at the wake boundary and that it is associated with strong magnetic field fluctuations. A synthetic energy spectrum of accelerated ions was obtained, combining the measurements of these ions on different revolutions of the Venera 10 spacecraft (Fig. 23) assuming the acceleration is steady-state. This procedure was used because the time of an energy-spectrum scan is longer than the time of the boundary crossing.

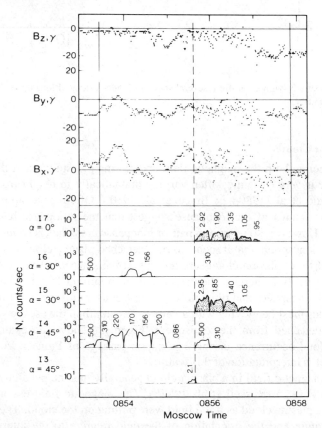

Fig. 22. Crossing of the wake boundary on 3 December 1975. Vertical dashed line marks the wake boundary. Magnetic field components are shown on the upper panels. The lower five panels show the measurements of the five electrostatic analyzers of RIEP: I3, (2.0-20 keV); I4, I6, (0.05-0.5 keV); and I5, I7, (0.3-3.0 keV).

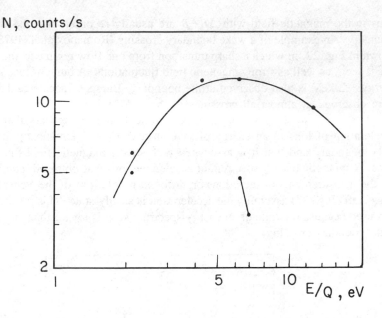

Fig. 23. Spectrum synthesized from several spectra of the accelerated ions upon crossing the wake boundary.

E. Plasma Clouds

As seen in both the magnetic field and plasma measurements, the Venus ionopause is very dynamic, often varying dramatically in height from day to day (Elphic et al. 1980a, b; Brace et al. 1980). Often what appear to be multiple crossings of the ionopause suggest that the ionopause has a wavy structure. However, sometimes a pair of ionopause crossings is so far from the rest of the ionosphere that it seems to be a detached plasma cloud (Brace et al. 1982a). If these plasma clouds are moving at the solar wind velocity or any significant fraction thereof, they represent an important loss mechanism for the Venus ionosphere and ultimately for the Venus atmosphere. At minimum these clouds must be moving with a velocity greater than the escape velocity if they are detached from the ionosphere. Figure 24 shows an example of a plasma cloud occurrence during a periapsis passage of Pioneer Venus, and Brace et al.'s interpretation of this feature.

The magnetic field in a plasma cloud generally has the peculiar variation shown in Fig. 25 (Russell et al. 1982b). The magnetic field essentially reverses in a plasma cloud as if the field were pulling on the cloud. The location of this feature and the orientation of the field relative to the interplanetary magnetic field orientation is that which would be expected if the field lines were hung up near the subsolar point and bent into a hairpin shape. This feature is shown projected on the surface of the planet in Fig. 26. The plasma

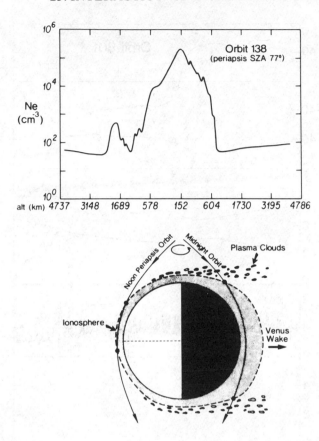

Fig. 24. Electron density profile through periapsis (SZA 77°) on Pioneer Venus orbit 138 containing a plasma cloud. Lower panel shows interpretation of these density enhancements (Brace et al. 1982a).

cloud and the bend in the field are along the noon-midnight magnetic meridian, the great circle through the subsolar point which is perpendicular to the interplanetary magnetic field. This observation suggests that the plasma cloud is, instead, a ridge of plasma on the ionosphere, presumably of flowing plasma. The plasma is accelerated and prevented from spreading by the magnetic field. Note the strong 100 Hz plasma waves occurring in the cloud. Increases in the plasma wave amplitudes are common at the ionopause but these have a different spectrum.

The Maxwell stress on this plasma cloud is $\sim 4 \times 10^{-8}$ dyne cm^{-2}. If we approximate the plasma cloud with a vertical slab 3° thick assumed to be oxygen, it should be accelerated by this stress at 0.7 km s^{-2}. If the plasma started at rest at the subsolar point, it would reach velocity 90 km s^{-1} at the

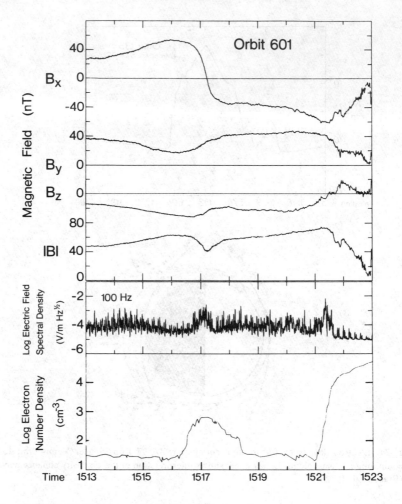

Fig. 25. Magnetic field in solar ecliptic coordinates, plasma wave amplitude and electron density through a plasma cloud (orbit 601) (Russell et al. 1982b).

point of observation with transit time 130 s if the acceleration were constant. If we assume that this accelerating cloud has height 500 km, the distance down to the ionopause as observed later on this pass, and that there is a similar cloud in the south, we would obtain a loss rate of 2×10^{25} ions s^{-1}. We can also estimate the plasma temperature from the diamagnetic depression observed at the center of the cloud in Fig. 25. The sum of electron and ion temperatures is 7.2 eV. Since the electron temperature was observed to be 1.2 eV at this time, the ion temperature must be 6 eV. If the ions are O$^+$, this temperature corresponds to a thermal speed of 8.5 km s^{-1} close to the escape velocity. This low temperature compared to that expected for creation in the

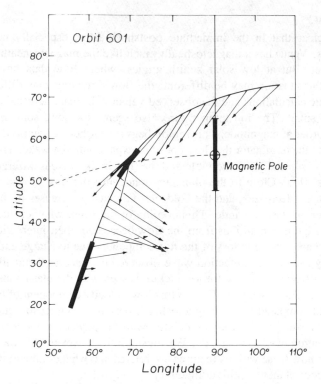

Fig. 26. Magnetic field on pass (orbit 601) shown in Fig. 25 projected on to surface of planet. Black bars on trajectory show plasma cloud and entry into ionosphere. Magnetic pole is intersection of noon-midnight magnetic meridian with terminator. Noon-midnight magnetic meridian is perpendicular of Y-Z ecliptic projection of interplanetary magnetic field, and passes through subsolar point.

solar wind electric field indicates that the ions were created either in the ionosphere or in a low-velocity zone outside the ionopause.

The frequent occurrence of these clouds in the Pioneer data suggests that they are an important contributor to the loss of ions from the Venus ionosphere. The magnetic structure suggests that the dayside ionopause is not quasi hemispherical but rather ridged along a noon-midnight magnetic meridian. The two possible sources of these clouds or ridges are the ionosphere itself and the mass accretion region or mantle outside the ionopause. The first possibility leaves unexplained how the field penetrated the plasma; likewise, the second possible source does not explain how the field lines gathered so much plasma in their passage from the subsolar region. On the other hand, the cloud loss rate estimated here is not much different from the expected loss due to photoionization and charge exchange above the ionopause, and the field lines in the cloud must have swept across the entire dayside ionosphere before being observed. Nevertheless, the cloud formation is still somewhat of a mystery.

F. Summary

It is clear that in the immediate postshock flow, especially near the terminators, Venus has a magnetosheath much like the magnetosheath of any other planet. But at low solar zenith angles, where little data have been gathered, the situation may be different; the flow directions may differ from gas-dynamic calculations and the observed velocity decrease across the shock seems too small. The ion temperature also seems too low, sometimes by almost an order of magnitude. This region begs for further exploration.

Beyond the terminator the flow does not expand into the wake. The wake boundary is essentially impenetrable and the flow in the wake is different from the external flow. Close to the planetary ionopause the flow slows, the hot plasma density decreases, and the cold plasma density increases in the magnetic barrier or plasma mantle. This cooling of the flow, whether due to the loss of the high gyroradii ions from the flow because of their interaction with the atmosphere, or due to loss of the hot component due to charge exchange, presumably causes the rarefaction wave observed by Veneras 9 and 10 (Vaisberg et al. 1976b). Evidence for ion pickup is seen with the plasma analyzers in this region. As is evident in Fig. 13 this low-velocity component of the flow is very cold compared to the magnetosheath protons. Another ion phenomenon perhaps even more closely associated with the physics of the ionopause, is the formation of plasma clouds. We must admit, however, that the understanding of how these various features are related and what dominant physical processes control them is still rudimentary.

III. THE MAGNETOTAIL

All planets with an intrinsic magnetic field have a magnetotail, which arises mainly because of tangential stresses applied by the solar wind. If there were only normal stresses on a planetary magnetosphere, it would assume a teardrop shape and would not have a long tail. Comets also have long tails; one type consists of dust, but the other consists of flowing ions presumably constrained by the magnetic field of a cometary magnetotail. A comet is not thought to have an intrinsic magnetic field; instead, the cometary magnetotail is thought to arise by capture of interplanetary magnetic flux by mass-loading of field lines which enter the cometary atmosphere (Alfvén 1957). Processes such as charge exchange, photoionization, and possibly even the critical ionization velocity process (Alfvén 1954) will add mass to the field lines in the region of closest approach to the comet. This mass accretion slows the motion of the magnetic flux tube. The plasma near the comet moves more slowly while that farther away moves at the solar wind velocity. The field line then bends around the comet and stretches out in the antisolar direction. Eventually these draped field lines will slip through or around the comet. In analogy with Dungey's (1965) estimate of the length of the terrestrial magnetotail, we can estimate the length of a cometary tail as the distance the solar wind travels as the field line drifts through or around the comet. If this slippage past the comet takes about half a day, the resulting tail would be ~ 0.1 AU long. However, the situation is not perfectly analogous. Once the field line has left the mass-loading region it will attempt to straighten itself, but can only do so slowly because of the added mass at the crook of the field line.

We would expect this same cometary process to occur at Venus, though we cannot see any visible evidence for a venusian tail. The expected atmospheric loss rates at Venus, on the order of 10^{25} ions s^{-1}, are many orders of magnitude below the loss rates for a large comet. However, our discussion in earlier sections demonstrates that processes analogous to those occurring in comets occur at Venus. We expect the field lines to be slowed as they move across the front of the planet. The plasma clouds discussed in Sec. II. E seem to be a manifestation of this process. The field lines then slip over the poles, or more precisely over the magnetic poles, of the planet, all the while trying to straighten up and thereby accelerating the plasma contained in the crook of the field line. This is what we expect, but do the observations support this picture?

A. First Observations

As shown in Figs. 3 and 4 Mariner 5 made a close approach to the axis of the Venus wake although it did not pass directly behind the planet. Close to point 3 of Fig. 3 the magnetic field became steady and was roughly aligned with the solar wind flow. This behavior suggests that Mariner 5 entered a Venus magenetotail though it could not be distinguished whether this tail was due to intrinsic field sources or to 'hung up' interplanetary field lines. Figure 4 illustrates the alignment of the field more clearly.

The Venera 9 and 10 orbiters probed the wake region often and at various distances. The properties of the tail region as seen in the magnetometer and plasma data much resembled those of Earth's magnetotail. Figure 27 shows ion fluxes recorded by the narrow angle plasma detectors in the distant tail (Romanov et al. 1978). The top panel shows magnetic field strength; the bottom panels, ion flux versus time and energy. The instrument does not sample all energies simultaneously, but samples one energy for 20 s, and then steps to the next lowest energy for 20 s, etc. Thus, time runs both upwards and to the left in this diagram. The feature of interest here is the periodic increases and decreases in the magnetic field strength accompanied by energizations of the plasma and sometimes followed by dropouts of the flux. Romanov et al. (1978) interpret these as substorms.

At this point we should give credit to the guidance of laboratory simulations in our interpretation of the Venus data. While the solar wind interaction with Venus cannot be duplicated in the laboratory, qualitative scaling can be performed in which ratios of scale lengths have the right ordering if not the proper size. Venus has been simulated with a wax sphere having a conducting core in a magnetized supersonically flowing plasma (Dubinin et al. 1978).

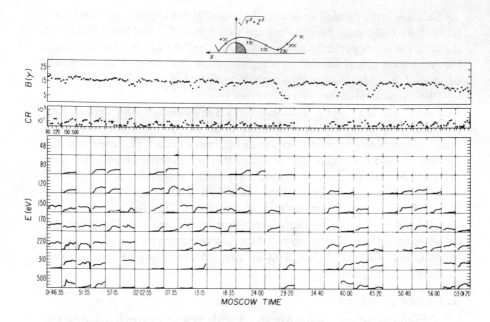

Fig. 27. Ion fluxes as a function of energy and time recorded by the narrow angle detectors in the distant tail on 17 April 1976 (Romanov et al. 1978). The top panel shows 20 s averages of the magnetic field; the bottom panels show the ion fluxes. The instrument does not sample all energies simultaneously but rather stays at each energy for 20 s and then steps to the next lowest energy; after 160 s it repeats the cycle.

The results of this simulation are shown in Fig. 28. The interplanetary magnetic field, as we expect from the cometary analogy, becomes draped around the obstacle. It is not clear, of course, whether the hanging up of the field lines is due to unipolar induction in the planetary ionosphere or mass-loading ahead of the obstacle. Dubinin et al. (1978) point out that these results show that the mere existence of a tail is no proof of an intrinsic magnetic field. The dependence of tail field polarity on the interplanetary magnetic orientation must be investigated.

The Venera 9 and 10 wake data, however, appear so much like similar observations in the Earth's magnetotail that Dolginov et al. (1978) interpreted them in terms of an intrinsic magnetic field. The surprising aspect of this interpretation is that the field lines do not radiate from the polar regions of the planet, but are confined to a conical region expanding away from the core. As we discuss below, the tail observed by PV flares slightly but shows no evidence of converging on to the planet this rapidly. On the other hand, there are small regions of radial field lines on the night side of Venus associated with holes in the nightside ionospheric density. Perhaps observations of the magnetic field in these nightside holes led to Dolginov et al.'s interpretation.

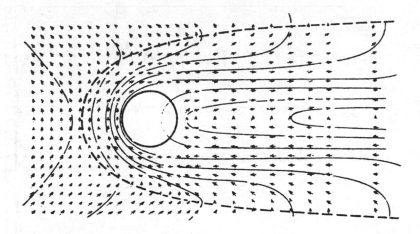

Fig. 28. Distribution of magnetic field vectors near a laboratory model of Venus (wax sphere 6 cm in diameter) (Dubinin et al. 1978).

The interpretation is, of course, incorrect; Venus does not have an intrinsic field large enough to supply the magnetic flux in such a tail. Yeroshenko (1979) correctly pointed out that the Venus tail polarity was controlled by the interplanetary magnetic field. Currently, none of the various principals involved are holding out for an intrinsic magnetotail at Venus.

B. Pioneer Observations

Because of the high inclination of the Pioneer Venus orbit, the spacecraft cuts through the wake in two regions of limited radial extent: one near the planet at $\lesssim 2000$ km altitude; and one far down the wake from ~ 7 to $12 \, R_{\varphi}$. Figure 29 shows the magnetic field on a pass through the distant wake. In the center of this period there is an interval of enhanced though fluctuating field strength, during which the field is mainly oriented along the solar wind direction (the X-direction). This interval has been defined as the magnetotail (Russell et al. 1981). The frequent reversals in the X-component signal tail current sheet crossings. Their coincidence with the low-field regions imbedded in the high-field region suggests the current sheet has an associated plasma sheet, in analogy to the Earth's plasma sheet and tail current sheet. Since the interplanetary magnetic field is usually close to the ecliptic plane, the tail current sheet should be mainly in the north-south plane, rather than the ecliptic plane as the Earth's current sheet is. The high inclination of the Pioneer orbit causes the spacecraft to fly roughly parallel to this sheet. Perhaps flapping of the tail causes the multiple current sheets observed. However, it is also possible that the Venus tail is actually striated.

Figure 30 shows another passage through the tail as seen in the Pioneer magnetic field data. In this figure we have added the amplitude of plasma

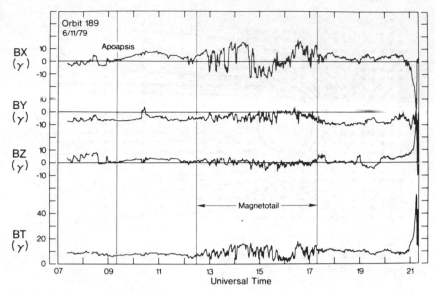

Fig. 29. Pioneer magnetic field measurements in the Venus wake region (orbit 189, 6/11/79). Coordinate system is spacecraft coordinates, which is very close to solar ecliptic.

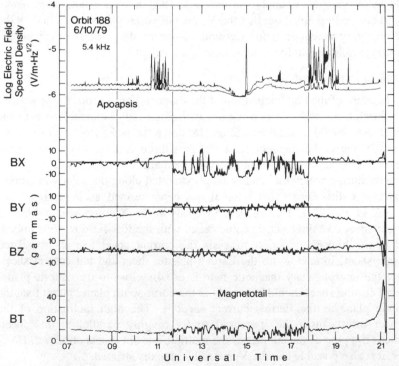

Fig. 30. Pioneer magnetic field measurements in the Venus wake together with the plasma wave amplitudes at 5.4 kHz (orbit 188, 6/10/79) (Russell et al. 1981).

waves seen at 5.4 kHz. The striking observation here is that the region immediately outside that identified as magnetotail has strong plasma wave emissions. Some special process occurs here in the plasma, but at this time we do not know what it is.

Figure 31 shows the location of these tail crossing intervals as projected in the plane perpendicular to the solar wind. The region is much larger than the optical shadow of Venus, ~ 4 R_φ across, and it is shifted from the optical shadow because of the 35 km s^{-1} planetary motion of Venus perpendicular to the solar wind. The circle we have drawn does not order the data perfectly, probably because of the variability of solar wind direction relative to the solar direction.

Figure 32 shows these locations as a function of distance along the Venus-Sun line, together with the distant bow shock locations. Also shown are corresponding observations with Mariner 5 and Veneras 4, 9, and 10. The tail is seen to expand slightly behind the planet and is consistent with the source of field lines being in the dayside magnetic barrier. However, some of the tail field lines certainly penetrate the night ionosphere. The nighttime ionosphere field is generally weak, but there are two magnetic features which

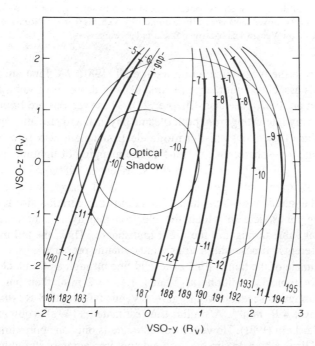

Fig. 31. Terminator plane projection of orbit during Pioneer tail crossings (heavy lines). Orbit numbers are given at the start of each pass (bottom of panel). Numbers along trajectory give the x distance at tail entry and exit (Russell et al. 1981).

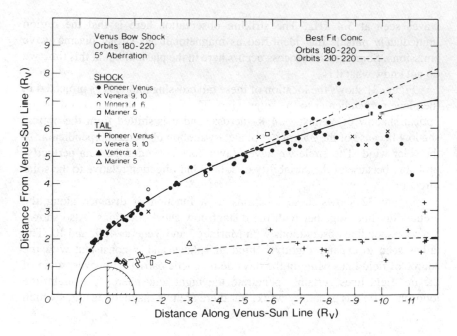

Fig. 32. Location of distant bow shocks and tail encounters in solar cylindrical coordinates (Russell et al. 1981). Boxes show near tail encounters by Veneras 9 and 10 (Romanov et al. 1978); triangles show Venera 4 and Mariner 5 near tail encounters.

may be related to the magnetotail (Luhmann et al. 1981a,b). First are radial magnetic fields associated with large plasma density dropouts or ionospheric holes (Brace et al. 1982b; see also Chapter 23). These regions are limited in extent but are very interesting from both plasma physical and also atmospheric viewpoints, since they allow whistler mode electromagnetic waves to propagate through the ionosphere, thus permitting the mapping of lightning source locations (W. W. L. Taylor et al. 1979b; Scarf et al. 1980a). The second feature is the occurrence of large horizontal low-altitude magnetic fields when the solar wind dynamic pressure is high. This field configuration also leads to a disappearing ionosphere in which the electron density is suppressed far below its usual value throughout the night ionosphere. Thus, the tail may be somewhat different at times of high solar wind dynamic pressure.

If our estimate of the transit time of a field line through a plasma cloud is appropriate for typical conditions at Venus, i.e. $\sim$ 2 min, then our introductory tail formation scenario would give a Venus tail length 8 R_φ since the solar wind moves 4 R_φ min^{-1}. A similar time estimate ($\sim$140 s) is obtained by Zeleny and Vaisberg (1981). However this estimate is only an approximation, since the field lines which do slip by Venus do not immediately straighten up. Thus we are not surprised to see a well-developed magnetotail at 12 R_φ. Another implication of this slippage is that all the field lines in this tail do not close across the day side of Venus, nor in the night ionosphere. Instead, many field lines close across the tail current sheet. We note that the magnetic flux in the tail is on the order of 3 megaweber. Even if we assume that the field in the magnetic barrier is 100 nT and 500 km thick, which are large estimates, there is only 1 megaweber across the front of the planet. Holes have < 0.5 megaweber. Again, this calculation suggests that there must be much closure across the current sheet.

Examination of the magnetometer data at high resolution across a current sheet in the center of the tail is shown in Fig. 33. The crossing seems to be that of a MHD tangential discontinuity with little normal component at this time. The direction of the normal to this current sheet was in the solar ecliptic Y-direction, as expected (Russell et al. 1982c). Figure 34 shows similar data across the outer boundary of the magnetotail; it shows the behavior of a rotational discontinuity. The direction of the normal to the boundary here was mainly north-south, as expected for a magnetopause crossing at the position of the spacecraft. Here there is a measureable normal magnetic field, as if the magnetosheath magnetic field and the tail field were interconnected.

The comparison of the Pioneer plasma and magnetic field data in this region is still at an early stage. Heavy ions are supposedly seen in this region (Mihalov et al. 1980) but sometimes there are no heavy ions, and sometimes there is no detectable plasma at all. Much work remains to be done.

C. Summary

There is agreement now that Venus has a magnetic and plasma tail. However, much of that field does not originate in the ionosphere but comes from the magnetic barrier and some of it, the older field lines, close across the center of the tail itself. It is perhaps incorrect to call such a tail induced, because it does not arise from either of the processes usually called induction. It is mainly an accreted tail. The tail in many respects seems similar to the Earth's tail, but this may be deceptive. It will be interesting to discover the exact interrelationship of the proton and oxygen flow regions to the magnetic structure. Furthermore, much work remains to be done on the microphysics of the tail. What is the cause of the plasma wave emissions seen bounding the tail? What is the usual hydromagnetic character of the tail current sheet and magnetopause? Is the formation of holes in the night ionosphere related to tail formation?

IV. THE BOW SHOCK

It is clear from the observations reported above that Venus presents a fairly hard obstacle to the solar wind flow. Only a small fraction of the solar wind appears to be absorbed. Since the solar wind flows much faster than the speed of pressure waves, i.e. magnetosonic waves, a shock wave must develop in the flow. This shock wave both heats and deflects the flow. One might not expect the strength of this shock or its dependence on solar wind

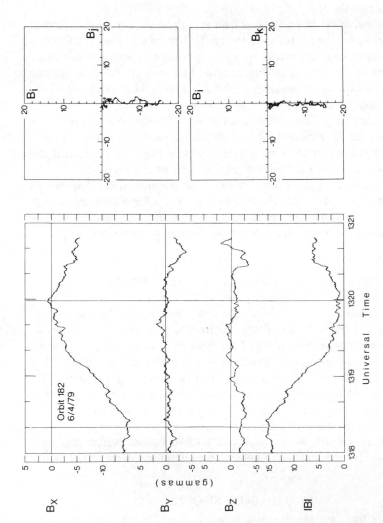

Fig. 33. Magnetic field in the Venus tail at current sheet crossing. Left, time series in spacecraft coordinates. Right, hodogram of tip of field vector between intervals marked by vertical lines on right, expressed in minimum variance coordinates (orbit 182, 6/4/79) (Russell et al. 1982c).

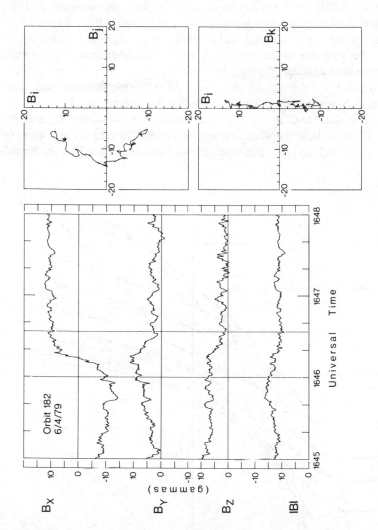

Fig. 34. Magnetic field in Venus tail at magnetopause crossing (orbit 182, 6/4/79) (Russell et al. 1982c).

conditions to differ markedly from the terrestrial bow shock, since the plasma conditions in the solar wind at Venus are similar to those at Earth. However, it is of interest to see how the Venus shock differs from the terrestrial shock. Disparities that could affect the shock are: the scale size of the Venus shock, one tenth of the size of the Earth's; and momentum removal and mass addition by such non-MHD processes as charge exchange and photoionization. There are a variety of tests we may employ. We can examine the position of the shock, the jump in field strength and number density at the shock, the dependence of the structure of the bow shock on plasma conditions, and the wave phenomena seen upstream from the bow shock.

As a point of reference for our observations we use the numerical computations of Spreiter et al. (1966) who studied the interaction of a supersonic gas-dynamic flow with a nonabsorbing nonviscous boundary. Figures 35, 36, and 37 show the streamlines and wave patterns, the velocity and temperature contours, and the mass flux contours for supersonic flow past the Earth's

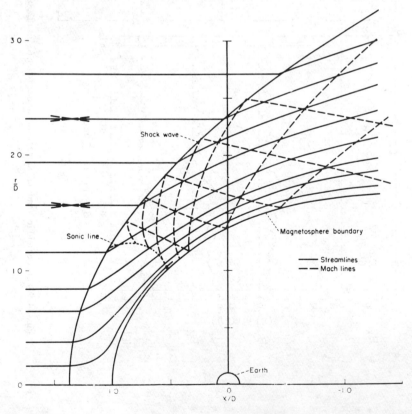

Fig. 35. Gas dynamic computer simulation of interaction of solar wind with an impenetrable object (Spreiter et al. 1966). Streamlines and wave patterns for supersonic flow past the magnetosphere are shown ($M_\infty = 8$, $\gamma = 2$).

magnetosphere for Mach number 8 and ratio of specific heats 2. We use a
specific heat ratio of 2 because such a ratio has been repeatedly observed to
give the best fit in studies of the bow shock position (Fairfield 1971; Zhuang
and Russell 1981) even though $\gamma = {}^{5}/_{3}$ is most frequently used in simulations.
We also note that a slightly lower Mach number would be more appropriate; it
would result in a bow shock slightly farther from the planet.

Spreiter et al (1970) have presented similar diagrams specifically for a
nonmagnetic planet. However, all the figures are given for a ratio of specific
heats of ${}^{5}/_{3}$ rather than 2. Their paper presents a correspondence rule that
relates diagrams derived for the Earth's magnetospheric interaction to ionospheric interactions. In particular, if the scale height of the Venus ionosphere
is 1500 km, then the curves are identical with the subsolar ionopause coincident with the subsolar magnetopause. If the scale height is 750 km, the same
curves can be used by shifting the center of the nonmagnetic planet away from
the Sun by 840 km. If the scale height is 75 km, the same curves can be used

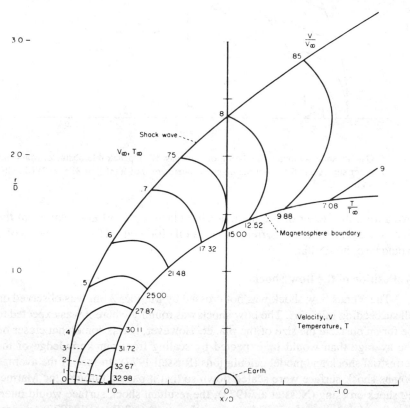

Fig. 36. Gas dynamic computer simulation of flow past an impenetrable obstacle. Velocity and
temperature contours for supersonic flow past the magnetosphere are shown ($M_{\infty} = 8$, $\gamma = 2$) (Spreiter et al. 1966).

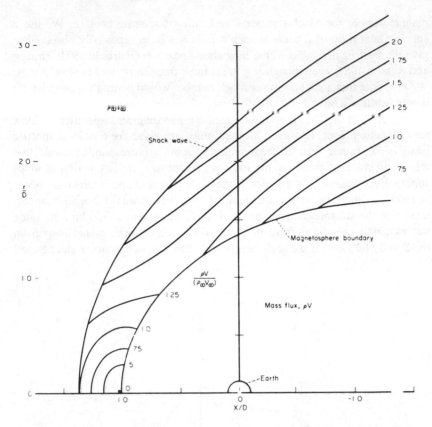

Fig. 37. Gas dynamic computer simulation of flow past an impenetrable obstacle. Mass flux contours for supersonic flow past the magnetosphere are shown ($M_\infty = 8$, $\gamma = 2$) (Spreiter et al. 1966).

by shifting the center of the nonmagnetic planet by 2430 km away from the Sun. We assume that the correspondence rule for a ratio of specific heats of 2 would be quite similar.

A. Position of the Bow Shock

The Venus bow shock was not crossed by Mariner 2 but was observed on all succeeding missions. The bow shock was roughly where it was expected to be for an obstacle the size of the planet. However, it was somewhat closer on the average than would be expected by scaling from our knowledge of the terrestrial shock and model calculations (Russell 1977). In fact, if the average Venus shock surface were scaled down so that it passed through the Mariner 10 shock crossing (Ness et al. 1974), the resultant shock surface would intersect the planetary surface. This observation prompted Russell (1977) to question whether, on occasion, the Venus shock could become attached to Venus as shown in Fig. 38. Even if the anomalous Mariner 10 crossing could be

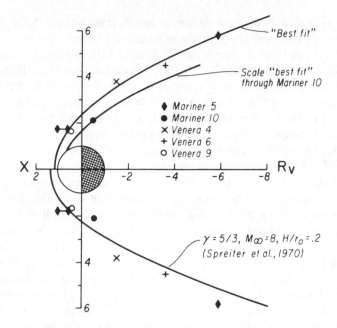

Fig. 38. Location of Venus bow shock up to initial Venera 9 reports (Russell 1977).

explained, there is still the implication that on the average 30% of the solar wind streamlines incident on Venus passed into the ionosphere, if we naively use the gas dynamic computations of Spreiter et al. (1966) to map the flow. This suggestion was criticized by Verigin et al. (1978) who claimed that there was nothing unusual about the Venus bow shock position as measured by Veneras 9 and 10. However, later analyses of an extended Venera 9 and 10 data base showed unequivocally that the subsolar bow shock, as well as the bow shock at the terminators, was well below the expected location (Smirnov et al. 1980). The subsolar shock is expected to occur at an altitude of $\sim$ 3000 km, but the subsolar altitude of the bow shock is close to half this value. It lies only slightly $>$ 1300 km above the surface of Venus, in surprising agreement with the initial estimate of 1450 km altitude, using only 7 data points and an approximate aberration correction (Russell 1977).

A further surprise in the study of the shock location came with the availability of Pioneer Venus observations. The bow shock near the terminator (where the Pioneer Venus observations occurred because of the orbital inclination) was farther from the planet than it was during the Venera 9 and 10 mission (Slavin et al. 1979a). The extrapolated subsolar point was also higher, $\sim$ 2270 km rather than $\sim$ 1500 km, but still much closer to the planet than would be expected for complete deflection of the solar wind. However, when aberration corrections are made this difference is much less than originally reported (Smirnov et al. 1981).

Slavin et al. attributed the difference between the Venera 9 and 10 positions and the Pioneer orbiter positions to solar cycle effects. Veneras 9 and 10 obtained data at solar minimum and Pioneer at solar maximum. However, latitudinal asymmetries controlled by the interplanetary magnetic field (IMF) direction had been predicted by Cloutier (1976) and reported in the Venera data by Romanov (1970). A search for these asymmetries revealed no such effect in the Pioneer data (Slavin et al. 1979b). In order to reconcile the differences between the Venera and Pioneer data, Smirnov et al. (1981) proposed a slightly different mechanism: the additional aberration of the shocked solar wind caused by magnetic stresses in the magnetosheath (Walters 1964) tilted the shock nose in the direction of planetary motion, moving the shock in, on the average, above the evening ($-Y$ solar ecliptic) terminator and out, on the average, above the morning terminator. The Venera 9 and 10 shock crossings were recorded mainly on the west side of the planet and were strongly affected by any shock tilt, whereas the Pioneer data at high latitudes would be little affected by this process. The available data support this interpretation, as shown in Fig. 39. Note that the terminator projection of the shocks as determined in this analysis has only a 5% asymmetry, presumably due to either solar cycle effects or latitudinal differences as proposed by Romanov. We may be able to test the possibility that the difference is related to the solar cycle with the Pioneer data as the solar cycle declines.

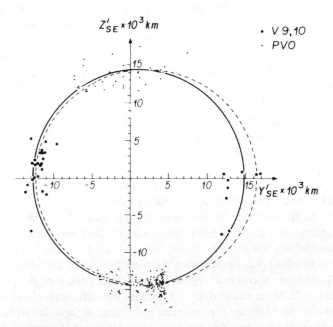

Fig. 39. Location of Venus bow shock in terminator plane as observed by Pioneer Venus and Veneras 9 and 10 (Smirnov et al. 1981).

Thus, the study of the location of the Venus bow shock has led to two surprises: (1) the bow shock is closer to the planet than expected for complete deflection of the flow; and (2), while the shock may show additional aberration by the Walters effect, it shows little asymmetry about this aberrated flow direction. We examine the implications of these two observations, taking the second one first. There are two reasons we might expect an asymmetry about the solar wind flow which is ordered by the IMF (cf. Cloutier 1976). First, the effective obstacle may be asymmetric. Ions created in the postshock solar wind flow will have large gyroradii of several thousand km, because they feel the solar wind electric field. Depending on the orientation of the IMF these ions will either initially spiral away from the planet and escape, or spiral into the ionosphere and be lost. If this mass-addition process is an important contributor to the deflection of the solar wind, then there should be a north-south asymmetry. However, the magnetic barrier is a region of weak solar wind and slow flow. The electric field is not expected to be large near the ionopause, where the rate of production of these ions would be greatest. The second proposed source of asymmetry is the asymmetric propagation of magnetosonic waves in a magnetized plasma. This effect may be detectable far downstream from Venus, but the shock shape over the front of the shock is principally determined by the shape of the obstacle. This fact can be appreciated by noting the relative insensitivity of the shock stand-off distance to changes in Mach number for usual solar-wind Mach numbers (Spreiter et al. 1966), and by referring to Fig. 35 in which the Mach lines indicate what portion of the shock is sensitive to what portions of the ionopause. In short, the shock is quite symmetric because it is only weakly sensitive to obstacle shape and furthermore, because the obstacle is quite symmetric.

The unexpected closeness of the Venus bow shock is less easy to explain. The Venus ionopause generally appears to behave as an equipotential surface. If only 10% of the solar wind flow directly interacted with the ionosphere, a potential drop of ~ 5 kV would be applied across the planet (400 km s^{-1} times 10 nT times 1200 km). This is to be compared with the 40 V of the model of Daniell and Cloutier (1977) and the ~ 1 V of Zeleny and Vaisberg (1981). The solution of this paradox must lie in the nature of the interaction. If momentum is removed from the flow via charge exchange, then the direct interaction with the neutral atmosphere occurs via fast charge-neutral particles. The charged particles can flow around the planet in the usual way, in which the single streamline intersecting the stagnation point coats the entire ionopause. The direct loss of momentum to the atmosphere obviates the need to bend the flow as sharply around the planet, and the shock can turn more normal to the incident flow throughout the subsolar region. This blunter shock will be closer to the subsolar ionopause than the normal shock. The loss of momentum to the planet will also alter the mass flux contours seen in Fig. 37. Charge exchange does not alter the number density (although it could alter mass density) but it will drastically reduce the velocity of the plasma and thus reduce the mass flux crossing the terminators. This will lead to further reductions in the height of the shock in the terminator region, even though the shape might not be affected here. With recent advances in numerical simulations these ideas could be readily tested, but this has not yet been done; until it has, we must treat these ideas only as a plausible scenario. On the other hand, observations of shock strength and calculations of processes that could lead to weakening the shock do support these ideas. Furthermore, studies of mass addition at the subflow point, a process which differs in sign from the effect of charge exchange, show the shock to move away from the obstacle (Cresci and Libby 1962).

Finally, Slavin et al. (1980) have studied the dependence of the terminator crossing of the bow shock on the ionopause altitude, the Alfvénic Mach number, and the solar wind pressure. They find that the position of the shock near the terminators depends little on the ionopause altitude, and weakly on solar wind pressure. The correlation with Alfvénic Mach number is much higher (0.7), corresponding to a more distant shock at low Mach numbers. This correlation is presumably due to two factors: a change of stand-off distance, especially at the lowest Mach numbers; and a change in the flaring angle of the shock.

B. The Strength of the Bow Shock

Strictly speaking, the strength of the bow shock is measured in terms of its Mach number. However, we operationally determine how strong a shock is by measuring the jumps in magnetic field density and temperature or the decrease in velocity across the bow shock. At low Mach numbers $\leqslant 3$ the jumps in magnetic field and density are very sensitive to the Mach number but they asymptotically approach upper limits as the Mach number increases above 3 (Tidman and Krall 1971). Mach number 3 is also the approximate dividing line between laminar and supercritical shocks. Supercritical shocks have abundant downstream ion heating, are more turbulent, and are presumably associated with upstream ion beams although this latter conjecture has not yet been adequately tested with terrestrial data. We would naively expect that since the bow shock stands in the flow, and therefore its velocity is equal to that of the solar wind but oppositely directed, the Mach number would be specified by knowledge of the upstream solar wind conditions. However, mass addition and charge exchange complicate the picture, so we cannot make this simple assumption, as we will demonstrate from observations.

The first investigator to point out something strange about the Venus bow shock was Wallis (1972), who proposed that because of mass accretion due to photoionization there might be no shock directly in front of Venus. He felt that the failure to detect upstream energetic particles at Venus supported this view. However, as we will discuss below, the structure of the Mariner 5 shock has another explanation and there certainly are upstream particle beams at Venus even though the evidence is indirect. Wallis (1972) proposed that this neutral exosphere from which the solar wind accreted mass through photo-ionization was constituted of helium, whereas we now know there is a hot oxygen exosphere. In a later paper Wallis (1973) admits to the existence of a Venus bow shock but proposes that it is weakened by the mass accretion process. Wallis also discusses the effects of symmetric charge exchange (proton-hydrogen) which has zero mass accretion, and points out that this process reduces the required expansion and so weakens the shock. In fact, the immediate downstream cooling rate can be so severe that the flow does not need to expand laterally. Although there is much to criticize in these two papers based on our present observational knowledge, the physics is sound and was very innovative for its time.

Wallis's evidence that the Venus shock was weak was the supposed lack of upstream particles and the behavior of the Mariner 5 shock. Other explanations exist for these observations, but it is difficult to dismiss the fact that the magnetic shock jump at Venus is less than expected from theory, as shown in Fig. 40 (Russell et al. 1979b). The theoretical number would be lowered somewhat by using a ratio of specific heats of 2, but not as far as the observed ratio. Nine terrestrial shocks were examined at median solar zenith angle 72° and found to have average shock jump of 3.1, i.e., 25% greater than their Venus counterparts. Further evidence for a weak bow shock lies in the attempts to model the interaction with a gas-dynamic code. Spreiter and Stahara (1980) have attempted to duplicate the parameters along two Pioneer Venus periapsis passes with their computer code. Using a specific heat ratio of 2 they obtain good qualitative agreement with the field and plasma behavior, but the observed field jump is less than the predicted jump in every case. In one case for which ion temperatures were available downstream, the temperature was ~½ the expected value. We note that when a ratio of specific heats of $5/3$ was used, an unrealistically low Mach number was required to get the predicted shock position out to where it was observed, and the predicted shock jump became too low. Mihalov (1981) followed this work with a study of six additional examples; he measured the solar wind conditions and hence knew what the Mach number should be. However, he allowed the Mach number to vary to get the best fit in the shock jump and shock position. This best fit Mach number was close to ½ the observed solar wind Mach number. Even choosing a Mach number much lower than expected for the calculations, Mihalov was unable to approximate the postshock ion temperatures. In every case examined the ions were factors of 2 to 10 colder than calculated. The missing heat could be in the electrons, but if so the electrons would have to be much hotter than the ions and much hotter than electrons observed in the terrestrial magnetosheath.

Gombosi et al. (1980) proposed and tested a new idea of removing solar wind from the postshock magnetosheath flow, that is, scattering of ions into the ionosphere by magnetic fluctuations. However, the dayside magnetic barrier has high field strengths and is generally quite steady. Computed loss rates

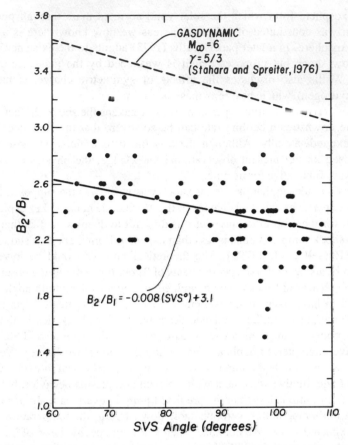

Fig. 40. Ratio of downstream to upstream magnetic field strength as a function of solar zenith angle at Venus, compared to computer simulation (Russell et al. 1979b).

using measured fluctuation amplitudes indicated that only 0.3% of the solar wind would be absorbed on the average, an entirely negligible amount. Since the same code could be used to calculate the charge exchange loss rate, Gombosi et al. (1980) also treated this problem and found that the loss rate was much greater, from 1% to 7% for ionopause altitudes from 1000 to 200 km respectively. The loss rate depends greatly on the atmospheric model used, and when the existence of a hot oxygen exosphere was realized by Nagy et al. (1981) the calculations were immediately redone because protons have an accidental charge exchange resonance with oxygen atoms. As shown in Fig. 41 the inclusion of hot oxygen doubled the loss rates (Gombosi et al. 1981).

In summary, the observed inward displacement of the bow shock from its expected location, the observed weakness of the bow shock and the theoretical

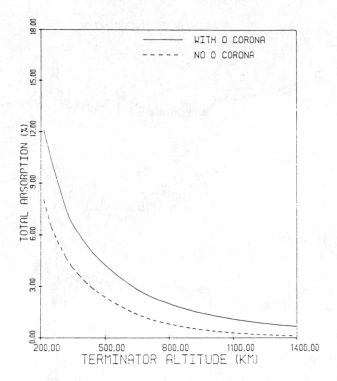

Fig. 41. Absorption of solar wind as a function of distance of streamline from planet at terminator. Top curve includes hot oxygen exosphere in calculation (Gombosi et al. 1981).

expectation of solar wind loss by charge exchange consistently show non-MHD processes to play a significant role in the postshock flow. However, we emphasize that to date these processes have not been incorporated in a self-consistent model which calculates shock position and strength in a charge exchanging flow.

C. Shock Structure

The structure of the Earth's bow shock depends on the Mach number of the solar wind flow, the ratio of solar wind flow velocity to the magnetosonic velocity, the β of the solar wind (the ratio of thermal to magnetic energy density) and the local angle between the upstream magnetic field and the shock normal (cf. Greenstadt and Fredricks 1979). The dependence of the structure of the Venus bow shock on the IMF direction was first pointed out by Greenstadt (1970) in the Mariner 5 data, the inbound shock being quasi perpendicular and the outbound shock quasi parallel. This outbound shock crossing is shown in the magnetic field as displayed in solar cylindrical coordinates in Fig. 42. The interplanetary magnetic field in the region labeled

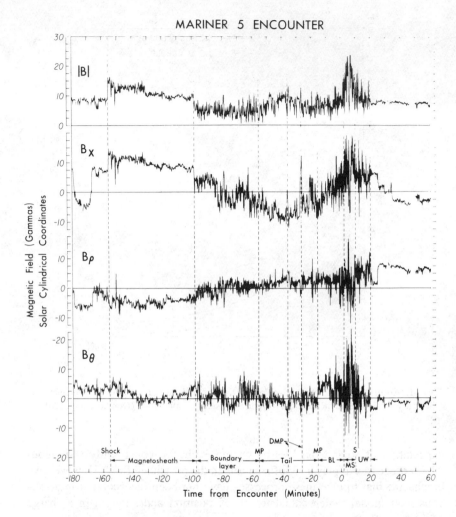

Fig. 42. Magnetic field data during Mariner 5 encounter, plotted in solar cylindrical coordinates.

UW (upstream waves) is closely aligned with the shock normal. When the field direction changes abruptly at the end of the UW interval to a direction not so nearly aligned with the shock normal, the upstream waves cease. This behavior is very similar to that observed at the terrestrial shock. The Mariner 10 outbound shock crossing shown in Fig. 43 was also a quasi parallel shock and was similarly disturbed (Ness et al. 1974). Due to the very turbulent nature of the quasi parallel shock it is difficult to characterize these shocks or to look for subtle differences between them and their terrestrial counterparts; in the following discussion we will therefore examine mainly quasi perpendicular shocks in which the interplanetary field lies mainly in the shock plane.

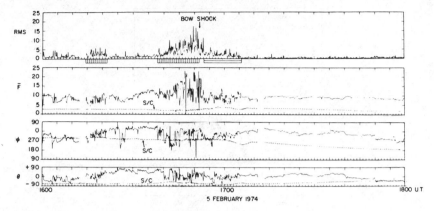

Fig. 43. Magnetic field data (6 s averages) during Mariner 10 encounter 5 February 1974 plotted in solar ecliptic coordinates (Ness et al. 1974). ----- S/C magnetic field; ▥ **ΔB** larger amplitude, lower frequency: ▤ **ΔB** smaller amplitude, higher frequency.

One of the features of supercritical quasi perpendicular shocks evident in terrestrial data is an overshoot in field magnitude immediately behind the shock. After a few tens of seconds the magnitude returns to an equilibrium value, often by passing through an undershoot (Russell and Greenstadt 1979). The plasma also participates in this phenomenon with electrons following the field closely (Bame et al. 1979). The overshoot appears to be associated with ion gyration at the shock front; the ions do not become thermalized until downstream of the overshoot.

Figure 44 shows a set of Venus shock crossings as measured in the magnetic field strength (Russell et al. 1982a). Well-developed overshoots such as those seen in the second and third panels from the bottom are somewhat rare in the Venus data. The overshoots also seem somewhat weaker at Venus than would be expected under the same solar wind conditions at other planets. Figure 45 shows the overshoot magnitude as a function of magnetosonic Mach number for Venus, Earth, Jupiter, and Saturn, where the Mach number is derived from measured solar wind conditions. The Venus points are lower than those from the other planets. If, as we have speculated above, charge exchange in the postshock flow indeed weakens the Venus shock so that its effective Mach number is less than its nominal Mach number, then the Venus points on Fig. 45 would move to the left and move into agreement with the measurements at other planets.

Note that there are significant differences in the spectrum of plasma waves as seen at Venus, Earth, Jupiter, and Saturn, shown in Fig. 46 (Scarf et al. 1981). The waves at the Venus shock are stronger at low frequencies and have a monotonic decrease with increasing frequency. Scarf et al. (1980b, 1981) attribute this to the difference in Mach number with heliocentric

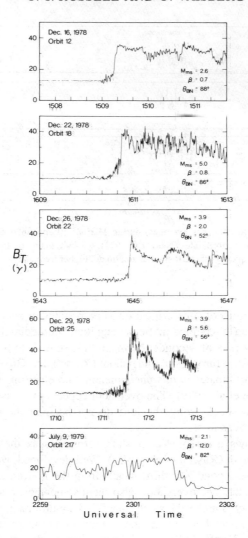

Fig. 44. Magnetic field magnitude across bow shock at Venus for varying solar wind conditions (Russell et al. 1982a).

radius. This may be true, but another apparent difference is the ratio of ion to electron temperature, which seems much lower for the Venus magnetosheath than the other planets. From these differences Scarf et al. (1980b) conclude that the plasma instability conditions at Venus are not the same as those in the Earth's bow shock region.

D. Upstream Waves and Particles

None of the Venus missions have included particle measurements with either sufficient sensitivity or the proper look directions to see particles emit-

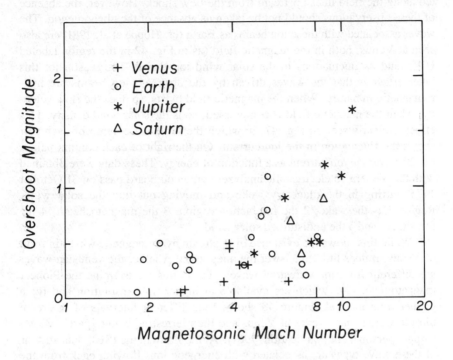

Fig. 45. Magnitude of the overshoot at the bow shocks of Venus, Earth, Jupiter, and Saturn as a function of magnetosonic Mach number (Russell et al. 1982*b*).

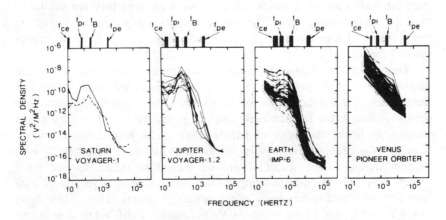

Fig. 46. Very low frequency electric plasma wave spectrum at the bow shocks of Saturn, Jupiter, Earth, and Venus (Scarf et al. 1981).

ted along the field lines upstream from the bow shock. However, the absence of these observations should not be taken as absence of the phenomenon. The waves associated with these ion beams as Earth (cf. Hoppe et al. 1981) are also seen at Venus, both in the magnetic field (as in Fig. 42 in the region labeled UW), and as fluctuations in the solar wind ion fluxes. The reason for this latter effect is that the waves driven by the upstream ion beams are hydromagnetic in nature. When the magnetic field twists, so does the flow velocity; when the magnetic field is compressed, so is the solar wind density. This effect is clearly seen in Fig. 47, in which the left-hand column of each pair shows the fluctuation in the ion current. On the right of each column pair is the 20 s average ion current as a function of energy. These data were obtained with the Venera 10 electrostatic analyzers on an outboard pass on 31 October 1975 starting in the planetary wake and moving out into the solar wind. Region 1 is the wake, 2 the rarefaction region, 3 the magnetosheath, 4 the foreshock, and 5 the undisturbed solar wind.

While this shows the existence of upstream hydromagnetic waves in front of Venus, it does little to answer the question of whether the venusian waves are different from the terrestrial waves. To do this we examine the Pioneer magnetometer data which are available at higher time resolution and for a greater time interval. Figure 48 shows four different intervals of upstream ultra-low-frequency waves at Venus and their terrestrial counterparts (M. M. Hoppe, personal communication, 1981). At the Earth in the ISEE data all four of these wave types are associated with energetic ions flowing back from the bow shock. These ion beams have properties that differ as a function of position. At the leading edge of the foreshock region, the region in which the ion beams are seen, the ion beams are narrow in both energy and pitch angle. Here the waves are weak or not seen. If waves are seen they are similar to those in the left-hand panels. The high-frequency waves seen at Venus are larger than at Earth; this is a general relationship, of which the examples shown here are typical.

Deeper into the foreshock region low-frequency transverse waves are often seen with high-frequency ($\sim$ 1 Hz) waves superimposed on the low frequencies, as in the second pair of panels. Still deeper into the foreshock the beams become spread in pitch angle and energy and the low-frequency waves steepen, as in the third pair of panels. Finally, far behind the foreshock boundary the beams become very diffuse and the waves very steep with leading (trailing in the spacecraft frame) wavetrains called discrete wave packets. We caution that this comparison was made from a very large data base without regard to frequency of occurrence of events. Thus, while these examples are typical of the waves at Venus and the Earth when they occur, this does not imply that the occurrence rates are the same; an occurrence rate study has not been done.

The upstream waves occur at slightly higher wave frequencies at Venus than at Earth, but slightly lower than at Mercury. In Fig. 49 we show the wave

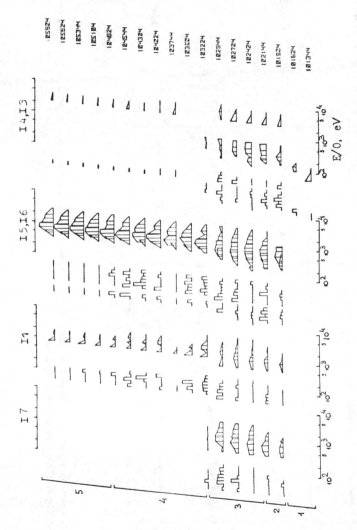

Fig. 47. Successive energy spectra (right hand columns in each pair) measured by the electrostatic analyzers of the RIEP plasma spectrometer on Venera 10 during a near planetary pass on 31 October 1975, through: (1) the wake, (2) rarefaction region, (3) magnetosheath, (4) foreshock, and (5) undisturbed solar wind. Left hand columns are the mean square fluctuations of the counting rate at each energy step during the 20 s measurement interval. Scale on the bottom is energy; scale on the right is time.

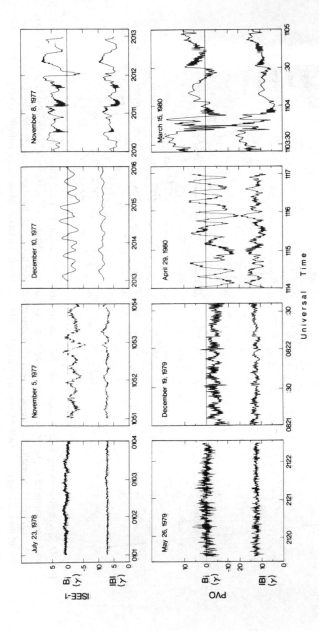

Fig. 48. Upstream waves at low frequencies observed by the Pioneer Venus magnetometer, compared with similar waves detected by the ISEE-1 magnetometer at Earth.

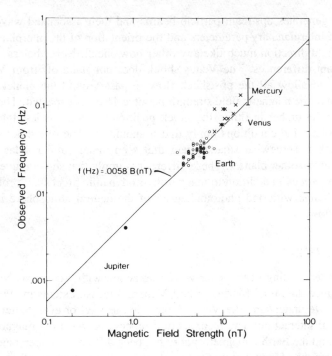

Fig. 49. Observed frequency of upstream low-frequency waves upstream of the bow shocks of Mercury, Venus, Earth, and Jupiter as a function of interplanetary field strength (Hoppe and Russell 1982).

period of the low-frequency waves, shown in the middle two panels at Mercury, Venus, Earth, and Jupiter as a function of the interplanetary field strength (Hoppe and Russell 1982). The linear dependence of wave frequency on field strength is as would be expected if the generating ion beams had the same energy at each planet. This relationship says nothing about the flux in the ion beams, only that the generation mechanism generates roughly the same energy at each planet. Note that the calculated energy of the beams is slightly but not significantly less for Venus and Mercury than for Earth and Jupiter.

Finally, we note that in the upstream solar wind VLF plasma waves associated with the electron and ion beams frequently occur (Scarf et al. 1980b). These have potential to help probe the properties of the beams, but as yet a morphological survey of the waves in this region has not been performed.

E. Summary

The observations clearly show that Venus has a well-developed bow shock, in many respects similar to the bow shocks of the other planets. It apparently generates upstream particle beams and their associated waves. It depends on interplanetary parameters and the orientation of the interplanetary magnetic field direction much like any other bow shock. Nevertheless, there are important differences. The Venus shock does not stand off from Venus sufficiently to allow all the postshock flow to pass around the planet. The shock jump in the magnetic field strength is not as large as expected. The best inferred Mach number in fitting the shock position and jump is less than the observed solar wind conditions imply that it should be. The overshoot at the Venus shock is somewhat smaller than that seen under similar solar wind conditions at the other planets. These differences imply that something else is occurring at Venus in addition to magnetohydrodynamic processes, probably charge exchange with and photoionization of the neutral atmosphere in the postshock flow.

V. THE FUTURE

Our understanding of the solar wind interaction with Venus has come a long way since the initial Mariner 5 and Venera 4 measurements in 1967. In many ways we understand the Venus interaction as well or even better than the terrestrial interaction. The reconnection process between the magnetized solar wind and the Earth's magnetosphere has been difficult to understand and has been a source of controversy for two decades. The deflection of the solar wind by Venus seems much more straightforward, although our initial guesses were incorrect. We also might claim to understand the origin of the magnetic field in the Venus ionosphere better than in the terrestrial ionosphere. The model of Cloutier et al. (1981), even if it does not work when the ionopause is at high altitudes, seems to replicate the observational data when the ionopause is at low altitudes. On the Earth we still do not understand the dynamo mechanism which generates the terrestrial field. Nevertheless, there is still much we do not know.

Figure 50 is an update of a summary diagram of Vaisberg et al. (1976c). At that time only the existence of upstream disturbances was known, whereas we now know that these disturbances closely resemble those at Earth. However, the amplitudes are different, as is the frequency of occurrence of the various wave types. The bow shock is closer to Venus than expected and the velocity decrease and the ion temperature increase are smaller than expected. As measured in terms of the jump in field strength across the shock, the jump is weaker than at similar locations on Earth. We might be able to understand these differences better if we either had more observations near the subsolar shock, or if the problem were treated with a self-consistent computer simulation taking into account charge exchange, photoionization, and the Zwan and Wolf effect.

The geometry of the distant tail seems well defined both in the magnetic and plasma data. The Pioneer Venus, Venera 9 and 10, and Mariner 5 data

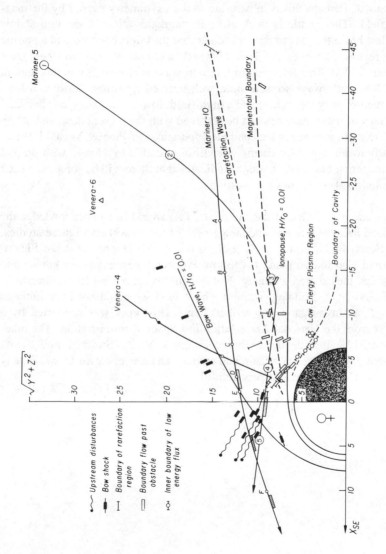

Fig. 50. Overview of solar wind interaction with Venus.

provide a similar picture. However, much work needs to be done in determining the relative morphology of the field and plasma regions and the variation of the tail structure with solar wind dynamic pressure, etc.

The structure of the near tail and wake region is still not completely understood. Perhaps this is in part due to the asymmetry caused by the magnetic field. The mantle is thick at high magnetic latitudes and thin at low latitudes. From the plasma data it is clear that the tail is connected to a plasma filling region. However, the 3-dimensional field and flow patterns here have not been settled. The definition of the ionopause is also murky in this region: there does not always seem to be a well-defined transition between a low-field, high-density ionosphere and a high-field, low-density magnetic barrier.

Many of these questions can be resolved with the present data, and others with data to be gathered in the continued operations of Pioneer Venus. It is still too early to tell which problems we will not be able to address with present measurements, but past experience indicates that there will be some, and each new solution also spawns new questions.

Acknowledgments. One of the authors (CTR) would like to acknowledge the continued help, advice and encouragement of his coworkers in these studies, F. L. Scarf, R. C. Elphic, J. G. Luhmann, M. M. Hoppe and J. A. Slavin. He would also like to thank his fellow Pioneer investigators for advice and data which they shared willingly. Notable in this regard are L. H. Brace, A. Barnes and J. D. Mihalov, without whose help we could not have initiated many of the investigations described here. This work was supported by a contract from the National Aeronautics and Space Administration. The other author (OLV) is indebted to his colleagues V. N. Smirnov and S. A. Romanov for their constant help and advice. Discussions with L. M. Zeleny and A. A. Galeev were also very helpful.

26. PHYSICS OF THE INTERACTION OF THE SOLAR WIND WITH THE IONOSPHERE OF VENUS: FLOW/FIELD MODELS

P. A. CLOUTIER and T. F. TASCIONE
Rice University

R. E. DANIELL, JR. and H. A. TAYLOR
Goddard Space Flight Center

R. S. WOLFF
Jet Propulsion Laboratory

Lacking an intrinsic magnetic field, Venus deflects the incident solar wind by thermal pressure and by magnetic pressure produced by ionospheric currents. These currents are driven by pressure gradients and electric fields induced in the ionosphere by the flowing magnetized solar wind plasma. The ionospheric currents connect across the ionopause into the ionosheath region and close on and behind the bow shock. The average volume distribution of the ionospheric currents is determined by minimizing the volume Joule heating by a variational technique. The computed ionospheric plasma velocity distribution is found to be spatially asymmetric with the fastest flow velocities perpendicular to the ionospheric magnetic fields. Since the magnetic fields are closely coupled with the plasma flow, the ionospheric magnetic field geometry exhibits as well a spatial asymmetry. Computations also show that the ionosphere is susceptible to a magnetohydrodynamic shear instability. The location and extent of this instability is a function of the external solar wind conditions. This proposed model of the Venus ionosphere compares very well with in situ observations by the Pioneer Venus Orbiter.

The solar wind interacts in various ways with the planets, comets, satellites, and other bodies of the solar system. In this chapter, we investigate the physics of the solar wind interaction with Venus, and attempt to show that the myriad details of the interaction revealed by planetary fly-bys, orbiters, and entry vehicles may be explained as the result of very fundamental and simple physical processes acting concurrently. The first sections of our chapter describe the theoretical basis necessary to understand the obstacle Venus presents to the solar wind, the resulting flow configuration, the direct effects on the ionosphere of Venus, and the connections between observed physical effects within the Venus ionosphere and the variability of the solar wind and interplanetary field. In these treatments, we describe various magnetohydrodynamic (MHD) processes and macroscopic low-frequency plasma instabilities, but make no attempt to theoretically model other aspects of macroscopic plasma physics involving distribution-function instabilities and high-frequency wave modes. We then compare observations of the simultaneous behavior of physical parameters in the Venus ionosphere with model calculations.

I. DEFINITION OF THE IONOSPHERIC OBSTACLE

A. Requirement for Shock Formation

We begin by considering the nature of the obstacle presented by Venus to the solar wind flow. Data from entry probes and orbiters (Dolginov et al. 1978; Russell et al. 1980b) have shown that Venus apparently has no significant magnetic dipole moment. The state upper limit $4.3 \pm 2.0 \times 10^{21}$ Gauss cm^3 precludes the possibility that the solar wind is deviated directly by a planetary field to form an intrinsic magnetospheric cavity as in the cases of Mercury, Earth, Jupiter, Saturn, and possibly Mars. Instead, the solar wind directly impinges on the atmosphere and ionosphere of Venus.

One might expect that a dense atmosphere could readily absorb and neutralize the incident solar wind plasma, in analogy to the process at the lunar surface. If the solar wind particles could be neutralized rapidly within the atmosphere and their momentum and energy transferred to the neutral atmosphere through collisions, then no deviation of the incident flow would occur, and no shock wave would form in the supersonic super-Alfvénic ($v^2 > B^2/\mu_0\rho$) flow upstream of the planet. However, observations by Mariner 5, later verified by numerous spacecraft measurements, showed a strong MHD shock wave upstream of Venus.

To understand why such a shock wave forms, we review a simple gas flow analogy by Cloutier et al. (1969). Consider a collision-dominated gas assumed to flow at velocity v_1 with density and pressure ρ_1 and P_1, respectively. This flow is normally incident on a thin region in which fresh gas particles are created at rest and with zero temperature. The creation rate of new gas particles is S (mass per unit area per unit time). After the initial flow accelerates and thermalizes the new gas particles in the source region, the combined flow emerges with velocity v_2, density ρ_2, and pressure P_2, as shown in Fig. 1a. Conservation of mass, momentum, and energy for steady-state flow in one dimension requires

$$\rho_2 v_2 = \rho_1 v_1 + S \tag{1}$$

$$\rho_2 v_2^2 + P_2 = \rho_1 v_1^2 + P_1 \tag{2}$$

$$\left(\frac{1}{2} \rho_2 v_2^2 + \frac{\gamma}{\gamma - 1} P_2 \right) v_2 = \left(\frac{1}{2} \rho_1 v_1^2 + \frac{\gamma}{\gamma - 1} P_1 \right) v_1 \tag{3}$$

where γ is the ratio of specific heats, c_p/c_v.

Simultaneous solution of these equations for the ratio of velocities $\eta = v_1/v_2$ gives

$$\eta = \frac{1}{2} \left[(1 + \eta_0) \pm \sqrt{(1 - \eta_0)^2 - 4\sigma^2_0} \right] \tag{4}$$

where $\eta_0 = (\gamma+1)M^2/2 + (\gamma-1)M^2$ is the velocity ratio across a standing shock transition obtained with $S = 0$; $M^2 = \rho_1 v_1^2/\gamma P_1$ is the incident flow Mach number; and $\sigma = S/\rho_1 v_1$.

Real solutions for Eq. (4) exist for $\sigma > 0$ only if $\sigma < (1-\eta_0)^2/4\eta_0$, which for highly supersonic incident flow means $\sigma < 1/(\gamma^2-1)$. No real steady-state solution to Eqs. (1), (2), and (3) exists for $\sigma > (1-\eta_0)^2/4\eta_0$. Physically, this means that when the mass addition rate exceeds a certain fraction of the incident mass flow rate $\rho_1 v_1$ (~9/16 for $\gamma = 5/3$ and $M \to \infty$), the flow can no longer accommodate the added mass by adjusting the downstream parameters ρ_2, v_2, and P_2. A transition (time-dependent) solution becomes necessary, with a transient wave propagating upstream to modify the incident flow parameters ρ_1, v_1 and P_1. Figure 1b illustrates an analogy to a stream emptying into a river. For a given river flow velocity upstream of v_1 and water level h_1, a maximum stream flow rate S exists for which the river simply carries the stream outflow downstream. For higher rates, the stream outflow spreads out in both directions from the entry point, modifying the upstream flow parameters v_1 and h_1 as well as the downstream flow parameters v_2 and h_2. In the original gas flow problem, modification of the incident flow paramenters $\rho_1 v_1$ and P_1 can occur via an ordinary acoustic pressure wave for $M < 1$ (subsonic flow); but if $M > 1$, only a shock wave can propagate upstream to modify the incident flow. This gas-flow model corresponds in many ways to the solar wind incident on the Venus atmosphere. Within the atmosphere, photoions and electrons are created by solar ultraviolet radiation in a region that is thin compared to the planetary radius. These ions and electrons, if created within the direct solar wind flow, would be accelerated to the local flow velocity by the electric field induced in their rest frame by the flowing magnetized solar

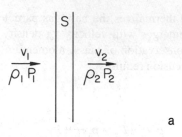

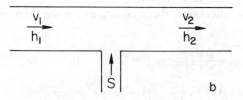

Fig. 1. (a) Jump conditions (with source); (b) stream model.

wind plasma ($E = - v_1 \times B$) rather than by direct collisional coupling. The incident solar wind is ordinarily highly supersonic and super-Alfvénic, and the column photoion mass addition rate within the Venus ionosphere greatly exceeds the incident mass flux of the solar wind (σ is roughly 10^2 for Venus). Thus there can be no steady-state solution in which the supersonic solar wind simply flows into the planetary atmosphere, picking up photoions and carrying them along toward the planetary surface. Instead, a shock wave must form and move out to modify the incident flow conditions. In the one-dimensional gas analog, the shock wave would continue to propagate upstream to infinity (or at least to the origin of the flow ρ_1, v_1, P_1). Venus's finite radius of curvature causes the shock to form as a standing bow shock in front of the planetary ionosphere, deflecting enough of the incident solar wind around the ionosphere to keep σ below its maximum allowable value. Since σ_{max} is $\sim$ 9/16 for highly supersonic flow, and σ is on the order of 10^2 for direct solar wind entry, this implies that the shock must deflect all but roughly 10^{-2} of the incident solar wind plasma around the planetary ionosphere. Thus the ionosphere of Venus is a very "hard" obstacle to the solar wind, allowing on the order of 1% of the incident flow to be absorbed by the atmosphere.

Of course, this treatment has not included possible effects of chemical and charge exchange reactions between solar wind plasma and constituents of the planetary atmosphere. Gombosi et al. (1981) have investigated the effects of charge exchange reactions and conclude that absorptions between 2% and 5% may occur for typical solar wind conditions. Thus the Venus ionosphere appears virtually impenetrable to the solar wind, with a standing bow shock deflecting most of the solar wind plasma around it. In the next section, we describe the flow pattern around this ''hard'' obstacle and examine the effects of this flow pattern on the ionosphere.

B. Flow Around the Obstacle

Ionopause and ionosheath. The problem of supersonic gas flow around a blunt nonabsorbing obstacle has been exhaustively treated in the aerodynamics literature, but with the shape of the obstacle usually predefined and not dependent on the incident flow conditions. In the case of Venus, the shape of the ionopause (the transition region between tangential solar wind flow around the ionosphere and the planetary ionosphere itself) depends on pressure balance between the solar wind and the planetary ionosphere. Assuming Newtonian pressure balance, Spreiter et al. (1970) calculated the shape of the ionopause boundary and the shock, for various ratios of ionospheric scale height to obstacle radius. Spreiter and Stahara (1980) have extended these calculations to Venus using Pioneer Venus data to fix model parameters. Their calculations yield the detailed shape of streamlines, density and temperature maps, velocity maps, field lines, and magnetic field magnitudes between the shock surface and the ionopause surface, assuming that the ionopause is completely impenetrable. The configuration of streamlines is shown schematically in Fig. 2. In this figure (and in Figs. 3–6) the atmospheric thickness is highly exaggerated to illustrate detail below the ionopause, and the ionopause altitude is artifically high relative to the planetary

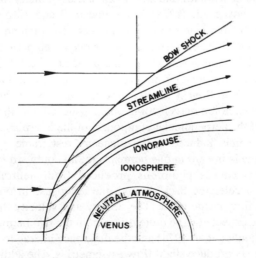

Fig. 2. Ionosheath streamlines.

surface and shock position. The ionopause in Fig. 2 is taken as an impenetrable (nonabsorbing) obstacle, and all flow lines entering the shock extend around the ionopause. In Spreiter and Stahara's (1980) treatment, the streamline geometry is axisymmetric about the planet-Sun line, and is independent of the magnitude or the orientation of the interplanetary magnetic field (parallel or perpendicular to the solar wind velocity vector). This result is obtainable from the assumption that gas dynamic pressure ρv^2 and static pressure P dominate the magnetic pressure $B^2/2\mu_0$ everywhere. However, several effects may contribute to asymmetries in the flow around Venus, and produce a slight modification to Spreiter and Stahara's (1980) results; these effects are associated with the perpendicular component of the interplanetary magnetic field. The first effect, treated theoretically by Zwan and Wolf (1976), results in enhancement of the magnetic field magnitude just above the ionopause due to the squeezing out of plasma along the magnetic field lines draped around the ionosphere by the flow. This depletes the plasma density there, requiring larger magnetic pressure to maintain pressure balance. This effect is apparently the cause of the large magnetic fields observed just above the ionopause by Pioneer Venus (Russell and Elphic 1979). The calculations of Zwan and Wolf (1976) also predict that the enhancement should be accompanied by a velocity asymmetry about the planet-Sun line, with larger velocities in the direction perpendicular to both the interplanetary magnetic field vector and the solar wind velocity vector. The compressibility of the plasma may also be different in directions parallel and perpendicular to the magnetic field vector, resulting in an asymmetry in shock altitude as a function of angle about the planet-Sun line. Cloutier (1976) calculated an asymmetry ranging from 0 to 12% as a function of the perpendicular noise component $\Delta B_\perp$ of the interplanetary magnetic field, and the maximum asymmetry was apparently verified by Romanov et al. (1978) using Venera 9 and 10 data. However, Slavin et al. (1980) have searched for but not detected such an asymmetry in the Pioneer Venus data. The difference may be due to temporal changes in the perpendicular noise component of the interplanetary field (asymmetry decreases with increasing noise) from Veneras 9 and 10 to Pioneer Venus. Another departure from flow symmetry may be caused by asymmetries in the drag produced by photoion mass loading on opposite sides of the planet. For a given direction of the perpendicular component of the interplanetary magnetic field, the electric field induced in the planetary rest frame by the flowing magnetized plasma is inward in one hemisphere, and outward in the opposite hemisphere. This causes photoions produced from neutrals above the ionopause to be accelerated back through the ionopause in one hemisphere and out into the flow in the other. The density and velocity distributions of planetary ions are expected to be quite different in the two hemispheres above the ionopause (Cloutier et al. 1974) and the difference in drag to the postshock solar wind flow may produce slight flow asymmetries. The addition of photoions in the postshock region above the ionopause may also change the strength of the shock transition and thus slightly alter the postshock flow conditions.

Examination of the effects described above leads us to conclude that the calculations of Spreiter and Stahara (1980), under the assumptions of flow symmetry about the planet-Sun line and an impenetrable ionopause, are a reasonable approximation if we include the enhancement of the magnetic field in the stagnation region due to the Zwan-Wolf effect. We now examine this model of the flow around Venus in more detail.

Impenetrable (superconducting) ionospheric model. We restrict our discussion to the case of the interplanetary magnetic field **B** perpendicular to the solar wind velocity vector. (This assumption is not very constraining, since a parallel component of **B** does not seriously affect our conclusions as long as $B_\perp$ is nontrivial.) The magnetic field lines, frozen in the preshock solar wind plasma, are compressed as they flow through the shock, and are draped around the ionopause as the flow diverges around the obstacle. Since the magnetic field is also assumed to be frozen in the postshock plasma flow, and since the ionopause of Spreiter and Stahara's model is assumed to be impenetrable to the plasma, no magnetic field can penetrate the ionopause. This requires a boundary current to flow in a thin layer along the ionopause to produce the change in magnetic field strength across the ionopause boundary. All the current that flows along and behind the shock to produce the increase in magnetic field in the postshock region must return in the ionopause current layer, producing a solenoidal current configuration as shown schematically in Fig. 3. The interplanetary magnetic field is perpendicular to the plane of the figure, and the contours represent current lines, with one tenth of the total

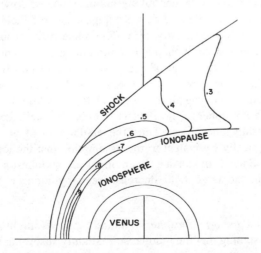

Fig. 3. Ionosheath contours of constant magnetic field strength for perfect conducting ionopause. Magnitudes are relative to stagnation point maximum.

current flowing between adjacent contours. The contours are labeled according to the fraction of total current flowing outside each contour, which is equivalent to the fraction of total magnetic field along each contour. The current flow direction is clockwise for an interplanetary magnetic field directed into Fig. 3, and the progressive field increase toward the nose of the ionopause may be thought of as caused by a nested set of solenoids having the shapes of the field contours. The field at the ionopause nose is largest because that point lies inside all of the solenoids. Current thus flows upward in Fig. 3 along and behind the shock, is distributed between contours in the region between the shock and the ionopause, and then returns in a thin current layer along the ionopause (downward in Fig. 3). The contours shown in this figure (and in Figs. 4 and 5) are adopted from Spreiter and Stahara (1980) with a field enhancement factor of 2 due to the Zwan-Wolf effect. The currents described in Fig. 3 are not due to electric fields, but are composed of magnetization and drift currents. In this case, it may be shown that if we assume all of the solar wind is deviated around the ionopause, it is equivalent to there being no electric field impressed across the ionosphere. In the solar wind flow upstream of the shock, the electric field induced in a reference frame stationary with respect to Venus is $E = - v \times B$. This means that the flow streamlines are electrical equipotentials (there is no component of E parallel to the velocity vector) and the streamlines must remain equipotentials in the postshock region for steady-state flow, if $\nabla \times E = 0$. For an impenetrable obstacle, the central flow line (along the equatorial axis in Fig. 2) splits at the nose of the obstacle to diverge around it. Since the potential difference between streamlines is linearly related to their separation distance before the shock, the central streamline deviated upward over one hemisphere is at the same electrical potential as the central streamline deviated downward around the other hemisphere, and the electric field imposed across the dayside ionosphere must therefore be identically zero. This same result would have been obtained if we had started with the assumption that the ionosphere were electrically superconducting: that is, the postshock magnetic field would be excluded from the ionosphere by a zero-thickness boundary current flowing along the ionopause. Since the plasma is frozen to the field, it is also excluded from entering the ionosphere. The electric field is identically zero within the perfectly conducting ionosphere. Thus the assumption of an impenetrable ionopause is physically equivalent to the assumption that the ionosphere is a superconductor. Such a model has been invoked by Johnson and Hanson (1979) to explain the large field decrease seen by Pioneer Venus at the ionopause.

Absorbing (conducting) ionospheric model. As we might expect from the foregoing, assuming that the ionopause is slightly absorbing is also physically equivalent to allowing the ionosphere to have large but finite conductivity. If the ionopause is only slightly absorbing, then to first order, the flow configuration will be unchanged from that of an impenetrable ionopause.

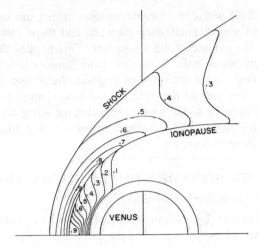

Fig. 4. Ionosheath and ionospheric contours of constant magnetic field strength (low field).

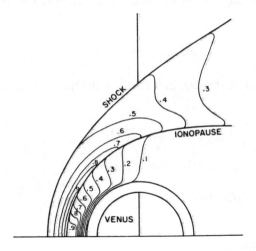

Fig. 5. Ionosheath and ionospheric contours of constant magnetic field strength (high field).

However, the absorption must take place by having the innermost stream-lines along the ionopause terminate inside the ionosphere. This means that the last streamlines to continue around the ionopause without termination must originate ahead of the shock some distance apart (above and below the planet-Sun line), and thus these streamlines are at different electrical potentials due to the electric field $\mathbf{E} = -\mathbf{v} \times \mathbf{B}$. This results in an electric field being impressed across the planetary ionosphere from pole to pole of magnitude $\Phi/2R_o$, where Φ is the potential difference between these two streamlines and R_o is the effective radius of the ionopause.

This electric field will drive Ohmic currents within the ionosphere in addition to the drift and magnetization currents, and these currents will be distributed within the volume of the ionosphere. This implies that the magnetic field in the postshock solar wind flow is no longer excluded from the ionosphere, in analogy to the diffusion of magnetic fields into a medium of finite conductivity. The sum of the drift, magnetization, and Ohmic currents must be equal and opposite to the total current flowing along and behind the shock to produce the postshock field enhancement. We will now investigate the distribution of these currents.

II. ELECTRODYNAMICS OF THE INTERACTION

A. Distribution of Ionospheric Currents

First we consider the distribution of currents within a conductor obeying the generalized Ohm's Law ($\mathbf{j}_E = \boldsymbol{\sigma} \cdot \mathbf{E}$) in the absence of drift and magnetization currents. Here $\boldsymbol{\sigma}$ is the conductivity tensor. In this case, the actual distribution of current will be that which produces the least Joule heating over the entire volume of the conductor. This is equivalent to a variational principle in which

$$\delta \int_{vol} \mathbf{j}_E \cdot \mathbf{E} \, dV = 0 \tag{5}$$

which may be shown (Daniell and Cloutier 1977) to correspond to the two Euler equations

$$\nabla \cdot (\boldsymbol{\sigma} \cdot \nabla \lambda) = 0 \tag{6}$$

$$\nabla \cdot (\boldsymbol{\sigma}^t \cdot \nabla \eta) = 0 \tag{7}$$

where λ and η are defined by

$$\mathbf{E} = \nabla \lambda \tag{8}$$

$$\mathbf{j}_E = \boldsymbol{\sigma}^t \cdot \nabla \eta \tag{9}$$

and $\boldsymbol{\sigma}^t$ is the transpose of $\boldsymbol{\sigma}$. Equations (8) and (9) are consistent with Ohm's Law as long as the magnetic field $\mathbf{B}$ used to calculate $\boldsymbol{\sigma}$ and the current density $\mathbf{j}_E$ obtained as a solution satisfy Ampere's Law. Two potentials, λ and η, are required because $\mathbf{j}_E$ and $\mathbf{E}$ must be treated as independent variables in carrying out the variation in Eq. (5). Solutions to Eq. (5) are straightforward for the case in which σ is a scalar whose magnitude is dependent on position within the conductor but independent of $\mathbf{B}$, and hence $\mathbf{j}_E$. As one might expect, current lines tend to follow regions of high conductivity within the conductor to reduce the volume Joule heating j_E^2/σ. If the applied electric field is transverse to the direction in which σ varies most rapidly, then the current will focus in the region of highest conductivity and be greatly reduced elsewhere.

Returning to the problem of Ohmic currents in combination with drift and magnetization currents, it is possible under certain assumptions to include drift and magnetization currents in the variational principle. For isotropic pressure and negligible acceleration drifts, the sum of the drift and magnetization currents may be written (Spitzer 1962) as

$$\mathbf{j}_D = \frac{\mathbf{B} \times \nabla P}{B^2} \; . \tag{10}$$

If $\mathbf{B}$ and P are treated as known quantities, then it may be shown (Cloutier and Daniell 1979) that the variational principle may be written using Eqs. (8) and (9) as

$$\delta\!\int \left[\left(\boldsymbol{\sigma}^t \cdot \nabla \eta \right) \cdot \nabla \lambda + \left(\nabla \eta + \nabla \lambda \right) \cdot \mathbf{j}_D \right] \mathrm{d}V = 0 \tag{11}$$

with corresponding Euler equations

$$\nabla \cdot \left(\boldsymbol{\sigma} \cdot \nabla \lambda + \mathbf{j}_D \right) = 0 \tag{12}$$

$$\nabla \cdot \left(\boldsymbol{\sigma}^t \cdot \nabla \eta + \mathbf{j}_D \right) = 0. \tag{13}$$

As before, these equations are consistent with Ohm's Law whenever the magnetic field $\mathbf{B}$ (used to calculate σ and $\mathbf{j}_D$) and the total current density obtained satisfy Ampere's Law.

We now consider the problem of obtaining self-consistent solutions to Eqs. (8) through (13); solutions of these equations would ordinarily be straightforward if it were not for two complicating factors. First, the conductivity tensor σ depends on the value of $\mathbf{B}$ locally, and second, the charge density which specified both σ and the plasma pressure P may be locally altered by the drifts we are trying to determine (even in the absence of pressure gradients and collisions, the plasma would convect due to $\mathbf{E} \times \mathbf{B}$ drifts at a velocity $\mathbf{v} = \mathbf{E} \times \mathbf{B}/B^2$). These complications require that we start with the following set of boundary conditions at the ionopause: $\mathbf{B}$ and P specified by Newtonian pressure balance (Spreiter and Stahara 1980); total current magnitude equal to that required to cancel the postshock $\mathbf{B}$ field enhancement; assumed model of $\mathbf{B}$ and P in two dimensions (latitude and altitude) in the noon meridian plane within the atmosphere to obtain starting values of σ and ∇P, and then to solve Eq. (11) for the total current $\mathbf{j} = \boldsymbol{\sigma} \cdot \mathbf{E} +$ $\mathbf{B} \times \nabla P/B^2$. The starting values of P can be obtained by a photochemical equilibrium model starting with, for example, a specified neutral atmosphere. It may be shown that the initial assumption on the behavior of $\mathbf{B}$ is not crucial to the final result (Daniell and Cloutier 1977); a model in which $\mathbf{B}$ varies in latitude according to Newtonian pressure balance (cos ψ, where ψ is solar zenith angle) and is constant with altitude, may be used as an initial estimate of **B**. Having obtained the total current **j** from Eq. (11), we must then iterate using Ampere's Law to obtain new values of **B** everywhere, and using the total plasma convection velocity to recalculate the plasma concentration. For the last step, we use the macroscopic drift relation (Spitzer 1962)

$$\mathbf{u} = \frac{1}{B^2}\left(\mathbf{E} - \frac{1}{n_e e}\nabla P_i\right) \times \mathbf{B} \tag{14}$$

where P_i is the pressure due to ions alone and n_e is the total electron concentration. From Eq. (14) we determine the convection streamlines within the ionosphere, and then integrate the ion conservation relations for each species along streamlines from the point of ionopause penetration

$$\nabla \cdot \left(n_s \mathbf{u}\right) = p_s - l_s \tag{15}$$

where p_s and l_s are the photochemical production and loss rates for each ion species. By "stacking" streamlines, we obtain a revised set of values of P_i. We then recalculate σ using the new values of **B** and n_e, recalculate ∇P_i from the values of P_i, and solve again Eq. (11). At each iteration we compare our solutions to their last values, and stop the iteration when all parameters are consistent with Eqs. (8) through (15). The final values so obtained represent the equilibrium solution for the ionospheric dynamic configuration of ion flow and electric and magnetic fields (hence the name "flow/field model").

If the initial assumed conditions in the model were photochemical equilibrium and constant **B** with altitude, the iterative procedure is analogous to the transients that would occur if the solar wind were suddenly "turned on" to the atmosphere. In this case, the rapid removal of ionization at high altitudes by convection would result in progressive depression of the topside ionosphere until the total photochemical production rate of ionization was in equilibrium with convective and chemical losses. Since the chemical loss rate is very slow at high altitudes, the equilibrium solution for total electron concentration is greatly depressed at high altitudes to form the ionopause. Equilibrium solutions have been obtained by Cloutier and Daniell (1979) for an assumed neutral atmospheric model of Venus (and Mars); the structure of the upper ionosphere was shown to agree with observations of Mariners 5 and 10 and Venera 9. Although iteration to convergence may be quite time-consuming even on a fast computer because of the complexities of both the variational calculation and the chemical processes entering into Eq. (15), this method is extremely useful in determining the convective equilibrium ionosphere from a neutral atmosphere without initial information on actual electron concentration distributions.

In practice, the variational technique described above becomes much more practical and useful in the solution of a much simpler problem in which the structure and composition of both the neutral atmosphere and ionosphere are known experimentally (as from Pioneer Venus observations). In this case, the only remaining complexity is the dependence of σ on $\mathbf{B}$, since n_e and ∇P are now given. (The reader is referred to Chapters 13, 23, and 24 in this book, for a complete description of the neutral atmosphere and ionosphere.) As discussed in these chapters, the dayside ionosphere generally shows the following structural features: a low altitude electron concentration peak (E peak) formed by a balance between O_2^+ photoproduction and ion-electron recombination; a topside region characterized by a slow decrease in concentrations of atomic ions (principally O^+) with altitude; and a sharp decrease in electron concentration at the ionopause owing to the large velocity shear there. Calculations of the conductivity tensor σ for such an ionospheric structure show that for any reasonable assumed variations of $\mathbf{B}$, the conductivity reaches a maximum in the vicinity of the low-altitude E peak. We thus expect most of the current driven by $\mathbf{E}$ to concentrate at the E peak. Since it can be shown that drift and magnetization currents are small compared to Ohmic currents at high altitudes, we may assume that within the ionosphere $\mathbf{B}$ remains relatively constant with altitude at a given latitude and longitude between the ionopause and the E peak, and that it then decreases rapidly in the vicinity of the peak. The conductivity near the peak is only weakly dependent on $\mathbf{B}$, and hence we may start the iteration for solution of Eq. (11) with $\mathbf{B}$ constant with altitude over the entire vertical extent of the ionosphere. The variation of $\mathbf{B}$ with latitude for our initial guess is given by Newtonian pressure balance, as before. We then iterate until convergence is reached, thereby establishing the ionospheric distribution of the various parameters. Using Eg. (11) with Eqs. (8) and (9) we obtain $\mathbf{E}$ and $\mathbf{j}_E$; from Ampere's Law with $\mathbf{j}_E$ (and with $\mathbf{j}_D$ given by Eq. [10]) we obtain $\mathbf{B}$; and using Eq. (14) we obtain $\mathbf{u}$, the plasma convection velocity. Using a model of the ionosphere and neutral atmosphere derived from Pioneer Venus measurements of the ion and neutral composition and structure (H. Taylor et al. 1979a,c; Niemann et al. 1979b), Cloutier et al. (1981) performed these calculations; the magnetic field results are shown schematically in Figs. 4 and 5. In these calculations, the ionopause is defined as the region where the thermal and superthermal planetary ions decrease rapidly with increasing altitude, going from 10^3 or 10^4 cm^{-3} down to 10 or 10^2 cm^{-3}, representing a large discontinuity in electrical conductivity between the postshock solar wind and the planetary ionosphere. As demonstrated below, this region may or may not coincide with large changes in magnetic field. It will also be evident that this definition of the ionopause, in terms of planetary ion gradients, is the physically meaningful definition in the physical description of ionospheric dynamics and structure. In Figs. 3 through 6, the magnetic field is perpendicular to the plane of the figure, and is

assumed to be draped around the planet so that the geometry applies locally
for a range of longitudes about local noon. Figure 4 shows the magnetic field
contours (current lines) for a relatively weak postshock interplanetary magnetic field (IMF) of ~60 γ maximum at the ionopause while Fig. 5 shows
the corresponding contours for a relatively strong postshock IMF of ~ 120 γ
maximum at the ionopause. The contours are labeled according to fraction of
maximum magnetic field strength, and the ionospheric thickness is highly
exaggerated for clarity. The results shown in these two figures are exactly
what our intuitive arguments led us to expect, namely, that the current which
would have flowed along the ionopause boundary for the impenetrable
ionopause distributes itself within the ionosphere for finite conductivity, with
the current concentrated in the vicinity of the low altitude conductivity
maximum. In effect, the current takes the shortest (vertical) path through the
low-conductivity upper atmosphere to reach the high-conductivity lower
ionosphere before flowing large horizontal distances, thereby minimizing the
Joule heating. The differences between Figs. 4 and 5 reflect the fact that the
Pederson and Hall conductivites vary differently with changing magnetic field
strength. Since the Pederson conductivity which dominates at high altitudes
varies inversely with B^2 for low collision frequencies and the conductivity of
the collision-dominated lower ionosphere is only weakly dependent (inversely) on **B**, this tends to distribute the current pattern to higher latitudes as
B increases. In either case, these current distributions illustrate the important
result that the total current is concentrated at the ionopause at high latitudes,
and in the lower ionosphere (E peak) at low and equatorial latitudes. Thus a
satellite penetrating the ionopause would detect a large field decrease at the
ionopause at high latitudes, but would remain in the large field region until the
E peak at the equator. This implies that the ionopause is coincident with a
large magnetic field transient at high latitudes, but that such a transient does
not mark the low-latitude ionopause. The total potential drop across the
diameter of Venus required to drive these currents is roughly 100 volts, which
is 0.2% of the 50 kilovolts impressed by the solar wind across the same
distance. By the arguments presented in the previous section, this implies that
only 0.2% of the solar wind plasma incident on the geometric disk of the
planet is absorbed by the ionosphere, showing (in agreement with the simple
gas flow analogy) that the ionopause of Venus represents a very hard obstacle
to the solar wind. If the flow around Venus is assumed to be axisymmetric,
then all the solar wind plasma absorbed by Venus flows through a circle only
320 km in radius at the nose of the shock. This is shown schematically in Fig.
6 by the shaded region just above the planet-Sun axis. The dashed lines below
the ionopause show the convection streamlines within the ionosphere described by Eq. (14). Convection is antisunward, with the plasma constrained
gravitationally to follow the curvature of the planet toward the antisolar point.
Because of the complex magnetic-field geometry, the model calculations of
Cloutier et al. (1981) are not expected to yield quantitative results past the

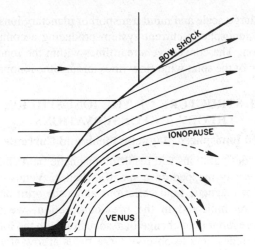

Fig. 6. Ionospheric streamlines. The shaded region designates the plasma absorption area.

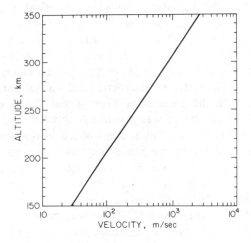

Fig. 7. Computed plasma flow velocity profile from electrodynamic model.

terminator, and the extrapolation into the night side is not based on exact calculations. On the day side, the model calculations of the convection velocity altitude profiles give results similar to those shown in Fig. 7, calculated at 30° from Eq. (14). The velocity decreases exponentially away from the ionopause, where it reaches several km s^{-1}. Because of the complexity of including inelastic collisions < 200 km altitude, the calculations were stopped at 200 km and the velocity profile was extrapolated to lower altitudes. The flow direction is approximately radially away from the subsolar point. Thus the absorption of a small amount of solar wind plasma by the ionosphere of

Venus results in large-scale and rapid transport of planetary ions antisunward, with an induced ionospheric current system producing a complex magnetic field configuration. The convective streamlines within the ionosphere represent the extension of the solar wind flow lines through the ionopause.

III. STRUCTURE OF THE IONOSPHERE: THEORY AND OBSERVATIONS

A. Ionopause and Ionosphere Magnetic Fields and Currents

The current configurations and resulting magnetic field contours of Figs. 4 and 5 may be easily mapped on a specific Pioneer Venus orbit to obtain a time sequence of magnetic field variations along the orbit passing through the ionosphere. As indicated in the previous section, we would expect to encounter a large rapid field change in association with the ionopause penetration at high latitudes, and to observe large fields at low altitudes near the equator. Pioneer Venus observations (Russell et al., 1979a; Elphic et al. 1980b) revealed an unexpected result. In a small number of cases, the field variations along the orbit agreed well with the theoretical predictions, but in most cases, the average field was much weaker inside the ionosphere than theory predicted, and within these low-field regions were localized regions of very intense fields. At times, isolated low-altitude regions with intense fields were observed. However, an interesting pattern emerged from these observations. When the postshock magnetic field was largest, the field profiles agreed best with the patterns of Figs. 4 and 5 and when the field was weakest, the disagreement was greatest (Luhmann et al. 1981a). Furthermore, the field magnitude within the isolated low-altitude strong-field regions tended to agree with the model calculations even when the surrounding areas did not. The model calculations completely failed to reproduce the extremely localized intense field regions, which were reported by Russell and Elphic (1979) to contain a helical magnetic field structure with greatest helicity on the outside. These structures, called flux ropes, indicated that some mechanism not encompassed in the original theory was required to explain the magnetic field observations. The helicity of the field led Wolff et al. (1980) to suggest that flux ropes were produced at the ionopause by the Kelvin-Helmholtz instability (driven by the large velocity shear there), and that the flux ropes were subsequently pulled down into the ionosphere by magnetic tension. However, the size distribution of flux ropes with altitude (Elphic et al. 1980b) argued against vertical transport from the ionopause (changing pressure with altitude would produce a different size variation with altitude due to compression than that observed). Furthermore, the flux ropes would be frozen in the ionospheric plasma at high altitudes, and would thus convect mainly horizontally rather than vertically. The same mechanism, the Kelvin-Helmholtz instability, was invoked by Cloutier et al. (1981) but the driving mechanism in this case is the velocity shear of the convection within the ionosphere (cf. Fig. 7). The following treatment, taken from Cloutier et al. describes the production of flux ropes by this mechanism.

In the simplest case of this instability, a plasma having mass density ρ flows at velocity $\mathbf{v}_0$ with respect to adjacent plasma of the same density, with a uniform magnetic field $\mathbf{B}$ over the entire region containing the flowing and stationary plasma. If $\mathbf{v}_0$ and $\mathbf{B}$ are parallel to the plane across which the velocity shear is supported, a growing wave mode (instability) with wave vector $\mathbf{k}$ exists whenever

$$\rho \left(\mathbf{k} \cdot \mathbf{v}_0 \right)^2 > \frac{(\mathbf{k} \cdot \mathbf{B})^2}{\mu_0} \quad \text{where } \mu_0 = 4\pi \times 10^{-7} \tag{16}$$

in mks units (see Hasegawa [1975] for a formal derivation of the result). The condition for maximum stability for all wave modes is $\mathbf{v}_0$ parallel to $\mathbf{k}$ and $\mathbf{B}$, and greatest instability occurs when $\mathbf{v}_0$ is parallel to $\mathbf{k}$ and perpendicular to $\mathbf{B}$. As shown by Eq. (16), the instability is driven by the kinetic energy of the flowing plasma, and stabilized by magnetic tension. Generalization of Eq. (16) for a finite thickness boundary with continuous velocity shear and density gradients in the direction of the velocity gradient results in an instability criterion similar to Eq. (16), but the thickness of the layer tends to stabilize short-wavelength perturbations (Ong and Roderick 1972).

The mechanism whereby magnetic field lines become twisted by this instability is straightforward for $\mathbf{v}_0 \parallel \mathbf{k} \perp \mathbf{B}$. In this case, the wave at the shear boundary pushes the boundary into the region of increasing v_0 causing the wave crest to advance in the direction of v_0 faster than the trough, with the wave breaking and wrapping up the field lines into discrete ropes spaced at intervals of $2\pi k^{-1}$, as shown schematically in Fig. 8. Since the total flux within a large volume containing the flux ropes must be conserved, the average field between flux ropes must be very small if the field within flux ropes is very large. This implies that the final thickness of the flux ropes formed must be small compared to the wavelength of the growing wave mode. Flux ropes thus form in a curtain lying in the (horizontal) plane perpendicular to the velocity gradient, and convect in the direction of v_0 parallel to that plane. For the case of $\mathbf{v}_0 \parallel \mathbf{k} \perp \mathbf{B}$, the flux ropes will be straight, but for $\mathbf{v}_0 \parallel \mathbf{k}$ but not $\perp \mathbf{B}$, the flux ropes, spaced at intervals $2\pi k^{-1}$, will also exhibit wave oscillations corresponding to a transverse Alfvén wave propagating along $\mathbf{B}$. The net flux along a flux rope is also a function of the relative angles between $\mathbf{v}_0$, $\mathbf{B}$, and $\mathbf{k}$. For $\mathbf{v}_0 \parallel \mathbf{k} \perp \mathbf{B}$, the net flux may be large, while for $\mathbf{v}_0 \parallel \mathbf{k} \parallel \mathbf{B}$, it is zero. It should be recognized in the first case that twisting of field lines requires a gradient in the velocity shear along the flux rope. For any geometry of $\mathbf{v}_0$ and $\mathbf{B}$, the most favorable condition for arbitrary wave mode instability is $\mathbf{v}_0 \parallel \mathbf{k}$, and we define the condition for instability as

$$\rho v_0^2 > \frac{B^2}{\mu_0} \cos^2\theta \tag{17}$$

where θ is the angle between $\mathbf{B}$ and $\mathbf{v}_0$.

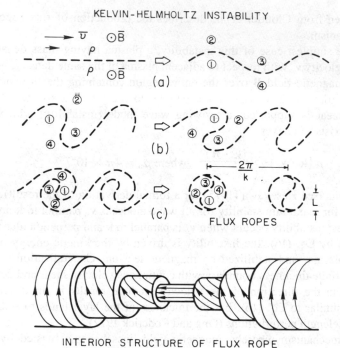

Fig. 8. The top sequence (a–c) shows schematically the formation of flux ropes by the Kelvin-Helmholtz instability. Sample field lines are labeled 1–4. Panel (d) shows the interior structure of a flux rope.

In the previous analysis, both the flowing and stationary plasma were assumed to have perfect electrical conductivity which forced the field lines to follow the plasma motions exactly (frozen-in flux). However, the finite conductivity of the atmosphere of Venus presents an additional complication to the formation of flux ropes by the Kelvin-Helmholtz instability, that of diffusion of the magnetic field within the flux ropes. In effect, the convection velocity of the plasma wrapping up the field must exceed the diffusion velocity of the field lines through the plasma in order for the wrapping up to occur. The ratio of plasma convection velocity v to magnetic diffusion velocity v_m is the magnetic Reynolds number

$$R_m = \mu_0 \sigma_3 L v \tag{18}$$

where σ_3 is the Cowling electrical conductivity and L is the scale length of the magnetic perturbation (radius of the flux rope). Using the model velocity distribution of Fig. 7 and the model of the ionosphere and neutral atmosphere derived from Pioneer Venus measurements, altitude profiles of the magnetic Reynolds number R_m were calculated by Cloutier et al. (1981) for an assumed magnetic field strength at the ionopause of 100 γ and an assumed flux rope scale size of 10 km. R_m was found to exhibit a monotonic decrease with increasing altitude, implying that flux ropes at high altitudes must be larger than those at low altitudes to give the same magnetic Reynolds number. Since least-action favors formation of the smallest possible flux ropes subject to the short-wavelength constraints of Ong and Roderick (1972), the size distribution of the flux ropes should vary from large flux ropes at high altitudes to small flux ropes at low altitudes, in agreement with the observations of Elphic et al. (1980b). In the region where the instability occurs, the average magnetic field between flux ropes will be greatly reduced relative to the values indicated by the magnetic contours of Figs. 4 and 5. Within flux ropes, magnetic diffusion will tend to reduce the field helicity toward the center relative to the exterior (see Fig. 8d), because the central core of a flux rope was formed earliest in the progression of Fig. 8, and diffusion has operated there for the longest time.

B. Global Flow and Field Patterns of the Day side Ionosphere

In principle, knowledge of the local values of the convection velocity and magnetic field vectors as a function of latitude, longitude, and altitude should allow direct evaluation of Eq. (17) to determine regions of flux-rope formation. However, the calculations performed thus far were based on the assumption that $\mathbf{v}$ (or $\mathbf{u}$ in Eq. [14]) was locally orthogonal to $\mathbf{B}$, which was satisfied in the noon meridian plane. Furthermore, since the components of $\mathbf{B}$ parallel and perpendicular to the velocity vector change with angle about the planet-Sun line (thus changing the contributions of magnetic pressure and tension to the flow momentum as a function of the angle about the planet-Sun line), we cannot assume that the flow velocity will be axially symmetric. This problem has already been considered in the postshock solar wind flow region by Zwan and Wolf (1976) who solved the momentum equations with the changing magnetic field components as a function of angle about the planet-Sun line. Their results indicated a velocity asymmetry about the planet-Sun line, with the largest velocity perpendicular to $\mathbf{B}$, and the smallest velocity parallel to $\mathbf{B}$. Physically, the largest velocity in the perpendicular direction occurs because the plasma is accelerated by the compression of both the plasma and magnetic field, whereas the plasma flow in the parallel direction is accelerated by gas compression alone. The resulting velocity asymmetry produced in the ionosheath will be carried into the ionosphere by the small fraction of solar wind plasma absorbed by the ionosphere. Since the same equations with ion production and loss included apply to the convective flow within the ionosphere, we may calculate the velocity asymmetries within the ionosphere as well.

The general equations describing the steady-state three-dimensional flow are

$$\nabla \cdot \rho \mathbf{v} = P_s - l_s \tag{19}$$

$$\rho (\mathbf{v} \cdot \nabla) \mathbf{v} = -\nabla P + \rho \mathbf{g} + \mathbf{j} \times \mathbf{B} \tag{20}$$

plus Maxwell's equations

$$\nabla \cdot \mathbf{E} = \rho_c / \epsilon_0 \tag{21}$$

$$\nabla \cdot \mathbf{B} = 0 \tag{22}$$

$$\nabla \times \mathbf{E} = 0 \tag{23}$$

$$\nabla \times \mathbf{B} = \mu_0 \mathbf{j} \tag{24}$$

where P_s and l_s are the ion production and loss rates, respectively; ρ_c is the free charge density; and $\mathbf{g}$ is the acceleration of gravity. In general, these equations cannot be solved analytically in three dimensions with arbitrary $\mathbf{B}$ field, but since draping produces horizontal $\mathbf{B}$ fields locally and $\mathbf{v}$ is horizontal as well, analytic solutions may be found for this geometry. Tascione (1982) notes that the velocity asymmetry decreases with decreasing altitude, and that the flow lines are radial with velocity increasing slowly along a given flow line. As in the ionosheath, the asymmetry is oriented with the largest velocity perpendicular to $\mathbf{B}$, in agreement with the results of Zwan and Wolf. Since the magnetic field is coupled to the flow, the velocity asymmetry also produces a field asymmetry, with field lines becoming nearly elliptical in the vicinity of the noon meridian, as shown in perspective in Fig. 9 by the solid lines.

Since the velocity asymmetry of Zwan and Wolf is a function of the interplanetary magentic field strength, the magnetic field line asymmetry also depends on the interplanetary field strength. Thus, specification of a magnetic field magnitude and direction outside the ionopause for a fixed solar wind dynamic pressure allows the values of $\mathbf{v}$, $\mathbf{B}$, and θ in Eq. (17) to be determined over the dayside ionosphere as a function of latitude, longitude, and altitude. Since we have assumed ρ to be independently provided by observations, we may determine the region(s) in which instability occurs (i.e., in which Eq. [17] is satisfied). The boundary separating regions of stability and instability is shown by the dashed contour in Fig. 9 (at the ionopause surface). The region of stability lies inside the dashed contour, and the region of flux-rope formation lies outside. The peculiarly shaped lobes near the center of the stability region are due to the geometry of $\mathbf{v}$ and $\mathbf{B}$ having three regions in which $\theta = 90°$ ($\mathbf{v} \perp \mathbf{B}$) in each hemisphere, one along the central meridian and one on either side where the field lines turn to become parallel to the equator. In these three

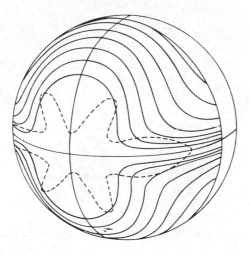

Fig. 9. Projection of region of stability (dashed lines) on to ionospheric magnetic field lines (solid curved lines).

regions (in each hemisphere), the instability will occur regardless of the magnitude of **B**. These regions are not shown extending inward to the subflow point (central meridian at the equator) because near the subflow point, the horizontal velocity is small and we assume that the regions of instability there become vanishingly small.

Below the ionopause, the region of stability becomes smaller and the stable lobes become less pronounced; this effect is mainly due to the velocity asymmetry decreasing with decreasing altitude. The variation in the shape and extent of the stability region with altitude is shown schematically in Fig. 10 where the ionopause surface is shown for simplicity to be flat rather than flared, according to the calculations of Spreiter and Stahara (1980). The lines of constant altitude at intervals below the ionopause represent 50 km altitude steps. The magnetic equator is taken to lie in the plane containing the IMF vector and the solar wind velocity vector. The subflow point is the point of symmetry for ionospheric convection, and may differ from the subsolar point or the point of normal solar wind velocity due to "garden-hose-angle" effects of the interplanetary field and real ionospheric density asymmetries. The figure is plotted on a "flat" Venus, so that the horizontal axes are latitude and longitude and radial vectors are represented by vertical lines. The equatorial axis of the figure covers roughly 150° longitude (magnetic) and the lobes span roughly 60° lat. The sides of the contour at constant longitude are concave.

The stability region described in Fig. 10 is based on the velocity profile shown in Fig. 7, which was arbitrarily extrapolated downward below 200 km. This extrapolation is questionable owing to several factors, and thus the exact shape of the stability region below 200 km is uncertain. For instance, the

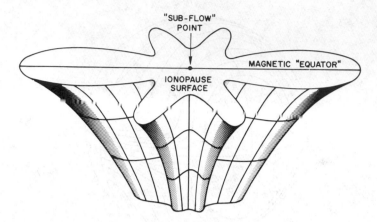

Fig. 10. Idealized stable region. The ionopause is at 350 km and the levels are separated by 50 km.

momentum loss (drag) to the neutral atmosphere by the convecting ionospheric plasma increases rapidly below 180 km due to inelastic collision processes. This tends to decrease ρv^2 compared to the magnetic tension $B^2 \, \mu_o^{-1}$, and increases the area of the stable region at altitudes below 180 km. This effect is somewhat compensated in the region just above the E peak by the rapid increase in ion concentration with decreasing altitude. However, as the interplanetary field magnitude increases, producing greater stabilizing influence over the entire ionosphere, the stable region below 180 km may greatly expand in area compared to the region above 180 km, producing a very wide base to Fig. 10 in the region between 180 and 150 km. This effect, which we have not attempted to portray quantitatively in Fig. 10, would result in a tendency for stable regions to occur just below 180 km altitude on medium- and high-field days over a greater range of solar zenith angles than shown in Fig. 10

The dimensions of the stable regions shown in Figs. 9 and 10 are appropriate for postshock magnetic field magnitude of 100 γ at the stagnation point of the ionopause. The volume of the stable region decreases with decreasing magnetic field strength and increases with increasing field strength. Also, as the field strength increases, the off-axis lobes tend to become more elongated. The entire pattern of field lines and stable regions described in Figs. 9 and 10 is therefore controlled by solar wind dynamic pressure, which is converted to magnetic pressure at the ionopause, and by IMF orientation, which causes the entire pattern to rotate about the subflow point as the interplanetary field rotates about the planet-Sun line.

Having described the model predictions concerning the shape, extent, and orientation of the regions of stability against the Kelvin-Helmholtz instability, we now consider the passage of a spacecraft (e.g., Pioneer Venus) through the dayside ionosphere, and examine the expected magnetic field profiles with time for various orientations of the spacecraft orbit and the interplanetary field as a function of magnetic field strength. A comprehensive study of possible geometric combinations with fixed angle between **v** and **B** along the orbit was performed by Cloutier et al. (1981), and it was shown that reasonable variations in solar wind parameters (dynamic pressure, interplanetary field magnitude and orientation) produce a rich variety of magnetic profiles along Pioneer Venus orbits which showed excellent agreement with observations. In the next section, we compare theoretical calculations to Pioneer Venus data for several selected orbits. Before examining these comparisons, however, it is instructive to consider several limiting cases:

Case 1. Weak interplanetary magnetic field lying in ecliptic plane, orthogonal to solar wind velocity vector. In this case we expect the stability region to be a very narrow equatorial belt with relatively short side lobes. A satellite such as Pioneer Venus in nearly polar orbit will intersect the stable region only if the entry and exit points of the orbit are on opposite sides of the magnetic equator. If periapsis is at low latitudes, the greatest chance of encountering the stable region is near the subflow point, which is near the subsolar point. Such an orbit geometry will produce a magnetic profile showing flux ropes in the region extending from below the ionopause to the stable region (which produces a single, isolated island of stability) and then beyond the stable region to the ionopause again. Ionopause passage coincides in this case with a large change in average field strength (strong magnetic gradient).

Case 2. Medium strength interplanetary magnetic field lying in ecliptic plane, orthogonal to solar wind velocity vector. In this case the stability region is a broad equatorial belt with side lobes more pronounced and elongated. Near the subflow region, the orbit may intersect a single broad stable region; or in the case where the orbit crosses one or more lobes, there may be several regions of stability along the orbit. (This latter case requires that the orbital path be quite long within the ionosphere.) If periapsis occurs near the center of the equatorial stable region, the magnetometer will detect a low-altitude, large-field region. If periapsis occurs outside the equatorial belt, then ionopause penetration may occur within the stable region and will not coincide with a large magnetic gradient, since the field magnitude increased upon entry to the stable region. This illustrates an important point (Cloutier et al. 1981), namely, that large magnetic field gradients are not reliable indicators of ionopause passage.

Case 3. Large interplanetary magnetic field lying in the ecliptic plane, orthogonal to the solar wind velocity vector. In this case the equatorial stability region may become so large that it encompasses the entire ionospheric passage of the satellite, producing complete stability everywhere. The variation in magnetic field strength along the orbit within the ionosphere should be described by the contours of Fig. 5 mapped onto the orbit plane. Thus large fields will penetrate the ionosphere and the ionopause is not clearly marked by large-field gradients.

Case 4. Arbitrary strength magnetic field at high inclination to ecliptic plane, orthogonal to solar wind velocity vector. In this case a satellite passage near the subflow region will show some region(s) of stability, but orbits removed from the subflow region may show a variety of magnetic signatures ranging from complete instability to isolated stable regions. If the field magnitude is large, the satellite is more likely to penetrate stable regions at low latitudes and low altitudes.

Day-to-night transport and inferred magnetic fields. As previously described, the ionospheric convection pattern sweeps ionization from the subsolar (subflow) region toward the terminator, and we expect the flow of the gravitationally bound plasma to continue around the planet toward the antisolar point, carrying the impressed magnetic field configuration to the night side. The continuation of the flow behind the terminator has been observed experimentally by H. Taylor et al. (1980) and Knudsen et al. (1980a), and the resulting nightside magnetic field configuration has been reported by Luhmann et al. (1981b). The observed magnetic field configuration is very similar to that deduced by H. Taylor et al. (1980) on the basis of ion spectrometer measurements alone, as shown in Fig. 11. The exact details of the flow and field configuration have not yet been quantitatively modeled, and large uncertainties exist concerning the physics of nightside transport and currents.

Viscous interaction processes. The treatment of the dayside ionosphere-solar wind interaction has assumed that no direct coupling of momentum exists between the solar wind flow outside the ionopause and the ionospheric convection except for the momentum carried in by the small amount of solar wind plasma absorbed by the ionosphere. This assumption of inviscid flow is apparently justified by the presence of the large tangential magnetic field at the ionopause (Perez de Tejada 1980b). However, this assumption may be invalid in the region of the polar terminator, and Perez de Tejada (1980a, b) has presented theoretical arguments that the shocked solar plasma behind the terminator is deflected into the planetary umbra, producing an observed nightside ionospheric bulge and convergence of the planetary wake. It thus appears that the inclusion of viscous drag effects, while unimportant on the day side, is necessary to explain the terminator and near-wake effects observed.

IV. CORRELATIVE STUDIES AND MODELING ATTEMPTS

A. Comprehensive Analysis of Selected Orbit Data Sets

In this section, we apply the theoretical model calculations to several sets of Pioneer Venus data, and show that the model is able to reproduce in fine

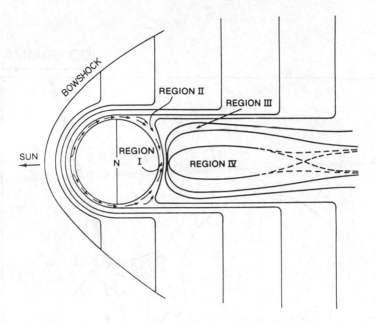

Fig. 11. Characteristic plasma regions. The ion flow direction is shown by the arrows.

detail many of the observed magnetic field features on these orbits. The orbits selected for analysis are 170, 176 and 177, chosen because they give a representative sampling of the wide variety of ionospheric conditions observed by Pioneer Venus. We give three diagrams for each orbit. The first diagram shows the projection of the computed magnetic field lines and stable regions on the surface of the planet; also plotted on this figure are the observed magnetic field vectors. The second diagram is an ionospheric cross section taken along the orbital plane, and the third diagram compares the computed total magnetic field strength and field direction with observations. The following discussion is, in part, a summary of the results of the authors.

Figure 12 shows the horizontal projection of the computed magnetic field lines (solid curved lines) for orbit 170. The field line spacing is not a measure of magnetic field strength; the observed magnetic field vectors are shown by the arrows along the orbit line. The varying thickness of the orbit line is used to distinguish the time spent in the ionosheath (heavy line) from the time the Orbiter is within the ionosphere. Once inside the ionosphere, orbital altitudes (in km) are shown by the sidebar; the ionopause crossings are indicated by the arrows at the edges of the sidebar. The center of symmetry for the convection pattern is shown by the solid circle, and the subsolar point is denoted by the large cross.

The region of stability is outlined by a set of dashed lines; the lowest altitude contour is distinguished by shading. The height of each contour is

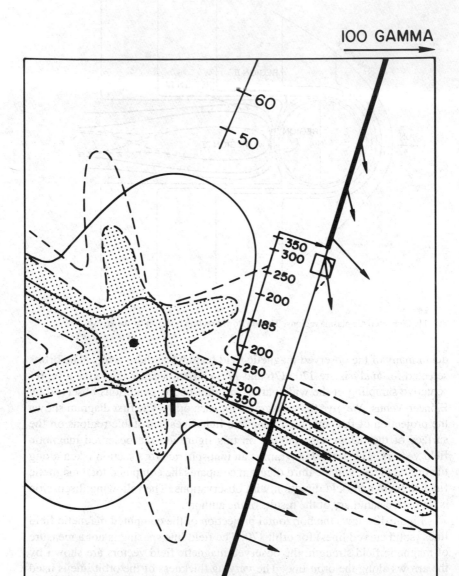

Fig. 12. Orbit 170: horizontal projection of the magnetic field parameters. Observed magnetic field vectors are shown by arrows. Model parameters are described in the text.

listed at the bottom of the figure. Within the ionosphere, stable areas encountered by the Orbiter are enclosed by a box. If the boxed region is shaded, it means that the stable area was encountered at (or below) the level of the lower (shaded) contour. The observed magnetic field vectors are shown only for stable ionospheric regions and the entire ionosheath. As shown in Fig. 12, a stable area is encountered shortly after the inbound ionopause crossing. A large unstable region extends from ~ 250 km inbound to ~ 250 km outbound; another stable area is encountered just before the outbound ionopause crossing.

The model parameters used for Orbit 170 are:

1. IMF direction is east-southeast.
2. Maximum magnetic field strength is 160 γ.
3. Inbound ionopause crossing is at 370 km.
4. Outbound ionopause crossing is at 350 km.
5. Magnetic field asymmetry factor is 2.0.

The ionopause crossings are determined from the atomic oxygen ion measurements by the Orbiter ion mass spectrometer (OIMS); these measurements are discussed in more detail later in this section. The outbound ionopause crossing is not clearly defined; Hartle et al. (1980) report that a suprathermal layer of O^+ ions was encountered just above 300 km during the outbound portion of orbit 170. These suprathermal ions are thought to be due to a nonequilibrium plasma region just above the relatively unperturbed ambient thermal ionosphere. Since the ionopause is defined as the boundary of the thermal ion envelope, the outbound ionopause crossing might lie slightly lower than the altitude used in the orbit 170 analysis. However, a small change in the ionopause altitude would not change the salient features of the analysis.

Figure 13 is an ionospheric cross section taken within the plane of orbit 170. The vertical axis gives the height of the ionosphere (in km) while the horizontal axis shows the distance from the magnetic equator (in deg). The horizontal scale is centered at the intersection of the orbital plane with the magnetic equator; positive distance is above the magnetic equator, negative distance below the equator. The solid lines are contours of constant ionospheric magnetic field strength, and dashed lines the analogous contours for the ionosheath. The contours are labeled relative to the maximum magnetic field strength (stagnation point). The ionosheath contours are modifications of the aerodynamic model of Spreiter and Stahara (1980), and the ionospheric contours are from the flow/field model of the authors. The crossmarks along the orbit line show one minute time intervals relative to periapsis; the orbit direction is shown by the arrows. The sawtooth pattern outlines the region of stability within the orbital plane. In Fig. 13, the orbit intersects the region of stability twice: the first intersection is with the stable lobe region (see Fig. 12), which appears as an unattached sawtooth pattern near 30°, and the second intersection occurs as the orbit crosses

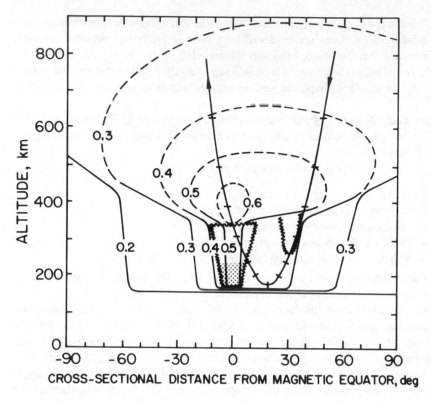

Fig. 13. Orbit 170: orbital plane cross section. Ionospheric contours of constant magnetic field
strengths are given by solid lines. Ionosheath contours are shown by dashed lines.

the magnetic equator, just above 250 km. The central stable region is
shaded below 250 km in order to coincide with the stability contour shading
used in Fig. 12.

In the top panel of Fig. 14, the total magnetic field strength for orbit 170
is plotted versus time. The solid line represents observations, and the dashed
line represents the model; the model results are shown only for the stable
regions. Conversely, the model predicts that a large unstable area begins near
-2.5 min (end of dashed line) and extends to ~ 2.1 min (start of dashed line).
The spike-like field strength features in the unstable areas are twisted magnetic field filaments or flux ropes (see Fig. 8). The magnetic field structure
within the flux ropes produces the rapid field direction changes shown in the
bottom panel of Fig. 14. The asterisks (*) are the OIMS measurements of
atomic oxygen. Only the atomic oxygen ion data are plotted because it predominates above 200 km; the loss of ion data near periapsis indicates that the
Orbiter has passed beneath the region of measurable atomic oxygen ion
concentration. The ion data are used to identify the top of the thermal ion
envelope (ionopause), and the ionopause locations are maked by the arrows
beneath the time axis.

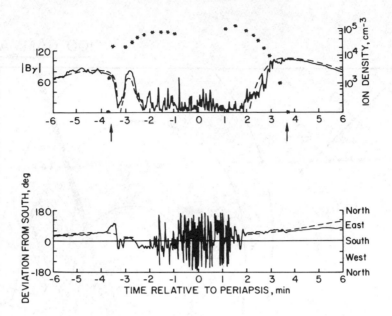

Fig. 14. Orbit 170: comparison of model with observations. Model results are shown by dashed lines. A break in the dashed line identifies a region of instability predicted by the model.

Orbit 176 is an example of a completely stable ionospheric observation; i.e., the ionosphere is not completely stable, but rather the Orbiter remains within the region of stability during the entire ionospheric transit (see Fig. 15). The model parameters for orbit 176 are:

1. IMF direction is nearly eastward.
2. Field strength maximum is approximately 190 γ.
3. Inbound ionopause crossing is at 280 km.
4. Outbound ionopause crossing is at 300 km.
5. Magnetic asymmetry factor is 2.5.

Figure 16 shows a cross-sectional view of the orbit trajectory through the region of stability. The observed total magnetic field strength (solid line), plotted in Fig. 17, has two areas of data dropout: between 1 and 2 min, and then again between 3 and 4 min. The model results (dashed line) closely match the observations.

Variability in the solar wind dynamic pressure is responsible for the change in the size and shape of the region of stability for orbit 176 as compared to orbit 170; i.e., increased solar wind dynamic pressure enhances the plasma depletion mechanism which, in turn, increases both the stagnation point magnetic field strength and the velocity asymmetry. Since the ionospheric magnetic field is closely coupled with the plasma flow, enhancing the plasma depletion also produces a larger magnetic field asymmetry. The ultimate result of all of these changes is the broadening of the region of stability

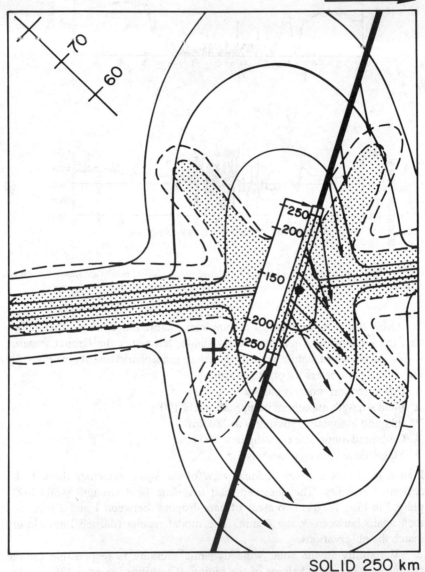

100 GAMMA

70

60

250
200
150
200
250

SOLID 250 km
DASH 300 km

Fig. 15. Orbit 176: horizontal projection of the magnetic field parameters. Observed magnetic field vectors are shown by arrows. Model parameters are described in the text.

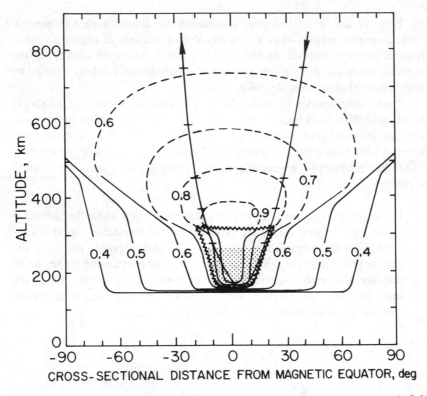

Fig. 16. Orbit 176: orbital plane cross section. Ionospheric contours of constant magnetic field strengths are shown by solid lines. Ionosheath contours are shown by dashed lines.

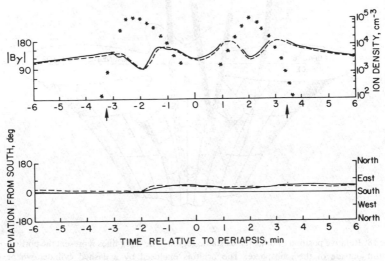

Fig. 17. Orbit 176: comparison of model with observations. Model results are shown by dashed lines.

(cf. Figs. 12 and 15). Therefore, statistically we would expect to observe stable magnetic regions more frequently during periods of high solar wind dynamic pressure because the Orbiter has a better chance of intersecting the larger region of stability. In fact, this statistical prediction has been verified by observations (Luhmann et al., 1980, 1981a).

The relative position of orbits 170 and 177 versus the region of stability is shown qualitatively in Fig. 18. Both orbits are plotted together for convenience; as discussed above, the region of stability for orbit 176 is larger than that for orbit 170. Notice that the principal difference between these two orbits is where they intersect the region of stability. Three factors control the location of this intersection.

1. The magnetic longitude of periapsis determines how close the Orbiter is to the stable region. Since the broadest part of the stable region is centered about the magnetic coordinate origin (subflow point), the closer the Orbiter is to the origin, the better the chance for intersecting the region of stability. Generally, the magnetic coordinate origin is within 15° of the subsolar point and therefore, orbits passing near the noon meridian are most likely to observe areas of stable magnetic fields.

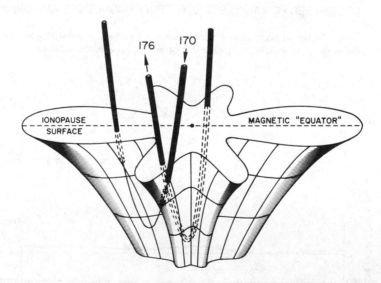

Fig. 18. Relative position of orbit versus stable region. The thick lines represent the portion of the orbit outside of the ionosphere. The orbit is enclosed by a dashed cylinder whenever it intersects the stable region.

2. The ionospheric magnetic field orientation is determined by the IMF. Since the IMF orientation is highly variable, the orientation of the ionospheric magnetic equator usually changes from orbit to orbit. As seen in Fig. 9, the orientation of the stable region is tied to the magnetic equator. Therefore, even if periapsis remains at a fixed magnetic longitude, the Orbiter could intersect different portions of the stable region from day to day because of changes in the IMF orientation.

3. The size of the region of stability is controlled by the solar wind dynamic pressure.

Orbit 177 is an example in which the IMF orientation is considerably different from the orientation for orbits 176 and 170 (see Fig. 19). The model parameters for orbit 177 are:

1. IMF direction northeast.
2. Maximum magnetic field strength is 110 γ.
3. Inbound ionopause crossing is at 350 km.
4. Outbound ionopause crossing is also at 350 km.
5. Magnetic field asymmetry factor is 1.5.

As seen in Figs. 19, 20, and 22, the Orbiter enters the stable region immediately after the inbound ionopause crossing, and remains within this region for $\sim$ 100 km. The Orbiter reenters the region of stability at an altitude of $\sim$ 200 km. This pattern is repeated during the outbound portion of the orbit. Figure 21 shows once again that there is a good match between the model and observations.

B. Dynamic Phenomena

Solar wind transients. In the previous sections, we have considered time-independent steady-state models of the solar wind interaction with Venus. However, a variety of transients are known to occur in the solar wind, and we now consider the effects of some of these transients.

Case 1. Sector boundary crossings and rotational discontinuities. If a sector boundary crossing results in a short period of very weak fields before field reversal takes place, then we might expect that large solar wind absorption could take place during that period. However, if the period of weak fields is shorter than the magnetic diffusion times (estimated to be on the order of several hr) over the thickness of the ionosphere, then the large remnant field within the ionosphere could provide the necessary coupling to produce a strong interaction without large absorption. After the field reversal, magnetic merging might result in a momentary increase of solar wind plasma entry through the ionopause. However, at present, no quantitative model of this process has been formulated, and no ionospheric effects attributable to such an event have been reported.

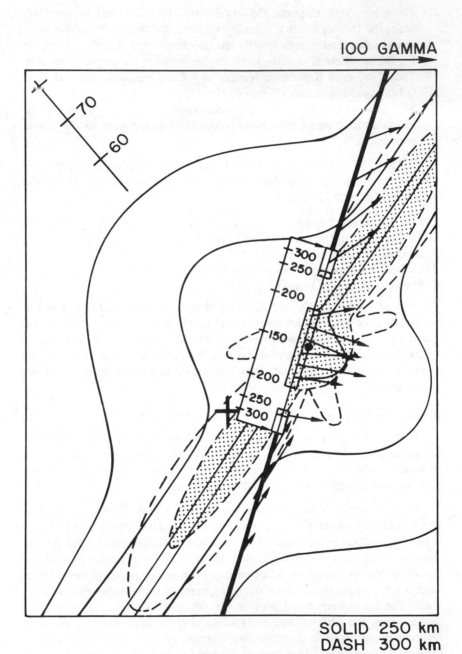

Fig. 19. Orbit 177: horizontal projection of the magnetic field parameters. Observed magnetic field vectors are shown by arrows. Model parameters are described in the text.

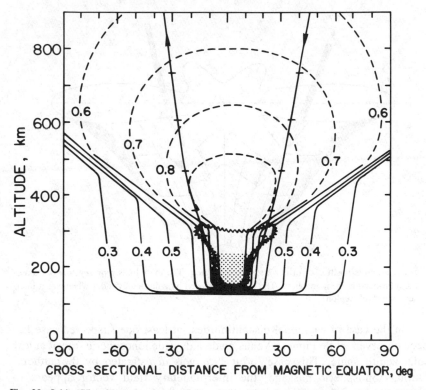

Fig. 20. Orbit 177: orbital plane cross section. Ionospheric contours of constant magnetic field strength are shown by solid lines. Ionosheath contours are shown by dashed lines.

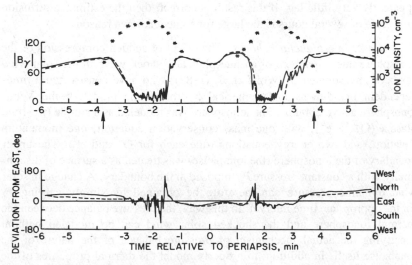

Fig. 21. Orbit 177: comparison of model with observations. Model results are shown by dashed line. A break in the dashed line indentifies a region of instability predicted by the model.

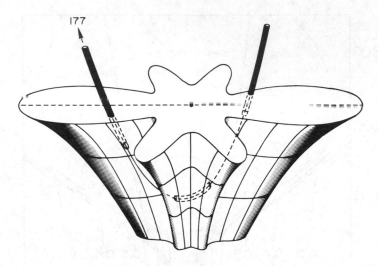

Fig. 22. Relative position of orbit versus stable region. The thick lines represent the portion of the orbit outside the ionosphere. The orbit is enclosed by a dashed cylinder whenever it intersects the stable region.

In the case of rotational discontinuities, at least one suspected case has been observed while Pioneer Venus made a dayside ionospheric transit at low solar zenith angle. This case, orbit 175, was investigated by the authors. During ionospheric passage, the interplanetary field vector appears to rotate about the planet-Sun line at a rate of $2°.5$ min^{-1}, and this rotation rate was found to be matched by the current pattern within the ionosphere with apparently very little lag. If this result is correct, then the estimated diffusion time scale of several hours is too large for some unknown reason.

Case 2. Interplanetary shocks. The effect of sudden compression of the ionosphere due to passage of an interplanetary shock wave in the solar wind has been investigated by Wolff et al. (1982), who have constructed a time-dependent one-dimensional numerical hydrodynamic model of the Venus ionosphere above 200 km. The ionosphere was modeled as a simple two-fluid plasma ($O^+ + e^-$), with one mass conservation equation, one momentum equation, and two energy equations (one each for O^+ and e^-). The upper boundary of the ionosphere (the ionopause) was treated as a surface of discontinuity with a constant pressure P_b imposed at the boundary. A change in solar wind dynamic pressure can therefore be obtained by simply varying P_b for the appropriate time interval. In this way, the pressure balance between the thermal ionosphere and the shocked solar wind can be modeled without specifying the actual mechanism of pressure balance or the nature of the ionopause itself. In addition, to properly model the thermal properties of the ionosphere, a constant conductive heat flux of 3×10^{10} eV cm^{-2} s^{-1} at the ionopause is specified. This is the electron heat flux used by Cravens et al. (1980) to model the mean electron temperature from 120 to 400 km and corresponds roughly to the upper limit for heating of the ionosphere by Landau damping of whistler waves (W. W. L. Taylor et al. 1979*a*). The ion conductive flux is chosen to be zero at the ionopause.

The results of the modeling indicate that the ionosphere can be compressed considerably as a result of a sudden change in solar wind dynamic pressure. For example, from an initial ionopause altitude of ~ 1000 km, an increase in solar wind dynamic pressure by a factor of 5 would initially compress the ionopause to ~ 400 km in ~ 300 s. In addition, sudden increases in solar wind dynamic pressure generate shock waves at the ionopause which propagate at ~ 2-5 km s^{-1}, ahead of the (downwardly moving) ionopause. These shock waves appear as density ledges in the ionosphere and exhibit characteristic temperature and velocity profiles as well. Finally, a sudden decrease in solar wind dynamic pressure produces rarefaction waves in the ionosphere which propagate downward from the ionopause. Examples of these two cases are shown in Figs. 23 and 24.

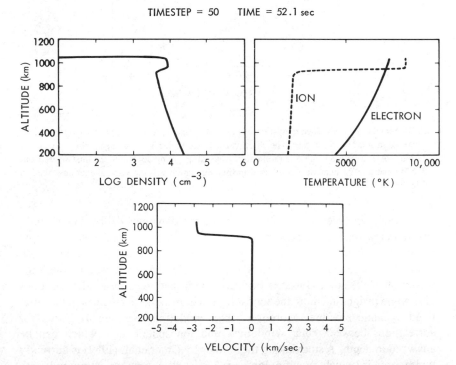

Fig. 23. Simulation of a compression of the Venus ionosphere as a result of a sudden 5-fold increase in solar wind pressure. As a result of the sudden compression, a shock wave is generated at the ionopause and propagates into the ionosphere ahead of the (downward moving) ionopause. In this simulation, only O$^+$ ions and electrons were considered, and the ionosphere was assumed to have no magnetic field. Note also, the dramatic increase in ion temperature in the shock region. The numerical simulation was performed at Michigan State University using a 1-dimensional Lagrangian hydrodynamic code developed by R. F. Stein of Michigan State University.

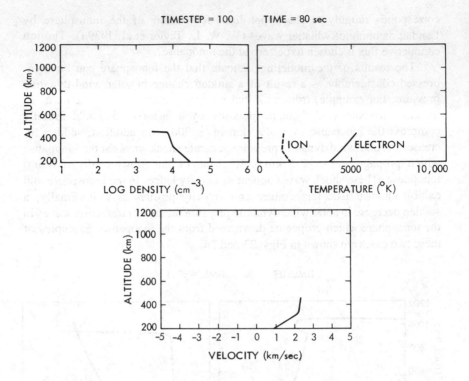

Fig. 24. Simulation of a sudden relaxation of the Venus ionopause as a result of a sudden 10-fold decrease in solar wind pressure. The sharp vertical rise in density, accompanied by a decrease in ion temperature is a result of a rarefaction wave propagating down into the ionosphere from the ionopause. The numerical code used in this simulation was the same as that described in Fig. 23.

Clearly, in order to correctly simulate the dynamics of the Venus ionosphere in response to changes in solar wind conditions, the effects of multiple ion species, neutral molecules, and magnetic fields must be considered as well as processes due to horizontal transport of species. In this sense, although the model of Wolff et al. (1982) is primitive with respect to these ideas, it does give some insight into how the ionosphere will respond to changes in the solar wind dynamic pressure. Moreover, their model can be used to search for correlations between solar wind data and ionospheric data which can be analyzed in depth. A study along these lines by Dryer et al. (1982) is currently under way in which specific Pioneer Venus data sets are compared with observed increases in solar wind dynamic pressure due to interplanetary shocks caused by solar flares.

Solar ultraviolet variations. It has been shown by Bauer and Taylor (1981) that the ionospheric density of Venus varies in response to changes in solar ultraviolet fluxes. This variability shows periodicities at the solar rotation rate, indicating that the effect is due to persistence of active ultraviolet emission region on the Sun.

V. SUMMARY AND CONCLUSIONS

In this chapter we have described the physical mechanisms believed to control the dynamical behavior of the ionopause and ionosphere of Venus, and have attempted to demonstrate that a number of simple mechanisms operate in concert to produce a seemingly complex flow and field configuration. We have described the basic mechanism of shock formation due to mass loading by photoions, and have shown that Venus is practically nonabsorbing to solar wind plasma in spite of the absence of intrinsic magnetic field. The small ($\leq$ 1%) amount of absorption results in electric currents distributed within the ionosphere. The electric fields and pressure gradients within the ionosphere drive ionospheric plasma convection patterns from the subsolar point through the terminator to the antisolar point, and the velocity shear in this convection pattern results in disruption of the steady-state magnetic field produced by the ionospheric currents owing to the Kelvin-Helmholtz instability. This instability occurs within the ionosphere where the magnetic fields are weakest and the field is oriented with largest angles to the convection velocity vector. Calculations of the dayside convection pattern and field configuration allowed contours of the stable regions to be determined for given values of interplanetary field magnitude and direction. We have shown that geometric mapping of selected Pioneer Venus orbits on these contours produced excellent agreement with the observed magnetic field configuration along these orbits, yielding field magnitude and direction within regions of stability, and predicting the regions in which the Kelvin-Helmholtz instability should wrap up the field lines into the observed flux ropes. The model is also able to explain the lack of correlation at times in the positions of magnetic gradients and the ionopause, and the occurrence at times of a low-altitude, low-latitude belt of stable large magnetic field. Although the effects of solar wind transients have not been considered in detail, we have discussed evidence that the ionosphere is able to respond rapidly to changing solar wind conditions.

A further point of interest is the comparison of these mechanisms with those expected in the case of comets. Except for the absence of gravity and the mass outflow, we might expect exactly the same mechanisms to operate in the case of a comet as in the case of Venus, and the resulting frontside flow and field configuration could be very similar. Future work should clearly be directed toward the comparison.

Acknowledgments. The authors are indebted to C. T. Russell, J. G. Luhmann, and R. C. Elphic for providing the Pioneer Venus magnetometer data and also for their helpful comments, discussions, and suggestions. This work was supported by a NASA Contract.

27. THE BOW SHOCK AND THE MAGNETOSPHERE OF VENUS ACCORDING TO MEASUREMENTS FROM VENERA 9 AND 10 ORBITERS

K. I. GRINGAUZ

USSR Academy of Sciences

Some features of the magnetosphere of Venus which were revealed by magnetic and plasma measurements aboard Veneras 9 and 10 are similar to the well-known peculiarities of the Earth's magnetosphere. They are: a well-developed outgoing shock wave; a current of plasma immediately behind the shock wave front, which perfectly satisfies the hydromagnetic theory; a wake of the magnetosphere protracted slightly along the Sun-Venus axis; and a plasma layer in the middle of the magnetic wake, which separates the two strands of antiparallel lines of force.

Although measurements of the magnetic field and plasma near Venus have been conducted since 1968 by Venera 4 (Gringauz et al. 1968; Dolginov et al. 1968), Mariner 5 (Bridge et al. 1967), Venera 6 (Gringauz et al. 1970), and Mariner 10 (Bridge et al. 1974; Ness et al. 1974), prior to 1975 some authors did not regard the findings as convincing. For example, Wallis (1972) believed that the bow shock does not exist near Venus, while Russell (1977) conceded that the shock wave near the planet is not an outgoing wave. These opinions changed after the Venera 9 and 10 magnetic and plasma measurements in 1975–1976, which gave information on many interactions of the shock wave and magnetosphere of Venus.

This chapter employs the following definition of the magnetosphere: a limited region of space with a magnetic field which, due to the existence of the planet within it, differs in magnitude, direction, and regularity from both the interplanetary field and the magnetic field in the transitional zone behind the front of the near-planetary shock wave, if this exists (Gringauz 1981). This definition is somewhat different from earlier ones, but it can be used for

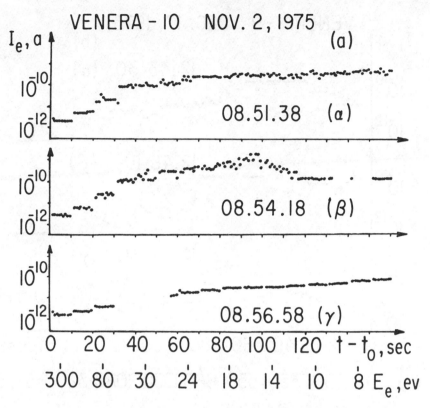

Fig. 1. Charged particle spectra obtained when crossing the near-planetary shockwave front: electron retardation curves from Venera 10, 2 October 1976.

all planets, and it avoids such terms as pseudomagnetosphere and quasi magnetosphere.

I. THE SHOCK WAVE NEAR VENUS

Measurements of the magnetic field near Venus by Veneras 9 and 10 were conducted by means of 3-component fluxgate magnetometers (Dolginov et al. 1976; Dolginov 1977), while measurements of the plasma were made by wide-angle plasma detectors, Faraday cylinder and planar electron analyzer with deceleration potential (Gringauz et al. 1976a,b) and by narrow-angle electrostatic analyzers (Vaisberg et al. 1976b,c).

All these measurements demonstrated the existence of a distinct and permanent shock wave around Venus. Figure 1 shows sample curves of electron deceleration from data obtained by Venera 10 on 2 October 1976, when the spacecraft was entering the unperturbed solar wind from an area near the planet. Curve α was obtained in a transitional region (the deceleration poten-

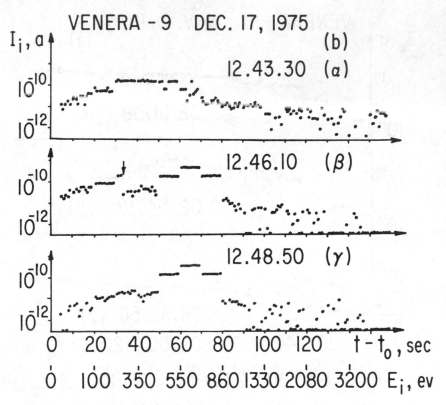

Fig. 2. Charged particle spectra obtained when crossing the near-planetary shockwave front: energy spectra of ions from data of the Faraday cylinder on board Venera 9, 17 December 1976.

tial diminishes, taking on fixed values, at each of which the collector current is measured 10 times). The deceleration curve γ was obtained in the solar wind. Curve β was obtained by intersecting the front of the shock wave; the transition from a deceleration curve of type α to type γ type takes place on segment S (see Fig. 2), corresponding to the shock wave front. This lasts $\sim$ 20 s, corresponding to a thickness of $\sim$ 150 km for the front, if the front was not moving during the measurements.

Figure 2 shows the energy spectra of ions, found by a Faraday cylinder mounted on board Venera 9 on 17 December 1976. Here the transition from the spectrum of type α to that of type γ is virtually a jump lasting 1 s, and the thickness of the shock wave front may be estimated at $\sim$ 8 km, if the front is immobile. However, cases were observed in which the changes in the energy spectra and particles of the magnetic field in the space around the planet took place very slowly and gradually, with large fluctuations, and the thickness of the shock wave front could then be estimated as 2000–3000 km. The broad

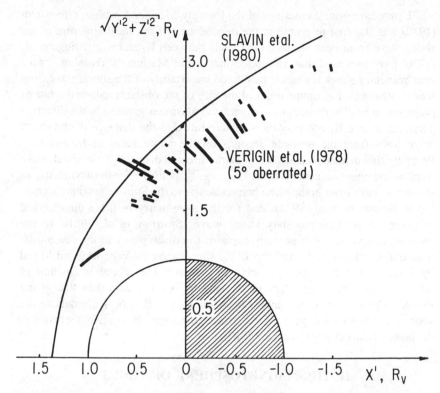

Fig. 3. Positions of crossings of the bow shock according to data of wide-angle plasma detectors on board Veneras 9 and 10 (end of 1975, 1976), and mean position of the front of the bow wave according to data of the PVO (end of 1978, 1979). From Verigin et al. (1978).

diffuse fronts correspond to the orientation $\mathbf{B}_I$ of the interplanetary magnetic field, near the local normals to the surface of the front $\mathbf{n}_{\mathrm{loc}}$, while the sharp fronts correspond to angles between $\mathbf{B}_I$ and $\mathbf{n}_{\mathrm{loc}}$ which are closer to 90°.

The results of measurements of the position and width of the near-planetary shock wave front obtained by Veneras 9 and 10 have been summarized by Verigin et al. (1978), the authors of the experiments with wide-angle plasma detectors in Fig. 3 in which the broad fronts are shown to occur at ~ 30% of the intersections. It is evident from Fig. 3 that the position of the shock wave during the measurement period by Veneras 9 and 10 (end of 1975, and 1976) was rather stable. This figure also shows the mean position of the near-planet shock wave from Pioneer Venus orbiter data, according to a paper by Salvin et al. (1980), who emphasized that the position of the shock wave is highly variable compared with the data of Verigin et al. (1978). Slavin et al. believe that this discrepancy can be explained by the different phases of solar activity during which the Venera and Pioneer measurements were conducted.

Using data from 2 crossings of the front by Mariner 5 orbiter, Greenstadt (1970) was the first to point out the dependence of the local structure of the shock wave front near Venus on the angle between $\mathbf{B}_I$ and $\mathbf{n}_{\mathrm{loc}}$. Bridge et al. (1976) have indicated that, in the Mariner 5 and Mariner 10 crossings of the near-planetary shock wave, the positional uncertainty of the shock wave front was ~ 3000 km, i.e. on the order of the size of the obstacle. Since the rate of propagation of disturbances in a magnetized plasma is greatest in the direction perpendicular to $\mathbf{B}$, not only the structure but also the distance of the shock wave front from the obstacle depends on the angle between $\mathbf{B}_I$ and $\mathbf{n}_{\mathrm{loc}}$. Precisely this dependence of the local structure and distance of the shock wave front on the angle defined by $\mathbf{B}_I$ and $\mathbf{n}_{\mathrm{loc}}$ was responsible for the asymmetry of the shock wave front in the plane perpendicular to the Sun-Venus line (according to Romanov et al. 1978), and for the asymmetry in the 3-dimensional structure of the near-planetary shock wave (Smirnov et al. 1981). In my opinion, this same effect partially explains the discrepancy in the determinations of the position and variability of the shock wave by Veneras 9 and 10 and by Pioneer Venus orbiter (cf. Fig. 3); because the orbital inclination of Veneras 9 and 10 to the plane of the ecliptic is much less than that of the Pioneer Venus orbiter, the angles between $\mathbf{B}_I$ and the local normals to the shock wave front are larger for Pioneer, i.e. the vector $\mathbf{B}_I$ is mainly situated in the plane of the ecliptic (Gringauz 1981).

II. THE MAGNETOSPHERE OF VENUS

At present, all researchers agree on the absence of a currently measurable internal magnetic field at Venus (see Chapter 25). This problem may be of interest for planetary dynamo processes, but not for the physics of the Venus magnetosphere.

Prior to Veneras 9 and 10, no spacecraft had explored the optical shadow of Venus. Indicators of the penetration of the spacecraft into the magnetosphere of Venus were considered to be (as in the case of the Earth's magnetosphere): abrupt increase in the magnetic field B, and a reduction in its fluctuations; and reduction of the fluxes of plasma, accompanying the growth of B, when the magnetic pressure exceeds the plasma pressure. In the wake of the magnetosphere, an additional indicator is the extension of B along the Sun-Venus line.

The closest that Veneras 9 and 10 came to the planet's surface was 1500 km. To date (1981) the richest data on the magnetic field and plasma at low altitude (< 1 $R_{\venus}$) has been obtained from the Pioneer Venus orbiter. On the other hand, information on the structure of the magnetic field and plasma at great height above the night side of the planet has come largely from Veneras 9 and 10. The Venera measurements are summarized on the basis of the experiment with wide-angle plasma detectors, Fig. 4 (Gringauz 1981; Verigin et al. 1978). Region A is the transitional zone behind the shock wave front;

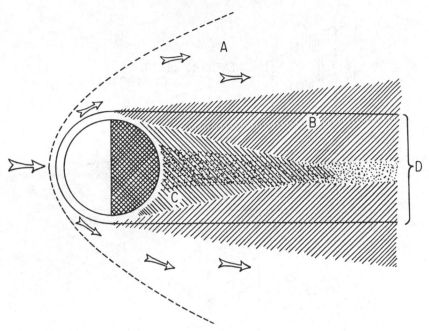

Fig. 4. Plasma-magnetic formations in the nightside portion of space around Venus: A, transitional region; B, boundary layer (particulate penumbra); C, particulate shadow; D, wake of the magnetosphere.

in it, the speed of the plasma is rather well described by hydrodynamic analyses made under the assumption that the obstacle is impervious to the solar wind (Spreiter 1976). Region B is a particulate penumbra or boundary layer (Gringauz et al. 1976b; Verigin et al. 1978), in which the plasma fluxes are reduced and the speed of the plasma V_i is less than that in the nondissipative model of Spreiter ($V_i \leq 0.75\ V_o$, where V_o is the speed of the undisturbed solar wind). This region, as is seen from Fig. 4, extends into the optical shadow of the planet. Verigin et al. (1978) noted that the retardation of the plasma flow in region B is caused by the quasi-viscous interaction between the solar plasma of the transitional region and the obstacle (i.e. the ionospheric plasma). As early as 1970 Spreiter et al. pointed out the possibility of formation of a broad layer by this cause. Verigin et al. (1978) have shown that the experimental data obtained in this region by the wind-angle detectors are in qualitative agreement with boundary-layer analyses (Perez-de-Tejada et al. 1977).

A segment of region B adjoining the terminator was apparently observed by the retarding-potential analyzer on board the Pioneer Venus orbiter (Spenner et al. 1980). For future study of the physics of region B, energy-mass ion analyzers would be very useful, but unfortunately these instruments were not on board Veneras 9 and 10, or the Pioneer Venus orbiter.

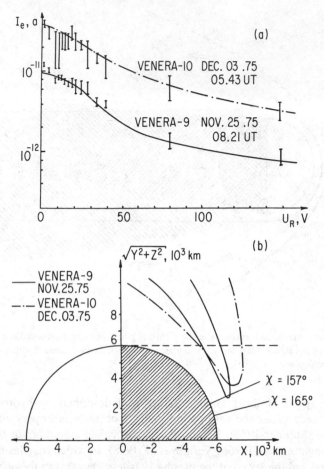

Fig. 5. (a) Examples of retardation curves for the electrons directed at the planet in the optical shadow of the latter, and (b) coordinates of Veneras 9 and 10 where the above retardation curves were obtained.

Region C (Fig. 4) is a particulate shadow, in which the Faraday cylinders on board Veneras 9 and 10 did not record regular ion fluxes from the transitional zone within the limits of their sensitivity (Gringauz et al. 1976a; Verigin et al. 1978). This region is roughly the shape of a cone with a base of radius $R_♀ + H_{ipt}$, where $R_♀$ is the radius of Venus and H_{ipt} is the height of the ionopause at the terminator, and the height of the cone is ~ 4 to 5 $R_♀$. A similar evaluation of the dimensions of the particulate shadow (called a cavity by the authors) is given by Intriligator et al. (1979), using the Pioneer plasma analyzer data. Despite the absence of regular ion fluxes (Gringauz et al. 1976a; Verigin et al. 1978), electron fluxes always exist in the particulate shadow, though they are extremely variable; cf. the integrated spectra (retardation curves in Fig. 5). Furthermore, in the particulate shadow sporadic ion fluxes

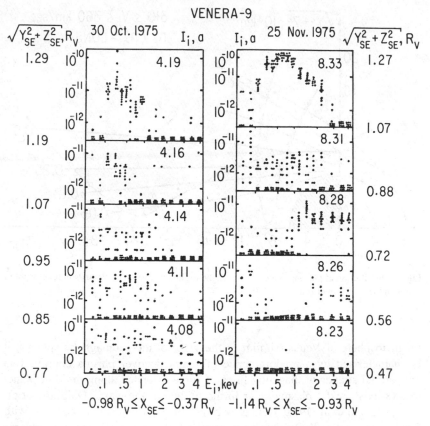

Fig. 6. Energy spectra of ions obtained when the Venera 9 spacecraft crossed from the particulate shadow into the transitional zone on 30 September 1975 and 25 October 1975.

are observed with energies E up to E_{max}, the maximum that can be registered by the Faraday cylinders (i.e. up to 4.5 keV); there may be sporadic fluxes of ions with $E > E_{max}$.

Figure 6 shows sample energy spectra of the ions according to the data of a Faraday cylinder oriented toward the planet along the Sun-Venus line, when Venera 9 was moving from the particulate shadow into the transitional layer (Gringauz et al. 1976a). In each energy interval, 10 measurements per second of the total current were made. Many points lie on the abscissa, although in ~10% of the measurements there were significant fluxes of ions with energies of 2 to 4 keV in the particulate shadow. Ions of such energy are not observed in the transitional layer, and apparently are a consequence of acceleration processes occurring in the particulate shadow or at great distances in the wake of the magnetosphere, in region D. Region D (Fig. 4) is the wake of the

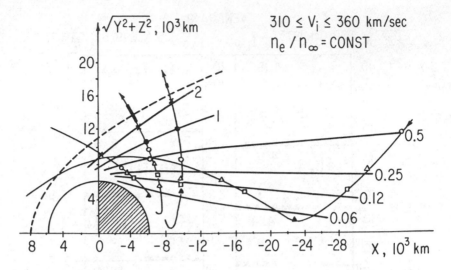

Fig. 7. Distribution of plasma concentration in the nightside portion of Venus's magnetosphere and in the transitional zone.

magnetosphere of Venus (similar to the wake of the geomagnetic sphere); it is characterized by a predominence of magnetic over plasma energy, two antiparallel strands of magnetic lines of force extended along the Sun-planet axis (x-axis), and a plasma layer in the middle of the wake (Dolginov et al. 1970; Verigin et al. 1978; Romanov et al. 1977). However, in contrast with Earth's magnetosphere, the orientation of the neutral layer and plasma layer varies with the change in component B_l in the plane YZ. The diameter of the wake of Venus's magnetosphere somewhat exceeds the diameter of the planet.

Figure 7 (Verigin et al. 1978) shows the distribution of the plasma concentration n_e in the nightside space around the planet, according to the data of retarding-potential electron analyzers; $n_{e\infty}$ is assumed the same as in interplanetary space. Figure 8 shows two consecutive measurements of ion spectra by Venera 10 passing from the transitional zone into the wake of the magnetosphere. The decrease in plasma flux at the wake boundary corresponds to an increase in B, while B_y and B_z are reduced. Figure 9 (Verigin et al. 1978) shows ion spectra and vectors B, as measured within the magnetosphere of Venus near the point where the x-component of B changes sign. The change in sign of B_x corresponds to the appearance of particles with energies of 1 to 3 keV in the ion spectrum. These are absent from the neighboring spectra within the magnetosphere, and evidently are related to a plasma layer of the venusian magnetosphere qualitatively similar to the plasma layer of the Earth's magnetosphere.

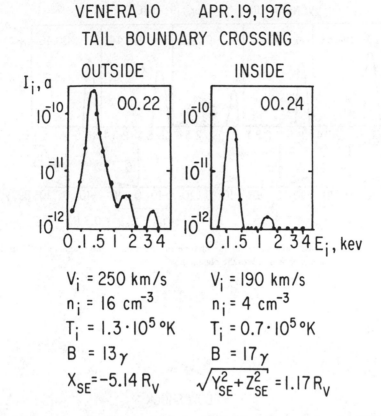

Fig. 8. Ion spectra obtained on crossing the boundary of the magnetospheric wake of Venus.

III. THE CORRELATION BETWEEN THE MAGNETOSPHERE AND IONOSPHERE OF VENUS

Figure 10 shows a simplified current system corresponding to the magnetic fields near Venus (Gringauz 1981). The dayside portion of this current system creates a magnetic barrier, discovered above the dayside ionosphere in Pioneer measurements, and injects current into the upper layer of the dayside ionosphere, which is closed across the plasma of the transitional zone (Elphic et al. 1980a). The current structure in the nightside portion is similar to that in the cross section of the geomagnetosphere wake (Θ-structure).

Part of the fluxes of electrons with energy <300 eV directed toward the planet, and measured in the particulate shadow (Fig. 5), evidently strongly influence the formation of the nightside ionosphere of Venus. The electron concentration n_e in the ionosphere of Venus was measured by the Venera 9

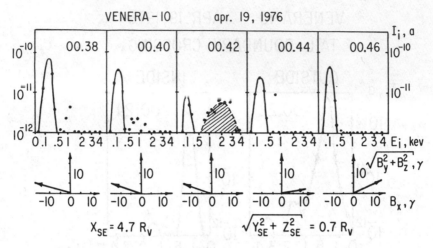

Fig. 9. Variation of the ion spectra for the change in direction of the magnetic field in the magnetospheric wake: appearance of the plasma layer.

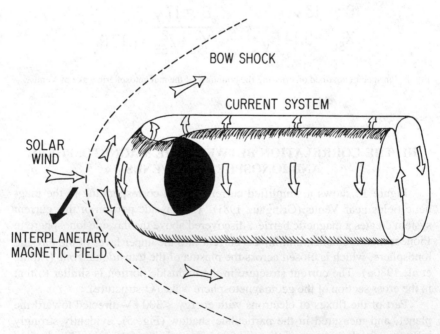

Fig. 10. Simplified current system in the magnetosphere of Venus constructed on the basis of available observations.

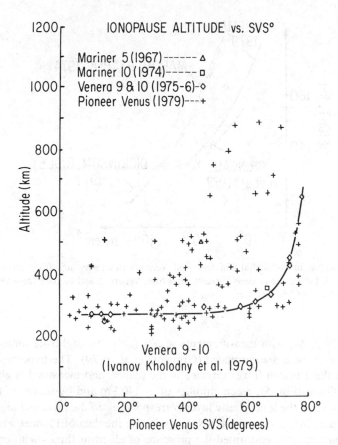

Fig. 11. Diurnal ionopause, from data of a radio occultation experiment on board Veneras 9 and 10, and data from the Pioneer Venus orbiter.

and 10 radio occultation experiment (Aleksandrov et al. 1976b; Ivanov-Kholodnyy et al. 1977). The measurements confirmed the existence of the ionopause as a characteristic and permanent feature of the dayside ionosphere of Venus; it was previously observed by the Mariner 5 flyby (Mariner Stanford group 1967) and the Mariner 10 flyby (Fjeldbo et al. 1975). The measurements from Veneras 9 and 10 established that the height of the ionopause increases with solar zenith angle; this tendency was confirmed by Pioneer Venus observations (cf. Figs. 11 and 12; Elphic et al. 1980a).

The nightside ionosphere of Venus, according to radio occultation data from Veneras 9 and 10, has a principal maximum n_e at the fixed height ~140 km. The value of $n_{e(max)}$ varies considerably with time. Often, but not always, this principal maximum is attended by another smaller maximum. A comparison of the time variations in the electron fluxes measured in the particulate

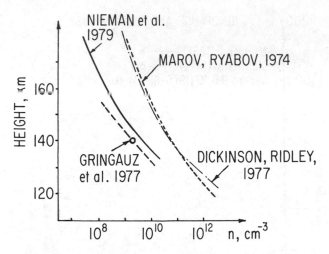

Fig. 12. Models of the neutral atmosphere of Venus with density at 140 km altitude from analyses in 1977 using electron measurements from Veneras 9 and 10 and experimental data from the Pioneer Venus orbiter.

shadow (Fig. 5), with measurements of $n_{e(max)}$ in the nightside ionosphere, revealed that these are correlated (Gringauz et al. 1979). The hypothesis was used that the electron fluxes directed at the planet, and measured at altitudes of > 1500 km (Fig. 5), reach altitudes of ~ 140 km and by means of impact ionization form the ionospheric layer corresponding to the principal nighttime maximum n_e in the ionosphere. (Note again that the data of Pioneer's electrostatic analyzer have confirmed the presence of electron fluxes with energies on the order of tens or hundreds of eV in the nightside ionosphere.) Estimates have shown that in order for these fluxes of ionizing electrons to reach the magnitude of $n_{e(max)}$, corresponding to observations in the nightside ionosphere, the concentration of neutral substances at an altitude of ~ 140 km must be 10^9 cm^{-3}, two orders of magnitude lower than that assumed in 1977 for the models of Venus's neutral atmosphere. The temperature of the neutral particles T_n in this case should also be much lower than that in the 1977 models.

The results of the Pioneer Venus orbiter mass spectrometer measurements published by Niemann et al. (1979a) revealed that the 1977 models were incorrect; instead, estimates of nighttime n_n and T_n at an altitude of ~ 140 km based on the hypothesis of impact ionization as the source of $n_{e(max)}$ and using data of the Venera 9 and 10 electron analyzers (cf. Fig. 11) were right. We regard this as convincing support for the theory that impact ionization by electrons from the magnetosphere wake is a mechanism, if not the main one, for formation of the principal ionization maximum in the nighttime magnetosphere of Venus.

IV. CONCLUSION

Several features of the Venus magnetosphere, as revealed by magnetic and plasma measurements on board Veneras 9 and 10, are similar to the well-known peculiarities of the Earth's magnetosphere:

1. A well-developed outgoing shock wave;
2. A current of plasma immediately behind the shock wave front, which satisfies the hydromagnetic theory;
3. A magnetosphere wake which extends slightly along the Sun-planet axis;
4. A plasma layer in the middle of the magnetic wake which separates the two strands of antiparallel lines of force (directed away from the Sun and toward the Sun).

These common features of magnetospheres, despite the substantially different nature of the obstacles to the solar wind, are one of the proofs that magnetosphere-like structures with long wakes may be formed in supersonic fluxes of magnetized plasma near widely differing obstacles, from comets to galaxies.

28. LABORATORY SIMULATION OF THE INTERACTION BETWEEN THE SOLAR WIND AND VENUS

I. M. PODGORNY
USSR Academy of Sciences

In experiments with an artificial solar wind, a laboratory model was created in which the interaction of the solar wind with Venus and the consequent magnetic wake were simulated.

In a series of projects carried out at the Space Research Institute of the USSR Academy of Sciences it was demonstrated that a number of phenomena observed in the interaction between the solar wind and planets can be successfully reproduced and studied under laboratory conditions (Podgorny 1978). For example, in experiments with an artificial solar wind, a laboratory model was created for the magnetosphere of Earth, with all the basic features of the actual magnetosphere. From this model of the terrestrial magnetosphere a number of geophysical processes could be understood. It was shown, for the first time, that there exist two independent mechanisms of convection: viscous, and that associated with the interconnection of lines of force of the fields of Earth and interplanetary space. It was demonstrated in the laboratory that, under certain conditions, it is possible to realize the well-known Dungey (1961) configuration, as well as other methods of changing the interconnection of the magnetic field lines. These, and many other results, have instilled confidence in the capabilities of laboratory simulation.

The confusing situation, which resulted from attempts to interpret the data obtained on the first flights to Venus, rendered the undertaking of a laboratory simulation experiment especially attractive. The data of a number of spacecraft distinctly pointed to the existence of a magnetic wake of Venus.

[994]

The similarity of this wake to the geomagnetic wake, as well as the existence of a bow shock at Venus similar to Earth's, made it tempting to postulate that Venus possesses its own magnetic moment, with a magnetosphere where general features are similar to those of Earth. By scaling the magnetosphere of Earth to the "magnetosphere" of Venus, one could estimate the magnetic moment of Venus using measurements in the wake. In this way, on several occasions, different values of magnetic moment were ascribed to this planet.

A second, initially less popular possibility of explaining the observed effects is based on the generation of unipolar-induction currents by the interaction between the magnetized solar wind and the ionosphere of a planet. The possible formation of this type of magnetosphere had been postulated by Alfvén (1957) for comets. However no one had yet observed such an induced magnetosphere, and it had not been possible to determine whether a shock wave would be created around it. In such a situation, it was advisable to conduct simulation experiments in order to answer the following questions:

1. Does an induced magnetosphere form in the interaction between an artificial solar wind and a body which has a plasma envelope?
2. Does the induced magnetosphere have an extensive magnetic wake?
3. Does the induced magnetosphere form an obstacle, responsible for the formation of a collisionless bow shock wave?

As in our other experiments on the simulation of space phenomena, the plasma parameters were chosen by the principle of a limited simulation. In limited simulation, nondimensional parameters, which characterize particular phenomena and which are on the order of unity in space, are chosen to be the same in the laboratory. Parameters which are much larger or smaller than unity in space are chosen to be respectively much larger or smaller than unity in the laboratory, although their order of magnitude may be different. Table I shows the chief properties of the artificial solar wind in this particular experiment. The flux of hydrogen plasma was collision-free, supersonic, and trans-Alfvènic. The ratio between the gas-kinetic and the magnetic pressure is

$$\beta = 8\pi nkT/B^2 \sim 1 \tag{1}$$

with the magnetic Reynolds number

$$\mathrm{Re}_m = 4\pi\sigma\mathrm{v}L/c^2 \sim 10^3 \ . \tag{2}$$

Here, n is the density of plasma, T the temperature, B the frozen-in magnetic field, σ the conductance, v the directional velocity of the plasma, L the characteristic dimension of the obstacle. The length of the stream of plasma was $\sim 20 \ \mu$ s.

To model Venus, we used wax balls of 5 cm and 10 cm diameter. The wax is sublimed when the plasma arrives at the surface of the ball. The

TABLE I

Plasma Parameters in the Experiment

Velocity of the flux	$1-3 \times 10^7$ cm s^{-1}
Plasma concentration	10^{13} cm^{-3}
Electron temperature	15 20 eV
Magnetic field frozen in the plasma	10–40 Gauss

sublimation products are ionized, basically by the ultraviolet radiation of the plasma source and the electron collisions. If the stream of plasma does not contain a frozen-in magnetic field, the glow of the wax ball is diffuse in nature (Fig. 1a) (Podgorny and Andriyanov 1978). The chief contribution to the glow is provided by the lines of ionized carbon. When the plasma contains a magnetic field of $B \gtrsim 20$ Gauss, the glow becomes localized near the ball (Fig. 1b). The boundary of the glow is very sharp, and in this case the formation of a bow shock may be seen. The collisionless bow shock has been detected by magnetic and electric directional probes. Typical measurement results are shown in Fig. 2. A directed electric probe, oriented perpendicular to the flux and facing the plasma accelerator, measured a practically constant current along the z-axis, directed toward the plasma source (Fig. 2a). This indicates that the flux $n\mathbf{v}$, is conserved and shows that the particular method is correct. The probe, oriented with its collecting surface parallel to the velocity vector, registers the current which is produced by the random motion of ions (Fig. 2b). Its readings begin to increase at a distance of 6 to 7 cm from the surface of the ball. At a distance of 3 cm, a second increase is evident, coinciding with the boundary of the plasma glow. The probe current in front of the second increase, in the area where the curve almost reaches saturation, is a factor of ~ 3 larger than the probe current in the undisturbed flux. Since the electron temperature increases here by only a factor of 1.5, this increase suggests a rise in the density. Within the limits of measurement precision, this increase corresponds to a shock with Mach number 3, which is appropriate for this series of measurements.

The existence of a bow shock is also suggested by the behavior of the transverse and longitudinal component of the magnetic field (Figs. 2c and 2d). In the region of the glow boundary a second increase in the transverse component $B_\perp$ and a drop in the component $B_\parallel$ parallel to the flow is observed. The field lines here become practically parallel to the glow boundary. These data suggest that an induced magnetosphere is formed by the interaction between a magnetized plasma stream and the plasma envelope of a body. This magnetosphere in turn acts as an obstacle, around which there forms a bow wave. However, the peak magnetic field pressure of this obstacle, $B^2/8\pi$, is somewhat lower than the dynamic pressure of the plasma stream. In a typical case, the plasma envelope acquires roughly 50% of the pressure. The mechanism of

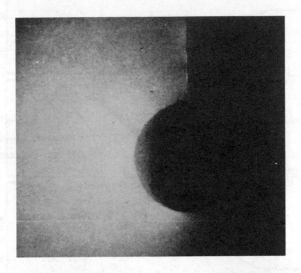

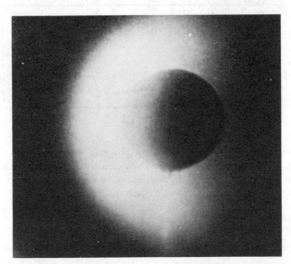

Fig. 1. Photographs of the glow of the plasma shell of a wax ball, placed in a stream of artificial solar wind: (a) without; (b) with a frozen-in magnetic field.

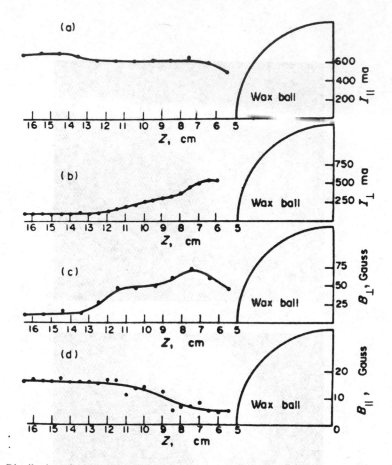

Fig. 2. Distribution of the plasma parameters along the z-axis on the day side of the laboratory
model of Venus: (a) plasma flux $ne\mathbf{v}$, measured by a directional electric probe, the collecting
surface of which was oriented opposite the velocity vector; (b) readings of a directional
probe with collecting surface parallel to the velocity vector. Probe current is proportional to
$ne\sqrt{T/M}$; (c) distribution of the transverse component of the magnetic field; (d) distribution
of the component of the magnetic field parallel to the flow.

transfer of the momentum has not been studied in detail, although it has been
established that the momentum is transferred over a length 3 to 5 times smaller
than the ion Larmor radius.

The wrapping of the obstacle by the magnetic field lines and the formation of an extensive magnetic wake are illustrated by the map of the magnetic
field, drawn in a plane which passes through the center of the ball (Fig. 3a)
(Dubinin et al. 1978). The magnetic field component normal to the current
layer always has the same sign as the vertical component of field in the plasma
stream. There is not yet sufficient data available to construct a similar map

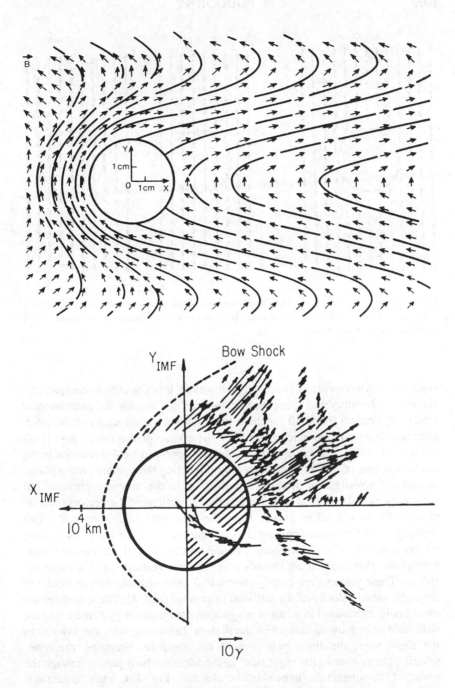

Fig. 3. (a) Configuration of the field in an artificially induced magnetosphere, observed in a simulation experiment; (b) distribution of the magnetic field vectors Venus's induced magnetosphere according to data from Veneras 9 and 10 spacecraft.

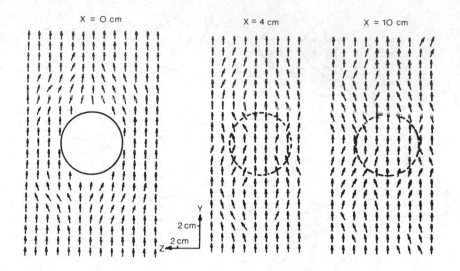

Fig. 4. Distribution of the projections of the magnetic field vectors in cross sections perpendicular to the axis of the wake of the artificially induced magnetosphere, at various distances
from the center of the model.

from the measurements of spacecraft. However, it is possible to compare the
results of laboratory and space measurements if we employ the measurement
results of Veneras 9 and 10 for those passes when the direction of the interplanetary field remains unchanged over the course of the entire pass (Dolginov et al. 1981). Figure 3b shows a chart of observed field directions in the
magnetosphere of Venus, constructed by projecting the vectors onto a plane.
A detailed investigation of the magnetic field in the magnetosphere of our
laboratory model of Venus revealed that the projections of the vectors onto the
plane (**vB**) do not differ greatly from the directions of the vectors which
actually lie in that plane, passing through the center of the body. Moreover,
the construction of a 3-dimensional model of the lines of the induced magnetosphere also enabled an identification of new features of this magnetosphere. These features are clearly shown in 3 cross sections perpendicular to
the undisturbed velocity of the artificial solar wind (Fig. 4). These features are
most easily understood in terms of magnetohydrodynamics by considering the
field lines to be moving along with the plasma. In moving from the day side to
the night side, the lines pass around the obstacle, dragging the ionospheric plasma over to the night side. In the section which passes through the
center of the sphere, the lines skirt the obstacle (Fig. 4a), while in sections
behind the sphere they converge into the plasma wake. A 3-dimensional
laboratory model of the magnetosphere of Venus in the form of a wire shell is
shown in Fig. 5 (Podgorny et al. 1980).

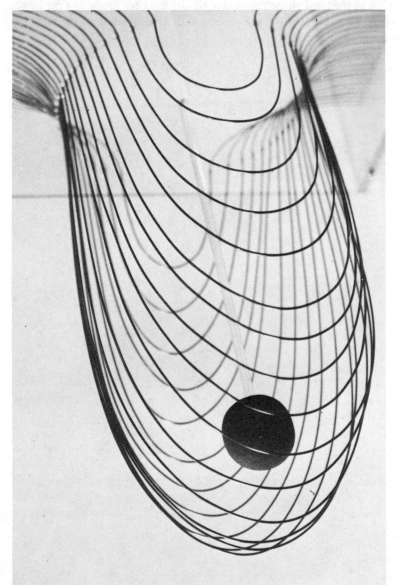

Fig. 5. A 3-dimensional wire model of the induced magnetosphere.

This model is valid when the body does not have its own field. If the induced field is similar in strength to the internal magnetic field, then a combined magnetosphere should be formed, with rather complex configuration. One of the most elementary cases ($\mathbf{B} \perp \mu$) has been studied in the laboratory (Dubinin et al. 1980) in connection with an evaluation of the combined magnetosphere of Russell (1978).

29. ORIGIN AND EVOLUTION OF THE
ATMOSPHERE OF VENUS

T. M. DONAHUE
University of Michigan

and

J. B. POLLACK
NASA Ames Research Center

The volatile inventories of Venus, Earth and Mars are compared and the impli-cations for the origin and evolution of the terrestrial planets assessed. The large gradient in noble gas abundance between 0.7 and 1.5 AU places severe con-straints on mechanisms for planetogenesis. Solar wind augmentation theories and grain accretion theories according to which the materials for the three planets were processed in regions or times of greatly different pressures are currently favored. Both have problems: the low ^{20}Ne to ^{36}Ar ratio on all three planets for the first class of theories; and confinement of planetesimals to re-stricted regions of the solar system for both. The reported high ratio of ^{36}Ar to Kr and Xe must be verified; if this ratio is correct even more severe restrictions will be placed on theories. The possible loss of appreciable water from Venus is discussed, as it is affected by recent evidence that deuterium has been enriched 100-fold. The ^{4}He and ^{40}Ar abundances suggest that outgassing has been ineffi-cient for much of Venus's lifetime in accordance with evidence that this planet is less active tectonically than Earth. The geochemistry of CO_2 on Venus is discussed. Finally, weak evidence is adduced that the climate on Venus may once have been benign.

Volatiles like H, C, N, and the noble gases are depleted relative to Si on the terrestrial planets, the lighter noble gases most severely. For this reason it is conceded that the atmospheres of these planets are not simple unaltered remnants of primordial protoatmospheres. Instead it is believed that the bulk of their atmospheres developed after or during planet formation. Heating associated with accretion, with differentiation, and with radioactivity is thought to have caused outgassing of volatiles that had somehow been acquired by the materials from which planetesimals and subsequently the planets were made; the gases thus produced form the present outgassed volatile inventory. Some of these gases are now atmospheric, some have condensed to form oceans and ice caps, some have been reincorporated chemically or physically in crusts and mantles and some have escaped into space. In the case of Venus the temperature at the surface is so high that it is usually assumed that, with some interesting exceptions (notably H_2O), volatiles of the sort partially locked in the Earth's crust are in the atmosphere of Venus. On the other hand, it is known that almost all of the CO_2 outgassed on Earth has reacted with the crust in the presence of water to form carbonates that have been layered down in the lithosphere. Since Venus and Earth are similar in size, mass, and distance from the sun, it is not surprising to find that the volatile inventories of C and N on the two planets are very similar. According to the measurements of CO_2 and N_2 abundances discussed in Chapter 13, the ratios of Cytherean to terrestrial C and N are between 0.6 and 1.7 (cf. Tables III and IV in Sec. I). Here the comparison is between atmospheric CO_2 and N_2 on Venus and atmospheric plus crustal C and N on Earth (Turekian and Clark 1975). On the other hand, while there is some water vapor in the atmosphere of Venus, there is certainly no ocean. The abundance ratio for water is $< 2 \times 10^{-4}$ and perhaps as small as 3×10^{-5} when the atmospheric inventory of Venus is compared with the terrestrial ocean. This situation poses one of the classical problems in understanding the origin and evolution of Venus. Was water outgassed in large amounts on Venus or was Venus formed essentially free of water? And if there once was much water, where is it now?

The problem of water is not only important for the atmosphere of Venus; it is probably crucial also for the geophysics of the solid planet (Chapters 6 and 10). If Venus was originally richly endowed with water, there was probably a strong relationship between the permanent loss of water from the interior, the rise in surface temperature associated with the attendant runaway greenhouse, and the development of a thick plastic lithosphere, inhibition of subduction, and ultimately a choking up of crustal material. A comparison of volatile inventories on the terrestrial planets must be carried out in parallel with a comparison of global tectonic properties of these planets for a full understanding of the mechanisms by which these planets were formed and have evolved.

I. COMPARATIVE VOLATILE INVENTORIES

On Earth and in meteorites, the noble gases are not only more severely depleted than other volatiles but they are fractionated as well, i.e., the relative abundances of the lighter noble gases are much smaller than in the solar atmosphere and presumably in the solar nebula. The ^{20}Ne to ^{84}Kr ratio on the

Sun is $\sim 10^5$; on Earth it is 23. The extension of a comparison of the volatile abundances on the terrestrial planets beyond CO_2, N_2 and H_2O to the noble gases has produced some major surprises. A premonitory signal of troubles to come was sounded by the results obtained on Mars during the Viking mission. The nonradiogenic rare gas abundances were found to be lower than the terrestrial by a factor of ~ 150. A similar depletion could be deduced for CO_2, N_2, and H_2O, if allowance was made for escape of some of these volatiles and sequestration of others in the present-day crust. The depletion of radiogenic ^{40}Ar was, however, less severe by an order of magnitude. On the other hand, the relative abundance pattern for the primordial rare gases (with the exception of Xe) was similar to the pattern for Earth and some chondrites. This situation led naturally to proposals that Mars and Earth received their volatiles by accretion of essentially the same kind of primitive matter but that for some reason Mars received a much less rich endowment than Earth did and was less thoroughly outgassed. The similarity of the CO_2 and N_2 abundances on Venus and Earth combined with the picture of the two planets as virtual twins led to the expectation, based on this mechanism, that the nonradiogenic rare gas inventory on Venus would be similar to that of Earth. Clearly a special treatment was needed for the anomaly of water. Anomalies in ^4He and ^{40}Ar were not unexpected because of possible differences in tectonism, volcanism and escape processes. These expectations were seriously upset, first by the discovery that neon and nonradiogenic argon appear to be more abundant on Venus than on Earth by a large factor (between 50 and 100) and later by a less well-documented indication that Kr and Xe are not greatly in excess on Venus. Cytherean ^{40}Ar and ^4He appear to be slightly in deficit unless appreciable helium has escaped from Venus as well as from Earth. Not only is the budget different on Venus but the distribution among species is different as well. This combination of results puts severe difficulties in the way of an effort to identify a unique mechanism by which the terrestrial planets (not to mention certain classes of meteorites) acquired their volatiles. It is no longer only the problem of water that forces us to search for specific mechanisms to account for volatile inventories on Venus.

A. Composition Measurements on Venus

An analysis of the present knowledge of the composition of Venus's atmosphere is presented in Chapter 13. For the purposes of this discussion the results of measurements of noble gas, CO_2, N_2 and H_2O mixing ratios for Venus, Earth, and Mars are drawn together in Table I and II. Isotope ratios are given in Table III. Some comments on these measurements are in order; they can be found in the Appendix.

B. Comparison of Terrestrial Planets

Meaningful comparison of volatile abundances on the three terrestrial planets can be made if they are converted to give the mass of each constituent

TABLE I
Mixing Ratios: Noble Gases

	^{20}Ne	^{36}Ar	Kr	^{84}Kr	Xe	^{129}Xe	^{132}Xe	^{4}He	^{40}Ar
Venus	ppm	ppm	ppb	ppb		ppb		ppm	ppm
LNMS	9^{+20}_{-6}	30^{+20}_{-10}	47^{+22}_{-35}	25^{+3}_{-18}	<40 (5)		<10		33^{+22}_{-11}
BNMS		<9						12^{+24}_{-6}	
LGC	4.3±0.7	33.0± 2.6	<2000						34.0± 2.6
VNMS	13 ±4	48 ± 8.5	1050±350	600±200				250?	57 ±10
VGC		21 ±15							25 ± 5
Earth	ppm	ppm	ppm	ppm		ppb		ppm	
	18.2	32	1.14	0.65	87	22.5	23.5	5.24	9.3(3)
Mars	ppm	ppm	ppm	ppm		ppb			
	$2.3^{+3.5}_{-1.5}$	5.3± 1.5	0.3±.06	0.17±.09	80±20	48±15	19±5		(1.6±.3) (4)

TABLE II

Mixing Ratios

	CO_2	N_2	H_2O
Venus	%	%	ppm
LNMS	96±2	4 ±2	<1000
BNMS			
LGC	96.4 ±1	3.4±0.1	1350±100
VNMS	96±0.3	4.5±0.5	50−120
VGC	97.5 ±0.5	2.5±0.3	<100
VSP			20−200
Earth	3.14 (−2)	78.1	
Mars	95.32±1.0	2.7±0.5	

per unit mass of planetary material. A ratio to the crustal mass would also be interesting but is clearly impossible to obtain at this stage of planetary exploration. The comparison is made in Table IV and in graphic form in Fig. 1.

Significant trends apparent in these data are highlighted in Tables V and VI. Earth is generally richer in outgassed volatiles than Mars by a factor of ~ 125. For nitrogen and carbon this is true, if Mars originally outgassed 10.6 times as much nitrogen as that presently found in the atmosphere, and if the ratio of outgassed carbon to nitrogen is the same as on Earth (McElroy et al. 1977). This is the lower part of the range suggested by the enrichment of ^{15}N in the atmosphere. The highest listings of martian nitrogen and carbon depend on the existence of surface chemical sinks for nitrogen whose strength is difficult to define. The enrichment is an order of magnitude smaller for ^{40}Ar and there is a xenon anomaly. The isotopic ratio of ^{129}Xe to ^{132}Xe is 2.5^{+2}_{-1} on Mars and 0.95 on Earth. The ratio of terrestrial to martian ^{132}Xe in the atmosphere is a factor of 4 smaller than the ratio for ^{36}Ar.

The situation for Venus is quite different. In Table IV two values are given for Earth's carbon inventory. The lower figure for the atmosphere and crust is the lower bound for the *outgassed* inventory (Walker 1977; Holland 1978); the larger value includes the upper mantle estimate of Turekian and Clark (1975) and is the upper bound for the outgassed inventory. Similarly for nitrogen, the atmospheric contribution, the atmospheric plus crustal and the

TABLE III

Isotopic Ratios

	Venus	Earth	Mars
$^{22}Ne/^{20}Ne$	0.07±0.02	0.097	0.1 ±0.03
$^{38}Ar/^{36}Ar$	0.18±0.02	0.187	0.19±0.02
$^{13}C/^{12}C$	≤1.19 (−2)	1.11 (−2)	(1.1 ±0.1) (−2)
$^{18}O/^{16}O$	2.0 ±0.1 (−3)	2.04 (−3)	(2 ±0.2) (−3)
$^2H/H$	1.6 ±0.2 (−2)	1.6 (−4)	—

TABLE IV

Volatile Abundances in gm/gm

	Venus[a]	Earth[b]	Mars[c]
C	$(2.67 \pm 0.30)\,(-5)$	$(1.5-4.5)\,(-5)$	$(1.2 \pm 0.1)\,(-8)$
			$1.1(-7) -? ? (-6)$
N	$(2.49 \pm 0.30)\,(-6)$	$6.66\,(-7)$	$(8.0 \pm 1.5)\,(-10)$
		$(1.45-2.43)\,(-6)$	$8.5(-9)-1.8\,(-7)$
^{4}He	$(1.1^{+2}_{-0.5})\,(-10)$	$1.65\,(-9)$	
	$(23 \pm 5)\,(-10)$		
^{20}Ne	$(2.3 \pm 0.5)\,(-10)$	$1.08\,(-11)$	$(4.9^{+7.4}_{-3.2})\,(-14)$
^{36}Ar	$(2.5 \pm 0.4)\,(-9)$	$3.52\,(-11)$	$(2.1 \pm 0.6)\,(-13)$
^{40}Ar	$(2.9 \pm 0.4)\,(-9)$	$1.14\,(-8)$	$(6.8 \pm 1.3)\,(-10)$
Kr	$(9.2^{+4.3}_{-6.9})\,(-12)$	$2.93\,(-12)$	$(2.7 \pm .5)\,(-14)$
^{84}Kr	$(4.9^{+0.6}_{-2.4})\,(-12)$	$1.66\,(-12)$	$(1.4 \pm .3)\,(-14)$
Xe	$<1.25\,(-11)$	$3.51\,(-13)$	$(1.1 \pm .25)\,(-14)$
^{129}Xe	$<2.9\,(-12)$	$8.92\,(-14)$	$(6.7 \pm 2)\,(-15)$
^{132}Xe	$<3.1\,(-12)$	$9.54\,(-14)$	$(2.7 \pm 0.7)\,(-15)$

[a]The Venus column is based on PV LNMS results.

[b]Two values are given for terrestrial C: crustal and crustral plus upper mantle; three are given for N: atmospheric, atmospheric plus crustal, and total. Terrestrial He is the amount that would be outgassed by a flux of 2×10^6 cm^{-2} s^{-1} maintained for 4.6×10^9 yr.

[c]Atmospheric abundances and the upper and lower bounds for outgassed C and N for Mars are those of McElroy et al. (1977).

upper bound, including the upper mantle (Turekian and Clark 1975) are listed separately. The atmospheric inventories of C and N on Venus are in approximate balance with the upper bound for the outgassed inventories on Earth (Walker 1977; Holland 1978). This result suggests that the total N and C volatile inventories are about the same on the two planets and that the mantle of Venus is largely outgassed. This also agrees with the proposal of Phillips et al. (1981) that the mantle of Venus is now dry, that there is no asthenosphere, and that large-scale plate movement has been arrested. On the other hand, if the mantle of Venus has not been highly degassed and contains a gas component comparable to that of Earth, the proper comparison is between the atmospheric inventory on Venus and the sum of the crustal and atmospheric inventories on Earth. In this case, the implication is that Venus has about twice as much C and N as Earth.

However, the nonradiogenic noble gases ^{36}Ar and ^{20}Ne are enhanced by factors of 72 and 43. If the PV LNMS results are accepted, this is in contrast with only a 3-fold enhancement of ^{84}Kr and $<$30-fold enrichment of the xenon isotopes, but very likely less than half that much. The Venera data show a krypton enrichment on Venus of 70 compared to a ^{36}Ar enrichment of 120. The atmosphere of Venus has only one fourth the terrestrial endowment

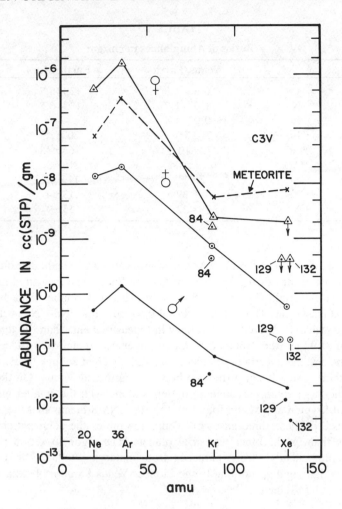

Fig. 1. Abundances of noble gases on Venus, Earth, and Mars and in class 3V carbonaceous chondrites.

of radiogenic ^{40}Ar, but between 175 and 3700 as much ^{4}He. However, when the comparison is between atmospheric helium on Venus and outgassed helium on Earth, it appears that <10% as much ^{4}He has entered the Cytherean atmosphere as the total that has been outgassed on Earth, if the outgassing rate of helium on Earth is 2 x 10^{6} cm^{-2}s^{-1} and escape of helium from Venus has been negligible.

Table VI, where the ratios of ^{36}Ar to other species are tabulated, shows these trends in a different way. All three atmospheres (like meteorites) are poor in neon relative to argon. The sun has 40 times as much ^{20}Ne as ^{36}Ar (Cameron 1973). Despite the large spread in the measured ^{20}Ne mixing ratios

TABLE V

Ratios of Abundances in gm/gm

	Venus/Earth	Earth/Mars
C	0.6 –1.8	133– 6.7
N	1.02– 1.7	133– 6.7
^{4}He	0.07– 1.4	—
^{20}Ne	21 ±5	200^{+280}_{-130}
^{36}Ar	72 ±10	165 ±45
^{40}Ar	0.25 ±.04	16 ±3
^{84}Kr	$2.9^{+3}_{-1.5}$	115^{+185}_{-75}
^{129}Xe	<32	13 ±4
^{132}Xe	<32	35 ±9

for Venus, the ratio of ^{20}Ne to ^{36}Ar abundances appears to be higher on Venus and in most classes of meteorites than it is on Earth. The measured ratio for Mars is so uncertain that it is of little help. That it is smaller than the solar ratio is at least clear. The isotopic ratio of 0.07 ± 0.2 for ^{22}Ne/^{20}Ne is closer to the solar value as derived from solar wind measurements than the terrestrial ratio of 0.97 or the "planetary" component in meteorites of ~ 0.13. Nevertheless, the uncertainty in the Venus ratio is great enough to discourage establishing a significant difference between Earth and Venus. On the other hand, the large argon enrichment on Venus along with the smaller enhancements of krypton and xenon found by the PV LNMS means that the relative abundance of these three gases on Venus resembles the solar pattern more strongly than on the other terrestrial planets. The argon to xenon ratio increases dramatically as one proceeds from meteorites to Mars to Earth to Venus to Sun, again according to the Pioneer Venus Large probe mass spectrometer (LNMS) data.

TABLE VI

Ratio of Abundances in cm³/gm of ^{36}Ar to Other Species

	^{20}Ne	^{36}Ar	^{84}Kr	Xe	^{129}Xe	^{132}Xe	^{12}C
Sun	.025	1	2500	6700	2.5(4)	2.5(4)	4.8 (−2)
Venus	0.16 ±0.02	1	80 ±25	<750	<3200	<3000	(3.1 ±0.5) (−5)
	0.3 ±0.15		1200^{+3000}_{-130}				
Earth	0.57 ±0.003	1	48 ±0.5	368 ±5	1420 ±15	1360 ±15	(1.5 ±0.7) (−6)
Mars	$0.4^{+.6}_{-.25}$	1	30^{+60}_{-20}	66 ±15	110 ±40	280 ±70	(0.31–6.4) (−7)
C1	0.27	1	109	28	120	114	1.7 (−8)
C3V	0.19	1	99	30	130	126	4.7 (−8)
C30	0.03	1	154	60	260	242	9.7 (−8)

II. ORIGIN OF THE ATMOSPHERE

A. Description of Theories

Theories that describe the formation of the terrestrial planets from the solar nebula can be separated several ways into classes. One classification contrasts environments where the planets were formed as gas-rich or gas-poor. A gas-rich environment would have prevailed if giant gaseous proto-planets (Cameron 1978a,b; Cameron et al. 1980; Slattery et al. 1980) were precursors of the inner planets. In this theory, as infall accretion proceeds, a massive nebula (~ 1 $M_\odot$) is believed to become unstable against perturbations leading to the formation of rings that then collapse into protoplanets. The protoplanets grow hot during the process of contraction. Solids gather toward the center and the gaseous envelope is eventually lost by planets in the inner solar system because of action of tidal forces. In another kind of gas-rich formation theory, all stages of planetary formation from the accretion of grains forming planetesimals to the coalescence of planetesimals into planets are considered to take place in the presence of the solar nebula. Subsequently the nebular gas and the protoatmospheres of these planets are at least partially swept away by the enhanced solar wind or by extreme ultraviolet (EUV) heating (Hayashi et al. 1979; Horedt 1979).

In the other class of theories, the later stages of planetary formation are assumed to have taken place after the solar nebula has been dissipated. It is again assumed that planets grow in stages starting with dust grains and proceeding through formation of colliding planetesimals. The grains may be of several types; one, interstellar in nature, was part of the initial population of dust and gas that collapsed into the nebula; another may have originated from a nearby supernova explosion if such an event triggered the formation of the nebula; a third formed from gases in the solar nebula as it cooled. Volatiles in this case are considered to be (a) supplied from substances originally incorporated in the grains or acquired by a selected set of grains by chemical reactions or physical absorption from the gas of the solar nebula, or (b) implanted in grains and larger objects by the solar wind. Combinations of these processes are also possible.

A version of this second set of theories postulates that the "bare" primitive terrestrial planets, generated in a region where the temperature of the solar nebula was high, were made of material that was volatile-poor. Most of their volatiles are supposed to have been incorporated in these planets in a late accreting "veneer" supplied by impacts with volatile-rich objects formed elsewhere in the solar system (Lewis 1974; Turekian and Clark 1975; Anders and Owen 1977; Owen et al. 1982). These volatile-rich objects were presumably born in places like the asteroid belt or Oort's cloud where low nebular temperatures favored development of solids well-endowed with volatiles. Orbital perturbations sent some of these bodies into orbits crossing those of the terrestrial planets and collisions with those planets during the final stages of the period of planetogenesis endowed the planets with their volatiles. Since meteorites may be the relics of such volatile-rich bodies, attempts have been made to reproduce a reasonable approximation to the present planetary inventories by assembling a suitable mixture of various types of meteorites. It is then proposed that a certain fraction of today's planets is composed of that mixture. Ideally this procedure would explain all three planets from a unique mixture at least in so far as the *relative* abundance of volatiles is concerned. Thus, the inelegant requirement that radically different classes of orbit crossing objects would have to be envisioned for Venus, Earth, and Mars separately would be avoided. It should be noted that in models of this class, the late accreting veneer need not necessarily be supplied by a stray body from the outer solar system. Owen et al. (1982) also propose the possibility that a late forming local condensate might be trapped in a gaseous protoplanet and contribute volatiles to some of the planets—notably to Venus.

Another version of the "gas-poor" class of theories is similar to the first but with an essential difference. It dispenses with the stray late arriving objects as the prime source of volatiles; instead, each planet's endowment of volatiles is taken to be supplied by dust grains that were involved from the beginning in the various local steps of planetary formation. As in the previous theory, volatiles are acquired by grains from the ambient nebular gas. Different conditions, particularly of pressure and temperature in the regions of the nebula from which each planet acquired the bulk of its volatile-rich grains are invoked to account for systematic differences between one planet and another. This class of theory is known as a "grain accretion" theory.

A variation on the grain accretion theory is one proposing that an important source of some volatiles (particularly the noble gases) on some planets (e.g. Venus) was bombardment of grains and larger bodies of planetesimal size by an enhanced flux of solar wind during the period of planet formation (Wetherill 1980, 1981; Donahue et al. 1981; McElroy and Prather 1981). The three planets are supposed to have received their basic endowment of volatiles by some common mechanism, say that of grain accretion. The material from which Venus was formed was exposed to an enhanced solar wind and acquired a significant fraction of its noble gases from this source. The objects destined to be incorporated in Earth and Mars were shielded from this intense solar wind and hence did not receive such a large dose of volatiles rich in inert gases. As planetesimals grew larger, the ratio of surface area to mass decreased and the nebular disk became more and more transparent. Hence an appropriate phasing of the period of enhanced solar wind and the time of planetesimal growth is required in this model to prevent Mars and Earth from acquiring too much noble gas. Either the clearing and transformation from small to large planetesimals must have occurred abruptly or the strength of the solar wind must have decreased after a sufficiently long time but before the inner part of the disk became transparent. Another necessary condition is that a very large fraction of the strongly irradiated planetesimals stayed in the region where Venus was formed and did not cross into the provinces of telluric or martian parent bodies. Because the basic planetary material was seriously depleted in noble gases relative to other volatiles, this mechanism can selectively enrich a planetary atmosphere in the rare gases.

If the formation of the solar nebula was triggered by a nearby supernova explosion, the timing of planet formation can also be crucial. McElroy and Prather (1981) have suggested that Mars grew to its present size before Earth and Venus did and that it formed while appreciable amounts of ^{26}Al left over from the supernova event were still present. Consequent radioactive heating of the planetesimals from which Mars was formed might then account for early loss of volatiles from that planet. The enhanced solar wind mechanism is also assumed in their model to account for the special component of noble gases on Venus. Supplemental sources of volatiles are the solar wind or the occasional chance comet or asteroid encountering the planet after the atmosphere was formed.

B. Comparison of Models with Observations

None of the theories discussed above is free of serious deficiencies in explaining the origin of Venus's atmosphere, particularly when called upon to account as well for the volatiles on Earth and Mars. Models in which the planets grew in a gas-rich environment naturally account for the gradient in Ne and Ar abundance with distance from the Sun, although this is not clearly the case for gaseous protoplanets. However, they have a fundamental problem in explaining the overall depletion of noble gases relative to other volatiles and the departure of the abundance ratios of the noble gases and, in some cases, of isotopic ratios from the solar pattern on all the planets. Escape of gases after accretion and outgassing is selective for light gases and hence qualitatively capable of accounting for the depletion pattern of the noble gases compared to the solar abundance spectrum. On the other hand, the fact that the ratio of ^{36}Ar to ^{20}Ne is similar for the terrestrial planets and carbonaceous chondrites (~ 100 times the solar ratio) despite the great differences in parent body and planet masses as well as location in the solar system, confronts this model with a serious difficulty. It also has a problem in explaining how argon can be more efficiently lost than carbon and nitrogen, or why the argon to krypton and argon to xenon ratios become more nearly solar the closer the planet is to the Sun.

The external source, or late accreting veneer model discussed by Turekian and Clark (1975) was further developed after Viking to explain the terrestrial and martian volatile inventories (Rasool and Le Sergeant 1977; Anders and Owen 1977). In the detailed treatment of Anders and Owen, a carbonaceous chondrite of type 3V was chosen as the archetype of the volatile-rich material. There is some question of whether or not the ^{36}Ar to C ratio used in this model is correct (Bogard and Gibson 1978). In any event the discovery that the noble gas pattern on Venus is radically different from that on Earth, while the C and N inventories are similar, is fatal for this particular version of the veneer model. Since stray bodies have similar probabilities for colliding with Venus and Earth, the model predicts similar Ar and Ne as well as C and N abundances for the two planets. Owen et al. (1982) have modified their model by proposing that along with the 7% of type C3V material appropriate to Earth and Mars, an additional 7% of a hypothetical type E3 chondritic material was incorporated in Venus. The supposed E3 material is presumed to be adequately rich in ^{36}Ar and neon, and it is proposed that it was also somehow processed by a gaseous protoplanet precursor of Venus. The authors point out that some peculiar chondritic meteorites have been found with ^{36}Ar to ^{84}Kr ratios between 260 and 450. Such ratios, even if imputed to the supposed E3 component of Venus, when combined with those appropriate to a C3V component, would fall far short of the ratio of 1200 reported for Venus by the PV LNMS. This problem, and the need to invoke a special source for Venus without a specific meteoritic analogue, limits the appeal of this mechanism.

A grain accretion theory was developed (Pollack and Black 1979; Oyama et al. 1980a) to account for the orders of magnitude increase in the abundances of atmospheric ^{36}Ar and ^{20}Ne from Mars to Earth to Venus (cf. Table V). According to the model this systematic trend with heliocentric distance is due to correspondingly large pressure differences but small temperature contrasts in the regions where rare gases were incorporated into certain types of grains, i.e. those that ultimately were involved in forming the terrestrial planets and meteorite parent bodies. Rare gases were incorporated into grains by adsorption and occlusion while C, N, and H_2O were incorporated chemically. The former set of processes is very sensitive to the nebular pressure while the latter is not.

The trends in the volatile abundances of rare gases and N_2 and CO_2 among the three outer terrestrial planets (cf. Table V) were reproduced by the grain accretion theory by appropriately adjusting three parameters: solar nebula pressure p and the efficiency of early and late outgassing e_1 and e_2. These parameters were found to have the following relative values from Venus to Earth to Mars: p:65/1/0.05; e_1:1/1/0.1; e_2:.2/1/0.05. The smaller values of e_2 for Venus and Mars, as required by their lower abundances of atmospheric ^{40}Ar, imply that these planets had a much less active *global* tectonic style during their middle and later history than did the Earth. Thus, during this interval, much less of the ^{40}Ar produced in their interiors from the radioactive decay of ^{40}K was able to escape into their atmospheres. The low value of e_2 for Venus also implies that a smaller fraction of radiogenic ^{4}He was vented into its atmosphere as compared to that of the Earth.

The grain accretion theory is not without problems. First, there is the objection (Wetherill 1981) that large relative velocities would develop between large and small objects in the presence of large pressure gradient forces. Subsequent collisions would result in fragmentation rather than coalescence.

However, it is possible that the pressure differences refer to different times as well as places at which the host grains grew and acquired rare gases. Next, it is not clear that it is physically plausible for the host grain to grow and incorporate rare gases at a narrow enough range of temperature that the amounts incorporated were dominated by nebular pressure and not temperature. Third, the relative proportion of rare gases varies somewhat (perhaps dramatically if the LNMS data on Kr and Xe are correct) among meteorites, Mars, Earth and Venus (cf., Table VI). Such variation is not consistent with the original form of the grain accretion model, although Pollack and Black (1981, 1982) have proposed a modified form of this model in which relative proportion can vary, in accord with data from various meteorites. Finally, as discussed more fully below, exchange of planetesimals among the feeding zones for the various planets could significantly reduce the ultimate differences in rare gas abundances among the planets.

Wetherill (1981), Donahue et al. (1981), and McElroy and Prather (1981) have proposed a "solar wind augmentation" model, according to which rare gases were derived from two sources: occlusion of rare gases in grains, analogous to the process invoked by Pollack and Black, and implantation of solar wind gases into planetesimals after the nebula had dissipated. The first source is supposed to provide a common component of rare gases for all three planets while the second changes by a large factor between Venus and the Earth because of the large opacity of the Venus planetesimals for the solar wind. The great change in relative abundances of rare gases between Earth and Venus, as implied by the LNMS results (but not the Venera 11–12 MS data) is easily explained by these models since the two components have greatly differing relative proportions of rare gases (planetary or solar) and since Venus received much more of the solar wind component than did Earth. In particular, if solar wind implantation results in a 75-fold enhancement of ^{36}Ar for Venus, it would lead to only a factor of 2 increment in ^{84}Kr and 2.5 in Xe.

An important problem facing the solar wind augmentation theories is one advanced by Wetherill himself. This is confinement of a sufficiently large number of the enriched planetesimals to the region where Venus is being formed despite the effect of perturbations tending to send them into the feeding ground for embryonic Earth and Mars. Indeed such mixing of orbits appears to be necessary to avoid premature termination of planetesimal growth before planetary-size objects are achieved. The latest calculations by Wetherill deal only with planetesimals of equal size but predict that the fraction of enriched bodies remaining in the Venus feeding region would not be able to account for the 75-fold enhancement of ^{36}Ar (and the ratio of 1200 in ^{36}Ar/^{84}Kr that is a corollary of this enhancement).

Another difficulty the augmented solar wind theory must face is accounting for the reduction in the ^{20}Ne to ^{36}Ar ratio from 40 in the solar wind to ≤ 0.3 on Venus. This low value for ^{20}Ne/^{36}Ar lies within the range that characterizes the trapped planetary component of meteorites and is distinctly smaller than values manifested by known samples of implanted solar wind gas. In particular, Brownlee dust (presumably cometary grains), lunar soils, and gas-rich meteorites all have $^{20}Ne/^{36}Ar$ ratios of ~ 10 (Hudson et al. 1981). While these values are somewhat smaller than that characterizing the solar wind, they are similar to one another, despite the rather different astrophysical sites to which they pertain, and they are ~ 30 times larger than the values for Venus. Advocates of the solar wind implantation theory have suggested that a more severe fractionation may have occurred for the solar wind implanted into the Venus planetesimals, due perhaps to enhanced diffusional losses caused by higher temperatures and/or an enhanced meteoroid bombardment flux. But diffusional processes do not appear to be responsible for the subsolar $^{20}Ne/$ ^{36}Ar values that characterize implanted solar wind gas (Pollack and Black 1981, 1982), and some lunar soils are quite mature, i.e., individual grains in the soil have remained at the surface for the maximum amount of time possible before being stirred to deeper depths by continuing impacts. It is not clear how neon could be fractionated with respect to argon to the degree required in solar wind implanted gas for Venus while the $^{36}Ar/^{84}Kr$ and $^{36}Ar/^{132}Xe$ ratios remain close to solar. Finally, the large fractionation of neon relative to argon might result in a nonsolar ^{20}Ne to ^{22}Ne ratio.

Another enigma is xenon. The already noted increase in the ^{36}Ar to Xe ratio from meteorites to Mars to Earth to Venus is striking. This suggests that there is 25 times as much terrestrial xenon in shales as in the atmosphere (Canalas et al. 1968; Fanale and Cannon 1971). That would reduce the ^{36}Ar to Xe ratio to ~ 60 and produce fair agreement with the ratio for meteorites and Mars. A recent more extensive survey of the amount of Xe trapped in shales and sedimentary rocks (Podosek 1980) has produced a drastic reduction in inventory. If xenon is imprisoned on Earth, it surely would have been released at Cytherean temperatures. And yet the ^{36}Ar to Xe ratio on Venus appears to be much larger than 750 according to the PV LNMS results listed in Table VI.

Theories assuming that the atmospheres of the terrestrial planets were essentially derived by outgassing of solid, volatile-rich components accreted from colliding planetesimals share a common problem. This is the difficulty of maintaining the observed large differences in noble gas inventories of the planets in both absolute and relative abundance, despite the imperatives of gravitational mixing of planetesimal orbits. This problem is aggravated by the requirement that the ^{20}Ne to ^{36}Ar ratio be similar on these planets and on meteorite parent bodies. For these reasons it would probably be advisable to examine carefully the possibility that these are important remnants of protoatmospheres present on at least some of the planets, notably on Venus. This is another way of giving to Venus's noble gases a solar signature and of explaining the large argon-krypton and argon-xenon ratios measured by Pioneer Venus. In this case, it might be possible to solve the neon problem by appealing to fractionation during the blow-off phase of the protoatmosphere and perhaps a modest gradient in the degree of fractionation in the component of noble gases contributed by accretion and outgassing of solid material. An approach to the problem that has some of these features has been made by Isakov (1979,1980,1981). Unfortunately, there are quantitative deficiencies in Isakov's models; they overestimate the inventory of martian volatiles by two orders of magnitude and do not offer a quantitative explanation for the fractionation in the protoatmospheric contribution.

III. OUTGASSING

A. Tectonic and Thermal History

Some gases in Venus's atmosphere, like radiogenic ^{40}Ar, are almost certainly derived by outgassing from the planet's interior. According to the discussion in Sec. II, it seems likely that the primordial rare gases were incorporated into the material that formed the planet and later partially outgassed. A similar situation may hold for other atmospheric constituents, like N_2, CO_2, and H_2O, although an asteroidal or cometary impact source cannot be ruled out. This section discusses the timing and degree of outgassing, as well as the composition of the outgassed volatiles. It also comments on the timing of the addition of volatiles brought in by volatile-rich meteoroids.

Juvenile gases are probably vented into the atmosphere in association with interior magmas that reach the surface. As a result of interior melting, gases are released or produced within the magma and can escape into the atmosphere if the magma can rise to the surface or if the layer overlying the magma is sufficiently porous. Thus, to help determine the timing of outgassing, it is useful to assess the timing of surface-reaching magmatic events.

Information about the occurrence of near-surface melting on Venus can be obtained from lunar chronology, thermal history models for the terrestrial planets, and relevant geologic data for Venus. Analyses of dated lunar samples indicate that there have been two major magmatic periods on the Moon (Wasserberg et al. 1977). A 60 km thick, sialic crust formed from a several hundred km deep magma ocean during the first several hundred million years. Then, over the next billion or more years, local basaltic volcanism occurred. The occurrence of the magma ocean is usually attributed to accretional energy of formation, while the basaltic volcanism is due chiefly to radioactive heating (Toksöz et al. 1978). For objects much more massive than the Moon, like the Earth and Venus, an even larger fraction of their outer portions can be expected to be melted due to accretional heat and hence a large amount of outgassing can be expected during their first billion or so years of history. Additional melting and outgassing for the Earth and Venus should occur during this time interval due to core formation. Finally, because of the larger size of these objects, lithospheric thickening occurs more slowly than for the Moon (presently the Earth's lithosphere is $\sim$ 100 km thick compared to $\sim$ 800 km for the Moon), and the localized magmatic events may be spread over much, if not all, of their history.

Topographic and gravity data obtained from the Pioneer Venus spacecraft, in conjunction with groundbased and spacecraft radar images of Venus's surface have provided a first order understanding of the geologic history of Venus that is consistent with the inferences drawn above from thermal history models (Masursky et al. 1980; Philips et al. 1981; and Chapters 6 and 10). About 65% of the surface consists of an "upland rolling plains" province. Based upon the abundance of crater-like forms and a granitic composition inferred from Venera gamma-ray spectrometry, this province is thought to represent the planet's ancient crust. Evidence for later, more localized volcanism is provided by Beta Regio, which is thought to consist of several large shield volcanoes, and the lowland province which covers 27% of the surface and which may be covered by basaltic lavas. Venus seems to have been tectonically active sometime during its history, as evidenced by at least one small continent that may be dynamically supported and by complex ridge and trough regions (Chapter 6). However, there is no evidence for the occurrence of an Earth-type plate tectonics now, though such a style of tectonism may have been realized in the past.

B. Timing of Outgassing

A useful upper bound on the duration of the bulk of the outgassing from Venus's interior can be obtained from the abundance of ^{40}Ar in the atmosphere (Oyama et al. 1980a; Pollack and Black 1981, 1982). ^{40}Ar in the atmospheres of the terrestrial planets is derived almost entirely from the decay of radiogenic ^{40}K in their interiors and subsequent outgassing. There is about four times as much ^{40}Ar in the Earth's atmosphere as in the atmosphere of Venus. This difference can be attributed to a combination of a shorter duration of outgassing for Venus, a shallower depth to which Venus has been outgassed, and/or a lower potassium abundance for Venus. It is likely that the first factor is the dominant one. First, based on our understanding of the thermal history of the terrestrial bodies (Toksöz et al. 1978), there is no reason for as much as a factor of 4 difference in the depth to which the two planets have been melted. Second, the amount of potassium in Venus's crust is very similar to that in the Earth's crust and it is distinctly larger than the corresponding value for the Moon and the average potassium abundance for these objects' interiors (Anders and Owen 1977). The concentration of potassium in the crust is a result of chemical differentiation that accompanies interior melting, with the degree of concentration providing a measure of the amount of melting. Finally, the very similar K to U ratios for the surface material of Venus and Earth imply very similar bulk K values (Pollack and Black 1981, 1982). Therefore it can be assumed that the difference in the abundance of ^{40}Kr between the atmospheres of the Earth and Venus is due chiefly to a difference in the duration of the bulk of the outgassing. The half life for the decay of ^{40}K to ^{40}Ar then implies that most of Venus's outgassing took place during its first billion years, a result in accord with the earlier discussion of thermal history models.

The vast majority of volatiles brought to Venus through impacts with volatile-rich comets and asteroids probably entered its atmosphere during its early history. In particular, studies of lunar chronology indicate that the Moon was bombarded by a flux of meteoroids that was orders of magnitude larger during its first 700 Myr than during later times (Wasserburg et al. 1977). Since all the objects in the inner solar system are subjected to essentially the same time history of bombardment (Wetherill 1975), almost all the impacts Venus experienced occurred during the first 700 Myr. Because of the high velocity of these impacts, the impacting object was almost entirely vaporized and hence any volatiles it contained were incorporated immediately and directly into Venus's atmosphere.

The degree to which Venus has been outgassed can be derived from estimates of the corresponding fraction for the Earth and the apparent similarity in this fraction for the two planets, as argued above. Estimates of this fraction for the Earth range from ~ 0.25 (Hart et al. 1979) to 0.5 (Stacey 1979) and therefore Venus has presumably outgassed to a similar extent.

C. Composition of Outgassed Volatiles

The decay of radionuclides in the interior results in the production of several gas species. The most abundant of these are ^{40}Ar, produced by the decay of ^{40}K, and ^{4}He, produced chiefly by the decay of isotopes of U and Th. Venus's atmosphere contains about 200 times as much He as the Earth's according to von Zahn et al. (1979a). In part, this difference is due to the rapid escape of He from the Earth's atmosphere, with its residence time being a mere 2 Myr (Holland 1978). The much larger amount of He in Venus's atmosphere implies that its residence time is much longer compared to that on Earth. First, suppose that the current outgassing rates are the same for the two planets ($\sim 2 \times 10^6$ cm^{-2} s^{-1}). Then He has a residence time of ~ 280 Myr for Venus. But the Earth very likely has a larger outgassing rate at present than does Venus. First, little of the ^{40}Ar produced after the planet's first billion years of history has been vented into the atmosphere, as argued above and second, the absence of plate tectonics on Venus at the current epoch implies a smaller outgassing rate now. Thus, 280 Myr is a gross lower bound on the residence time of He in Venus's atmosphere.

On the other hand, it would take 10.5×10^9 yr for the mixing ratio of ^{4}He to grow to 450 ppm as suggested by the PV LNMS, provided that no helium has escaped. For this amount of helium to accumulate in 4.5×10^9 yr the flux would have to be 4.7×10^6 cm^{-2} s^{-1}. With such a flux the lifetime of helium on Earth would only be 750,000 yr. According to Pollack and Black (1982), 450 ppm is about an order of magnitude greater than the total radiogenic helium production during the life of the planet. If both measurement and model are correct, the implication is that most of Venus's helium is primordial.

Useful estimates of the oxidation state of outgassed volatiles can be obtained from thermodynamic equilibrium considerations, once the oxygen fugacity is specified (Holland 1978). The oxygen fugacity, in turn, may be largely determined by the oxidation state of iron in the parent magma. Volcanic gases on present-day Earth typically have H_2O to H_2 and CO_2 to CO ratios of ~ 60 and 20, respectively, with almost all the N-containing species being N_2 and with much smaller amounts of the more highly reduced C and N gases present. In this case the oxygen fugacity is controlled by magmas having no metallic iron and Fe_2O_3 to FeO ratios of ~ 0.5.

Gases vented from Venus's interior during its early history, when much of the outgassing occurred, probably had a somewhat more reducing character due to a lower oxygen fugacity. This oxygen fugacity in turn reflects the lower oxidation state of the magma iron that is due to several factors. First, in its earlier history, before core formation was completed, metallic iron may have been present in the outer layers of the planet. Note that core formation and outgassing may have partially overlapped one another in time. Second, the Fe_2O_3 to FeO ratio may have been lower because Venus may have formed in a somewhat hotter part of the solar nebula (Lewis 1972) and also because the current ratio for the Earth may, to some degree, be determined by the cumulative oxidation of the crust and upper mantle by atmospheric oxygen. Conceivably, the H_2O to H_2 and CO_2 to CO ratios for Venus could have approached unity for the early outgassed volatiles, though N_2 would have remained the dominant N-containing volatile and little CH_4 would have been expected.

IV. LOSS PROCESSES

A. Escape: Thermal and Nonthermal

Some of the volatiles originally present on Venus may have been lost by escape from the upper atmosphere. Escape may have occurred by a variety of processes including hydrodynamic flow, Jeans escape, or by nonthermal mechanisms such as charge exchange or momentum transfer from the solar wind. The treatment of escape will be brief because of the existence of other exhaustive treatments of the subject (Hunten and Donahue 1976; Hunten 1982). Hydrodynamic loss involves mass outflow of gases into interplanetary space and may involve a substantial fraction of the entire energy budget of the planet's thermosphere. The criterion for hydrodynamic loss is that the internal energy of the gas be greater than the depth of the gravitational potential well in which it is situated. Quantitatively this means that the escape parameter, $\lambda_c = GMm/kT_cr_c = r_c/H_c$, be less than 2 or 3/2 depending upon whether the gas is monatomic or diatomic. Here G is the gravitational constant, M the mass of the planet, m the mass of a gas molecule, k Boltzmann's constant, T_c, H_c and r_c the temperature, scale height, at and planetocentric distance to the critical or escape level. Watson et al. (1982) have recently studied this process in some detail. The energy to power hydrodynamic flow is supplied by solar EUV.

There is a limit to the rate at which gas can escape from a planet for a given solar flux. The limit is set because the cooling of a rapidly flowing atmosphere puts a limit on the radial extent of that atmosphere, no matter how much hydrogen it may contain. The global external energy that can be absorbed is therefore also limited. For a Venus atmosphere dominated by hydrogen but containing a negligible amount of dust, the energy limited H_2 flux is 1.1×10^{12} cm^{-2} s^{-1}. Loss of the equivalent of an ocean of water would occur in ~ 300 Myr at this rate.

If the mixing ratio of total hydrogen becomes small enough, escape may be limited by the rate at which hydrogen compounds can diffuse through the background gas (Hunten 1973). On Venus this condition is reached when the total hydrogen mixing ratio in the bulk atmosphere drops to $\sim 2\%$ (Watson et al. 1982). In the more recent atmosphere of Venus, the exospheric temperature is so low that if the Jeans mechanism alone were responsible for escape, the loss of hydrogen would be at a rate even lower than that allowed by diffusion limited flux.

Watson et al. (1982) have also determined the conditions under which rapidly flowing hydrogen undergoing hydrodynamic escape will drag along with it heavier atmospheric constituents. In the rest frame of the escaping hydrogen, the issue is determined by whether or not the heavier gas can diffuse downward fast enough—a rate set by limiting flux. As a result, when hydrogen is escaping at its energy limited rate, molecules > 7 times as massive as the escaping hydrogen will not be entrained and gases lighter than nitrogen will be lost, including deuterium in the form of HD. Hence the escape of the hydrogen belonging to an ocean of water would not leave a large residue of deuterium behind during the early phases of escape. However, once the mixing ratio of hydrogen drops below 2% the HD also will diffuse downward rapidly enough so that the flowing hydrogen will no longer carry HD along with it. Since the mixing ratio of total hydrogen is now on the order of 200 ppm, a 100-fold (but not much more) enrichment of deuterium with respect to hydrogen is possible. (If such an enrichment does not exist, it seems unlikely that Venus ever had a large reservoir of water.)

In fact, the enrichment of D, according to the measurement of a D to H ratio of 1.6% (Donahue et al. 1982), is 100 if the original ratio had the terrestrial value of 1.6×10^{-4}. Since escaping H_2 stops carrying HD along with it when the mixing ratio of H_2 drops below 2%, an enrichment by a factor of 100 would call for a present mixing ratio of hydrogen in all of its forms of 200×10^{-6} v/v. If the hydrogen is predominantly in the form of H_2O between 50 km and the cloud tops, this value is in good agreement with the Venera spectrophotmetric measurement. Because the enrichment appears to have its maximum allowable value, it can be concluded only that the initial abundance of Venus water was at least 2% of the gas in the atmosphere. If the atmospheric mass had the present value of 5×10^{23} g, Venus had at least 4×10^{21} g of H_2O, 2.5×10^{-3} times a terrestrial ocean. A larger initial reservoir is possible but the exact quantity cannot be inferred from the D/H measurement, given the present state of the escape theories; however, it is suggestive that the measured enrichment is about as large as the simple model of escape allows.

The exospheric temperature of Venus today is so low that Jeans escape of hydrogen is negligible. Kumar et al. (1981) have proposed that escape of fast H atoms produced by dissociative recombination of OH^+ is the dominant mechanism in the present atmosphere. However, the presence of a large amount of H_2 is required to produce the OH^+ and there are no known sources of H_2 to balance the sinks. In fact, 20 ppm in the mixed atmosphere is needed and an upper limit of 10 ppm is set by the PV LGC. A large supply of H would be produced from the H_2 and it would have to flow out of the daytime thermosphere. The escape flux is not large enough to accommodate such a large outward flow. Kumar and Hunten proposed transport to the nightside but no quantitative model has been developed. McElroy et al. (1982) have proposed instead that the ion of mass 2 in the ionosphere is D^+, an interpretation that requires the ratio of deuterium to hydrogen to be $\sim 10^{-2}$ in the mixed atmosphere. This result is in reasonable agreement with the value measured in the lower atmosphere by the LNMS (Donahue et al. 1982). They have also developed a model of Venus hydrogen in which there is a large escape flux driven by collisions of hydrogen atoms with fast oxygen atoms produced by dissociative recombination of O_2^+. Deuterium atoms do not acquire escape energy in such collisions. McElroy et al. calculate that this mechanism would exhaust the hydrogen associated with one third of a terrestrial ocean. However, according to the discussion of hydrodynamic escape by Watson et al. (1982), it is doubtful that escape would proceed at this relatively slow rate once the H_2 mixing ratio exceeded $\sim 2\%$.

Once hydrodynamic escape ceases, hydrogen, on the average, must escape at a rate substantially higher than the present rate if the present low abundance of water is to be reached in a time interval comparable to or smaller than the age of the solar system. During this transition period, hydrogen escape is expected to occur chiefly through nonthermal processes, such as charge exchange and dissociation recombination, just as it does at the present epoch. Preliminary sensitivity experiments indicate that the hydrogen escape rate does increase as the water vapor abundance increases above its present value (S. Kumar, personal communication; Pollack et al. 1980). Hence, it may be possible to realize the present water content of Venus's atmosphere, starting from a much more water-rich state, but additional calculations are needed to confirm this.

B. Loss of Water

The current amount of water in the atmosphere of Venus and hence the volatile inventory is four to five orders of magnitude smaller than that in the Earth's volatile inventory (Sec. I). Conceivably, this difference reflects a huge deficiency in Venus's initial endowment of water, perhaps resulting from

Venus forming in a portion of the solar nebula where it was too hot for hydrated minerals to be stable (Lewis 1972). Alternatively, Venus may have formed with much more water, perhaps not very different from that of the Earth, but subsequently lost most of this water. The principal argument against the first alternative has been that even if the planetesimals that formed Venus are totally dry, gravitational stirring would be expected to promote a significant exchange of planetesimals between the Venus and Earth zones of accretion (Wetherill 1981). An unresolved paradox is involved in this argument, because an inhibition of this kind of exchange appears to be necessary if the gradient in noble gas abundance is to be understood. The recent finding of a 100-fold enrichment of D on Venus (Donahue et al. 1982) is persuasive evidence that in fact Venus was once much wetter than it is now. In this section, we explore the feasability of the hypothesis that Venus initially had much more water than it does today by reviewing calculations of water loss rates.

There are at least four mechanisms in which water can be eliminated from Venus's atmosphere:

1. Water can be photodissociated by solar ultraviolet radition, with the resulting hydrogen escaping from the top of the atmosphere and the oxygen combining with crustal material (Walker 1975).
2. Water can react with FeO in surface rocks, with magnetite and hydrogen being produced and hydrogen escaping to space (Watson et al. 1983).
3. Water vapor can react with reduced gas species, especially carbon monoxide, that are contained in the initial outgassed volatiles (Sec. III), with carbon dioxide and hydrogen being generated and the hydrogen escaping to space (Pollack and Yung 1980).
4. Water can be recycled into the interior, where it is partially dissociated thermally; the resulting oxygen reacts with the magma and the hydrogen is vented into the atmosphere and eventually escapes to space (Pollack and Yung 1980).

A common feature to all four loss mechanisms is the implicit requirement that molecular hydrogen readily escape. Otherwise, they would be incompatible with the trace amount of molecular hydrogen that characterizes Venus's current atmosphere of at most 10 ppm v/v. This requirement on the hydrogen loss rate is nontrivial in view of its current escape rate of only $\sim 10^7$ H cm^{-2} s^{-1} (McElroy et al. 1982). At this rate, only $\sim$ 9 m of water could be eliminated from Venus over its lifetime. As was discussed in the preceding subsection, much larger amounts of hydrogen can be lost by hydrodynamic escape when the mixing ratio of hydrogen is much larger than its current value. Still, it is unclear whether or not a combination of other loss mechanisms (Jeans escape, exothermic chemical reactions, solar wind sweeping, etc.) can operate at rates substantially above their present rates in order to provide a smooth transition between the current state and the one that pertains when hydrodynamic escape can no longer take place.

In discussing the magnitude of the water losses that can be realized with each of the four loss processes it will be assumed that hydrogen can readliy escape to space, although as we have seen above, this remains to be demonstrated. Walker (1975) discussed several stages through which mechanism 1 (photodissociation) could result in the loss of a full terrestrial ocean of water from Venus. Beginning at a time when the runaway greenhouse occurred (Sec. V) water vapor would be the major atmospheric constituent. The tropopause cold trap would be ineffective in limiting water abundance at high altitudes. Water would be photolyzed at the photon-limited rate of 10^{13} $cm^{-2}s^{-1}$, hydrogen lost rapidly to space, and the oxygen gradually become a major component of the atmosphere. Later, the loss of hydrogen would be limited by diffusion and by the development of a cold trap. Crustal oxidation would dispose of the oxygen.

This scenario runs into a number of serious problems. Solar EUV energy limitations would hold the H_2 escape flux below 10^{12} cm^{-2} s^{-1} during the runaway phase (Watson et al. 1982). A mechanism is lacking for the deep crustal oxidation required in the absence of Earth-like erosional processes and plate tectonism. Shielding of water vapor from solar ultraviolet by the oxygen that builds up in the atmosphere would become important, and a cold trap may have developed when the water vapor mixing ratio dropped to $\sim 10\%$.

Due to an enhanced greenhouse, the surface temperature may have approached the basalt solidus at times when the water vapor partial pressure exceeded ~ 10 bar (Watson et al. 1983) (see Fig. 2). Under these circumstances, water could be lost rapidly through a redox reaction with the FeO component of the surface material (ibid). In addition, since the lithosphere was very thin or nonexistent at such times, new material would be constantly brought to the surface and much water could be lost in this manner. Exactly the same type of thermodynamic equilibrium considerations apply to this situation as to the buffering of gases in a magma chamber during the initial outgassing, as described in Sec. III. Thus, the redox reaction is reversible, with the system tending towards an equilibrium ratio of H_2/H_2O that depends on the state of oxidation of the iron in the surface material. As H_2 escapes to space, more water would be consumed in attempting to reestablish the equilibrium ratio. This mechanism may also be useful for quickly eliminating oxygen created by the photodissociation mechanism. As the partial pressure of water drops, so too would the surface temperature. Thus, this loss mechanism may have been particularly effective in eliminating the first 90% of a full terrestrial equivalent of water, but would have become inoperative in eliminating the remaining 10%. Nevertheless, as discussed in Sec. III, localized basaltic volcanism has probably occurred over much of Venus's history. A limited amount of water could have been eliminated through chemical interactions with these surface magmas.

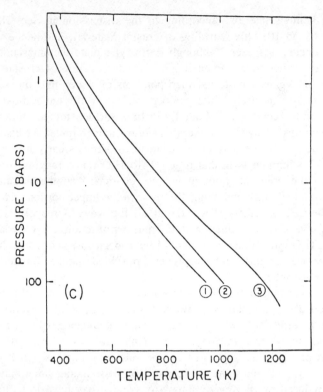

Fig. 2. Temperature profiles: ① 10 bars H_2O, 90% cloud albedo; ② 10 bars H_2O, 55% albedo; ③ 90 bars H_2O, 90% albedo.

The importance of mechanism 3 (redox reactions with CO) depends upon the CO to CO_2 ratio in the initial outgassed volatiles and the inital abundance of water. Reasonable bounds on the CO to CO_2 ratio are 1/40 to 4 (Pollack and Yung 1980), with values towards the upper end perhaps being more likely, as discussed in Sec. III. Thus, an amount of water ranging from $\sim 1\%$ to almost 100% of a terrestrial ocean equivalent could be eliminated by this mechanism. However, since it is highly improbable that the amount of outgassed water and carbon monoxide would be close to stoichometric balance to give the current low abundance of both gases, the redox mechanism, while probably important, is not the only key loss process for water.

One expression of plate tectonics on present-day Earth is recycling of volatiles among atmospheric, surface, and subsurface reservoirs on a time scale of a few times 10^8 yr (Walker 1977). Water that is incorporated into sedimentary rocks is eventually freed in zones of melting, where it is partially thermally dissociated. The resulting water vapor and hydrogen are outgassed and the oxygen is taken up by the molten rock. The importance of this process for eliminating water on Venus depends upon the degree to which water has been recycled and the H_2 to H_2O ratio that characterizes thermally dissociated water in Venus's interior. According to the discussion in Sec. III and in Chapters 6 and 10, little recycling of crustal material, and hence volatiles, probably occurs at present, although Earth-type plate tectonics might have occurred at earlier times. Even if a vigorous recycling of crustal material did occur in Venus's past, the incorporation of water into the subducted material is not assured Such incorporation could have most readily occurred under conditions of an Earth-like surface temperature because of the instability of hydrous minerals at high temperatures. As discussed in Sec. V, such a climate may have occurred in Venus's early history due to a lower solar luminosity at that time. H_2 to H_2O ratios ranging from a few percent to the order of unity may have characterized the thermally dissociated water (Pollack and Yung 1980), with the upper portion of this range perhaps being more likely (Sec. III). If all the water is recycled anywhere from 1 to several tens of times, with the number of required cycles depending on the H_2/H_2O ratio, much water could have been lost. Thus, mechanism 4 could have been an important loss process, provided that the climate of Venus was once Earth-like.

In conclusion, all four loss mechanisms may have contributed to the elimination of large amounts of water from Venus. While several of these operate quite effectively when the water vapor abundance is high, the transition to the current low abundance on a time scale of several billion years remains a problem; the hydrogen escape rate may decrease greatly in this intermediate regime and some mechanisms for dissociating water may operate at much reduced rates. Key reasons why Venus may have lost most of its water while the Earth did not, may include:

a. the occurrence of a runaway greenhouse and the attendant transfer of all the water into the atmosphere (mechanisms 1 and 2);
b. a somewhat reduced inventory of water for Venus (mechanisms 3 and 4);
c. a higher FeO/Fe_2O_3 and hence higher CO/CO_2 and H_2/H_2O ratios in outgassed volatiles for Venus (mechanisms 3 and 4).

A plausible sequence by which Venus could even have lost the equivalent of a terrestrial ocean is that mechanism 2 disposed of the water when its partial pressure was > 20 bar and mechanism 3 (the reaction with CO) accounted for the remaining water. Any CO in excess of the 20 bar needed to react with water could have reacted with the large concentrations of surface Fe_2O_3 and been converted to CO_2. The "end game" would account for the present low abundance of CO and O_2 without requiring excessive overturning of the crust. Again the possibility that Venus never had more than 100 times as much water as it does today cannot be excluded, although the likelihood seems small.

C. CO_2 Geochemical Cycle

As illustrated in Fig. 3, CO_2 on Earth is partitioned between various atmospheric, surface, and subsurface reservoirs, there being fluxes of CO_2

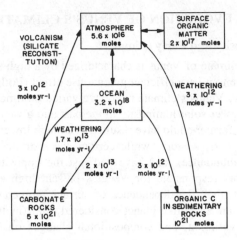

Fig. 3. Carbon cycles on Earth.

among the reservoirs that lead to the establishment of quasi-equilibrium amounts in the various reservoirs (Walker 1977). Almost all the outgassed CO_2 of the Earth currently resides in subsurface rock reservoirs. Most of the subsurface CO_2 is locked into carbonate rocks that form when running water dissolves Ca and Mg cations from continental silicate rocks. These cations combine with carbonate anions in seawater, resulting in carbonate minerals that precipitate to the ocean bottoms; these minerals are later buried and subducted into the interior. There the carbonate rocks are eventually metamorphosed, freeing CO_2 gas, which is vented back into the atmosphere to complete the cycle. One complete cycle takes several hundred million years at the current rates of exchange between reservoirs.

The quasi-equilibrium abundance of CO_2 in the atmosphere is determined by the fluxes of CO_2 among the various reservoirs and their dependence upon geochemical, geophysical, and atmospheric factors. For example, if the outgassing rate is increased, say due to a thinner lithosphere or higher interior heat flux, the atmospheric CO_2 content increases (Pollack and Yung 1980; Walker et al. 1981). Also an increased atmospheric CO_2 abundance and/or temperature may lead to an enhanced weathering rate and thus act as a negative feedback on the quasi-steady state abundance of atmospheric CO_2 (Walker et al. 1981).

The applicability of the above CO_2 geochemical cycle to Venus is uncertain. If Venus had an Earth-like climate in its past, then the equilibrium abundance of CO_2 in its atmosphere may have been much smaller than the amount in its volatile inventory, with this abundance depending upon fluxes among various CO_2 reservoirs. For present conditions, essentially all the CO_2 resides in the atmosphere because it is very difficult to form carbonate rocks at Venus's high surface temperature in the absence of running water and because little subduction appears to be occurring now (Sec. III).

V. EVOLUTION OF VENUS'S CLIMATE

A. Solar Luminosity and Early Climate

The current climate of Venus is characterized by a high surface temperature of 730 K, due to a very efficient greenhouse effect (Pollack et al. 1980). Quite conceivably, Venus's climate may have varied on time scales of $\sim 10^9$ yr due to both a lower solar luminosity in the past and a varying atmospheric composition. The former could have resulted in a much lower surface temperature in Venus's early history, while certain cases of the latter, e.g., an enhanced water abundance, may have had just the opposite effect. In this section, the factors responsible for Venus's present high surface temperature are reviewed, then the dependence of surface temperature on distance from the Sun for an Earth-like planet considered, and finally the impact of solar luminosity variation, and compositional changes on Venus's climate discussed.

Using constraints on atmospheric composition and solar energy deposition provided by the Pioneer Venus and Venera spacecraft missions, Pollack et al. (1980) have been able to closely reproduce Venus's observed surface temperature and the lapse rate profile in the lower atmosphere with a one-dimensional greenhouse model. In order for the greenhouse effect to result in a 500 K elevation of surface temperature above the temperature at which the planet radiates to space, it is necessary that there be significant amounts of infrared opacity at *all* thermal wavelengths at which there are nontrivial quantities of blackbody radiation. This opacity arises primarily from the massive amount of CO_2 in Venus's atmosphere; the CO_2 provides most of the infrared opacity and also pressure broadens the transitions of other key optically active constituents, namely H_2O and SO_2.

Some insights into the effects of a time varying solar luminosity can be obtained by considering the dependence of surface temperature on distance from the Sun for an Earth-like planet. Calculations performed by Ingersoll (1969), Rasool and de Bergh (1970), and Pollack (1971) show a dramatic increase in surface temperature as the solar flux incident on a water covered planet is raised from that at the Earth's orbit to that at Venus's orbit. Figure 4 illustrates the dependence of surface temperature on incident solar flux for atmospheres containing only water vapor (H_2O) and those having water vapor, 1 bar of N_2, and a prescribed lapse rate in the troposphere, Γ (Pollack 1971). The arrows labelled "Venus now" and "Earth now" show the values of the current incident solar flux for Venus and the Earth, respectively. The sharp increase in surface temperature as the flux increases between these two points is due to the strong feedback effect between surface temperature and water vapor abundance.

When the flux rises to a critical value lying between its present values for the Earth and Venus for the calculations of Fig. 4, all the water enters the atmosphere and a "runaway" greenhouse is said to occur. This critical value

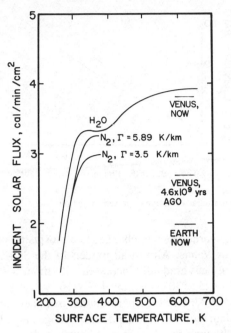

Fig. 4. Runaway greenhouse: surface temperatures on Venus and Earth as functions of solar flux.

occurs in the figure at the places where the slope of the curves undergo a large change. In the Rasool and de Bergh calculations, an additional positive feedback between surface temperature and atmospheric composition is provided by the partitioning of CO_2 between the atmosphere and carbonate rocks. Thermodynamic equilibrium is assumed to hold, in which case the partial pressure of CO_2 depends exponentially on surface temperature.

While the above calculations are pedagogically interesting, their usefulness in fully understanding the difference in surface temperature between Venus and Earth may be somewhat limited. First, Venus might not have had much more water than currently resides in its atmosphere, although this now seems unlikely. Since this amount is comparable to that in the Earth's atmosphere, water alone would not have significantly raised Venus's surface temperature above that of the Earth, although strictly speaking, a runaway greenhouse would have occurred according to the calculations of Fig. 4. Second, in view of the discussion in Sec. IV, it is a gross oversimplification to use thermodynamic equilibrium considerations to relate the amount of CO_2 in the atmosphere to the surface temperature. Kinetic factors play an important if not dominant role in this regard. Third, the current global heat balance of Venus implies that the current state of the climate depends on its past history. Venus absorbs less sunlight than the Earth does (Tomasko et al. 1980*a*), yet it has a much higher surface temperature.

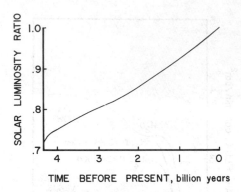

Fig. 5. Change in solar luminosity relative to present with time.

Solar luminosity variations and changes in atmospheric composition also impact the climate of Venus. Almost all models of the Sun's evolution indicate that its luminosity has gradually increased with time, with the magnitude of the change being ~ 25 to 30% over the 4.6×10^9 yr age of the solar system (Newman and Rood 1977; R. Stothers as quoted in Canuto and Hsieh 1978). A representative calculation of these long-term changes in solar luminosity is shown in Fig. 5. Although the discrepancy between observed solar neutrino fluxes and those predicted by most models (Rowley et al. 1980) indicates either an imperfect understanding of the interior structure of the Sun, and/or the nature of neutrinos (e.g., they may have a nonzero mass), the predicted variation in solar luminosity is almost identical for a wide range of solar models (Newman and Rood 1977). Only cosmological models, in which the universal constant of gravitation is allowed to be time varying, display significantly different variations of solar luminosity with time than that given in Fig. 5 (Canuto and Hsieh 1978).

A lower solar luminosity in the past leads to a paradox for both the Earth and Mars. According to greenhouse calculations for models with a fixed atmospheric composition and constant relative humidity, the mean surface temperature of the Earth should have been below the freezing point of water and hence the Earth should have been entirely ice covered between 4.6 and 2.3×10^9 yr ago (Sagan and Mullen 1972). Yet life and liquid oceans are known to have been present much further back in time than 2.3 billion years ago: 3.5 billion for life, 3.8 billion for liquid water. Similarly, certain types of channels on Mars suggest that liquid water flowed across the Martian surface several billion years ago, at a time when the temperatures were apparently much warmer than they are today (Pollack 1979). But a lower solar luminosity in the past implies a colder, not a warmer climate. A common resolution to the above solar luminosity paradox has been to invoke atmospheres with altered compositions for the Earth and Mars, ones capable of generating enhanced greenhouses in the past. Because of difficulties encountered with the stability of reducing atmospheres, a popular altered atmosphere of this type is one containing much more CO_2 than the current amount (Pollack 1979; Owen et al. 1979; Cess et al. 1980). The enhancement in atmospheric CO_2 is attributed to a combination of geophysical factors, such as increased outgassing, and atmospheric factors that result in a smaller fraction of the volatile inventory of CO_2 being contained in subsurface reservoirs, like carbonate rocks (Walker et al. 1981). In assessing the implications of a lower solar luminosity for Venus's past climate, the above resolution of the solar luminosity paradox for the Earth and Mars should be kept in mind.

At first glance, a lower solar luminosity raises the possibility that Venus may have had an Earth-like climate in its early history (Pollack 1971). For example, surface temperatures only slightly in excess of the Earth's present mean value are realized in greenhouse models dominated by water vapor for a solar flux equal to that incident on Venus 4.6×10^9 yr ago (Fig. 4). A similar conclusion follows when amounts of atmospheric CO_2 determined from thermodynamic equilibrium considerations are also included in the greenhouse calculations (Pollack 1979).

Whether Venus ever had a moderate surface temperature over a significant fraction of its history depends on the initial amount of water in its volatile inventory, the time history of outgassing, and especially the partitioning of CO_2 between its atmospheric and surface reservoirs. Suppose first that Venus never had much more water than is now contained in its atmosphere. Then, even for the lower early solar luminosity, the surface temperature may have been high enough to drive all the water into the atmosphere during early times; it would have been difficult to produce large quantities of carbonate rocks from atmospheric CO_2 without the presence of liquid water on the surface. In this case, a moderate surface temperature would be difficult to realize.

Next suppose that Venus had much more water in its past than it does today. If outgassing occurred only very gradually much lower surface temperatures than today's value must have occurred at early times when the amount of outgassed CO_2 was much less than the amount now in the atmosphere, regardless of the exact partitioning of CO_2 between reservoirs. If the outgassing of juvenile gases was largely completed at a time when the solar luminosity was still much lower than its present value, then the value of the surface temperature would have depended sensitively on the partitioning of CO_2 between its various reservoirs. To the extent that an enhanced level of CO_2 is the solution to the solar luminosity paradox for the Earth and Mars, and to the extent that this solution has relevance for Venus, a moderate value of surface temperature in Venus's past may have been hard to achieve. An early divergence in the tectonic styles of Venus and Earth is implied by the 4-fold deficiency of ^{40}Ar in Venus's atmosphere. This divergence is probably associated with the establishment of a runaway greenhouse.

There may have been epochs in Venus's past when its surface temperature was significantly higher than its present value. Such a situation would

have occurred at times of enhanced atmospheric water, especially when the
solar luminosity had values not much less than the current value. In fact, the
surface temperature may have even exceeded the solidus of basalt when the
water vapor partial pressure exceeded 10 bar (Sec. IV).

VI. SYNTHESIS. OUTSTANDING PROBLEMS

The noble gas inventory of Venus as it is currently reported has not been
explained in a completely satisfactory way by any theory of planetary origin
so far developed. On the other hand, the observational data presently available
are far from incontrovertible. In particular, the need to resolve the disagreement concerning the krypton mixing ratio is urgent and a good measurement
of xenon abundance and isotope mixing ratios would be very useful. Clearly,
confirmation of a large ^{36}Ar to ^{84}Kr abundance ratio on Venus would put a
strong constraint on theories for planetogenesis. In any case, the resolution of
the neon paradox will be called for. This paradox is the persistence in the
Venus atmosphere of a low ratio of ^{20}Ne to ^{36}Ar, similar to the ratio found
elsewhere in the inner solar system and in chondrites. Precise measurements
of noble gas abundances and isotopic ratios in the atmosphere of Venus (and
Mars as well) would be welcome. It is possible that some enlightenment
would follow from a study of the xenon isotope ratios and absolute abun-
dances. The mixing ratios of ^{128}Xe, ^{129}Xe, ^{130}Xe, ^{131}Xe, and ^{132}Xe would be
~ 0.06, 0.07, 0.1, 0.6, and 0.7 ppb v/v if the atmosphere had the Earth's
allotment of xenon plus a solar contribution consistent with the 75-fold enrichment of argon; 60% of the xenon would be solar. Models that attempt to
account for the large gradient in noble gas abundance (and perhaps water)
between 1.4 and 0.7 AU also must dispose of the tendency of planetesimals to
be strongly scattered gravitationally throughout this region and indeed of the
apparent necessity that they do so in order that development of embryo planets
will not be arrested. Hybrid models in which a portion of today's atmospheres, particularly that of Venus, are remnants of massive protoatmospheres
have enough promise to warrant careful study (Walker 1982).

The recent measurements of the deuterium to hydrogen ratio indicate that
the mixing ratio of H_2O was once 100 times greater than it is today. Obviously
the actual value deduced for the earlier mixing ratio depends on an accurate
measurement of the present mixing ratio. The current level of uncertainty
regarding this crucial quantity must be reduced. A careful study of the nature
of hydrogen loss in two transition regimes is also required. One of these
regimes involves the change from gas kinetic to hydrodynamic escape—the
other, the transition from a flow which fully entrains deuterium in the atmosphere to one in which deuterium is retained. With these issues settled it will
be possible to assert with greater confidence the early abundance of water—
whether it was no more than the lower limit implied by the D/H measurement
or perhaps a much larger fraction of a terrestrial ocean. The present D to H

ratio probably can never reveal the extent of the original water endowment. Evidence for loss of a large amount of water probably can be found in the oxidation state of the lithosphere. A determination of the oxidation state of the planet at depth could do so if a way could be found to make the measurement.

It would also be interesting to explore the tectonic implications of surface temperatures possibly above the basalt solidus for a prolonged period (300 Myr) while Venus was disposing of its ocean. How does the existence of one small continent like Ishtar fit in such a picture? Was the climate on Venus benign in the beginning, its surface mostly covered by an ocean, and enough water present in the mantle to permit the development of an asthenosphere? If so, did plate tectonics develop and proceed to the production of the one continent that is now Ishtar? If the ancient shoreline was at the 6053 km level on Ishtar $\sim$ 90% of the planet's area would have been covered by an ocean whose volume would have been $\sim$ 70% of the terrestrial ocean—without taking into account the depression of the surface that would have been caused by the weight of such an ocean. Radar mapping of Ishtar might be able to disclose traces of ancient fluvial features terminating at a recognizable shoreline. Aphrodite Regio, Beta Regio and Phoebe Regio may be much younger manifestations of intraplate-like plume activity occuring after a runaway greenhouse had led to the loss of the ocean and cessation of plate movement (Phillips et al. 1981). If so, no evidence of ancient river beds would be found on their surfaces.

The picture now emerging is one of a Venus that began to evolve along a path similar to that of Earth but suffered a catastrophe in the form of a runaway greenhouse early in its lifetime. How early may be suggested by the present low inventories of radiogenic argon and helium in the atmosphere. Clearly this story carries a lesson with it for humanity lest it trigger a similar catastrophe on Earth. The chances seem to be very small that such an event can happen, but prudence suggests that the probability be evaluated with the case of Venus in mind.

Acknowledgments. The authors are grateful to J. Walker and D. Black for helpful comments. The work reported here was supported in part by grants from the National Aeronautics and Space Administration.

APPENDIX

Too many problems developed during the Pioneer Venus sounder probe descent and too many discrepancies have turned up among results obtained by the probe instruments, the bus mass spectrometer, and the Venera instruments to allow a confident discussion of the implications of these measurements without strict caveats. There are important lacunae in the Viking measurements also.

First of all, there is a measurement of helium by the Pioneer Venus bus neutral mass spectrometer (BNMS). Obtained in the upper atmosphere, the measured concentration must be extrapolated to give a mixing ratio for the bulk atmosphere, a model dependent procedure. The result obtained was 12^{+24}_{-6} ppm by volume. Data at the ^{4}He mass number were also obtained by the Pioneer Venus Large probe mass spectrometer (PV LNMS); these have not been published because of problems in the interpretation and because helium used in other instruments may have contaminated the mass spectrometer. However, the data do suggest an extraordinarily high mixing ratio of ~ 450 ppm for helium, and would present another problem of reconciliation and interpretation, if they should be verified.

The principal difficulty with the argon data is the disagreement between the BNMS upper limit and all probe results from Pioneer Venus and Veneras 11 and 12. The upper limit is placed by an extrapolation to the mixed atmosphere from an absence of data clearly ascribable to ^{36}Ar and ^{40}Ar in the upper atmosphere. Modeling is therefore involved. Furthermore, there were problems with sensor gain values during the crucial period of measurement. None of the other measurements is perfectly straightforward. The gas chromatographs detect only total argon; the isotope ratios are assumed to be those obtained by the mass spectrometers. The Venera neutral mass spectrometer (VNMS) measurements were obtained during an enrichment cycle and the LNMS mixing ratios depend on a post-flight analysis of pumping speeds which are very different for CO_2 and argon. Nevertheless, the consensus among the probe results is sufficiently strong to bestow an aura of credibility to them and for the purpose of this discussion, the ^{36}Ar mixing ratio will be taken to be 30 ppm. Only the mass spectrometers obtained isotope ratios. The values obtained are shown in Table III.

For neon the agreement among the three instruments obtaining data is not as good as for argon, the Pioneer Venus Large probe gas chromotograph (PV LGC) value being a factor of 2 smaller than the LNMS and VNMS figures. In any event a large enrichment of Cytherean neon and a large ratio of argon to neon abundance is indicated. In this chapter, a mixing ratio of 5 ppm will be adopted for neon. This choice reflects a preference for the gas chromatograph value because of the large probable errors assigned to the mass spectrometer neon measurements. The ^{22}Ne/^{20}Ne isotope ratio was obtained only by the LNMS. It was measured in an inert gas enrichment cell where the mass 22 signal was not contaminated by CO_2^{++}. However, it should be noted that the degree of enrichment of rare gases in the cell fell far short of expectations and the neon isotopes were the only ones detected with sufficient accuracy for a ratio measurement. The quoted uncertainty in the ratio reflects this problem.

A serious disagreement exists for the heavier noble gases. Unambiguous detection of the Kr isotopes was obtained by the VNMS; the same cannot be said for the LNMS measurements. Although altitude dependent signals were detected in the krypton isotope channels (and not in other neighboring high-mass channels), the variation with altitude was not the same as that of ^{36}Ar (Donahue et al. 1981). Analysis of the results on the basis of partial contributions from atmospheric krypton, and from krypton released from pumps by chemical reaction with contaminants, leads to the central values listed in the tables. The upper limits of ~ 70 ppb for total Kr and 30 ppb for ^{84}Kr are obtained when all counts are ascribed to atmospheric Kr. Even these limits are 15 to 20 times lower than the VNMS results. The number of counts per interval associated with the LNMS measurements range from 2 to 10; they would lie between 30 and 400 if the mixing ratio were that obtained by the VNMS, provided that no mass dependent deterioration of spectrometer sensitivity occurred between Earth and Venus. Donahue et al. (1981) present evidence that no such problem existed. On the other hand the VNMS ion sputter pumps were exposed to terrestrial krypton. There is a possibility that the gas detected was carried to Venus in the pump as discussed in Chapter 13. The Venera investigators argue against this suggestion on the basis of the similar behavior of krypton and other noble gases during the enrichment cycle when their measurements were made. The situation is clearly unsatisfactory and a careful repetition of these measurements is definitely in order.

For xenon the problems are even more serious. The LNMS measurements (Donahue et al. 1981) are not only complicated by the presence of xenon presumably produced inside the instruments in a manner similar to the non-atmospheric krypton component but also by the fact that there is clearly some contribution to the signals from xenon isotopes introduced into the spectrometer along with the 136 amu isotope which was used for calibration purposes. The data are consistent with the presence of no detectable xenon. (Even if the apparatus had been functioning perfectly, the limit of detectability for xenon was ~ 5 ppb.) The upper limits shown in Table III were obtained after subtraction of a background caused by the marker gas contamination. Without such subtraction, the upper limits would be a factor of 3 higher. It should be noted that all krypton and xenon abundances were measured relative to ^{36}Ar and a value of 30 ppm is taken for the ^{36}Ar mixing ratio.

There is a considerable spread in values obtained for water vapor. It is notoriously difficult to measure low concentrations of water vapor properly. Mass spectrometers must be treated with scrupulous care even in the Earth's atmosphere if measurements are not to be plagued by outgassing from the apparatus. This problem is not as serious for a gas chromatograph with a large volume to area ratio which was the case for the PV LGC. Nevertheless, there is a presumption in favor of the lowest values measured provided that there are no reasons to suspect systematic errors. For the measurements in the atmosphere of Venus, the spectrophotometer has the advantage of sensing the atmosphere of Venus remote from the descent probe; these measurements are probably very clean and should be preferred.

Donahue et al. (1982) have reported measurements of the HDO to H_2O

abundance with the PV LNMS. The ratio given is $(1.6 \pm 0.2)\%$. As is discussed in Chapter 13, the PV LNMS observed a large mass peak at 2.016 amu that was certainly H_2. However, the source was probably CH_4, introduced deliberately into the mass spectrometer or reactions involving H_2SO_4 as well as atmospheric H_2. In several mass channels near 3 amu, there were also high counting rates. However, they must be ascribed to HD with an abundance ratio of 3×10^{-4} to the H_2 (hence terrestrial) and to H_3^+ produced in the ion chamber, as well as to atmospheric HD. So far, efforts to extract the contributions of Venus H_2 and HD from the data have been thwarted.

30. THE PROBLEM OF RARE GASES IN THE VENUS ATMOSPHERE

L. M. MUKHIN
USSR Academy of Sciences

A summary is given of isotopic and elemental ratios among the noble gases. Theories attempting to account for these results, along with the near equality of C and N on Earth and Venus, are summarized. None of them is as yet fully satisfactory.

The mass-spectrometric and gas-chromatographic data on the chemical and isotopic composition of the Venus atmosphere were among the most interesting results obtained by the Pioneer Venus and Venera 11 and 12 missions (Pollack and Black 1979). Moreover, the data now being processed and additional calibration of the instruments give rise to unexpected results that can affect significantly the cosmochemical and planetological interpretation of the experiments. For example, the data recently published by Donahue et al. (1981) on the krypton content in the Venus atmosphere differ sharply from the results of Istomin et al. (1980*b*). We shall show that this discrepancy is important for the general problem of genesis of inert gases in the Venus atmosphere. The main unresolved questions are:

1. Why is the Venus atmosphere enriched markedly in nonradiogenic isotopes of inert gases, ^{20}Ne, ^{36}Ar, Kr and probably Xe as compared with the Earth's atmosphere? The absolute amounts of these gases in the atmosphere of any of the terrestrial planets is a measure of degassing of its solid matter and, therefore, the next question immediately arises:

2. Why are the contents of nitrogen and carbon dioxide in the Earth and Venus atmospheres similar (if we take into account carbon in Earth sediments) in contrast to this excess of inert gases in the Venus atmosphere?

[1037]

Two possible sources of volatile components for planets of the solar system can be singled out: the gaseous component of a nebula and the solar wind. Volatile components of a nebula, on the one hand, can be directly accreted into massive bodies, as is seen in the case of the giant planets, and on the other hand, they may be somehow included (Fanale and Cannon 1972; Lancet and Anders 1973; Niemeyer and Marti 1980) into the dust component of the nebula. Later on, it is the dust component with a certain amount of volatiles that may serve as initial material for the solid matter of the terrestrial planets.

The solar wind can be an additional source of volatiles in the terrestrial planets. The effect of this factor for different planets can be estimated from the assumption that the wind intensity varies as the inverse square of the distance from the Sun.

We shall now analyze the available attempts to solve the questions mentioned above. In so doing, we will use the data on the volatile content and isotopic composition of inert gases in the meteorites and in the atmospheres of the Earth, Venus and Sun (Table I). From examination of this table it can be concluded that the Venus atmosphere has high neon ratios similar to those of the primitive solar wind (Black 1970). This fact offers additional difficulties in solving the problem of the origin of the Venus atmosphere. Even the phase Q (Alaerts 1977), which is richest in inert gases in meteorites has a maximum ratio from 8.6 to 11 and, therefore, could never provide Venus with neon of such a high isotopic ratio $^{20}Ne/^{22}Ne$.

This ratio is even lower in the other mineral phases of carbonaceous chondrites and, for example, the silicate fraction of Orgueil varies between 4.5 and 10.3 (Herzog and Anders 1974). Tolstikhin (1981) concentrated on the fact that meteorite matter could lose a significant amount of volatiles due to thermal processes. If so, the protomatter from which Venus was formed contained a higher amount of inert gases than that from which the Earth was formed. But if this hypothetical fraction of the protomatter of Venus (at least to some extent) were similar to the phase Q, then in this case the ratio $^{36}Ar/^{20}Ne$ in the Venus atmosphere should depend strongly on the content of ^{36}Ar (Alaerts 1977), and due to this they should differ sharply from $^{36}Ar/^{20}Ne$ in the Earth's atmosphere. However, as is seen from Table I these ratios for the Earth and Venus are approximately the same. In addition, it is difficult to explain the equality of C and N concentrations for Earth and Venus within the frame work of Tolstikhin's hypothesis since diffusive losses of carbon and nitrogen and those of inert gases ^{36}Ar, Kr, Xe are of the same order (Herzog et al. 1979). Thus, the explanation of the excess of inert gases suggested by Tolstikhin (1981) is somewhat speculative.

One would think that the high ratio $^{20}Ne/^{22}Ne$ in the Venus atmosphere confirms the Wetherill (1980) hypothesis about the irradiation of planetesimals by the solar wind at the Venus orbit. Wetherill assumed that the Earth ratio $^{36}Ar/^{20}Ne$ can be explained by the diffusion loss of H, He, and Ne following

TABLE I[a]

Ratios of Noble Gases in Various Objects of the Solar System[d]

Body of the Solar system	$^{20}Ne/^{22}Ne$	$^{36}Ar/^{20}Ne$	$^{36}Ar/^{84}Kr$	$^{36}Ar/^{132}Xe$
Earth	9.8	~2	46	~10^3
Venus	12 ± 0.7	4	~50(3)	
			500(2)	≲3 × 10^2
Meteorites				
Carbonaceous Chondrites	4–13	~5[c]	45	30–50
Normal Chondrites (phase Q)		~0.5	100	100
Sun	13.4			
	13.7	0.05	~5 × 10^3	~10^5
	14.5[b]			

[a] The data used here were taken from Pollack and Black (1979), Donahue et al. (1981), Istomin et al. (1980b), and Geiss (1972).

[b] The primitive solar wind.

[c] $^{36}Ar/^{20}Ne$ for meteorites varies widely (from 1 to 100) depending on concentration of ^{36}Ar.

[d] Also, Venus/Earth ratios are: ^{20}Ne, ~ 59 to 43; ^{36}Ar, ~ 100; ^{84}Kr, ~ 3.5 (Donahue et al. 1981) or ~48 (Istomin et al. 1980b) or ~84 (Mukhin et al. 1982); ^{132}Xe, ≤ 30. Primary attention should be given to the neon ratios.

implantation by the solar wind. But the Wetherill hypothesis is not free from internal contradictions. One of the most serious contradictions is the high ratio $^{20}Ne/^{22}Ne$ which at the same time seems to be in favor of this hypothesis. But if Wetherill explains the non-solar ratio $^{36}Ar/^{20}Ne$ by diffusion losses of neon, it cannot be also true for Ar and Kr; therefore the ratio $^{36}Ar/Kr$ in the Venus atmosphere should be solar, rather than close to the Earth ratio (Istomin et al. 1980b). The Wetherill hypothesis may be used as a working model only if the results of Donahue et al. (1981) are valid. According to the data of their paper $^{36}Ar/Kr \simeq 6.3 \times 10^2$. However, even this ratio is less than the solar one. The essential difference of diffusion losses between ^{36}Ar, on the one hand, and Kr, Xe on the other could save this model, but the experimental data available on step-by-step annealing does not confirm this possibility (Herzog and Anders 1974).

Another attempt to explain the excess of inert gases in the Venus atmosphere was made by Isakov (1979, 1980) using the concept of gas accretion from the nebula to the planet. It was assumed, since according to the majority of cosmogonic models (Safronov 1969; Weidenschilling 1976) Venus was accreted at a rate several times higher than the Earth, that there could be a gas-dynamic accretion from the nebula; by the time of Earth's formation the nebula had dissipated and only Venus has the excess of inert gases. As the gas-dynamic accretion occurs (if it does) from a nebula of solar composition, the ratio ^{36}Ar/Kr would have been close to solar in the Venus atmosphere and therefore the objections against the Wetherill hypothesis remain applicable to this model. Certainly, it can be assumed that the nebula has fractionated itself by the time of the gas-dynamic accretion. But if the fractionation of Ne/Ar seems to be possible, the fractionation of Ar/Kr and Ar/Xe requires additional study. Actually the degree of fractionation can be defined, for example, for Ne and Ar as $K = (Ne/Ar)_{\oplus}/(Ne/Ar)_{\odot}$. Then K for Ne and Ar will be $\simeq 60$ and for Ar and Kr $\simeq 100$ (for Istomin's data) and $\simeq 10$ for the data of Donahue et al. (1981). It is immediately seen from the comparison of these data how important is such a parameter as the degree of the fractionation for estimating the validity of either of these models.

The main concepts of the Pollack and Black (1979) model are the following. The amount of rare gases captured by the nebular dust component is proportional to pressure and decreases exponentially with temperature. Pollack and Black use an approximately isothermal model of the nebula with high radial gradients of pressure. In this case two aspects can be questioned. First, it is not clear why for high gradients of pressure the amount of chemically active volatiles should be approximately equal for Venus and the Earth. On the other hand, the pressure gradients in the nebula used in this paper do not agree with any cosmogonic model.

The same objections, which have been made against the ideas of Isakov (1979) can be applied to the model of Mukhin and Gerasimov (1981). They envision sorption of the active gases on grains, together with a much smaller amount of the noble gases. The grains are accreted into the planets, followed by some fraction of the residual nebular gas. The required fractionation is possible over almost the entire likely range of nebular temperatures during sorption. We come to the conclusion that the search for the real physical mechanism responsible for the peculiarities of the chemical composition of the Venus atmosphere is still far from successful.

In spite of critical remarks made about the Pollack and Black paper, the idea of a pressure gradient in the nebula remains interesting. The physical reality of this model can be provided by adding one more physical process—nebular dissipation. In other words, let us consider the pressure as a function not of radial distance but of time. For this explanation there is no need to use Shimizu's (1978) radical assumption; only a comparatively small range of pressure (by 2–4 orders) is needed. Let us consider the processes which could occur in the cold nebula in detail. Some estimates by Cameron (1962) give a rather small time scale ($\leqslant 10^3$ yr) for cooling the protoplanetary nebula. Some additional data of Fisher (1975) also favor the cold accretion of planets. Safronov's (1969) calculations of the thermal regime of a dust layer show that the temperature of dust particles in the zone of terrestrial planets could be very low with a slight difference of temperatures in the Earth and Venus zones.

During cooling of the nebula through the temperature range ~ 1000 to $\sim$ 400 K, the processes of capturing inert gases by various mineral phases (Lancet and Anders 1973) and chemical sorption on growing grains of such molecules as nitrogen, ammonia, water, carbon dioxide, oxygen, hydrogen prevailed. Therefore, at this stage in the sufficiently isothermal nebula with the low radial gradient of pressure, the reserve of volatiles in grains in the zones of the Earth and Venus should have been almost the same. In this case attention should be paid to an important circumstance which is of prime importance for a comparative analysis of the composition of Venus and Earth atmospheres. The ratio $^{36}Ar/C$ in carbonaceous chondrites is $\sim 10^{-8}$ and the same ratio for the Sun is $\sim 10^{-2}$. These data show a very strong fractionation of the nebular elements in condensation processes and yield a conclusion that most of the carbon mass was in the solid substance of the nebula from which planets were formed. Consequently, neither the implantation by the solar wind nor the gas-dynamic accretion could strongly change absolute amounts of carbon and nitrogen on Venus. Since carbon and nitrogen constitute the main content of the organic substance of carbonaceous chondrites (carbon as elemental carbon and carbynes) these elements must have been included in the matrix of the solid component at the 300–500 K stage. This is the required temperature range for synthesis of hydrocarbons by the Fischer-Tropsch process and for formation of carbynes.

Then, as temperatures dropped to 100–50 K the processes of physical sorption predominated. Thus, two types of volatile captured by solid matter of the nebula can be considered in any cosmogonic model. The division is somewhat arbitrary but is convenient. The first are the volatiles incorporated to the crystal lattice of mineral phases at higher temperatures. Among them there are both chemically active volatiles (water, for example, can take part in hydration reactions) and noble gases. The fractionation of a captured gas component relative to the nebular composition will start already at this stage. The second type are the sorbed gases on the surface of dust grains. In this case fractionation also takes place during the processes of sorption followed by desorption, as the nebular pressure drops.

Different workers on planetary cosmogony assume different physical conditions in the disk, primarily associated with the choice of solar system formation model. The two main possibilities are simultaneous formation of the nebula and the Sun and the hypothesis of gas-dust cloud capture by the Sun. The first scenario is associated with high temperature in the nebula, followed by cooling, the second one with an initial low temperature, which then slightly increases. The proposed model of rare gas enrichment in the Venus atmosphere is sensitive to only one parameter: temperature. The use of sufficiently low temperatures is physically reasonable.

Let us try to make some semi-quantitative estimates using the Freundlich isotherm (Freundlich 1930) and the experimental data obtained by Peters and Weil (1930). The amount of sorbed gas can be found from

$$v = kP^{1/n} \tag{1}$$

where k and n are constants, and P is the sorbed gas pressure. Carbon is chosen as a sorbent because the carbon-carrying phase in meteorites is richest in noble gases (Lewis et al. 1980).

A reasonable pressure range in the nebula is $10^{-6} - 10^{-2}$ atm (Lancet and Anders 1973) and assuming, as a first approximation, that hydrogen is responsible for this pressure, we can estimate partial pressures of noble gases by using solar elemental abundances in the protoplanetary nebula (Voitkevich and Zakrutkin 1976). For a total pressure of 10^{-2} atm and $T = 200$ K, $P_{Ar} = 2 \times 10^{-4}$ torr, and for 10^{-4} atm $P_{Ar} \simeq 2 \times 10^{-6}$ torr. Then using the Freundlich isotherm parameters (Fisher 1975) we obtain an amount of sorbed argon near 2×10^{-6} cm^3 g^{-1} at 193 K and $P = 2 \times 10^{-6}$ torr, close to the content of ^{36}Ar in the Venus atmosphere. Thus, with a reasonable set of physical parameters for the nebula we can explain the observed excess concentration of noble gases in the Venus atmosphere even at relatively high temperatures. We have demonstrated the applicability of our model only for argon. For krypton and xenon the Freundlich isotherm parameters provide noticable fractionation in sorption (Earth ratios). Unfortunately, a comparison can be made only for krypton, xenon and argon, since in Peters and Weil (1930) there are no Freundlich isotherm constants for neon.

Finally, it is worthwhile to discuss some kinetic factors. It will be shown that chemically-active molecules remain on the grains as the nebular pressure falls, while the amount of noble gases sorbed tends to follow the pressure. The time that a gas molecule stays on the surface is approximately

$$\tau = \tau_0 \, e^{Q/RT} \tag{2}$$

where $\tau_0 \sim 10^{-13}$ s, Q is the heat of sorption, and R the gas constant (Adamson 1976). The characteristic times of nebular dissipation and planetary accretion are both $\sim 10^8$ yr (Vityazev and Pechernikova 1980).

At low temperatures, the lifetime of chemically active molecules with large heats of sorption will be comparable to the nebular time scales. Noble gases have much smaller values of Q, and their sorbed amounts will follow the ambient pressure. The hundredfold larger gas quantities on Venus, relative to Earth, therefore require the same ratio of nebular gas pressures during accretion. Because of the tighter sorption of active molecules, both planets get the same amounts of C and N.

In conclusion, the recent data (Istomin et al. 1982; Mukhin et al. 1982) on the content of noble gases and their isotopes in the Venus atmosphere still cannot give a final answer as to which mechanism is responsible for the enrichment of Venus's atmosphere by noble gases. Post-flight calibrations of the gas chromatograph and mass spectrometer should be decisive here, followed by the updating of the krypton results. It would then be possible to make a final correction for all the models considered above.

Appendices

APPENDIX A
MODELS OF VENUS'S ATMOSPHERIC STRUCTURE

A. SEIFF

NASA Ames Research Center

The models of the lower, middle, and upper atmosphere of Venus which are given below are based on the analysis of the currently available data presented in Chapter 11. The rationale for the models is discussed in the Appendix to Chapter 11. It is recommended that users of the tables read that Appendix to familiarize themselves with the assumptions and limitations inherent in the models, the degree of uncertainty, and such matters as the expected amplitude of temporal variability. The tables were selected to represent the temporal mean state, and below 100 km, the diurnal mean as well.

The altitudes are all referred to a planet radius of 6052 ± 0.5 km. This is 0.5 km above the mean planetary radius (6051.5 ± 0.1 km) defined by Pettengill et al. (1980). Thus, for reference to the mean radius of Venus, all altitudes may be increased by 0.5 km. At the mean radius, $p = 95.0$ bar and $T = 737$ K.

The tables are generally given to 3 significant figures for the sake of internal consistency in satisfying hydrostatic equilibrium and the equation of state. At best, they could approach an accuracy of 3 significant figures at selected levels in the lower atmosphere. However, there are regions where temporal variability is ~ 5 to 10 K, and model uncertainties increase progressively above the clouds to perhaps 20% in p and ρ, and ~ 10 K in the dayside upper atmosphere.

TABLE AI

Models for the Lower Atmosphere, Below Cloud Top

z, km[a]	Latitudes below 40			60° latitude		
	T, K	p, bar	ρ, kg m^{-3}	T, K	p, bar	ρ, kg m^{-3}
0	733	92.1	65.0	737	92.1	64.6
2	717	81.1	58.7	722	81.2	58.4
4	701	71.2	52.9	706	71.3	52.6
6	685	62.4	47.6	691	62.5	47.2
8	669	54.4	42.5	675	54.6	42.3
10	653	47.3	37.9	659	47.5	37.7
12	637	41.0	33.8	644	41.2	33.6
14	623	35.4	29.8	627	35.6	29.9
16	609	30.5	26.3	611	30.7	26.4
18	595	26.2	23.2	594	26.3	23.3
20	579	22.4	20.4	577	22.5	20.5
22	562	19.0	17.8	561	19.1	17.9
24	545	16.1	15.5	544	16.1	15.6
26	529	13.55	13.48	527	13.56	13.56
28	512	11.35	11.67	509	11.34	11.72
30	495	9.45	10.05	493	9.43	10.07
32	478	7.82	8.61	477	7.80	8.60
34	461	6.43	7.34	461	6.40	7.30
36	446	5.25	6.19	445	5.22	6.17
38	430	4.25	5.20	429	4.23	5.18
40	416	3.42	4.32	415	3.40	4.31
42	403	2.73	3.56	400	2.71	3.56
44	391	2.17	2.91	388	2.15	2.91
46	379	1.71	2.37	375	1.69	2.36
48	366	1.336	1.91	359	1.316	1.92
50	349	1.034	1.553	344	1.014	1.545
52	332	0.791	1.248	324	0.772	1.248
54	312	0.595	0.999	303	0.577	0.995
56	291	0.440	0.792	282	0.422	0.783
58	274	0.318	0.608	262	0.302	0.602
60	263	0.227	0.451	243	0.210	0.452
62	252	0.159	0.330			
64	244	0.110	0.236			
66	236	0.076	0.168			

[a] Values relative to a radius of 6052.0 km.

TABLE AII

Models for the Middle Atmosphere, 60 to 100 km[a]

z, km	0° to 30° latitude			45° latitude			60° latitude			75° latitude			85° latitude			g, m s⁻²	R, J kg⁻¹ K⁻¹
	T, K	p, mb	ρ, kg m⁻³	T, K	p, mb	ρ, kg m⁻³	T, K	p, mb	ρ, kg m⁻³	T, K	p, mb	ρ, kg m⁻³	T, K	p, mb	ρ, kg m⁻³	g, m s⁻²	R, J kg⁻¹ K⁻¹
60	263	227	0.451	258	221	0.447	243	210	0.452	235	190	0.422	243	173	0.372	8.696	191.40
62	252	159	0.330	247	154	0.326	228	143	0.327	213	127	0.311	243	119	0.256	8.690	191.40
64	244	110	0.236	240	106	0.231	217	95	0.229	220	82.9	0.197	243	81.9	0.176	8.684	191.40
66	236	76	0.168	235	73	0.161	219	62.6	0.149	232	55.5	0.125	243	56.4	0.121	8.679	191.40
68	237	51.8	0.114	240	49.5	0.108	240	42.2	.0918	238	37.7	.0829	242	38.8	.0838	8.673	191.40
70	233	35.2	.0789	236	33.8	.0749	237	28.9	.0636	239	25.8	.0564	242	26.7	.0576	8.667	191.40
72	228	23.8	.0545	231	23.0	.0519	236	19.7	.0436	238	17.7	.0388	241	18.3	.0398	8.662	191.40
74	223	15.9	.0373	226	15.4	.0357	234	13.4	.0299	235	12.0	.0268	238	12.6	.0276	8.656	191.40
76	217	10.54	.0254	221	10.3	.0244	227	9.06	.0208	230	8.16	.0185	235	8.58	.0191	8.650	191.40
78	211	6.91	.0171	214	6.80	.0166	218	6.04	.0145	224	5.48	.0128	230	5.82	.0132	8.645	191.40
80	203	4.47	.0115	207	4.43	.0112	210	3.96	9.85−3	217	3.57	8.60−3	224	3.91	9.11−3	8.639	191.40
82	194	2.84	7.65−3	198	2.84	7.49−3	202	2.56	6.61	206	2.33	5.91	215	2.59	6.29	8.634	191.40
84	184	1.76	5.00	189	1.78	4.92	194	1.62	4.30	195	1.49	3.98	203	1.68	4.33	8.628	191.44
86	176	1.07	3.17	181	1.09	3.16	185	1.01	2.84	187	0.927	2.59	192	1.07	2.90	8.622	191.46
88	171	0.637	1.95	175	0.660	1.97	177	0.613	1.81	180	0.568	1.65	183	0.660	1.88	8.617	191.53
90	167	0.374	1.17	169	0.391	1.21	170	0.365	1.12	174	0.342	1.02	176	0.400	1.19	8.611	191.57
92	164	0.217	6.90−4	165	0.228	7.22−4	165	0.213	6.75−4	169	0.202	6.25−4	170	0.238	7.30−4	8.605	191.66
94	162	0.125	4.02	163	0.132	4.23	163	0.123	3.95	166	0.118	3.72	168	0.140	4.34	8.600	191.80
96	162	.0719	2.31	164	.0764	2.43	165	.0715	2.26	166	.0690	2.17	168	.0820	2.54	8.594	191.93
98	163	.0415	1.32	167	.0445	1.39	173	.0421	1.27	168	.0404	1.25	170	.0483	1.48	8.589	192.06
100	170	.0242	0.740	173	.0263	0.791	177	.0253	0.742	173	.0239	0.719	173	.0287	0.862	8.583	192.24

[a] A notation such as 5.42 −3 is to be read 5.42×10^{-3}.

TABLE AIII

Models for the Upper Atmosphere, 100 to 180 km[a]

z, km	Antisolar or Midnight Model SZA > 120°				Subsolar or Noon Model SZA < 50°				
	T, K	p, mb	ρ, kg m^{-3}	R, J kg^{-1} K^{-1}	T, K	p, mb	ρ, kg m^{-3}	R, J kg^{-1} K^{-1}	g, m s^{-2}
100	170	2.50 −2	7.64 −5	192.2	170	2.50 −2	7.64 −5	192.2	8.583
104	160	8.48 −3	2.75	192.6	172	8.81 −3	2.66	192.6	8.572
108	150	2.70	9.31 −6	193.1	175	3.16	9.36 −6	193.1	8.561
112	140	7.95 −4	2.93	193.6	179	1.16	3.36	193.6	8.550
116	129	2.15	8.56 −7	194.3	184	4.41 −4	1.23	194.3	8.539
120	117	5.17 −5	2.26	195.3	190	1.73	4.66 −7	195.1	8.527
124	125	1.23	4.99 −8	197.1	196	7.00 −5	1.82	196.3	8.517
128	135	3.30 −6	1.22	200.6	204	2.95	7.30 −8	198.3	8.506
132	142	9.91 −7	3.35 −9	208.2	212	1.31	3.05	202.1	8.494
136	146	3.30	1.02	220.1	220	6.07 −6	1.33	208.1	8.483
140	147	1.26	3.26 −10	263	230	2.98	6.01 −9	215.4	8.472
144	140	5.65 −8	1.25	324	240	1.55	2.87	224.4	8.461
148	134	2.83	5.41 −11	390	250	8.51 −7	1.43	238	8.450
152	131	1.53	2.63	442	260	4.96	7.54 −10	253	8.440
156	129	8.68 −9	1.42	475	272	3.06	4.16	270	8.429
160	127	5.07	7.91 −12	504	289	1.99	2.35	293	8.418
164	126	3.03	4.57	526	299	1.37	1.44	319	8.407
168	126	1.84	2.65	551	300	9.78 −8	9.42 −11	346	8.397
172	125	1.13	1.55	581	300	7.16	6.43	371	8.386
176	125	7.19 −10	9.33 −13	616	300	5.35	4.48	398	8.375
180	125	4.73	5.69	665	300	4.08	3.22	422	8.364

[a] A notation such as 2.50 −2 is to be read 2.50×10^{-2}.

APPENDIX B

DATA TABLE INDEX

APPENDIX C
NASA Technical Memorandum Numbers

Most of the material submitted by Soviet authors was in Russian and was translated courtesy of NASA. The following list of Technical Memorandum numbers may be helpful in the tracking down of original wording or of papers that were not included in the book.

Our Number, Author, Title	Number	Book Chapter
1. V. I. Moroz	TM-76780	3
2. N. N. Izakov, Noble Gases...	TM-76774	—
3. B. Ye. Moshkin, A. P. Ekonomov, Yu. M. Golovin, Dust on the Venus Surface	TM-76771	—
4. Yu. M. Golovin	TM-76778	7
5. V. I. Moroz	TM-76772	Part of 13
6. A. P. Ekonomov	TM-76773	19
7. K. P. Florenskiy	TM-76777	8
8. T. K. Breus, The Venus Night Ionosphere	TM-76776	—
9. V. A. Krasnopol'sky and V. A. Parshev	TM-76770	14
10. V. A. Krasnopol'sky	TM-76775	15
11. O. Vaisberg	TM-76769	Part of 25
12. L. V. Ksanfomality, ... Cloud and Haze ...	TM-76768	—
13. L. V. Ksanfomality	TM-76767	Part of 20
14. V. N. Zharkov, Interior Structure	TM-76766	—
15. Yu. A. Surkov	—	9
16. A. D. Kuz'min	—	4
17. O. I. Yakovlev and S. S. Matyugov, Neutral Atmosphere by Radio Occultation	—	—
18. K. I. Gringauz	—	27
19. N. A. Savich and L. N. Samoznayev, Ionosphere by Radio Occultation	—	—
20. M. Ya. Marov	—	Part of 16
21. I. M. Podgorny	TM-76817	28
22. V. V. Kerzhanovich and M. Ya. Marov	TM-76818	22
23. V. S. Avduevskiy	TM-76819	12
24. G. S. Golitsyn, Atmospheric Dynamics	TM-76816	—
25. L. V. Ksanfomality	TM-76820	Part of 17
26. Yu. A. Surkov et al., Aerosol Composition	TM-76781	—
27. V. I. Moroz	—	5
28. L. M. Mukhin	—	30

Color Section

CONTENTS OF ILLUSTRATIONS

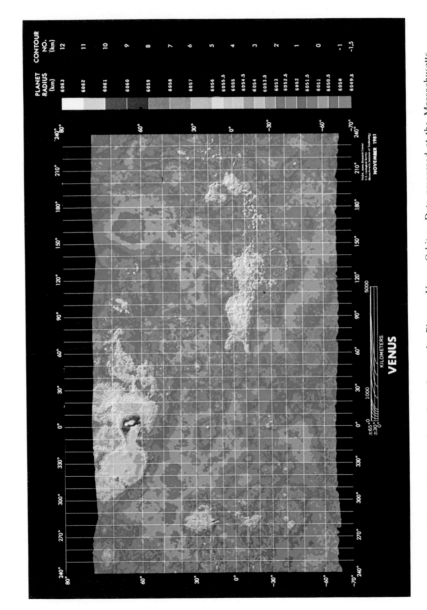

Plate 1. Topographic map from the radar on the Pioneer Venus Orbiter. Data processed at the Massachusetts Institute of Technology and the United States Geological Survey, Flagstaff. For names see Plate 3.

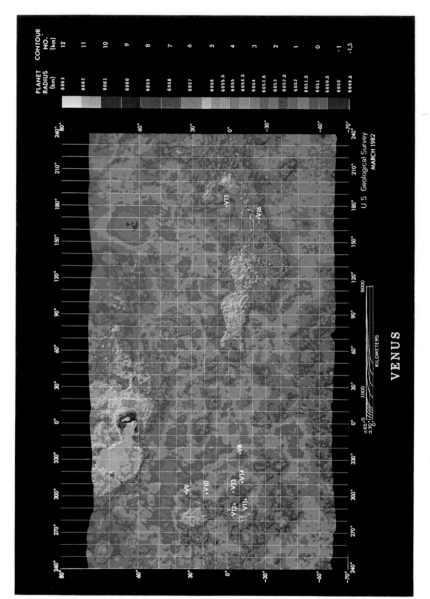

Plate 2. The map of Plate 1 showing the sites of the Soviet Venera landers.

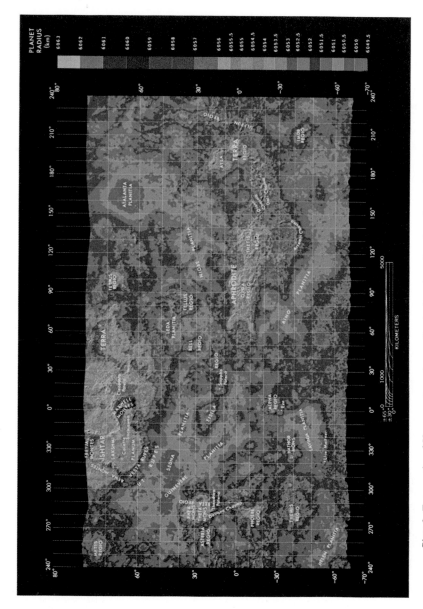

Plate 3. Topography of Venus contoured according to the prismatic color scheme shown in the column on the right. The Mercator projection exaggerates the areas of features at high latitudes.

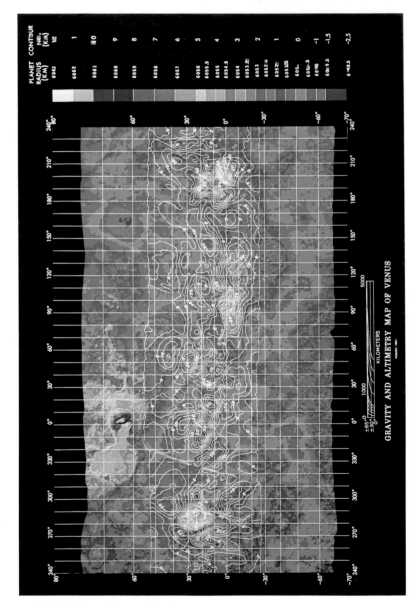

Plate 4. Line-of-sight acceleration of Pioneer Venus orbiter contoured onto topography. Gravity contour interval is 5 mgal.

Plate 5. Panorama of the Venus surface taken by Venera 13.

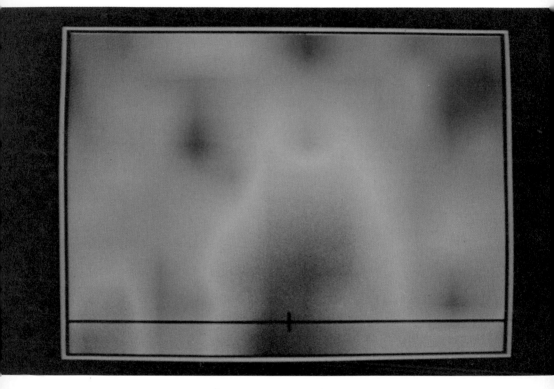

Plate 6. The mean temperture of the 95–110 km layer as a function of solar-fixed latitude and longitude. The highest temperature shown (red) is 184 K; the lowest (blue) is 170 K (Taylor et al. 1981).

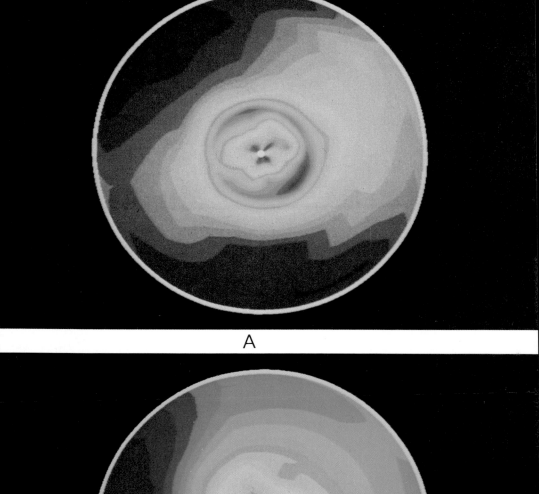

A

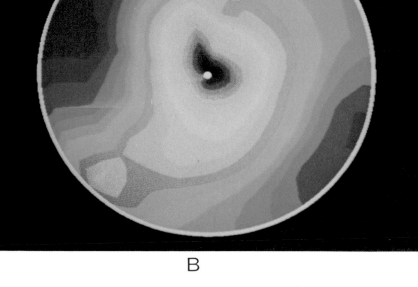

B

Plate 7. Map of the northern hemisphere in polar coordinates showing the mean temperature of (a) the 75 mbar surface, and (b) the 1 mbar surface reconstructed from Pioneer Venus temperature data. The subsolar point is at the bottom of the picture.

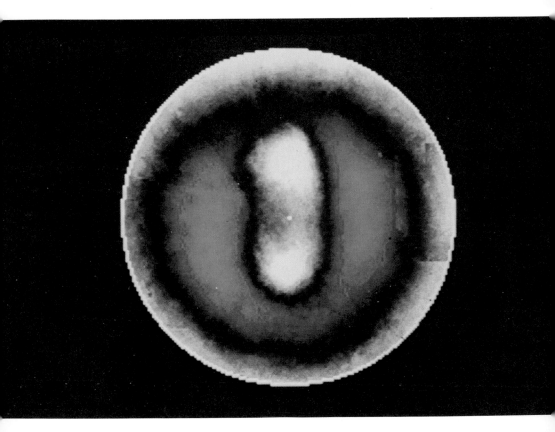

Plate 8. False color image of the polar dipole feature at a wavelength of 11.5 μm. Pioneer Venus orbiter infrared data from several different orbits were combined in a coordinate frame fixed with respect to the axis of the dipole to produce this picture. The north pole is at the center (color plate courtesy of D. J. Diner).

REFERENCES

Abell, G. 1969. *Exploration of the Universe* (New York: Holt, Rinehart and Winston).

Abramovich, S. K., Ageyeva, P. D., and Akim, E. L. 1979. Ballistics and navigation of the automatic interplanetary stations Venera 11 and Venera 12. *Kosmich. Issled.* 17:670–677.

Adams, J. B., and Filice, A. L. 1967. Spectral reflectance 0.4 to 2.0 microns of silicate rock powders. *J. Geophys. Res.* 72:5705–5715.

Adams, J. B., and McCord, T. B. 1969. Mars: Interpretation of spectral reflectivity of light and dark regions. *J. Geophys. Res.* 74:4851–4856.

Adams, W. S., and Dunham, T. 1932. Absorption bands in the infra-red spectrum of Venus. *Publ. Astron. Soc. Pacific* 44:243–247.

Adamson, A. W. 1976. *Physical Chemistry of Surfaces* (New York: Wiley).

Adel, A., and Dennison, D. M. 1933. The infrared spectrum of carbon dioxide. Part I. *Phys. Rev.* 43:716–723.

Adel, A., and Slipher, V. M. 1934. Concerning the carbon dioxide content of the atmosphere of the planet Venus. *Phys. Rev.* 46:240.

Adel, A., and Lampland, C. O. 1941. Planetary atmospheres and water-cell temperatures. *Astrophys. J.* 93:391–396.

Alaerts, L., Lewis, R. S., and Anders, E. 1977. Primordial noble gases in chondrites: The abundance pattern was established in the solar nebula. *Science* 198:927–929.

Albritton, D. L. 1978. Ion-neutral reaction-rate constants measured in flow reactors through 1977. *Atomic Data and Nucl. Data Tables* 22:1–101.

Aleksandrov, Yu. N., Vasil'ev, M. B., Vyshlof, A. S., Dolbezhev, G. G., Dubrovin, V. M., Zaitsev, A. L., Kolosov, M. A., Petrov, G. M., Savich, N. A., Samovol, V. Z., Samoznaev, L. N., Siborenko, A. I., Khasyanov, A. F., and Shtern, D. Ya. 1976a. Nighttime ionosphere of Venus from the results of two-frequency radioscopy from Venera 9 and Venera 10. *Kosmich. Issled.* 14:824–827.

Aleksandrov, Yu. N., Vasil'ev, M. B., Vyshlof, A. S., Dubrovin, V. M., Zaitsev, A. L., Kolosov, M. A., Krymov, A. A., Makovoz, G. I., Petrov, G. M., Savich, N. A., Samovol, V. Z., Samoznaev, L. N., Siborenko, A. I., Khasyanov, A. F., and Shtern, D. Ya. 1976b. Preliminary results of two-frequency radioscopy of the daytime ionosphere of Venus from Venera 9 and Venera 10. *Kosmich. Issled.* 14:819–823.

Aleksandrov, Yu. N., Golovkov, V. K., Dubrovin, V. M., Zaitsev, A. L., Kaevitser, V. I., Krymov, A. A., Petrov, G. M., Khasyanov, A. F., Shakhovskoi, A. M., and Shamparova, O. S. 1980. The reflection properties and surface relief of Venus: Radar surveys at 39-cm wavelength. *Soviet Astron. AJ* 24:139–146; (*Astron. Zh.* 57:237–249).

Alfvén, H. 1954. *On the Origin of the Solar System* (Oxford: Clarendon).

Alfvén, H. 1957. On the theory of comet tails. *Tellus* 9:92–96.

Allen, C. W. 1973. *Astrophysical Quantities* (London: Athlone).

Anders, E., and Owen, T. 1977. Origin and abundances of volatiles. *Science* 198:453–465.

Anderson, D. E. Jr. 1976. The Mariner 5 ultraviolet photometer experiment: Analysis of hydrogen Lyman-alpha data. *J. Geophys. Res.* 81:1213–1216.

Anderson, D. L. 1980. Tectonics and composition of Venus. *Geophys. Res. Letters* 7:101–102.

Anderson, D. L. 1981. Plate tectonics on Venus. *Geophys. Res. Letters* 8:309–311.

Anderson, J. L., Belton, M. J. S., Danielson, G. E., Evans, N., and Soha, J. M. 1978. Venus in motion. *Astrophys. J. Suppl.* 36:275–284.

Anderson, L. G. 1976. Atmospheric chemical kinetics data survey. *Rev. Geophys. Space Phys.* 14:151–171.

Anderson, R. C., Pipes, J. G., Broadfoot, A. L., and Wallace, L. 1969. Spectra of Venus and Jupiter from 1800 to 3200 Å. *J. Atmos. Sci.* 26:874–888.

Andreichikov, B. M. 1978. Distribution of water vapor in upper troposphere of Venus based on the data Venera 4, 5, and 6 probes. *Geok.* 1:11.

Andreyev, B. N., Guslyakov, V. T., Kerzhanovich, V. V., Druglov, Yu. M., Lysov, V. P., Marov, M. Ya., Onishekenleo, L. V., Rozhdestvensky, M. K., Sorokin, V. P., and Shnygin, Yu. N. 1974. Measurements of the wind speed in the atmosphere of Venus by means of the automatic interplanetary station Venera 8. *Kosmich. Issled.* 12:419–429.

Antsibor, N. M., Bakit'ko, R. V., Ginzburg, A. L., Gudlyakov, V. T., Kerzhanovich, V. V., Makarov, E. P., Marov, M. Ya., Molotov, E. P., Rogal'skii, V. I., Rozhdestvensky, M. K., Sopokin, V. P., and Shnzgin, Yu. N. 1976. Estimates of wind velocity by instruments dropped from Venera 9 and Venera 10. *Cosmic Res.* 14;625–631; (*Kosmich. Issled.* 14:714–721).

Apt, J., and Goody, R. M. 1979a. Anomalous features in thermal radiance maps of Venus. *J. Geophys. Res.* 84:2529–2538.

Apt, J., and Goody, R. M. 1979b. Infrared image of Venus at the time of the Pioneer Venus encounter. *Science* 203:785–787.

Apt, J., and Leung, J. 1982. Thermal periodicities in the Venus atmosphere. *Icarus* 49: 423–427.

Apt, J., Brown, R. A., and Goody, R. M. 1980. The character of the thermal emission from Venus. *J. Geophys. Res.* 85:7934–7940.

Arvidson, R. E. 1981. Effects of lateral resolution on the interpretability of geologic features sampled by the Pioneer-Venus altimeter. *Lunar Planet. Sci.* 12:31–33.

Arvidson, R. E., Batiza, R., and Guiness, E. A. 1980. Simulation of Pioneer-Venus altimetry using northern Pacific ocean floor bathymetry. *NASA Tech. Mem.* 82385:77–78.

Arvidson, R. E., and Davies, G. F. 1981. Effects of lateral resolution on the identification of volcano-tectonic provinces on Earth and Venus. *Geophys. Res. Letters* 8:741–744.

J. Atmospheric Science Vol. 32 1975. "Conference on the Atmosphere of Venus" held on 15–17 October 1974 at GISS. Papers published in *J. Atmos. Sci.* 32, No. 6.

Avduevsky, V. S., Marov, M. Ya., and Rozhdestvensky, M. K. 1968. Model of the planet Venus based on results of the measurements made by the Soviet automatic interplanetary station Venera-4. *J. Atmos. Sci.* 25:537–545.

Avduevsky, V. S., Marov, M. Ya., and Rozhdestvensky, M. K. 1970. A tentative model of the Venus atmosphere based on the measurements of Veneras 5 and 6. *J. Atmos. Sci.* 27:561–568.

Avduevsky, V. S. et al. 1973a. *Doklady AN SSSR* 210:799.

Avduevsky, V. S., Marov, M. Ya., Moshkin, B. E., and Ekonomov, A. P. 1973b. Measurement of solar illumination through the atmosphere of Venus. *J. Atmos. Sci.* 30:1215–1218.

Avduevsky, V. S., Borodin, N. F., Burtsev, V. P., Malkov, Ya. V., Marov, M. Ya., Morozov, S. F., Rozhdestvenskii, M. K., Romanov, R. S., Sokolov, S. S., Fokin, V. G., Cheremukhina, Z. P., and Shkirina, V. I. 1976a. Automatic stations Venera 9 and Venera 10: Functioning of descent vehicles and measurement of atmospheric parameters. *Cosmic Res.* 14:577–586.

Avduevsky, V. S., Vishnevetskii, S. L., Golov, I. A., Karpeishkii, Yu. Ya., Lavrov, A. D., Kikhuskin, V. Ya., Marov, M. Ya., Melnikov, D. A., Pomagin, N. I., Pronina, N. N., Razin, K. A., and Fokin, V. G. 1976b. Measurement of wind velocity on the surface of Venus during the operation of stations Venera 9 and Venera 10. *Cosmic Res.* 14:622–625.

Avduevsky, V. S. et al. 1976c. *Doklady AN SSSR* 229:579.

Avduevsky, V. S., Golovin, Yu. M., Kavelevich, F. S., Likhushin, V. Ya., Marov, M. Ya., Mel'nikov, D. A., Merson, Ya. I., Moshkin, B. E., Razin, K. A., Chernoshchekov, L. I., and Ekonomov, A. P. 1976d. Preliminary results of the investigation of the light regime in the atmosphere and on the surface of Venus. *Kosmich. Issled.* 14:735.

Avduevsky, V. S., Vishnevetsky, S. L., Golov, I. Z., Karpeisky, Yu. Ya., Lavrov, A. D., Likhashin, Y. Ya., Marov, M. Ya., Mel'nikov, D. A., Pomogin, N. N., Razin, K. A., and Fokin, V. G. 1976e. *Sov. Space Res.* 14:710.

Avduevsky, V. S., Borodin, N. F., Vasil'ev, V. N., Godnev, A. G., Karyagin, V. P., Kover'yanov, V. A., Kovtunenko, V. M., Kremnev, R. S., Pavlova, V. M., Rozhdestvenskii, M. K., Serbin, V. I., Sukhanov, K. G., Uspenskii, G. R., and Cheremukhina, Z. P. 1979. Parameters of Venus' atmosphere at Venera 11 and 12 landing sites (an analysis of the results of measurements made by these automatic interplanetary stations). *Cosmic Res.* 17:539–544.

Aveni, A. F. 1979. May–June 1979 , Venus and the Maya. *Amer. Sci.* 67:274–285.

Ayers, G. P., Gillett, R. W., and Gros, J. L. 1980. On the vapor pressure of sulfuric acid. *Geophys. Res. Letters* 7:433–436.

Baer, A. J. 1977. Speculations on the evolution of the lithosphere. *Precambrian Res.* 5:249–260.

Bagnold, R. A. 1941. *The Physics of Blown Sand and Desert Dunes* (New York: Wm. Marrow).

Baker, R. H. 1958. *Astronomy* (Princeton: Van Nostrand).

Bame, S. J., Asbridge, J. R., Gosling, J. T., Halbig, W., Paschmann, G., Sckopke, N., and Rosenbauer, H. 1979. High temporal resolution observations of electron heating at the bow shock. *Space Sci. Rev.* 23:75–92.

Banks, P. M., and Axford, W. I. 1970. Origin and dynamical behavior of thermal protons in the Venusian ionosphere. *Nature* 225:924–926.

Banks, P. M., and Kockarts, G. 1973. *Aeronomy*, 2 Vols., Parts A, B (New York: Academic).

Barakat, A. R., and Schunk, R. W. 1981. Transport equations for multi-component space plasmas: A review. *Plasma Phys.* 24:389–418.

Barath, F. T. 1964. Symposium on radar and radiometric observations of Venus during the 1962 conjunction. *Astron. J.* 69:1–2.

Barath, F. T., Barrett, A. H. Copeland, J., Jones, D. C., and Lilley, A. E. 1963. Microwave radiometers, part of Mariner II: Preliminary reports on measurements of Venus. *Science* 139:908–909.

Barath, F. T., Barrett, A. H., Copeland, J., Jones, D. E., and Lilley, A. E. 1964. Mariner 2 microwave radiometer experiment results. *Astron. J.* 69:49–58.

Barin, I., and Knacke, O. 1973. *Thermodynamic Properties of Inorganic Substances* (New York: Springer Verlag).

Barker, E. S. 1975a. Comparison of simultaneous CO_2 and H_2O observations of Venus. *J. Atmos. Sci.* 32:1071–1075.

Barker, E. S. 1975b. Observations of Venus water vapor over the disk of Venus: The 1972–74 data using the H_2O lines at 8197 Å and 8176 Å. *Icarus* 25:268–281.

Barker, E. S. 1979. Detection of SO_2 in the UV spectrum of Venus. *Geophys. Res. Letter* 6:117–120.

Barker, E. S., Woodman, J. H., Perry, M. A., Hapke, B. A., and Nelson, R. 1975. Relative spectrophotometry of Venus from 3067 to 5960 Å. *J. Atmos. Sci.* 32:1205–1211.

Bar-Nun, A. 1979. Acetylene formation on Jupiter: Photolysis or thunderstorms: *Icarus* 38:180–191.

Bar-Nun, A. 1980. Production of nitrogen and carbon species by thunderstorms on Venus. *Icarus* 42:338–342.

Barsukov, V. L., Khodakovsky, I. L., Volkov, V. P., and Florensky, C. P. 1980a. The geochemical model of the troposphere and lithosphere of Venus based on new data. *Space Res.* 20:197–203.

Barsukov, V. L., Volkov, V. P., and Khodakovsky, I. L. 1980b. The mineral composition of surface rocks: A preliminary prediction. *Proc. Lunar Planet. Sci. Conf.* 11:765–773.

Barth, C. A. 1964. Three-body reactions. *Ann. Geophys.* 20:182–196.

Barth, C. A. 1968. Interpretation of the Mariner 5 Lyman alpha measurements. *J. Atmos. Sci.* 25:564–567.

Barth, C. A., Pearce, J. B., Kelly, K. K., Wallace, L., and Fastie, W. G. 1967. Ultraviolet emission observed near Venus from Mariner 5. *Science* 158:1675–1678.

Barth, C. A., Wallace, L., and Pearce, J. B. 1968. Mariner 5 measurement of Lyman-alpha radiation near Venus. *J. Geophys. Res.* 73:2541–2545.

Basaltic Volcanism Study Project 1981. *Basaltic Volcanism on the Terrestrial Planets* (New York: Pergamon), pp. 633–699.

Bauer, S. J., and Hartle, R. E. 1974. Venus ionosphere: An interpretation of Mariner 10 observations. *Geophys. Res. Letters* 1:7–9.

Bauer, S. J., and Taylor, H. A. 1981. Modulation of Venus ion densities associated with solar variations. *Geophys. Res. Letters* 8:840–842.

Bauer, S. J., Brace, L. H., Hunten, D. M., Intriligator, D. S., Knudsen, W. C., Nagy, A. F., Russell, C. T., Scarf, F. L., and Wolfe, J. H. 1977. The Venus ionosphere and solar wind interaction. *Space Sci. Rev.* 20:413–430.

Bauer, S. J., Donahue, T. M., Hartle, R. E., and Taylor, H. A. 1979. Venus ionosphere: Photochemical and thermal diffusion control of ion composition. *Science* 205:109–112.

Baulch, D. L., Drysdale, D. D., Duxbury, J., and Grant, S. 1976. *Evaluated Kinetic Data for High Temperature Reactions* (London: Butterworths).

Baulch, D. L., Cox, R. A., Hampson, R. F. Jr., Kerr, J. A., Troe, J., and Watson, R. T. 1980. Kinetic and photochemical data for atmospheric chemistry. *J. Phys. Chem. Ref. Data* 9:295–471.

Beatty, J. K., O'Leary, B., and Chaikin, A., eds. 1981. *The New Solar System* (New York: Cambridge Univ. Press and Sky Publ.).

Beaumont, C. 1978. The evolution of sedimentary basins on a viscoelastic lithosphere: Theory and examples. *Geophys. J. Roy. Astron. Soc.* 55:471–497.

Beebe, R. F. 1972. Ultraviolet clouds on Venus: An observational bias. *Icarus* 17:602–607.

Beljajev, W. N., Kurt, W. G., Melioranski, A. S., Smirnov, A. S., Sorokin, L. S., and Tijt, W. M. 1970. Measurements of ultraviolet radiation from the interplanetary spacecraft Venera 5 and Venera 6. *Kosmich. Issled.* 8:740.

Belton, M. J. S. 1968. Theory of the curve of growth and phase effects in a cloudy atmosphere: Application to Venus. *J. Atmos. Sci.* 25:596–609.

Belton, M. J. S., and Hunten, D. M. 1966. Water vapor in the atmosphere of Venus. *Astrophys. J.* 146:307–308.

Belton, M. J. S., and Hunten, D. M. 1968. A search for O_2 on Mars and Venus: A possible detection of oxygen in the atmosphere of Mars. *Astrophys. J.* 153:963–974. Also Erratum 1969. *Astrophys. J.* 156:797.

Belton, M. J. S., Hunten, D. M., and Goody, R. M. 1968. Quantitative spectroscopy of Venus in the region 8,000–11,000 Å. In *The Atmospheres of Venus and Mars,* eds. J. C. Brandt and M. B. McElroy (New York: Gordon and Breach), pp. 69–98.

Belton, M. J. S., Smith, G., Elliott, D. A., Klaasen, K., and Danielson, G. E. 1976a. Space-time relationships in UV markings on Venus. *J. Atmos. Sci.* 33:1383–1393.

Belton, M. J. S., Smith, G., Schubert, G., and Del Genio, A. D. 1976b. Cloud patterns, waves, and convection in the Venus atmosphere. *J. Atmos. Sci.* 33:1394–1417.

Bertaux, J. L., Blamont, J., Marcelin, M., Kurt, V. G., Romanova, N. N., and Smirnov, A. S. 1978. Lyman-alpha observations of Venera 9 and 10. I. The nonthermal hydrogen population in the exosphere of Venus. *Planet. Space Sci.* 26:817–831.

Bertaux, J. L., Blamont, J. E., Lepine, V. M., Kurt, V. G., Romanova, N. N., and Smirnov, A. S. 1981. Venera 11 and Venera 12 observations of EUV emissions from the upper atmosphere of Venus. *Planet. Space Sci.* 29:149–166.

Bertaux, J. L., Lepine, V. M., Kurt, V. G., and Smirnov, A. S. 1982. Altitude profile of H in the atmosphere of Venus from Lyman-alpha observations of Venera 11 and Venera 12 and the origins of the hot exospheric component. *Icarus* Vol. 52, No. 2.

Betz, A. L., Johnson, M. A., McLaren, R. A., and Sutton, E. C. 1976. Heterodyne detection of CO_2 emission lines and wind velocities in the atmosphere of Venus. *Astrophys. J.* 208:L141–L144.

Betz, A. L., Sutton, E. C., McLaren, R. A., and McClary, C. W. 1977. Laser heterodyne spectroscopy. In *Proc. of the Symposium on Planetary Atmospheres,* ed. A. V. Jones (Ottawa, Ontario: Roy. Soc. Canada), pp. 29–33.

Bickle, M. J. 1978. Heat loss from the Earth: A constraint on Archaean tectonics from the relation between geothermal gradients and the rate of plate production. *Earth Planet. Sci. Letters* 40:301–315.

Black, D. C. 1970. Trapped helium-neon isotopic correlations in gas-rich meteorites and carbonaceous chondrites. *Geochim. Cosmochim.* 34:132–139.

Blamont, J., and Ragent, B. 1979. Further results of the Pioneer Venus nephelometer experiment. *Science* 205:67–70.

Boese, R. W., Pollack, J. B., and Silvaggio, P. M. 1979. First results of the large probe infrared radiometer experiment. *Science* 203:797–800.

Bogard, D., and Gibson, E. K. Jr. 1978. The origin and relative abundance of C, N, and the noble gases on the terrestrial planets and in meteorites. *Nature* 271:150–153.

Boishot, A., Ginat, M., and Kazes, I. 1963. Observation du rayonnement de planetes Venus et Jupiter sur 13 cm et 21 cm de longueur d'onde. *Astrophys. J.* 26:385.

Borucki, W. J. 1982. Comparison of venusian lightning observations. *Icarus* Vol. 52, No. 2.

Borucki, W. J., Dyer, J. W., Thomas, G. Z., Jordon, J. C., and Comstock, D. A. 1981. Optical search for lightning on Venus. *Geophys. Res. Letters* 8:233–236.

Borucki, W. J., Levine, Z., Whitten, R. C. Keesee, R. G., Capone, L. A., Toon, O. B., and Dubach, J. 1982. Predicted electrical conductivity between 0 and 80 km in the venusian atmosphere. *Icarus* 51:302–321.

Boyer, C. 1973. The 4-day rotation of the upper atmosphere of Venus. *Planet. Space Sci.* 21:1559–1561.

Boyer, C., and Camichel, H. 1961. Observation photographique de la planete Venus. *Ann. Astrophys.* 24:531–535.

Boyer, C., and Guerin, P. 1966. Mise en evidence directe, par la photographie d'une rotation retrograde de Venus en 4 jours. *Compt. Rend. Acad. Sci.* 263:253–255.

Boyer, C., and Guerin, P. 1969. Etude de la rotation retrograde, en 4 jours de la couche exterieure nuageuse de Venus. *Icarus* 11:338–355.

Brace, L. H., Taylor, H. A., Cloutier, P. A., Daniell, R. E., and Nagy, A. F. 1979a. On the configuration of the nightside Venus ionopause. *Geophys. Res. Letters* 6:346–349.

Brace, L. H., Theis, R. F., Krehbiel, J. P., Nagy, A. F., Donahue, T. M., McElroy, M. B., and Pedersen, A. 1979b. Electron temperatures and densities in the Venus ionosphere: Pioneer Venus orbiter electron temperature probe results. *Science* 203:763–764.

Brace, L. H., Theis, R. F., Niemann, H. B., Mayr, H. G., Hoegy, W. R., and Nagy, A. F. 1979c. Empirical models of the electron temperature and density in the nightside Venus ionosphere. *Science* 205:102–104.

Brace, L. H., Theis, R. F., Hoegy, W. R., Wolfe, J. H., Mihalov, J. D., Russell, C. T., Elphic, R. C., and Nagy, A. F. 1980. The dynamic behavior of the Venus ionosphere in response to solar wind interactions. *J. Geophys. Res.* 85:7663–7678.

Brace, L. H., Theis, R. F., and Hoegy, W. R. 1982a. Plasma clouds above the ionopause of Venus and their implications. *Planet. Space Sci.* 30:29–37.

Brace, L. H., Theis, R. F., Mayr, H. G., Curtis, S. A., and Luhmann, J. G. 1982b. Holes in the nightside ionosphere of Venus. *J. Geophys. Res.* 87:199–211.

Braham, R. R. 1975. Overview of urban climate. "Conf. on Urban Physical Environment," held at Syracuse, NY.

Brass, G. W., and Harrison, C. G. A. 1982. On the possibility of plate tectonics on Venus. *Icarus* 49:86–96.

Brenning, N. 1981. Experiments on the critical ionization velocity interaction in weak magnetic fields. *Plasma Phys.* 23:967–977.

Breus, T. K. 1979. Solar wind interaction. *Space Sci. Rev.* 23:253–275.

Bridge, H. S., Lazarus, A. J., Snyder, C. W., Smith, E. J., Davis, L., Coleman, P. J., and Jones, D. E. 1967. Plasma and magnetic fields near Venus. *Science* 158:1669–1673.

Bridge, H. S., Lazarus, A. J., Scudder, J. D., Ogilvie, K. W., Hartle, R. E., Asbridge, J. R., Bame, S. J., and Siscoe, G. L. 1974. Observations at Venus encounter by plasma science experiments on Mariner 10. *Science* 183:1293–1296.

Bridge, H. S., Lazarus, A. J., Siscoe, G. L., Hartle, R. E., Ogilvie, K. W., Scudder, J. D., and Yeates, C. M. 1976. Interaction of the solar wind with Venus. In *Solar Wind Interaction with the Planets, Mercury, Venus and Mars,* ed. N. F. Ness (Washington, D. C.: NASA SP-397), pp. 63–79.

Brinton, H. C., Taylor, H. A. Jr., Niemann, H. B., and Mayr, H. G. 1979. Venus nighttime hydrogen bulge. *Bull. Amer. Astron. Soc.* 11:538 (abstract).

Brinton, H. C., Taylor, H. A. Jr., Niemann, H. B., Mayr, H. G., Nagy, A. F., Cravens, T. E., and Strobel, D. F. 1980. Venus nighttime hydrogen bulge. *Geophys. Res. Letters* 7:865–868.

Broadfoot, A. L., and Kendall, K. R. 1968. The airglow spectrum, 3100–10,000 Å. *J. Geophys. Res.* 73:426–428.

Broadfoot, A. L., Kumar, S., Belton, M. J. S., and McElroy, M. B. 1974. Ultraviolet observations of Venus from Mariner 10: Preliminary results. *Science* 183:1315–1318.

Broadfoot, A. L., Clapp, S. S., and Stuart, F. E. 1977a. Mariner 10 ultraviolet spectrometer: Airglow experiment. *Space Sci. Instrum.* 3:199–208.

Broadfoot, A. L., Clapp, S. S., and Stuart, F. E. 1977b. Mariner 10 ultraviolet spectrometer: Occultation experiment. *Space Sci. Instrum.* 3:209–218.

Bronshten, V. A. 1972. *The Planet Venus*. NASA Technical Translation TT-F-14,519 of *Planet Venera* (Moscow: Znaniye Press).

Bukowinski, M. S. T. 1976. The effect of pressure on the physics and chemistry of potassium. *Geophys. Res. Letters* 3:491–494.

Burke, K., and Wilson, J. T. 1972. Is the African plate stationary? *Nature* 239:387–389.

Burnett, D. S. 1975. Lunar science: The Apollo legacy. *Rev. Geophys. Space Phys.* 13: 13–34.

Busse, F. H. 1976. Generation of planetary magnetism by convection. *Phys. Earth Planet. Interiors* 12:350–358.

Butler, D. M. 1975. The Ionosphere of Venus. Ph.D. thesis, Rice Univ., Houston, TX.

Butler, D. M., and Chamberlain, J. W. 1976. Venus' nightside ionosphere: Its origin and maintenance. *J. Geophys. Res.* 81:4757–4760.

Butler, D. M., and Stolarski, R. S. 1978. Photoelectrons and electron temperatures in the Venus ionosphere. *J. Geophys. Res.* 83:2057–2065.

Cameron, A. G. W. 1962. The formation of the sun and planets. *Icarus* 1:13–69.

Cameron, A. G. W. 1973. Abundances of elements in the solar system. *Space Sci. Rev.* 15:121–128.

Cameron, A. G. W. 1978a. Physics of the primitive solar nebula and of giant gaseous protoplanets. In *Protostars and Planets,* ed. T. Gehrels (Tucson: Univ. Arizona Press), pp. 453–487.

Cameron, A. G. W. 1978b. The primitive solar accretion disk and the formation of the planets. In *The Origin of the Solar System,* ed. S. F. Dermott (New York: Wiley), pp. 49–74.

Cameron, A. G. W., De Campli, W. M., and Bodenheimer, P. H. 1980. Numerical experiments with giant gaseous protoplanets embedded in the primitive solar nebula. *Lunar Planet. Sci.* 11:122–124 (abstract).

Campbell, D. B., and Burns, B. A. 1980. Earth-based radar imagery of Venus. *J. Geophys. Res.* 85:8271–8281.

Campbell, D. B., Dyce , R. B., and Pettengill, G. H. 1976. New radar image of Venus. *Science* 193:1123–1124.

Campbell, D. B., Burns, B. A., and Boriakoff, V. 1979. Venus: Further evidence of impact cratering and tectonic activity from radar observations. *Science* 204:1424–1427.

Campbell, D. B., Burns, B. A., Harmon, J. K., and Hine, A. 1981. Radar reflectivity images of large topographic features on the surface of Venus. *Trans. Amer. Geophys. Union* 62:386 (abstract).

Campbell, I. M., and Gray, C. N. 1973. Rate constants for $O(^3P)$ recombination and association with $N(^4S)$. *Chem. Phys. Letters* 18:607.

Campbell, M. J., and Ulrichs, J. 1969. Electrical properties of rocks and their significance for lunar radar observations. *J. Geophys. Res.* 74:5867–5881.

Canalas, R. A., Alexander, E. C., and Manuel, O. K. 1968. Terrestrial abundances of noble gases. *J. Geophys. Res.* 73:3331–3334.

Canuto, V., and Hsieh, S. H. 1978. Scale co-variant cosmology and the temperature of the Earth. *Astron. Astrophys.* 65:389–391.

Carpenter, R. L. 1964. Study of Venus by CW radar. *Astron. J.* 69:2–11.

Carpenter, R. L. 1966. Study of Venus by CW radar—1964 results. *Astron. J.* 71:142–152.

Carr, M., Wilhelms, D., Greeley, R., and Guest, J. 1976. Stratigraphy and structural geology. In *A Geologic Basis for the Exploration of the Terrestrial Planets,* eds. R. Greeley and M. Carr (Washington, D. C.: NASA SP-417), pp. 13–32.

Cazenave, A., and Dominh, K. 1981. Elastic thickness of the Venus lithosphere estimated from topography and gravity. *Geophys. Res. Letters* 8:1039–1042.

Cess, R. D., Ramanathan, V., and Owen, T. 1980. The Martian paleoclimate and enhanced atmospheric carbon dioxide. *Icarus* 41:159–165.

Chalmers, J. A. 1967. *Atmospheric Electricity* (New York: Pergamon).

Chamberlain, J. W. 1965. The atmosphere of Venus near her cloud tops. *Astrophys. J.* 141:1184–1205.

Chamberlain, J. W. 1977. Charge exchange in a planetary corona: Its effect on the distribution and escape of hydrogen. *J. Geophys. Res.* 82:1–9.

Chamberlain, J. W., and Kuiper, G. P. 1956. Rotational temperature and phase variation of the carbon dioxide bands of Venus. *Astrophys. J.* 124:339–405.

Chamberlain, J. W., and Smith, G. R. 1971. Comments on the rate of evaporation of a non-maxwellian atmosphere. *Planet. Space Sci.* 19:675–684.

Chameides, W. L., Walker, J. C. G., and Nagy, A. F. 1979. Possible chemical impact of planetary lightning in the atmospheres of Mars and Venus. *Nature* 280:820–822.

Chandrasekhar, S. 1960. *Radiative Transfer* (New York: Dover).

Chapman, S. 1931. The absorption and dissociative or ionizing effect of monochromatic radiation in an atmosphere of a rotating earth. II. Grazing incidence. *Proc. Phys. Soc. London* 43:483–501.

Chapman, S., and Cowling, T. G. 1970. *The Mathematical Theory of Nonuniform Gases* (New York: Cambridge Univ. Press).

Chapman, S., and Lindzen, R. S. 1970. *Atmospheric Tides* (Dordrecht:Reidel).

Charney, J. G., and Stern, M. E. 1962. On the stability of internal baroclinic jets in a rotating atmosphere. *J. Atmos. Sci.* 19:159–172.

Chase, S. C., Kaplan, L. D., and Neugebauer, G. 1963. The Mariner 2 infrared radiometer experiment. *J. Geophys. Res.* 68:6157–6169.

Chase, S. C., Miner, E. D., Morrison, D., Munch, G., and Neugebauer, G. 1974. Preliminary infrared radiometry of Venus from Mariner 10. *Science* 183:1291–1292.

Chen, R. H. 1977. The Venus Ionosphere. Ph. D. thesis, Univ. of Michigan, Ann Arbor, MI.

Chen, R. H., and Nagy, A. F. 1978. A comprehensive model of the Venus ionosphere. *J. Geophys. Res.* 83:1133–1140.

Chen, R. H., Cravens, T. E., and Nagy, A. F. 1978. The Martian ionosphere in light of the Viking observations. *J. Geophys. Res.* 83:3871–3876.

Cheremukhina, Z. P., Morozov, S. F., and Borodin, N. F. 1974. Estimate of the temperature of the stratosphere of Venus by drag data of the Venera automatic station. *Kosmich. Issled.* 12:264.

Chernyshov, V. P., and Sheinman, D. N. 1973. *Radio Wave Propagation and Antenna Feed Structure* (Moscow: Svyaz'). In Russian.

Chiu, V. T., and Ching, B. K. 1978. The response of atmospheric and lower ionospheric layer structures to gravity waves. *Geophys. Res. Letters* 5:529–542.

Chub, E. V., and Yakovlev, O. I. 1980. Temperature and zonal circulation of Venus based on the data of radio probe experiments. *Cosmic Res.* 18:331–336.

Cicerone, R. J., Swartz, W. E., Stolarski, R. S., Nagy, A. F., and Nisbet, J. S. 1973. Thermalization and transport of photoelectrons: A comparison of theoretical approaches. *J. Geophys. Res.* 78:6709–6728.

Cimino, J. B., Elachi, C., Kliore, A. J., McCleese, D. J., and Patel, I. R. 1980. Polar cloud structure as derived from the Pioneer Venus orbiter. *J. Geophys. Res.* 85:8082–8088.

Clancy, R. T., Muhleman, D. O., and Berge, G. L. 1981. Mesospheric photochemistry and circulation from CO microwave measurements. "Inter. Conf. on the Venus Environment," held at Palo Alto, CA.

Clark, B. G., and Spencer, C. L. 1964. Some decimeter observations of Venus during the 1962 conjunction. *Astron. J.* 69:59–61.

Cloutier, P. A. 1970. Dynamics of the interaction of the solar wind with a planetary atmosphere. *Radio Sci.* 5:387–389.

Cloutier, P. A. 1976. Solar wind interaction with planetary ionospheres. In *Solar Wind Interaction With the Planets Mercury, Venus and Mars,* ed. N. F. Ness (Washington, D. C: NASA SP-397), pp. 111–119.

Cloutier, P. A., and Daniell, R. E. Jr. 1973. Ionospheric currents induced by solar wind interaction with planetary atmospheres. *Planet. Space Sci.* 21:463–474.

Cloutier, P. A., and Daniell, R. E. Jr. 1979. An electrodynamic model of the solar wind interaction with the ionosphere of Mars and Venus. *Planet. Space Sci.* 27:1111–1112.

Cloutier, P. A., McElroy, M. B., and Michel, F. C. 1969. Modification of the Martian ionosphere by the solar wind. *J. Geophys. Res.* 74:6215–6228.

Cloutier, P. A., Daniell, R. E. Jr., and Butler, D. M. 1974. Ionospheric currents induced by solar wind interaction with planetary atmospheres. *Planet. Space Sci.* 21:463–474.

Cloutier, P. A., Tascione, R. F., and Daniell, R. E. Jr. 1981. An electrodynamic model of electric currents and magnetic fields in the dayside ionosphere of Venus. *Planet. Space Sci.* 29:635–652.

Coblentz, W. W., and Lampland, C. O. 1925. Some measurements of the spectral components of planetary radiation and planetary temperature. *J. Franklin Inst.* 200:103–126.

Coffeen, D. L. 1968. A Polarimetric Study of the Atmosphere of Venus. Ph.D. thesis, Univ. Arizona, Tucson, AZ.

Coffeen, D. L. 1969. Wavelength dependence of polarization. XVI. Atmosphere of Venus. *Astron. J.* 74:446–460.

Colin, L. 1980. The Pioneer Venus program. *J. Geophys. Res.* 85:7575–7598.

Colin, L., and Hunten, D. M. 1977. Pioneer Venus experiment descriptions. *Space Sci. Rev.* 20:451–525.

Connes, J., and Connes, P. 1966. Near-infrared planetary spectra by Fourier spectroscopy. I. Instruments and results. *J. Opt. Soc. Amer.* 56:896–910.

Connes, J., Connes, P., and Maillard, J. P. 1969. *Atlas des Spectres dans le Proche Infrarouge de Venus, Mars, Jupiter et Saturn* (Paris: Centre Nat. de la Recherche Sci.).

Connes, P., and Michel, G. 1975. Astronomical Fourier spectrometer. *Appl. Optics* 14:2067–2084.

Connes, P., Connes, J., Benedict, W. S., and Kaplan, L. D. 1967. Traces of HCl and HF in the atmosphere of Venus. *Astrophys. J.* 147:1230–1237.

Connes, P., Connes, J., Kaplan, L. D., and Benedict, W. S. 1968. Carbon monoxide in the Venus atmosphere. *Astrophys. J.* 152:731–743.

Connes, P., Noxon, J. F., Traub, W. A., and Carleton, N. P. 1979. O_2 ($^1\Delta$) emission in the day and night airglow of Venus. *Astrophys. J.* 233:L29–L32.

Conrad, J., and Schunk, R. W. 1979. Diffusion and heat flow equations with allowance for large temperature differences between interacting species. *J. Geophys. Res.* 84:811–822.

Conway, R. R., McCoy, R. P., Barth, C. A., and Lane, A. L. 1979. IUE detection of sulfur dioxide in the atmosphere of Venus. *Geophys. Res. Letters* 6:629–631.

Cordell, B. M. 1981. Thermal relaxation of craters on Venus. "Inter. Conf. on the Venus Environment," held at Palo Alto, CA.

COSPAR Information Bull. No. 73 1973. p. 23.

Counselman, C. C. III, Gourevich, S. A., King, R. W., Loriot, G. B., and Prinn, R. G. 1979. Venus winds are zonal and retrograde below the clouds. *Science* 205:85–87.

Counselman, C. C. III, Gourevitch, S. A., King, R. W., and Loriot, G. B. 1980. Zonal and meridional circulation of the lower atmosphere of Venus determined by radio interferometry. *J. Geophys. Res.* 85:8026–8030.

Courtin, R., Lena, P., de Muizon, M., Rouan, D., Nicollier, C., and Wijnbergen, J. 1979. Far infrared photometry of planets Saturn and Venus. *Icarus* 38:411–419.

Courtoy, C. P. 1959. Spectre infra-rouge a grande dispersion et constantes moleculaires du CO_2. *Ann. Soc. Sci. Bruxelles* 73:5–203.

Covey, C. C., and Schubert, G. 1981a. Mesoscale convection in the clouds of Venus. *Nature* 290:17–20.

Covey, C. C., and Schubert, G. 1981b. 4-day waves in the Venus atmosphere. *Icarus* 47:130–138.

Covey, C. C., and Schubert, G. 1982a. Planetary scale waves in the Venus atmosphere. *J. Atmos. Sci.* In press.

Covey, C. C., and Schubert, G. 1982b. Venus atmospheric waves: A challenge for nonlinear dynamics. *Physica D.* In press.

Craig, R. A., Reynolds, R. T., Ragent, B., Carle, G. C., Woeller, F., and Pollack, J. B. 1981. Sulphur trioxide in the lower atmosphere of Venus? "Internat. Conf. on the Venus Environment," held at Palo Alto, CA.

Cravens, T. E., Nagy, A. F., Chen, R. H., and Stewart, A. I. 1978. The ionosphere and airglow of Venus: Prospects for Pioneer Venus. *Geophys. Res. Letters* 5:613–616.

Cravens, T. E., Nagy, A. F., Brace, L. H., Chen, R. H., and Knudsen, W. C. 1979. The energetics of the ionosphere of Venus: A preliminary model based on Pioneer Venus observations. *Geophys. Res. Letters* 6:341–344.

Cravens, T. E., Gombosi, T. I., Kozyra, J. U., and Nagy, A. F. 1980a. Model calculations of the dayside ionosphere of Venus: Energetics. *J. Geophys. Res.* 85:7778–7786.

Cravens, T. E., Gombosi, T. I., and Nagy, A. F. 1980b. Hot hydrogen in the exosphere of Venus. *Nature* 283:178–180.

Cravens, T. E., Kliore, A. J., Kozyra, J. U., and Nagy, A. F. 1981a. The ionospheric peak on the Venus dayside. *J. Geophys. Res.* 86:11323–11329.

Cravens, T. E., Nagy, A. F., and Crawford, S. 1981b. A two-dimensional model of the ionosphere of Venus. "4th IAGA Scientific Assembly," held at Edinburgh, Scotland.

Cresci, R. J., and Libby, P. A. 1962. The downstream influence of mass transfer at the nose of a slender cone. *J. Aerospace Sci.* 29:815–826.

Croft, T. A. and Price, G. H. 1983. Evidence for a low altitude origin of lightning on Venus 1980. Submitted to *Icarus*.

Cruikshank, D. P., and Kuiper, G. P. 1967. An upper limit for the abundance of SO_2 in the Venus atmosphere. *Comm. Lunar. Planet. Lab.* 97:195–197.

Crutzen, P. J. 1976. The possible importance of CSO for the sulfate layer of the stratosphere. *Geophys. Res. Letters* 3:73–76.

Daniell, R. E., and Cloutier, P. A. 1977. Distribution of ionospheric currents induced by the solar wind interaction with Venus. *Planet. Space Sci.* 25:621–628.

Danjon, A. 1949. *Bull. Astron.* 14:315.

Davey, A. 1967. The motion of a fluid due to a moving source of heat at the boundary. *J. Fluid. Mech.* 29:137–150.

Delaney, J. R., Muenow, D. W., and Graham, D. G. 1978. Abundance and distribution of water, carbon, and sulfur in the glassy rims of submarine pillow basalts. *Geochim. Cosmochim.* 42:581–594.

Del Genio, A. D., and Rossow, W. B. 1982. Temporal variability of ultraviolet cloud features in the Venus stratosphere. *Icarus* 51:391–415.

Del Genio, A. D., Schubert, G., and Straus, J. M. 1979. Acoustic-gravity waves in the thermosphere of Venus. *Icarus* 39:401–417.

De More, W. B. 1970. The efficiency of CO_2 formation in the Mars and Venus atmospheres. *J. Geophys. Res.* 75:4898–4899.

Dessler, A. J. 1968. Ionozing plasma flux in the Martian upper atmosphere. In *The Atmosphere of Venus and Mars,* ed. J. C. Brandt and M. B. McElroy (New York: Gordon and Breach), pp. 241–250.

Detrick, R. S., von Herzen, R. P., Crough, S. T., Epps, D., and Fehn, U. 1981. Heat flow on the Hawaiian swell and lithospheric reheating. *Nature* 292:142–143.

Dickel, J. R., Warnock, W. W., and Medol, W. J. 1968. Lack of phase variation of Venus. *Nature* 220:1183–1185.

Dickinson, R. E. 1972. Infrared radiative heating and cooling in the Venusian mesosphere. I. Global mean radiative equilibrium. *J. Atmos. Sci.* 29:1531–1556.

Dickinson, R. E., 1976. Venus mesosphere and thermosphere temperature structure. I. Global mean radiative and conductive equilibrium. *Icarus* 27:479–493.

Dickinson, R. E., and Ridley, E. C. 1972. Numerical solution for the composition of a thermosphere in the presence of a steady subsolar-to-antisolar circulation with application to Venus. *J. Atmos. Sci.* 29:1557–1570.

Dickinson, R. E., and Ridley, E. C. 1975. A numerical model for the dynamics and composition of the Venusian thermosphere. *J. Atmos. Sci.* 32:1219–1231.

Dickinson, R. E., and Ridley, E. C. 1977. Venus mesosphere and thermosphere temperature structure. II. Day-night variations. *Icarus* 30:163–178.

Diner, D. J., and Westphal, J. A. 1978. Phase coverage of Venus during the 1975 apparition. *Icarus* 36:119–126.

Dlugach, J. M., and Yanovitskij, E. G. 1978. Interpretation of Venera 10 optical measurements. *Pisma Astron. Zh.* 4:471–476.

Dobrovolskis, A. R., and Ingersoll, A. P. 1980. Atmospheric tides and the rotation of Venus. I. Tidal theory and balance of torques. *Icarus* 41:1–17.

Dolginov, Sh. Sh. 1977. The magnetism of planets. *Geomagnetizm i Aeronomiya* 17:569.

Dolginov, Sh. Sh., Yeroshenko, Ye. G., and Zhuzgov, L. N. 1968. The study of the magnetic field from the interplanetary station Venera-4. *Kosmich. Issled.* 6:561–575.

Dolginov, Sh. Sh., Yeroshenko, Ye. G., and Davis, L. 1969. On the nature of the magnetic field near Venus. *Kosmich. Issled.* 7:747–752.

Dolginov, Sh. Sh., Zhuzgov, L. N., Sharova, V. A., Buzin, V. B., and Yeroshenko, Ye. G. 1970. The magnetosphere of the planet Venus. Preprint 1Z MIR AN SSSR, No. 19/193, Moscow.

Dolginov, Sh. Sh., Yeroshenko, Ye. G., Zhuzgov, L. N., Buzin, V. B., and Sharova, V. A. 1976. Preliminary results of magnetic field measurements at pericenter region of "Venera-9 satellite." *Pisma Astron. Zh.* 2:88–90.

Dolginov, Sh. Sh., Zhuzgov, L. N., Sharova, V. A., and Buzin, V. B. 1978. Magnetic field and magnetosphere of the planet Venus. *Cosmic Res.* 16:657–687.

Dolginov, Sh. Sh., Duburin, E. M., Yeroshenko, Ye. G., Israelevich, P. L., Podgorny, I. M., and Shkol'nikova, S. I. 1981. On the configuration of the field in the magnetic tail of Venus. *Kosmich. Issled.* 19:624.

Dollfus, A. 1975. Venus: Evolution of the upper atmospheric clouds. *J. Atmos. Sci.* 32:1060–1070.

Dollfus, A., and Coffeen, D. L. 1970. Polarization of Venus. I. Disk observations. *Astron. Astrophys.* 8:251–266.

Dollfus, A., Auriee, M., and Santer, R. 1979. Wavelength dependence of polarization. XXXVII. Regional observations of Venus. *Astron. J.* 84:1419–1436.

Donahoe, F. J. 1970. Is Venus a polywater planet? *Icarus* 12:424–430.

Donahue, T. M. 1968. The upper atmospherre of Venus: A review. *J. Atmos. Sci.* 25:568–573.

Donahue, T. M. 1969. Deuterium in the upper atmosphere of Venus and Earth. *J. Geophys. Res.* 74:1128–1137.

Donahue, T. M., Hoffman, J. H., and Hodges, R. R. Jr. 1981. Krypton and xenon in the atmosphere of Venus. *Geophys. Res. Letters* 8:513–516.

Donahue, T. M., Hoffman, J. H., Hodges, R. R. Jr., and Watson, A. J. 1982. Venus was wet: A measurement of the ratio of D to H. *Science* 216:630–633.

Dostal, J., and Capedri, S. 1975. Partition coefficients of uranium for some rock-forming minerals. *Chem. Geol.* 15:285–294.

Douglas, A. E., Moller, C. K., and Stoicheff, B. P. 1963. The absorption spectra of $^{35}O_2$ from 4780–6000 Å. *Can. J. Phys.* 41:1174–1192.

Douglas, H. A., Mason, P. J., and Hinch, E. J. 1972. Motion due to a moving internal heat source. *J. Fluid Mech.* 54:469–480.

Drake, F. D. 1962a. 10-cm observations of Venus near superior conjunction. *Publ. NRAO* 1:165.

Drake, F. D. 1962b. 10-cm observations of Venus near superior conjunction. *Nature* 195:894.

Drake, F. D. 1964. Microwave observations of Venus in 1962–1963. *Astron. J.* 69:62–64.

Drake, L. D. 1970. Rock texture: An important factor for clast shape studies. *J. Sediment. Petrol.* 40:1356–1361.

Dryer, M., Perez de Tejada, H., Taylor, H. A., Intriligator, D. S., Mihalov, J. D., and Rompolt, B. 1982. Compression of the Venusian ionopause on 10 May 1979 by the interplanetary shock generated by the eruption of 8 May 1979. Submitted to *J. Geophys. Res.*

Dubinin, E. M., Podgorny, I. M., Potanin, Ya. N., and Shkol'nikova, S. I. 1978. Determining the magnetic moment of Venus by magnetic measurements in the tail. *Kosmich. Issled.* 16:870–876.

Dubinin, E. M., Israelevich, P. L., and Podgorny, I. M. 1980. The combination magnetosphere. *Kosmich. Issled.* 18:470.

Dungey, J. W. 1961. Interplanetary magnetic field and the auroral zones. *Phys. Rev. Letters* 6:47–48.

Dungey, J. W. 1965. The length of the magnetospheric tail. *J. Geophys. Res.* 70:1753 (letter).

Dunham, T. Jr. 1952. Spectroscopic observations of the planets at Mount Wilson. In *Atmospheres of the Earth and Planets,* ed. G. P. Kuiper (Chicago, IL: Univ. Chicago Press), pp. 288–305.

Durrance, S. T. 1981. The carbon monoxide fourth positive bands in the Venus day glow. I. Synthetic spectra. *J. Geophys. Res.* 86:9115–9124.

Durrance, S. T., Barth, C. A., and Stewart, A. I. F. 1980. Pioneer Venus observations of the Venus dayglow spectrum 1250–1430 Å. *Geophys. Res. Letters* 7:222–224.

Durrance, S. T., Stewart, A. I. F., and Swansen, R. 1981. Concentration of CO in Venus' lower thermosphere (120–150 km). "Internat. Conf. on the Venus Environment," held at Palo Alto, CA.

Dyce, R. B., and Pettengill, G. H. 1967. Radar determination of the rotation of Venus and Mercury. *Astron. J.* 72:351–359.

Egan, W. G., and Becker, J. F. 1969. Determination of the complex index of refraction of rocks and minerals. *Appl. Optics* 8:720–721.

Eggler, D. H. 1976. Does CO_2 cause partial melting in the low-velocity layer of the mantle? *Geology* 4:69–72.

Eggler, D. H., and Holloway, J. R. 1977. Partial melting of the peridotite in the presence of H_2O and CO_2: Principles and review. In *Magma Genesis,* ed. H. J. B. Dick (Oregon Dept. Geol. Mining Ind. Bull.), 96:15–36.

Ekonomov, A. P. et al. 1978.*Kosmich. Issled.* 16:901.

Ekonomov, A. P., Moshkin, B. E., Golovin, Yu. M., Parfentev, N. A., and Sanks, N. F. 1979. Spectrophotometric experiment on the Venera 11 and Venera 12 landers. *Kosmich. Issled.* 17:714–726.

Ekonomov, A. P., Golovin, Yu. M., and Moshkin, B. E. 1980. Visible radiation near the surface of Venus: Results and their interpretation. *Icarus* 41:65–75.

Elachi, C. 1980. Spaceborne imaging: Geologic and oceanographic applications. *Science* 209:1073–1082.

Eliasen, E., and Machenhauer, B. 1965. On the observed large-scale atmospheric wave motions. *Tellus* 21:149–165.

Elphic, R. C., Russell, C. T., Slavin, J. A., and Brace, L. H. 1980*a*. Observations of the dayside ionopause and ionosphere of Venus. *J. Geophys. Res.* 85:7679–7696.

Elphic, R. C., Russell, C. T., Slavin, J. A., Brace, L. H., and Nagy, A. F. 1980*b*. The location of the dayside ionopause of Venus: Pioneer Venus orbiter magnetometer observations. *Geophys. Res. Letters* 7:561–564.

Elphic, R. C., Russell, C. T., Luhmann, J. G., Scarf, F. L., and Brace, L. H. 1981. The Venus ionopause current sheet: Thickness, length, scale, and controlling factors. *J. Geophys. Res.* 86:11430–11438.

Elson, L. S. 1977. A primitive equation, solar driven, perturbation model of the thermosphere of Mars and Venus. *Geophys. Astrophys. Fluid Dyn.* 8:57–90.

Elson, L. S. 1978. Barotropic instability in the upper atmosphere of Venus. *Geophys. Res. Letters* 5:603–605.

Elson, L. S. 1979. Preliminary results from the Pioneer Venus orbiter infrared radiometer: Temperature and dynamics in the upper atmosphere. *Geophys. Res. Letters* 6:720–722.

Elson, L. S. 1982. Circulation modelling of the Venusian atmosphere from the cloud tops to 100 km. In preparation.

Epstein, E. E., Oliver, J. P., Schorn, R. A., Wilson, W. J., and Soter, S. J. 1968. Venus: On an inverse variation with phase in the 3–4 mm emission during 1965 through 1967. *Astron. J.* 73:271–274.

Esposito, L. W. 1980. Ultraviolet contrasts and the absorbers near the Venus cloud tops. *J. Geophys. Res.* 85:8151–8157.

Esposito, L. W. 1981. Absorber seen near the Venus cloud tops from Pioneer Venus. *Adv. Space. Res.* 1:163–166.

Esposito, L. W., and Gates, L. J. 1981. Horizontal and vertical distribution of sulfur dioxide on Venus. *Trans. Amer. Geophys. Union (EOS)* 62:321 (abstract).

Esposito, L. W., and Travis, L. D. 1982. Polarization studies of the Venus UV contrasts: Cloud height and haze variability. *Icarus* 51:374–390.

Esposito, L. W., Winick, J. R., and Stewart, A. I. F. 1979. Sulfur dioxide in the Venus atmosphere: Distribution and implications. *Geophys. Res. Letters* 6:601–604.

Esposito, P. B., Sjogren, W. L., Mottinger, N. A., Bills, B. G., and Abbott, E. A. 1982. Venus gravity: Analysis of Beta Regio. *Icarus* 51:448–459.

Evans, D. C. 1966. Ultraviolet reflectivity of Venus and Jupiter. *NASA/GSFC Rept.* X613–66–72; 1967. *The Moon and Planets* (Amsterdam: North-Holland), pp. 135–149.

Evans, J. V. 1969. Radar studies of planetary surfaces. *Ann. Rev. Astron. Astrophys.* 7:201–249.

Evans, J. V., and Hagfors, T., eds. 1968. *Radar Astronomy* (New York: McGraw-Hill).

Evans, W. F. J., Llewellyn, E. J., and Jones, A. V. 1972. Altitude distribution of the O_2 ($^1\Delta_g$) nightglow emission. *J. Geophys. Res.* 77:4899–4901.

Fairfield, D. H. 1971. Average and unusual locations of the Earth's magnetopause and bow shock. *J. Geophys. Res.* 76:6700–6716.

Fanale, F. P., and Cannon, W. A. 1971. Physical adsorption of rare gas on terriginous sediments. *Earth Planet. Sci. Letters* 11:362–368.

Fanale, F. P., and Cannon, W. A. 1972. Origin of planetary primordial rare gas: The possible role of adsorption. *Geochim. Cosmochim.* 36:319–328.

Farlow, N. H., Oberbeck, V. R., Colburn, D. S., Ferry, G. V., Lem, H. Y., and Hayes, D. D. 1981. Comparison of stratospheric aerosol measurements over Poker Flat, Alaska, July 1979. *Geophys. Res. Letters* 8:15–17.

Feinberg, Je. L. 1961. *Propagation of Radiowaves Along the Earth Surface* (Moscow: USSR Publ. House). In Russian.

Feldman, P. D., Moos, H. W., Clarke, J. T., and Lane, A. L. 1979. Identifications of the UV nightglow from Venus. *Nature* 279:221–222.

Fels, S., and Lindzen, R. S. 1974. The interaction of thermally excited gravity waves with mean flows. *Geophys. Fluid Dyn.* 6:149–192.

Ferrin, I. R. 1976. The Distribution of Hot Hydrogen Atoms in the Upper Atmosphere of Venus. Ph.D. thesis, Univ. Colorado, Boulder, CO.

Fink, U., Larson, H. P., Kuiper, G. P., and Poppen, R. F. 1972. Water vapor in the atmosphere of Venus. *Icarus* 17:617–631.

Fisher, D. E. 1975. The early history of the Earth. In *Proceedings of a NATO Advanced Study Institute*.

Fjeldbo, G., and Eshleman, V. R. 1969. Atmosphere of Venus as studied with the Mariner 5 dual radio-frequency occultation experiment. *Radio Sci.* 4:879–897.

Fjeldbo, G., Seidel, B., Sweetnam, D., and Howard, T. 1975. The Mariner 10 radio occultation measurements of the ionosphere of Venus. *J. Atmos. Sci.* 32:1232–1236.

Flammarion, C. 1897. *The Planet Venus* (Paris: Gauthier-Villars). (English trans. available from D. Cruikshank.)

Flasar, F. M., Samuelson, R. E., and Conrath, B. J. 1981. Titan's atmosphere: Temperature and dynamics. *Nature* 292:693–698.

Florenskiy, C. P., Basilevsky, A. T., Burba, G. A., Nikolaeva, O. V., Pronin, A. A., and Volkov, V. P. 1977a. First panoramas of the Venusian surface. *Proc. Lunar Planet. Sci. Conf.* 8:2655–2664.

Florenskiy, C. P., Ronca, L. B., and Basilevsky, A. T. 1977b. Geomorphic degradations on the surface of Venus: An analysis of Venera 9 and Venera 10 data. *Science* 196:869–871.

Florenskiy, C. P., Ronca, L. B., Basilevsky, A. T., Burba, G. A., Nikolaeva, O. V., Pronin, A. A., Trakhtman, A. M., Volkov, V. P., and Zazetsky, V. V. 1977c. The surface of Venus as revealed by Soviet Venera 9 and 10. *Geol. Soc. Amer. Bull.* 88:1537–1545.

Florenskiy, C. P., Volkov, V. P., and Nikolaeva, O. V. 1978. A geochemical model of the Venus troposphere. *Icarus* 33:537–553.

Florenskiy, C. P., Basilevsky, A. T., Kryuchkov, V. P., Kuz'min, R. O., Nikolaeva, O. V., Pronin, A. A., Chernaya, I. M., Selivanov, A. S., Naraeva, M. K., and Tyuflin, Yu. S. 1982. Analysis of panoramas of the Venera 13 and Venera 14 landing sites. *Pisma Astron. Zh.* 8:429–432.

Formisano, V., Galeev, A. A., and Sagdeev, R. Z. 1982. The role of the critical ionization velocity phenomenon in production of inner coma cometary plasma. *Planet. Space Sci.* 30:491–497.

Forsyth, D. W. 1979. Lithospheric flexure. *Rev. Geophys. Space Phys.* 17:1109–1114.

Forsyth, D. W., and Uyeda, S. 1975. On the relative importance of the driving forces of plate motion. *Geophys. J. R. Astron. Soc.* 43:163–200.

Fox, J. L. 1982. The chemistry of metastable species in the Venusian ionosphere. *Icarus* Vol. 51, No. 2.

Fox, J. L., and Dalgarno, A. 1979. Electron energy deposition in carbon dioxide. *Planet. Space Sci.* 27:491–502.

Fox, J. L., and Victor, G. A. 1981. O^{++} in the Venusian ionosphere. *J. Geophys. Res.* 86:2438–2442.

Frank, L. A., Van Allen, J. A., and Hills, H. K. 1963. Mariner 2: Preliminary reports on measurements of Venus charged particles. *Science* 139:905–907.

Freundlich, H. 1930. *Kapallarchemie*. Leipzig 1:153.

Froidevaux, C., and Schubert, G. 1975. Plate motion and structure of the continental asthenosphere: A realistic model of the upper mantle. *J. Geophys. Res.* 80:2553–2564.

Fuchs, N. A. 1964. *The Mechanics of Aerosols* (New York: Pergamon).

Fuchs, N. A., and Sutugin, A. G. 1971. Highly dispersed aerosols. In *Topics in Current Aerosol Research* Vol. II., eds. G. M. Hidy and J. R. Brock (New York: Pergamon).

Fultz, D., Long, R., Owens, G., Bohan, W., Kaylor, R., and Weil, J. 1959. Studies of thermal convection on a rotating cylinder with some implications for large-scale atmospheric motions. *Meteor. Monogr.* 4:36–39.

Galilei, G. 1610. Note sent to G. de Medici who forwarded it to J. Kepler. In *Le Opere de Galileo Galilei* Vol. 19, p. 612 (Edizione Nationale, 1890–1939).

Gallouet, L. 1964. Magnitude stellaire du soleil. *Ann. d'Astrophys.* 27:423–428.

Garvin, J. B. 1981. Dust clouds observed in Venera 10 panorama of Venusian surface: Inferred surface processes. *Proc. Lunar Planet. Sci.* 12:1493–1505.

Garvin, J. B., Mouginis-Mark, R. J., and Head, J. M. 1981. Characterization of rock populations on planetary surfaces, techniques and a preliminary analysis of Mars and Venus. *Moon Planets* 24:355–387.

Gavrik, A. L., Ivanov-Kholodny, G. S., Mihalov, A. V., Savich, N. A., and Samoznaev, L. N. 1980. The formation of the daytime Venusian ionosphere: The results of dual-frequency occultation experiments. In *Space Research XX* (Oxford: Pergamon), pp. 231–235.

Gebbie, H. A., Delbouille, L., and Roland, G. 1962. The use of a Michelson interferometer to obtain infrared spectra of Venus. *Mon. Not. R. Astron. Soc.* 123:497–500.

Geiss, J. P., Bucler, F., Curutti, H., Eberhardt, R., and Felleuch, Ch. 1972. Apollo 16 Preliminary Science Report, NASA.

Gektin, Yu. M., and Panilov, A. S. 1978. Statistics of the contrasts and relief features of the locality on the panoramas of the surface of Venus. *Kosmich. Issled.* 16:557–562.

Gel'man, B. G., Zolotukhin, V. G., Lamonov, N. I., Levchuk, B. V., Lipatov, A. N., Mukhin, L. M., Nenarokov, D. F., Rotin, V. A., and Okhotnikov, B. P. 1979a. An analysis of the chemical composition of the atmosphere of Venus on an AMS of the Venera-12 using a gas chromatograph. *Kosmich. Issled.* 17:708; *Cosmic Res.* 17:585, 1980.

Gel'man, B. G., Zolotukhin, V. G., Lamonov, N. I., Levchuk, B. V., Mukhin, L. M., Nenarokov, D. F., Okhotnikov, B. P., Rotin, V. A., and Lipatov, A. N. 1979b. Gaschromatographical analysis of chemical composition of the atmosphere of Venus done by the Venera 12 probes. *Pisma Astron. Zh.* 5:217; *NASA Tech. Mem.* TM-75476.

Gel'man, B. G., Zolotukhin, V. G., Mukhin, L. M., Lamonov, N. I., Levchuk, B. V., Nenarokov, D. F., Okhothikov, B. P., Rotin, V. A., and Lipatov, A. N. 1980. Gas chromatograph analysis of the chemical composition of the Venus atmosphere. *Space Res.* 20:219.

Gerard, J.-C. 1981. Observations optiques de Venus a l'aide des sondes spatiales et leur interpretation. *Bull. Acad. Roy. Belgique* (Classe des Sciences) 67:151–170.

Gerard, J.-C., Stewart, A. I. F., and Bougher, S. W. 1981. The altitude distribution of the Venus ultraviolet nightglow and implications on vertical transport. *Geophys. Res. Letters* 8:633–636.

Germogenova, T. A. et al. 1977. *Kosmich. Issled.* 15:5.

Gibson, J. E., and McEwan, R. J. 1959. The brightness temperature of Venus at 8.6 μm. *Paris Sym. on Radio Astronomy.*

Gierasch, P. J. 1970a. The four-day rotation in the stratosphere of Venus: A study of radiative driving. *Icarus* 13:25–33.

Gierasch, P. J. 1970b. The fourth Arizona conference on planetary atmospheres: Motions in planetary atmospheres. *Earth Extrater. Sci.* 1:171–184.

Gierasch, P. J. 1975. Meridional circulation and the maintenance of the Venus atmospheric rotation. *J. Atmos. Sci.* 32:1038–1044.

Gierasch, P. J., Goody, R., and Stone, P. 1970. The energy balance of planetary atmospheres. *Geophys. Fluid Dyn.* 1:1–18.

Gierasch, P. J., Ingersoll, A. P., and Williams, R. T. 1973. Radiative instability of a cloudy planetary atmosphere. *Icarus* 19:473–481.

Gillett, F. C., Low, F. J., and Stein, W. A. 1968. Absolute spectrum of Venus from 2.8 to 14 microns. *J. Atmos. Sci.* 25:594–595.

Gmitro, J. I., and Vermeulen, T. 1964. Vapor-liquid equilibrium for aqueous sulfuric acid. *Amer. Inst. Chem. Eng. J.* 10:740–746.

Goettel, K. A. 1981. Density constraints on the composition of Venus. *Proc. Lunar Planet. Sci. Conf.* 12:347–349.

Goettel, K. A., and Lewis, J. S. 1974. Ammonia in the atmosphere of Venus. *J. Atmos. Sci.* 31:828–830.

Goettel, K. A., Shields, J. A., and Decker, D. A. 1981. Density constraints on the composition of Venus. *Proc. Lunar Planet. Sci. Conf.* 12:1507–1516.

Gold, T., and Soter, S. 1969. Atmospheric tides and the resonant rotation of Venus. *Icarus* 11:356–366.

Gold, T., and Soter, S. 1971. Atmospheric tides and the 4-day circulation on Venus. *Icarus* 14:16–20.

Gold, T., and Soter, S. 1979. Theory of the Earth-synchronous rotation of Venus. *Nature* 277:280–281.

Goldreich, P., and Peale, S. 1970. The obliquity of Venus. *Astron. J.* 75:273–284.

Goldstein, R. M. 1964. Venus characteristics by Earth-based radar. *Astron. J.* 69:12–18.

Goldstein, R. M. 1965. Preliminary Venus radar results. *Radio Sci.* 69D:1623–1625.

Goldstein, R. M., and Rumsey, H. Jr. 1970. A radar snapshot of Venus. *Science* 169: 974–977.

Goldstein, R. M., Green, R. R., and Rumsey, H. C. 1976. Venus radar images. *J. Geophys. Res.* 81:4807–4817.

Goldstein, R. M., Green, R. R., and Rumsey, H. C. 1978. Venus radar brightness and altitude images. *Icarus* 36:334–352.

Golitsyn, G. S. 1970. A similarity approach to the general circulation of planetary atmospheres. *Icarus* 13:1–24.

Golitsyn, G. S. 1973. *Introduction to the Dynamics of Planetary Atmospheres* (Leningrad: Gidrometeoizdat).

Golitsyn, G. S. 1982. Problems of the Venus atmospheric dynamics. *NASA Tech. Mem.* 76816.

Golovin, Yu. M. 1979. The optical properties of the surface of Venus: Dependence of the albedo on the wavelength. *Kosmich. Issled.* 17:473–476.

Golovin, Yu. M. and Ustinov, E. A. 1981. *Kosmich. Issled.* 19.

Golovin, Yu. M. et al. 1977. Preprint Pr-341 K AN SSR.

Golovin, Yu. M. et al. 1978. *Kosmich. Issled.* 16:563.

Golovin, Yu. M. et al. 1981. *Kosmich. Issled* 19:421.

Gombosi, T. I., Cravens, T. E., Nagy, A. F., Elphic, R. C., and Russell, C. T. 1980. Solar wind absorption by Venus. *J. Geophys. Res.* 85:7747–7753.

Gombosi, T. I., Horanyi, M., Cravens, T. E., Nagy, A. F., and Russell, C. T. 1981. The role of charge exchange in the solar wind absorption by Venus. *Geophys. Res. Letters* 8:1265–1268.

Goody, R. M. 1964. *Atmospheric Radiation* (London: Oxford Univ. Press).

Goody, R. M., and Robinson, A. R. 1966. A discussion of the deep circulation of the atmosphere of Venus. *Astrophys. J.* 146:339–355.

Goody, R. M., and Belton, M. J. S. 1967. Radiative relaxation times for Mars: A discussion of Martian atmospheric dynamics. *Planet. Space Sci.* 15:247–256.

Goody, R. M., and Walker, J. C. G. 1972. *Atmospheres* (Englewood Cliffs, N. J.: Prentice Hall).

Gordon, J. J., Jauncey, D. L., and Yerbury, M. J. 1973. The brightness temperature of Venus at 70 cm. *Astrophys. J.* 183:1035–1080.

Grebowsky, J. M., and Curtis, S. A. 1981. Venus nightside ionospheric holes: The signatures of parallel electric field acceleration regions? *Geophys. Res. Letters* 8:1273–1276.

Grechnev, K. V., Istomin, V. G., Ozerov, L. N., and Klimovitskiiy, V. G. 1979. Mass-spectrometer for Venera-11 and -12. *Kosmich. Issled.* 17:697.

Greeley, R. J., Leach, R., and Pollack, J. B. 1980a. Venus: Consideration of aeolian (windblown) processes. *Lunar Planet. Sci.* 11:360–361 (abstract).

Greeley, R. J., White, B. R., Leach, R., Leonard, R., Pollack, J. B., and Iverson, J. D. 1980b. Venus aeolian processes: Saltation studies and the Venusian wind tunnel. *NASA Tech. Mem.* 82385:275–277.

Green, D. H. 1973. Experimental melting studies on a model upper mantle composition at high pressure under water-saturated and water-undersaturated conditions. *Earth. Planet. Sci. Letters* 19:37–53.

Greenstadt, E. W. 1970. Dependence of shock structure at Venus and Mars on the orientation of the interplanetary magnetic field. *Cosmic Electrodyn.* 1:380–388.

Greenstadt, E. W., and Fredricks, R. W. 1979. Shock system in collisionless plasmas. In *Solar System Plasma Physics Vol. III.*, eds. C. F. Kennel, L. J. Lanzerotti, and E. N. Parker (Amsterdam: North-Holland), pp. 3–43.

Greer, R. G. H., Llewellyn, E. J., Solheim, B. H., and Witt, G. 1981. The excitation of O_2 ($b^1\Sigma_g^+$) in the nightglow. *Planet. Space Sci.* 29:383–390.

Griffin, P. H., Thornton, D. D., and Welch, W. L. 1967. The microwave spectrum of Venus in the frequency range 18–36 Gc/sec. *Icarus* 6:175–188.

Gringauz, K. I. 1981. A comparison of the magnetospheres of Mars, Venus and Earth. *Adv. Space Res.* 1:5–24.

Gringauz, K. I., and Breus, T. K. 1969. Comparison of ionospheric features of planets in the terrestrial group: Mars, Venus, and Earth. *Kosmich. Issled.* 7:871–890.

Gringauz, K. I., Bezrukikh, V. V., Musatov, L. S., and Breus, T. K. 1968. Plasma measurements conducted in the viscinity of Venus by the spacecraft Venera 4. *Kosmich. Issled.* 5:411–419.

Gringauz, K. I., Bezrukikh, V. V., Volkov, G. I., Musatov, L. S., and Breus, T. K. 1970. Interplanctary plasma disturbances near Venus determined from Venera 4 and Venera 6 data. *Kosmich. Issled.* 8:431–436.

Gringauz, K. I., Bezrukikh, V. V., Breus, T. K., Gombosi, T., Remizov, A. P., Verigin, M. I., and Volkov, G. I. 1976a. Plasma observations near Venus onboard the Venera 9 and 10 satellites by means of wide angle plasma detectors. In *Physics of Solar Planetary Environments,* ed. D. J. Williams (Washington, D. C.: Amer. Geophys. Union), pp. 918–932.

Gringauz, K. I., Bezrukikh, V. V., Breus, T. K., Verigin, M. I., Volkov, G. I., Remizov, A. P., and Sluchendov, G. F. 1976b. Preliminary results of plasma measurements near Venus using the Venera-9 orbiter. *Pisma Astron. Zh.* 2:82–87.

Gringauz, K. I., Verigin, M. I., Breus, T. K., and Gombosi, T. 1977. Elektronnye potoki izmerennye v opticheskoi teni Venery na sputnikakh "Venera-9" i "Venera-10"—osnovnoi istochnik ionizatsii v nochnoi ionosfere Venery. *Doklady Adademii Nauk SSR* 232:1039–1042.

Gringauz, K. I., Verigin, M. I., Breus, T. K., and Gombosi, T. 1979. The interaction of electrons in the optical umbra of Venus with the planetary atmosphere—the origin of the nightside ionosphere. *J. Geophys. Res.* 84:2123–2127.

Guinot, B., and Feissel, M. 1968. Spectrographic measurement of movements in the atmosphere of Venus. *J. Observat.* (Marseille) 51, N1:13–20.

Gulkis, S., Kakar, R. K., Klein, M. J., Olsen E. T., and Wilson, W. J. 1977. Venus: Detection of variations in stratospheric carbon monoxide. In *Proc. of the Symp. on Planetary Atmospheres,* ed. A. V. Jones (Ottawa, Ontario: Roy. Soc. Canada), pp. 61–65.

Gull, T. R., O'Dell, C. R., and Parker, R. A. R. 1974. Water vapor in Venus determined by airborne observations of the 8200 Å band. *Icarus* 21:213–218.

Hagfors, T. 1970. Remote probing of the moon by infrared and microwave emissions and by radar. *Radio Sci.* 5:219–227.

Hagfors, T., and Evans, J. V., eds. 1968. Radar studies of the moon. In *Radar Astronomy* (New York: McGraw-Hill), pp. 219–273.

Halley, E. 1686. An historical account for the trade winds and monsoons, etc. *Phil. Trans. Roy. Soc.* 16:153–158.

Halley, E. 1716. Methodus singularis qua solis parallaxis sive distantia a terra, ope Veneris intra solem conspiciendae, tuto determinari poterit. *Phil. Trans.* 29:454–464.

Hamill, P. Turco, R. P., Toon, O. B., Kiang, C. S., and Whitten, R. C. 1982. On the formation of sulfate aerosol particles in the stratosphere. *J. Aerosol Sci.* In press.

Handbook of Basics of Radioelectronics Theory 1977. eds. B. H. Krwizky and V. N. Dubin (Moscow: Energy Publ. House). In Russian.

Handbook of Geophysics 1960 (New York: McMillan).

Hanel, R., Conrath, B., Flasar, M., Kunde, V., Lowman, P., Maguire, W., Pearl, J., Pirraglia, J., Samuelson, R., Gautier, D., Gierasch, P., Kumar, S., and Ponnamperuma, C. 1979. Infrared observations of the Jovian system from Voyager 1. *Science* 204: 972–976.

Hansen, J. E., and Matsushima, S. 1967. The atmosphere and surface temperature of Venus: A dust insulation model. *Astrophys. J.* 150:1139–1157.

Hansen, J. E., and Arking, A. 1971. Clouds of Venus: Evidence for their nature. *Science* 171:669–672.

Hansen, J. E., and Hovenier, J. W. 1974. Interpretation of the polarization of Venus. *J. Atmos. Sci.* 31:1137–1160.

Hansen, J. E., and Travis, L. D. 1974. Light scattering in planetary atmospheres. *Space Sci. Rev.* 16:527–610.

Hapke, B., and Van Horn, H. 1963. Photometric studies of complex surfaces with applications to the moon. *J. Geophys. Res.* 68:4545–4570.

Hapke, B., and Nelson, R. 1975. Evidence for an elemental sulfur component of the clouds from Venus spectroscopy. *J. Atmos. Sci.* 32:1212–1218.

Hara, T. 1976. Lightning in the planetary atmospheres and in the primordial solar nebula. Preprint, Kyoto Univ.

Hargraves, R. B. 1978. Punctuated evolution of tectonic style. *Nature* 276.159 161

Harker, R. I., and Tuttle, O. F. 1956. Experimental data of the P_{CO_2}-T curve for the reaction: Calcite + quartz $\rightleftarrows$ wollastonite + CO_2. *Amer. J. Sci.* 254:239–249.

Hart, R., Dymond, J., and Hogan, L. 1979. Preferential formation of the atmosphere-sialic crust system from the upper mantle. *Nature* 278:156–159.

Hartle, R. E. 1981. Ion composition of the nightside ionospheric holes of Venus. "Internat. Conf. on the Venus Environment," held at Palo Alto, CA.

Hartle, R. E., Taylor, H. A. Jr., Bauer, S. J., Brace, L. H., Russell, C. T., and Daniell, R. E. 1980. Dynamical response of the dayside ionosphere of Venus to the solar wind. *J. Geophys. Res.* 85:7739–7746.

Hasegawa, A. 1975. *Plasma Instabilities and Nonlinear Effects* (New York: Springer-Verlag).

Haurwitz, B. 1964. Atmospheric tides. *Science* 144:1415–1422.

Hayashi, Y., and Golder, D. G. 1978. The generation of equatorial transient planetary waves: Control experiments with a GFDL general circulation model. *J. Atmos. Sci.* 35:2068–2082.

Hayashi, C., Nakazawa, K., and Mizuno, H. 1979. Earth's melting due to the blanketing effect of the primordial dense atmosphere. *Earth Planet. Sci. Letters* 43:22–28.

Head, J. W., Yuter, S. E., and Solomon, S. G. 1981. Topography of Venus and Earth: A test for the presence of plate tectonics. *Amer. Sci.* 69:614–623.

Hedin, A. E., Niemann, H. B., Kasprzak, W. T., and Seiff, A. 1983. Global empirical model of the Venus thermosphere. *J. Geophys. Res.* In press.

Heestand, R. L., and Crough, S. T. 1981. The effect of hot spots on the oceanic age-depth relation. *J. Geophys. Res.* 86:6107–6114.

Heicklen, J. 1976. *Atmospheric Chemistry* (New York: Academic).

Held, I. M., and Hou, A. Y. 1980. Nonlinear axially symmetric circulation in a nearly inviscid atmosphere. *J. Atmos. Sci.* 37:515–533.

Herman, J. R., Hartle, R. E., and Bauer, S. J. 1971. The dayside ionosphere of Venus. *Planet. Space Sci.* 19:443–460.

Herzberg, G. 1949. Laboratory absorption spectra obtained with long paths. In *The Atmospheres of the Earth and Planets,* ed. G. P. Kuiper (Chicago, IL.: Univ. Chicago Press), pp. 406–414.

Herzog, G. F., and Anders, E. 1974. Primordial noble gases in separated meteoritic minerals, II. *Earth Planet. Sci. Letters* 24:173–181.

Herzog, G. F., Gibson, E. K., and Lipschutz, M. E. 1979. Thermal metamorphism of primitive meteorites—VIII. Noble gases, carbon and sulfur in Allende (C3) meteorite heated at 400–1000°C. *Geochim. Cosmochim.* 43:395–404.

Hess, H. H. 1962. History of ocean basins. In *Petrologic Studies: A volume in Honor of A. F. Buddington,* eds. A. E. J. Engel, H. L. James, and B. L. Leonard (Geol. Soc. Amer.), pp. 599–620.

Hess, S. L. 1968. The hydrodynamics of Mars and Venus. In *The Atmospheres of Mars and Venus,* eds. J. C. Brandt and M. B. McElroy (New York: Gordon and Breach), pp. 109–131.

Hide, R. 1969. Dynamics of the atmospheres of the major planets with an appendix on the viscous boundary layer at the rigid bounding surface of an electrically-conducting rotating fluid in the presence of a magnetic field. *J. Atmos. Sci.* 26:841–853.

Hide, R., Birch, N. T., Morrison, L. V., Shea, D. J., and White, A. A. 1980. Atmospheric angular momentum fluctuations and changes in the length of the day. *Nature* 286:114–117.

Hinch, E. J., and Schubert, G. 1971. Strong streaming induced by a moving thermal wave. *J. Fluid Mech.* 47:291–304.

Hines, C. O. 1960. Internal atmospheric gravity waves at ionospheric heights. *Can. J. Phys.* 38:1441–1481.

Hines, C. O. 1965. Dynamical heating of the upper atmosphere. *J. Geophys. Res.* 70: 177–183.

Hinteregger, H. E., Bedo, D. E., and Manson, J. E. 1973. The EUV spectrophotometer on Atmosphere Explorer. *Radio Sci.* 8:349–359.

Hinteregger, H. E., Bedo, D. E., Manson, J. E., and Skillman, D. R. 1977. EUV flux variations with solar rotation observed during 1974–1976 from the AE-C satellite. *Space Res.* 17:533–544.

Hinteregger, H. E., Fukui, K., and Gilson, B. R. 1982. Observational, reference, and model data on solar EUV from measurements of AE-E. *Geophys. Res. Letters.* In press.

Hochstim, A. R., and Massel, G. D. 1969. Calculations of transport coefficients in ionized gases. In *Kinetic Processes in Gases and Plasmas,* ed. A. R. Hochstim (New York: Academic), pp. 141–255.

Hodges, R. R. Jr., and Tinsley, B. A. 1981. Charge exchange in the Venus ionosphere as the source of the hot exospheric hydrogen. *J. Geophys. Res.* 86:7649–7656.

Hoegy, W. R., Brace, L. H., Theis, R. F., and Mayr, H. G. 1980. Electron temperature and heat flow in the nightside Venus ionosphere. *J. Geophys. Res.* 85:7811–7816.

Hoffman, J. H., Hodges, R. R. Jr., and Duerksen, K. D. 1979a. Pioneer Venus large probe neutral mass spectrometer. *J. Vacuum Sci. Tech.* 16:692–694.

Hoffman, J. H., Hodges, R. R. Jr., McElroy, M. B., and Donahue, T. M. 1979b. Venus lower atmospheric composition: Preliminary results from Pioneer Venus. *Science* 203:800–802.

Hoffman, J. H., Hodges, R. R. Jr., McElroy, M. B., Donahue, T. M., and Kolpin, M. 1979c. Composition and structure of the Venus atmosphere: Results from Pioneer Venus. *Science* 205:49–52.

Hoffman, J. H., Hodges, R. R. Jr., Donahue, T. M., and McElroy, M. B. 1980a. Composition of the Venus lower atmosphere from the Pioneer Venus mass spectrometer. *J. Geophys. Res.* 85:7882–7890.

Hoffman, J. H., Hodges, R. R. Jr., Wright, W. W., Blevins, V. A., Duerksen, K. D., and Brooks, L. D. 1980b. Pioneer Venus sounder probe neutral gas mass spectrometer. *IEEE Trans. Geosci. and Remote Sensing* GE-18:80–84.

Hoffman, J. H., Oyama, V. I., and von Zahn, U. 1980c. Measurements of the Venus lower atmosphere composition: A comparsion of results. *J. Geophys. Res.* 85:7871–7881.

Holland, H. D. 1978. *The Chemistry of the Atmosphere and Oceans* (New York: Wiley).

Holton, J. R. 1972. Waves in the equatorial stratosphere generated by tropospheric heat sources. *J. Atmos. Sci.* 29:368–375.

Holton, J. R. 1973. A note on the frequency distribution of atmospheric Kelvin waves. *J. Atmos. Sci.* 30:499–501.

Holton, J. R. 1975. *The Dynamic Meteorology of the Stratosphere and Mesosphere,* Meteorol. Monogr. 15, No. 3, (Boston, MA: Amer. Meteorol. Soc.).

Holton, J. R. 1979. *An Introduction to Dynamic Meteorology,* 2nd Ed. (New York: Academic).

Holton, J. R., and Lindzen, R. S. 1972. An updated theory for the quasi-biennial cycle of the tropical stratosphere. *J. Atmos. Sci.* 29:1076–1080.

Hoppe, M. M., and Russell, C. T. 1982. Particle acceleration at planetary bow shock waves. *Nature* 295:41–42.

Hoppe, M. M., Russell, C. T., Frank, L. A., Eastman, T. E., and Greenstadt, E. W. 1981. Upstream hydromagnetic waves and their association with backstreaming populations: ISEE 1 and 2 observations. *J. Geophys. Res.* 86:4471–4492.

Horak, H. G. 1950. Diffuse reflection by planetary atmospheres. *Astrophys. J.* 112:445–463.

Horedt, G. P. 1979. Cosmogony of the solar system. *Moon Planets* 21:63–121.

Houben, H. C. 1982. Radiative-dynamic instability in the atmosphere of Venus. *Icarus.* In press.

Houghton, J. T. 1977. *The Physics of Atmospheres* (New York: Cambridge Univ. Press).

Houghton, J. T., and Taylor, F. W. 1975. On remote sounding of the upper atmosphere of Venus. *J. Atmos. Sci.* 32:620–629.

Hovis, W. A. Jr. 1965. Infrared reflectivity of iron oxide minerals. *Icarus* 4:425–430.

Howard, H. T., Tyler, G. L., Fjeldbo, G., Kliore, A. J., Levy, G. S., Brunn, D. L., Dickinson, R., Edelson, R. E., Martin, W. L., Postal, R. B., Seidel, B., Sesplaukis, T. T., Shirley, D. L., Stelzried, C. T., Sweetnam, D. N., Zygielbaum, A. E., Esposito,

P. B., Anderson, J. D., Shapiro, I. I., and Reasenberg, R. D. 1974. Venus mass, gravity field, atmosphere and ionosphere as measured by Mariner 10 dual-frequency radio system. *Science* 183:1297–1301.

Huber, P. J. 1977. Early cuneiform evidence for the existence of the planet Venus. In *Scientists Confront Velikovsky,* ed. D. Goldsmith (Ithaca, NY: Cornell Univ. Press), pp. 117–144.

Hudson, D., Flynn, G. J., Froundorf, P., Hohenberg, C. M., and Shirck, J. 1981. Noble gases in stratospheric dust particles: Confirmation of extraterrestrial origin. *Science* 211:383–386.

Hughes, M. P. 1965. Planetary observations at a wavelength of 6 cm. *Planet. Space Sci.* 14:1017–1022.

Hunt, G. R., and Ross, H. P. 1967. A bidirectional reflectance accessory for spectroscopic measurements. *Appl. Optics* 6:1687–1690.

Hunt, G. R., and Moore, P. 1981. *Jupiter* (Chicago: Rand McNally).

Hunten, D. M. 1971. Composition and structure of planetary atmospheres. *Space Sci. Rev.* 12:539–599.

Hunten, D. M. 1973. The escape of light gases from planetary atmospheres. *J. Atmos. Sci.* 30:1481–1494.

Hunten, D. M. 1974. Energetics of thermospheric eddy transport. *J. Geophys. Res.* 79:2533–2534.

Hunten, D. M. 1975a. Vertical transport in atmospheres. In *Atmospheres of Earth and Planets,* ed. B. M McCormac (Dordrecht: D. Reidel), pp. 59–72.

Hunten, D. M. 1975b. The atmosphere of Venus: Conference review. *J. Atmos. Sci.* 32:1262–1265.

Hunten, D. M. 1982. Thermal and nonthermal escape mechanisms for terrestrial bodies. *Planet. Space Sci.* 30:773–783.

Hunten, D. M., and Donahue, T. M. 1976. Hydrogen loss from the terrestrial planets. *Ann. Rev. Earth Planet. Sci.* 4:265–292.

Hunten, D. M., and Dessler, A. J. 1977. Soft electrons as a possible heat source for Jupiter's thermosphere. *Planet. Space Sci.* 25:817–821.

Hunten, D. M., McGill, G. E., and Nagy, A. F. 1977. Current knowledge of Venus. *Space Sci. Rev.* 20:265–282.

Huygens, C. 1967. Observations astronomiques systeme d'Saturne. Trauxaux astronomique, 1658–1666. In *Complete Works of Christian Huygens,* Vol. 15. Societe Hollandaise des Sciences.

IEEE Transactions on Geosciences and Remote Sensing, GE-18, 1980. PV Issue, January.

Imyonitov, I. M., Chubarina, Je. V., and Shwartz, Ja. M. 1971. *Electricity of Clouds* (Leningrad: Hydrometeorology Publ. House). In Russian.

Ingersoll, A. P. 1969. The runaway greenhouse: A history of water on Venus. *J. Atmos. Sci.* 26:1191–1198.

Ingersoll, A. P., and Orton, G. S. 1974. Lateral inhomogeneities in the Venus atmosphere: Analysis of thermal infrared maps. *Icarus* 21:121–128.

Ingersoll, A. P., and Dobrovolskis, A. R. 1978. Venus rotation and atmospheric tides. *Nature* 275:37–38.

Intriligator, D. S., Collard, H. R., Mihalov, J. D., Whitten, R. C., and Wolfe, J. H. 1979. Electron observations and ion flows from the Pioneer Venus Orbiter plasma analyzer experiment. *Science* 205:116–119.

Irvine, W. M. 1968. Monochromatic phase curves and albedos for Venus. *J. Atmos. Sci.* 25:610–616.

Isakov, M. N. 1979. Inertnye gasy v atmosferakh Venery, Zemli i Marsa i vopros o proiskhozhdenii planetnykh atmosfer. *Kosmich. Issled.* 17:602.

Isakov, M. N. 1980. Letuchie veshchestva v meteoritakh v protoplanetnoi tumannosti i obrazovanie planetnykh atmosfer. *Kosmich. Issled.* 18:918.

Isakov, M. N. 1981.*Kosmich. Issled.* 19:N6.

Istomin, V. G., Grechnev, K. V., and Kochnev, V. A. 1979a. Mass spectrometer measurements of the composition of the lower atmosphere of Venus. *Pisma Astron. Zh.* 5:211–216.

Istomin, V. G., Ksanfomality, L. V., Moroz, V. I., Mukhiu, L. M., Narimanov, G. S., and Perminov, V. G. 1979b. Venera-11 and -12 new scientific results. "COSPAR Meeting," held at Bangalore, India.

Istomin, V. G., Grechnev, K. V., Kochnev, V. A., and Ozerov, L. N. 1979c. Composition of low atmosphere of Venus according to mass-spectrometer data. Kosmich. Issled. 17:703.

Istomin, V. G., Grechnev, K. V., and Kochnev, V. A. 1980a. Mass spectrometer measurements of the lower atmosphere of Venus. Space Res. 20:215–218.

Istomin, V. G., Grechnev, K. V., and Kochnev, V. A. 1980b. Mass spectrometry of the lower atmosphere of Venus: Krypton isotopes and other recent results of the Venera-11 and -12 data processing. "23rd COSPAR meeting," held at Budapest, Hungary; also preprint D-298, Space Res. Inst., Acad. of Sci., USSR.

Istomin, V. G., Grechnev, K. V., and Kochnev, V. A. 1982. Preliminary results of mass-spectrometric measurements on board the Venera 13 and Venera 14 probes. Pisma Astron. Zh. 8:391–398.

Ivanov-Kholodny, G. S., Kolosov, M. A., Savich, N. A., Aleksandrov, Yu. N., Vasil'yev, M. B., Vyshlov, A. S., Dubrovin, V. M., Zaytsev, A. L., Samovol, V. A., Samoznayev, L. N., Sidorenko, I. A., Khasyanov, A. F., and Shtern, D. Ya. 1977. The ionosphere of Venus according to the data of the orbiters Venera 9 and Venera 10 and certain peculiarities in its formation. Uspekhi. Fizich. Nank. 123.

Ivanov-Kholodny, G. S., Kolosov, M. A., Savich, N. A., Aleksandrov, Yu. N., Vasil'yev, M. B., Vyshlov, A. S., Dubrovin, V. M., Zaytsev, A. L., Michailov, A. V., Petrov, G. M., Samovol, V. A., Samoznayev, L. N., Sidorenko, I. A., and Hasyznov, A. F. 1979. Daytime ionosphere of Venus as studied with Venera 9 and 10 dual-frequency radio occultation experiments. Icarus 39:209–213.

Iversen, J. D., Greeley, R. J., and Pollack, J. B. 1976a. Windblown dust on Earth, Mars, and Venus. J. Atmos. Sci. 33:2425–2429.

Iversen, J. D., Pollack, J. B., Greeley, R. J., and White, B. R. 1976b. Saltation threshold on Mars: The effect of interparticle force, surface roughness and low atmospheric density. Icarus 29:381–393.

Izakov, M. N. 1978. Influence of turbulence on the thermal conditions of planetary thermospheres. Cosmic Res. 16:324–331.

Jacchia, L. G. 1977. Thermospheric temperature, density, and composition: New models. Smithsonian Astrophys. Obs. Special Rept. No. 375.

Jackson, G. E., and Calvert, J. G. 1971. The triplet sulfur dioxide-carbon monoxide reaction excited within the $SO_2(^1A_1)-SO_2(^3B_1)$ forbidden band. J. Amer. Chem. Soc. 93:2593–2599.

JANAF Thermochemical Tables 1978 (Midland, MI: Dow Chemical Co.).

Janssen, M. A., and Klein, M. J. 1981. Constraints on the composition of the Venus atmosphere from microwave measurements near 1.35 cm wavelengths. Icarus 46:58–69.

Janssen, M. A., Hills, R. E., Thornton, D. D., and Welch, W. J. 1973. Venus: New microwave measurements show no atmospheric water vapor. Science 179:994–997.

Jastrow, R., and Rasool, S. I. 1969. The Venus Atmosphere (New York: Gordon and Breach).

Johnson, F. S. 1975. Transport processes in the upper atmosphere. J. Atmos. Sci. 32:1658–1662.

Johnson, F. S., and Hanson, W. B. 1979. A new concept for the daytime magnetosphere of Venus. Geophys. Res. Letters 6:581–584.

Johnson, R. E. 1978. Comment on ion and electron temperatures in the Martian upper atmosphere. Geophys. Res. Letters 5:989–992.

Jones, D. E. 1961. The microwave temperature of Venus. Planet. Space Sci. 5:166–167.

Journal of Geophysical Research 1980 Vol. 85 (Special Venus issue).

Jurgens, R. F., Goldstein, R. M., Rumsey, H. R., and Green, R. R. 1980. Images of Venus by three-station radar interferometry: 1977 results. J. Geophys. Res. 85:8282–8294.

Kakar, R. K., Waters, J. W., and Wilson, W. J. 1976. Venus: Microwave detection of carbon monoxide. Science 191:379–380.

Kalaghan, P. M., Wulfsberg, K. N., and Telford, L. E. 1969. Observations of the phase effect of Venus at 8.6 mm wavelength. Astron. J. 73:970–971.

Kalnay de Rivas, E. 1973. Numerical models of the circulation of the atmosphere of Venus. *J. Atmos. Sci.* 31:763–779.

Kalnay de Rivas, E. 1975. Further numerical calculations of the circulation of the atmosphere of Venus. *J. Atmos. Sci.* 32:1017–1024.

Karjagina, Z. V. 1955. *Izv. Astrophys. Inst. AN Kaz. SSR* 1–2:85.

Kasprzak, W. T., Hedin, A. E., Niemann, H. B., and Spencer, N. W. 1980. Atomic nitrogen in the upper atmosphere of Venus. *Geophys. Res. Letters* 7:106–108.

Kassal, T. 1975. Resonant fluorescent scattering of solar radiation by the fourth positive band system of CO. *Appl. Optics* 14:1513–1515.

Kaula, W. M. 1975. The seven ages of a planet. *Icarus* 26:1–15.

Kaula, W. M. 1981. The beginning of the Earth's thermal evolution. In *The Continental Crust and its Mineral Deposits,* ed. D. W. Strangway (Geol. Assoc. Canada), Special Paper 20:25–34.

Kaula, W. M., and Phillips, R. J. 1981. Quantitative tests for plate tectonics on Venus. *Geophys. Res. Letters* 8:1187–1190.

Kaula, W. M., and Muradian, L. 1982. Is plate tectonics on Venus concealed by volcanism? *Geophys. Res. Letters* 9:1021–1024.

Kawabata, K. D., Coffeen, D. L., Hansen, J. E., Lane, W. A., Sato, M., and Travis, L. D. 1980. Cloud and haze properties from Pioneer Venus polarimetry. *J. Geophys. Res.* 85:8129–8140.

Kawabata, K. D., Sato, M., and Travis, L. D. 1982. Polarimetric determination of aerosol properties and variation of haze and cloud structure on Venus. Submitted to *Icarus.*

Keating, G. M., and Prior, E. J. 1968. The winter helium bulge. *Space Res.* 8:982–992.

Keating, G. M., Mullins, J. A., and Prior, E. J. 1971. Simultaneous measurements of exospheric densities near opposite poles. *Space Res.* 11:987–994.

Keating, G. M., Prior, E. J., Levine, J. S., and Mullins, J. A. 1972. Seasonal variations in the thermosphere and exosphere: 1968–1970. *Space Res.* 12:765–775.

Keating, G. M., Taylor, F. W., Nicholson, J. Y., and Hinson, E. W. 1979a. Short-term cyclic variations and diurnal variations of the Venus upper atmosphere. *Science* 205:62–64.

Keating, G. M., Tolson, R. H., and Hinson, E. W. 1979b. Venus thermosphere and exosphere: First satellite drag measurements of an extraterrestrial atmosphere. *Science* 203:772–774.

Keating, G. M., Nicholson, J. Y., and Lake, L. R. 1980. Venus upper atmosphere structure. *J. Geophys. Res.* 85:7941–7956.

Keldysh, M. Y. 1977. Venus exploration with the Venera 9 and Venera 10 spacecraft. *Icarus* 30:605–625.

Keller, G. V. 1966. Electrical properties of rocks and minerals. In *Handbook of Physical Constants,* ed. S. P. Clark, Jr. (Geol. Soc. Amer. Mem.) 97:553–577.

Kellerman, K. I. 1966. The thermal radio emission from Mercury, Venus, Mars, Saturn, and Uranus. *Icarus* 5:487–490.

Kellerman, K. I., Pauliny-Toth, I. I. K., and Williams, P. J. S. 1969. The spectra of radio sources in the revised 3C Catalogue. *Astrophys. J.* 157:1–34.

Kellogg, W. W., and Sagan, C. 1961. The atmospheres of Mars and Venus. *Nat. Acad. Sci.* Publ. No. 944.

Kemurdzhian, A. L., Gromov, V. V., and Shvarev, V. V. 1978. Investigation of the physico-mechanical properties of extraterrestrial grounds. In *The Achievements of Soviet Science in the Exploration of Outer Space* (Moscow: Nauka), pp. 352–380.

Kenner, R. D., Ogryzlo, E. A., and Turley, S. 1979. On the excitation of the night airglow on Earth, Venus, and Mars. *J. Photochem.* 10:199–203.

Kerzhanovich, V. V., and Marov, M. Ya. 1977. On the wind velocity measurements from Venera spacecraft data. *Icarus* 30:320–325.

Kerzhanovich, V. V., and Marov, M. Ya. 1982. The atmospheric dynamics of Venus according to Doppler measurements aboard the Venera automatic interplanetary stations. *NASA Tech. Mem.* 76818.

Kerzhanovich, V. V., Gotlib, V. M., Chetyrkin, N. V., and Andreyev, B. N. 1969. Results of the determination of the atmospheric dynamics of Venus from measurement data on the radial speed of the automatic interplanetary station Venera 4. *Kosmich. Issled.* 7:592–596.

Kerzhanovich, V. V., Marov, M. Ya., and Rozhdestvensky, M. K. 1972. Data on dynamics of the subcloud Venus atmosphere from Venera space probes measurements. *Icarus* 17:659–674.

Kerzhanovich, V. V., Kolzov, F. I., Selivanov, A. S., Tyuflin, Yu. S., and Khizhnichenko, V. I. 1979a. Structure of venusian cloud layer according to Venera 9 televised pictures. *Cosmic Res.* 17:57–66.

Kerzhanovich, V. V., Makarov, Yu. F., Marov, M. Ya., Rozhdestvensky, M. K., and Sorokin, V. P. 1979b. Venera 11 and Venera 12: Preliminary estimates for the wind speed and turbulence in the atmosphere of Venus. *NASA Tech. Note* N80-23240; *Moon Planets* 23:261–270.

Kerzhanovich, V. V., Makarov, Yu. F., Marov, M. Ya., Molotov, E. P., Rozhdestvensky, M. K., Sorokin, V. P., Antsibor, N. M., Kustodiev, V. D., and Puchkov, V. I. 1979c. An estimate of the wind velocity and turbulence in the atmosphere of Venus on the basis of reciprocal Doppler measurements by the Venera 11 and Venera 12 spacecraft. *Cosmic Res.* 17:569–575.

Kerzhanovich, V. V., Makarov, Yu. F., Molotov, E. P., Sorokin, V. P., Sukhanov, K. G., Antsibor, N. M., Kustodiev, V. D., Matsygorin, I. A., and Tikhonov, V. F. 1982. Estimates of wind velocity from Doppler measurements on Venera 13 and Venera 14 spaceprobes. First results. *Pisma Astron. Zh.* 8:414–418.

Kharitonov, V. A. et al. 1978. *Svodny Spektrophotometricheskiji Kataloy Zvezd.* (Alma-Ata: Nauka Kaz SSR).

Khodakovsky, I. L. 1982. Atmospheric and surface interactions on Venus and implications for atmospheric evolution. *Planet. Space Sci.* 30:803–817.

Khodakovsky, I. L., Volkov, V. P., Sidorov, I. Yu., Borisov, M. V., and Lomonosov, M. V. 1979. Venus: Preliminary prediction of the mineral composition of surface rocks. *Icarus* 39:352–363.

King, J. I. F. 1960. Probe observations of Venus at close range. *Johns Hopkins Univ. Rept.* ed. J. Strong.

Kliore, A. J., and Patel, I. R. 1980. The vertical structure of the atmosphere of Venus from Pioneer Venus orbiter radio occultations. *J. Geophys. Res.* 85:7957–7962.

Kliore, A. J., and Patel, I. R. 1982. Thermal structure of the atmosphere of Venus from Pioneer Venus radio occultations. *Icarus* Vol. 52, No. 2.

Kliore, A. J., Levy, G. S., Cain, D. L., Fjeldbo, G., and Rasool, S. I. 1967. Atmosphere and ionosphere of Venus from the Mariner 5 S-band radio occultation measurement. *Science* 158:1683–1688. Reproduced in *The Venus Atmosphere*, eds. R. Jastrow and S. I. Rasool (New York: Gordon and Breach), pp. 105–124.

Kliore, A. J., Cain, D. L., Levy, G. S., Fjeldbo, G., and Rasool, S. I. 1969. Structure of the atmosphere of Venus derived from Mariner 5 S-band measurements. In *Space Research IX* (Amsterdam: North-Holland), pp. 712–729.

Kliore, A. J., Elachi, C., Patel, I. R., and Cimino, J. B. 1979a. Liquid content of the lower clouds of Venus as determined from Mariner 10 radio occultation. *Icarus* 37:51–72.

Kliore, A. J., Patel, I. R., Nagy, A. F., Cravens, T. E., and Gombosi, T. I. 1979b. Initial observations of the nightside ionosphere of Venus from Pioneer Venus orbiter radio occultations. *Science* 205:99–102.

Kliore, A. J., Woo, R., Armstrong, J. W., and Patel, I. R. 1979c. The polar ionosphere of Venus near the terminator from early Pioneer Venus orbiter occultations. *Science* 203:765–768.

Knollenberg, R. G., and Hunten, D. M. 1979a. Clouds of Venus: Particle size distribution measurements. *Science* 203:792–795.

Knollenberg, R. G., and Hunten, D. M. 1979b. Clouds of Venus: A preliminary assessment of microstructure. *Science* 205:70–74.

Knollenberg, R. G., and Gilland, J. R. 1980. Pioneer Venus sounder probe particle size spectrometer. *IEEE* GE-18: 100–104.

Knollenberg, R. G., and Hunten, D. M. 1980. Microphysics of the clouds of Venus: Results of the Pioneer Venus particle size spectrometer experiment. *J. Geophys. Res.* 85:8039–8058.

Knollenberg, R. G., Hansen, J., Ragent, B., Martonchik, J., and Tomasko, M. 1977. The clouds of Venus. *Space Sci. Rev.* 20:329–354.

Knollenberg, R. G., Travis, L., Tomasko, M., Smith, P., Ragent, B., Espositio, L., McCleese, D., Martonchik, J., and Beer, R. 1980. The clouds of Venus: A synthesis report. *J. Geophys. Res.* 85:8059–8081.

Knollenberg, R. G., Dasher, A. J., and Huffman, D. 1982. Measurement of the aerosol and ice crystral populations in tropical stratospheric cumulonimbus anvils. *Geophys. Res. Letters* 9:613–616.

Knuckles, C. F., Sinton, M. K., and Sinton, W. M. 1961. UBV photometry of Venus. *Lowell Obs. Bull.* 5:153–156.

Knudsen, W. C., Spenner, K., Whitten, R. C., Spreiter, J. R., Miller, K. L., and Novak, V. 1979a. Thermal structure and energy influx to the day and nightside Venus ionosphere. *Science* 205:105–107.

Knudsen, W. C., Spenner, K., Whitten, R. C., Spreiter, J. R., Miller, K. L., and Novak, V. 1979b. Thermal structure and major ion composition of the Venus ionosphere: First RPA results from Venus orbiter. *Science* 203:757–763.

Knudsen, W. C., Spenner, K., Michelson, P. F., Whitten, R. C., Miller, K. L., and Novak, V. 1980a. Suprathermal electron energy distribution within the dayside Venus ionosphere. *J. Geophys. Res.* 85:7754–7758.

Knudsen, W. C., Spenner, K., Miller, K. L., and Novak, V. 1980b. Transport of ionospheric O^+ ions across the Venus terminator and implications. *J. Geophys. Res.* 85:7803–7810.

Knudsen, W. C., Spenner, K., Whitten, R. C., and Miller, K. L. 1980c. Ion energetics in the Venus nightside ionosphere. *Geophys. Res. Letters* 7:1045–1048.

Knudsen, W. C., Spenner, K., and Miller, K. L. 1981. Antisolar acceleration of ionospheric plasma across the Venus terminator. *Geophys. Res. Letters* 8:241–244.

Koenig, L. R., Murray, F. W., Michaux, C. M., and Hyatt, H. A. 1967. *Handbook of the Physical Properties of the Planet Venus* (Washington, D. C.: NASA SP-3029).

Kohlstedt, D. L., and Goetze, C. 1974. Low-stress high-temperature creep in olivine single crystals. *J. Geophys. Res.* 79:2045–2051.

Kolosov, M. A., Yakovlev, O. I., Efimov, A. I., Matyugov, S. S., Timofeeva, T. S., Chub, E. V., Pavelyev, A. G., Kucheriavenkov, A. I., Kalashnikov, I. E., and Milekhin, O. E. 1980. Investigation of the Venus atmosphere and surface by the method of radio sounding using Venera 9 and 10 satellites. *Acta Astron.* 7:219–234.

Kolosov, M. A., Yakovlev, O. I., Pavelyev, A. G., Kucheriavenkov, A. I., and Milekhin, O. E. 1981. Characteristics of the surface and features of the propagation of radio waves in the atmosphere of Venus from data of bistatic radiolocation experiments using Venera 9 and 10 satellites. *Icarus* 48:188–200.

Kondrat'yev, V. N. 1971. *The Rate Constants of Gas-Phase Reactions* (Moscow: Nanka).

Kong, T. Y., and McElroy, M. B. 1977. Photochemistry of the martian atmosphere. *Icarus* 32:168–189.

Korol'kov, D. V., Pariyskiy, Yu. N., Timofeeva, Yu. N., and Kharykiu, S. E. 1963. Higher resolution radio astronomy observations of Venus. *Dokl. AN SSSR* 149:65–67.

Kozyrev, N. A. 1954. On the light of the night sky of Venus. *Izv. Krymskoi. Astrofiz. Obs.* 12:169–176.

Krasnopol'sky, V. A. 1978. Upper limits on some components of Mars' and Venus' atmospheres from the results of twilight airglow spectroscopy onboard Mars 5 and Venera 9, 10 orbiters. *Cosmic Res.* 16:895–900.

Krasnopol'sky, V. A. 1979a. Nightside ionosphere of Venus. *Planet. Space Sci.* 27:1403–1408.

Krasnopol'sky, V. A. 1979b. Venera 9, 10: Spectroscopy of scattered radiation in overcloud atmosphere. *Cosmic Res.* 18:899–906.

Krasnopol'sky, V. A. 1980. On lightnings in the Venus' atmosphere according to the Venera 9 and 10 data. *Cosmic Res.* 18:429–434.

Krasnopol'sky, V. A. 1981. Excitation of oxygen emissions in the night airglow of the terrestrial planets. *Planet. Space Sci.* 29:925–929.

Krasnopol'sky, V. A., and Krysko, A. A. 1976. On the night airglow of the Martian atmosphere. *Space Res.* 16:1005–1008.

Krasnopol'sky, V. A., and Krysko, A. A. 1979. Venera 9, 10: Is there a dust ring around Venus? *Planet. Space Sci.* 27:951–958.

Krasnopol'sky, V. A., and Parshev, V. A. 1979a. On the chemical composition of the Venus troposphere and the cloud layer based on Venera 11, 12 and Pioneer Venus measurements. *Space Res. Inst. Acad. Sci.* USSR, PR-514; *Kosmich. Issled.* 17:763–771.

Krasnopol'sky, V. A., and Parshev, V. A. 1979b. Ozone and photochemistry of the Martian lower atmosphere. *Planet. Space Sci.* 27:113–120.

Krasnopol'sky, V. A., and Parshev, V. A. 1980a. Chemical composition of Venus' troposphere and cloud layer based on Venera 11, Venera 12, and Pioneer Venus measurements. *Cosmic Res.* 17:630.

Krasnopol'sky, V. A., and Parshev, V. A. 1980b. Initial data for calculation of Venus' atmospheric photochemistry at heights and down to 50 km. *Cosmic Res.* 19:87–103.

Krasnopol'sky, V. A., and Parshev, V. A. 1980c. Photochemistry of the Venus atmosphere down to 50 km (results of calculations). *Space Res. Inst. Acad. Sci.* USSR, PR-591; *Kosmich. Issled.* 19:261.

Krasnopol'sky, V. A., and Tomashova, G. V. 1980. Venus nightglow variations. *Kosmich. Issled.* 18:766.

Krasnopol'sky, V. A., and Parshev, V. A. 1981a. Chemical composition of the atmosphere of Venus. *Nature* 292:610–613.

Krasnopol'sky, V. A., and Parshev, V. A. 1981b. Photochemistry of the atmosphere of Venus at altitudes above 50 km. Part 2. Results of calculations. *Cosmic Res.* 19:261–278.

Krasnopol'sky, V. A., and Parshev, V. A. 1981c. Photochemistry of Venus at altitudes above 50 km. Part 1. Initial data of the calculation. *Kosmich. Issled.* 19:87.

Krasnopol'sky, V. A., Krysko, A. A., Rogachev, V. N., and Parshev, V. A. 1976a. Observations of Venus from the interplanetary spacecraft Venera 9 and Venera 10. *Kosmich. Issled.* 14:789.

Krasnopol'sky, V. A., Krysko, A. A., Rogachev, V. N., and Parshev, V. A. 1976b. Spectroscopy of the Venus night airglow from the Venera-9, 10 orbiters. *Cosmic Res.* 14:789–795.

Kraus, J. 1956. Impulsive radio signals from the planet Venus. *Nature* 178:33.

Kraus, J. 1957. Recent observations of radio signals from Venus at 11 meters wavelength. *Astron. J.* 62:21 (abstract).

Krotikov, V. D. 1962. Several electrical characteristics of the Earth's rocks and a comparison with the surface layer of the moon. *Izv. Vuzov. Radiofiz.* 5/6:1057–1061.

Krupenie, P. H. 1972. The spectrum of molecular oxygen. *J. Phys. Chem. Ref. Data* 1:423–534.

Ksanfomality, L. V. 1977. Some features of the global thermal emission and radiometric albedo of Venus. *Pisma Astron. Zh.* 8:368–372.

Ksanfomality, L. V. 1979a. Lightning in Venus' cloud layer. *Cosmic Res.* 17:747–762.

Ksanfomality, L. V. 1979b. Soviet probes explore Venus. *Space World* P-8-190:28.

Ksanfomality, L. V. 1980a. Discovery of frequent lightning discharges in clouds on Venus. *Nature* 284:244–246.

Ksanfomality, L. V. 1980b. Venera 9 and 10 thermal radiometry. *Icarus* 41:36–64.

Ksanfomality, L. V., and Petrova, E. V. 1978. Photometry in the 3500 Å-band from Venera-9 and -10 spacecraft. *Kosmich. Issled.* 16:886.

Ksanfomality, L. V., and Selivanov, A. S. 1978. The clouds of Venus: A comparsion of altitude of UV-contrast markings. *Kosmich. Issled.* 16:959–961.

Ksanfomality, L. V., Vasilchikov, N. M., Ganpantzerova, O. F., Petrova, E. V., Suvorov, A. P., Filippov, G. F., Yablons-Kaya, O. V., and Yabrova, L. V. 1979. Electrical discharges in the Venus atmosphere. *Pisma Astron. Zh.* 5:229–236.

Ksanfomality, L. V., Goroshkova, N. V., Naraeva, M. K., Suvorov, A. P., Khondyrev, V. K., and Yabrova, L. V. 1982a. The wind velocity in the Venera 13 and Venera 14 landing sites from acoustic measurements. *Pisma Astron. Zh.* 8:419–423.

Ksanfomality, L. V., Vasil'chikov, N. M., Goroshkova, N. V., Petrova, E. V., Suvorov, A. P., and Khondyrev, V. K. 1982b. Low-frequency electromagnetic field in the atmosphere of Venus from Venera 13 and Venera 14 measurements. *Pisma Astron. Zh.* 8:424–428.

Ksanfomality, L. V., Zubkova, V. M., Morozov, N. A., and Petrova, E. V. 1982c. Microseims in the landing sites of Venera 13 and Venera 14. *Pisma Astron. Zh.* 8:444–447.

Kuenen, P. H. 1939. Quantitative estimations relating to eustatic movements. *Geol. Mijnbouw* 18:194–201.

Kuiper, G. P. 1947. Infrared spectra of planets. *Astrophys. J.* 106:251–254.

Kuiper, G. P. 1949. Survey of planetary atmospheres. In *The Atmospheres of the Earth and Planets,* ed. G. P. Kuiper (Chicago, IL: Univ. Chicago Press), pp. 306–405.

Kuiper, G. P. 1954. Determination of the pole of rotation of Venus. *Astrophys. J.* 120:603–605.

Kuiper, G. P. 1962. Infrared spectra of stars and planets. I. Photometry of the infrared spectrum of Venus, 1–2.5 microns. *Comm. Lunar Planet. Lab.* 1·83–117.

Kuiper, G. P. 1969. Identification of the Venus cloud layers. *Comm. Lunar Planet. Lab.* 6:229–250.

Kuiper, G. P. 1971. On the nature of the Venus clouds. In *Planetary Atmospheres,* eds. C. Sagan, T. C. Owen, and H. J. Smith (Dordrecht:D. Reidel). pp. 91–109.

Kuiper, G. P., and Forbes, F. F. 1967. High altitude spectra from NASA CV 990 Jet. I: Venus, 1–2.5 microns, resolution 20 per cm. *Comm. Lunar Planet. Lab.* 6:177–189.

Kuiper, G. P., Forbes, F. F., and Johnson, J. L. 1967. Program of astronomical spectroscopy from aircraft. *Comm. Lunar Planet. Lab.* 6:155–170.

Kuiper, G. P., Forbes, F. F., Steinmetz, D. L., and Mitchell, R. I. 1969. High altitude spectra from NASA CV 990 Jet. II: Water vapor on Venus. *Comm. Lunar Planet. Lab.* 6:209–228.

Kumar, S., and Hunten, D. M. 1974. Venus: An ionospheric model with an exospheric temperature of 350°K. *J. Geophys. Res.* 79:2529–2532.

Kumar, S., and Broadfoot, A. L. 1975. Helium 584 Å airglow emission from Venus: Mariner 10 observations. *Geophys. Res. Letters* 2:357–360.

Kumar, S., and Taylor, H. A. Jr. 1981. Predawn H_2^+ bulge: Evidence of electron impact ionization on the nightside of Venus. *Trans. Amer. Geophys. Union. (EOS)* 62:320 (abstract).

Kumar, S., Hunten, D. M., and Broadfoot, A. L. 1978. Non-thermal hydrogen in the Venus exopshere: The ionospheric source and the hydrogen budget. *Planet. Space Sci.* 26:1063–1075.

Kumar, S., Hunten, D. M., and Taylor, H. A. Jr. 1981. H_2 abundance in the atmosphere of Venus. *Geophys. Res. Letters* 8:237–239.

Kunde, V. G., Hanel, R. A., and Herath, L. W. 1977. High spectral resolution ground based observations of Venus in the 450–1250 cm^{-1} region. *Icarus* 32:210–224.

Kundt, W. 1977. Spin and atmospheric tides of Venus. *Astron. Astrophys.* 60:85–91.

Kurt, V. G., Dostovalow, S. B., and Sheffer, E. K. 1968. The Venus far ultraviolet observations with Venera 4. *J. Atmos. Sci.* 25:668–671.

Kuz'min, A. D. 1965. Measurements of the brightness temperature of the illuminated side of Venus at wavelength 10.6 cm. *Astron. J.* 42:1281–1286.

Kuz'min, A. D. 1967. *Radio Physical Studies of Venus* (Moscow: VINITI).

Kuz'min, A. D., and Salomonivich, A. E. 1960. The radio emission of Venus in the 8 cm range of wavelengths. *Astron. J.* 37:297–300.

Kuz'min, A. D., and Salomonivich, A. E. 1962. Observations of the radio emission of Venus and Jupiter at wavelength 8 cm 1962. *Astron. J.* 39:660–668.

Kuz'min, A. D., and Clark, B. G. 1965. The polarization measurement and the distribution of the brightness temperature of Venus at wavelength 10.6 cm. *Astron. J.* 43:595–617.

Kuz'min, A. D., and Dent, W. 1966. Measurement of brightness temperature and the polarization of the radio emission of Venus at wavelength 3.75 cm. *Astron. J.* 43:692–694.

Kuz'min, A. D., and Marov, M. Ya. 1974. *Physics of the Planet Venus* (Nauka: Moscow). (In Russian)

Kuz'min, A. D., and Marov, M. Ya. 1975. Physics of the Planet Venus. *NASA Tech. Trans.* TT F-16, 226; [*Fizika Planety Venera* 1974 (Moscow: Nauka Press).] (transl. of above)

Kuz'min, A. D., Nanmov, A. P., Smirnova, T. V., and Vetukhnooskaya, Yu. N. 1971. Lower atmosphere of Venus from radio astronomical and space measurements. *Space Res.* 11:141–145.

Lacis, A. A. 1975. Cloud structure and heating rates in the atmosphere of Venus. *J. Atmos. Sci.* 32:1107–1124.

Lacis, A. A., and Hansen, J. E. 1974a. A parameterization for the absorption of solar radiation in the Earth's atmosphere. *J. Atmos. Sci.* 31:118–133.

Lacis, A. A., and Hansen, J. E. 1974b. Atmosphere of Venus: Implications of Venera 8 sunlight measurements. *Science* 184:979–982.

Lambert, R. St. J. 1976. Archean thermal regimes, crustal and upper mantle temperatures, and a progressive evolutionary model for the Earth. In *The Early History of the Earth,* ed. B. F. Windley (London: Wiley), pp. 363–373.

Lancet, M. S., and Anders, E. 1973. Solubilities of noble gases in magnetite: Implications for planetary gases in meteorities. *Geochim. Cosmochim.* 37:1371–1388.

Lane, W. A., and Opstbaum, R. 1981. High altitude Venus haze from limb scans at 365 nm and 690 nm. "Internat. Conf. on the Venus Environment," held at Palo Alto, CA.

Latham, J., and Stow, C. D. 1969. Airborne studies of the electrical properties of large convective clouds. *Quart. J. Roy. Met. Soc.* 95:486.

Lawrence, G. M., Barth, C. A., and Argabright, V. 1977. Excitation of the Venus night airglow. *Science* 195:573–574.

Leonovich, A. K., Gromov, V. V., Semyonov, P. S., Surkov, Yu. A., Perminov, V. G., and Dmitriev, A. D. 1977. On the physical and mechanical properties of venusian soil. *Int. Astron. Fed. XXVIII Congress* 1AF 77–75, 1–15.

Leovy, C. B. 1964. Simple models of thermally driven mesospheric circulation. *J. Atmos. Sci.* 21:327–341.

Leovy, C. B. 1966. Photochemical destabilization of gravity waves near the mesopause. *J. Atmos. Sci.* 23:223–232.

Leovy, C. B. 1973. Rotation of the upper atmosphere of Venus. *J. Atmos. Sci.* 30:1218–1220.

Leovy, C. B. 1979. Martian meteorology. *Ann. Rev. Astron. Astrophys.* 17:387–413.

Lestrade, J. P. 1979. The ultraviolet markings on Venus: An analysis of the Venera 9 data. *Icarus* 39:418–432.

Lettau, H. 1951. Diffusion in the upper atmosphere. In *Compendium of Meteorology,* ed. T. F. Malone (New York: Amer. Meteorology Soc.), pp. 320–333.

Levin, Z., and Tzur, I. 1981. Atmosphere electric field near the Venusian surface. "Internat. Conf. on the Venus Environment," held at Palo Alto, CA.

Levine, J. S., Hughes, R. E., Chameides, W. L., and Howell, W. E. 1980. N_2O and CO production by electric discharge: Atmospheric implications. *Geophys. Res. Letters* 6:557–559.

Levine, J. S., Gregory, G. L., Harvey, G. A., Howell, W. E., Buroki, W. J., and Orville, R. E. 1982. Production of nitric oxide by lightning on Venus. *Geophys. Res. Letters* 9:893–896.

Levy, J. B. 1962. The thermal decomposition of perchloric acid vapor. *J. Phys. Chem.* 66:1092–1097.

Lewis, J. S. 1968a. An estimate of the surface conditions on Venus. *Icarus* 8:434–456.

Lewis, J. S. 1968b. Composition and structure of the clouds of Venus. *Astrophys. J.* 152:L79–L83.

Lewis, J. S. 1969. Geochemistry of the volatile elements on Venus. *Icarus* 11:367–385.

Lewis, J. S. 1970. Venus: Atmospheric and lithospheric composition. *Earth Planet. Sci. Letters* 10:73–80.

Lewis, J. S. 1971. The atmosphere, clouds, and surface of Venus. *Amer. Sci.* 59:557–566.

Lewis, J. S. 1972. Metal/silicate fractionation in the solar system. *Earth Planet. Sci. Letters* 15:286–290.

Lewis, J. S. 1973. Chemistry of the planets. *Ann. Rev. Phys. Chem.* 24:339–351.

Lewis, J. S. 1974. The temperature gradient in the solar nebula. *Science* 186:440–443.

Lewis, J. S., and Kreimendahl, F. A. 1980. Oxidation state of the atmosphere and crust of Venus from Pioneer Venus observations. *Icarus* 42:330–337.

Lewis, R. S., Matsuda, J., Whittaker, A. G., Watts, E. J., and Anders, E. 1980. Carbynes: Carriers of primordial noble gases in meteorites. *Proc. Lunar Planet. Sci. Conf.* 11: 624–625.

Limaye, S. S. 1977. Venus Stratospheric Circulation: A Diagnostic Study. Ph.D. thesis, Univ. Wisconsin, Madison, WI.

Limaye, S. S., and Suomi, V. E. 1977a. A normalized view of Venus. *J. Atmos. Sci.* 34:205–215.

Limaye, S. S., and Suomi, V. E. 1977b. Possible detection of an atmospheric wave on Venus. *Bull. Amer. Astron. Soc.* 9:509 (abstract).

Limaye, S. S., and Suomi, V. E. 1981. Cloud motions on Venus: Global structure and organization. *J. Atmos. Sci.* 38:1220–1235.

Limaye, S. S., Grund, C. J., and Burre, S. P. 1982. Zonal mean circulation at the cloud level on Venus: Spring and Fall 1979 OCPP observations. *Icarus* 51:416–439.

Lindzen, R. S. 1970. The application and applicability of terrestrial atmospheric tidal theory to Venus and Mars. *J. Atmos. Sci.* 27:536–549.

Lindzen, R. S. 1971a. Equatorial planetary waves in shear: Part I. *J. Atmos. Sci.* 28:609–622.

Lindzen, R. S. 1971b. Tides and gravity waves in the upper atmosphere. In *Mesospheric Models and Related Experiments,* ed. G. Fiocco (Dordrecht: D. Reidel), pp. 122–130.

Lindzen, R. S. 1981. Turbulence and stress owing to gravity wave and tidal breakdown. *J. Geophys. Res.* 86:9707–9714.

Liu, S. C., and Donahue, T. M. 1975. The aeronomy of the upper atmosphere of Venus. *Icarus* 24:148–156.

Llewellyn, E. J., and Solheim, B. H. 1980. Optical emissions in the polar auroral E-region. In *Exploration of the Polar Upper Atmosphere*, eds. C. S. Deehr and J. A. Holtet (Dordrecht: D. Reidel), pp. 165–174.

Llewellyn, E. J., Evans, W. F. J., and Wood, H. C. 1973. O_2 ($^1\Delta$) in the atmosphere. In *Physics and Chemistry of the Upper Atmospheres,* ed. B. M. McCormac (Dordrecht: Reidel), pp. 193–202.

Llewellyn, E. J., Solheim, B. H., Stegman, J., and Witt, G. 1979. A measurement of the O_2 ultraviolet nightglow emission. *Planet. Space Sci.* 27:1507–1511.

Loeb, L. B. 1953. Experimental contributions to the knowledge of charge generation. In *Thunderstorm Electricity,* ed. H. R. Byers (Chicago, IL: Univ. Chicago Press), pp. 150–192.

Logan, L. M., Hunt, G. R., Long, D. A., and Dybwad, J. P. 1974. Absolute infrared radiance measurements of Venus and Jupiter. *Air Force Cambridge Res. Lab.* TR 74-0573.

Lomonosov, M. V. 1955. Phenomena of Venus on the Sun, observed in the St. Petersburg imperial academy of sciences, May 26, 1761. In *Polnoye Sobraniye Sochinenii* 4:363–376, (Moscow: USSR Acad. Sci.).

Longuet-Higgins, M. S. 1968. The eigenfunctions of Laplace's tidal equations over a sphere. *Phil. Trans. R. Soc. London* A262:511–607.

Lowell, P. 1897. Determination of the rotation period and surface character of the planet Venus. *Mon. Not. R. Astron. Soc.* 57:148–149.

Luhmann, J. G., Elphic, R. C., Russell, C. T., Mihalov, J. D., and Wolfe, J. H. 1980. Observations of large scale steady magnetic fields in the dayside Venus ionosphere. *Geophys. Res. Letters* 7:917–920.

Luhmann, J. G., Elphic, R. C., Russell, C. T., and Brace, L. H. 1981a. On the role of the magnetic field in the solar wind interaction with Venus: Expectations versus observations. *Adv. Space Res.* 1:123–128.

Luhmann, J. G., Elphic, R. C. Russell, C. T., Slavin, J. A., and Mihalov, J. D. 1981b. Observations of large scale steady magnetic fields in the nightside Venus ionosphere and near wake. *Geophys. Res. Letters* 8:517–520.

Luhmann, J. G., Elphic, R. C. and Brace, L. H. 1981c. Large-scale current systems in the dayside Venus ionosphere. *J. Geophys. Res.* 86:3509–3513.

Lukashevich, N. A. et al. 1974. *Kosmich. Issled.* 12:272.

Madden, R. A. 1978. Further evidence of traveling planetary waves. *J. Atmos. Sci.* 35:1605–1618.

Madden, R. A. 1979. Observations of large-scale traveling Rossby waves. *Rev. Geophys. Space Phys.* 17:1935–1949.

Makarova, E. A., and Kharitonov, A. V. 1972. *Energy Distribution in Solar Spectrum and Solar Constant* (Moscow: Nauka).

Malin, M. C. 1978. Surfaces of Mercury and the moon: Effects of resolution and lighting conditions on the discrimination of volcanic features. *Proc. Lunar Planet. Sci. Conf.* 9:3395–3409.

Malin, M. C. 1981. Speculations on the geology of Venus. *Trans. Amer. Geophys. Union (EOS)* 62:386 (abstract).

Malin, M. C., and Saunders, R. S. 1977. Surface of Venus. Evidence of diverse landforms from radar observations. *Science* 196:987–990.

Malin, M. C., and Paluzzi, P. R. 1980. Topography of Earth and Venus. *Trans. Amer. Geophys. Union (EOS)* 61:1019 (abstract).

Mange, P. 1957. The theory of molecular diffusion in the atmosphere. *J. Geophys. Res.* 62:279–296.

Mantas, G. P., and Hanson, W. B. 1979. Photoelectron fluxes in the martian ionosphere. *J. Geophys. Res.* 84:369–385.

Mariner Stanford Group 1967. Venus: Ionosphere and atmosphere as measured by dual frequency occultation on Mariner 5. *Science* 158:1678–1683.

Marov, M. Ya. 1972. Venus: A perspective at the beginning of planetary exploration. *Icarus* 16:415–461.

Marov, M. Ya. 1978. Results of Venus missions. *Ann. Rev. Astron. Astrophys.* 16:141–170.

Marov, M. Ya. 1979. The atmosphere of Venus: Venera data. *Fund. of Cosmic Phys.* 5:1–46.

Marov, M. Ya., and Ryabov, O. L. 1974. A model atmosphere of Venus. Preprint No. 112. Keldysh Inst. Applied Math., USSR Acad. Sci.

Marov, M. Ya., Avduevsky, V. S., Borodin, N. F., Ekonomov, A. P., Kerzhanovich, V. V., Lysov, V. P., Moshkin, B. Ye., Rozhdestvensky, M. K., and Ryabov, O. L. 1973a. Preliminary results on the Venus atmosphere from the Venera 8 descent module. *Icarus* 20:407–421.

Marov, M. Ya., Avduevsky, V. S., Kerzhanovich, V. V., Rozhdestvensky, M. K., Borodin, N. F., and Ryabov, O. L. 1973b. Venera 8: Measurements of temperature, pressure, and wind velocity on the illuminated side of Venus. *J. Atmos. Sci.* 30:1210–1214.

Marov, M. Ya., Avduevsky, V. S., Kerzhanovich, V. V., Rozhdestvensky, M. K., Borodin, N. F., and Ryabov, O. L. 1973c. Measurements of temperature, pressure, and wind velocity in the venusian atmosphere on the AIS Venera-8. *Dokl. AN SSSR* 210:559–562.

Marov, M. Ya., Byvshev, B. V., Manuilov, K. N., Baramov, Yu. P., Kuznetsov, I. S., Lebedev, V. N., Lystev, V. E., Maksimov, A. V., Popandopulo, G. K., Razdolin, V. A., Sandimirov, V. A., and Frolov, A. M. 1976a. Nephelometric investigations on the stations of Venera 9 and Venera 10. *Kosmich. Issled.* 14:729.

Marov, M. Ya., Lebedev, V. N., and Lystsev, V. E. 1976b. Some preliminary estimates of aerosol component in the Venus atmosphere. *Pisma Astron. Zh.* 2:251–256.

Marov, M. Ya., Lystsev, V. E., and Lebedev, V. N. 1978. Structura i mikrophizicheskie svoistra oblakov Veneray. Preprinted No. 144. Inst. of Applied Math., USSR Academy of Sci.

Marov, M. Ya., Bivshev, B. V., Baranov, Ju. P., Lebedev, V. N., Lystsev, V. E., Maximov, A. V., Manuylov, K. N., and Frolov, A. M. 1979a. Aerosol component of atmosphere of Venus from data of Venera-11. *Kosmich. Issled.* 17:743–746.

Marov, M. Ya., Lebedev, V. N., Lystev, V. E., and Maknylov, K. K. 1979b. The structure and microphysical properties of the clouds of Venus. *Acad. Sci. USSR* 5:5.

Marov, M. Ya., Lystev, V. E., Lebedev, V. N., Lukashevich, N. L., and Shari, V. P. 1980. The structure and microphysical properties of the Venus clouds: Venera 9, 10, and 11 data. *Icarus* 44:608–639.

Martonchik, J., and Beer, R. 1975. Analysis of spectrophotometer observations of Venus in the 3–4 micron range. *J. Atmos. Sci.* 32:1151–1156.

Masursky, H., Kaula, W. M., McGill, G. E., Pettengill, G. H., Phillips, R. J., Russell, C. T., Schubert, G., and Shapiro, I. I. 1977. The surface and interior of Venus. *Space Sci. Rev.* 20:431–449.

Masursky, H., Eliason, E., Ford, P. G., McGill, G. E., Pettengill, G. H., Schaber, G. G., and Schubert, G. 1980. Pioneer Venus radar results: Geology from images and altimetry. *J. Geophys. Res.* 85:8232–8260.

Masursky, H., Dial, A. L. Jr., Schaber, G. G., and Strobell, M. E. 1981a. Venus: A first geologic map based on radar altimetric and image data. *Proc. Lunar Planet. Sci. Conf.* 12:661–663.

Masursky, H., Schaber, G. G., Dial, A. L. Jr., and Strobell, M. E. 1981b. Tectonism and volcanism on Venus deduced from Pioneer Venus images and altimetry. *Trans. Amer. Geophys. Union (EOS)* 62:385 (abstract).

Masursky, H., Dial, A. L. Jr., Eliason, E. M., and Strobell, M. E. 1982. Venus: Volcanism and tectonism. To be submitted to *Icarus*.

Mauersberger, K., Engebretson, M. J., Potter, W. E., Kayser, D. C., and Nier, A. D. 1975. Atomic nitrogen measurements in the upper atmosphere. *Geophys. Res. Letters* 2:337–340.

Mauersberger, K., von Zahn, U., and Krankowsky, D. 1979. Upper limits on argon isotope abundances in the Venus thermosphere. *Geophys. Res. Letters* 6:671–674.

Mayer, C. H., McCullough, T. P., and Sloanaker, R. M. 1958. Observations of Venus at 3.16 cm wavelength. *Astrophys. J.* 127:1–10.

Mayer, C. H., McCullough, T. P., and Sloanaker, R. M. 1960. Observations of Venus at 10.2-cm wavelength. *Astron. J.* 65:349–350.

Mayer, C. H., McCullough, T. P., and Sloanaker, R. M. 1962. 3.15-cm observations of Venus in 1961. *Mem. Soc. Roy. Sci. Liège* 7:357–363.

Mayr, H. G., Harris, I., Hartle, R. E., and Hoegy, W. R. 1978. Diffusion model for the upper atmosphere of Venus. *J. Geophys. Res.* 83:4411–4416.

Mayr, H. G., Harris, I., Niemann, H. B., Brinton, H. C., Spencer, N. W., Taylor, H. A., Hartle, R. E., Hoegy, W. R., and Hunten, D. M. 1980. Dynamic properties of the thermosphere inferred from Pioneer Venus mass spectrometer measurements. *J. Geophys. Res.* 85:7841–7847.

McCleese, D. J., Cimino, J. B., and Diner, D. J. 1981. The nature of the cloud and temperature morphologies in the mid-latitude collar and north polar regions of Venus. "Internat. Conf. on the Venus Environment," held at Palo Alto, CA.

McCullough, T. P. 1972. Phase dependence of the 2.7 cm wavelength radiation of Venus. *Icarus* 16:310–313.

McCullough, T. P., and Boland, J. W. 1964. Observations of Venus at 2.07 cm. *Astron. J.* 69:68.

McElroy, M. B. 1968. The upper atmosphere of Venus. *J. Geophys. Res.* 73:1513–1521.

McElroy, M. B. 1969. Structure of the Venus and Mars atmospheres. *J. Geophys. Res.* 74:29–41.

McElroy, M. B., and Hunten, D. M. 1969. The ratio of deuterium to hydrogen in the Venus atmosphere. *J. Geophys. Res.* 74:1720–1739.

McElroy, M. B., and Strobel, D. F. 1969. Models for the nighttime Venus ionosphere. *J. Geophys. Res.* 74:1118–1127.

McElroy, M. B., and Donahue, T. M. 1972. Stability of the Martian atmosphere. *Science* 177:986–988.

McElroy, M. B., and Prather, M. J. 1981. Noble gases in the terrestrial planets: Clues to evolution. *Nature* 293:535–539.

McElroy, M. B., Sze, N. D., and Yung, Y. L. 1973. Photochemistry of the Venus atmosphere. *J. Atmos. Sci.* 30:1437–1447.

McElroy, M. B., Kong, T. Y., and Yung, Y. L. 1977. Photochemistry and evolution of Mars' atmosphere: A Viking perspective. *J. Geophys. Res.* 82:4379–4388.

McElroy, M. B., Prather, M. J., and Rodríguez, J. M. 1982. Escape of hydrogen from Venus. *Science* 215:1614–1615.

McGill, G. E. 1979a. Tectonics of Venus. *NASA Tech. Mem.* 80339:39–41.

McGill, G. E. 1979b. Venus tectonics: Another Earth or another Mars? *Geophys. Res. Letters* 6:739–741.

McGill, G. E. 1980. Evidence for continental-style rifting from the Beta region of Venus. *NASA Tech. Mem.* 82385:79–80.

McGill, G. E., Steenstrup, S. J., Barton, C., and Ford, P. G. 1981. Continental rifting and the origin of Beta Regio, Venus. *Geophys. Res. Letters* 8:737–740.

McKenzie, D. P. 1977. The initiation of trenches: A finite amplitude instability. In *Island Arcs, Deep Sea Trenches and Back-Arc Basins*, eds. M. Talwani and W. C. Pitman, III (Washington D. C.: Amer. Geophys. Union), pp. 57–61.

McKenzie, D. P., and Weiss, N. O. 1975. Speculations on the thermal and tectonic history of the Earth. *Geophys. J. R. Astron. Soc.* 42:131–174.

McKenzie, D. P., and Bowin, C. 1976. The relationship between bathymetry and gravity in the Atlantic ocean. *J. Geophys. Res.* 81:1903–1915.

McNeal, R. J., and Durana, S. C. 1969. Absolute chemiluminescent reaction rates for emission of the O_2 Herzberg bands in oxygen and oxygen-inert-gas afterglows. *J. Chem. Phys.* 51:2955–2960.

Merritt, D., and Thompson, K. 1980. Thermal energy transport in the Venus ionosphere: Classical and saturated electron temperature profiles. *J. Geophys. Res.* 85:6778–6782.

Meyer, B. T., Stroyer-Hansen, T., and Oommen, T. V. 1972. The visible spectrum of S_3 and S_4. *J. Molec. Spec.* 42:335–343.

Michel, F. C. 1971. Solar wind interaction with planetary atmospheres. *Rev. Geophys. Space Phys.* 9:427–435

Mihalov, J. D. 1981. Comparison of Gas Dynamic Model for Solar Wind Flow Around Venus with Pioneer Venus Orbiter Data. Ph.D. thesis, Stanford Univ., Palo Alto, CA.

Mihalov, J. D., and Barnes, A. 1982. Evidence for the acceleration of ionospheric O^+ in the magnetosphere of Venus. *Geophys. Res. Letters.* In press.

Mihalov, J. D., Wolfe, J. H., and Intriligator, D. S. 1980. Pioneer Venus plasma observations of the solar-wind Venus interaction. *J. Geophys. Res.* 85:7613–7624.

Miller, K. L., Knudsen, W. C., Spenner, K., Whitten, R. C., and Novak, V. 1980. Solar zenith angle dependence of ionospheric ion and electron temperatures and density on Venus. *J. Geophys. Res.* 85:7759–7764.

Mills, K. C. 1974. *Thermodynamic Data for the Inorganic Sulphides, Selenides, and Tellurides* (London: Butterworth).

Monin, A. S., and Yaglom, A. M. 1967. *Statistical Hydromechanics, Part 2* (Moscow: Nauka).

Moore, P. 1959. *The Planet Venus* (New York: MacMillan).

Moore, P. 1961. *The Planet Venus* 3rd Ed. (London: Faber and Faber).

Moos, H. W., and Rottman, G. J. 1971. OI and HI emissions from the upper atmosphere of Venus. *Astrophys. J.* 169:L127–L130.

Morgan, J. W., and Anders, E. 1980. Chemical composition of Earth, Venus, and Mercury. *Proc. Nat. Acad. Sci.* 77:6973–6977.

Morgan, P., and Phillips, R. J. 1983. Hot spot heat transfer and its application to Venus. *J. Geophys. Res.* In press.

Morgan, W. J. 1971. Convection plumes in the lower mantle. *Nature* 230:42–43.

Moroz, V. I. 1964. *Astron. J. Zh.* 41:711.

Moroz, V. I. 1968. The CO_2 bands and some optical properties of the atmosphere of Venus. *Sov. Astron.–A. J.* 11:653–661.

Moroz, V. I. 1971. The atmosphere of Venus. *Uspekhi Fizichi Nauk* 104:255–296.

Moroz, V. I. 1976. Panorama of the venusian surface. Some conclusions concerning boundary layer of atmosphere. *Kosmich. Issled.* 14:691.

Moroz, V. I. 1979. Several conclusions as to the structure of the boundary layer of the atmosphere of Venus. In *First Panoramas of the Surface of Venus* (Moscow: Nauka), pp. 128–130.

Moroz, V. I. 1981. The atmosphere of Venus. *Space Sci. Rev.* 29:3–127.

Moroz, V. I., Parfent'ev, N. A., San'ko, N. F., Zhegulev, V. S., Zasova, L. V., and Ustinov, E. A. 1976. Preliminary results of a narrow band photometer investigation of the Venus cloud layers in the spectral region 0.80–0.87 μm from the descent modules Venera 9 and Venera 10. *Kosmich. Issled.* 14:743–757.

Moroz, V. I., Moshkin, B. E., Ekonomov, A. P., San'ko, N. F., Parfent'ev, N. A., and Golovin, Yu. M. 1978. Spectrophotometric experiment on board the Venera-11, -12 descenders: Some results of the analysis of the Venus day-sky spectrum. *Space Res. Inst. Acad. Sci. Leningrad Publ.* 270.

Moroz, V. I., Moshkin, B. E., Ekonomov, A. P., San'ko, N. F., Parfent'ev, N. A., and Golovin, Yu. M. 1979a. Spectroscopic experiment aboard descent probes Venera 11 and Venera 12: Some results of the venusian day sky spectrum. *Pisma Astron. Zh.* 5:222–228.

Moroz, V. I., Parfent'ev, N. A., and San'ko, N. F. 1979b. Spectrometric experiment on the Venera 11 and 12 landers. *Kosmich. Issled.* 17:727–742.

Moroz, V. I., Moshkin, B. E., Ekonomov, A. P., San'ko, N. F., Parfent'ev, N. A., and Golovin, Yu. M. 1979c. Spectrophotometric experiment on-board the Venera 11 and 12 descenders: Some results of the analysis of Venus day-sky spectrum. *Space Res. Inst. Publ.* 117.

Moroz, V. I., Golovin, Yu. M., Ekonomov, A. P., Moshkin, B. E., Parfent'ev, N. A., and San'ko, N. F. 1980a. Spectrum of the Venus day sky. *Nature* 284:243–244.

Moroz, V. I., Moshkin, B. E., Ekonomov, A. P., San'ko, N. F., Parfent'ev, N. A., and Golovin, Yu. M. 1980b. Venera 11 and 12 lander results on the Venus day-sky spectrum. *Space Res.* 20:209.

Moroz, V. I., Parfent'ev, N. A., and San'ko, N. F. 1980c. Spectrophotometric experiment in the Venera 11 and Venera 12 modules 2: Analysis of Venera 11 spectra by layer-addition method. *Cosmic Res.* 17:601.

Moroz, V. I., Golovin, Yu. M., Moshkin, B. E., and Ekonomov, A. P. 1981. Spectrophotometric experiment on the Venera 11 and 12 descent modules. 3: Results of the photometric measurements. *Kosmich. Issled.* 19:599.

Moroz, V. I., Moshkin, B. E., Ekonomov, A. P., Golovin, Yu. M., Gnedykh, V. I., and Grigor'ev, A. V. 1982. Spectrophotometrical experiment on Venera 13 and Venera 14 *Pisma Astron. Zh.* 8:404–410.

Morrison, D. 1969. Venus: Absence of a phase effect at 2-cm wavelength. *Science* 163:815–817.

Moshkin, B. E., Ekonomov, A. P., and Golovin, Yu. M. 1978. Spectral composition of solar radiation in Venus' atmosphere according to light intensity measurements on Venera 9 and Venera 10. *Cosmic Res.* 16:412–418.

Mueller, G. 1893. *Publ. Potsdam Astrophys. Obs.* 8:197–209. In German.

Mueller, R. F. 1963. Chemistry and petrology of Venus: Preliminary deductions. *Science* 141:1046–1047.

Mueller, R. F., 1969. Planetary problems: Origin of the Venus atmosphere. *Science* 163:1322–1324.

Mueller, R. F., and Saxena, S. K. 1977. *Chemical Petrology* (New York: Springer-Verlag).

Muhleman, D. O. 1969. Microwave opacity of the Venus atmosphere. *Astron. J.* 74:57–69.

Muhleman, D. O., Berge, G. L., and Orton, G. S. 1973. The brightness temperature of Venus and the absolute flux-density scale at 608 MHz. *Astrophys. J.* 183:1081–1085.

Mukhin, L. M., and Gerasimov, M. V. 1981. On the mechanism of the enrichment of the Venus atmosphere with noble gases. *Proc. Lunar Planet. Sci. Conf.* 12:738–740.

Mukhin, L. M., Gel'man, B. G., Lamonov, N. I., Mel'nikov, V. V., Nenarokov, D. F., Okhotnikov, B. P., Rotin, V. A., and Khokhlov, V. N. 1982. Gas-chromatographical analysis of chemical composition of the atmosphere of Venus done by Venera 13 and Venera 14 probe. *Pisma Astron. Zh.* 8:399–403.

Murphy, R. 1982. *Graphic Timetable of the Heavens* (Baltimore, MD: Maryland Acad. of Science and Scientia).

Murray, B. C., Wildey, R. L., and Westphal, J. A. 1963. Infrared photometric mapping of Venus through the 8- to 14-micron atmospheric window. *J. Geophys. Res.* 68:4813–4818.

Murray, B. C., Belton, J. S., Danielson, G. E., Davies, M. E., Gault, D., Hapke, B., O'Leary, B., Strom, R. G., Suomi, V., and Trask, N. 1974. Venus: Atmospheric motion and structure from Mariner 10 pictures. *Science* 183:1307–1315.

Mysen, B. O., and Boettcher, A. L. 1975. Melting of a hydrous mantle. I. Phase relations of natural peridotite at high pressure and temperatures with controlled activities of water, carbon dioxide, and hydrogen. *J. Petrology* 16:520–548.

Nagy, A. F., and Banks, P. M. 1970. Photoelectron fluxes in the ionosphere. *J. Geophys. Res.* 75:6260–6270.

Nagy, A. F., Donahue, T. M., Liu, S. C., Atreya, S. K., and Banks, P. M. 1975. A model of the Venus ionosphere. *Geophys. Res. Letters* 2:83–86.

Nagy, A. F., Cravens, T. E., Chen, R. H., Taylor, H. A. Jr., Brace, L. H., and Brinton, H. C. 1979. Comparison of calculated and measured ion densities on the dayside of Venus. *Science* 205:107–109.

Nagy, A. F., Cravens, T. E., Smith, S. G., Taylor, H. A. Jr., and Brinton, H. C. 1980. Model calculations of the dayside ionosphere of Venus: Ionic composition. *J. Geophys. Res.* 85:7795–7801.

Nagy, A. F., Cravens, T. E., Yee, J.-H., and Stewart, A. I. F. 1981. Hot oxygen atoms in the upper atmosphere of Venus. *Geophys. Res. Letters* 8:629–632.

NASA SP-8011 1972. Models of Venus atmosphere; revised Sept. 1972. NASA Space Vehicle Design Criteria (Environment).

Nepoklonov, B. V., Leykin, G. A., and Selivanov, A. S. 1979. The processing and topographical interpretations of the television panoramas obtained from the descent capsules of Venera 9 and Venera 10. In *First Panoramas of the Surface of Venus* (Moscow: Nauka), pp. 80–106.

Ness, N. F., Behannon, K. W., Lepping, R. P., Whang, Y. C., and Schatten, K. H. 1974. Magnetic field observations near Venus: Preliminary results from Mariner 10. *Science* 183:1301–1306.

Neugebauer, M., and Snyder, C. W. 1965. Solar-wind measurements near Venus. *J. Geophys. Res.* 70:1587–1591.

Newkirk, G. 1959. The airglow of Venus. *Planet. Space Sci.* 1:32–36.

Newman, M. J., and Rood, R. T. 1977. Implications of solar evolution for the Earth's early atmosphere. *Science* 198:1035–1037.

Niemann, H. B. 1977. Orbiter neutral mass spectrometer (ONMS). *Space Sci. Rev.* 20:489–491.

Niemann, H. B., Hartle, R. E., Hedin, A. E., Kasprzak, W. T., Spencer, N. W., Hunten, D. M., and Carignan, G. R. 1979a. Venus upper atmospheric neutral gas composition: First observations of the diurnal variations. *Science* 205:54–56.

Niemann, H. B., Hartle, R. E., Kasprzak, W. T., Spencer, N. W., Hunten, D. M., and Carignan, G. R. 1979b. Venus upper atmosphere neutral composition: Preliminary results from the Pioneer Venus orbiter. *Science* 203:770–772.

Niemann, H. B., Hedin, A. E., Kasprzak, W. T., Spencer, N. W., and Hunten, D. M. 1979c. The neutral gas composition in the thermosphere and exosphere of Venus. *Bull. Amer. Astron. Soc.* 11:537 (abstract).

Niemann, H. B., Booth, J. R., Cooley, J. E., Hartle, R. E., Kasprzak, W. T., Spencer, N. W., Way, S. H., Hunten, D. M., and Carignan, G. R. 1980a. Pioneer Venus orbiter neutral gas mass spectrometer experiment. *IEEE Trans. Geosci. Remote Sensing* GE-18:60–65.

Niemann, H. B., Kasprzak, W. T., Hedin, A. E., Hunten, D. M., and Spencer, N. W. 1980b. Mass spectrometric measurements of the neutral gas composition of the thermosphere and exosphere of Venus. *J. Geophys. Res.* 85:7817–7827.

Niemeyer, S., and Marti, K. 1980. Planetary noble gases: Rare-gas trapping by laboratory carbon condensates. *Proc. Lunar Planet. Sci. Conf.* 11:812–814.

Nikonova, E. K. 1949. *Astrophys. Obs.* 114. In Russian.

Nisbet, J. S. 1968. Photoelectron escape from the ionosphere. *J. Atmos. Terr. Phys.* 30:1257–1278.

Noll, R. B., and McElroy, M. B. 1972. *Models of Venus Atmosphere* (Washington, D.C.:NASA SP-8011).

Noxon, J. F. 1968. Day airglow. *Space Sci. Rev.* 8:92–134.

Noxon, J. F., Traub, W. A., Carleton, N. P., and Connes, P. 1976. Detection of O_2 dayglow emission for Mars and the martian ozone abundance. *Astrophys. J.* 207:1025–1035.

Nozette, S., and Lewis, J. S. 1982. Venus: Chemical weathering of igneous rocks and buffering of atmospheric composition. *Science* 216:181–183.

Nur, A. 1971. Viscous phase in rocks and the low-velocity zone. *J. Geophys. Res.* 76:1270–1277.

Offermann, D., and Drescher, A. 1973. Atomic oxygen densities in the lower thermosphere as derived from in situ 5577 Å night airglow and mass spectrometer measurements. *J. Geophys. Res.* 78:6690–6700.

O'Leary, B. 1975. Venus: Vertical structure of stratospheric hazes from Mariner 10 pictures. *J. Atmos. Sci.* 32:1091–1100.

Ong, R. S. B., and Roderick, N. 1972. On the Kelvin-Helmholtz instability of the Earth's magnetopause. *Planet. Space Sci.* 20:1–10.

Öpik, E. J. 1961. The aeolosphere and atmosphere of Venus. *J. Geophys. Res.* 66:2807–2819.

Oran, E. S. 1983. Photoelectron theory and measurements. *Rev. Geophys. Space Phys.* In press.

Oran, E. S., and Strickland, D. J. 1978. Photoelectron flux in the Earth's ionosphere. *Planet. Space Sci.* 26:1161–1177.

O'Reilly, T. C., and Davies, G. F. 1981. Magma transport of heat on Io: A mechanism allowing a thick lithosphere. *Geophys. Res. Letters* 8:313–316.

Orville, P. 1975. Crust-atmosphere interactions. In *The Atmosphere of Venus*, ed. J. E. Hansen (Washington, D. C.: NASA SP 382) pp. 190–195.

Otsuka, K., and Calvert, J. G. 1971. Decay mechanism of triplet sulfur dioxide molecules found by intersystem crossing in the flash photolysis of sulfur dioxide (2400–3200 Å). *J. Amer. Chem. Soc.* 93:2581–2586.

Owen, T. 1967. Water vapor on Venus—a dissent and clarification. *Astrophys. J.* 150:L121–L123.

Owen, T., and Sagan, C. 1972. Minor constituents in planetary atmospheres: Ultraviolet spectroscopy from the Orbiting Astronomical Observatory. *Icarus* 16:557–568.

Owen, T., Biemann, K., Rushneck, D. R., Biller, J. E., Howarth, D. W., and Lafleur, A. L. 1977. The composition of the atmosphere at the surface of Mars. *J. Geophys. Res.* 82:4635–4639.

Owen, T., Cess, R. D., and Ramanathan, V. 1979. Early Earth: An enhanced carbon dioxide greenhouse to compensate for reduced solar luminosity. *Nature* 277:640–642.

Owen, T., Crabb, J., and Anders, E. 1982. Venus: A chondritic source for atmospheric noble gas? Submitted to *Icarus*.

Onburgh, D. R., and Parmentier, E. M. 1977. Compositional and density stratification in oceanic lithosphere: Causes and consequences. *J. Geol. Soc. London* 133:343–355.

Oyama, V. I., Carle, G. C., Woeller, F., and Pollack, J. B. 1979. Venus lower atmosphere composition: Analysis by gas chromatography. *Science* 203:802–804.

Oyama, V. I., Carle, G. C., Woeller, F., Pollack, J. B., Reynolds, R. T., and Craig, R. A. 1980*a*. Pioneer Venus gas chromatography of the lower atmosphere of Venus. *J. Geophys. Res.* 85:7891–7902.

Oyama, V. I., Carle, G. C., Woeller, F., Rocklin, S., Vogrin, J., Potter, W., Rosiak, G., and Reichwein, C. 1980*b*. Pioneer Venus sounder probe gas chromatograph. *IEEE Trans. Geosci. and Remote Sensing* GE-18:85–93.

Palmen', E., and Newton, C. W. 1969. *Atmospheric Circulation Systems* (New York: Academic).

Palmer, K. F., and Williams, D. 1973. Optical constants of sulfuric acid; Application to the clouds of Venus? *Appl. Optics* 14:208–219.

Panfilov, A. S., and Selivanov, A. S. 1979. Model representations and initial information used in the preparation for the survey of the surface of Venus. In *First Panoramas of the Surface of Venus* (Moscow: Nauka), pp. 24–37.

Parisot, J. P., and Moreels, G. 1981. Fluorine chemistry in the upper atmosphere of Venus. "Internat. Conf. on the Venus Environment," held at Palo Alto, CA.

Parkinson, T. D., and Hunten, D. M. 1972. Spectroscopy and aeronomy of O_2 on Mars. *J. Atmos. Sci.* 29:1380–1390.

Parsons, B. 1982. Causes and consequences of the relation between area and age of the ocean floor. *J. Geophys. Res.* 87:289–302.

Parsons, B., and Sclater, J. G. 1977. An analysis of the variation of ocean floor bathymetry and heat flow with age. *J. Geophys. Res.* 82:803–827.

Perez-de-Tejada, H. 1979. On the viscous flow character of the shocked solar wind at the venusian ionopause. *J. Geophys. Res.* 84:1555–1558.

Perez-de-Tejada, H. 1980*a*. Evidence for a viscous boundary layer at the Venus ionopause from preliminary Pioneer Venus results. *J. Geophys. Res.* 85:7709–7714.

Perez-de-Tejada, H. 1980*b*. Viscous flow circulation of the shocked solar wind behind Venus. *Science* 207:981–983.

Perez-de-Tejada, H., and Dryer, M. 1976. Viscous boundary layer for the venusian ionopause. *J. Geophys. Res.* 81:2023–2029.

Perez-de-Tejada, H., Dryer, M., and Vaisberg, O. L. 1977. Viscous flow in the near-venusian plasma wake. *J. Geophys. Res.* 82:2837–2841.

Peters, K., and Weil, K. 1930. Adsorption experiments of the heavy noble gases. *Z. Phys. Chem.* 148:1.

Pettengill, G. H., Dyce, R. B., and Campbell, D. B. 1967. Radar measurements at 70 cm of Venus and Mercury. *Astron. J.* 72:330–337.

Pettengill, G. H., Ford, P. G., Brown, W. E., Kaula, W. M., Keller, C. H., Masursky, H., and McGill, G. E. 1979*a*. Pioneer Venus radar mapper experiment. *Science* 203:806–808.

Pettengill, G. H., Ford, P. G., Brown, W. E., Kaula, W. M., Masursky, H., Eliason, E., and McGill, G. E. 1979*b*. Venus: Preliminary topographic and surface imaging results from the Pioneer Orbiter. *Science* 205:91–93.

Pettengill, G. H., Eliason, E., Ford, P. G., Loriot, G. B., Masursky, H., and McGill, G. E. 1980*a*. Pioneer Venus radar results: Altimetry and surface properties. *J. Geophys. Res.* 85:8261–8270.

Pettengill, G. H., Ford, P. G., Masursky, H., and Eliason, E. 1980*b*. The surface of Venus. In *Plenary Meeting COSPAR* p. 16 (abstract).

Pettengill, G. H., Nozette, S., and Ford, P. G. 1980*c*. The radar reflectivity of Venus. *Bull. Amer. Astron. Soc.* 12:690 (abstract).

Pettengill, G. H., Ford, P. G., and Nozette, S. 1982. Venus: Global surface radar reflectivity. *Science* 217:640–642.

Pettit, E. 1961. Planetary temperature measurements. In *Planets and Satellites,* eds. G. P. Kuiper and B. M. Middlehurst (Chicago, IL: Univ. Chicago), pp. 400–428.

Pettit, E., and Nicholson, S. B. 1924. Radiation measures on the planet Mars. *Publ. Astron. Soc. Pacific* 36:269–272.

Pettit, E., and Nicholson, S. B. 1955. Temperatures on the bright and dark sides of Venus. *Publ. Astron. Soc. Pacific* 67:293–303.

Phillips, R. J., and Lambeck, K. 1980. Gravity fields of the terrestrial planets: Long-wavelength anomalies and tectonics. *Rev. Geophys. Space Phys.* 18:27–76.

Phillips, R. J., Sjogren, W. L., Abbott, E. A., and Zisk, S. H. 1978. Simulation gravity modeling to spacecraft-tracking data: Analysis and application. *J. Geophys. Res.* 83:5455–5464.

Phillips, R. J., Sjogren, W. L., Abbott, E. A., Smith, J. C., Wimberly, R. N., and Wagner, C. A. 1979. Gravity field of Venus: A preliminary analysis. *Science* 205:93–96.

Phillips, R. J., Kaula, W. M., McGill, G. E., and Malin, M. C. 1981. Tectonics and evolution of Venus. *Science* 212:879–887.

Philpotts, J. A., and Schnetzler, C. C. 1970. Apollo 11 lunar samples: K, Rb, Sr, Ba and rare Earth concentrations in some rocks and separated phases. In *Proc. Apollo 11 Lunar Science Conf.* (New York: Pergamon), pp. 1471–1486.

Podgorny, I. M. 1978. Simulation studies of space. *Fund. Cosmic Phys.* 4:1–72.

Podgorny, I. M., and Andrijanov, Yu. V. 1978. Simulation of the solar wind interaction with non-magnetic celestial bodies. *Planet. Space Sci.* 26:99–109.

Podgorny, I. M., Dubinin, E. M., and Israelevich, P. L. 1980. Laboratory simulation of induced magnetospheres of comets and Venus. *Moon Planets* 23:323–338.

Podosek, F. A., Honda, M., and Ozima, M. 1980. Sedimentary noble gases. *Geochim. Cosmochim.* 44:1875–1884.

Polaski, L. J. 1979. Anomalous behavior of the Pioneer Venus entry probes during lower descent. *NASA Tech. Rept.*

Pollack, J. B. 1969a. A non-gray CO_2–H_2O greenhouse model of Venus. *Icarus* 10:314–341.

Pollack, J. B. 1969b. Temperature structure of nongray planetary atmospheres. *Icarus* 10:301–313.

Pollack, J. B. 1971. A non-gray calculation of the runaway greenhouse: Implications for Venus' past and present. *Icarus* 14:295–306.

Pollack, J. B. 1979. Climatic change on the terrestrial planets. *Icarus* 37:479–553.

Pollack, J. B., and Sagan, C. 1965a. The infrared limb darkening of Venus. *J. Geophys. Res.* 70:4403–4426.

Pollack, J. B., and Sagan, C. 1965b. The microwave phase effect of Venus. *Icarus* 4:62–103.

Pollack, J. B., and Morrison, D. 1970. Venus: Determination of atmospheric parameters from the microwave spectrum. *Icarus* 12:376–390.

Pollack, J. B., and Morrison, D. 1971. Venus: Determination of atmospheric parameters from the microwave spectrum. In *Planetary Atmospheres,* eds. C. Sagan, T. C. Owen, and H. H. Smith (New York: Springer-Verlag), pp. 28–31.

Pollack, J. B., and Young, R. 1975. Calculations of the radiative and dynamical state of the Venus atmosphere. *J. Atmos. Sci.* 32:1025–1037.

Pollack, J. B., and Black, D. C. 1979. Implications of the gas compositional measurements of Pioneer Venus for the origin of planetary atmospheres. *Science* 205:56–59.

Pollack, J. B., and Cuzzi, J. 1979. Scattering by nonspherical particles of size comparable to the wavelength: A new semi-empirical theory and its application to tropospheric aerosols. *J. Atmos. Sci.* 37:868–881.

Pollack, J. B., and Yung, Y. L. 1980. Origin and evolution of planetary atmospheres. *Ann. Rev. Earth Planet. Sci.* 8:425–487.

Pollack, J. B., and Black, D. C. 1981. Origin of rare gases on Venus. "Internat. Conf. on the Venus Environment," held at Palo Alto, CA.

Pollack, J. B., and Black, D. C. 1982. Noble gases in planetary atmospheres: Implications for the origin and evolution of atmospheres. *Icarus* 51:169–198.

Pollack, J. B., Erickson, E. F., Goorvitch, D., Baldwin, B. J., Stecker, D. W., Witteborn, F. C., and Augason, G. C. 1975. A determination of the composition of the Venus clouds from aircraft observations in the near infrared. *J. Atmos. Sci.* 32:1140–1150.

Pollack, J. B., Strecker, D. W., Witteborn, F. C., Erickson, E. F., and Baldwin, B. J. 1978. Properties of the clouds of Venus, as inferred from airborne observations of its near-infrared reflectivity spectrum. *Icarus* 34:28–45.

Pollack, J. B., Ragent, B., Boese, R., Tomasko, M. G., Blamont, J., Knollenberg, R. G., Esposito, L. W., Travis, L., and Wiedman, D. 1980*b*. Distribution and source of the UV absorption in Venus' atmosphere. *J. Geophys. Res.* 85:8141–8150.

Pollack, J. B., Toon, O. B., and Boese, R. 1980*a*. Greenhouse models of Venus' high surface temperature, as constrained by Pioneer Venus measurements. *J. Geophys. Res.* 85:8223–8231.

Pollack, J. B., Toon, O. B., Whitten, R. C., Boese, R., Ragent, B., Tomasko, M. G., Esposito, L. W., Travis, and L. Wiedman, D. 1980*b*. Distribution and source of the UV absorption in Venus' atmosphere. *J. Geophys. Res.* 85:8141–8150.

Prinn, R. G. 1971. Photochemistry of HCl and other minor constituents in the atmosphere of Venus. *J. Atmos. Sci.* 28:1058–1068.

Prinn, R. G. 1972. Venus atmosphere: Structure and stability of the ClOO radical. *J. Atmos. Sci.* 29:1004–1007.

Prinn, R. G. 1973*a*. The upper atmosphere of Venus: A review. In *Physics and Chemistry of the Upper Atmosphere*, ed. B. M. McCormac (Dordrecht: Reidel), pp. 334–335.

Prinn, R. G. 1973*b*. Venus: Composition and structure of the visible clouds. *Science* 182:1132–1135.

Prinn, R. G. 1974. Venus: Vertical transport rates in the visible atmosphere. *J. Atmos. Sci.* 31:1691–1697.

Prinn, R. G. 1975. Venus: Chemical and dynamical processes in the stratosphere and mesosphere. *J. Atmos. Sci.* 32:1237–1247.

Prinn, R. G. 1978. Venus: Chemistry of the lower atmosphere prior to the Pioneer Venus mission. *Geophys. Res. Letters* 5:973–976.

Prinn, R. G. 1979. On the possible roles of gaseous sulfur and sulfanes in the atmosphere of Venus. *Geophys. Res. Letters* 6:807–810.

Prinn, R. G. 1982. Perchloric acid clouds on Venus? To be submitted to *Icarus*.

Prokofyev, V. K. 1964. On the presence of oxygen in the atmosphere on Venus II. *Izv. Krysmkoi Astrofiz. Obs.* 31:276–280.

Pruppacher, H. R., and Klett, J. D. 1978. *Microphysics of Clouds and Precipitation* (Hingham, MA: D. Reidel).

Quenisset, F. 1921. Photographies de la planete Venus. *Compt. Rend. Acad. Sci. Paris* 172:1645–1647.

Ragent, B., and Blamont, J. 1979. Preliminary results of the Pioneer Venus nephelometer experiment. *Science* 203:790–793.

Ragent, B., and Blamont, J. 1980. Structure of the clouds of Venus: Results of the Pioneer Venus nephelometer experiment. *J. Geophys. Res.* 85:8089–8105.

Rao, T. N., Collier, S. S., and Calvert, J. G. 1969*a*. Primary photophysical processes in the photochemistry of sulfur dioxide. *J. Amer. Chem. Soc.* 91:1609–1615.

Rao, T. N., Collier, S. S., and Calvert, J. G. 1969*b*. The quenching reactions of the first excited singlet and triplet states of sulfur dioxide with oxygen and carbon dioxide. *J. Amer. Chem. Soc.* 91:1616–1621.

Rasool, S. I. 1970. The structure of Venus clouds—summary. *Radio Sci.* 5:367–368.

Rasool, S. I., and de Bergh, C. 1970. The runaway greenhouse and the accumulation of CO_2 in the Venus atmosphere. *Nature* 226:1037–1039.

Rasool, S. I., and Le Sergeant, L. 1977. Implications of the Viking results for volatile outgassing from Earth and Mars. *Nature* 266:822–823.

Reasenberg, R. D., Goldberg, Z. M., MacNeil, P. E., and Shapiro, I. I. 1981. Venus gravity: A high-resolution map. *J. Geophys. Res.* 86:7173–7179.

Reasenberg, R. D., Goldberg, Z. M., and Shapiro, I. I. 1982. Venus: Comparison of gravity and topography in the vicinity of Beta Regio. *Geophys. Res. Letters* 9:637–640.

Revercomb, H. E., Sromovsky, L. A., and Suomi, V. E. 1982. Reassessment of net radiation measurements in the atmosphere of Venus. *Icarus* Vol. 52, No. 2

Richardson, R. S. 1958. Spectrographic observations of Venus for rotation made at Mount Wilson in 1956. *J. Brit. Interplanet. Soc.* 16, N9 (82):517–524.

Richter, F. M., and Parsons, B. 1975. On the interaction of two scales of convection in the mantle. *J. Geophys. Res.* 80:2529–2541.

The Ring Group 1981. Gulf stream cold-core rings: Their physics, chemistry, and biology. *Science* 212:1091–1100.

Ringwood, A. E. 1975. *Composition and Petrology of the Earth's Mantle* (New York: McGraw-Hill).

Ringwood, A. E. 1979. *Origin of the Earth and Moon* (New York: Springer-Verlag).

Ringwood, A. E., and Major, A. 1970. The system Mg_2SiO_4-Fe_2SiO_4 at high pressures and temperatures. *Phys. Earth Planet. Interior* 3:89–108.

Ringwood, A. E., and Anderson, D. L. 1977. Earth and Venus: A comparative study. *Icarus* 30:243–253.

Rizzi, A. 1971. Solar Wind Flow Past the Planets Earth, Mars, and Venus, Ch. XI. Ph.D. thesis, Stanford Univ. (Univ. Michigan: Univ. Microfilms, 72–5982).

Robie, R. A., Hemingway, B. S., and Fisher, J. R. 1979. Thermodynamic properties of minerals and related substances at 298.15° K and 1 bar (10^5 pascals) pressure and at higher temperatures. *U. S. Geol. Surv. Bull.* No. 1452.

Rodgers, C. D. 1976. Evidence for the 5-day wave in the upper stratosphere. *J. Atmos. Sci.* 33:710–711.

Rodgers, C. D., and Walshaw, C. D. 1966. The computation of infrared cooling rate in planetary atmospheres. *Quart. J. Roy. Meteor. Soc.* 92:67–92.

Rogers, A. E. E., and Ingalls, R. P. 1969. Venus: Mapping the surface reflectivity by radar interferometry. *Science* 165:797–799.

Rohrbaugh, R. P., Nisbet, J. S., Bleuler, E., and Herman, J. R. 1979. The effect of energetically produced O_2^+ on the ion temperatures of the Martian thermosphere. *J. Geophys. Res.* 84:3327–3338.

Romanov, S. A. 1978. Asymmetry of the region of the solar wind interaction with Venus according to data of Venera 9 and Venera 10. *Kosmich. Issled.* 16:318.

Romanov, S. A., Smirnov, V. N., and Vaisberg, O. L. 1977. On the nature of solar-wind-Venus interaction. Preprint D-253, Space Res. Inst., USSR Acad. Sci.

Romanov, S. A., Smirnov, V. N., and Vaisberg, O. L. 1978. Interaction of the solar wind with Venus. *Cosmic Res.* 16:603–611.

Rosen, J. M., Hofmann, D. J., and Knollenberg, R. G. 1982. *Results of Instrument Comparisons for the July 1981 Ace Mission.* Univ. Wyoming, Dept. of Physics and Astronomy, Rept. No. AP-69.

Ross, F. E. 1928. Photographs of Venus. *Astrophys. J.* 68:57–92.

Ross, H. P., Adler, J. E. M., and Hunt, G. R. 1969. A statistical analysis of the reflectance of igneous rocks from 0.2 to 2.65 microns. *Icarus* 11:46–54.

Rossow, W. B. 1977. The clouds of Venus: II. An investigation of the influence of coagulation on the observed droplet size distribution. *J. Atmos. Sci.* 34:417–431.

Rossow, W. B. 1978. Cloud microphysics: Analysis of the clouds of Earth, Venus, Mars, and Jupiter. *Icarus* 36:1–5.

Rossow, W. B. 1982. A general circulation model of a Venus-like atmosphere. Submitted to *J. Atmos. Sci.*

Rossow, W. B., and Sagan, C. 1975. Microwave boundary conditions on the atmosphere and clouds of Venus. *J. Atmos. Sci.* 32:1164–1176.

Rossow, W. B., and Williams, G. P. 1979. Large-scale motion in the Venus stratosphere. *J. Atmos. Sci.* 36:377–389.

Rossow, W. B., and Kinsella, E. 1982. Variation of winds on Venus. Submitted to *Science*.

Rossow, W. B., Del Genio, A. D., Limaye, S. S., Travis, L. D., and Stone, P. H. 1980a. Cloud morphology and motions from Pioneer Venus images. *J. Geophys. Res.* 85:8107–8128.

Rossow, W. B., Fels, S. B., and Stone, P. H. 1980b. Comments on "A three-dimensional model of dynamical processes in the Venus atmosphere." *J. Atmos. Sci.* 37:250–252.

Rottman, G. J., and Moos, H. W. 1973. The ultraviolet (1200–1900 Angström) spectrum of Venus. *J. Geophys. Res.* 78:8033–8084.

Rowley, J. K., Cleveland, B. T., Davis, R., Hampsel, W., and Kursten, T. 1980. The present and past neutrino luminosity of the sun. *Geochim. Cosmochim. Acta Suppl.* 13:45–62.

Rumsey, H. C., Morris, G. A., Green, R. R., and Goldstein, R. M. 1974. A radar brightness and altitude image of a portion of Venus. *Icarus* 23:1–7.

Rusch, D. W., and Cravens, T. E. 1979. A model of the neutral and ion nitrogen chemistry in the daytime thermosphere of Venus. *Geophys. Res. Letters* 6:791–794.

Russell, C. T. 1976a. The magnetosphere of Venus: Evidence for a boundary layer and a magnetotail. *Geophys. Res. Letters* 3:589–592.

Russell, C. T. 1976b. The magnetic moment of Venus: Venera-4 measurements reinterpreted. *Geophys. Res. Letters* 3:125–128.

Russell, C. T. 1976c. Venera-9 magnetic field measurements in the Venus wake: Evidence for an Earth-like interaction. *Geophys. Res. Letters* 3:413–416.

Russell, C. T. 1977. The Venus bow shock: Detached or attached? *J. Geophys. Res.* 82:625–631.

Russell, C. T. 1978. Some comments on the role of induced and intrinsic planetary magnetic fields at Venus. Preprint.

Russell, C. T., and Elphic, R. C. 1979. Observations of flux ropes of the Venus ionosphere. *Nature* 279:616–618.

Russell, C. T., and Greenstadt, E. W. 1979. Initial ISEE magnetometer results: Shock observations. *Space Sci. Rev.* 23:3–37.

Russell, C. T., Elphic, R. C., and Slavin, J. A. 1979a. Initial Pioneer Venus magnetic field results: Dayside observations. *Science* 203:745–748.

Russell, C. T., Elphic, R. C., and Slavin, J. A. 1979b. Initial Pioneer Venus Venus magnetic field results: Nightside observations. *Science* 205:114–116.

Russell, C. T., Elphic, R. C., and Slavin, J. A. 1979c. Pioneer magnetometer observations of the Venus bow shock. *Nature* 282:815–816.

Russell, C. T., Elphic, R. C., Luhmann, J. G., and Slavin, J. A. 1980a. On the search for an intrinsic magnetic field at Venus. *Proc. Lunar Planet. Sci. Conf.* 11:1897–1906.

Russell, C. T., Elphic, R. C., and Slavin, J. A. 1980b. Limits on the possible intrinsic magnetic field of Venus. *J. Geophys. Res.* 85:8319–8332.

Russell, C. T., Siscoe, G. L., and Smith, E. J. 1980c. Comparison of ISEE 1 and 3 interplanetary magnetic field observations. *Geophys. Res. Letters* 7:381–384.

Russell, C. T., Luhmann, J. G., Elphic, R. C., and Scarf, F. L. 1981. The distant bow shock and magnetotail of Venus: Magnetic field and plasma wave observations. *Geophys. Res. Letters* 8:843–846.

Russell, C. T., Hoppe, M. M., and Livesey, W. A. 1982a. Overshoots in planetary bow shocks. *Nature* 296:45–48.

Russell, C. T., Luhmann, J. G., Elphic, R. C., and Brace, L. H. 1982b. Magnetic field and plasma wave observations in a plasma cloud at Venus. *Geophys. Res. Letters* 9:45–48.

Russell, C. T., Luhmann, J. G., Elphic, R. C., and Neugebauer, M. 1982c. Solar wind interaction with comets: Lessons from Venus. In *Comets*, ed. L. L. Wilkening (Tucson: Univ. Arizona Press), pp. 561–585.

Russell, H. N. 1899. The atmosphere of Venus. *Astrophys. J.* 9:284–299.

Safronov, V. S. 1969. Evolutsiya doplanetnogo oblaka i obrazovanie Zemli i planet. *M. Nauka* 7.

Sagan, C. 1960a. The radiation balance of Venus. *JPL Tech. Rept.* 32–34.

Sagan, C. 1960b. The surface temperature of Venus. *Astron. J.* 65:352–353.

Sagan, C. 1961. The planet Venus. *Science* 133:849–858.

Sagan, C. 1962. Structure of the lower atmosphere of Venus. *Icarus* 1:151–169.

Sagan, C., and Pollack, J. B. 1969. On the structure of the Venus atmosphere. *Icarus* 10:274–281.

Sagan, C., and Mullen, G. 1972. Earth and Mars: Evolution of atmospheres and surface temperature. *Science* 177:52–56.

Sagdeev, R. Z., and Moroz, V. I. 1982a. *Pravda,* March 12 (In Russian).

Sagdeev, R. Z., and Moroz, V. I. 1982. Venera 13 and Venera 14. *Pisma Astron. Zh.* 8:387–390.

Samuelson, R. E. 1967. Greenhouse effect in semi-infinite scattering atmospheres: Application to Venus. *Astrophys. J.* 147:782–798.

Sanko, N. F. 1980. Gaseous sulfur in the atmosphere of Venus. *Kosmich. Issled.* 18:600.

Santer, R., and Herman, M. 1979. Wavelength dependence of polarization, 38. Analysis of ground-based observations of Venus. *Astron. J.* 84:1802–1810.

Satterfield, C. N., Reid, R. C., and Briggs, D. R. 1954. Rate of oxidation of hydrogen sulfide by hydrogen peroxide. *J. Amer. Chem. Soc.* 76:3922–3923.

Saunders, R. S., and Malin, M. C. 1976. Venus: Geologic analysis of radar images. *Geologica Romana* 15:507–515.

Saunders, R. S., and Malin, M. C. 1977. Geologic interpretation of new observations of the surface of Venus. *Geophys. Res. Letters* 4:542–550.

Saxon, R. P., and Liu, B. 1977. Ab initio configuration interaction study of the valence states of O_2. *J. Chem. Phys.* 67:5432–5441.

Scarf, F. L. 1977. Orbiter electric field detector. *Space Sci. Rev.* 20:501–502.

Scarf, F. L., Taylor, W. W. L., and Green, I. M. 1979. Plasma waves near Venus: Initial observations. *Science* 203:748–750.

Scarf, F. L., Taylor, W. W. L., Russell, C. T., and Brace, L. H. 1980a. Lightning on Venus: Orbiter detection of whistler signals. *J. Geophys. Res.* 85:8158–8166.

Scarf, F. L., Taylor, W. W. L., Russell, C. T., and Elphic, R. C. 1980b. Pioneer Venus plasma wave observations: The solar wind-Venus interaction. *J. Geophys. Res.* 85: 7599–7612.

Scarf, F. L., Taylor, W. W. L., and Virobik, P. F. 1980c. The Pioneer Venus orbiter plasma wave investigation. *IEEE Trans. Geosci. Remote Sensing* 18:36–38.

Scarf, F. L., Gurnett, D. A., and Kurth, W. S. 1981. Plasma wave turbulence at planetary bow shocks: Saturn, Jupiter, Earth, and Venus. *Nature* 292:747–750.

Scarf, F. L., Taylor, W. W. L., Russell, C. T., Elphic, R. C., and Luhmann, J. G. 1982. Venus lightning: A review of the Pioneer Orbiter whistler measurements. Submitted to *Geophys. Res. Letters.*

Schaber, G. G. 1976. Geologic application of radar data to planetary exploration. *Geologica Romana* 15:363–365.

Schaber, G. G. 1982. Venus: Limited extension and volcanism along zones of lithospheric weakness. *Geophys. Res. Letters* 9:499–502.

Schaber, G. G., and Boyce, J. M. 1977. Probable distribution of large impact basins on Venus: Comparison with Mercury and the moon. In *Impact and Explosion Cratering,* eds. D. J. Roddy, R. O. Pepin, and R. B. Morrill (New York: Pergamon), pp. 603–612.

Schaber, G. G., and Masursky, H. 1981. Large-scale equatorial and mid-latitude surface disruptions on Venus: Preliminary geologic assessment. *Lunar Planet. Sci.* 11:929–931 (abstract).

Schloerb, F. P., Robinson, S. E., and Irvine, W. M. 1980. Observation of CO in the stratosphere of Venus via its $J=0-1$ rotation transition. *Icarus* 43:121–127.

Schofield, J. T., and Taylor, F. W. 1982a. The mean, retrieved, solar fixed temperature and cloud top structure of the middle atmospheres of Venus. Submitted to *J. Quart. J. Roy. Met. Soc.*

Schofield, J. T., and Taylor, F. W. 1982b. Net global thermal emission for the venusian upper atmosphere. *Icarus* Vol. 52. No. 2.

Schofield, J. T., Taylor, F. W., and McCleese, D. J. 1982. The global distribution of water vapor in the middle atmosphere of Venus. *Icarus* Vol. 52, No. 2

Schorn, R. A., Barker, E. S., Gray, L. D., and Moore, R. C. 1969. High-precision spectroscopic studies of Venus II. The water vapor variations. *Icarus* 10:98–104.

Schubert, G. 1979. Subsolidus convection of the mantles of terrestrial planets. In *Annual Review of Earth and Planetary Sciences,* eds. F. A. Donath, F. G. Stehli, and G. W. Wetherill (Palo Alto, CA: Annual Review Inc.), 7:289–342.

Schubert, G., and Whitehead, J. 1969. Moving flame experiment with liquid mercury: Possible implications for the Venus atmosphere. *Science* 163:71–72.

Schubert, G., and Young, R. E. 1970. The 4-day Venus circulation driven by periodic thermal forcing. *J. Atmos. Sci.* 27:523–528.

Schubert, G., and Covey, C. 1981. The atmosphere of Venus. *Sci. Amer.* 245:66–74.

Schubert, G., Turcotte, D. L., and Oxburgh, E. R. 1969. Stability of planetary interiors. *Geophys. J. Roy. Astron. Soc.* 18:441–460.

Schubert, G., Counselman, C. C. III, Hansen, J., Limaye, S., Pettengill, G., Seiff, A., Shapiro, I. I., Suomi, V. E., Taylor, F., Travis, L., Woo, R., and Young, R. E. 1977. Dynamics, winds, circulation and turbulence in the atmosphere of Venus. *Space Sci. Rev.* 20:357–387.

Schubert, G., Cassen, P., and Young, R. E. 1979. Subsolidus convective cooling histories of terrestrial planets. *Icarus* 38:192–211.

Schubert, G., Covey, C., Del Genio, A., Elson, L. S., Keating, G., Seiff, A., Young, R. E., Apt, J., Counselman, C. C. III, Kliore, A. J., Limaye, S. S., Revercomb, H. E.,

Sromovsky, L. A., Suomi, V. E., Taylor, F., Woo, R., and von Zahn, U. 1980*a*. Structure and circulation of the Venus atmosphere. *J. Geophys. Res.* 85:8007–8025.

Schubert, G., Stevenson, D., and Cassen, P. 1980*b*. Whole planet cooling and the radiogenic heat source contents of the Earth and Moon. *J. Geophys. Res.* 85:2531–2538.

Schunk, R. W. 1975. Transport equations for aeronomy. *Planet. Space Sci.* 23:437–485.

Schunk, R. W. 1977. Mathematical structure of transport equations for multispecies flows. *Rev. Geophys. Space Phys.* 15:429–445

Schunk, R. W., and Walker, J. C. G. 1969. Thermal diffusions of the topside ionosphere for mixtures which include multiply-charged ions. *Planet. Space Sci.* 17:853–868.

Schunk, R. W., and St. Maurice, J.–P. 1977. Plasma transport in the topside Venus ionosphere. *Planet. Space Sci.* 25:921–930.

Schunk, R. W., and Nagy, A. F. 1978. Electron temperatures in the F region of the ionosphere: Theory and observations. *Rev. Geophys. Space Phys.* 16:355–399.

Schunk, R. W., and Nagy, A. F. 1980. Ionospheres of the terrestrial planets. *Rev. Geophys. Space Phys.* 18:813–852.

Schunk, R. W., and Raitt, W. J. 1980. Atomic nitrogen and oxygen ions of the daytime high-latitude F region. *J. Geophys. Res.* 85:1255–1272.

Schunk, R. W., and St. Maurice, J.–P. 1981. Ion temperature anisotropy and heat flow in the Venus lower ionosphere. *J. Geophys. Res.* 86:4823–4827.

Science 1979*a*. Vol. 203, Pioneer Venus issue, February 23.

Science 1979*b*. Vol. 295, Pioneer Venus issue, July 6.

Sclater, J. G., Jaupart, C., and Galson, D. 1980. The heat flow through oceanic and continental crust and the heat loss of the Earth. *Rev. Geophys. Space. Phys.* 18:269–311.

Scott, A. H., and Reese, E. J. 1972. Venus: Atmosphere rotation. *Icarus* 17:602–607.

Seiff, A. 1982. Dynamical implications of observed thermal contrasts in Venus' upper atmosphere. *Icarus* 51:574–592.

Seiff, A., and Kirk, D. B. 1982. Structure of the Venus mesosphere and lower thermosphere from measurements during entry of the Pioneer Venus probes. *Icarus* 49:49–70.

Seiff, A., Kirk, D. B., Sommer, S. C., Young, R. E., Blanchard, R. C., Juergens, D. W., Lepetich, J. E., Intrieri, P. F., Findlay, J. T., and Derr, J. S. 1979*a*. Structure of the atmosphere of Venus up to 110 kilometers: Preliminary results from the four Venus entry probes. *Science* 203:787–790.

Seiff, A., Kirk, D. B., Young, R. E., Sommer, S. C., Blanchard, R. C., Findlay, J. T., and Kelly, G. M. 1979*b*. Thermal contrast in the atmosphere of Venus: Initial appraisal from Pioneer Venus probe data. *Science* 205:46–49.

Seiff, A., Kirk, D. B., Young, R. E., Blanchard, R. C., Findlay, J. T., Kelly, G. M., and Sommer, S. C. 1980. Measurements of thermal structure and thermal contrasts in the atmosphere of Venus and related dynamical observations: Results from the four Pioneer Venus probes. *J. Geophys. Res.* 85:7903–7933.

Selivanov, A. S., Panfilov, A. S., Narayeva, M. K., Chemodanov, V. P., Bokhonov, M. I., and Gerasimov, M. A. 1976*a*. Photometric analysis of panoramas on the surface of Venus. *Cosmic Res.* 14:596–603.

Selivanov, A. S., Panfilov, A. S., and Narayeva, M. K. 1976*b*. Photometric processing of the panoramas of the surface of Venus. *Kosmich. Issled.* 14:678–686.

Selivanov, A. S., Chemodanov, V. P., and Narayeva, M. K. 1976*c*. The television devices for transmission of panoramic images aboard Venera 9 and Venera 10. *Tekhnika kino i televedeniya* 5:26–31.

Selivanov, A. S., Gektin, Yu. M., Narayeva, M. K., Panfilov, A. S., and Fokin, A. B. 1982. Evolution of the images received from Venera 13. *Pisma Astron. Zh.* 8:433–436.

Selivanov, A. S., Gektin, Yu. M., Kerhanovich, V. V., Narayeva, M. K., Panfilov, A. S., Chemodanov, V. P., and Chochia, P. A. 1978. Survey of the Venus cloud layer from the Venera 9 orbiter. *Cosmic Res.* 16:698–706.

Serry, D. J., and Britton, D. 1964. The continuous absorption spectra of chlorine, bromne, bromne chloride, iodine chloride and iodine bromide. *J. Phys. Chem.* 68:2263–2266.

Shapiro, I. I. 1968. Spin and orbital motions of the planets. In *Radar Astronomy*, eds. J. V. Evans and T. Hagfors (New York: McGraw-Hill), pp. 143–185.

Shapiro, I. I., Campbell, D., and De Campli, W. 1979. Nonresonance rotation of Venus? *Astrophys. J. Letters* 230:123–126.

Shari, V. P. 1976. *Kosmich. Issled.* 14:97.

Shari, V. P., and Lukashevich, N. L. 1977. Preprint N105. Inst. Applied Mathematics, USSR Academy of Sci.

Sharonov, V. V. 1965. *Planeta Venera* (Moscow: Nauka).

Sharpe, H. N., and Peltier, W. R. 1978. Parameterized mantle convection and the Earth's thermal history. *Geophys. Res. Letters* 5:737–740.

Sharpe, H. N., and Peltier, W. R. 1979. A thermal history model for the Earth with parameterized convection. *Geophys. J. R. Astron. Soc.* 58:171–203.

Shaya, E., and Caldwell, J. 1976. Photometry of Venus below 2200 Å from the Orbiting Astronomical Observatory-2. *Icarus* 27:255–264.

Shefer, R. A., Lazarus, A. J., and Bridge, H. S. 1979. A reexamination of plasma measurements from the Mariner 5 Venus encounter. *J. Geophys. Res.* 84:2109–2114.

Sheldon, C. S. II. 1971. *Soviet Space Programs, 1966–1970.* Staff Report for Comm. on Aeron. and Space Sci., U. S. Senate Doc. No. 92–51.

Shimizu, M. 1963. Vertical distribution of neutral gases on Venus. *Planet. Space Sci.* 11:269–273.

Shimizu, M. 1968. The recombination mechanism of CO and O in the upper atmosphere of Venus and Mars. *Icarus* 9:593–597.

Shimizu, M. 1969. A model calculation of the cytherean upper atmosphere. *Icarus* 10:11–25.

Shimizu, M. 1971. The effect of atmospheric dynamics on the upper atmosphere phenomena of Mars and Venus. In *Planetary Atmospheres*, eds. C. Sagan, T. C. Owen, and H. J. Smith (New York: Springer-Verlag), pp. 331–336.

Shimizu, M. 1973. Atmospheric mixing in the upper atmosphere of Mars and Venus. *J. Geophys. Res.* 78:6780–6783.

Shimizu, M. 1978. Implication of ^{36}Ar excess on Venus. *ISAS RN* 69.

Sidebottom, H. W., Babcock, C. C., Calvert, J. G., Reinhardt, G. W., Rabe, B. R., and Damon, E. K. 1971. A study of decay processes in the triplet sulfur dioxide molecules excited at 3828.8 Å. *J. Amer. Chem. Soc.* 93:2587–2592.

Sill, G. T. 1972. Sulfuric acid in the Venus clouds. *Comm. Lunar Planet. Lab.* 9:191–198.

Sill, G. T. 1975. The composition of the ultraviolet dark markings on Venus. *J. Atmos. Sci.* 32:1201–1204.

Sill, G. T. 1981. The clouds of Venus: Sulfuric acid by the lead chamber process. Submitted to *Icarus.*

Sinton, W. M. 1954. Distribution of Temperature and Spectra of Venus and Other Planets. Ph.D. thesis, Johns Hopkins Univ., Baltimore, MD.

Sinton, W. M. 1961. Recent radiometric studies of the planets and the moon. In *Planets and Satellites*, eds. G. P. Kuiper and B. M. Middlehurst (Chicago, IL: Univ. Chicago Press), pp. 429–441.

Sinton, W. M. 1963*a*. Infrared observations of Venus. *Mem. Soc. R. Sci. Liège* 7:300–310.

Sinton, W. M. 1963*b*. Recent infrared spectra of Mars and Venus. *J. Quant. Spect. Rad. Trans.* 3:551–558.

Sinton, W. M., and Strong, J. 1960. Radiometric observations of Venus. *Astrophys. J.* 131:470–490.

Sjogren, W. L., Phillips, R. J., Birkeland, P. W., and Wimberly, R. N. 1980. Gravity anomalies on Venus. *J. Geophys. Res.* 85:8295–8302.

Skinner, B. J. 1962. Thermal expansion of ten minerals. *U. S. Geol. Survey Prof. Paper* 450D:109–112 .

Slanger, T. G. 1978*a*. Generation of O_2 ($c^1\Sigma_u^-$, $c^3\Delta_u$, $A^3\Sigma_u^+$) from oxygen atom recombination. *J. Chem. Phys.* 69:4779–4791.

Slanger, T. G. 1978*b*. Metastable oxygen emission bands. *Science* 202:751–753.

Slanger, T. G. 1981*a*. O(^{1}S) quenching by O_2($^1\Delta_g$). *Report 28th Congress IUPAC*, Vancouver, Canada.

Slanger, T. G. 1981*b*. On the three-body recombination of O(^{3}P)+CO. *J. Geophys. Res.* 86:7795 (abstract).

Slanger, T. G., and Black, G. 1976. O(^{1}S) production from oxygen atom recombination. *J. Chem. Phys.* 64:3767–3773.

Slanger, T. G., and Black, G. 1978. The O_2 ($C^3\Delta_u \rightarrow a^1\Delta_g$) bands in the nightglow spectrum of Venus. *Geophys. Res. Letters* 5:947–948.

Slanger, T. G., and Black, G. 1979. Interactions of $O_2(b^1\Delta)$ with $O(^3P)$ and O_3. *J. Chem. Phys.* 70:3434–3438.

Slanger, T. G., and Huestis, D. L. 1981. O_2 ($c^1\Sigma_u^- \rightarrow x^3\Sigma_g^-$) emission in the terrestrial nightglow. *J. Geophys. Res.* 86:3551–3554.

Slattery, W. L., De Campli, W. M., and Cameron, A. G. W. 1980. Protoplanetary core formation by rain-out of minerals. *Moon Planets* 23:381–390.

Slavin, J. A., Elphic, R. C., and Russell, C. T. 1979a. A comparison of Pioneer Venus and Venera bow shock observations: Evidence for a solar cycle variation. *Geophys. Res. Letters* 6:905–908.

Slavin, J. A., Elphic, R. C., Russell, C. T., Wolfe, J. H., and Intriligator, D. S. 1979b. Position and shape of the Venus bow shock: Pioneer Venus Orbiter magnetometer observations. *Geophys. Res. Letters* 6:901–904.

Slavin, J. A., Elphic, R. C., Russell, C. T., Scarf, F. L., Wolfe, J. H., Mihalov, J. D., Intriligator, D. S., Brace, L. H., Taylor, H. A., and Daniell, R. E. 1980. The solar-wind interaction with Venus: Pioneer Venus observations of bow shock location and structure. *J. Geophys. Res.* 85:7625–7641.

Smirnov, V. N. 1981. *Kosmich. Issled.* In press.

Smirnov, V. N., Vaisberg, O. L., and Intriligator, D. S. 1980. An empirical model of the venusian outer environment. 2. The shape and location of the bow shock. *J. Geophys. Res.* 85:7651–7654.

Smirnov, V. N., Vaisberg, O. L., Romanov, S. A., Slavin, J. A., Russell, C. T., and Intriligator, D. S. 1981. Three-dimensional shape and position of the Venus bow shock. *Kosmich. Issled.* 19:613.

Smirnova, T. V. 1973. Radio Astronomy Studies of the Atmospheres of Venus, Jupiter, and Saturn. Dissertation, MFTI.

Smirnova, T. V., and Kuz'min, A. D. 1974. The evaluation of water vapor and ammonia content in the Venus atmosphere from radar measurements. *Astron. J. Zh.* 51:607.

Smith, B. A. 1967. Rotation of Venus: Continuing contradictions. *Science* 158:114–116.

Smith, B. A., Soderblom, L. A., Beebe, R. F., Boyce, J., Briggs, G. A., Carr, M. H., Collins, S. A., Cook, A. F. II, Danielson, G. E., Davies, M. E., Hunt, G. E., Ingersoll, A. P., Johnson, T. V., Masursky, H., McCauley, J. F., Morrison, D., Owen, T., Sagan, C., Shoemaker, E. M., Strom, R. G., Suomi, V. E., and Veverka, J. 1979a. Voyager 2 encounter with the Jovian system. *Science* 206:925–950.

Smith, B. A., Soderblom, L. A., Johnson, T. V., Ingersoll, A. P., Collins, S. A., Shoemaker, E. M., Hunt, G. E., Masursky, H., Carr, M. H., Davies, M. E., Cook, A. F. II, Boyce, J., Danielson, G. E., Owen, T., Sagan, C., Beebe, R. F., Veverka, J., Strom, R. G., McCauley, J. F., Morrison, D., Briggs, G. A., and Suomi, V. E. 1979b. The Jupiter system through the eyes of Voyager 1. *Science* 204:952–972.

Smith, B. A., Shoemaker, E. M., Kieffer, S. K., and Cook, A. F. II 1979c. Volcanism on Io—the role of SO_2. *Nature* 280:738–743.

Smith, B. A., Soderblom, L. A., Beebe, R. F., Boyce, J., Briggs, G. A., Bunker, A., Collins, S. A., Hansen, C. J., Johnson, T. V., Mitchell, J. L., Terrile, R. J., Carr, M. H., Cook, A. F. II, Cuzzi, J., Pollack, J. B., Danielson, G. E., Ingersoll, A. P., Davies, M. E., Hunt, G. E., Masursky, H., Shoemaker, E. M., Morrison, D., Owen, T., Sagan, C., Veverka, J., Strom, R. G., and Suomi, V. E. 1981. Encounter with Saturn: Voyager 1 imaging results. *Science* 212:163–191.

Smith, B. A., Soderblom, L. A., Batson, R., Bridges, P., Inge, J., Masursky, H., Shoemaker, E. M., Beebe, R. F., Boyce, J., Briggs, G. A., Bunker, A., Collins, S. A., Hansen, C. J., Johnson, T. V., Mitchell, J. L., Terrile, R. J., Cook, A. F. II, Cuzzi, J., Pollack, J. B., Danielson, G. E., Ingersoll, A. P., Davies, M. E., Hunt, G. E., Morrison, D., Owen, T., Sagan, C., Veverka, J., Strom, R. G., and Suomi, V. E. 1982. A new look at the Saturn system: The Voyager 2 images. *Science* 215:504–537.

Smith, E. J., Davis, L. Jr., Coleman, P. J., and Sonett, C. P. 1963. Mariner II: Preliminary reports on measurements of Venus magnetic field. *Science* 139:909–910.

Smith, E. J., Davis, L. Jr., Coleman, P. J., and Sonett, C. P. 1965. Magnetic measurements near Venus. *J. Geophys. Res.* 70:1571–1586.

Smith, F. L., and Smith, C. 1972. Numerical evaluation of Chapman's grazing incidence integral ch (X,χ). *J. Geophys. Res.* 77:3592–3597.

Smith, J. V. 1980. Planetary crusts: A comparative review. *Geochim. Cosmochim. Acta Suppl.* 12:441–456.

Smith, R. N., Young, D. A., Smith, E. N., and Carter, C. C. 1963. The structure and properties of carbon suboxide polyus. *Inorg. Chem.* 2:829–838.

Sobolev, V. V. 1944. *Astron. Zh.* 21:244 (in Russian).

Sobolev, V. V. 1968. *Astron. Zh.* 45:169 (in Russian).

Sobolev, V. V. 1972. *Rassejanie Sveta v Atmospherakh Planet* (Moscow: Nauka).

Solar Data 1978, No. 12, in Russian.

Solomon, S. C., and Head, J. W. 1982. Mechanisms for lithospheric heat transport on Venus: Implications for tectonic style and volcanism. *J. Geophys. Res.* 87:9236–9246.

Solomon, S. C., Stephens, S. K., and Head, J. W. 1982. On Venus impact basins: Viscous relaxation of topographic relief. *J. Geophys. Res.* 87:7763–7771.

Sonett, C. P., and Colburn, D. S. 1968. The principle of solar wind induced planetary dynamos. *Phys. Earth Planet. Inter.* 1:326–346.

Space Science Reviews 1977 Vol. 20 (PV Special Issue, May 3).

Space Science Reviews 1977 Vol. 20 (PV Special Issue, June 4).

Spenner, K., Knudsen, W. C., Miller, K. L., Novak, V., Russell, C. T., and Elphic, R. C. 1980. Observations of the Venus mantle, the boundary region between solar wind and ionosphere. *J. Geophys. Res.* 85:7655–7662.

Spenner, K., Knudsen, W. C., Whitten, R. C., Michalson, P. F., Miller, K. L., and Novak, V. 1981. On the maintenance of the Venus nighttime ionosphere: Electron precipitation and plasma transport. *J. Geophys. Res.* 86:9170–9178.

Spinrad, H., and Richardson, E. H. 1965. An upper limit to the molecular oxygen content of the Venus atmosphere. *Astrophys. J.* 141:282–286.

Spinrad, H., and Shawl, S. J. 1966. Water vapor on Venus—a confirmation. *Astrophys. J.* 146:L328–L:329.

Spitzer, L. 1962. *Physics of Fully Ionized Gases* (New York: Interscience).

Spreiter, J. R. 1976. Magnetohydrodynamic and gas dynamic aspects of solar wind flow around terrestrial planets: A critical review. In *Solar-Wind Interaction with the Planets Mercury, Venus, and Mars,* ed. N. F. Ness, (Washington, D. C.: NASA SP-397), pp. 135–149.

Sprieter, J. R., and Stahara, S. S. 1980. Solar wind flow past Venus: Theory and comparisons. *J. Geophys. Res.* 85:7715–7738.

Spreiter J. R., Summers, A. L., and Alksne, A. Y. 1966. Hydromagnetic flow around the magnetosphere. *Planet. Space Sci.* 14:223–253.

Spreiter, J. R., Summers, A. L., and Rizzi, A. W. 1970. Solar wind flow past nonmagnetic planets—Venus and Mars. *Planet. Space Sci.* 18:1281–1299.

Stacey, F. D. 1977. *Physics of the Earth* (New York: Wiley).

Stacey, F. D. 1980. The cooling Earth: A reappraisal. *Phys. Earth Planet. Inter.* 22:89–96.

Stankevich, K. S. 1970. Observations of Mars and Venus at 11.1 cm. *Austral. J. Phys.* 23:111–112.

Stebbins, J., and Kron, G. P. 1957. Six-color photometry of stars. X. The stellar magnitude and color index of the sun. *Astrophys. J.* 126:266–280.

Steffes, P. G., and Eshleman, V. R. 1982. Sulfuric acid vapor and other cloud-related gases in the Venus atmosphere: Abundances inferred from observed radio opacity. *Icarus* 51:322–333.

Stern, M. E. 1959. The moving flame experiment. *Tellus* 11:175–179.

Stevenson, D. J., Schubert, G., Cassen, P., and Reynolds, R. T. 1980. Core evolution and magnetism of the terrestrial planets. *Lunar and Planet. Sci.* 11:1091–1093.

Stewart, A. I. F. 1968. "Second Arizona Conf. on Planetary Atmospheres," held at Tucson, AZ.

Stewart, A. I. F. 1980. Design and operation of the Pioneer Venus orbiter ultraviolet spectrometer. *IEEE Trans. Geosci. and Remote Sensing* 18:65–70.

Stewart, A. I. F., and Barth, C. A. 1979. Ultraviolet night airglow of Venus. *Science* 205:59–62.

Stewart, A. I. F., Anderson, D. E., Esposito, L. W., and Barth, C. A. 1979. Ultraviolet spectroscopy of Venus: Initial results from the Pioneer Venus orbiter. *Science* 203:777–779.

Stewart, A. I. F., Gerard, J.-C., Rusch, D. W., and Bougher, S. W. 1980. Morphology of the Venus ultraviolet night airglow. *J. Geophys. Res.* 85:7861–7870.

Stewart, R. W. 1968. Interpretation of Mariner 5 and Venera 4 data on the upper atmosphere of Venus. *J. Atmos. Sci.* 25:578–582.

Stolarski, R. S. 1972. Analytic approach to photoelectron transport. *J. Geophys. Res.* 77:2862–2870.

Stone, P. H. 1968. Some properties of Hadley regimes on rotating and non-rotating planets. *J. Atmos. Sci.* 25:644–657.

Stone, P. H. 1974. The structure and circulation of the deep Venus atmosphere. *J. Atmos. Sci.* 31:1681–1690.

Stone, P. H. 1975. The dynamics of the atmosphere of Venus. *J. Atmos. Sci.* 32:1005–1016.

Strickland, D. J. 1973. The OI 1304 and 1356-Å emissions from the atmosphere of Venus. *J. Geophys. Res.* 78:2827–2836.

Strickler, S. J., Vikesland, J. P., and Bier, H. D. 1974. $^3B_1 - {}^1A_1$ transition of SO_2 gas. II. Radiative lifetime and radiationless processes. *J. Chem. Phys.* 60:664–667.

Suomi, V. E. 1974. Cloud motions on Venus. In *The Atmosphere of Venus*, ed. J. Hansen (New York: Goddard Inst. Space Studies), pp. 42–58.

Suomi, V. E., and Limaye, S. S. 1978. Venus: Further evidence of vortex circulation. *Science* 201:1009–1011.

Suomi, V. E., Sromovsky, L. A., and Revercomb, H. E. 1979. Preliminary results of the Pioneer Venus small probe net flux radiometer experiment. *Science* 205:82–85.

Suomi, V. E., Sromovsky, L. A., and Revercomb, H. E. 1980. Net radiation in the atmosphere of Venus: Measurements and interpretation. *J. Geophys. Res.* 85:8200–8218.

Surkov, Yu. A. 1977. Geochemical studies of Venus by Venera 9 and 10 automatic interplanetary stations. *Proc. Lunar Planet. Sci. Conf.* 8:2665–2689.

Surkov, Yu. A. 1979a. Paper presented at 22nd COSPAR Meeting, held at Bangalore, India.

Surkov, Yu. A. 1979b. *Pisma Astron. Zh.* 5:7.

Surkov, Yu. A., Andreichikov, B. M., and Kalinkina, O. M. 1974. Composition and structure of the cloud layer of Venus. *Space Res.* 14:673–687.

Surkov, Yu. A., Kirnozov, F. F., Glazov, V. N., Dunchenko, A. G., and Tatsil, L. P. 1976a. The content of natural radioactive elements in venusian rock as determined by Venera 9 and Venera 10. *Kosmich. Issled.* 14:704–709.

Surkov, Yu. A., Kirnozov, F. F., Khristianov, V. K., Korchuganov, B. N., Glazer, V. N., and Ivanov, V. F. 1976b. Density of surface rock on Venus from data obtained by the Venera 10 automatic interplanetary station. *Kosmich. Issled.* 14:697–703.

Surkov, Yu. A., Andreichikov, B. M., and Kalinkina, O. M. 1977. Gas-analysis equipment of the automatic interplanetary stations Venera 4, 5, 6, and 8. *Space Sci. Instrum.* 3:301–310.

Surkov, Yu. A., Ivanova, V. F., Pudov, A. N., Verkin, B. I., Bagrov, N. N., and Pilipenko, A. P. 1978. Mass-spectral study of the chemical composition of the Venus atmosphere by the unmanned space probes Venera 9 and Venera 10. *Geokhimiya* 4:506.

Surkov, Yu. A., Kurnozov, F., Guryanov, V., Glazov, V., Dunchenko, V., Kurochkin, V., Raspontny, E., Kharitonova, L., Tatsiy, L., and Gimadov, V. 1981. Venus cloud aerosol investigation on the Venera 12 space probe (preliminary data). *Geokhimiya* 7:3–9.

Surkov, Yu. A., Ivanova, V. F., Pudov, A. N., Pavlenko, V. A., Davydov, N. A., and Shejnin, D. M. 1982a. The measurements of water vapor concentration in the Venus atmosphere by Venera 13 and Venera 14. *Pisma Astron. Zh.* 8:411–413.

Surkov, Yu. A., Moskaleva, L. P., Shcheglov, O. P., Kharyukova, V. P., Manvelyan, O. S., and Smirnov, G. G. 1982b. Preliminary results of the determination of rock elemental composition on Venus by Venera 13 and Venera 14. *Pisma Astron. Zh.* 8:437–443.

Sykes, L. R., and Sbar, M. L. 1973. Intraplate earthquakes, lithospheric stresses, and the driving mechanism of plate tectonics. *Nature* 245:298–302.

Sze, N. D., and McElroy, M. B. 1975. Some problems in Venus aeronomy. *Planet. Space Sci.* 23:763–786.

Takacs, P. Z., Broadfoot, A. L., Smith, G. R., and Kumar, S. 1980. Mariner 10 observations of hydrogen Lyman-alpha emission from the Venus exosphere: Evidence of complex structure. *Planet. Space Sci.* 28:687–701.

Taranova, O. G. 1977. *Astron. Tsirk.* 950:1 (in Russian).

Tascione, T. F. 1982. A Three Dimensional Model of the Plasma Flow and Magnetic Field in the Dayside Ionosphere of Venus. Ph.D. Thesis, Rice Univ., Houston, TX.

Tauber, M. E., and Kirk, D. B. 1976. Impact craters on Venus. *Icarus* 28:351–357.

Taylor, F. W. 1975. Interpretation of Mariner 10 infrared observations of Venus. *J. Atmos. Sci.* 32:1101–1106.

Taylor, F. W., Diner, D. J., Elson, L. S., Hanner, M. S., McCleese, D. J., Martonchik, J. V., Reichley, P. E., Houghton, J. T., Delderfield, J., Schofield, J. T., Bradley, S. E., and Ingersoll, A. P. 1979a. Infrared remote sounding of the middle atmosphere of Venus from the Pioneer orbiter. *Science* 203:779–781.

Taylor, F. W., Diner, D. J., Elson, L. S., McCleese, D. J., Martonchik, J. V., Delderfield, J., Bradley, S. P., Schofield, J. T., Gille, J. C., and Coffey, M. T. 1979b. Temperature, cloud structure, and dynamics of Venus middle atmosphere by infrared remote sensing from Pioneer orbiter. *Science* 205:65–67.

Taylor, F. W., Vescelus, F. E., Locke, J. R., Foster, G. T., Forney, P. B., Beer, R., Houghton, J. T., Delderfield, J., and Schofield, J. T. 1979c. Infrared radiometer for the Pioneer Venus orbiter. I. Instrument description. *Appl. Optics* 18:3893–3900.

Taylor, F. W., Beer, R., Chahine, M. T., Diner, D. J., Elson, L. S., Haskins, R. D., McCleese, D. J. Martonchik, J. V., Reichley, P. E., Bradley, S. P., Delderfield, J., Schofield, J. T., Farmer, C. B., Froidevaux, L., Leung, J., Coffey, M. T., and Gille, J. C. 1980. Structure and meteorology of the middle atmosphere of Venus: Infrared remote sensing from the Pioneer orbiter. *J. Geophys. Res.* 85:7963–8006.

Taylor, F. W., Schofield, J. T., and Bradley, S. P. 1981. Pioneer Venus atmospheric observations. *Phil. Trans. Roy. Soc. London* A303:215–2233.

Taylor, H. A. Jr., and Walsh, W. J. 1972. The light ion trough, the main trough, and the plasmapause. *J. Geophys. Res.* 77:6716–6723.

Taylor, H. A. Jr., Brinton, H. C., Bauer, S. J., Hartle, R. E., Cloutier, P. A., Daniell, R. E., and Donahue, T. M. 1979a. Ionosphere of Venus: First observations of day-night variation of the ion composition. *Science* 205:96–99.

Taylor, H. A. Jr., Brinton, H. C., Bauer, S. J., Hartle, R. E., Cloutier, P. A., Michel, F. C., Daniell, R. E., Donahue, T. M., and Maehl, R. C. 1979b. Ionosphere of Venus: First observations of the effects of dynamics on the dayside ion composition. *Science* 203: 755–757.

Taylor, H. A. Jr., Brinton, H. C., Bauer, S. J., Hartle, R. E., Donahue, T. M., Cloutier, P. A., Michel, F. C., Daniell, R. E., and Blackwell, B. H. 1979c. Ionosphere of Venus: First observations of the dayside ion composition near dawn and dusk. *Science* 203:752–754.

Taylor, H. A. Jr., Brinton, H. C., Bauer, S. J., Hartle, R. E., Cloutier, P. A., and Daniell, R. E. 1980. Global observations of the composition and dynamics of the ionosphere of Venus: Implications for the solar wind interaction. *J. Geophys. Res.* 85:7765–7777.

Taylor, H. A. Jr., Bauer, S. J., Daniell, R. E., Brinton, H. C., Mayr, H. G. and Hartle, R. E. 1981a. Temporal and spatial variations observed on the ionospheric composition on Venus: Implication for empirical modeling. *Adv. Space Res.* 1:37–51.

Taylor, H. A. Jr., Daniell, R. E., Hartle, R. E., Brinton, H. C., Bauer, S. J., and Scarf, F. L. 1981b. Dynamic variations observed in thermal and superthermal ion distributions in the dayside ionosphere of Venus. *Adv. Space Res.* 1:247–258.

Taylor, H. A. Jr., Hartle, R. E., Niemann, H. B., Brace, L. H., Daniell, R. E., Bauer, S. J., and Kliore, A. J. 1982. Observed composition of the ionosphere of Venus: Implications for the ionization peak and the maintenance of the nightside ionosphere. *Icarus* 51:283–295.

Taylor, K. E. 1980. The roles of mean meridional motions and large-scale eddies in zonally averaged circulations. *J. Atmos. Sci.* 37:1–19.

Taylor, S. R. 1982. *Planetary Science: A Lunar Perspective* (Houston, TX: Lunar Planet. Sci. Inst.).

Taylor, W. W. L., Scarf, F. L., Russell, C. T., and Brace, L. H. 1979a. Absorption of whistler mode waves in the ionosphere of Venus. *Science* 205:112–114.

Taylor, W. W. L., Scarf, F. L., Russell, C. T., and Brace, L. H. 1979b. Evidence for lightning on Venus. *Nature* 282:614–616.

Teitelbaum, H., and Cot, C. 1981. Calculation of the solar gravitational torque on the Venus thermal tide. *Astron. Astrophys.* 97:265–268.

Theis, R. F., Brace, L. H., and Mayr, H. G. 1980. Empirical models of the electron temperature and density in the Venus ionosphere. *J. Geophys. Res.* 85:7787–7794.

Thomas, L., Greer, R. G. H., and Dickinson, P. H. G. 1979. The excitation of the 557.7 nm
line and Herzberg bands in the nightglow. *Planet. Space Sci.* 27:925–931.

Thompson, R. 1970. Venus' general circulation is a merry-go-round. *J. Atmos. Sci.*
27:1107–1116.

Thornton, D. D., and Welch, W. L. 1964. Radio emission from Venus at 8.35 mm. *Astron. J.*
69:71–72.

Tidman, D. A., and Krall, N. A. 1971. *Shock Waves in Collisionless Plasmas* (New York:
Wiley-Interscience).

Timofeeva, T. S., Yakovlev, O. I., and Efimov, A. I. 1978. Radio wave fluctuation and
turbulence on the nighttime venusian atmosphere from radioscopy data of the space probe
Venera 9. *Cosmic Res.* 16:226–232.

Toksoz, M. N., Hsui, A. T., and Johnson, D. H. 1978. Thermal evolutions of the terrestrial
planets. *Moon Planets* 18:281–320.

Tolbert, C. W., and Straiton, A. W. 1964. 35 Gc/s, 70 Gc/s and 94 Gc/s cytherean radiation.
Nature 204:1242–1245.

Tolchel'nikov, Yu. S. 1974. *The Optical Properties of a Landscape* (Leningrad: Nauka).

Tolstikhin, I. N. 1981. K voprosu o rasprostranennosti planetarnykh pervichnykh gazov v
atmosferakh Venery, Zemli i Marsa. *Geokhimia* 6:803.

Tomasko, M. G., Boese, R., Ingersoll, A. P., Lacis, A. A., Limaye, S. S., Pollack, J. B.,
Seiff, A., Stewart, A. I. F., Suomi, V. E., and Taylor, F. W. 1977. The thermal balance of
the atmosphere of Venus. *Space Sci. Rev.* 20:389–412.

Tomasko, M. G., Doose, L. R., Palmer, J., Holmes, A., Wolfe, W., Castillo, N. D., and
Smith, P. H. 1979*a*. Preliminary results of the solar flux radiometer experiment aboard the
Pioneer Venus multiprobe mission. *Science* 203:795–797.

Tomasko, M. G., Doose, L. R., and Smith, P. H. 1979*b*. Absorption of sunlight in the
atmosphere of Venus. *Science* 205:80–82.

Tomasko, M. G., Doose, L. R., Smith, P. H., and Odell, A. P. 1980*a*. Measurements of the
flux of sunlight in the atmosphere of Venus. *J. Geophys. Res.* 85:8167–8186.

Tomasko, M. G., Smith, P. H., Suomi, V. E., Sromovsky, L. A., Revercomb, H. E.,
Taylor, F. W., Martonchik, D. J., Seiff, A., Boese, R., Pollack, J. B., Ingersoll, A. P.,
Schubert, G., and Covey, C. C. 1980*b*. The thermal balance of Venus in light of the
Pioneer Venus mission. *J. Geophys. Res.* 85:8187–8199.

Toon, O. B., and Farlow, N. H. 1981. Particles above the tropopause: Measurement models
of stratospheric aerosols, meteoric debris, nacreous clouds, and noctilucent clouds. *Ann.
Rev. Earth and Planet. Sci.* 9:19–58.

Toon, O. B., and Blamont, J. 1982. Large solid particles in the clouds of Venus: Do they
exist? Submitted to *Icarus*.

Toon, O. B., and Turco, R. P. 1982. The ultraviolet absorber on Venus: Amorphous sulphur.
Icarus 51:358–373.

Toon, O. B., Turco, R. P., Keesee, R., and Whitten, R. 1982. Physics of the Venus clouds.
Submitted to *Icarus*.

Torr, D. G., and Torr, M. R. 1978. Review of rate coefficients of ionic reactions determined
from measurements made by the Atmosphere Explorer satellites. *Rev. Geophys. Space
Phys.* 16:327–340.

Torr, M. R., and Torr, D. G. 1982. The role of metastable species in the thermosphere. *Rev.
Geophys. Space Phys.* 20:91–144.

Torr, M. R., Torr, D. G., Ong, R. A., and Hinteregger, H. E. 1979. Ionization frequencies
for major thermospheric constituents as a function of solar cycle 21. *Geophys. Res. Letters*
6:771–774.

Torr, M. R., Richards, P. G., and Torr, D. G. 1980. A new determination of the ultraviolet
heating efficiency of the thermosphere. *J. Geophys. Res.* 85:6819–6826.

Transactions of the IAU 1971. Vol. 14B (Dordrecht: Reidel).

Traub, W. A., and Carleton, N. P. 1973. A search for H_2 and O_2 on Venus. *Bull. Amer.
Astron. Soc.* 5:299–300 (abstract).

Traub, W. A., and Carleton, N. P. 1974. Observations of O_2, H_2O and HD in planetary atmospheres. In *Exploration of the Planetary System,* eds. A. Woszczyk and C.
Iwaniszewska (Dordrecht: Reidel), pp. 223–228.

Traub, W. A., and Carleton, N. P. 1975. Spectroscopic observations of winds on Venus. *J. Atmos. Sci.* 32:1045–1059.

Traub, W. A., and Carleton, N. P. 1979. Retrograde winds on Venus: Possible periodic variations. *Astrophys. J.* 227:329–333.

Traub, W. A., Carleton, N. P., Connes, P., and Noxon, J. R. 1979. The latitude variations of O_2 dayglow and O_3 abundance on Mars. *Astrophys. J.* 229:846–850.

Travis, L. D. 1975. On the origin of ultraviolet contrasts on Venus. *J. Atmos. Sci.* 32: 1190–1200.

Travis, L. D. 1978. Nature of the atmospheric dynamics on Venus from power spectrum analysis of Mariner 10 images. *J. Atmos. Sci.* 35:1584–1595.

Travis, L. D., Coffeen, D. L., Del Genio, A. D., Hansen, J. E., Kawabata, K., Lacis, A. A., Lane, W. A., Limaye, S. S., Rossow, W. B., and Stone, P. H. 1979a. Cloud images from the Pioneer Venus orbiter. *Science* 205:74–76.

Travis, L. D., Coffeen, D. L., Hansen, J. E., Kawabata, K., Lacis, A. A., Lane, W. A., Limaye, S. S., and Stone, P. H. 1979b. Orbiter cloud photopolarimeter investigation. *Science* 203:781–785.

Tritton, D. J. 1975. Internally heated convection in the atmosphere of Venus and in the laboratory. *Nature* 257:110–112.

Tullis, J. A. 1979. High temperature deformation of rocks and minerals. *Rev. Geophys. Space Phys.* 17:1137–1154.

Turco, R. P, 1982. Models of stratospheric aerosols and dust. In *The Stratospheric Aerosol Layer,* ed. R. C. Whitten (Heidelberg: Springer-Verlag), Ch. 4.

Turco, R. P., Toon, O. B., Keesee, R., and Whitten, R. 1982a. Chemistry of the Venus clouds. Submitted to *Icarus.*

Turco, R. P., Toon, O. B., Whitten, R. C., and Keesee, R. G. 1982b. Noctilucent and nacreous clouds on Venus. *Icarus.* In press.

Turcotte, D. L. 1979. Convection. *Rev. Geophys. Space Phys.* 17:1090–1098.

Turcotte, D. L., Cooke, F. A., and Willeman, R. J. 1979. Parameterized convection within the moon and the terrestrial planets. *Proc. Lunar Planet. Sci. Conf.* 10:2375–2392.

Turcotte, D. L., Willeman, R. J., Haxby, W. F., and Norberry, J. 1981. Role of membrane stresses in the support of planetary topography. *J. Geophys. Res.* 86:3951–3959.

Turekian, K. K., and Clark, S. P. 1975. The nonhomogeneous accumulation model for terrestrial planet formation. *J. Atmos. Sci.* 32:1257–1261.

Tverskoy, P. N. 1962. *Meteorology* (Leningrad: Gidrometeoizdat).

Ulich, B. L., Cogdell, J. R., and Davies, J. H. 1972. Planetary observations at millimeter wavelength. *NASA Tech. Rept.* No. NGL-006-72-1.

Uman, M. A. 1969. *Lightning* (New York: McGraw-Hill).

Urey, H. C. 1952. *The Planets, Their Origin, and Development* (New Haven: Yale Univ. Press).

U. S. Standard Atmosphere 1976. Nat. Oceanic Atmos. Admin. S/T 76-1562 (Washington, D. C.: U. S. Gov't. Printing Office).

Usikov, A. Ya., Korniyenko, Yu. V., and Shkwratov, Yu. G. 1982. Analysis of the relation between the altitude and the surface roughness of Venus according to radar data from the Pioneer/Venus space vehicle. *Doklady AN SSSR.* In press.

Ustinov, E. A. 1977. *Kosmich. Issled.* 15:768.

Ustinov, E. A. 1979. *Kosmich. Issled.* 17:81.

Vaisberg, O. L. 1980. On the asymmetry of the internal flow in the wake of Venus. *Kosmich. Issled.* 18:809.

Vaisberg, O. L., Bogdanov, A. V., Smirnov, V. N., and Romanov, S. A. 1976a. On the nature of the solar wind Mars interaction. In *Solar-Wind Interaction with the Planets Mercury, Venus, and Mars,* ed. E. F. Ness (Washington, D. C.: NASA SP-397), pp. 21–40.

Vaisberg, O. L., Romanov, S. A., Smirnov, V. N., Karpinsky, I. P., Khazanov, B. I., Polenov, B. V., Bogdanov, A. V., and Antonova, N. M. 1976b. The structure of the field of interaction between the solar wind and Venus according to measurements of the characteristics of the ion flux aboard the automatic stations Venera-9 and Venera-10. *Kosmich. Issled.* 14:827.

Vaisberg, O. L., Smirnov, V. N., Karpinsky, I. P., Khazanov, B. I., Polenov, B. N., Bogdanov, A. V., and Antonova, N. M. 1976c. Ion flux parameters in the solar-wind-

Venus interaction region according to Venera-9 and Venera-10 data. In *Physics of Solar Planetary Environments*, ed. D. J. Williams (Boulder, CO: Amer. Geophys. Union), pp. 904–917.

Vaisberg, O. L., Intriligator, D. S., and Smirnov, V. N. 1980. An empirical model of the Venusian outer environment. I. The shape of the dayside solar wind atmosphere interface. *J. Geophys. Res.* 85:7642–7650.

Van Allen, J. A., Krimigis, S. M., Frank, L. A., and Armstrong, T. P. 1967. An upper limit on intrinsic dipole moment based on absence of a radiation belt. *Science* 158:1673–1675.

de Vaucouleurs, G. 1964. Geometric and photometric parameters of the terrestrial planets. *Icarus* 3:187–235.

de Vaucouleurs, G., and Menzel, D. H. 1960. Results of the occultation of Regulus by Venus, July 7, 1959. *Nature* 188:28–33.

Verigin, M. I., Gringauz, K. I., Gombosi, T., Breus, T. K., Bezrukikh, V. V., Remizov, A. P., and Volkov, G. I. 1978. Plasma near Venus from the Venera-9 and Venera-10 wide angle analyzer data. *J. Geophys. Res.* 83:3721–3728.

Vetukhnovskaya, Yu. N. 1969. Radio Astronomy Measurements of Venus and Several Results from an Interpretation of these Measurements on the Basis of AIS Venera-4 Data. Ph.D. thesis, FIAN. In Russian.

Vetukhnovskaya, Yu. N., Kuz'min, A. D., Kutusa, B. G., Losovskiy, B. Ya., and Salomonovich, A. E. 1963. Measurements of the spectra of the radio emission of the nightside of Venus. *Radiofisika* 6:1054–1056.

Vetukhnovskaya, Yu. N., Kuz'min, A. D., and Losovskiy, B. Ya. 1971. Measurements of the phase change of the radio emission of Venus at wavelength 8.2 mm. *Astron. J.* 48:1033–1037.

Vinogradov, A. P., Surkov, Yu. A., and Florensky, C. P. 1968. The chemical composition of Venus atmosphere based on the data of the interplanetary station Venera 4. *J. Atmos. Sci.* 25:535–536.

Vinogradov, A. P., Surkov, Yu. A., Andreichikov, B. M., Kalinkina, O. M., and Grechishcheva, I. M. 1971. Chemical composition of the Venus atmosphere. In *Planetary Atmospheres*, eds. C. Sagan, T. C. Owen, and H. J. Smith (New York: Springer-Verlag), pp 3–16; *Space Res.* 11:129.

Vinogradov, A. P., Surkov, Yu. A., and Kirnozov, F. F. 1973. The contents of uranium, thorium, and potassium in the rocks of Venus as measured by Venera 8. *Icarus* 20:253–259.

Vinogradov, A. P., Florenskiy, K. P., Bazilevskiy, A. T., and Selivanov, A. S. 1976. The first panoramas of the surface of Venus. (Preliminary analysis of the images.) *Doklady AN SSSR* 228:570–573.

Vityazev, A. V., and Pechernikova, G. V. 1980. Thermal dissipation of gas from the protoplanetary cloud. In *Proc. COSPAR XXIII*. Planetary meeting held at Budapest, Hungary, p. 244 (abstract).

Voitkevich, G. V., and Zakrutkin, V. V. 1976. Osnovy Geokhimii. M.: Vyshaya shkola, 136.

Vonder Haar, T. H., and Suomi, V. E. 1971. Measurements of the Earth's radiation budget from satellites during a five year period. I. Extended time and space means. *J. Atmos. Sci.* 28:305–314.

Vorontsov-Velyaminov, B. A. 1956. *Essays on the History of Astronomy in Russia* (Moscow: State Publ. House for Tech. Theor. Literature). In Russian.

Walker, J. C. G. 1975. Evolution of the atmosphere of Venus. *J. Atmos. Sci.* 32:1248–1256.

Walker, J. C. G. 1977. *Evolution of the Atmosphere* (New York: McMillan).

Walker, J. C. G. 1982. The earliest atmosphere of the Earth. *Precambrian Res.* 17:147–171.

Walker, J. C. G., Hays, P. B., and Kasting, J. F. 1981. A negative feedback mechanism for the long-term stabilization of Earth's surface temperature. *J. Geophys. Res.* 86:9776–9782.

Wallace, J. M. 1971. Spectral studies of tropospheric wave disturbances in the tropical western Pacific. *Rev. Geophys. Space Phys.* 9:557–612.

Wallace, J. M. 1973. General circulation of the tropical lower stratosphere. *Rev. Geophys. Space Phys.* 11:191–222.

Wallace, J. M., and Kousky, V. E. 1968. Observational evidence of Kelvin waves in the tropical stratosphere. *J. Atmos. Sci.* 25:900–907.

Wallace, L. 1969. Analysis of the Lyman alpha observations of Venus made from Mariner 5. *J. Geophys. Res.* 74:115–131.

Wallace, L., and Hunten, D. M. 1978. The Jovian spectrum in the region 0.4–1.1 μm. The C/H ratio. *Rev. Geophys. Space Phys.* 16:289–319.

Wallace, L., Stuart, F. E., Nagel, R. H., and Larson, M. D. 1971. A search for deuterium on Venus. *Astrophys. J.* 168:L29–L31.

Wallace, L., Caldwell, J. J., and Savage, B. D. 1972. Ultraviolet photometry from the Orbiting Astronomical Observatory. III. Observations of Venus, Mars, Jupiter, and Saturn longward of 2000 Å. *Astrophys. J.* 172:755–769.

Wallis, M. K. 1972. Comet-like interaction of Venus with the solar wind. I. *Cosmic Electrodyn.* 3:45–59.

Wallis, M. K. 1973. Weakly-shocked flows of the solar wind plasma through atmospheres of comets and planets. *Planet. Space Sci.* 21:1647–1660.

Walters, G. K. 1964. Effect of oblique interplanetary magnetic field on shape and behavior of the magnetopause. *J. Geophys. Res.* 69:1769–1783.

Walters, G. K. 1966. On the existence of a second standing shockwave attached to the magnetopause. *J. Geophys. Res.* 71:1341–1344.

Walterscheid, R. L. 1980. Traveling planetary waves in the stratosphere. *Pure Appl. Geophys.* 118:239–265.

Walterscheid, R. L. 1981. Dynamical cooling induced by dissipating internal gravity waves. *Geophys. Res. Letters* 8:1235–1238.

Ward, D. B., Gill, G. E., and Harwit, M. 1977. Far infrared spectral observations of Venus, Mars, and Jupiter. *Icarus* 30:295–300.

Ward, W. R., and DeCampli, W. M. 1979. Comments on the Venus rotation pole. *Astrophys. J.* 230:L117–L121.

Warner, J. L. 1980. Venus: Do sediments cover low-lands? *Bull. Amer. Astron. Soc.* 12:691 (abstract).

Warner, J. L. 1982. Sedimentary processes and crustal cycling on Venus. *Proc. Lunar Planet. Sci. Conf.* 12. In press.

Warner, J. L. and Morrison, D. A. 1978. Planetary tectonics. I. The role of water. *Proc. Lunar Planet. Sci. Conf.* 9:1217–1219 (abstract).

Warnock, W. W., and Dickel, J. R. 1972. Venus: Measurements of brightness temperatures in the 7–15 cm wavelength range and theoretical radio and radar spectra for a two layer subsurface model. *Icarus* 17:682–691.

Wasserburg, G. J., MacDonald, G. J. F., Hoyle, F., and Fowler, W. A. 1964. Relative contributions of uranium, thorium, and potassium to heat production in the Earth. *Science* 143:465–467.

Wasserburg, G. J., Papanastassiou, D. A., Tera, F., and Huneke, J. C. 1977. Outline of a lunar chronology. In *The Moon, A New Appraisal* (London: Royal Society).

Watson, A. J., Donahue, T. M., Stedman, D. H., Knollenberg, R. G., Ragent, B., and Blamont, J. 1979. Oxides of nitrogen and the clouds of Venus. *Geophys. Res. Letters* 6:743–746.

Watson, A. J., Donahue, T. M. and Walker, J. C. G. 1982. The dynamics of a rapidly escaping atmosphere: Applications to the evolution of Earth and Venus. *Icarus* 48:150–166.

Watson, A. J., Donahue, T. M., and Walker, J. C. G. 1983. Temperature in a high pressure steam atmosphere of Venus. Submitted to *Icarus*.

Watson, R., Machado, G., Fischer, S., and Davis, D. D. 1976. A temperature dependence kinetics study of the reactions of Cl and O_3, CH_4, and H_2O_2. *J. Chem. Phys.* 65:2126–2138.

Watts, A. B., Bodine, J. H., and Steckler, M. S. 1980. Observations of flexure and the state of stress in the oceanic lithosphere. *J. Geophys. Res.* 85:6369–6376.

Wayne, R. P. 1973. Reaction involving excited state of O and O_2. In *Physics and Chemistry of the Upper Atmosphere*, ed. B. M. McCormac (Dordrecht: D. Reidel), pp. 125–132.

Weertman, J. 1970. The creep strength of the Earth's mantle. *Rev. Geophys. Space Phys.* 8:145–168.

Weertman, J. 1979. Height of mountains on Venus and the creep properties of rock. *Phys. Earth Planet. Int.* 19:197–207.

Weertman, J., and Weertman, J. R. 1975. High temperature creep of rock and mantle viscosity. *Ann. Rev. Earth Planet. Sci.* 3:293–315.

Weidenshilling, S. J. 1976. Accretion of the terrestrial planets. *Icarus* 27:161–170.

Weidenschilling, S. J. 1978. Iron/silicate fractionation and origin of Mercury. *Icarus* 35:99–111.

Weinberg, J. L., and Newkirk, G. 1961. Airglow of Venus: A reexamination *Planet. Space Sci.* 5:163–164.

Welch, W. L., and Thornton, D. D. 1965. Recent planetary observations at wavelength near 1 cm. *Astron. J.* 70:149–150.

Weller, C. S., and Biondi, M. A. 1967. Measurements of dissociative recombination of CO_2^+ ions with electrons. *Phys. Rev. Letters* 19:59–61.

Wendlandt, R. F., and Morgan, P. 1982. Lithospheric thinning associated with rifting in East Africa. *Nature* 298:734–736.

Werner, M. W., Neugebauer, G., Houck, J. R., and Hauser, M. G. 1978. One millimeter brightness temperatures of the planets. *Icarus* 35:289–296.

Wetherill, G. W. 1975. Late heavy bombardment of the moon and terrestrial planets. *Proc. Lunar Planet. Sci. Conf.* 4:1539–1559 (abstract).

Wetherill, G. W. 1980. ^{36}Ar on Venus. *Proc. Lunar Planet. Sci. Conf.* 11:1239–1242 (abstract).

Wetherill, G. W. 1981. Solar wind origin of ^{36}Ar on Venus. *Icarus* 46:70–80.

Whitcomb, S. E., Hildebrand, R. H., Keene, J., Stiening, R. F., and Harper, D. A. 1979. Submillimetre brightness temperatures of Venus, Jupiter, Uranus, and Neptune. *Icarus* 38:75–80.

White, R. E. 1978. On the detectability of millimeter wavelength spectral lines from planetary atmospheres. *Astron. J.* 83:1122–1138.

Whitehead, J. A. 1972. Observations of rapid mean flow produced in mercury by a moving heater. *Geophys. Fluid Dyn.* 3:161–180.

Whitten, R. C. 1969. Thermal structure of the ionosphere of Venus. *J. Geophys. Res.* 74:5623–5628.

Whitten, R. C. 1970. The daytime upper ionosphere of Venus. *J. Geophys. Res.* 75:3707–3714.

Whitten, R. C., Baldwin, B., Knudsen, W. C., Miller, K. L., and Spenner, K. 1982. The Venus ionosphere at grazing incidence of solar radiation: Transport of plasma to and in the nightside. *Icarus* 51:261–270.

Widing, K. G., Purcell, J. D., and Saudlin, G. D. 1970. The UV continuum solar temperature minimum. *Solar Phys.* 12:52–62.

Wildt, R. 1940a. Note on the surface temperature of Venus. *Astrophys. J.* 91:266–268.

Wildt, R. 1940b. On the possible existence of formaldehyde in the atmosphere of Venus. *Astrophys. J.* 92:247–255.

Williams, B. G., Mottinger, N. A., and Birkeland, P. W. 1982. Venus gravity: Seventh degree and order field. *Geophys. Res. Letters.* In press.

Williams, M. A., Thomason, L. W., and Hunten, D. M. 1982. The transmission to space of the light produced by lightning in the clouds of Venus. *Icarus.* In press.

Wilson, W. J., and Klein, M. J. 1981. Venus: Observed variations in the carbon monoxide distribution. "Internat. Conf. on the Venus Environment," held at Palo Alto, CA.

Wilson, W. J., Klein, M. J., Kakar, R. K., Gulkis, S., Olsen, E. T., and Ho, P. T. P. 1981. Venus. I. Carbon monoxide distribution and molecular-line searches. *Icarus* 45:624–637.

Winick, J. R., and Stewart, A. I. F. 1980. Photochemistry of SO_2 in Venus' upper cloud layers. *J. Geophys. Res.* 85:7849–7860.

Wise, D. U. 1972. Freeboard of continents through time. In *Studies in Earth and Space Sciences* (Geol. Soc. Amer.), GSA Memoir 132:87–100.

Witt, G., Stegman, J., Solheim, B., and Llewellyn, E. J. 1979. A measurement of the O_2 (b $^1\Sigma_g^+ \rightarrow X^3\Sigma_g^-$) atmospheric band and $O(^1S)$ green line in the nightglow. *Planet. Space Sci.* 27:341–350.

Wofsy, S. C., and McElroy, M. B. 1974. HO_x, NO_x, and ClO_x: Their role in atmospheric photochemistry. *Can. J. Chem.* 52:1582–1591.

Wofsy, S. C., and Sze, N. D. 1975. Venus cloud models. In *Atmospheres of Earth and the Planets,* ed. B. M. McCormac (Dordrecht: D. Reidel), pp. 369–384.

Wofsy, S. C., McElroy, M. B., and Yung, Y. L. 1975. The chemistry of atmospheric bromine. *Geophys. Res. Letters* 2:215–218.

Wolff, R. S., Stein, R. F., and Taylor, H. A. 1979. Initial Pioneer Venus magnetic field results: Nightside observations. *Science* 205:114–116.

Wolff, R. S., Goldstein, B. E., and Yeates, C. M. 1980. The onset and development of Kelvin-Helmholtz instability at the Venus ionopause. *J. Geophys. Res.* 85:7697–7707.

Wolff, R. S., Stein, R. F., and Taylor, H. A. 1982. Dynamics of Venus ionosphere. I. Simulation of compression of upper dayside ionosphere. *J. Geophys. Res.* In press.

Woo, R. 1975. Observations of turbulence in the atmosphere of Venus using Mariner 10 radio occultation measurements. *J. Atmos. Sci.* 32:1084–1090.

Woo, R., and Ishimaru, A. 1981. Eddy diffusion coefficient for the atmosphere of Venus from radio scintillation measurements. *Nature* 289:383–384.

Woo, R., Ishimaru, A., and Kendall, W. B. 1974. Observations of small-scale turbulence in the atmosphere of Venus by Mariner 5. *J. Atmos. Sci.* 31:1698–1706.

Woo, R., Armstrong, J. W., and Ishimaru, A. 1980. Radio occultation measurements of turbulence in the Venus atmosphere by Pioneer Venus. *J. Geophys. Res.* 85:8031–8038.

Woo, R., Armstrong, J. W., and Kliore, A. J. 1982. Small scale turbulence in the atmosphere of Venus. *Icarus* Vol. 52, No. 2.

Wood, C. 1979. Venusian volcanism: Environmental effects of style and landforms. *NASA Tech. Mem.* 80339:244–276

Wood, J. A. 1962. Chondrules and the origin of the terrestrial planets. *Nature* 194:127–130.

Wraight, P. C. 1982. Association of atomic oxygen and airglow excitation mechanism. *Planet. Space Sci.* 30:251–259.

Wright, W. H. 1927. Photographs of Venus made by infrared and by violet light. *Publ. Astron. Soc. Pacific* 39:220–221.

Wrixon, G. T., Welch, W. J., and Thornton, D. D. 1971. The spectrum of Jupiter at millimeter wavelength. *Astrophys. J.* 169:171–183.

Wu, C. Y. R., and Judge, D. L. 1981. SO_2 and CS_2 cross section data in the ultraviolet region. *Geophys. Res. Letters* 8:769–771.

Wyllie, P. J. 1971. *The Dynamic Earth* (New York: Wiley).

Yakovlev, O. I., and Matyugov, S. S. 1982. Neutral atmosphere of Venus as measured by radio occultation by Venera 9 and Venera 10 orbiters. Submitted to *Icarus*.

Yakovlev, O. I., Efimov, A. I., Timofeeva, T. S., Yakovleva, G. D., Chub, E. V., Tikhonov, V. F., and Shtrykov. V. K. 1976. Preliminary radio transmission data and the venusian atmosphere from Venera 9 and Venera 10. *Cosmic Res.* 14:632–637.

Yakovlev, O. I., Efimov, A. I., Matyugov, S. S., Timofeeva, T. S., Chub, E. V., and Yakovleva, G. D. 1978. Radioscopy of the nighttime atmosphere of Venus by probes Venera 9 and Venera 10. *Cosmic Res.* 16:88–93.

Yanai, M. 1975. Tropical meteorology. *Rev. Geophys. Space Phys.* 13:685–710.

Yanai, M., and Maruyama, T. 1966. Stratospheric wave disturbance propagating over the equatorial Pacific. *J. Meteor. Soc. Japan* 44:291–294.

Yanovitskiy, E. G. 1972. *Astron. J. Zh.* 49:845.

Yefanov, V. A., Kislyakov, A. G., Moiseyev, I. G., and Naurnov, A. I. 1969. The radio emission of Venus at wavelengths 2.25 and 8 cm. *Astron. J.* 46:147–151.

Yeroshenko, E. G. 1979. Unipolar induction effects in the magnetic tail of Venus. *Kosmich. Issled.* 17:93–105.

Yoder, H. S. Jr. 1976. *Generation of Basaltic Magma* (Washington D. C.: Nat. Acad. Sci.).

Young, A. T. 1973. Are the clouds of Venus sulfuric acid? *Icarus* 18:564–582.

Young, A. T. 1975a. Is the four-day "rotation" of Venus illusory? *Icarus* 24:1–10.

Young, A. T. 1975b. The clouds of Venus. *J. Atmos. Sci.* 32:1125–1132.

Young, A. T. 1977. An improved Venus cloud model. *Icarus* 32:1–26.

Young, A. T. 1979. Chemistry and thermodynamics of sulfur on Venus. *Geophys. Res. Letters* 6:49–50.

Young, A. T. 1980. Possible explanation of large thermal fluxes measured from Pioneer Venus entry probes. *Bull. Amer. Astron. Soc.* 12:717 (abstract).

Young, A. T., and Young, L. D. G. 1973. Comments on the composition of the Venus cloud tops in light of recent spectroscopic data. *Astrophys. J.* 179:39–43.

Young, L. D. G. 1972. High resolution spectra of Venus: A review. *Icarus* 17:632–658.

Young, L. D. G. 1974. Infrared spectra of Venus. In *Exploration of Planetary Systems*, eds. A. Woszczyk and A. Iwaniszewska (Dordrecht: Reidel), pp. 77–160.

Young, L. D. G. 1981. A new interpretation of the Venera 11 spectra of Venus. "Internat. Conf. on the Venus Environment," held at Palo Alto, CA.

Young, L. D. G., Schorn, R. A. J., and Smith, J. J. 1970. High-dispersion spectroscopic observations of Venus. IX. The carbon dioxide bands at 12030 Å and 12177 Å. *Icarus* 13:74–81.

Young, L. D. G., Young, A. T., Sanko, N. F., and Zasova, L. V. 1982. A new interpretation of the Venera 11 spectra of Venus. Submitted to *Icarus*.

Young, R. A., and Black, G. 1966. Excited state formation and destruction in mixtures of atomic oxygen and nitrogen. *J. Chem. Phys.* 44:3741–3751.

Young, R. E. 1981. Models of dynamic instability in the Venus atmosphere. "Internat. Conf. on the Venus Environment," held at Palo Alto, CA.

Young, R. E., and Schubert, G. 1973. Dynamical aspects of the Venus 4-day circulation. *Planet. Space Sci.* 21:1563–1580.

Young, R. E., and Pollack, J. B. 1977. A three-dimensional model of dynamical processes in the Venus atmosphere. *J. Atmos. Sci.* 34:1315–1351.

Young, R. E., and Pollack, J. B. 1980. Reply. *J. Atmos. Sci.* 37:253–255.

Young, R. E., Schubert, G., and Torrance, K. E. 1972. Nonlinear motions induced by moving thermal waves. *J. Fluid Mech.* 54:163–187.

Young, R. E., Pfister, L., Seiff, A., and Houben, H. 1982. Baroclinic instability in the Venus atmosphere. In *Proc. of the NASA Fourth Annual Meeting of Planetary Atmospheres Principal Investigators*.

Yung, Y. L., and McElroy, M. B. 1979. Fixation of nitrogen in the prebiotic atmosphere. *Science* 203:1002–1004.

Yung, Y. L., and DeMore, W. B. 1982. Photochemistry of the stratosphere of Venus: Implications for atmospheric evolution. *Icarus* Vol. 51, No. 2.

von Zahn, U. 1963. Monopole spectrometer, a new electric field mass spectrometer. *Rev. Sci. Instrum.* 34:1–4.

von Zahn, U. 1977. Bus neutral mass spectrometer (BNMS). *Space Sci. Rev.* 20:451–458.

von Zahn, U., and Mauersberger, K. 1978. Small mass spectrometer with extended capabilities at high pressures. *Rev. Sci. Instrum.* 49:1539–1542.

von Zahn, U., Fricke, K. H., Hoffman, H. J., and Pelka, K. 1979a. Venus: Eddy coefficient in the thermosphere and the inferred helium content of the lower atmosphere. *Geophys. Res. Letters* 6:337–340.

von Zahn, U., Krankowsky, D., Mauersberger, K., Nier, A. O., and Hunten, D. M. 1979b. Venus thermosphere: In-situ composition measurements, the temperature profile, and the homopause altitude. *Science* 203:768–770.

von Zahn, U., Fricke, K. H., Hunten, D. M., Krankowsky, D., Mauersberger, K., and Nier, A. O. 1980. The upper atmosphere of Venus during morning conditions. *J. Geophys. Res.* 85:7829–7840.

Zahniser, M. S., and Kaufman, F. 1977. Kinetics of the reactions of ClO with O and with NO. *L. Chem. Phys.* 66:3673–3681.

Zaitsev, A. V., Karyagin, V. P., Kovtunenko, V. M., Kremnev, R. S., Fokin, V. G., Pichkhadze, K. M., Rodin, A. L., Sukhanov, K. G., Turchaninov, V. N., and Fedorov, W. S. 1979. Venera 13 and Venera 14 probes: The description of descent in the atmosphere of planets. *Kosmich. Issled.* 17:655.

Zasetokiy, V. V., Nesterenko, V. V., Trakhtman, A. M., and Chemodanov, V. P. 1979. Preliminary digital processing of the images of the surface of Venus. In *First Panoramas of the Surface of Venus.* (Moscow: Nauka), pp. 57–62.

Zasova, L. V. et al. 1980. In *Proc. COSPAR XXIII.* Planetary Meeting held at Budapest, Hungary.

Zasova, L. V., Krasnopolsky, V.A., and Moroz, V. I. 1981. Vertical distribution of SO_2 in upper cloud layer of Venus and origin of UV-absorption. *Adv. Space Res.* 1:13–16.

Zeldovich, Y. B., and Raizer, Y. B. 1966. *Physics of Shockwaves and High Temperature Hydrodynamic Phenomena* (New York: Academic).

Zeleny, L. M., and Vaisberg, O. L. 1981. Formation of the plasma mantle in the magnetosphere of Venus. *Kosmich. Issled.* In press.

Zel'manovich, I. L., and Shifrin, K. S. 1968. *Tablitsy po Sveetoparsejuniju*, t. III. (Leningrad: Gidrometeoizdat).

Zharkov, V. N. 1981. Models of the interior structure of Venus. In *Modeli Vnutrennego Stroyeniya Venery* (Moscow: USSR Academy of Sciences), pp. 1–66.

Zhuang, H.-C., and Russell, C. T. 1981. An analytic treatment of the structure of the bow shock and magnetosheath. *J. Geophys. Res.* 86:2191–2205.

Zohar, S., and Goldstein, R. M. 1968. Venus map: A detailed look at the feature β. *Nature* 219:357–358.

Zohar, S., Goldstein, R. M., and Rumsey, H. C. 1980. A new radar determination of the spin vector of Venus. *Astron. J.* 85:1103–1111.

Zwan, B. J., and Wolf, R. A. 1976. Depletion of solar wind plasma near a planetary boundary. *J. Geophys. Res.* 81:1636–1648.

Glossary

GLOSSARY

adiabatic lapse rate
lapse rate: negative of vertical temperature gradient, according to most authors. Adiabatic lapse rate: value taken up in rapid vertical motion (see Chap. 11, Eq. 6).

albedo
(Latin, whiteness): ratio of reflected to incoming energy. Bond or spherical albedo: total over all directions. Geometric albedo: in backward direction or referred to this (see Chap. 3). Monochromatic albedo: in some specific narrow wavelength band.

amu
atomic mass unit.

anemopause
inner boundary of the solar wind at Venus.

ashen light
light from the dark side of Venus, occasionally reported by visual observers.

asteroid belt
the region between the orbits of Mars and Jupiter.

asthenosphere
the mobile region of a planet just below the lithosphere (q.v.).

atmosphere
1.013 bars, standard Earth surface pressure unit.

AU
astronomical unit, mean Earth-Sun distance, 1.496×10^8 km.

β
in plasma physics, the ratio of plasma and magnetic field energy densities (see Chap. 25, p. 939).

bar
10^6 dyne cm^{-2} or 10^5 Pa (N m^{-2}).

BIMS
Pioneer Venus multiprobe bus ion mass spectrometer.

BNMS
Pioneer Venus multiprobe bus neutral mass spectrometer.

bolometric temperature
the temperature corresponding to the total thermal flux from an object, also called effective temperature.

Bond albedo
(spherical albedo); see albedo and Chap. 3.

Boussinesq approximation
in atmospheric fluid dynamics, the neglect of the basic density variation with height, while changes about the mean density are retained.

Brownian motion
kinetic-theory motion of microscopic particles.

Brunt-Väisäla   frequency of buoyancy oscillations in an atmosphere (see
  frequency, N Chap. 21, Eq. 6).

carbonaceous   a class of meteorite, regarded as the least subject to
  chondrite   changes since formation.

Chapman grazing   generalization of the air mass function (secant of zenith
  incidence   angle) to a spherical atmosphere; also see Chapman
  function   (1931).

coagulation   growth of cloud particles by collision due to their random
                       (Brownian) motion.

coalescence   growth of large particles by sweeping up of smaller ones
                       due to their different sedimentation speeds.

cold trap   the concept that a vapor (such as water) is kept out of
                       higher regions by condensation at some cold level.

conductivity,   tensor components of the electrical conductivity of a
  Cowling, Hall,   collisional, magnetized plasma. Roughly speaking,
  Pedersen σ_P is along the electric field and σ_H is transverse
                       to it. The Cowling conductivity $\sigma_C = \sigma_P + \sigma_H^2/\sigma_P$
                       and is related to the power dissipation in currents trans-
                       verse to the magnetic field.

Coriolis force   a pseudo force that appears in the dynamical equations
                       when set up for a rotating coordinate system. In the
                       Earth's northern hemisphere, it tends to turn all velocity
                       vectors to the right (see Chap. 21, Eq. 2).

Coulomb   collisions between charged particles, mediated by the
  collisions   electrostatic force between them.

cryosphere   on Venus, a term used to describe the cold upper atmo-
                       sphere on the night side.

ctn   cotangent.

cyclostrophic   a condition in which the meridional atmospheric pres-
  balance   sure gradient provides the horizontal component of cen-
                       tripetal force required to constrain the zonal flow circling
                       the planet (see Chap. 11, Eqs. 3–5; Chap. 21, Sec. III).

dB or db   decibel, $10 \log_{10} (P_2/P_1)$ where P_1, P_2 are power levels.

diabatic   used to describe heating or cooling of an air parcel by
                       radiation.

diffusive   a state, found in upper atmospheres, where each con-
  equilibrium   stituent takes up its own scale height.

effective temperature, T_e	same as bolometric temperature (q.v.).
exobase	base of the exosphere (q.v.).
exosphere	outermost region of an atmosphere, where atoms and molecules are on ballistic or escaping trajectories and collisions can usually be neglected.
FWHM	full width of a spectral line or other feature at half-maximum intensity.
Gauss	CGS unit of magnetic field strength, 10^{-4} Tesla.
geometric albedo	see albedo and Chap. 3.
geostrophic balance	a balance between inertial and Coriolis forces, approximately valid for the atmosphere of a rapidly rotating planet.
Gyr	10^9 yr.
Hadley cell	a simple convective cell, rising in warm regions and descending in cold ones.
Hagfors constant	used in describing angular distribution of radar echoes (C in Chap. 6, Eq. 1).
homopause	level in an atmosphere at which gases cease being uniformly mixed and pass over to diffusive equilibrium (q.v.). Often called turbopause.
IMF	interplanetary magnetic field.
ionopause	abrupt termination of the ionosphere (see Chap. 23, Figs. 1, 4, 8a, 21, and 22).
ionosheath	see Chap. 23, Fig. 28.
ionosphere	the ionized region of an upper atmosphere.
ISEE	International Sun-Earth Explorer, a system of Earth satellites.
IUE	International Ultraviolet Explorer, a telescope and spectrometer.
Jeans escape	thermal escape of light atoms from an upper atmosphere.
Joule heating	heating by dissipation of electric currents.
Junge layer	a layer of H_2SO_4 particles, typically 0.1 μm radius, on the Earth near 20 km.

Kelvin effect the tendency for small cloud particles to vaporize while the vapor is acquired by larger ones (see Chap. 16).

Kelvin-Helmholtz instability tendency of waves to grow on a shear boundary between two regions in relative motion parallel to the boundary.

Knudsen number, Kn ratio of mean free path to particle radius.

Kohler effect effect of dissolved material on vapor pressure of a drop (see Chap. 16).

KR kilorayleigh (see R).

laminar flow smooth, nonturbulent flow.

Landau damping a form of plasma-wave absorption due to nonthermal electrons.

Larmor radius also gyroradius. Radius of gyration of a charged particle in the ambient magnetic field.

LAS PV Large probe atmospheric structure experiment.

LCPS PV Large probe cloud particle size spectrometer.

LGC PV Large probe gas chromatograph.

LIR PV Large probe infrared radiometer.

lithosphere, elastic a cool, relatively strong boundary layer lying on top of the asthenosphere, a slowly convecting, weaker, less viscous region (see Chap. 6, Sec. IV).

LN PV Large probe nephelometer.

LNMS PV Large probe neutral mass spectrometer.

LSFR PV Large probe solar (net) flux radiometer.

LST local solar time.

LTE local thermodynamic equilibrium, the approximation that a radiation field can be computed as if all atomic and molecular energy states are in thermal equilibrium.

LVT local Venus time.

magnetohydro-dynamics (MHD) dynamics of a conducting medium coupled to a magnetic field.

mbar millibar, 10^{-3} bar (q.v.).

megaweber, MWb 10^6 webers.

Meissner effect tendency of a conductor to exclude the ambient magnetic field.

mesopause upper boundary of mesosphere (see the Preface).

mgal an acceleration (usually gravitational) of 10^{-3} cm s^{-2}.

Mie theory Electromagnetic theory of scattering by a homogeneous spherical particle.

monochromatic geometric albedo see albedo.

MR megarayleigh (see R).

nadir the downward direction; intensity seen looking downward.

nanotesla, nT 10^{-9} T (q.v.).

neper $\log_e (X_2/X_1)$ where X_1, X_2 are two values of some quantity.

nephelometer instrument that measures cloud by light scattering.

noble gases He, Ne, Ar, Kr, Xe (also Rn). Also called rare or inert gases.

nT unit of magnetic field strength, also called gamma; 10^{-9} Tesla.

Nusselt number, Nu dimensionless constant used in describing convection (see Chap. 10, Eq. 9).

Nyquist limit requirement that a waveform must be sampled at least twice per cycle for that frequency to be properly recorded.

OCPP PV orbiter cloud photopolarimeter (and imager).

OEFD PV orbiter electric field detector (plasma wave analyzer).

OETP PV orbiter electron temperature (Langmuir) probe.

OIMS PV orbiter ion mass spectrometer.

OIR PV orbiter infrared radiometer.

OMAG PV orbiter magnetometer.

ONMS PV orbiter neutral mass spectrometer.

Oort's cloud	a cloud of proto-comets believed to be present between 20,000 and 100,000 AU.
OPA	PV orbiter (solar wind) plasma analyzer.
ORAD	PV orbiter radar.
ORO	PV orbiter radio occultation experiment.
ORPA	PV orbiter retarding potential analyzer.
OUVS	PV orbiter ultraviolet spectrometer.
PEPSIOS	a high-resolution spectrometer based on a cascade of three Fabry-Perot etalons.
phase angle	angle between Sun and observer, taken at the object observed.
Planck function	energy distribution of blackbody radiation, usually in intensity units.
pm	picometer, 10^{-12} m or 10^{-6} μm.
Porapak-N	a molecular sieve used as the medium in a gas chromatograph.
ppb	parts per billion (10^9).
ppm	parts per million (10^6).
prograde	rotating in the same sense as the orbital motions of the planets (also called direct).
PV	Pioneer Venus; PVO, orbiter.
R	Rayleigh, $10^6/4\pi$ photons cm^{-2} steradian^{-1}.
radiogenic	produced by radioactive decay (e. g. ^{4}He, ^{40}Ar).
Raoult's law	the property that the vapor pressure of a component in a solution is proportional to its mole fraction.
Rayleigh number, Ra	dimensionless constant used in describing convection (see Chap. 10, Eqs. 10 and 26).
Rayleigh-Taylor instability	a "fingering" instability common when a dense fluid overlies (or is at higher pressure than) a less dense one.
retrograde	opposition of prograde (q.v.).
Reynolds criterion	if the Reynolds number Re (q.v.) exceeds a critical value $\sim$ 6000, a flow is likely to become turbulent.

Reynolds number, Re	$Re = vL/\nu$; ν is kinematic viscosity, L distance along a flow, and v velocity.
Richardson number, Ri	number used to describe the onset of turbulence due to wind shear (see Chap. 11, Eq. 8; Chap. 21, Eq. 29).
RIEP	a plasma sensor used on Venera orbiters and flybys.
Rossby number, Ro	a measure of the importance of planetary rotation on atmospheric motion (small for large Ro) (see Chap. 21, Eq. 1).
SAS	PV Small probe atmospheric structure experiment.
shock front	a sudden jump in properties at a dynamical interface (e. g. a ''sonic boom'').
sial, sialic	rock rich in Si, Al, K, and Na (generally granitic).
sima	rock rich in Mg, Fe, and Ca (generally basaltic).
SN	PV Small probe nephelometer.
SNFR	PV Small probe net flux radiometer.
S/N ratio	signal-to-noise ratio.
solar cylindrical coordinates	a cylindrical coordinate system with the axis along the planet-Sun line.
Sounder probe	PV Large probe.
SPGC	same as LGC.
Stefan-Boltzmann (or Stefan) constant	σ in the relation (thermal radiative flux) $= \sigma T^4$ for a black body, where T is temperature.
stellar magnitude	$-2.5 \log_{10} (F_2/F_1)$, where F_1, F_2 are fluxes from two objects, usually in some standard wavelength interval. Often F_1 represents a standard object.
stilb	a unit of surface brightness, 1 candle cm^{-2} or 1 lumen cm^{-2} sr^{-1}.
SVS angle	Sun-Venus-spacecraft angle, or phase angle.
synodic year	the time for the relative positions of Earth and a planet to repeat (584 d for Venus).
terrane (variant of terrain)	an area of ground considered as to its extent and natural features.

Tesla, T	MKS unit of magnetic field, 10^4 Gauss.
thermal tides	atmospheric variations with periods of 1 solar day and harmonics, induced by solar heating.
thermopause	the upper boundary of the thermosphere.
thermosphere	region of temperature rise due to ionospheric heating (see the Preface).
tidal structure	any atmospheric variation having a period of 1 solar day.
turbopause	see homopause.
VGC	Venera gas chromatograph.
visual magnitude	stellar magnitude (q.v.) in the visual band (around 5500 Å).
volatiles	materials that accrete into a planet but subsequently degas as atmosphere or hydrosphere.
wavenumber	(1) for radiation, reciprocal of wavelength, (2) for planetary waves, the number of waves in one global circuit.
weighting function	the kernel of an integral representing relative contributions of different layers to outgoing radiation (see Chap. 20, Eq. 1 and Fig. 2).
whistler-mode waves	a kind of plasma wave, thought to be responsible for heating of Venus's dayside ionosphere. Also, the mode in which electromagnetic waves from lightning can penetrate the nightside ionosphere.
Zwan-Wolf effect	tendency of plasma to be squeezed out of a magnetic field tube in directions parallel to the field (see Chap. 26).

List of Contributors
with Acknowledgments
to Funding Agencies

LIST OF CONTRIBUTORS

The following people helped to make this book possible, in organizing, writing, refereeing, or otherwise. The members of the Organizing Committee and Chairpersons are indicated with an asterisk (*).

L. K. Acheson Jr., Hughes Aircraft Laboratory, Los Angeles, CA.

V. Adams, Lunar and Planetary Laboratory, University of Arizona, Tucson, AZ.

G. Andlauer, CPR, Strasburg, France.

J. Apt, Jet Propulsion Laboratory, Pasadena, CA.

J. W. Armstrong, Jet Propulsion Laboratory, Pasadena, CA.

R. Arvidson, Department of Earth and Planetary Sciences, Washington University, St. Louis, MO.

M. G. Austin, Physics International Company, San Leandro, CA.

V. S. Avduevskiy, USSR Academy of Sciences, Moscow, USSR.

F. D. Baker, Fullerton College, Fullerton, CA.

W. F. Ballhaus Jr., NASA Ames Research Center, Moffett Field, CA.

A. Barnes, NASA Ames Research Center, Moffett Field, CA.

V. L. Barsukov, Vernadsky Institute, USSR Academy of Science, Moscow, USSR.

J. Bateman, Hughes Aircraft Company, Los Angeles, CA.

S. J. Bauer, NASA Goddard Space Flight Center, Greenbelt, MD.

A. T. Bazilevsky, Vernadsky Institute, USSR Academy of Science, Moscow, USSR.

G. L. Berge, California Institute of Technology, Pasadena, CA.

J. L. Bertaux, Service d'Aeronomie, Centre Nationale de al Recherche Scientifique, Verrieres, France.

K. Bhatki, Jet Propulsion Laboratory, Pasadena, CA.

*D. C. Black, NASA Ames Research Center, Moffett Field, CA.

J. Blamont, Service d'Aeronomie, Centre Nationale de la Recherche Scientifique, Verrieres, France.

W. J. Borucki, NASA Ames Research Center, Moffett Field, CA.

C. Bowin, Woods Hole Oceanographic Institution, Woods Hole, MA.

L. H. Brace, NASA Goddard Space Flight Center, Greenbelt, MD.

H. C. Brinton, NASA Goddard Space Flight Center, Greenbelt, MD.

W. E. Brown, Jet Propulsion Laboratory, Pasadena, CA.

J. R. Bruman, Jet Propulsion Laboratory, Pasadena, CA.

G. A. Burba, USSR Academy of Sciences, Moscow, USSR.

D. B. Campbell, Arecibo Observatory, Arecibo, Puerto Rico.

G. C. Carle, NASA Ames Research Center, Moffett Field, CA.

M. H. Carr, U.S. Geological Survey, Flagstaff, AZ.

P. Cassen, NASA Ames Research Center, Moffett Field, CA.

K. R. Chan, NASA Ames Research Center, Moffett Field, CA.

S. Chang, NASA Ames Research Center, Moffett Field, CA.

V. P. Chemodanov, State Scientific Research Center, Moscow, USSR.

R. H. Chen, Center for Radar Astronomy, Stanford University, Stanford, CA.

Z. P. Cheremukhina, USSR Academy of Sciences, Moscow, USSR.

J. B. Cimino, California Institute of Technology, Pasadena, CA.

R. T. Clancy, California Institute of Technology, Pasadena, CA.

P. A. Cloutier, Department of Space Physics and Astronomy, Rice University, Houston, TX.

G. Clow, U.S. Geological Survey, Flagstaff, AZ.

D. S. Colburn, NASA Ames Research Center, Moffett Field, CA.

*L. Colin, NASA Ames Research Center, Moffett Field, CA.

D. L. Compton, NASA Ames Research Center, Moffett Field, CA.

B. J. Conrath, NASA Goddard Space Flight Center, Greenbelt, MD.

B. M. Cordell, Department of Physics, California State College, Bakersfield, CA.

C. Cot, Service d'Aeronomie, Centre Nationale de la Recherche Scientifique, Verrieres, France.

R. A. Craig, NASA Ames Research Center, Moffett Field, CA.

T. E. Cravens, Department of Atmospheric and Oceanic Sciences, University of Michigan, Ann Arbor, MI.

D. Crisp, Geophysical Fluid Dynamics Program, Princeton University, Princeton, NJ.

D. P. Cruikshank, Institute for Astronomy, Honolulu, HI.

S. A. Curtis, NASA Goddard Space Flight Center, Greenbelt, MD.

J. A. Cutts, Planetary Science Institute, Science Applications Inc., Pasadena, CA.

J. Cuzzi, NASA Ames Research Center, Moffett Field, CA.

R. E. Daniell Jr., NASA Goddard Space Flight Center, Greenbelt, MD.

A. D. Del Genio, NASA Goddard Institute for Space Studies, New York, NY.

W. B. DeMore, Jet Propulsion Laboratory, Pasadena, CA.

K. Denomy, Lunar and Planetary Laboratory, University of Arizona, Tucson, AZ.

A. L. Dial, U.S. Geological Survey, Flagstaff, AZ.

D. J. Diner, Jet Propulsion Laboratory, Pasadena, CA.

A. R. Dobrovolskis, Jet Propulsion Laboratory, Pasadena, CA.

T. M. Donahue, Department of Atmospheric and Oceanic Sciences, University of Michigan, Ann Arbor, MI.

L. Doose, Lunar and Planetary Laboratory, University of Arizona, Tucson, AZ.

D. Doreo, Edmonds College, Lynnwood, WA.

M. Dryer, Space Environment Laboratory, National Oceanic and Atmospheric Administration, Boulder, CO.

D. Dugan, NASA Ames Research Center, Moffett Field, CA.

S. T. Durrance, Department of Physics, Johns Hopkins University, Baltimore, MD.

P. Dyal, NASA Ames Research Center, Moffett Field, CA.

J. Dyer, NASA Ames Research Center, Moffett Field, CA.

L. Edsinger, NASA Ames Research Center, Moffett Field, CA.

A. P. Ekonomov, USSR Academy of Sciences, Moscow, USSR.

E. M. Eliason, U. S. Geological Survey, Flagstaff, AZ.

R. C. Elphic, Institute of Geophysics and Planetary Physics, University of California, Los Angeles, CA.

L. S. Elson, Jet Propulsion Laboratory, Pasadena, CA.

V. R. Eshleman, Center for Radar Astronomy, Stanford University, Stanford, CA.

L. W. Esposito, Laboratory for Atmospheric and Space Physics, University of Colorado, Boulder, CO.

P. B. Esposito, Jet Propulsion Laboratory, Pasadena, CA.

A. Esquivel, Tucson Typographic Service, Inc., Tucson, AZ.

S. Fels, Geophysical Fluid Dynamics Laboratory, Princeton University, Princeton, NJ.

R. O. Fimmel, NASA Ames Research Center, Moffett Field, CA.

F. M. Flasar, NASA Goddard Space Flight Center, Greenbelt, MD.

C. P. Florenskiy, Vernadsky Institute, USSR Academy of Science, Moscow, USSR.

P. G. Ford, Department of Earth and Planetary Sciences, Massachusetts Institute of Technology, Cambridge, MA.

J. Fortuno, Tucson Typographic Service, Inc., Tucson, AZ.

J. L. Fox, Center for Astrophysics, Harvard University, Cambridge, MA.

J. A. Gardner, Jet Propulsion Laboratory, Pasadena, CA.

J. B. Garvin, Department of Geological Sciences, Brown University, Providence, RI.

J.-C. Gerard, Université de Liège, Liège, Belgium.

W. Gilbreath, NASA Ames Research Center, Moffett Field, CA.

L. Giver, NASA Ames Research Center, Moffett Field, CA.

C. M. Glass, Goodyear Aerospace Corporation, Akron, OH.

K. A. Goettel, Department of Earth and Planetary Sciences, Washington University, St. Louis, MO.

Z. M. Goldberg, Department of Earth and Planetary Sciences, Massachusetts Institute of Technology, Cambridge, MA.

Yu. M. Golovin, USSR Academy of Sciences, Moscow, USSR.

T. I. Gombosi, Central Research Institute for Physics, Budapest, Hungary.

*R. M. Goody, Center for Earth and Planetary Physics, Harvard University, Cambridge, MA.

D. Goorvitch, NASA Ames Research Center, Moffett Field, CA.

A. Ya. Gorbachevskiy, USSR Academy of Sciences, Moscow, USSR.

D. Gordon, Tucson Typographic Service, Inc., Tucson, AZ.

F. Gray, NASA Ames Research Center, Moffett Field, CA.

R. Greeley, Geology Department, Arizona State University, Tempe, AZ.

S. Gulkis, Jet Propulsion Laboratory, Pasadena, CA.

*J. E. Hansen, NASA Goddard Institute for Space Studies, New York, NY.

W. B. Hansen, Center for Space Science, University of Texas at Dallas, Richardson, TX.

B. W. Hapke, Department of Geology, University of Pittsburgh, Pittsburgh, PA.

I. Harris, NASA Goddard Space Flight Center, Greenbelt, MD.

R. E. Hartle, NASA Goddard Space Flight Center, Greenbelt, MD.

W. K. Hartmann, Planetary Sciences Institute, Tucson, AZ.

G. A. Harvey, NASA Langley Research Center, Hampton, VA.

J. W. Head, Department of Geological Sciences, Brown University, Providence, RI.

A. E. Hedin, NASA Goddard Space Flight Center, Greenbelt, MD.

R. R. Hodges Jr., Center for Space Science, University of Texas at Dallas,
Richardson, TX.

J. H. Hoffman, Center for Space Science, University of Texas at Dallas, Richardson, TX.

J. L. Holloway, Geophysical Fluid Dynamics Laboratory, Princeton University,
Princeton, NJ.

R. E. Holzer, Institute of Geophysics and Planetary Physics, University of California,
Los Angeles, CA.

H. Houben, Mycol, Inc., Sunnyvale, CA.

W. E. Howell, NASA Langley Research Center, Hampton, VA.

*D. M. Hunten, Department of Planetary Sciences, University of Arizona, Tucson, AZ.

A. P. Ingersoll, California Institute of Technology, Pasadena, CA.

E. C. Y. Inn, NASA Ames Research Center, Moffett Field, CA.

D. S. Intriligator, Carmel Research Center, Santa Monica, CA.

G. Israel, Service d'Aeronomie, Centre Nationale de la Recherche Scientifique,
Verrieres, France.

J. Iversen, NASA Ames Research Center, Moffett Field, CA.

R. W. Jackson, NASA Ames Research Center, Moffett Field, CA.

S. Jaffe, Jet Propulsion Laboratory, Pasadena, CA.

W. W. James, Jet Propulsion Laboratory, Pasadena, CA.

M. A. Janssen, Jet Propulsion Laboratory, Pasadena, CA.

W. T. Kasprzak, NASA Goddard Space Flight Center, Greenbelt, MD.

D. P. Kastel, National Research Council, Washington, DC.

J. F. Kasting, NASA Ames Research Center, Moffett Field, CA.

*W. M. Kaula, Department of Earth and Space Sciences, University of California,
Los Angeles, CA.

K. Kawabata, NASA Goddard Institute for Space Studies, New York, NY.

G. M. Keating, NASA Langley Research Center, Hampton, VA.

R. G. Keesee, NASA Ames Research Center, Moffett Field, CA.

V. V. Kerzhanovich, USSR Academy of Sciences, Moscow, USSR.

E. A. King, Geology Department, University of Houston, Houston, TX.

D. B. Kirk, NASA Ames Research Center, Moffett Field, CA.

M. J. Klein, Jet Propulsion Laboratory, Pasadena, CA.

A. J. Kliore, Jet Propulsion Laboratory, Pasadena, CA.

R. G. Knollenberg, Particle Measuring Systems, Inc., Boulder, CO.

W. C. Knudsen, Lockheed Palo Alto Research Laboratory, Palo Alto, CA.

M. Kobrick, Jet Propulsion Laboratory, Pasadena, CA.

D. Kojiro, NASA Ames Research Center, Moffett Field, CA.

Yu. V. Kornienko, Institute of Radiophysics Electronics, Ukrainian SSR Academy of
Science, Kharkov, USSR.

V. A. Krasnopol'sky, USSR Academy of Sciences, Moscow, USSR.

Ksanfomality, USSR Academy of Sciences, Moscow, USSR.

W. R. Kuhn, Department of Atmospheric and Oceanic Sciences, University of Michigan,
Ann Arbor, MI.

Yu. N. Kulikov, USSR Academy of Sciences, Moscow, USSR.

S. Kumar, Space Sciences Institute, University of Southern California, Tucson, AZ.
V. G. Kurt, USSR Academy of Sciences, Moscow, USSR.
A. D. Kuz'min, USSR Academy of Sciences, Moscow, USSR.
L. Lake, Systems and Applied Sciences Corporation, Hampton, VA.
A. I. Lane, Jet Propulsion Laboratory, Pasadena, CA.
W. A. Lane, NASA Goddard Institute for Space Studies, New York, NY.
P. Layer, Astronomy Department, Stanford University, Stanford, CA.
R. Leach, NASA Ames Research Center, Moffett Field, CA.
M. Lehwalt, NASA Ames Research Center, Moffett Field, CA.
C. Leidich, NASA Ames Research Center, Moffett Field, CA.
*C. Leovy, Department of Astronomy, University of Washington, Seattle, WA.
V. Lepine, Service d'Aeronomie, Centre Nationale de la Recherche Scientifique, Verrieres, France.
L. Lerman, Astronomy Department, Stanford University, Stanford, CA.
J. Leung, California Institute of Technology, Pasadena, CA.
Z. Levin, Department of Geophysics and Planetary Sciences, Tel-Aviv University, Tel-Aviv, Israel.
J. S. Levine, NASA Langley Research Center, Hampton, VA.
S. S. Limaye, Department of Physics and Astronomy, University of Wisconsin, Madison, WI.
J. J. Lissauer, NASA Ames Research Center, Moffett Field, CA.
D. Lozier, NASA Ames Research Center, Moffett Field, CA.
J. G. Luhmann, Institute of Geophysics and Planetary Physics, University of California, Los Angeles, CA.
M. C. Malin, Department of Geology, Arizona State University, Tempe, AZ.
M. Ya. Marov, USSR Academy of Sciences, Moscow, USSR.
F. Marschak, Santa Barbara City College, Santa Barbara, CA.
J. Martin, Marietta Observatory, State University of New York, Morrisville, NY.
J. Martonchik, Jet Propulsion Laboratory, Pasadena, CA.
H. Masursky, U.S. Geological Survey, Flagstaff, AZ.
M. A. Matthews, Lunar and Planetary Laboratory, University of Arizona, Tucson, AZ.
M. S. Matthews, Lunar and Planetary Laboratory, University of Arizona, Tucson, AZ.
H. G. Mayr, NASA Goddard Space Flight Center, Greenbelt, MD.
D. J. McCleese, Jet Propulsion Laboratory, Pasadena, CA.
G. E. McGill, Department of Geology and Geography, University of Massachusetts, Amherst, MA.
G. McLaughlin, Lunar and Planetary Laboratory, University of Arizona, Tucson, AZ.
A. E. Metzger, Jet Propulsion Laboratory, Pasadena, CA.
P. F. Michelson, Department of Physics, Stanford University, Stanford, CA.
J. D. Mihalov, NASA Ames Research Center, Moffett Field, CA.
K. L. Miller, Lockheed Palo Alto Research Laboratory, Palo Alto, CA.
G. Moreels, Observatoire de Besancon, Besancon, France.
V. I. Moroz, USSR Academy of Sciences, Moscow, USSR.
P. Morris, Department of Space Physics and Astronomy, Rice University, Houston, TX.
D. Morrison, Institute for Astronomy, Honolulu, HI.
B. Ye. Moshkin, USSR Academy of Sciences, Moscow, USSR.
N. Mottinger, Jet Propulsion Laboratory, Pasadena, CA.
P. Mouginis-Mark, Department of Geological Sciences, Brown University, Providence, RI.
D. O. Muhleman, California Institute of Technology, Pasadena, CA.
L. Mukhin, USSR Academy of Sciences, Moscow, USSR.
J. Murphy, NASA Ames Research Center, Moffett Field, CA.
R. E. Murphy, Solar System Exploration Division, NASA Headquarters, Washington, DC.
A. F. Nagy, Space Physics Research Laboratory, University of Michigan, Ann Arbor, MI.
M. K. Narayeva, State Scientific Research Center, Moscow, USSR.
H. Nebel, Alfred University, Alfred, NY.
D. M. Newlands, Hughes Aircraft Company, Los Angeles, CA.
J. Nicholson, Systems and Applied Sciences Corporation, Hampton, VA.
H. B. Niemann, NASA Goddard Space Flight Center, Greenbelt, MD.

O. V. Nikolaeva, Vernadsky Institute, USSR Academy of Science, Moscow, USSR.

V. Novak, Fraunhofer Institut für Physikalische Messtechnik, Freiburg, West Germany.

S. Nozette, Department of Earth and Planetary Sciences, Massachusetts Institute of Technology, Cambridge, MA.

R. R. Nunamaker, NASA Ames Research Center, Moffett Field, CA.

R. Opstbaum, NASA Goddard Institute for Space Studies, New York, NY.

R. E. Orville, Department of Atmospheric Physics, State University of New York, Albany, NY.

A. S. Panfilov, State Scientific Research Center, Moscow, USSR.

J. P. Parisot, Observatoire de Besancon, Besancon, France.

V. A. Parshev, USSR Academy of Sciences, Moscow, USSR.

M. Patapoff, Jet Propulsion Laboratory, Pasadena, CA.

I. R. Patel, Jet Propulsion Laboratory, Pasadena, CA.

J. B. Pechmann, California Institute of Technology, Pasadena, CA.

H. Perez-de-Tejada, Instituto de Astronomia, UNAM, Baja, Mexico.

G. H. Pettengill, Department of Earth and Planetary Sciences, Massachusetts Institute of Technology, Cambridge, MA.

L. Pfister, NASA Ames Research Center, Moffett Field, CA.

B. Phillips, Tucson Typographic Service, Inc., Tucson, AZ.

R. J. Phillips, Department of Geological Sciences, Southern Methodist University, Dallas, TX.

J. Pingree, Massachusetts Institute of Technology, Cambridge, MA.

P. Pitluga, Adler Planetarium, Chicago, IL.

W. Pittle, NASA Ames Research Center, Moffett Field, CA.

I. M. Podgorny, USSR Academy of Sciences, Moscow, USSR.

L. Polaski, NASA Ames Research Center, Moffett Field, CA.

J. B. Pollack, NASA Ames Research Center, Moffett Field, CA.

S. M. Pompea, Marietta Observatory, State University of New York, Morrisville, NY.

F. Prevost, NASA Ames Research Center, Moffett Field, CA.

*R. G. Prinn, Department of Meteorology and Physical Oceanography, Massachusetts Institute of Technology, Cambridge, MA.

A. A. Pronin, Vernadsky Institute, USSR Academy of Science, Moscow, USSR.

B. Ragent, NASA Ames Research Center, Moffett Field, CA.

R. Ramos, NASA Ames Research Center, Moffett Field, CA.

A. S. P. Rao, Department of Geology, Osmania University, Hyderabad, India.

S. Rathjen, NASA Ames Research Center, Moffett Field, CA.

R. D. Reasenberg, Department of Earth and Planetary Sciences, Massachusetts Institute of Technology, Cambridge, MA.

R. T. Reynolds, NASA Ames Research Center, Moffett Field, CA.

P. Rigaud, Laboratoire de Physique et Dynamique de l'Atmosphere, Université Paris, Paris, France.

S. Ritke, Jet Propulsion Laboratory, Pasadena, CA.

W. B. Rossow, NASA Goddard Institute for Space Studies, New York, NY.

C. T. Russell, Institute of Geophysics and Planetary Physics, University of California, Los Angeles, CA.

R. E. Samuelson, NASA Goddard Space Flight Center, Greenbelt, MD.

N. F. Sanko, Institute of Cosmic Research, Moscow, USSR.

M. Sato, NASA Goddard Institute for Space Studies, New York, NY.

S. R. Saunders, Jet Propulsion Laboratory, Pasadena, CA.

F. L. Scarf, Space Sciences Department, TRW Defense and Space Systems Group, Redondo Beach, CA.

J. T. Schofield, Department of Atmospheric Physics, Clarendon Laboratory, Oxford, England.

G. Schubert, Department of Earth and Space Sciences, University of California, Los Angeles, CA.

P. H. Schultz, Lunar and Planetary Institute, Houston, TX.

D. R. Schumaker, SLU Planetarium, Felton, CA.

A. Seiff, NASA Ames Research Center, Moffett Field, CA.

A. S. Selivanov, State Scientific Research Center, Moscow, USSR.

I. I. Shapiro, Department of Earth and Planetary Sciences, Massachusetts Institute of Technology, Cambridge, MA.

V. P. Shari, USSR Academy of Sciences, Moscow, USSR.

M. Sheehan, NASA Ames Research Center, Moffett Field, CA.

Yu. G. Shkuratov, Kharkov State University, Kharkov, USSR.

G. Sill, Lunar and Planetary Laboratory, University of Arizona, Tucson, AZ.

R. A. Simpson, Center for Radar Astronomy, Stanford University, Stanford, CA.

W. L. Sjogren, Jet Propulsion Laboratory, Pasadena, CA.

J. A. Slavin, Institute of Geophysics and Planetary Physics, University of California, Los Angeles, CA.

D. A. Slezak, Goodyear Aerospace Corporation, Akron, OH.

V. N. Smirnov, USSR Academy of Sciences, Moscow, USSR.

P. H. Smith, Lunar and Planetary Laboratory, University of Arizona, Tucson, AZ.

S. M. Smith, NASA Ames Research Center, Moffett Field, CA.

T. S. Smith, Lunar and Planetary Laboratory, University of Arizona, Tucson, AZ.

*L. A. Soderblom, U.S. Geological Survey, Flagstaff, AZ.

S. Sommer, NASA Ames Research Center, Moffett Field, CA.

*N. W. Spencer, NASA Goddard Space Flight Center, Greenbelt, MD.

K. Spenner, Fraunhofer-Institut für Messtechnik, Freiburg, West Germany.

J. R. Spreiter, Division of Applied Mechanics, Stanford University, Stanford, CA.

S. Squyres, NASA Ames Research Center, Moffett Field, CA.

S. S. Stahara, Nielsen Engineering and Research, Inc., Mountain View, CA.

W. L. Starr, NASA Ames Research Center, Moffett Field, CA.

P. G. Steffes, Center for Radar Astronomy, Stanford University, Stanford, CA.

A. I. F. Stewart, Laboratory for Atmospheric and Space Physics, University of Colorado, Boulder, CO.

E. L. Strickland, Department of Earth and Planetary Sciences, Washington University, St. Louis, MO.

M. E. Strobell, U.S. Geological Survey, Flagstaff, AZ.

Yu. A. Surkov, USSR Academy of Science, Moscow, USSR.

R. Swansen, Laboratory for Atmospheric and Space Physics, University of Colorado, Boulder, CO.

C. A. Syvertson, NASA Ames Research Center, Moffett Field, CA.

M. A. Talman, Icarus, Cornell University, Ithaca, NY.

C. Tang, Jet Propulsion Laboratory, Pasadena, CA.

T. Tascione, Department of Space Physics and Astronomy, Rice University, Houston, TX.

F. W. Taylor, Department of Atmospheric Physics, Oxford University, Oxford, England.

H. A. Taylor Jr., NASA Goddard Space Flight Center, Greenbelt, MD.

W. W. L. Taylor, Space Sciences Department, TRW Defense and Space Systems Group, Redondo Beach, CA.

J. Terhune, NASA Ames Research Center, Moffett Field, CA.

P. Thompson, Lunar and Planetary Institute, Houston, TX.

T. W. Thompson, Planetary Science Institute, Science Applications Inc., Pasadena, CA.

B. A. Tinsley, Center for Space Science, University of Texas at Dallas, Richardson, TX.

M. G. Tomasko, Lunar and Planetary Laboratory, University of Arizona, Tucson, AZ.

O. B. Toon, NASA Ames Research Center, Moffett Field, CA.

L. D. Travis, NASA Goddard Institute for Space Studies, New York, NY.

R. P. Turco, R and D Associates, Marina del Rey, CA.

I. Tzur, National Center for Atmospheric Research, Boulder, CO.

A. Ya. Usikov, Institute of Radiophysics Electronics, Ukrainian SSR Academy of Science, Kharkov, USS.

G. R. Uspenskiy, USSR Academy of Sciences, Moscow, USSR.

O. Vaisberg, Space Research Institute, Moscow, USSR.

P. J. Valdes, Department of Atmospheric Physics, Oxford University, Oxford, England.

J. Valentine, NASA Ames Research Center, Moffett Field, CA.

G. Visconti, Instituto di Fisica, Universita dell'Aquila, L'Aquila, Italy.

N. Vojvodich, NASA Ames Research Center, Moffett Field, CA.

D. L. Wallach, Department of Astronomy, Pennsylvania State University,
    University Park, PA.
J. L. Warner, NASA Johnson Space Center, Houston, TX.
A. J. Watson, Department of Atmospheric and Oceanic Sciences, University of Michigan,
    Ann Arbor, MI.
B. White, NASA Ames Research Center, Moffett Field, CA.
R. C. Whitten, NASA Ames Research Center, Moffett Field, CA.
D. Wilhelms, U.S. Geological Survey, Flagstaff, AZ.
G. P. Williams, Geophysical Fluid Dynamics Laboratory, Princeton University,
    Princeton, NJ.
S. Williams, NASA Ames Research Center, Moffett Field, CA.
L. Wilson, Department of Geological Sciences, Brown University, Providence, RI.
W. J. Wilson, Jet Propulsion Laboratory, Pasadena, CA.
F. Woeller, NASA Ames Research Center, Moffett Field, CA.
R. S. Wolff, Jet Propulsion Laboratory, Pasadena, CA.
R. Woo, Jet Propulsion Laboratory, Pasadena, CA.
A. T. Young, Department of Astronomy, San Diego State University, San Diego, CA.
D. Young, Systems and Applied Sciences Corporation, Hampton, VA.
L. D. G. Young, Department of Astronomy, San Diego State University, San Diego, CA.
R. E. Young, NASA Ames Research Center, Moffett Field, CA.
K. Youngman, Lunar and Planetary Institute, Houston, TX.
Y. L. Yung, California Institute of Technology, Pasadena, CA.
U. von Zahn, Physikalisches Institut, Universität Bonn, Bonn, West Germany.
L. V. Zasova, Institute of Cosmic Research, Moscow, USSR.
L. Zeleny, USSR Academy of Sciences, Moscow, USSR.
J. M. Zucconi, Observatoire de Besancon, Besancon, France.
R. W. Zurek, Jet Propulsion Laboratory, Pasadena, CA.

*The Editors acknowledge the support of NASA Grant NAGW-149 and NASA Ames Research
Center. The following authors wish to acknowledge specific funds involved in supporting the
preparation of their chapters.*

Arvidson, R. E.; NASA Grant NSG-7087.
Cloutier, P. A.; NASA Contract NAS-23338.
Cravens, T. E.; NASA Grants NAGW-15, NGR 23-005-015 and NAS 2-9130, and NSF INT
    8026015.
Cruikshank, D. P.; NASA Grant NGL 12-001-057.
Daniell, R. E. Jr.; NASA Contract NAS-23338.
Esposito, L. W.; NASA Grants NAGW-197 and NAS 29477.
Gombosi, T. I.; NASA Grants NAGW-15 and NGR 23-005-015, and NSF INT 8026015.
Knudsen, W. C.; NASA Grant NAS2-9481.
Kumar, S.; NASA Grant NAGW-57.
Malin, M. C.; NASA Grant NAGW-1.
McGill, G. E.; NASA Grant NAG2-138.
Nagy, A. F.; NASA Grants NAGW-15, NGR 23-005-015 and NAS2-9130, and NSF INT
    8026015.
Phillips, R. J.; NASA Contract NASW-3389.
Russell, C. T.; NASA Contract NAS 2-9491.
Scarf, F. L.; NASA Contract NAS 2-9842.
Schubert, G.; NASA Grant NSG-7164 and Contract NAS 2-9131.
Tascione, T. F.; NASA Contract NAS-23338.
Taylor, H. A. Jr.; NASA Contract NAS-23338.
Tomasko, M. G.; NASA Grant NAGW-55.
Toon, O. B.; NASA Grants NAS-9126 and NSG-7404.
Turco, R. P.; NASA Grants NAS-9126 and NSG-7404.
Wolff, R. S.; NASA Contract NAS-7-100 and Pioneer Venus Guest Investigator Program.

Index

INDEX[a]

Absorption cross sections, 849, 850 (tab)
Acid rain, 555
Adiabatic lapse rate, 241 (def), 651
Aeronomy, 7, 333, 431–457
Aerosols
  general, 473–478
  physical processes, 542–548
Airglow
  excitation, 467–473
  general, 412–419
Akna Montes, 101, 105, 107
Albedo. *See also* Bond albedo *and under*
  Spherical albedo
  data, 27–35
  general, 144, 149, 150, 656–658, 659 (fig),
    666–668, 667–668 (figs)
Alpha Regio, 101
Ambipolar diffusion, 845
Ammonia, 364
Anemopause, 885, 889, 898
Angular momentum, 700–702, 701, (fig),
  712–729, 777
Aphrodite Terra, 100 (fig), 101, 105, 109,
  115 (fig), 126–127 (figs), 129, 1056
  (Color Plate)
Argon, 123, 303, 315, 409, 410, 1006–1008
  (tab), 1010 (tab), 1015, 1016, 1018, 1019,
  1032, 1037, 1039, 1039 (tab), 1040
Artemis Chasma, 111
Ashen light, 8, 19, 592, 593
Asteroid belt, 1011
Asthenosphere, 121, 122
Atmosphere
  composition, 47–58, 299–430, 325 (tab),
    1004–1010, 1033–1036
  discovery, 2
  history, 217, 218
  lower, 218–246, 301
  middle, 246–255, 301
  models, 269–279, 1045, 1045–1048 (tab)
  superrotation, 694–705
  surface interaction, 85–96
  upper, 255–266, 301, 347
  vertical transport, 326–333. *See also*
    Vertical motions
  waves. *See under* Waves

Barotropic instability, 750
Beta, 155
Beta-Phoebe region, 130
Beta Regio, 101, 105, 111, 113–118, 116–118
  (figs), 129, 1018, 1056 (Color Plate)
Bolometric temperature. *See under*
  Temperature
Bond albedo, 20 (tab)
Bow shock. *See also* structure *under* Shock
  general, 917–938, 980–993, 996–1001
  position, 922–926
  strength, 926–929
Brunt-Vaisala frequency, 691, 692, 693 (fig)

C, 315
Carbon and oxygen compounds, 333. *See also*
  CO, CO_2 *and* COS
Carbonaceous chondrite, 1013
Chapman layer, 784
Charge exchange, 890, 920
Chemical thermodynamics, 86–90
Chlorides, 404, 500
Chondritic composition, 163, 164
Circulation, 693, 694
Cloud
  composition, 532–538, 666
  dynamics, 529–531
  general, 59–62, 243–246, 447–449,
    484–564, 624–629, 637–643
  layers, 506–519
  models, photochemical models, 538–553
  morphology, 492–498
  nature, 3
  optical thickness, 642, 643
  particle modes, 514–516
  particles (Mode 0), 617, 618, 628, 629
  particles (Mode 3), 524–526
  polar structure. *See* Polar cloud structure
  properties, 487 (tab)
  spectra, 639 (fig)
  upper, 486–504
Cloud top
  conditions, 605
  height, 659
CO, 5, 302, 310, 311, 311 (fig), 315, 334–347,
  349–352, 614

---

[a]See also table of Contents.